Y0-EDW-232

TECHNICAL PHYSICS

Clarence R. Green
Davidson County Community College

PRENTICE-HALL, INC., Englewood Cliffs, New Jersey 07632

Library of Congress Cataloging in Publication Data

GREEN, CLARENCE R. (date)
Technical physics.

Includes bibliographies and index.
1. Physics. I. Title.
QC212.G728 1984 530 83-16025
ISBN 0-13-898387-9

Editorial/production supervision: Karen Skrable
Interior design: Karen Skrable and Anne T. Bonanno
Cover design: Anne T. Bonanno
Cover photo: Geoffrey Gove, The Image Bank
Manufacturing buyer: Anthony Caruso

Printed in the United States of America

10 9 8 7 6 5 4 3 2 1

ISBN 0-13-898387-9

Prentice-Hall International, Inc., *London*
Prentice-Hall of Australia Pty. Limited, *Sydney*
Editora Prentice-Hall do Brasil, Ltda., *Rio de Janeiro*
Prentice-Hall Canada Inc., *Toronto*
Prentice-Hall of India Private Limited, *New Delhi*
Prentice-Hall of Japan, Inc., *Tokyo*
Prentice-Hall of Southeast Asia Pte. Ltd., *Singapore*
Whitehall Books Limited, *Wellington, New Zealand*

CONTENTS

PREFACE

The term *technical physics* is used by educators to refer to an introductory course that is not as advanced as *college physics* or *university physics* and that requires no mathematics beyond algebra and trigonometry. In technical physics, the ideas are presented in terms of verbal exposition rather than by the rigorous mathematical developments that characterize advanced physics.

This text is intended for a one-year course in technical physics. All of the topics traditionally covered are included, and several areas have been extended. In thermodynamics, for example, complete working equations are given for the analysis of cyclic processes, and in electronics an entire chapter is devoted to applied solid-state physics. Both metric and nonmetric units are used, especially in the beginning, but metric (SI) units are used almost exclusively in the latter chapters. An entire chapter (Chap. 2) is devoted to units and unit conversions so that this aspect of studying physics may be mastered early.

A great effort has been made to present the ideas clearly and concisely. From the outset, my goal has been to make this text easy, inasmuch as the subject permits, for students to read. But no compromise of depth of treatment or selection of topics has been made just to facilitate easy reading. Rather, a greater effort has been made to express the more difficult concepts in language that is straightforward and comprehensible.

At this point, in order to comply with several polite requests to "put my name in the book," I take great pride in including

Seth, Wes, Julie, and Christie,

and of course, I must not forget

Faye.

Finally, I would like to remember my father and mother, **C. A.** and **Opha Green,** of Butler, Tennessee, who taught me the importance of education and thereby led me to one of the greatest joys—that of learning and of sharing it with others.

Clarence R. Green

CONVERSION FACTORS

LENGTH

1 m = 39.37 in. = 3.281 ft
1 ft = 30.48 cm = 0.3048 m
1 in. = 2.54 cm = 25.4 mm
1 mi = 5280 ft = 1609 m
1 nautical mile = 1.1508 mi
  = 6076.1 ft = 1852 m
1 angstrom (Å) = 10^{-10} m = 10^{-8} cm
1 micron (μ) = 10^{-6} m = 10^{-4} cm
1 light-year = 9.461×10^{12} km
  = 5.880×10^{12} mi
1 astronomical unit (A.U.)
  = 1.49598×10^{11} m
  = 93,000,000 mi
1 parsec = 206,265 A.U.
  = 3.262 light-years
1 mil = 0.001 in.
1 rod = 16.5 ft
1 fathom = 6 ft

VOLUME

1 m^3 = 1000 L = 10^6 cm^3
  = 264.2 gal = 35.313 ft^3
  = 1.308 yd^3
1 ft^3 = 1728 $in.^3$ = 7.481 gal
  = 0.02832 m^3 = 28.32 L
1 $in.^3$ = 16.387 cm^3
1 L = 1000 cm^3 = 0.001 m^3
  = 1.0576 qt = 61.03 in^3
  = 0.0353 ft^3 = 0.242 gal
1 mL = 1 cm^3
1 qt = 57.75 $in.^3$ = 945.54 cm^3
1 gal = 231 $in.^3$ = 3.7854 L
  = 0.1337 ft^3 = 3786 cm^3

PRESSURE

1 N/m^2 = 1 Pascal (Pa) = 1.450×10^{-4} $lb/in.^2$
  = 10 dyn/cm^2 = 10^{-5} bar = 9.869×10^{-6} atm
  = 7.501×10^{-3} mm Hg
1 bar = 10^5 Pa = 14.50 $lb/in.^2$ = 0.9869 atm
  = 750 mm Hg
1 mbar = 10^{-3} bar = 0.750 mm Hg
1 $lb/in.^2$ (psi) = 6895 N/m^2 = 51.71 mm Hg
  = 27.68 in. Hg = 6.895 kPa = 68.96 mbar
1 atm = 1.01325×10^5 N/m^2 = 101.3 kPa
  = 76 cm Hg = 760 mm Hg = 760 torr
  = 1.013 bar = 14.7 $lb/in.^2$ = 34 ft H_2O
1 in. H_2O = 0.03613 $lb/in.^2$ = 1.868 mm Hg
  = 249.1 Pa = 2.491 mbar
1 torr = 1 mm Hg

AREA

1 $in.^2$ = 6.452 cm^2 = 645.2 mm^2
1 ft^2 = 144 $in.^2$ = 929.0 cm^2
  = 0.09290 m^2
1 m^2 = 10.764 ft^2 = 1550 $in.^2$
  = 10^4 cm^2 = 10^6 mm^2
1 acre = 43,560 ft^2 = 160 rod^2
  = 4047 m^2 = 4840 yd^2
1 square mile = 640 acres
1 circular-mil = 7.854×10^{-7} $in.^2$
1 $in.^2$ = 1.273×10^6 circular-mils
1 barn = 10^{-28} m^2 = 10^{-24} cm^2

HOUSEHOLD

1 teaspoon = 5 milliliters (mL)
3 teaspoons = 1 tablespoon
1 fluid ounce (oz) = 2 tablespoons
1 jigger = 3 tablespoons
  = 1.5 fluid ounces
16 tablespoons = 1 cup = 8 oz
1 cup = 236.4 mL = 14.43 $in.^3$
1 qt = 32 oz = 2 pints
8 qt = 1 peck = 2 gal
4 pecks = 1 bushel = 8 gal
1 bushel = 32 qt = 1.07 ft^3

ENERGY

1 joule (J) = 0.7376 ft-lb = 10^7 ergs
  = 6.242×10^{18} eV = 2.778×10^{-7} kW · h
1 ft-lb = 1.356 J = 0.3241 cal
1 BTU = 1054 J = 777.3 ft-lb = 252 cal
1 cal = 4.184 J = 3.086 ft-lb
1 kW · h = 3.6×10^6 J = 3415.6 BTU = 8.604×10^5 cal
1 watt-second (W · s) = 1 J
1 L · atm = 101.3 J

POWER

1 watt (W) = 0.7376 ft-lb/s
1 HP = 746 W = 550 ft-lb/s

CONVERSION FACTORS

MASS

1 kg = 2.2046 lb-mass
= 0.06852 slug
1 lb-mass = 0.4536 kg
= 0.03108 slug
1 slug = 32.17 lb-mass
= 14.59 kg
1 metric ton = 1000 kg

SPEED

60 mi/h = 88 ft/s = 96.5 km/h
= 26.82 m/s = 52.13 knots
1 m/s = 3.60 km/h = 2.237 mi/h
= 3.281 ft/s = 1.944 knots
1 mi/h = 0.4470 m/s = 1.609 km/h
= 1.466 ft/s = 0.8689 knot
1 ft/s = 0.3048 m/s = 0.682 mi/h
1 km/h = 0.9113 ft/s = 0.6214 mi/h
1 knot = 1 nautical mile per hour
= 1.151 mi/h

FORCE

1 Newton (N) = 0.2248 lb
= 0.1020 kg-wt = 102 g-wt
= 10^5 dynes
1 lb = 4.448 N = 0.4536 kg-wt
= 453.6 g-wt
1 gram mass = 980 dynes
1 U.S. ton = 2000 lb

DENSITY

1 g/cm^3 = 62.43 lb/ft^3
= 1000 kg/m^3
1 lb/ft^3 = 0.01602 g/cm^3
= 16.02 kg/m^3

SELECTED PHYSICAL CONSTANTS AND DATA

Speed of light, c	2.997925×10^8 m/s
Acceleration of gravity, g	9.8 m/s^2 = 32 ft/s^2
Gravitational constant, G	6.672×10^{-11} N · m^2/kg^2
Avogadro's number, N_a	6.023×10^{23} molecules/g · mol
Universal gas constant, R	8.3143×10^3 J/kg · mol · K
	= 0.821 L · atm/g · mol · K
	= 0.0237 psi-ft^3/g · mol · R
Boltzmann's constant, k	1.381×10^{-23} J/K
	= 8.617×10^{-5} eV/K
Stefan-Boltzmann constant,	5.6696×10^{-8} W/m^2 · K^4
Nominal molecular weight of air	29 g/g · mol
Electronic charge, e	1.602×10^{-19} C
Coulomb's law constant, k	$1/4\pi\epsilon_0 = 8.988 \times 10^9$ N · m^2/C^2
Permittivity of free space, ϵ_0	8.854×10^{-12} C^2/N · m^2
Permeability of free space, μ_0	$4\pi \times 10^{-7}$ Wb/A · m
Planck's constant, h	6.626×10^{-34} J · s
	= 6.626×10^{-27} erg · s
	= 4.136×10^{-15} eV · s
Mass unit, u	1.6606×10^{-27} kg $\leftrightarrow$ 931.5 MeV/c^2
Electron rest mass	9.1091×10^{-31} kg = 0.0005486 u
Proton rest mass	1.6725×10^{-27} kg = 1.0072766 u
Neutron rest mass	1.6748×10^{-27} kg = 1.0086654 u
Hydrogen atom rest mass	1.673×10^{-27} kg = 1.007825 u
Helium atom, $^4_2$He, rest mass	6.647×10^{-27} kg = 4.002603 u

Radius of Earth	
Equatorial	6.378×10^6 m = 3963 mi
Polar	6.357×10^6 m = 3950 mi
Nominal value	6.37×10^6 m = 3957 mi
Radius of moon	1.738×10^6 m = 887 mi
Radius of sun	6.960×10^8 m = 4.33×10^5 mi
Mass of sun	1.99×10^{30} kg
Mass of Earth	5.98×10^{24} kg
Mass of moon	7.36×10^{22} kg
Earth-sun distance (average)	1.496×10^{11} m
Earth-moon distance (average)	3.84×10^8 m

CHAPTER 1

INTRODUCTION

The first part of this introductory chapter consists of general comments about physics and the relationship of physics to the other sciences, engineering in particular. Also, some benefits of studying physics are proposed.

It appears that physics, as a subject to be studied, learned, and passed, requires a significantly different approach than chemistry, biology, or pure mathematics. Many good students who do well elsewhere have difficulty with physics. Why this happens is not because physics is just naturally more difficult—it is not—but *physics is different*. Perhaps if we can identify the differences, we can also relieve some of the difficulty.

The second portion of this chapter is devoted to tips on how to study physics. Almost a dozen tips are given that range from how to read a physics book to steps in problem solving. Among these items, which deal mostly with organization, technique, and procedure, are suggestions about how to think about things in physics. This brings up a good question.

Is thinking something that can be learned or improved, or is it something that a person just does, whether good or bad? Can a person improve his or her ability to "think physics"? Keep this question in mind as you read the second part of this chapter.

1-1 WHAT IS PHYSICS?

A good idea of what physics is all about can be obtained by flipping through the pages of this book, if you have not done so already. It is obvious that many things are studied under the broad general heading of *physics*. Some of these are the laws of motion, forces, gravitational phenomena, the structure of matter, heat, light, sound, electricity, magnetism, electromagnetic waves, the structure of the atom, and elementary particles.

The companion sciences to physics are chemistry and biology. Chemistry deals with the chemical composition of substances and with the various chemical reactions and transformations that substances undergo. Speaking in the broadest terms, biology deals with living things. There is much overlap between these broad areas so that sharp lines cannot be drawn. Indeed, we hear of physical chemistry, chemical physics, biophysics, and biochemistry. Within the broad areas there are the fields of medicine, pharmacology, bacteriology, botany, astronomy, meteorology, mineralogy, geology, and oceanography—to name just a few. And within each of these areas there are dozens—if not hundreds—of smaller, specialized fields, each of which is big enough to absorb entire careers.

Mathematics is the foremost tool of the scientist, and the importance of mathematics to the sciences is well put in the statement made by Lord Kelvin [William Thompson (1824–1907)]:

> I often say that when you can measure what you are speaking about, and express it in numbers, you know something about it; but when you cannot express it in numbers, your knowledge is of a meagre and unsatisfactory kind; it may be the beginning of knowledge, but you have scarcely, in your thoughts, advanced to the stage of Science, whatever the matter may be.

This oft-quoted statement has had many opponents over the years who point out that certain things like art, history, beauty, happiness, and love are neither capable of being expressed in numbers nor stand in need of such expression. A less severe and stoic view of science is reflected in the statement made by the American astronomer and educator, Maria Mitchell (1818–1889):

> We especially need imagination in science. It is not all mathematics, nor all logic, but it is somewhat beauty and poetry.

The significance of both statements is better appreciated after a considerable expenditure of time and effort in the study of a science such as physics.

Physics vs. engineering

At the introductory level, the most apparent difference between physics and engineering is that physics covers a wide spectrum of topics whereas engineering almost always represents a specialization. The broadness of physics is indicated in part by the variety of topics in this book. On the other hand, several fields of engineering—aerodynamic, automotive, civil, electrical—each a specialty, come to mind. Beginning engineers study physics in order to become familiar with the areas of physics outside the area of their specialization.

Traditionally, physics has been called a *pure* science; engineering is considered to be an *applied* science. In a pure science, there is little concern for practical applications of the principles; the major thrust is toward understanding. In an applied science, the objective is to put the principles to use. It is sometimes said that physicists develop the formulas and engineers find ways to use them.

At the highest level, the difference between an engineer and a physicist becomes less apparent. It is primarily a matter of background. Both will be specialists, and both may be engaged in research aimed at a better understanding of new principles or directed toward the development of new equipment, techniques, or products.

The rewards of studying physics

There are many benefits derived from studying physics aside from vocational pursuits and meeting graduation requirements. A good knowledge of physics gives one an advantage in dealing to maximum advantage with day-to-day experiences. For example, . . . well, no specific examples come to mind at the moment, but the principles of physics are continually in effect all around us—in the kitchen, in the car, on the ski slopes, and so on, ad infinitum. Everything and everybody must obey physical laws. Knowledge of those laws surely must be of benefit!

An understanding of physics increases one's appreciation of a wide variety of things—rainbows, echos, snowflakes, icicles, ocean waves, the sound of a violin, the artistry of an ice skater, the flight of a bird, and even the sound of the wind. Furthermore, physics helps us to know more about the universe and to better appreciate its magnitude. Via physics we know quite a lot about galaxies that are millions of light-years away, and we know something about the inner workings of atoms and of the particles of which they are made.

These benefits, however, come with a price. That price is time and effort devoted to *the study of physics*, the topic of the next section.

1-2 TIPS ON HOW TO STUDY PHYSICS

Why is physics so hard?

If a survey were taken, it is likely that physics would be indicated as one of the more difficult subjects. Many students who excel in other areas have difficulty with physics. In considering why this is often the case, we see that physics is actually a combination of two areas: physical concepts on the one hand, mathematics on the other. A lack of strength in mathematics will invariably lead to difficulties because of the quantitative nature of physics. Furthermore, the exactitude that is characteristic of physical laws requires precise and unambiguous thinking. Vague notions and general feelings are not sufficient. You have to know it cold.

To do well in physics requires that you be able to take a concept or a physical principle and apply it to a situation not previously encountered. A good memory alone is not sufficient for the needs and requirements of physics. Some people say that physics requires an analytical mind, referring to the fact that physical phenomena and physical laws first must be analyzed and understood as a prerequisite to the application of these laws and principles to problem sets or actual physical situations. This demand for a person to know the material and then be able to apply it with mathematical precision is one of the most prominent reasons why physics is difficult. This is not intended to be frightening, however,

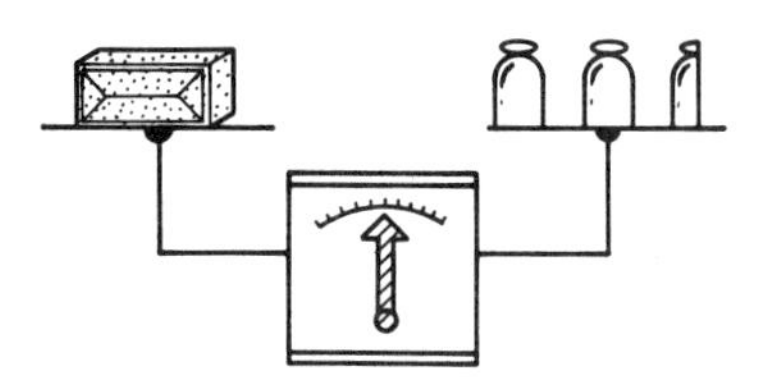

Figure 1-1 A possible mental image stemming from the following statement: The mass of a common brick is 2.5 kg.

because we shall see that there are many specific things that can make problem solving easier.

How to read a physics book

Obviously, a physics book or other technical material cannot be read with the same speed with which you read a novel, history, or other literary work, but few students new to physics realize the full extent of this. Technical material must be taken slowly and deliberately; speed-reading is definitely out. Your exact reading rate will depend upon your ability and upon the density of the material, but 10 minutes or more per page is not too long if that much time is required. The important point is that you must understand and reflect upon the material as you read. You cannot rush.

Making notes, sketches, and diagrams in a notebook as you read helps fix your attention; if carefully done, the sketches and notes will provide for a quick review of the material as test time approaches. All but the simplest examples should be worked out on paper.

Do not skip the math! Would you believe that some students read the words but skip right over the mathematical parts. This leads to vague and imprecise thinking that will only cause greater difficulties later on. Because most physics tests and examinations consist mainly of problems to solve, we must pay particular attention to the mathematics. After the first reading, it always helps to come back later for a second, more-rapid reading to refresh and reinforce your understanding of the material.

Visualize objects in detail

As you read, think of the details of the objects mentioned in the text to make the mental picture more vivid. Consider this simple statement: The mass of a common brick is about 2.5 kilograms. What mental pictures might make this statement come alive?

First, what kind of brick do you imagine? Does it have holes in it? What color is it, exactly? How would you describe its shape? What is the nature of its surface texture? Is it smooth or rough? And, when you have visualized the brick adequately, how will you determine its mass? Imagine that you place the brick on a balance and add the proper number of 1-kilogram masses to the opposite pan to bring it to balance. Your final image might be something like Fig. 1-1.

Be careful not to clutter up your mental picture with too many irrelevant details. For example, in Fig. 1-1, the brick and the masses are rendered in considerable detail, but only the barest essentials are provided for the balance.

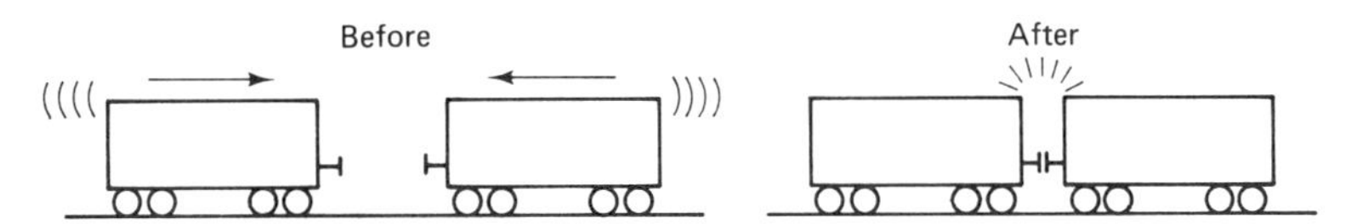

Figure 1-2 A sketch made as an aid to visualization of the following statement: "Two railroad cars traveling toward each other on a level track collide and become coupled."

This is proper because the important point about the balance is the fact that it *is balanced*. It is not necessary to visualize the mechanical or ornamental details of the balance; these details are irrelevant, and if a truly fine and beautiful balance were visualized, it might detract from the brick.

Draw diagrams and sketches

Draw diagrams and make sketches as you read as an aid to visualization. This is extremely important because a sketch provides a visual impression and reinforcement of an idea that otherwise would remain largely on the verbal level. The sketches need not be elaborate, artistic renditions that take a lot of time to do; they should be simple and to the point. For example, Fig. 1-2 illustrates the following statement: Two railroad cars traveling toward each other on a level track collide and become coupled. Note that the sketch illustrates the conditions both before and after the collision.

Think in the extreme

Often, a question or concept can be clarified by imagining one aspect of the situation to be either increased or decreased to an extreme or greatly exaggerated extent that is more easily understood. Here is an example.

Archimedes (287–212 B.C.), after discovering the principle of the lever and its properties of multiplying forces, is reported to have said that if he were given a place to stand and were provided with a lever sufficiently long, then he could move the earth. This is a rather profound statement, and one naturally wonders how such a conclusion might have been reached. Maybe it was reached in the following manner.

First, imagine a small earth, about the size of a pumpkin. Given a place to stand and a fairly short lever, this earth could be moved with little effort. Now increase the size of the earth to the size of a physics professor's desk. The earth can still be moved, but it requires a longer lever to do it with the same applied force. Let the earth now swell to the size of a house. An earth this size is not going to be moved easily, but with some ingenuity and with a considerably longer lever, the possibility still is very real that the house-sized earth could be moved. The basic principle still holds true even though technical difficulties (where to get a lever that long?) become more of a problem. Because we see no basic difficulty with the underlying principle, we can extend the concept to apply to the earth itself, speculating that if we could find a lever sufficiently long—given a place to stand, of course—we *could* move the earth. Nothing to it! There is no historical basis for this line of reasoning, but it does illustrate the idea of thinking in the extreme.

Should you memorize formulas?

This question is a standard because beginning students invariably comment, "There are so many formulas to memorize!" The true physicist will say, "No, thou shalt *not* memorize the formulas. Physics is more than that." But what are you going to do on the tests if you do not know the formulas?

The idea here is that the student will work so many homework problems that the formulas will become second nature and will become known without a purposeful act of conscious memorization. However, life being what it is, few of us ever work as many homework problems as we should. Therefore, the recommendation is to memorize the formulas if it becomes necessary to do so. It should be evident which formulas are most important; it is not necessary to memorize every single line of mathematics in the book!

A common misconception is that an appropriate formula exists for every conceivable problem. In other words, for every problem there is a formula, and if you know enough formulas, you should have no difficulty. This simply is not the case. Half of being able to work a problem consists of being able to decide whether a given formula is applicable. This calls for a firm understanding of the theory behind the formulas. For many problems, there may be no specific formula readily available that is directly applicable. Other problems may require more than one formula or more than one application of the same formula.

Think with timewise accuracy

Physical events such as a ball bouncing or a large airliner taking off are easily visualized in terms of general mental images, but it takes a greater effort to visualize the events at the physically correct rate of speed. Our minds seem to be like movie projectors that can be speeded up or slowed down, and we do not automatically visualize action scenes at the correct rate of speed. The value of this may be arguable, but the ability to think with timewise accuracy lends an extra dimension of realism to a person's thought processes, and this would appear to be of benefit to a physicist. Consequently, the following is offered to illustrate some possibilities for mental games that might enliven an otherwise uneventful day.

It takes about 1 second for most people to say the word *hippopotamus*. This provides a convenient and readily available time reference because of the popularity of the *second* as a unit of time in physics. Now, let us pay attention to the time required for various events to happen. Suppose a ball is dropped from a basketball goal, the height of which is 10 feet. Can you estimate the time required for it to fall to the floor? The calculated value is 0.79 seconds. Can you visualize this event, paying attention to the time element?

Extend this thought process to a 10-story building, which we assume to be 120-feet tall. How many seconds does it take for a brick to fall from a height of 120 feet, i.e., from the roof of the building to the ground? Can you make a good estimate based on your physical intuition? The calculated answer is about 2.83 seconds.

A reasonable altitude from which a parachutist might jump is 3000 feet above the surface of the earth. This is high enough to make it all seem worthwhile, yet it is not so high that undue concern has to be given to winds and other factors. Now suppose an unlucky parachutist has a bad day—a really bad day—and the chute does not open. How long will it take to fall from 3000 feet?

Air resistance complicates this problem somewhat, but a good estimate (assuming a terminal velocity of 120 miles per hour) is that it will take the unfortunate being about 20 seconds to fall to the ground.

To make this a bit more meaningful, look at a watch for 20 seconds while imagining that *you* are the jumper! You jump. Five seconds later, you know you have a problem. After five more seconds, the situation does not seem to be improving very much. The ground does not seem so far away now. The wind is blowing. The ground is getting closer. Your life begins to pass before your eyes . . . all this in 20 seconds!

Quantify common experience

As your study of physics progresses, make it a habit to associate numbers and quantities with everyday experiences. Your ability to do this may be somewhat limited at first, depending upon your background, but the ability will come quickly if you pay attention to it. Chapter 2 is devoted to units of measurement, but a major thrust is toward the increased awareness and appreciation of numbers. All this will greatly enhance your "feel" for the subject and it will be of great benefit because you will be able to tell if an answer looks right just by thinking about it. This goes hand in hand with the idea of visualization. The following paragraph illustrates how a fairly common experience can be *quantified*.

Suppose we are at an airport watching a helicopter start up in anticipation of taking off. The engine starts and the rotor slowly begins to turn. We watch, noting that we can say *hippopotamus* twice for each time the rotor goes around one time. (This gives a rate of rotation of the rotor of 0.5 revolutions per second.) Soon, as the rotor speeds up, we can say *hippopotamus* only one time for each revolution of the rotor (1.0 revolutions per second). As the downwash from the rotor begins to blow around us, we count two revolutions per *hippopotamus*. As the rotor speed increases to about 3 revolutions per second, we find that we can no longer keep our eyes focused on a single rotor blade. The roar of the engine increases, and when the rotor speed reaches about 5 revolutions per second, the helicopter lifts off.

Perform thought experiments

A *thought experiment* is a mental exercise; it is an experiment done completely within the mind. No apparatus is used, but mathematics may be used to aid the thinking process. A carefully specified set of circumstances or initial conditions is given, and the result of the experiment is determined simply by thinking about it.

However, the thinking process must be carefully guided by physical laws—one wrong assumption or one wrong thought and the entire experiment can go awry. The thought experiment is *not* to be an exercise in fantasy, and the end result is not supposed to be *science fiction.*

Thought experiments are most useful in situations where the actual physical experiment would be difficult or impossible to perform. For example, the idea of moving the earth with a lever sufficiently long might be termed a thought experiment.

Here is another experiment for thought. Suppose a hole were bored all the way through the center of the earth to the other side. Then suppose a bag were filled with mail, and tossed in. What will happen? The bag will obviously fall into the hole toward the center of the earth and toward the other side of the earth. It is easy to visualize the bag falling *down* into the hole, but what will happen when the bag gets to the exact center of the earth? Will it stop? Or, will it be traveling at such high speed that it will keep on going and eventually fall *up* and out of the hole on the other side of the earth?

Can you imagine yourself standing beside a hole in the ground, waiting for a bag of mail to arrive? And if mail can travel through the center of the earth, why would it not be possible to design a vehicle so that people could make the trip also? Is there something wrong with this line of reasoning, something more than simple technical problems? This is a thought experiment, illustrated in Fig. 1-3.

Predict the future!

Students often ask, "What is the benefit of studying physics?" This is a good question. Skipping over the obvious items such as the benefits of increased awareness and understanding and the fact that this course is required anyway, a reply often given is that with physics you can predict the future! And somehow, predicting the future always seems to be a worthwhile endeavor.

Speaking more precisely, if we carefully specify the conditions at the beginning of an experiment and if we know what physical laws are applicable, we can calculate or predict the outcome of the experiment ahead of time. The following examples illustrate this point.

In Fig. 1-4, a big box is resting on the floor; a rope and scale are attached. The question is: What will be the reading on the scale when the box begins to move?

If, ahead of time, we know the weight of the box and the coefficient of friction between the floor and the box, we can then calculate the force required to start the box moving. (We study the details of this in Chap. 4.) The scale should indicate this force just as the box begins to move across the floor. Now, is it not predicting the future when you can point to a mark on the scale and say that the box will move when the indicator gets there?

Obviously, such predictions as these cannot be made without the benefit of prior knowledge of the situation. First, the applicable physical principle must be known. Second, appropriate data to permit the required calculations must be available for the particular situation. To predict the point at which the box will

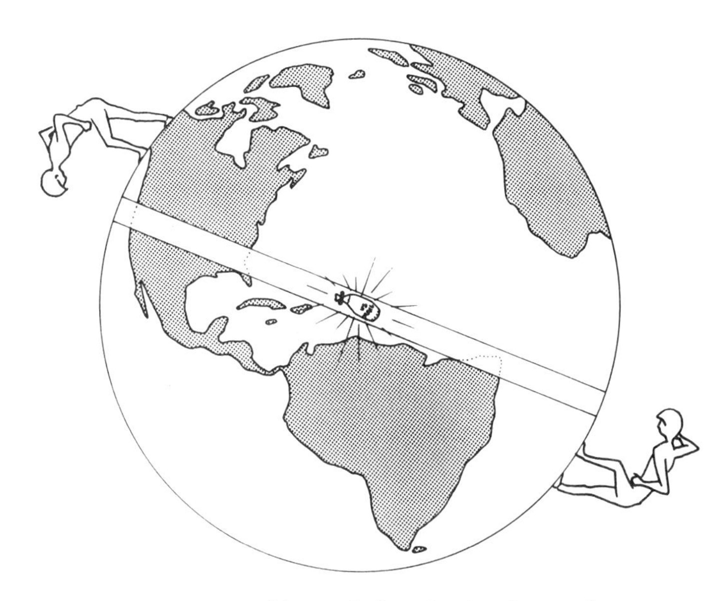

Figure 1-3 A possible conclusion of a thought experiment.

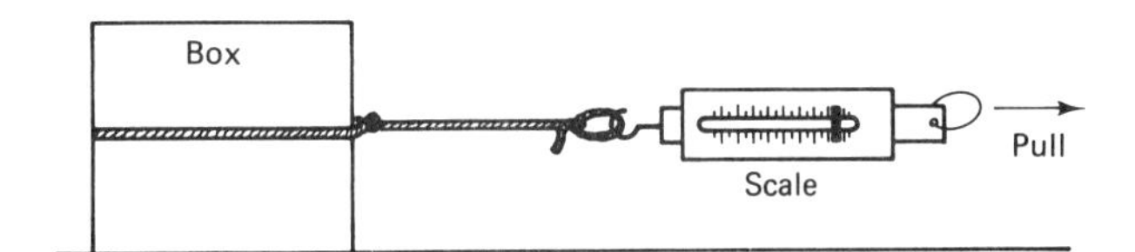

Figure 1-4 An illustration of predicting the future.

move requires the use and knowledge of the coefficient of friction between the box and the floor. Further, an experiment had to have been done previously in order to determine this coefficient. Consequently, we must admit that the future can be predicted only when both the physical principles and the required data are available.

The benefits of working problems

The benefits of working problems are often recited, but the subject always has been and continues to be a bit unpleasant—nobody likes to work problems. However, because it is almost a certainty that you will be asked to work a problem or two, the following benefits are enumerated.

Working problems forces a detailed study of the material. You can be under the impression that you understand it completely, only to find that you cannot work the problems without again consulting the text. Moreover, the text should be read and studied prior to attempting the problems. To do otherwise will probably result in a greater expenditure of time and a fragmented coverage of the material.

Working problems is good mental exercise and is extremely helpful in developing a feel for the subject. The problems almost always require visualization of a situation, and the quantitative nature of most problems forces an association of numbers with the subject of the problem. Further, there is but one thing that will cement an idea into your mind more securely than for you to struggle with the idea in working a problem. That one thing is to miss a problem on an exam for lack of the idea!

Working problems gives an immediate indication of the degree of your mastery of the material. If you can work the problems, things are probably going well. If you cannot work the problems, further study of the text is appropriate.

Working problems (assuming moderate success) gives you confidence in your ability to handle the new material. Further, there is a cumulative effect in manipulating ideas and mathematical formulas to solve problems. All the effort and experience contributes to your total maturity as a mathematician, physicist, and problem solver, where the term *maturity* is used to denote confidence, experience, and capability.

Steps in problem solving

Now that we have firmly established the benefits of working problems, let us briefly consider the various steps that may be recognized in the process of solving a problem. At this point, we must speak in generalities; details and special techniques for specific types of problems are given within each chapter where appropriate. Here are eight suggested steps, with brief comments:

1. Understand the applicable principles; pay attention to the conditions imposed on the principles. In other words, read the text very carefully before trying to work the problems.
2. Sketch the physical situation as an aid to visualization. A sketch also helps you to identify and organize the data.
3. Determine precisely what the problem asks. Most of the time, it will be one thing and one thing only, even though a problem with several parts may ask for "one thing only" several times.
4. Locate or identify the appropriate mathematics; i.e., find the formula. Often, algebraic manipulation may be required in order to solve for an unknown. This step may also demand clever application of the formulas and principles involved.
5. Check the units on all numerical quantities to ensure compatibility. Generally, all units should belong to the same system. All necessary conversions of units should be done systematically prior to doing the main calculation. (This is described in the next chapter.)
6. If you haven't already, write the applicable formula in the form in which it is to be used; then insert the numerical quantities and check the units. Do the computations and once again check the units.
7. Examine the result. Does it seem to be correct from an intuitive point of view? Check the answer if it is provided. Quickly review the entire problem and then reflect, weigh, and consider as deemed appropriate.
8. File the solution away for future reference. Be sure the problem is clearly identified by number, page, and so on.

These steps are designed to clarify the problem-solving process and ease the burden somewhat. However, it is likely that the problems will continue to be a laborious part of studying physics, as they have always been. Therefore, as a special favor, no problems are included at the end of this chapter.

Topics for Discussion/Questions

1. Do you believe a person can improve upon his or her ability to think by practicing? How does a person practice thinking?
2. How many apples can you clearly visualize at the same time? Does arranging the apples into patterns help you to visualize more apples? What pattern do you use for 5 apples? for 7? for 9? (The author finds that most people begin having difficulty seeing *all* the apples simultaneously with 6 or 7 apples. However, practice seems to help. The record, unverifiable of course, is 10 or 11 apples, and that person made an A in physics.)
3. What might a student do whose mind wanders while reading a chapter in the physics book? What can be done to improve one's ability to concentrate?
4. Discuss the benefits of studying physics, if any, to the following: an auto mechanic; an accountant; a mathematician; an airplane pilot; a cosmetologist; an English major; a musician; a psychologist. Are there any disadvantages to studying physics?

CHAPTER 2

UNITS AND CONVERSIONS

People just beginning to study physics often seem to have as much trouble manipulating units as dealing with the concepts of physics. This chapter provides a study of units and unit conversions before we settle into the mainstream of physics in the next chapter.

Three systems of units are introduced, the SI (MKS), the CGS, and the British, or foot-pound-second (fps), system. Additionally, many units of the household variety—teaspoons, bushels, acres—are included simply because they are useful and familiar.

Perhaps the most important idea of this chapter is that of a *unit bracket*. Unit brackets provide a systematic approach to unit conversions, such as from miles per hour to feet per second. Further, many commonsense problems can be solved just by manipulating unit brackets; several examples are given. This provides our initial introduction to problem solving and working with numbers.

Scientific calculators are capable of providing answers with eight or more digits, and the initial impulse might be to copy all available digits in order not to be wasteful of the calculator's accuracy. Therefore, a brief discussion of significant figures is included to serve as a guide as to how many digits should be retained. Also, the distinction between accuracy and precision is given.

Many problems are included at the end of the chapter to provide practice in manipulating units and to provide a feel for the subject. A conscientious effort should be made to master unit conversions at this point.

2-1 UNITS AND DIMENSIONS

What are units?

At the grocery store we buy sugar by the pound, eggs by the dozen, milk by the gallon, and cigarettes by the carton. These specifications—pounds, dozens, gallons, and cartons—are called units and are essential for specifying quantities of the materials. We say that the unit of measure of sugar is the pound, for example. Units are not unique; cigarettes are often bought by the pack rather than by the carton. A pack is simply a smaller unit than a carton.

In physics, great importance is placed upon the units associated with numbers. Not only must the number be correct, but it must be accompanied by the proper unit. This is reasonable because a number without the proper unit is meaningless. A man who is 6 *feet* tall will not attract much attention because of his height, but a Lilliputian man 6 *inches* tall is a great curiosity. Units are important.

Dimensional quantities

Three fundamental *dimensional quantities* of physics are *length, mass*, and *time*. A dimensional quantity is a quantity associated with a physical attribute such as length or mass, and which requires units as well as numbers for complete specification. Dimensionless quantities are pure numbers which do not require associated units.

A variety of units may exist for each dimension. *Time*, for example, is measured in such units as seconds, minutes, hours, and days; *mass* is measured in grams, kilograms, and slugs; and *length* is measured in inches, feet, meters, and miles.

A more complicated dimensional quantity is *area*, such as the surface area of the top of a professor's desk. The area of a rectangular object is obtained by multiplying the *length* by the *width*, both of which have the dimensions of length. Consequently, the dimension of area is *length squared*.

Square brackets, [ ], are used to denote the dimensions of the quantity enclosed. The dimension of volume V is length cubed, which may be written

$$[V] = [L^3]$$

More complicated physical quantities, such as velocity, may be written in terms of the fundamental dimensions. The dimension of velocity is that of a length (distance) divided by a time. This may be determined from the units of velocity—e.g., meters per second, feet per minute, or miles per hour. Therefore,

$$[\text{Velocity}] = \frac{[L]}{[T]}$$

The rate at which a housepainter can paint a surface may be expressed as square feet per hour—so much area per so much time. Hence, the dimensions of the rate of painting of surface are

$$[\text{Painting rate}_{\text{surface}}] = \frac{[\text{area}]}{[\text{time}]} = \frac{[L^2]}{[T]}$$

On the other hand, the painter may wish to express the rate of painting by giving the number of gallons of paint used per day. *Gallons* is a volume unit, so in terms of volume of paint consumed, the dimensions of the rate of painting are

$$[\text{Painting rate}_{\text{volume}}] = \frac{[\text{volume}]}{[\text{time}]} = \frac{[L^3]}{[T]}$$

A painter who paints picket fences provides a good illustration of how dimensional quantities appearing as fractions may cancel as if they were algebraic quantities. Suppose the painter wishes to express the paint requirements of a certain type of fence: e.g., 1 gal per 100 ft. Dimensionally, this is a ratio of a volume to a length:

$$[\text{Paint requirement}] = \frac{[\text{volume}]}{[\text{length}]} = \frac{[L^3]}{[L]} = [L^2]$$

Thus, the dimensions of "paint requirement per length of picket fence" turns out to be the dimension of an area. Do you suppose this has anything to do with the fact that the paint-consuming properties of a fence are related to its surface area?

Only quantities having the same dimensional form may be added; you cannot add peaches to pears nor dollars to doughnuts. This requirement of *dimensional consistency* within an equation is the basis of *dimensional analysis*, which can provide a check for the possible validity of a newly derived equation. Dimensional consistency by itself does not guarantee the validity of an equation, but any equation that is *not* dimensionally consistent cannot be valid.

Systems of units

Scientists have developed self-consistent *systems of units* so that no difficulties will arise as long as all units in a given problem are chosen from the same system. This is of great benefit, but unfortunately several systems have been used so that sometimes the systems themselves tend to complicate matters. Consequently, we must pay careful attention to the *systems*, as well as to the units themselves.

The basic units in any system relate to the fundamental dimensions, length, mass, and time. Units for other quantities, such as work and force, may then be expressed in terms of the basic units and may be given names in their own right, such as the joule and the newton.

As we mentioned, there are three systems with which we should be familiar: (1) the now-preferred Système International (SI), also known as the modern MKS system; (2) the older CGS system; and (3) the British system, which is no longer encouraged. The fundamental units for SI (modern MKS) are the meter, the kilogram, and the second. For the CGS system, the fundamental units are the centimeter, the gram, and the second. In the British system, the force unit (the pound) rather than the mass unit is taken as basic; the system is sometimes called the *fps system*. In that system the slug is the functional unit of mass even though it is not formally recognized. A mass of one slug weighs 32.17 lb on the surface of the earth.

In this text, we shall initially use all three systems in order to become familiar with them, but later on, SI units will be used almost exclusively.

Dimensionless numbers

Not all quantities in physics have dimensions and units; such *dimensionless quantities* are called *pure numbers*. All ratios of like quantities are pure numbers because the dimensions (and units) cancel. Consider the number π, which is the ratio of the circumference of a circle to its diameter. The circumference is measured in units of length, as is the diameter; when one is divided by the other, the units of length cancel to give a pure number.

2-2 A SYSTEMATIC METHOD FOR DOING UNIT CONVERSIONS

What are unit conversions?

Suppose we wish to calculate the time required for a car traveling at 55 mi/h to travel a distance of 100 ft. We are immediately confronted with *mixed units* of length, namely, miles and feet. We cannot solve the problem until this difference is reconciled. However, 55 mi/h is the equivalent of 80.67 ft/s, as determined by a *units conversion*, and a simple calculation gives 1.24 s as the required time. Most real physics or engineering problems will require unit conversions in the course of working the problems. Consequently, it is important to be able to do unit conversions quickly and accurately.

Obviously, we must have a base of information from which to work. In the above example, we needed to know that 1 mi is the equivalent of 5280 ft, and 1 h is the equivalent of 3600 s. Such information is contained in unit conversion charts or tables. Such a chart is included in the frontmatter of this book. The more common items should be committed to memory.

Unit brackets

We propose the name *unit bracket* (sometimes called *unit fraction*) for the following mathematical construction obtained by writing two different representations of the same physical measure as a fraction:

$$\left(\frac{5280\text{ ft}}{1\text{ mi}}\right)$$

The correspondence between inches and centimeters can be seen in this photo of two meter sticks.

The name is appropriate because the bracket contains units and because the quantity as a whole is the equivalent of unity and is dimensionless. We shall find unit brackets extremely useful in doing unit conversions. Other examples of unit brackets are shown below. In particular, note that the reciprocal of the unit bracket above is given and is perfectly valid in its own right.

$$\left(\frac{1\text{ mi}}{5280\text{ ft}}\right) \quad \left(\frac{12\text{ in}}{1\text{ ft}}\right) \quad \left(\frac{4\text{ qt}}{1\text{ gal}}\right) \quad \left(\frac{1\text{ year}}{365\text{ days}}\right)$$

You can make your own unit brackets from the information contained in unit conversion charts.

As a small sample of the usefulness of unit brackets, let us examine the effects of multiplying them together. Here is the first example:

$$\left(\frac{2\text{ pt}}{1\text{ qt}}\right)\left(\frac{4\text{ qt}}{1\text{ gal}}\right) = \left(\frac{2 \times 4\text{ pt}}{1 \times 1\text{ gal}}\right) = \frac{8\text{ pt}}{1\text{ gal}}$$

Observe that the numbers carry through as in simple fraction arithmetic, and the units (quarts in this case) cancel as if they were algebraic quantities. Did you know there are 8 pt in 1 gal?

The next example is quite similar.

$$\left(\frac{1\text{ ft}}{12\text{ in.}}\right)\left(\frac{1\text{ in.}}{2.54\text{ cm}}\right) = \frac{1\text{ ft}}{30.48\text{ cm}} = \frac{0.0328\text{ ft}}{1\text{ cm}}$$

Here we have that a 1-ft ruler is 30.48 cm long. The last, optional step was taken simply to illustrate that 1 could be obtained in the denominator by simply taking the reciprocal.

What would happen if we take a unit bracket and multiply it by itself?

$$\left(\frac{1\text{ ft}}{12\text{ in.}}\right)\left(\frac{1\text{ ft}}{12\text{ in.}}\right) = \left(\frac{1 \cdot 1\text{ ft} \cdot \text{ft}}{12 \cdot 12\text{ in.} \cdot \text{in.}}\right) = \frac{1\text{ ft}^2}{144\text{ in.}^2}$$

This is worthy of note—an area of 1 ft^2 is equivalent to an area of 144 in.^2! This may be extended to determine that a volume of 1 ft^3 is the equivalent of 1728 in.^3.

The simplest case

The simplest case of unit conversion is that in which only one unit is involved. For example, suppose we wish to convert 47 mi to kilometers. How do we do it?

Consulting a conversion table, we find that 1 mi = 1.609 km. This is the essential information. Then write the original quantity:

$$(47\text{ mi})$$

Next, form a unit bracket using the information obtained from the conversion table. There are two possibilities, depending upon whether the miles or the kilometers go on top in the unit bracket:

$$\left(\frac{1\text{ mi}}{1.609\text{ km}}\right) \quad \text{or} \quad \left(\frac{1.609\text{ km}}{1\text{ mi}}\right)$$

What we propose to do is to multiply the original quantity (47 mi) by a unit bracket so that the miles will cancel. Clearly, the rightmost unit bracket above is the one required. Form the product—

$$(47\ \cancel{\text{mi}})\left(\frac{1.609\ \text{km}}{1\ \cancel{\text{mi}}}\right) = 4.7 \cdot 1.609\ \text{km}$$

$$= 75.62\ \text{km}$$

The miles cancel, leaving kilometers, and multiplication of the numerical parts gives the desired result: 47 mi = 75.62 km.

Suppose we wish to determine the number of centimeters in 1 yd using the information that 1 yd = 36 in. and 1 in. = 2.54 cm. The first step is to write the starting quantity; then follow with unit brackets, as shown here:

$$(1\ \text{yd})\left(\frac{36\ \text{in.}}{1\ \text{yd}}\right)\left(\frac{2.54\ \text{cm}}{1\ \text{in.}}\right) = 91.44\ \text{cm}$$

Final result.

Converts inches to centimeters.

Converts yards to inches.

Starting quantity.

In order to get rid of a particular unit, that unit must appear in the denominator of the following unit bracket.

Units appearing as fractions

Units of velocity, for example, occur in fractional form because the dimensions are length divided by time. Here are typical velocities, complete with units:

$$\left(60\ \frac{\text{mi}}{\text{h}}\right) \quad \left(88\ \frac{\text{ft}}{\text{s}}\right) \quad \left(26.8\ \frac{\text{m}}{\text{s}}\right) \quad \left(10\ \frac{\text{cm}}{\text{s}}\right) \quad \left(1\ \frac{\text{mm}}{\text{s}}\right)$$

In doing unit conversions with fractional units, we may desire to change either the top unit, the bottom unit, or both. A unit bracket is required for each unit changed.

The cruising speed of a jet airliner is typically 600 mi/h. This velocity is hard for us to visualize, so let us convert the velocity to miles per minute. Here is the conversion:

$$\left(600\ \frac{\text{mi}}{\text{h}}\right)\left(\frac{1\ \text{h}}{60\ \text{min}}\right) = \left(\frac{600\ \text{mi}}{60\ \text{min}}\right) = 10\ \frac{\text{mi}}{\text{min}}$$

Here we wanted to get rid of the unit *hour* in the denominator of 600 mi/h. Consequently, hour had to appear in the numerator of the following unit bracket.

We often need to convert miles per hour to feet per second to work a problem. This conversion requires that both the miles and the hours be changed, as shown by the following:

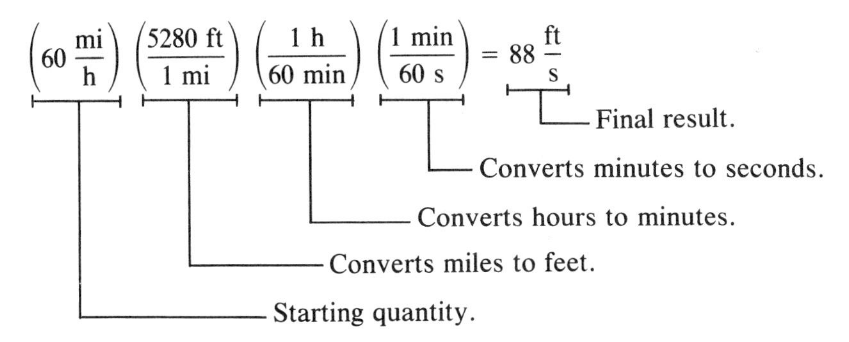

You should remember this result—60 mi/h equals 88 ft/s.

Units appearing as reciprocals

In order to save space in typesetting, a unit may be written as a reciprocal, such as min^{-1}, rather than placed in the denominator of a fraction. This should cause no confusion:

$$(360\ \text{ft}\cdot\text{min}^{-1}) \quad \text{means the same as} \quad \left(360\ \frac{\text{ft}}{\text{min}}\right)$$

However, care must be taken not to make an error in constructing the unit brackets by putting the reciprocated units in the denominator:

$$\underbrace{(360\ \text{ft}\cdot\text{min}^{-1})}_{\text{start}}\underbrace{\left(\frac{1\ \text{min}}{60\ \text{s}}\right)}_{\text{Note that min are on top.}} = \left(6\ \frac{\text{ft}}{\text{s}}\right) = (6\ \text{ft}\cdot\text{s}^{-1})$$

equivalent

The natural tendency while under the pressure of an exam might be to do it this way:

WRONG WAY!

$$(360\ \text{ft}\cdot\cancel{\text{min}}^{-1})\left(\frac{60\ \text{s}}{1\ \cancel{\text{min}}}\right) = 21{,}600\ \text{ft}\cdot\text{s}$$

This looks good—until you see the -1 atop the minutes!

Units appearing as a product

A physicist claims that *work* is done when a force exerted against a box succeeds in pushing the box across the floor in the direction of the force. The amount of work done depends upon the distance the box is pushed and upon the magnitude of the force required to push it. The formula for work is

$$\text{Work} = \text{force} \times \text{distance}$$

We now ask what the units are for work. This can be determined from the formula, provided we know what units are used for the force and the distance. If the force is measured in pounds and if the distance is measured in feet, the work done will be calculated by multiplying the pounds of force by feet of distance. Out of this comes pound · feet as the natural unit. It is commonly known as the foot-pound.

In the MKS system (SI), the unit of force is the newton and the unit of distance is the meter. Consequently, the MKS unit of work is the newton · meter. This unit is given a name in its own right, the joule. In the CGS system force is measured in dynes and distance in centimeters. The unit of work is the dyne · centimeter. One dyne · centimeter is called an erg.

Getting back to the subject of unit conversions, let us determine the relationship between a foot · pound and a newton · meter (a joule):

$$1 \text{ joule} = (1 \text{ N} \cdot \text{m}) \left(\frac{1 \text{ lb}}{4.448 \text{ N}}\right) \left(\frac{3.281 \text{ ft}}{1 \text{ m}}\right) = 0.7376 \text{ ft} \cdot \text{lb}$$

Thus, we see that the procedure is essentially the same as before.

Units appearing as powers

So far, all the unit conversions we have considered have dealt with the first power of the unit (neglecting reciprocals). We now examine the conversion of units of area and volume, examples of which are square feet (ft^2), square meters (m^2), cubic feet (ft^3), and cubic meters (m^3). Of course, units in higher powers appear in many places other than in conjunction with areas and volumes.

To begin, let us convert 1 ft^2 to square inches. This is tantamount to asking how many square inches it takes to make 1 ft^2. Note that (1 ft^2) is the same as (1 ft · ft). Consequently,

$$(1 \text{ ft}^2) = (1 \text{ ft} \cdot \text{ft}) \left(\frac{12 \text{ in.}}{1 \text{ ft}}\right) \left(\frac{12 \text{ in.}}{1 \text{ ft}}\right) = 144 \text{ in} \cdot \text{in.} = 144 \text{ in.}^2$$

Observe that in order to remove square feet, the unit bracket (in./ft) had to appear twice. This is the key to dealing with powers of units.

Visualize a million

A cubic meter (1 m^3) is a volume comparable in size to a physics professor's desk, but a cubic centimeter (1 cm^3) is even smaller than a die (one of a pair of dice). Sometimes it is necessary to know how many of the smaller cubes (1 cm^3) will fit inside the larger cube (1 m^3). We can determine this by doing a unit conversion. Note that the unit bracket (cm/m) is repeated three times to get rid of the unit bracket (m^3).

$$(1 \text{ m}^3) = (1 \text{ m} \cdot \text{m} \cdot \text{m}) \left(\frac{100 \text{ cm}}{1 \text{ m}}\right) \left(\frac{100 \text{ cm}}{1 \text{ m}}\right) \left(\frac{100 \text{ cm}}{1 \text{ m}}\right)$$

$$= 1{,}000{,}000 \text{ cm}^3 = \text{one million cubic centimeters}$$

This is an impressively large number of little cubes to fit into the bigger one!

Now, imagine that we have an abundant supply of the small cubes, and suppose we set out to build a stack of them in the form of a cube that is 1 m on a side (and whose volume is 1 m^3). The first layer will consist of 100 lines of cubes, and each line will contain 100 cubes. Thus, there will be 10,000 cubes in the first layer of the stack. Furthermore, a total of 100 layers are required, and you may verify that 100 times 10,000 is 1,000,000, as we obtained earlier. To what extent can you visualize this stack of small cubes? Can you visualize one complete layer of 10,000 cubes?

This is the best mental picture of a million that comes to mind, but it is still quite unimaginable. Try this. Suppose we put one little cube on the stack per second, every second. How long would it take to complete the first layer? The answer is 2 h 46 min 40 s. And if you worked diligently 8 h a day, 5 days a week, it would take just about 2 h 15 min short of 7 weeks to complete the stack.

Manipulating unit combinations

We end this section on unit conversions with two examples that illustrate how combinations of units can be manipulated as if they were one unit alone. This often results in a great saving of time and effort.

Earlier we established that a velocity of 60 mi/h is equivalent to 88 ft/s. Now suppose we wish to convert a velocity of 44 ft/s to miles per hour. It might go like this:

$$\left(44\ \frac{\text{ft}}{\text{s}}\right)\left(\frac{60\ \text{mi/h}}{88\ \text{ft/s}}\right) = 30\ \text{mi/h}$$

Note that the unit bracket is used exactly as before, with the exception that combinations of units now appear where individual units were before.

Suppose we know the density of mercury (Hg) is 13.6 g/cm^3, but suppose we need the density of mercury in pounds per cubic foot. To go through the basic conversion process would be long and tedious, requiring many steps. However, our conversion chart tells us that (1 g/cm^3) = (62.4 lb/ft^3). We use this to construct a unit bracket that simplifies the problem greatly. Here is the conversion done both ways:

$$\left(13.6\ \frac{\text{g}}{\text{cm}^3}\right)\left(\frac{62.43\ \text{lb/ft}^3}{1\ \text{g/cm}^3}\right) = 849.0\ \text{lb/ft}^3$$

Compare to:

$$\left(13.6\ \frac{\text{g}}{\text{cm}^3}\right)\left(\frac{1\ \text{lb}}{453.59\ \text{g}}\right)\left(30.48\ \frac{\text{cm}}{\text{ft}}\right)\left(30.48\ \frac{\text{cm}}{\text{ft}}\right)\left(30.48\ \frac{\text{cm}}{\text{ft}}\right) = 849.0\ \text{lb/ft}^3$$

The advantages are obvious.

Conversion factors

Consider the following conversion from feet per second to miles per hour, which should be old hat:

$$\underbrace{\left(100\ \frac{ft}{s}\right)}_{\text{start}}\underbrace{\left(60\ \frac{s}{min}\right)\left(60\ \frac{min}{h}\right)\left(\frac{1\ mi}{5280\ ft}\right)}_{\text{conversion factor}} = \underbrace{68.182\ \frac{mi}{h}}_{\text{final result}}$$

All the unit brackets may be multiplied together to obtain a *conversion factor*:

$$\text{Conversion factor} = \left(0.68182\ \frac{s \cdot mi}{h \cdot ft}\right) = \left(0.68182\ \frac{mi/h}{ft/s}\right)$$

The conversion factor is the number (including units) that is multiplied by the starting quantity to obtain the final result. The advantage of using conversion factors is that only one calculation is required for the conversion—a "quick and dirty" way to do the job. Disadvantages are (1) the conversion factors are less easily remembered than the more familiar information required for unit brackets, and (2) the question often arises about whether to multiply or divide when a number only (without the units) is remembered. In any event, whatever the method, unit conversions must be done with confidence.

2-3 MEASURED QUANTITIES AND SIGNIFICANT FIGURES

Measured vs. unmeasured quantities

Certain numbers encountered in science and engineering are exact and totally free from error. These numbers are either derived or defined, and therefore they are mathematically exact. For example, there are 12 eggs in a dozen; the number 12 is exact because it arises from the definition of a dozen. There are 144 $in.^2$ in 1 ft^2; the number 144 is exact because it is derived from the fact that there are (defined to be) 12 in. in 1 ft.

On the other hand, the number 2.997925×10^8 m/s—the speed of light—is not exact because it is a measured quantity. All measurements necessarily involve some degree of error, and great efforts are expended in the course of sophisticated experiments to reduce the error and to estimate the magnitude of the possible error.

Because all measured quantities involve some degree of error, two words—accuracy and precision—are frequently used in referring to the nature of the possible error. These may seem to have very similar meanings, but such is not the case. *Precision* refers to the capability of making fine distinctions in the reading of an instrument, such as a thermometer. For example, a thermometer with marks for each 0.1 degree is more precise than one with marks only for each whole degree. *Accuracy* refers to "freedom from error" in a measurement. The closer a given measurement is to the correct value, the more accurate it is said to be. The distinction between accuracy and precision is illustrated in Fig. 2-1.

Accuracy and precision do not necessarily go hand in hand. It is possible for an instrument to be very precise, but inaccurate at the same time. Suppose the water in a certain pot is known to be at a temperature of 76.824°F. A certain thermometer might measure it to be 76.53°F, which is wrong. Further, the

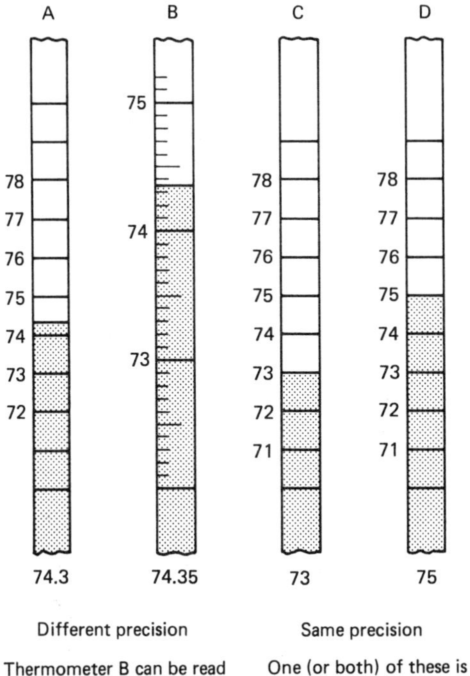

Figure 2-1 The distinction between accuracy and precision.

thermometer might indicate that the temperature is neither 76.54°F nor 76.52°F. In describing this thermometer, we might say that it is capable of a precision of 0.01°, but it is inaccurate because the reading is too low by about 0.29°. In less formal language, we say that accuracy is "how close we are to being right," while precision refers to the ability of an instrument to "split hairs."

Significant figures

Suppose you are asked to calculate the gasoline required for a car that gets 23 mi/gal to go a distance of 27 mi. The following calculation might occur:

$$\frac{27 \text{ mi}}{23 \text{ mi/gal}} = 1.17391304 \text{ gal}$$

You are apt to smile and be proud that your calculator gives such precise answers. But what is the significance of each digit in this answer that your calculator so dutifully provided? An illustration of the significance of the various digits is given in Fig. 2-2 to the fifth place to the right of the decimal. In the number above, the 4 in the eighth place to the right of the decimal represents such a small quantity of gasoline (40-billionths of a gallon) that a magnifying glass would be required in order to see it. In the realm of measuring gasoline, this truly is splitting hairs.

The number above is the result of dividing two 2-digit numbers representing measured quantities, 27 mi and 23 mi/gal. Because only two digits are given, we

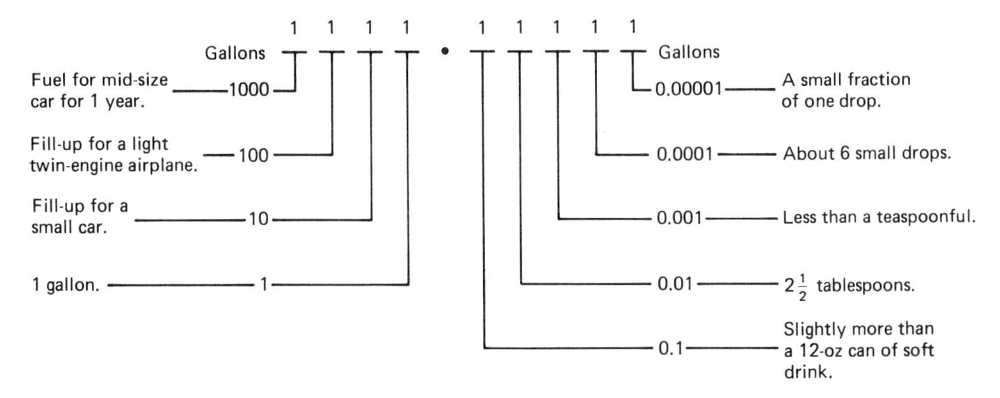

Figure 2-2 The significance of 1-digit changes.

conclude that the intended precision is not great. How precisely were the measures 27 mi and 23 mi/gal determined? We may assume the distance is closer to 27 mi than to either 26 or 28 mi, but we cannot say that the distance is definitely not 27.1 mi nor 26.9 mi. Nor can we assume the distance is 27.0 mi. In other

In addition and subtraction, do not include more than one column that contains a doubtful digit:

```
 6.234        3.47652
 4.1         -2.13
 5.426       -------
------        1.346
15.75
```

In multiplication and division, retain the same number of significant figures in the answer as there are in the factor having the least number of significant figures:

$2.3(4.694) = 11$

$2.31(4.694) = 10.8$

$2.311(4.694) = 10.84$

$\frac{57.952}{8.1} = 7.2$

$\frac{57.952}{8.13} = 7.13$

Zeros included to establish place value are not necessarily significant figures (SF):

93,000,000	2 SF
96,005	5 SF
186,000	3 SF
186,300	4 SF
0.12	2 SF
0.120	3 SF
0.012	2 SF

A bar, or tag, may be placed above a zero to indicate that it is significant when otherwise it would not be interpreted as significant:

$49{,}\bar{0}00$	3 SF
$49{,}0\bar{0}0$	4 SF

An example of rounding:

3.43655	→ 3.4366	426,532	→ 426,530
	→ 3.437		→ 426,500
	→ 3.44		→ 427,000
	→ 3.4		→ 430,000

Figure 2-3 Significant figures in arithmetic calculations.

words, we are dealing with a number that is accurate to two significant figures. Similar considerations apply to the value 23 mi/gal.

In arithmetic calculations, it is customary to retain only as many significant digits in the answer as are contained in the least precise of the numbers used in the calculation. Consequently, we write

$$\frac{27 \text{ mi}}{23 \text{ mi/gal}} = 1.2 \text{ gal}$$

where the answer, 1.2, is obtained by rounding 1.1739 to two significant digits. General guidelines for handling significant figures in arithmetic calculations are given in Fig. 2-3.

Rounding numbers

Numbers displayed on a calculator must be rounded so that only the proper number of significant digits are retained. This involves changing the nonsignificant digits to zeros. Here are the rules for rounding numbers.

1. If the first digit changed to zero is less than 5, the last digit retained is unchanged:

$$186{,}432 \rightarrow 186{,}000$$
$$3.743 \rightarrow 3.740 \rightarrow 3.74$$

2. If the first digit changed to zero is greater than 5, the last digit retained is increased by 1:

$$45{,}562 \rightarrow 45{,}600$$
$$4.1666 \rightarrow 4.1700 \rightarrow 4.17$$

3. If the first digit changed to zero is exactly 5, the last digit retained is increased by 1 unless all digits following the 5 are zero:

$$426{,}532 \rightarrow 427{,}000$$
$$6.3452 \rightarrow 6.3500 \rightarrow 6.35$$

When all digits following the 5 are zero, the last digit retained is left unchanged if it is even, but it is increased by one if it is odd:

$$426{,}500 \rightarrow 426{,}000$$
$$425{,}500 \rightarrow 426{,}000$$
$$5.36500 \rightarrow 5.36000 \rightarrow 5.36$$
$$5.35500 \rightarrow 5.36000 \rightarrow 5.36$$

As a concluding remark, we admit that the rules and suggestions given for handling significant figures are not always followed in the examples and exercises included in this text. More figures or digits are usually written than is merited by the data. This is done consciously so that the answers given in this text might agree more closely with what appears on your calculator; thus you can more easily judge the correctness of your methods.

2-4 EXAMPLES AND ILLUSTRATIONS OF UNIT CONVERSION

This section presents several illustrations of the practical use of unit conversions. The objective is to provide a bit of initial experience with numbers and units prior to putting numbers, units, and physical laws all together in the next chapter.

EXAMPLE 2-1 A certain water glass holds 1 c (liquid measure) of water. To compute the volume of the glass in cubic centimeters, we perform a unit conversion:

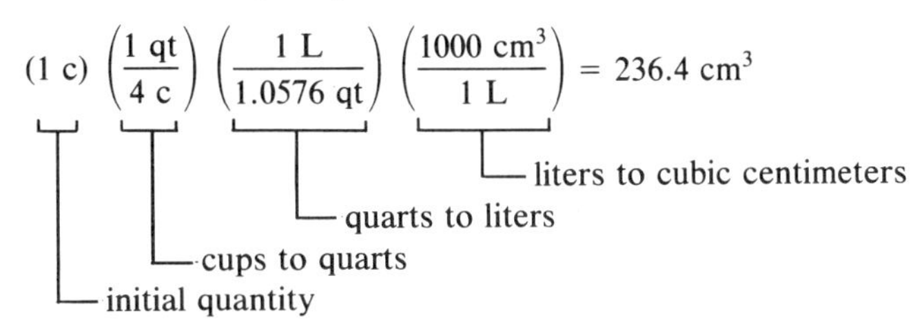

EXAMPLE 2-2 **WALKING SPEED**

A moderate rate of walking is 90 steps per minute with an average stride of 3 ft/step. Let us compute the time required to walk 1 mi at this rate. We write the following:

$$\left(\frac{90\ \text{steps}}{1\ \text{min}}\right)\left(\frac{3\ \text{ft}}{1\ \text{step}}\right)\left(\frac{1\ \text{mi}}{5280\ \text{ft}}\right) = \left(0.0511\ \frac{\text{mi}}{\text{min}}\right)$$

- initial information: $\left(\frac{90\ \text{steps}}{1\ \text{min}}\right)$
- steps in terms of distance: $\left(\frac{3\ \text{ft}}{1\ \text{step}}\right)$
- feet to miles: $\left(\frac{1\ \text{mi}}{5280\ \text{ft}}\right)$
- intermediate result: $\left(0.0511\ \frac{\text{mi}}{\text{min}}\right)$

The intermediate result is given appropriately in terms of miles and minutes, but the information conveyed is the number of miles the walker can walk in 1 min. We want just the opposite of this, namely, the number of minutes required to walk 1 mi. Therefore, take the reciprocal of the intermediate result obtained above:

$$\left(0.0511\ \frac{\text{mi}}{\text{min}}\right)^{-1} = 19.555\ \frac{\text{min}}{\text{mi}}$$

This is the desired result—it takes almost 20 min to walk a mile. To look at this from a different point of view, take the intermediate result and do an additional conversion to obtain the walking speed in miles per hour:

$$\left(0.0511\ \frac{\text{mi}}{\text{min}}\right)\left(60\ \frac{\text{min}}{\text{hr}}\right) = 3.06\ \text{mi/h}$$

EXAMPLE 2-3 **WEIGHT OF A CLOUD**

Suppose an afternoon thunderstorm produces a rainfall of 1 in. over an area of 1 mi^2. This is not a large storm, although a rainfall of 1 in. is moderate. Let us calculate the total volume and weight of the water that falls.

By squaring 5280, we find the area of 1 mi^2:

$$\text{Area of 1 mi}^2 = 2.788 \times 10^7\ \text{ft}^2$$

If this area is covered with water to a depth of 1 in. ($\frac{1}{12}$ ft), the volume of water in cubic feet will be:

$$\text{Volume of water} = 2.323 \times 10^6 \text{ ft}^3$$

Since 1 ft^3 of water weighs 62.4 lb and because 1 ton is 2000 lb,

$$(2.323 \times 10^6 \text{ ft}^3)\left(62.4 \frac{\text{lb}}{\text{ft}^3}\right)\left(\frac{1 \text{ ton}}{2000 \text{ lb}}\right) = 72{,}483 \text{ tons}$$

This is a lot of water, but an area of 1 mi^2 is fairly large also. To bring this closer to home, calculate the weight of water that falls on the roof of a house during a rainfall of 1 in. Assume the roof to have an area of 2000 ft^2, corresponding to a medium-sized home. The volume of water is 166.67 ft^3, and a unit conversion as above indicates that 5.2 tons of water falls upon the house. It is fortunate that all the water does not fall at one time.

Perhaps we should reflect upon this for a moment. It rains because the water in the cloud condenses to form drops, and the drops fall. All the water that is rained was initially contained in the white, billowy cloud. Here is a question: Because the cloud must have weighed *at least* 72,483 tons, what holds it up and how did it get up there in the first place?

EXAMPLE 2-4

THE DRIPPING SINK

It has been determined that 3.64 drips from the faucet of a kitchen sink is equivalent to 1 milliliter (mL) of water. Further, a fairly rapid and obnoxious drip occurs every 2 s, or 30 times per minute. Let us calculate the amount of water lost overnight—for example, in 8 h:

$$\left(30 \frac{\text{drips}}{\text{min}}\right)\left(\frac{60 \text{ min}}{1 \text{ h}}\right)\left(\frac{8 \text{ h}}{1 \text{ night}}\right)\left(\frac{1 \text{ mL}}{3.64 \text{ drips}}\right)\left(\frac{1 \text{ qt}}{945.54 \text{ mL}}\right) = 4.18 \frac{\text{qt}}{\text{night}}$$

This is just a bit over 1 gal.

EXAMPLE 2-5

GAS MILEAGE PER SPOONFUL

If a small compact car gets 35 mi/gal, how far can it go on 1 tbsp of gasoline?

First, convert a gallon to tablespoons:

$$(1 \text{ gal})\left(\frac{4 \text{ qt}}{1 \text{ gal}}\right)\left(\frac{32 \text{ oz}}{1 \text{ qt}}\right)\left(\frac{2 \text{ tbsp}}{1 \text{ oz}}\right) = 256 \text{ tbsp}$$

Use this to convert miles per gallon to feet per tablespoon:

$$\left(35 \frac{\text{mi}}{\text{gal}}\right)\left(\frac{1 \text{ gal}}{256 \text{ tbsp}}\right)\left(\frac{5280 \text{ ft}}{1 \text{ mi}}\right) = 721.88 \frac{\text{ft}}{\text{tbsp}}$$

Thus, 1 tbsp of gasoline will power the car for 721.88 ft.

Assuming the car to be traveling at 55 mi/h, calculate the number of seconds for which 1 tbsp of gasoline will provide the necessary fuel. Using information from above,

$$\left(\frac{1 \text{ tbsp}}{721.88 \text{ ft}}\right)\left(55 \frac{\text{mi}}{\text{h}}\right)\left(\frac{88 \text{ ft/s}}{60 \text{ mi/h}}\right) = 0.112 \frac{\text{tbsp}}{\text{s}}$$

Take the reciprocal of this to obtain the result in seconds per tablespoon:

$$\left(0.112\ \frac{\text{tbsp}}{\text{s}}\right)^{-1} = 8.949\ \frac{\text{s}}{\text{tbsp}}$$

This means that if you were spoon-feeding the engine, you would have to give it one spoonful about every 9 s.

EXAMPLE 2-6 **THE ATOMS OF A PENNY ARE MANY**

The mass of an ordinary penny is on the order of 3.1 g, and it contains about 2.9×10^{22} atoms. We assume all of them to be copper, even though a penny is actually an alloy of copper. The diameter of a copper atom is very nearly 2.6×10^{-8} cm (2.6 Å). In order to obtain a better appreciation for the large number of atoms in a penny, let us imagine the atoms to be taken from the penny and placed side by side in a line, in single file. How long will the line of atoms be?

By multiplying the atomic diameter by the total number of atoms, the length of the line is found to be

$$\left(2.6 \times 10^{-8}\ \frac{\text{cm}}{\text{atom}}\right)(2.9 \times 10^{22}\ \text{atoms}) = 7.54 \times 10^{14}\ \text{cm}$$

$$= 7.54 \times 10^{12}\ \text{m}$$

This is an extremely long line. Converting it to miles gives

$$(7.54 \times 10^{12}\ \text{m})\left(\frac{3.28\ \text{ft}}{1\ \text{m}}\right)\left(\frac{1\ \text{mi}}{5280\ \text{ft}}\right) = 4.68 \times 10^{9}\ \text{mi}$$

This is almost 5 billion miles! In comparison, the average distance from the earth to the sun is 93 million miles. Dividing the length of the line by the earth-to-sun distance gives

$$\frac{4.68 \times 10^{9}\ \text{mi}}{93 \times 10^{6}\ \text{mi}} = 50.4$$

which means that the line of atoms would reach from the earth to the sun *and back* more than 25 times.

The speed of light in empty space is 3×10^{8} m/s. Dividing the length of the line by the velocity of light gives the time required for a ray of light to travel from one end of the line to the other:

$$\frac{7.54 \times 10^{12}\ \text{m}}{3 \times 10^{8}\ \text{m/s}} = (25{,}133\ \text{s})\left(\frac{1\ \text{h}}{3600\ \text{s}}\right) = 6.98\ \text{h}$$

Thus, almost 7 h would be required for the ray of light to go from one end of the line of atoms to the other.

Questions

1. What is the difference between a unit and a dimension?
2. Name the three systems of units commonly encountered. Which is the preferred system?
3. What are the disadvantages, if any, of the English (fps) system of units?
4. Discuss some of the difficulties and things that must be considered in changing units in

the United States from the English (fps) to the metric system. What is to be gained by switching to metric?

5. What is a conversion factor? What are the advantages and disadvantages of using conversion factors as opposed to unit brackets?

6. Briefly explain the difference between accuracy and precision.

7. After doing a calculation, what should be kept in mind as the result is copied from the calculator?

8. Give several examples of measured and unmeasured quantities.

9. Give a brief summary of the rules for rounding numbers.

Problems

1. **(a)** What is the length of a football field (100 yd) in meters?
(b) One hundred meters is how many yards?

2. **(a)** How many centimeters are in 1 yd?
(b) What length in inches equals 10 cm?

3. **(a)** One stick of butter weighs ¼ lb. How many newtons is this?
(b) What fraction (approximately) of one stick of butter weighs 1 N?

4. Express the weight of a 90-lb weakling in newtons.

5. A "fifth" is ⅕ gal. How many jiggers are contained in a fifth?

6. How many fluid ounces are in 1 gal?

7. The gasoline tank of a certain car requires 14 gal for a typical fill-up. How many liters is this?

8. The radius of the sun is 696,000 km. The radius of the earth is 6370 km.
(a) How many times larger is the radius of the sun than the radius of the earth?
(b) How many times greater is the volume of the sun than the volume of the earth?

9. Suppose the gasoline mileage rating of a car is given as 12 km/L. Convert this to miles per gallon.

10. Obtain a conversion factor for converting the price of gasoline in dollars per gallon to cents per cup.

11. Obtain a conversion factor for converting miles per gallon to feet per teaspoon.

12. The typical engine revolutions per minute (rpm) of a small plane in level flight is 2400 rpm. At this rate, what period of time is required for the engine to turn through **(a)** 1 million revolutions; **(b)** 1 billion revolutions?

13. Suppose a person lives the Biblically allotted 3 score and 10 years upon this earth (70 years). Suppose further that the average heartbeat rate is 80 beats/min. Calculate the number of heartbeats that occur **(a)** per hour; **(b)** per day; **(c)** per year; **(d)** per 70 years.
(e) By substracting your age from 70 years, compute the number of heartbeats you have left.

14. Calculate the annual wage of a person who is paid 1¢/s. Assume the person works 8 hours a day, 5 days a week, and 50 weeks per year.

***15.** Calculate the weight of a water bed whose water compartment is 6.5 ft wide, 7.0 ft long, and 10 in. deep.

***16.** A cylindrical tank 5 ft in diameter and 7 ft long is installed on the back of a truck. A pump installed on the truck pumps water to and from the tank at the rate of 20 gal/min.
(a) Calculate the volume of the tank in cubic feet.
(b) Calculate the volume of the tank in gallons.
(c) Compute the time required to fill the tank with water using the pump on the truck.
(d) What is the weight of the water required to fill the tank?

***17.** A woman watering her lawn turns and squirts the entire stream of water from the

*Throughout this text, the asterisk will be used to denote problems of somewhat greater difficulty.

hose into a 2.5-gal bucket that happens to be nearby. The bucket fills with water in 1 min. The lawn is small, 20×30 ft.

(a) If the woman sprinkles the water uniformly over the lawn, how long will it take to obtain the equivalent of 1 in. of rain?

(b) How many gallons of water are required?

18. Determine the surface area of a penny, considering that the penny is a short cylinder 19 mm in diameter and 1 mm long. Express your answer in square meters.

***19.** Suppose the atoms of a penny were arranged in a square layer one atom thick and with adjacent atoms touching. Assume there are 2.9×10^{22} atoms (assumed to be spherical) in a penny, and assume the diameter of each atom is 2.6×10^{-8} cm.

(a) Calculate the length of one side of the square in meters and in feet.

(b) Calculate the surface area (top only) of the square in square meters.

***20.** The line of atoms formed by arranging the atoms of a penny into a single line may be considered to form a cylinder 2.6×10^{-10} m in diameter and 7.54×10^{12} m long. Calculate the surface area of the cylinder and compare with the result of Problems 18 and 19 above.

21. The torque rating of electric motors may be expressed in foot-pounds or, in the case of very small motors, in inch-ounces. How many inch-ounces is the equivalent of a torque of 1 ft-lb? (An ounce of force is $\frac{1}{16}$ of a pound.)

22. One ångström unit (Å) (often used to express the wavelength of light) is a length of 1×10^{-8} cm. One micron (μ) is a millionth of a meter, 1×10^{-6} m.

(a) What fraction of a meter is 1 Å?

(b) How many ångströms equal a length of 1 μ?

(c) One nanometer (nm), now the preferred unit in spectroscopy, is 1×10^{-9} m. How many ångströms equal 1 nm?

23. The yellow light of a sodium vapor lamp has a wavelength of about 5890 Å. Express this length in **(a)** centimeters; **(b)** nanometers.

24. In nuclear physics, a barn is a unit of area equal to 10^{-28} m^2, or 10^{-24} cm^2. Calculate the number of barns in an area of 1 Å^2.

CHAPTER 3

POSITION, VELOCITY, AND ACCELERATION

This chapter marks the official beginning of our study of physics. It deals with position, velocity, and acceleration, the objective being first to define these quantities in mathematical terms and then use the definitions to develop a mathematical description of uniformly accelerated motion.

Because objects in free fall accelerate uniformly until the effects of air resistance come into play, the equations for uniformly accelerated motion apply. With only slight modification, we obtain a set of equations for describing falling bodies. The independence of vertical and horizontal motions allows us to consider the motion of a projectile fired from a vertical and from a horizontal cannon.

Many physics professors prefer to introduce vectors before treating the topics of this chapter. Consequently, you may be asked to study the first four sections of Chapter 5 before proceeding. Such a study is not mandatory, however.

3-1 SPECIFYING POSITION

How do you tell someone where a certain house is located? If you think about this for a moment, you will realize that the position, or location, of an object is always given relative to another object or landmark, which serves as a reference. To emphasize this, just try to describe the position of some nearby object without referring to some other object that serves as the reference.

The physical universe is three-dimensional; three different specifications, or *coordinates*, are required in order to specify the position of an object. For example, the position of an airplane may be specified by giving its north-south coordinate (latitude), its east-west coordinate (longitude), and its altitude above the surface of the earth.

Certain objects, such as a boat on a lake, are naturally constrained to move in only *two* dimensions. The boat can move either north and south or east and west, but unless the boat leaves the surface of the lake (up or down), it is not necessary for us to specify the third coordinate, its altitude. The same idea holds

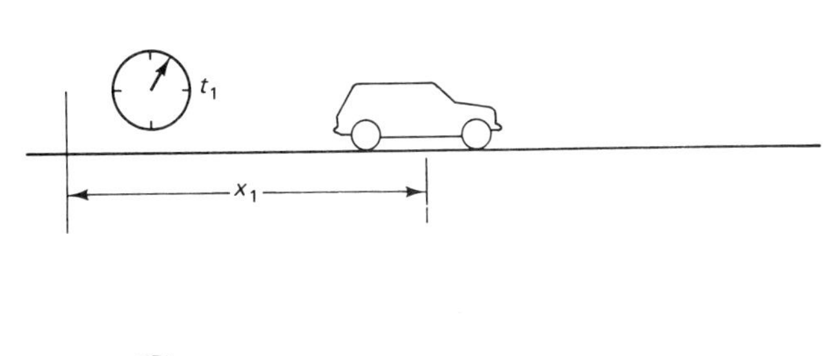

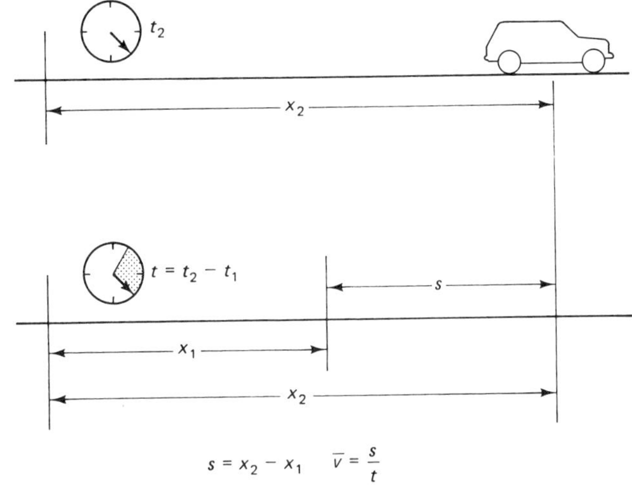

Figure 3-1 Position, time, and velocity.

true for cars, horses, bicycles, and other devices that move over the surface of the earth while remaining in contact with it. The position of these devices can be specified by giving only *two* coordinates.

Because a train must follow the track, the position of a train may be specified by giving only one coordinate. The train can move to the front or to the rear, but it cannot move from side to side or up and down. Thus, the position of the train is specified completely by giving its distance along the track from a known landmark. This assumes that the route followed by the track is known, but this point is not important to us here. Given a train and a track to put it on, we can specify the position of the train by giving only *one* coordinate.

In the following sections concerning velocity and acceleration, we consider motion in only one dimension so that the position can be specified using only one coordinate. This keeps the mathematics as simple as possible, but it requires that we consider objects that travel along a predetermined line. For example, we may consider the motion of a train down a track or the motion of a car being driven along a road. In all cases, however, the position of the object or vehicle will be specified by giving a single coordinate. This coordinate, denoted by x, is the distance to the object, measured from a *reference point*. Thus, *position* is specified by giving a number whose unit is length. This is illustrated and clarified in Fig. 3-1, where the meanings of x_1 and x_2 are made clear.

Figure 3-1 also shows that *time* is often associated with position when the position of an object is changing. Thus, x_1 and t_1 go together, as do x_2 and t_2.

Further, note that s is the distance between positions x_1 and x_2. The quantity s is sometimes called *displacement*, especially when a direction is involved, but we shall simply call it *distance* at this point. We denote the interval of time between t_1 and t_2 as t. Hence, t is the time required for a vehicle to travel the distance s from position x_1 to position x_2.

3-2 VELOCITY

Velocity is the rate of change of position with time. Obviously, velocity and speed are similar in meaning, but the two are not quite the same. The term *velocity* implies that the direction of motion is being considered, as well as the rapidity of the change in position. *Speed*, on the other hand, refers only to the rapidity of the motion without regard to the direction. We see in Chap. 5 that velocity is a *vector* quantity, whereas speed is a *scalar* quantity.

In this chapter and in many places throughout this text, the terms *speed* and *velocity* are used interchangeably. When the vector (directional) attributes of velocity are important, we shall refer to the velocity vector if there is any chance of confusion. Speed is never used to denote vector properties.

In mathematical terms, using the symbols of Fig. 3-1, *average velocity* is defined as

$$\bar{v} = \frac{x_2 - x_1}{t_2 - t_1} \tag{3-1}$$

This is average velocity, which is designated by placing a bar over the v, read "vee bar." Because $x_2 - x_1 = s$ and $t_2 - t_1 = t$, we may write

$$\bar{v} = \frac{s}{t}\,. \tag{3-2}$$

If a shoe salesperson leaves town A at 3:00 P.M. and arrives at town B, 45 mi away, at 4:00 P.M., the average speed of the shoe salesperson from 3 o'clock until 4 o'clock is 45 mi/h. However, this does not imply that the salesperson drove at exactly 45 mi/h for the whole trip. Half the distance could have been covered at 60 mi/h and the other half at 30 mi/h to give an average speed of 45 mi/h.

The speedometer of a car indicates *instantaneous* speed because the indicated speed is the speed at any instant. As the time interval t of Eq. (3-1) is made smaller and smaller, the average velocity and the instantaneous velocity become more nearly the same. In brief, average velocity refers to long trips made at varying speeds, while instantaneous velocity is the velocity at any instant.

EXAMPLE 3-1 A runner crosses the 30 yard line of a football field at the same time the sweep hand of a stopwatch passes the 14-s mark. The runner crosses the 80 yard line as the stopwatch passes the 20-s mark. Calculate the average speed of the runner.

Solution. First, we make a sketch to aid in visualizing the problem, as in Fig. 3-2. The applicable formula is

$$\bar{v} = \frac{x_2 - x_1}{t_2 - t_1}$$

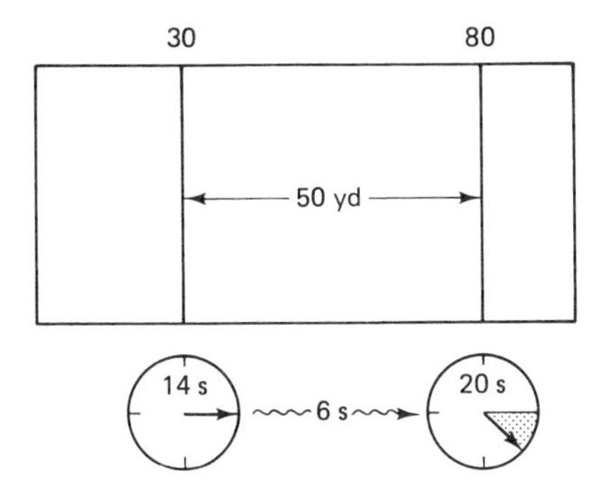

Figure 3-2

Hence,

$$\bar{v} = \frac{80 - 30 \text{ yd}}{20 - 14 \text{ s}}$$

$$\bar{v} = \frac{50 \text{ yd}}{6 \text{ s}} = 8.33 \text{ yd/s}$$

Converting this to feet per second gives

$$\left(8.33 \frac{\text{yd}}{\text{s}}\right)\left(\frac{3 \text{ ft}}{1 \text{ yd}}\right) = 25 \text{ ft/s}$$

which, as you may verify, corresponds to 17 mi/h.

3-3 ACCELERATION

In Sect. 3-2, velocity was defined to be the rate of change of position. We now consider the rate of change of velocity (or speed), which is *acceleration*. The most familiar example of acceleration is probably that of an automobile starting from rest. The velocity obviously changes as the vehicle gets under way, and the speedometer indicates this as the needle moves upward. It is clear that the velocity is changing as long as the speedometer needle is moving toward higher and higher speeds—upward past 10 mi/h, 12 mi/h, to 15 mi/h, and beyond. The car is *accelerating* because the velocity is increasing. If the speed were to increase to 30 mi/h and then remain *constant* at 30 mi/h, the car would no longer be accelerating.

Greater accelerations cause the speedometer needle to move up the scale more rapidly. Also, we observe that a greater effort seems to be required of the engine in order to produce larger accelerations. Going in the other direction, if the car is traveling at a high rate of speed—for example, 55 mi/h—and we jam on the brakes, the speedometer needle will begin moving downward as the car slows down. This may be called a *negative acceleration*, or a *deceleration*. The harder we apply the brakes, the greater the deceleration and the more rapidly the vehicle slows down.

Units

At this point, even before we give the mathematical definition of acceleration, let us see what units we might expect an acceleration to have. We consider the changing speedometer needle, and we use a unit of time of 1 s—the time to say *hippopotamus*. A moderate acceleration is 2 mi/h per second. That is, every time we say *hippopotamus*, the needle increases by 2 mi/h. The acceleration is 2 mi/h *per second*. Writing this in terms of units, we get

$$\text{Units of acceleration} = \frac{\text{mi/h}}{\text{s}} = \frac{\text{mi}}{\text{h} \cdot \text{s}}$$

Here we have both hours and seconds within the same expression, which would normally cause great alarm, but in this case we are interested only in the dimensions of acceleration:

$$[\text{Acceleration}] = \frac{[L]}{[T]\,[T]} = \frac{[L]}{[T^2]}$$

where [ ] stands for *the dimension of* the enclosed quantity. Thus, the dimensions of acceleration are that of a length divided by a time squared. Consequently, the units expected would be of the form ft/s^2. The units of acceleration are usually a source of mystery for beginning physicists, but a unit like ft/s^2, when written as

$$\frac{\text{ft/s}}{\text{s}}$$

becomes meaningful as we envision a speedometer moving upward at one division per second on a speedometer calibrated in feet per second.

Mathematical definition of acceleration

Having now been forewarned of the units to be expected, let us consider the mathematical definition of acceleration. It is the time rate-of-change of velocity. Refer to the speedometer of Fig. 3-3. Suppose the needle passes V_1 when a

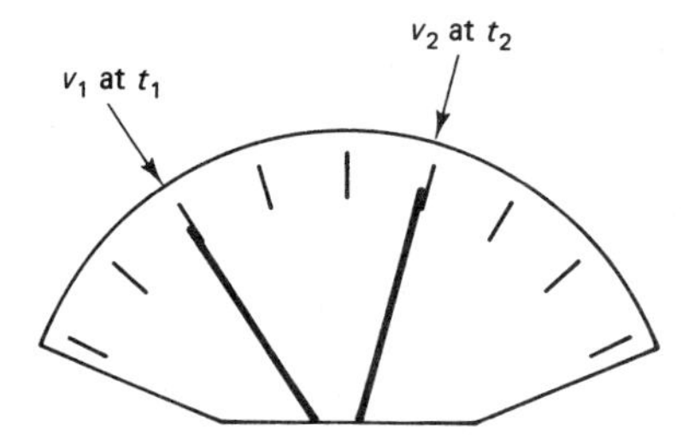

Figure 3-3 Use of a speedometer of an automobile in visualizing the meaning of acceleration.

stopwatch passes time t_1 and suppose it passes V_2 when the stopwatch passes time t_2. We can then write the time-rate of change of velocity in mathematical terms:

$$a = \frac{v_2 - v_1}{t_2 - t_1} \tag{3-3}$$

If, as before, we denote the time interval by t, where $t = t_2 - t_1$, we obtain

$$a = \frac{v_2 - v_1}{t}$$

This parallels the definition of average velocity given earlier, so that it might appear that this should be an *average* acceleration. Such is the case, but we omit the bar because all accelerations we consider are constant or *uniform accelerations* so that the average and instantaneous accelerations are equivalent.

A change in notation

At this point we propose a change in notation that tacitly assumes a slight change in the method of recording the time for the time intervals previously bounded by t_2 and t_1. Two formulas are involved:

$$\bar{v} = \frac{x_2 - x_1}{t_2 - t_1} \quad \text{and} \quad a = \frac{v_2 - v_1}{t_2 - t_1}$$

If we agree to set our stopwatch initially to zero so that $t_1 = 0$, the time interval $t = t_2 - t_1$ becomes simply $t = t_2$. It is then unnecessary to distinguish between t and t_2. We denote time simply as t, the number of seconds since the time interval of interest began.

The other factors we wish to modify are $x_2 - x_1$ and $v_2 - v_1$. The quantities x_1 and v_1 are the values of position and velocity at the beginning of the time interval. We designate these *original* values as x_0 and v_0. Continuing, we drop the subscripts from x_2 and v_2, which gives simply x and v. The resulting formulas are

$$\bar{v} = \frac{x - x_0}{t} \quad \text{and} \quad a = \frac{v - v_0}{t}$$

These formulas stem directly from the definition of velocity and acceleration. Algebraic rearrangement gives

$$x = x_0 + \bar{v}t \quad \text{and} \quad v = v_0 + at \tag{3-4}$$

In these equations, x and v are the values of position and velocity that occur at time t (seconds, perhaps) after the beginning of the time interval when the position and velocity were x_0 and v_0.

3-4 AVERAGE VELOCITY (UNIFORM ACCELERATION)

Suppose a car is traveling down a long stretch of road at 40 mi/h. As it passes a certain mailbox, the driver suddenly depresses the accelerator so that the car accelerates uniformly at 2 mi/h/s. This acceleration is maintained for 10 s, and at the end of the interval the car is traveling at 60 mi/h, as calculated by Eq. (3-4). The car then proceeds at 60 mi/h, with no acceleration. This situation is shown in Fig. 3-4. We need to obtain an equation for the average velocity of the car as it travels from the mailbox to the fire hydrant.

From the point of view of mathematical rigor, this problem is not trivial;

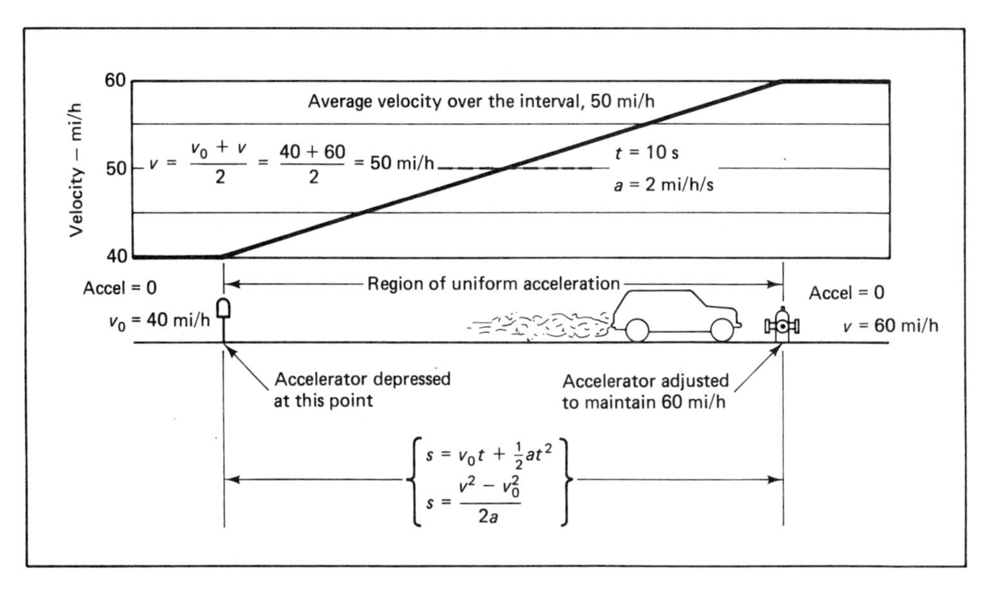

Figure 3-4 An example of uniformly accelerated motion.

calculus is required because the velocity is constantly changing. Therefore, we simply write the result and note that it appears intuitively obvious. Thus, the average velocity of the car while it is traveling over the region of acceleration is given by

$$\bar{v} = \frac{v_0 + v}{2} \tag{3-5}$$

This equation is *derived* from the earlier definition of average velocity. The equation is *not* a definition, and it applies only to the case of uniformly accelerated motion. We use this equation in the next section to derive other formulas.

3-5 DERIVATION OF TWO IMPORTANT FORMULAS

Thus far, we have obtained the following formulas pertaining to velocity and acceleration:

1. $\bar{v} = \dfrac{x - x_0}{t}$ definition of average velocity
2. $s = \bar{v}t$ distance in terms of $\bar{v}$ and t
3. $a = \dfrac{v - v_0}{t}$ definition of acceleration (3-6)
4. $v = v_0 + at$ velocity in terms of v_0 and a
5. $\bar{v} = \dfrac{v_0 + v}{2}$ derived equation for $\bar{v}$ for uniform acceleration

We have not yet inquired about how far a vehicle or other object accelerating uniformly for a given time will travel. In Fig. 3-4, this is equivalent to asking the distance from the mailbox to the fire hydrant.

In any event, the distance is given by the general equation

$$s = \bar{v}t$$

This equation is of benefit to us here only because we already have an applicable equation for the average velocity, namely,

$$\bar{v} = \frac{v_0 + v}{2}$$

By substituting this into the equation above, we obtain

$$s = \left(\frac{v_0 + v}{2}\right) t$$

We can eliminate the final velocity v by substituting $v = v_0 + at$, so that the resulting formula contains only the initial velocity, the acceleration, and the time. The result is the desired equation for the distance traveled during the period of acceleration:

$$(\textit{the } at^2 \textit{ equation}) \quad s = v_0 t + \tfrac{1}{2} at^2 \qquad (3\text{-}7)$$

Details of the algebra are given in Fig. 3-5.

The at^2 equation	
General equation.	$s = \bar{v}t$
Substitute $v = \dfrac{v + v_0}{2}$.	$s = \left(\dfrac{v + v_0}{2}\right) t$
Expand.	$s = \frac{1}{2} vt + \frac{1}{2} v_0 t$
Substitute $v = v_0 + at$.	$s = \frac{1}{2}(v_0 + at)t + \frac{1}{2} v_0 t$
Expand.	$s = \frac{1}{2} v_0 t + \frac{1}{2} at^2 + \frac{1}{2} v_0 t$
Collect terms.	$s = v_0 t + \frac{1}{2} at^2$

The $2as$ equation	
Previous result	$s = v_0 t + \frac{1}{2} at^2$
Eliminate t via $t = (v - v_0)/a$.	$s = v_0\left(\dfrac{v - v_0}{a}\right) + \frac{1}{2} a \left(\dfrac{v - v_0}{a}\right)\left(\dfrac{v - v_0}{a}\right)$
Multiply by a and expand.	$as = v_0(v - v_0) + \frac{1}{2}(v - v_0)(v - v_0)$ $as = v_0 v - v_0^2 + \frac{1}{2}(v^2 - 2v_0 v + v_0^2)$
Multiply by 2 and expand further.	$2as = 2v_0 v - 2v_0^2 + v^2 - 2v_0 v + v_0^2$
End result.	$2as = v^2 - v_0^2$

Figure 3-5 Mathematical details of two derivations.

One other equation that is very useful in problem solving can be obtained by eliminating t from the at^2 equation. This is done by substituting

$$t = \frac{v - v_0}{a} \quad \text{into} \quad s = v_0 t + \frac{1}{2} at^2$$

and simplifying to obtain a relationship between the velocities, acceleration, and distance traveled:

$$(\textit{the } 2as \textit{ equation}) \quad 2as = v^2 - v_0^2 \qquad (3\text{-}8)$$

Algebraic details are given in Fig. 3-5.

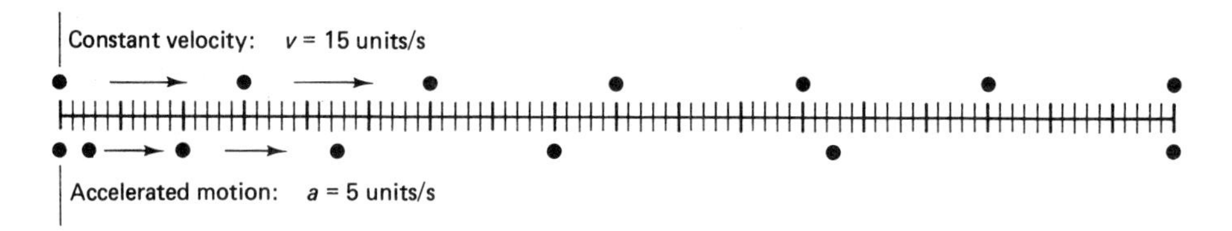

Figure 3-6 A comparison of accelerated and unaccelerated motion. The dots mark the positions of the objects at 1-s intervals.

We are now well supplied with equations for dealing with problems of uniformly accelerated motion. Several examples and illustrations are provided in the next section. Accelerated and unaccelerated motion are compared in Fig. 3-6.

3-6 ILLUSTRATIVE PROBLEMS (UNIFORMLY ACCELERATED MOTION)

At this point you may wish to review the general procedure for solving problems, which was given in the introduction.

EXAMPLE 3-2 An automobile starts from rest and accelerates to 30 mi/h in 8 s.
(a) What is its acceleration in ft/s^2?
(b) How far does the car travel during the period of acceleration?
(c) What is the average velocity during the 8-s period?

Solution. First, sketch the problem and indicate the items given and those that are to be calculated (see Fig. 3-7).

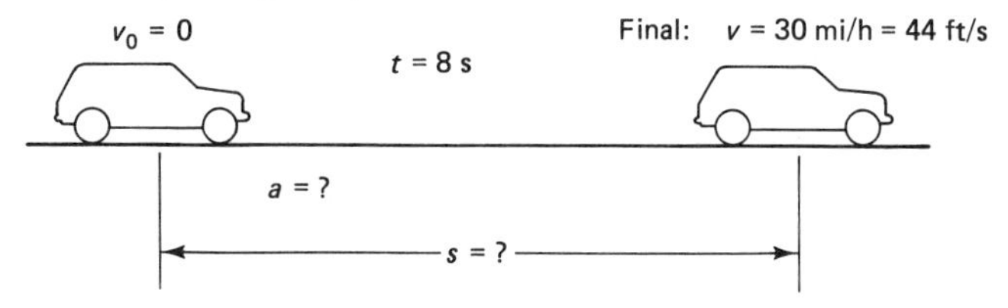

Figure 3-7

(a) Noting that the acceleration is requested in feet per second, we convert 30 mi/h to 44 ft/s (this conversion is not shown here). Place the result on the sketch. The starting velocity and the final velocity are given, along with the time required for the change in velocity to occur. Thus, the applicable formula is the definition of acceleration given in Eq. (3-6):

$$a = \frac{v - v_0}{t}$$

Substituting gives

$$a = \frac{44 - 0 \text{ ft/s}}{8 \text{ s}} = \frac{44 \text{ ft/s}}{8 \text{ s}}$$

$$a = 5.5 \text{ ft/s}^2$$

(b) Now that we have calculated the acceleration, we consider it to be one of the known quantities of the problem. And if we know the initial velocity, the acceleration, and the time, it is a simple matter to calculate the distance traveled during the period of acceleration. This calls for the at^2 formula, Eq. (3-7):

$$s = v_0 t + \frac{1}{2} at^2$$

Because $v_0 = 0$, this reduces to

$$s = \frac{1}{2} at^2$$

Substituting,

$$s = \frac{1}{2}\left(5.5 \frac{\text{ft}}{\text{s}^2}\right)(8 \text{ s})^2 = \frac{1}{2}\left(5.5 \frac{\text{ft}}{\text{s}^2}\right)(64 \text{ s}^2)$$

$$s = 176 \text{ ft}$$

(c) For objects starting from rest and accelerating with uniform velocity for a certain interval of time, the average velocity is simply one-half the maximum velocity attained. Consequently, the average velocity in this example is 22 ft/s, as can be verified by Part 2 or 5 of Eq. (3-6).

EXAMPLE 3-3 A car traveling at the fairly slow speed of 3 m/s begins to accelerate as it passes a certain tree. It accelerates for 5 s before passing a blue mailbox at a velocity of 10 m/s.
(a) What is the acceleration of the car?
(b) How far is it from the tree to the mailbox?
(c) Calculate the average velocity of the car while traveling from the tree to the mailbox.

Solution. First, we sketch the problem (Fig. 3-8) and collect the data. This sketch is very simple, but it serves the purpose and demonstrates that a sketch does not need to be complicated.

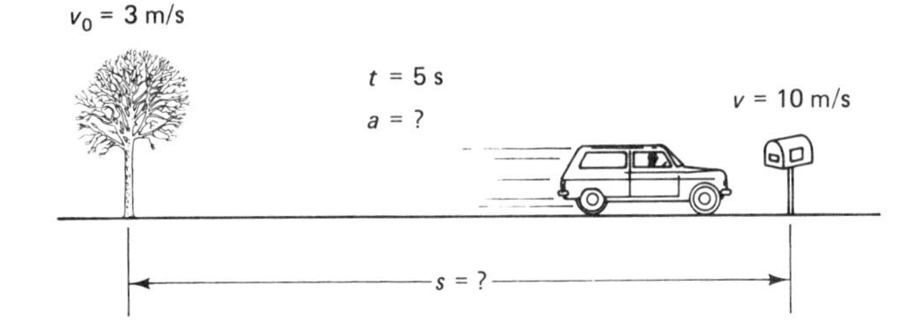

Figure 3-8

(a) By definition of acceleration, part 3 of Eq. (3-6):

$$a = \frac{v - v_0}{t} = \frac{10 - 3 \text{ m/s}}{5 \text{ s}} = 1.4 \text{ m/s}^2$$

(b) Use the at^2 formula (Eq. (3-7)) for relating s to v_0, a, and t:

$$s = v_0 t + \frac{1}{2} at^2$$

Substituting,

$$s = \left(3\ \frac{\text{m}}{\text{s}}\right)(5\ \text{s}) + \frac{1}{2}\left(1.4\ \frac{\text{m}}{\text{s}^2}\right)(5\ \text{s})^2$$

$$= 15\ \text{m} + 17.5\ \text{m}$$

$$= 32.5\ \text{m}$$

This result could also have been obtained with the $2as$ formula.

(c) For uniform acceleration, the average velocity is given by Eq. (3-5):

$$\bar{v} = \frac{v_0 + v}{2} = \frac{3 + 10\ \text{m/s}}{2} = 6.5\ \text{m/s}$$

Substituting the data and doing the arithmetic gives

$$\bar{v} = \frac{3 + 10\ \text{m/s}}{2} = 6.5\ \text{m/s}$$

EXAMPLE 3-4 Suppose a small trainer airplane accelerates uniformly from rest to a takeoff speed of 65 mi/h. The airplane runs 1100 feet down the runway before becoming airborn.

(a) Calculate the acceleration in feet per second per second.

(b) Calculate the time required for the takeoff run.

(c) What is the velocity of the plane 6 s after it begins to roll down the runway?

Solution. As usual, the sketch is made first (see Fig. 3-9). Note that the result of the units conversion (65 mi/h = 95.33 ft/s) is given on the sketch.

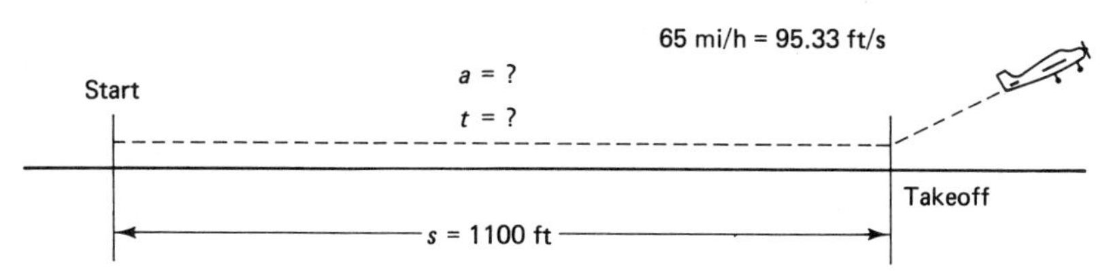

Figure 3-9

(a) We know the initial velocity ($v_0 = 0$), the distance traveled while accelerating uniformly, and the final velocity. We are looking for the acceleration. The $2as$ formula applies:

$$2as = v^2 - v_0^2 \xrightarrow[v_0 = 0]{} 2as = v^2 \rightarrow a = \frac{v^2}{2s}$$

Substituting,

$$a = \frac{(95.33\ \text{ft/s})^2}{2\ (1100\ \text{ft})} = \frac{9087.8\ \text{ft}^2/\text{s}^2}{2200\ \text{ft}}$$

$$a = 4.13\ \text{ft/s}^2$$

(b) The time required for the takeoff run may be found from part 4 of Eq. (3-6) by rearranging it to read

$$t = \frac{v - v_0}{a} \xrightarrow[v_0 = 0]{} t = \frac{v}{a}$$

Substituting,

$$t = \frac{95.33 \text{ ft/s}}{4.13 \text{ ft/s}^2} = 23.08 \text{ s}$$

(c) Use part 4 of Eq. (3-6) with v_0 equal to zero:

$$v = v_0 + at \xrightarrow[v_0 = 0]{} v = at$$

Substituting $t = 6$ s with a as defined above gives

$$v = \left(4.13 \frac{\text{ft}}{\text{s}^2}\right)(6 \text{ s}) = 24.78 \frac{\text{ft}}{\text{s}}, \quad \text{or} \quad 16.9 \text{ mi/h}$$

This example may be checked by using the at^2 formula with the values of t and a calculated above to determine that the calculated length of the takeoff roll is 1100 feet, as it should be.

EXAMPLE 3-5 The driver of a car traveling at 60 mi/h sees a small child run into the road 300 ft ahead of the point at which he jams on the brakes. If the car decelerates uniformly at 18 ft/s^2, will the car strike the child?

Solution. The problem involves a deceleration to a stop from a given initial velocity utilizing a known acceleration (see Fig. 3-10). We must calculate the distance to stop. No *time* information is provided, so we ignore the at^2 equation in favor of the $2as$ formula with $v = 0$:

$$2as = v^2 - v_0^2 \xrightarrow[v = 0]{} 2as = -v_0^2 \rightarrow s = -\frac{v_0^2}{2a}$$

Substituting, after converting units and remembering that the acceleration is negative because the car is coming to a stop,

$$s = \frac{(88 \text{ ft/s})^2}{2(-18 \text{ ft/s}^2)} = -\frac{7744 \text{ ft}^{\not{2}} \cdot \not{s}^2}{-36 \not{\text{ft}} \cdot \not{s}^2}$$

$$s = 215 \text{ ft} \qquad \text{stopping distance}$$

Thus, the car could stop before reaching the child.

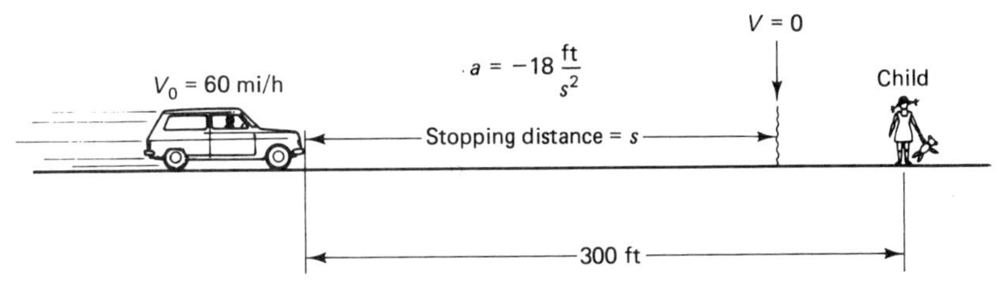

Figure 3-10

3-7 THE ACCELERATION OF GRAVITY AND FALLING BODIES

Acceleration of gravity

When a small, heavy object is released and allowed to fall toward the earth, it does not fall with a constant velocity. The velocity steadily increases with a uniform acceleration. Further, all objects accelerate downward with the *same* acceleration, as long as air resistance is negligible. The numerical value of the *acceleration of gravity* near the earth's surface is determined by three things: (1) the mass of the earth; (2) the radius of the earth; and (3) the fundamental properties of gravitational attraction. Other factors, such as the rotation of the earth, give rise to small secondary effects, but the most notable exception is that the acceleration of gravity does *not* depend upon the mass of the falling body. The standard symbol for the acceleration of gravity is g, and numerical values are

$$g = 32 \text{ ft/s}^2 = 9.8 \text{ m/s}^2 = 980 \text{ cm/s}^2$$

Falling-body formulas

Because falling bodies undergo uniformly accelerated motion, the formulas derived in Sect. 3-4 are applicable. It is customary to modify the notation somewhat, but the formulas are the same as before. For example, let us consider the at^2 formula and see how it is modified for falling bodies.

$$s = v_0 t + \tfrac{1}{2} at^2 \rightarrow h = v_0 t + \tfrac{1}{2} gt^2 \qquad (3\text{-}9)$$

Observe two changes: (1) s is changed to h, which might stand for *height*; and (2) the symbol g is substituted for a because g, the acceleration of gravity, is the applicable acceleration.

Usually, *upward* velocities and distances (or heights) above a reference level are taken as positive. Thus, if an object is calculated to have a positive velocity, it is assumed to be traveling upward. The acceleration of gravity, however, causes objects to fall *downward*, which is opposite to the direction for positive velocities. Therefore, in order for the formulas to work, the acceleration of gravity, g, must be inserted as a negative quantity. That is, we must attach a minus sign to the numerical value of g, giving -32 ft/s^2, for example. We must not forget to do this.

Two other formulas are modified as shown here:

$$v = v_0 + at \quad \rightarrow \quad v = v_0 + gt \qquad (3\text{-}10)$$

$$2as = v^2 - v_0^2 \quad \rightarrow \quad 2gh = v^2 - v_0^2 \qquad (3\text{-}11)$$

The formula for the average velocity is unchanged. Now that the basic formulas are available, we can investigate a few specific situations that are often encountered.

A body falling from rest

When we say a body is *at rest*, we mean that its velocity is zero. A body falling *from rest* is one that has been simply held up (somehow) and let fall without giving it an initial velocity either upward or downward. Because the initial velocity v_0 is then zero, all terms containing v_0 in the falling body formulas drop out. The result is:

[formulas for the special case of a body falling from rest ($v_0 = 0$)]

$$h = \tfrac{1}{2} gt^2 \qquad [3\text{-}12(a)]$$

$$v = gt \qquad [3\text{-}12(b)]$$

$$2gh = v^2 \qquad [3\text{-}12(c)]$$

Note that h is a negative quantity in these equations. By rearranging these equations a bit, we can obtain equations that answer two commonly asked questions:

1. How much time is required for an object to fall from a height h above the ground?

$$h = \frac{1}{2} gt^2 \quad \rightarrow \quad \frac{2\,h}{g} = t^2 \quad \rightarrow \quad t = \sqrt{\frac{2\,h}{g}} \qquad (3\text{-}13)$$

2. How fast will an object be traveling after falling a height h?

$$2gh = v^2 \quad \rightarrow \quad v = \sqrt{2gh} \qquad (3\text{-}14)$$

EXAMPLE 3-6

In high-diving competition, contestants frequently dive from a height of 80 feet or more.

(a) Assuming the diver rises a negligible distance above the board, calculate the time required for a diver to fall 80 ft.

(b) Calculate the speed with which the diver enters the water.

Solution. (a) The applicable formula is $t = \sqrt{2h/g}$. Taking h as a negative quantity and using $g = -32$ ft/s^2:

$$\text{Time} = t = \sqrt{\frac{2h}{g}} = \sqrt{\frac{2\,(-80\text{ ft})}{-32\text{ ft/s}^2}} = 2.24\text{ s}$$

(b) Using $v = \sqrt{2gh}$ gives

$$v = \sqrt{2\,(-32\text{ ft/s}^2)(-80\text{ ft})} = \sqrt{5120\text{ ft}^2/\text{s}^2} = 71.55\text{ ft/s}$$

Converting this to miles per hour gives 71.55 ft/s = 48.78 mi/h.

Note in regard to units, that both parts of this example required that a square root be taken of a combination of units:

$$\sqrt{\text{s}^2} = \text{s} \quad \text{and} \quad \sqrt{\text{ft}^2/\text{s}^2} = \text{ft/s}$$

EXAMPLE 3-7

An object falls from rest.

(a) How far (in feet) will it fall during the first second?

(b) How far will it fall during the second 1-s interval?

Solution. The formula for distance fallen from rest is $h = \tfrac{1}{2} gt^2$. If we take $g = -32$ ft/s^2, the distance fallen in t seconds is given simply by

$$h = 16t^2 \qquad (3\text{-}15)$$

Time of fall, *s*	Velocity ft/s; mi/h	Distance fallen, ft
0	0; 0	0
1	32; 21.8	16
2	64; 43.6	64
3	96; 65.4	144
4	128; 87.3	256
5	160; 109	400
6	192; 130	576
7	224; 153	784
8	256; 174	1024
9	288; 196	1296
10	320; 218	1600

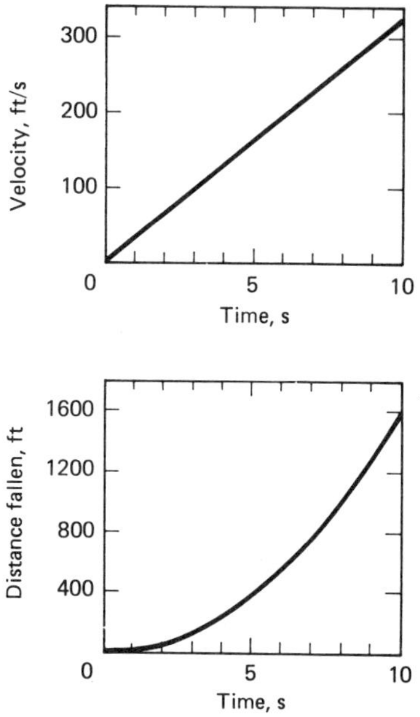

Figure 3-11 Distance fallen and velocity as a function of time. The acceleration of gravity is taken as 32 ft/s^2 and any effect of air resistance is not considered.

Consequently, after 1 s the object will have fallen 16 ft; after 2 s it will have fallen 64 ft, and so on. Thus, for part (a), the answer is 16 ft. For part (b), the answer is 64 ft minus 16 ft = 48 ft (see Fig. 3-11).

3-8 THE VERTICAL CANNON

Suppose a toy cannon shoots a marble straight up with an initial velocity v_0. The following are easily calculated: (1) the maximum height reached by the marble; (2) the time required for the marble to travel from the cannon to the highest point; (3) the time required for the marble to fall from the highest point back to the cannon; (4) the total time of flight of the marble; and (5) the velocity of the marble when it returns to the cannon.

This problem embodies several important ideas, as we shall see. This same treatment would be applicable to a real cannon (a rifle, or arrow shot from a bow) except that the resistance of the air must be taken into account when the bullets, cannon balls, or other objects travel at high velocities. For this reason, we consider a *toy* cannon, whose ball never travels at high velocities.

To find the *maximum height* reached by the marble, we consider the motion of the marble as it leaves the cannon and travels to the highest point. The initial condition is that of the marble traveling upward at a velocity v_0 just as it leaves the cannon. The final condition is that of the marble at the highest point, where its velocity, v, is zero. While the marble travels upward, its velocity steadily decreases because the acceleration of gravity is in the direction opposite to its initial velocity. Thus, our problem of finding the maximum height reduces to the familiar problem of finding the distance required for a decelerating body to come to a stop from an initial velocity v_0.

The applicable formula is the falling-body version of the $2as$ formula, further simplified by setting $v = 0$ because the motion stops at the highest point.

(maximum height reached by a projectile fired vertically upward with an initial velocity v_0)

$$2gh = v^2 - v_0^2 \xrightarrow[v=0]{} 2gh = v_0^2$$
$$h = \frac{-v_0^2}{2g} \tag{3-16}$$

To find the *time* required for the marble to reach the highest point, we must find the time required for a decelerating body to come to a stop ($v = 0$) from a given initial velocity. The applicable formula is

$$v = v_0 + gt$$

A bit of manipulation and setting $v = 0$ gives

(time required for a projectile fired vertically upward to reach its highest point)

$$t = \frac{-v_0}{g} \tag{3-17}$$

The problem of finding the time for the marble to fall from the highest point involves nothing new. This is exactly the same problem as that of finding the time required for a body to fall from a height h, starting from rest. The marble does indeed fall from rest because its velocity at the top of its path is zero. In an earlier section, we obtained:

(time required for a body initially at rest to fall from a height h)

$$t = \sqrt{\frac{2h}{g}} \tag{3-18}$$

Earlier in this section we obtained an expression for the maximum height obtained by the marble, namely,

$$h = \frac{-v_0^2}{2g}$$

When this height is substituted into the preceding equation, we obtain an alternative expression for the time of fall:

$$t = \sqrt{\frac{2h}{g}} = \sqrt{\frac{2\,(v_0^2/2g)}{g}} = \sqrt{\frac{v_0^2}{g^2}} = \frac{v_0}{g} \tag{3-19}$$

In comparing this with Eq. (3-16), we see that:

The time required for a projectile fired vertically with initial velocity v_0 to fall from its maximum height is exactly the same as the time required for it to reach its highest point, v_0/g. (This assumes the effects of air resistance to be negligible.)

Because equal times are required for the upward and downward motion, the total time of flight of the projectile is simply twice the time to fall from the maximum height:

(total time of flight of a projectile fired vertically upward with initial velocity v_0)

$$\text{Total time} = \sqrt{\frac{2h}{g}} = \frac{2v_0}{g} \tag{3-20}$$

h is maximum height

To find the velocity the marble will have when it returns to the cannon, we simply use information already available. The problem is that of finding the velocity attained by a body undergoing acceleration g for a given amount of time after starting from rest. The general formula is, after dropping the initial velocity term,

$$v = at$$

Here, the acceleration is $-g$ and the time [from Eq. (3-16)] is v_0/g. Consequently,

$$v = -g\left(\frac{v_0}{g}\right) = -v_0$$

Thus, we find that the marble returns *to* the cannon with the same velocity as when it was fired *from* the cannon. This idea enjoys considerable generality, and we may write:

Neglecting air resistance, any projectile fired at any angle from the surface of the earth (as from a cannon) will return to the earth with the same speed as that with which it was fired.

This result is far-reaching in its significance. To wit: What child of the country is there who, upon receiving a new bow and several arrows, has not soon thereafter aimed the device as nearly vertical as the eye will provide and fired the arrow with full force to see the height that it would attain?

To appreciate this, you must realize that the arrow almost always is lost from view as soon as it is released, and it will appear to have vanished entirely until it suddenly is heard to strike the ground nearby.

3-9 TWO COMPONENTS OF VELOCITY

If an object located in midair is not supported by a force at least equal to its weight, it will begin to fall to the ground. Furthermore, the motion toward the ground is not affected by any sideways, or horizontal, movement that may be imparted to the object. Consequently, objects fall at the same rate (with the same acceleration) irrespective of their horizontal motion. Moreover, the horizontal motion is not affected by any vertical motion that might occur, assuming that the objects in question have no aerodynamic qualities.

A 10-year-old girl discovers that while standing, she can drop a penny from eye level into a styrofoam cup placed on the floor near her toe. Further, she is able to do the trick on board a train while the train is standing stationary at the station. She also finds that she can also do the trick while the train is moving at high speed down the track. As long as the train is moving at constant speed in a straight line, the motion of the train does not influence the trick of dropping the penny into the cup; that is, the vertical component of the penny's motion is not affected by the horizontal motion of the train.

Most people have seen, either on television or at the movies, strings of bombs being dropped from high-altitude aircraft of the World War II era. The bombs are observed to remain under the aircraft as they fall, and it is not until the bombs have fallen a great distance that they begin to drift backwards and lag behind the plane. This provides a vivid illustration that the horizontal motion of the bomb is not affected by the fact that it has been dropped from the plane and is falling freely. After a while, of course, air resistance will reduce the horizontal velocity, causing the bombs to lag behind the aircraft. The vertical and horizontal components of the motion are independent.

3-10 A HORIZONTAL CANNON

In regard to a cannon fired horizontally, as shown in Fig. 3-12, we might wish to compute the range R and the time of flight of the bullet. These calculations are not difficult, because we realize that the vertical and the horizontal components of the bullet's motion are independent.

The muzzle velocity of the cannon is denoted by v_m, and the cannon is fired exactly horizontally. Thus, no vertical component of velocity is imparted to the bullet by the discharge of the powder; any vertical velocity is due to gravitational attraction. Further, the magnitude of the horizontal velocity is always v_m (ignoring air resistance).

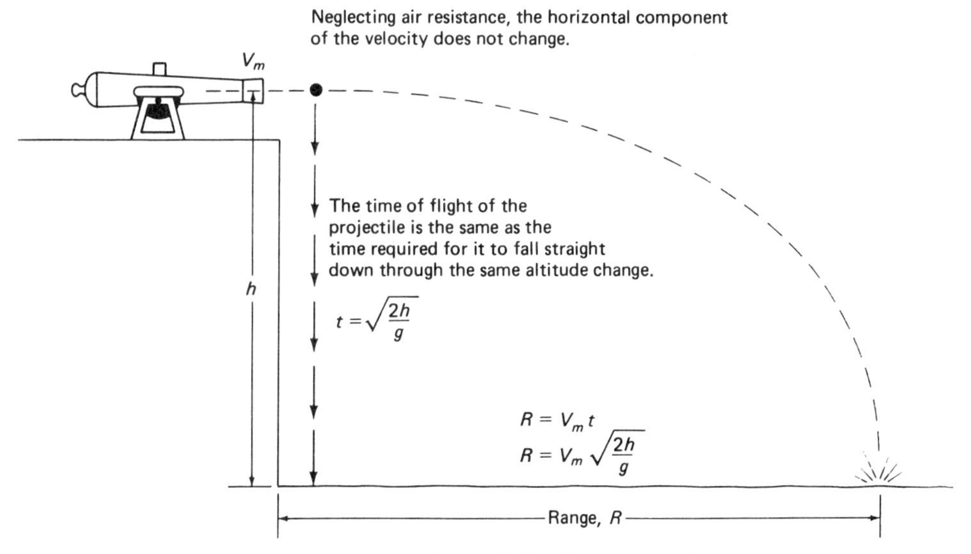

Figure 3-12 The horizontal cannon.

The time of flight of the bullet is exactly the same as the time required for a bullet to fall from the level of the cannon to the sea, starting from rest. This result was obtained earlier as Eq. (3-18), namely,

$$t = \sqrt{\frac{2h}{g}}$$

The range R is the horizontal distance from the cannon that the bullet strikes the sea. Because the horizontal component, v_m, remains constant, the range R is given by

$$R = v_m t = v_m \sqrt{\frac{2h}{g}} \tag{3-21}$$

In passing, we note that the total velocity of the bullet is not always v_m. (It is the horizontal component that is v_m.) In fact, the total velocity will be greater than v_m. In Chap. 5 we see that the velocity with which the bullet hits the water is given by

$$\text{Impact velocity} = \sqrt{2gh + v_m^2} \tag{3-22}$$

3-11 EXAMPLES AND ILLUSTRATIONS (FALLING-BODY PROBLEMS)

EXAMPLE 3-8

A toy cannon shoots a marble straight up with an initial velocity of 7.67 m/s.
(a) Calculate the maximum height to which the marble will rise above the cannon.
(b) What is the time required for the marble to reach its maximum height?
(c) What is the velocity of the marble 0.6 s after the cannon is fired?
(d) What is the velocity of the marble 0.9 s after the cannon is fired?

Solution. All parts of this problem utilize equations that are available in the text without modification.
(a) The maximum height is given by Eq. (3-16),

$$h = \frac{-v_0^2}{2g} \rightarrow h = \frac{-(7.67 \text{ m/s})^2}{2(-9.8 \text{ m/s}^2)} = 3.0 \text{ m}$$

We note that 3.0 m is about 10 ft.
(b) The time to reach the maximum height is given by Eq. (3-17),

$$t = \frac{-v_0}{g} \rightarrow t = \frac{-7.67 \text{ m/s}}{-9.8 \text{ m/s}^2} = 0.783 \text{ s}$$

(c) The velocity of the marble at any time t is given by Eq. (3-10). For $t = 0.6$ s,

$$v = v_0 + gt \rightarrow v = 7.67 \frac{\text{m}}{\text{s}} + \left(-9.8 \frac{\text{m}}{\text{s}^2}\right)(0.6 \text{ s}) = 1.79 \text{ m/s}$$

(d) The velocity of the marble after 0.9 s is

$$v = v_0 + gt \rightarrow v = 7.67 \frac{\text{m}}{\text{s}} + \left(-9.8 \frac{\text{m}}{\text{s}^2}\right)(0.9 \text{ s}) = -1.15 \text{ m/s}$$

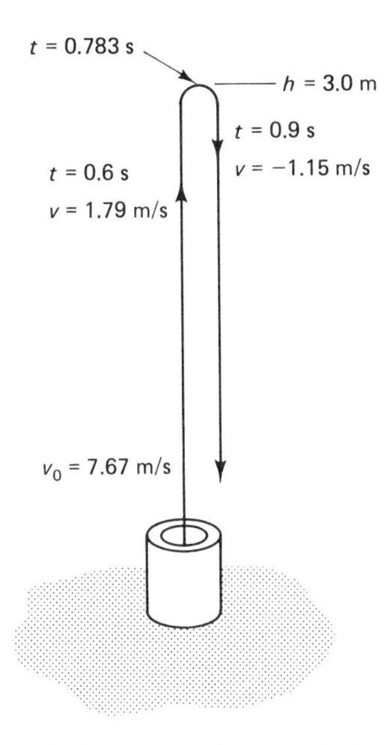

Figure 3-13

Note that at $t = 0.6$ s, the velocity calculated is positive, which implies upward motion; the marble has not yet reached the top. However, at $t = 0.9$ s the velocity is negative; the marble has reached the top and is falling back. This is illustrated in Fig. 3-13.

EXAMPLE 3-9 An archer shoots an arrow horizontally with an initial velocity of 156 m/s. If the arrow leaves the bow at a height of 1.5 m above the ground, how far from the archer will the arrow strike the ground (neglect air resistance)? See Fig. 3-14.

Solution. The time of flight equals the time for an arrow to fall 1.5 m. Using Eq. (3-18),

$$\text{Time of flight} = \sqrt{\frac{2h}{g}} = \sqrt{\frac{2(1.5\text{m})}{9.8\ \text{m/s}^2}} = 0.55\ \text{s}$$

Because the horizontal velocity is constant at 156 m/s, the distance traveled horizontally is

$$\begin{aligned}\text{Horizontal distance} &= \text{horizontal velocity} \times \text{time of flight}\\ &= 156\ \text{m/s} \times 0.55\ \text{s}\\ &= 85.8\ \text{m}\end{aligned}$$

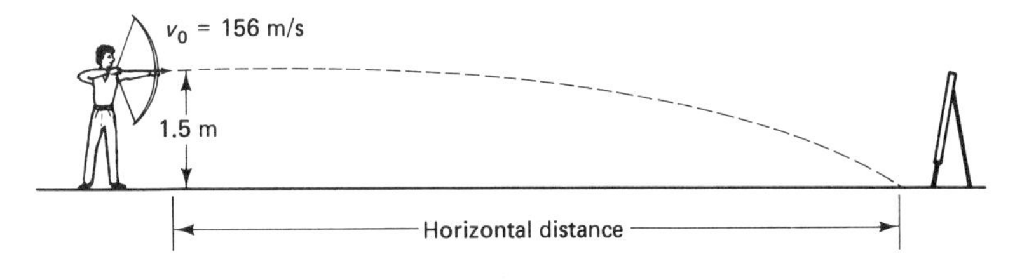

Figure 3-14

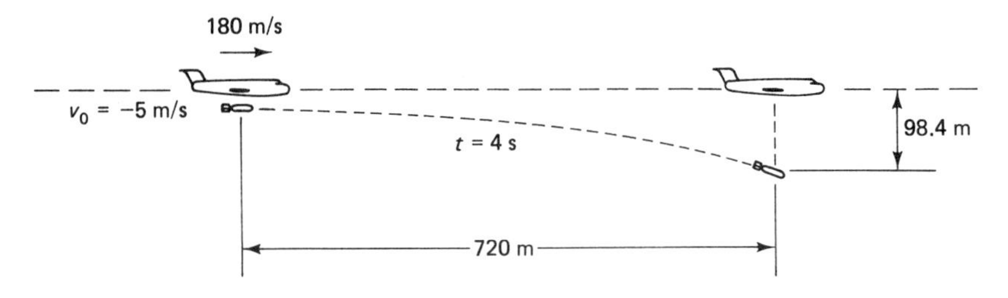

Figure 3-15

EXAMPLE 3-10 The bomb-release mechanism of an aircraft hurls the bomb downward with an initial velocity of 5 m/s in order to effect a clean separation between the bomb and the plane.

(a) At the end of 4 s, how far will the bomb have fallen?

(b) How far will it have traveled horizontally if the plane was flying level at 180 m/s when the bomb was released? See Fig. 3-15.

Solution. (a) Use the at^2 equation for falling bodies, Eq. (3-9). The initial velocity is downward and is therefore negative:

$$s = v_0 t + \frac{1}{2} g t^2 \rightarrow s = \left(-5 \frac{\text{m}}{\text{s}}\right)(4 \text{ s}) + \frac{1}{2}\left(-9.8 \frac{\text{m}}{\text{s}^2}\right)(4 \text{ s})^2$$

$$= -20 \text{ m} - 78.4 \text{ m}$$

$$= -98.4 \text{ m}$$

The negative distance indicates the bomb's position is below the plane, which seems proper. Note that if the bomb had fallen from rest, it would have fallen only 78.4 m in 4 s.

(b) The horizontal distance traversed in 4 s is the horizontal velocity multiplied by the elapsed time, $s = \bar{v}t$:

$$\text{Horizontal distance} = \left(180 \frac{\text{m}}{\text{s}}\right)(4 \text{ s})$$

$$= 720 \text{ m}$$

EXAMPLE 3-11 Most rambunctious children do not care to jump from objects higher than about 8 ft. From 8 ft, the impact velocity is 22.6 ft/s, and the time of flight is 0.71 s. On the moon, however, the acceleration of gravity is only about $\frac{1}{6}$ the earth value, i.e., $g_m = 5.33$ ft/s^2.

(a) Calculate the maximum height from which a rambunctious child might jump on the moon.

(b) Calculate the time of flight.

Solution. (a) We can calculate the height by turning the problem around and calculating the maximum altitude the child would achieve if fired vertically with an initial velocity of 22.6 ft/s. We can do this because of the symmetric nature of upward and downward motion. Consequently, by Eq. (3-16),

$$h = \frac{-v^2}{2g_m} = \frac{-(22.6 \text{ ft/s})^2}{2(-5.33 \text{ ft/s}^2)} = 47.9 \text{ ft} \quad (\approx 48 \text{ ft})$$

(b) The time of flight, by Eq. (3-18), is:

$$t = \frac{-v_0}{g_m} = \frac{-22.6 \text{ ft/s}}{-5.33 \text{ ft/s}^2} = 4.24 \text{ s}$$

Note that if rounding error is neglected, the maximum height and the time of flight is six times the earth value. This is because $g_m = \frac{1}{6} g_{\text{earth}}$.

Questions

1. What distinction is made between *velocity* and *speed*?
2. What are the dimensions and typical units of acceleration?
3. Can an object have a nonzero velocity when its acceleration is zero?
4. Can an object traveling upward be accelerating downward at the same time? Is this at all unusual?
5. Near the surface of the earth, how far will an object released from rest fall in 1 s?
6. Suppose a heavy object falls off a tall building. How much does its velocity increase per second?
7. Suppose an object is fired upward from a vertical cannon. When it reaches its highest point, **(a)** what is its speed; **(b)** what is its acceleration?
8. Can you suggest a method for demonstrating terminal velocity in the lab?
9. Suppose a bulldozer is driving down a road at 4 ft/s. Consider the track.
   (a) Relative to the ground, what is the speed of the track that is in contact with the ground?
   (b) What is the speed of the top track relative to the ground? relative to the driver?
10. An automobile is traveling along a highway at 55 mi/h. What is the speed, relative to the ground, of the points on the tires that are in contact with the pavement? What is the speed of the tops of the tires, relative to the pavement?

Problems

1. Compute the average velocity in miles per hour of a runner who can run a mile in 4 min.
2. Do a unit conversion to obtain a conversion factor for changing an acceleration given in mi/h/s to ft/s/s.
3. In a time interval of 8 s, a car speeds up, going from 30 mi/h to 50 mi/h. What is the average acceleration of the car in ft/s^2?
4. If a car originally traveling at 20 mi/h accelerates at 3 mi/h/s for 10 s, what is the final velocity of the car?
5. A car starts from rest and accelerates at 5 ft/s^2 for 8 s.
   (a) What is the final velocity of the car?
   (b) How far does the car travel during the acceleration period?
   (c) What is the average velocity of the car during the acceleration period?

* 6. The radius of the earth's orbit around the sun is about 93×10^6 mi, with one revolution being completed every 365 days. Calculate the average velocity of the earth as it progresses along the orbital path. (Answer in miles/second.)

* 7. The radius of the moon's orbit around the earth is about 240,000 mi, and the moon completes one orbit every 27.3 days (one sidereal month). Compute the velocity of the moon in its orbital path around the earth.

8. Traveling at 55 mi/h, how long would it take to drive 240,000 miles?
9. Radio waves travel at the speed of light, namely 186,000 mi/s. Calculate the time required for a radio signal to travel from the earth to the moon.

10. An astronaut is being rocketed aloft with an acceleration of $5g$.
  (a) What is the acceleration in miles per hour per second?
  (b) At this rate, how much will his velocity increase in 1 min?

11. In leaving a speed zone, a freight train accelerates uniformly from a speed of 25 mi/h to 45 mi/h at the rate of 0.4 ft/s^2.
  (a) How far does the train travel during the acceleration period?
  (b) How much time was required for the change in velocity to occur?

***12.** Suppose a small child runs into the road ahead of a car traveling at 55 mi/h. The driver jams on the brakes 0.6 s *after* seeing the child, and the car decelerates at 22 ft/s^2.
  (a) How far does the car travel in the time interval between seeing the child and applying the brakes?
  (b) What is the total distance required for the vehicle to come to a stop?

13. Repeat Problem 12 if the car is traveling at 70 mi/h and if, at the same time, the reaction time of the driver is slowed to 1.0 s.

14. An apple, falling freely, is falling at 50 ft/s. How fast will it be falling 1 s later?

15. A brick falls from the top of a building 122 ft high.
  (a) How fast will the brick be traveling as it hits the ground?
  (b) How much time is required for the brick to fall the first 61 ft?
  (c) How much time is required for the brick to fall the last (lower) 61 ft?

***16.** A bomb is hurled downward from an airplane with an initial vertical velocity of 3 m/s.
  (a) How far will the bomb fall during the first second of its flight?
  (b) How far below the plane will the bomb be after 10 s?
  (c) If the plane is traveling horizontally at 350 mi/h, how far will the bomb travel horizontally during the first 10 s of its flight?

17. A stone is dropped from rest from a bridge, and it strikes the water after 3.8 s. How high is the bridge above the river below?

18. A waterhose is pointed straight up. The stream of water rises to a height of 10 ft above the level of the nozzle before falling back. With what velocity does the water leave the nozzle?

19. In the preceding problem, if the nozzle is aimed horizontally while being held at a level 6 ft above the ground, how far from the point below the nozzle will the water strike the ground?

20. A cannonball is fired directly upward with an initial velocity of 55 mi/h.
  (a) Compute the time for the cannonball to reach its highest point.
  (b) What is the total time of flight?
  (c) What is the maximum height attained?
  (d) Ignoring air resistance, with what velocity will it strike the ground?

21. **(a)** Compute the height (in feet) from which an object must be dropped in order for it to strike the ground with a velocity of 30 mi/h.
  (b) Repeat **(a)** for a velocity of 60 mi/h.

***22.** **(a)** A hobo atop a train doing 12 mi/h jumps to the ground 9 ft below. With what velocity (total) does he strike the ground?
  (b) From what height could he jump from a stationary train and obtain the same impact velocity?

23. A horizontal cannon 20 ft above a level parking lot fires a cannonball which strikes the ground 150 ft from the point underneath the cannon.
  (a) Compute the time of flight of the cannonball.
  (b) Compute the muzzle velocity of the cannon.
  (c) With what velocity does the cannonball strike the ground?

***24.** Suppose a monkey wishes to drop a coconut onto the head of an evil tiger, who passes underneath the monkey's tree every day. The monkey plans to release the coconut from a height of 72 ft as the tiger crosses a mark on the trail a calculated distance from the point under the monkey. If the tiger always travels at 18 mi/h, how far from the impact point should the monkey make the mark on the trail?

CHAPTER 4

FORCE, MASS, AND NEWTON'S LAWS

In Chap. 3, we developed formulas to describe the motion of objects undergoing uniform acceleration, but we did not consider the cause of the acceleration. The primary objective of this chapter is to present the three laws of motion set forth by Sir Isaac Newton. In so doing we see that an object accelerates in direct proportion to the net force exerted upon it, with the mass of the object being a measure of its reluctance to be accelerated (inertia). Also, we establish the important distinction between mass and weight, and the basic concepts of friction and frictional forces are presented.

Historically, concepts of motion and its cause date back to Aristotle (384–322 B.C.). Aristotle's system holds that bodies are imbued with "natural" tendencies that cause an object to move toward its natural resting place, either toward or away from the earth. Further, any motion that is not natural is "forced," and the force must be applied continually for the motion to be maintained. Aristotle believed that objects always fall at constant speed and that heavier objects fall faster.

Galileo (1564–1642), Kepler (1571–1630), and Newton (1642–1727) demonstrated the invalidity of Aristotle's system through experiment, analysis, and mathematical description (the scientific method). Newton's laws of motion now form the basis of classical mechanics, but Newton stood heavily upon the shoulders of Galileo and Kepler. A famous experiment conducted by Galileo at (or at least near) the Leaning Tower of Pisa played an important role in understanding these basic concepts. That experiment demonstrated that heavy and light objects fall at the same rate (provided air resistance is negligible) and that Aristotle was wrong.

4-1 A FRICTIONLESS SURFACE

To push a book across a table, a force must be applied continually in order to maintain the motion. If you stop pushing, the book comes to a stop. Frictional forces developed between the book and the table are always directed opposite to the direction of motion. If you give the book a shove, it will proceed on its own for awhile, but it gradually slows down and finally stops as a result of the frictional forces.

One way to reduce the friction between the book and the table is to spread tiny beads over the surface of the table so that the book rolls on top of the beads. The effect of reducing the friction is to make the book easier to push. When given a shove, the book will go farther before coming to a stop. Also, for a given applied force, the book moves more readily, with greater acceleration. Reducing the friction seems to give the book greater freedom to move.

Suppose we could find some way to eliminate the friction entirely so that objects placed on the table would move without any frictional effects whatever. How would objects placed on the frictionless surface behave under the influence of an applied force? We shall see that motion does not take place spontaneously or with perfect ease even on a frictionless surface. Large forces are involved in starting, stopping, and in changing the direction of motion of the object placed on the surface. This is the topic of Newton's laws, which are described in later sections. However, we can make some observations about the motion of an object on a frictionless surface.

Once an object is set in motion on the frictionless surface, it will continue to move with uniform velocity until it reaches the edge of the surface. It does not slow down at all, even if the object is moving slower than a snail crawls. No motion, no matter how slow, will ever slow down and stop.

A small windup racecar placed in the center of the table would be hopelessly stuck. The absence of friction between the wheels and the table would cause the wheels to spin without budging the car. A walking Snoopy toy would not walk; his feet would just slide back and forth in the same place without giving him any forward motion. A live bug placed on the surface would be doomed because its feet, like Snoopy's, would slip and slide so that it could go nowhere.

There is no way to achieve a net horizontal motion by pushing against a horizontal, frictionless surface. On the other hand, external forces—such as a puff of air blown against the toy racecar—would start the racecar moving in the direction of the "wind." If the wind happened to blow against the side of the car, the car would move sideways in a skid. There would be no way for the car to recover from the skid because the steering of a car depends upon a frictional contact between the wheels and the road. The little car would skid out of control in whatever direction the wind might blow. And, of course, the brakes would be useless.

A person would find it impossible to walk on a frictionless surface enlarged to the size of a skating rink. The best ice skater would be equally frustrated because he or she could move only in a straight line from one side of the rink to the other. Recall that ice skates move very easily either forward or backward, but the blade cuts into the ice so that sideways motion of the skate is impossible. This gives a skater control. We could *not* "skate without skates" on a frictionless surface.

4-2 FORCE

Intuitively, a force is simply a push or a pull. More scientifically, however, we say that a force is any action on a material body that tends to change the velocity of the body. Because a change in velocity is related to acceleration, a force tends to cause a body to accelerate. Further, if a body is accelerating, we say with certainty that a force is acting on it.

A force has a directional characteristic. A force exerted on an object tends to cause the object to accelerate in the direction of the force. In the next chapter, we shall see that a force is a *vector quantity*.

Net force

Chances are that few objects you see in your immediate surroundings are accelerating, although each object is being acted on by at least two forces. The lack of acceleration in the presence of several forces is due to the fact that the forces acting on most objects balance out to give a net force of zero. Thus, we define *net force* to be the unbalanced component that results when many forces act on a given body. Consider a chandalier, for example. Gravity pulls *down*; the chain pulls *up* an equal amount. The net force is zero, and the acceleration is zero. But if the chain breaks so that the downward force of gravity is no longer canceled by the upward pull of the chain, the chandalier will accelerate downward impressively. Almost always, it is the unbalanced forces (the net force) that produce the most dramatic effects.

Remember the unfortunate person whose parachute failed to open and who fell 3000 feet in 20 s at a terminal velocity of 120 mi/h? What was the net force acting on the person after he or she reached terminal velocity? The answer: zero! If the acceleration is zero, the net force applied must be zero. After leaving the plane, the jumper's velocity increases (the person accelerates) until the air resistance (upward force) equals weight (downward force), so that the net force (and acceleration) is zero. Hence, the person's velocity no longer increases and he or she then falls at the constant terminal velocity.

4-3 MASS AND INERTIA: NEWTON'S FIRST LAW

An experiment

Suppose we place a 200-lb anvil in the center of a frictionless surface constructed as a circular platform about 20 ft in diameter, as shown in Fig. 4-1. A string capable of withstanding a 5-lb pull is attached to the anvil, and a scale at the other end of the string indicates the tension (force) in the string. Our objective is to determine how the anvil will respond to forces exerted on it by the string, i.e., the velocity or change in velocity a given force will produce. We observe the following:

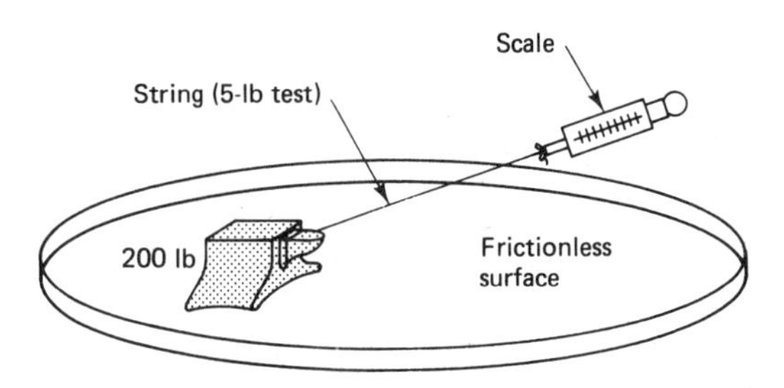

Figure 4-1 An anvil on a hypothetical frictionless surface.

1. When we pull on the string, we feel a resistance and see a reading on the scale. At the same time, the anvil begins to move, but not very rapidly; the motion does not stop when we stop pulling. The resistance to the pull and the slowness of the motion surprises us because we expected the anvil to zip along with only the slightest pull because of the frictionless surface.
2. The slightest pull exerted on the string causes the anvil to accelerate for as long as the pull is maintained. If the pull is very weak, the resulting acceleration is small and the resulting velocities are small. The acceleration is always in proportion to the applied force.
3. After pulling and after relieving the tension, the anvil keeps on moving at a constant velocity in a straight line. If the motion is allowed to continue, the anvil will eventually hit the sidewall of the frictionless surface.
4. If the anvil is moving, a force must be applied to the anvil in the direction opposite to its motion in order to bring it to a stop. We note that it is very difficult to achieve a velocity of zero, e.g., a stationary anvil.
5. When the anvil is in motion, it can be made to follow a curved path by pulling on it at an angle to its direction of movement. In the absence of a sideways pull, however, the anvil always moves in a straight line.
6. The anvil can be made to move rapidly by pulling on the string with greater force and by sustaining the force for longer periods of time. Greater forces produce greater accelerations, and accelerations maintained for longer periods result in greater velocities.
7. In an effort to move the anvil very quickly, a sharp jerk is given on the string. The string breaks, and the anvil moves slowly in the direction of the jerk.

These observations will help us understand the three laws of motion, set forth by Sir Isaac Newton more than 300 years ago. We now consider the first of these laws, which deals with inertia.

Inertia

Inertia is the tendency of a body to resist a change in its state of motion. A noticeable pull was required in order to get the anvil to give up its state of motion of zero velocity (point 1). Further, once it was moving, a force had to be exerted on it to cause it to stop (point 4), and a force was required just to get the anvil to change directions (point 5). These ideas are embodied in Newton's first law, the law of inertia, which may be stated as follows:

If no net force acts on a material body, the body will maintain its present state of motion: If it is stationary, it will remain stationary; if it is moving at a particular speed in a particular direction, it will maintain both the speed and the direction until acted upon by an unbalanced external force.

At the time this law was first put forth, the idea that motion could exist without a continuous driving force was a new concept that seemed to contradict everyday experience. When horses stopped pulling, the sleds and wagons behind them also stopped. We now understand that the purpose of the horse in such endeavors is to balance out the frictional forces so that a condition of balanced forces could prevail.

Mass—A measure of inertia

Physicists do not measure inertia directly, nor do they often speak of it, because the *mass* of a body is directly related to its inertia and is more readily observable. A common definition of mass is that it is a quantitative measure of the amount of inertia a body has. A more commonly held notion of mass is that it is the amount of *matter* contained in an object. But this idea runs into trouble when an object is accelerated to velocities approaching the speed of light. At such high velocities, the inertia increases as evidenced by the fact that objects become increasingly difficult to accelerate. Because mass is a measure of inertia, we then say that the mass increases at high velocities. It is hard to imagine that the amount of *matter* in a body could increase just because it is speeded up.

4-4 NEWTON'S SECOND LAW: $F = ma$

Newton's second law establishes the relationship between net force, mass, and acceleration. Thus, this law is one of the most important in all of physics. We state it as follows:

1. When a net, unbalanced force, F_{net}, acts upon an object of mass m, the object will experience an acceleration as given by $a = F_{net}/m$ in the direction of the force.

An alternative statement is the following:

2. If an object of mass m is undergoing an acceleration a, a net, unbalanced force, F_{net}—given by $F_{net} = ma$—must be acting on the body in the direction of the acceleration in order to produce the acceleration.

Mathematically, the law is very simple:

$$F_{net} = ma \tag{4-1}$$

From these statements we see that a given force produces accelerations that are inversely proportional to the mass; *larger* masses are accelerated *less* by the same net force. Furthermore, the force required to produce a given acceleration is directly proportional to the mass; *large* masses require *large* forces.

Units

Equation (4-1) is a simple equation mathematically, but many students have difficulty with it because of the units of the quantities involved: force, mass, and acceleration. The proper units for three systems are shown in Fig. 4-2.

System	Length	Mass	Time	Force	Acceleration
SI (MKS)	meter (m)	kilogram (kg)	second (s)	newton (N)	m/s^2
CGS	centimeter (cm)	gram (g)	second (s)	dyne	cm/s^2
British	foot (ft)	slug*	second (s)	pound (lb)	ft/s^2

*Instead of the slug, the pound is formally defined.

Figure 4-2 Three systems of units.

Special care must be taken to avoid confusing units of force and mass. A pound is a unit of *force*, not *mass*. A gram is a unit of *mass*, not *force*. Pounds must never be used as units of mass, and grams or kilograms must never be used as units of force in Eq. (4-1).

The following relationships stem from $F = ma$:

$$1 \text{ newton} = (1 \text{ kg})\ (1 \text{ m/s}^2);\ 1 \text{ dyne} = (1 \text{ g})\ (1 \text{ cm/s}^2);\ 1 \text{ pound} = (1 \text{ slug})\ (1 \text{ ft/s}^2)$$

From these, we obtain the following statements:

1. One newton is the force required to give a mass of 1 kg an acceleration of 1 m/s^2.
2. One dyne is the force required to give a mass of 1 g an acceleration of 1 cm/s^2.
3. One pound is the force required to give a mass of 1 slug an acceleration of 1 ft/s^2.

The difference between mass and weight, $W = mg$

The mass of any object is a measure of its inertia and is a constant property of the object (with the exception that the mass of a given object increases when it is accelerated to velocities approaching the speed of light). At velocities within the realm of common experience, we may assume mass to be a constant property, whether the body is located on the moon, Earth, Mars, Jupiter, or anywhere else.

Weight, on the other hand, is the gravitational attraction of the earth for a body. It so happens, because of the nature of gravitational forces, that weight and mass are proportional. More massive bodies weigh more. Because of small variations in the acceleration of gravity over the surface of the earth, an object will weigh slightly different amounts at different places; hence, the weight of a body is *not* a constant property of the body. An object obviously has weight on the moon because of the moon's gravitational attraction. However, objects on the moon weigh only about one-sixth as much as on the earth.

The mathematical relationship between mass and weight is that of a simple proportionality, with the acceleration of gravity being the proportionality constant:

$$W = mg \tag{4-2}$$

This equation is of the same dimensional form as $F = ma$, namely, a force equals the product of a mass and an acceleration. Often, in problems where mass units are being avoided, the equivalent expression W/g is substituted for m. Doing this to $F = ma$ yields

$$F = \frac{W}{g}a \tag{4-3}$$

We shall see the usefulness of this in a later section.

EXAMPLE 4-1 Consider the 200-lb anvil initially at rest on the frictionless surface, and suppose we pull on the string with a force of 4 lb. What acceleration will result?

Solution. We first calculate the mass of the anvil, which weighs 200 lb. We divide by the acceleration of gravity, $g = 32\ \text{ft/s}^2$:

$$m = \frac{200\ \text{lb}}{32\ \text{ft/s}^2} = 6.25\ \text{slugs}$$

The acceleration is given by

$$a = \frac{F_{\text{net}}}{m} = \frac{4\ \text{lb}}{6.25\ \text{slugs}} = 0.64\ \text{ft/s}^2$$

This is not a large acceleration, but the anvil will begin to move fairly rapidly if the force is maintained for a period of 5 s, for instance.

$$v = v_0 + at$$
$$v = 0 + (0.64\ \text{ft/s}^2)\ (5\ \text{s}) = 3.2\ \text{ft/s}$$

At this velocity, the anvil will go 10 ft in about 3 s.

EXAMPLE 4-2 What force is required to give the anvil an acceleration of 4 ft/s^2?

Solution. We use the mass calculated in Example 4-1 (6.25 slugs) and $F = ma$:

$$F = ma$$

$$F = (6.25\ \text{slugs})\ (4\ \text{ft/s}^2) = 25\ \frac{\text{slug-ft}}{\text{s}^2} = 25\ \text{lb}$$

4-5 NEWTON'S THIRD LAW: ACTION-REACTION

Perhaps you have heard that for every action there is an equal and opposite reaction. This principle has been applied to many things by many people, but in the context of Newton's third law, it applies to the fact that forces always occur in pairs. When one object exerts a force on a second object, the second object also exerts an equal force, oppositely directed, on the first. Two large boxes on a frictionless surface connected by a long, flexible spring illustrate this idea (Fig. 4-3). A force is exerted on each box by the spring, and it is impossible for the force exerted on one box to be different from the force exerted on the other.

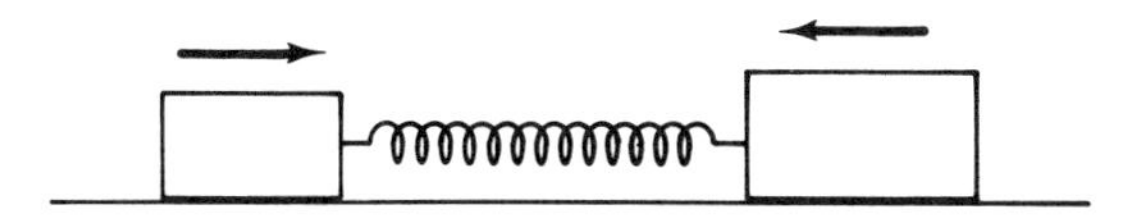

Figure 4-3 A spring stretched between two boxes exerts an equal force on each box. This is an example of the action-reaction principle.

Consider the chair on which you are now sitting. Are you pushing down on it? Is it pushing up on you? This is an illustration of the third law.

A rocket engine exerts great forces toward the rear to accelerate the exhaust gases to high velocities. The opposing force is exerted toward the front and constitutes the thrust of the rocket. It is *not* true that the exhaust gases of a rocket must push on something (against the ground or against the atmosphere) in order to provide thrust. Consequently, rockets function equally well in the nearly perfect vacuum of space.

As the final example of the action-reaction principle, here is the classic: An angler wishes to take the outboard motor home. The person drives the boat alongside the dock and ties up the front of the boat, removes the motor from the back of the boat, and prepares to lift the motor up and over onto the dock. All is well as the person lifts the motor upward, but suddenly a considerable shove is required to move the motor horizontally toward the dock.

A force on the motor toward the dock is accompanied by a force on the boat away from the dock. Because the back of the boat is free to move, it does. The distance from boat to dock increases unexpectedly, throwing the person off balance and causing the motor to drop into the water.

The elevator problem

An elevator provides an interesting illustration of $F = ma$ acting in conjunction with the acceleration of gravity. In Fig. 4-4 an expression is derived for the tension in the supporting cable of an elevator of mass m that is accelerating with acceleration a. The elevator is assumed to run on frictionless tracks.

The net force on the elevator is the upward tension in the cable minus the weight (mg) of the elevator. This is substituted into $F = ma$ and rearranged to obtain

$$T = m(g + a) \qquad (4\text{-}4)$$

This shows that the tension in the supporting cable depends upon the acceleration of the elevator. The tension in the cable equals the weight of the elevator only

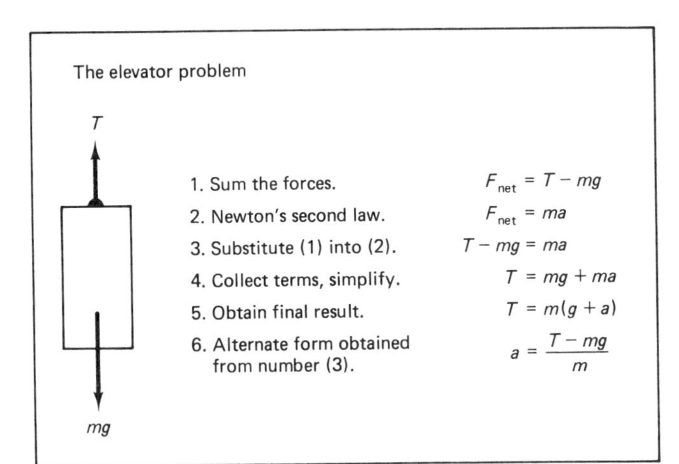

Figure 4-4 Derivation of a formula for the tension in a cable supporting an elevator of mass m accelerating upward with acceleration a.

when the acceleration is zero, i.e., when the elevator is moving with uniform velocity or when it is stationary. When the elevator is accelerating upward, the tension in the cable is greater than the weight of the elevator. When the acceleration is downward, the tension is less than the weight of the elevator.

By substituting W/g for m in Eq. (4-4), we obtain

$$T = W\left(\frac{g + a}{g}\right) \quad \text{or} \quad T = W\left(1 + \frac{a}{g}\right) \tag{4-5}$$

This equation is more convenient in the British system where the weight in pounds is more commonly given than the mass.

EXAMPLE 4-3

Calculate the tension in the supporting cable of a 1000-lb elevator: (a) accelerating upward at 4 ft/s^2; (b) accelerating downward at 4 ft/s^2.

Solution. (a) For upward acceleration,

$$T = W\left(1 + \frac{a}{g}\right) = (1000 \text{ lb})\left(1 + \frac{4 \text{ ft/s}^2}{32 \text{ ft/s}^2}\right)$$

$$= 1125 \text{ lb}$$

(b) For downward acceleration, we use the same formula, substituting a negative value for the acceleration as it is downward.

$$T = (1000 \text{ lb})\left(1 + \frac{(-4) \text{ ft/s}^2}{32 \text{ ft/s}^2}\right)$$

$$= 875 \text{ lb}$$

EXAMPLE 4-4

The mass of a certain elevator is 850 kg, and the supporting cable can withstand a maximum tension of 11,000 N. Calculate the maximum upward acceleration the cable can impart to the elevator.

Solution. We use the equation that expresses the acceleration in terms of the tension in the cable:

$$a = \frac{T - mg}{m} = \frac{11{,}000 \text{ N} - (850 \text{ kg})(9.8 \text{ m/s}^2)}{850 \text{ kg}}$$

Maximum acceleration = 3.14 m/s^2 upward

EXAMPLE 4-5 In Example 4-4, calculate the tension in the supporting cable when the elevator has the same acceleration, except downward.

Solution. Use the formula for the tension in terms of acceleration, but attach a negative sign to the downward acceleration.

$$\begin{aligned} T &= m(g + a) \\ &= (850 \text{ kg})(9.8 \text{ m/s}^2 - 3.14 \text{ m/s}^2) \\ &= 5661 \text{ N} \end{aligned}$$

Atwood's machine

Atwood's machine consists of two masses suspended by a string and a light, frictionless pulley, as shown in Fig. 4-5. If the masses are unequal, the heavier mass will descend, pulling the other one up. The acceleration of the two-mass system is proportional to the difference in magnitude of the masses. Accelerations resulting in clockwise rotation of the pulley are taken as positive. Because the string is assumed to be perfectly flexible and the pulley very light and frictionless, the same tension will exist on the string on both sides of the pulley. In general, we want to calculate the acceleration with which the masses move and the tension in the string.

The expression for the acceleration is derived in Fig. 4-5. The net force tending to accelerate the masses is the difference in weight of the masses. The mass being accelerated, represented by m in $F = ma$, is the sum of the two masses because both masses execute the same motion, with the exception of direction. Substitution of these quantities into $F = ma$ gives the following formula for the acceleration.

$$(\textit{Atwood's machine}) \quad a = \frac{(m_2 - m_1)g}{(m_1 + m_2)} \tag{4-6}$$

An expression for the tension in the string may be obtained by isolating m_1 and determining the force required to produce the acceleration that it is known to have. Details are given in the second part of Fig. 4-5. The final result is

$$(\textit{string tension}) \quad T = 2\left(\frac{m_1 m_2}{m_1 + m_2}\right)g \tag{4-7}$$

Examination of Eqs. 4-6 and 4-7 shows that as long as m_2 is greater than m_1, the acceleration will be positive. Also, the acceleration will be zero when

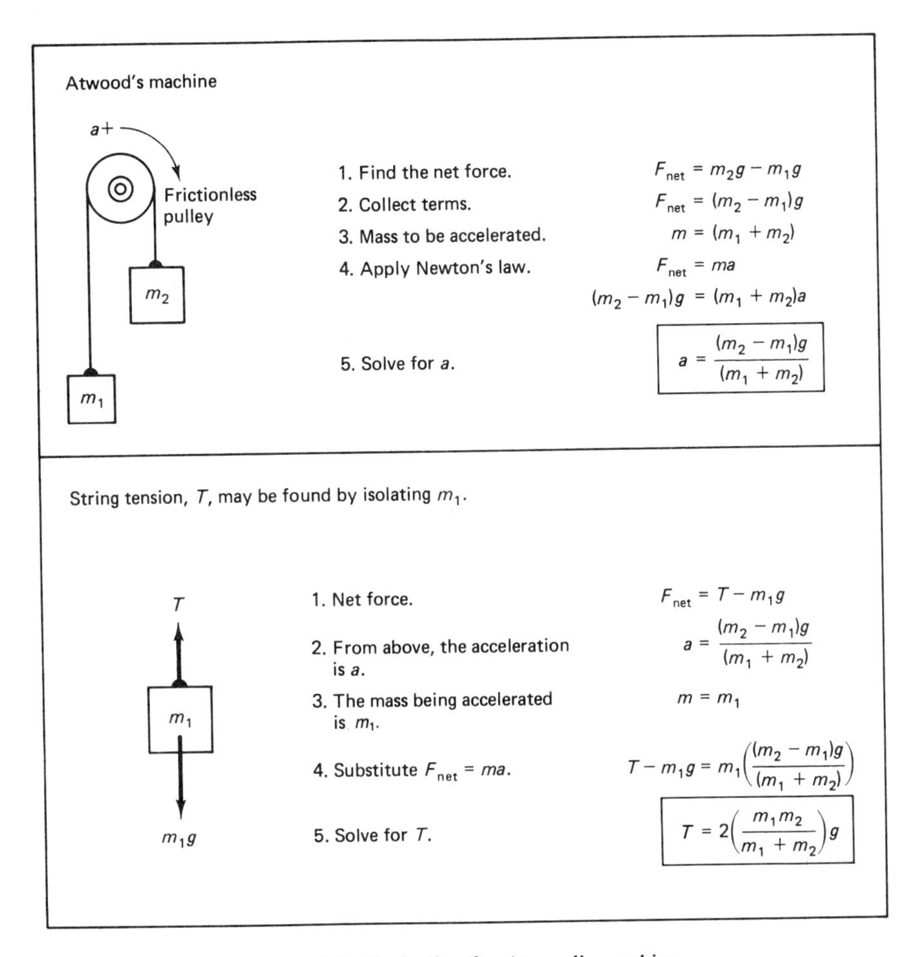

Figure 4-5 Derivation for Atwood's machine.

$m_1 = m_2$; the system will be in a state of balance and $T = m_1g = m_2g$. Further, if m_1 were zero, m_2 would fall unimpeded. In this case, T would be zero.

Another configuration of accelerating masses is shown in Fig. 4-6. Analysis is left as an exercise. Consider the special cases when either m_1 or m_2 is zero.

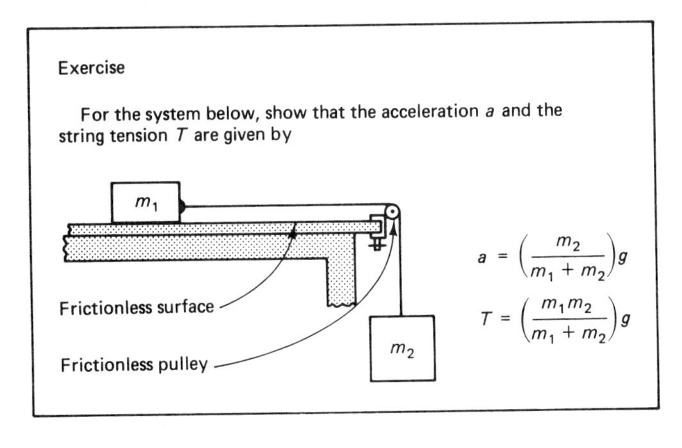

Figure 4-6 Another system of accelerating masses.

4-6 MORE ABOUT FRICTION

Suppose we place a book on the surface of an ordinary table and equip the book with a string and scale, as in Fig. 4-7, so that we can measure the force required to set the book in motion. The pull on the string is gradually increased from zero until the book moves. If the book weighs 24 oz (1.5 lb), we might find that a force of 12 oz is required to start the book moving from rest. No motion occurs until the applied force reaches 12 oz; at this time, the book "breaks loose" and accelerates across the table.

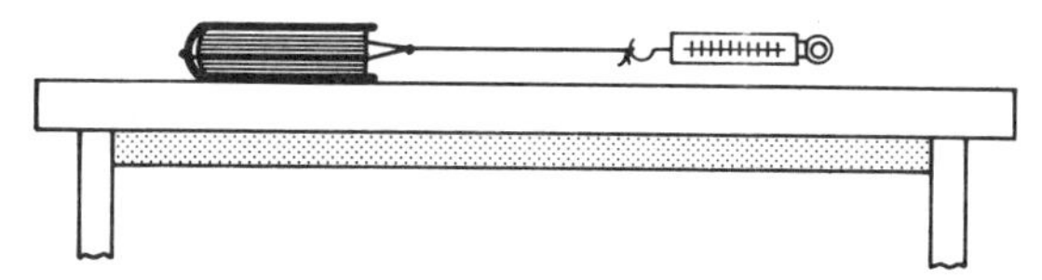

Figure 4-7 The scale permits the force required to pull the book to be determined.

Suppose now that we tap the book gently in the direction of the force as the force is gradually increased from zero. Each tap sets the book in motion for just an instant, and we might find that an applied force of only 8 oz is sufficient to sustain uniform (unaccelerated) motion of the book once it is started by tapping. Thus, we see that less force is required to *sustain* the motion than is required to start the motion.

In both cases, the frictional forces must be overcome before motion can ensue. The frictional force is given by

$$F = \mu N \tag{4-8}$$

where F is the frictional force, μ is the *coefficient of friction*, and N, the *normal force*, is the force with which the sliding surfaces are pressed together in the direction perpendicular to the plane of the surfaces (see Fig. 4-8). The term *normal* implies *perpendicularity*. In this case, N is the weight of the book, 24 oz,

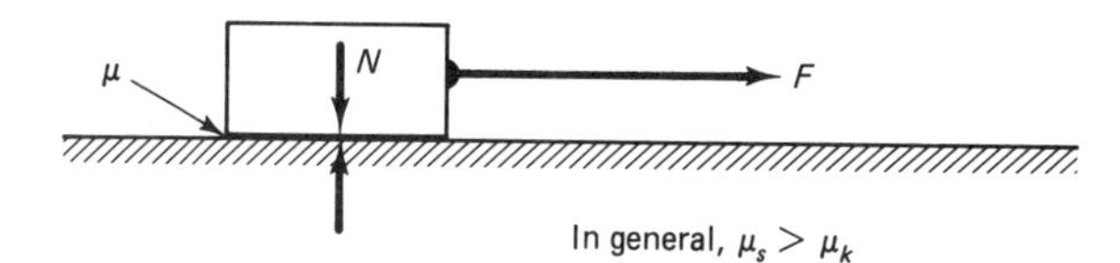

Figure 4-8 The normal force N is the force pushing the two surfaces together. For an object resting on a horizontal surface, the normal force is simply the weight of the object, $N = W = mg$. The frictional forces are parallel to the surfaces of contact.

because the surfaces are horizontal. The value of the coefficient of friction μ depends upon the nature of the sliding surfaces and may be calculated from

$$\mu = \frac{F}{N} \tag{4-9}$$

Note, however, that two different situations are involved in the example above. One involves *static friction*, where the book was started from rest; the other involves *kinetic friction*, where the applied force simply maintains a uniform motion that had already been started. Thus, we define a coefficient of static friction μ_s and a coefficient of kinetic friction μ_k:

Static Friction	*Kinetic Friction*
$F_s = \mu_s N$	$F_k = \mu_k N$
$\mu_s = \dfrac{F_s}{N}$	$\mu_k = \dfrac{F_k}{N}$

In all cases μ_k is less than μ_s, and for surfaces that are not "sticky" or otherwise unusual, both coefficients are less than 1.0. Thus, to put an object in motion by pulling on it horizontally requires an applied force less than the weight of the object. In the case of the book described above, the coefficient of static friction is $\frac{12}{24} = 0.5$ and the coefficient of kinetic friction is $\frac{8}{24} = 0.33$.

Frictional forces are thought to be due to a combination of two phenomena, namely, (1) molecular attractions between portions of the opposing surfaces that come into intimate molecular contact, and (2) interlocking of irregularities on the opposing surfaces. Molecular attractions (miniature "welds") occur between "roughness peaks," as illustrated in Fig. 4-9, where the roughness is greatly exaggerated for purposes of illustration. Oil or other lubricant reduces friction by forming a film between the two surfaces, thereby preventing the formation of the miniature welds. The microscopic details of frictional forces are not well understood.

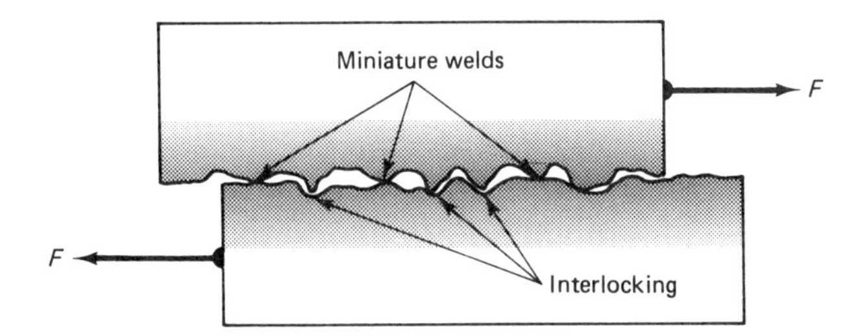

Figure 4-9 A microscopic view of the cause of frictional forces.

Here are two properties observed of frictional forces between dry, sliding surfaces:

1. Within the realm of moderate forces, the coefficient of friction does not depend upon the area of contact between the two surfaces.
2. At low velocities, the frictional force does not depend upon the relative velocity between the two surfaces. At higher velocities, the friction decreases.

If the surfaces are lubricated with an ample supply of oil, the properties of friction are altered significantly from those properties given above. For example, the frictional force depends more upon the properties of the oil and less upon the nature of the two surfaces in contact. More information may be found in the various mechanical engineering handbooks and in books addressed solely to friction and lubrication.

Reference

Excellent historical and philosophical discussions are presented in *The Nature of Physics* by Peter J. Brancazio. New York: Macmillan, 1975.

Questions

1. A book that weighs 1 lb is at rest on a table. Describe the two forces that act on the book. What is the net force on the book, assuming nothing is visibly pushing or pulling the book?
2. What is the net force acting on a 180-lb person falling through the air at terminal velocity?
3. Why is a performer able to snatch a tablecloth from under the dishes resting on the cloth without disturbing the dishes? What would happen if the performer pulled the cloth very slowly instead of very quickly?
4. What is the relationship between mass and weight?
5. How is Newton's first law (inertia) responsible for whiplash injuries that occur when a car is struck from behind?
6. What weight in pounds (at the surface of the earth) is the equivalent of a mass of one slug?
7. If we see that an object is accelerating, what can we say about the net force acting on the object?
8. A small child stands flat-footed, with knees bent slightly, and then jumps straight up into the air. What was it that exerted a force on the child that propelled the child into the air?
9. Suppose a person somehow sits upon a bathroom scale placed on the floor of a recreational vehicle. Describe how the scale reading will vary as the vehicle passes over a speedbreaker (a bump) in a parking lot.
10. Is it possible for an airplane to be sustained by the air without having the plane exert a net downward force on the air? Does the air move downward because of this force? From what does the downwash of a helicopter result?

Problems

1. A 5-lb cat is sitting on a 200-lb crate at rest on the floor of a warehouse. What is the net force on the crate?
2. A 195-lb skydiver is falling at terminal velocity in free fall. What is the net force on the skydiver?
3. What is the acceleration of a 5-kg mass if a net force of 15 N acts upon it?
4. A toy car is accelerating at the rate of 3 m/s^2. If the mass of the car is 450 g, what net force must be acting upon it?
5. A force of 12 lb is applied to a 200-lb anvil on a frictionless surface. What will be the acceleration of the anvil?
6. A net force of 44 dynes is applied to a mass of 10 g. What is the resulting acceleration of the mass?
7. In terms of pounds, what is the force equivalent of 1 N?
8. How many dynes of force is equivalent to a pound?
9. When a net force of 10 N is applied to a certain object, the resulting acceleration is 3 m/s^2. What is the *weight* of the object?
10. An object weighing 96 lb is accelerating at the rate of 4 ft/s^2. What net force is acting on the object?
11. A car weighing 3000 lb is accelerating at the rate of 2 mi/h/s. What net force must be exerted on the car?

*12. A pitched baseball weighing 4 oz travels toward the batter at 78 mi/h. The ball strikes the catcher's mitt and is brought to a stop in a distance of 6 in. by the recoil of the mitt.

(a) Compute the average deceleration of the ball.
(b) What is the average force acting on the ball during the stopping process?
(c) What is the average force on the mitt?

13. If the anvil in Problem 5 is initially at rest, how far will it travel during the first 4 s that the force is applied?

14. The weight of an elevator and contents is 3000 lb. Assume that it runs on frictionless tracks. Compute the tension in the supporting cable if: **(a)** the elevator is moving upward at a constant velocity of 3 ft/s; **(b)** the elevator is moving downward at a constant velocity of 3 ft/s; **(c)** the elevator is accelerating upward at 3 ft/s^2; **(d)** the elevator is accelerating downward at 3 ft/s^2.

15. The mass of a certain elevator is 850 kg. The tension in the supporting cable is 7500 N. Describe (in quantitative terms) the motion of the elevator.

16. Suppose the elevator of Problem 15 is accelerating downward at the rate of 1 m/s^2. What will be the tension in the supporting cable?

17. The masses of an Atwood's machine as in Fig. 4-5 are 4 kg and 6 kg. If the pulley is frictionless and very light, compute: **(a)** the acceleration of the two masses, specifying the direction; **(b)** the tension in the string.

18. In Problem 17, how far will the heavier mass descend in the first 2 s after the system is released from rest?

19. Repeat Problem 17 for masses of 1 kg and 9 kg. Compare the results.

20. The heavier mass of an Atwood's machine descends 48 cm in 3 s after being released from rest.
(a) Compute the acceleration of the system.
(b) If the total mass of the two masses is 2000 g, compute the mass of each mass.

21. In Fig. 4-6, suppose $m_1 = 1$ kg and $m_2 = 2$ kg. **(a)** Compute the acceleration of the system, and **(b)** the tension in the string. **(c)** If the system starts from rest, how far will mass m_1 move during the first 1-s interval after the system is released?

***22.** A wooden box weighing 20 lb rests on a smooth floor. The coefficient of static friction between the box and the floor is 0.75; the coefficient of kinetic friction is 0.50.
(a) What horizontal force must be applied to the box to set it in motion?
(b) What force is required to maintain the box in motion at constant velocity?
(c) Calculate the acceleration of the box assuming that the force [calculated in part (a)] required to set the box in motion is maintained even after the box begins to move.
(d) How far will the box move in the first 5 s?
(e) What will be the velocity of the box at the end of the first 5 s?

***23.** A small box weighing 100 lb rests on top of a larger box, which weighs 200 lb. The small box is tied by a horizontal string to a wall behind the two boxes. The coefficient of static friction between the two boxes is 0.55 and between the lower box and the floor is 0.70. What force is required to pull the lower box from underneath the upper box?

CHAPTER 5

VECTORS AND APPLICATIONS

Many physical quantities, such as velocity and force, have an associated direction as well as a magnitude and are called *vector quantities*. Other quantities, such as mass and volume, have no associated direction and are called *scalar quantities*. A *vector* is a representation of a vector quantity. In this chapter we present the basic properties of vectors and several applications of vectors. We see, for example, how two forces at right angles are added to obtain a single resultant force. We give the three conditions for equilibrium of a rigid body, and we describe the circumstances under which forces on rigid bodies produce a twisting force, called *torque*. Some of the topics presented earlier without vectors are revisited to show how the use of vectors provides a more meaningful treatment of certain problems.

Because the first four sections of this chapter deal with the mathematics of vectors, your instructor may ask you to study these sections before you study either Chap. 3 or Chap. 4.

5-1 VECTOR AND SCALAR QUANTITIES

Vector quantities

A vector quantity is a physical quantity that has both a magnitude and a direction. Examples of vector quantities are displacement, velocity force, and acceleration. Each of these has both a magnitude and a direction. Taking a velocity as an example, 45 mi/h due north specifies a velocity vector of magnitude 45 mi/h in the direction due north.

A vector may be represented graphically by an arrow drawn in the appropriate direction whose length is proportional to the magnitude of the vector quantity. If the vector quantity is a velocity, the arrow is drawn in the direction of motion and is given a length that corresponds to the speed of the motion. The *line of action* of a vector is the line on which the vector is drawn, and it extends infinitely in both directions.

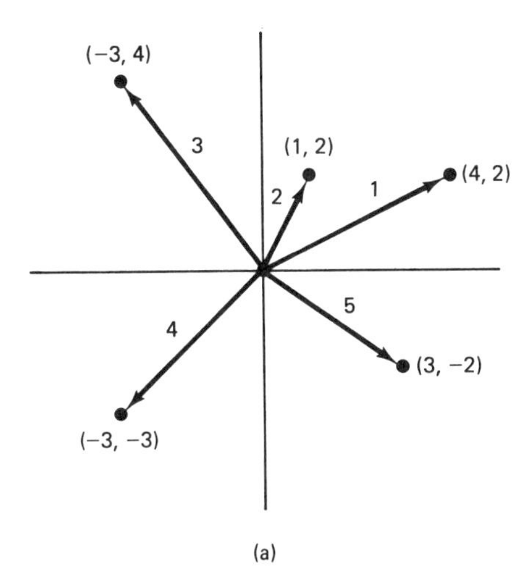

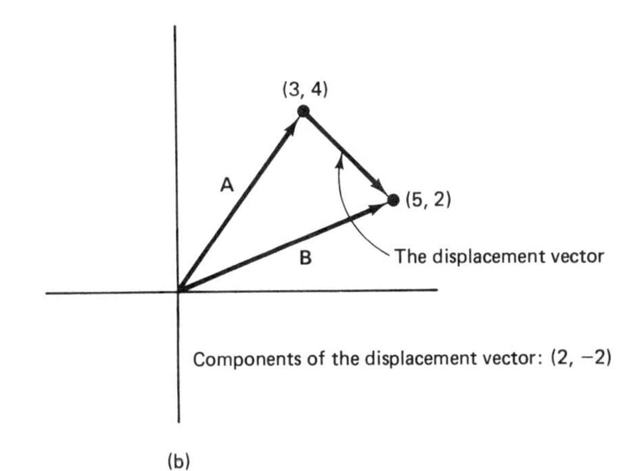

Figure 5-1 Position and displacement vectors.

It should be clear that force, velocity, and acceleration are vector quantities. Going further, a *position* can be represented by a *position vector*, which extends from a point of reference (the origin) to the point in question. The magnitude of the position vector is the distance from the origin to the point. A *displacement vector* is the vector drawn from the tip of an *initial* position vector to the tip of a *final* (later-occurring) position vector, indicating a change in position. Position vectors and a displacement vector are illustrated in Fig. 5-1(a) and (b).

Scalar quantities

Scalar quantities have only a magnitude; they have no associated direction. Examples are mass, volume, temperature, energy, speed, and time. Of these, speed is the only one that might appear to have a directional characteristic. However, speed is taken to be the magnitude of the velocity vector and does not imply a direction, as does velocity. For example, a car traveling at 80 mi/h around

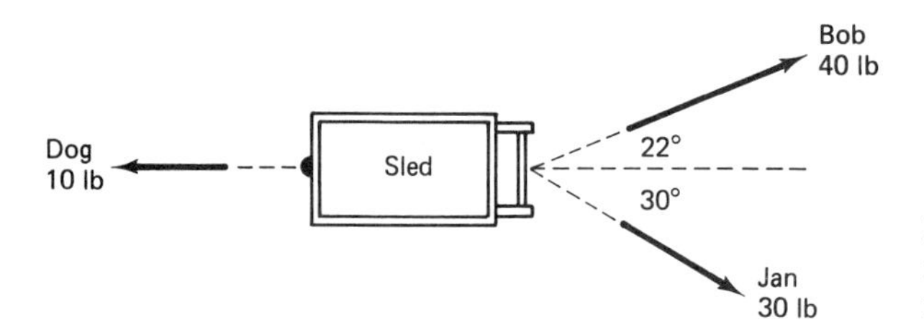

Figure 5-2 Vectors are useful in determining the net force on a sled when the forces do not all lie on the same line.

a circular racetrack is said to have a *constant speed*, but its velocity is continually changing because of its continually changing direction. Velocity implies both speed and direction; speed refers only to how fast something is happening. When something travels at a constant velocity, neither its speed nor its direction change.

Students sometimes insist that mass should be a vector quantity because mass has weight that is directed downward. Certainly a mass at the earth's surface has weight, and weight is a vector quantity, but here we are confusing an attribute of mass (weight) with mass itself. Mass has no directional characteristic, and it is therefore a scalar quantity.

A hint of the usefulness of vectors is given in Fig. 5-2, which shows a sled being pulled by two children and a dog. We might wish to know the net force on the sled so that we could determine its acceleration. Vectors are helpful because the forces do not all lie on the same line.

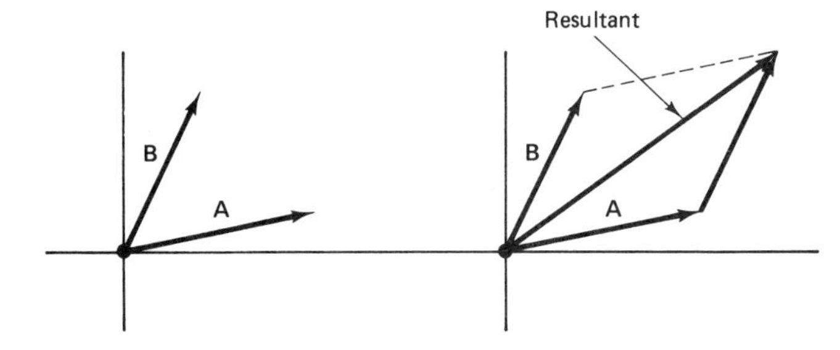

Figure 5-3 Vector addition. Two vectors are added by placing the tail of one to the head of the other to form a vector parallelogram.

5-2 VECTOR ADDITION

The graphical addition of two vectors is illustrated in Fig. 5-3. Initially, both vectors are drawn with their tails at the origin. Vector **B** is then moved so that the tail of **B** joins the tip of **A**. The vector sum, called the *resultant*, is the vector **R** that extends from the origin to the tip of **B**. Note that the direction of **B** is unchanged by the move, which results in the formation of a parallelogram. The order of addition is not important. The same resultant will be obtained if **A** is placed at the tip of **B**.

This procedure of connecting vectors head-to-tail is applicable to more than two vectors, as is illustrated in Fig. 5-4. In this case, a vector *polygon*, rather than a parallelogram, is formed. As before, the resultant is the vector drawn from the origin to the tip of the last vector, and the order of addition is unimportant.

Two vectors may be added analytically by using the law of cosines and the law of sines, as in Fig. 5-5. First, the magnitude of the resultant **R** is determined by using the law of cosines, paying attention to the sign of the term containing cos θ.

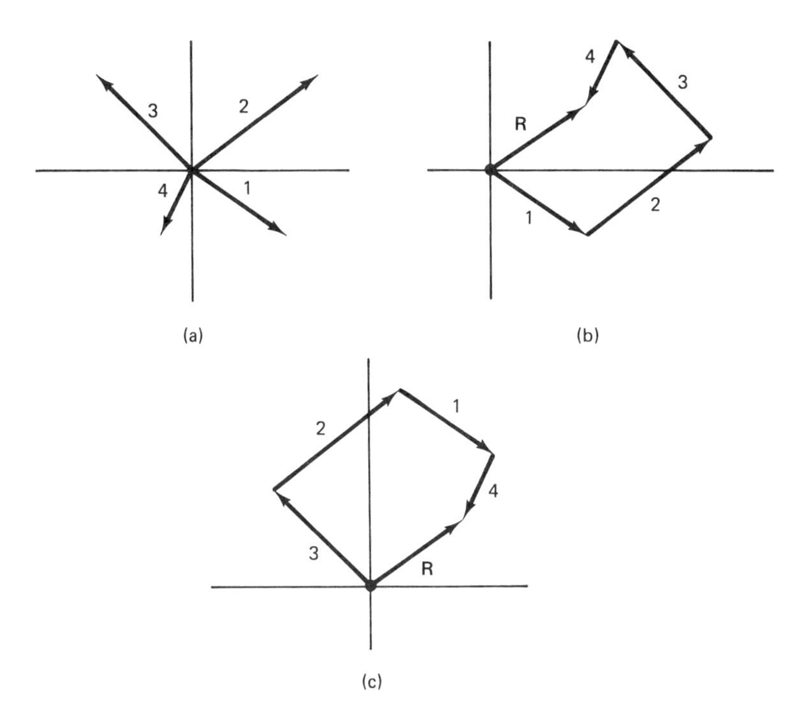

Figure 5-4 (a) Vectors to be added. (b) Vector polygon. (c) Same resultant obtained by connecting vectors in different order.

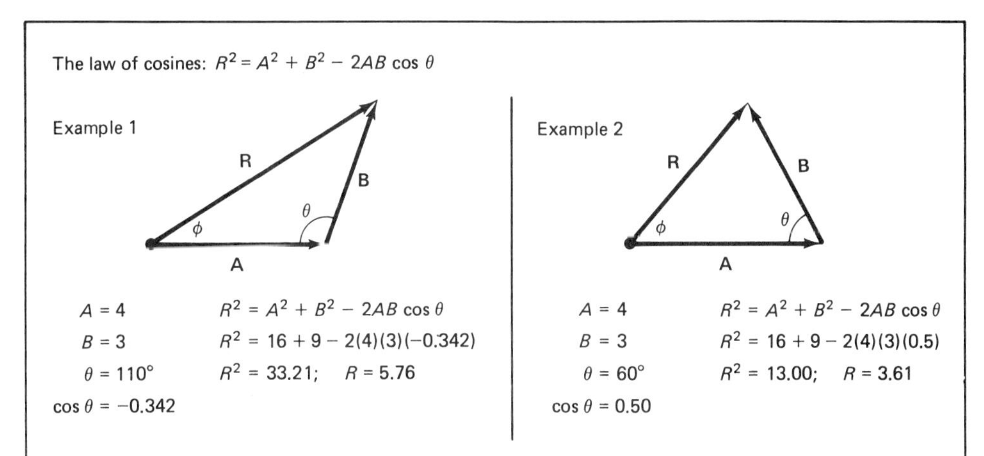

Figure 5-5 Mathematical addition of two vectors.

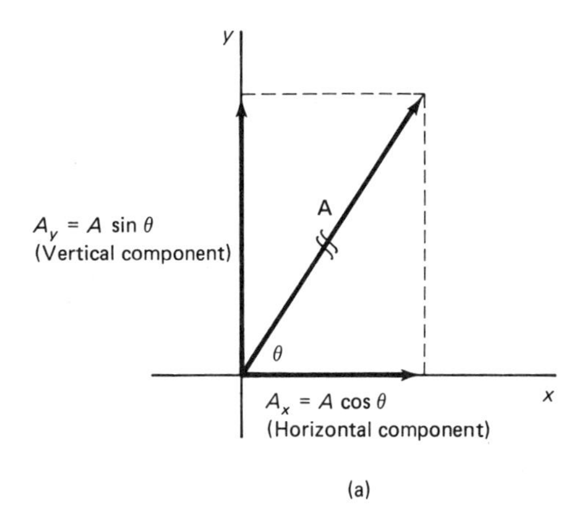

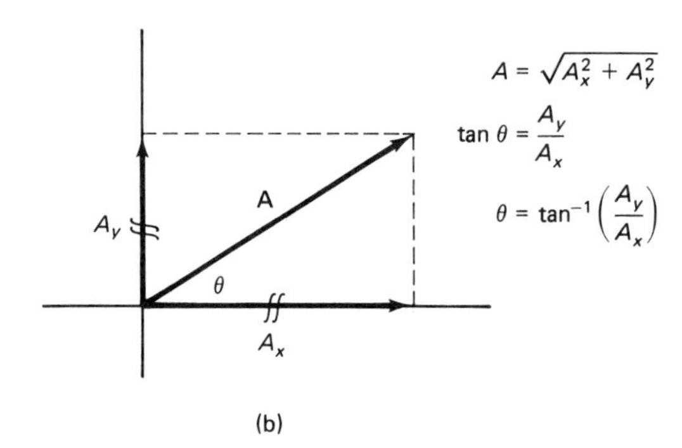

Figure 5-6 (a) Resolving a vector into components. (b) Reconstructing a vector from its components.

Then, the angle the resultant makes with the adjacent vector is found using the law of sines.

When more than two vectors are to be added analytically, each vector is first resolved into its *x*- and *y*-components. All the *x*-components and all the *y*-components are then added separately to obtain the *x*- and *y*-component of the resultant. Before this procedure can be described, however, we must first see how to resolve vectors into components.

5-3 RESOLVING VECTORS INTO COMPONENTS

A vector can be resolved into *x*- and *y*-components, as illustrated in Fig. 5-6(a). The two marks on vector **A** indicate that **A**, which makes an angle θ with the *x*-axis, is resolved into perpendicular components. Vector **A** is replaced by its components because the effect of the components is identical to the effect of the original vector **A**. Thus, a vector can be replaced by its components, and two components can be added to obtain a vector whose effect is in all respects identical to that of the components.

When a vector is resolved into components, the components are almost always at right angles; i.e., one is horizontal and the other is vertical, or one is the *x*-component while the other is the *y*-component. Consequently, the angle

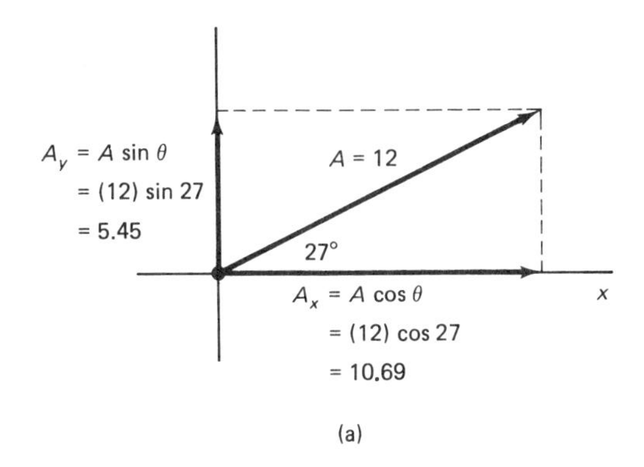

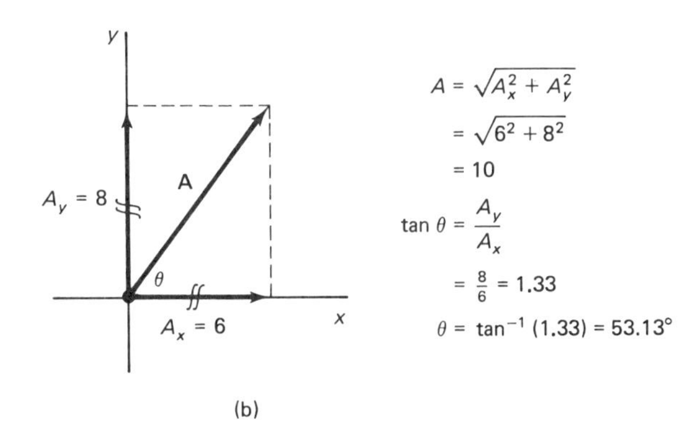

$$A = \sqrt{A_x^2 + A_y^2} = \sqrt{6^2 + 8^2} = 10$$

$$\tan\theta = \frac{A_y}{A_x} = \frac{8}{6} = 1.33$$

$$\theta = \tan^{-1}(1.33) = 53.13^\circ$$

Figure 5-7 (a) Resolving a vector into components. (b) Reconstructing a vector from its components.

involved in the vector addition of the components is 90°, which means we must deal only with right-angle trigonometry. The procedure for reconstructing a vector **A** from its components A_x and A_y is shown in Fig. 5-6(b). It is clear that if the components A_x and A_y are added, the result is **A**. The magnitude of **A** is obtained via the Pythagorean theorem, and the angle θ is found via the inverse tangent relationship. An example of resolving vectors into components is given in Fig. 5-7.

5-4 VECTOR ADDITION BY COMPONENTS

Any number of vectors may be added by resolving each vector into components and then summing all the x-components and all the y-components to obtain the x- and y-components of the resultant. The procedure is summarized in the following steps:

1. Compute the x- and y-components of each vector, paying careful attention to the algebraic sign. Vectors tending toward the left have negative x-components; vectors tending downward have negative y-components.
2. Organize the x- and y-components in a table.
3. Compute the sum of the x-components and the sum of the y-components. These are denoted by Σx and Σy, respectively.

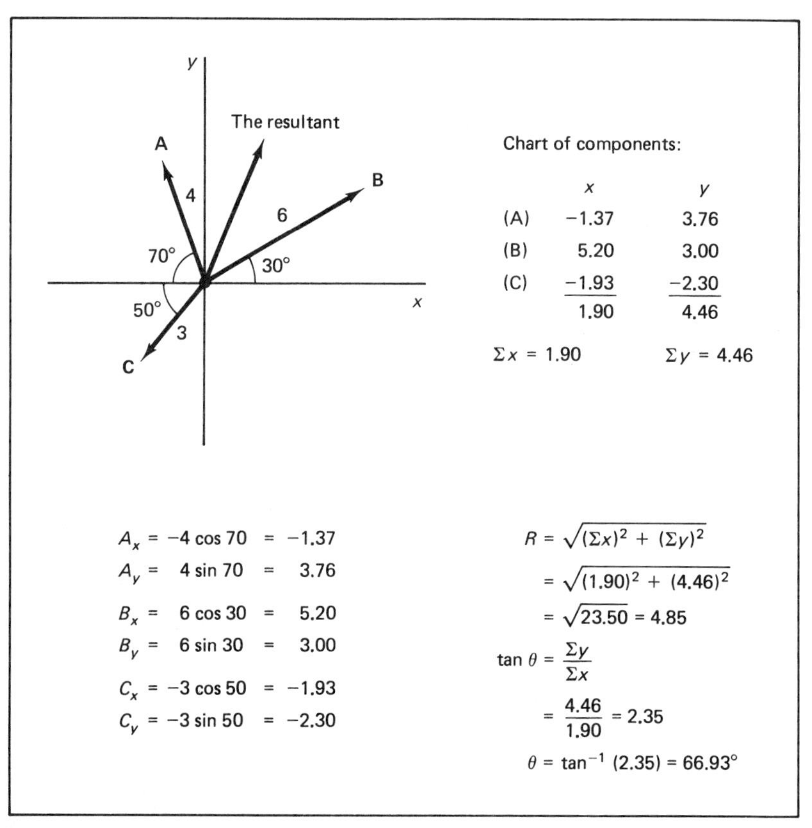

Figure 5-8 Analytical method of finding the resultant of several vectors.

4. Calculate the magnitude of the resultant **R** using the Pythagorean theorem.

$$R = \sqrt{(\Sigma x)^2 + (\Sigma y)^2} \tag{5-1}$$

5. Calculate the angle the resultant makes with the x-axis by using the tangent and inverse tangent:

$$\tan\theta = \frac{\Sigma y}{\Sigma x} \qquad \theta = \tan^{-1}\left(\frac{\Sigma y}{\Sigma x}\right) \tag{5-2}$$

An example is given in Fig. 5-8.

EXAMPLE 5-1 A force of 50 lb is applied to a sled at an angle of 30° to the horizontal. Calculate the force tending to pull the sled horizontally, and calculate the force that tends to lift the sled off the ground.

Solution. This problem amounts to finding the vertical and horizontal components of the force applied to the sled (see Fig. 5-9), which is drawn as a vector on a vector diagram. The resulting vector is resolved into vertical and horizontal components.

The horizontal component of 43.3 lb is the force tending to pull the sled forward along the ground, and the vertical component of 25 lb is the force tending to lift the sled.

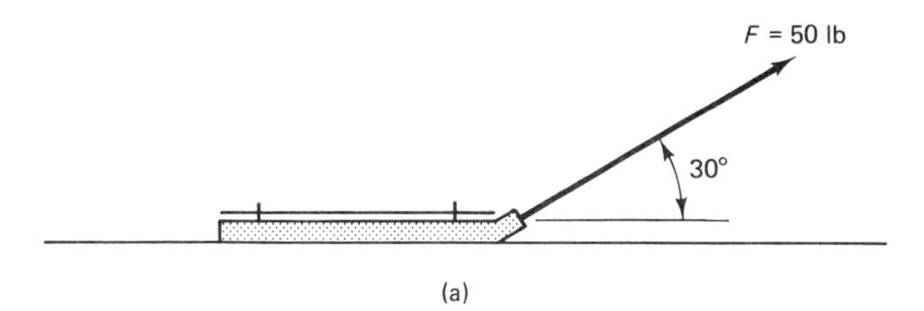

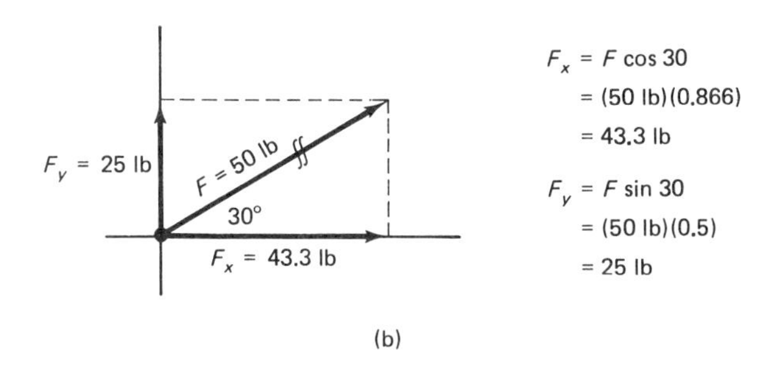

(b)

Figure 5-9

EXAMPLE 5-2 An airplane pilot sets a heading of due east at an indicated airspeed of 100 mi/h. A wind is blowing from the north at 20 mi/h. Find the direction of motion of the plane relative to the ground, and find the true ground speed.

Solution. This problem amounts to finding the resultant velocity vector of the easterly motion of the plane and the southward motion due to the wind. The vector diagram and computational details are given in Fig. 5-10.

You may confirm that if the pilot wishes to fly due east, the plane should be aimed 11.54° north of east. The ground speed will then be 97.98 mi/h in an easterly direction.

See if you can verify that if a pilot is to fly a round trip from one city to another and back, the entire trip can be made in the *least time* when there is *no wind*. Put another way, any wind from any direction at any speed will increase the

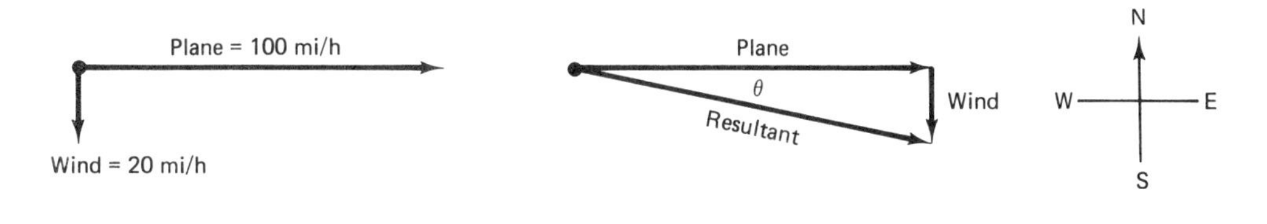

Find the resultant R:

$$R = \sqrt{100^2 + 20^2} = \sqrt{10{,}400} = 101.98 \text{ mi/h}$$

Find the angle θ:

$$\tan\theta = \frac{20 \text{ mi/h}}{100 \text{ mi/h}} = 0.2$$

$$\theta = \tan^{-1}(0.2) = 11.31°$$

The true ground speed is about 102 mi/h in a direction 11.31° south of east.

Figure 5-10

time required to make the trip. (We assume the speed and direction of the wind remain constant for the duration of the flight.)

5-5 EQUILIBRIUM

Definition

A body in *equilibrium* is in a state of balance. All forces on the body cancel; the net force is zero. Further, all forces that tend to cause the body to twist or turn are balanced by opposing forces so that no net effect is produced on any rotational motion the body might have.

A body in *static equilibrium* is stationary; it is at rest. It will remain at rest until some external force disturbs the equilibrium condition. However, a body in *dynamic equilibrium* is in motion; it is moving with constant speed, either in a straight line or in rotation about a point (or even in a combination of these motions). Any object in either dynamic or static equilibrium cannot be accelerating because the condition for equilibrium presumes that the net force is zero.

The world is full of examples of objects in static equilibrium. Any object about you that is at rest qualifies. Examples of dynamic equilibrium are less plentiful but are still quite common. The turntable of a phonograph is in dynamic equilibrium when a record is being played, and any object that is falling at terminal velocity is in dynamic equilibrium.

Three types of equilibrium states—*stable, unstable*, and *neutral*—may be distinguished in regard to the stability of the equilibrium. An object (such as a pendulum) in a state of stable equilibrium tends to return to the equilibrium position when slightly displaced from it. An object slightly displaced from a state of unstable equilibrium tends to depart further from the equilibrium position. An example is a pencil balanced on its point or an egg stood on its end on a flat surface. A marble placed on a smooth, level surface is in a state of neutral equilibrium because no force results from a slight displacement from the equilibrium position. Indeed, there is no *one* equilibrium position; the marble will be in equilibrium at any point on the smooth, level surface.

Mathematical definition of equilibrium

We stated earlier that no net force acts on a body in equilibrium. In relation to a vector polygon, this means that the polygon is *closed*—the resultant is *zero*. The tip of the last vector falls exactly on the origin. From this it follows that the sum of the x-components of all the forces and the sum of the y-components must each be zero. If the problem is three-dimensional, the sum of all the z-components is zero also. For a body in equilibrium, we may use the summation symbol to write that

$$\Sigma F_x = 0 \qquad \Sigma F_y = 0 \qquad \Sigma F_z = 0 \tag{5-3}$$

When a force acts on a body causing it to turn, we say that a *torque* is exerted on the body. Consequently, a torque is a *twisting force*. Torques are described later, but at this point we can state that for a body to be in equilibrium, the sum of the torques must equal zero. This implies that either no torque is exerted on the body, or that all torques tending to rotate the body clockwise are balanced by torques tending to rotate the body counterclockwise.

Suppose the lines of action of all the forces acting on a body intersect at a single point. In such case, the forces are said to be *concurrent*, and no torque is exerted on the body.

Force table

A *force table* provides a convenient means of testing the properties of vectors. Forces are applied to a small, freely movable ring at the center of a circular table equipped with pulleys and strings for applying the forces at various angles. A force table is shown in Fig. 5-11.

A typical problem for a force table might be to find the resultant of two force vectors that make an angle of 60° to each other, assuming the magnitude of one vector is 80 units and the other is 120 units. Incidentally, the unit usually employed in force table work is the gram because the mass suspended from the pulleys is measured in grams. A gram is a unit of mass, not force, but no confusion will arise if we remember that the force involved is actually the *weight* of 1 g of mass.

The procedure consists of first setting the pulleys for the two original vectors and suspending the proper weights to correspond to the forces given. The ring is held secure during this procedure. Then, another pulley is positioned and the force adjusted so that when released, the ring will remain stationary at the center of the table. When this is achieved, the ring is in equilibrium and it is clear that the third force balances the other two. The third force is called the *equilibrant* because it successfully holds the ring in equilibrium against the action of the other two forces. The equilibrant is equal in magnitude to the resultant, but it is directed

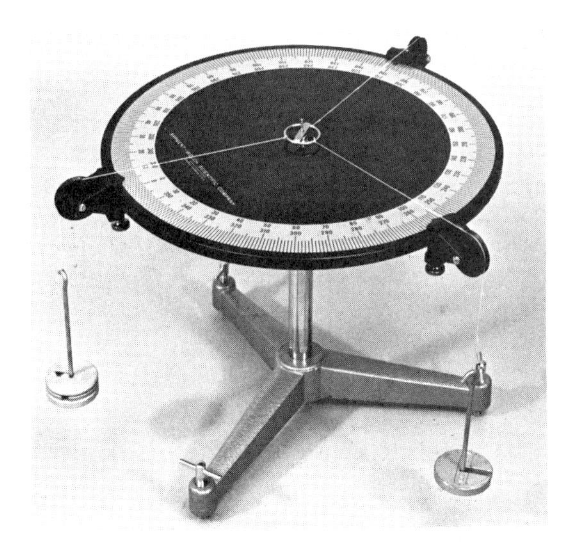

Figure 5-11 A force table.

in exactly the opposite direction. Thus, the resultant is obtained by simply reversing the direction of the equilibrant.

In a typical laboratory exercise, the vectors are taken from the force table and carefully plotted on graph paper to form a vector polygon. Ideally, the polygon should be closed. The results obtained on the force table may also be checked analytically by using the law of cosines or by resolving each vector into components.

5-6 SUSPENDED-BODY PROBLEMS

The simplest case

A class of problems that vividly illustrates the applications of vectors involves objects suspended by light, perfectly flexible strings. In such problems, the vector diagram is nearly identical to the sketch of the physical situation. The item usually sought is the tension in one or more of the strings. We begin with the simplest case, that of an object suspended by a vertical string, as in Fig. 5-12(a).

In this problem, it is obvious that the tension in the string simply equals the weight of the object. However, note that on the vector diagram the body is replaced by a point with the two force vectors acting on the point.

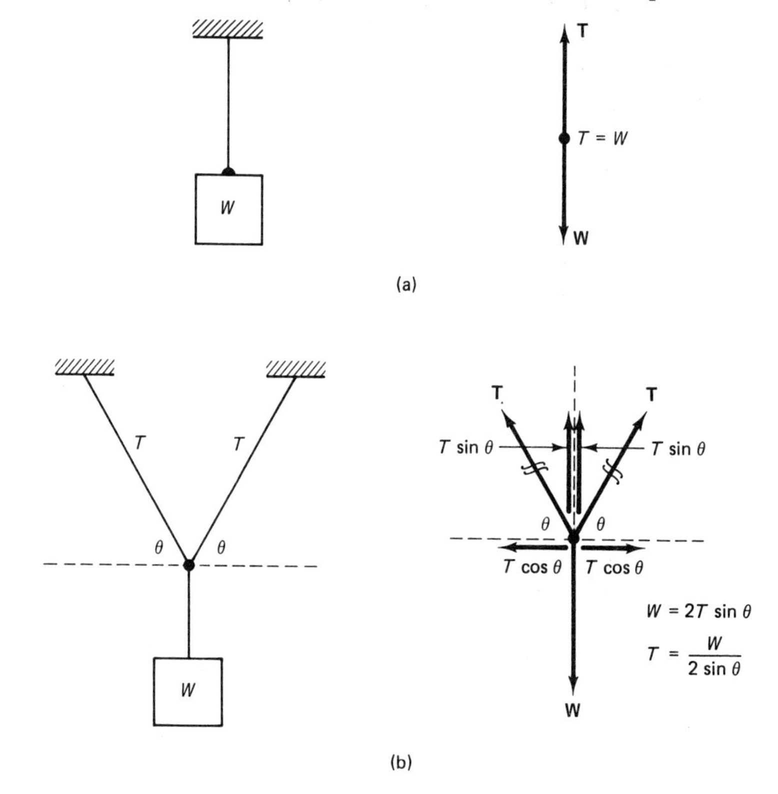

Figure 5-12 Two examples of suspended-body problems.

Two oblique but symmetrical supporting strings

In Fig. 5-12(b), a weight is supported by two strings that make equal angles to the horizontal. Because of the symmetry of the configuration, the same tension T exists in both strings. Further, because the weight is stationary, it must be in equilibrium. Consequently, the sum of the horizontal and of the vertical components must be zero.

Due to the symmetry about the vertical axis, it is obvious that the horizontal components of the tension in the two strings cancel. Adding the vertical components on the knot where the strings join gives

$$2T \sin \theta = W \tag{5-4}$$

from which we obtain

$$T = \frac{W}{2 \sin \theta} \tag{5-5}$$

The factor 2 arises because each string has a vertical component of $T \sin \theta$.

Suspended cables must have enough sag so that the weight of a few dozen gulls will not cause the cable to snap.

Three special cases are of interest, namely, the cases where $\theta = 90°$, $\theta = 30°$, and $\theta = 0°$. When $\theta = 90°$ ($\sin 90° = 1.0$), the two strings will be vertical and each string will support half the weight of the object. The tension in each string will be $T = W/2$. For $\theta = 30°$, the tension in each string equals the weight of the object ($\sin 30° = 0.5$). Practically, it is impossible for θ to equal zero because the tension would be theoretically infinite ($\sin 0 = 0$), and the strings would break.

One oblique and one horizontal string

The configuation of Fig. 5-13 involves both horizontal and vertical components in a nontrivial, yet simple, form. The vertical component of the oblique string is the only component that contributes to the lifting of the weight. The force F has no vertical component because it is exactly horizontal. The vector diagram and the physical configuration are almost identical, so we work directly from the drawing, omitting the vector diagram. We sum the forces on the knot where the strings come together. The result is given in Fig. 5-13.

As special cases, what happens to T and F when $\theta = 90°$? Could θ ever equal zero? What value of θ is required for F to be numerically equal to W?

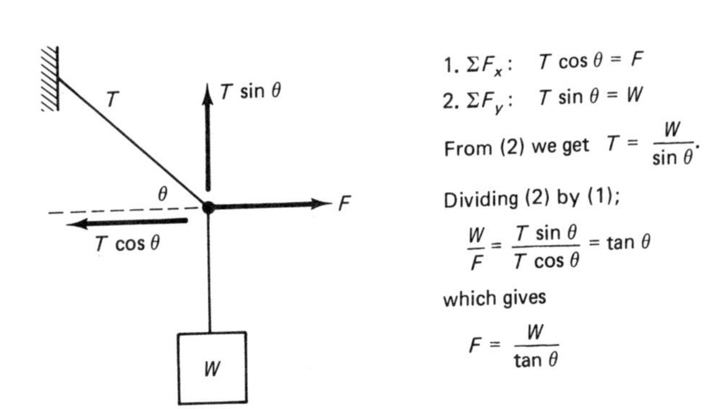

Figure 5-13 A body suspended by one oblique and one horizontal string.

Two oblique (but not symmetric) supporting strings

This example is more complicated, but the same procedure of summing the vertical and horizontal components on the knot is used. Refer to Fig. 5-14 as we describe the mathematical steps.

Summing the vertical components gives Eq. (1). Summing the horizontal components gives Eq. (2). Both T_1 and T_2 appear in both equations, giving two equations with two unknowns. We solve by substitution.

When Eq. (2) is solved for T_2, the result is Eq. (3). Equation (4) is obtained by substituting the expession for T_2 [Eq. (3)] into Eq. (1). Algebraic manipulation gives Eq. (5). We observe the following combination in the numerator of Eq. (5):

$$\cos \theta_1 \sin \theta_2 + \cos \theta_2 \sin \theta_1$$

We recognize this trigonometric identity to be equal to $\sin(\theta_1 + \theta_2)$. A math handbook aids the recognition process. Equation (6) results from substituting the identity, and the final expression for T_1 follows.

If the final expression for T_1 is substituted into Eq. (3), the expression for T_2 is obtained by a cancellation of $\cos \theta_2$. Thus, two similar expressions for the tensions T_1 and T_2 are obtained.

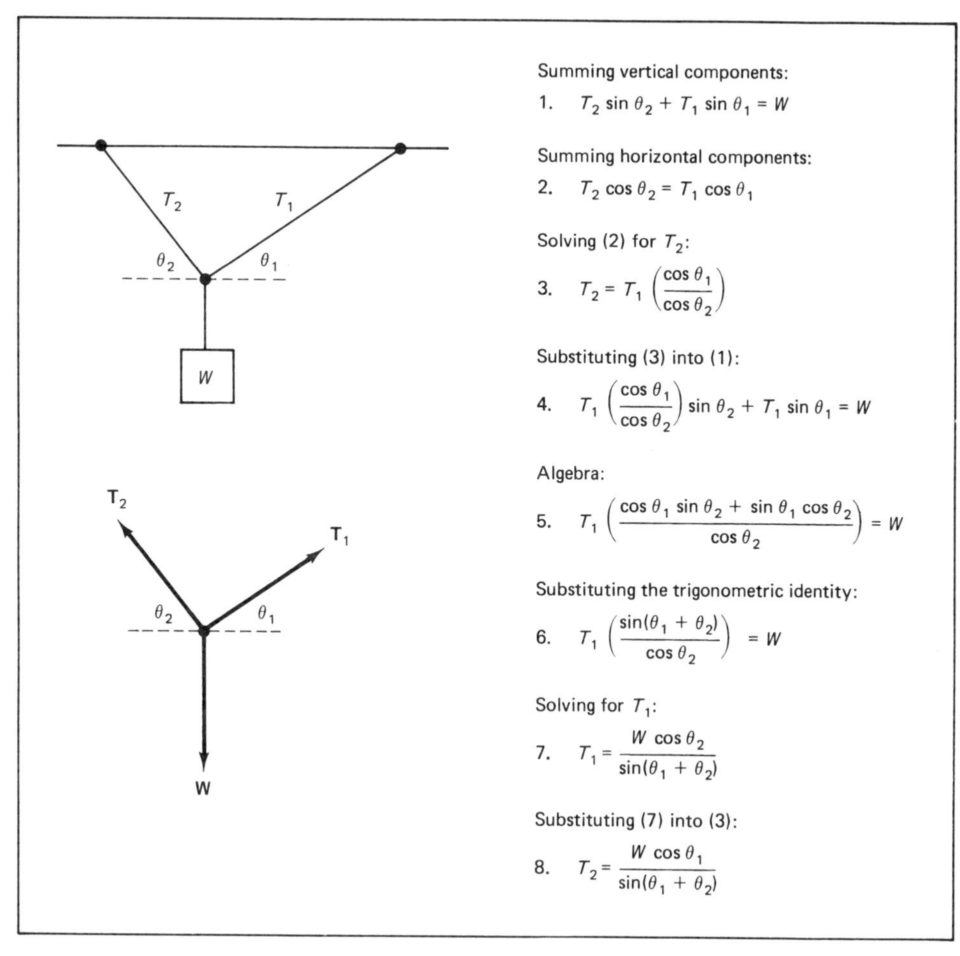

Figure 5-14 Derivation of string tensions in the unsymmetric suspended-body problem.

5-7 THE INCLINED PLANE

When an object is placed on an incline, the weight acts downward, as always, but a component of the weight tends to cause the object to slide down the plane. This can be seen more clearly if the weight vector is resolved into components parallel and perpendicular to the surface of the plane, as shown in Fig. 5-15. The component perpendicular to the surface is called the *normal* component. Note that the angle between the weight vector and the normal component is the same as the angle of the incline. The force tending to slide the object down the plane is given by $W \sin \theta$, and the force with which the object presses in the direction normal to the surface is $W \cos \theta$, as illustrated.

Let us suppose that the surface of the inclined plane is frictionless and assume the mass of the object on the incline is m. The weight is then mg, and the

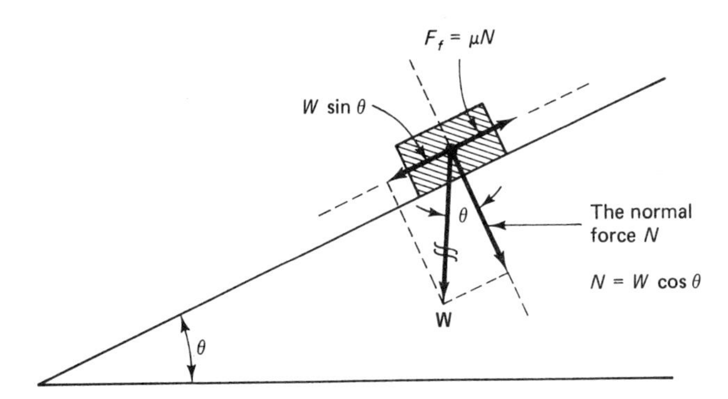

Figure 5-15 Forces exerted on an object sliding down an incline.

force component directed down the plane is $mg \sin \theta$. By Newton's second law, the object will tend to accelerate down the plane:

$$F = ma$$

$$\cancel{m}g \sin \theta = \cancel{m}a \rightarrow a = g \sin \theta \tag{5-6}$$

Note, however, that the acceleration of the object down the plane is reduced by the $\sin \theta$ term. If the plane is inclined at an angle of 5°, the acceleration down the plane is 2.79 ft/s^2, or only 8.7% of the acceleration of gravity. Consequently, the acceleration occurs slowly enough to be observed easily.

Frictional force on an inclined plane

Frictional forces developed between sliding surfaces are calculated by multiplying the coefficient of friction μ for the two surfaces by the normal force. For an object on an inclined plane, the normal force equals $W \cos \theta$; that is, the normal force varies with the angle of the incline. When the angle of inclination is 0°, corresponding to a horizontal surface, the normal force equals the weight of the object because $\cos 0° = 1$. In light of this, the frictional force on an inclined plane is given by

$$F_f = \mu N = \mu W \cos \theta \tag{5-7}$$

Of course, μ is either the coefficient of static or kinetic friction, as the case may be.

Frictional forces are directed to oppose the motion, or the impending motion if motion has not yet begun. Therefore, for an object about to slide *down* an incline, the frictional force will be directed up the plane, as in Fig. 5-15. However, if the object were being pulled *up* the plane by an external force, the frictional force would be directed *down* the plane.

A simple method for finding the coefficient of friction between an object and an inclined surface is shown in Fig. 5-16.

EXAMPLE 5-3 A large crate rests on a ramp inclined at an angle of 30° to the horizontal. If the crate weighs 200 lb and if the coefficient of friction between the crate and the ramp is 0.7, calculate the force F_a required to (a) pull the crate *down* the ramp; (b) pull the crate *up* the ramp. (c) What is the limiting angle of repose for the surface-ramp combination?

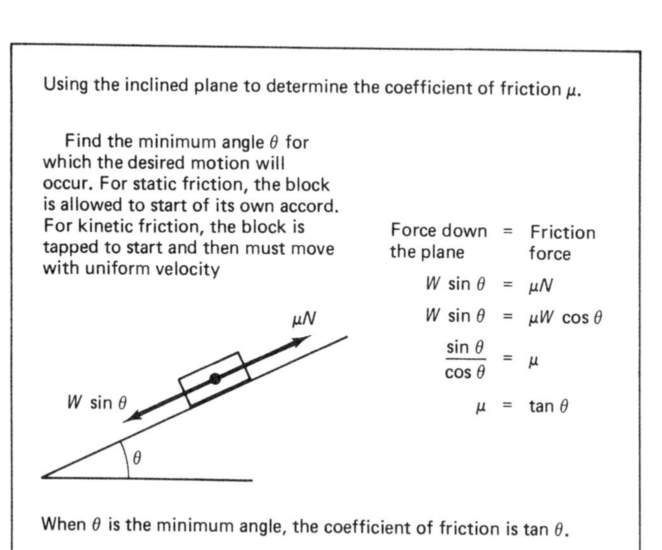

Figure 5-16 A simple method for determining the coefficient of friction μ.

Solution. First, sketch the physical situation and draw the vector diagram, as in Fig. 5-17. Then calculate the normal force, which is found to be 173.2 lb. The frictional force is then the product of the coefficient of friction and the normal force, namely, 121.2 lb. The frictional force may be directed either up or down the plane depending upon the impending motion of the crate.

(a) To pull the crate down the incline, the weight component and the applied force act in the same direction. Only the frictional force up the plane opposes the motion.

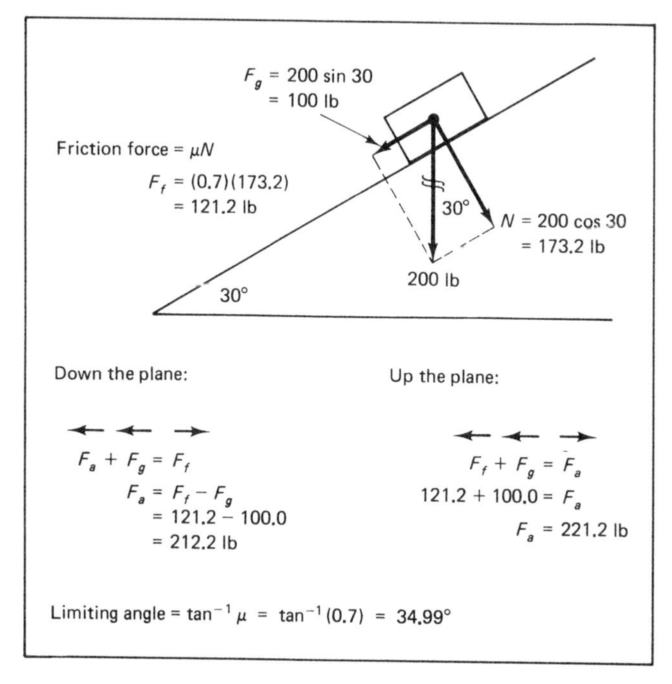

Figure 5-17

At the point where motion ensues, the forces down the plane equal the forces up the plane. This equality yields 21.2 lb as the force required to pull the crate down the plane.
(b) To pull the crate up the incline, the applied force must overcome both the frictional force and the weight component. When the motion is about to begin, the applied force will equal the frictional force plus the weight component. Thus, an applied force of 221.2 lb is required.
(c) The limiting angle of repose is simply the angle whose tangent equals the coefficient of friction, or 34.99°.

5-8 VECTORS IN PROJECTILE MOTION

Vectors are particularly suited to projectile motion because of the independence of the vertical and horizontal components of the motion. The case in which a projectile is fired at angle θ to the horizontal is a good example, as illustrated in Fig. 5-18. The initial velocity of the projectile is v_m, the muzzle velocity of the gun. In such problems, we usually want to calculate the range of the gun, the maximum height reached by the projectile, the time of flight, and so forth.

The initial velocity v_m is resolved into vertical and horizontal components v_{my} and v_{mx}, which are given by

$$\begin{aligned} (\textit{vertical}) \qquad v_{my} &= v_m \sin\theta \\ (\textit{horizontal}) \qquad v_{mx} &= v_m \cos\theta \end{aligned} \tag{5-8}$$

Once these components are computed, the vertical and horizontal elements of the problem can be considered separately. Neglecting air resistance, the horizontal

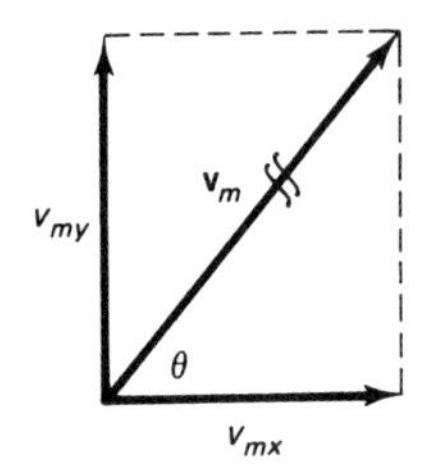

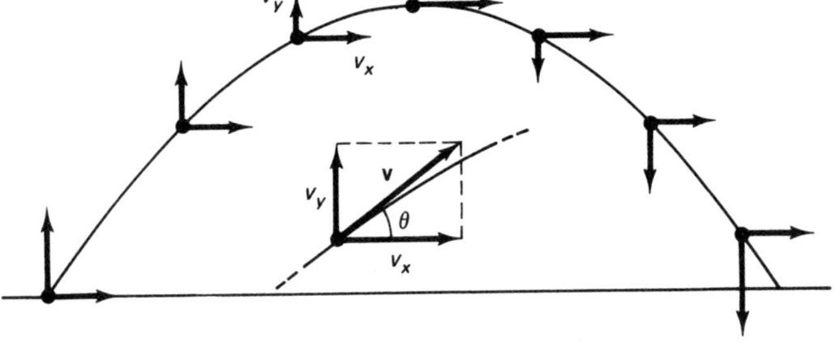

Figure 5-18 Position and velocity components of a projectile.

Water squirted from a hose traverses a parabolic trajectory. Surface tension causes the stream of water to break up.

component v_x remains constant and equal to v_{mx}. However, the vertical component v_y varies due to the acceleration of gravity and is treated as in the problem of the vertical cannon of Sect. 3-8. In terms of time t, the x- and y-components of the trajectory are given by

$$y = v_{my}t + \tfrac{1}{2}gt^2 \qquad \text{(obtained from } h = V_0t + \tfrac{1}{2}gt^2\text{)}$$
$$x = v_{mx}t \qquad \text{(obtained from } s = \bar{v}t\text{)} \tag{5-9}$$

where the projectile is assumed to be fired from the origin ($x = 0$, $y = 0$) at time $t = 0$ and the numerical value of g is entered as a negative quantity. The velocity components v_y and v_x are given by

$$v_y = v_{my} + gt \qquad \text{(obtained from } v = v_0 - gt\text{)}$$
$$v_x = v_{mx} \qquad \text{(a constant value)}$$

Note in these equations and in Fig. 5-18 that the x-component of the velocity remains constant, but that the vertical component of the velocity gradually decreases to zero and then turns downward. The total velocity at any point on the trajectory is obtained from the Pythagorean theorem and the tangent relationship:

$$v = \sqrt{v_x^2 + v_y^2} \qquad \theta = \tan^{-1}\left(\frac{v_y}{v_x}\right)$$

Making the proper substitutions into the formulas for the vertical cannon (Sect. 3-8) gives the following for the time of flight:

$$(Eq.\ 3\text{-}18) \quad t = \frac{-2v_0}{g} \rightarrow t_f = \frac{-2v_{my}}{g} = \frac{-2v_m \sin\theta}{g} \tag{5-10}$$

The maximum height is given by:

$$(Eq.\ 3\text{-}15) \quad h = \frac{-v_0^2}{2g} \rightarrow h = \frac{-v_{my}^2}{2g} = \frac{-v_m^2 \sin^2 \theta}{2g} \tag{5-11}$$

The range is given by the product of the horizontal component of the velocity v_{mx} and the time of flight:

$$\text{Range} = v_{mx} t_f = (v_m \cos \theta) \left(\frac{-2v_m \sin \theta}{g} \right)$$
$$\text{Range} = \frac{-2v_m^2 \sin \theta \cos \theta}{g} = \frac{-v_m^2 \sin 2\theta}{g} \tag{5-12}$$

You can verify that the range is a maximum when the angle of elevation θ is 45° above the horizon. Furthermore, a given range will result for *two* angles of elevation: If $\theta_1 = \alpha$ produces a given range, $\theta_2 = 90 - \alpha$ will produce the same range.

"Shoot-the-monkey" problem

This problem has been handed down through several generations of physicists. A hunter sees a monkey in a distant tree and aims a rifle directly at the monkey without making any adjustment to the sights to allow for distance. The monkey sees fire and smoke when the hunter fires, and instantly lets go and falls (from rest) toward the ground. Will the bullet hit the monkey?

We might at first think that the bullet will pass over the monkey, but this is not the case; the bullet will hit the monkey. Once the bullet is released, the same acceleration of gravity (g) acts on it as acts on the monkey. The monkey falls downward from the limb on the tree and the bullet falls downward an equal amount measured vertically from the line of sight.

We can verify mathematically that the bullet will hit the monkey as follows. First of all, the horizontal component of the bullet's velocity is $v_m \cos \theta$, which remains constant. Therefore, we can calculate t_c, the time after firing that the bullet arrives at the point of collision:

$$t_c = \frac{x}{v_m \cos \theta} \tag{5-13}$$

where x is the horizontal distance from the hunter to the tree in which the monkey was sitting.

We now calculate the height h_b of the bullet as it passes the tree. Substituting $v_m \sin \theta$ for v_0 in Eq. (5-9) and using t_c for t, we obtain

$$h_b = (v_m \sin \theta) t_c + \tfrac{1}{2} g t_c^2$$

When the expression for t_c [Eq. (5-13)] is substituted into the first term of this equation, we have

$$h_b = (v_m \sin \theta) \left(\frac{x}{v_m \cos \theta} \right) + \frac{1}{2} g t_c^2$$

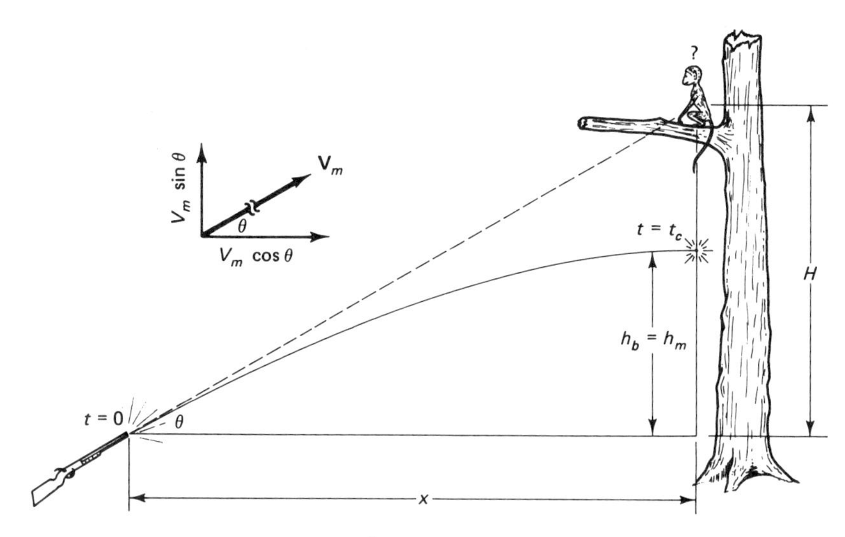

Figure 5-19 Shoot-the-monkey problem.

which simplifies to

$$h_b = x \tan \theta + \tfrac{1}{2} g t_c^2$$

because $\sin \theta / \cos \theta = \tan \theta$. From the sketch shown in Fig. 5-19, we see that $\tan \theta = H/x$, where H is the initial height of the monkey. Hence,

$$h_b = H + \tfrac{1}{2} g t_c^2 \tag{5-14}$$

This is the height of the bullet as it passes the monkey. (Recall that g is a negative quantity.)

We now consider the height of the monkey when the bullet passes. Because the monkey was at height H when the hunter fired, we add H to Eq. (5-9):

$$h_m = H + v_0 t + \tfrac{1}{2} g t^2$$

Because the monkey fell from rest, $v_0 = 0$, and we use t_c for t because we are interested in h_m at the instant the bullet passes. We then have

$$h_m = H + \tfrac{1}{2} g t_c^2 \tag{5-15}$$

This expression is identical to Eq. (5-14), so that the monkey and bullet will be at the same height when the bullet arrives. This implies that the bullet will hit the monkey.

5-9 TORQUE

Definition

When a force is applied to an object, a turning or twisting force is often produced, as illustrated in Fig. 5-20, where a force applied to the handle of the wrench tends to turn the nut. Note the *line of action* extending from the force vector. The

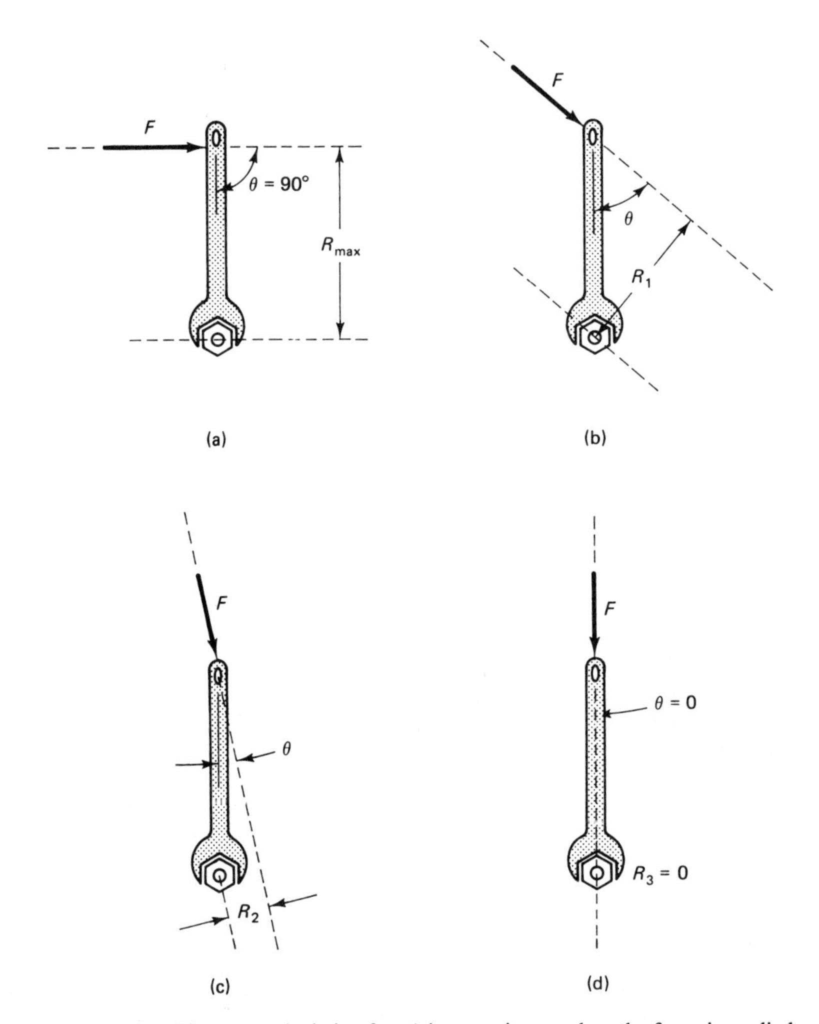

Figure 5-20 The torque (twisting force) is a maximum when the force is applied at right angles to the wrench, as in (a). In (b), moment arm R_1 is less than R_{max}, and the moment arm R_2 in part (c) is even smaller. In (d), the moment arm is zero, and the torque is zero because the line of action of the force passes through the pivot point.

moment arm R is the shortest distance from the line of action to the point of rotation, the pivot. The line representing the shortest distance is the perpendicular to the line of action that passes through the pivot point. The magnitude of the twisting force, called *torque* and denoted by the Greek letter tau τ, is the product of the force and moment arm.

The moment arm is greatest when the force vector is perpendicular to a line extending from the pivot to the point of application of the force. When the force is exerted in any other direction, the line of action of the force will pass closer to the pivot, reducing the moment arm. This is illustrated in Fig. 5-20, where it is obvious that R_1, R_2, and R_3 are progressively smaller; R_3 equals zero because the line of action passes through the pivot point.

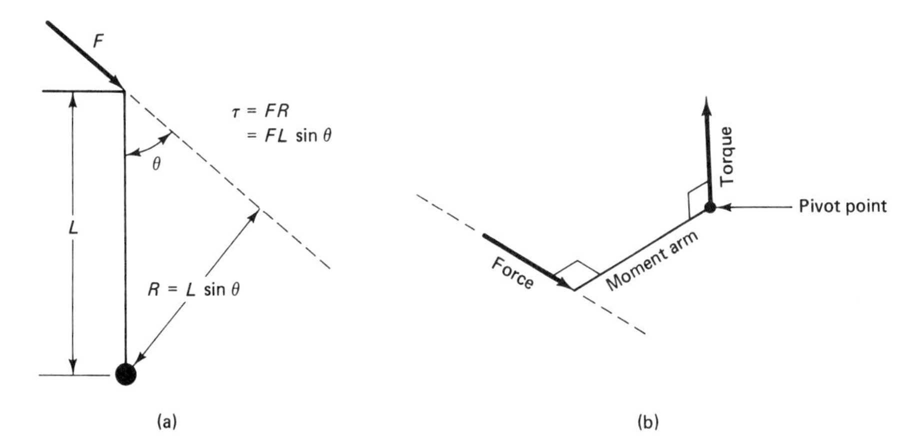

Figure 5-21 (a) The torque depends upon the sine of the angle between the force and moment arm. (b) Torque is a vector quantity whose direction is perpendicular to both the force and moment arm.

Because the moment arm varies with the direction of the applied force, the torque τ does also. Specifically, the torque varies with the sine of the angle θ between the line of action of the force and the handle of the wrench, $\tau = FL \sin \theta$, as shown in Fig. 5-21(a).

In the strictest sense, torque is a vector quantity whose direction is perpendicular to both the applied force and the moment arm, as shown in Fig. 5-21(b). However, in this text we need only to consider whether a torque produces a turning force that is *clockwise* or *counterclockwise*. For the sake of brevity, we refer to counterclockwise torques as τ_+ and to clockwise torques as τ_-.

The second condition for equilibrium

One condition for equilibrium is that all forces applied to an object must have a vector sum of zero; that is, the sums of the x-components and of the y-components must equal zero. The second condition for equilibrium is that all torques must have a vector sum that gives a net torque of zero. Put another way, the sum of the counterclockwise torques must equal the sum of the clockwise torques so that a condition of balance exists. If it is determined that a given object is in equilibrium, it may be assumed that the following conditions are met:

$$\Sigma F_x = 0 \qquad \Sigma F_y = 0 \qquad \Sigma F_z = 0 \qquad \Sigma \tau = 0 \tag{5-16}$$

The forces in the z-direction need be considered only if the problem is three-dimensional.

EXAMPLE 5-4 An example of torques in equilibrium is provided by the step-pulley arrangement of Fig. 5-22. A weight W is suspended from the pulley of smaller radius R_1, and a force F is applied to the larger pulley of radius R_2. We wish to find the relationship between the weight W and the applied force F.

Solution. If the system is in equilibrium, the torque to the left must equal the torque to the right. The moment arm for each torque is the radius of the respective pulley.

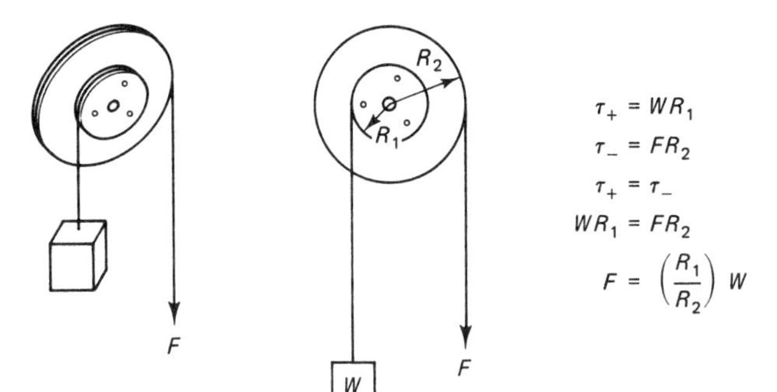

Figure 5-22 Step-pulley system in equilibrium.

Consequently, $\tau_+ = WR_1$ and $\tau_- = FR_2$. Setting these quantities equal and rearranging gives the relationship between F and W.

EXAMPLE 5-5 Examine the system of Fig. 5-23, in which a bucket of weight W is held up by force F_1 acting at the end of a light, rigid rod attached to a wall by a frictionless pivot. The weight of the rod is assumed to be negligible. We wish to find the relationship between F_1 and W.

Solution. Assuming, naturally, that the bar may rotate about the pivot, we sum the torques about the pivot. The weight of the bucket W and moment-arm distance L_2 give the torque to the right. The upward force F_1 and moment-arm distance L_1 give the torque to the left. The force acting upward through the pivot does not give rise to a torque about the pivot, because the moment-arm distance is zero. Consequently, the two torques involved are WL_2 and F_1L_1, as indicated. Setting these equal gives the force F_1 in terms of W and the distances L_1 and L_2.

If W is given, we can compute F_1, but as yet we have not considered the upward force on the pivot. This force is most easily determined by summing the vertical forces. The sum of the upward forces, F_p and F_1, must equal the downward force W. This gives $F_p = W - F_1$.

Examine the solution to Example 5-5 and determine, by examining the expression for F_1, what happens to F_1 as L_2 becomes smaller in relation to L_1. What happens to the force F_1 as the bucket is moved closer to the pivot? What will

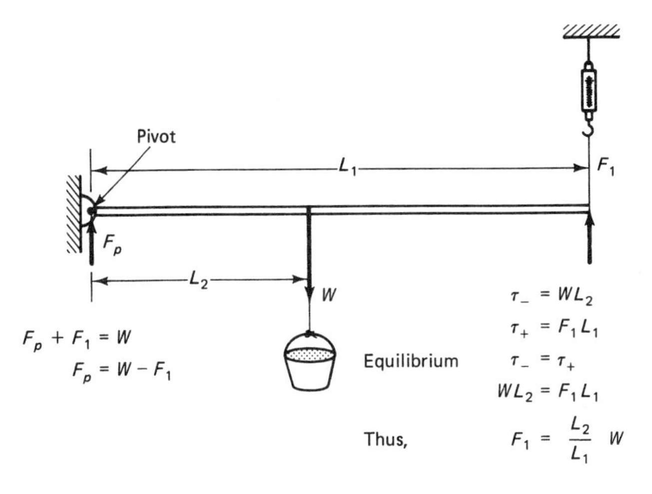

Figure 5-23

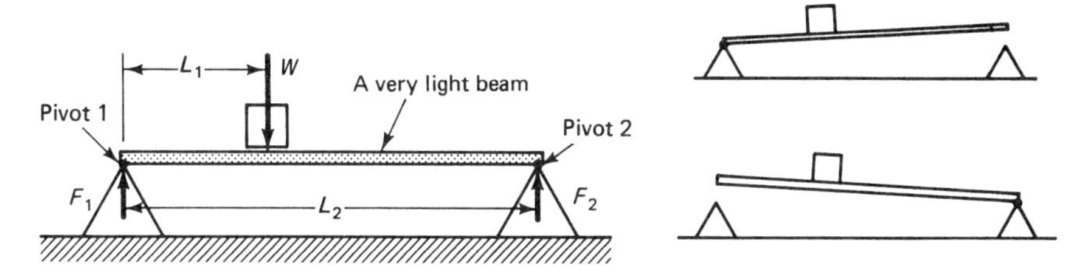

Figure 5-24

be the value of F_1 when the bucket is moved directly under the scale so that $L_1 = L_2$?

EXAMPLE 5-6 Torques may be summed about points in physical configurations that are not actual pivots or places where rotation is normally possible or expected. In Fig. 5-24, a weight W rests on a light beam supported by two triangular blocks. There are no points of rotation because this system normally does not rotate. However, for the purpose of summing torques, we may assume the beam to rotate about the apex of block 1; or, equivalently, we may assume it to rotate about the apex of block 2. These points are labeled pivot 1 and pivot 2 in the figure, even though no actual pivot is present.

Summing the torques about pivot 1 gives

$$\tau_- = WL_1; \qquad \tau_+ = F_2L_2; \qquad \tau_- = \tau_+ \rightarrow WL_1 = F_2L_2$$

$$F_2 = \left(\frac{L_1}{L_2}\right) W$$

Thus, summing the torques about pivot 1 yields an expression for the force F_2. In order to find the force F_1 in terms of the weight, we sum the torques about the other pivot, pivot 2.

$$\tau_+ = W(L_2 - L_1); \qquad \tau_- = F_1L_2; \qquad \tau_- = \tau_+ \rightarrow W(L_2 - L_1) = F_1L_2$$

$$F_1 = \left(\frac{L_2 - L_1}{L_2}\right) W$$

We now can calculate both forces F_1 and F_2 when the weight is given.

To illustrate that the torques can be summed about any point, we can sum the torques about the point directly under the weight in order to obtain an expression relating F_1 and F_2. The weight W will not appear in the expression because the line of action of the weight vector passes through the point about which the torques are summed, giving a moment arm of zero so that no torque results. The torque to the right is F_1L_1; the torque to the left is $F_2(L_2 - L_1)$. Equating these and solving for the desired unknown gives

$$F_1 = F_2\left(\frac{L_2}{L_1} - 1\right) \qquad F_2 = F_1\left(\frac{L_1}{L_2 - L_1}\right)$$

EXAMPLE 5-7 The technique of summing torques about properly chosen pivot points is well suited to a class of problems involving booms and cranes, as illustrated in Fig. 5-25. A horizontal string maintains the boom at an angle θ to the vertical. A weight W is

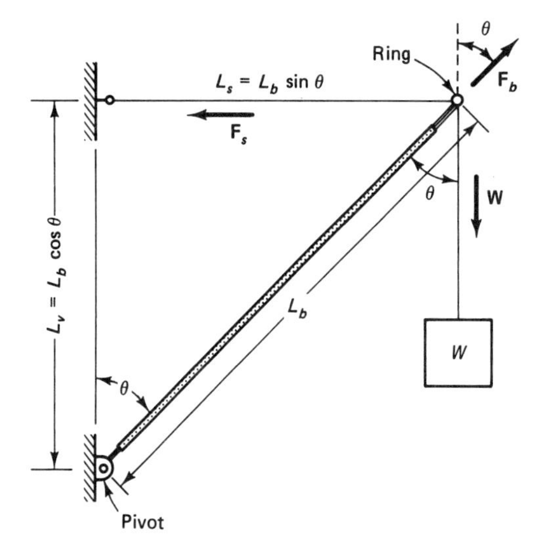

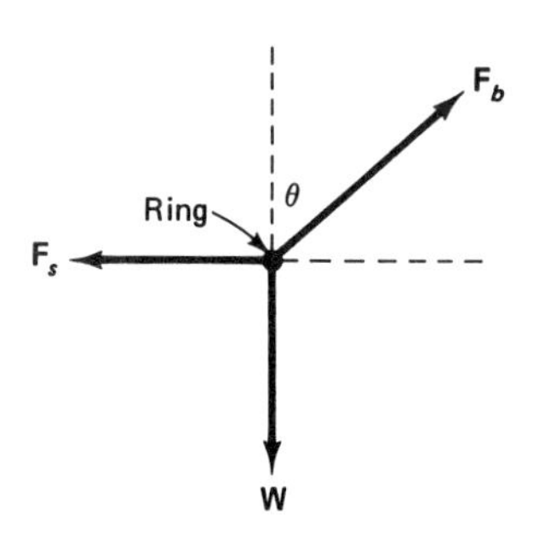

Sum torques about the pivot:

$$\tau_+ = \tau_-$$
$$F_s L_v = W L_s$$
$$F_s L_b \cos\theta = W L_b \sin\theta$$
$$F_s = W \frac{\sin\theta}{\cos\theta} = W \tan\theta$$

Sum vertical forces at ring:

$$F_b \cos\theta = W \qquad F_b = \frac{W}{\cos\theta}$$

Solution obtained by summing components of forces:

$$\Sigma F_x: \quad F_s = F_b \sin\theta$$
$$\Sigma F_y: \quad F_b \cos\theta = W$$
$$F_b = \frac{W}{\cos\theta}$$
$$F_s = F_b \sin\theta = \frac{W}{\cos\theta} \sin\theta$$
$$F_s = W \tan\theta$$

Figure 5-25

suspended from the boom, which is free to rotate in the pivot at the lower end. We want to calculate the tension F_s in the string and the compression force F_b in the boom.

Solution. Because the boom pivots at the bottom, the force vector of the boom will lie on line with the boom. The lines of action of the three forces involved intersect at the ring on the top of the boom, and this ring provides a convenient origin for the vector diagram of the forces shown in the figure.

The dimensions L_s and L_v constitute the moment arms for W and F_s, respectively, when the torque is summed about the pivot. These dimensions are obtained from the length of the boom and the angle θ by simple trigonometry. Summing the torques gives F_s in terms of W and the angle θ: $F_s = W \tan\theta$.

Summing the torques about the pivot cannot yield an expression for the force in the boom F_b because the line of action of the force passes through the pivot. Therefore, F_b is most easily determined by summing the vertical forces acting on the ring at the top of the boom. The upward component of the force is $F_b \cos\theta$, and this must equal the weight W. The expression for F_b follows.

This particular problem also may be solved quite readily by summing vertical and horizontal forces at the ring, as shown in the figure.

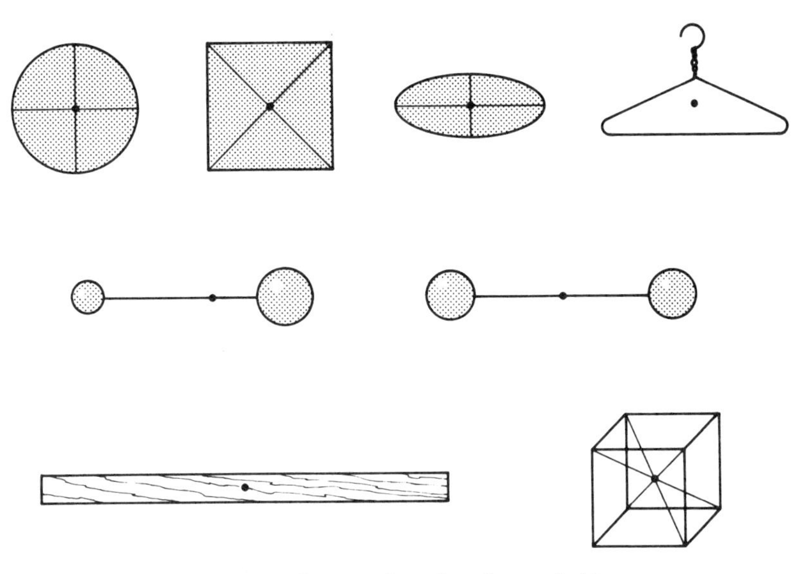

Figure 5-26 Centers of gravity of several objects.

5-10 CENTER OF GRAVITY

Definition

We define the *center of gravity* in a roundabout way by saying that if a body is suspended by or is supported at the center of gravity, the body will show no tendency to rotate—it will be perfectly balanced in all directions. For any particular body, there is only one point for which this will be the case.

Figure 5-26 shows several objects and the center of gravity of each. For flat, regularly shaped objects such as the circle, square, and ellipse, the center of gravity is at the geometric center of the object. For masses connected by a light, rigid rod, the center of gravity will lie closer to the larger mass. For a heavy beam that is uniform throughout its length, the center of gravity will lie at the geometric center of the beam. A coat hanger is unusual because its center of gravity lies in the space within the loop where there is no wire.

The center of gravity of a flat, irregularly shaped body can be found by suspending the body from at least two different pivot points and marking the intersection of the two vertical lines obtained, as illustrated in Fig. 5-27. An object freely suspended will come to rest with the center of gravity exactly underneath the pivot point.

The concept of center of gravity is useful in many aspects of physics. One use is illustrated in Fig. 5-28, where a heavy beam supported at two points holds two objects. The action of gravity is distributed over the entire length of the beam. However, the entire weight of the beam may be assumed to be concentrated at the center of gravity and may be represented by a downward-directed vector of magnitude equal to the weight of the beam. In a similar manner, the two objects on the beam may be concentrated at their respective centers of gravity, as indicated.

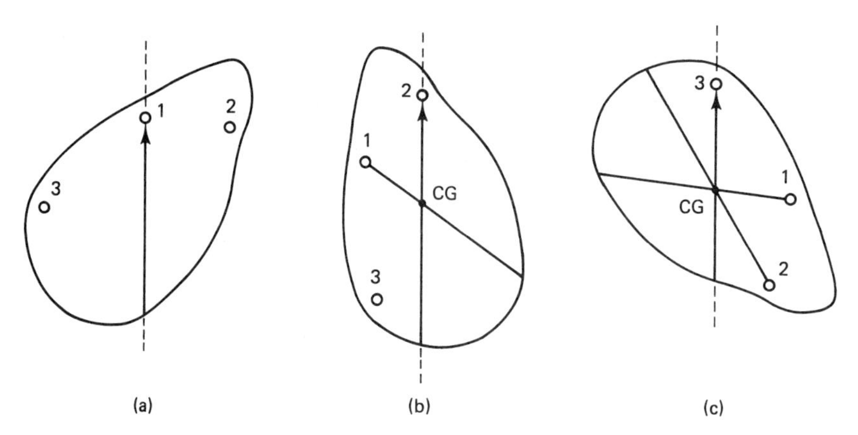

Figure 5-27 An object freely suspended comes to rest with the center of gravity (CG) directly underneath the pivot point. This fact can be used to locate the center of gravity of a planar object, as illustrated.

The resulting, simplified problem is shown in Fig. 5-29. The forces exerted on the two supports F_1 and F_2 may now be found by application of the conditions of equilibrium.

Center of gravity and equilibrium

Section 5-5 describes stable, neutral, and unstable equilibrium. We can now relate the stability of the equilibrium to the vertical movement, or shifting, of the center of gravity of the body when the body undergoes a small displacement from the equilibrium position.

1. *Stable equilibrium.* The center of gravity is raised to a higher level by a small displacement from the equilibrium position.
2. *Neutral equilibrium.* The vertical level of the center of gravity is not changed by a small displacement from the equilibrium position.
3. *Unstable equilibrium.* The center of gravity is shifted to a lower level by a small displacement from the equilibrium position.

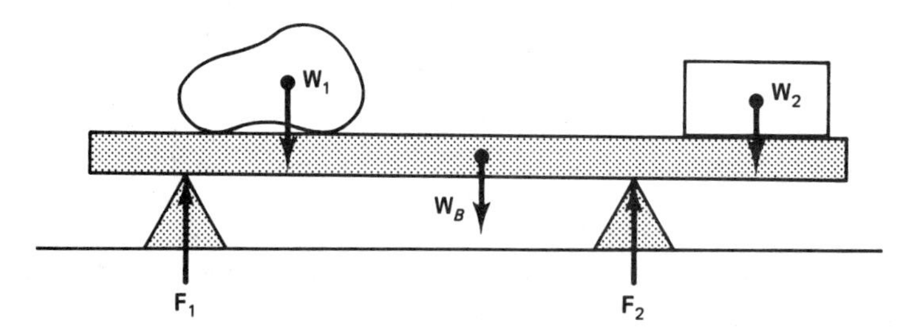

Figure 5-28 The weight of the objects can be represented as a single force acting downward through the center of gravity.

Total load → $L = W_1 + W_2 + W_B$

Sum of vertical forces: $F_1 + F_2 = L$

Sum torques about apex of the left-hand support:

$$F_2 L_2 = W_1 L_1 + W_2 L_3 + W_B L_B$$

$$F_2 = \frac{1}{L_2}(W_1 L_1 + W_2 L_3 + W_B L_B)$$

Then, using the result obtained above,

$$F_1 = L - F_2$$

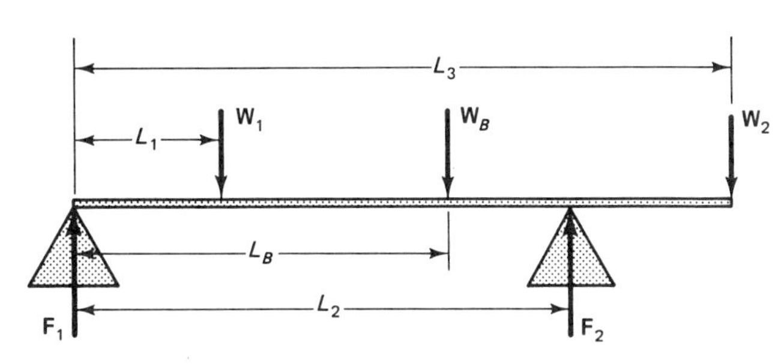

Figure 5-29 The simplified problem of Fig. 5-28.

From this it is clear that for a condition of stable equilibrium to exist, the center of gravity must be at its lowest possible position (within the confines or conditions set forth by the problem).

Thus, for stable equilibrium, the potential energy of the system must be a minimum at the equilibrium position. The higher the center of gravity, the greater the potential energy. Consequently, when the center of gravity is lowest, the potential energy is least. If a small displacement of the body from the equilibrium position increases the potential energy, the equilibrium is stable. If the potential energy is unchanged, the equilibrium is neutral; and, if the potential energy is decreased, the equilibrium is unstable. This will probably become clearer after the concept of energy is introduced in the next chapter.

Questions

1. Which of the following are vector and which are scalar quantities? **(a)** Mass **(b)** Velocity **(c)** Force **(d)** Speed **(e)** Time **(f)** Acceleration **(g)** Weight **(h)** Volume **(i)** Torque **(j)** Temperature
2. A car rounds a curve at 55 mi/h. Is its speed constant? Is its velocity constant?
3. Suppose vector **B** is to be subtracted from vector **A**. Describe the procedure for doing it graphically.
4. A pilot flies from town W due east to town E. If the wind is from the south, what will be the effect of the wind on the time required for the flight? (That is, will it require a longer time, a shorter time, or have no effect?)
5. Briefly describe the stable, unstable, and neutral states of equilibrium. Give one or more examples of each.
6. A 100-lb crate rests on a rough floor. First, suppose nothing pushes horizontally against the crate. What frictional force does the floor exert on the crate? Next, suppose a child pushes horizontally against the crate with a force of 10 lbs. What frictional force is exerted on the crate?
7. When a tire on an automobile begins to spin in the snow, should the driver apply more gas or less gas? Why?
8. Briefly explain how torque is used in solving problems related to booms and cranes even though these items do not rotate.
9. Suppose an object is displaced slightly from its equilibrium position. In such a case, its center of gravity must move upward slightly, move downward slightly, or remain at the same level. Describe the type of equilibrium of which each movement would be characteristic.

Problems

1. In Fig. 5-1(a), give the components of the displacement vector between each pair of vectors. **(a)** **1** and **3** **(b)** **1** and **5** **(c)** **3** and **5**
2. In Fig. 5-1(a), by summing components, determine the vector sum of each combination of vectors. **(a)** **1** and **2** **(b)** **2** and **3** **(c)** **4** and **5**
3. In Fig. 5-1(b), use the method of Fig. 5-6(b) to find the magnitude and direction of each vector.
   (a) **A**
   (b) **B**
4. Determine the magnitude and direction of the displacement vector in Fig. 5-1(b).
5. Use Fig. 5-3.
   (a) Determine the components of vectors **A** and **B**, and then, by summing components, determine the components of the resultant.
   (b) What angle does the resultant make with the horizontal?
6. Use Fig. 5-3.
   (a) Determine the magnitude of **A** and **B**.
   (b) What angle does each vector make with the horizontal?
   (c) What is the angle *between* **A** and **B**?
   (d) Use the law of cosines to find the magnitude of the resultant.
7. In Fig. 5-4, determine the components of the resultant **R** by summing the components of the constituent vectors.
8. What is the net force on the sled of Fig. 5-2?
9. In Fig. 5-6(a), compute the components of the vector **A** if the magnitude of **A** is 16 cm and θ is 64°.
10. Find the magnitude and direction of the resultant of the following combinations of vectors

(a)	*Magnitude*	*Angle*	(b)	*Magnitude*	*Angle*
A	12	20°	**A**	12	20°
B	6	30°	**B**	10	120°
C	8	70°	**C**	8	−45°
			D	6	0°

11. Repeat Example 5-1 for a force of 40 lb applied at an angle of 25°.
12. Suppose in Fig. 5-10 that a small plane cruises at only 65 mi/h. Compute the ground speed and direction of flight if the pilot aims the plane due east.
13. Repeat Problem 12 for a jet airliner traveling at 600 mi/h, and compare the effect of the wind on the two aircraft.
14. Suppose the wind is blowing from the south toward the north at 30 mi/h.
    (a) In what direction should a plane traveling at 140 mi/h be aimed (relative to due east) in order to travel due east?
    (b) What will be the true ground speed?
15. In Fig. 5-12(b), the strings make equal angles with the horizontal and support a weight of 25 lb. Find the tension in each string for each angle.
    (a) 60° **(b)** 45° **(c)** 90° **(d)** 5°
16. In Fig. 5-13, for a weight W of 25 lb, what force F is required to produce each angle θ?
    (a) 60° **(b)** 90° **(c)** 30° **(d)** 5°
17. Compute the string tension T for each angle in Problem 16.
18. In Fig. 5-13, for a weight of 25 lb, what angle θ results for each applied force?
    (a) 5 lb **(b)** 15 lb **(c)** 25 lb **(d)** 50 lb
    (e) 100 lb
19. Suppose the weight in Fig. 5-14 is 100 lb; assume $\theta_1 = 40°$ and $\theta_2 = 60°$
    (a) Compute the tension in each string.
    (b) Calculate the vertical component of the tension in each string.
    (c) What is the sum of the vertical components?
20. Repeat Problem 19 letting $\theta_1 = 30°$ and $\theta_2 = 70°$.
21. At what angle should a (frictionless) inclined plane be set so that the acceleration of an object down the plane will be one-half the acceleration of gravity?
22. The coefficient of static friction, μ_s, between a block of wood and an adjustable inclined plane is 0.5. As the angle of the incline is gradually increased, at what angle will the block begin to move?
23. A 200-g block of wood is stationary on a

plane inclined at 20°. The coefficient of static friction, μ_s, is 0.6.

(a) What is the normal force between the block and the plane?

(b) What component of the weight is directed down the plane?

(c) What frictional force is directed up the plane?

(d) What is the net force on the block?

24. Repeat the previous problem for a 30° incline.

25. A block at rest begins to move down an incline when the angle reaches 39°. What is the coefficient of static friction between the block and the incline?

26. Repeat Example 5-3 for a ramp inclined at 34°.

27. A physics demonstration cannon fires a 200-g brass ball from floor level with a muzzle velocity v_m of 750 cm/s at an angle of elevation of 50° to the horizontal. Calculate the **(a)** vertical and **(b)** horizontal components of v_m. **(c)** To what maximum height will the ball ascend? **(d)** Compute the **(e)** time of flight, and **(f)** range of the ball.

28. In the preceding problem, compute **(a)** the position of the ball and **(b)** its velocity (vertical component, horizontal component, and total) 0.25 s after firing the cannon.

29. A horizontal cannon fires a heavy ball with a muzzle velocity of 200 ft/s from the top of a seaside cliff 180 ft high.

(a) Determine the horizontal and vertical velocity components at a time t of 1 s, 2 s, and 3 s after firing.

(b) What is the total time of flight of the cannonball?

(c) With what velocity and at what angle does the ball strike the water?

30. Suppose a bomb is dropped from a plane traveling at 300 mi/h in level flight at an altitude of 1000 ft.

(a) Compute the time required for the bomb to fall to the ground.

(b) From what horizontal distance from the target should the plane release the bomb?

(c) With what velocity and at what angle will the bomb strike the ground? (Neglect air resistance.)

31. Suppose in Fig. 5-22 that $R_1 = 2$ and $R_2 = 6$ in. What force F is required to support a weight W of 300 lb?

32. Refer to example 5-5 and Fig. 5-23.

(a) Compute the force F_1, given that $L_1 = 8$ ft, $L_2 = 2$ ft, and $W = 100$ lb.

(b) What is the upward force F_p at the pivot?

(c) Repeat the parts **(a)** and **(b)** if the load is moved 1 ft farther from the pivot.

33. In Fig. 5-24, assume $L_2 = 8$ ft and let $W = 100$ lb. Compute the forces F_1 and F_2 for each value of L_1.

(a) $L_1 = 1$ ft **(b)** $L_1 = 2$ ft **(c)** $L_2 = 4$ ft
(d) $L_2 = 6$ ft

34. In Fig. 5-25, the boom is 6 ft long and θ is 40°. Assume W is 100 lb.

(a) Compute the tension in the horizontal string.

(b) Find the compression force in the boom.

35. Repeat Problem 34 if θ is increased to 50°.

***36.** In Fig. 5-28, suppose the uniform beam is 10 ft long and weighs 24 lb. The force F_1 is applied 1 ft from the left end of the beam, and the two supports are 7 ft apart. The center of gravity of W_1 (40 lbs) is 2 ft from the left end of the beam, while the center of gravity of W_2 (60 lbs) is 1 ft from the right end of the beam.

(a) Make a sketch to scale showing the points of application of the various forces, as in Fig. 5-29.

(b) Compute the forces F_1 and F_2.

***37.** Repeat Problem 36, but neglect the weight of the beam.

CHAPTER 6

WORK, ENERGY, AND POWER

The concept of energy and of the conservation of energy is one of the most pervasive and powerful in all of physics. Even so, it is difficult to define energy in a meaningful way; perhaps the best way to appreciate the concept of energy and its physical reality is by studying its properties.

In this chapter we see that energy occurs in several basic forms, and it may be transformed from one form to another. Indeed, the principle of the conservation of energy states that energy can only be transformed from one form to another; it can neither be created nor destroyed.

There is a close relationship between work and energy. Energy is sometimes defined to be the capacity for doing work. Therefore, we begin this chapter with a scientific definition of work.

In the last section of this chapter, we briefly discuss sources of energy that may help to satisfy the incessant demands of society for greater and greater quantities of energy.

6-1 WORK

Definition

The scientific definition of *work* is far more specialized in meaning than work as understood by nonscientific people. The scientific definition is very narrow in scope, but it is precise. We say that work is done when a force applied to an object causes the object to move in the direction of the applied force. This is illustrated in Fig. 6-1, where a force F pushes a box across the floor a distance s. Further, we can say exactly how much work is done and express it in numbers. The work W done by the force F in pushing the box a distance s is the product of the force and the distance:

$$\text{Work} = \text{force} \times \text{distance}$$
$$W = F \cdot s \tag{6-1}$$

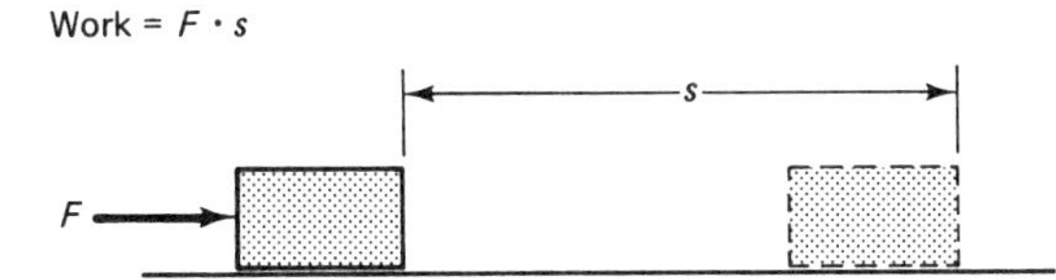

Figure 6-1 Work is done when a force F pushes a box a distance s across the floor.

The mere application of a force to an object does not necessarily mean that work is done in the process. The object must move, and a component of its motion must move in the direction of the force. It is the combination of force and *displacement* (the distance moved) that results in work being done. If either is zero, no work is done in the scientific sense.

Work and the units of work are first mentioned in Sect. 2-2. The unit of work is that of a force times a distance. Hence, we obtain newton-meters (joules), dyne-centimeters (ergs), and foot-pounds in the SI, CGS, and British systems, respectively, as illustrated in Fig. 6-2. If an applied force of 1 lb acts over a displacement of 1 ft, one ft-lb of work is done. A force of 1 N and a displacement of 1 m gives work in the amount of 1 N · m. The quantity 1 N · m, is called a joule in honor of James Prescott Joule (1818–1889). In a similar manner, 1 erg of work is done when a force of 1 dyn acts over a displacement of 1 cm.

System	Force	Distance	Work or Energy
British	pound (lb)	foot (ft)	foot-pound
SI (MKS)	newton (N)	meter (m)	newton-meter (joule, J)
CGS	dyne (dyn)	centimeter (cm)	dyne-cm (erg)

Figure 6-2 Units of work and energy. (The units of work and energy are the same.)

It often happens that the applied force and the resulting displacement are not in exactly the same direction, as illustrated in Fig. 6-3. In such case, the work done is computed by taking the component of the force in the direction of the displacement. Consequently, the work done is

$$W = F(\cos\theta) \cdot s \quad \text{or} \quad W = F \cdot s \cdot \cos\theta \tag{6-2}$$

When F and s are in line, θ is zero and $\cos\theta = 1$. Then Eq. (6-2) becomes the same as Eq. (6-1). If F and s are perpendicular so that $\theta = 90°$ ($\cos 90 = 0$), no work is done by the force.

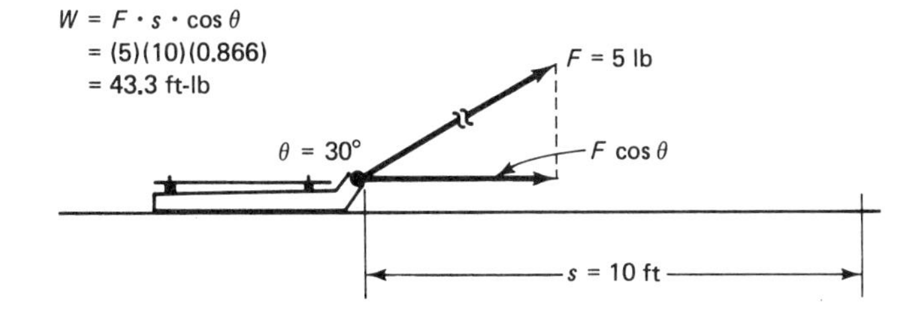

Figure 6-3 When the force is not in the same direction as that of the motion it produces, the component of the force in the direction of the motion is used in calculating the work done.

Work done against a gravitational force

When a load of mass M is lifted vertically by a crane, winch, forklift, elevator, or similar object, the work done is obtained by multiplying the weight of the load by the vertical distance the load is lifted, in accordance with Eq. (6-1). The applied force equals the weight of the load, Mg. If it is lifted a height h, the work done is

$$\textit{(to lift a load vertically)} \quad W = Mgh \tag{6-3}$$

If the weight of the load is given directly (instead of having to be calculated from the mass), the formula for calculating the work done is simply

$$\text{Work} = \text{weight} \times \text{height} \tag{6-4}$$

(*Note*: For convenience, use a *positive* numerical value for g.)

EXAMPLE 6-1 A constant force of 5 N moves an object 3 m in the direction of the force. How much work is done?

Solution. Apply Eq. (6-1).

$$W = F \cdot s = (5 \text{ N})\,(3 \text{ m}) = 15 \text{ N} \cdot \text{m} = 15 \text{ J}$$

EXAMPLE 6-2 A child pulls a sled a horizontal distance of 10 ft by exerting a 5-lb pull on a string at a 30° angle to the horizontal. How much work is done? (See Fig. 6-3.)

Solution. Use Eq. (6-2) because the force and the displacement are not in line. Only the component of the force in the direction of the displacement is effective in doing work.

$$\begin{aligned} W = F \cdot s \cdot \cos\theta &= (5 \text{ lb})\,(10 \text{ ft})\,(\cos 30^\circ) \\ &= 43.3 \text{ ft-lb} \end{aligned}$$

EXAMPLE 6-3 An unfortunate person, out of gas, pushes against the car with a force of 52 lb. However, the car does not budge. How much work was done on the car?

Solution. No work was done on the car because the car did not move. Therefore, the displacement was zero.

EXAMPLE 6-4 How much work is done when a dog carries a 1-lb bone 50 ft across a horizontal lawn?

Solution. Assuming the dog runs at a steady pace across the measured interval, no horizontal forces will be exerted on the bone. The weight of the bone (1 lb) is directed downward, at right angles to the horizontal displacement. Substituting into Eq. (6-2) reveals that no work is done because $\cos 90^\circ = 0$.

6-2 ENERGY: DEFINITION AND DESCRIPTION

What is energy?

There is a close relationship between work and energy. Energy is consumed (converted from one form to another) when work is done and is often described as the capacity for doing work. Energy is hard to define because it appears at first to be a rather abstract concept. It is an invisible, untouchable commodity, but it can be measured with great precision and is bought and sold at carefully negotiated prices. Energy is real, but we do not call it a substance or a material even though Einstein's relationship $E = mc^2$ implies a direct relationship between mass m and energy E (c is the speed of light).

The fundamental nature of energy is a mystery, but we can gain an appreciation of its properties by studying many examples of energy transformations. Energy can neither be created or destroyed; it can only be converted from one form to another. Because of the close relationship between energy and work, the same units that are used to measure work are also used to measure energy, namely, joules, ergs, and foot-pounds.

Forms of energy

Energy appears in several different forms. The following is a brief description of each.

1. *Potential.* The energy of position or configuration. Examples are a heavy weight suspended above the floor, a compressed string, and water at the top of a waterfall.
2. *Kinetic.* The energy of mass in motion. Any object of mass m moving with velocity v has kinetic energy $\frac{1}{2} mv^2$.
3. *Acoustical.* The energy contained in vibrations of the air. Any sound wave carries energy with it: music, noise, sonic booms.
4. *Radiant* (*light*). The energy contained in electromagnetic radiation. Examples are sunlight, the warmth from a fireplace, infrared and ultraviolet light.
5. *Thermal* (*heat*). The energy contained in objects that are hot. Thermal energy refers to the aggregate of the kinetic energy of individual molecules.
6. *Chemical.* The energy contained in chemicals. Obvious examples are nitroglycerine (for its explosive properties) and sulfuric acid, which—within a car battery—causes chemical energy to be converted to electrical energy.
7. *Electrical.* The energy associated with electrical fields, charges, and currents. Any separation of positive and negative electrical charges involves electrical energy.
8. *Nuclear.* The energy obtained from the nucleus of the atom by virtue of the conversion of mass to energy in accordance with $E = mc^2$. In nuclear reactors, the energy appears first as heat, and ultimately as electrical energy.

Potential energy

The work required to lift an object of mass m a distance h is given by mgh. Conversely, an object that has been lifted to a height h above a reference surface has been "worked on" in the amount of mgh. In lifting an object, the work done is converted to potential energy, and the amount of potential energy the object has is exactly equal to the amount of work done in lifting the object. Hence, we arrive at a formula for potential energy (PE):

$$\text{PE} = mgh \tag{6-5}$$

Because mg is the weight of the object, we can also write

$$\text{PE} = (\text{weight})(h) \tag{6-6}$$

That is, the potential energy of an object that has been elevated to a height h above a reference surface is the product of the weight and the height.

Because energy is the capacity for doing work, an object possessing potential energy can give up its energy and do work. However, the work that can be obtained from the elevated object can be no greater than its potential energy.

Other objects, such as stretched springs or clock springs that are wound up, also have potential energy because potential energy is the energy stored by virtue of position or configuration. A mousetrap provides a good example of energy stored in the form of a coiled spring.

Kinetic energy

Newton's second law ($F = ma$) states that a force is required in order to produce a change in the velocity of a body. Let us now consider the work done on a body initially at rest, which is acted on by a constant applied force. We know from Chap. 3 that the body will accelerate, but here our main interest lies in the amount of energy imparted to the body by the force.

Immediately after application of the force, the body will begin to move, producing a displacement in the direction of the force. Consequently, work is done on the body by the force, and the velocity of the body increases—the body gains *kinetic energy*.

The work done on the body by the force F while the body moves a distance s is

$$W = F \cdot s$$

If the force F is understood to be the net force, F can be replaced by ma to obtain

$$W = ma \cdot s$$

Recalling the physical situation, the body—initially at rest—is caused to accelerate. At any time after the acceleration begins, the following relationship [the $2as$ equation, Eq. (3-8)] relates the velocity, v, the acceleration, a, and the distance traveled, s:

$$2as = v^2 \qquad as = \tfrac{1}{2} v^2$$

Because the equations above apply to the same physical situation, we may substitute the expression for as into the expression for the work done on the body:

$$\begin{aligned} W &= mas = m(\tfrac{1}{2}v^2) \\ W &= \tfrac{1}{2} mv^2 \end{aligned} \tag{6-7}$$

Thus, in accelerating an object of mass m from rest to velocity v, an amount of work $\frac{1}{2} mv^2$ is done on the body. Because work has been done on the body, the body has energy; in this case, it has kinetic energy. Consequently, the formula for kinetic energy is

$$\text{KE} = \frac{1}{2} mv^2 \tag{6-8}$$

Here the formula was obtained assuming the body started from rest. A more general treatment shows that the work and kinetic energy involved in changing the velocity of a body from v_0 to v is given by

$$W = \Delta\text{KE} = \frac{1}{2} mv^2 - \frac{1}{2} mv_0^2 \tag{6-9}$$

This equation requires neither that the applied force be constant nor that the acceleration be uniform.

6-3 THE CONSERVATION OF ENERGY

The scientific meaning of the term *conservation* has nothing to do with frugality or in being economical in the use of resources. When something is *conserved*, the amount of it at hand in a given system does not change. In physics, conservation implies an unchanging amount.

One of the fundamental laws of physics is that energy is conserved within a *closed* system. That is, if no energy enters or leaves the system, the amount of energy within the system will always be the same, although the energy may be converted from one form to another.

This idea may seem contrary to popular experience. One might think that energy can be used up and that it must be continually resupplied. Indeed, it is true that energy can be converted from a usable form to a form where it is not usable, but the energy is not used up, consumed, or destroyed. Understanding that energy is conserved is complicated additionally by the definition and requirement of a closed system. Most systems commonly encountered are not closed. The following examples and discussions should clarify these ideas.

6-4 ENERGY TRANSFORMATIONS

Falling bodies

One easily observed energy transformation is the transformation of potential energy to kinetic energy, as in a brick falling from the roof of a tall building (Fig. 6-4). As the brick falls, its potential energy decreases. However, as the potential energy decreases, the kinetic energy increases in exactly the same proportion so that the sum total of the potential and kinetic energy is a constant. This is a consequence of the law of conservation of energy.

Before the brick begins to fall, its potential energy is mgh, and the kinetic energy is zero because the velocity of the brick is zero. At the instant of impact at the bottom, the potential energy is zero, and if the brick is moving downward at

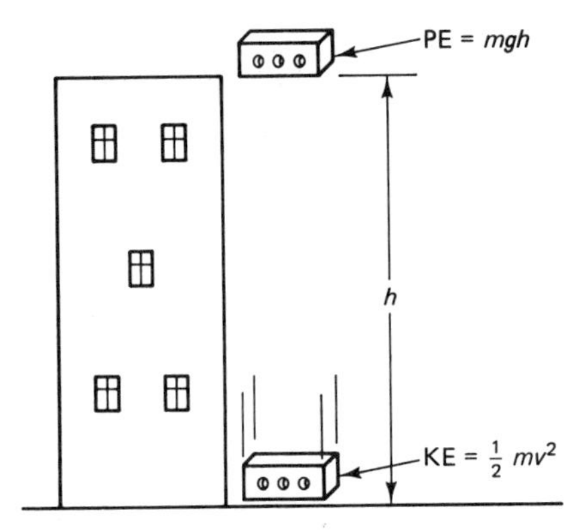

Figure 6-4 Potential energy is converted to kinetic energy as a brick falls from the top of a tall building.

velocity v, the kinetic energy is $\frac{1}{2}mv^2$. Thus, all the energy initially is potential; all the energy at the instant of impact is kinetic. If the resistance of the air is neglected, the conservation of energy law stipulates that the initial potential energy and the final kinetic energy must be equal:

$$\text{Initial potential energy of the brick at height } h = \text{Final kinetic energy of the brick at impact}$$

$$mgh = \tfrac{1}{2}mv^2$$

By canceling the mass m and rearranging, we obtain

$$v = \sqrt{2gh} \quad \text{or} \quad h = \frac{v^2}{2g} \qquad (6\text{-}10)$$

Thus, a relationship between v and h is obtained by utilizing the conservation of energy. These two equations were derived in Chap. 3 by considering the force acting on the body in conjunction with Newton's laws.

Similar problems are illustrated in Figs. 6-5 and 6-6. The maximum height attained by the projectile of a vertical cannon is found by appropriately equating potential and kinetic energy, and in Fig. 6-6, the velocity with which a pendulum

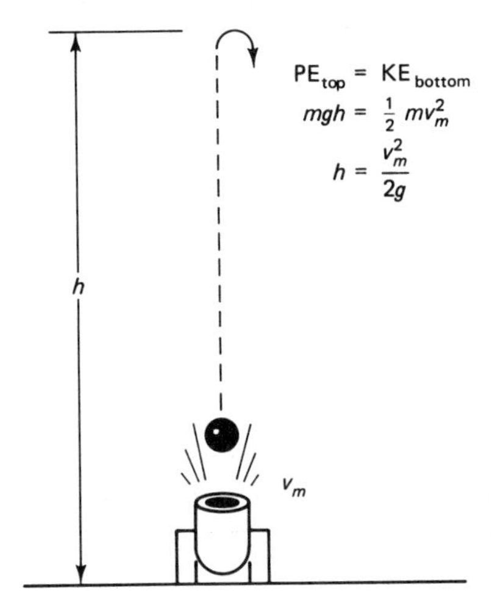

Figure 6-5 The maximum height attained by a projectile fired vertically can be calculated from energy considerations.

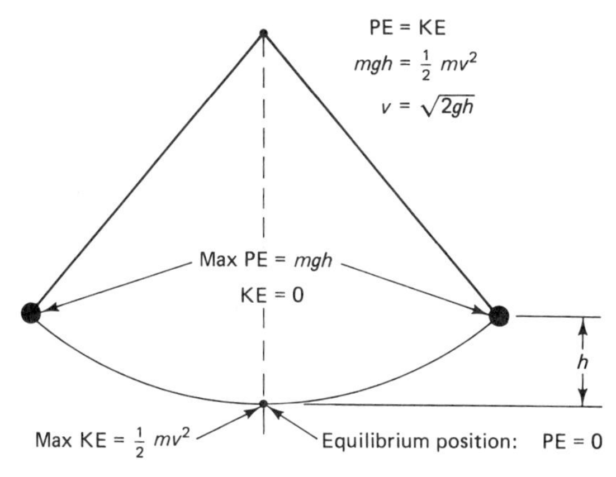

Figure 6-6 Calculating the velocity of the pendulum bob as it passes through the equilibrium position.

bob passes through the equilibrium position is found by considering that the kinetic energy of the bob at the equilibrium position equals the potential energy it has at the end points of the swing. As the pendulum swings, its energy is continually converted from potential to kinetic to potential to kinetic to potential . . . , cyclically, without end.

Object descending a curved track

A problem related to the falling-body problems is that in which an object descends a frictionless track of irregular curvature. Conservation of energy allows the velocity of the object at any point on the track to be determined if the height of the point is known in relation to another point where the velocity is given. Because of the irregular curvature of the track, such problems would be more difficult to solve by considering the forces acting on the descending object. By using energy considerations—namely, the conservation of energy—detailed knowledge of the forces involved is not necessary. Of course, the track must be frictionless; otherwise, a portion of the potential energy would be converted to heat rather than to kinetic energy.

Referring to Fig. 6-7, note that the velocity v_a is given at point A. We wish to compute the velocity of the object when it reaches point B, which is a vertical distance h lower than point A.

The key to the problem is that the kinetic energy at point B equals the kinetic energy at point A *plus* an increase equal to the change in potential energy between the two points. Thus, we may write

$$\text{KE}_b = \text{KE}_a + \Delta\text{PE}$$

Hence,

$$\tfrac{1}{2} mv_b^2 = \tfrac{1}{2} mv_a^2 + mgh$$

from which comes

$$v_b = \sqrt{v_a^2 + 2gh} \tag{6-11}$$

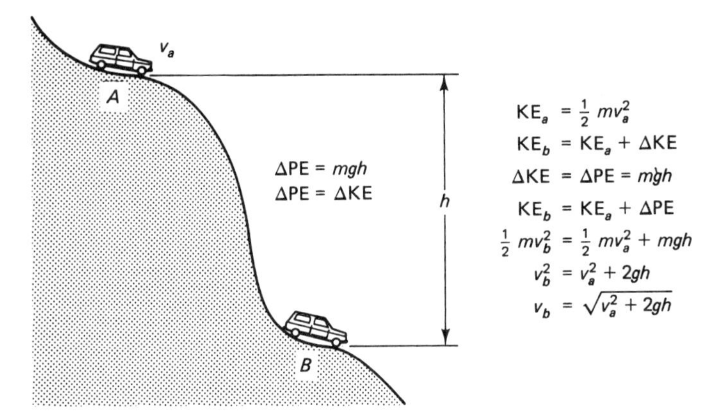

Figure 6-7 Vehicle descending a curved, frictionless track. (All kinetic energy of the vehicle is assumed to appear as translational kinetic energy.)

This formula is also applicable to an object projected up the track; in such a case, the height h would be entered as a negative quantity.

Energy transformations in an automobile

An automobile provides a good illustration of energy transformations because all forms of energy except nuclear are represented. The source of the automobile's energy is, of course, the gasoline (chemical energy) that goes into the tank. Energy for the lights, heater, horn, air conditioner, and all other accessories is derived from the gasoline.

Gasoline is burned in the cylinders to produce rotational motion of the engine crankshaft. Additionally, the burning gasoline gives off great amounts of heat (thermal energy) by way of the engine exhaust and the radiator. A portion of this heat may be discharged through the heater for warming the interior of the car, but the energy discharged as heat is of no benefit for the most part.

As mechanical power is taken from the crankshaft, the crankshaft becomes harder to turn and will slow down unless the flow of gasoline to the engine is increased. Thus, the rate of gasoline consumption is increased by anything that tends to make the crankshaft more difficult to turn.

Mechanical power delivered to the wheels produces kinetic energy as the vehicle comes up to highway speed. If the vehicle happens to be traveling up an incline, the entire vehicle will be gaining potential energy as it reaches higher elevations. The kinetic energy and the potential energy are obtained from the chemical energy of the gasoline.

The generator (or alternator), which supplies electrical energy for charging the battery and for operating the other electrical accessories, is driven by the engine crankshaft. When the generator is producing electricity, the generator becomes more difficult to turn and consequently tends to slow the crankshaft. This requires an additional expenditure of gasoline in order to maintain the crankshaft speed. From this we conclude that any accessory on an automobile that consumes electrical power results in increased gasoline consumption. Ultimately, this lowers the miles-per-gallon rating of the vehicle.

Thus far, we have seen that the original chemical energy of the gasoline is transformed to potential, kinetic, thermal, and electrical energy. Going further, energy is stored as chemical energy within the battery. When the headlights are turned on at night, radiant energy is produced from electrical energy, and when the horn is activated, acoustical energy is produced from electrical energy.

In summary, the chemical energy of the gasoline is transformed to all other forms of energy except nuclear. If the energy expended in all forms could be accurately measured, tabulated, and totaled, the sum of the energy expended would exactly equal the energy consumed in the form of gasoline. This is in accordance with the law of conservation of energy, and it is in this sense that the law is valid.

6-5 POWER

Definition and units

Power is the rate at which work is done. Or, because work always involves a transformation of energy, power may be defined as the rate at which energy is transformed from one form to another. In common usage, there is much confusion between the terms *energy* and *power*. The term power is often used where energy is intended. Scientifically, energy and power are entirely different; one is how fast the other is transformed, expended, or dissipated. Mathematically,

$$\text{Power} = \frac{\text{work}}{\text{time}} \quad \text{or} \quad \text{power} = \frac{\text{energy transformed}}{\text{time}} \tag{6-12}$$

Because power refers to work or energy performed, transformed, or expended per unit time, the units of power are, for example, joules per second or foot-pounds per second:

$$(\textit{units of power}) \quad \text{J/s}, \qquad \text{N} \cdot \text{m/s}, \qquad \text{ft-lb/s}, \qquad \text{ergs/s}$$

Two commonly used units of power are the horsepower (hp), which is 550 ft-lb/s, and the watt (W), which is 1 J/s. A units conversion reveals that

$$1 \text{ hp} = 746 \text{ W}$$

Also familiar is the kilowatt (kW), which is 1000 W.

EXAMPLE 6-5 Suppose a crane lifts a stack of bricks that weighs 600 lb to a height of 44 ft. If 22 s are required to lift the load, compute the power output of the motor driving the crane, assuming all the energy output of the motor goes into lifting the load.

Solution. First, we must calculate the work done by the crane in lifting the load:

$$\begin{aligned} \text{Work} &= \text{force} \times \text{distance} \\ &= 600 \text{ lb} \times 44 \text{ ft} = 26{,}400 \text{ ft-lb} \end{aligned}$$

We then use the definition of power:

$$\text{Power} = \frac{\text{work}}{\text{time}} = \frac{26{,}400 \text{ ft-lb}}{22 \text{ s}}$$

$$= 1200 \text{ ft-lb/s} = 2.18 \text{ hp}$$

Here, 1200 ft-lb/s was divided by 550 to convert to horsepower.

EXAMPLE 6-6 Suppose a 1-hp motor is being used to pump water to a height of 96 ft above the level of water in a reservoir. Calculate the maximum number of gallons per minute that can be pumped.

Solution. The amount of work done by the 1-hp motor in 1 min is

$$\begin{aligned} \text{Work} &= \text{power} \times \text{time} \\ \text{Work/min of a 1-hp motor} &= 1 \text{ hp} \times 60 \text{ s/min} \\ &= (550 \text{ ft-lb/s})\,(60 \text{ s/min}) \\ &= 33{,}000 \text{ ft-lb/min} \end{aligned} \tag{6-13}$$

One gallon of water weighs 62.4 lb. To lift it to a height of 96 ft requires work in the following amount.

$$\begin{aligned} \text{Work to lift 1 gal of water} &= (62.4 \text{ lb})\,(96 \text{ ft}) \\ &= 5990 \text{ ft-lb/gal} \end{aligned}$$

Dividing the work by the motor per minute by the work required per gallon of water gives the maximum water per minute that can be lifted 96 ft by a 1-hp motor:

$$\text{Gal/min} = \frac{33{,}000 \text{ ft-lb/min}}{5990 \text{ ft-lb/gal}} = 5.50 \text{ gal/min}$$

The relationship between energy, time, and power

It is often asked how much energy is dissipated when a certain gadget operates at a certain power level for a given length of time. Rearranging the definition of power gives

$$\text{Energy} = \text{power} \times \text{time} \tag{6-14}$$

Consequently, the product of the power and the time gives the total energy expended during the time period under consideration.

In the field of electricity and electrical power, the watt is the most commonly used unit of power. One watt is a power dissipation of 1 J/s. Because a joule is 1 N · m, a watt is also 1 N · m/s. To compute the energy expended in a given amount of time by a gadget operating at a power level given in watts, multiply the watts by the time, as in Eq. (6-14).

$$\begin{aligned} \text{Energy} &= \text{power} \times \text{time} \\ &= \text{watts} \times \text{seconds} \end{aligned}$$

In light of this, we are not surprised that a unit of energy is the watt-second ($W \cdot s$). This is the energy dissipated by a 1-W light bulb (a very small bulb) in burning for 1 s. A more familiar unit of energy is the watt-hour ($W \cdot h$) which is the energy consumed when a device operates with a power of 1 W for a time period of 1 h. A unit conversion gives

$$1 \text{ W} \cdot \text{h} = 3600 \text{ W} \cdot \text{s}$$

Finally, the standard unit of energy used by electrical power utilities is the kilowatt-hour ($kW \cdot h$), which is the energy expended by a device operating at a power level of 1000 W for 1 h. Hence,

$$1 \text{ kW} \cdot \text{h} = 1000 \text{ W} \cdot \text{h}$$

EXAMPLE 6-7 An electric heater used in a centrally heated residential home operates at a power level of 20 kW.

(a) Calculate the energy (in kilowatt-hours) dissipated by the heater in 15 min.
(b) If the heater is on 15 min out of every hour, calculate the energy dissipated in 24 h.
(c) Calculate the total energy consumption for 1 month (30 days) if conditions are as in part (b).
(d) If energy costs 5¢ per kilowatt-hour, compute the cost of heating the home for 1 month.

Solution.
(a) 15 min is $\frac{1}{4}$ h. Hence,

$$\begin{aligned} \text{kW} \cdot \text{h} &= \text{power} \times \text{time} \\ &= (20 \text{ kW})\,(0.25 \text{ h}) \\ &= 5 \text{ kW} \cdot \text{h} \end{aligned}$$

(b) At this rate, the energy dissipation for 24 h is 120 $kW \cdot h$.
(c) For 1 month, the total energy consumed is

$$(30 \text{ days/month})\,(120 \text{ kW} \cdot \text{h/day}) = 3600 \text{ kW} \cdot \text{h/month}$$

(d) At 5¢ (\$0.05) per kilowatt-hour, the monthly bill is

$$\left(0.05 \frac{\$}{\text{kW} \cdot \text{h}}\right)\left(3600 \frac{\text{kW} \cdot \text{h}}{\text{month}}\right) = \$180.00 \text{ per month}$$

The conclusion: Heating a home is expensive.

Power and velocity

Examine the definition of power:

$$\text{Power} = \frac{\text{work}}{\text{time}} = \frac{\text{force} \times \text{distance}}{\text{time}} = F\frac{s}{t}$$

Recalling that s/t is a velocity, we obtain the important result that

$$\text{Power} = \text{force} \times \text{velocity} \tag{6-15}$$

This assumes, of course, that the force and the velocity are in line. An example illustrating the significance of the force-velocity product is that of a person pushing a car on a level surface. A much greater physical exertion is required to move the car rapidly than is required to move it slowly.

EXAMPLE 6-8

Suppose a 100-lb force is required to move a car on a level roadway. Calculate the power required to push the car: (a) at slow walking speed of 3 mi/h, (b) at a faster rate of 10 mi/h; (c) at a motoring speed of 20 mi/h.

Solution. First, we do the unit conversion to find that 3 mi/h = 4.4 ft/s; 10 mi/h = 14.67 ft/s; 20 mi/h = 29.33 ft/s. Then use Eq. (6-15):

(a)
$$\begin{aligned}\text{Power} &= \text{force} \times \text{velocity}\\ &= (100\text{ lb})\,(4.4\text{ ft/s})\\ &= 440\text{ ft-lb/s} = 0.8\text{ hp}\end{aligned}$$

(b)
$$\begin{aligned}\text{Power} &= (100\text{ lb})\,(14.67\text{ ft/s})\\ &= 1467\text{ ft-lb/s} = 2.67\text{ hp}\end{aligned}$$

(c)
$$\begin{aligned}\text{Power} &= (100\text{ lb})\,(29.33\text{ ft/s})\\ &= 2933\text{ ft-lb/s} = 5.33\text{ hp}\end{aligned}$$

For comparison, Example 6-9 illustrates the power-output capability of a rather strong human being.

EXAMPLE 6-9

A strong young person from the country weighing 180 lb can carry a 100-lb sack of hog feed up a set of steps at the rate of 1 ft (vertical) per second for a short distance. Compute the power dissipation.

Solution. The weight being moved upward is 180 lb plus 100 lb, which is 280 lb. At 1 ft/s vertical velocity, the work done per second is 280 ft-lb. Converting this to horsepower yields 280/550, or 0.51 hp.

6-6 EFFICIENCY

No practical machine is perfectly efficient in its use of energy. Of the energy put into a machine such as an automobile, only a fraction (about one-third) of the input energy is delivered to the output in useful form. For example, a 1-hp electric motor is capable of delivering 746 W of mechanical power at the output. To do this, however, requires that about 1000 W of electrical power be applied to the terminals of the motor. The difference between the input energy and the output energy is lost due to heat generation within the motor.

We define *efficiency*, where we are referring to some system having an input and an output for energy in some form, as follows:

$$\text{Efficiency} = \frac{\text{work out}}{\text{work in}} \tag{6-16}$$

Efficiency is typically expressed as a percentage by multiplying the preceding fraction by 100.

An equivalent definition is the following, given in terms of power:

$$\text{Efficiency} = \frac{\text{power out}}{\text{power in}} \tag{6-17}$$

EXAMPLE 6-10 Compute the efficiency of an electric motor that consumes power at the rate of 1100 W and delivers 1 hp of mechanical power at the output.

Solution. By Eq. (6-17) and because 1 hp = 746 W,

$$\text{Efficiency} = \frac{746 \text{ W}}{1100 \text{ W}} = 0.68$$

$$= 68\%$$

EXAMPLE 6-11 Suppose power is applied to a small motor at the rate of 500 W. If the motor is 74% efficient, what power level is available at the output?

Solution. First, express the efficiency as a decimal rather than as a percentage. Then use the relationship

$$\text{Power output} = \text{power input} \times \text{efficiency}$$

$$= (500 \text{ W})\,(0.74) = 370 \text{ W}$$

Of the original 500 W, 130 W is being converted (mostly) to heat within the motor.

6-7 MASS-ENERGY TRANSFORMATIONS

The earlier statement of the law of conservation of energy had to be revised when it became apparent that a transformation could occur between mass and energy. The theoretical foundation for this was put forth by Albert Einstein in 1905 and 1907, and the first experimental evidence in support of the mass-energy transformation was obtained in 1932 by John Cockroff and Ernest Walton working at the Cavendish Laboratory in Cambridge.

The earlier statements of the conservation law often contained the phrase energy can neither be created nor destroyed, or something similar. This became somewhat bothersome in light of evidence that mass could be converted to energy by complex processes occurring within the nucleus of an atom. Rigorous statements of the law of the conservation of energy now include mass as a form of energy; some writers speak of the conservation of mass-energy.

Conversions of mass to energy occur only on a small scale insofar as the quantity of mass is concerned, but the quantities of energy given off are enormous in comparison. To make this more definite, we now consider the splitting (fission) of the nucleus of the U^{235} isotope of uranium. This material plays an important role in atomic bombs and nuclear reactors.

When an atom of U^{235} fissions, the sum total of the masses of the fission fragments (the pieces) is less than the mass of the original atom. The mass that is lost is converted to energy. When one gram of U^{235} fissions, about 8.9×10^{-4} gram of mass is converted to energy. By $E = mc^2$, this mass is converted to roughly 81×10^9 J of energy. About 83% of this (67×10^9 J) appears in the form of usable heat. This heat is then used to generate steam for driving turbines to produce electricity. If the conversion process is 33% efficient, about 22×10^9 J of usable, marketable energy can be derived from 1 g of U^{235}.

Converting this to kilowatt-hours ($1\ kW \cdot h = 3.6 \times 10^6$ J), we get 6111 kW · h per gram of U^{235}. At 5¢ per kilowatt-hour, this quantity of energy is worth about $305. And, it may be obtained from the "consumption" of less than 0.1 mg of mass, although 1 g of U^{235} was involved in the process.

6-8 ENERGY SOURCES

This section briefly describes several sources of energy, which emphasizes the fact that energy must be *obtained* from somewhere; it cannot simply be created on the spot. Energy sources of all types are becoming more important in light of ever increasing energy consumption.

Fossil fuels include coal, oil, and natural gas. The importance of these fuels is well publicized. As the name suggests, these carbonaceous fuels are products of vegetation of past eras, and because plants depend upon the sun for energy for photosyntheses, the energy of the fossil fuels originally came from the sun.

Nuclear fission is the fundamental process of nuclear reactors. The energy is obtained from the conversion of mass to energy, which occurs when the nucleus of an atom of uranium (U^{235}) or other fissionable material splits to form two lighter elements. While extremely large quantities of energy are obtained from impressively small amounts of nuclear fuel, the known reserves of suitable uranium are limited; hence, nuclear fission as presently implemented cannot be the ultimate source of energy for the future. However, *breeder reactors* are capable of producing *more* nuclear fuel than they use for producing energy. Breeder reactors are presently under development and construction on an experimental basis.

The problems associated with generation of nuclear power are well known and widely debated. Aside from the catastrophic nuclear accident that might threaten large populations, more mundane problems of thermal pollution, radioactive wastes, and economic feasibility make generation of nuclear power less attractive than it once appeared to be.

Nuclear fusion represents the energy source of the future. Fusion refers to the combining of two light nuclei to form a single heavier nucleus, with large amounts of energy being given off in the process. The most promising fuel is deuterium, an isotope of hydrogen found in abundance (relatively speaking) in sea water and easily recovered. Thus, we may regard the fuel for fusion as being unlimited. This sounds good, but there is a catch. The physical conditions required for a fusion reaction to occur are stringent indeed. The temperature of the fuel must be at least 50,000,000°C; simultaneously, the fuel must be confined or contained in a state sufficiently dense and for a period of time long enough for the fusion reaction to occur. Thus, the proper temperature, density, and time must be achieved simultaneously. As of 1980, no successful sustained fusion reaction had been achieved.

Hydroelectric power is derived from the potential energy of water backed up behind dams or naturally occurring at an elevated position, such as at Niagra

Falls. Presently, about 4% of the energy in the United States is obtained from hydroelectric sources. Hydroelectric generation is essentially nonpolluting; the only significant environmental consideration is the vast area that must be flooded by the reservoir. Generation of hydroelectric power is not expected to increase significantly relative to the total energy needs because few acceptable sites remain for the construction of dams. Many sites that are suitable from the engineering viewpoint are protected by legislation.

Solar energy is present in abundance, but it is spread rather thinly over the surface of the earth. The maximum solar power delivered to the earth's surface under optimum conditions is on the order of 100 W/ft^2. However, a salient feature of solar power is its variability. The energy received varies with cloud cover and atmospheric conditions, latitude, season, and the properties of the collecting device. Of course, the time of day (the angle of the sun) is important too; no energy at all is received at night. While the maximum of 100 W/ft^2 is an appreciable power level, a more meaningful figure is the energy delivered per day per square foot of surface. The maximum value is on the order of 0.7 kW $\cdot$ h/ft^2 per day. However, the efficiency of the solar collector must be taken into account, and the usable energy is only a fraction of this amount. Collectors with large surface areas are required.

Maximum efficiency is obtained if the solar energy is used in the form of heat. If a conversion to electrical power is contemplated, a steam generator or other "heat engine" is necessary for converting heat to rotary motion. Because heat engines are only about 33% efficient, nearly two-thirds of the collected energy is lost. Consequently, the generation of electricity by way of solar energy is an inefficient process. Solar cells, which convert sunlight directly to electrical energy, are only about 11% efficient and are rather expensive. Clearly, a technological breakthrough is needed before solar energy can become a viable and competitive energy source.

Winds contain large amounts of energy, but winds are neither steady nor dependable, and the energy is not concentrated. Experimentation with sophisticated windmills is in progress, but it is unlikely that a significant portion of our total energy requirement will be obtained from the winds. Incidentally, the energy of the wind is a secondary form of solar energy because the wind is caused by the solar heating of the atmosphere. About 19% of the solar radiation incident upon the atmosphere is absorbed by the atmosphere.

Tides may be used to generate energy by damming the high tide and capturing the potential energy of the water. However, the total energy capability of the tides is not large, and only a few suitable sites are available where such a project might be economically successful. Considerable variation exists in the levels of the tides, so that the energy production could not be steady. The ultimate source of tidal energy is the kinetic energy of rotation of the earth-moon system.

Geothermal energy refers to the thermal energy of the interior of the earth. The central core of the earth is molten and is kept in the molten state by energy given off by the disintegration of radioactive materials within the core. There are many areas where hot molten rock, called *magma*, is pushed up to within a few kilometers of the surface. The magma heats the rocks above it to as much as

700°F. If the rocks are porous and contain water, steam will be produced. If a layer of rock closer to the surface happens to be impervious to the steam, the steam will be trapped and often is under great pressure. By drilling through the impervious layer, the steam can be obtained for use by steam turbines in producing electricity. The steam requires minor processing to remove impurities before it is admitted to the turbines. The steam must lie within about 3 km of the earth's surface to be of practical use, and the steam must not be unduly contaminated with impurities.

Because the temperature of the earth increases by about 48°C per mile of depth, it might be proposed that a system of pipes be installed at a great depth so that water circulated through the pipes would be heated. This is presently not economically feasible due to the expense of the initial installation (digging or drilling) and because the earth is not a good thermal conductor. The earth would be cooled in the vicinity of the pipes, and this would reduce the output of the system.

Photosynthesis is the process by which plants utilize the energy of the sun to convert carbon dioxide and water into carbohydrates and oxygen. A good example of a solar energy converter is sugar cane, which may be used to produce alcohol at a rate of about 600 gal/acre. Alcohol may be added to gasoline in order to extend the petroleum product. As the scarcity of energy becomes more pronounced, the lesser energy sources—for example, photosynthesis, geothermal, and winds—will receive ever-increasing attention whereas in the past, they were readily discerned to be economically unfeasible.

Questions

1. For the SI, CGS, and British systems, give the unit of: **(a)** work; **(b)** energy; and **(c)** power.
2. When gasoline is used by a car, it is gone forever. Does this violate the principle of conservation of energy?
3. What is the relationship between time, power, and total work done or energy transformed?
4. The thrust of a turbojet engine is more or less independent of the velocity of the airplane to which it is attached. Assuming identical engine performance, does a turbojet engine develop more power when the plane is traveling at 600 mi/h or 300 mi/h?
5. An object dropped from a height of 30 ft will impact the ground with a speed of about 30 mi/h. From what height should an object be dropped so that it will impact at 60 mi/h? (Not 60 ft!)
6. Can an object be given energy without touching it or disturbing it in any manner? Does a cinder block have any potential energy while it is lying motionless on the ground? Now suppose a deep hole is dug adjacent to the cinder block. Does it now have any potential energy? What does this tell us about potential energy?
7. The possibility that mass could be converted to energy (and vice versa) mandated a revision of the law of conservation of energy. Explain.
8. Starting with the sun as the energy source, trace the path of a quantity of energy through the various forms it might have taken before it winds up as heat energy contained in a cup of coffee.
9. A large airliner coming in to land possesses *considerable* kinetic energy. Where does the kinetic energy go as the airliner lands and rolls to a stop?

Problems

1. A force of 25 lb is applied horizontally in pushing a box across a level floor a distance of 4 ft. How much work is done by the force?
2. A rope making an angle of 30° with the horizontal is used to pull a box 8 ft across a floor. How much work is done if a tension of 50 lb is maintained in the rope?
3. How much work is done in lifting a mass of 1 g a distance of 1 cm?
4. A mass of 50 kg is lifted vertically a distance of 6 m. How much work is done?
5. Which is the larger amount of work, 1 J or 1 ft-lb? (*Hint*: Do a unit conversion.)
6. One gallon of water weighs 8.34 lb. What is the potential energy of 1 gal of water in a reservoir 180 ft above a turbine for hydroelectric-power generation?
7. How much work is done by a constant net force exerted on a 2-kg mass in accelerating it from rest to a velocity of 5 m/s?
8. How much work must be done on a 3-kg object to cause it to accelerate from a velocity of: **(a)** 10 m/s to 20 m/s; **(b)** 20 m/s to 30 m/s?
9. Suppose a mass of 1 kg is dropped from a height of 100 m. Compute the kinetic energy and potential energy of the mass as it falls past the following heights, measured from the ground.
   (a) 100 m **(b)** 75 m **(c)** 50 m **(d)** 25 m
   (e) 0 m
10. With what velocity will the mass of Problem 9 strike the ground?
11. Suppose the small car in Fig. 6-7 is traveling at 3 m/s as it passes point A. How fast will the car be traveling as it passes point B, a vertical distance of 6 m below point A?
12. A pendulum bob is pulled aside until it is lifted vertically a distance of 6 cm above the level of the equilibrium position. If released, with what velocity will it pass through the equilibrium position?
13. A 1400-lb load is lifted a vertical distance of 10 ft by a crane in 5 s. What is the power output of the crane in lifting the load: **(a)** in foot-pounds per second; **(b)** in horsepower?
14. A 1-hp motor can do work at the rate of 550 ft-lb/s. What is the maximum rate (in feet per second) at which such a motor can lift a load weighing: **(a)** 100 lb; **(b)** 400 lb; **(c)** 1000 lb; **(d)** 1 ton (2000 lb)?
15. The gross weight of a certain fully loaded elevator is 4000 lb. If the elevator must be capable of ascending at a velocity of 5 ft/s, what is the minimum horsepower required for the drive mechanism?
16. A 1-hp motor actually consumes energy at the rate of about 1000 watts to deliver 746 watts of mechanical power.
    (a) How much energy in joules is required to operate such a motor for 1 h?
    (b) Express **(a)** in watt-hours.
    (c) Express **(a)** in kilowatt-hours.
17. **(a)** How many joules of energy per second are consumed by a 7-W nightlight?
    (b) If the light burns for 8 h, how many kilowatt-hours of energy are consumed?
    (c) If electrical energy costs 5¢ per kilowatt-hour, how much does it cost to operate the light for 8 h?
18. Repeat Problem 17 for a 100-W bulb.
19. A tow truck exerts a steady force of 300 lb against a vehicle in towing it along a level road at 30 mi/h. What is the power output of the tow truck that is used to pull the vehicle?
20. Repeat Problem 19 for a velocity of 55 mi/h.
21. Steam turbines and other present-day processes for converting thermal energy to electrical energy are at best only about 40% efficient. In turn, an electric motor may be only about 70% efficient, giving an overall efficiency for the process of 28%. Based on this, what thermal power dissipation is required at the generating station in order to provide energy for a 1-hp motor (746 W) at some distant point?
22. **(a)** Determine the number of joules in a kilowatt-hour.
    (b) At 5¢ per kilowatt-hour, how many joules of energy can be purchased for a penny?

***23.** **(a)** Use $E = mc^2$ to compute the energy produced in a nuclear reactor when 1 g of matter is converted to energy. (*Hint*: Use MKS units.)

(b) Assuming 33% efficiency, how many salable kilowatt-hours of electrical energy does the 1 g of matter produce?

(c) At 5¢ per kilowatt-hour, what is the value of the energy?

(d) How many kilowatt-hours of thermal energy must be discarded as a result of the 67% inefficiency of the conversion process?

***24.** At cruise power a certain two-place, light utility aircraft will travel at about 100 mi/h, while the engine develops about 75 hp. Compute the thrust developed by the propeller (assumed to be 100% efficient) during these conditions.

***25.** A certain small airplane can glide (engine at idle) 5 ft forward for every 1 ft of vertical descent. Its total weight is 1400 lb, and it glides at 70 mi/h. From these data, estimate the horsepower required of the engine to maintain level flight at 70 mi/h. What assumptions must be made?

CHAPTER 7

MOMENTUM, IMPULSE, AND COLLISIONS

Momentum is another basic and widely applicable concept of physics. As the product of the mass and the velocity of a moving object, it is, in a sense, a measure of the motion of the moving body. Momentum is a vector quantity because velocity is a vector quantity. The usefulness of the concept of momentum arises from the fact that momentum is conserved in all physical processes.

Whereas kinetic energy is related to the work done on an object in accelerating it from rest to a certain velocity, momentum is related to the product of force and time—the force acting on a body causing it to accelerate and the time the force is in effect This product is called *impulse*.

An understanding of momentum, impulse, and the conservation of momentum provides additional tools for solving problems, especially in situations (such as in collisions) where detailed knowledge of the forces involved may not be available.

7-1 MOMENTUM

A measure of motion

In looking for a mathematical formula to express the "amount of motion" involved when a body of mass m travels with a velocity v, two possibilities come to mind. One is the kinetic energy, $\frac{1}{2} mv^2$. The other is the product of the mass and the velocity, mv, called the *momentum*. Both quantities are greater for larger masses and for higher velocities, as intuitive reasoning says they should be. However, momentum increases with the *first power* of the velocity, while kinetic energy increases with the *square* of the velocity. We might ask which of these provides the best measure of the amount of motion a moving body possesses.

This is a good question, but it does not have a good answer because the term *motion* has no special scientific significance. The two quantities—momentum and the kinetic energy—are both extremely useful concepts. Further, the physical properties of the two are very different, even though they are mathematically similar. These differences will become apparent in the following sections. At this point, we define the momentum of a body of mass m moving at velocity v as

$$\text{(definition of momentum)} \quad P = mv \tag{7-1}$$

$$\mathbf{P} = m\mathbf{v} \tag{7-1(a)}$$

Momentum is a vector quantity, as indicated by the boldface type in Eq. [7-1(a)]. The direction of the momentum is that of the velocity, and its magnitude is mv.

Because momentum is a vector quantity, momenta can be added vectorially only; the direction of each momentum must be taken into account. This is not the case for kinetic energy, which is a scalar quantity and which can be added *algebraically*. In most cases, we consider only straight-line motion in one dimension so that the vector properties of momentum do not have to be considered explicitly. Hence, we shall use Eq. (7-1) more often than Eq. [7-1(a)].

Units of momentum

The unit of momentum is that of a mass times a velocity. In the three systems, the units are

(units of momentum) $\quad$ kg · m/s $\quad$ g · cm/s $\quad$ slug-ft/s

These units have not been given any special names and they appear to be a bit awkward, but no significant inconvenience results.

EXAMPLE 7-1 Compute the momentum and kinetic energy of a 4-kg mass traveling with a velocity of 6 m/s.

Solution. This requires a straightforward application of the definition of momentum and kinetic energy. Consequently,

$$\text{Momentum} = P = mv = (4 \text{ kg})\,(6 \text{ m/s})$$
$$= 24 \text{ kg} \cdot \text{m/s}$$
$$\text{Kinetic energy} = \tfrac{1}{2}\,mv^2 = (0.5)\,(4 \text{ kg})\,(6 \text{ m/s})^2$$
$$= 72 \text{ kg} \cdot \text{m}^2/\text{s}^2 = 72 \text{ J}$$

EXAMPLE 7-2 Compute the momentum and kinetic energy of a 4-g rifle bullet traveling at 900 ft/s.

Solution. A confusion of units exists here, and we have a choice of whether to convert to the British system or to SI. Preferring SI, 4 g = 0.004 kg and 900 ft/s = 274.39 m/s. Then,

$$\text{Momentum} = mv = (0.004 \text{ kg})\,(274.39 \text{ m/s})$$
$$= 1.10 \text{ kg} \cdot \text{m/s}$$
$$\text{Kinetic energy} = \tfrac{1}{2}\,mv^2 = (0.5)\,(0.004 \text{ kg})\,(274.39 \text{ m/s})^2$$
$$= 150.58 \text{ J}$$

7-2 FORCE AND MOMENTUM

There is an intimate relationship between *force* and *momentum*. Precisely, the rate at which momentum changes with time is equal to the net force applied to the body in question. This result comes directly from Newton's second law, $F = ma$. The connecting link is the fact that the acceleration is the rate at which velocity changes with time:

$$F = ma$$

Because $a = (v - v_0)/t$,

$$F = m\frac{v - v_0}{t}$$

Expanding,

$$F = \frac{mv - mv_0}{t} \tag{7-2}$$

and, if we let $P = mv$ and $P_0 = mv_0$,

$$F = \frac{P - P_0}{t}$$

Recognizing that $P - P_0$ is a change in momentum—$\Delta P = P - P_0$, we see that force F is the change in momentum per unit time:

$$F = \frac{\Delta P}{t} \tag{7-3}$$

Rearranging Eq. (7-2) gives

$$F \cdot t = mv - mv_0 \quad \text{or} \quad F \cdot t = m(v - v_0) \tag{7-4}$$

Here, for the first time, we see the product $F \cdot t$. This is the important physical quantity *impulse*, which is the topic of a later section.

EXAMPLE 7-3 People in a spaceship in free space determine that the ship's velocity must be increased by 49 m/s in order to achieve the desired trajectory. The thrust of the spaceship's rocket engine is 18,000 N, and the mass of the ship and contents is 3000 kg. Calculate the burn time of the engine required to produce the desired change in velocity.

Solution. Rearranging Eq. (7-4) and solving for t gives

$$t = \frac{m(v - v_0)}{F}$$

In this problem, we do not know v or v_0, but we are given that the velocity must change by 49 m/s. Hence, $(v - v_0) = 49$ m/s. Straightforward substitution gives

$$t = \frac{(3000\ \text{kg})\ (49\ \text{m/s})}{18{,}000\ \text{N}}$$

$$= 8.17\ \text{s}$$

EXAMPLE 7-4 An object of mass 7 kg is dropped from rest and falls freely for 8 s. Use momentum considerations to compute the velocity of the object at the end of the 8-s period.

Solution. The net force acting on the body is the weight:

$$w = mg = (7 \text{ kg})\ (9.8 \text{ m/s}^2)$$
$$= 68.6 \text{ N}$$

Because $v_0 = 0$, Eq. (7-4) reduces to

$$F \cdot t = mv \quad \text{or} \quad v = \frac{F \cdot t}{m}$$

Substituting,

$$v = \frac{(68.6 \text{ N})\ (8 \text{ s})}{7 \text{ kg}}$$
$$= 78.4 \text{ m/s}$$

This identical answer is obtained more directly by using the equation $v = v_0 + gt$ of an earlier chapter. However, the applicability of momentum considerations is made clear by this example.

7-3 CONSERVATION OF MOMENTUM

One of the remarkable properties of momentum is that it is conserved. That is, if no net external force acts upon a body (or a system of particles), the momentum of the body (or system) remains constant. This is illustrated by Eq. (7-3), where ΔP will be zero if $F = 0$. This means that once a certain "amount of motion" (momentum) is initiated in a certain direction, the same amount of motion will prevail until the moving object or system is acted upon by some external force.

A striking example of the conservation of momentum is provided by a space-walker who forgets to secure the tether before stepping out for a walk. Once the space-walker is in motion away from the spacecraft, he or she has no way to stop or reverse the motion and will continue to drift away from the ship. While floating in free space, no external forces act on the space-walker, and momentum (motion away from the ship) is preserved.

Conservation of momentum is evident when a large gun fires a projectile because the momentum of the recoiling gun is equal and opposite to the momentum of the projectile (Fig. 7-1). Because both the gun and the projectile are initially at rest, the total momentum before the gun is fired is zero. After firing the gun, the total momentum is still zero. Therefore, we may write

$$m_G v_G + m_B v_B = 0 \quad \text{or} \quad m_G v_G = -m_B v_B \tag{7-5}$$

where the subscripts G and B refer to the gun and the bullet, respectively. This equation may be rearranged to give

$$v_G = -\frac{m_B}{m_G} v_B \quad \text{and} \quad v_B = -\frac{m_G}{m_B} v_G$$

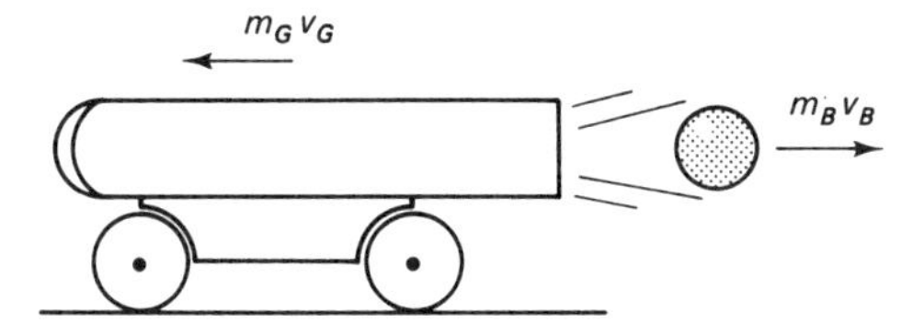

Figure 7-1 Conservation of momentum causes the momentum of the gun to be equal and opposite to that of the bullet.

which relates v_G and v_B. It is interesting to contemplate the result of firing a large bullet from a small gun.

Momentum is conserved in collisions as long as no net external forces act on any of the colliding objects. As an example, consider two railroad cars rolling without friction on a level track, as shown in Fig. 7-2. The cars collide, become coupled, and roll together with final velocity v_f. Velocities to the right are assumed to be positive. By equating the momenta before and after the collision, we obtain

$$\text{Momentum before} = \text{momentum after}$$
$$m_1 v_1 + m_2 v_2 = (m_1 + m_2) v_f \tag{7-6}$$

If the initial velocities v_1 and v_2 are known, the final velocity is given by

$$v_f = \frac{m_1 v_1 + m_2 v_2}{(m_1 + m_2)} \tag{7-7}$$

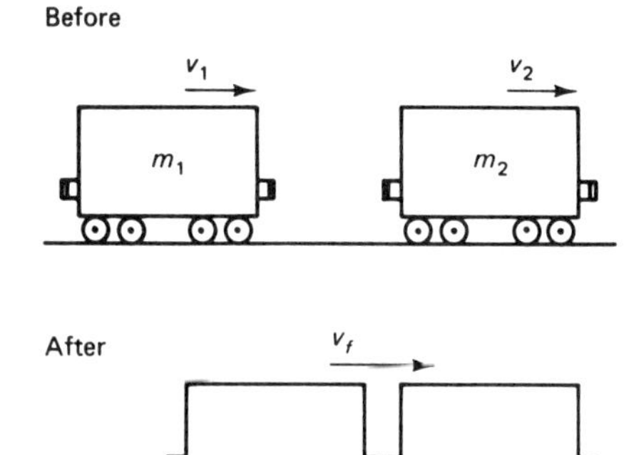

Figure 7-2 Momentum is conserved during the collision of two railroad cars.

Another example in which the conservation of momentum is put to use is that of a rocket engine. Fuel and oxidizer are pumped into the combustion chamber, where they combine and burn. The hot gases of combustion then pass through a nozzle, which causes them to be accelerated to exit velocities as great as 8000 ft/s. The gases constitute a flow of momentum toward the rear, and by the conservation of momentum, an equal momentum is imparted to the body of the rocket in the forward direction. Because the operation of a rocket engine does *not* depend upon pushing against an atmosphere and because a rocket carries its oxidizer with it, a rocket operates in the vacuum of space just as well as it does within the atmosphere of the earth.

7-4 IMPULSE

Definition and units

In Sect. 7-2 we saw that force is the time rate of change of momentum, and we derived the formula $F = \Delta P/t$ from Newton's second law. Rearranging gives

$$F \cdot t = \Delta P \qquad (7\text{-}8)$$

which was expanded to

$$F \cdot t = mv - mv_0$$

The product $F \cdot t$ is called the impulse. The force F is assumed to be constant during the time interval t. Here we have obtained a relationship between impulse and momentum: impulse equals the change in momentum.

The unit of impulse is that of force times time; consequently, we obtain these units for the three systems.

(*units of impulse*) N · s dyn · s lb-s

Because of the equality of impulse and the change in momentum, impulse and momentum must have the same units. That is, the following must be true:

$$\text{Units of impulse} = \text{units of momentum}$$
$$\text{N} \cdot \text{s} = \text{kg} \cdot \text{m/s}$$

To see that this is true, we recall (by remembering $F = ma$) that

$$N = \text{kg} \cdot \text{m/s}^2$$

Substituting into the impulse units to eliminate N and canceling gives

$$\text{N} \cdot \text{s} = \text{kg} \cdot \text{m/s}$$
$$(\text{kg} \cdot \text{m/s}^2) \cdot \text{s} = \text{kg} \cdot \text{m/s}$$
$$\text{kg} \cdot \text{m/s} = \text{kg} \cdot \text{m/s}$$

This identity illustrates that impulse and momentum have the same units.

Impulsive forces

In order to simplify the mathematics, we made the assumption that the force F in Eq. (7-8) was constant during the time t. Generally speaking, the force varies over a wide range, suddenly and sharply. Indeed, the term *impulsive force* implies a force occurring in a sharp blow such as a golf club striking a golf ball or a metal hammer striking a steel spike. In these cases, the duration of the force is extremely short and the magnitude of the force is extremely large. Other examples of impulsive forces are a marble dropped on a pane of glass, the collision of two billiard balls, and the encounter between a baseball and a baseball bat.

A bouncing ball illustrates many aspects of collision phenomena and the origin and nature of impulsive forces. The process is illustrated and summarized in Fig. 7-3. A force is first developed when initial contact is made between the ball

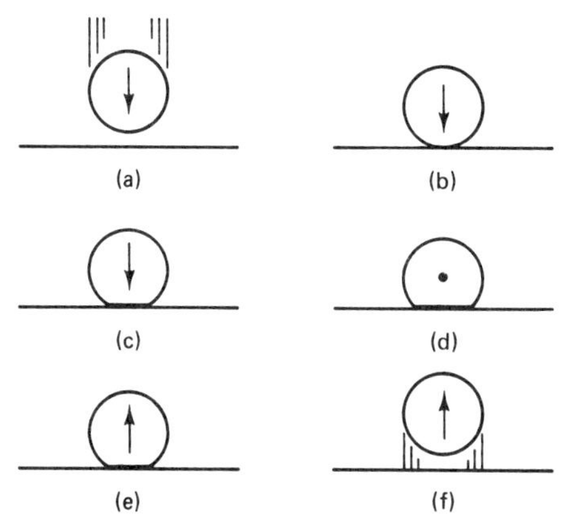

Figure 7-3 A bouncing ball. (a) Contact not yet made. (b) Initial contact. (c) Deformation of ball and surface is beginning; an upward force is being exerted on the ball. (d) Ball is stopped; upward motion begins. Force on ball is a maximum. (e) Ball accelerates upward as deformation diminishes. (f) Ball travels upward.

and the surface. Both the ball and the surface then undergo a process of deformation; both will be deformed, but not necessarily in equal amounts. The impulsive forces increase in proportion to the deformation, and the force on the ball causes it to decelerate until its forward motion ceases. At this point, the deformation and the force on the ball will be maximum. Consequently, the ball will accelerate in the opposite direction, away from the surface. As the deformation is relaxed, the forces diminish until they become zero at the instant of separation. If the nature of the ball and surface are such that no energy is absorbed in the process (a perfectly elastic collision), the ball will rebound from the surface at the same velocity with which it approached the surface.

Initially, the momentum of the ball is directed toward the surface. After the collision, the momentum is the same in magnitude (for a perfectly elastic collision), but its direction is reversed. Because momentum is a vector quantity, the directions of the two momenta must be considered. Therefore, the magnitude of the change in momentum is twice the magnitude of the original momentum of the ball. Because impulse equals the change in momentum, the change in momentum is an important physical quantity.

In many cases, as in the rebound of a ball thrown against a vertical wall, the change in momentum can be accurately stated, but the time of contact of the ball with the wall can be determined with much less accuracy. Furthermore, the force in effect between the ball and the wall can be determined with the least accuracy. Consequently, a calculation of the impulse made by using the product of force and time cannot be expected to be very accurate. However, because the impulse and the change in momentum are equal, the impulse can be obtained indirectly from the change in momentum.

For a given impulse (or change in momentum that occurs in a particular collision or impact under consideration), we can form an inverse proportion between F and t, namely,

$$F = \frac{\text{impulse}}{t} \tag{7-9}$$

We see that shorter times of interaction, t, produce greater forces. An example of a short interaction is a marble dropped on a pane of glass. A longer interaction is illustrated by a tennis ball dropped on the floor or colliding with a tennis racket.

7-5 ELASTIC AND INELASTIC COLLISIONS

Momentum is always conserved in collisions, but it is *not* true that kinetic energy is conserved. It is only in perfectly elastic collisions that kinetic energy and momentum are both conserved. Collisions between hardened steel balls, glass marbles, billiard balls, and other extremely hard objects most closely approximate perfectly elastic collisions on the observable level. Colliding bodies that are easily deformed absorb a portion of the original kinetic energy (converting it to heat) in the process of deformation. Many collision phenomena on the atomic level are perfectly elastic.

If a collision is not perfectly elastic, it is *inelastic*, meaning that some of the kinetic energy is converted to heat. Because kinetic energy is related to velocity, inelastic collisions are characterized by a loss of velocity. A ball that collides inelastically with a wall rebounds with less than the velocity of approach. In a perfectly inelastic collision, there is no rebound at all; the colliding bodies stick together. A ball of putty that sticks when thrown against a wall or two railroad cars that collide and become coupled are examples of inelastic collisions, in which momentum is conserved, but kinetic energy is not.

7-6 ELASTIC COLLISION IN ONE DIMENSION

Derivation

A one-dimensional collision is illustrated in Fig. 7-4. The bodies approach each other traveling at velocities u_1 and u_2. After the collision, they travel away from each other with velocities v_1 and v_2. We shall assume the collision is perfectly elastic so that we can use the conservation of kinetic energy as well as the conservation of momentum. Our objective is to find the velocity of each body after the collision, assuming we know the initial velocities and the masses. To derive the expressions for v_1 and v_2, we use the conservation of momentum and the conservation of kinetic energy.

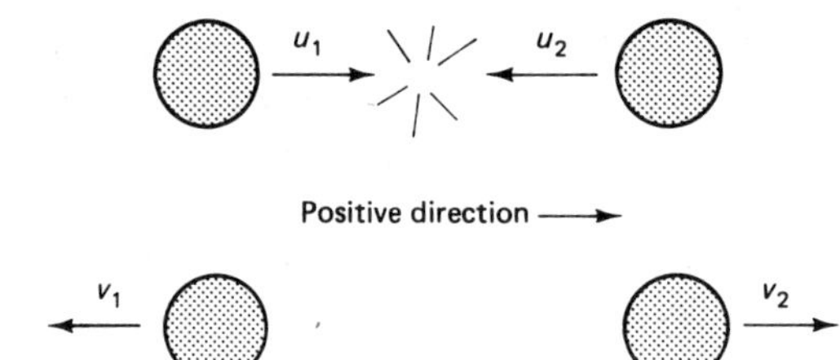

Figure 7-4 A collision in one dimension. Velocities to the left (v_1 and u_2) appear as negative quantities.

The conservation of momentum dictates that the momentum before the collision must equal the momentum after the collision:

$$\text{Momentum before} = \text{momentum after}$$

$$m_1u_1 + m_2u_2 = m_1v_1 + m_2v_2 \qquad [7\text{-}10(a)]$$

Rearranging gives

$$m_1(u_1 - v_1) = m_2(v_2 - u_2) \qquad [7\text{-}10(b)]$$

The collision is elastic, so the kinetic energy before the collision and after the collision is the same:

$$\text{Kinetic energy before} = \text{kinetic energy after}$$
$$\tfrac{1}{2}\, m_1u_1^2 + \tfrac{1}{2}\, m_2u_2^2 = \tfrac{1}{2}\, m_1v_1^2 + \tfrac{1}{2}\, m_2v_2^2 \qquad [7\text{-}10(c)]$$

Multiplying by 2 and factoring gives

$$m_1(u_1^2 - v_1^2) = m_2(v_2^2 - u_2^2) \qquad [7\text{-}10(d)]$$

Because the difference of two squares can be factored, Eq. [7-10(d)] can be written as

$$m_1(u_1 + v_1)(u_1 - v_1) = m_2(v_2 + u_2)(v_2 - u_2) \qquad [7\text{-}10(e)]$$

By dividing Eq. [7-10(e)] by Eq. [7-10(b)], we obtain

$$u_1 + v_1 = u_2 + v_2 \qquad [7\text{-}10(f)]$$

Solving this equation for v_1 and v_2 gives

$$v_1 = u_2 + v_2 - u_1 \qquad [7\text{-}10(g)]$$

$$v_2 = u_1 + v_1 - u_2 \qquad [7\text{-}10(h)]$$

These two equations are now substituted, one at a time, into Eq. [7-10(b)] to obtain expressions for v_1 and v_2 in terms of the initial velocities u_1 and u_2 and the masses m_1 and m_2 of the colliding bodies. The result is

$$v_1 = \left(\frac{m_1 - m_2}{m_1 + m_2}\right)u_1 + \left(\frac{2m_2}{m_1 + m_2}\right)u_2 \qquad (7\text{-}11)$$

$$v_2 = \left(\frac{2m_1}{m_1 + m_2}\right)u_1 + \left(\frac{m_2 - m_1}{m_1 + m_2}\right)u_2 \qquad (7\text{-}12)$$

EXAMPLE 7-5 In Fig. 7-5, an 8-kg hardened steel ball is about to collide with a similar ball whose mass is 3 kg. The smaller ball travels to the right at 10 m/s while the other ball travels to the left at 3 m/s. Calculate the velocity of the two balls after the collision, assuming the collision is perfectly elastic.

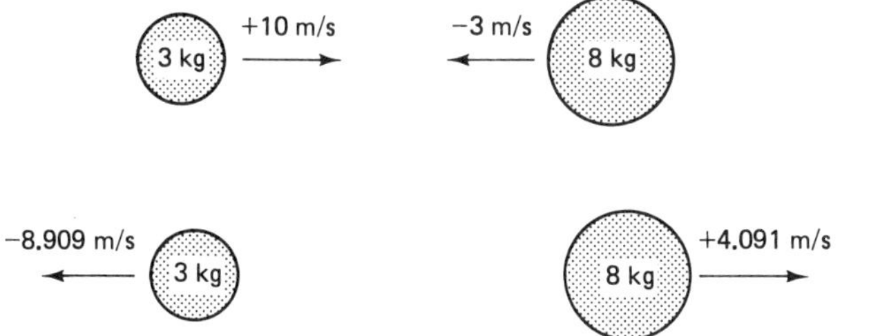

Figure 7-5

Solution. Equations 7-11 and 7-12 are applicable; only a direct substitution is required. However, we note that the initial velocity of the larger ball is negative because it is traveling to the left. Using Eqs. (7-11) and (7-12),

$$v_1 = \left(\frac{m_1 - m_2}{m_1 + m_2}\right)u_1 + \left(\frac{2m_2}{m_1 + m_2}\right)u_2$$

$$v_1 = \left(\frac{3 - 8 \text{ kg}}{3 + 8 \text{ kg}}\right)(10 \text{ m/s}) + \left(\frac{2(8 \text{ kg})}{3 + 8 \text{ kg}}\right)(-3 \text{ m/s})$$

$$v_1 = -8.909 \text{ m/s}$$

$$v_2 = \left(\frac{2m_1}{m_1 + m_2}\right)u_1 + \left(\frac{m_2 - m_1}{m_1 + m_2}\right)u_2$$

$$v_2 = \left(\frac{2(3 \text{ kg})}{3 + 8 \text{ kg}}\right)(10 \text{ m/s}) + \left(\frac{8 - 3 \text{ kg}}{3 + 8 \text{ kg}}\right)(-3 \text{ m/s})$$

$$v_2 = 4.091 \text{ m/s}$$

After the collision, the smaller ball travels to the left at 8.909 m/s while the larger ball rebounds to the right at 4.091 m/s.

Suggested Exercise Calculate and compare the total momentum and kinetic energy before and after the collision to verify that they are equal.

Special cases

Several special cases of the collision described by Eqs. (7-11) and (7-12) exist. For these cases, the equations for v_1 and v_2 may be simplified. First of all, when the masses are equal, the velocities of the two bodies are simply interchanged:

(*when* $m_1 = m_2$)

$$v_1 = u_2 \qquad v_2 = u_1 \tag{7-13}$$

This result is obtained by setting $m_1 = m_2$ in Eqs. (7-11) and (7-12) and simplifying.

When one body (m_2) is initially at rest so that $u_2 = 0$, we use Eqs. (7-11) and (7-12) to obtain:

(*when* $u_2 = 0$)

$$v_1 = \left(\frac{m_1 - m_2}{m_1 + m_2}\right)u_1 \qquad v_2 = \left(\frac{2m_1}{m_1 + m_2}\right)u_1 \tag{7-14}$$

If, in addition to stipulating that m_2 is initially at rest, we further assume that m_2 is much larger than m_1, we obtain the following, because $m_1/m_2 \approx 0$.

(*when* $u_2 = 0$ *and* $m_2 >> m_1$)

$$v_1 = \frac{(m_1/m_2 - 1)}{\left(\dfrac{m_1}{m_2} + 1\right)}u_1 \approx -u_1 \tag{7-15}$$

$$v_1 = \frac{2(m_1/m_2)}{\left(\frac{m_1}{m_2} + \frac{m_2}{m_2}\right)} u_1 \approx 2\left(\frac{m_1}{m_2}\right) u_1 \approx 0$$

Here we see that the smaller body simply bounces off the larger body without causing the larger body to move very much. An example of this collision would be a marble shot against a bowling ball initially at rest.

Finally, when m_2 is a very small body at rest, which is struck by a much larger body, we obtain

(when $u_2 = 0$ and $m_2 << m_1$)

$$v_1 = \frac{(1 - m_2/m_1)}{\left(1 + \frac{m_2}{m_1}\right)} u_1 \approx u_1$$

$$v_2 = \frac{2}{(1 + m_2/m_1)} u_1 \approx 2u_1 \tag{7-16}$$

This indicates that when a large body strikes a small body initially at rest, the small body takes off with twice the velocity of the larger body. Also, the velocity of the larger body remains essentially the same.

EXAMPLE 7-6 Consider a pitched baseball being struck by a bat whose effective mass is 10 times as great as that of the ball. We let body 1 be the bat and body 2 be the ball. Consequently, $m_1 = 10m_2$ in order to approximate the ball and bat. When this is substituted into Eq. (7-11), we obtain

$$v_1 = \left(\frac{10m_2 - m_2}{10m_2 + m_2}\right) u_1 + \left(\frac{2m_2}{10m_2 + m_2}\right) u_2$$

$$v_1 = 0.818u_1 + 0.182u_2$$

$$v_2 = \left(\frac{2(10m_2)}{10m_2 + m_2}\right) u_1 + \left(\frac{m_2 - 10m_2}{10m_2 + m_2}\right) u_2$$

$$v_2 = 1.818u_1 - 0.818u_2$$

As a numerical example, assume the ball is pitched at 85 mi/h (124 ft/s), which is a pretty fast ball, and suppose the batter swings the bat at 50 mi/h (73 ft/s). We can compute the final velocities of the ball and bat using these equations.

$$\begin{aligned} v_1 &= 0.818u_1 + 0.182u_2 \\ &= (0.818)(73 \text{ ft/s}) + (0.182)(-124 \text{ ft/s}) \\ &= 37.15 \text{ ft/s} \end{aligned}$$

$$\begin{aligned} v_2 &= 1.818u_1 - 0.818u_2 \\ &= (1.818)(73 \text{ ft/s}) - (0.818)(-124 \text{ ft/s}) \\ &= 234.15 \text{ ft/s} \end{aligned}$$

Thus, we see that the bat is slowed by hitting the ball (as expected), while the ball's velocity is increased dramatically, being almost doubled.

Fractional decrease of kinetic energy

When incident particle m_1 traveling at velocity u_1 strikes a stationary mass m_2, the incident particle loses a fraction of its kinetic energy. The fraction of the kinetic energy lost depends upon the relative masses of the two particles. To derive an expression for the fractional decrease of kinetic energy, we use Eq. (7-14),

$$v_1 = \left(\frac{m_1 - m_2}{m_1 + m_2}\right)u_1$$

which gives the velocity of the incident particle after the collison. If the initial and final kinetic energies are denoted by K_i and K_f, respectively, the fractional decrease is

$$\text{Fractional decrease} = \frac{K_i - K_f}{K_i}$$

where

$$K_i = \tfrac{1}{2}\, m_1 u_1^2 \quad \text{and} \quad K_f = \tfrac{1}{2}\, m_1 v_1^2$$

If the expression given above for v_1 is substituted into K_f and if the expressions for K_i and K_f are then substituted into the formula for the fractional decrease, the result is

$$\textit{(fractional decrease of kinetic energy of the incident particle)} = \frac{4m_1 m_2}{(m_1 + m_2)^2} \tag{7-17}$$

For the special case in which $m_1 = m_2$ (the masses of the particles are equal), the fractional decrease (as a percentage) is 100%. This is consistent with the fact that the incident particle is "stopped cold," while all its original velocity is given to the originally stationary particle.

7-7 INELASTIC COLLISIONS

When a collision is perfectly inelastic, no rebound occurs; the colliding bodies stick together. Kinetic energy is *not* conserved. The general situation is illustrated in Fig. 7-6. We wish to find v_f in terms of the masses and initial velocities, and we give an expression for the loss in kinetic energy that results from the collision.

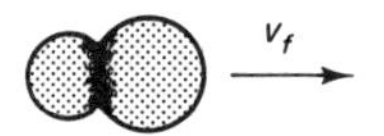

Figure 7-6 In a perfectly inelastic collision, the colliding bodies stick together.

Using the conservation of momentum, we write

$$\text{Momentum before} = \text{momentum after}$$

$$m_1v_1 + m_2v_2 = (m_1 + m_2)v_f$$

and we obtain

$$v_f = \frac{m_1v_1 + m_2v_2}{m_1 + m_2} \qquad (7\text{-}18)$$

This equation was obtained earlier as Eq. (7-7) as an illustration of the conservation of momentum.

To calculate the change in kinetic energy that occurs during the collision, we subtract the final from the initial kinetic energy. That is,

$$\Delta\text{KE} = K_i - K_f = \tfrac{1}{2}\,m_1v_1^2 + \tfrac{1}{2}\,m_2v_2^2 - \tfrac{1}{2}\,(m_1 + m_2)v_f^2$$

Substituting the expression for v_f [Eq. (7-18)] into the preceding equation and simplifying gives

$$\Delta\text{KE} = \frac{1}{2}\left(\frac{m_1m_2}{m_1 + m_2}\right)(v_1 - v_2)^2 \qquad (7\text{-}19)$$

This amount of kinetic energy is converted to other forms.

EXAMPLE 7-7 For the inelastic collision illustrated in Fig. 7-7, compute the final velocity and the loss of kinetic energy produced by the collision.

Solution. Noting that v_2 is negative, a straightforward application of Eq. (7-18) gives v_f:

$$\begin{aligned} v_f &= \frac{m_1v_1 + m_2v_2}{m_1 + m_2} \\ &= \frac{(5\text{ kg})(10\text{ m/s}) + (3\text{ kg})(-7\text{ m/s})}{(5\text{ kg} + 3\text{ kg})} \\ &= 3.62\text{ m/s} \end{aligned}$$

The change in kinetic energy is obtained from Eq. 7-19:

$$\begin{aligned} \Delta\text{KE} &= \frac{1}{2}\left(\frac{m_1m_2}{m_1 + m_2}\right)(v_1 - v_2)^2 \\ &= \frac{1}{2}\left(\frac{5(3)\text{ kg}}{5 + 3\text{ kg}}\right)[10 - (-7)\text{ m/s}]^2 \\ &= 0.5(1.875)(17)^2 \\ &= 270.9\text{ J} \end{aligned}$$

For the special case where $v_2 = 0$ in Eq. (7-18)—so that the incident particle collides inelastically with a stationary body—a simple expression can be obtained

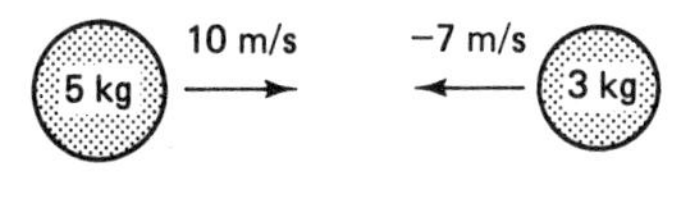

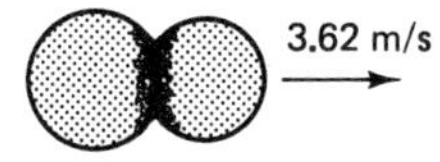

Figure 7-7

for the fraction of the original kinetic energy that is retained after the collision. Without proof, the expression is

$$\frac{K_f}{K_i} = \frac{m_1}{m_1 + m_2} \tag{7-20}$$

where m_2 is the mass of the stationary particle.

Coefficient of restitution

The coefficient of restitution is defined as the ratio of the relative velocity of separation to the relative velocity of approach for two bodies that collide. That is,

(*coefficient of restitution*) $$e = \frac{v_2 - v_1}{u_1 - u_2} \tag{7-21}$$

In the analysis of an elastic collision in one dimension, we obtained Eq. [7-10(f)]:

$$u_1 + v_1 = u_2 + v_2$$

This may be rearranged to give

$$u_1 - u_2 = v_2 - v_1$$

By considering the coefficient of restitution [Eq. (7-21)] in light of this equation, we see that $e = 1.0$ for a perfectly elastic collision, indicating that the relative velocity of separation equals the relative velocity of approach. On the other hand, the relative velocity of separation for a perfectly inelastic collision is zero because the colliding bodies stick together. Therefore, $e = 0$ for a perfectly inelastic collision. Collisions that are neither perfectly elastic nor perfectly inelastic are characterized by values of e between 0 and 1.

7-8 INELASTIC COLLISION IN ONE DIMENSION

Derivation

This section parallels the earlier treatment for a perfectly elastic collision, and we refer once again to Fig. 7-4. We wish to find expressions for v_1 and v_2 in terms of the initial velocities and the masses, comparable to Eqs. (7-11) and (7-12).

Kinetic energy is not conserved in an inelastic collision; a part of it is converted to heat, sound, and so forth as a result of the impact and deformation of the colliding bodies. Therefore, we cannot obtain an equation from the properties of kinetic energy as we did previously. Instead, we use the definition of the coefficient of restitution [Eq. (7-21)], which may be solved for v_1 and v_2 to give

$$v_1 = v_2 - e(u_1 - u_2) \qquad [7\text{-}22(a)]$$

$$v_2 = e(u_1 - u_2) + v_1 \qquad [7\text{-}22(b)]$$

Because momentum is conserved, we may write

$$m_1u_1 + m_2u_2 = m_1v_1 + m_2v_2 \qquad [7\text{-}22(c)]$$

When Eq. [7-22(b)] is substituted into Eq. [7-22(c)] in order to eliminate v_2, we obtain the desired expression for v_1. Likewise, when Eq. [7-22(a)] is substituted in order to eliminate v_1, we obtain the expression for v_2. Omitting algebraic details, the expressions are

$$v_1 = \frac{(m_1 - m_2e)}{(m_1 + m_2}u_1 + \frac{m_2(1 + e)}{(m_1 + m_2)}u_2 \qquad (7\text{-}23)$$

$$v_2 = \frac{m_1(1 + e)}{(m_1 + m_2)}u_1 + \frac{(m_2 - m_1e)}{(m_1 + m_2)}u_2 \qquad (7\text{-}24)$$

These equations apply to perfectly elastic collisions ($e = 1.0$), to completely inelastic collisions ($e = 0$), and to intermediate collisions ($0 < e < 1.0$). We note that when $e = 1.0$ is used, these equations become identical to the equations derived earlier [Eqs. (7-11) and (7-12)] for an elastic collision, and when $e = 0$ is used, both v_1 and v_2 become equivalent to Eq. (7-17) for inelastic collisions. This is not surprising because the defining equation for the coefficient of restitution embodies the same information that was obtained earlier from the conservation of kinetic energy, viz., the relationship between the initial and final velocities.

Bouncing ball

If body 2 is initially at rest ($u_2 = 0$) and is very large ($m_2 = \infty$), we arrive at a situation corresponding to the bouncing of a ball. By setting $u_2 = 0$ in Eqs. (7-23) and (7-24), we obtain

$$\begin{aligned} v_1 &= \frac{(m_1 - m_2e)}{(m_1 + m_2)}u_1 = \frac{(m_1/m_2 - e)}{(m_1/m_2 + 1)}u_1 \\ v_2 &= \frac{m_1(1 + e)}{(m_1 + m_2)}u_1 = \frac{(m_1/m_2)(1 + e)}{(m_1/m_2 + 1)}u_1 \end{aligned} \qquad (7\text{-}25)$$

The second step in these equations results from dividing both the numerator and the denominator by m_2. If m_2 becomes very large, as we stipulated above, the ratio m_1/m_2 becomes very small and may be neglected. Consequently, the preceding equations reduce to

$$v_1 = -eu_1 \quad \text{and} \quad v_2 = 0 \qquad (7\text{-}26)$$

Thus, a tennis ball bounced off the court will rebound with a fraction (e) of the initial velocity, while the court will not move. This is in accord with common experience.

Recall that if a ball is dropped from a height h, its velocity at impact is given by

$$v = \sqrt{2gh}$$

Therefore, we can say that if the impact velocity is u_1, the height h_i from which it dropped is given by

$$h_i = \frac{u_1^2}{2g}$$

Further, if the ball rebounds back to height h_f, h_f is given by

$$h_f = \frac{v_1^2}{2g} = \frac{e^2 u_1^2}{2g}$$

where we have used Eq. (7-26). Dividing the two preceding equations gives

$$\frac{h_f}{h_i} = e^2 \quad \text{or} \quad e = \sqrt{\frac{h_f}{h_i}} \tag{7-27}$$

which provides a convenient method of determining the coefficient of restitution. On the other hand, if the coefficient of restitution is already known, the height of the rebound is given by

$$h_f = e^2 h_i \tag{7-28}$$

These relationships are illustrated in Fig. 7-8.

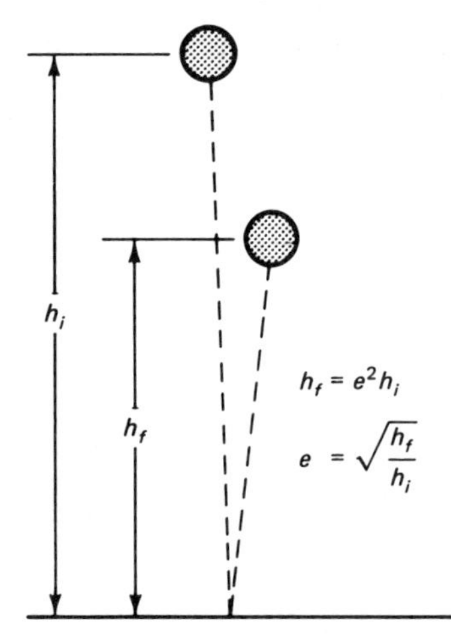

Figure 7-8 A bouncing ball is an example of a partially elastic collision. The coefficient of restitution, e, describes the degree of elasticity of the collision.

7-9 COLLISIONS IN TWO DIMENSIONS

The vector nature of momentum must be taken into account when colliding bodies are free to move in two dimensions. The conservation of momentum requires that both the x-component and the y-component be conserved. That is, the total x-component momentum before and after the collision must be the same, as must be the total y-component momentum. In general, two-dimensional collisions are

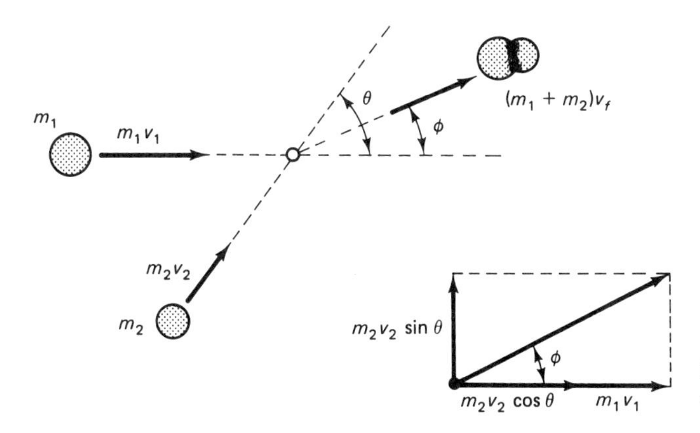

Figure 7-9 A collision in two dimensions.

sufficiently complicated to be beyond the scope of this text because the direction of the velocities, in addition to their magnitudes, come into play.

To illustrate the vector nature of momentum, consider the problem shown in Fig. 7-9, where two objects approach at an angle θ and collide inelastically, sticking together or becoming otherwise coupled together. We wish to find the magnitude and direction of the velocity with which the coupled masses move away from the point of impact.

Because the total y-momentum remains the same,

$$y\text{-momentum after} = y\text{-momentum before}$$
$$(m_1 + m_2)v_f \sin\phi = m_2v_2 \sin\theta \qquad [7\text{-}29(a)]$$

Similarly, for the x-momentum,

$$x\text{-momentum after} = x\text{-momentum before}$$
$$(m_1 + m_2)v_f \cos\phi = m_2v_2 \cos\theta + m_1v_1 \qquad [7\text{-}29(b)]$$

At this point, we note that the left-hand sides of these equations are similar, and we recall that the sine divided by the cosine gives the tangent. Therefore, we divide Eq. [7-29(a)] by Eq. [7-29(b)] to obtain

$$\tan\phi = \frac{m_2v_2 \sin\theta}{m_2v_2 \cos\theta + m_1v_1} \qquad (7\text{-}30)$$

Once the tangent of ϕ is known, the inverse function gives ϕ. We then can compute sine ϕ for use in the following expression for v_f, obtained by rearranging Eq. [7-29(a)]:

$$v_f = \frac{m_2v_2 \sin\theta}{(m_1 + m_2)\sin\phi} \qquad (7\text{-}31)$$

We now have algebraic expressions for computing both the magnitude and direction of the velocity of the coupled bodies.

EXAMPLE 7-8 Suppose two objects collide and become coupled together, as in Fig. 7-10. Compute the final velocity and the direction in which the entangled mass moves.

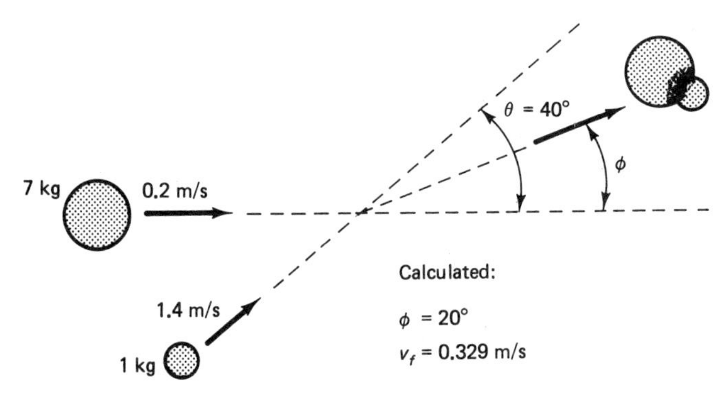

Figure 7-10

Solution. This requires straightforward application of Eqs. (7-30) and (7-31). Therefore,

$$\tan \phi = \frac{m_2 v_2 \sin \theta}{m_2 v_2 \cos \theta + m_1 v_1} = \frac{(1 \text{ kg})(1.4 \text{ m/s})\sin 40°}{(1 \text{ kg})(1.4 \text{ m/s})(\cos 40°) + (7 \text{ kg})(0.2 \text{ m/s})}$$

$$\tan \phi = 0.364$$

$$\phi = 20°$$

We then use a calculator to obtain: sin 20° = 0.342. Proceeding,

$$v_f = \frac{m_2 v_2 \sin \theta}{(m_1 + m_2)\sin \phi} = \frac{(1 \text{ kg})(1.4 \text{ m/s})(\sin 40°)}{(7 \text{ kg} + 1 \text{ kg})\sin 20°}$$

$$v_f = 0.329 \text{ m/s}$$

A collision such as this may be demonstrated on a frictionless air table.

7-10 THE BALLISTIC PENDULUM

Elaborate and sophisticated apparatus is required in order to determine the velocity of the bullet fired from a rifle or pistol by measuring the time required for the bullet to travel a known distance. However, a *ballistic pendulum*, illustrated in Fig. 7-11, allows the velocity of the bullet to be expressed in terms of easily measured quantities. A bullet of mass m_b is fired into a target of mass m_t that is suspended in order to be free to swing like a pendulum. A mechanical mechanism stops the target at its maximum height so that the vertical distance h that the target and bullet rise can be measured. This, in addition to the masses of the bullet and target, permits the velocity of the bullet to be determined. We now derive the appropriate expression.

The conservation of momentum during the collision gives

$$m_b v_b = (m_b + m_t)v_f \qquad [7\text{-}32(a)]$$

The kinetic energy that the target-bullet combination has immediately after the collision is

$$\tfrac{1}{2}(m_b + m_t)v_f^2 \qquad [7\text{-}32(b)]$$

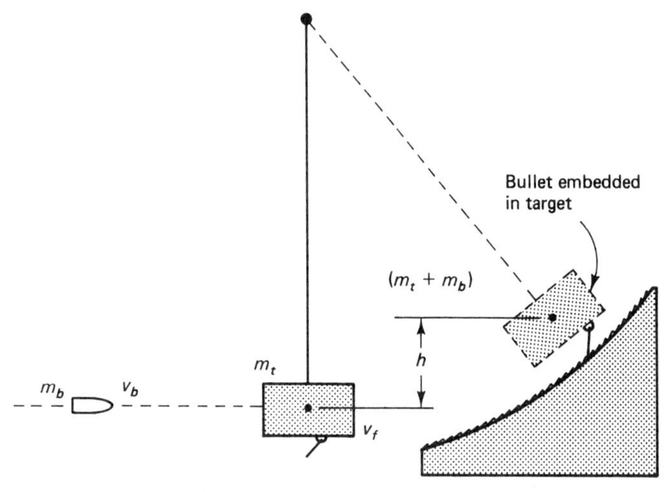

Figure 7-11 A ballistic pendulum.

and this will be converted completely to potential energy as the bullet-target mass rises a distance h. Hence,

$$\tfrac{1}{2}(m_b + m_t)v_f^2 = (m_b + m_t)gh \qquad [7\text{-}32(c)]$$

from which we obtain

$$v_f = \sqrt{2gh} \qquad [7\text{-}32(d)]$$

Substituting this into Eq. [7-32(a)] gives the desired result, the velocity of the bullet in terms of measurable quantities:

$$v_b = \left(\frac{m_b + m_t}{m_b}\right)\sqrt{2gh} \qquad (7\text{-}33)$$

We note that kinetic energy is not conserved during the collision; most of the original kinetic energy of the bullet is either converted to heat or is consumed in disrupting the target. After the collision, the kinetic energy of the target-bullet combination is converted to potential energy without a significant loss of energy. This fact allows us to write Eq. [7-32(c)].

Questions

1. Discuss the following collisions in terms of impulsive forces.
   (a) An egg is dropped onto the kitchen floor.
   (b) An egg is dropped onto a pillow.
   (c) A parachutist, when landing, must not lock his knees.

2. Which produces the most kick, a heavy shotgun or a lightweight shotgun?

3. During the launching of a rocket, the engines ignite and burn for a brief interval before the rocket is released from the pad. Explain how momentum is conserved during this interval.

4. Suppose a rocket hovers a short distance above the ground so that the rocket appears to stand on its tail of fire. Discuss this event in terms of the conservation of momentum.

5. The pendulum of a grandfather clock repeatedly swings back and forth. Does the pendulum have momentum? Is it conserved? Discuss.

6. Discuss the following in terms of the conservation of momentum.
   (a) Eruption of a volcano.
   (b) A tennis player bouncing a ball on the court prior to serving.
   (c) Landing of a large airliner.
   (d) Collision of two outfielders trying to catch a fly ball.

7. Why does a ball bounce? Why does a marble bounce when dropped onto a pane of glass? Is deformation essential to bouncing?

8. Why are several fire fighters (instead of just one) often observed to hold and aim a high-pressure hose?

9. Suppose an ultra-superball were invented that always rebounded with a velocity slightly greater than its impact velocity.
   (a) What principles of physics would be violated?
   (b) What would eventually happen if you threw it into an empty room and then quickly closed the door?

Problems

1. What is the momentum of a 1-kg mass moving at 1 m/s?

2. Compute the momentum of a truck whose total weight is 6 tons if it is moving at 60 mi/h.

3. In a time interval of 8 s, the momentum of a freely falling body changes from 88.2 kg · m/s to 323.4 kg · m/s.
   (a) What force is acting on the body?
   (b) What is the mass of the body ($g = 9.8$ m/s^2)?
   (c) What was the initial velocity of the body?

4. A net force of 16 N acts on an object whose mass is 8 kg for 6 s.
   (a) What is the impulse of the force?
   (b) By how much does the momentum of the object change?
   (c) If the initial velocity of the object was 24 m/s, what is the final velocity?

5. A 2-kg mass is allowed to fall freely from rest for 4 s.
   (a) What impulse is delivered to the mass by the gravitational attraction of the earth?
   (b) By how much does the velocity change?
   (c) By how much does the momentum change?

6. **(a)** Compute the recoil velocity of a rifle of mass 2.4 kg that fires a 4-g bullet at a muzzle velocity of 1200 m/s.
   (b) Compute the kinetic energy of the bullet as it leaves the barrel.
   (c) Compute the kinetic energy of the gun as it recoils.

7. Repeat Problem 6 for an 8-g bullet.

8. Suppose an extremely small and light handgun is developed whose mass is only four times as great as the bullet it fires.
   (a) If the gun is to fire a 4-g bullet at a muzzle velocity of 800 m/s, what will be the recoil velocity of the gun?
   (b) Compute the kinetic energy of the recoiling gun and compare with the kinetic energy of the bullet. Comment.

* **9.** Two identical railroad cars of mass 20 tons are about to collide on a level track. One car travels toward the right (assumed to be the positive direction) while the other travels toward the left. If the cars become coupled after the collision, compute the final velocity of the coupled cars for the following combinations of velocities: **(a)** (+5, −4 ft/s); **(b)** (+5, 0 ft/s); **(c)** (+5, +4 ft/s); **(d)** (+5, −5 ft/s).

***10.** Repeat Problem 9, but assume the car traveling to the right has a mass of 40 tons rather than 20 tons.

11. A small rubber ball of mass 0.1 kg bounces off a vertical wall. The velocity of the ball both before and after the collision is 4 m/s.
- **(a)** What are the momenta of the ball before and after the collision?
- **(b)** What is the change in momentum?
- **(c)** If the time of contact of the ball with the wall is 0.15 s, what average force does the wall exert on the ball?

12. Repeat part **(c)** of Problem 11 for the following times of contact, paying attention to the resulting forces: **(a)** 0.05 s; **(b)** 0.10 s; **(c)** 0.20 s; **(d)** 0.30 s.

13. A glass marble whose mass is 20 g is dropped from a height of 1 ft onto a horizontal pane of glass. Assume the marble bounces back to the same height from which it was dropped and the time of contact between the marble and the glass is 0.002 s.
- **(a)** With what velocity does the marble strike the glass?
- **(b)** With what velocity does it rebound from the glass?
- **(c)** What is the change of momentum of the marble?
- **(d)** Compute the average force exerted on the glass by the marble.

***14.** Repeat Problem 13, but assume the coefficient of restitution between the marble and the glass is 0.8.

15. Repeat Example 7-5, but for balls of different masses. Let the 8-kg ball become 6 kg, while the 3-kg ball is increased to 5 kg.

16. Two balls of mass 100 g are about to collide elastically. One ball travels to the right at 3 m/s, while the other travels to the left at 2 m/s. What will be their velocities after the collision?

17. Repeat Problem 16, but let the 2-m/s ball be initially at rest.

18. Suppose a bowling ball and a marble are suspended from the ceiling by light strings so that their centers are at the same height from the floor. If the bowling ball is pulled aside and released so that it strikes the marble (elastically) with a velocity of 1 m/s, what will be the resulting velocity of the marble?

***19.** Suppose a 2-kg hardened steel ball moving at 6 m/s collides elastically with a stationary ball whose mass is 4 g.
- **(a)** Calculate the final velocity of each ball.
- **(b)** What was the initial kinetic energy of each ball?
- **(c)** What is the final kinetic energy of each ball?
- **(d)** Compute the fractional decrease in kinetic energy of the smaller ball.
- **(e)** Compare the total kinetic energy of both balls with the initial kinetic energy of the smaller ball.

***20.** Repeat Problem 19, but let the 2-kg ball become 3 kg.

CHAPTER 8

CIRCULAR MOTION

Thus far we have been concerned primarily with objects in straight-line (translational) motion. In this chapter we consider objects in rotation.

It is noteworthy that every physical quantity developed for translational motion has a rotational analog. Thus, we encounter angular displacement, angular velocity, angular acceleration, and so forth; we see that torque is the rotational analog of force, and the moment of inertia is the rotational analog of mass (pertaining to inertia). We use the translation-rotation analogy to obtain a set of equations for describing rotational motion. We obtain expressions for rotational kinetic energy and angular momentum, and we see that angular momentum is conserved.

Newton's laws are applicable to particles or objects moving in a circular path. As a matter of fact, it is the tendency of a particle to move in a straight line that produces centripetal force and coriolis force. Moreover, we see that the precession of a gyroscope is caused by the same thing, that a body continues its motion in a straight line until a force acts to cause it to do otherwise.

8-1 RADIAN MEASURE

The familiar unit of angular measure is the *degree*; there are 360° in one complete circle. However, the disadvantage of the degree is that there is no simple relationship between the angle, the radius of the circle in which the angle is included, and the arc length subtended by the angle. This disadvantage may be avoided by defining a new unit of angular measure called the *radian* (rad). The included angle is defined to be 1 rad when the angle is such that the subtended arc equals the radius of the circle. Thus, the relationship between the angle and its subtended arc is inherent in the definition of a radian.

Because the circumference of a circle is $2\pi R$, there are 2π radians in a complete circle. Consequently, there are $360/2\pi$ (about 57.3) degrees in 1 rad. Conversion from degrees to radians and vice versa may be accomplished by

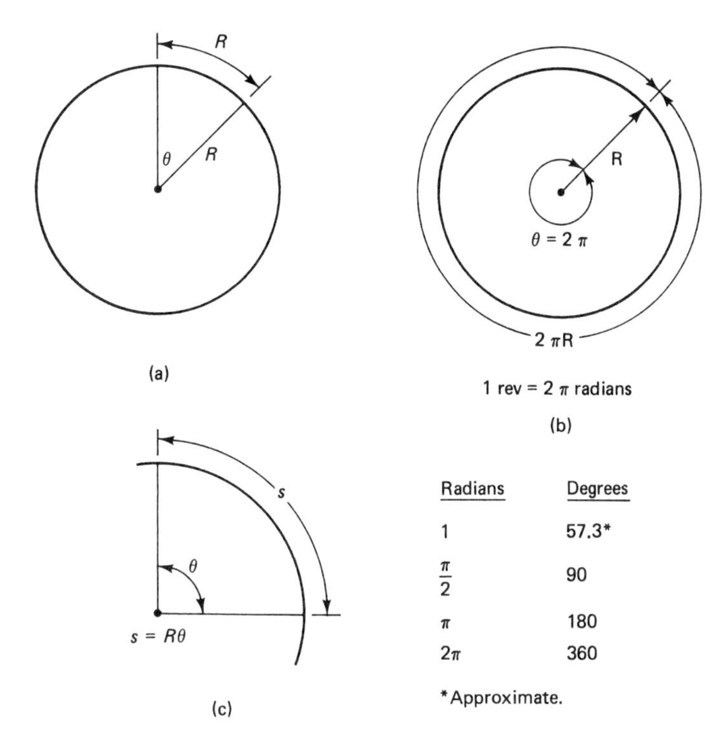

Radians	Degrees
1	57.3*
$\frac{\pi}{2}$	90
π	180
2π	360

Figure 8-1 Radian measure.

remembering that π radians is the equivalent of 180°. These relationships are illustrated in Fig. 8-1.

If the length of arc s is given relative to a circle of radius R, the angle subtended by the arc (in radians) is

$$\theta = \frac{s}{R} \tag{8-1}$$

which may be turned around to give

$$s = R\theta \tag{8-2}$$

Thus, if θ is given in radians, the arc length is simply the product of the radius and the angle.

When the angle is small—for instance, 10° or less—a useful simplification is obtained by assuming the straight vertical side of a right triangle to be an arc of equal length, as illustrated in Fig. 8-2. Adapting Eq. (8-2), the angle θ (in radians) is given approximately by

(for small angles)
$$\theta \approx \frac{h}{l} \quad \text{and} \quad h \approx l\theta \tag{8-3}$$

On the other hand, if the triangle is considered, the exact value of the angle θ is given by

(exact, for all angles)
$$\theta = \tan^{-1}\frac{h}{l} \quad \text{and} \quad h = l \tan\theta \tag{8-4}$$

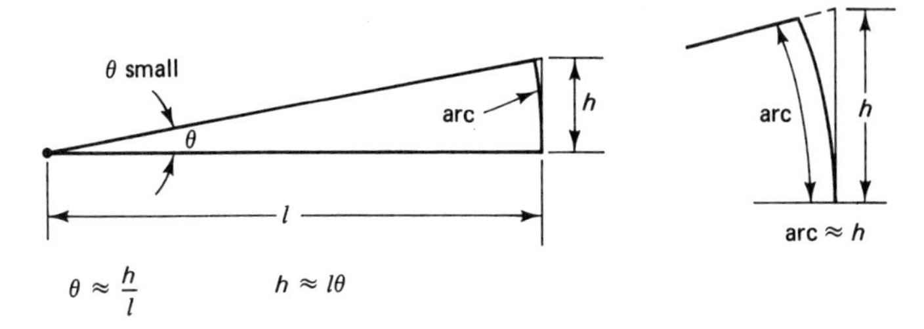

Figure 8-2 When θ is small, the length of the arc is very nearly the same as the height of the right triangle.

Although not widely used, another unit of angular measure, the *grad*, is found on many scientific calculators. By definition, a grad is one-hundredth of a right angle. Hence, 100 grads is the equivalent of 90° and 400 grads is the equivalent of 360°, or 2π radians. The grad suffers from the same disadvantages as the degree in regard to the arc length vs. angle relationship.

8-2 ANGULAR DISPLACEMENT

In describing the motion of rotating objects, it is frequently desirable to specify the amount of rotation that has occurred. This quantity is called the *angular displacement*, and it is simply the angle through which the object has rotated. This is illustrated in Fig. 8-3(a).

An object undergoing simple rotation will return to its original position when the angular displacement becomes 2π rad (one revolution). If the rotation continues, the object will again return to its original position when the angular displacement is 4π, 6π, 8π, . . . rad. Thus, for a uniform rotation, the angular

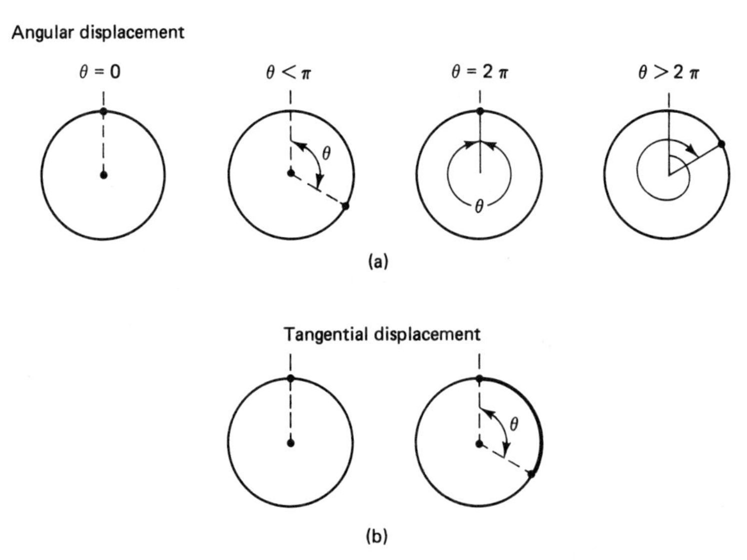

Figure 8-3 Angular and tangential displacement.

displacement steadily increases, even though the object periodically returns to its initial position.

Because angular displacement is an angle, the units may be degrees, radians, or (quite commonly) revolutions. Of course, one revolution is the equivalent of 2π rad or 360°.

Suppose a point is marked on the rim of a wheel of radius R. If the wheel rotates through an angle θ, the point will move along a curved path, the arc of a circle subtended by θ. Therefore, the distance the point moves is given by $R\theta$, according to Eq. (8-2). This distance, measured along the circumference of the wheel, is called the *tangential displacement* [see Fig. 8-3(b)].

8-3 ANGULAR VELOCITY

The speed of rotation is called the *angular velocity* and is denoted by the Greek letter omega ω. The angular velocity is the rate of change of angular displacement. In mathematical terms, following the definition of linear velocity in Sect. 3-2,

$$\omega = \frac{\theta_2 - \theta_1}{t_2 - t_1} \quad \text{or} \quad \omega = \frac{\theta - \theta_0}{t} \tag{8-5}$$

from which we obtain the rotational analogy of Eq. (3-4):

$$\theta = \theta_0 + \omega t \tag{8-6}$$

The units of angular velocity may be rad/s, deg/s, rev/s, or rev/min (rpm). Incidentally, because a radian is a ratio of an arc length to the radius (the ratio of two lengths), the units cancel so that a radian is actually a dimensionless quantity. However, in order to emphasize that the unit of angular measure is the radian, we carry the designation *rad* almost as if it were a legitimate unit.

An angular velocity may be ascribed to an object traveling in a circular path even though the object is not physically connected to a central pivot point. An example is a race car traveling around a circular track, as illustrated in Fig. 8-4. An observer at the center of the track views the car through the telescope which

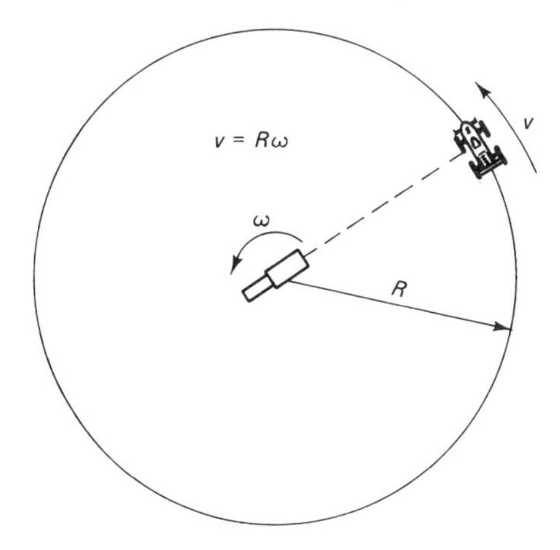

Figure 8-4 Tangential velocity and angular velocity.

must be continually rotated in order to keep the car in view. The *angular velocity* of the car is the same as that of the rotating telescope. The *tangential velocity* of the car is its speed around the track as indicated on the car's speedometer. We now look for a relationship between the angular velocity ω and the tangential velocity v.

In a given time period, the car will proceed a distance s around the track and the telescope will have been rotated an angle θ. From Eq. (8-2), $s = R\theta$. When both sides of this equation are divided by t, we obtain

$$\frac{s}{t} = R\,\frac{\theta}{t}$$

and it follows from basic definitions that

$$v = R\omega \tag{8-7}$$

which is the desired relationship.

8-4 ANGULAR ACCELERATION

The speeding up and slowing down of a rotating object may be described by specifying the *angular acceleration* α (Greek alpha), which is defined as the rate at which the angular velocity changes with time. Hence,

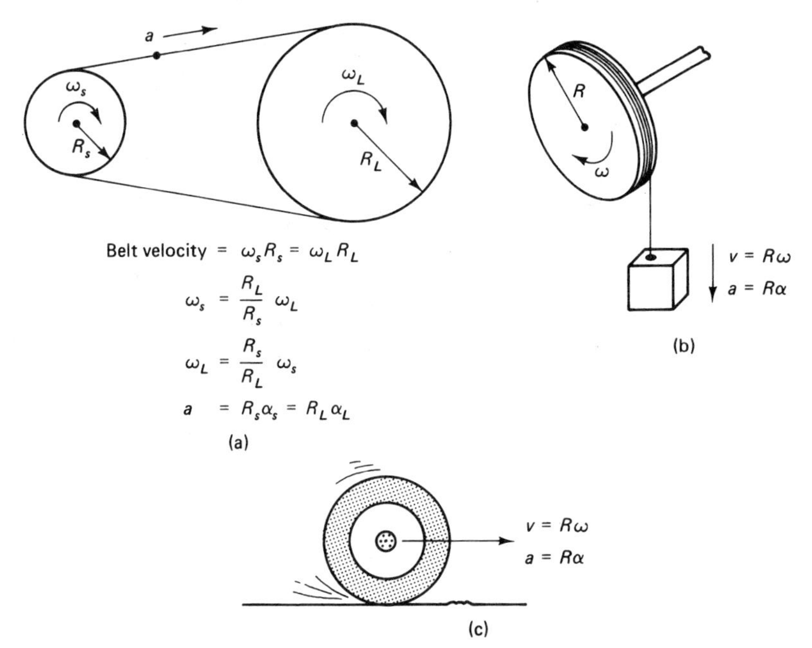

Figure 8-5 The relationship between linear and angular velocity and acceleration for three physical situations.
(a) Belt and pulley system.
(b) An object unwinding from a rotating drum.
(c) A rolling wheel.

$$\alpha = \frac{\omega_2 - \omega_1}{t_2 - t_1} \quad \text{or} \quad \alpha = \frac{\omega - \omega_0}{t} \tag{8-8}$$

which follows the same pattern as in Sect. 3-3 for linear acceleration. Rearranging the above equation gives

$$\omega = \omega_0 + \alpha t \tag{8-9}$$

which is the rotational counterpart of Eq. (3-4).

The units of angular acceleration are those of a change in angular velocity per unit time. Namely, the units are rad/s^2, deg/s^2, or perhaps rpm/s^2. Note that the unit of angular acceleration is characterized by a time-squared term in the same manner as for linear acceleration.

Angular acceleration α is related to linear acceleration a by the following equation which is obtained in a manner similar to Eq. (8-7).

$$a = R\alpha \tag{8-10}$$

This equation relates to physical situations such as those illustrated in Fig. 8-5.

8-5 ROTATIONAL EQUATIONS OF MOTION

At this point we have defined three rotational quantities corresponding to linear displacement, velocity, and acceleration. Therefore, it should be evident that a full complement of equations of motion can be developed for rotational motion as we do for linear motion in Chap. 3. Some of these we have already obtained; the others are given in Table 8-1.

TABLE 8-1 CORRESPONDENCE BETWEEN LINEAR AND ROTATIONAL EQUATIONS OF MOTION

Linear	Rotational	
$2as = v^2 - v_0^2$	$2\alpha\theta = \omega^2 - \omega_0^2$	(8-11)
$s = v_0t + \frac{1}{2}at^2$	$\theta = \omega_0 t + \frac{1}{2}\alpha t^2$	(8-12)
$v = v_0 + at$	$\omega = \omega_0 + \alpha t$	(8-9)
$\bar{v} = \frac{v + v_0}{2}$	$\bar{\omega} = \frac{\omega + \omega_0}{2}$	(8-13)

To relate linear rotational quantities:

$$s = R\theta \qquad v = R\omega \qquad a = R\alpha$$

EXAMPLE 8-1 When first turned on, a motor is observed to attain its standard speed of 1750 rpm in 4 s. Assuming the angular acceleration is uniform: (a) calculate the standard speed in radians per second; (b) calculate the angular acceleration α; (c) find how many revolutions the motor will turn as it comes up to speed.

Solution. (a) Using the fact that 2π radians equals 1 revolution,

$$1750 \text{ rpm} = \left(\frac{\text{rev}}{\text{min}}\right)\left(\frac{1 \text{ min}}{60 \text{ s}}\right)\left(\frac{2\pi \text{ rad}}{1 \text{ rev}}\right) = 183.3 \text{ rad/s}$$

(b) Making use of Eq. (8-9) with $\omega_0 = 0$, we obtain

$$\alpha = \frac{\omega}{t}$$

$$\alpha = \frac{183.3 \text{ rad/s}}{4 \text{ s}} = 45.83 \text{ rad/s}^2$$

(c) This can be worked two ways. (We first compute the angular displacement in radians.)

1. Using Eq. (8-11) with $\omega_0 = 0$:

$$2\alpha\theta = \omega^2$$

$$\theta = \frac{\omega^2}{2\alpha}$$

$$\theta = \frac{(183.3 \text{ rad/s})^2}{2(45.83 \text{ rad/s}^2)}$$

$$\theta = 366.6 \text{ rad}$$

2. Using Eq. (8-12) with ω_0,

$$\theta = \frac{1}{2}\alpha t^2$$

$$\theta = (0.5)(45.83 \text{ rad/s}^2)(4 \text{ s})^2$$

$$\theta = 366.6 \text{ rad}$$

Finally, converting 366.6 rad to revolutions,

$$366.6 \text{ rad} = (366.6 \text{ rad})\left(\frac{1 \text{ rev}}{2\pi \text{ rad}}\right) = 58.35 \text{ rev}$$

8-6 CENTRIPETAL ACCELERATION

An object traveling uniformly in a circular path must have a force constantly applied in the direction of the center of the circle. This force is called the *centripetal force* and it produces a *centripetal acceleration* of the object toward the center of the circle. The circular path represents the combined effect of the tangential velocity and the velocity resulting from the centripetal acceleration. If an object on the end of a string is swung above one's head, a force is felt that tends to pull the string outward toward the object. This reaction force to the centripetal force is called the *centrifugal force*.

We can obtain an expression for the centripetal acceleration by considering the rate of change of velocity as the object travels with angular velocity ω in the arc of a circle of radius R. We consider the change in velocity that occurs in a small time interval Δt and then divide by Δt to obtain $\Delta v/\Delta t$, which is the acceleration. We assume Δt is very small so that the angle θ in Fig. 8-6 is also

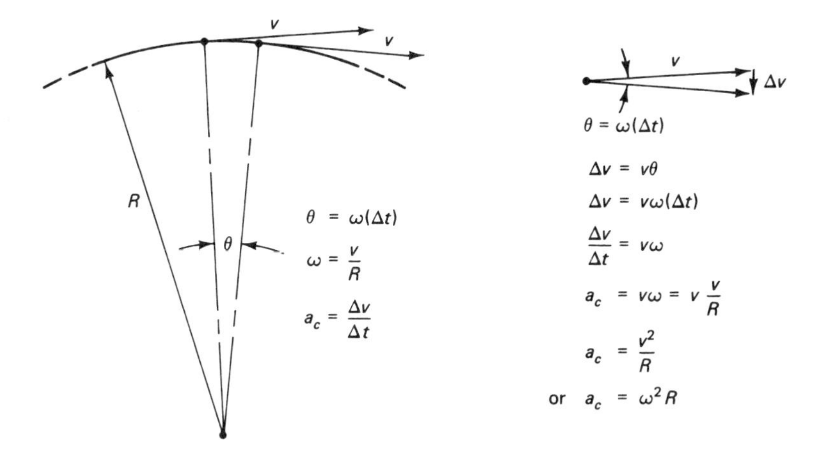

Figure 8-6 Derivation of the formula for centripetal acceleration.

small. The magnitude of the velocity is constant; only the direction changes as the object progresses around the circular path.

In Fig. 8-6, we note that the angle between the velocity vectors is θ because of the right angle between the radius and the tangential velocity. Because θ is small, we can use the approximation of Sect. 8-1 and write that $\Delta v = v\theta$. The angle θ is the angular displacement that the object undergoes in the time Δt, namely, $\omega(\Delta t)$. Algebra reveals $\Delta v/\Delta t = v\omega$, an interesting result because this shows that the centripetal acceleration, $\Delta v/\Delta t$, is the product of the tangential and angular velocities. Further algebra gives

$$a_c = \frac{v^2}{R} \quad \text{or} \quad a_c = \omega^2 R \qquad \text{(8-14)}$$

(*centripetal acceleration*)

The centripetal force is the product of the centripetal acceleration and the mass of the object:

$$F_c = m\frac{v^2}{R} \quad \text{or} \quad F_c = m\omega^2 R \qquad \text{(8-15)}$$

(*centripetal force*)

The centripetal force is directed toward the center of the circular path.

EXAMPLE 8-2 Compute the centripetal acceleration at the equator due to the rotation of the earth.

Solution. The earth rotates once every 24 h,[1] so its angular velocity is

$$\left(\frac{1\text{ rev}}{24\text{ h}}\right)\left(\frac{1\text{ h}}{3600\text{ s}}\right)\left(\frac{2\pi\text{ rad}}{1\text{ rev}}\right) = 7.27 \times 10^{-5}\text{ rad/s}$$

The average radius of the earth is 6.37×10^6 m.

$$\text{Centripetal acceleration} = \omega^2 R = (7.27 \times 10^{-5}\text{ rad/s})^2(6.37 \times 10^6\text{ m})$$
$$= 0.0337\text{ m/s}^2$$

[1]See problem 27 at the end of this chapter.

This is about 3.37 cm/s^2, which is 0.34% of 980 cm/s^2, the nominal acceleration of gravity.

Motion in a vertical circle

When an object travels around a circle in the vertical plane, the net force on the object is the resultant of the centrifugal force and the gravitational force of the earth on the object. Typical of such objects are a ball on a string, an airplane doing a loop, and a roller coaster going through a loop. At the top of the loop, the two forces are oppositely directed; hence, the net force is the difference between the centrifugal and the gravitational forces. But at the bottom of the loop, the forces are in the same direction and add together. Thus, we experience the minimum g-force at the top, and the maximum g-force at the bottom, as illustrated in Fig. 8-7.

A velocity greater than a certain minimum must be maintained at the top of the loop in order to prevent the object from falling out of the circular path. The minimum velocity is that velocity for which the net force at the top of the loop is exactly zero; or, it is the velocity that causes the centrifugal and gravitational forces to be exactly equal at the top of the loop. Formulas for the minimum velocity are given in Fig. 8-7.

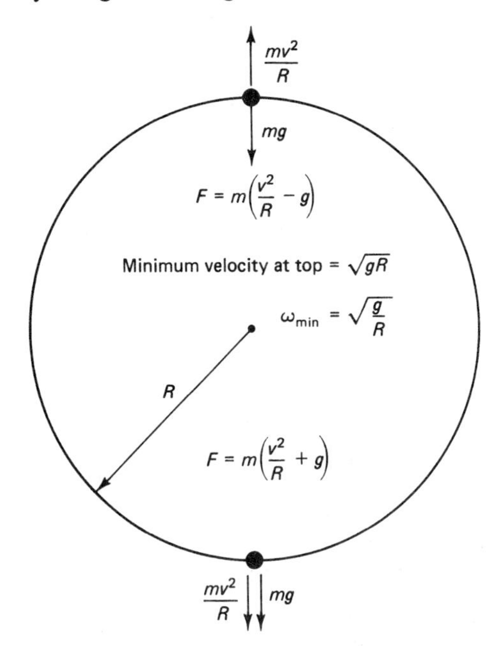

Figure 8-7 Motion in a vertical circle.

Rotating-ball governor

One type of rotational speed-control device involves a rotating-ball governor, shown schematically in Fig. 8-8. Centrifugal force causes the balls to be slung outward, which causes the output coupling to rise and fall as the rotational speed of the governor varies. The output coupling is linked mechanically to the throttle (or equivalent control) of the device whose speed is being controlled.

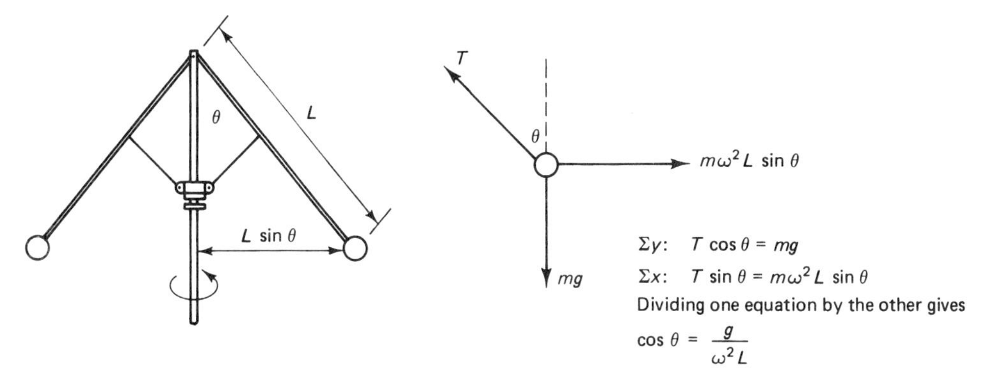

Figure 8-8 Rotating-ball governor.

If we ignore the effects of the coupling device (which cannot be done in practice), a simple formula can be derived that relates the angle θ to the angular velocity ω. This is a vector problem; details are given in Fig. 8-8. The relationship may be verified experimentally using a physics-lab governor that does not have the coupling.

Banking of curves

Curves of roadways and railroads are banked in order to compensate for the effects of centrifugal force. For similar reasons, an airplane banks when executing a turn. We now consider the situation where a vehicle rounds a curve of radius R that is banked at an angle θ.

For a curve with a particular radius and angle of bank, there is a single velocity (or speed) v at which a vehicle should round the curve so that there is no tendency to skid, either to the outside or to the inside. At greater speeds, the vehicle tends to skid to the outside; at lesser speeds, the vehicle tends to skid to the inside. Friction between the wheels and the road broadens the range of speeds for which no skid occurs. If there is no friction at all, there will be only one speed with which a vehicle may round a curve without skidding.

The forces acting on the vehicle are shown in Fig. 8-9. Note that the normal force is increased by one component of the centrifugal force. When the vehicle is on the verge of skidding up the plane (an outside skid), the forces down the plane equal the forces up the plane. In this case, the frictional force acts down the plane. The corresponding velocity is the maximum velocity v_{max} with which the vehicle may round the curve without skidding. The formula for v_{max} derived in the figure is

$$v_{max} = \sqrt{gR\left(\frac{\mu + \tan\theta}{1 - \mu\tan\theta}\right)} \tag{8-16}$$

As an alternative, we may solve for the minimum angle of bank required for a given velocity. The result is

$$\tan\theta_{min} = \frac{v^2 - \mu gR}{\mu v^2 + gR} \tag{8-17}$$

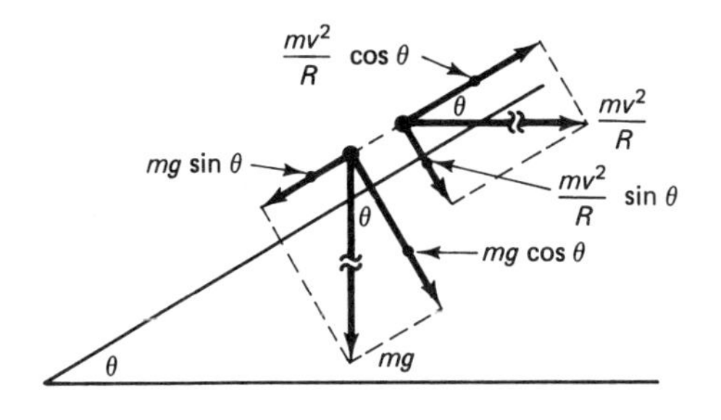

$$N = mg\cos\theta + \frac{mv^2}{R}\sin\theta$$

$$\text{Frictional force} = \mu N = \mu\left(mg\cos\theta + \frac{mv^2}{R}\sin\theta\right)$$

When on the verge of skidding up the plane,

Forces down the plane = forces up the plane

$$mg\sin\theta + \mu\left(mg\cos\theta + \frac{mv^2}{R}\sin\theta\right) = \frac{mv^2}{R}\cos$$

$$g\sin\theta + \mu g\cos\theta = \frac{v^2}{R}(\cos\theta - \mu\sin\theta)$$

Divide by $\cos\theta$ and then solve for v:

$$g\tan\theta + \mu g = \frac{v^2}{R}(1 - \mu\tan\theta)$$

$$v_{max} = \sqrt{gR\,\frac{\mu + \tan\theta}{1 - \mu\tan\theta}}$$

Figure 8-9 Derivation of the maximum velocity with which a vehicle may round a curve of radius R banked at angle θ; the coefficient of friction between the wheels and the road is μ.

In practical computations, this angle may turn out to be a negative quantity, which means that the coefficient of friction is sufficiently great to prevent a skid even if the road is banked to the outside.

If the vehicle rounds the curve too slowly, it will tend to skid to the inside. Picture a car creeping around an icy curve that is steeply banked. In this case, the car must maintain a velocity greater than a certain minimum velocity v_{min}. For inside skids the frictional force is directed *up* the plane. An analysis similar to that in Fig. 8-9 gives the minimum velocity required in order to avoid an inside skid:

$$v_{min} = \sqrt{gR\left(\frac{\tan\theta - \mu}{1 + \mu\tan\theta}\right)} \tag{8-18}$$

When a negative quantity is obtained under the square root sign, it means that the vehicle may come to a complete stop on the curve without skidding to the inside.

The ideal velocity for a vehicle rounding a curve is the velocity that would not produce a skid even in the total absence of friction. A formula for this velocity may be obtained by setting $\mu = 0$ in Eq. (8-16) or Eq. (8-18). Thus, in the absence of friction, the required velocity for no skid to occur is

$$v = \sqrt{gR\tan\theta} \tag{8-19}$$

If the velocity is given, the ideal angle of bank is given by

$$\tan \theta = \frac{v^2}{gR} \tag{8-20}$$

A graph of the no-skid velocity range for a curve banked at 30° is given in Fig. 8-10 as a function of the coefficient of friction. Observe that as the roadway becomes slicker, the velocity must be more carefully controlled.

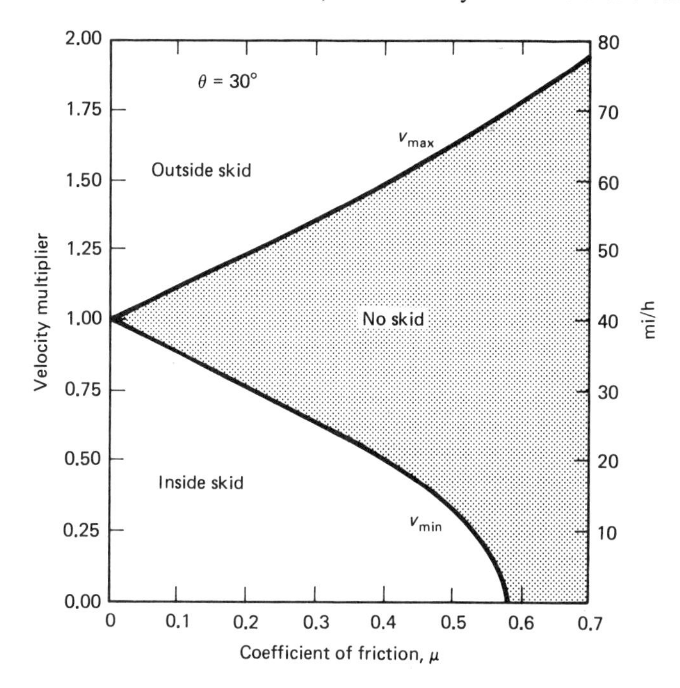

Figure 8-10 The no-skid velocity range for a curve banked at 30° plotted as a function of the coefficient of friction. The numerical example in miles per hour refers to a curve of radius 186 ft.

8-7 THE DYNAMICS OF ROTATION

Torque and angular acceleration

We first discussed torque in Sect. 5-9, where it is defined as a twisting force that results when a force F acts in conjunction with a moment arm R about a pivot point. In that section torque is considered in conjunction with the conditions for equilibrium, and actual rotation of the object under consideration never occurs. In this section we consider the effect of torque on rotating objects; a torque applied to a rotating object will cause the rotational speed of the body to change. That is, a net torque applied to a rotating body produces an angular acceleration of the body. Moreover, the acceleration may be positive or negative, depending upon the sense of the applied torque.

Rotational inertia refers to the reluctance of a rotating body to change its state of rotational motion. If a body is rotating with a certain angular velocity, the body will maintain that velocity until it is acted upon by an unbalanced torque. Or, if the body is at rest, it will tend to remain at rest until it is acted upon by an

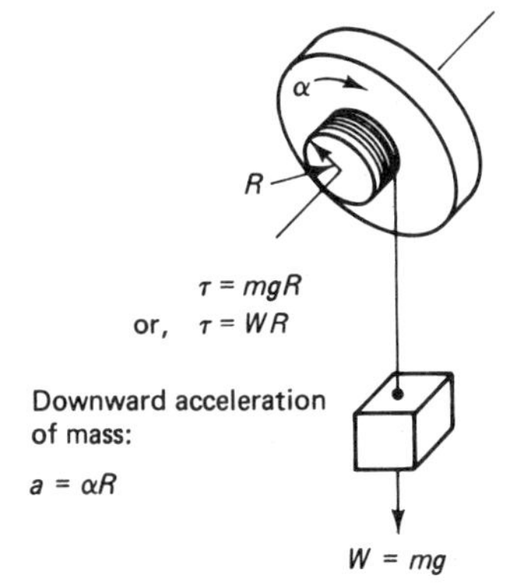

Figure 8-11 The weight of the descending mass exerts a torque on the flywheel causing it to experience an angular acceleration.

unbalanced torque. This parallels Newton's first law for linear motion (Sect. 4-3). In Fig. 8-11, a constant applied torque causes the flywheel to undergo an angular acceleration; at the same time, the reluctance of the flywheel to be accelerated causes the mass to be accelerated downward at a rate much less than the acceleration of gravity. Also, note the relationship between the angular acceleration of the flywheel and the linear (and downward) acceleration of the mass.

Moment of inertia

The rotational counterpart of $F = ma$ relates the resulting angular acceleration α to the applied torque τ. The relationship involves the moment of inertia (frequently abbreviated MOI), which is a measure of the rotational inertia of a rotating system in the same way that mass is a measure of inertia in a linear system.

To see the logic behind the MOI, consider the rotating system of Fig. 8-12, which consists of a single mass m rotating about a point. A force F acts on the mass at right angles to the pivot arm.

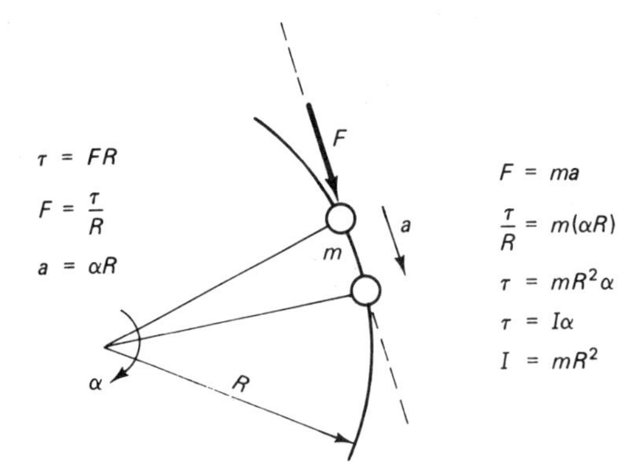

Figure 8-12 The moment of inertia I of a point mass traveling in a circular path of radius R is mR^2. This stems directly from Newton's second law, $F = ma$.

If the time interval is sufficiently small, the difference in position of the object at the beginning and end of the interval will be so small that the arc of the circular path will be essentially straight. Therefore, $F = ma$ may be applied to the motion of the body in a straightforward manner. When the appropriate angular quantities (τ/R and αR) are substituted for F and a, the result is

$$\tau = mR^2\alpha \tag{8-21}$$

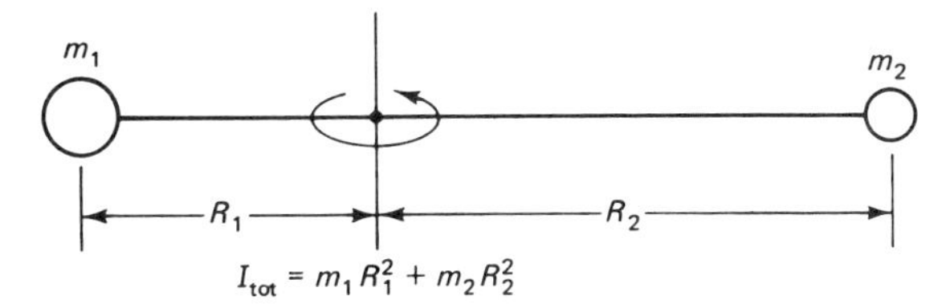

Figure 8-13 The moments of inertia of individual masses in a system are additive.

Thus, the constant of proportionality between τ and α is mR^2, and this is the MOI of a point mass (a small, concentrated mass) rotating in a circular path of radius R. The square of the distance R appears in the MOI, so the units of the MOI are that of mass times a distance squared.

If the rotating system consists of two masses, as in Fig. 8-13, the MOI of the system is the sum of the moments of the component masses. The MOI's are

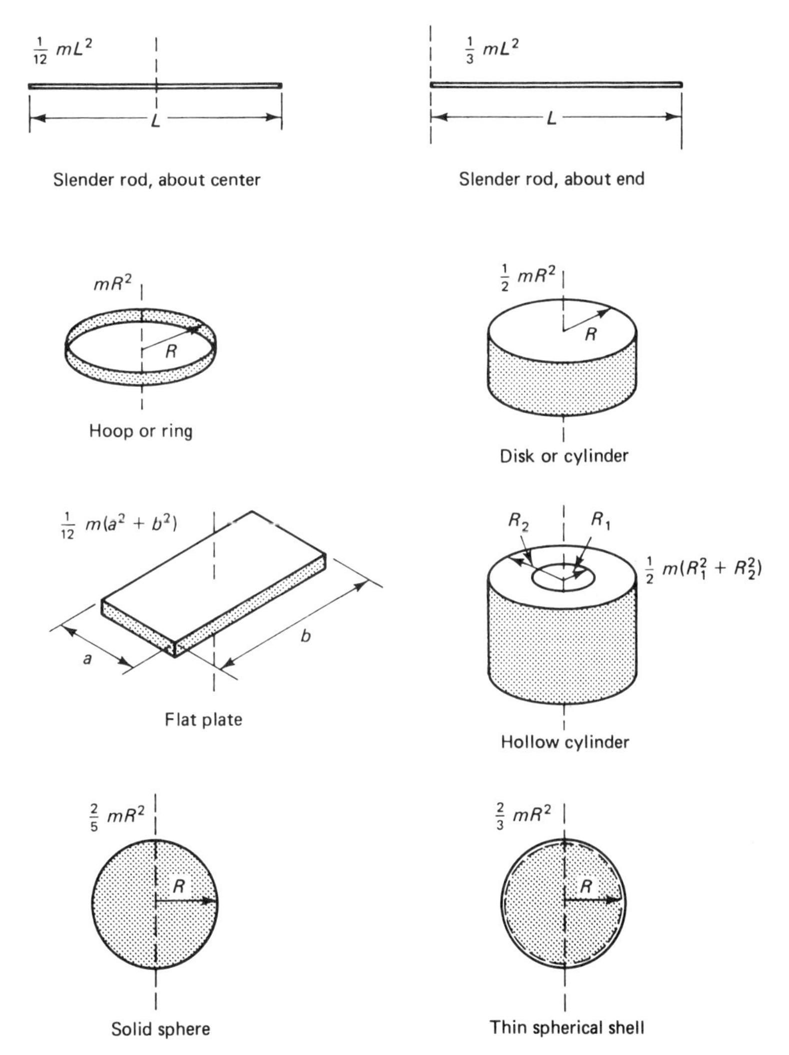

Figure 8-14 Moments of inertia.

additive, and a given system may include many masses with each contributing to the total MOI.

The MOI of an extended body, such as a rod or a flat circular disk, may be obtained by considering the body to consist of a large number of tiny constituent masses joined together to make the whole. The total MOI is the sum of the MOI of the constituent masses. For most objects of geometric shape, the sum is easily computed using calculus. The moments of inertia of several bodies are given in Fig. 8-14.

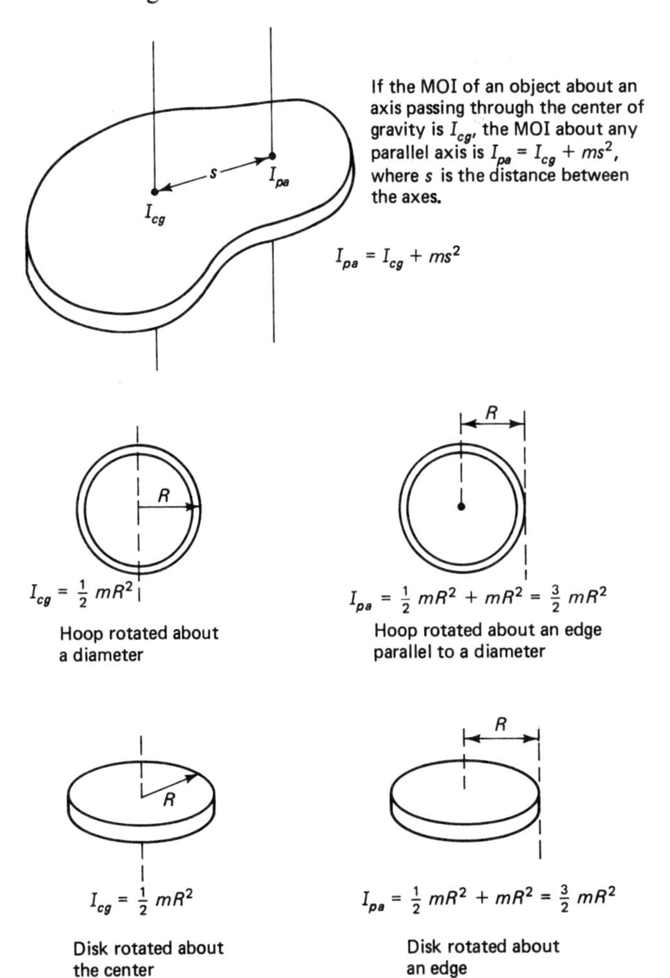

Figure 8-15 The parallel-axis theorem permits the calculation of the moment of inertia about an axis displaced from, but parallel to, the axis running through the center of gravity.

The MOI of a particular object depends upon the location of the axis of rotation. For example, if a rod of mass m and length L is rotated about its center, the MOI is $\frac{1}{12}mL^2$. But if it is rotated about one end, its MOI is increased by a factor of four, to $\frac{1}{3}mL^2$. The MOI is least when the axis of rotation passes through the center of gravity of the body. If the MOI is known for an object rotating about an axis passing through the center of gravity, the *parallel-axis theorem* may be used to calculate the MOI about an axis parallel to, but displaced from, the axis passing through the center of gravity. The parallel-axis theorem is illustrated in Fig. 8-15.

8-8 ROTATIONAL WORK, POWER, AND KINETIC ENERGY

When a torque τ is exerted on a body that rotates through an angle θ, work is done on the body by the agent producing the torque. The amount of work done is the product of the angular displacement and the torque, $\tau\theta$, which is justified in Fig. 8-16. The work done by the force F in producing the displacement s is Fs. But $s = R\theta$ and $\tau = FR$, so that rotational work is given by $\tau\theta$. Because power is the work done per unit time, it follows from basic definitions that rotational power is $\tau\omega$. This corresponds to the relationship for power in linear (straight-line) motion, Fv, obtained in Sect. 6-5.

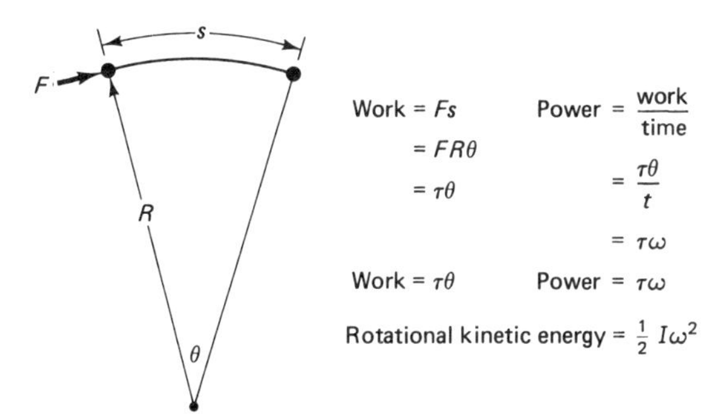

Figure 8-16 Rotational work, power, and kinetic energy.

The kinetic energy of a rotating body of angular velocity ω must be equal to the work done on the body in bringing it from rest up to the given angular velocity. An applied torque produces the angular acceleration, so

$$\tau = I\alpha \tag{8-22}$$

If, during the acceleration period, the body rotates through the angular displacement θ (which may represent many revolutions), the work done on the body is $\tau\theta$. Hence, we may write

$$\text{Work done} = \tau\theta = I\alpha\theta \tag{8-23}$$

From Sect. 8-5, the rotational analog of the $2as$ formula (with $\omega_0 = 0$) is

$$2\alpha\theta = \omega^2 \quad \text{or} \quad \alpha\theta = \frac{\omega^2}{2}$$

and we obtain

$$\begin{aligned}\text{Work done} &= I\alpha\theta \\ &= I\frac{\omega^2}{2} = \frac{1}{2}I\omega^2\end{aligned}$$

Thus, because the kinetic energy stored in the rotating body must have arisen from the work done in accelerating the body, it follows that

$$KE_{rot} = \frac{1}{2} I\omega^2 \tag{8-24}$$

8-9 ANGULAR MOMENTUM

Definition

The angular momentum L of a point mass rotating in a circle of radius R is defined to be

$$L = mvR \tag{8-25}$$

which is the product of the linear momentum mv and the radius R. Recalling that $v = R\omega$, we obtain an equivalent definition in terms of the angular velocity ω:

$$L = mR^2\omega \tag{8-26}$$

Because the moment of inertia I for a point mass rotating in a circle of radius R is mR^2, we may write

(*angular momentum*) $$L = I\omega \tag{8-27}$$

This definition is *not* limited to a point mass. If an extended body of any shape with moment of inertia I is rotating about an axis with angular velocity ω, the angular momentum is given by $I\omega$. Of course, the moment of inertia must correspond to the particular axis of rotation being considered.

The MKS units of angular momentum are

$$\frac{\text{kg} \cdot \text{m}^2}{\text{s}}$$

In reconciling these units with Eq. (8-27), remember that the units of I are kg · m^2 and the units of ω are "inverse seconds." For ω, we must use the rigorously correct unit 1/s rather than the more explicit rad/s. (See Sect. 8-3.)

Angular momentum is a measure of rotational motion in the same way that linear momentum is a measure of straight-line motion. Whereas *force* is the rate of change of linear momentum, *torque* can be shown to be the rate of change of angular momentum. To demonstrate this relationship, we use $L = mvR$ to write an expression for a change in angular momentum of a particle, as in Fig. 8-17. A tangential force acts on the particle, which we consider during a small time interval t; during the interval, the linear velocity of the particle will increase from v_0 to v. Moreover, a change in angular momentum, ΔL, will arise from the change in velocity.

$$\frac{\Delta L}{t} = \tau \tag{8-28}$$

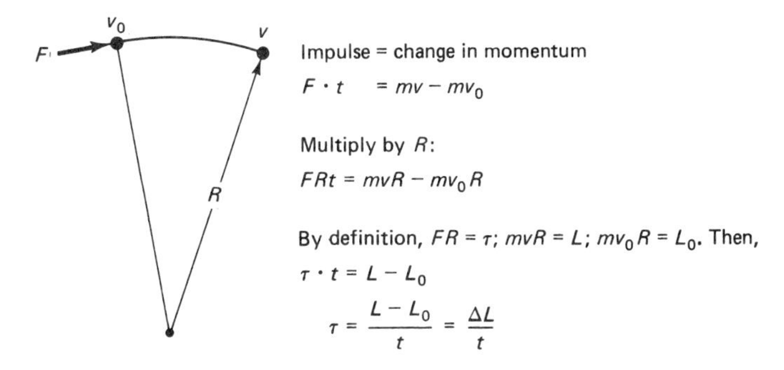

Figure 8-17 The change in angular momentum per unit time is equal to the torque applied to the rotating object.

as shown in Fig. 8-17. While this result is obtained in regard to a single particle, the result is general and may be applied to any rotating body.

Conservation of angular momentum

The result just obtained demonstrates that the angular momentum of a system will change only if the system is acted on by a torque τ. If no torque is allowed to act on the system, the angular momentum must remain constant. In other words, the amount of rotational motion will not change unless a torque is applied from an external source.

Special measures must be taken in order to control the rotation of satellites in orbit about the earth. Obviously, no external forces can be applied (unless small control rocket thrusters are used), and once a small rotational velocity is imparted to a satellite, it will continue to rotate indefinitely. Therefore, elaborate gyroscopes (described later) and control systems are employed to achieve rotational stability. Extreme care must be taken during the launch phase to ensure that appreciable rotation is not imparted to the satellite.

EXAMPLE 8-3 Suppose two disks are mounted on a frictionless horizontal axle, as shown in Fig. 8-18. Initially, the disks are separated and are rotating independently, perhaps at different angular velocities. They are then pushed together so that they rub (with friction between them) and finally become coupled together, rotating at the same speed. The conservation of angular momentum allows the final angular velocity of the combination to be calculated. Algebraic details are given in the figure.

Assume disk 1 to have a radius of 15 cm and a mass of 1.5 kg and disk 2 to have a radius of 20 cm and a mass of 2.5 kg. The initial angular velocity of disk 1 is 100 rad/s and of disk 2 is 200 rad/s. Calculate the angular velocity of the disks after they become coupled together.

Solution. Using the formulas given in the figure (with MKS units):

$$I_1 = \tfrac{1}{2}m_1R_1^2 = (0.5)(1.5\text{ kg})(0.15\text{ m})^2 = 0.0169\text{ kg}\cdot\text{m}^2$$

$$I_2 = \tfrac{1}{2}m_2R_2^2 = (0.5)(2.5\text{ kg})(0.2\text{ m})^2 = 0.0500\text{ kg}\cdot\text{m}^2$$

Then,

$$\omega_f = \frac{I_1\omega_1 + I_2\omega_2}{I_1 + I_2}$$

$$= \frac{(0.0169)(100) + (0.050)(200)}{0.0169 + 0.0500}$$

$$= 174.74 \text{ rad/s}$$

You may verify that the initial kinetic energies of the disks are 84.5 and 1000 J, respectively, a total initial energy of 1084.5 J. The final kinetic energy is 1021.36 J. Some 63.14 J of energy were dissipated due to frictional effects when the disks came together.

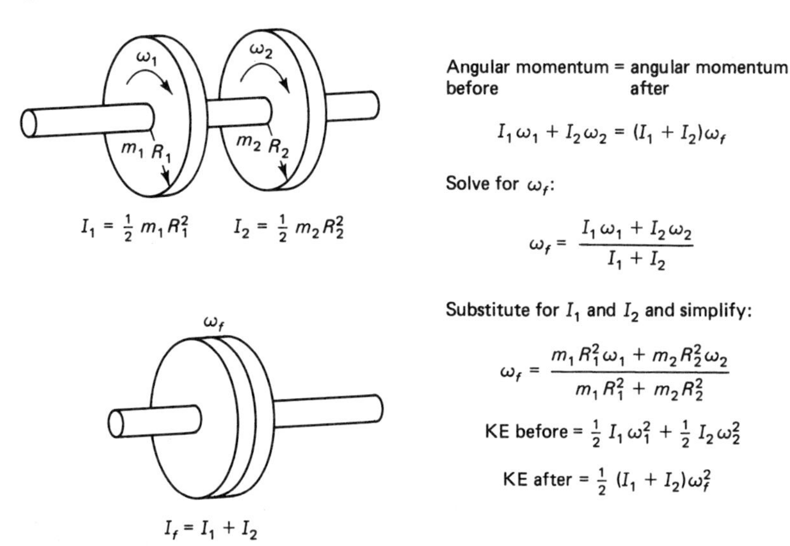

Figure 8-18 Conservation of angular momentum allows the final angular velocity of the disks to be calculated.

The effects of ΔI

If the mass distribution of a rotating body is changed or altered, it is possible for the moment of inertia to change also. This, in light of the conservation of angular momentum, provides some rather interesting effects. First of all, if mass is moved to a region farther from the axis of rotation, the moment of inertia will be increased. The moment of inertia will be decreased if the mass moves closer to the axis of rotation.

Because angular momentum is conserved, the product $I\omega$ remains constant. Therefore, if I decreases, ω must increase; and if I increases, ω must decrease accordingly. Thus, the rotational speed of an object can be changed simply by changing its moment of inertia. Note that no external forces are involved, and it is not necessary for any external torque to be applied to the rotating system. However, work may be required and energy expended in order to effect the mass redistribution. The kinetic energy of the rotating body does not remain constant as the moment of inertia changes.

In mathematical terms, because of the conservation of angular momentum,

$$I_0\omega_0 = \text{constant} = I_1\omega_1 \tag{8-29}$$

Therefore, because the products equal the same constant, they must equal each other. Hence,

$$I_0\omega_0 = I_1\omega_1$$

which yields

$$\frac{I_0}{I_1} = \frac{\omega_1}{\omega_0} \tag{8-30}$$

An ice skater can achieve high rotational speeds by starting the rotation with both arms and a leg outstretched and subsequently drawing the extended limbs in close to the body and close to the vertical axis of rotation. Conversely, the rotation is slowed by extending the limbs. Trampoline artists and high divers also make use of the principle; higher rates of rotation are achieved when the body is drawn into a ball.

A cat dropped from a reasonable height always lands upright, even if released upside down. The cat is able to turn itself in midair by extending and contracting its legs and tail, appropriately changing the moment of inertia so that the front and back parts of the cat become alternately more and less difficult to rotate. In this way, the cat rotates first one end and then the other until satisfied that conditions are proper for landing.

8-10 COMBINED ROTATIONAL AND TRANSLATIONAL MOTION

It is commonplace for many objects to engage in simultaneous rotational and translational (straight-line) motion, an immediate example being the rolling of a wheel on a car. The wheel rotates about the hub; at the same time, the wheel moves horizontally on the roadway. If the wheel is considered as a mass, without regard to its rotation, the kinetic energy of the translational motion is

$$\text{KE}_{\text{trans}} = \tfrac{1}{2}mv^2 \tag{8-31}$$

where m is the mass of the wheel and v is the velocity of its center of gravity. Because the wheel is rotating, it will possess rotational kinetic energy given by

$$\text{KE}_{\text{rot}} = \tfrac{1}{2}I\omega^2$$

Consequently, the total kinetic energy of the rolling wheel consists of two parts:

$$\begin{aligned} \text{KE}_{\text{total}} &= \text{KE}_{\text{trans}} + \text{KE}_{\text{rot}} \\ \text{KE}_{\text{total}} &= \tfrac{1}{2}mv^2 + \tfrac{1}{2}I\omega^2 \end{aligned} \tag{8-32}$$

If the wheel rolls without slipping, the familiar relationship $v = R\omega$ will exist between v and ω, and the total kinetic energy may be written in terms of the translational velocity as

$$\mathrm{KE}_{\mathrm{total}} = \frac{1}{2}mv^2 + \frac{1}{2}I\frac{v^2}{R^2} \tag{8-33}$$

where R is the radius of the wheel.

If the wheel under consideration happens to be a simple disk, so that its moment of inertia is $\frac{1}{2}mR^2$, we may substitute $I = \frac{1}{2}mR^2$ into Eq. (8-33) to obtain

(*rolling disk*)

$$\begin{aligned}\mathrm{KE}_{\mathrm{total}} &= \tfrac{1}{2}mv^2 + \tfrac{1}{2}(\tfrac{1}{2}mR^2)\frac{v^2}{R^2} \\ &= \tfrac{1}{2}mv^2 + \tfrac{1}{4}mv^2 \\ \mathrm{KE}_{\mathrm{total}} &= \tfrac{3}{4}mv^2\end{aligned} \tag{8-34}$$

Thus, when a disk rolls with translational velocity v, its total kinetic energy is $\frac{3}{4}mv^2$.

Motion down an incline

If a disk starts from rest and rolls without slipping down an incline—descending a vertical distance h in the process—the change in potential energy must equal the increase in kinetic energy. Hence,

(*disk rolling down incline*)

$$\begin{aligned}\Delta\mathrm{PE} = \Delta\mathrm{KE} &= \mathrm{KE}_{\mathrm{trans}} + \mathrm{KE}_{\mathrm{rot}} \\ mgh &= \tfrac{1}{2}mv^2 + \tfrac{1}{2}I\omega^2 \\ mgh &= \tfrac{3}{4}mv^2\end{aligned} \tag{8-35}$$

Solving for the velocity v, which is the translational velocity at the bottom of the incline, yields

(*disk rolling down incline*)

$$v = \sqrt{\tfrac{4}{3}gh} \tag{8-36}$$

For purposes of comparison, if the incline were perfectly frictionless so that the disk would slide down the incline without rotating, the velocity at the botom of the incline would be

$$v = \sqrt{2gh} \tag{8-37}$$

which is greater than the velocity of Eq. (8-36). It takes a disk (or any other body) longer to roll down an incline than to slide down a similar, frictionless incline.

If the incline is of uniform slope, the linear acceleration is constant so that the equations of uniformly accelerated motion are applicable. In this case, referring to Fig. 8-19(a), we can compute the time required for a body to start from rest and roll a distance L down the plane inclined at angle θ to the horizontal. Because the acceleration is uniform and because the object starts from rest, the average velocity will be one-half the final velocity, namely, $\bar{v} = v/2$, where v is the velocity of the body after it rolls a distance L down the plane. Therefore, the time required to roll a distance L is

$$t = \frac{L}{\bar{v}} = \frac{L}{v/2} = \frac{2L}{v} \tag{8-38}$$

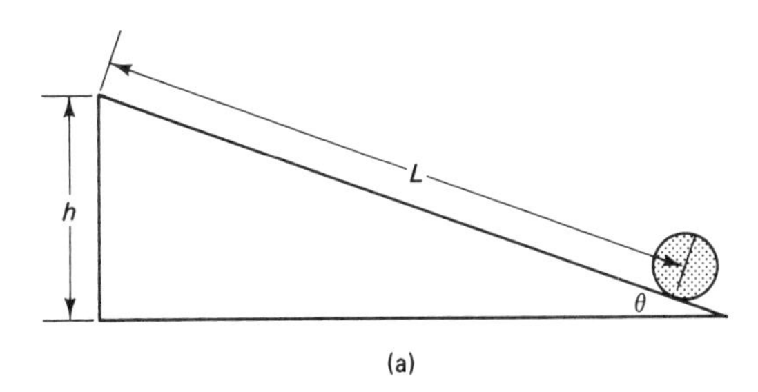

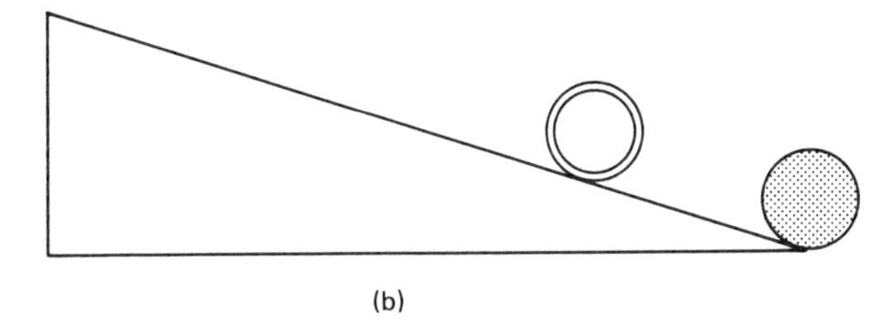

Figure 8-19 A ring and a disk are of equal diameters and have the same total mass. The moment of inertia of the ring will be greater than that of the disk and the disk will win a race down an incline.

For the disk, the velocity v is given by Eq. (8-36), which may be expressed in terms of L by using the fact that $h = L \sin \theta$. Consequently, the time required for a disk to start from rest and roll a distance L down an inclined plane is

(disk)
$$t = \frac{2L}{\sqrt{\frac{4}{3}gL \sin \theta}} \tag{8-39}$$

If a similar analysis is done for a ring ($I = mR^2$), we obtain

(ring)
$$t = \frac{2L}{\sqrt{gL \sin \theta}} \tag{8-40}$$

and we see that more time is required for the ring to descend the incline. For a sliding object on a frictionless incline,

(sliding object)
$$t = \frac{2L}{\sqrt{2gL \sin \theta}} \tag{8-41}$$

8-11 THE EFFECTS OF ROTATING REFERENCE SYSTEMS

Coriolis force

The captain of a ship has a dartboard in the main cabin. The dartboard is on the forward wall, so that the path of the darts is exactly in line with the forward motion of the ship. As long as the ship travels in a straight line, the captain hits the target with great precision. But when the ship is turning, the darts seem to curve to the right or left, causing the captain to miss the target by a small amount. When the ship is turning to the left, the darts are deflected to the right. When the ship is turning to the right, the darts are deflected to the left. This is illustrated in Fig. 8-20.

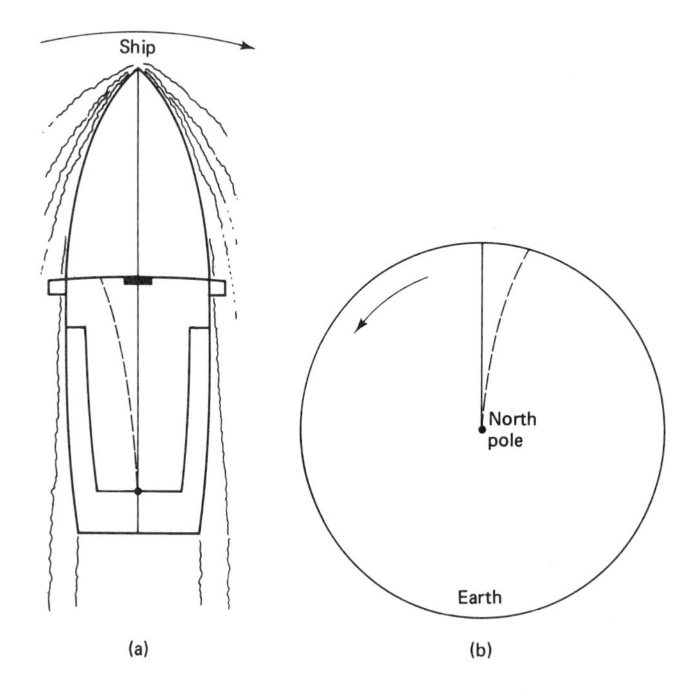

Figure 8-20 (a) Consequences of coriolis forces. Darts tossed in a ship that is turning to the right appear to be deflected to the left.
(b) A projectile fired from the north pole of the earth appears to be deflected to the right.

The deflection of the darts results from the turning motion of the ship, of which the captain may be unaware. The darts actually travel in a straight line in accordance with Newton's first law. But the captain, being very much at home and at ease in the cabin, perceives the darts to curve as if acted upon by an applied force. This *fictitious* force is an example of a *coriolis* force; it is only an apparent force which appears when objects move in rotating reference systems such as the room on the turning ship.

The earth rotates on its axis once a day, and coriolis effects come into play when objects move (in free flight) over the surface of the earth. Suppose a cannon situated exactly at the north pole fires a projectile in the direction of a polar bear 15 mi south of the pole. Because the earth is rotating toward the east, the projectile will curve to the right and fall to the west of the polar bear. A coriolis force is said to act on the projectile even though in actuality the earth rotates underneath the flight path of the projectile.

Because of the rotation of the earth, a point on the equator moves with a tangential velocity of about 1000 mi/h toward the east. The tangential velocity is less at points north and south of the equator, becoming zero at the poles. Suppose a cannon at the equator fires a projectile due north for a great distance. Initially, the projectile has a tangential velocity of about 1000 mi/h toward the east, and this velocity is maintained as the projectile travels north. The result is that the projectile appears to curve to the right because the tangential velocity of the projectile exceeds that of the terrain below. If the cannon were fired from the equator toward the south, the projectile would appear to curve to the left. This curvature to the right in the northern hemisphere and to the left in the southern hemisphere is a manifestation of coriolis force.

The weather report given on television in conjunction with the evening news usually includes a view of the national weather map on which is displayed the various low- and high-pressure centers. Speaking in broad terms, air flows inward toward the center of a low, and air flows out and away from the center of a high. However, because of the coriolis force that stems from the rotation of the earth, a circular motion is imparted to the winds as they are deflected to the right in the northern hemisphere. Thus, air flow is generally clockwise about a high and counterclockwise about a low. The directions are reversed in the southern hemisphere.

The vertical coriolis effect

An object at the top of a high tower is farther from the center of the earth and is therefore farther from the axis of rotation of the earth than a point at the base of the tower. Consequently, the tangential velocity of the object at the top of the tower will be greater than the tangential velocity of a point at the foot of the tower. If the object is allowed to fall freely to the ground, it will be deflected to the east (in the direction of the earth's motion) because the tangential velocity it had at the top of the tower will be maintained. The deflection will be toward the east in both the northern and southern hemispheres. If the object were dropped from a height of 100 ft, the deflection toward the east would amount to only about 0.15 in., a rather small amount that can easily be masked by the effects of air currents and other effects. Therefore, the vertical coriolis effect is difficult to observe directly.

In Sect. 1-2 we propose the construction of a mail chute through the center of the earth so that bags of mail would fall from one side of the earth to the other. It is now evident that the vertical coriolis effect would soon cause the bag of mail to collide with the east wall of the chute. This would produce frictional forces which would slow the bag until it would not fall all the way up the other side. Obviously, there are other problems too.

Suppose a projectile—a rocket, perhaps—is fired from the surface of the earth straight up to a great height. As it gains altitude, its tangential velocity (surface value) becomes insufficient to maintain its position over the same point on the earth below. It therefore falls behind the imaginary vertical line along which it was aimed, and it will return to the earth at a point to the west of the launch site. This, of course, assumes that no corrective measures are taken to compensate for the vertical coriolis effect.

The Foucault pendulum

If a pendulum consisting of a small mass suspended by a light, flexible string is constructed on a rotatable platform such as a lazy Susan, rotation of the platform does *not* cause the plane of the pendulum's swing to rotate also. If the pendulum swings in an east-west direction, that direction will be maintained even though the platform and pendulum support are rotated. This is in keeping with Newton's first law.

It should now be obvious that if a pendulum is located at the north pole, the plane of the pendulum will rotate through one complete revolution every 24 h. The plane of the pendulum remains fixed (relative to the stars) while the earth rotates underneath it. It is somewhat surprising, however, to learn that the time required for the plane to complete one revolution is *not* 24 h everywhere on the surface of the earth. As a matter of fact, if the pendulum is located at the equator, the plane of the pendulum's swing will not rotate at all! This is best understood with the use of a world globe, using a pencil as a pendulum. It is convenient to let the to-and-fro motion occur in a plane parallel to the support ring of the globe, always keeping the pencil very nearly perpendicular to the surface of the earth.

A pendulum specifically arranged for demonstrating the effect of the rotation of the earth is called a *Foucault pendulum*. A Foucault pendulum is typically

The Foucault pendulum at the Smithsonian Institution in Washington, D.C.

several stories tall and is suspended over a circle graduated in degrees, so that the rotation of the pendulum's swing can be noted. The supports and drive system of such a pendulum must be carefully designed or other effects come into play that cause the plane of the pendulum to rotate much faster than the effect produced by the rotation of the earth. The angular velocity and period of rotation of a Foucault pendulum are given by

$$\text{Angular velocity} = \omega_E \sin \lambda \qquad \text{period} = \frac{24\ \text{h}}{\sin \lambda} \tag{8-42}$$

where ω_E is 7.29×10^{-5} rad/s (angular velocity of the earth's rotation) and λ is the latitude. Note that when $\lambda = 90°$ (at the poles), the period is 24 h, as expected.

8-12 GYROSCOPES

The term *gyroscope* usually brings to mind a freely spinning wheel mounted in a set of gimbal rings, as shown in Fig. 8-21. The axis of the wheel is free to rotate in any direction due to the system of pivots supporting the gimbal rings. When the wheel rotates at high angular velocity, the plane of the wheel remains fixed in space even as the gimbal support is rotated, tilted, or inverted. That is, if the

Figure 8-21 Laboratory gyroscope and gimbal rings.

wheel is started spinning with its axis pointed toward the north, the axis will continue to point toward the north independent of the orientation of the gimbal support. This property makes a gyroscope useful as an orientation indicator for airplanes, rockets, spacecraft, and so forth, and a gyrocompass is a standard item on aircraft instrument panels.

The stability of a "gyro" is a consequence of Newton's first law, which says that a particle tends to travel in a straight line unless acted upon by an external force. If the wheel of a gyro is considered as an assembly of small particles of mass joined together, each particle will attempt to follow a straight line. However, the structure of the wheel exerts centripetal forces to cause each particle to follow a curved path in the plane of the wheel as the wheel rotates on its axis. Because the centripetal forces all lie in the plane of the wheel, no force will tend to cause any particle to move in a direction other than parallel to the plane of the wheel. Therefore, the wheel will maintain its orientation until it is acted upon by an external force.

From this discussion, it might appear that a stationary wheel would be just as stable as a rotating wheel. To see the importance of rotation, consider the rotating disk of Fig. 8-22(a), in which forces are applied in opposite directions at points A and B. The force at A tends to make the mass at point A move in the

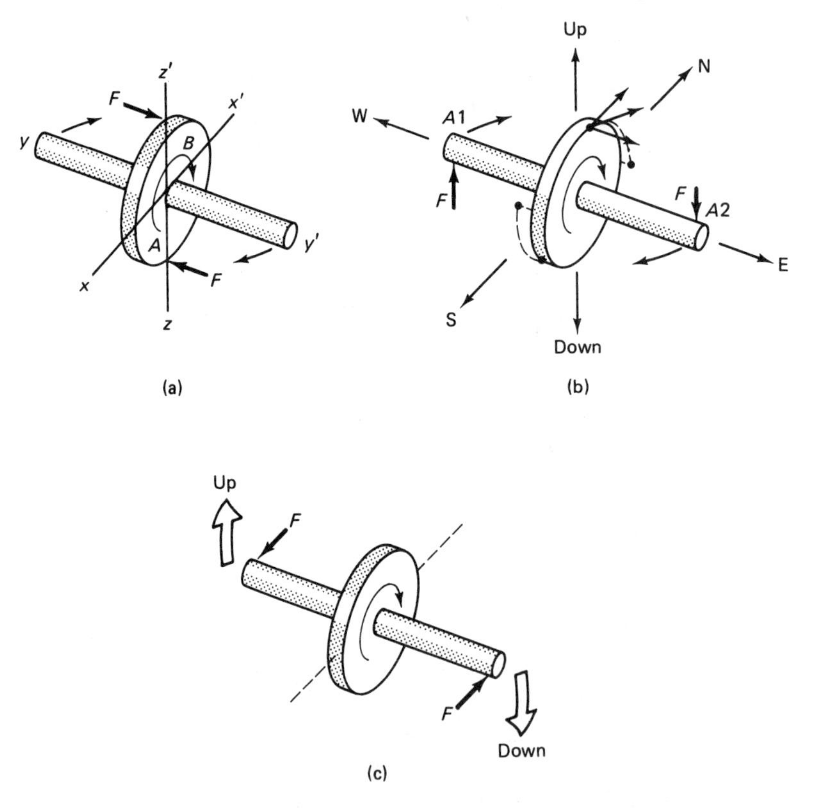

Figure 8-22 Precession of a gyroscope.

direction of y. The force at B tends to make the mass at B move toward y'. In the absence of rotation, the disk would tend to rotate about the line xx'. However, because the wheel is rotating rapidly, the mass particle initially at point A soon rotates around to point B, where the force is in the opposite direction. Thus, a given mass point sees a force first in one direction and then in the other so that the net effect of the alternating force is minimal. The result is that little, if any, rotation will occur about the line xx'. Instead, with the forces applied as in Fig. 8-22(a), the plane of the wheel will tend to slowly rotate about the vertical axis, zz'. This somewhat surprising result is called the *precession* of a gyroscope, and it is puzzling at first because it occurs about an axis at right angles to the axis anticipated.

Gyroscopic precession

When two forces are applied perpendicular to the axis of a rapidly spinning gyro, as in Fig. 8-22(b), the axis will remain horizontal, but it will turn slowly in the clockwise direction (viewed from above). This slow rotation (precession) of the axis at right angles to the applied force will continue at a uniform rate for as long as the forces are applied. The rate of precession is proportional to the magnitude of the applied forces. The direction of the precession depends upon the direction in which the gyro is spinning; if the direction of the spin is reversed, the direction of the precession will be reversed also.

To see why the precession occurs, consider a mass particle momentarily at the top of the spinning wheel while an upward force is applied to $A1$ and a downward force is applied to $A2$. Because of the rigidity of the wheel and axle, the initial force on the particle will be toward the east, and the particle will begin to move in that direction in accordance with Newton's law. But at the same time, the particle is moving toward the north due to the spinning motion of the gyro. Consequently, two components of velocity come into play, the spinning component toward the north and the smaller component toward the east, which arises from the applied forces. Therefore, the particle moves toward the northeast in the direction of the vector sum of the two velocities. This is possible only if the plane of the wheel rotates; $A2$ moves horizontally toward the south, while $A1$ moves horizontally toward the north.

Similarly, a particle momentarily at the bottom of the wheel will see an initial force toward the west, due to the applied forces at $A1$ and $A2$, and the particle will develop a velocity component toward the west. This causes the particle to move toward the southwest as it moves upward, and as the motion of the particle continues, it will pass across the top of the gyro from the southwest toward the northeast. Thus, the motion of the particle at the bottom is compatible with the motion of the particle at the top; they will follow very nearly the same path (but at slightly different times), and both motions are consistent with the same precessional motion. Figure 8-22(c) further illustrates the direction of precession that a given set of forces will produce.

Spinning top

The fundamental difference between a spinning top (a whirligig) and a gyro is that the gyro is mounted so that it pivots freely in any direction about the center of gravity. A top pivots about a point on the axis of rotation, where its "sharp point" touches the floor or tabletop. The fascinating aspect of a top is its ability to stand up when common sense says it should fall over. Obviously, the principles of the gyroscope are involved, and spinning tops that lean from vertical precess for the same reasons that gyroscopes precess. Figure 8-23 illustrates how the weight of a top gives rise to forces on the particles constituting the top. You should be able to predict the direction of the resulting precession.

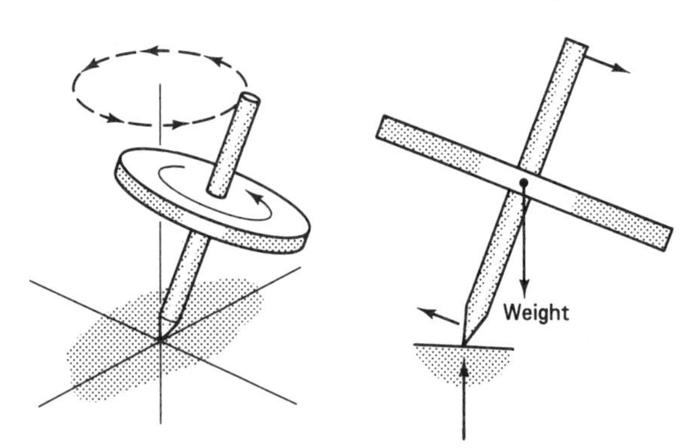

Figure 8-23 The weight of a spinning top produces a torque that tends to rotate the axis in the direction shown. However, gyroscopic action causes the top to precess rather than to topple over.

As a spinning top slows down, it wobbles more and more until finally it topples over. Obviously, the motion of the top is quite complicated. A meaningful quantitative treatment of tops and gyroscopes is quite involved mathematically because the motions can be far more complicated than the simple precession described here.

Questions

1. To the nearest half-dozen, how many radians are there in one complete circle?

2. What is the distinction between angular and tangential displacement?

3. Why should a motorist not drive too slowly when rounding a banked curve when the roadway is covered with ice?

4. The flywheel of a reciprocating engine "smooths out" the rotational motion of the crankshaft. Explain.

5. Would a lawnmower engine (for a small, push mower) be expected to operate properly when the blade is removed?

6. Describe the origin of the Coriolis force. Is it a real force? Give examples illustrating the coriolis effect for both horizontal and vertical motion.

7. Gun barrels are *rifled* with a spiral groove that causes the bullet to spin. Why is this desirable?

8. Observant motorcyclists know that to initiate a turn to the right, a slight pressure on the handlebars must be applied *to the left*. Can you explain this phenomenon in terms of the gyroscopic properties of the front wheel? (The same thing can be accomplished by consciously leaning to the right. In practice, a combination of the two effects is most common.)

* **9.** Viewed from the cockpit, the propeller of a

single-engine airplane rotates clockwise. Further, the spinning prop exhibits considerable gyroscopic effects, tending to precess noticably when the aircraft turns either right or left or changes its pitch attitude (nose up or nose down). What effect on the aircraft would the spinning prop have (turn left or turn right, climb or dive) when the airplane: **(a)** turns to the left; **(b)** turns to the right; **(c)** increases pitch attitude at the takeoff point; **(d)** decreases pitch attitude when leveling off from a climb?

10. Why must the conventional helicopter have a tail rotor?

11. Some dying, burned-out stars suffer gravitational collapse to form neutron stars only a few kilometers in diameter. Most rotate quite rapidly, one rotating about 30 revolutions per second. Another object, recently discovered, rotates at the astonishing rate of 642 revolutions per second. What accounts for the high rate of rotation? (You might wish to read about *pulsars* in an astronomy book.)

Problems

1. Express the following angles in radians: **(a)** 90°; **(b)** 180°; **(c)** 360°; **(d)** 45°; **(e)** 30°.

2. Convert the following angles given in radians to degrees: **(a)** π; **(b)** $\pi/6$; **(c)** $\pi/2$; **(d)** 1.52; **(e)** 3.4.

3. What arc length is subtended by the following angles in a circle of radius 3 cm: **(a)** 1 rad; **(b)** 2 rad; **(c)** 45°; **(d)** 60°?

4. Repeat Problem 3 for a circle of radius 6 cm.

5. **(a)** What angle is subtended by a microwave tower known to be 1 mi away if the tower is 100 ft tall? (Express the answer in degrees and in radians.)
(b) How tall is an adjacent tower that subtends an angle of 2°?

6. A bicycle wheel 26 in. in diameter rolls through one-half revolution.
(a) What is the tangential displacement of a point initially at the top of the wheel?
(b) What is its angular displacement?

7. **(a)** What is the angular velocity of a racecar on a circular track $\frac{1}{2}$ mi in diameter if the speedometer indicates 80 mi/h?
(b) What is the tangential velocity of the car?

8. A 26-in. diameter tire on a truck rotates with an angular velocity of 74.5 rad/s. How fast is the truck traveling?

9. **(a)** What is the angular velocity in radians per second of a stereo turntable platter rotating at $33\frac{1}{3}$ rpm?
(b) What is the tangential velocity of a point on the rim of a standard record 12 in. in diameter?

10. A flywheel 4 ft in diameter starting from rest accelerates for 1 min with an angular acceleration of 0.3 rad/s^2.
(a) What will be the final angular velocity of the wheel?
(b) What will be the angular displacement of a point on the rim at the end of the 1-min interval?
(c) What will be the linear (or tangential) acceleration of a point on the rim of the flywheel?

11. Power is disconnected from a 1750-rpm motor, and the motor comes to a stop in exactly 1 min.
(a) Calculate the angular acceleration (deceleration).
(b) Through how many revolutions will the motor shaft turn in coming to a stop?

12. A 1-kg mass makes 1 revolution per second about a circular path. What centripetal force is required to produce a circular path whose radius is: **(a)** 1 m; **(b)** 2 m; **(c)** 3 m?

13. A toy car of mass 1 kg always runs at a constant speed of 2 m/s. What centripetal force is required for the car to stay on a circular track of radius: **(a)** 1 m; **(b)** 2 m; **(c)** 3 m?

14. A flywheel of mass 500 kg is constructed in the shape of a uniform disk of radius 0.75 m. If the flywheel is rotating at 100 rpm, calculate its: **(a)** rotational kinetic energy; **(b)** its angular momentum.

15. **(a)** What net torque is required to give a flywheel whose moment of inertia is 750 kg · m^2 an angular acceleration of 5 rad/s^2?
(b) Verify that the units obtained are valid units for torque.

16. A flywheel (MOI = 400 kg · m^2) initially at rest is acted upon for 14 s by a net torque of 600 N · m.
(a) What angular acceleration of the flywheel results?
(b) Through how many revolutions will the flywheel turn during the 14-s period?
(c) What is the final rotational velocity obtained?
(d) What is its final kinetic energy?
(e) What is its final angular momentum?

17. Show that the expression $\Delta L/t$, the change in angular momentum per unit time, has the units of torque, namely, newton-meters.

18. In Fig. 8-18, suppose m_1 = 10 kg, m_2 = 20 kg, R_1 = 10 cm, and R_2 = 14 cm. Compute the moment of inertia of each disk. For each combination of ω_1 and ω_2, calculate the final angular velocity and the loss of kinetic energy that occurs as the disks are pushed together and become coupled: **(a)** $\omega_1 = 30$ rad/s, $\omega_2 = 0$; **(b)** $\omega_1 = 30$, $\omega_2 = 15$ rad/s; **(c)** $\omega_1 = 0$, $\omega_2 = 30$ rad/s; **(d)** $\omega_1 = +20$, $\omega_2 = -20$ rad/s.

19. Repeat Problem 18 for two identical disks, each of mass 10 kg and radius 10 cm.

20. Simplify the formula for ω_F in Fig. 8-18 for the special case of identical disks. (*Hint*: $m_1 = m_2$; $R_1 = R_2$.)

21. Suppose an earth satellite with MOI of 1200 kg · m^2 is rotating slowly at 0.2 rad/s. On command, two solar panels are extended, increasing the MOI to 1400 kg · m^2.
(a) What will be the new rate of rotation?
(b) Compute the change in rotational kinetic energy of the satellite as the panels are extended.
(c) If the panels are later retracted to their initial position, what will be the final rate of rotation?
(d) What will be the final rotational kinetic energy?

***22.** A sphere of solid brass 10 cm in diameter has a mass of 4.5 kg. Two such spheres are mounted on a light rod so that their centers are 1 m apart, and the rod is pivoted at its midpoint.
(a) What is the MOI of one of the spheres about an axis through its center?
(b) What is the MOI of one of the spheres about an axis 50 cm from the center of the sphere?
(c) What is the MOI of the brass spheres mounted on the rod?
(d) What would be the MOI of the system if the spheres were treated as point masses?
(e) What percent error is made in computing the system MOI if the spheres are treated as point masses?

***23.** A metal ring (hoop) is 6 in. in diameter and has a mass of 0.5 kg. It is placed on an inclined plane 5 ft from the bottom. The plane makes an angle of 20° with the horizontal. **(a)** Calculate the time required for the ring to roll down the incline. **(b)** Calculate the rotational and **(c)** the translational kinetic energy of the ring when it gets to the bottom of the incline. **(d)** Compare the total kinetic energy with the change in potential energy of the ring.

***24.** Repeat Problem 23 for a sphere of uniform density of the same diameter and mass as the ring.

***25.** A ring and a disk, made of different materials, have the same overall dimensions (16 cm in diameter, 3 cm thick, mass 1 kg). They are allowed to race down a 3-m-long, 15° incline. Obviously, the disk will win the race.
(a) Compute the time required for the disk to roll down the incline.
(b) How far will the ring have traveled when the disk reaches the bottom?
(c) How far will the disk be ahead of the ring as the disk crosses the finish line?
(d) Make a sketch to scale showing the positions of the ring and disk at the end of the race.

**26. Repeat Problem 25 for two spheres of the same diameter (16 cm) and mass (1 kg), but assume that one sphere is solid (of uniform density) while the other is a thin-walled sphere made of heavier material, but with the same mass as the solid sphere.

27. Relative to the "fixed stars," the earth rotates once every 23 h 56 min. However, because of the orbital motion of the earth relative to the sun, the sun appears directly overhead once every 24 h.

(a) Compute ω_E, the angular velocity of rotation of the earth, using the two time intervals, and compare.

(b) In calculating the effects of centripetal force on objects rotating with the earth, which value of ω_E should be used?

*__28.__ Identical masses of 4 kg each are attached to the ends of a light rod 1 m long. Considering the masses as point masses, compute the MOI of the system when the rod is pivoted: **(a)** about the midpoint, 50 cm from each mass; **(b)** about a point 40 cm from one mass; **(c)** about a point 20 cm from one mass; **(d)** about an axis that passes through the center of one of the masses.

*__29.__ Assuming the earth to be spherical with a radius of 3960 mi, compute the tangential velocity of a point at latitude: **(a)** 30°; **(b)** 45°; **(c)** 60°; **(d)** 90°.

CHAPTER 9

GRAVITATION AND KEPLER'S LAWS

In this chapter we present another fundamental law of nature, the universal law of gravitation discovered by Sir Isaac Newton. This law enables us to calculate the mutual gravitational force between two masses, and we shall use the law to obtain a formula for calculating g, the acceleration of gravity at the surface of the earth.

Also presented are the three laws of planetary motion discovered by Johannes Kepler. These laws, and the universal law of gravitation, arose as part of an effort to understand the solar system. The same laws govern the motion of satellites. We obtain formulas for calculating orbital parameters of satellites in circular orbit about the earth.

The possibility of space travel stirs the imagination. In the last section of this chapter we briefly discuss a major problem involved in traveling, for example, to the nearest star.

9-1 THE UNIVERSAL LAW OF GRAVITATION

The difficulties of early astronomers trying to understand the movements of the planets were compounded by the fact that the nature of the forces acting between celestial objects was not understood. However, one man—Sir Isaac Newton—provided major advances in three areas, which essentially solved the problem. We have already studied the three laws of motion put forth by Newton relating the concepts of inertia (mass), force, and acceleration. These laws, obviously, are essential to understanding the response of a planet to an applied force. The origin and the characteristics of the forces between celestial objects was established in Newton's universal law of gravitation. Newton recognized that the ability to produce and to respond to gravitational attraction was an inherent property of all mass, whether the mass was of the earth or of celestial origins. It was a major conceptual advance to realize that the same type of force that caused an apple to fall from a tree was also responsible for maintaining the moon in its orbit about the earth. Thus, a connection was made between terrestrial and celestial events, at the time a point of major philosophical impact. Furthermore, in the absence of mathematics appropriate for dealing with planetary motion, Newton developed the calculus. Using it, he demonstrated that the (closed) orbit resulting from an inverse-square force field (such as gravity) was an ellipse with the larger body located at a focus of the ellipse. The concept of elliptical orbits had been put forth earlier in qualitative form by Johannes Kepler as Kepler's first law, which we describe in Sect. 9-5.

His own genius notwithstanding, Newton was able to deduce the universal law of gravitation largely because of the work of scientists and astronomers who preceded him. The history of the developments that led to our present-day understanding of the solar system is a fascinating story in which the names Nicolaus Copernicus, Tycho Brahe, Johannes Kepler, Galileo Galilei, and Isaac Newton are prominent. Interested readers are referred to the references at the end of the chapter.

In 1666, at age 23, Newton put forth the universal law of gravitation, which may be stated as follows:

> Each and every mass in the universe exerts a mutual, attractive gravitational force on every other mass in the universe. For any two masses, the force is directly proportional to the product of the two masses and is inversely proportional to the square of the distance between them.

In mathematical terms, the law may be written as

$$F_G = G\frac{m_1 m_2}{r^2} \tag{9-1}$$

where m_1 and m_2 are the two masses considered and r is the distance between the centers of the masses (which we assume to be spherical or spherically symmetric). (See Fig. 9-1.) The proportionality constant G is commonly called "big G" to distinguish it from "little g," the *acceleration* of gravity. Big G is the *universal gravitational constant*, and it has the following value throughout the universe (to the best of our knowledge):

$$G = 6.673 \times 10^{-11}\ \text{N} \cdot \text{m}^2/\text{kg}^2 \qquad (\text{m}^3/\text{s}^2 \cdot \text{kg})$$

The alternate set of units arises from substituting kg · m/s^2 for N.

The numerical value of G was first measured in 1798 by Lord Cavendish in an experiment that first demonstrated that the same type of forces which act

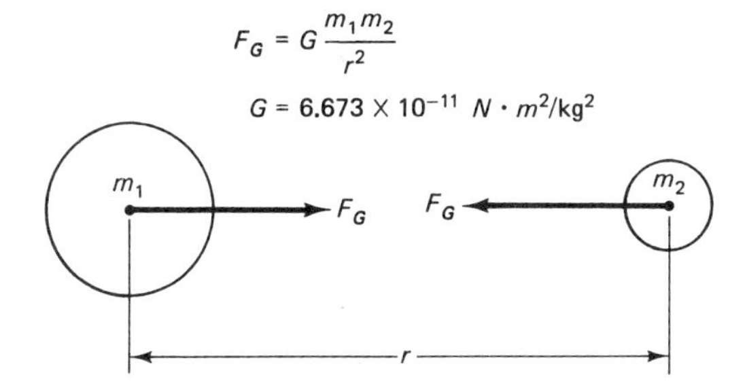

Figure 9-1 Universal law of gravitation.

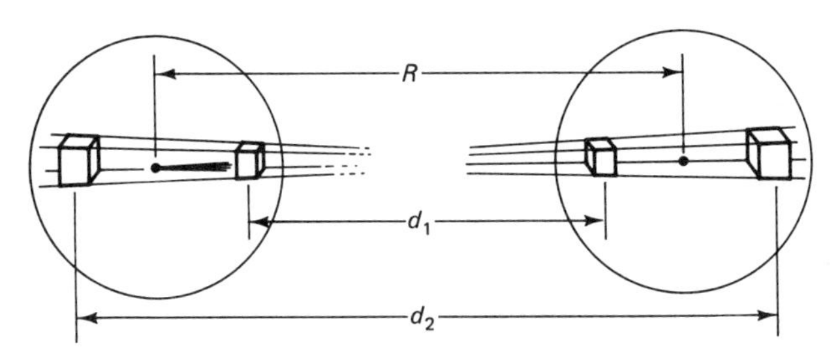

Figure 9-2 The greater attraction of masses of separation d_1 is exactly offset by the lesser attraction of symmetrically located (but unequal) masses of separation d_2. The result is that the total attraction between the spheres is the same as if the masses were concentrated into points at the center of each sphere.

between celestial bodies also act between objects on the earth. This occurred 71 years after the death of Newton in 1727. The Cavendish experiment is described in the next section.

Spherically symmetric mass distributions

It is extremely significant that gravitational forces vary as the square of the distance between the centers of the masses. This makes possible the stable orbits of planets and comets about the sun, of moons about the planets, and of man-made satellites about the earth, the moon, the planets, and the sun. Another consequence of the inverse-square law is the following simplification: Any spherically symmetric distribution of mass can be treated, for purposes of calculating gravitational forces, as if all the mass were concentrated in a point at the center of the original mass. This is illustrated in Fig. 9-2, where the added attraction between closer portions of the two spherical masses exactly balances the weaker attraction between more distant, symmetrically located portions. Therefore, in calculating the gravitational attraction between the earth and a cannonball at the earth's surface, both the earth and the cannonball may be shrunk to two points of mass separated by a distance equal to the radius of the earth (neglecting the radius of the cannonball in comparison to that of the earth). The same principle applies to spherically symmetric masses such as basketballs and Ping-Pong balls, which are hollow inside. Replace the thin spherical shell of mass by an equivalent mass located at the center of the ball.

It also follows from the inverse square law that a spherical shell of mass will exert no gravitational force on another smaller mass located inside the shell. The gravitational forces are always balanced in all directions, irrespective of the position of the small mass inside the spherical shell. Thus, if a marble and a basketball were in free space, a gravitational force would tend to pull them together if the marble were outside the ball. But if the marble were inside the ball, it would float about freely and would experience no net force from the basketball.

No method has so far been devised of shielding against a gravitational field. If three masses are in line, the gravitational attraction of the end masses for each other is not affected by the presence of the middle mass.

9-2 THE CAVENDISH EXPERIMENT

If two basketball-size lead spheres, each of mass 100 kg, are suspended so that the closest points of the two spheres are about 2 in. apart (30 cm between centers), the gravitational force on each sphere is about 7.4×10^{-6} N, an extremely small

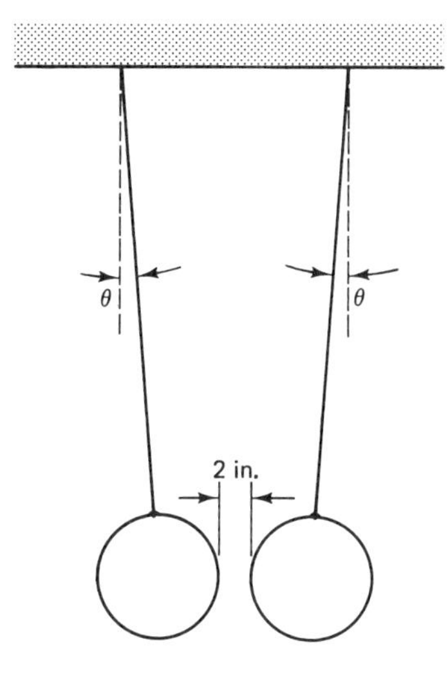

Figure 9-3 Two lead spheres suspended 30 cm between centers near the surface of the earth will attract each other, but the angle from the vertical of the supporting strings will be less than 0.000001°.

force. At the surface of the earth, each sphere would weigh about 220 lb. With the spheres positioned as described, the weight of one sphere is about 100 million times as great as the attractive force between the two spheres. If the masses were suspended by a strong, flexible string, the supporting string would be pulled aside from the vertical by an amount less than one millionth of a degree, as illustrated in Fig. 9-3. Clearly, such a small force as the gravitational force between objects of moderate size requires a delicate apparatus for its measurement.

An apparatus similar to that used by Lord Cavendish in 1798 is illustrated in Fig. 9-4. Two small spherical masses on a light rod are suspended by a fine fiber to form a *torsion pendulum* which will swing back and forth very slowly in the horizontal plane. Larger spherical masses, located as shown on a platform near the small masses, attract the smaller masses and cause the light-beam indicator to deflect to one side of the scale. When the large masses are then rotated almost

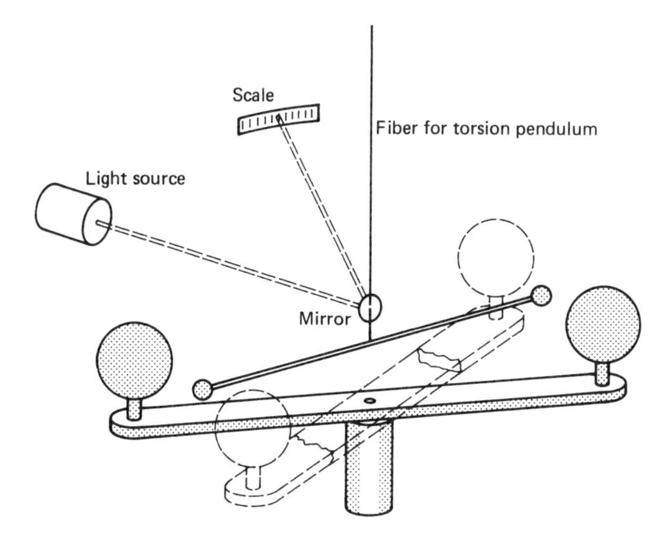

Figure 9-4 A diagrammatic representation of the apparatus used by Lord Cavendish in 1798 to measure G.

180° to the position indicated by the dotted lines, the light beam will move in the other direction. The average of the two deflections is proportional to the gravitational force between the large and small masses.

The force required to produce a given deflection of the light beam remains to be determined. This amounts to finding the torque required to twist the fiber supporting the two small masses. This is achieved by using the relationship between the period of oscillation of a torsion pendulum and the torsional constant of the fiber. In brief, the force required to produce a given deflection can be determined from a measurement of the time required for the small masses to move back and forth through one complete cycle (with the larger masses having been removed). Thus, by knowing the mass of each body, the distances separating the bodies, and the resulting gravitational force, the gravitational constant can be calculated.

The mass of the earth

Once the value of G is at hand, it is a simple matter to determine the mass of the earth, M_e, providng the radius R_e of the earth is known. For any given object of mass m_s at the surface of the earth, the gravitational attraction F_G is simply the weight $m_s g$ of the object, which is easily determined. Applying the gravitation formula to this situation yields

$$F_G = G\frac{M_e m_s}{R_e^2} = m_s g \tag{9-2}$$

in which every factor except M_e is known (the mass m_s of the specimen cancels out). Solving for M_e,

$$M_e = \frac{gR_e^2}{G} \tag{9-3}$$

Substituting numerical values with $g = 9.8$ m/s^2, $R_e = 6.367 \times 10^6$ m, and $G = 6.673 \times 10^{-11}$ m^3/kg $\cdot$ s^2 gives

$$M_e = 5.95 \times 10^{24} \text{ kg}$$

which agrees closely with the accepted value of 5.983×10^{24} kg.

The volume of the spherical earth can be calculated from its radius. If the mass just obtained is divided by the volume, we obtain the average density of the earth. It is about 5500 kg/m^3, corresponding to a specific gravity of 5.5. The average density of rocks and soil near the surface of the earth is about 2700 kg/m^3 (specific gravity = 2.7), which is only about half the average density of the earth. From this we conclude two things: (1) the mass of the earth as revealed by the Cavendish experiment appears reasonable; and (2) the density of the central core of the earth must be several times greater than the density of materials near the surface. Thus, we marvel that an experiment performed on the surface of the earth has revealed information about the interior of the earth.

9-3 RELATION OF g TO G

An object of mass m_s located near the surface of the earth will be attracted toward the center of the earth by a force given by

$$F_G = G\frac{M_e m_s}{R_e^2}$$

and the mass will accelerate downward with an acceleration a given by Newton's second law. Consequently, we write

$$F_G = G\frac{M_e m_s}{R_e^2} = m_s a$$

Solving for a (noting that m_s cancels out) gives

$$a = G\frac{M_e}{R_e^2}$$

Because m_s cancels out of the equation, we conclude that the acceleration is independent of the mass of the falling body. Large and small masses accelerate downward with the same acceleration, g, which we call the acceleration of gravity. Thus,

$$g = G\frac{M_e}{R_e^2} \tag{9-4}$$

Note that g depends upon the mass of the earth, M_e, and the radius of the earth, R_e.

This expression can also be used to compute the acceleration of gravity at the surface of a planet or other celestial body (assumed to be spherical) of mass m_p and radius R_p. We denote the acceleration as g_p to distinguish it from g and from g_a in the following. Hence,

(at surface of planet or other celestial body)

$$g_p = G\frac{M_p}{R_p^2} \tag{9-5}$$

EXAMPLE 9-1 Calculate g_j, the acceleration of gravity at the surface of the planet Jupiter.

Solution. We need the radius of Jupiter and the mass of Jupiter: $R_j = 7.15 \times 10^7$ m and $M_j = 1.898 \times 10^{27}$ kg. The universal constant is $G = 6.673 \times 10^{-11}$ m³/kg · s². Substituting numerical values into the expression above gives

$$g_j = (6.673 \times 10^{-11})\frac{1.898 \times 10^{27}}{(7.15 \times 10^7)^2} \qquad \text{(MKS units)}$$

$$= 24.77 \text{ m/s}^2$$

This value is about 2.5 times the acceleration of gravity of the earth.

Variation of g with altitude

The acceleration of gravity is less at high altitudes than it is at the surface of the earth because of the greater distance to the center of the earth. Recall that for points above ground, the earth may be considered as a point mass at its center. Therefore, the acceleration of gravity, g, at the surface of the earth is unique only in that it is the acceleration peculiar to the distance R_e from the center of the earth. From Eq. (9-5),

$$\text{(at surface of the earth)} \qquad g = \mathrm{GM}_e \frac{1}{R_e^2} \qquad (9\text{-}6)$$

At a greater distance R_a from the center of the earth, which corresponds to an altitude (distance from the surface) of $R_a - R_e$, the acceleration of gravity g_a is

$$\text{(at altitude)} \qquad g_a = GM_e \left(\frac{1}{R_a^2}\right) \qquad (9\text{-}7)$$

Dividing Eq. (9-6) by Eq. (9-7) allows us to eliminate GM_e:

$$\frac{g}{g_a} = \frac{GM_e\,(1/R_e^2)}{GM_e\,(1/R_a^2)} \qquad g_a = g\left(\frac{R_e}{R_a}\right)^2 \qquad (9\text{-}8)$$

Because R_a is greater than R_e, g_a is obviously less than g.

TABLE 9-1 VARIATION OF g WITH LATITUDE AT THE EARTH'S SURFACE

Latitude	g, m/s^2
0	9.78039
10	9.78195
20	9.78641
30	9.79329
40	9.80171
50	9.81071
60	9.81918
70	9.82608
80	9.83059
90	9.83217

TABLE 9-2 VARIATION OF g WITH ALTITUDE AT 45° LATITUDE

Altitude, km	g, m/s^2
0	9.806
1	9.803
4	9.794
8	9.782
16	9.757
32	9.708
100	9.589

EXAMPLE 9-2 Compute the acceleration of gravity, g_a, at an altitude of 1000 mi above the surface of the earth (1000 mi = 1.609×10^6 m; $R_e = 6.367 \times 10^6$ m).

Solution. We first calculate R_a by adding R_e to the altitude of 1000 mi. In MKS units, $R_a = 7.976 \times 10^6$ m. Then, using Eq. (9-8),

$$g_a = g\left(\frac{R_e}{R_a}\right)$$

$$= \left(9.80\ \frac{\text{m}}{\text{s}^2}\right)\left(\frac{6.367 \times 10^6\ \text{m}}{7.976 \times 10^6\ \text{m}}\right)^2$$

$$= 6.24\ \text{m/s}^2$$

9-4 CIRCULAR ORBIT OF A SATELLITE

A satellite in orbit is constantly pulled toward the center of the earth by the gravitational attraction of the earth. However, a stable orbit will result if the velocity and altitude of the satellite are such that the centripetal force required to maintain the circular orbit exactly equals the mutual gravitational attraction between the earth and the satellite (Fig. 9-5). In mathematical terms,

$$\text{Gravitational force} = \text{centripetal force}$$

$$G\frac{M_e m_s}{R^2} = \frac{m_s v^2}{R}$$

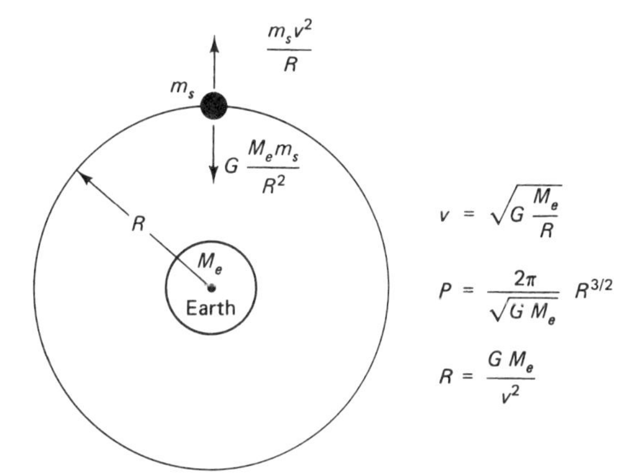

Figure 9-5 For a satellite in orbit, the gravitational force and the centripetal force arising from the circular motion are equal.

We see immediately that m_s, the mass of the satellite, appears on both sides of the equation and therefore drops out. The relationship simplifies to

$$G\frac{M_e}{R} = v^2 \quad \text{or} \quad v = \sqrt{G\frac{M_e}{R}} \tag{9-9}$$

Equation (9-9) gives the tangential velocity required for a satellite to orbit the earth at a distance R from the earth's center. Conversely, we may rearrange the expression to obtain R in terms of v:

$$R = G\frac{M_e}{v^2} \tag{9-10}$$

EXAMPLE 9-3 Calculate the orbital velocity required for a satellite to orbit the earth in a circular orbit 100 mi above the surface of the earth.

Solution. First, we compute the radius of the orbit, taking 3960 mi as the mean radius of the earth. The distance of the satellite above the surface is added to the radius of the earth to obtain 4060 mi as the orbital radius. To convert from miles to meters, multiply by 1609 m/mi, which yields an orbital radius of 6.53×10^6 m.

The product GM_e appears frequently in earth-orbit calculations; it is convenient to remember its numerical value of 3.99×10^{14} in MKS units. Using Eq. (9-9) to calculate the orbital velocity,

$$v = \sqrt{\frac{GM_e}{R}}$$

$$= \sqrt{\frac{3.99 \times 10^{14}}{6.53 \times 10^6}} = 7817 \text{ m/s}$$

Because we were careful to convert all quantities to the MKS system, we attach the appropriate MKS unit (m/s) to the numerical answer even though the units were not explicitly carried through with the arithmetic. A units conversion may be used to convert 7817 m/s to 17,490 mi/h.

EXAMPLE 9-4 Repeat the calculation of Example 9-3, above, but let the satellite be at an altitude of 200 mi above the surface rather than 100 mi. By what percentage relative to the 100-mi orbit does the required velocity change?

Solution. The new orbital radius is 4160 mi, or 6.69×10^6 m. Consequently,

$$v = \sqrt{\frac{3.99 \times 10^{14}}{6.69 \times 10^6}} = 7723 \text{ m/s} = 17{,}279 \text{ mi/h}$$

Thus, we find that the required velocity is less for the higher orbit, but by only 1.22%. This indicates that the satellite velocity must be controlled very precisely in order to achieve an orbit of a given radius.

Orbital period of a satellite

The *orbital period* (or *period*) of a satellite is the time required for the satellite to make one complete revolution around the earth. An expression for the period can be derived for a circular orbit by dividing the circumference of the orbit ($2\pi R$) by the velocity of the satellite in the oribt. The velocity is given by Eq. (9-9). Hence,

$$\textit{(orbital period)} \qquad P = \frac{2\pi R}{\sqrt{G\dfrac{M_e}{R}}} = \frac{2\pi R}{\sqrt{GM_e}\left(\dfrac{1}{R^{1/2}}\right)}$$

$$\textit{(The famous } \tfrac{3}{2} \textit{ law)} \quad P = \frac{2\pi}{\sqrt{GM_e}} R^{3/2} \quad \text{or} \quad P^2 = \frac{4\pi^2}{GM_e} R^3 \qquad (9\text{-}11)$$

For satellites orbiting the earth, because $GM_e = 3.99 \times 10^{14}$,

$$P = (3.15 \times 10^{-7})\, R^{3/2} \qquad \text{(MKS units)} \tag{9-12}$$

For an orbit 200 mi above the surface of the earth (6.69×10^6 m), the period is

$$P = (3.15 \times 10^{-7})(6.69 \times 10^6)^3 \qquad \text{(MKS units)}$$
$$= 5{,}451 \text{ s} \quad \text{or 90 min 51 s}$$

If the orbit were 300 mi above the surface of the earth, the period would be slightly more than 94 min.

For convenience, we express R in terms of P to obtain

$$\textit{(for earth orbit only)} \qquad R = 21{,}620\, P^{2/3} \qquad \text{(MKS units)} \tag{9-13}$$

which allows us to calculate the orbital radius required to produce a given period.

Orbital energies of circular orbits

The total energy that must be given to a satellite in launching it into orbit is a matter of interest because that energy must be delivered to the satellite by the launch vehicle, and the energy requirements will have a bearing upon the total fuel consumption. Work is done on a satellite in lifting it to a height above the surface of the earth; this work manifests itself as the potential energy of the satellite. Also, work is done in accelerating the satellite to a high velocity, thereby giving it kinetic energy. Thus, a satellite in orbit has both potential energy and kinetic energy. The kinetic energy is easily computed from the orbital velocity, but the potential energy is somewhat more difficult to calculate.

Near the surface of the earth, the potential energy of an object that has been lifted a distance h to an elevated position is mgh. The weight of the object, mg, is assumed to be constant over the short distance h. If, however, an object is lifted to large distances from the surface of the earth, the gravitational acceleration varies with altitude so that the weight of the object is not constant. This variation of weight with altitude must be taken into account in computing the potential energy of the object at high altitudes.

This problem is easily solved using calculus which effectively considers the varying weight by breaking up the total distance, h, into many short sections. Over each short section, the weight is assumed to be constant. The final result is obtained by summing the contributions from the short sections. This mathematical process is called *integration*. When used to compute the work done in lifting a mass m from distance R_1 to a greater distance R_2 from a larger mass M [as shown in Fig. 9-6(a)], the result is

$$W = GMm\left(\frac{1}{R_1} - \frac{2}{R_2}\right) \tag{9-14}$$

Consequently, the work that must be done in lifting a satellite from the surface of the earth to the height of an orbit of radius R_0 is

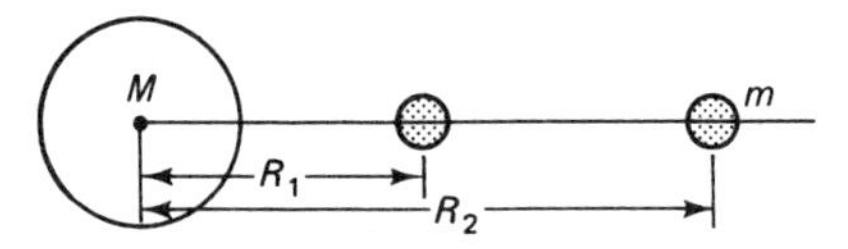

$$W = GMm\left(\frac{1}{R_1} - \frac{1}{R_2}\right)$$

(a) Work that must be done to lift a stationary object (not orbiting) from R_1 to R_2:

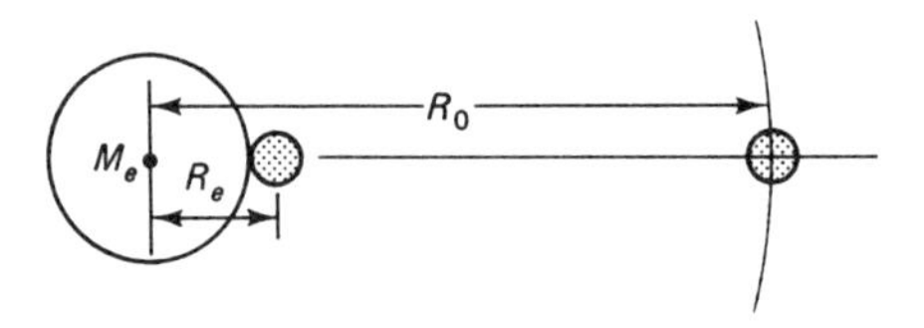

$$PE = GM_em_s\left(\frac{1}{R_e} - \frac{1}{R_0}\right)$$

$$KE = \frac{GM_em_s}{2R_0}$$

$$E_{tot} = GM_em_s\left(\frac{1}{R_e} - \frac{1}{2R_0}\right)$$

(b) The potential energy, kinetic energy, and total energy of a satellite in circular orbit about the earth:

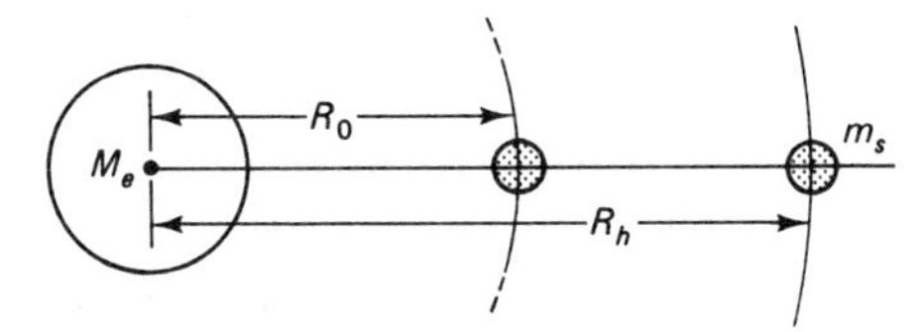

$$\Delta E = \frac{GM_em_s}{2}\left(\frac{1}{R_0} - \frac{1}{R_h}\right)$$

(c) The energy required to boost a satellite from a lower to a higher orbit:

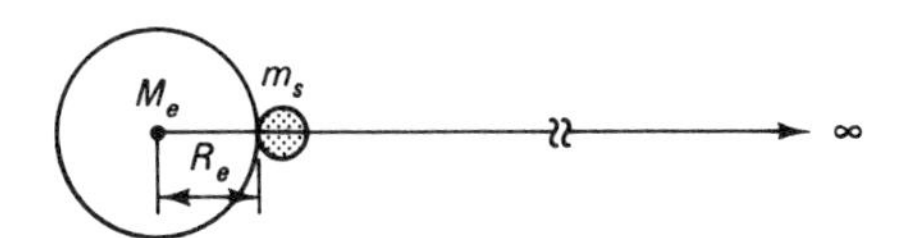

$$E_{tot} = \frac{GM_em_s}{R_e}$$

$$v_{escape} = \sqrt{2\frac{GM_e}{R_e}}$$

Additional energy required to escape from any orbit $= GM_e\frac{m_s}{2R_0}$

(d) Escape velocity:

Figure 9-6 Summary of formulas for various energy parameters.

(potential energy of a satellite in a circular orbit of radius R_0 about the earth)

$$\text{Work} = \text{PE} = GM_em_s\left(\frac{1}{R_e} - \frac{1}{R_0}\right) \tag{9-15}$$

The preceding is also designated as the potential energy, PE, of the satellite because the PE of a satellite in orbit is simply the work done in lifting it to the orbit. This expression does *not* include the kinetic energy, which we now consider.

The kinetic energy of any object of mass m moving with velocity v is given by $\frac{1}{2}mv^2$. We have already obtained an expression for the orbital velocity of a satellite, namely, Eq. (9-9):

(orbital velocity)

$$v = \sqrt{G\frac{M_e}{R}}$$

When this expression for the velocity is substituted into $\frac{1}{2}mv^2$, we obtain

(*Kinetic energy of a satellite in a circular orbit*)

$$\text{KE} = \tfrac{1}{2}m_s v^2 = \tfrac{1}{2}m_s G \frac{M_e}{R}$$

$$\text{KE} = GM_e m_s \left(\frac{1}{2R}\right) \qquad (9\text{-}16)$$

Note that the kinetic energy is inversely proportional to R, the radius of the orbit. This is consistent with the fact that satellites in higher orbits have lower velocities than satellites orbiting closer to the earth.

Now that we have expressions for both the kinetic and the potential energy, we can add the two expressions to obtain the total energy that must be given to a satellite in order to lift it from the surface of the earth to the level of the orbit and to accelerate it from rest to the orbital velocity. Thus, the total energy of the satellite orbiting the earth is

(*total energy of a satellite*)

$$E_{\text{tot}} = \text{PE} + \text{KE}$$

$$E_{\text{tot}} = GM_e m_s \left(\frac{1}{R_e} - \frac{1}{R}\right) + GM_e m_s \left(\frac{1}{2R}\right)$$

$$E_{\text{tot}} = GM_e m_s \left(\frac{1}{R_e} - \frac{1}{R} + \frac{1}{2R}\right) \qquad (9\text{-}17)$$

$$E_{\text{tot}} = GM_e m_s \left(\frac{1}{R_e} - \frac{1}{2R}\right)$$

For a particular satellite, G, M_e, m_s, and R_e are constant factors, leaving R as the only variable in the preceding expression. Therefore, the total energy of a satellite depends upon the radius of the orbit and, of course, the mass m_s of the satellite. As R is made larger in Eq. (9-17), E_{tot} becomes larger; more energy is required to launch a satellite into a higher orbit.

Escape velocity

An object thrown upward with moderate velocity slows down, stops, and returns to earth a short time later. If the initial upward velocity is increased, the object will attain a greater altitude before beginning its descent. We might expect this same sequence of events to occur even for extremely large velocities, with the object always returning to earth eventually. However, such is not the case because the gravitational attraction of the earth diminishes as the object gets farther from the center of the earth. An object hurled upward with a velocity greater than the *escape velocity* will never return to the earth even though the object is always being decelerated by the gravitational attraction of the earth. If the initial velocity is exactly equal to the escape velocity, the magnitude of the deceleration is such that an *infinite* distance is required for the object to come to rest. Such objects are said to have escaped from the gravitational field of the earth. If the object is given a velocity somewhat greater than the escape velocity, the object will "arrive at infinity" with a velocity equal to the velocity it had initially *in excess* of the escape velocity.

Mathematically, we may consider that an object escapes from the earth by being launched into an orbit of infinite radius. Therefore, setting R equal to infinity in Eq. (9-17), we obtain an expression for the energy required to move an object from the surface of the earth to infinity (at which point its velocity will be zero):

$$E_\infty = G\frac{M_e m_s}{R_e}$$

If this amount of energy is initially given to the object in the form of kinetic energy—as in blasting the object from a cannon, for example—we may write

$$\text{Initial KE} = E_\infty$$

$$\tfrac{1}{2}m_s v^2 = \frac{GM_e m_s}{R_e}$$

Canceling m_s, the mass of the object, gives

(escape velocity)
$$v_{\text{escape}} = \sqrt{2\frac{GM_e}{R_e}} \qquad (9\text{-}18)$$

Substituting numerical values yields about 11,200 m/s, which is on the order of 25,000 mi/h, or 6.94 mi/s. The significance of the escape velocity to space exploration is obvious.

Note that the escape velocity is independent of the mass of the escaping object. Moreover, if an object is at rest an infinite (almost) distance from the earth and if it "falls" to the earth, it will reach the earth with a velocity equal to the escape velocity. This, of course, does not consider any effect of the atmosphere upon the object.

It is apparent that a high escape velocity is essential for celestial bodies, such as the planets or the moon, to retain an atmosphere. Rapidly moving molecules of a celestial atmosphere may exceed the escape velocity and escape. This process could continue until the entire atmosphere is gone. The temperature of the atmosphere under consideration is important because atoms and molecules move faster at high temperatures. Further, at a given temperature, the lighter elements such as hydrogen and helium move faster so that they are the most likely to escape. The escape velocity of the moon, which has no atmosphere, is but 1.5 mi/s. Mercury, the planet closest to the sun, has an escape velocity of 2.5 mi/s and has no atmosphere of any consequence. The escape velocity of 7 mi/s maintains the familiar atmosphere of the earth, while Jupiter, which has an extensive atmosphere of hydrogen and helium, has an escape velocity of almost 36 mi/s. (Actually, Jupiter and Saturn are gaseous planets that have no sharply defined surface.)

Synchronous satellite

Because the orbital period of a satellite increases with the radius of the orbit, the interesting possibility arises of launching a satellite whose period is exactly equal to the period of rotation of the earth. If launched into the equatorial plane of the earth, the satellite would *hover* over the same point on the earth at all times, as shown in Fig. 9-7. Such satellites are widely used for communications, navigation, and monitoring weather systems.

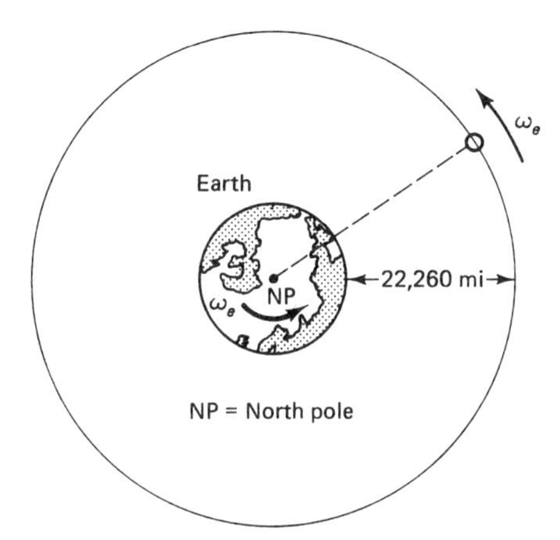

Figure 9-7 A synchronous satellite.

An expression for the radius of a *synchronous orbit* may be obtained by equating the centripetal and gravitational forces, using $m\omega^2R$ for the centripetal force, where ω_e is the angular velocity of rotation of the earth:

$$m_s\omega_e^2R = G\frac{M_em_s}{R^2}$$

Canceling m_s and solving for R yields

$$R^3 = \frac{GM_e}{\omega_e^2}$$

$$R = \sqrt[3]{\frac{GM_e}{\omega_e^2}} \tag{9-19}$$

Substituting, $GM_e = 3.99 \times 10^{14}$ m^3/s^2 and $\omega_e = 7.29 \times 10^{-5}$ rad/s:

$$R = \sqrt[3]{7.508 \times 10^{22}}$$

$$= 4.219 \times 10^7 \text{ m} = 26{,}220 \text{ mi}$$

Subtracting the radius of the earth (3960 mi) to obtain the height of a synchronous satellite above the surface of the earth gives 22,260 mi.

9-5 KEPLER'S LAWS—ELLIPTICAL ORBITS

Historical

Based upon a lengthy analysis of observational data collected by the Danish astronomer Tycho Brahe, Johannes Kepler put forth three laws that effectively describe the motions of the planets. These laws, although initially formulated in regard to the planetary orbits about the sun, are equally applicable to the orbits of satellites, comets, asteroids, and other celestial bodies. For the sake of definiteness, however, we shall speak of a planet and the sun with the understanding that the point in question is equally applicable to a satellite in orbit about the earth.

In passing, we note that Kepler died in 1630, some 12 years before the birth of Newton in 1642. Thus, Kepler's work preceded Newton's laws of motion and gravitation. Kepler's laws state *how* the planets move without giving the reasons *why*. It remained for Newton to pull the picture together in mathematical terms, but recall that it was in 1798—long after the death of Newton—that Cavendish measured G, the proportionality constant in Newton's universal law of gravitation.

Kepler's first law

The orbit of each planet around the sun is an ellipse with the sun located at one of the two focal points of the ellipse.

At this point we note that a circular orbit, as described in the previous section, is just a particular type of an elliptical orbit. A circle may be considered to be an ellipse whose *eccentricity* is zero. Thus, there is no conflict between circular orbits and Kepler's first law.

An in-depth analysis of an elliptical orbit is mathematically complex, but a good appreciation of an elliptical orbit can be achieved simply by studying the geometric properties of an ellipse. Figure 9-8(a) illustrates the semimajor axis, the semiminor axis, and the two focal points. Also, the eccentricity is the ratio of the distance between focal points to the length of the major axis. A well-known method of constructing an ellipse is shown in Fig. 9-8(b).

Terminology and mathematical relationships pertaining to the orbit of a planet about the sun are illustrated in Fig. 9-9. The point on the ellipse closest to the sun is the *perihelion*, while the most distant point is the *aphelion*. (In case of an orbit about the earth, corresponding points are called the *perigee* and *apogee*, respectively.) At perihelion and at aphelion, the orbital velocity of the planet is at right angles to a line (called the *position vector*) connecting the planet and the sun. An orbit is often described or specified by giving the perihelion and aphelion distances, or by giving the eccentricity of the orbit and either the perihelion or aphelion.

As a planet proceeds from perihelion toward aphelion, it is essentially moving uphill away from the sun, and it will gradually slow down to a minimum

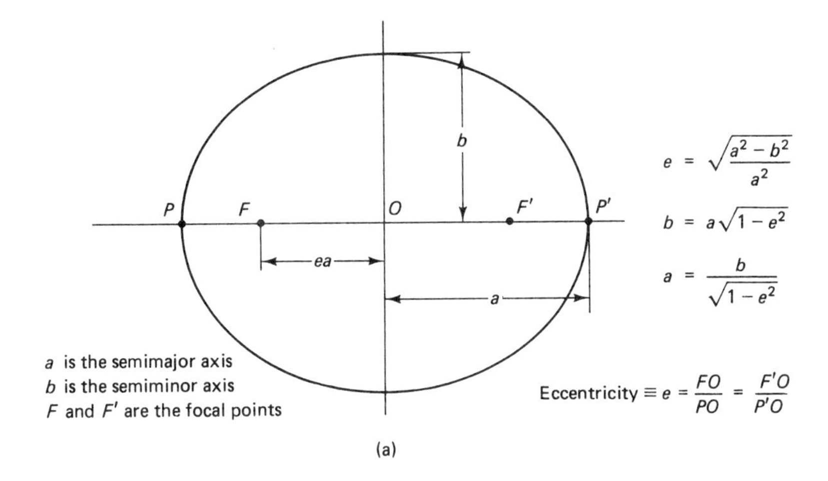

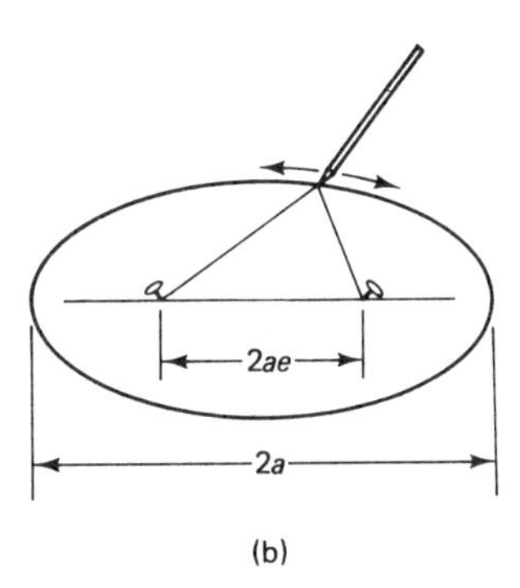

Figure 9-8 (a) Details of an ellipse. (b) In using the familiar pencil and string method to construct an ellipse, the distance between the foci (pins) is $2ae$ and the length of string required is $2a$, where e is the eccentricity and a is the semimajor axis.

velocity at aphelion. On the other side of the orbit, the planet speeds up to a maximum velocity at perihelion. Thus, the orbital velocity of a planet in an elliptical orbit is not constant. Expressions are given in Fig. 9-9 (without proof or derivation) for the velocity at perihelion and aphelion.

Kepler's second law

Because the gravitational force exerted on a planet by the sun is always directed through the center of the sun, the moment arm of the force about the center of the sun is zero. Therefore, no torque is developed by the gravitational force to change the angular momentum of the planet in its motion about the sun. Consequently, the momentum of the planet in its orbit around the sun is a constant, and it is this fact that gives rise to Kepler's second law. Forces always directed toward the same point within an orbit are called *central forces*.

For a circular orbit, the tangential velocity v_t of the planet is always perpendicular to the position vector R, so that the angular momentum is given by mv_tR. For elliptical orbits, this situation occurs only at aphelion and perihelion,

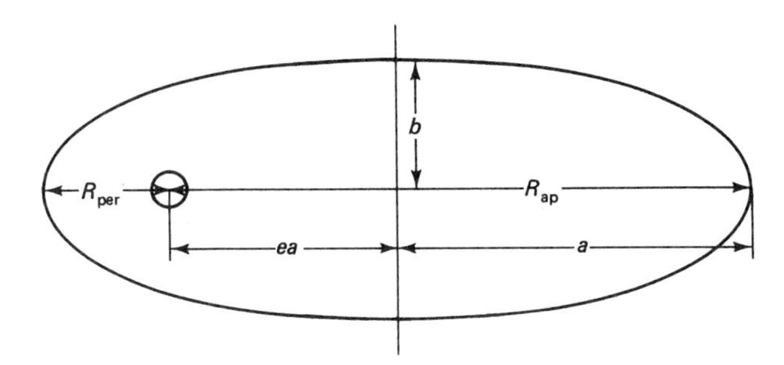

$$a = \frac{1}{2}(R_{ap} + R_{per}) \qquad R_{ap} = \left(\frac{1+e}{1-e}\right) R_{per}$$

$$e = \frac{R_{ap} - a}{a} \qquad R_{per} = \left(\frac{1-e}{1+e}\right) R_{ap}$$

$$b = a\sqrt{1-e^2} \qquad a = \frac{R_{ap}}{1+e}$$

Orbital velocity at aphelion and perihelion:

$$v_{per} = \sqrt{\frac{GM}{a}\left(\frac{1+e}{1-e}\right)} \qquad v_{ap} = \sqrt{\frac{GM}{a}\,\frac{1-e}{1+e}}$$

$G = 6.673 \times 10^{-11}\ \mathrm{m^3/kg \cdot s^2}$; a = semimajor axis; e = eccentricity of orbit.

Figure 9-9 Relationships between selected orbital parameters.

and a more complicated expression is required for the angular momentum. Specifically, the angular momentum is given by $mv_tR_\perp$, where $R_\perp$ is the length of the line drawn through the sun and perpendicular to the line of action of the tangential velocity vector. For a given planet in a given orbit, the angular momentum is the same at all points of the orbit. Hence, considering two points on an orbit (one denoted by primes), we may write

$$m_p v_t R_\perp = m_p v_t' R_\perp'$$

which gives

$$v_t R_\perp = v_t' R_\perp' \tag{9-20}$$

and we see that an inverse proportion exists between $R_\perp$ and v_t.

When the planet is near aphelion or perihelion, $R_\perp$ is very nearly the same as R, so that

$$v_{ap}R_{ap} = v_{per}R_{per} \quad \text{or} \quad R_{ap}^2\omega_{ap} = R_{per}^2\omega_{per} \tag{9-21}$$

where we have substituted $R_{ap}\omega_{ap} = v_{ap}$ and $R_{per}\omega_{per} = v_{per}$ in order to eliminate the tangential velocity.

Let us now consider the area swept out in a short time interval Δt by the position vector as the planet passes perihelion or aphelion (see Fig. 9-10). By approximating the portion of the orbit of interest by a circle of radius R (which may be either R_{ap} or R_{per}), a simple expression for the area swept per unit time is obtained from the proportional relationship between arc lengths and sector areas of a circle. The result is that in a given time interval, the position vector sweeps out the same area at aphelion as it does at perihelion. In fact, this relationship is

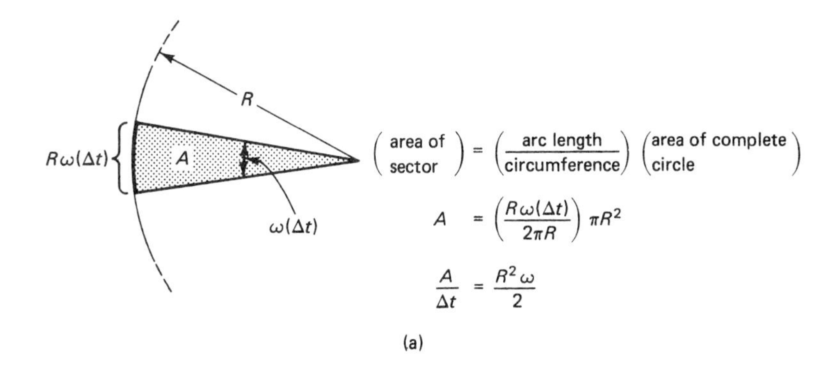

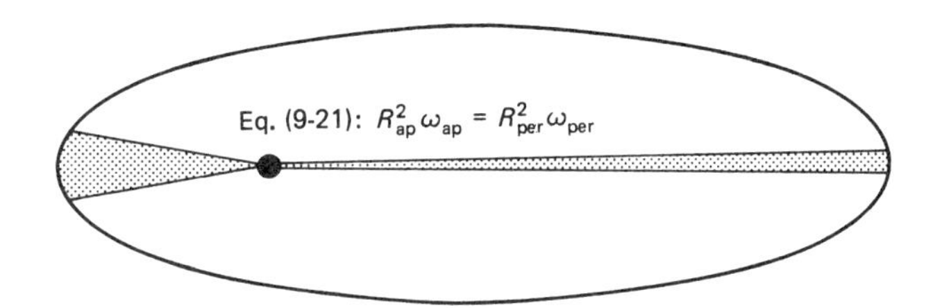

(b) Dividing Eq. (9-21) by 2 yields $\frac{R_{ap}^2\omega_{ap}}{2} = \frac{R_{per}^2\omega_{per}}{2}$. Thus, $\left(\frac{A}{\Delta t}\right)_{ap} = \left(\frac{A}{\Delta t}\right)_{per}$

Figure 9-10 (a) Area swept out per unit time for a particle in a circular orbit. (b) Equal areas are swept out at aphelion and perihelion for a particle in an elliptical orbit.

true for all portions of the orbit, as shown in Fig. 9-11. We limited this discussion to the aphelion and perihelion just to avoid the mathematical complexities of dealing with an ellipse. Kepler arrived at this result from analysis of observational data; it is now known as Kepler's second law:

The area swept out per unit time by the position vector of a planet is constant for all portions of the orbit.

This law also applies to the orbits of comets or other objects that may pass near the sun in an unbounded (parabolic or hyperbolic) orbit. It is therefore possible to calculate ahead of time the velocity with which a newly discovered comet will swing around the sun on its way to the outer reaches of the solar system.

Kepler's third law

Kepler's third law relates the period of revolution of a planet to its mean radius from the sun.

The square of the period of revolution of a planet is directly proportional to the third power of the average distance of the planet from the sun.

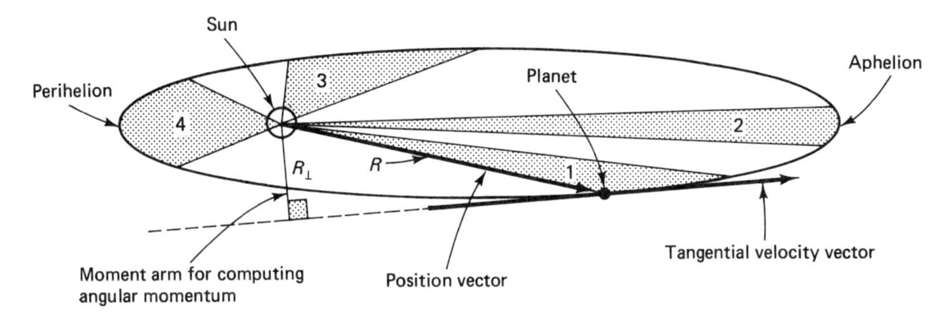

Figure 9-11 The position vector sweeps out equal areas in equal time intervals for all portions of an elliptical orbit.

Actually, we have already obtained this result from the rather simple case of a circular orbit. Equation 9-11 gives the period in terms of the orbital radius:

(orbital period of planet in orbit of radius R about the sun)

$$P^2 = \frac{4\pi^2}{GM_s} R^3 \qquad (9\text{-}22)$$

A similar result can be derived from the general case of an elliptical orbit. Note that this equation involves the mass of the sun ($M_s = 1.99 \times 10^{30}$ kg), but not the mass of the planet. Substituting numerical values for the constants gives:

(orbital period of a planet about the sun)

$$P^2 = (2.974 \times 10^{-19})R^3 \quad \text{(MKS units)} \qquad (9\text{-}23)$$

A unit conversion that is somewhat intricate gives the relationship

$$P = 0.408\, R^{3/2} \qquad (9\text{-}24)$$

where P is in days when R is expressed in millions of miles. Substituting R = 93,000,000 mi gives a period of revolution of 365.9 days for the earth.

Going further, if P is measured in earth years and R is given in terms of astronomical units, then, to a good approximation,

$$P^2 = R^3 \qquad (9\text{-}25)$$

TABLE 9-3 ORBITAL PROPERTIES OF THE PLANETS

	Semimajor axis		Length of year		Inclination
	A. U.	10^6 km	(in Earth days)	Eccentricity	to plane of Earth's orbit
Mercury	0.387	57.9	87.96	0.2056	7°00′
Venus	0.723	108.2	224.68	0.0068	3°24′
Earth	1.000	149.6	365.26	0.0167	0°00′
Mars	1.524	227.9	686.95	0.0934	1°51′
Jupiter	5.203	778.3	4337	0.0483	1°18′
Saturn	9.539	1427	10,760	0.0560	2°29′
Uranus	19.191	2871	30,700	0.0461	0°48′
Neptune	30.061	4497	60,200	0.0100	1°46′
Pluto	39.529	5913	90,780	0.2484	17°09′

9-6 THE PRACTICALITY OF SPACE TRAVEL

In studying the solar system and astronomy in general, one is inevitably impressed with the "astronomical" distances involved. So large are the distances that special units have been adopted for their measure. The mean distance from the earth to the sun is the astronomical unit (A.U.), while the distance that a ray of light will travel in one year is a light-year (l-y). A parsec is a much larger unit of distance that is based on the annual parallax of a nearby star as measured against a fixed field of more distant stars. The relationships of these units are as follows.

$$1 \text{ A.U.} = 1.5 \times 10^8 \text{ km} = 93 \times 10^6 \text{ mi}$$

$$1 \text{ l-y} = 9.46 \times 10^{12} \text{ km} = 5.88 \times 10^{12} \text{ mi} = 6.31 \times 10^4 \text{ A.U.}$$

$$1 \text{ parsec} = 3.08 \times 10^{13} \text{ km} = 1.92 \times 10^{13} \text{ mi} = 206{,}265 \text{ A.U.} = 3.26 \text{ l-y}$$

The moon is the celestial body closest to the earth, being a mere 238,000 mi away. The closest planet is Venus, which may approach us to within about 25,000,000 mi. The most distant planet is Pluto, whose orbital radius about the sun is 39.4 A.U. If we take the orbit of Pluto as being the outer limit of the solar system, then the distance across the system is more than 7 billion miles. While these distances are extremely large relative to our terrestrial experience, the distances to the other systems of the universe are far greater.

The distance to the nearest star, Alpha Centauri, is 4.24 l-y. The Milky Way galaxy, which includes our sun as one of its approximately 100 billion stars, is thought to be about 110,000 l-y in diameter. The nearest galaxy to the Milky Way is the Large Magellanic Cloud, which is about 160,000 l-y away. The nearest galaxy similar to our own Milky Way is the Great Galaxy in Andromeda, at a distance of 2.2 million l-y. While distances as these are already quite unimaginable, distances of more remote objects are given in terms of millions or even billions of light years.

The significance of astronomical distances to space travel is obvious in that longer journeys require more time. Travel times may be reduced by traveling at higher velocities, but achieving higher velocities requires greater expenditures of energy. And in any event, according to Einstein's theory of relativity, the maximum possible velocity that any object might attain is the speed of light. Thus, the minimum possible time required for an astronomical journey is the time required for a light wave to make the same trip. Therefore, in terms of time requirements, a distance of 1 l-y requires a travel time of *at least* 1 year.

If we ignore technological limitations and assume we have the ultimate spaceship which can travel at very nearly the speed of light, we see that a round trip to Alpha Centauri would require at least 8.5 years. While this is not clearly out of reach, it is an appreciable fraction of a human lifetime, the time scale against which such travel times must be considered. This in itself may not be a problem, however, because according to relativity theory, time and physical processes slow down for high-velocity travelers—they age more slowly than their colleagues at home. However, there is another problem. Energy requirements for such a venture are formidable. Even if an engine were developed that could convert matter directly into energy with 100% efficiency according to $E = Mc^2$, not enough matter for fuel could be carried along for even a modest intergalactic voyage. In short, irrespective of technological developments, it appears that extended space travel is out of the question. We are confined by the fundamental laws of physics to a very small region of the universe in the vicinity of the solar system. Furthermore, admitting technological limitations—such as not having an ideal speed-of-light spaceship with a 100% efficient matter engine at present—we see that we are truly doomed to stay at home until a breakthrough is made in fundamental physics that will provide a way around the exceedingly large distances involved. Whether this will occur is a matter of speculation.

9-7 THE TIDES

We now consider an effect of the gravitational interaction between the moon and the earth, the tides of the oceans. It is easy to understand how water, being fluid and free to flow, will be drawn by the gravitational attraction of the moon to create a slight bulge in the ocean on the side of the earth adjacent to the moon. Then, as the daily rotation of the earth moves a point of land underneath the bulge, a high tide will be experienced. While this sounds plausible at first, it can account for only *one* high tide per day, when in fact there are *two*. Something more is involved.

It is commonly said that the moon revolves about the earth. This is true, but the earth also revolves around the moon. Actually, the two bodies revolve around their common center of mass, called the *barycenter*, as illustrated in Fig. 9-12. Thus, the earth travels in a comparatively small circle as a result of the orbital motion of the moon. This circular motion requires a centripetal force, which, of course, is provided by the moon.

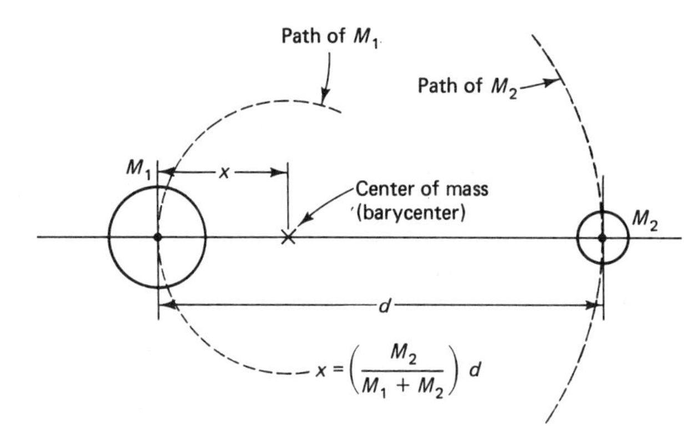

Figure 9-12 Two bodies in mutual orbit revolve about their common center of mass, called the barycenter.

If the attractive force of the moon were the same at all points on the earth, there would be no tides—even though the moon pulls on the water of the oceans, as well as the rest of the earth. All the earth, including the water, would accelerate toward the moon with the same (centripetal) acceleration. However, the moon's attraction is greater on the near side of the earth than it is on the more distant far

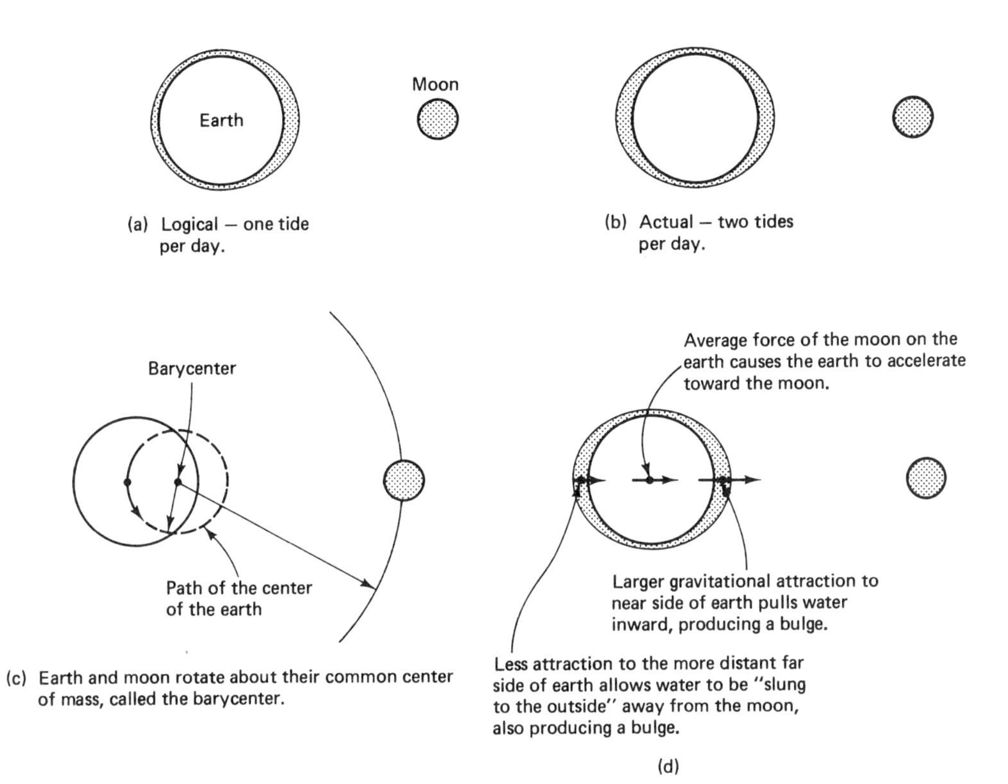

Figure 9-13 The relative motion of the moon and earth produces two tides per day instead of one.

side (a difference of about 6%), and this unbalance is responsible for the *two* tides that occur each day (Fig. 9-13).

Let us accept the moon's pull on the center of the earth as being the "average" force of attraction. The total net attraction provides the centripetal force to maintain the circular motion of the earth about the earth-moon barycenter. However, the side of the earth facing the moon experiences a greater than average force, and a tidal bulge develops on the surface of the earth underneath the moon (very nearly). On the other side of the earth, the water (and other objects) receives a slightly smaller force than is required to maintain the circular motion around the barycenter. Consequently, water flows to the outside of the circular path—to the side of the earth *away* from the moon.

Because the earth rotates on its axis once a day, a given point on the beach will pass each of the two bulges of water once a day and will, therefore, experience two high tides per day. Actually, the tidal cycle is somewhat longer than 24 h (24 h, 50 min) because the moon moves toward the east by about 13° each day in traversing its (almost) monthly cycle around the earth.

To Go Further

Read about the following in an encyclopedia:

Nicolaus Copernicus (1473–1543), Polish astronomer and scholar
Tycho Brahe (1546–1601), Danish astronomer

Galileo Galilei (1564–1642), Italian mathematician and astronomer
Johannes Kepler (1571–1630), German mathematician
Isaac Newton (1642–1727), English scholar
Gottfried Leibniz (1646–1716), German philosopher-mathematician
Edmund Halley (1656–1742), English astronomer

References

BRANCAZIO, PETER J., *The Nature of Physics*, New York: Macmillan Publishing Co., Inc., 1975. An introductory physics text that includes an entire chapter on the development of astronomy.

PASACHOFF, JAY M., *Contemporary Astronomy*. 2nd ed. New York: Saunders College Publishing, 1981. A magnificent text on introductory astronomy that includes much historical background.

SAGAN, CARL, *COSMOS*. New York: Random House, 1980. A bestseller with a bias toward astronomy that is noteworthy for the variety of topics covered and its detailed historical accounts.

VON BRAUN, W. and F. I. ORDWAY III, *History of Rocketry & Space Travel*, New York: Thomas Y. Crowell Co., 1969. A profusely illustrated history of the development of the rocket and space travel.

Questions

1. In your own language, state the universal law of gravitation. Why is it called *universal*?
2. How did the Cavendish experiment amount to "weighing the earth"?
3. Why does g vary with altitude? Why does g vary with latitude?
4. Strictly speaking, is the orbit of a satellite a circle? Can it ever be a circle?
5. Must a satellite in a high orbit have a larger or smaller velocity than a satellite in a lower orbit?
6. What role does the escape velocity play in determining whether a celestial body will retain an atmosphere?
7. Describe a synchronous satellite. Does it really hover over the earth?
8. What is an astronomical unit? a light-year? a parsec?
9. What role does centrifugal force play in the development of tides on the earth?
10. If the gravitational field of the moon were uniform over the region of space occupied by the earth, would there be none, one, or two tides per day?
11. Why is the tidal cycle very nearly 25 h instead of 24 h, the number of hours in 1 day?

Problems

1. Suppose the value of G to be 1 $N \cdot m^2/kg^2$ instead of its normal value. Using this value, compute the gravitational force of attraction between two 20 kg spherical masses when the distance between the centers is: **(a)** 1 m; **(b)** 2 m; **(c)** 3 m; **(d)** 4 m.
2. Repeat Problem 1 using the correct value of G.
3. Show how the unit $N \cdot m^2/kg^2$ is the equivalent of $m^3/(s^2 \cdot kg)$.
4. A 100-kg lead ball and a smaller 0.1-kg ball

are located so that their centers are separated by 0.40 m.

(a) What gravitational force is exerted on the larger ball?

(b) What gravitational force is exerted on the smaller ball?

(c) If both balls are perfectly free to move, what would be the initial acceleration of the larger ball?

(d) Repeat (c) for the smaller ball.

5. What is the mass of the moon, given that the lunar acceleration of gravity is 1.63 m/s^2 and the radius of the moon is 1.738×10^6 m?

6. Calculate the average density of the moon and compare it with the value for the earth.

7. Use Eq. (9-4) to calculate g for the earth.

8. The radius of the planet Saturn is 6.0×10^7 m and its mass is 5.69×10^{26} kg. Compute the acceleration of gravity at the surface of Saturn.

9. Use the data of Problem 8 to compute the average density of Saturn. (Would Saturn float in a large-enough body of water?)

10. Calculate the acceleration of gravity at a distance of $3R_e$ from the surface of the earth.

11. (a) What tangential velocity is required for a satellite to orbit the earth at an altitude of 2000 mi?

(b) What would be the orbital period of the satellite?

12. What (a) tangential velocity and (b) orbital radius are required for an earth satellite to have an orbital period of 12 h?

13. Can a satellite be placed in orbit around a satellite? Suppose a spherical satellite with a mass of 1000 kg is in a high orbit about the earth. A small lead shot of mass 0.5 gram is placed in orbit around the satellite so that the orbital radius is 5 m. Calculate (a) the tangential velocity and (b) the orbital period of the lead shot as it orbits the satellite.

14. (a) What is the total energy (KE + PE) of a small satellite of mass 1 kg in orbit 100 mi above the earth?

(b) What is the ratio of the KE to the PE?

15. Repeat Problem 14 for an orbit 1000 miles above the surface of the earth. Compare the results of the two problems.

16. In Example 9-4, we found only a 1.22% difference in tangential velocity between a 100-mi and a 200-mi (above the surface) orbit. By what percentage is the total energy of a 200-mi orbit greater than for a 100-mi orbit?

17. Calculate the escape velocity of the moon.

18. Compute the escape velocity of the satellite of Problem 13.

19. What additional energy in joules must be given to a 1000-kg synchronous satellite to have it escape from the earth?

20. Compute the orbital period of the planet Venus, given that Venus is—on the average—67 million miles from the sun.

21. The planet Mercury orbits the sun once every 88 days. From this, compute the radius of the orbit.

22. Perform the unit conversion that leads to Eq. (9-24).

23. Taking the velocity of light to be 3×10^8 m/s, compute the time required for a ray of light to travel: (a) from the earth to the moon; (b) from the sun to the earth; (c) from the earth to Alpha Centauri, a distance of 4.24 light-years.

24. At points within the interior of the earth, the acceleration of gravity g_i is given by

$$g_i = G\frac{M_e}{R_e^3}\, r$$

where R_e is the radius of the earth and where r is the distance from the center of the earth. (This assumes the earth to be of uniform density, which is not exactly true.) Use this expression to calculate g_i: (a) at the center of the earth; (b) 1000 mi from the center of the earth; (c) 100 mi below the surface of the earth; (d) at the surface of the earth.

25. Suppose the earth rotated once every 12 h instead of once every 24. What would be the required altitude (measured from the earth's surface) of a synchronous satellite?

26. Given that the mass of Venus is 4.9×10^{24} kg and its radius is 6.054×10^6 m, compute the acceleration of gravity at the surface of Venus.

27. By writing Eq. (9-9) for two different orbits (1 and 2) and then dividing the resulting two expressions, the following relationship is obtained:

$$\frac{V_1^2}{V_2^2} = \frac{R_2}{R_1}$$

(a) Test the validity of this relationship using the results of Examples 9-3 and 9-4.
(b) Use the relationship to compute the required tangential velocity of an orbit 500 mi above the surface of the earth.

28. When two spherical masses, m_1 and m_2, are separated a distance d between centers, there is a point between the masses a distance xd from m_1 at which the net gravitational field is zero because of the "equal and opposite" attractions of the two masses. The value of x is given by

$$x = \frac{1}{\sqrt{m_2/m_1} + 1}$$

Use this expression to show that the point of zero gravity between the earth and the moon is located about 218,000 mi from the center of the earth.

29. An approximate formula for the acceleration of gravity g_a at an altitude h above the surface of the earth is

$$g_a = g\left(1 - 2\frac{h}{R_e}\right)$$

This approximation is good as long as h is small in comparison to R_e, the radius of the earth. Use this expression to compute g_a for altitudes of: **(a)** 100 mi; **(b)** 1000 mi; **(c)** 2000 mi. Compare the result with the *exact* value computed from Eq. (9-8).

30. It is desired to use the pin-and-string method of Fig. 9-8 to construct an ellipse whose major axis is 4 in. and whose eccentricity is 0.5.

(a) How long should the string be?
(b) How far apart should the pins be placed?

31. A string 6 in. long attached to two pins 4 in. apart is used to construct an ellipse. What are: **(a)** the major axis; **(b)** the minor axis; **(c)** the eccentricity of the resulting ellipse?

CHAPTER 10

SIMPLE MACHINES

A simple machine is a simple system such as a lever, system of pulleys, or an inclined plane that might multiply a force, change the direction of a force, change straight-line motion to rotary motion, effect a change in speed, or simply transmit energy from one place to another. In this chapter we consider several simple machines and use the principles elaborated in earlier chapters to obtain mathematical expressions for the ideal mechanical advantage of each.

Most people do not find simple machines to be as exciting as rockets, gravitational phenomena, or even the Cavendish balance, an elegant device in its own right. But the simple machines presented here constitute the ingredients of more complex machines that we do find exciting: typewriters, line printers for computers, automobiles, grandfather clocks, sewing machines, ski lifts, chain saws, and fancy corkscrews for wine bottles, to name a few. To see the variety of simple machines utilized in even a modest factory causes one to stand agape and marvel at the capabilities of physics. Even better is a tour of the aerospace and technology facilities at the Smithsonian Institution in Washington, D. C.

10-1 FUNDAMENTAL PRINCIPLES OF SIMPLE MACHINES

Simple machines may be divided into two broad categories, namely, *translation* machines and *rotation* machines. Translation machines involve motion of shafts or other force-bearing members in a straight line. Rotation machines involve rotating shafts, gears, or pulleys. These two types are illustrated in Fig. 10-1.

The major concept for machines of either type is that the movement of the output shaft is related to and is controlled by the movement of the input shaft. In all the machines we shall consider, the output shaft moves in direct proportion to the input shaft. Certain machines, such as the inclined plane and the wedge, do not involve shafts or any other moving parts, but the same principles are applicable.

We refer to an *ideal* machine as one that is frictionless and therefore not

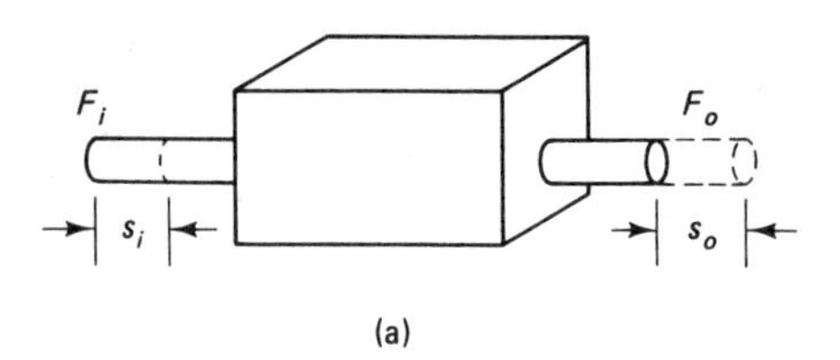

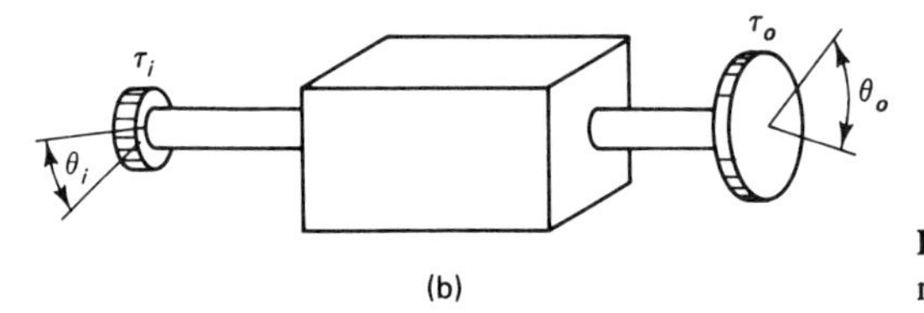

Figure 10-1 Two categories of simple machines; (a) translational; (b) rotational.

subject to frictional losses. For such a machine, all the energy that goes in at the input shaft will come out at the output shaft. Hence, for an ideal machine, the work input must equal the work output, and the power input must equal the power output. Further, because the rate of movement of the input and output shafts are different—while the energy input and output are the same—a simple machine can serve as a force or torque *multiplier*. Using a simple machine, a powerful force moving through a short distance can be converted to a lesser force that moves through a greater distance. The converse is also true, and the same possibilities exist for torque and rotational motion.

Mechanical advantage

The factor by which force or torque is multiplied by a machine is called the *mechanical advantage* of the machine. The output force of a machine having a mechanical advantage of two is twice as great as the input force. The *ideal mechanical advantage* (IMA) is the mechanical advantage that would result if the machine were ideal. The *actual mechanical advantage* (AMA) is the mechanical advantage actually obtained by a real, working machine. The IMA is typically obtained via a simple calculation, whereas the AMA usually is obtained experimentally from measurements made on a working machine. In the following discussion, simple expressions are derived for the IMA of a simple machine, and the *efficiency* of a machine is given as the ratio of the AMA to the IMA.

Consider the general machines of Fig. 10-1. The work done by the force or torque at the input is

$$F_i s_i \quad \text{or} \quad \tau_i \theta_i$$

and the work done at the output is

$$F_o s_o \quad \text{or} \quad \tau_o \theta_o$$

Because of the conservation of energy principle, these must be equal.

$$F_i s_i = F_o s_o \qquad \tau_i \theta_i = \tau_o \theta_o \tag{10-1}$$

These may be rearranged to give

$$F_o = \frac{s_i}{s_o} F_i \qquad \tau_o = \frac{\theta_i}{\theta_o} \tau_i \tag{10-2}$$

and

$$\frac{F_o}{F_i} = \frac{s_i}{s_o} \qquad \frac{\tau_o}{\tau_i} = \frac{\theta_i}{\theta_o} \tag{10-3}$$

If we define the IMA as

$$\text{IMA} = \frac{F_o}{F_i} \quad \text{or} \quad \text{IMA} = \frac{\tau_o}{\tau_i} \tag{10-4}$$

for the translational and rotational cases, respectively, then

$$\text{IMA} = \frac{s_i}{s_o} \quad \text{or} \quad \text{IMA} = \frac{\theta_i}{\theta_o} \tag{10-5}$$

Thus, the IMA is simply the ratio of the input distance s_i to the corresponding output distance s_o. If the input shaft moves more than the output shaft, the output force (or torque) will be greater than that at the input.

Efficiency

All practical machines are subject to frictional forces that cause energy to be wasted in the course of operating the machine. Consequently, the useful work output is less than the work input, and the actual mechanical advantage is less than the ideal mechanical advantage. The AMA may be defined as

$$\text{AMA} = \frac{F_{oA}}{F_i} \quad \text{or} \quad \text{AMA} = \frac{\tau_{oA}}{\tau_i} \tag{10-6}$$

where F_{oA} and τ_{oA} are the actual force and torque, respectively, obtained at the output of the machine in question.

The efficiency is defined as the ratio of the work output to the work input, expressed as a percentage:

$$\text{Efficiency (\%)} = \frac{\text{work output}}{\text{work input}} \times 100$$

and because

$$\frac{\text{Work output}}{\text{Work input}} = \frac{F_{oA} s_o}{F_i s_i} = \frac{F_{oA}/F_i}{s_i/s_o} = \frac{\text{AMA}}{\text{IMA}}$$

we obtain

$$\text{Eff (\%)} = \frac{\text{AMA}}{\text{IMA}} \times 100 \tag{10-7}$$

If the efficiency (eff) of a machine is given, the AMA is calculated from

$$\text{AMA} = \text{eff} \times \text{IMA}$$

where the efficiency is written as a decimal fraction. The actual output force or torque is given by

$$F_{oA} = \text{AMA} \times F_i \qquad \tau_{oA} = \text{AMA} \times \tau_i \tag{10-8}$$

10-2 LEVERS

Three types, or classes, of levers may be distinguished as in Fig. 10-2 according to the relative positions of the load L, applied force F, and the pivot point, called the *fulcrum* of the lever. In each case, the force is applied a distance r_i from the pivot point, while the load is situated a distance r_o from the pivot. When r_i is greater than r_o, the IMA of the lever (which equals r_i/r_o) is greater than one. The IMA of a

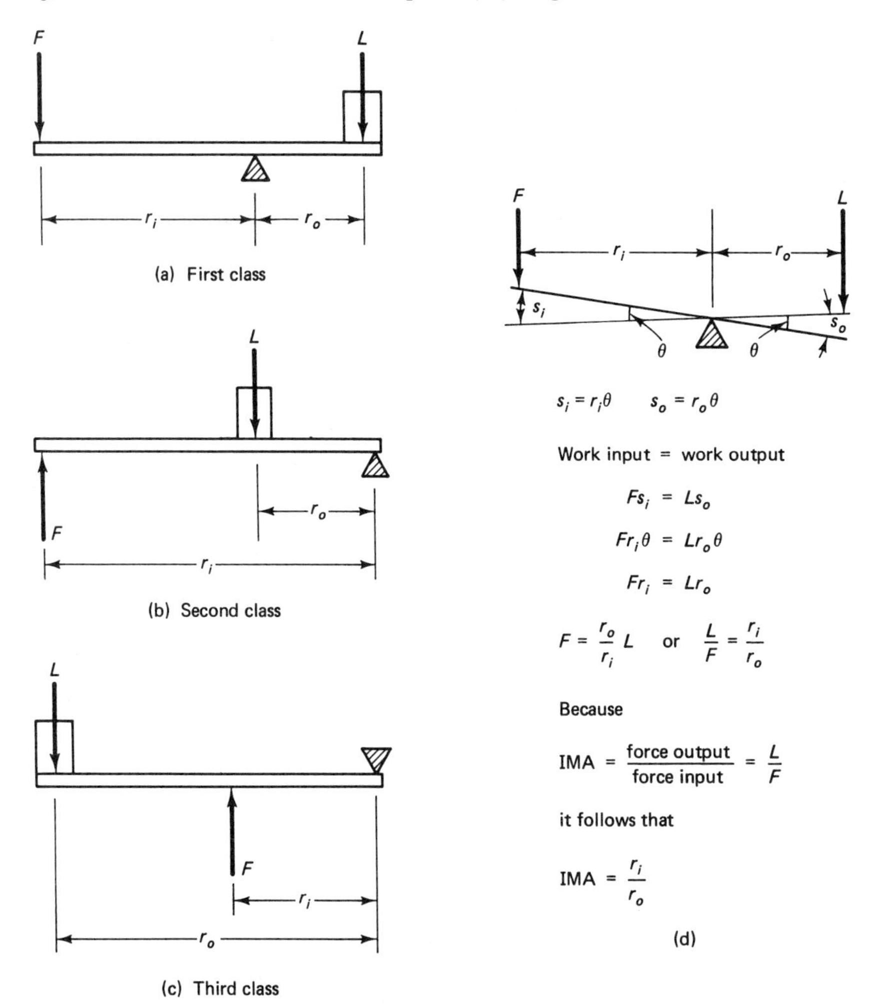

Figure 10-2 Three types of levers are shown in (a), (b), and (c). The formulas obtained in (d) are applicable to all three cases.

given lever may be obtained either by summing torques about the pivot point or by equating the work input and the work output. In this section, we take the latter approach (see Sect. 5-9).

In Fig. 10-2(d), the input distance s_i and the output distance s_o are related to the angle of rotation θ of the lever. We assume θ is small so that F and L are very nearly in line with s_i and s_o. By equating work input (Fs_i) to the work output (Ls_o) and writing the distances as $r_i\theta$ and $r_o\theta$, respectively, the force F is related to the load L in terms of r_i and r_o. It then follows that the IMA is r_i/r_o. The other two types of levers may be treated in a similar manner.

Three examples of lever systems are shown in Fig. 10-3. The calculations relevant to each case should be obvious.

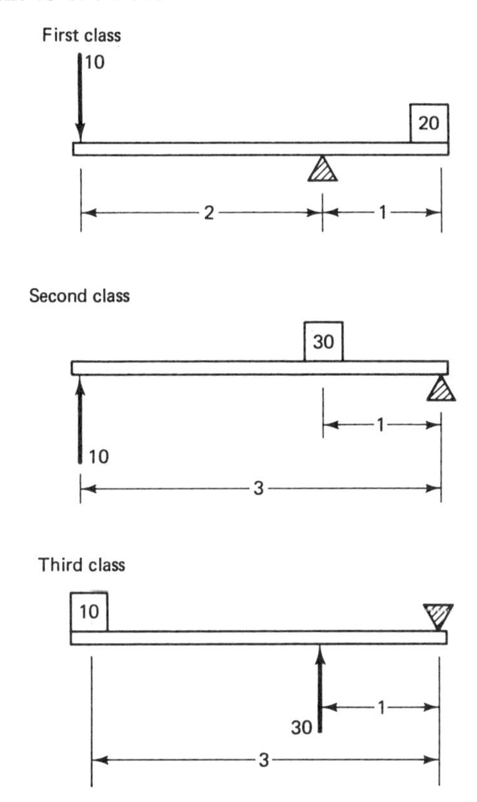

Figure 10-3 Three examples of problems involving levers.

10-3 THE INCLINED PLANE AND THE WEDGE

An inclined plane may be considered as a simple machine even though it has no moving parts. A heavy load can be elevated a vertical distance $h = s_o$ by applying a lesser force over the greater distance s_i up the incline. In practical situations, frictional effects are extremely important; these were considered in Sect. 5-7. However, if the load is mounted on high-quality rollers or wheels that roll on a hard surface, frictional effects are minimized. In any case, the IMA represents an idealization, and the AMA of any practical situation is naturally expected to be less. We refer to Fig. 10-4.

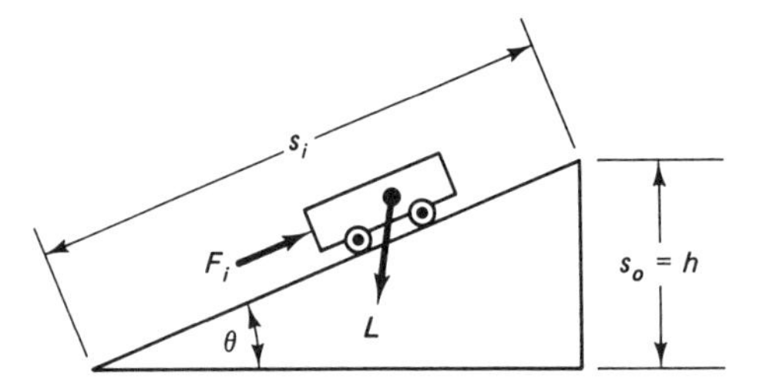

Work done = potential energy

$$F_i s_i = L s_o$$

$$F_i = \frac{s_o}{s_i} L \quad \text{or} \quad \frac{L}{F_i} = \frac{s_i}{s_o}$$

$$\text{IMA} = \frac{s_i}{s_o} = \frac{s_i}{s_i \sin\theta}$$

$$\text{IMA} = \frac{1}{\sin\theta}$$

Figure 10-4 Frictionless inclined plane.

Ignoring frictional effects, the work done in pushing the load up the incline ($F_i s_i$) must equal the potential energy of the load at the elevated position. The potential energy is mgh which can be written as Ls_o. From this it follows that the IMA is s_i/s_o. Because $s_o = s_i \sin\theta$, the IMA of a frictionless inclined plane, as illustrated in Fig. 10-4, is found to be

(frictionless inclined plane)

$$\text{IMA} = \frac{1}{\sin\theta} \tag{10-9}$$

where θ is the angle of the incline.

The wedge

Referring to Fig. 10-5, we see that when a wedge is driven a distance s_i into a body, a lateral displacement s_o results. Thus, we consider a wedge to be a simple machine similar to an inclined plane. An analysis of the geometry involved yields the expression for the IMA given in the figure.

The overall efficiency of a wedge is rather low due to frictional effects, but a wedge is very useful in converting large, impulsive forces delivered to the wedge by a heavy hammer to sustained forces directed at right angles to the penetration of the wedge. Because the impulsive forces are quite large, the lateral forces are large also, even in light of frictional losses and the relatively small IMA of a wedge. This is illustrated in Example 10-1.

EXAMPLE 10-1 An 8-lb hammer strikes a wedge with a velocity of 34 ft/s and causes the wedge to penetrate a block of wood a distance of $\frac{1}{2}$ in. (0.0417 ft). The wedge angle is 20°.
(a) Compute the average force exerted on the wedge by the hammer.
(b) What is the IMA of the wedge?
(c) What lateral force is exerted on the wood by the wedge?

Solution. (a) Assuming the hammer to be uniformly decelerated from $v_0 = 34$ ft/s to

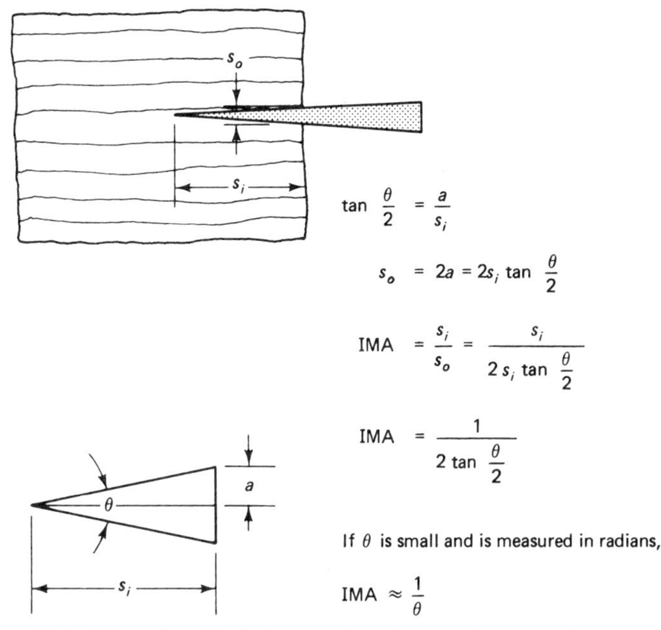

Figure 10-5 IMA of a frictionless wedge. In practice, a wedge must have sufficient friction to keep it from being expelled from the block.

a final velocity of zero ($v_0 = 0$), we apply the $2as$ formula of Sect. 3-5 to calculate the average deceleration:

$$2as = v^2 - v_0^2$$

$$a = -\frac{(34 \text{ ft/s})^2}{2(0.0417 \text{ ft})}$$

$$= -13{,}872 \text{ ft/s}^2 \quad \text{(minus sign implies deceleration)}$$

Using $F = ma$ with a mass of 0.25 slug (8 lb), the force exerted on the hammer to stop its motion is

$$F = ma$$

$$= (0.25 \text{ slug})(13{,}872 \text{ ft/s}^2)$$

$$= 3468 \text{ lb}$$

This is also the force exerted on the wedge, by virtue of the action-reaction principle.
(b) The IMA of the wedge is calculated using the formula given in Fig. 10-4(b).

$$\text{IMA} = \frac{1}{2 \tan(\theta/2)} = \frac{1}{2 \tan 10°}$$

$$= 2.84$$

(c) The force exerted on the wood (ignoring friction) is

$$F_{\text{wood}} = \text{IMA} \times F$$

$$= 2.84 \times 3468 \text{ lb}$$

$$= 9849 \text{ lb}$$

This force is almost 5 tons! Even if the wedge is only 20% effective, the force applied to the wood will be on the order of 1 ton—a rather sizable force.

10-4 SYSTEMS OF PULLEYS

Pulleys may be arranged in various combinations to provide a fairly wide range of IMA's with moderately good efficiency. In such systems, pulleys are described as being either *fixed* or *movable*. All pulleys rotate freely on their axles, of course, but the axle of a *fixed* pulley is attached to a rigid support and does not move in a straight line as the system is operated. On the other hand, a *movable* pulley moves in a straight line in proportion to the movement of the load. In Fig. 10-6, the axles of all movable pulleys are connected directly to the load.

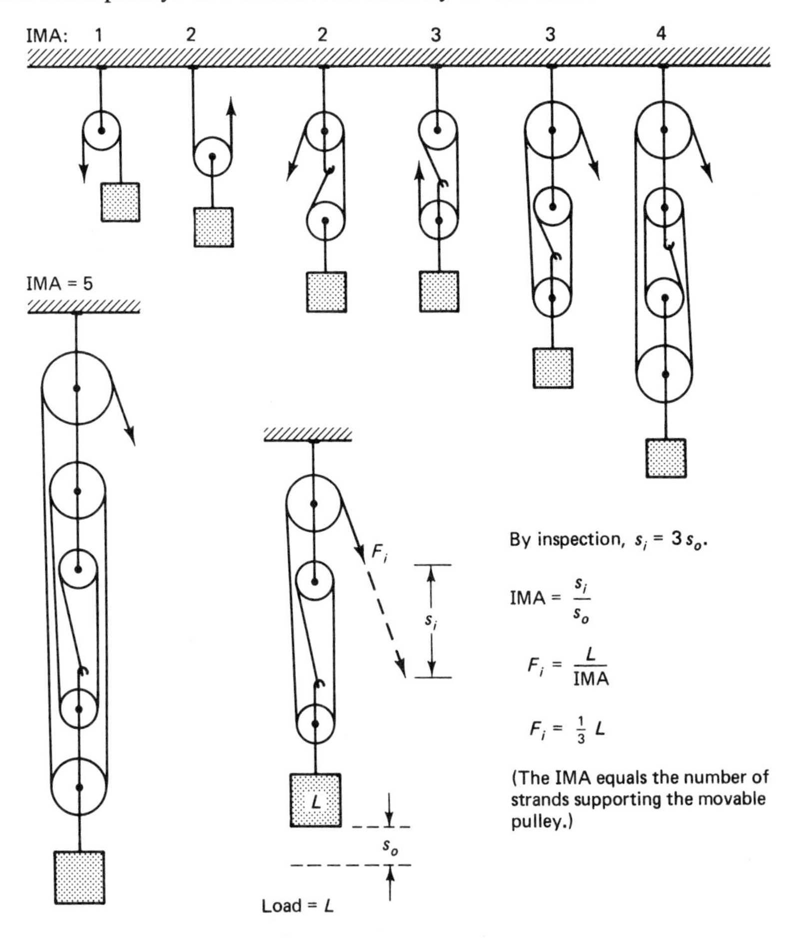

Figure 10-6 Systems of pulleys.

The IMA of a given combination of pulleys may be determined by relating the movement s_o of the load to the movement s_i of the applied force. Obviously, the number of strands connecting the fixed and movable pulleys determines the relationship between s_i and s_o. If the load is lifted a distance s_o, each strand of rope supporting the movable pulleys must be shortened by that amount. Consequently, the distance s_i equals $N \times s_o$, where N is the number of strands supporting the movable pulleys. It follows that the IMA of a simple combination of pulleys equals the number of strands supporting the movable pulleys, N.

The pulleys of Fig. 10-6 are considered to be frictionless and very light. If this is the case the tension in the string will be the same at all points in a particular system.

A compound system

Counting the strands to determine the IMA is *not* applicable to the compound system of Fig. 10-7. The system is best considered as a cascaded system of four single-movable pulleys. The tension in each strand of rope can be determined by inspection (starting at the load and working upward), and the input force F_i is found to be $\frac{1}{16}$ of the load. Consequently, the IMA of the system is 16 even though far fewer than 16 strands of rope support the load. The fixed pulley simply reverses the direction of F_i and does not enter into the determination of the IMA.

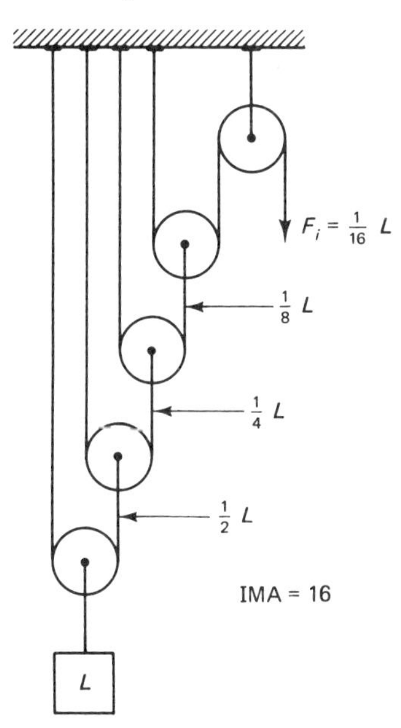

Figure 10-7 A compound system of pulleys. The tensions in the ropes are not all the same.

Differential hoist

The system shown in Fig. 10-8 is a modification of a pulley system that achieves a far greater IMA than is possible with a conventional pulley system, using the same number of pulleys. A double pulley at the top has unequal radii, and the two sections are joined so that they rotate as one unit. The bottom pulley is a conventional single pulley. A chain in often used instead of a rope, and the upper pulleys are constructed so that the chain cannot slip or slide across the tops of the pulleys.

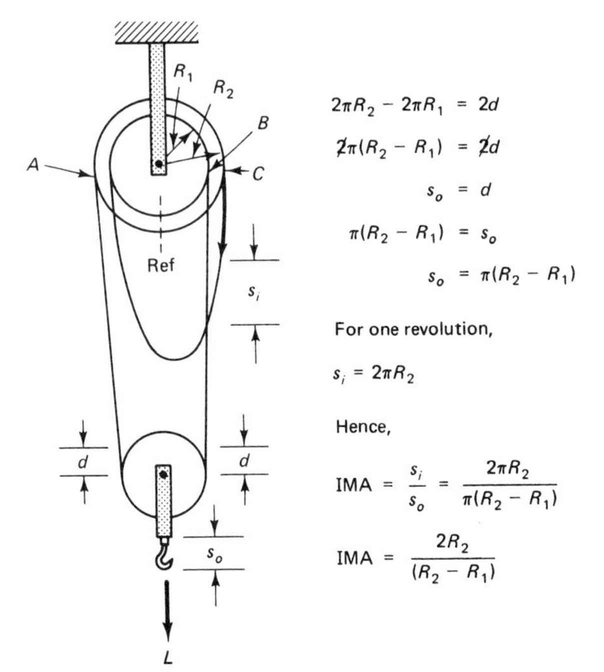

Figure 10-8 A differential hoist.

As the chain is pulled down a distance s_i at point C, chain is taken up at point A, but chain is let out at point B. Because R_2 is greater than R_1, chain is taken up at point A somewhat faster than it is let out at B. Therefore, the total length of chain connecting the fixed and movable pulleys is shortened, and the load is lifted a distance s_o.

The net chain taken up by one revolution of the top pulley is the difference in circumference of the two sections of the top pulley. This must equal twice the distance d illustrated in the figure because two strands support the movable pulley. This equality, and the fact that s_o equals d, gives the following value for s_o, the distance the load is lifted by one revolution of the top pulley:

$$s_o = \pi(R_2 - R_1)$$

To produce one revolution of the top pulley, however, the chain at point C must be pulled down a distance equal to the circumference of the larger top pulley; hence, for one revolution,

$$s_i = \pi R_2$$

Then, from basic principles, the IMA is the ratio s_i/s_o, which gives the IMA of the device:

(differential hoist)
$$\text{IMA} = \frac{2R_2}{(R_2 - R_1)} \qquad (10\text{-}10)$$

Note that as the two radii become more nearly equal, the IMA becomes larger. If R_2 is 6 in. while R_1 is 5 in., the IMA is 12. A conventional pulley system would require 12 strands supporting the movable pulley to give the same IMA.

10-5 GEAR AND BELT SYSTEMS

When gears with different numbers of teeth are meshed, as in Fig. 10-9, the rates of rotation of the two gears will be different. The rotation of the smaller gear will be greater by a factor equal to the ratio of the number of teeth on the two gears, namely, N_1/N_2. This result can be obtained by assuming a certain rather large number of teeth on each gear (denoted by N_t) pass the point of contact between the two gears. Obviously, N_t will be the same for each gear. Therefore, we compute the revolutions made by each gear:

$$\text{rev}_1 = \frac{N_t}{N_1} \qquad \text{rev}_2 = \frac{N_t}{N_2} \tag{10-11}$$

By solving each expression for N_t and setting the resulting expressions equal, we obtain

$$N_1\text{rev}_1 = N_2\text{rev}_2 \quad \text{or} \quad \text{rev}_2 = \frac{N_1}{N_2}\text{rev}_1 \tag{10-12}$$

Because the angular displacement θ (measured in radians) is proportional to the revolutions turned through, we may also write

$$N_1\theta_1 = N_2\theta_2 \quad \text{or} \quad \theta_2 = \frac{N_1}{N_2}\theta_1 \tag{10-13}$$

If we consider the amount of time taken for the revolutions to occur, we obtain, by dividing by the time and invoking basic definitions,

$$N_1\text{rpm}_1 = N_2\text{rpm}_2 \quad \text{or} \quad N_1\omega_1 = N_2\omega_2 \tag{10-14}$$

These relationships are extremely useful in analyzing systems of gears.

In a two-gear system, one gear is driven by the other, which receives power

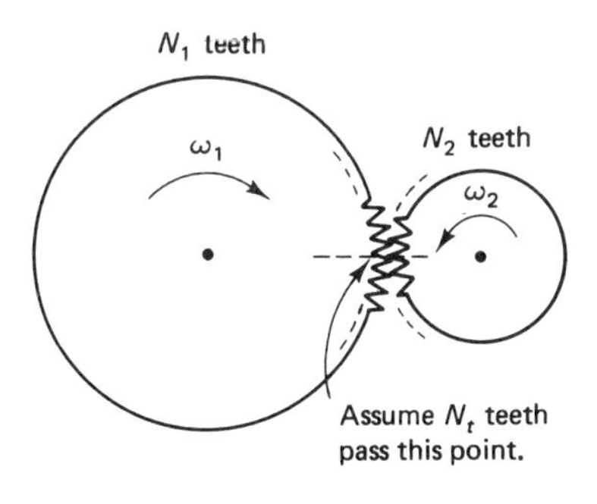

$$N_t = N_1 \text{ rev}_1 = N_2 \text{ rev}_2$$

$$\text{rev}_2 = \frac{N_1}{N_2} \text{ rev}_1$$

$$\text{rpm}_2 = \frac{N_1}{N_2} \text{ rpm}_1$$

$$N_1\omega_1 = N_2\omega_2$$

Figure 10-9 Gear relationships for revolutions and rpm.

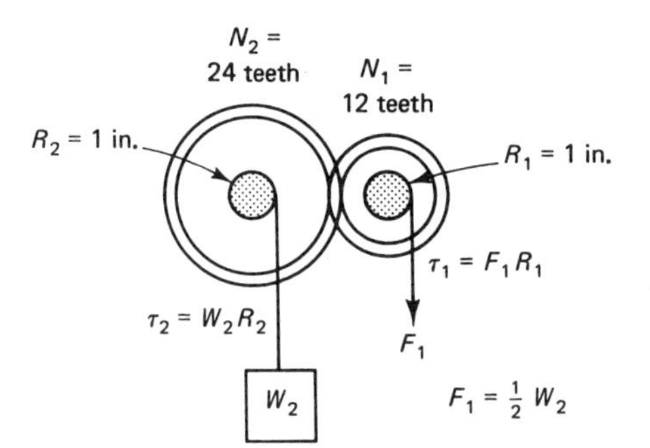

Figure 10-10 System of gears in equilibrium.

from an external source. The gear to which external power is applied is called the *driver* because it drives the other gear, called the *driven* gear.

If an external agent does an amount of work on the driver, the driven gear will do the same amount of work on whatever is connected to the output of the system. In other words, the work input equals the work output, a familiar principle. Recalling that rotational work is the product of the torque and angular displacement, we write

$$\text{Work input} = \text{work output}$$
$$\tau_1\theta_1 = \tau_2\theta_2 \tag{10-15}$$

If we substitute the expression for θ_2 given as Eq. (10-13) and then cancel θ_1, which appears on both sides of the resulting equation, we obtain

$$\tau_1 = \frac{N_1}{N_2}\tau_2 \quad \text{or} \quad \tau_2 = \frac{N_2}{N_1}\tau_1 \tag{10-16}$$

These expressions define the torque conversion properties of a gear system, and we see that the torque is transformed according to the ratio of the teeth on the respective gears.

Return to Eq. (10-15). Considering that the angular displacements θ_1 and θ_2 must have occurred in the same time interval t, we divide both sides of the equation by t to obtain

$$\tau_1\frac{\theta_1}{t} = \tau_2\frac{\theta_2}{t}$$

which is equivalent to

$$\tau_1\omega_1 = \tau_2\omega_2 \tag{10-17}$$

This expression relates the torques and angular velocities. It is actually a statement that the power input equals the power output because $\tau\omega$ is the expression for rotational power.

EXAMPLE 10-2 Compute the force F_1 required to hold the system of Fig. 10-10 in equilibrium.

Solution. Applying Eq. (10-16), we obtain

$$\tau_2 = \frac{N_2}{N_1}\tau_1 = \frac{24}{12}\tau_1$$
$$= 2\tau_1$$

Substituting the expressions for the torque produced by W_2 and F_1 yields

$$W_2 R_2 = 2F_1 R_1$$

Solving for F_1 and noting that R_2/R_1 equals 1 because the shafts of the two gears are of the same diameter, we have

$$\begin{aligned} F_1 &= \frac{1}{2}\left(\frac{R_2}{R_1}\right)W_2 \\ &= \tfrac{1}{2}(1)W_2 \\ &= \tfrac{1}{2}W_2 \end{aligned}$$

Consequently, if W_2 is 16 lb, the force F_1 must be 8 lb to maintain equilibrium.

EXAMPLE 10-3 A $\frac{1}{4}$-HP motor runs at the standard 1750 rpm.
(a) Use the formula for rotational power, $\tau\omega$, to compute the maximum torque available at the output shaft of the motor.
(b) If a speed reducer is used to slow the rotational speed down to 100 rpm, what torque would be available at the output of the speed reducer?

Solution. (a) Multiply by $2\pi \times 60$ to convert 1750 rpm to 377 rad/s. Then, recalling that 1 HP = 550 ft-lb/s,

$$\tfrac{1}{4}(550 \text{ ft-lb/s}) = \tau(377 \text{ rad/s})$$

Hence, the maximum torque is

$$\tau = \frac{550 \text{ ft-lb/s}}{4(377 \text{ lb/s})} = 0.365 \text{ ft-lb} \qquad (\text{at 1750 RPM})$$

(b) If the output of the speed reducer is 100 rather than 1750 rpm, the output torque is 17.5 times as large (1750/100 = 17.5). Thus, the maximum torque is 6.39 ft-lb at 100 rpm.

Belt systems

The key idea in the analysis of a belt system, as shown in Fig. 10-11(a), is that the tengential velocity v_t of a point on the belt is the same as the tangential velocity of points on the rim of both pulleys. Consequently, we write

$$(\textit{tangential velocity}) \qquad R_1\omega_1 = v_t = R_2\omega_2$$

which may be arranged to give

$$\omega_2 = \left(\frac{R_1}{R_2}\right)\omega_1 \quad \text{or} \quad \omega_1 = \left(\frac{R_2}{R_1}\right)\omega_2 \qquad (10\text{-}18)$$

In terms of revolutions per minute instead of the angular velocity, ω,

$$\text{rpm}_2 = \left(\frac{R_1}{R_2}\right)\text{rpm}_1 \quad \text{and} \quad \text{rpm}_1 = \left(\frac{R_2}{R_1}\right)\text{rpm}_2 \qquad (10\text{-}19)$$

This is the familiar result that, for such systems, the smaller pulley rotates faster than the larger pulley.

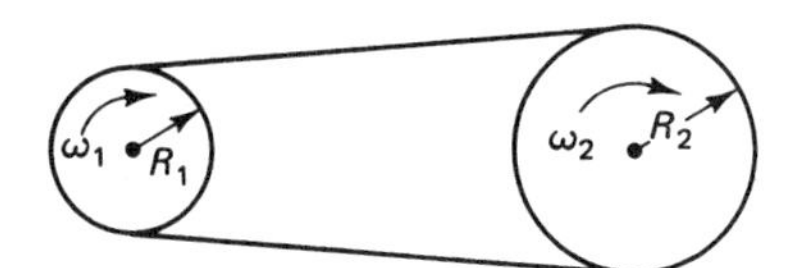

Tangential velocity:

$$R_1\omega_1 = R_2\omega_2$$

$$\omega_1 = \left(\frac{R_2}{R_1}\right)\omega_2$$

$$\frac{\omega_1}{\omega_2} = \frac{R_2}{R_1} = \frac{\text{rpm}_1}{\text{rpm}_2}$$

Power:

$$\tau_1\omega_1 = \tau_2\omega_2$$

$$\tau_1 = \left(\frac{\omega_2}{\omega_1}\right)\tau_2$$

$$\tau_1 = \left(\frac{R_1}{R_2}\right)\tau_2$$

(a)

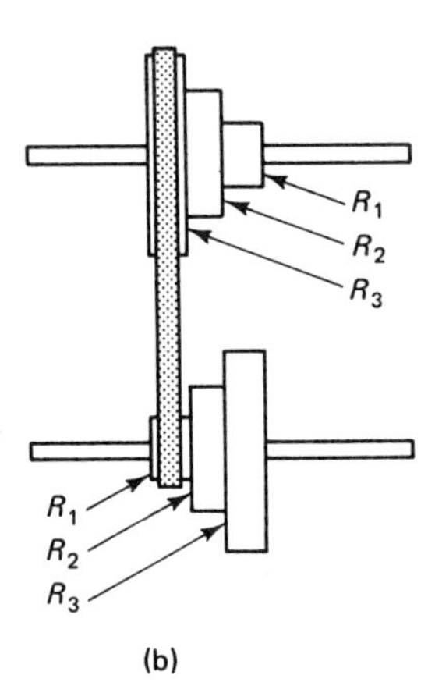

(b)

Figure 10-11 (a) Radius, rpm, and torque relationships for a belt system. (b) Step pulleys.

Because of the equality of the power input and power output, we may write

$$\tau_1\omega_1 = \tau_2\omega_2 \tag{10-20}$$

from which we obtain

$$\tau_1 = \left(\frac{\omega_2}{\omega_1}\right)\tau_2 \qquad \tau_2 = \left(\frac{\omega_1}{\omega_2}\right)\tau_1 \tag{10-21}$$

Using Eq. (10-18), these expressions can be written in terms of R_1 and R_2 as

$$\tau_1 = \left(\frac{R_1}{R_2}\right)\tau_2 \qquad \tau_2 = \left(\frac{R_2}{R_1}\right)\tau_1 \tag{10-22}$$

These expressions are quite similar to those obtained for the gears.

Step pulleys, as shown in Fig. 10-11(b), are frequently used to vary the relative rpm of two shafts to which the pulleys are attached. With the belt at a given position, the rpm of each shaft and the torque transformation can be calculated using the formulas given above. The pulley diameters are chosen so that only a minimal variation in belt tension occurs when the belt is switched from one position to another.

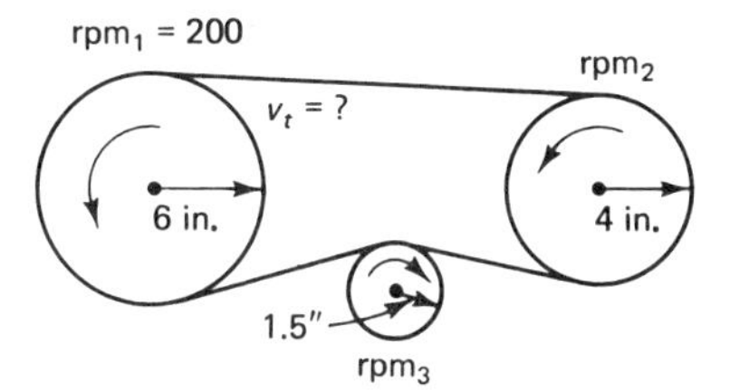

Figure 10-12

EXAMPLE 10-4 Two pulleys are connected by a belt, as shown in Fig. 10-12. An idler pulley maintains the tension in the lower portion of the belt. If the large pulley is driven at the rate of 150 rpm: (a) calculate the rpm of the other two pulleys; (b) calculate the tangential velocity of the belt.

Solution. (a) Use Eq. (10-19) to find the rpm of the other two pulleys:

$$\text{rpm}_2 = \left(\frac{R_1}{R_2}\right)\text{rpm}_1 \qquad \text{rpm}_3 = \left(\frac{R_1}{R_3}\right)\text{rpm}_1$$

$$\text{rpm}_2 = \left(\frac{6}{4}\right)(200) \qquad \text{rpm}_3 = \left(\frac{6}{1.5}\right)(200)$$

$$\text{rpm}_2 = 300 \qquad \text{rpm}_3 = 800$$

(b) Because all three pulleys are driven by the same belt, the tangential velocity of all three pulleys is the same, namely, that of the belt. Consequently, we can compute the tangential velocity of the belt knowing the radius and rpm of any one of the three pulleys. Let us use the 4-in. pulley. Its angular velocity is 300 rpm, which is 31.42 rad/s. Hence, the tangential velocity is

$$\begin{aligned} v_t &= R\omega \\ &= (4 \text{ in.})(31.42 \text{ rad/s}) \\ &= 125.7 \text{ in./s} \\ &= 10.47 \text{ ft/s} \end{aligned}$$

This same result is obtained if either of the other pulleys is used.

Worm gear

A worm gear is often used whenever a large speed reduction is required in going from one rotating shaft to another. The mechanism is illustrated in Fig. 10-13. The

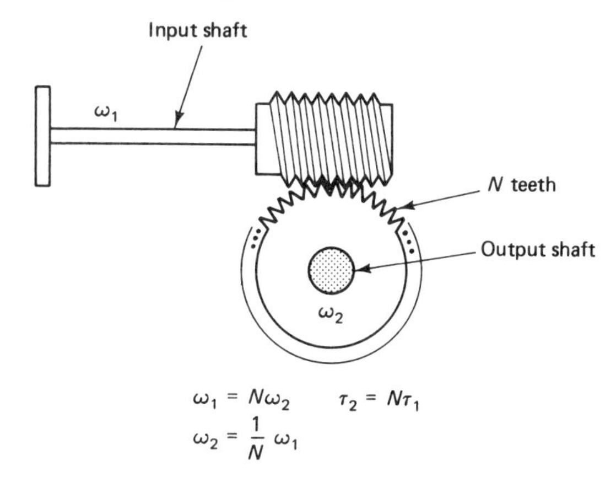

Figure 10-13 Worm gear.

A speed reducer with the housing cut away to show the worm gear.

large output gear advances one tooth for each revolution of the input shaft. Therefore, in order for the output gear to turn through one revolution, the input shaft must turn through as many turns as there are teeth on the output gear. The number of turns of the spiral thread on the input shaft has nothing to do with the speed reduction ratio. The output torque is N times as great as the input torque.

Screw jack

A screw jack is illustrated in Fig. 10-14. When the handle is turned through one revolution ($S_i = 2\pi R$), the load is raised an amount (S_o) equal to the pitch P of the screw thread. Consequently, the IMA is the ratio of the circumference of the circle traversed by the handle to the pitch of the thread. Details are given in the figure.

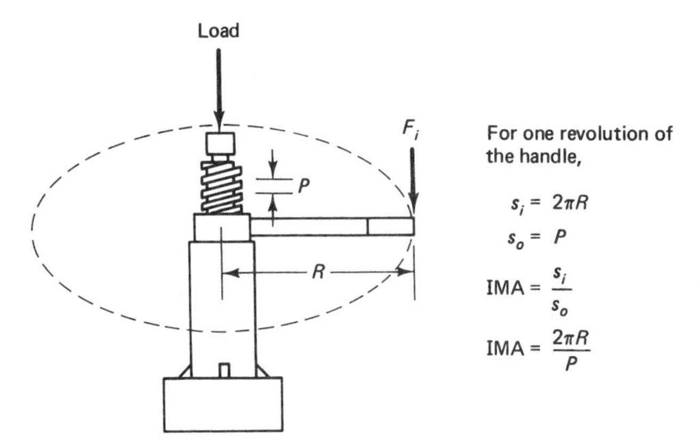

Figure 10-14 Screw jack.

To Go Further

Find out about: (a) planetary gear systems; (b) elliptical gears; (c) bevel gears; (d) helical gears; (e) the differential of an automobile; (f) cams and camshafts; (g) variable-speed drives used on lathes and drill presses.

Questions

1. Discuss or describe these items in terms of simple machines: a pair of scissors, a pair of wire cutters, a nutcracker, a claw hammer used to withdraw a nail, a knife, tweezers, a wheelbarrow.
2. Give a simple procedure for determining the IMA of a system of pulleys.
3. For simple systems of gears, how is the IMA determined?
4. What determines the IMA of a belt and pulley system?
5. How is the linear velocity of a belt related to the angular velocity of the pulleys over which the belt runs?
6. What is the primary advantage of a worm gear? In what applications are they most often used? Can the output and input be interchanged in regard to which is the driven member?
7. Why is it necessary for an automobile to have a transmission?

Problems

1. In Fig. 10-2(a) and (b), suppose the load is 100 lb. If $r_i = 6$ ft while $r_o = 1$ ft, what force F is required to lift the load?
2. In Fig. 10-2(c), let $r_o = 6$ ft, while $r_i = 1$ ft. What force is required to lift a 100-lb load?
3. What is the IMA of an inclined plane that makes an angle of 10° with the horizontal?
4. What is the IMA of a wedge if the angle between the faces is 12°?
5. In a conventional pulley system—as in Fig. 10-6—five strands of rope support the movable pulleys.
   (a) What is the IMA of the system?
   (b) Through what distance must the input force act to lift the load 1 ft?
   (c) What applied force is required to lift a 200-lb load?
   (d) What amount of work is done in lifting the load 1 ft?
   (e) What amount of work is done by the input force to lift the load 1 ft?
6. Repeat Problem 5 for a system in which four strands support the movable pulley. Compare the results.
7. The upper pulleys of a differential hoist (Fig. 10-8) have radii of 8 in. and 9 in.
   (a) What is the IMA of the device?
   (b) What length of chain must be pulled past point C (Fig. 10-8) to lift the load 1 ft?
   (c) If the applied force is 100 lb, what is the weight of the load?
8. Repeat Problem 7, but let the 8-in. pulley be changed to 8.5 in. Compare the results.
9. A 400-lb load is lifted by a conventional pulley system with friction by an applied force of 150 lbs.
   (a) What is the AMA of the system?
   (b) What is the probable IMA?
   (c) How many strands support the movable pulley?
   (d) What work must be done at the input to lift the load 1 ft?
   (e) What is the efficiency of the system?
10. Repeat Problem 10, but suppose an applied force of 110 lb is required to lift the load.
11. A small gear with 14 teeth on the output shaft of an electric motor meshes with a larger gear with 56 teeth. The small gear rotates at 1750 rpm.
    (a) What is the rpm of the larger gear?
    (b) What is the angular velocity of the two gears, in radians per second?
    (c) If the motor delivers a torque of 2.7 ft-lb to the small gear, what torque is avail-

able at the output shaft of the larger gear?

(d) What is the horsepower rating of the motor?

12. Repeat parts (a), (b), and (c) of Problem 11, but assume the larger gear has 42 teeth instead of 56.

13. A simple worm-gear speed reducer provides a reduction ratio of 11:1.

(a) How many times greater is the output torque than the input torque?

(b) If the *worm* is driven at 1750 rpm, what will be the rpm of the output shaft of the speed reducer?

14. A belt-and-pulley system, as in Fig. 10-11(a), consists of a 2-in. diameter pulley and a 4-in. pulley.

(a) If the small pulley is driven at 800 rpm by an external source of power, what will be the rpm of the larger pulley?

(b) What will be the tangential velocity of the belt?

15. A step-pulley system, as in Fig. 10-11(b), consists of pulleys of radius 3, 4, and 5 in. If the input pulley is driven at 100 rpm, what rotational speeds are available at the output?

16. The linear speed of a conveyor belt is 4 ft/s. The belt runs across a supporting roller 6 in. in diameter. What is the rpm of the roller when the conveyor is in operation?

17. The handle of a screw jack is 2 ft long (measured from the center) and the pitch of the screw thread is $\frac{1}{4}$ in. (Fig. 10–14).

(a) What is the IMA of the jack?

(b) What force applied to the handle will lift a load of 2000 lb, assuming 100% efficiency?

(c) What force will be required if the jack is only 50% efficient?

18. Suppose an extension is used to make the handle in Problem 17 3 ft long instead of 2. **(a)** What will be the new IMA, and **(b)** what load can be lifted by the jack if the same force as before is applied to the handle (assume 100% efficiency)?

19. A pulley system whose IMA is 3 is found to be 80% efficient.

(a) What input force is required to lift a load of 200 lb?

(b) If the load is lifted 10 ft, how much work must be done by the input force?

***20.** Design a step-pulley system that will deliver output shaft rpm's of 600, 700, and 800, when the rpm of the input shaft is 700. Can you design it so that the same belt will work for all three sets of pulleys?

***21.** Suppose a 1750-rpm motor is to supply power to a conveyor belt that is to move at nearly 25 in./s. The driven roller of the conveyor belt is 4 in. in diameter. Design the drive system to interface the motor to the drive roller. Make a sketch.

***22** An engineer in a cornflake factory needs a pusher to push the boxes a distance of 20 in. from one conveyor to another. A hydraulic cylinder is available that moves back and forth a distance of 4 in. Design a pusher using the hydraulic cylinder as the source of power. Make a sketch.

CHAPTER 11

PROPERTIES OF MATERIALS; HYDROSTATICS

This chapter begins with a quick review of the basics of the structure of matter, which should be familiar. We then consider the density and specific gravity of materials, concepts that are needed for the sections that follow.

The middle portion of this chapter is devoted to pressure in liquids, Archimedes' principle of buoyancy, Pascal's principle and hydraulics, and surface tension. The final sections deal with elasticity (Hooke's law) and elastic constants of materials.

This material will tremendously broaden the range of physical situations that we can address with the aim of making meaningful calculations. And you may be gratified to observe that the material is actually less difficult (or at least no more difficult) than that in several earlier chapters.

11-1 FUNDAMENTALS OF THE STRUCTURE OF MATTER

Three fundamental particles

It was once thought that all matter was composed of three *fundamental particles*, namely, protons, neutrons, and electrons. These particles were envisioned as the smallest, simplest, and most basic form in which matter could appear. More complex forms of matter were thought to consist of various combinations of the basic particles. The fundamental particles represented a sort of lower limit to the smallest unit in which matter could exist and were considered to be the basic building blocks of the universe.

This view is still held to be true, but with one exception. Many other particles have been found that are comparable to the so-called fundamental particles. Thus, it must be concluded that the fundamental particles—the protons, neutrons, and electrons—are not really fundamental at all. Hundreds of additional particles have been discovered and current thinking is that there may be no particles truly fundamental in the sense of particles representing a lower limit to the manifestation of matter. Protons, neutrons, and electrons are the most popular particles, of course, and we briefly describe the structure of the atom in terms of these.

Protons carry the fundamental unit of positive charge; electrons carry the fundamental unit of negative charge. Neutrons have no net electrical charge. They are thought to be a combination of a proton and an electron, so that the positive and negative charges cancel. Electrical charges are exceedingly important to the constitution of an atom; indeed, it is the attraction between unlike charges that holds the negative electrons in orbit around the positive nuclei.

The atom

An atom consists of a heavy central part, called the *nucleus*, which is orbited by electons in a manner somewhat reminiscent of the sun and planets of the solar system. The nucleus is composed of protons and neutrons and has a positive charge because of the protons. Most of the mass of an atom is concentrated in the nucleus because the mass of either a proton or a neutron is roughly 1840 times the mass of an electron. Even so, the nucleus is exceedingly small in comparison to the diameters of the electron orbits. Thus, an atom might be pictured as a heavy, pinpoint central core with electrons orbiting the core in orbits situated at various distances and orientations.

The feature that distinguishes one type of atom from another is the number of protons in the nucleus, called the *atomic number* of the atom. The simplest atom is hydrogen, whose nucleus consists of only one proton. The next atom in the sequence is helium, which has two protons in the nucleus. Lithium has three protons, beryllium has four, and so on, as shown in the chart of Appendix B. All together, a few more than 100 different types of atoms have been discovered. For an atom to be electrically neutral, the total number of electrons orbiting the nucleus must equal the number of protons within the nucleus.

The total number of particles in the nucleus is called the *atomic weight* of the atom. The number of neutrons present may be computed by subtracting the atomic number from the atomic weight. The number of neutrons associated with a nucleus of a given type of atom is not a unique feature of that type of atom. For example, carbon atoms typically have 6 protons and 6 neutrons in the nucleus. However, it is possible for a carbon atom to have 6 protons and 8 neutrons. The fact that this nucleus is a nucleus of carbon is determined by the fact that there are 6 protons present. Atoms having the same number of protons (atomic number) but which have different numbers of neutrons (atomic weight) are called *isotopes*.

While the atom is often compared with the solar system, the electrons simply do not behave as miniature planets governed by the laws of Kepler. Extremely small particles such as electrons do not always follow the laws of motion as set forth by Newton. Small particles must be treated by quantum mechanics, the branch of physics devoted to the behavior of small particles. In many situations, small particles act differently from what one would rationally expect of larger particles. This is the case for the electron orbits about the nucleus.

A satellite can be placed in orbit at any distance from the earth, but an electron can exist only in certain *allowed* orbits about the nucleus. This gives rise to imaginary shells surrounding the nucleus, each of which may accommodate a certain maximum number of electrons. For example, the first shell will hold no more than 2 electrons. The second shell may contain as many as 8; the third shell 18, the fourth 32, and so on. Atoms with a large atomic number have more electrons and will have more filled shells (beginning with the shell closest to the nucleus) than an atom of smaller atomic number. Thus, the number of electrons available determines the *electronic structure* of the atom, which, in turn, determines many of its physical and chemical properties. Electrons existing within completely filled shells have less effect upon the atom's properties than the outermost *valence* electrons, which exist in partially filled shells. Thus, whether an atom is a metal—like gold—or a nonmetal—such as sulfur—is determined by its electronic structure.

Elements and compounds

Substances such as hydrogen, carbon, sulfur, or gold, whose smallest integral unit is an atom are called *elements*. Because there are about 100 different types of atoms, there are, accordingly, about 100 different elements. These are listed in Appendix B.

One or more atoms of different types may join together to form a molecule. Substances whose smallest unit is a molecule are called *compounds*. Two atoms of hydrogen may combine with one atom of oxygen to form a molecule of water, H_2O. Familiar compounds include sodium chloride (salt), NaCl, sulfuric acid, H_2SO_4, and carbon dioxide, CO_2. While only about 100 elements are known, thousands of compounds are available, and the number is increasing as research continues.

Atoms of the same type frequently join to form a *diatomic* molecule, of which hydrogen, H_2, and oxygen, O_2, are examples. Ozone is a gaseous compound of oxygen; an ozone molecule is three atoms of oxygen joined together to form O_3. Sometimes the term *molecule* may be used in a broad sense to refer to the smallest unit of either an element or a compound. Consequently, a single atom may be called a molecule, as in speaking of the molecular motion due to heat in a column of mercury. The smallest unit of mercury is an *atom* of mercury. No confusion should arise from this usage.

Molecular weight; definition of a mole

The *molecular weight* of a compound is the sum of the atomic weights of all the atoms in one molecule of the compound. Because the atomic weight of carbon is 12 and of oxygen is 16 (dealing in round numbers), the molecular weight of CO is 12 plus 16, or 28; that of CO_2 is 12 plus 2 times 16, or 44. For sulfuric acid, H_2SO_4, the molecular weight is 2 plus 32 plus 4 times 16, which gives 98. For *diatomic* hydrogen, H_2, the molecular weight is 2, whereas for *monatomic* helium, He, the molecular weight is 4, the same as the atomic weight.

The molecular weight is often used indirectly in specifying quantities of compounds in terms of *gram-moles* or *kilogram-moles*. One gram-mole (g-mole) of a substance is the mass in grams numerically equal to the molecular weight. Similarly, one kilogram-mole (kg-mole) is a mass in kilograms equal to the molecular weight. The value in using a g-mole (or kg-mole) as a unit of measure lies in the fact that a g-mole of one substance (such as H_2O) will contain the same number of molecules as a g-mole of any other substance (such as CO_2). Specifically, the number of molecules in 1 g-mole of *any* compound is Avogadro's number, 6.02×10^{23} molecules/g-mole. (Avogadro's number is 6.02×10^{26} molecules/kg-mole.) We make use of this in Chapter 15 in our study of an ideal gas.

Three states of matter

Three states of matter are *solids, liquids*, and *gases*. The distinguishing feature of each state is the degree of freedom of the molecules comprising the substance. In a solid, the molecules are confined to a localized region of the solid; they are not free to roam around, but they can and do vibrate back and forth about their average position in the solid. In liquids, the molecules are less tightly bound together so that they can move around at random through the entire volume of the liquid. The molecules of a liquid exert significant forces of cohesion on each other (producing surface tension as one effect of this), but the forces are not sufficiently strong for a liquid to hold its shape without external support. In gases, the molecules are only weakly attracted to each other. To a first approximation, the molecules of a gas may be assumed to be completely free of each other. Consequently, a gas must be enclosed in a container, and any amount of gas put into a container will spread out and fill the entire volume of the container. Several aspects of solids and liquids are considered in this and the following chapter. Gases are considered in Chap. 15.

Crystalline vs. amorphous solids

In most solids, the atoms or molecules are arranged in an orderly fashion in a three-dimensional array, as opposed to being piled together in a random pattern like the grains of sand in a sandpile. Materials whose molecular structure forms a regular pattern are called *crystals*, and the pattern is called a *crystal lattice*. According to scientific definition, a crystal is any material whose atoms or molecules are arranged in an orderly, geometric pattern in three dimensions. Solids whose molecules are *not* so arranged are called *amorphous* solids. Many substances that are not commonly regarded as crystals are indeed crystalline. All metals, for example, are crystalline.

Polycrystalline solids

Almost all crystalline solids exhibit a *polycrystalline* structure consisting of a large number of small crystals, called *grains*, joined together to form the bulk of the solid, as illustrated in Fig. 11-1. Although the boundries of a grain may be irregular

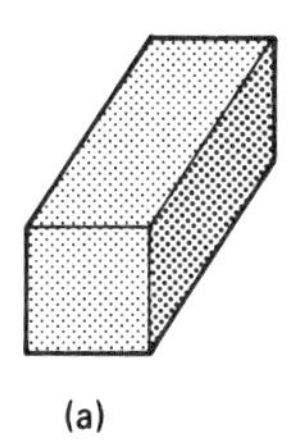

Figure 11-1 Etching a metal specimen with a suitable acid solution reveals the grain structure of the specimen. Specimen (a) is a single crystal, whereas (b) is polycrystalline.

in shape, the atoms or molecules comprising the grain are arranged in an orderly crystal lattice. The atoms form lines and planes that provide suitable reference for describing the orientation of the lattice. Within a particular grain, the lines and planes are continuous, but their direction and orientation may change at the *grain boundry* between grains.

A solid that consists of only one very large grain, so that the entire bulk of the solid is a part of one crystal structure, is called a *single crystal*. Single crystals of metals may be quite large, having dimensions on the order of an inch or more. Large single crystals are obtained by a process of slow solidification in a laboratory. On the other hand, mineralogy and gemnology are very much involved with large single crystals that occur naturally within the earth. Almost all gemstones—rubies, diamonds, or sapphires—are single crystals. The occurrence in nature of very beautiful and almost perfectly formed geometrical crystals is a source of wonder.

A major concern of metallurgy, the science of metals, is the manner in which grain size affects the physical properties of the metal. In broad terms, large grains produce metals that are soft and pliable, whereas small grains produce metals that are hard and brittle. Grain size may be controlled; a slow process of cooling and solidification from the molten state produces large grains. Rapid cooling produces small grains. Grain size may be altered by heat treating even after a metal object is formed into its final shape. Thus, two objects made of steel may have the same chemical composition and outward appearance, but one may be hard while the other is soft.

11-2 DENSITY AND SPECIFIC GRAVITY

Some materials are obviously more dense than others: Wood is more dense than cork or styrofoam, water is more dense than oil or gasoline, mercury is more dense than water, and lead is more dense than iron or steel. Intuitively, the density of a substance is the amount of material existing within a unit volume of the substance. However, the amount of material may be given in terms of either the weight or the mass of the material. Thus, there are two densities, between which we must be very careful to distinguish:

$$\begin{aligned} &(\textit{weight density}) \qquad D = \frac{\text{weight}}{\text{volume}} \qquad (\text{lb/ft}^3,\ \text{N/m}^3,\ \text{dyn/cm}^3) \\ &(\textit{mass density}) \qquad \rho = \frac{\text{mass}}{\text{volume}} \qquad (\text{slug/ft}^3,\ \text{kg/m}^3,\ \text{g/cm}^3) \end{aligned} \tag{11-1}$$

Because the weight of a particular amount of material may vary, depending upon the acceleration of gravity at a particular point, the weight density may vary accordingly. However, the mass density is a constant. There is a relationship between D and ρ, which should not be surprising since $W = mg$:

$$D = \rho g \tag{11-2}$$

Table 11-1 lists the densities of several common materials.

TABLE 11-1 DENSITY OF SELECTED MATERIALS

Material	kg/m^3	S_{gr} (g/cm^3)	lb/ft^3
Aluminum	2,650	2.65	164
Brass	8,600	8.60	535
Copper	8,930	8.93	555
Gold	19,320	19.32	1,200
Iron, cast	7,200	7.20	450
Lead	11,370	11.37	710
Silver	10,500	10.5	655
Tin	7,290	7.29	455
Uranium	18,680	18.68	1,165
Zinc	7,100	7.10	443
Alcohol	789	0.789	49.3
Gasoline	790	0.790	49.4
Glycerin	1,260	1.260	78.7
Kerosene	820	0.820	51.2
Mercury	13,600	13.60	840
Water			
fresh	1,000	1.00	62.43
sea	1,030	1.03	64.4
Ice (0°C)	917	0.917	57.2
Cork	240	0.24	15
Glass, common	2,600	2.60	162
Wood,			
balsa	120	0.12	7.5
oak	800	0.8	50
pine	500	0.5	31
lignum vitae	1,250	1.25	78

The *specific gravity* of a substance is defined as the ratio of the density of the substance to the density of water:

$$(\textit{specific gravity}) \qquad S_{gr} = \frac{\text{density of substance}}{\text{density of water}} \qquad (S_{gr} \text{ has no units}) \tag{11-3}$$

It makes no difference whether the mass density or the weight density is used to compute the specific gravity of a substance. The density of water is very nearly 1.00 g/cm^3 at 4°C, the temperature at which water is most dense. In another system of units, the density of water is 62.4 lb/ft^3.

If the specific gravity of a substance is known, its density may be computed from a rearrangement of Eq. (11-3):

$$\text{Density of substance} = S_{gr} \times \text{density of water} \tag{11-4}$$

Similarly, if the weight (or mass) of a certain volume of water is known, the weight (or mass) of an equal volume of a particular substance (such as mercury) is the product of the specific gravity of the substance and the weight of the water.

EXAMPLE 11-1 Calculate the weight of a block of aluminum in the shape of a cube 1 ft on a side. The specific gravity of aluminum is 2.70.

Solution. The volume of the block is obviously 1 ft^3, and we know that 1 ft^3 of water weighs 62.4 lb. We can find the weight of an equal volume of aluminum by multiplying the specific gravity of aluminum (2.70) by the weight of the volume of water, 62.4 lb. Consequently, the weight of the block of aluminum is 2.70×62.4, which is 168.5 lb.

Range of densities

The densities of matter that occur naturally in the differing environments of the universe span a wide range, extending from the rarified regions of interstellar space on the one hand to the extreme densities of very old stars on the other. In interstellar space, hydrogen is the most abundant material and occurs at the rate of about 0.7 atoms/cm^3. In more dense regions, hydrogen may occur at the rate of 1000 to 10,000 atoms/cm^3, as in the emission nebula in the sword of the constellation Orion. This concentration of matter is far less than that within a good laboratory high-vacuum system, where (at 10^{-8} mm Hg) there may be about 10^7 molecules/cm^3. In the earth's atmosphere at an altitude of 100 mi, there are about 10^{10} molecules/cm^3, and this figure increases rapidly with decreasing altitude until at sea level there are about 10^{19} molecules/cm^3. In 1 cm^3 of water, there are about 10^{22} water molecules.

The density of water is 1 g/cm^3. The average density of the earth is 5.5 g/cm^3, and the density of the central portion of the sun is thought to be between 50 and 400 g/cm^3. One type of very old stars (white dwarfs) that are nearing the end of their evolutionary cycle exhibit densities ranging to over 1,000,000 g/cm^3. Another form of an old star is a *neutron star* in which matter exists as a crystalline form of neutrons. The densities of neutron stars (also known as *pulsars* because they emit rapidly pulsating light) may be on the order of 10^{15} g/cm^3, a billion times that of a white dwarf. Some old stars are thought to collapse through the neutron star stage to form *black holes*, which—according to present theories—may have densities approaching infinity. Elementary texts on astronomy provide fascinating reading in this area.

11-3 PRESSURE; PRESSURE IN LIQUIDS

The commonly held concept of pressure is invariably linked to the force, or exertion, that tends to make water squirt. More scientifically, pressure deals with forces that are distributed over an area. We are familiar with atmospheric pressure which exerts a force of about 14.7 lb against every square inch of surface area of objects near the earth's surface. Mathematically, pressure is force per unit area:

$$P = \frac{F}{A} \qquad (11\text{-}5)$$

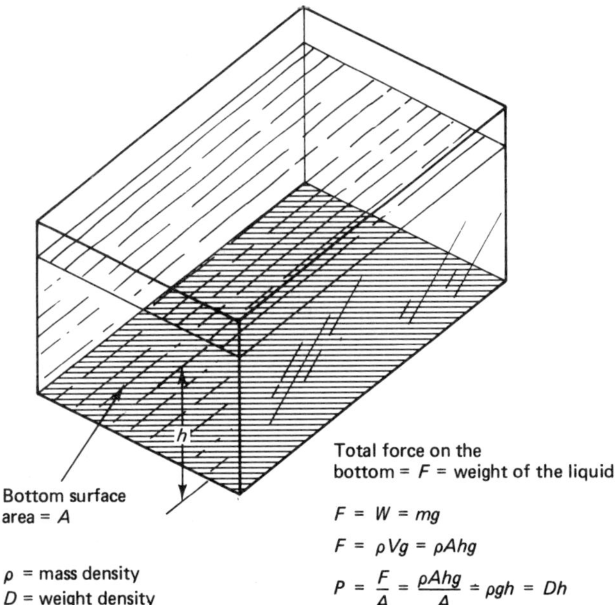

Figure 11-2 Hydrostatic pressure at the bottom of an aquarium.

The units of pressure are lb/ft^2, lb/in^2, N/m^2, dyn/cm^2 and so forth. In a following section, we study other units for measuring pressure.

Pressure in stationary liquids

Let us now consider the pressure at the bottom of an aquarium of bottom surface area A filled with water to a depth h, as shown in Fig. 11-2. The total downward force on the bottom of the tank obviously equals the weight of the water in the tank. In terms of the weight density D and the volume of water V, the weight of the water in the tank is DV. Because the volume can be expressed as hA, the total downward force on the bottom of the tank is DhA. Pressure is the force per unit area, so division by A gives the pressure at the bottom of the tank:

$$P = Dh \tag{11-6}$$

In terms of the mass density ρ, this can be written as

$$P = \rho g h \tag{11-7}$$

because $D = \rho g$. Thus, we see that the pressure depends upon the density of the liquid and the depth of the liquid, being directly proportional to both quantities.

This expression was obtained by considering conditions at the bottom of the tank, but it is also applicable to intermediate points between the bottom and the top surfaces.

Thus far we have ignored any pressure, such as atmospheric pressure, that may be applied to the top surface of the liquid. If such a pressure is present, the external pressure P_{ext} is added to the pressure $\rho g h$ due to the liquid alone:

$$P_{total} = P_{ext} + \rho g h \tag{11-8}$$

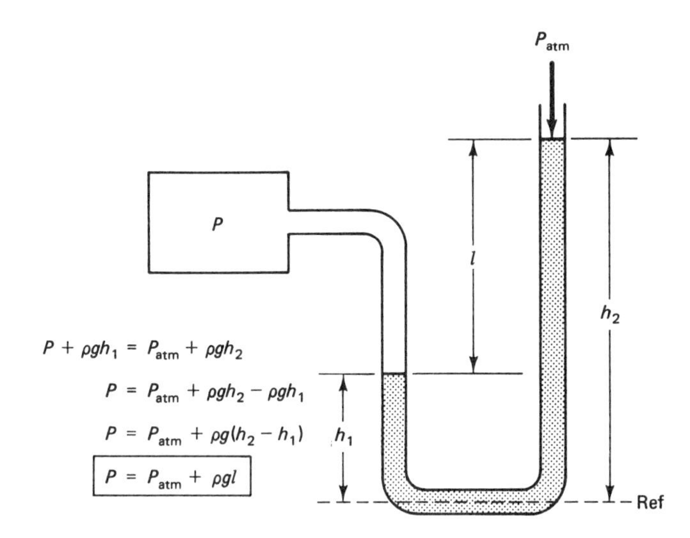

Figure 11-3 U-tube manometer used to measure the pressure in the tank.

For a liquid whose top surface is exposed to the atmosphere, the pressure within the liquid at a distance h below the top surface is

$$P_{total} = P_{atm} + \rho g h \tag{11-9}$$

where P_{atm} is atmospheric pressure.

U-tube manometer

A U-tube manometer may be used to determine the pressure P in a tank, as shown in Fig. 11-3. One side of the manometer is connected to the tank; the other is open to the atmosphere. The liquid may be a special oil, water, or mercury, depending upon the application. In practice, a means must be provided for determining the distance l, the difference in height of the left and right columns of the manometer.

Because the liquid is in equilibrium, the pressure at all points on the horizontal reference line must be the same. Therefore, the pressure at the bottom of the left-hand column must equal the pressure at the bottom of the right-hand column. Thus, in light of the preceding discussion, we write expressions for the pressure at the bottom of the two columns of liquid and set the expressions equal to obtain

$$P = P_{atm} + \rho g l \tag{11-10}$$

as shown in Fig. 11-3. The atmospheric pressure P_{atm} may be obtained from a barometer.

Liquid column heights as pressure units

Figures 11-2 and 11-3 show that pressure differences can be unambiguously expressed in terms of the difference in heights l of two columns of liquid. In Fig.

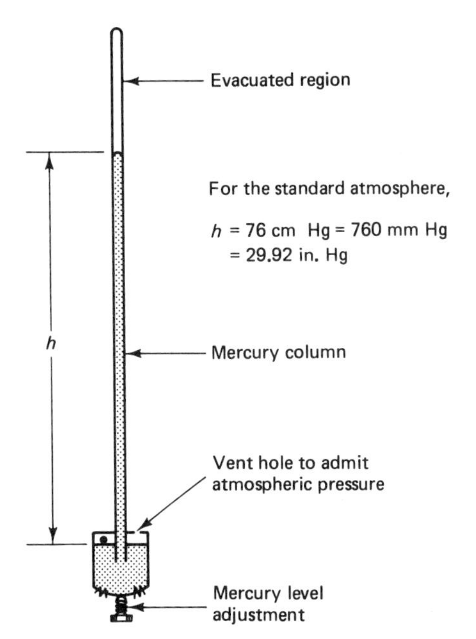

Figure 11-4 Barometer.

11-3, the gauge pressure in the tank may be given in inches of water, inches of mercury (in. Hg), or perhaps centimeters of mercury (cm Hg). Atmospheric pressure is commonly given in terms of the distance a column of mercury will rise in an evacuated tube (a barometer) as a result of the air pressing down on the mercury in the cup at the bottom of the tube. A barometer is illustrated in Fig. 11-4.

The standard atmospheric pressure is 76 cm Hg, which corresponds to 29.92 in. Hg, or about 14.7 lb/in^2. A pressure of 1 mm of mercury (mm Hg) is called a torr in honor of Evangelista Torricelli, who invented the barometer in 1643.

A unit such as cm Hg is not a valid unit of pressure in terms of pressure as a force per unit area even though such units are quite useful. The SI unit of pressure is the pascal (Pa), a pressure of 1 N/m^2, a rather small pressure. In meteorology, the bar (or millibar, mbar) is a commonly used pressure unit equivalent to 10^5 Pa or 10^6 dyn/cm^2. A kilopascal (kPa) is also frequently encountered and is equal to 10 mbar. Table 11-2 gives several equivalents that may be used to convert from one pressure unit to another.

TABLE 11-2 PRESSURE EQUIVALENTS

1 N/m^2	= 1 Pa = 1.450×10^{-4} lb/in^2 = 10 dyn/cm^2
1 Pa	= 10^{-5} bar = 9.869×10^{-6} atm = 7.501×10^{-3} mm Hg
1 bar	= 10^5 Pa = 14.50 lb/in^2 = 0.9869 atm = 750.0 mm Hg
1 mbar	= 10^{-3} bar = 0.750 mm Hg
1 lb/in^2	= 6895 N/m^2 = 51.71 mm Hg = 27.68 in. H_2O
	= 6.895 kPa = 68.96 mbar
1 atm	= 1.01325×10^5 N/m^2 = 101.3 kPa = 76 cm Hg = 760 mm Hg
	= 760 torr = 1.013 bar = 14.7 lb/in^2 34 ft H_2O
1 in. H_2O	= 0.03613 lb/in^2 = 1.868 mm Hg = 249.1 Pa = 2.491 mbar
1 torr	= 1 mm Hg

We are usually oblivious to atmospheric pressure because of its constancy and uniformity. Therefore, the pressure P just determined is called the *absolute*

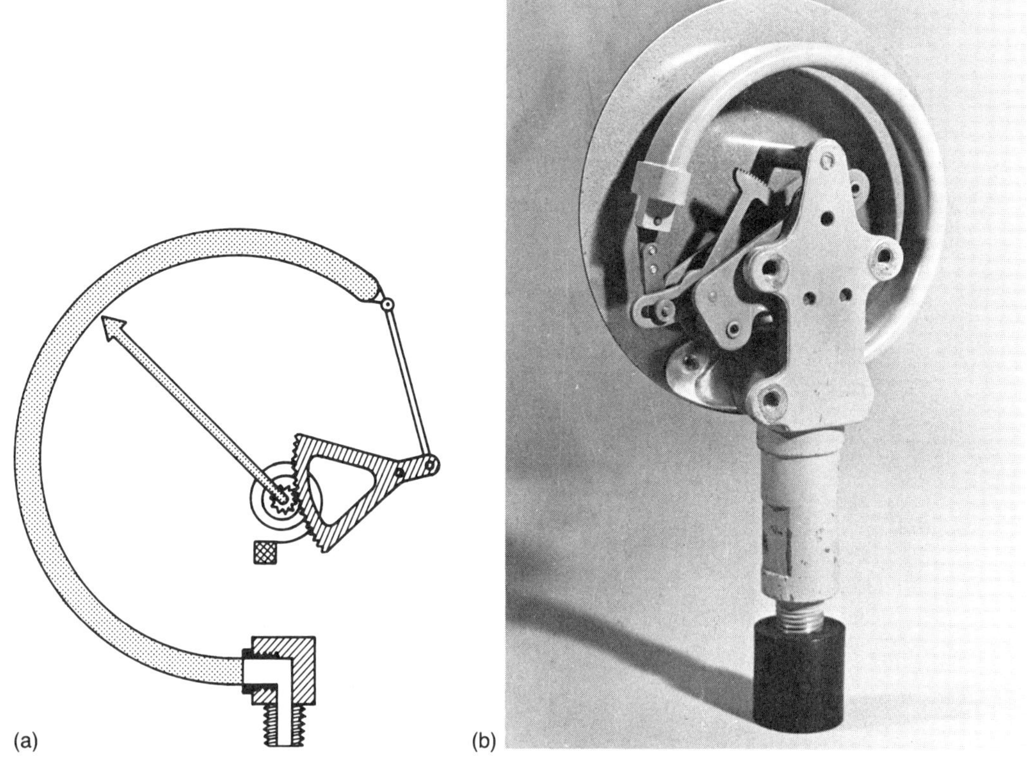

Figure 11-5 (a) When a fluid under pressure is admitted to the interior of the C-shaped Bourdon tube, the tube tends to straighten slightly. The small movement at the tip of the tube is conveyed through a sector-gear linkage to a rotating pointer.
(b) Internal view of a pressure gauge that uses a bourdon tube.

pressure to distinguish it from the *gauge pressure*. The absolute pressure is the force per unit area exerted on the interior walls of the tank. On the other hand, gauge pressure is the pressure in the tank above and beyond the ever-present, essentially constant atmospheric pressure. Gauge pressure and absolute pressure are related as follows:

$$\begin{aligned} P_{\text{gauge}} &= P_{\text{abs}} - P_{\text{atm}} \\ P_{\text{abs}} &= P_{\text{gauge}} + P_{\text{atm}} \end{aligned} \tag{11-11}$$

Almost all pressure-indicating devices used in industrial applications are calibrated to read gauge pressure. A common pressure gauge utilizing a *Bourdon tube* is illustrated in Fig. 11-5.

11-4 ARCHIMEDES' PRINCIPLE

When an object is either partially or completely immersed in a liquid such as water, oil, or mercury, an upward force is exerted on the object that tends to make the object float. The object will float if its average density is less than that of the

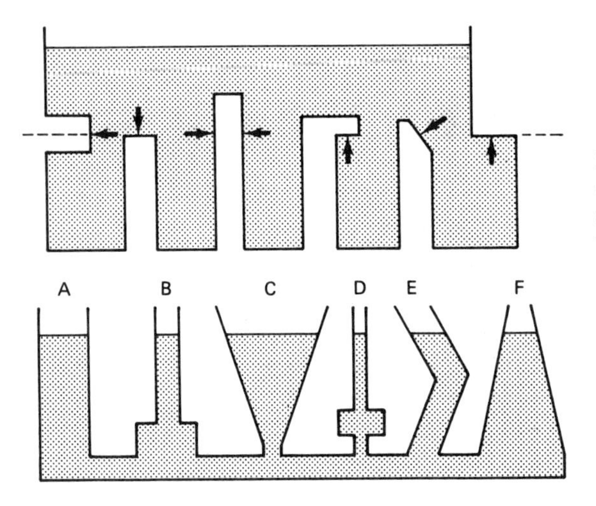

Figure 11-6 Pressure in a fluid is the same in all directions at the same level in the fluid, as illustrated by this container of rather unusual cross section.

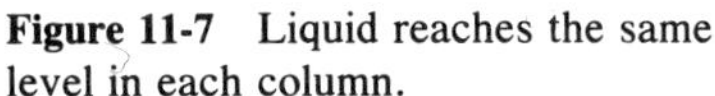

Figure 11-7 Liquid reaches the same level in each column.

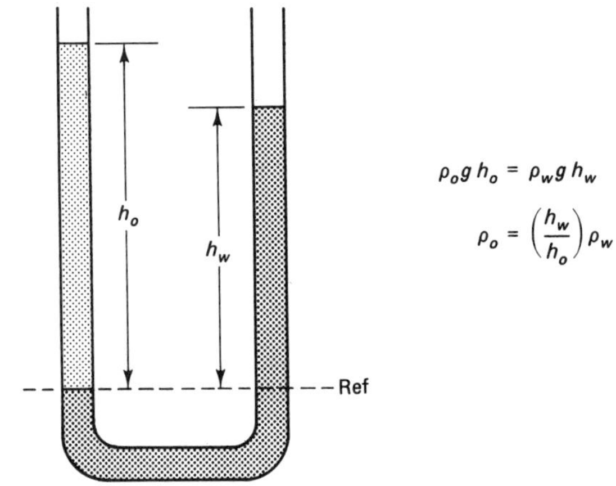

Figure 11-8 The density of a liquid such as oil or gasoline, which does not mix with water, can be determined with a U-tube manometer. The pressure in both columns is the same at the reference level.

liquid; it will sink if its average density is greater than that of the liquid. Even though a stone sinks to the bottom when placed in water, the upward force mentioned above is still exerted upon the stone and tends to make it weigh less in the water than in air. We next describe *Archimedes' principle*, which states that the upward force (the buoyant force) exerted on an object placed in a liquid equals the weight of the liquid *displaced* by the object.

Suppose a thin-walled clear plastic cube is placed in a fluid so that the top and bottom surfaces are at distances h_t and h_b below the top surface of the fluid, as shown in Fig. 11-9. The top and bottom surfaces have equal areas, A. The fluid will exert a downward force on the top surface, and it will exert an upward force on the bottom surface. Of these, the upward force will be greater so that the cube tends to float. The upward force is greater because the pressure at the bottom surface is greater than the pressure at the top surface. This difference in pressure produces the buoyant force F_B. The mathematical details are given in Fig. 11-9, where we find that

$$\text{Buoyant force} = \text{weight of displaced fluid} \qquad (11\text{-}12)$$
$$F_B = \rho g V = DV$$

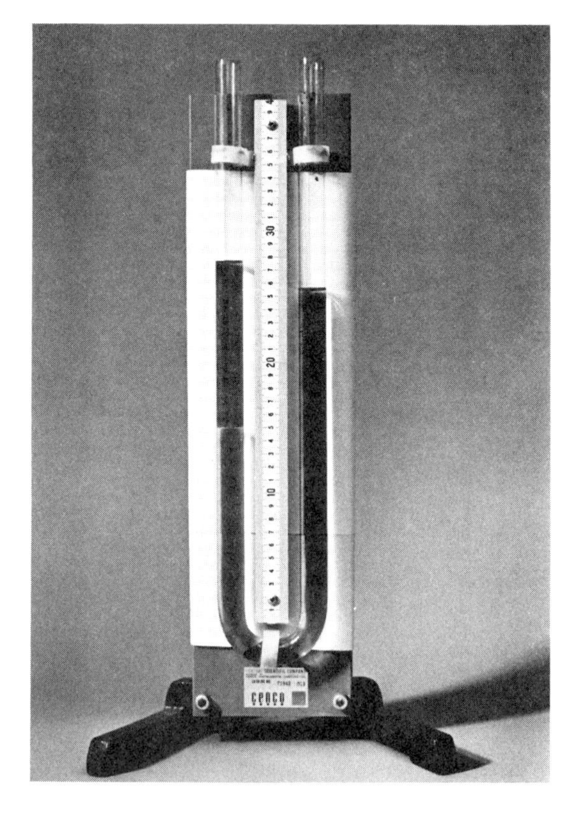

A U-tube manometer. In the left-hand tube, a 12-cm column of oil rises above the inky water in the other tube.

where ρ is the mass density of the fluid and D is its weight density. Also, V is the volume of the submerged object.

In the foregoing, we assumed the cube was empty and very light so that we

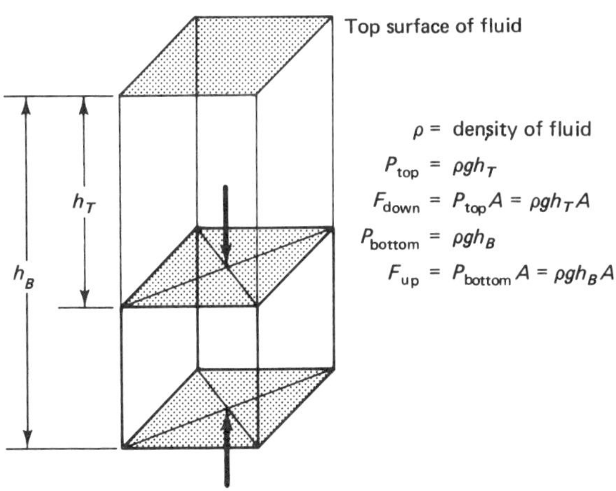

Figure 11-9 Archimedes' principle states that the buoyant force exerted on an object immersed in a fluid equals the weight of the displaced fluid.

could ignore its weight. However, in practice, the weight of the cube, W, must be subtracted from the buoyant force, F_B, to obtain the net vertical force on the cube:

$$\text{Vertical force} = \text{buoyant force} - \text{weight of object}$$
$$F_{\text{vert}} = F_B - W \tag{11-13}$$

By writing the weight of the cubical specimen as $\rho_S gV$ and the buoyant force as $\rho_F gV$, we obtain

$$F_{\text{vert}} = gV(\rho_F - \rho_S) \tag{11-14}$$

or, in terms of the weight density,

$$F_{\text{vert}} = V(D_F - D_S) \tag{11-15}$$

where ρ_F is the mass density of the fluid, ρ_S is the average density of the material of which the cube is made, and D_F and D_S are the weight density of the fluid and the cube, respectively. These results apply equally well to objects of irregular shape; we consider a cube only because of its simplicity.

EXAMPLE 11-2 Compute the vertical force on a brass sphere immersed in gasoline if the volume of the sphere is 24 cm^3.

Solution. We apply Eq. (11-14) and use density data from Table 11-1 on page 218:

$$\begin{aligned} F_{\text{vert}} &= gV(\rho_F - \rho_S) \\ &= (980 \text{ cm/s}^2)(24 \text{ cm}^3)(0.79 - 8.60 \text{ g/cm}^3) \\ &= -183{,}691 \text{ dyn} \end{aligned}$$

The negative sign indicates that the force is downward; the sphere tends to sink. We may convert from dynes to gram equivalent weight by dividing by $g = 980$. The result is 187.7 g, which may be compared with the calculated mass of the sphere of 206.4 g.

Volume of an irregular object

The volumes of objects of regular geometric shapes (for example, rectangular solids, cubes, spheres, or cones) can be determined from a few fairly simple measurements and the application of a formula. The volumes of irregular objects, such as stones, mineral specimens, and crowns, are not so easily determined. However, the principles of bouyancy can be used to calculate the volume of an irregular object, requiring only that the object be weighed in air and in a liquid of known density. This stems from the fact that the weight of an object suspended in a liquid, W_{liq}, equals the weight of the object in air, W_{air}, minus the buoyant force, F_B. Hence, in mathematical terms,

$$W_{\text{liq}} = W_{\text{air}} - F_B$$
$$W_{\text{liq}} = W_{\text{air}} - \rho gV \tag{11-16}$$

Solving for the volume of the object yields

$$V = \frac{W_{\text{air}} - W_{\text{liq}}}{\rho_F g} \tag{11-17}$$

Here, ρ_F is the mass density of the fluid. Because the weight density $D_F = \rho_F g$, we may also write

$$V = \frac{W_{air} - W_{liq}}{D_F} \qquad (11\text{-}18)$$

EXAMPLE 11-3 An irregular chunk of feldspar, a pale white mineral, weighs 12.5 lb in air and 7.6 lb when suspended in water. Compute: (a) the volume of the chunk; (b) the weight density; (c) the specific gravity.

Solution. (a) By Eq. (11-18)

$$V = \frac{W_{air} - W_{liq}}{D_F}$$

$$= \frac{12.5 - 7.6 \text{ lb}}{62.4 \text{ lb/ft}^3}$$

$$= 0.079 \text{ ft}^3$$

(b)

$$D = \frac{\text{weight of specimen}}{\text{volume of specimen}}$$

$$= \frac{12.5 \text{ lb}}{0.079 \text{ ft}^3}$$

$$= 158 \text{ lb/ft}^3$$

(c)

$$S_{gr} = \frac{\text{density of specimen}}{\text{density of water}}$$

$$= \frac{158 \text{ lb/ft}^3}{62.4 \text{ lb/ft}^3}$$

$$= 2.53$$

Density determination

The average density ρ_S of an irregular specimen can be determined by weighing the specimen in air and then in a fluid (such as water) of known density ρ_F. The weight of the specimen in air is given by

$$W_{air} = \rho_S g V \qquad V = \frac{W_{air}}{\rho_S g} \qquad (11\text{-}19)$$

If this expression for the volume V of the specimen is substituted into Eq. (11-18), we obtain

$$\frac{W_{air}}{\rho_S g} = \frac{W_{air} - W_{liq}}{\rho_F g}$$

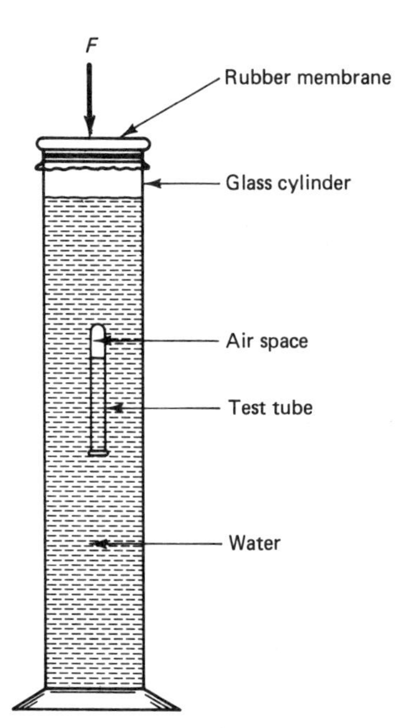

Figure 11-10 Demonstration of Archimedes' principle. A force exerted on the membrane increases the pressure within the cylinder, causing additional water to flow into the test tube. Then, because the displaced water is less, the buoyant force is less, and the test tube descends to the bottom.

Rearranging gives two useful possibilities:

$$\rho_S = \left(\frac{W_{\text{air}}}{W_{\text{air}} - W_{\text{liq}}}\right)\rho_F \quad \text{or} \quad \rho_F = \left(\frac{W_{\text{air}} - W_{\text{liq}}}{W_{\text{air}}}\right)\rho_S \tag{11-20}$$

Thus, by weighing a specimen in air and then in a fluid, we can use a fluid of known density to determine the density of an unknown specimen, or we can use a specimen of known density to determine the density of an unknown fluid. If the fluid is water, a very simple expression results for determining the specific gravity of a solid specimen:

$$S_{gr} = \frac{W_{\text{air}}}{W_{\text{air}} - W_{\text{water}}} \tag{11-21}$$

Archimedes was a Greek philosopher (287–212 B.C.) who reportedly ran, sparsely clad, down the streets of Syracuse, Sicily, shouting ''Eureka'' after deducing the principle of buoyancy. This principle enabled him to verify that the crown of King Hiero had been made of impure gold. Any encyclopedia will provide an account of this and the many other achievements of Archimedes.

The law of flotation

Objects that float, such as pieces of wood, cork, empty bottles, and ships, do not rest precisely upon the top surface of the water (or other fluid). An object will sink into the water until the weight of the displaced water equals the weight of the floating object. This principle allows us to compute the fraction of a floating body that will be submerged. The waterline of a ship can be computed prior to launching. The procedure is illustrated in Fig. 11-11 for a rectangular object of cross-sectional area A. From the figure, we see that any one of four quantities (x, L, D, or D_F) can be computed if the other three are known. This is the principle of the hydrometer, described in the following section.

Hydrometer

Consider a rod of length L and uniform cross section A whose density ρ_R is known. If the rod is forced to float upright in a liquid of unknown density ρ_F, the density of the liquid is given by

$$\rho_F = \frac{L}{x}\rho_R \tag{11-22}$$

This is the operating principle of the hydrometer, a simple floatation device for determining the density (specific gravity) of liquids. The density of the fluid is inversely proportional to the length of rod submerged. A hydrometer is shown in Fig. 11-12.

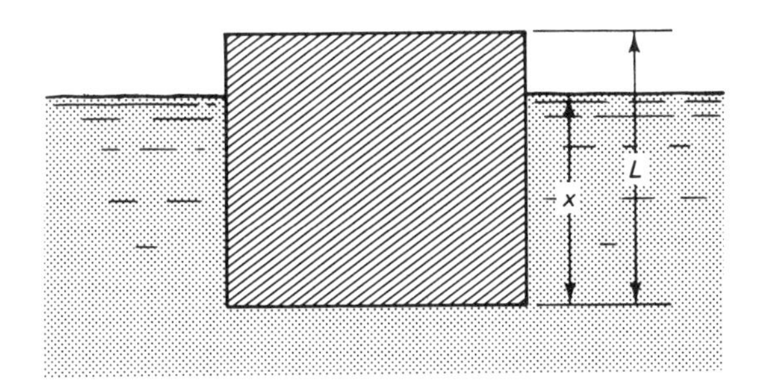

Weight density of body: D

Cross sectional area of body: A

Volume of body: AL

Weight of body: DAL

Weight density of fluid: D_F

Fluid displaced: Ax

Weight of displaced fluid: $D_F Ax$

Law of floatation: Weight of body = weight of displaced fluid

$$DAL = D_F Ax$$

Hence,

$$x = \left(\frac{D}{D_F}\right) L \quad \text{or} \quad D = \left(\frac{x}{L}\right) D_F$$

If the fluid is water, the specific gravity S_{gr} of the body is given by

$$S_{gr} = \frac{x}{L}$$

Or, x can be calculated by using

$$x = S_{gr} L.$$

Figure 11-11 Finding the waterline of a partially submerged object.

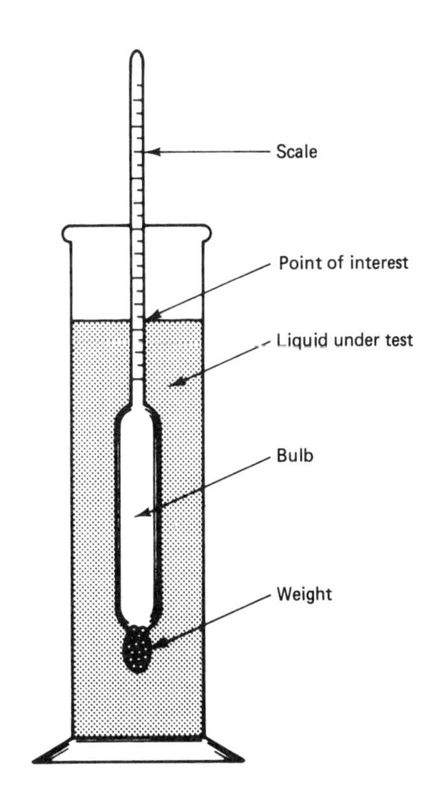

Figure 11-12 Hydrometer being used to determine the density (or specific gravity) of a liquid.

11-5 PASCAL'S PRINCIPLE—HYDRAULICS

When a fluid such as the water in an aquarium is at rest, the difference in pressure between two points within the fluid depends only upon the vertical distance between the points and upon the density of the fluid in question. This holds true even if the fluid is subjected to an external pressure applied to its free surface. When atmospheric pressure increases at the top surface of the fluid, the pressure at all points within the fluid increases by the same amount. This is Pascal's principle: A change in pressure imparted to one portion of a confined fluid is transmitted equally to all parts of the fluid.

Pascal's principle is commonly applied to hydraulic systems in which the system pressure is very great in comparison to the pressure differences within the system caused by differences in level. If gravitational effects may be ignored, we may simplify Pascal's principle and say that the fluid pressure at all points in a system (as in Fig. 11-13) is the same. Thus, the pressure at the small piston is the same as the pressure at the larger piston. However, because the surface areas of the pistons are unequal, different forces are applied to the pistons. Hence, a small force applied to the small piston creates an increased pressure, which—because it is applied to the larger area—produces a much greater force on the larger piston. This is the operating principle of the hydraulic jack; the IMA is given in Fig. 11-13.

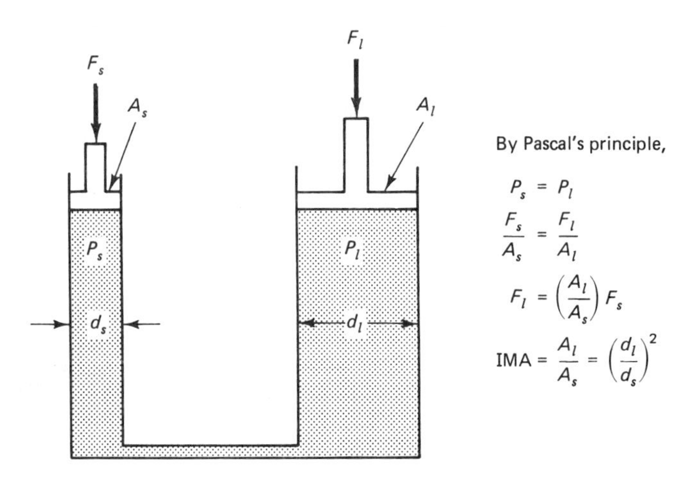

$$P_s = P_l$$

$$\frac{F_s}{A_s} = \frac{F_l}{A_l}$$

$$F_l = \left(\frac{A_l}{A_s}\right) F_s$$

$$\text{IMA} = \frac{A_l}{A_s} = \left(\frac{d_l}{d_s}\right)^2$$

Figure 11-13 Hydraulic jack.

11-6 SURFACE TENSION

Atoms or molecules situated closely together, as in a liquid, exert attractive forces on each other. Ultimately, these forces stem from the attraction of the negative electrons of one atom to the positive nuclei of nearby atoms. Forces between molecules of the same type are called *cohesive* forces, and forces between atoms or molecules of a different type are *adhesive* forces. Both types are attractive. In a solid, the forces are strong enough to bond the molecules into a rigid structure. In

A grading machine with a hydraulic cylinder prominently visible.

a gas, the molecules are so far apart that few observable effects result from the internal forces. The effects are most easily observed in liquids and are responsible for the familiar phenomenon of surface tension.

Within the interior of a liquid, the forces on a given molecule are uniformly distributed in all directions, so that no net force results. At the surface, however, an unbalance, which pulls the molecules toward the interior of the liquid, occurs. The net effect is that a liquid tends to "pull itself together" into a shape that minimizes the surface area. Of course, gravitational forces also come into play and most of the time tend to overshadow the relatively weaker forces of surface tension. Even so, the effects of surface tension are readily observed.

Two simple soap film experiments shown in Fig. 11-14 illustrate many aspects of surface tension. Note the film formed on a wire support with a loop of string included in the film. The loop of string hangs limp until the soap film inside the loop is ruptured. The loop is then drawn into a circular shape, indicating that the film applies forces uniformly to all parts of the loop. We assume the forces inside and outside the loop were initially balanced, allowing the loop to assume an arbitrary shape. Rupturing the enclosed film produced the unbalance which caused the circular loop to be formed. This demonstrates that tension forces are present in a surface film and that the forces ordinarily are balanced.

In Fig. 11-14(c), a film is caught in a wire frame with one freely movable slider of length L, as shown. A force F must be maintained on the movable slider or else the surface tension of the film will pull the slider upward, allowing the film to collapse. A novel feature of surface tension, however, is that the force F required to maintain the film does not increase if the film is "stretched" an additional amount to position D. The same force F will maintain the film at positions A, B, C, or D. This behavior is quite different from that of a thin rubber membrane which requires an ever-increasing force to produce additional stretching. As the surface area of the films is increased, more molecules move from the interior of the film (from within the bulk liquid between the *two* surface films) to take up positions as part of the surface.

The strength of the forces of surface tension for a particular liquid is given by a coefficient of surface tension, T. Often, T is referred to simply as the *surface tension*. It is defined as the force per unit length required to hold a surface in equilibrium, as in Fig. 11-14(c). If F is the total force exerted on the film (including

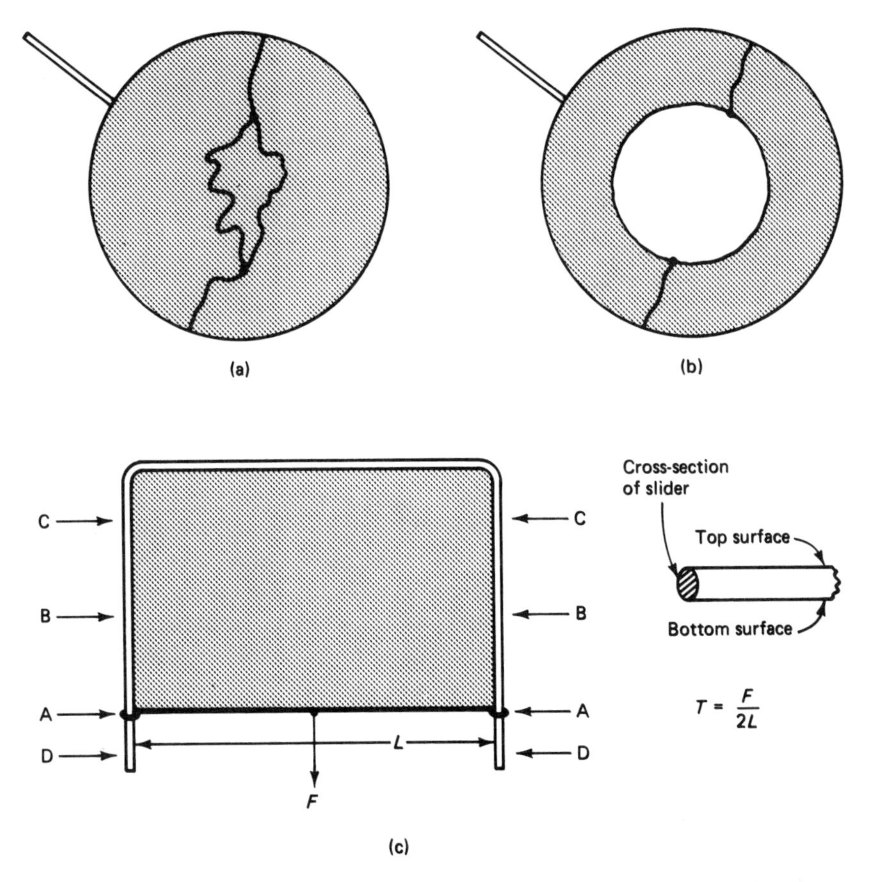

Figure 11-14 (a) String embedded within a soap film. (b) Film within the loop is ruptured; surface tension pulls the loop into a circle. (c) A means of measuring surface tension.

the weight of the slider if the frame is held vertically), the coefficient of surface tension, T, is given by

$$T = \frac{F}{2L} \tag{11-23}$$

The factor 2 appears in the denominator because the one film has two surfaces in contact with the slider. Thus, the total length of the edge of the surface is $2L$ instead of just L. Even though the film is very thin, it is still extremely thick in comparison to the dimensions of a single molecule. On the other hand, the surface layer producing the surface tension may be only a few molecules thick. The units of surface tension are force per unit area, typically, dyn/cm^2. Values of surface tension for common liquids are given in Table 11-3.

From the work-energy point of view, the surface tension is the work per unit area that must be done to increase the area of the surface. If the force F in Fig. 11-14(c) pulls the slider down a distance x to position D, the work done is Fx, and the increase in surface area is $2Lx$. Because the force F is constant (and equal to $2LT$),

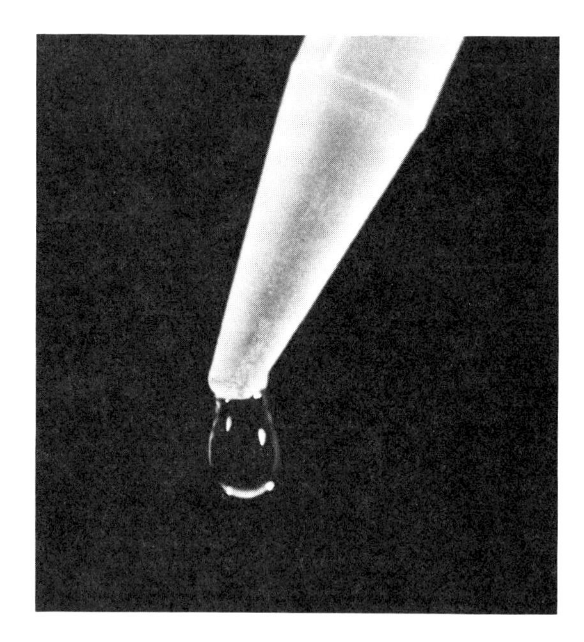

Surface tension tends to pull a drop of water into the shape of a sphere.

the ratio of the work done to the increase in surface area is

$$\frac{\text{Work done}}{\text{Area increase}} = \frac{Fx}{2Lx} = \frac{2LTx}{2Lx} = T \tag{11-24}$$

as stated above. Also, from the relationship between work and energy, the surface tension is the *surface energy per unit area* that the surface contains. Consequently, the surface tension is often given in units of ergs/cm^2.

Because surface energy is a type of potential energy, the fact that a liquid tends to assume a shape that minimizes the surface area is consistent with the general principle that any physical system tends toward a configuration that minimizes the potential energy.

Now that surface tension has been mathematically defined, consider the experiment depicted in Fig. 11-15 in which a solid disk and a thin ring are being lifted from a liquid. The surface tension of the liquid exerts a downward force on the object (in addition to the weight of the object) that must be overcome before the object is lifted clear. The important difference between the ring and the disk is

TABLE 11-3 SURFACE TENSION OF SELECTED LIQUIDS (IN CONTACT WITH AIR AT 20°C)

Liquid	T, dyn/cm[a]
Acetone	23.7
Benzene	28.9
Ethyl alcohol	22.3
Glycerin	63.4
Mercury	54.5
Water	72.8
Soap solution	25.0

[a]Multiply by 10^{-3} to convert to N/m.

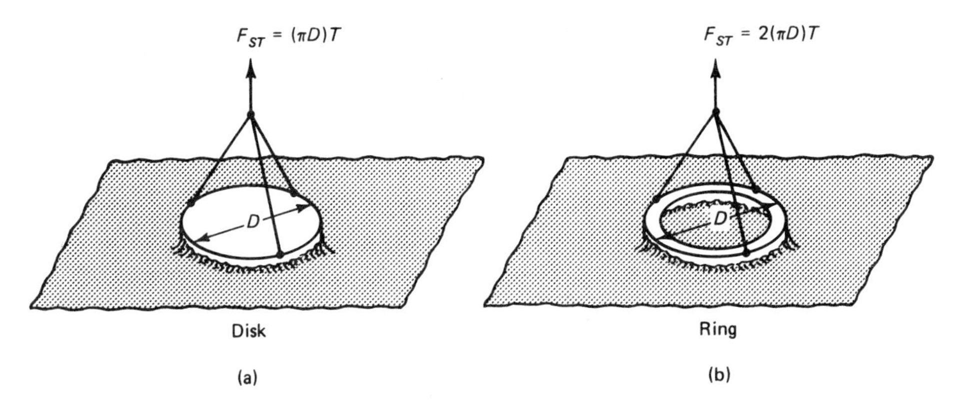

Figure 11-15 Practical method of determining the surface tension of a liquid, using (a) a disk and (b) a ring. The total upward force required to lift the object from the surface is the sum of the force of the surface tension and the weight of the object.

that there are *two* surfaces acting on the ring (one inside, one outside), whereas only *one* acts upon the disk. By using a sensitive scale to measure the force, either object may be used to experimentally determine the surface tension of the liquid. On the other hand, if the surface tension is known, the force required to lift the object can be computed.

Contact angle, wetability, and capillarity

When a liquid comes in contact with a solid, the liquid may or may not *wet* the solid, as shown in Fig. 11-16, where a drop of water is placed on a block of paraffin and on a piece of glass. Note the angle of contact between the liquid and the solid. If the *cohesive* forces within the liquid are greater than the *adhesive* forces between the liquid and the solid, the angle of contact will be more than 90°, and the liquid is said *not* to wet the solid. But if the adhesive forces are stronger, the angle of contact will be less than 90°, and the liquid *wets* the solid. A liquid poured into a test tube will have a top surface (called the *meniscus*) that is curved, and the curvature will be concave upward if the liquid wets the walls of the tube. If, as

TABLE 11-4 CONTACT ANGLES

Water–glass	0°
–paraffin	107°
–silver	90°
Mercury–glass	140°
Alcohol–glass	0°
Kerosene–glass	26°

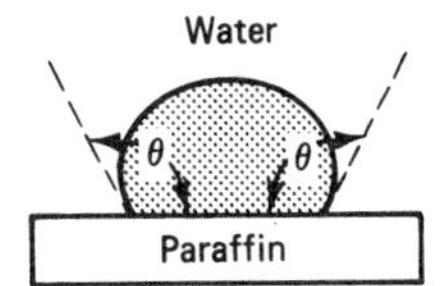

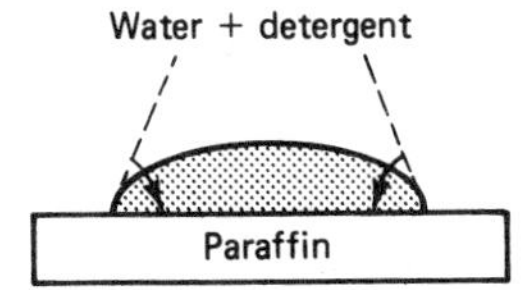

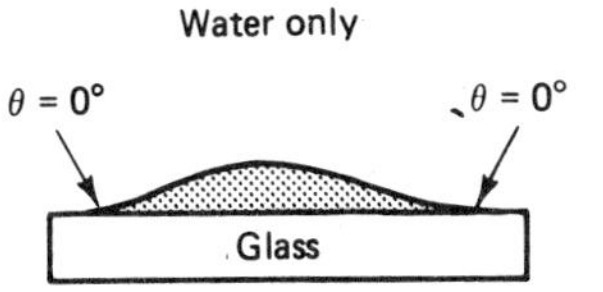

Figure 11-16 Surface tension determines the contact angle θ.

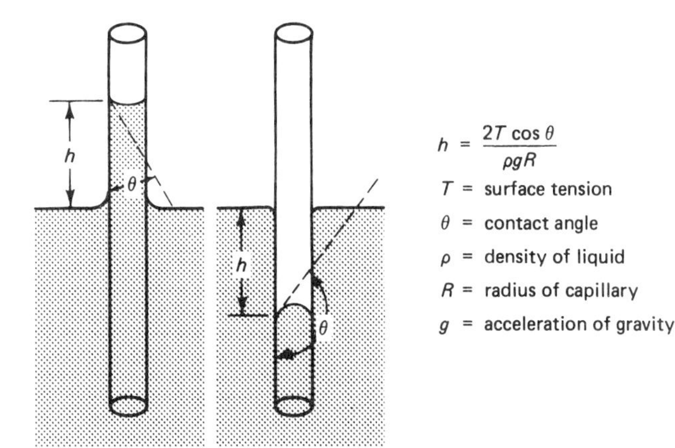

Figure 11-17 A liquid will either rise or be depressed in a capillary tube, depending upon whether the liquid wets the tube.

with the case of mercury in glass, the liquid does not wet the walls of the tube, the meniscus will be concave downward. Detergents are *wetting agents*, which reduce the angle of contact between a liquid and a solid.

The relative strengths of the adhesive and cohesive forces (as indicated by the contact angle) is important to *capillarity*, another effect of surface tension. When an open glass tube with a very small inside diameter (bore), called a *capillary*, is inserted into a liquid, the liquid will be pulled up into the capillary if the liquid wets the glass. Or, if the liquid does not wet the glass, the liquid level in the capillary will be depressed. These effects are illustrated in Fig. 11-17, together with a formula for computing the resulting rise or depression. Note that the rise or depression is given by the same formula and that capillary effects are greater in tubes of smaller diameter.

11-7 ELASTICITY; HOOKE'S LAW

Because the molecular bonds holding a solid together are not absolutely rigid, all solids may be deformed to a certain degree by an external applied force, and the solid will return to its original shape when the force is removed. The term *elasticity* refers to the property of a material to be deformed and then return to its original shape. If the solid is deformed beyond the *elastic limit* so that its original shape is not regained after the deforming force is removed, the solid is said to have been *plastically* deformed. Thus, a distinction is made between *elastic deformation* and *plastic deformation*. In the following, we consider the fundamentals of the elastic properties of solids.

Hooke's law

The English physicist Robert Hooke (1635–1703) discovered a relationship between the extension x of a spring or other elastic body and the force F required to maintain the extension, as illustrated in Fig. 11-18. As long as the elastic limit is not exceeded, the required force is directly proportional to the amount the spring is stretched. Because the direction of the force exerted by the spring on the stretching device is always opposite to the extension, a negative sign is customarily incorporated into the mathematical statement of Hooke's law:

$$F = -kx \tag{11-25}$$

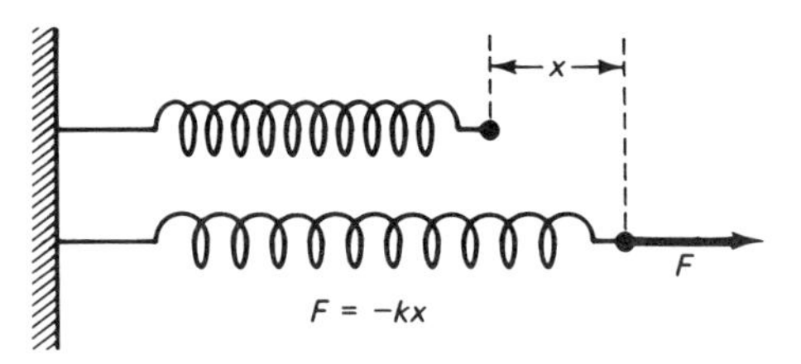

Figure 11-18 The force required to maintain the extension x of a spring is given by Hooke's law.

For springs, the proportionality constant k is called the *spring constant* and has the units of force per unit of distance. The spring constant is simply the additional force required to increase the extension by one distance unit. It depends upon the material of which the spring is made, the length and diameter of the spring, the diameter of the wire used to wind the spring, and the number of turns per unit length of the spring. Springs are frequently wound of wire that does not have a circular cross section, and the diameter of the spring may vary along its length in order to have the spring obey Hooke's law over greater extensions.

Potential energy of a stretched spring

Work is done on a spring in stretching it a distance x beyond its normal unstretched length. The energy expended in doing the work is stored in the spring as potential energy. By considering the force applied and the resulting displacement, the work done in stretching the spring can be computed. Calculus is required because the force varies as the spring is stretched. The resulting expression is

(*Work done in stretching a spring, or the potential energy stored in a spring*)

$$W = \frac{1}{2} kx^2 \tag{11-26}$$

where k is the spring constant and x is the elongation of the spring.

11-8 STRESS, STRAIN, AND YOUNG'S MODULUS

In the preceding section, we considered springs which are designed to be stretched or compressed in order to exert a force on some part. In this section, we turn our attention to structural members such as wires, beams, and rods, whose primary purpose is to support a load. Because of their elastic properties, such load-bearing members are stretched or compressed a certain amount when the load is initially applied or when it is altered. If a thin wire is used to support a load, the weight of the load will stretch the wire.

Tensile stress is defined as the force per unit area that tends to stretch or compress an object, as in Fig. 11-19. In most cases where the tensile stress is applicable, the major dimension of the load-bearing member is its length, as for a thin wire, a rod, or a beam. The area considered is situated at right angles to the length, and the force is applied perpendicular to the area.

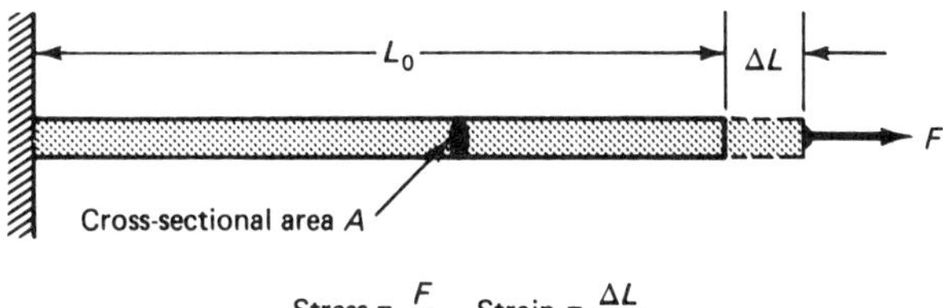

Stress = $\frac{F}{A}$ Strain = $\frac{\Delta L}{L_0}$

Figure 11-19 Tensile stress and strain.

Tensile strain is the amount the length of the member changes and is given as a fraction of the original length, as illustrated in Fig. 11-19.

As long as the applied stresses and the resulting strains do not exceed the elastic limit of the material, a type of relationship similar to Hooke's law exists between the stress and the strain. The proportionality constant is called *Young's modulus*, and is defined as

(*Young's modulus*) $$Y = \frac{\text{stress}}{\text{strain}} = \frac{F/A}{\Delta L/L_0} \tag{11-27}$$

This relationship may be rearranged to give the following expressions, which are quite useful:

$$\Delta L = \left(\frac{L_0}{Y}\right)\left(\frac{F}{A}\right) \qquad F = AY\left(\frac{\Delta L}{L_0}\right) \qquad Y = \left(\frac{L_0}{A}\right)\left(\frac{F}{\Delta L}\right) \tag{11-28}$$

Values of Young's modulus for various materials are given in Table 11-5.

TABLE 11-5 ELASTIC MODULI[(a)]

Material	Young's modulus,[(b)] N/m^2	Shear modulus, N/m^2	Bulk modulus, N/m^2
Aluminum	7.0×10^{10}	2.5×10^{10}	7.5×10^{10}
Brass	10	3.5	11
Copper	13	4.8	14
Steel	20	8.0	14
Tungsten	36	15	20
Fused quartz	5.6	2.5	2.7
Ethyl alcohol			0.11
Glycerin			0.40
Mercury			2.80
Water			0.21

[(a)]Elastic moduli of solids are subject to considerable variation because they are strongly dependent upon the heat treatment and work-hardening history of the material.
[(b)]Multiply by 1.45×10^{-4} to convert from N/m^2 to lb/in^2.

EXAMPLE 11-4 A steel wire 1 mm in diameter is to be used as the support wire for a Foucault pendulum. The pendulum bob has a mass of 4 kg, and the support wire is 12 m long.
(a) Compute the stress in the wire.
(b) How much will the support wire be stretched by the weight of the pendulum bob?

Solution. (a) The cross-sectional area of the wire ($\pi D^2/4$) is 7.85×10^{-7} m^2. The weight of the bob is 39.2 N. Thus, the stress is F/A, which gives a 4.99×10^7 N/m^2.

(b) Using Eq. (11-28) and Young's modulus from Table 11-5,

$$\Delta L = \left(\frac{L_0}{Y}\right)\left(\frac{F}{A}\right) = \frac{(12\ \text{m})(39.2\ \text{N})}{(20 \times 10^{10}\ \text{N/m}^2)(7.85 \times 10^{-7}\ \text{m}^2)}$$

$$= 3 \times 10^{-3}\ \text{m} = 3\ \text{mm}$$

11-9 SHEAR MODULUS

Another frequently occurring type of stress is one in which the applied forces tend to cause one part of a body to slide across an adjacent part, producing a sideways disruption if the stress is great enough. This type of stress, illustrated in Fig. 11-20, is called *shearing stress* and is defined as the force per unit area across which the shearing action occurs. In this case, the forces are applied "edge on" to the area; for tensile stress, the forces are applied at right angles to the area.

Shearing strain is the deformation that occurs as a result of a shearing stress. It is defined in reference to a rectangular block, as shown in Fig. 11-20. The distance Δx is much exaggerated for purposes of illustration. The shearing strain can also be specified by giving the angle θ. In radians, $\theta = \Delta x/h$.

The shear modulus n, sometimes called the *modulus of rigidity*, is defined as the ratio of shearing stress to shearing strain:

$$n = \frac{\text{shearing stress}}{\text{shearing strain}} = \frac{F/A}{\Delta x/h} \quad \text{or} \quad n = \frac{F/A}{\theta} \tag{11-29}$$

Rearranging yields

$$\Delta x = \left(\frac{1}{n}\right)\left(\frac{h}{A}\right)F \qquad F = \left(\frac{nA}{h}\right)\Delta x \qquad \theta = \frac{F}{nA} \tag{11-30}$$

The units of the shear modulus n are force per area.

Torsion constant

Suppose a rod of length l and radius r is clamped at one end and subjected to a twisting force, a torque, applied to the other end, as shown in Fig. 11-21. The rod

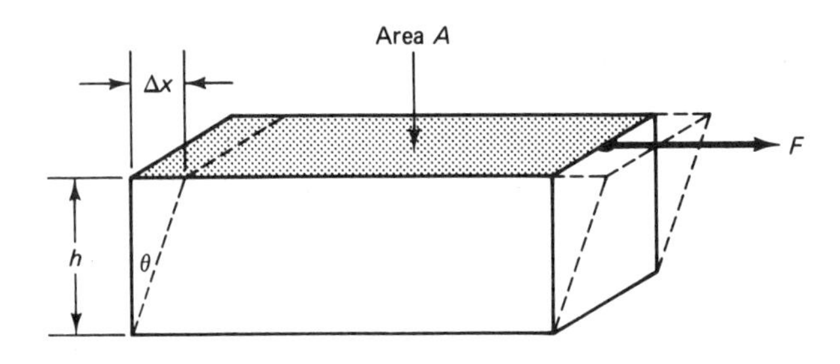

Figure 11-20 Shearing stress and strain.

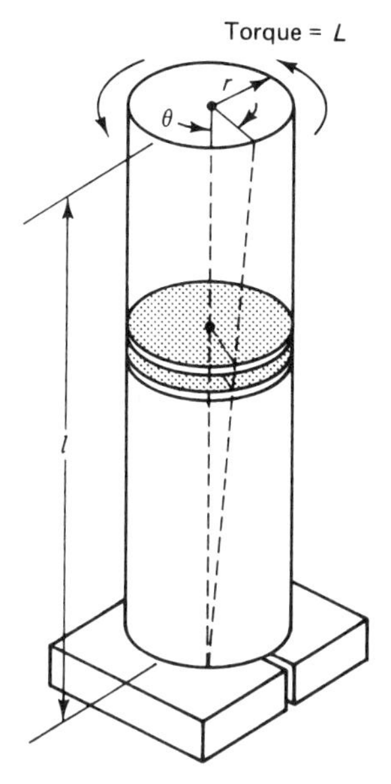

Figure 11-21 Rod clamped at one end is twisted by a torque applied to the other end. Shear forces come into play as adjacent layers of the rod are forced to rotate relative to each other by the twisting action.

will be twisted through an angle θ. The amount of twist resulting from an applied torque L depends upon the length of the rod, the fourth power of the radius, and the shear modulus n of the material. The relationship between L and θ is

$$\theta = \frac{2l}{\pi r^4 n} L \quad \text{or} \quad L = \left(\frac{\pi r^4 n}{2l}\right)\theta \tag{11-31}$$

For a particular rod, this expression may be written as

$$L = k\theta \quad \text{where} \quad k = \frac{\pi r^4 n}{2l} \tag{11-32}$$

which illustrates that a direct proportion exists between the angle of twist and the applied torque. The constant k is sometimes called the *moment of torsion* for the particular rod to which it applies.

EXAMPLE 11-5 A steel rod 0.25 in. in diameter and 4 ft long is subjected to a torque of 50 ft-lb. How much does the rod twist?

Solution. From Table 11-5, n for steel is 12×10^6 lb/in^2. In order to express all lengths in inches, we convert 50 ft-lb to 600 in.-lb. Then, using Eq. (11-31),

$$\theta = \frac{2(48 \text{ in.})(600 \text{ in.-lb})}{(3.14)(0.125)^4(12 \times 10^6 \text{ lb/in.}^2)}$$

$$= 6.26 \text{ radians}$$

This is slightly less than one complete revolution. The same answer is obtained (with a bit more work, however) if the foot is taken as the unit of length. The moment of torsion for this rod is $k = 95.87$ in.-lb/rad, from Eq. (11-32). This corresponds to 1.67 in.-lb/degree.

11-10 VOLUME ELASTICITY; BULK MODULUS

Suppose a rectangular block of steel is lowered to a great depth in the ocean so that a large pressure is exerted on all sides of the block. The block will be compressed, and its volume will decrease slightly. This phenomenon is an example of *volume elasticity*, and the pertinent elastic modulus is the *bulk modulus*. The reciprocal of the bulk modulus is the *compressibility*. Volume elasticity is actually a three-dimensional manifestation of tensile stress and strain. The definition of bulk modulus parallels that of Young's modulus, being the ratio of a stress to strain:

$$\text{Volume stress} = \frac{F}{A} \qquad \text{(a pressure)}$$

$$\text{Volume strain} = \frac{\Delta V}{V} \qquad \text{(fractional change in volume)}$$

$$\text{Bulk modulus} = B = -\frac{\text{volume stress}}{\text{volume strain}} = -\frac{F/A}{\Delta V/V} \tag{11-33}$$

Rearranging gives

$$\Delta V = -\left(\frac{1}{B}\right)V\left(\frac{F}{A}\right) \tag{11-34}$$

Because an increase in pressure (F/A) always produces a decrease in volume, a negative sign is customarily included in the definition of the bulk modulus. The bulk modulus is also applicable to liquids because all liquids are compressible. The treatment of gases is more involved and is given in a later chapter.

EXAMPLE 11-6 The bulk modulus of steel is 17×10^{10} N/m^2. Compute the change in volume of a hugh block of steel in the shape of a cube 1 m on a side if the block is lowered to a depth of 1000 ft in the ocean, where the pressure is on the order of 30 atm.

Solution. In MKS units, a pressure of 30 atm is $30 \times 1.013 \times 10^5$ N/m^2, which is about 3×10^6 N/m^2. Substituting this for F/A in Eq. (11-34) gives

$$\Delta V = -\left(\frac{1}{B}\right)V\left(\frac{F}{A}\right) = -\frac{(1\text{ m}^3)(3 \times 10^6\text{ N/m}^2)}{17 \times 10^{10}\text{ N/m}^2}$$

$$= -1.76 \times 10^{-5}\text{ m}^3 = -17.6\text{ cm}^3$$

This is a very small change in volume (17 parts per million) in comparison to the total volume of the steel block.

To Go Further

Read about the following in an encyclopedia:

- Archimedes (287–212 B.C.), Greek philosopher
- Amadeo Avogadro (1776–1856), Italian scientist
- Robert Hooke (1635–1703), English scientist
- Blaise Pascal (1623–1662), French scientist and philosopher

Reference

SULLIVAN, WALTER, *Black Holes*. New York: Anchor Press/Doubleday, 1979. A readable account of the various possibilities for dying stars, one of which leads to the formation of black holes.

Questions

1. What property of an atom uniquely distinguishes it from an atom of a different element?
2. Is it correct to say that an atom is a miniature solar system in which the electrons play the role of planets? Explain.
3. What is it that determines the chemical properties of an atom?
4. In addition to solid, liquid, and gas, another state of matter is a plasma, which resembles a "soup" of electrons and ions of various sorts. Could we also say that the matter in a neutron star and the matter in a black hole represent two other states of matter?
5. Describe crystalline and amorphous solids. Does transparency have anything to do with whether a material is crystalline?
6. Distinguish between absolute pressure and gauge pressure.
7. Why is it easier to float in sea water?
8. Why does water on the hood of a newly waxed car puddle in strange patterns?
9. Might it be possible for an astronaut in orbit to "construct" a spherical blob of water 6 in. in diameter? What difficulties might be encountered?
10. When a glass capillary tube is inserted in mercury, the level is depressed. Does the tube push the level down or does the mercury pull itself down?
11. What is the distinction made between elastic and plastic deformation?
12. A rock hound finds a rock and wonders what it is made of. Describe a procedure for finding the specific gravity of the rock.
13. An ice cube floats in a brimful glass of water. The tip of the ice is above the water and is therefore above the rim of the glass. When the ice melts, will the water spill over the rim, stay at the same level, or go down slightly?
*14. How could we measure the density of water at the bottom of the ocean to see if and how much it is compressed?
15. A gas-station attendent uses a hydrometer to test the freezing point of the radiator coolant. Explain. (The specific gravity of ethylene glycol is 1.1.)

Problems

1. The diameter of the nucleus of a hydrogen atom is about 2.4×10^{-13} cm. The diameter of the electron orbit (in the lowest energy state) is about 1×10^{-8} cm. If the atom were enlarged until the nucleus was as large as a tennis ball (diameter ≈ 6 cm), what would be the diameter of the electron orbit?
2. The mass of a proton and of a neutron is 1.67×10^{-27} kg. One of the heavier atoms is uranium, whose atomic number is 92 and atomic weight (of the most common isotope) is 238.
   (a) How many neutrons are in the nucleus of this isotope of uranium?
   (b) What is the mass of one atom of this isotope?
   (c) How many electrons does one atom of uranium have, assuming the atom is uncharged?
   (d) What is the total mass of the electrons?
   (e) What percentage of the total mass of the atom do the electrons represent?
3. Two semiconductor materials widely used in solid state electronics are silicon (atomic number 14) and germanium (atomic number 32).
   (a) How many electrons are in each shell of each atom?

(b) Why is it likely that the two elements should have similar physical and chemical properties?

4. What is the weight density of a material whose mass density is 2650 kg/m^3?

5. Suppose a balloon is blown up with water rather than air. What are: (a) the weight density; (b) the mass density of the water inside the balloon at sea level? If the balloon is placed aboard a satellite in earth orbit, what will be: (c) the weight density; (d) the mass density of the water?

6. The weight density of carbon tetrachloride is 99.6 lb/ft^3. What is its specific gravity?

7. Calculate the weight of 1 gal of chloroform, whose specific gravity is 1.489.

8. Compute the weight in pounds of (a) a brass sphere and (b) a lead sphere each 10 cm in diameter.

9. What volume in cubic centimeters of (a) mercury and (b) alcohol weighs 1 lb?

10. Compute the pressure in pounds per square inch in a vat of mercury 1 ft below the surface.

11. An observation window on a diving bell is 6 in. in diameter.
    (a) Compute the pressure outside the window when the bell is at a depth of 2000 ft.
    (b) What total force will be exerted on the window at that depth?

12. Suppose, in Fig. 11-3, that the manometer is filled with mercury and that the difference in height of the two columns is 6 cm.
    (a) Calculate the gauge pressure in the tank.
    (b) Calculate the absolute pressure in the tank, assuming the atmospheric pressure is 760 mm Hg.

13. Repeat Problem 12, but assume the manometer is filled with oil of specific gravity 0.8 instead of mercury.

14. In Fig. 11-8, the column of oil is 12 cm in length, whereas the column of water, measured from the same reference level, is only 10 cm. What is the specific gravity of the oil?

15. Compute the weight of 1 gal of the oil in Problem 14.

16. A cylindrical sample of aluminum 3 cm in diameter and 5 cm tall is suspended in water. For the sample, compute: (a) its volume; (b) its mass; (c) its weight. Further, calculate: (d) the weight of the displaced water; (e) the tension in the supporting string.

17. A block of cast iron weighs 14 lb in air. How much will it weigh suspended in water?

18. A mineral specimen of irregular shape weighs 72.8 g in air (using a gram as a unit of weight), and it weighs 44.8 g when suspended in water. Use this information to determine its: (a) volume; (b) density; (c) specific gravity.

19. An aluminum cylinder of mass 100 g weighs 69.1 g while suspended in a light oil. Compute the specific gravity of the oil.

20. The "solid gold" crown of a king weighs 650 g in air and 608 g when suspended in water. For the crown, compute: (a) the volume; (b) the density; (c) the specific gravity. (d) Refer to Table 11-1 and comment on the purity of the gold used to fabricate the crown.

21. A flat block of wood 4.0 cm thick and coated lightly with wax floats on water, as in Fig. 11-11. The waterline is 2.5 cm from the bottom surface of the wood. Compute the specific gravity of the wood.

22. A test tube containing a bit of sand is 1.5 cm in (outside) diameter and is 16 cm tall. The total mass of the test tube and sand is 21 grams. When floated in (a) water and (b) alcohol, what length of the tube will remain above the surface?

23. A hydraulic jack has a large cylinder that is 4.75 in. in diameter, while the small cylinder is only 0.6 in. in diameter.
    (a) Compute the IMA of the jack.
    (b) What will be the pressure in the fluid when the jack lifts a load of 2000 lb?

24. What should be the ratio of the cylinder diameters of a hydraulic jack in order for the jack to have an IMA of 100?

25. A thin aluminum ring 5 cm in diameter weighs 2.5 g (of force) while suspended in air. What total force is required to lift it from the surface of water, as illustrated in Fig. 11-

15? Express the answer in grams. (*Hint*: Compute F_{st} in dynes and then divide by 980 to convert to gram equivalent weights.)

26. Repeat Problem 25, but assume a bit of soap is added to the water.

27. Suppose the aluminum ring of the preceding problems is used to determine the surface tension of a liquid. If a total force of 5.76 g is required to lift it from the surface, what is the surface tension of the liquid?

28. A D student finds a smooth, shiny disk of copper 7 cm in diameter and 0.6 cm thick, which the student proposes to use to determine the surface tension of ethyl alcohol, as in Fig. 11-15. By what percentage will the total force required to lift the disk from the surface be greater than the weight of the disk hanging free in the air?

29. Calculate the distance a column of mercury will be depressed by a glass capillary tube 0.5 mm in diameter, as in Fig. 11-17.

30. What distance will water be drawn up into a glass capillary tube 2 mm in diameter inserted into the water?

31. Repeat Problem 30 for alcohol instead of water.

32. When steel floats on mercury, what fraction of the volume is submerged?

33. By what percentage does the volume of water increase as it freezes into ice?

34. A mass of 50 g is hung on a spring with a spring constant of 19.6×10^3 dyn/cm. Compute: **(a)** the extension of the spring; **(b)** the energy stored in the extended spring.

35. When 350 g is hung on a certain spring, the spring is stretched by 10 cm. Compute the spring constant in: **(a)** dynes per centimeter; **(b)** grams (of force) per centimeter.

36. A spring has a spring constant of 5000 N/m. What mass must be hung on the spring in order to stretch it 0.1 m?

37. What distance must a spring of force constant 5000 N/m be stretched in order to store 25 J of energy?

38. A mass of 2 kg is suspended by a brass wire 2 m long and 2 mm in diameter. Compute the change in length of the wire when the load is first applied.

39. A steel rod of cross-sectional area 1 cm^2 and nominal length 3 m is subjected to a tensile force of 20,000 N. Compute the resulting: **(a)** stress; **(b)** strain.

40. A rectangular solid of brass is subjected to a shearing force of 1000 N, as in Fig. 11-20. The block is 20 cm long, 3 cm wide, and 6 cm high. Compute: **(a)** the shearing stress, in Newtons per square meter; **(b)** the shearing strain, expressed as a percentage; **(c)** the angle θ, expressed in degrees.

41. Repeat Problem 40, but assume the block of brass is rotated about its length so that it becomes 6 cm wide and 3 cm high.

42. **(a)** Compute the moment of torsion for an aluminum rod 1 m long and 5 mm in diameter.

(b) What torque is required to twist this rod by 1° (0.0175 rad)?

43. A torque of 1.5 N · m applied to a certain rod 0.5 m long and 4 mm in diameter twists the rod through an angle of 0.7 rad.

(a) Compute the moment of torsion of the rod.

(b) Find the shear modulus of the material of which the rod is made.

44. **(a)** Compute the pressure existing at the maximum depth of the ocean, 35,597 ft, which occurs in the Mariana Trench near Guam in the Pacific. (Neglect the effects of temperature and salinity variations.)

(b) Compute the percent increase in the density of water at the maximum depth due to pressure, using the bulk modulus for water given in Table 11-5.

CHAPTER 12

FLUIDS

The study of fluid flow and the design of fluid-flow systems is one of the most important engineering aspects of a technical society. The wide range of applications includes pipeline systems for providing water for urban areas, for transporting oil and natural gas over large distances, and for transporting sewage to treatment plants outside metropolitan areas. Another aspect of fluid flow deals with the motion of a ship or submarine through water, and the field of aerodynamics is concerned with the flow of air around an aircraft. Yet another aspect deals with the design of turbines and compressors in aircraft engines, steam turbines for electric power generation, and even windmills.

In this chapter we present several basic concepts of fluid flow. We consider laminar and turbulent flow, streamlines, boundary layers, the peculiar phenomenon of a vortex, and viscosity. We derive Bernoulli's equation and present several applications including a venturi tube, pitot tube, and a curve ball. A discussion of Reynolds' number and the calculation of drag coefficients for a sphere and a cylinder is also included.

Obviously, the subject of fluid flow is quite large and complex. Therefore, in a chapter such as this, we can hope only to hit the high points and perhaps convey the most important ideas.

12-1 FLUIDS; DEFINITION AND BASIC CONCEPTS

Definition

The term *fluid* brings to mind liquids like water, molasses, or perhaps hydraulic fluid, but the term is also used to refer to air and other gases when their fluidlike properties are being considered. Two determining properties of a fluid are that (1) the substance forms a *continuum* (no voids exist within the substance) and (2) the substance cannot support static shear stresses. Clearly, any void that occurs in water soon disappears, and it is impossible to place a body of water in static shear by forcing the top surface toward the right while forcing the bottom of the water toward the left. Thus, water is a fluid according to these criteria.

Obviously, fluids exist with densities that span a wide range, from light gases, such as hydrogen or helium, to very heavy liquids, like mercury. Density is important in determining the properties of a fluid. Another important property that varies widely from one fluid to another is the *viscosity*, which may be thought of as the frictional resistance that occurs when one elemental volume of a fluid tries to move relative to an adjacent volume. Molasses is a *viscous* liquid whereas water is much less viscous. The minute viscosity of gases must be considered in many practical situations.

All fluids are compressible. Liquids are only slightly compressible under tremendous pressures, but gases are readily compressible even at small to moderate pressures. The study of compressible fluid flow is exceedingly complex. We concern ourselves only with fluids in situations where appreciable compression does not occur. Compression effects must be considered in regard to supersonic flight and in the study of shock waves.

Fluid flow through a pipe

We begin our study of fluid flow by considering the flow through a circular pipe of diameter D and area A ($\pi D^2/4$). An assumption we make, for purposes of simplification, is that all parts of the fluid across any particular cross section of the pipe move at the same velocity. In practice, this is not quite true, but it is a good approximation where the fluid is not particularly viscous and where the velocity is fairly large in a pipe of moderate diameter.

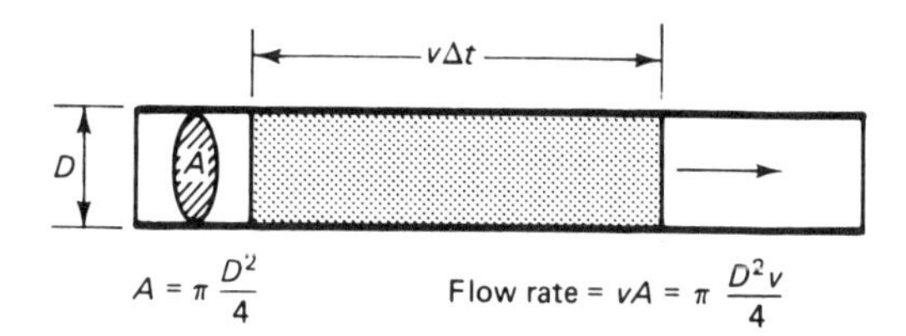

Figure 12-1 An element of fluid flowing in a pipe.

In Fig. 12-1, the fluid moves through a horizontal pipe with velocity v. In a short time interval Δt, a given element of fluid moves down the pipe a distance $v\Delta t$. Accordingly, a volume of fluid given by $V = v(\Delta t)A$ will move past a given cross section of the pipe. Hence, the flow rate (volume per unit time) of fluid through the pipe is obtained by dividing the time interval, Δt:

(*flow rate*) $$\frac{V}{\Delta t} = vA \quad \text{or} \quad \frac{V}{\Delta t} = \frac{\pi D^2 v}{4} \qquad (12\text{-}1)$$

This expresssion gives the quantity of fluid discharged from the end of an open pipe of area A in which the fluid moves with an average velocity v.

A rather unusual situation is shown in Fig. 12-2, in which fluid flows through a pipe having three different diameters at various sections of its length. Because the same volume of fluid per unit time must flow through each section of the pipe, the flow rate must be the same for each. Consequently, from Eq. (12-1), we write

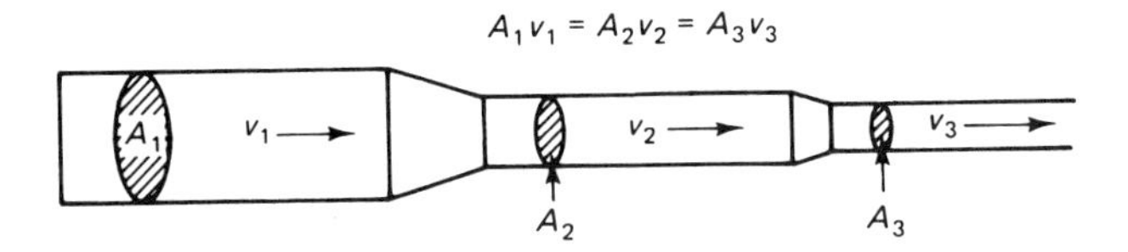

Figure 12-2 Fluid moves with greater velocity in the smaller pipe.

$$\frac{V}{\Delta t} = v_1A_1 = v_2A_2 = v_3A_3$$

and for any two sections of the pipe we obtain

$$v_2 = \left(\frac{A_1}{A_2}\right)v_1 \quad \text{or} \quad v_2 = \left(\frac{D_1}{D_2}\right)^2 v_1 \tag{12-2}$$

The fluid moves with lower velocity in the larger pipe. If the diameter of one pipe is twice that of the other, the velocity in the larger pipe is only one-fourth the velocity in the smaller pipe.

EXAMPLE 12-1 Water flows in a pipe with a 4-in. inside diameter (ID) at the average velocity of 3 ft/s. Compute the flow rate, and express the answer in gallons per minute.

Solution. We first calculate the cross-sectional area of the pipe to be 12.57 in.^{2} Then, by Eq. (12-1),

$$\frac{V}{\Delta t} = vA = (36 \text{ in./s})(12.57 \text{ in.}^2) = 452.5 \text{ in.}^3\text{/s}$$

where the velocity has been converted to units of inches per second. A unit conversion converts this to gallons per minute:

$$(452.5 \text{ in.}^3\text{/s})(60 \text{ s/min})(1 \text{ gal/231 in.}^3) = 117.5 \text{ gal/min}$$

EXAMPLE 12-2 A 4-in. ID pipe connects through a reducer to a 3-in. ID pipe. If water flows at an average velocity of 5 ft/s in the larger pipe, what will be the average velocity of the water in the smaller pipe?

Solution. This requires a straightforward application of Eq. (12-2).

$$v_2 = \left(\frac{D_1}{D_2}\right)^2 v_1 = \left(\frac{4}{3}\right)^2 (5 \text{ ft/s})$$

$$= 8.89 \text{ ft/s}$$

Measuring fluid flow within a pipe

Generally speaking, the velocity of the fluid within a pipe is of less interest than the *flow rate* in volume per unit time. Consequently, flowmeters to determine the flow rate are more common than velocity meters. One type of flow meter, called a *rotameter*, is shown in Fig. 12-3. A rotameter utilizes the phenomenon of terminal velocity in conjunction with the principle embodied in Eq. (12-2), that fluids flow slower in larger diameter pipes. The tube of the rotameter increases in diameter from the bottom to the top so that the fluid gradually slows down as it rises. The rotor is lifted upward by the rising fluid until the rate of fall of the rotor relative to

Laminar flow of water. Water streams silently from the finger hole of a biscuit cover to the bucket three feet below.

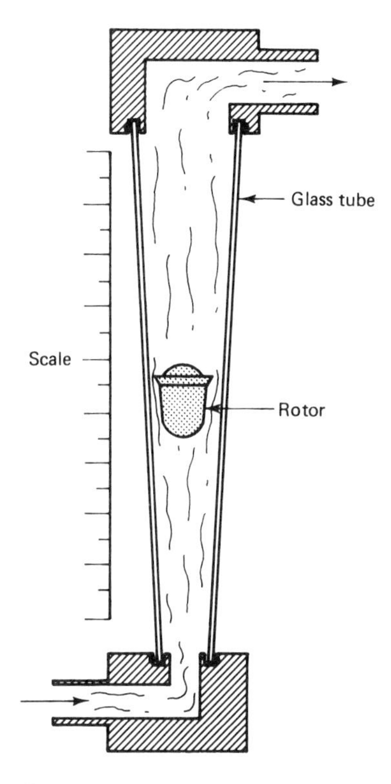

Figure 12-3 A rotameter for determining flow rates of liquids and gases.

the fluid exactly equals the rate of rise of the fluid. High flow rates produce greater velocities within the tube, and the rotor rises to a higher level. The scale is calibrated in units of volume per unit of time. A particular rotameter is intended for use in a particular fluid; the calibrations for water are slightly different from the calibrations for gasoline, for example. Rotameters are also useful for measuring flow rates of gases such as air, nitrogen, and oxygen hydrogen.

12-2 FLUID FLOW: LAMINAR AND TURBULENT, STEADY AND UNSTEADY

The type of fluid flow that occurs when we pour a thick syrup from a bottle onto a pancake is noticeably different from the flow occurring when we pour water from an identical bottle. The flow of syrup is a smooth, orderly flow called *laminar* flow. On the other hand, the water might initially exhibit such an orderly flow, but it will break up into a chaotic, *turbulent* flow only a short distance from the bottle. Turbulent flow is characterized by randomness; miniature swirls, each of which is called a *vortex*, are present in large numbers even though they may not be visible.

Often, smoke rising from a lighted cigarette is observed to exhibit laminar flow for several inches before the flow breaks up into turbulent flow. Steam rising from a stewpot is an obvious example of tubulent flow. Later, we investigate some of the factors that determine whether a given flow will be laminar or turbulent.

Streamlines

If a snapshot could be taken so that the direction of flow at all points within a fluid could be determined, we would find that the motion of the fluid is such that lines called *streamlines* may be traced throughout the region of flow. In other words, the fluid flows along the lines of the stream. The streamlines are not necessarily straight, and they may not appear at the same place in a second snapshot taken only an instant after the first. The streamlines provide a graphical method of illustrating fluid flow. Fluid flows parallel to the streamlines, and it is assumed that fluid flows faster where the streamlines are closer together. If the streamlines remain fixed, always appearing at the same place in snapshop after snapshot, the flow is said to be *steady*. Otherwise, the flow is *unsteady*. A streamline may be visualized as being the centerline of an imaginary tube of flow that surrounds the streamline. The diameter of the tube is not necessarily constant; the diameters of adjacent tubes are imagined to increase if the streamlines diverge. Streamlines never cross each other. If they did, how would a particular particle of fluid approaching the crossing know whether to go left or to go right?

When a fluid flows past an object, or equivalently, when an object passes through a fluid, the object produces a *wake* on the downstream side of the object. The wake is a region of eddy currents and vortices that appear as spirals and whirlpools representing turbulence. Energy is contained in this region of turbulence, and in the case of a car, ship, or airplane, the energy represents a waste because it must come from the fuel used to power the vehicle.

The condition of zero slip

When a fluid flows by a solid surface, the thin layer of fluid in actual contact with the surface does not move. This thin layer of fluid at the fluid-solid interface remains perfectly stationary. We might initially expect the fluid somehow to *slip by* the surface, but this does not occur. We refer to this phenomenon, which arises as a consequence of viscosity, as *the condition of zero slip*. This condition prevails for such situations as water flowing in a pipe, for a ship passing through a body of water, and for air flowing over the surface of an airplane wing, and it is responsible for the formation of the *boundary layer* described in the following section. A *velocity profile* is useful in illustrating the variation of velocity of fluid flow near a solid surface. Several velocity profiles are illustrated in Fig. 12-4.

The Karman vortex street

Suppose a fluid (air or water) flows past a long cylindrical rod or pipe with a velocity that initially is very low and gradually increases to very high values. The nature of the fluid flow past the cylinder is rather different for various ranges of

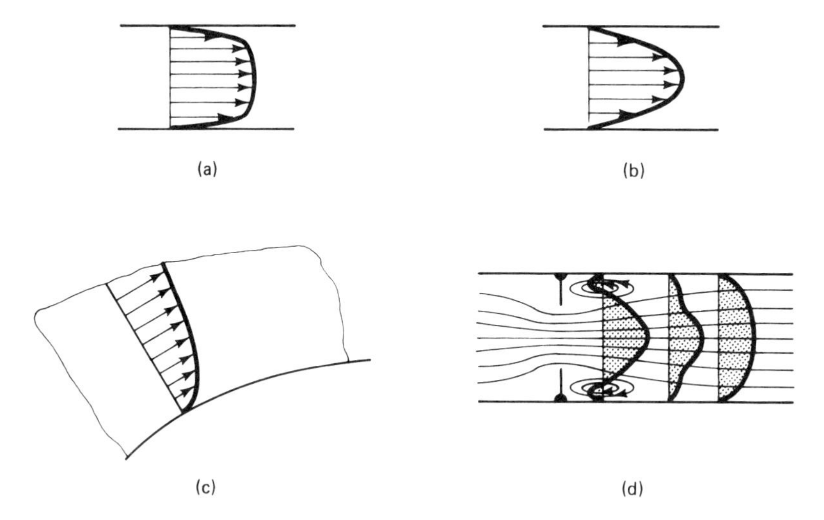

Figure 12-4 Velocity profiles illustrate variations in the fluid velocity.
(a) Turbulent flow of fluid in a pipe.
(b) Laminar flow in a pipe.
(c) Fluid flow near a curved boundary.
(d) Flow near an orifice in a pipe with back-streaming near the orifice.

the velocity; at least five different types of flow past the cylinder can be distinguished. These are shown in Fig. 12-5 in order of increasing velocity; the rod is $\frac{1}{4}$ in. in diameter, and the fluid is assumed to be water.

In the following, we assume that a pencil-sized rod is inserted into a pan of water, as shown in Fig. 12-5(a), and is then moved horizontally through the water at various speeds. We are interested in the flow patterns behind the moving rod. The patterns can be made visible by sprinkling fine powder or other material onto the surface of the water.

When the rod is moved extremely slowly—so slowly, in fact, that the movement is barely perceptible—the flow pattern of Fig. 12-5(b) is obtained. No vortices are present, and the streamlines downstream return to their original configuration. When the rod is speeded up to about $\frac{1}{4}$ in./per s, the pattern of part (c) is obtained. Two symmetric vortices are present, but the downstream streamlines are left undisturbed. At the slightly higher velocity of $\frac{3}{4}$ in./per s, two symmetric vortices are present, but the downstream streamlines are left curved, as shown. The streamlines do not oscillate back and forth at a given point; the curve is like that obtained by having a snake crawl across a dusty tabletop.

As the velocity is increased still further, the curvature of the downstream track becomes increasingly great until vortices are formed at the extreme positions of the curve, as shown in part (e). A trail of miniature whirlpools is left behind the rod, each remaining essentially stationary. The overall features of the pattern of vortices are mathematically predictable. This trail is called a *Karman vortex street* in honor of Theodore von Karman, who first gave a mathematical account of the behavior in 1911.

Finally, as velocities on the order of several inches per second are reached, a turbulent wake is formed downstream from the rod. Higher velocities may

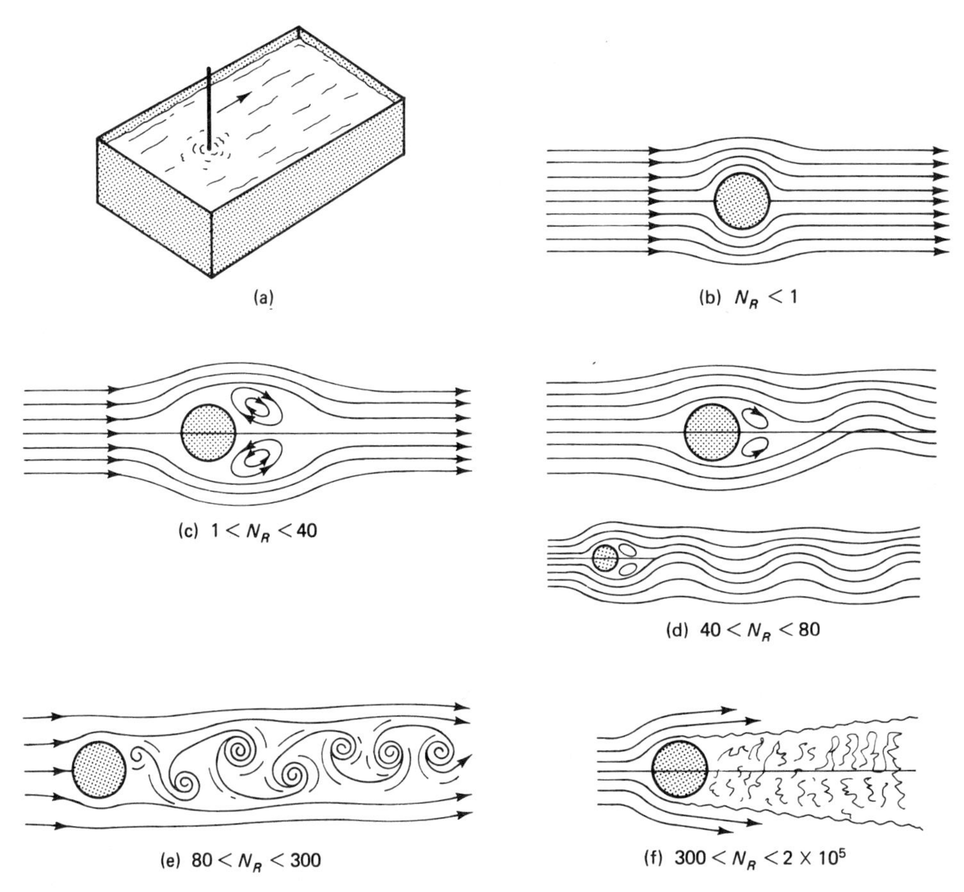

Figure 12-5 Different types of flow past a cylindrical rod.
(a) A vertical rod is moved through the water.
(b) Laminar flow around the rod.
(c) Two symmetrical vortices.
(d) Streamlines are left-curved.
(e) The Karman vortex street.
(f) A turbulent wake.

produce fairly minor differences in the nature of the wake, but the flow pattern does not change drastically after the transition from the Karman region to that of the turbulent wake.

If we take a closer look at the development of the Karman vortex trail, we see that a vortex begins to grow near the upper portion (in the figure) of the cylinder. It stays close to the cylinder as it increases in size. At some point, however, it will begin to detach itself from the cylinder and will snap away to become a part of the downstream trail. Simultaneously, as vortex shedding occurs at the top of the cylinder, a new vortex begins to grow near the bottom and the process repeats itself in a rhythmic, oscillatory manner. This is extremely significant because each time a vortex is shed, a vertical force is exerted on the rod by the departing vortex. If the rhythmic shedding of the vortices happens to be close in frequency to a natural resonant frequency of the rod, a vibration of large amplitude can develop which can be potentially damaging. This effect is responsible for the humming sound made by wires in a strong wind.

The most famous result of this phenomenon is the 1940 collapse of the Tacoma Narrows bridge because a wind blowing against large steel plates on the side of the bridge caused vortices to be shed at very nearly the resonant frequency of oscillation of the bridge. The amplitude grew over a period of several hours, until, finally, a twisting mode of oscillation developed that was particularly damaging to the structure of the bridge. Soon afterward, the bridge fell into the river below. Subsequent bridges have been given the benefit of extensive and very careful aerodynamic analysis.

Boundary layer

An important aspect of this high-speed motion of the rod through the water is the formation of a thin boundary layer on the forward surface of the rod. The boundary layer is a region of laminar flow that is characterized by marked differences in fluid-flow velocity as one proceeds through the layer in a direction at right angles to the flow. At the surface of the rod, the zero slip condition requires the fluid to move along with the rod. At a short lateral distance from the rod, the fluid remains stationary as the rod passes by. Between these two points, the velocity of the fluid must vary from zero at the point away from the rod to a velocity equal to that of the rod, which occurs at the surface of the rod. This change in velocity occurs within the boundary layer. A boundary layer is illustrated in Fig. 12-6; such a layer can also be observed around a boat or ship traveling at reasonable speed.

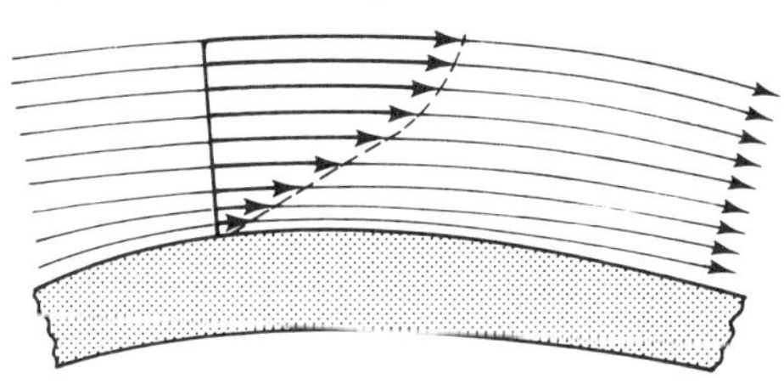

Figure 12-6 Boundary layer near the surface of an airplane wing. The vertical scale is greatly exaggerated.

At a certain distance from the most forward point, the boundary layer becomes detached from the surface of the rod to become part of the downstream wake. In this case, the boundary layer delineates between the wake and the surrounding fluid. The point at which the boundary layer becomes detached is important in aerodynamic design because the point of detachment plays a large part in determining the width of the wake. A wide wake entails greater energy losses than a narrow wake. If the angle of attack of an airfoil is increased excessively, boundary-layer separation can induce a stall, a condition of greatly increased drag and dramatic loss of lift.

The boundary layer for an airplane wing may be no thicker than the thickness of a dime. Consequently, the surface finish of the wing is very important. Flush-headed rivets are used in wing fabrication to minimize obstruction of the boundary layer. Pilots of small aircraft are warned about trying to take off when even a thin layer of frost covers the wings, even though the layer is hardly visible.

12-3 VISCOSITY

Definition and units

We are now ready to give a more precise definition of viscosity; refer to the rudimentary viscosimeter shown in Fig. 12-7(a). A light cylinder is arranged to

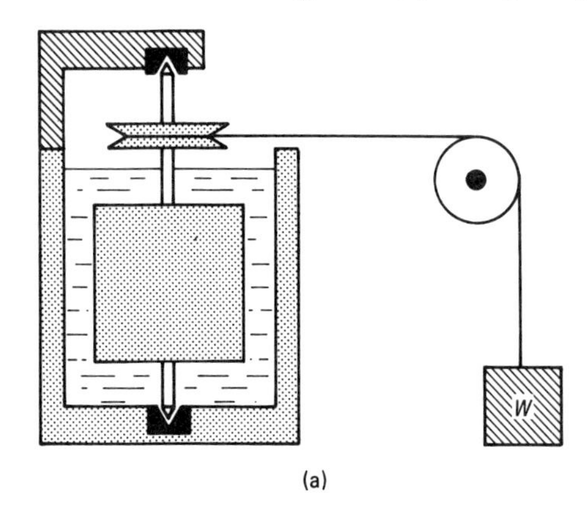

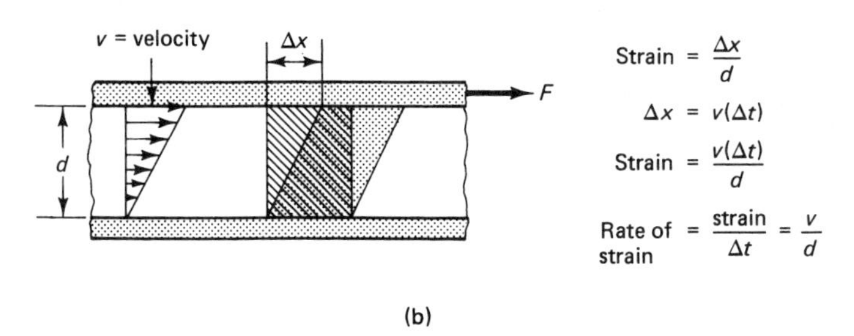

Figure 12-7 Viscosimeter.
(a) The liquid whose viscosity is to be determined fills the annular space between the cylinders.
(b) Shear forces are developed in a liquid confined between two surfaces moving relative to each other.

turn inside a stationary concentric cylinder of slightly larger diameter holding the liquid whose viscosity is to be determined. When the weight is released, the rotating cylinder will quickly come to a terminal velocity, which depends upon the viscosity of the liquid filling the annular space. The viscosity of the liquid may be determined from the final rotational velocity.

Because the annular space between the cylinders is fairly thin, we can imagine the curved surfaces of the cylinders to form parallel surfaces, as shown in Fig. 12-7(b). The liquid between the surfaces is placed in shear because the no-slip condition prevails at both surfaces. Laminar flow occurs as the upper layers of the liquid are forced to slide over the lower layers. A shearing action takes place in a manner similar to the action that occurs when a deck of cards is placed in shear—the upper cards in the deck slide over the lower cards. Resistance to the motion of one layer of liquid across a lower layer arises because of the viscosity of the liquid.

If the area of the moving surfaces is A and if the force F is applied in the direction of the motion, the shearing stress is F/A, as defined in the preceding chapter for the shear stress developed in solids. Recall, however, that a defining property of a fluid is that it cannot sustain a *static* shearing stress. On the other hand, a shearing stress will be present as long as there is relative motion between the two surfaces because a force is required to produce the sliding of the layers of liquid over each other. Furthermore, the faster the sliding action, the greater the force required. The shearing stress is proportional to the rate at which the shearing strain occurs. Figure 12-7 shows that

$$\text{Rate of shearing strain} = \frac{v}{d}$$

We now define the viscosity of the liquid as the ratio of the shearing stress to the *rate* of shearing strain:

$$\text{Viscosity} \equiv \eta = \frac{\text{shearing stress}}{\text{rate of shearing strain}} = \frac{F/A}{v/d} \tag{12-3}$$

This can be rearranged to yield

$$\frac{F}{A} = \eta\frac{v}{d} \quad \text{or} \quad F = \frac{\eta A v}{d} \tag{12-4}$$

From the defining equation, Eq. (12-3), it is easy to determine that the natural MKS unit of viscosity is $\text{N} \cdot \text{s/m}^2$. This unit is sometimes referred to as the *poiseuille*. An older unit based on the CGS system is $\text{dyn} \cdot \text{s/cm}^2$; it is called a *poise*. A unit conversion shows that 1 poiseuille is the equivalent of 10 poise. Viscosities of common liquids and gases are given in Table 12-1. The viscosity exhibits a marked dependence upon temperature. Liquids become less viscous with an increase in temperature, whereas gases exhibit a greater viscosity at high temperatures.

Kinematic viscosity is frequently encountered in *fluid mechanics*, the study of the motion and properties of fluids. Kinematic viscosity is the ratio of the viscosity as just defined to the mass density of the fluid:

$$\text{Kinematic viscosity} = \frac{\text{viscosity}}{\text{density}} = \frac{\eta}{\rho} \tag{12-5}$$

Stokes' law

In 1845, Sir George Stokes derived the following formula for the viscous force exerted on a sphere of radius r falling *slowly* at a relative velocity v through a fluid of viscosity η:

(*viscous force*) $$F = 6\pi\eta r v \tag{12-6}$$

If a small sphere is dropped into a viscous liquid, such as glycerin or motor oil, the sphere will soon attain a terminal velocity and thereby achieve a state of equilibrium so that the upward forces must equal the weight of the sphere. There

TABLE 12-1 VISCOSITY OF SELECTED LIQUIDS AND GASES AT 20°C.

Material	$(N \cdot s/m^2)$
Acetone	0.0003
Castor oil	0.98
Ethyl alcohol	0.0012
Ethylene glycol	0.019
Gasoline	0.0004
Glycerin	1.48
Kerosene	0.0018
Mercury	0.0015
Water	0.001
Air	18×10^{-6}
Argon	22×10^{-6}
Carbon dioxide	15×10^{-6}
Hydrogen	9×10^{-6}
Nitrogen	17×10^{-6}
Oxygen	20×10^{-6}

are two upward forces in effect: the viscous force, as given by Stokes' law, and the bouyant force, as described by Archimedes' principle. Thus, at terminal velocity,

$$\begin{aligned} \text{Bouyant force} &+ \text{viscous force} = \text{weight of sphere} \\ \tfrac{4}{3}\pi r^3 \rho_F g &+ 6\pi\eta r v_T = \tfrac{4}{3}\pi r^3 \rho_s g \end{aligned} \tag{12-7}$$

ρ_s = mass density of the sphere $\qquad \eta$ = the viscosity of the liquid

ρ_F = mass density of the fluid $\qquad g$ = the acceleration of gravity

The above relationship may be solved for the terminal velocity v_T, or the viscosity η can be determined if v_T is known:

$$v_T = \frac{2}{9}\frac{r^2 g}{\eta}(\rho_s - \rho_F) \qquad \eta = \frac{2}{9}\frac{r^2 g}{v_T}(\rho_s - \rho_F) \tag{12-8}$$

Timing the descent of a sphere in a column of liquid provides a convenient method of determining the viscosity of the liquid.

EXAMPLE 12-3 Compute the terminal velocity with which a small glass sphere 0.5 cm in diameter and specific gravity 2.6 will descend through ethylene glycol at 20°C. The specific gravity of ethylene glycol is 1.11, and its viscosity is $0.019\ N \cdot s/m^2$.

Solution. First, we use the specific gravities and the known density of water to determine that the mass densities of the glass sphere and ethylene glycol are 2600 and 1110 kg/m^3, respectively. Then, with all quantities in MKS units, we use Eq. (12-8):

$$\begin{aligned} v &= \frac{2}{9}\frac{r^2 g}{\eta}(\rho_s - \rho_F) \\ &= \frac{2(0.0025)^2(9.8)(2600 - 1110)}{9(0.019)} \\ &= 1.07 \text{ m/s} \end{aligned}$$

12-4 BERNOULLI'S PRINCIPLE

Derivation of Bernoulli's equation

Now that several aspects of fluid flow have been described, we turn our attention to the mathematics of laminar flow within a tube of flow that may vary both in cross-sectional area and elevation, as shown in Fig. 12-8. The fluid may be either a liquid or a gas, but it is assumed to be *nonviscous* and *incompressible*.

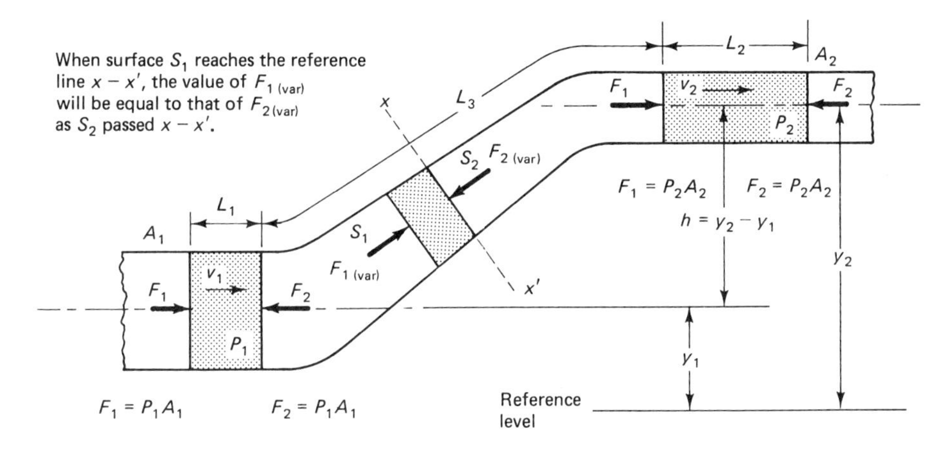

Figure 12-8 Derivation of Bernoulli's equation.

In the first section of this chapter, we found that the velocity of a fluid had to increase if the cross-sectional area of the tube of flow became smaller. Because each small element of fluid has mass, it follows that each fluid element must be acted upon by a force in the direction of motion in order to produce the requisite acceleration, giving rise to the increase in velocity. What is the origin of this force? What produces it? We shall see that any change in fluid velocity along a streamline is accompanied by either a change in *pressure*, a change in *elevation* of the streamline, or both. Thus, the force acting to accelerate the fluid must arise either from a pressure difference along the streamline or from gravitational effects.

Referring to Fig. 12-8, we propose to calculate the net work done on the volume of fluid in the lower horizontal section to move it to the upper section. We imagine forces F_1 and F_2 to act on the volume element, as shown. The magnitude of each force is the product of the pressure and the cross-sectional area of the tube. Because F_2 is directed opposite to the motion, the work done by F_2 is negative, and it is *subtracted* in the calculation of the net work done on the small volume of fluid. We consider movement of the fluid element over three distances (L_1, L_2, and L_3), but each force (F_1 or F_2) moves over only *two* of the three.

Also, F_1 and F_2 will be constant only in the horizontal sections of constant cross-sectional area. Further, the initial value of F_1 (at elevation y_1) and the final value of F_1 (at elevation y_2) will be different, and a smooth variation of F_1 and F_2 will occur along the inclined path, which we emphasize by the subscript *var*.

The work done by force F_1 over the distances L_1 and L_3 is

$$F_1L_1 + [F_{1(\text{var})}L_3]$$

and the work done by force F_2 over the distances L_2 and L_3 is

$$-F_2L_2 - [F_{2(\text{var})}L_3]$$

The net work is the sum of these, namely,

$$\text{Net work} = F_1L_1 + [F_{1(\text{var})}L_3] - F_2L_2 - [F_{2(\text{var})}L_3] \tag{12-9}$$

Over the distance L_3, $F_{1(\text{var})}$ and $F_{2(\text{var})}$ act over the same region (although at slightly different times), so that at corresponding points they have the same magnitude. Therefore, the quantity of work represented by $F_{1(\text{var})}L_3$ and $F_{2(\text{var})}L_3$ must be the same. Furthermore, because they differ in sign, they cancel out of Eq. (12-9), leaving, as the net work,

$$F_1L_1 - F_2L_2 = \text{net work} \tag{12-10}$$

At this point we express the forces F_1 and F_2 in terms of the pressure and cross-sectional area in the respective sections. At the same time, we reason that because the fluid is nonviscous, the energy contributed to the fluid by doing work on it must equal the change in energy that it contains in the form of kinetic energy and potential energy. We thus obtain

$$P_1A_1L_1 - P_2A_2L_2 = \Delta\text{KE} + \Delta\text{PE} \tag{12-11}$$

Because the fluid is incompressible, the volume V of fluid in the two sections must be equal, so that $V = A_1L_1$ and $V = A_2L_2$. Moreover, the mass of the volume of fluid is given by ρV. Bringing all this together gives us

$$P_1V - P_2V = \tfrac{1}{2}\rho V(v_2^2 - v_1^2) + \rho Vg(y_2 - y_1) \tag{12-12}$$

The volume V appears in all terms and therefore drops out of the equation. Rearrangement gives

$$P_1 + \tfrac{1}{2}\rho v_1^2 + \rho g y_1 = P_2 + \tfrac{1}{2}\rho v_2^2 + \rho g y_2 \tag{12-13}$$

This is *Bernoulli's equation*, first derived by Daniel Bernoulli in 1738. Each term in the equation represents an *energy density*, the energy per unit volume of fluid. Each *side* of the equation represents the *total* energy density, the total arising from contributions from the pressure, the kinetic, and the potential energy. In effect, Bernoulli's equation is a statement that the total energy density remains constant along the streamlines in laminar flow.

Special cases

If gravitational effects can be ignored within a given system, Eq. (12-13) can be simplified by dropping the PE term $\rho Vg(y_2 - y_1)$. The result is

$$P_1 - P_2 = \tfrac{1}{2}\rho(v_2^2 - v_1^2) \quad \text{or} \quad \Delta P = \tfrac{1}{2}\rho v_2^2 - \tfrac{1}{2}\rho v_1^2 \tag{12-14}$$

Here we see that a change in pressure along a streamline is directly related to a change in kinetic energy density, and it is apparent that the force required to accelerate the fluid elements arises from a pressure difference, as was stated earlier.

If no velocity change occurs along a streamline (if, for example, v_1 and v_2 are both zero), we obtain from Eq. (12-12) that

$$P_1 - P_2 = \rho g(y_2 - y_1) \quad \text{or} \quad \Delta P = \rho g h \tag{12-15}$$

which is the same result we obtained in our study of static fluids in the preceding chapter.

12-5 APPLICATIONS OF BERNOULLI'S PRINCIPLE

Venturi tube

The Venturi tube shown in Fig. 12-9 can be used to determine the velocity of fluid flow within a pipe through direct application of Bernoulli's equation. The approach to and departure from the section of reduced diameter is tapered, or streamlined, to avoid turbulence. A manometer may be used to measure the pressure difference between the two sections. Once the velocity of fluid is known, the flow rate can be calculated. The tube is horizontal, so the simplified form of Bernoulli's equation [Eq. (12-14)] may be used. The mathematics is given in the figure.

At this point we can make the following generalization in regard to laminar flow: When the streamlines are somehow compressed together, the fluid velocity increases and a smaller lateral pressure is exerted. On the other hand, diverging streamlines give rise to lower velocities and greater pressures.

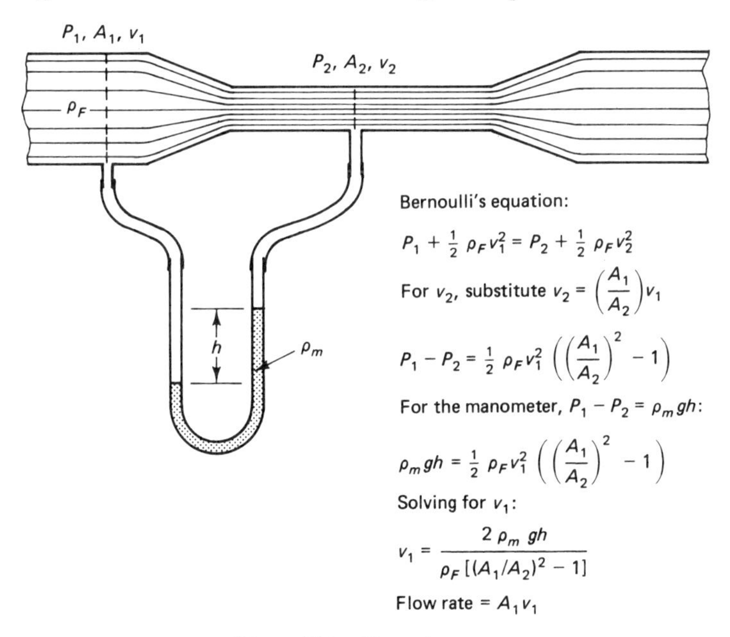

Figure 12-9 Venturi tube.

Pitot tube

An airspeed indicator for an airplane consists of a Pitot tube and a sensitive pressure-measuring device calibrated to read airspeed directly. The system is illustrated schematically in Fig. 12-10. The open end of the Pitot tube faces

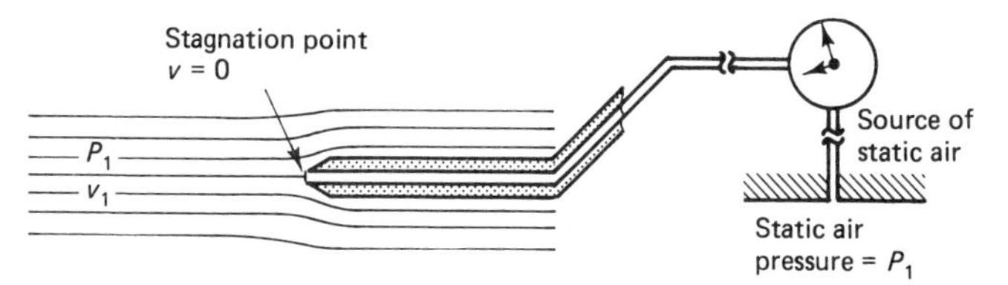

Figure 12-10 Pitot tube used as an airspeed indicator on an aircraft.

directly into the airstream so that a pressure is developed inside the tube. The pressure so obtained is compared with the static pressure from a small *static air opening* usually located somewhere on the side of the plane. The static pressure is the pressure of the surrounding air, and it is used for both the airspeed indicator and the altimeter.

The Pitot tube is designed to produce minimal disruption of the streamlines, but one streamline of the relative air will impinge directly upon the open end of the tube to form a *stagnation point* at which the air velocity relative to the tube is zero. Applying Bernoulli's principle to the streamline striking the tube head on:

$$P_1 + \tfrac{1}{2}\rho v_1^2 = P_s \qquad (12\text{-}16)$$

where P_s is the pressure within the tube and at the stagnation point. Solving for v_1 gives the airspeed in terms of the pressure difference and the density of the air:

$$v_1 = \sqrt{\frac{2(P_s - P_1)}{\rho}} \qquad (12\text{-}17)$$

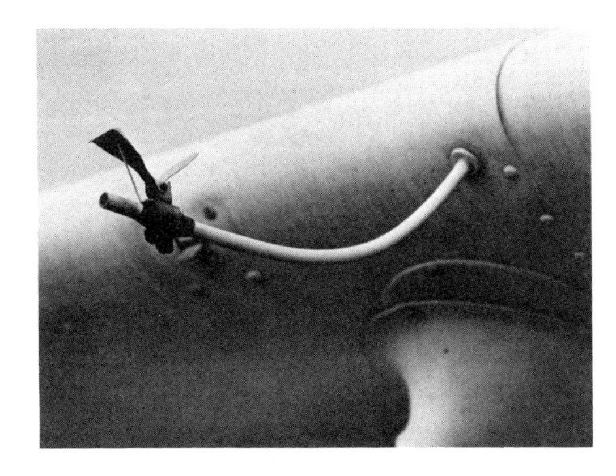

An older version of a pitot tube consists only of a metal tube extending in front of the leading edge of the wing of a small airplane. The pitot tube is the sensor for the airspeed indicator.

A modern pitot tube contains a heater to prevent it from icing up in freezing conditions. (A small twig was used by the photographer to hold the cover in the open position.)

Speed of efflux

Bernoulli's principle can be used to calculate the velocity with which a liquid will issue from an opening near the bottom of a tank, as shown in Fig. 12-11. The top surface of the liquid is assumed to be a distance h above the opening, and the opening is assumed to be frictionless so that it does not affect the velocity of

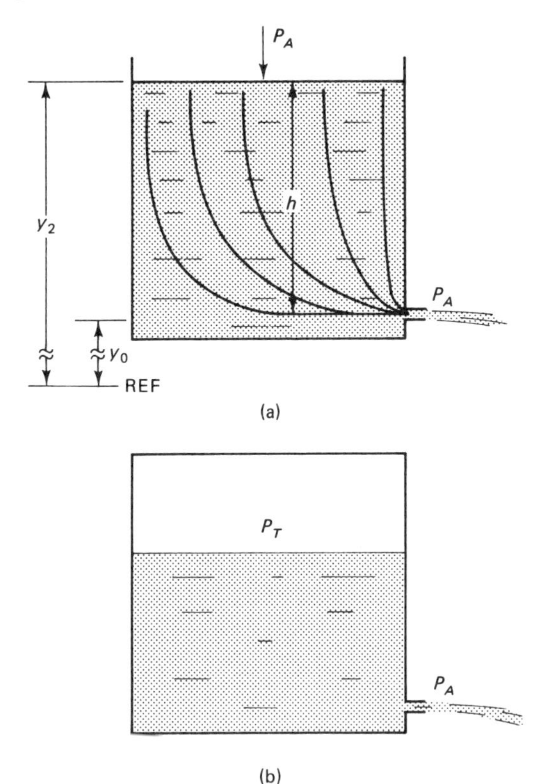

Figure 12-11 Bernoulli's principle can be used to determine the speed of efflux from an orifice on the side of a tank.
(a) Open tank.
(b) Closed tank.

the liquid flowing through it, the *speed of efflux*. Two possibilities exist for the pressure above the liquid. If the tank is open, atmospheric pressure will exist both at the top surface and at the opening so that the pressure effects cancel in Bernoulli's equation. However, if the tank is closed and pressurized to P_T, the pressure will serve to increase the speed of efflux. We assume the diameter of the tank is large in comparison to the diameter of the opening so that the velocity at the top surface of the liquid is small enough to neglect. This gives a further simplification of Bernoulli's equation, which is written in regard to the top surface and in regard to the opening. In its entirety, then,

$$P_T + \tfrac{1}{2}\rho v_T^2 + \rho g y_T = P_A + \tfrac{1}{2}\rho v_0^2 + \rho g y_0 \qquad (12\text{-}18)$$

If the tank is open, $P_T = P_A$, and we further assume $v_T \approx 0$. This gives

$$\rho g y_T = \tfrac{1}{2}\rho v_0^2 + \rho g y_0$$

Noting that the height h of the top surface above the opening is $y_T - y_0$, we obtain

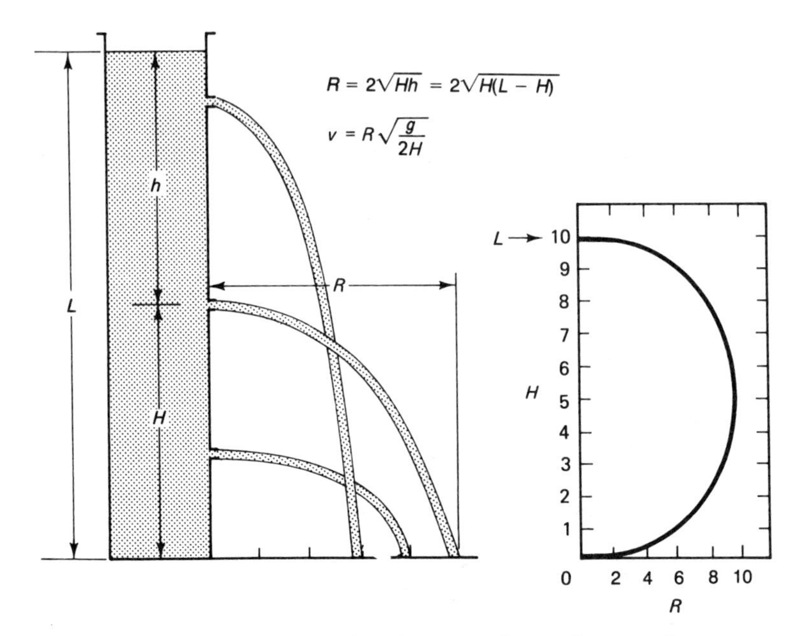

Figure 12-12 Trajectories of streams of water from a tank.

$$\tfrac{1}{2}\rho v_0^2 = \rho g h \quad \text{or} \quad v_0 = \sqrt{2gh} \tag{12-19}$$

We observe (1) that the speed of efflux does not depend upon the density of the liquid, and (2) by comparing with Eq. (3-14), the speed with which an element of fluid passes through the opening is the same as the speed that would be attained by a particle falling freely through a distance h. This latter result is known as *Torricelli's theorem.*

Similar reasoning for a pressurized tank gives the following result for the speed of efflux:

$$v_0 = \sqrt{\frac{2(P_T - P_A)}{\rho} + 2gh} \tag{12-20}$$

Three trajectories of issuing streams are shown in Fig. 12-12, along with a formula for computing the range of the projecting liquid. This problem is quite similar to that of the horizontal cannon of Sect. 3-10. The range is obtained by multiplying the speed of efflux, $\sqrt{2gh}$, by the time of fall of a particle from a height H. The result is surprisingly simple, and the range does *not* depend upon the acceleration of gravity! It is possible to measure the range as a part of a laboratory exercise and work backward to determine the speed of efflux. This provides a means of determining the effect of the opening upon the speed of the liquid flowing through it. A formula for computing the velocity from the range is given in the figure.

Curve ball

A good pitcher can make a baseball curve by causing it to spin rapidly as it approaches the batter. To understand this phenomenon, first consider the ball to be spinning without any forward motion through the air, as shown in Fig. 12-13(a).

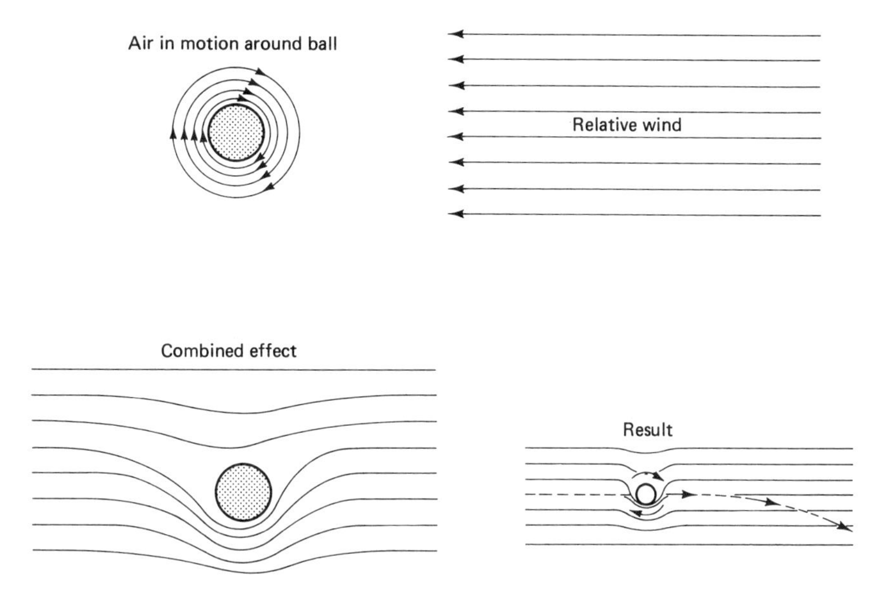

Figure 12-13 A curve ball curves as a result of the combined effect of the relative wind and a spinning field of air near the spinning ball.

The air adjacent to the spinning ball will also be set in motion, due to the viscosity of the air. This effect in itself produces no force on the ball because of the symmetry of the configuration, but when the spinning ball is exposed to a "wind" flowing past the ball, the flow of the wind is combined with the circular flow of the spinning ball to produce an unsymmetrical flow pattern, as shown in Fig. 12-13(c). The streamlines are compressed more on one side of the ball than on the other, and this causes a difference in pressure to occur on opposite sides of the ball. The pressure will be less on the side of the ball where the streamlines are closest together, and a force will be exerted on the ball in that direction. The result is shown in Fig. 12-13(d).

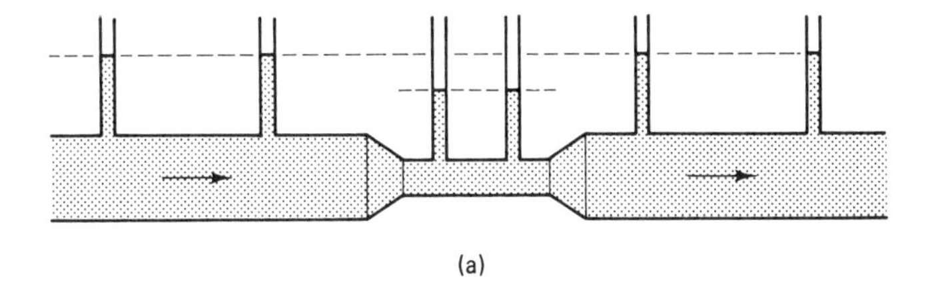

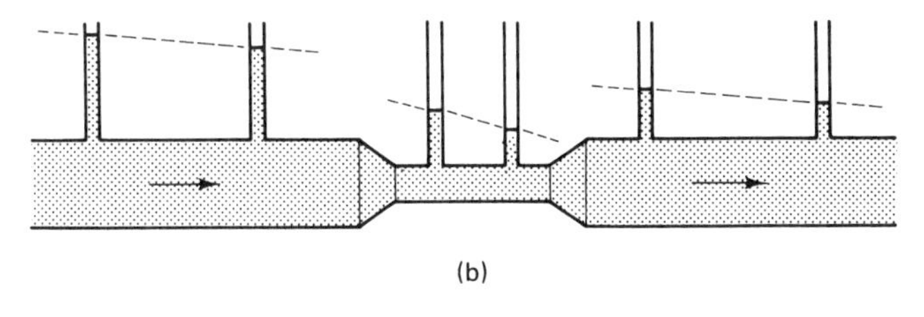

Figure 12-14 Effects of friction on the flow of a liquid through a pipe.
(a) In the absence of friction, the pressure does not vary along a straight section.
(b) Frictional effects cause a reduction in pressure along straight sections.

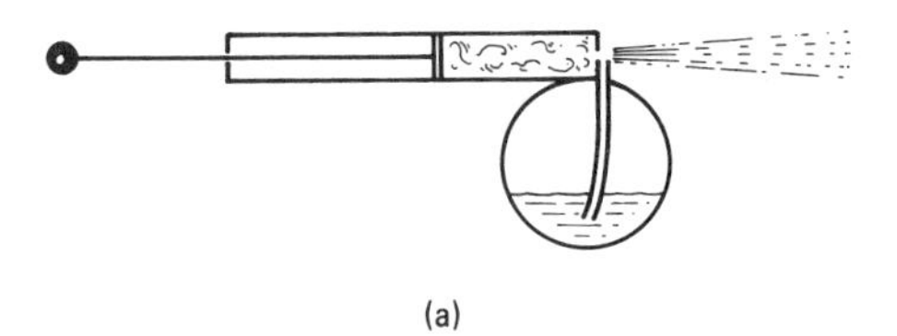

(a)

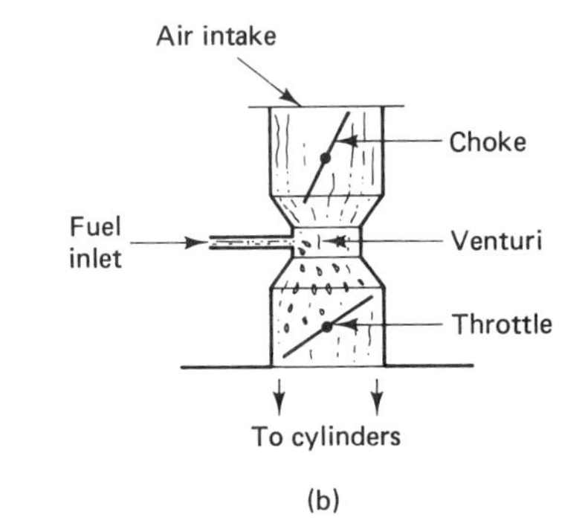

(b)

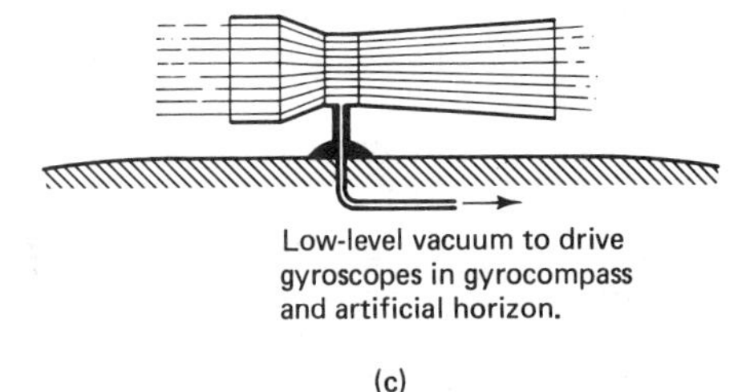

(c)

Figure 12-15 (a) Air rushing past the top of the tube produces a reduction in pressure that draws the liquid upward and into the spray of a common household sprayer. (b) The increased speed of air through the Venturi of a carburetor produces a reduction in pressure that pulls the fuel into the airstream. (c) Small aircraft of the 1940s derived a "suction" from a Venturi tube mounted on the side of the aircraft, which was used to drive the gyroscopic instruments.

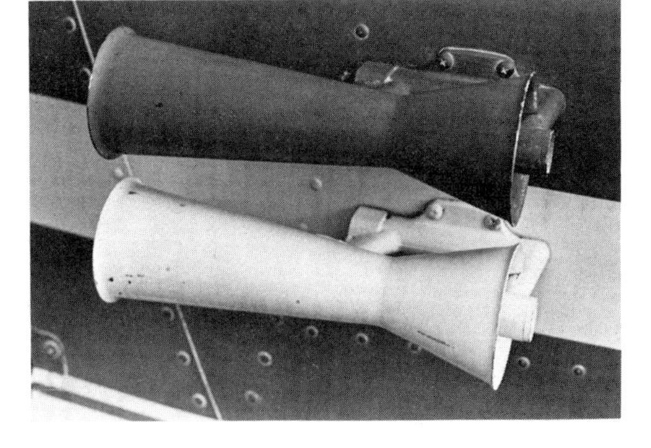

Two Venturi tubes mounted on the side of a small airplane. The tubes provide suction for the air-driven gryoscopic instruments.

Airfoils

The cross section of an airplane wing is called an *airfoil*; it is designed to produce lift by virtue of its motion through the air. The shape of the airfoil is such that the air flowing across the top surface has a greater velocity than the air flowing past the bottom surface. By Bernoulli's principle, the greater velocity at the top of the wing produces a smaller pressure than occurs underneath the wing. Hence, the upward force acting on the bottom of the wing is greater than the downward force acting on the top. The unbalance is the lift of the wing.

The velocity difference depends upon the shape and *angle of attack* of the airfoil (Fig. 12-16) and upon the speed with which the wing travels through the air.

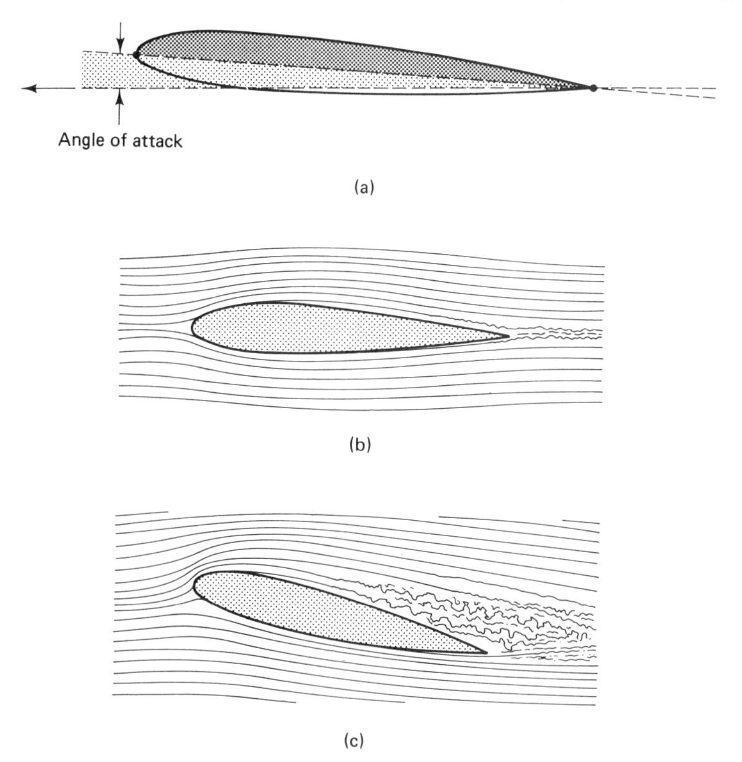

Figure 12-16 (a) Angle of attack of an airfoil. (b) Airflow pattern for a low angle of attack. (c) Airflow pattern for a greater angle of attack.

Lift is greater at high speeds and at high angles of attack. But if the angle of attack is increased beyond a certain critical value, the laminar flow around the airfoil may be disrupted so that turbulent flow ensues. The resulting condition of decreased lift and increased drag is known as a *stall*.

Calculation of the velocity difference between the top and bottom surface produced by a given airfoil is a complicated problem in fluid flow. We can see the nature of the effect, however, by making the simplifying assumptions that the velocity and pressure have the same values at all points of the respective surfaces and then applying Bernoulli's equation. A simplified expression for the lift of an airplane wing is

$$\text{Lift} = \tfrac{1}{2}\rho A(v_t^2 - v_b^2) \tag{12-21}$$

where A is the area of the wing, ρ is the density of the air, and v_t and v_b are the velocities at the top and bottom of the wing, respectively. By factoring the term involving a difference of squares and recognizing that $(v_t + v_b)/2$ is very nearly equal to the speed of the aircraft, which we denote as $\bar{v}$, we obtain the following expression for the lift:

$$\text{Lift} = \rho A \bar{v}(v_t - v_b) \tag{12-22}$$

From this we see that the lift depends upon the aircraft velocity, the density of the air, the area of the wing, and the velocity difference, $(v_t - v_b)$.

In an earlier section we described the boundary layer that forms on the surface of an airfoil. The behavior of this thin layer is very important in determining the properties of the wing. The drag is the force that must be overcome to maintain the motion of the wing through the air, and it depends strongly upon the manner in which the boundary layer becomes detached from the wing. Premature detachment may result in a thick wake and increased drag. These matters are for the specialist in the field of aerodynamics.

12-6 REYNOLDS NUMBER

One of the most important parameters in fluid flow is the Reynolds number, named in honor of the British engineer, Osborne Reynolds (1842–1912). It is given by

$$N_R = \frac{\rho v D}{\eta} \tag{12-23}$$

where ρ and η are the density and viscosity of the fluid, v is the velocity of the fluid flow, and D is the characteristic dimension of the particular physical configuration being considered. By and large, the numerical value of this dimensionless (unitless) parameter gives a good indication of whether the flow will be laminar or turbulent. Higher values of N_R indicate tendencies toward turbulent flow. Refer to Fig. 12-5 where the range of N_R for each type of flow is given.

The Reynolds number increases in direct proportion to the fluid density, fluid velocity, and size parameter D. It is inversely proportional to the viscosity. Viscous fluids, such as syrup or molasses, give rise to smaller values of N_R. The minute viscosities of gases typically yield fairly high values of N_R. Because N_R is dimensionless, the same numerical value will be obtained irrespective of the system of units employed.

Conceptually, viscosity and density may be easily confused, the tendency being to consider viscous fields to be naturally more dense, or vice versa. However, motor oil is *less dense*, but *more viscous*, than water. Warm oil is less viscous than cold oil even though a change of temperature has a negligible (in comparison) effect upon the density of the oil.

Reynolds discovered that fluid flow through a pipe will be laminar when N_R is less than about 2000. The flow will be turbulent when N_R is greater than about 3000. For values of N_R between 2000 and 3000, the flow may be either laminar or turbulent depending upon the other factors involved. Whether the flow is laminar or turbulent will have an effect upon the velocity profile across the diameter of the pipe (See Fig. 12-4).

EXAMPLE 12-4 Water flows through a 1-in. ID pipe at an average velocity of 2 ft/s. Calculate the Reynolds number and predict whether the flow is likely to be laminar or turbulent.

Solution. From Table 12-1, the viscosity of water is 0.001 N · s/m^2, and we recall that the density of water is 1000 kg/m^3. Converting to the MKS system (2 ft/s = 0.6096 m/s; 1 in. = 0.0254 m),

$$N_R = \frac{(1000 \text{ kg/m}^3)(0.6096 \text{ m/s})(0.0254 \text{ m})}{0.001 \text{ N} \cdot \text{s/m}^2}$$

$$= 15{,}484$$

The flow is definitely turbulent because N_R is far greater than 2000. This is reasonable because the corresponding flow rate is about 4.9 gal/min, which is rather large for a 1-in. diameter pipe. The water is obviously traveling at high velocity.

EXAMPLE 12-5 Compute the maximum velocity with which water could flow through a 1-in. ID pipe without producing a Reynolds number over 2000.

Solution. Rearranging the formula for N_R and utilizing the information of the preceding example, we have

$$v = \frac{\eta N_R}{\rho D} = \frac{(0.001 \text{ N} \cdot \text{s/m}^2)(2000)}{(1000 \text{ kg/m}^3)(0.0254 \text{ m})}$$

$$= 0.079 \text{ m/s} = 7.9 \text{ cm/s} = 3.10 \text{ in./s}$$

This velocity corresponds to a flow rate of about 0.6 gal/min.

Dynamic similarity

Suppose we have two identical 1-in. ID pipes, where water flows through one pipe and air flows through the other. Further, suppose that the average flow velocity of the air and water are the same. It is a simple matter to show that the resulting Reynolds number for the water is about 14 times as great as that of the air. Thus, the water is much more apt to be turbulent than the air. However, if the velocity of the air is increased by a factor of 14, giving an air velocity 14 times as great as the water velocity, the Reynolds number for the two systems would be the same. Under such conditions, the two systems are said to be *dynamically similar*, and the principle of *dynamic similitude* states that when the Reynolds number is the same for two similar systems of the same shape, the flow patterns in the two systems will be the same.

Dynamic similarity forms the basis for wind-tunnel testing in which a small model airplane is used to derive vital information about the performance characteristics of a full-size airplane of identical shape. Thus, engineers are able to predict the behavior of an aircraft before the test pilot ever takes it aloft. Similarly, a bridge can be tested for aerodynamic stability by placing a model of the bridge in a wind tunnel and selecting the proper wind velocity to give dynamic similarity. The value of such testing for aircraft, bridges, tall buildings, parachutes, and so forth needs no elaboration.

Suppose an airfoil is to be tested in a wind tunnel that is too small to accommodate a full-size section of the airfoil. A one-quarter-scale model is constructed for test purposes. If the airspeed of interest on the full-scale airfoil is 100 mi/h, what should be the air velocity used in the wind tunnel test to produce a dynamically similar situation?

Dynamic similitude requires equal Reynolds numbers for the two cases. Therefore, we write

$$\left(\frac{\rho}{\eta}\right)v_m D_m = \left(\frac{\rho}{\eta}\right)v_f D_f$$

where the subscript m relates to the model, while f denotes the full-scale airfoil. Because the density and viscosity are the same in both cases, the expression above can be simplified to give

$$v_m D_m = v_f D_f \quad \text{or} \quad v_m = \frac{D_f}{D_m}v_f \tag{12-24}$$

Consequently, for a one-quarter-scale model, D_f/D_m equals 4, and the wind velocity for the test should be

$$v_m = (4)v_f = 4(100 \text{ mi/h}) = 400 \text{ mi/h}$$

This velocity is well within reach even though at first it appears to be rather high. These same ideas can be used to design a test in liquid for an object ultimately intended to operate in air. Generally speaking, the design and testing of models is a complicated endeavor requiring the best efforts of experienced engineers.

Drag coefficient

It is often of interest to calculate the force of drag, F_D, exerted on an object either exposed to the wind or traveling through the air (or other fluid). This is a complicated problem for which there is no simple formula. The force of drag depends upon the shape of the object, the properties of the fluid, and the Reynolds number (which, in turn, depends on the velocity). By defining a *drag coefficient*—whose value is determined experimentally and expressed in graphical form—the problem of computing the drag can be negotiated in a straightforward manner for geometric shapes for which experimental values of C_D are available. The working equation is

$$F_D = C_D\left(\frac{\rho v^2}{2}\right)A \tag{12-25}$$

where

C_D = drag coefficient

ρ = density of the fluid

v = velocity of the fluid

A = projected area of the object

Note that $A = DL$ for a circular cylinder of diameter D and length L and $A = \pi D^2/4$ for a sphere of diameter D.

To use the above formula in a given situation, the Reynolds number is first calculated and is then used to determine C_D from the graph of Fig. 12-17. The variation of C_D with Reynolds number provides an indication of the overall complexity of the problem.

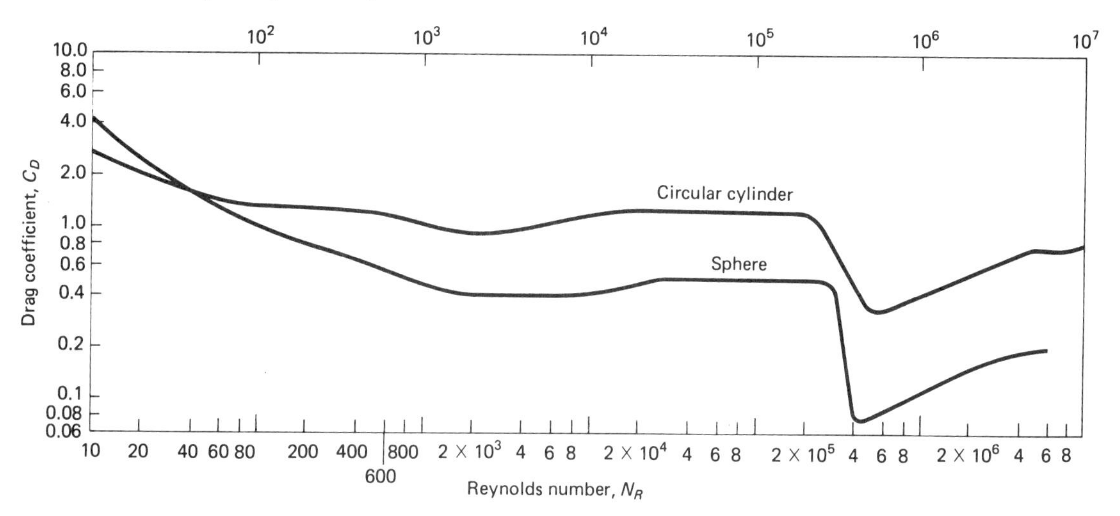

Figure 12-17 Variation of the drag coefficient for a sphere and a circular cylinder.

Dimples on a golf ball decrease aerodynamic drag. Without dimples, most par-four holes would be par sixes

EXAMPLE 12-6 Determine the drag on a sphere 1 cm in diameter traveling through air at a velocity of 60 m/s.

Solution. We first compute the Reynolds number:

$$N_R = \frac{\rho v D}{\eta} = \frac{(1.189)(60)(0.01)}{18 \times 10^{-6}} = 39{,}600 \qquad \text{(MKS units)}$$

From Fig. 12-17, the drag coefficient for a sphere at Reynolds number 39,600 is about 0.5. The projected area for the sphere in question is 7.85×10^{-5} m^2. Hence,

$$F_D = C_D\left(\frac{\rho v^2}{2}\right)A$$

$$= (0.5)\left(\frac{(1.189)(60)^2}{2}\right)(7.85 \times 10^{-5})$$

$$= 0.084 \text{ N}$$

If the sphere is made of brass of mass density 8500 kg/m^3, it will weigh 0.0436 N. Therefore, the drag on the sphere is nearly twice its weight.

12-7 SUPERSONIC FLOW

When fluid flow velocities are increased until they approach the speed of sound, many factors come into play which are negligible at lower velocities. Air, for example, can no longer be considered incompressible. Shock waves (sonic booms) are produced as an airplane or projectile traveling at supersonic speeds outruns the sound that it produces. Bernoulli's equation, in the form of Eq. (12-13), can no longer be used because its derivation is based upon the assumption that the fluid is incompressible. Obviously, mathematical complexities preclude any simple analysis. At this point, we briefly describe what is perhaps the most apparent effect of supersonic flight, the sonic boom.

The formation of a shock wave that produces a sonic boom is illustrated in Fig. 12-18. In part (a), the current position of the plane is shown, along with its position at five previous times. The circles represent cross sections of spherical sound waves produced by the plane at the various positions. Note that the plane punctures through each wavefront as it is produced, and observe in part (b) that the leading edges of the waves join to form a conical shock wave, called the *Mach cone*, with the plane at the apex. A sonic boom is experienced where the Mach cone interesects the surface of the earth, as illustrated in part (c).

The ratio of the plane's velocity v_p to the speed of sound in the surrounding air is called the *Mach number*:

$$\text{Mach number} = \frac{v_p}{v_s} \tag{12-26}$$

Any speed having Mach number of 1.0 or greater is supersonic. The apex angle (called the *Mach angle*) of the Mach cone depends upon the Mach number. It is given by

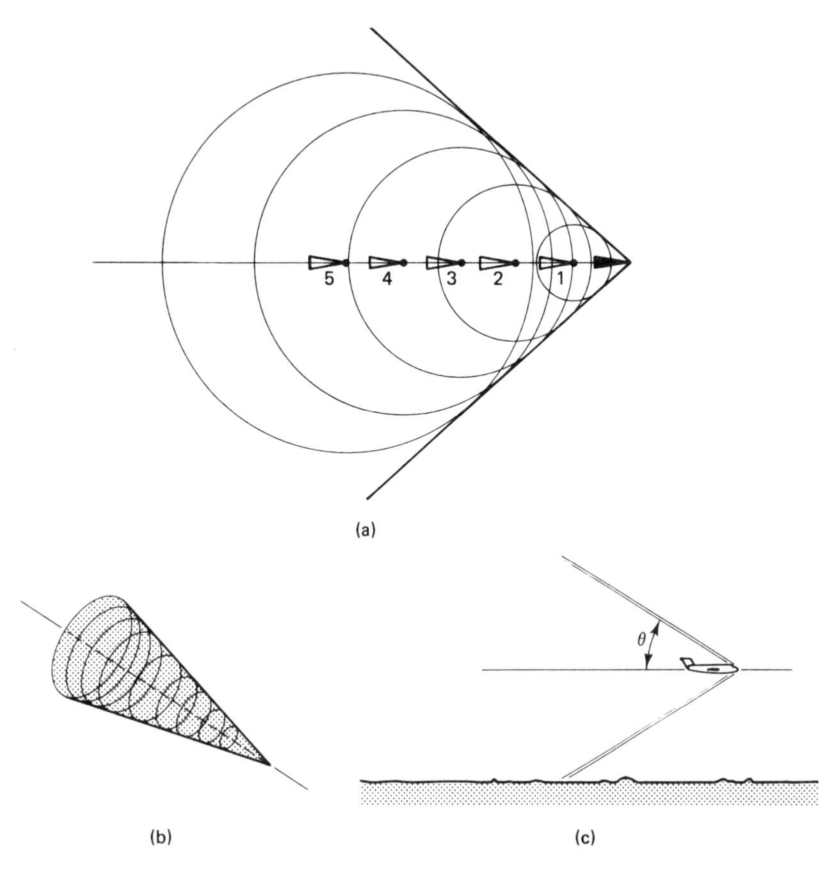

Figure 12-18 Mach cone.

$$\sin\theta = \frac{1}{\text{Mach number}} = \frac{v_s}{v_p} \qquad \theta = \sin^{-1}\left(\frac{v_s}{v_p}\right) \tag{12-27}$$

A person being approached by a subsonic projectile (tennis ball, artillery shell, or ordinary airplane) will hear the sound of the oncoming object before it passes by. On the other hand, a supersonic projectile will not be heard until it arrives, and the initial sound is the blast of a sonic boom. The passing of a supersonic plane produces an unnerving blast (the sonic boom) followed by the relatively faint (if even audible) normal sound of the plane as the Mach cone passes.

The Mach number, cone, and so on are named in honor of Ernst Mach (1838–1916), Austrian physicist, psychologist, and philosopher.

To Go Further

In an encyclopedia, read about:

Daniel Bernoulli (1700–1782), Swiss mathematician
Jean Poiseuille (1799–1869), French physicist
Ernst Mach (1838–1916), Austrian physicist
Evangelista Torricelli (1608–1647), Italian scientist

References

COMBS, HARRY, with MARTIN CAIDIN, *Kill Devil Hill*. Boston: Houghton Mifflin Company, 1979. The story of the Wright Brothers and their development of the first airplane; highly recommended.

SHAPIRO, ASCHER H., *Shape and Flow*. New York: Anchor Books Doubleday & Co, Inc., 1961. A small volume that presents the basic concepts of fluid flow with many illustrations and photographs.

VON KARMAN, T., with LEE EDSON, *The Wind and Beyond: Theodore von Karman*. Boston: Little, Brown, 1967. A biography of the noted aerodynamicist, Theodore von Karman.

Questions

1. Water flows through a large pipe that necks down to a smaller pipe one-half the diameter of the large pipe.
   (a) In which pipe is the fluid velocity greater?
   (b) How does the flow rate vary from one pipe to the other?
2. What does a rotameter have in common with a parachute?
3. Describe the Karman vortex street. When a vortex is formed and snaps off, a force is exerted on the member. What are some of the practical implications of these forces?
4. Early aviators judged airspeed by listening to the hum of the wires that constituted a large portion of their craft. What causes the wires to hum? Why does a gentle breeze through pine trees sound like a gentle breeze through pine trees?
5. Explain how the condition of zero slip gives rise to a boundary layer.
6. Distinguish carefully between density and viscosity. Give examples of liquids that are:
   (a) very dense but not very viscous;
   (b) very viscous but not very dense.
7. What supplies the force that causes the acceleration when a flowing fluid is speeded up?
8. What principle makes it possible to use scale models of large aircraft for wind-tunnel tests instead of having to build a wind tunnel large enough to accommodate the full-sized aircraft?
9. Do smooth objects always create less drag than rough objects? Why is a golf ball dimpled?
10. Certain spidery-looking crawley things have the much revered ability to walk on water and without getting their feet wet. How are they able to do this?
11. Under certain conditions the flow from a water faucet can be laminar. In such case, the stream is observed to shrink in diameter with distance from the faucet until the stream breaks up. Why does the diameter become smaller, and why does it eventually break up?

Problems

1. What is the flow rate in gallons per minute through a 1-in. ID pipe if the average velocity of the flowing liquid is 4 ft/s?
2. Water passes from a 2-in. ID pipe through a coupling to a 4-in. ID pipe. The flow velocity is 6 ft/s in the smaller pipe. Calculate: **(a)** the flow velocity in the larger pipe; **(b)** the flow rate (gallons/minute) in the smaller pipe; **(c)** the flow rate in the larger pipe.
3. The average speed of the current in a river is 2 ft/s. The river is 100 ft wide, and its average depth is 4 ft. Calculate the flow rate in gallons/second of water down the river.

4. The stream of water issuing from a certain faucet is 0.75 in. in diameter. The stream will fill a 1-gal milk jug in 30 s. Compute the velocity of the stream as it leaves the faucet.

5. Suppose two parallel plates of area 0.25 m^2 are immersed in glycerin at 20°C, as in Fig. 12-7(b). The separation between the plates is 0.02 m, and the top plate moves relative to the bottom plate at a velocity of 0.05 m/s. Compute: **(a)** the rate of shearing strain; **(b)** the force that must be exerted on the top plate to overcome the effects of viscosity.

6. Calculate the kinetic viscosity of: **(a)** water; **(b)** mercury.

7. A glass sphere of radius 0.005 m and specific gravity 2.6 attains a terminal velocity of 4.93 cm/s while falling through glycerin at 20°C (S_{gr} = 1.26). Use this data to calculate the viscosity of glycerin.

8. A tiny red sphere intended for viscosity determinations is 3 mm in diameter and has a specific gravity of 0.850. Compute the terminal velocity of this sphere in kerosene of specific gravity 0.820 and viscosity 0.0018 $N \cdot s/m^2$.

9. Refer to Fig. 12-8 and verify that Bernoulli's equation is satisfied by the following set of parameters. The fluid is water, and its flows from a 4-in. pipe at the lower level into a 2-in. pipe 3 meters higher.

$$y_1 = 2 \text{ m} \qquad v_1 = 0.25 \text{ m/s}$$

$$y_2 = 5 \text{ m} \qquad v_2 = 1.00 \text{ m/s}$$

$$P_1 = 405{,}300 \text{ N/m}^2 \qquad A_1 = 4A_2$$

$$P_2 = 375{,}431 \text{ N/m}^2 \qquad \rho = 1000 \text{ kg/m}^3$$

10. A horizontal pipe 3 in. in diameter connects via a sloping, tapered fitting to a 4-in. diameter pipe 1.5 m lower. Water flows in the 3-in. pipe with a velocity of 2 m/s under a pressure of 2 atm.
    (a) Compute the velocity of water in the 4-in. pipe.
    (b) Use Bernoulli's equation to calculate the pressure in the 4-in. pipe.
    (c) Suppose the flow ceases so that the water is stationary in the pipes. Calculate the difference in pressure between the two levels under static conditions and compare with the pressure difference obtained in part (b).

11. A Venturi tube (Fig. 12-9) is used to determine the velocity with which dry air at 20°C (density = 1.20 kg/m^3) flows through a pipe. The throat area of the Venturi is one-fourth the area of the pipe, and oil of specific gravity 0.8 is used as the manometer fluid. With the flow in progress, the difference in height of the manometer columns is 1 cm. Compute the flow velocity; **(a)** in the pipe; **(b)** in the throat of the Venturi. **(c)** For a 4 in. ID pipe, compute the flow rate of air through the pipe in cubic feet per minute.

12. Assume the Venturi of Problem 11 is used to measure the flow of water, with mercury used as the manometer fluid.
    (a) Compute the flow velocity of the water when the manometer columns differ in height by 1 cm.
    (b) Calculate the flow rate in gallons per minute through the 4-in. ID pipe.

13. A Venturi mounted on the side of a small plane provides a vacuum for driving the gyroscopes of the gyrocompass, artificial horizon, and turn indicator. The ratio of inlet area to throat area of the Venturi tube is 4:1. Compute the pressure difference in pounds per square inch existing between inlet and throat of the Venturi at the cruise speed of 100 mi/h.

14. Compute the airspeed of a plane flying in air of density 1.2 kg/m^3 when the pressure within the Pitot tube is **(a)** 1000 N/m^2; **(b)** 1200 N/m^2; **(c)** 1400 N/m^2.

15. Suppose a Pitot tube is mounted on the front of a car, and a homemade manometer filled with water is used to measure the Pitot pressure. What will be the difference in height of the water columns (in inches) at: **(a)** 30 mi/h; **(b)** 60 mi/h; **(c)** 90 mi/h?

16. A tank, pressurized as in Fig. 12-11(b), contains water whose top surface is 8 ft above a small hole near the bottom of the tank. Calculate the speed of efflux of water from the hole if the air pressure (gauge) is: **(a)** 40 lb/in^2; **(b)** 20 lb/in^2; **(c)** zero; **(d)** -5 lb/in^2.

17. The water level in a tall tank is 12 ft above

ground level, and a small hole is drilled in the side of the tank 5 ft above ground level. What horizontal distance from the bottom of the tank does the issuing stream of water strike the ground?

18. A stream of water from a hole in an elevated water tank strikes the ground a horizontal distance 2 m from a point directly underneath the hole. The hole is 3 m above the ground.
- **(a)** Calculate the speed of efflux of water from the hole.
- **(b)** What distance above the hole is the water level in the tank?

19. Calculate the Reynolds number resulting when glycerin flows through a 1-in. diameter pipe at a velocity of 1 m/s.

20. A stream of water from a faucet exhibits laminar flow. The diameter of the stream at a certain point is 4 mm, and the flow rate is 350 cm³/min.
- **(a)** Calcuate the flow velocity at the point where the diameter of the stream is 4 mm.
- **(b)** Verify that this situation meets the general rule that N_R must be less than 2000 for laminar flow to occur.

21. A vertical rod 0.05 m in diameter is drawn through a bath of glycerin (as in Fig. 12-5) at a velocity of 0.1 m/s. Compute the Reynolds number, and predict the type of trail that will be left behind the rod.

22. Compute the maximum velocity with which a glass rod 4 mm in diameter can be drawn through water without having the Reynolds number exceed 1.0.

23. With what velocity should the rod of Problem 22 be drawn through water if a Karman vortex trail is to be formed?

24. Compute the force of drag on a sphere 10 cm in diameter moving through the air at 60 mi/h.

25. A cylindrical strut on an airplane is 5 ft long and 1 in. in diameter.
- **(a)** Calculate the drag on the strut at 120 mi/h.
- **(b)** Recalling that power equals force times velocity, compute the horsepower expended in moving the strut through the air at 120 mi/h.

***26.** A fiberglass exhaust stack is 2 ft in diameter and extends 25 ft above a flat roof. Calculate the lateral force on the stack in a wind of 30 mi/h.

***27.** By trial and error, determine the approximate terminal velocity in air of a sphere 6.5 cm in diameter that weighs 0.54 N (a tennis ball). Remember that terminal velocity results when the drag equals the weight of the object.

28. The velocity of a sound in a certain region is 340 m/s. Determine the apex angle of the Mach cone for a plane in the same region traveling at: **(a)** 344 m/s; **(b)** 354 m/s; **(c)** 380 m/s; and **(d)** 500 m/s.

29. Compute the Mach number for each velocity in Problem 28.

CHAPTER 13

HEAT AND TEMPERATURE

This chapter begins with a description of heat and temperature and the difference between the two. Four temperature scales and the concept of absolute zero are presented.

The second section deals with heat capacity, the ability of objects to store thermal energy. A method is given for determining the heat capacity of a solid object by using a simple calorimeter.

We shall see how to calculate the thermal expansion of a rod that occurs when its temperature is increased, and we shall see how to design a thermometer.

Other topics include phase changes such as melting and freezing, and we see under what conditions it is possible for a substance to absorb heat without experiencing a rise in temperature. Procedures are given for determining the heat of fusion and heat of vaporization of water.

The last sections include discussions of humidity, dew point, boiling, *P-V-T* diagrams, heat of combustion, and the mechanical equivalent of heat. This is a full chapter, which should provide many new insights into familiar experience.

13-1 HEAT AND TEMPERATURE

If we could somehow examine a substance and actually see the atoms and molecules, we would observe one noteworthy feature of all substances—whether solid, liquid, or gas. That common feature is molecular motion. The atoms and molecules of all materials are continually moving; a given molecule will almost never find itself sitting still. And on the rare occasion that it should happen, a bump, jar, or other excitation from a neighboring molecule will soon set the stationary molecule in motion again.

In a crystalline solid, the molecular motion takes the form of a fervent vibration back and forth around the lattice site. In a liquid, a particular molecule is not bound to one locality; the molecules move around at random throughout the entire volume of the liquid. In a gas, the molecules enjoy an almost perfect freedom, being constrained only by the walls of the containing vessel and a very weak attraction for each other. Thus, there is ample opportunity for molecular motion to occur in all three states of matter.

A rifle bullet has kinetic energy because of its high velocity. Likewise, each atom or molecule of a solid, liquid, or gas has kinetic energy because of its motion. The total collective kinetic energy of all the atoms and molecules is called *heat*, or *thermal energy*. Consequently, heat energy is really kinetic energy. Can an object sitting still have kinetic energy? The answer is yes if we consider the kinetic energy of the individual atoms and molecules, but it is better to refer to this energy as either heat or thermal energy to avoid confusion.

The kinetic energy of a rifle bullet is sometimes referred to as *ordered* kinetic energy because all the atoms or molecules of the bullet travel in the same direction. The kinetic energy of a hot cup of water, however, is *disordered* kinetic energy because of the randomly directed molecular velocities. Furthermore, a molecule may rotate as well as travel in a straight line, and the atoms comprising a particular molecule may vibrate back and forth relative to each other. Consequently, the heat energy of a given material may consist of the kinetic energy of translation of the molecule as a whole, the rotation of the molecule as a whole, and the vibration of the atoms relative to each other, as shown in Fig. 13-1.

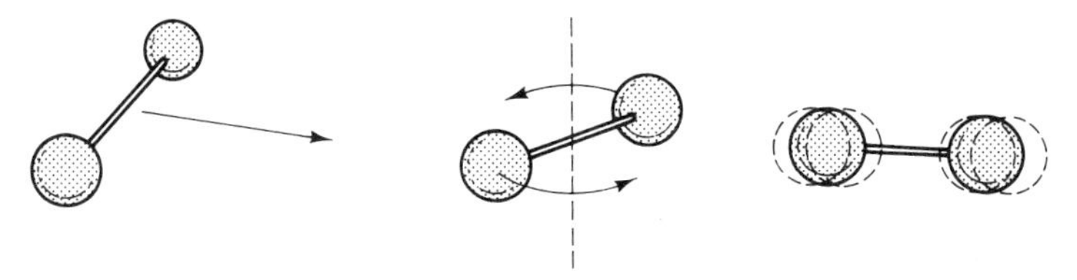

Figure 13-1 A molecule may possess energy by virtue of its translation, rotation, or vibration.

Thus far, nothing has been said about temperature. It is well known that temperature has to do with *hot* and *cold*, but what is the significance of temperature on the atomic or molecular level? As it turns out, the answer is fairly simple. The temperature of a substance is a measure of the average kinetic energy possessed by the molecules of the substance. The higher the temperature, the more pronounced the molecular motion. The term *hot* refers to high temperatures and *cold* refers to low temperatures. Cold is merely the absence or lack of heat; it is simply the opposite of hot and has no rigorous scientific meaning. We often speak of the flow of heat, but not the flow of cold.

Absolute zero

Suppose we devise some means of slowing the molecular motion so that we diminish the average kinetic energy of a given collection of molecules. As we do this, the temperature becomes lower because the temperature is a measure of the average kinetic energy per molecule. Therefore, as the molecules begin to move slower and slower, the temperature becomes lower and lower. Can this process be continued indefinitely so that we could achieve temperatures as low as we wished? The answer is a definite no because after a while, the molecules will be moving so slowly that they will have been stopped completely. A molecule cannot possibly move more slowly than when it is sitting perfectly still. Therefore, the lowest possible temperature is that temperature at which all molecular motion ceases. This temperature is called *absolute zero*; it is equivalent, approximately, to −460°F or −273°C.

It is theoretically impossible to achieve a temperature of absolute zero. It would be far easier to arrange for all the leaves on a huge oak tree to be perfectly motionless than to arrange for all the molecules of even a tiny sample of material to be perfectly still. Even so, temperatures have been achieved in rigorously controlled laboratory experiments that come to within 0.000001°C of absolute zero. The field of study of very low temperatures is called *cryogenics*.

On the other end of the temperature scale, there is no physical principle that limits the maximum temperature possible. Temperatures on the order of millions of degrees occur within the explosion of a hydrogen bomb and in the interior of stars.

Temperature scales

Four temperature scales are commonly encountered in scientific literature, and all are named in honor of pioneers in the field of heat and thermodynamics. These are Fahrenheit, F; Celsius, C; Rankine, R; and Kelvin, K. Of these scales, two (R, K) are absolute, and two (F, C) are relative. The Fahrenheit and Rankine scales use the Fahrenheit degree while the Celsius and Kelvin scales use the Celsius degree. The scales are illustrated in Fig. 13-2, and formulas for converting from one scale to any other are given.

Temperature scales are defined relative to certain reproducible fixed points, the most familiar being the *ice point* and the *boiling point* of water. The ice point is the temperature at which ice and air-saturated water exist in equilibrium. The size of the Celsius and Fahrenheit degrees results from dividing the temperature difference between these two points into 100 Celsius degrees and into 180 Fahrenheit degrees. Consequently, a change in temperature of 1°C is the equivalent of a change of 1.8°F. Current practice utilizes only one fixed point, the *triple point of water*, which is the temperature at which ice, pure water, and water vapor exist in equilibrium. The construction of a *triple-point cell* is shown in Fig. 13-3. The triple point is arbitrarily assignd the temperature of 0.01°C and 273.16 K. This assignment results in a temperature for the ice point of 0.0°C. Note that the ice point is 0.01°C lower than the triple point. This definition also gives the *steam point* a value of exactly 100.00°C. The steam point is the temperature at which pure water and steam exist in equilibrium at exactly 1 atm of pressure.

A temperature of zero on the absolute scales corresponds to the point at which the thermal energy of the molecules is zero. Because kinetic energy cannot be negative, there can be no negative absolute temperatures. On the other hand, the assignment of a temperature of zero on the Celsius and Fahrenheit scales is a

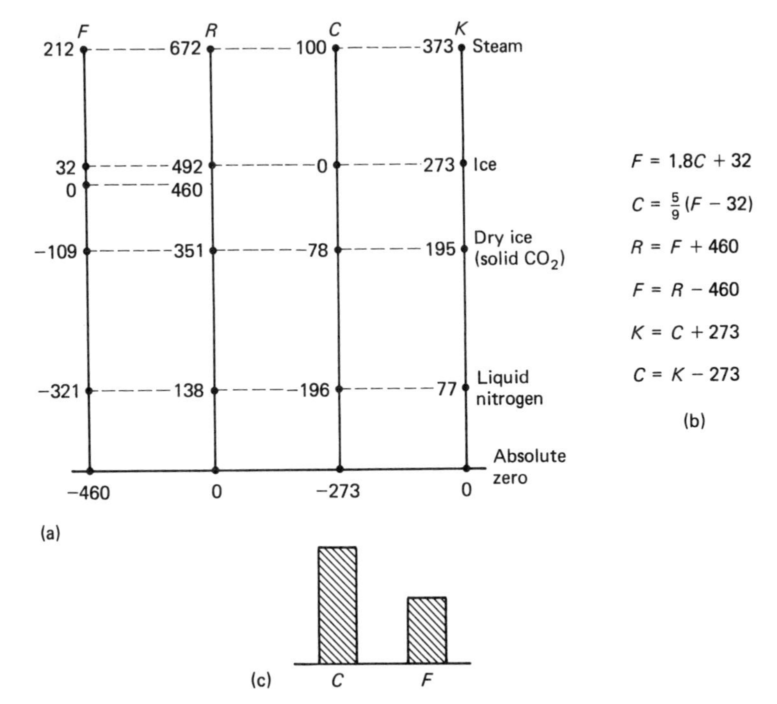

Figure 13-2 (a) Comparison of the four temperature scales.
(b) Conversion formulas.
(c) Comparison of the Fahrenheit and Celsius degree.

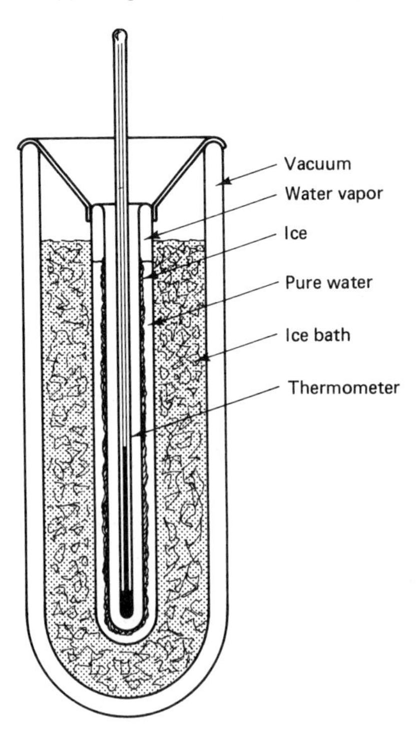

Figure 13-3 A triple-point cell for establishing a temperature of 273.16 K.

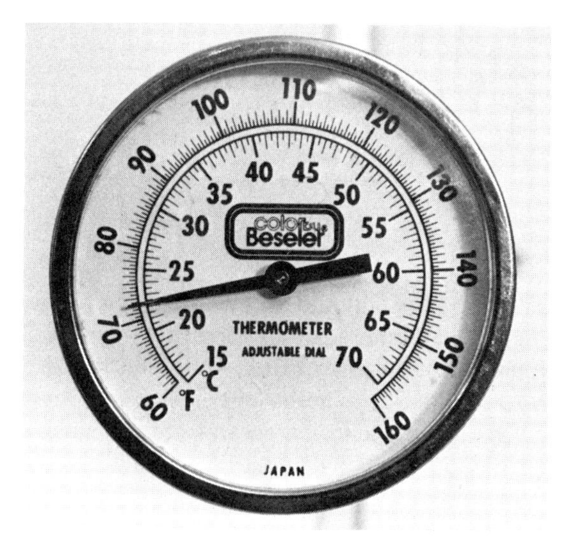

The correspondence between the Celsius and Fahrenheit scales is evident in this photo of a dial thermometer.

rather arbitrary matter. Consequently, negative temperatures on these scales are commonplace but have no particular physical significance.

Units of heat energy

Because heat is just another form of energy, there is no need to introduce new units in order to measure quantities of heat. Joules and ergs are perfectly valid units for the measurement of heat. Historically, however, two other units, the calorie and the BTU (for British thermal unit) have been defined and are still widely used. Hence, we define these units here. Conversion factors are given in Table 13-1.

A calorie is that quantity of heat required to raise the temperature of 1 g of water 1°C between 14.5 and 15.5°C.

A BTU is that quantity of heat required to raise the temperature of 1 lb of water 1°F between 58.5 and 59.5°F.

The BTU is obviously the larger quantity of heat because 1 lb of water is much greater than 1 g of water. A unit conversion reveals that one BTU is the equivalent of 252 cal.

Obviously, a kilocalorie (kcal) is a quantity of heat 1000 times as large as 1 cal. However, an equivalent unit known as a large calorie is sometimes used without explanation and is often simply called a calorie. This creates confusion. The most familiar example is in expressing the energy content of foods. The food calorie is a large calorie, the equivalent of 1000 of the smaller calories just defined.

TABLE 13-1 CONVERSION FACTORS FOR THERMAL ENERGY UNITS

1 BTU	= 252 cal = 1055 J = 778 ft-lb
1 kcal	= 4186 J = 3087 ft-lb = 3.968 BTU
1 cal	= 4.186 J = 3.087 ft-lb
1 J	= 0.7376 ft-lb = 10^7 ergs
1 W · h	= 3600 J = 860 cal = 3.413 BTU

13-2 HEAT CAPACITY AND SPECIFIC HEAT

The *heat capacity* of an object, a quantity of liquid, or a volume of gas refers to the amount of heat required to raise the temperature of the system by 1°. The heat capacity is directly proportional to the mass of the system, and it also depends upon the type of material involved. A large quantity of material will obviously be able to absorb more heat for a given change in temperature than a small quantity of the same material. It is less obvious, however, that the same mass of different materials is able to absorb differing quantities of heat for the same change in temperature.

The experimental arrangement of Fig. 13-4 provides a vivid demonstration of heat capacity. A calorimeter made of material that is a good thermal insulator (styrofoam, for example) contains a known mass of liquid. A battery provides electrical power for an electric heater immersed in the liquid. A thermometer indicates temperature changes, and a stirrer is provided for keeping the temperature of the liquid the same at all points.

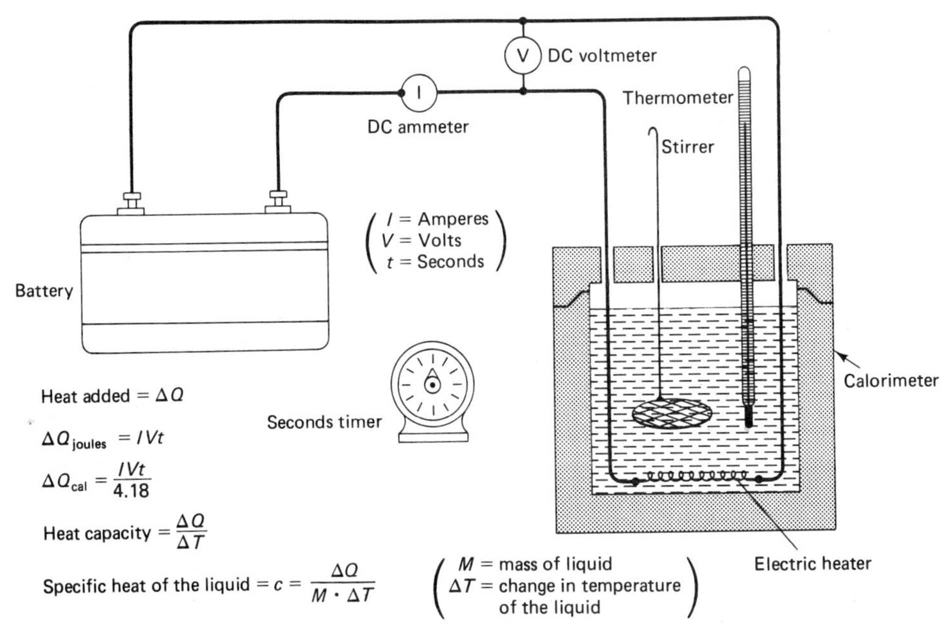

Figure 13-4 An experimental arrangement for determining the specific heat of a liquid.

When the electrical connection is made to the battery, energy begins to flow to the heater at the rate of $I \cdot V$ joules per second (watts). If the circuit operates for t seconds, the total number of joules added to the liquid inside the calorimeter is IVt. Because the electrical energy is converted to thermal energy, the heat absorbed by the liquid, ΔQ, is given by IVt. The absorption of the heat produces a temperature rise ΔT of the liquid, which is indicated on the thermometer. Hence, we can compute the heat capacity directly:

$$\text{Heat capacity} = \frac{\Delta Q}{\Delta T} \qquad \text{(units: J/degree or cal/degree)} \tag{13-1}$$

In this experiment, we have assumed that the heat absorbed by the walls of the calorimeter is small enough to ignore.

EXAMPLE 13-1 A calorimeter equipped with an electric heater, as in Fig. 13-4, contains 325 g of water at an initial temperature of 15°C. When the electrical circuit is energized, the voltmeter indicates a voltage of 12 V and the ammeter indicates a current flow of 4 A. After 4 min 40 s, the heater is switched off, and the thermometer soon stabilizes at a final temperature of 25°C. Compute the heat capacity of the water in the calorimeter.

Solution. The heat energy ΔQ added to the water because of the heater is given by

$$\Delta Q = IVt = (4\text{ A})(12\text{ V})(280\text{ s})$$
$$= 13{,}440\text{ J}$$

Using the definition of heat capacity [Eq. (13-1)] and noting that the temperature of the water changed 10°C, we get

$$\text{Heat capacity} = \frac{\Delta Q}{\Delta T} = \frac{13{,}440\text{ J}}{10°\text{C}}$$
$$= 1344\text{ J/°C}$$

Heat capacity is commonly given in terms of calories per degree Celsius. Because 1 cal is the equivalent of 4.18 J, a unit conversion gives

$$\text{Heat capacity} = \left(1344\,\frac{\text{J}}{°\text{C}}\right)\left(\frac{1\text{ cal}}{4.18\text{ J}}\right)$$
$$= 321\text{ cal/°C}$$

Notice that this is very nearly equal to the mass of the water in grams. Had the experiment been done with perfect accuracy, we would have obtained a value of 325 cal/°C.

Specific heat

The specific heat of a substance is the heat required to produce a 1° change in temperature of unit mass of the material. It is the heat capacity per unit mass of the material. If a mass M of the material in question has a heat capacity given by

$$\text{Heat capacity} = \frac{\Delta Q}{\Delta T}$$

the specific heat capacity c is obtained by dividing this quantity by the mass of material:

$$c = \frac{1}{M}\left(\frac{\Delta Q}{\Delta T}\right) \quad \text{or} \quad c = \frac{\Delta Q}{M(\Delta T)} \tag{13-2}$$

Typical units of specific heat are

$$\frac{\text{J}}{\text{g-°C}} \quad \text{or} \quad \frac{\text{cal}}{\text{g-°C}}$$

TABLE 13-2 SPECIFIC HEATS OF SELECTED MATERIALS

Material	cal/g-°C	J/kg-°C
Aluminum	0.22	920
Brass	0.94	393
Copper	0.093	390
Iron	0.11	460
Lead	0.031	130
Mercury	0.033	138
Silver	0.056	234
Tin	0.054	226
Zinc	0.093	389
Ice	0.55	2302
Glass	0.12	502
Granite	0.19	803
Acetone	0.53	2210
Ethyl alcohol	0.58	2428
Glycerin	0.56	2344
Olive oil	0.47	1967
Paraffin	0.69	2888
Sugar	0.27	1130
Wood	0.42	1758
Water	1.00	4186

EXAMPLE 13-2 Determine the specific heat of water from the information given in Example 13-1.

Solution. In Example 13-1, we determined the heat capacity of 325 g of water to be 321 cal/°C. Dividing this by 325 to put it in terms of unit mass gives us

$$\text{Specific heat} = \frac{321 \text{ cal/°C}}{325 \text{ g}} = 0.988 \frac{\text{cal}}{\text{g-°C}}$$

If we examine the definition of a calorie given earlier, we see that the specific heat of water must be exactly 1.0 cal/g-°C, by definition. Thus, we conclude that an error was made somewhere in the experiment of Example 13-1. Hence,

$$\textit{Specific heat of water} = 1.0 \frac{cal}{g\text{-}°C}$$

In this discussion, we have ignored the slight variation of the specific heat of water with temperature that occurs, as shown graphically in Fig. 13-5.

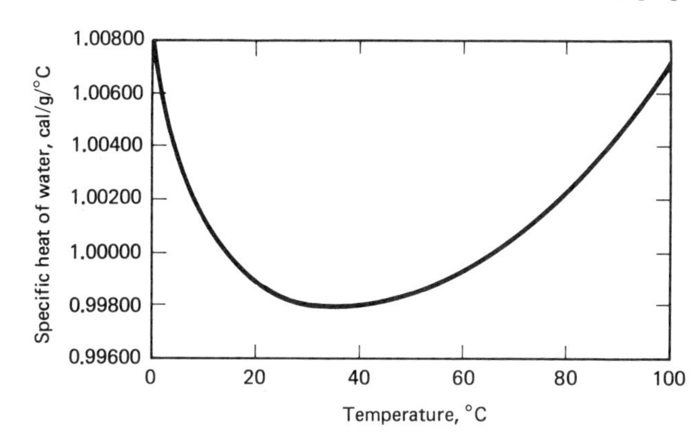

Figure 13-5 Specific heat of water vs. temperature.

Heat transfer vs. temperature change

If we know the mass M of a substance of specific heat c, we can compute the heat lost or gained by the substance when it undergoes a temperature change ΔT. The heat transferred to or from the substance is given by

$$\Delta Q = Mc \cdot \Delta T \tag{13-3}$$

This is a very important relationship, which is obtained directly from a simple rearrangement of Eq. (13-2). Example 13-3 illustrates the use of this relationship.

EXAMPLE 13-3 Compute the heat that must be removed from a 320-g sample of copper in order to cool it from a temperature of 98°C to 20°C. The specific heat of copper is 0.093 cal/g-°C.

Solution. A direct substitution into Eq. (13-3) gives

$$\Delta Q = Mc \cdot \Delta T = (320 \text{ g})(0.093 \text{ cal/g-°C})(98\text{°C} - 20\text{°C})$$
$$= 2321 \text{ cal}$$

To illustrate the fairly low specific heat of copper, 24,960 cal would have to be removed from an equal mass of water to effect the same temperature change.

Basic calorimetry

Suppose two containers of an identical liquid are poured together, as illustrated in Fig. 13-6. If the two masses are different and if the temperatures are different, we might wish to calculate the final temperature of the combined liquid. For purposes of simplification, we assume that no heat is lost to the surroundings and we neglect the heat capacity of the containers.

We attack the problem from the point of view of heat exchange between the warm and the cool liquid. Because no heat is lost to the surroundings (an assumption) and because heat is energy and simply cannot disappear, the heat lost by the warm liquid must equal the heat gained by the cool liquid as it is warmed by mixing with the warmer liquid. Thus, we use Eq. (13-3) to write

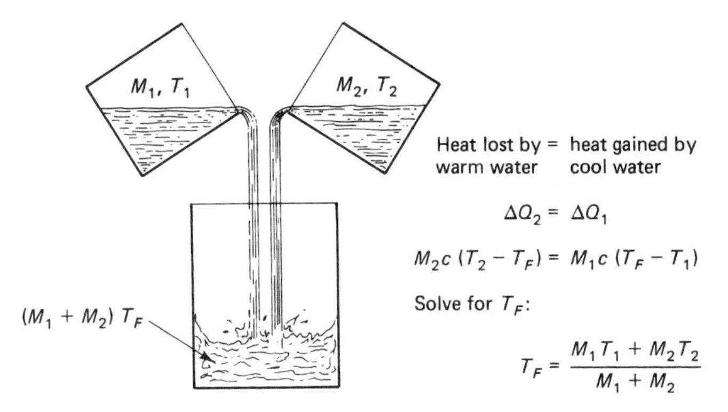

Figure 13-6 Calculation of the final temperature resulting from mixing two quantities of water at different temperatures.

$$\text{Heat lost by warm liquid} = \text{heat gained by cool liquid}$$

$$M_2\not{c}(T_2 - T_f) = M_1\not{c}(T_f - T_1)$$

Because the specific heat c appears on both sides of this expression, it drops out. The resulting expression may be solved for the final temperature:

$$T_f = \frac{M_1T_1 + M_2T_2}{M_1 + M_2} \tag{13-4}$$

The volume of a liquid is directly proportional to its mass, so this expression for the final temperature may be written in terms of volumes by simply writing V_1 for M_1 and V_2 for M_2.

Determination of specific heat

It is a simple matter to determine the specific heat of a metal specimen by heating the specimen to a known temperature and then placing it into a calorimeter containing water, as shown in Fig. 13-7. The specimen may be conveniently heated to 100°C by placing it in boiling water. The water in the calorimeter will be warmed when the specimen is inserted, and the thermometer will indicate the temperature change that occurs. The *heat balance equation* is

$$\begin{array}{c}\text{Heat lost by}\\ \text{specimen}\end{array} = \begin{array}{c}\text{heat gained}\\ \text{by water}\end{array} + \begin{array}{c}\text{heat gained by}\\ \text{calorimeter can}\end{array} \tag{13-5}$$
$$M_sc_s(T_s - T_f) = M_wc_w(T_f - T_w) + M_cc_c(T_f - T_w)$$

Note that we are including the heat-absorbing properties of the calorimeter can. When the above expression is solved for the specific heat of the specimen, the expression given in Fig. 13-7 results.

13-3 THERMAL EXPANSION

Because the amplitude of molecular vibration increases with the increasing temperature of a solid, it is not surprising that the average distance between molecules increases with an increase in temperature. A solid expands as it is

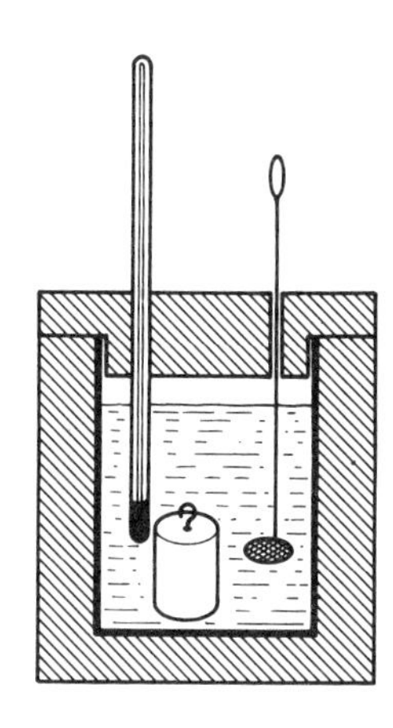

T_w = initial temperature of water

T_f = final temperature of water

T_s = initial temperature of specimen

M_w = mass of water

M_c = mass of calorimeter can

M_s = mass of specimen

c_w, c_s, c_c = specific heat of water, specimen, and calorimeter can

$$c_s = \frac{(M_w c_w + M_c c_c)(T_f - T_w)}{M_s(T_s - T_f)}$$

Figure 13-7 Determination of specific heat.

heated. A similar effect occurs in liquids. The expansion of liquids per degree rise in temperature is generally about ten times as great as the expansion of solids. We study the expansion of gases in a following chapter.

Effects of thermal expansion are exceedingly important. Building materials, structural components, machine parts—all materials change in size when they experience a change in temperature. Serious problems can result if these dimensional changes are not anticipated. Expansion joints must be included in the design of bridges, sidewalks, and long sections of pipelines. Sufficient clearance must be provided between moving machine parts so that the parts will not seize when a change in temperature occurs. The steel used to fabricate steel-reinforced concrete structures must have thermal expansion properties identical to those of the concrete in order to avoid cracking and crumbling caused by uneven expansion of the two components. If the temperature of an object must be changed over a wide temperature range, it is important that the temperature be changed gradually so that uneven shrinkage or expansion does not fracture the object. Boiling water poured into a heavy glass pitcher can easily crack the pitcher. The boiling water causes the interior surface to expand while the exterior surface is still cool, and the pitcher literally prys itself apart.

Water is a striking exception to the general rule of thermal expansion because in the temperature region from 0°C to about 4°C, water *shrinks* as the temperature is increased. The density of water as a function of temperature is shown in Fig. 13-8. As a consequence of this peculiar property of water, water at very nearly 0°C will be lighter (less dense) than slightly warmer water, and the colder water will rise to the surface. Hence, ice forms first at the top surface of a body of water, an exceedingly important ecological consideration.

Coefficient of linear expansion

The coefficient of linear expansion α describes the tendency of a solid material to expand or contract with a change in temperature. The term *linear* refers to any straight-line dimension of a body, as opposed to an *area* or a *volume*, which we consider later. We consider the expansion of a rod, shown in Fig. 13-9(a), as the temperature changes from an initial temperature T_0 to a final temperature T_f. We

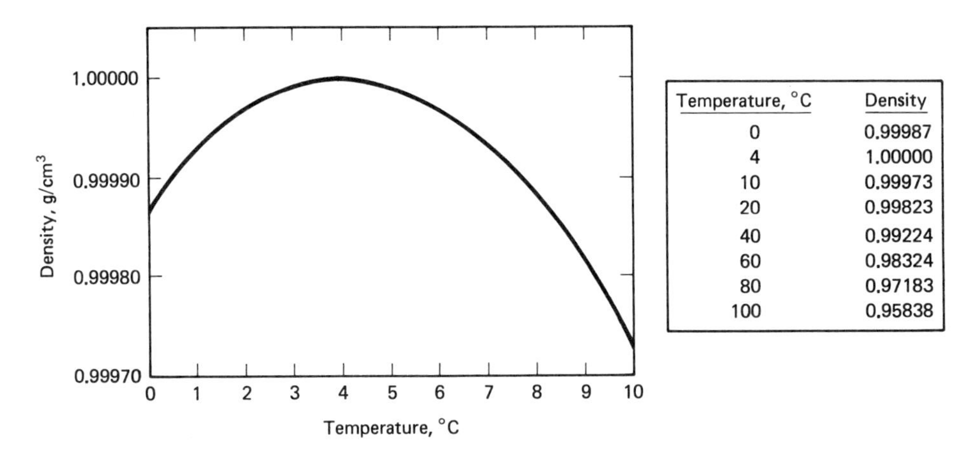

Temperature, °C	Density
0	0.99987
4	1.00000
10	0.99973
20	0.99823
40	0.99224
60	0.98324
80	0.97183
100	0.95838

Figure 13-8 Variation of the density of water with temperature.

denote the change in temperature as ΔT, and the change in length as Δl, as shown in the figure.

The coefficient of linear expansion is the change in length per degree change in temperature of a rod whose original length is one unit. Put another way, it is the fractional change in length of the rod per degree change in temperature. Mathematically, the coefficient of linear expansion is given by

$$\alpha = \frac{\Delta l}{l_0(\Delta T)} \tag{13-6}$$

The change in length of a rod depends directly upon the original length of the rod and upon the temperature change, ΔT. From this point of view, the coefficient of linear expansion, α, is simply the proportionality constant relating Δl to the product of the original length l_0 and the temperature change:

$$\Delta l = \alpha l_0(\Delta T) \tag{13-7}$$

From this expression we can obtain a formula for the length of the rod when it is at any temperature T_f provided we know the length l_0 of the rod at any reference temperature T_0:

$$l = l_0 + \alpha l_0(T_f - T_0) \tag{13-8}$$

Variations of this relationship are given in Fig. 13-9(a).

Coefficient of area and volume expansion

Figure 13-9(b) shows a flat plate of area A_0 at a temperature T_0. If the temperature changes to a higher temperature T_f, both the length and the width will increase according to the relationship given previously for linear expansion. The new area A will be the product of the new length and the new width. When this notion is put in mathematical terms and carried through as shown in the figure, the term $\alpha^2(\Delta T)^2$ appears in the expanded form of an expression for the new area. Because

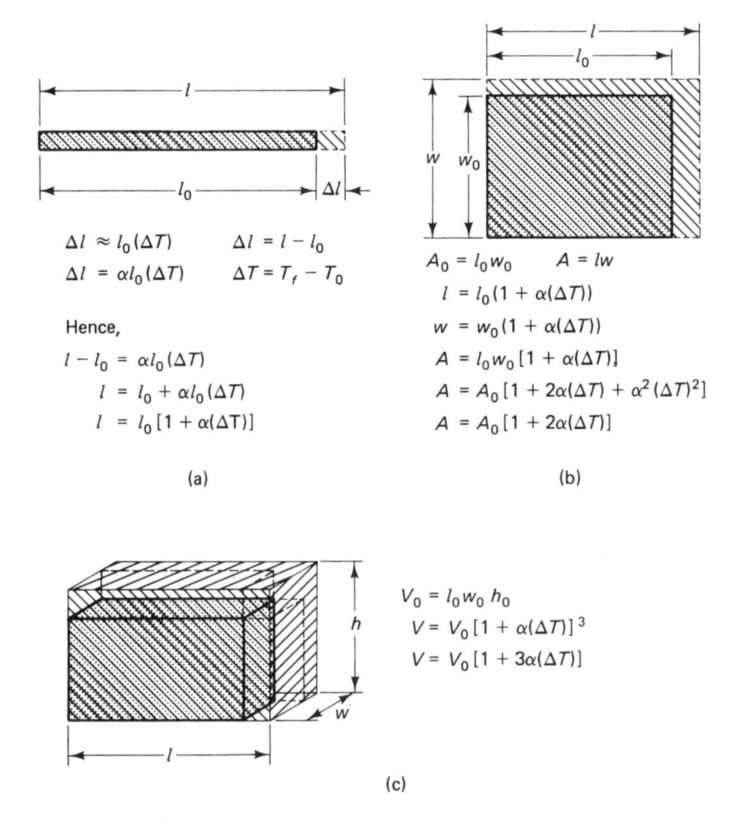

Figure 13-9 Thermal expansion of a rod, flat plate, and rectangular solid.

α is very small (on the order of 10^{-6}), α^2 is even smaller (on the order of 10^{-12}) so the term containing α^2 may be neglected in comparison to the other terms in the equation. Therefore, we drop it to obtain a simpler expression for the new area A in terms of the original area A_0 at T_0. The final expression is

$$A = A_0[1 + 2\alpha(\Delta T)] \tag{13-9}$$

Note the factor of 2 appearing in this equation. A comparison with the corresponding expression for linear expansion reveals that the coefficient of area expansion is simply twice as large as the coefficient of linear expansion.

By a similar line of reasoning, it is easy to show that the volume of a solid or a liquid will expand with a temperature change according to the following expression, which contains a conspicuous factor of 3.

$$V = V_0[1 + 3\alpha(\Delta T)] \tag{13-10}$$

Because of the impossibility of determining the coefficient of linear expansion for a liquid (why?), a coefficient β of volume expansion is defined so that Eq. (3-30) becomes

$$V = V_0[1 + \beta(\Delta T)] \tag{13-11}$$

EXAMPLE 13-4 In a laboratory experiment, a brass rod 60 cm long at 5°C is observed to increase in length by 0.108 cm as the rod is heated to 100°C by passing live steam over the rod. Compute the coefficient of thermal expansion of brass.

Solution. We use Eq. (13-6), noting that the temperature change is 95°C:

$$\alpha = \frac{\Delta l}{l_0(\Delta T)} = \frac{0.108 \text{ cm}}{(60 \text{ cm})(95°\text{C})}$$

$$= 18.9 \times 10^{-6}°\text{C}^{-1}$$

Note that the units of the coefficient of thermal expansion is reciprocal degrees (or *per degree*, if you prefer); note also that it was not necessary to convert from centimeters to any other unit of length.

EXAMPLE 13-5 The coefficient of volume expansion β of gasoline is about $960 \times 10^{-6}°\text{C}^{-1}$. If 1000 gal of gasoline at a temperature of 85°F is pumped from a tank truck to a storage tank in the ground where the temperature is 60°F, compute the quantity of gasoline "lost" due to thermal contraction as it cools.

Solution. We first convert the change in temperature (25°F) to its equivalent of 13.89°C. Then, using a variation of Eq. (13-11),

$$\Delta V = V_0\beta(\Delta T) = (1000 \text{ gal})(960 \times 10^{-6}°\text{C}^{-1})(13.89°\text{C})$$

$$= 13.33 \text{ gal}$$

Thus, for every 1000 gal of gasoline in a tank, the volume changes nearly 1 gal for every change in temperature of 1°C.

Suppose a metal plate has internal openings, as shown in Fig. 13-10. When the plate is heated, will the hole become larger or smaller? Will the distance l between the points become larger or smaller? This question can be easily answered if it is realized that thermal expansion takes place in the same manner as a photographic enlargement. That is, all dimensions of the object are increased in the same proportion. Consequently, the diameter of the hole must increase as the temperature of the plate rises, and the distance between the points will increase also. (It might at first be tempting to argue that the points would expand inward as each point becomes longer, but this is not the case.) To compute the dimensional change of an internal void, simply treat the void as if it were made of the same

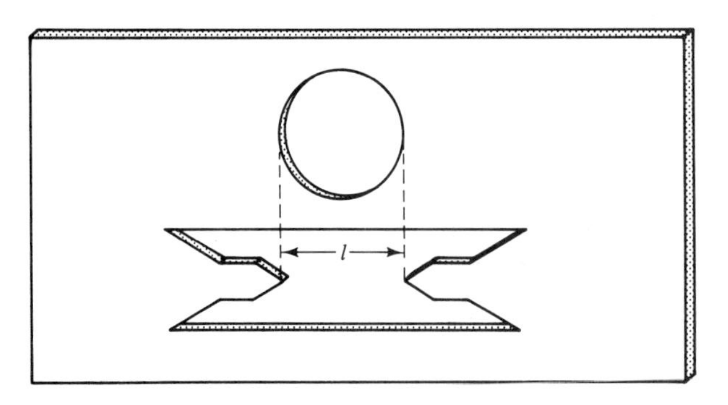

Figure 13-10 What is the effect of heating on this flat metal plate? Will the hole become larger, or will it become smaller? Will the distance between the points increase or decrease?

material as the enclosing body—assign a coefficient of linear expansion to the void equal to that of the surrounding material.

Thermal stresses

Suppose a steel beam just fits between two immovable supports, as illustrated in Fig. 13-11. If the temperature of the beam is then increased an amount ΔT, the

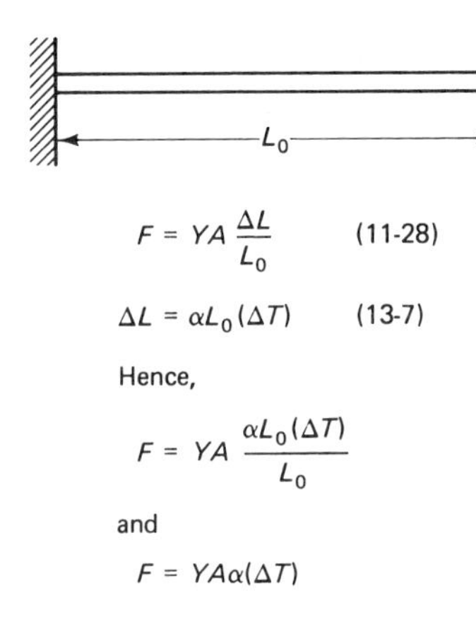

Figure 13-11 F is the force exerted on each support by the rod when it is heated an amount ΔT. Y is Young's modulus, A is the cross-sectional area of the rod, and α is the coefficient of linear expansion.

beam will try to expand, but it will be constrained by the supports. The beam will be placed in compression, and large forces will be exerted on the supports. Indeed, if the supports are not firmly anchored, the beam will push them apart because the forces exerted are quite large.

To calculate the magnitude of the forces, we first determine how much the beam would have expanded in the *absence* of the supports. Equation (13-7) is the appropriate expression; the calculation is the same as for any linear expansion. Then we appeal to Young's modulus and Eq. (11-28) in order to calculate the force that would be required to compress the beam back to its original length. The result of this calculation is the force exerted on the supports by the beam. The length of the rod does not appear in the final formula. An example should make this clear.

EXAMPLE 13-6 Suppose a steel rod 1 m long and 5 cm in diameter is constrained so that it cannot expand as its temperature is increased from 32°F to 98°F. Calculate the force that must be exerted on each end of the rod in order to prevent the rod from expanding.

Solution. We first convert the temperature change of 66°F to 36.67°C. From Table 13-3, the coefficient of linear expansion of steel is $11 \times 10^{-6}\,°\text{C}^{-1}$. Applying Eq. (13-7) to calculate the free expansion of the rod gives

$$\Delta l = \alpha l_0(\Delta T) = (11 \times 10^{-6}\,°\text{C}^{-1})(1\ \text{m})(36.67°\text{C})$$
$$= 4.03 \times 10^{-4}\ \text{m}$$

Once the amount the rod would have expanded is known, we use Eq. (11-28) to compute the force required to return the rod to its original length. The calculation is straightforward, but we must be careful with units—especially in computing the cross-sectional area of the rod.

TABLE 13-3 COEFFICIENTS OF THERMAL EXPANSION

Substance	Linear α, °C$^{-1(a)}$	Volume β, °C$^{-1(a)}$
Aluminum	26×10^{-6}	77×10^{-6}
Brass	19	56
Copper	17	42
Iron (steel)	11	33
Invar (64% Fe + 36% Ni)	1	2.7
Lead	29	87
Silver	19	58
Quartz	0.42	1.2
Glass, soft	8.5	26
Glass, Pyrex	3.3	10
Ice	—	112
Concrete	12	36
Ethyl alcohol		750
Gasoline		960
Glycerin		490
Mercury		180
Turpentine		970
Water		210

[a]Divide by 1.8 to convert to °F^{-1}.

$$F = YA\frac{\Delta L}{L_0} = (20 \times 10^{10}\ \text{N/m}^2)(1.96 \times 10^{-3}\ \text{m}^2)\left(\frac{4.03 \times 10^{-4}\ \text{m}}{1\ \text{m}}\right)$$

$$= 1.58 \times 10^5\ \text{N}$$

$$= 35{,}500\ \text{lb} = 17.7\ \text{tons}$$

This is a rather large force. We obtained Young's modulus for steel from Table 11-5.

Liquid-in-glass thermometers

The common thermometer is a liquid-in-glass thermometer with the liquid being either red-colored alcohol or mercury, which has a silvery appearance. We lead up to a working equation for a thermometer by first considering the expansion of a volume of liquid in a container that also expands or contracts as the temperature changes. Refer to Figure 13-12(a), which illustrates a beaker containing a volume V_0 of liquid at the initial temperature. When a rise in temperature occurs, both the volume of the beaker and the volume of the liquid will increase. But whether the liquid will rise in the beaker depends upon whether the volume expansion of the beaker is greater than or less than the volume expansion of the liquid.

After the temperature change, the new volumes of the beaker, V_B, and the liquid, V_L, are given by the expressions in Fig. 13-12(a). The change in volume, ΔV, is obtained by computing the difference $V_L - V_B$, The result is

$$\Delta V = V_0(\beta_L - \beta_B)(\Delta T) \tag{13-12}$$

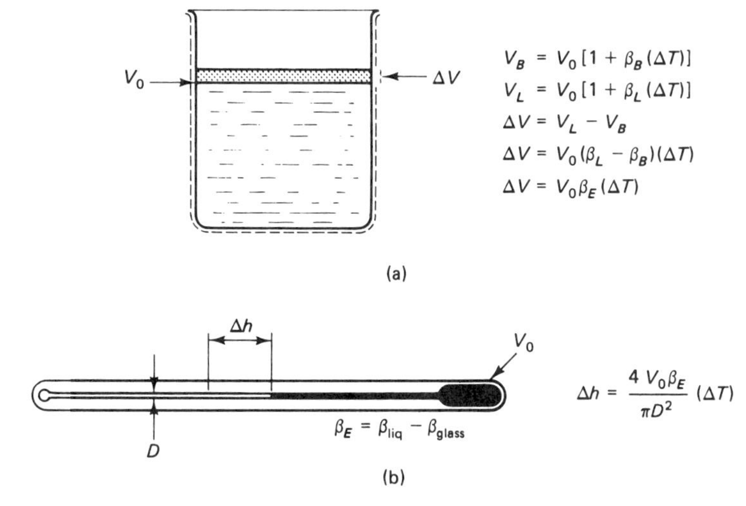

Figure 13-12 (a) Differential expansion between a beaker and liquid. (b) Liquid-in-glass thermometer.

where β_L and β_B are the coefficients of volume expansion of the liquid and beaker, respectively. We may define an effective volume expansion β_E for the particular liquid contained in the particular type of container. The expression for the change in volume then becomes

$$\Delta V = V_0\beta_E(\Delta T) \qquad \beta_E = \beta_L - \beta_B \tag{13-13}$$

A liquid-in-glass thermometer consists of a bulb containing a volume of liquid which expands into the bore of a fine capillary tube. The change in volume, ΔV, occuring with a rise in temperature is accommodated by having the liquid expand a greater distance into the bore. If the bore diameter is D, the increase in length, Δh, of the liquid column is obtained as follows:

$$\begin{aligned} \Delta V &= V_0\beta_E(\Delta T) = \left(\frac{\pi D^2}{4}\right)(\Delta h) \\ \Delta h &= \frac{4V_0\beta_E}{\pi D^2}(\Delta T) \end{aligned} \tag{13-14}$$

If we define the *sensitivity* of a thermometer to be the change in column length per unit change in temperature, $\Delta h/\Delta T$, then the sensitivity is given by

$$\text{Thermometer sensitivity} = \Delta h/\Delta T = \frac{4V_0\beta_E}{\pi D^2} \tag{13-15}$$

Factors resulting in a more sensitive thermometer are a greater bulb volume, a smaller capillary-bore diameter, and a greater coefficient of expansion, β_E.

13-4 PHASES AND PHASE CHANGES

The three states of matter—solid, liquid, and gas—are sometimes referred to as *phases*. The three phases of water are ice, liquid water, and steam, or water vapor. Generally speaking, the term *phase* refers to one of perhaps several distinct forms in which a substance can occur. Ice, for example, can exist in at least seven different forms if the pressure is great enough. Liquid helium changes phase from helium I to helium II at a temperature of 2.18 K, producing a dramatic change in the properties of the liquid. Liquid helium II, often called a *superfluid*, exhibits no measurable viscosity and an infinite thermal conductivity. On the other hand, helium I behaves much more like an ordinary liquid, even though its boiling point is very low, namely, 4.2 K. In the following, we consider changes of phase from solid to liquid and from liquid to a vapor. The discussion is addressed to water because of its familiarity, but the same principles apply to most other substances.

Ice melts and water freezes at 32°F. What is the difference between ice at 32°F and water at 32°F? Obviously, one is a liquid and the other a solid, but what does this mean on the molecular level? The difference is in the bonding between the water molecules. In ice, a solid, the water molecules are rigidly bound to their positions in the crystal lattice. In liquid water, the bonds have been broken, so that molecules can move about with greater freedom. In ice and in water at 32°F, the average kinetic energy per molecule is the same; therefore, the temperature is the same. However, to convert ice to water, energy must be added to the ice in order to disrupt the bonds and set the molecules free to become a liquid. Heat energy is absorbed when ice melts, but no change in temperature occurs during the melting.

Going in the other direction, water at 32°F must give up heat energy in order to freeze into ice at the same temperature. Solidification can occur only as rapidly as the removal of heat will allow. The heat that is transferred during the freezing or melting process is called *latent heat* because the transfer of heat is not accompanied by a temperature change. Recall that in the absence of a phase change, any transfer of heat to or from a body is accompanied by a temperature change according to Eq. (13-3).

The latent heat associated with freezing (solidification) and melting is called the *heat of fusion*. When water evaporates (changing phase from a liquid to a vapor), the latent heat involved is called the *heat of vaporization*. For water, the heat of fusion is 80 cal/g and the heat of vaporization is 540 cal/g. The value of the latent heat is the same whether water is freezing to become solid or whether ice is melting to become liquid. The same is true for evaporization and condensation; when steam condenses, heat is given up at the rate of 540 cal/g.

The phase change from ice to water tends to hold an ice-water mixture at 32°F. If heat is added to the system, which would tend to make the temperature rise, a bit of ice will melt, absorb the heat, and prevent the rise in temperature. If the system is cooled, a bit of water will freeze and liberate its latent heat, which prevents a further lowering of the temperature. As long as both ice and water are present, the temperature will remain at 32°F, provided the system is well mixed so that all parts of the system are at the same temperature.

A similar effect occurs at 212°F, the boiling point of water. Water can be readily heated to 212°, but further addition of heat results in the evaporation of water rather than heating of water to temperatures above 212°. The evaporating water absorbs the heat that is added. A pressure cooker or other pressure vessel must be used in order to heat water (and steam) to temperatures higher than 212°F. (See Fig. 13-13.)

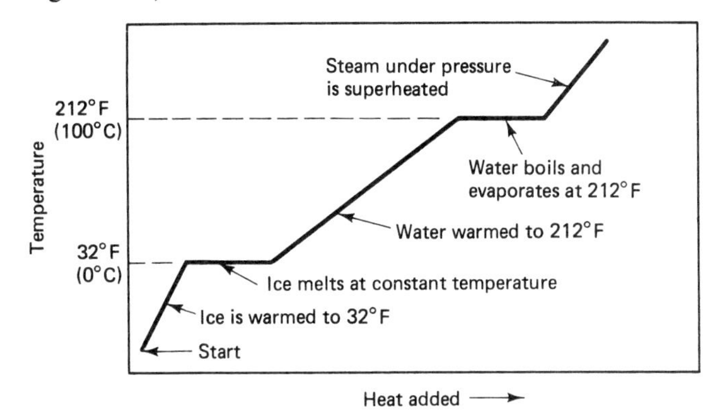

Figure 13-13 Variation of temperature as heat is added to a quantity of ice that is gradually warmed until it is finally converted to steam.

Latent heat calculations

The latent heat exchanged during a phase change is simply the product of the mass of the substance undergoing the phase change and the appropriate latent heat. Latent heats for various substances are tabulated in handbooks so that if the mass M of the substance is known, the heat exchanged is given by

$$\text{For melting:} \quad \Delta Q = h_f M \qquad \text{For evaporation:} \quad \Delta Q = h_v M \tag{13-16}$$

where h_f is the heat of fusion and h_v is the heat of vaporization. Latent heats of several substances are given in Table 13-4.

EXAMPLE 13-7 An ice cube of mass 36 g is dropped into a glass of ice tea. How much heat is absorbed from the tea by the melting of the ice?

Solution. The heat of fusion of water is 80 cal/g. Therefore, the melting of the ice will absorb heat given by

$$\Delta Q = h_f M = (80 \text{ cal/g})(36 \text{ g}) = 2880 \text{ cal}$$

This quantity of heat is absorbed simply by the melting of the ice. We did not consider that 36 g of water at 32°F resulted from the ice melting. This water absorbs additional heat as it is warmed by the tea.

EXAMPLE 13-8 How many calories are required to boil away 250 g of water in a saucepan at 212°F?

Solution. Using the heat of vaporization of water, 540 cal/g, the heat required to evaporate the water is

$$Q = h_v M = (540 \text{ cal/g})(250 \text{ g})$$
$$= 135{,}000 \text{ cal}$$

TABLE 13-4 LATENT HEATS OF SELECTED SUBSTANCES

Substance	Melting point, °C	Heat of fusion, cal/g	Boiling point, °C	Heat of vaporization, cal/g
Water	0	80	100	540
Mercury	−39	2.8	357	70
Lead	327	5.9	1753	206
Ammonia	−75	108	−33	327
Nitrogen	−210	6.1	−196	48
Oxygen	−218	3.3	−183	51
Helium	—	—	4.2 K	5
Hydrogen	—	—	20 K	108

Heat of fusion determination

We can determine the heat of fusion of water by doing a simple calorimeter experiment, as follows. Suppose the calorimeter consists of an aluminum can surrounded by styrofoam. The mass of the can is 46 g, and the specific heat of aluminum is 0.22. A thermometer and stirrer are provided, as for the calorimeter of Fig. 13-4.

Initially, the calorimeter contains 130 g of water at 24°C. Chunks of ice are then removed from a nearby container of ice water and wrapped in a towel so that any water on the surface of the ice will be removed. Several pieces of the dry ice (not solid CO_2) are then put into the calorimeter, cooling the calorimeter to a final temperature of 14°C. After reading the final temperature, the calorimeter can is carefully removed for weighing to determine the mass of the ice that was added. The water in the calorimeter now has a mass of 145 g, and it follows that 15 g of ice was added. We now can compute the heat of fusion of water from this information.

The heat balance equation is as follows:

Heat loss by calorimeter water	+	heat loss by calorimeter can	=	heat absorbed by melting of ice	+	Heat absorbed in heating icewater from 0°C to T_f

$$M_w c_w (T_w - T_f) + M_c c_c (T_w - T_f) = h_f M_{\text{ice}} + M_{\text{ice}} c_w (T_f - 0^\circ\text{C}) \quad (13\text{-}17)$$

Inserting the numerical values gives the following (units omitted for brevity):

$$(130)(1.0)(24 - 14) + (46)(0.22)(24 - 14) = h_f(15) + (15)(1.0)(14)$$

$$1300 + 101 = h_f(15) + 210$$

$$1191 = h_f(15)$$

(for water-ice) $$h_f = 79.4 \text{ cal/g}$$

This is very close to the accepted value of 80 cal/g.

Heat of vaporization determination

A method similar to that used for the determination of the heat of fusion can be used to determine the heat of vaporization of water. The calorimetry method is indicated in Fig. 13-14; steam from a steam generator is bubbled into the cool

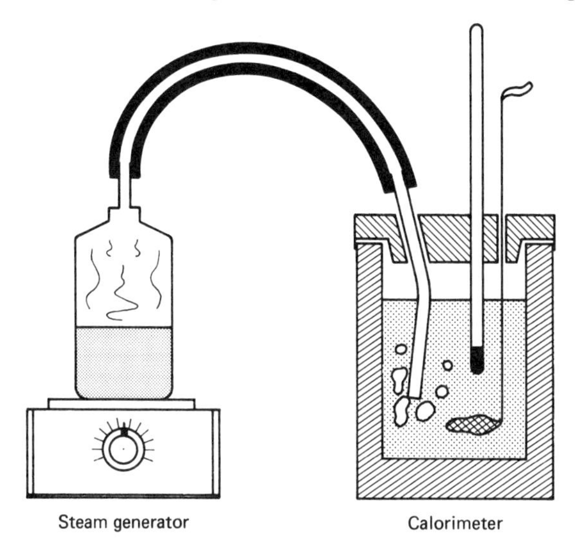

Figure 13-14 Determining the heat of vaporization of water.

water of a calorimeter where it condenses. The amount of steam condensed can be determined by weighing the water before and after the addition of the steam. The heat balance equation is

Heat contributed by condensation of steam	+	heat contributed by cooling the steam-water from 100°C to T_f	=	heat absorbed by calorimeter water	+	heat absorbed by calorimeter can

$$h_v M_s + M_s c_w(100°C - T_f) = M_w c_w(T_f - T_w) + M_c c_c(T_f - T_w) \quad (18\text{-}18)$$

where M_s is the mass of steam condensed, M_w and M_c are the masses of the calorimeter water and can, respectively, and T_w T_f are the initial and final temperatures of the water in the calorimeter. If temperatures are given in degrees Celsius and if the masses are in grams, then c_w, the specific heat of water, will be 1.0 cal/g-°C.

Sublimation

Another type of phase change occurs when molecules pass directly from the solid to the vapor phase without having the solid melt first. A vivid example is that of dry ice (CO_2). A chunk of dry ice exposed to room temperature will produce a visible vapor as sublimation occurs, but no liquid CO_2 is observed in the process. In frost-free freezers, the ice cubes, if left there for extended periods, will literally disappear, leaving behind an empty tray. Sublimation is the culprit.

13-5 THE VAPOR PRESSURE OF WATER; HUMIDITY

Suppose a heavy-walled pressure vessel from which the air has been removed is partially filled with water, as shown in Fig. 13-15. Water vapor will exist in the region above the water, and the pressure gauge will read the absolute pressure

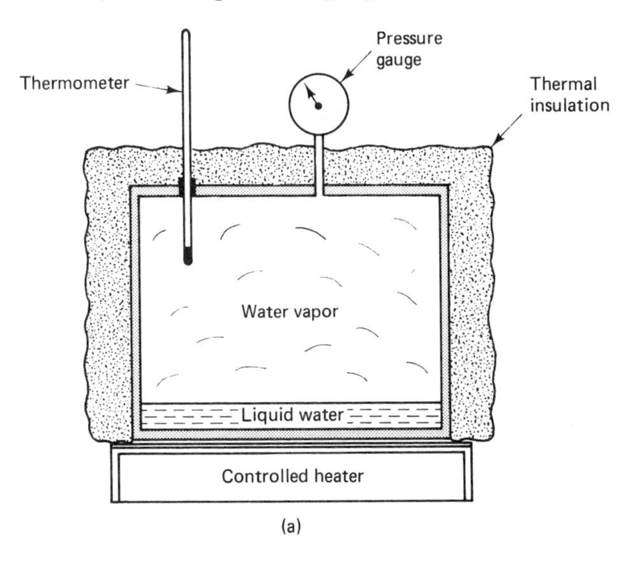

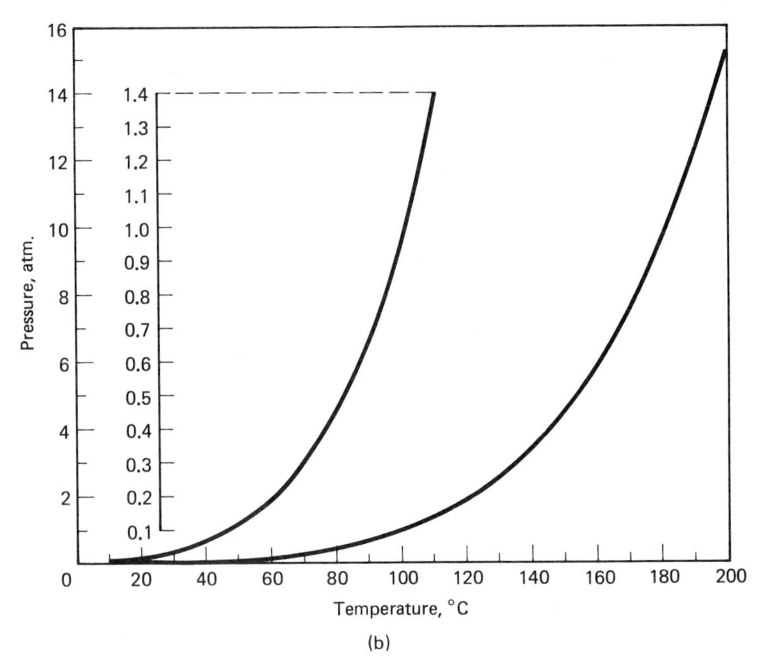

Figure 13-15 (a) Pressure vessel for determining the saturated vapor pressure of water at various temperatures.
(b) Graph of the result.

exerted on the walls of the vessel by the water vapor. The pressure so obtained is called the *saturated vapor pressure* of water at the particular temperature. Let us consider this process in more detail and see how the pressure varies as we gradually raise the temperature from 0°C to 200°C. In this experiment, heat from the temperature-controlled heater is supplied to the pressure vessel only as required to maintain the temperature.

The system comprised of the water and water vapor is in a state of dynamic equilibrium (at a given temperature). Water molecules continually escape from the surface of the water to become a part of the vapor, but at the same time, water molecules from the vapor collide with the surface of the water and recondense to become a part of the liquid. The rate of evaportaion is proportional to the temperature; the rate of condensation is proportional to the pressure of the vapor. If a small increase in temperature occurs, the evaporation will occur at a greater rate than the condensation until the pressure increases to the point where the rate of condensation equals the rate of evaporation, restoring the state of equilibrium. If the temperature decreases, condensation will exceed evaporation until the pressure falls appropriately.

The variation of pressure with temperature is shown in the graph of Fig. 13-15 and in Table 13-5. Note that nothing dramatic occurs when the temperature of the water reaches 100°C.

Humidity

The term *humidity* refers to the quantity of water vapor in the air. The capacity of air to hold water vapor varies with temperature; warm air is able to hold more water vapor than cool air. When air contains the maximum amount of water vapor

TABLE 13-5 SATURATED VAPOR PRESSURE OF WATER AT VARIOUS TEMPERATURES

Temperature		Pressure	
°C	°F	atm	mm Hg
0	32	0.00603	4.579
10	50	0.0121	9.205
20	68	0.0230	17.51
30	86	0.0417	31.71
40	104	0.0725	55.13
50	122	0.121	92.30
60	140	0.196	149.2
70	158	0.307	233.5
80	176	0.467	355.1
90	194	0.691	525.8
100	212	1.000	760.0
110	230	1.413	1,074
120	248	1.959	1,489
140	284	3.564	2,709
160	320	6.096	4,633
180	356	9.887	7,514
200	392	15.325	11,647

that it can hold, the air is said to be *saturated*. Moreover, saturated air at a given temperature contains the same quantity of water vapor per unit volume as the vapor region of the pressure vessel of Fig. 13-15 at the same temperature. This holds true even though, at the outset, all air was removed from the pressure vessel.

Absolute humidity refers to the mass of water vapor present per unit volume of air. It is usually expressed in terms of grams per cubic meter. The maximum quantity of water vapor that 1 m^3 of air can hold at various temperatures is given in the graph and chart of Fig. 13-16.

Unsaturated air contains less than the maximum possible amount of water vapor at a given temperature. That is, more water vapor may be evaporated into the same volume of air without causing condensation. The *relative humidity* gives the degree of saturation of the air. It is the ratio, expressed as a percentage, of the water vapor present in a volume of air to the maximum quantity that the same volume would contain if it were saturated:

$$\text{Relative humidity} = \frac{\text{water vapor present}}{\text{saturation level of water vapor}} \times 100\% \qquad (13\text{-}19)$$

Accordingly, the relative humidity of saturated air is 100%.

EXAMPLE 13-9 What is the relative humidity of a mass of air if the air contains 10.9 g of water vapor per cubic meter? The temperature of the air is 68°F.

Solution. From Fig. 13-16, we find that saturated air at 68°F contains 17.3 g/m^3 of

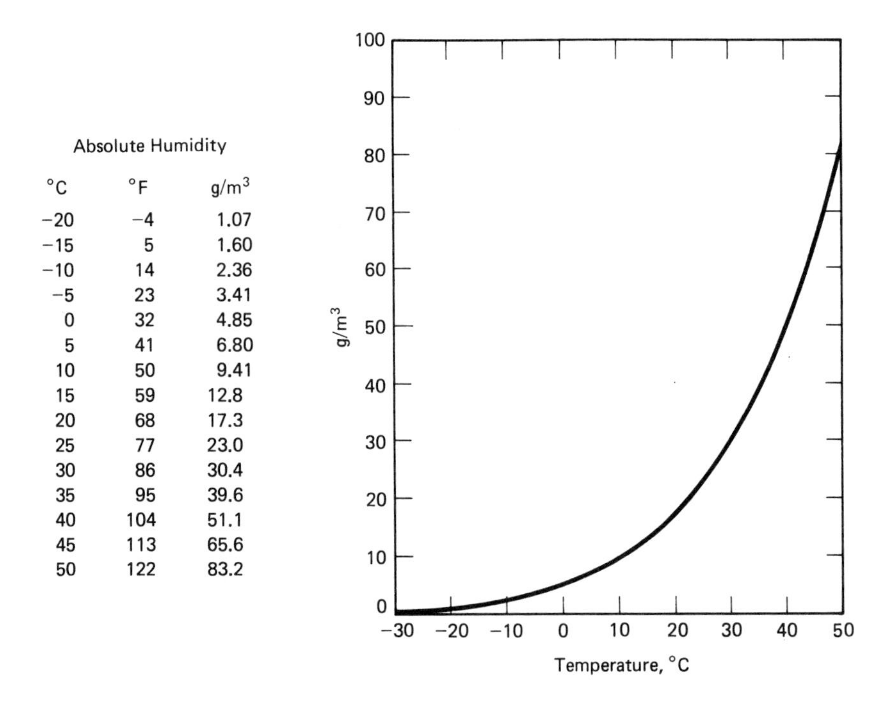

Absolute Humidity

°C	°F	g/m³
−20	−4	1.07
−15	5	1.60
−10	14	2.36
−5	23	3.41
0	32	4.85
5	41	6.80
10	50	9.41
15	59	12.8
20	68	17.3
25	77	23.0
30	86	30.4
35	95	39.6
40	104	51.1
45	113	65.6
50	122	83.2

Figure 13-16 Absolute humidity as a function of temperature.

water vapor. Therefore, using the definition of relative humidity, the relative humidity (RH) of the air in question is

$$\text{RH} = \frac{10.9 \text{ g/m}^3}{17.3 \text{ g/m}^3} \times 100\%$$
$$= 63\%$$

EXAMPLE 13-10 The relative humidity of the air in a classroom is 54% and the temperature is 68°F. Calculate the quantity of water that is suspended in the air as water vapor if the dimensions of the classroom are 9 × 20 × 30 ft.

Solution. We first calculate the volume of the classroom in cubic feet (5400) and convert it to cubic meters: 5400/35.31 = 152.9 m^3. From Fig. 13-16, each cubic meter would contain 17.3 g of water if the air were completely saturated. However, because the relative humidity is only 54%, we multiply the saturation value of the water by 0.54. Hence, the quantity of water in the room is given by

$$\text{Water} = (0.54)(17.3 \text{ g/m}^3)(152.9 \text{ m}^3)$$
$$= 1428 \text{ g}$$

This is 1.428 L of water, or about 1.5 qt.

EXAMPLE 13-11 Assuming no air enters or leaves the classroom of Example 13-10, compute the resulting relative humidity if an additional quart (946 g) of water is evaporated into the air. The temperature is assumed to remain constant at 68°F.

Solution. Using the result of the previous example, the total quantity of water in the air will be 2374 g after the additional quart is evaporated. Dividing by the volume of the room (152.9 m^3) gives a water density of 15.53 g/m^3. The saturation level at 68°F is 17.3 g/m^3, so the relative humidity is:

$$\text{RH} = \frac{15.53}{17.3} \times 100\% = 89.75\%$$

Thus, evaporation of an additional quart of water raised the humidity by more than 35%. This result is misleading in practice, however, because ventilation air entering and leaving a room would carry away a significant portion of the newly evaporated water. Humidity control in the interior of large auditoriums, theaters, or manufacturing facilities is a major engineering concern.

Reduced humidity of warmed air

The air inside a residential dwelling must circulate so that new air from the outside periodically replaces the inside air. This must be the case because otherwise all the oxygen in the inside air would be depleted so that the occupants could not survive. Thus, the inside air must have initially been outside air. We now ask what happens to the relative humidity as the (possibly) cold outside air is brought inside and warmed.

Two effects serve to reduce the humidity as the air is warmed. First, warm air can hold a greater quantity of water vapor than cold air. Second, air expands when heated, so 1 m^3 of outside air will occupy a greater volume when brought inside and warmed. This causes the same amount of water vapor to be distributed through a greater volume, reducing the absolute humidity of the inside air. We have not yet studied the expansion of gases, so we shall simply introduce a correction factor into the following formula to give an approximate value of the relative humidity on the inside of a dwelling.

If the relative humidity of outside air of temperature T_{out} is RH_{out}, the inside humidity, RH_{in}, at temperature T_{in} is given approximately by

$$RH_{in} = \left(1 - \frac{\Delta T}{490}\right)\left(\frac{D_{out}}{D_{in}}\right)RH_{out} \tag{13-20}$$

where

ΔT = the change in temperature $T_{in} - T_{out}$, in degrees Fahrenheit

D_{out} = the density of saturated water vapor at T_{out}

D_{in} = the density of saturated water vapor at T_{in}

Numerical values of D_{out} and D_{in} are obtained from Fig. 13-16. The factor $1 - \Delta T/490$ is the correction factor to account for the expansion of the outside air as it is heated.

EXAMPLE 13-12 The outside air temperature on a winter day is 41°F at the same time the relative humidity of the outside air is 80%. What will be the relative humidity of the air after it is brought inside and warmed to 68°F?

Solution. A direct application of Eq. (13-16) is in order, but we must find the saturation densities of water vapor at 41°F and 68°F from Fig. 13-16. Hence, D_{out} is 6.8 g/m^3 and D_{in} is 17.3 g/m^3. The change in temperature is 27°F. Substituting gives

$$\begin{aligned} RH_{in} &= \left(1 - \frac{\Delta T}{490}\right)\left(\frac{D_{out}}{D_{in}}\right)RH_{out} \\ &= \left(1 - \frac{27}{490}\right)\left(\frac{6.8}{17.3}\right)(80\%) \\ &= 29.7\% \end{aligned}$$

Dew point

When warm, unsaturated air is cooled, its relative humidity increases because cool air can hold less water vapor than warm air. As the temperature falls, a temperature is eventually reached at which the air becomes saturated; the relative humidity becomes 100%. A further lowering of temperature causes condensation to occur as fog or dew. The temperature at which the condensation begins is called the *dew point*. Frost is formed when dew freezes. Fog can be predicted when the air temperature begins to fall and approach the dew point.

Boiling

Boiling occurs when the vapor pressure of a liquid equals atmospheric pressure (or the pressure exerted on the liquid if different from atmospheric) and is evidenced by the familiar bubbling of the liquid as pockets of vapor form and rise

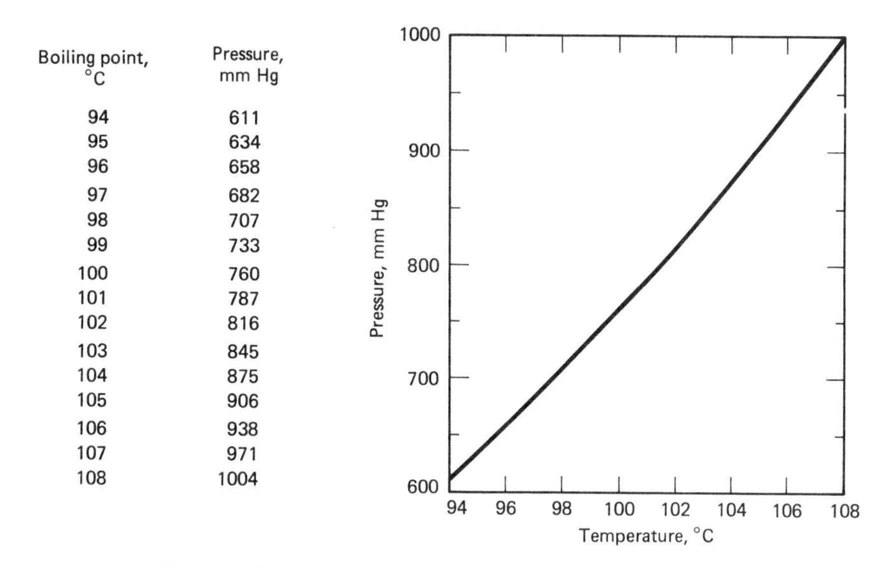

Boiling point, °C	Pressure, mm Hg
94	611
95	634
96	658
97	682
98	707
99	733
100	760
101	787
102	816
103	845
104	875
105	906
106	938
107	971
108	1004

Figure 13-17 Boiling point of water as a function of pressure.

toward the surface. The *boiling point* of a liquid is the temperature at which the vapor pressure exactly equals standard atmospheric pressure. The temperature at which water boils depends upon the local atmospheric pressure; cooking times must be extended at high altitudes because water boils at a lower temperature. A pressure cooker allows food to be cooked at higher temperatures because the pressure elevates the boiling point. The effect of pressure variations on the boiling point can be discerned from the graph of Fig. 13-17.

By "linearizing" the data found in a table of standard atmospheric pressures vs. altitude, the following equation is found to give a good approximation of the atmospheric pressure P (in millimeters of mercury) at an elevation h (in feet) above sea level, up to about 10,000 ft:

$$P = (-0.0256)h + 760 \text{ mm Hg} \tag{13-21}$$

By doing the same thing to the data of Fig. 13-17 for pressures ranging from 611 to 760 mm Hg, we find that the boiling point of water is given by

$$T_{\text{bpw}} = (0.0403)P + 69.4°\text{C} \tag{13-22}$$

Substituting Eq. (13-17) into Eq. (13-18) gives an approximate formula for calculating the boiling point of water (degrees celsius) at elevations h (in feet) above sea level:

$$T_{\text{bpw}} = 100°\text{C} - (0.00103)h \tag{13-23}$$

Thus, for every 1000 ft of elevation, the boiling point is lowered by about 1°C. For elevations up to 10,000 ft, these formulas are accurate to better than 3%.

13-6 MISCELLANEOUS TOPICS

P-V-T diagram

The state of a substance under various conditions of pressure, volume, and temperature is shown in a *P-V-T diagram*, as illustrated in Fig. 13-18. The *P-V-T* surface is actually a graph in three dimensions. If any two variables are given, the

1. A solid under high pressure below the freezing point; no vapor present.
2. Volume increased so that a vapor forms around the solid; temperature below freezing.
3. The triple line; solid, liquid, and vapor can coexist in equilibrium at the temperature of the triple point.
4. Region of liquid and vapor; temperature above melting point (the triple point).

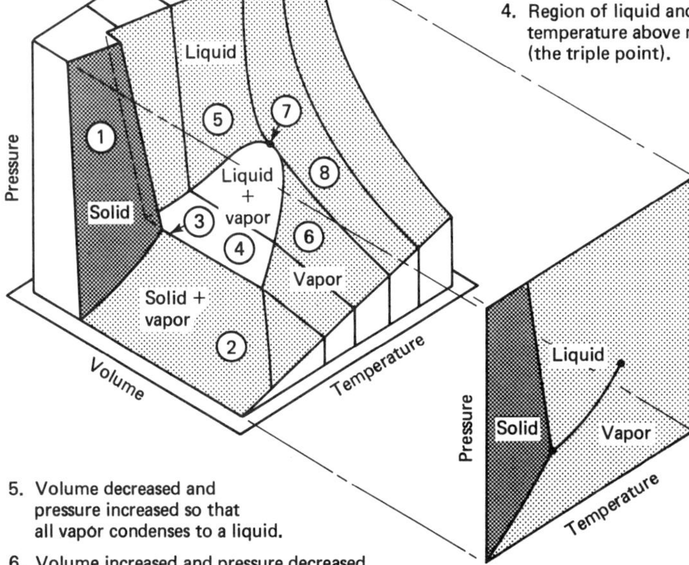

5. Volume decreased and pressure increased so that all vapor condenses to a liquid.
6. Volume increased and pressure decreased so that all liquid evaporates to a vapor.
7. The critical point; this is the highest temperature at which a liquid can exist.
8. High temperature, high pressure vapor.

Figure 13-18 *P-V-T* surface for a substance, such as water, that expands upon freezing.

value (or possible values) of the remaining variable can be determined along with the corresponding state of the substance.

Figure 13-18 pertains to a substance—for example, water—that expands upon freezing. The expansion produces the break in the surface at the solid-liquid interface. Further, we can see that the melting point is slightly lower at high pressures.

One variable (such as the volume) may be eliminated by projecting the *P-V-T* surface as illustrated to show the pressure-temperature relationships on a *P-T diagram*. In this case, the *triple line* appears as a *triple point*. Obviously, no volume information can be determined from the *P-T* diagram. The *P-V-T* surface can be projected in a different direction to obtain a *P-V diagram*. *P-V* diagrams are widely used in the study of thermodynamic cycles involving the compression and expansion of gases.

Heat of combustion

An important engineering consideration is the heat of combustion of various fuels such as coal, natural gas, and fuel oil. The heat of combustion is the heat produced by the complete oxidation (burning) of one unit of mass of the substance in question. Units in which the heat of combustion are given include cal/g, BTU/lb, and cal/g-mole. One g-mole (gram-mole, after gram-molecular weight) is the mass in grams of a substance numerically equal to its molecular weight. For example, one g-mole of water, H_2O, is 18 g (see Sect. 11-1). Heats of combustion of a large number of substances are tabulated in handbooks of chemical and physical data.

Heats of combustion are determined in specially designed combustion calorimeters. A bomb calorimeter for use with solid or liquid fuels is shown in Fig. 13-19, and a continuous-flow calorimeter is shown in Fig. 13-20. To use the bomb calorimeter, a small quantity of the test fuel is placed in the crucible in contact with the electric heater. After closing the bomb, it is filled with oxygen to a pressure of several atmospheres. The oxygen inlet valve is then closed so that the bomb is completely sealed. The fuel is ignited by an electric current which heats the heater wire to incandescence, and the heat evolved from the combustion produces a rise in temperature of the bomb and water bath, from which the heat of combustion of the fuel can be determined.

The continuous-flow calorimeter depends, for its operation, upon the accurate metering (measuring) of the fuel flow to the burner and of the cooling water that flows uniformly through the calorimeter. The rate of fuel flow gives the mass of fuel consumed per unit time, and the difference in temperatures of the

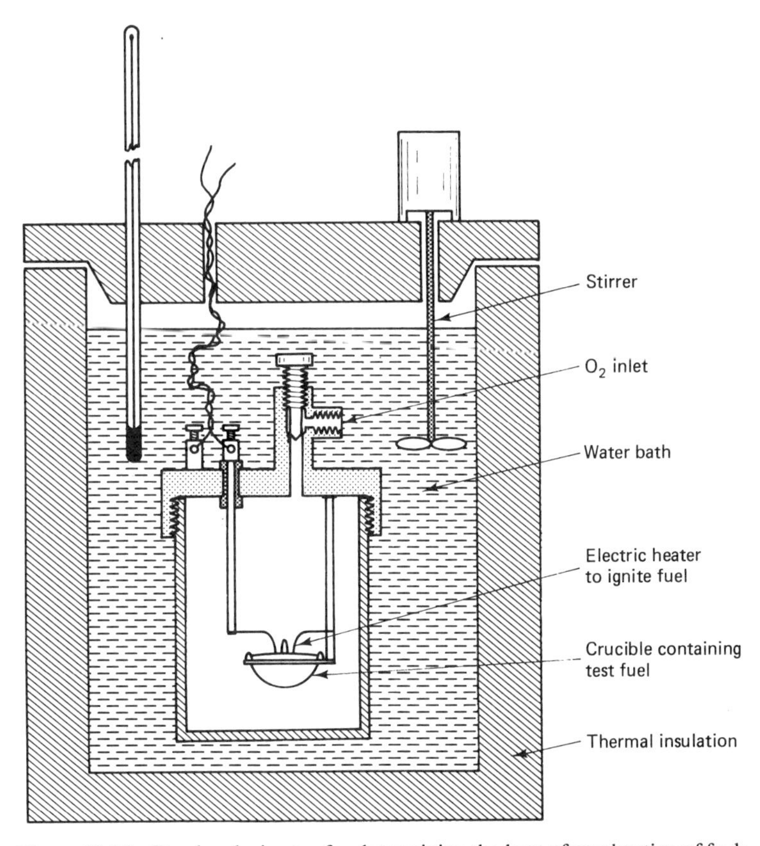

Figure 13-19 Bomb calorimeter for determining the heat of combustion of fuels.

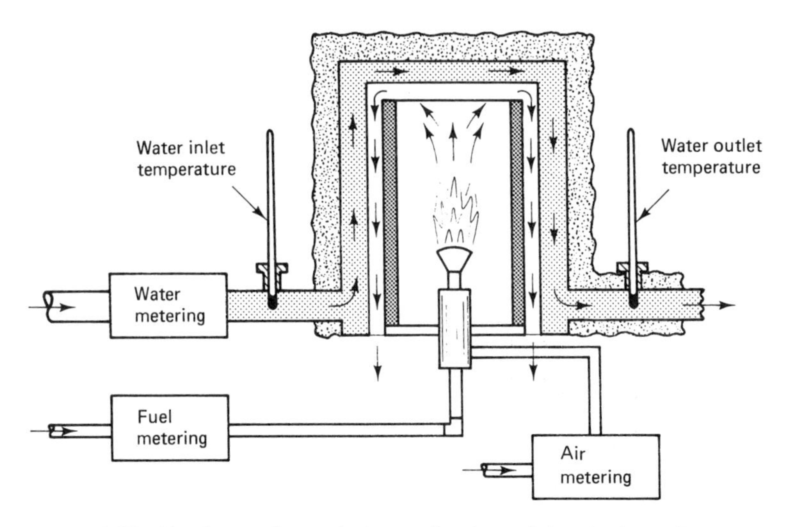

Figure 13-20 Continuous-flow calorimeter for determining the heat of combustion of gaseous or liquid fuels.

inlet and outlet water gives the quantity of heat carried away. From this, the heat of combustion of the fuel can be determined. A continuous-flow calorimeter is described in more detail in Sect. 14-1, in conjunction with determinations of thermal conductivity.

Mechanical equivalent of heat

The *mechanical equivalent of heat* is an outdated concept that arose from asking a question such as, How many joules of work must be done in order to produce one calorie of heat? It is well known nowadays that heat is just another form in which energy may appear, but the *mechanical equivalent* is still found in many older texts, so we briefly discuss it here. It emphasizes the point that there is a direct relationship between mechanical work or mechanical potential energy on the one hand, and heat or thermal energy on the other.

The mechanical system of Fig. 13-21 uses the potential energy of the weight to stir the liquid contained in the calorimeter. Neglecting losses due to friction in

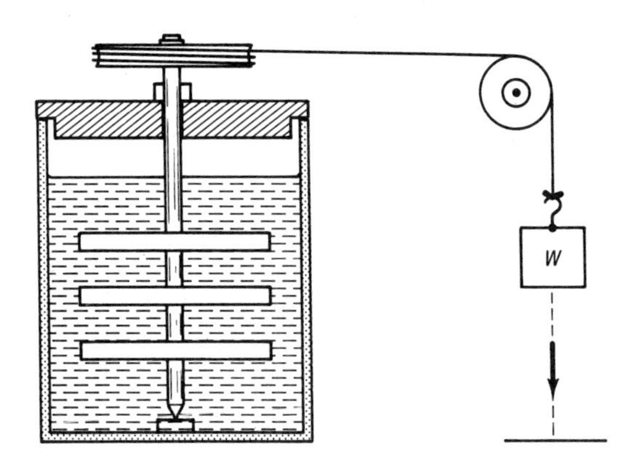

Figure 13-21 Apparatus for determining the mechanical equivalent of heat.

the pulley system, the original potential energy of the elevated weight is converted to heat by stirring the viscous liquid. If the potential energy is calculated in joules, the heat produced in calories may be calculated by using the mechanical equivalent of heat: 4.186 J is the equivalent of 1 cal.

To Go Further

Read about the early theories of heat (caloric) and the development of the various temperature scales. Also:

Anders Celsius (1701–1744), Swedish astronomer
Antoine Laurent Lavoisier (1743–1794), French chemist
Count Rumford [Benjamin Thompson] (1753–1814), a spy
Gabriel Daniel Fahrenheit (1686–1736), German scientist
James Prescott Joule (1818–1889), English brewery owner and physicist
William Rankine (1820–1872), Scottish engineer
Lord Kelvin [William Thomson], (1824–1907), British scientist

Questions

1. What is the distinction between ordered and disordered kinetic energy?

2. Why is there an absolute zero but no absolute maximum of temperature?

3. Which represents a greater change in temperature, 1°F or 1°C?

4. Briefly explain the units used to measure quantities of heat.

5. Suppose ice did not float and that water was most dense at 0°C rather than at 4°C. What would be the implications of this to marine life and to the polar ice cap?

6. For a solid, what is the relationship between its coefficient of linear expansion and of volume expansion?

7. When a rod constrained at both ends is heated, very large forces of compression are developed (Fig. 13-11). Why do the forces not depend upon the length of the rod?

8. Suppose one spoke of a cast-iron wheel is heated (as by welding) while the rest of the wheel and the rim in particular is allowed to remain cool. What is likely to happen?

9. Suppose the rim (Question 8) is heated rather than the spokes. What is apt to occur?

10. If all parts of any solid object are heated uniformly, do any thermal stresses develop?

11. Discuss the advisability of using warm water to melt the frost off the windshield of an automobile.

12. Why is the latent heat called *latent heat*? Where does the heat come from that must be removed from 0°C water in order for it to freeze?

***13.** How could one determine if there is a latent heat asssociated with the sublimation of dry ice (CO_2)?

14. Why is a pressure cooker able to cook meats and potatoes faster than an open pot?

15. Certain recipes call for additional cooking time at elevations above 3000 ft. Why?

16. Why does water condense on the outside of an iced-tea glass?

17. Why do pilots of noninstrument airplanes become concerned when the temperature falls toward the dew point?

18. Why are burns from live steam often more extensive and severe than those resulting from comparable exposures to boiling water alone?

Problems

1. Liquid oxygen boils at −183°C. Express this temperature in degrees Fahrenheit, Rankine, and Kelvin.

2. The optimum temperature for many photographic processes is 20°C. Express this in degrees Fahrenheit, Rankine, and Kelvin.

3. Biscuits bake best at 450°F. What is this temperature in degrees Celsius, Rankine, and Kelvin?
4. Calcined petroleum coke leaves a rotary kiln at 2400°F. Express this in degrees Celsius and Kelvin.
5. Grain size in black-and-white photography is affected by temperature variations during the processing of the film. A variation of no more than 0.5°F from one bath to the next is recommended. Express this in degrees Celsius.
6. What is normal body temperature (98.6°F) in degrees Celsius?
7. The mass of 1 gal of water is 3785 g. Compute the heat required to raise the temperature of 1 gal of water from 52°F to 140°F. Express the result in: **(a)** joules; **(b)** calories; **(c)** BTU.
8. The mass of 1 gal of mercury is 51,476 g. Compute the heat required to raise the temperature of 1 gal of mercury from 52°F to 140°F.
9. A styrofoam bucket contains 3 kg of water at 80°F. To this is added 1 kg of water at a temperature of 40°F. What is the final temperature of the mixture?
10. One quart of ethyl alcohol at 70°F is poured into a plastic bucket containing 1 gal of water at 80°F. Compute the final temperature of the mixture. (Ignore chemical effects.)
11. Five gallons of water at 60°F are mixed with 2 gal of water at 120°F in an insulated tank. What is the final temperature of the mixture?
12. A 100-g specimen of a copper alloy is heated to 212°F by a boiling water bath. It is then placed into a calorimeter containing 250 g of water at 65.0°F, and the temperature of the water rises to 69.2°F. The calorimeter can is made of aluminum and has a mass of 50 g. Compute the specific heat of the specimen.
13. What is the water equivalent of the calorimeter can in the preceding problem?
14. To determine the temperature of the top surface of a woodstove, 300 g of lead shot is placed on the stove; the lead is subsequently dumped into a styrofoam cup containing 150 g of water at 54°F. The water temperature rises to 71°F. What is the temperature of the stove?
15. An electric heater immersed in 500 g of glycerin draws 3 A at 18 V. It is turned on for 5 min. If no heat is lost from the glycerin, compute the resulting change in temperature of the glycerin.
16. A rod 60 cm long at 0°C is heated from 0°C to 100°C. Compute the change in length of the rod due to thermal expansion, if the rod is made of: **(a)** aluminum; **(b)** brass; **(c)** steel; **(d)** invar.
17. A steel sewer pipe is 24 ft long and has an ID of 12 in. when the temperature is 60°F. Compute **(a)** the length of the pipe and **(b)** the ID of the pipe when the temperature falls to 15°F.
18. On a hot day, gasoline is pumped from a truck into an underground tank. The temperature of the gasoline as it leaves the truck is 85°F, and it subsequently cools to 55°F. If 10,000 gal are pumped from the truck, compute the loss in volume of gasoline due to thermal contraction as it cools.
19. The length of a concrete section of a bridge is 100 ft long. Compute the change in length of the section as the temperature rises from 0°F to 100°F over the 6-month period from January to July.
20. A steel beam 12 ft long is installed between two immovable brick walls when the temperature is 50°F. Each end of the beam is embedded in concrete. Compute the force exerted on each wall by the beam when the temperature reaches 95°F.
21. The capillary tube of a certain thermometer made of Pyrex glass and containing mercury is 0.5 mm in diameter, and the bulb volume is 0.6 cm^3. How much does the length of the mercury column change per degree Celsius?
22. Repeat Problem 21 if the capillary diameter is 0.25 mm instead of 0.5 mm.
23. Repeat Problem 21 if ethyl alcohol is used instead of mercury.
24. One hundred grams of ice are contained in a calorimeter equipped with an electric heat-

er. The initial temperature of the ice is −12°C. Calculate the heat energy that must be added to: **(a)** raise the temperature of the ice to 0°C; **(b)** melt the ice at 0°C; **(c)** heat the resulting water to 100°C; **(d)** evaporate the water by boiling it away.

25. If the electric heater in the preceding problem supplies energy at the constant rate of 8 cal/s, calculate the time required for each step.

26. A calorimeter whose water equivalent is 22 g contains 300.0 g of water at 16°C. Steam is bubbled into the water until its temperature rises to 28°C. The calorimeter is then found to contain 306.3 g of water. Determine the heat of vaporization of water from this data.

27. In an experiment to determine the heat of fusion of water, 75 g of ice at 0°C is added to 300 g of water contained in a calorimeter whose water equivalent is 35 g. The initial temperature of the water is 32°C; the final temperature is 12°C. Compute the heat of fusion of water from this data.

28. The specific gravity of liquid helium is 0.122, and the heat of vaporization is 5 cal/g.

(a) Calculate the heat of vaporization of liquid helium in cal/cm^3.

(b) How many calories of heat are absorbed when 1 L of liquid helium evaporates at 4.2 K?

29. The experimental chamber of a liquid helium cryostat is charged with 1 L of liquid helium. Left undisturbed, the helium is found to evaporate completely in 4 h. Calculate the rate at which heat leaks into the cryostat in: **(a)** calories per minute; **(b)** Joules per second.

***30.** Suppose the experimental chamber of the cryostat in Problem 29 is filled with liquid nitrogen instead of helium. If the heat leak remains the same (an unlikely event), how much time would be required for 1 L of liquid nitrogen to boil away? (The density of liquid nitrogen is 0.808 g/cm^3.)

31. A 4-g bullet traveling at 500 m/s strikes and becomes embedded in a 2-kg block of wood covered with thermal insulation. Assuming that the kinetic energy of the bullet is converted to heat within the block, calculate the increase in temperature of the block.

***32.** An aluminum mass of 1 kg is cooled by 1°C.

(a) Calculate the energy in joules that must be extracted from the mass.

(b) If this quantity of energy is given back to the mass in the form of kinetic energy, calculate the resulting velocity of the mass.

(c) Assume the mass to be hurled upward with an initial velocity equal to that calculated in part **(b)**. To what maximum altitude would the mass rise?

***33.** A 20-kg mass is hoisted by a thin line to the top of a tower 80 m tall. The line is part of a gear-pulley system that connects to a small electrical generator. Electrical energy from the generator is converted to heat inside a calorimeter containing 500 g of water. If 80% of the original potential energy of the hoisted mass is converted to heat inside the calorimeter, compute the resulting rise in temperature of the water as the mass slowly descends.

34. The relative humidity of the air in a room is 65%, and the temperature is 77°F. How much water is contained in each cubic meter of the air?

35. What is the maximum absolute humidity of air at 86°F?

36. What is the relative humidity in a room at 15°C if the air contains 9 g of water per cubic meter?

37. On a winter day when the outside air is 32°F, the relative humidity outside is 74%. The inside air temperature is a cozy 77°F. Calculate the approximate relative humidity inside.

CHAPTER 14

HEAT TRANSFER

Heat transfer is another subject of major engineering importance. In this chapter we describe the three basic mechanisms of heat transfer: conduction, convection, and radiation.

We can consider only the simplest systems because in practice, heat-transfer calculations tend to be very complex. Convection, for example, involves the movement of a heated fluid, and this fact alone brings us face to face with the complexities of fluid flow. Radiative heat transfer calculations require detailed knowledge of the surface of the radiating and absorbing bodies. The process of conduction is probably the simplest, but this process quickly becomes complicated in light of often complicated geometries of structures through which heat is being conducted. In short, a person can devote an entire career to the study and application of heat transfer processes.

14-1 THERMAL CONDUCTION

If one end of a long, thin rod is very hot while the other end is very cold, a flow of heat occurs from the hot to the cold end of the rod. The heat is said to be *conducted* through the rod, and the process is called *thermal conduction.* If we consider the rod on the molecular level, remembering that heat is related to atomic or molecular motion, the atoms at the hot end will be in a more agitated state of motion than the atoms at the cool end. Because the atoms are coupled together by electrical forces that we may picture as tiny springs, it is easy to see that the thermal agitation at the hot end of the rod will soon cause an increase in thermal motion at the cool end. The hot end of the rod becomes cooler while the cool end becomes warmer. Figure 14-1 illustrates a model that is useful in picturing the process of thermal conduction. The vibratory motion of the pendulums is analagous to the vibratory motion of the atoms within the rod.

If molecules are imagined to "jiggle," a hot molecule obviously jiggles more than a cool molecule. Consequently, a transfer of heat is a transfer or redistribu-

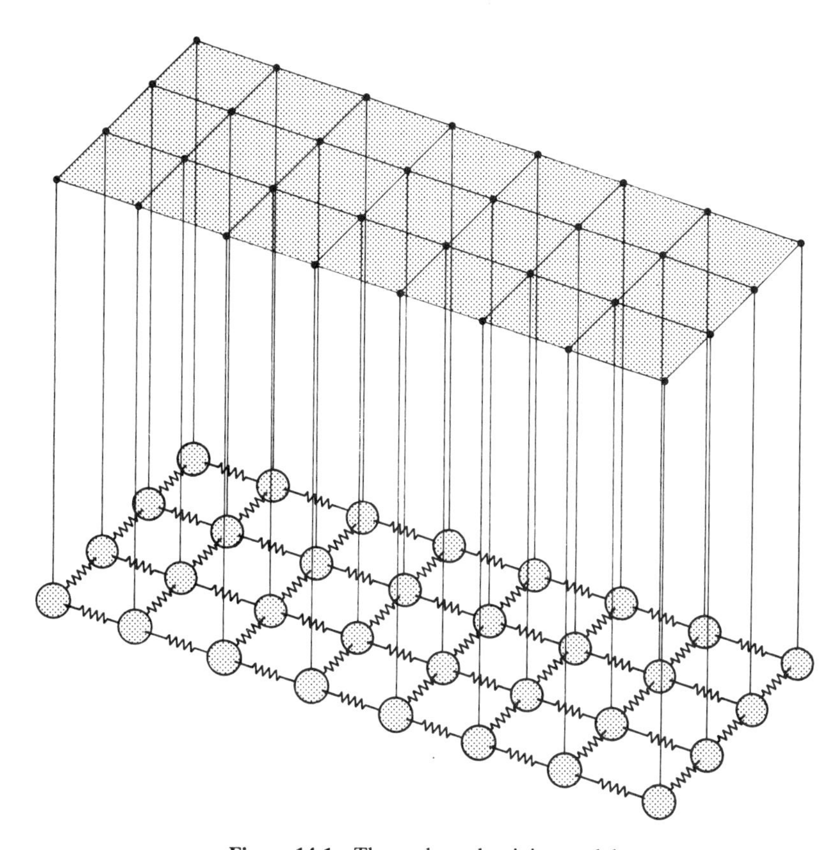

Figure 14-1 Thermal conductivity model.

tion of jiggling motion. It is hard to imagine that the atoms or molecules at one end of an isolated rod would exist permanently in a state of fervent jiggling while the molecules at the other end of the rod hardly jiggle at all.

The *thermal conductivity* of a substance is a measure of its ability to conduct heat. Not all materials conduct heat equally well because the strength of the molecular coupling mechanism (the tiny springs) varies from one substance to another. A method for using a continuous-flow calorimeter to determine the thermal conductivity of a material is described in the following section.

Determination of thermal conductivity

The apparatus shown in Fig. 14-2 illustrates how a heat flow of known proportions can be established in a rod. Hot water or steam circulates around the left end of the rod while cold water circulates on the right. Thermal insulation prevents loss of heat from the sides of the rod so that all heat entering at the hot end travels the full length of the rod. A steady flow of cold water is maintained about the rightmost end of the rod. The amount of heat flowing through the rod can be determined from the flow rate and rise in temperature of the cold water as it flows past the end of the rod.

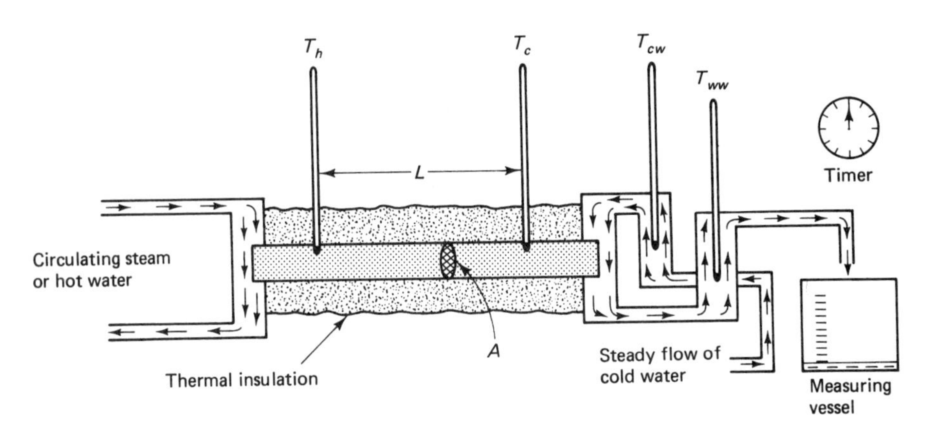

Figure 14-2 Continuous-flow calorimeter.

If M is the mass of cold water flowing through the calorimeter each second, the heat carried away by the water in time interval t is

$$Q = Mc(T_{ww} - T_{cw})t \tag{14-1}$$

where c is the specific heat of water. This quantity of heat must be the same as the heat flowing through the rod.

The heat flow Q through a rod of cross-sectional area A, length L, and thermal conductivity k in a time interval t is

$$Q = kA\frac{(T_h - T_c)}{L}t \tag{14-2}$$

where T_h and T_c are the hot and cool temperatures, respectively, at the ends of the rod. (In Fig. 14-2, we consider only the length of rod between the thermometers inserted into small holes drilled into the rod.) Setting Eqs. (14-1) and (14-2) equal, we obtain

$$kA\frac{(T_h - T_c)}{L}t = Mc(T_{ww} - T_{cs})t$$

which may be solved for the thermal conductivity k:

$$k = \frac{Mc(T_{ww} - T_{cw})L}{A(T_h - T_c)} \tag{14-3}$$

Thus, we can compute the thermal conductivity of the rod. Incidentally, if the water flow rate is given in terms of gallons per minute, the flow in grams per second can be obtained by multiplying by 63.0. Hence, a flow of 2 gal/min is equivalent to 126.0 g/s.

Thermal conduction through thin walls

It is often of interest to calculate the heat conducted through a thin layer of material, such as through a window glass or through a flat metal plate. The general situation is shown in Fig. 14-3, where heat is conducted through a slab of area A,

TABLE 14-1 APPROXIMATE THERMAL CONDUCTIVITIES OF SELECTED MATERIALS

Material	k cal/m · °C · s	k BTU · in./ft^2 · °F · h
Aluminum	49	1400
Brass	26	750
Copper	92	2700
Lead	8.3	240
Silver	97	2800
Steel	11	320
Air	0.0053	0.15
Asbestos	0.14	4
Brick	0.15	4.4
Concrete	0.4	12
Dry soil	0.03	0.9
Glass	0.25	7.2
Fiberglass	0.009	0.3
Ice	0.5	15
Water	0.14	4.0
Vacuum	0	0

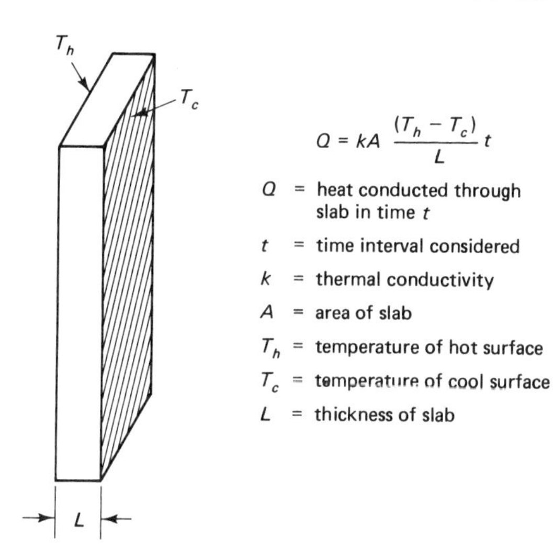

Figure 14-3 Thermal conduction through a slab.

thickness L, and thermal conductivity k. The temperature of the hot and cold surfaces of the slab are T_h and T_c, respectively. The same formula that we applied to the long, thin rod can be applied to the slab, tacitly assuming that a slab is simply a very short rod of large cross-sectional area. Consequently, the applicable formula is Eq. (14-2).

At this point, note that we were careful to specify the temperatures of the hot and cold surfaces of the slab rather than the temperature of the air, water, or whatever substance happens to be in contact with the surface. A thin film of material near the surface (similar to a boundary layer) tends to insulate the surface from the rest of the material, so the surface temperatures of the slab can be significantly different from the ambient temperatures. Consequently, the problems of computing the heat loss through a window pane is not as simple as it might first appear. The principles of fluid flow come into play as convection currents cause the material adjacent to the surface to move. This is an engineering problem. *Heat-transfer coefficients* are tabulated in mechanical engineering handbooks to enable the effects of surface films to be computed. A typical temperature profile is illustrated in Fig. 14-4.

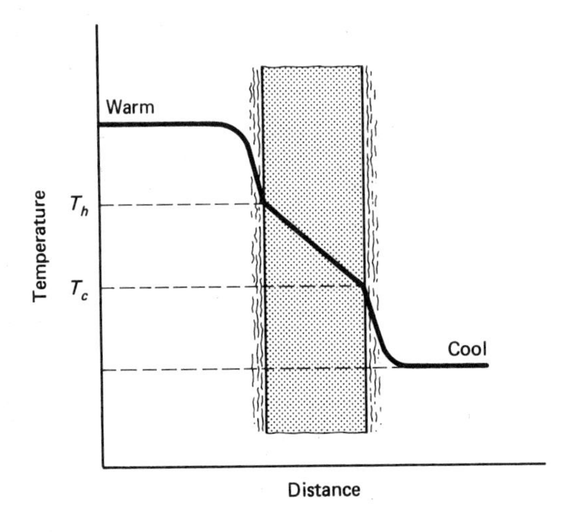

Figure 14-4 Temperature profile for thermal conduction through a wall.

Units of thermal conductivity

Various combinations of units have been utilized for the thermal conductivity, each of which seems to have a particular advantage for particular areas of engineering. The immediate concern in looking up thermal conductivity data is to understand the units being used. Generally speaking, this is no problem because a unit conversion is easily done. However, we must be wary of certain mixtures of units that are frequently encountered.

To determine the units of k, we examine the basic heat-conduction equation, Eq. (14-2), that is used to define the thermal conductivity. Solving the equation for k and inserting units gives

$$k = \frac{QL}{A(T_h - T_c)t} \qquad [k] = \frac{\text{cal} \cdot \text{m}}{\text{m}^2 \cdot {}^\circ\text{C} \cdot \text{s}}$$

We note that a length unit m appears in both the numerator and the denominator, so a cancellation is possible:

$$[k] = \frac{\text{cal}}{\text{m} \cdot {}^\circ\text{C} \cdot \text{s}}$$

By examining the units of k, we see that the heat Q is given in calories, the area is in square meters, the temperature is in degrees Celsius, the time is in seconds, and the thickness of the slab (or the length of the rod) is in meters. Many alternative forms are possible: Heat might be measured in joules or British thermal units, degrees Fahrenheit might be used instead of degrees Celsius, and so forth. However, a peculiar situation arises when a length unit is used to measure the slab thickness that is different from the length unit used to give the area. For example, the area is frequently given in square feet while the slab thickness is given in inches. Then, the units of thermal conductivity appear as

$$[k] = \frac{\text{BTU-in.}}{\text{ft}^2\text{-°F-h}}$$

and we note that no cancellation is possible between the numerator and the denominator as was done earlier. Unit brackets may be used to convert the mixed units to the more familiar unmixed form preferred by physicists.

EXAMPLE 14-1 A window 4 ft wide and 9 ft tall is made of glass 0.375 in. thick with a thermal conductivity of 5.5 BTU-in./ft^2-°F-h. Calculate the heat transfer by conduction through the window per hour if the cold surface of the glass is 40°F while the warm surface is 60°F.

Solution. We first examine the units of the thermal conductivity and the statement of the problem to see that they are consistent. Then, using Eq. (14-2),

$$\begin{aligned} Q &= kA\frac{(T_h - T_c)}{L}t \\ &= \left(5.5\ \frac{\text{BTU-in.}}{\text{ft}^2\text{-°F-h}}\right)(36\ \text{ft}^2)\frac{(60 - 40\text{°F})}{(0.0375\ \text{in.})}(1\ \text{h}) \\ &= 10{,}560\ \text{BTU} \end{aligned}$$

Note that a mixture of units was used in this example, and the unit of time is the hour.

14-2 THERMAL CONVECTION

Thermal convection is the process of heat transfer that occurs when a liquid or gas heated at one location is subsequently caused to flow to another location, carrying the heat with it. Figure 14-5(a) illustrates convection currents surrounding a candle in a box. Hot air rises because it is buoyed up by heavier, cooler air. Above a candle, hot air rises in a column of surprisingly small diameter until it is forced to spread out by the ceiling of the box or until it gradually breaks up by mixing with cooler air. The hot gases from an open fire in a room will stream to the ceiling in a narrow column, filling the room with hot, poisonous gases from the top down—as if the room were inverted and being filled with water.

On a sunny day, fields of grain or newly mowed hay become warmer more quickly than a boundary of pine trees. Consequently, warm air rises over the fields and cooler air falls over the trees. A small airplane flying alternately over fields and trees will experience a bumpy ride as the plane passes from regions of rising air to regions of falling air. A column of rising air is called a *thermal*. Soaring birds and glider pilots seek out thermals as sources of lift in order to gain altitude with minimal effort. The bumpy ride of the small plane is often erroneously

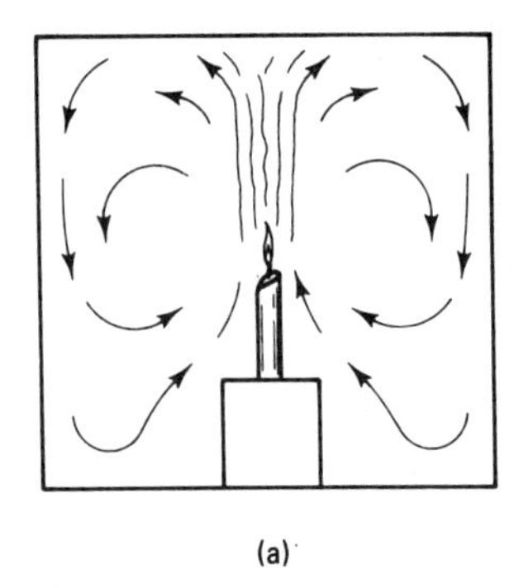

(a)

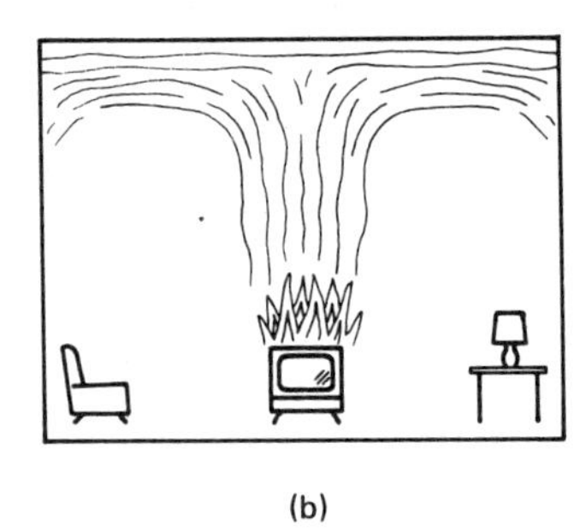

(b)

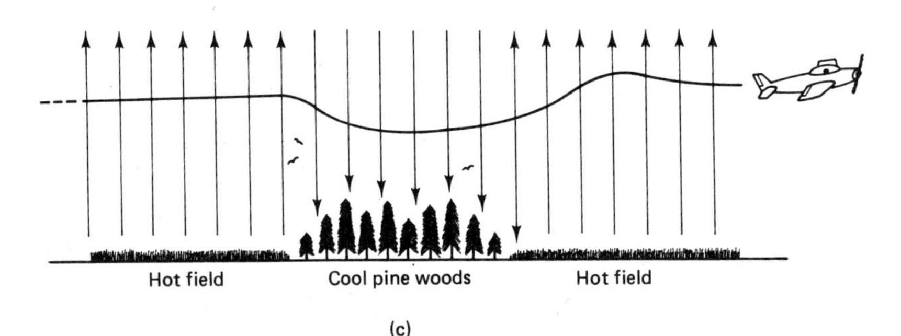

(c)

Figure 14-5 (a) Convection currents around a candle in a box.
(b) A fire in a room causes the room to fill from the top down with hot, poisonous gases and smoke.
(c) Convection currents provide a bumpy ride for a low-flying plane.

attributed to air pockets, an expression which brings up mental images of voids and empty spaces in the air into which the plane periodically plunges.

The convection occurring as described above is called *natural convection*. It arises by virtue of the earth's gravitational field and the fact that air expands when heated. *Forced convection* refers to the movement of a fluid by pumps, blowers, or fans. The forced-air, central-heating system of a home provides an example of forced convection.

Computation of heat transfer by convection occurring at a surface is a complicated problem involving the geometric shape of the surface plus the density, viscosity, thermal conductivity, and specific heat of the medium surrounding the surface. An expression of the following form is generally used:

$$\frac{Q}{t} = hA(T_s - T_a) \tag{14-4}$$

where Q/t is the heat transferred per unit time, A is the area of the surface, $(T_s - T_a)$ is the temperature difference between the surface and the ambient environment, and h is a heat-transfer coefficient typically determined (or verified) experimentally. The value of h is different for various configurations, such as vertical pipes, horizontal pipes, vertical plates, heated plates facing upward, or cooled plates facing downward. Typical values of h range from less than 1 up to several BTU/(ft^2-h-°F) for moderate temperature differences. Interested readers may find more information in books on heat transfer or in a handbook for mechanical engineers.

EXAMPLE 14-2 A 3-in. diameter horizontal pipe, exposed to the air, carries water at a temperature of 180°F. The pipe is 20 ft long; the air temperature is 32°F. How much heat per hour is lost from the pipe?

Solution. The surface area of the pipe (πDL) is 15.7 ft^2 and ΔT is 148°F. From Table 14-2, the heat-transfer coefficient is 1.2 BTU/ft^2-h-°F. Hence, using Eq. (14-4),

$$\frac{Q}{t} = hA(T_s - T_a)$$

$$= (1.2 \text{ BTU/ft}^2\text{-h-°F})(15.7 \text{ ft}^2)(180 - 32\text{°F})$$

$$= 2790 \text{ BTU/h}$$

TABLE 14-2 NOMINAL VALUES OF CONVECTIVE HEAT-TRANSFER COEFFICIENTS FOR SELECTED CONFIGURATIONS (BTU/Ft^2-h-°F)

$\Delta T^{(a)}$	50	100	150
Vertical plate			
Height = 2 ft	0.6	0.7	1.0
Height = 4 ft	0.7	0.9	1.0
Horizontal cylinder			
Diameter = 3 in.	0.9	1.1	1.2
Diameter = 6 in.	0.8	0.9	1.0
Diameter = 12 in.	0.7	0.8	1.0
Horizontal surface (area ≈ 16 ft^2)			
Facing upward	0.8	1.0	1.2
Facing downward	0.23	0.27	0.3

[a] $\Delta T = T_{\text{surface}} - T_{\text{ambient}}$

14-3 THERMAL RADIATION

Electromagnetic radiation

Before jumping headlong into a discussion of thermal radiation, we should first consider *electromagnetic* radiation, which is most frequently related to radio waves, television waves, radar waves, and so forth. Electromagnetic radiation (or electromagnetic waves) encompasses much more than the waves associated with radio, television, and radar. Visible light, infrared radiation, ultraviolet radiation, and X-rays also are forms of electromagnetic radiation; the difference between one form and another is the *frequency* of the radiation, the number of waves emitted per second.

As a consequence of the frequency difference, the *wavelength* associated with each form is different also, and the energy content is different because the basic unit of energy of each form is directly proportional to the frequency. All forms of electromagnetic radiation travel at the same velocity through a vacuum, namely, at the velocity of light—very nearly 3×10^8 m/s. The different forms have different properties because of the differences in frequency and wavelength; but, fundamentally, all forms of electromagnetic radiation are various manifestations of the same physical phenomenon.

The electromagnetic spectrum is illustrated in Fig. 14-6. It is noteworthy that the visible spectrum occupies such a small portion of the total spectrum; our eyes

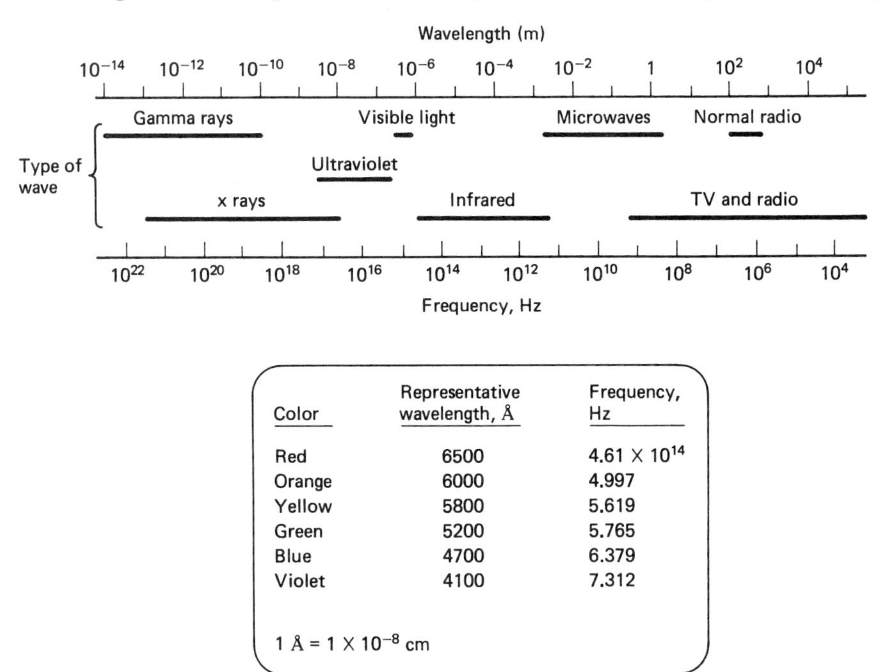

Color	Representative wavelength, Å	Frequency, Hz
Red	6500	4.61×10^{14}
Orange	6000	4.997
Yellow	5800	5.619
Green	5200	5.765
Blue	4700	6.379
Violet	4100	7.312

1 Å = 1×10^{-8} cm

Figure 14-6 The electromagnetic spectrum.

do not respond to a wide range of frequencies. The sensation of color is due to the mental response arising from a frequency difference; blue light is of a higher frequency than red. Because the rest of the spectrum is invisible, we tend either to ignore it or to treat it as if it were somewhat mysterious. *Black light* is simply ultraviolet radiation which we cannot see. It is possible to take photographs using infrared or ultraviolet radiation.

Radio and television waves are produced at the transmitter by having electrons (electrical charges) oscillate or vibrate back and forth along the transmitting antenna. The vibration involves accelerations because the velocity of the electrons is constantly changing. It is the acceleration of electrical charges that is responsible for the production of the electromagnetic waves. Any charge that undergoes an acceleration will emit an electromagnetic wave. The term *electromagnetic* stems from the fact that both an electric field and a magnetic field are inseparably associated with an electromagnetic wave.

There is no clear-cut dividing line between the various regions of the spectrum. Radio waves behave more like light waves as the frequency of the radio waves is increased. The beam from a microwave tower can be envisioned as a beam from a searchlight, and the shape of the reflector dish sending out the beam is of the same shape (parabolic) as that of a searchlight. However, the frequency of visible light is on the order of 100,000 times as high as the highest frequency used in electronics applications.

Thermal radiation

Because of the electrical charge associated with electrons and protons, it is not surprising that a substance should emit electromagnetic radiation because of the thermally induced atomic and molecular motion. *Thermal radiation* is the energy given off from a solid, liquid, or gas in the form of electromagnetic waves due to the temperature (thermal agitation) of the emitting substance. Furthermore, the intensity and frequency distribution of the emitted radiation depends strongly upon the temperature. A body heated to about 600°C glows a dull red, indicating that energy is being emitted in the visible spectrum—even though most of the energy is emitted as infrared radiation. A body at room temperature emits also, but at a much smaller rate and at correspondingly lower frequencies. In theory, thermal radiation from a body ceases only at a temperature of absolute zero, the temperature at which all molecular motion ceases.

The physical sensations produced by thermal radiation are familiar. The warmth felt on the face when standing in front of a fireplace or a hot stove is due to thermal radiation, as is (in part) the warmth of sunlight. The heat received from a bonfire is primarily radiant, and a person standing nearby may be warmed by thermal radiation even though the temperature of the surrounding air is rather low. Hence, because thermal radiation from a fire can "shine" only on one side of a person at a time, one roasts on one side while freezing on the other.

We allude to the thermal radiation of sunlight in the preceding paragraph, but sunlight contains electromagnetic radiation from all portions of the spectrum, including infrared, visible, and ultraviolet. Actually, the warmth of sunlight is due to the combined effects of the radiation from the different portions of the spectrum. Ultraviolet radiation is especially significant in producing a sunburn or in causing eye damage if one looks directly at the sun.

The Stefan-Boltzmann law

Even though the physical processes involved in radiation are extremely complex, the energy radiated from a body may be computed by a simple expression known as the Stefan-Boltzmann law. The energy I radiated per unit area per unit time is given by

$$I = \epsilon\sigma T^4 \tag{14-5}$$

where T is the absolute temperature. The radiation constant σ (sigma) is a universal constant, meaning that it applies to all substances. On the other hand, the *emissivity* ϵ (epsilon) depends upon the surface characteristics of the radiating body. Careful measurement gives

$$\sigma = 5.67 \times 10^{-8}\ \text{J/s}\cdot\text{m}^2\cdot\text{K}^4 = 5.67 \times 10^{-12}\ \text{J/s}\cdot\text{cm}^2\cdot\text{K}^4 = 1.36 \times 10^{-8}\ \text{cal/s}\cdot\text{m}^2\cdot\text{K}^4$$

It is helpful to visualize thermal radiation as being generated within the interior volume of a body so that it must pass through the body surface in order to

TABLE 14-3 EMISSIVITY OF SELECTED MATERIALS[a]

Material	Emissivity
Aluminum, unoxidized	0.03
oxidized	0.15
Brass, unoxidized	0.03
oxidized	0.6
Copper, unoxidized	0.02
oxidized	0.6
Iron, rusted	0.6
wrought	0.94
Steel, unoxidized	0.08
oxidized	0.8
Steel plate, rough	0.95

[a]Because the emissivity depends strongly upon the surface condition and somewhat upon the temperature, these values are subject to considerable variation.

escape. A highly polished surface will act like a mirror and reflect most of the radiation back into the body. But a rough, dull, black surface will allow almost all the radiation to pass through the surface and be radiated. The emissivity describes the surface condition; rough, dull, black surfaces have emissivities very nearly 1.0, while shiny, highly polished surfaces have emissivities near zero. Thus, the emissivity ranges between 0.0 and 1.0. The emissivity of highly polished pure gold may be as low as 0.018 while rough steel plate may have an emissivity as high as 0.97. Incidentally, the emissivity has no units.

Thus far, we have considered only the ability of a body to emit radiation, but a body absorbs a fraction of any radiation that falls on its surface. Furthermore, a body that is a good emitter is also a good absorber, and a body that emits poorly will absorb poorly. Thus, the emissivity applies to the absorbing qualities of a body as well as to its radiating properties. A body with an emissivity of 1.0 absorbs all the radiation falling on it; a body with an emissivity of 0.0 is a perfect reflector and does not absorb any of the incident radiation.

Ordinary objects are made visible by light reflected from their surfaces. Consequently, an object that reflects absolutely no light would appear completely black and would therefore be invisible. Such objects do not exist in reality, but it is often helpful to refer to such an idealized perfect absorber, called a *blackbody*. A blackbody has an emissivity of 1.0 and is a perfect absorber and a perfect emitter. A close approximation to a blackbody is illustrated in Fig. 14-7.

All bodies radiate and absorb energy simultaneously. If a body is completely enclosed—as shown in Fig. 14-8, where T_b is the temperature (absolute) of the body and T_w is the temperature of the walls of the enclosure—then the heat radiated and absorbed by the body of area A in time t are given by

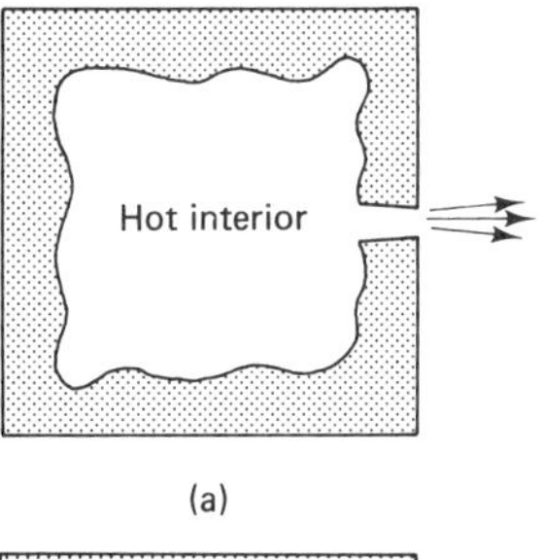

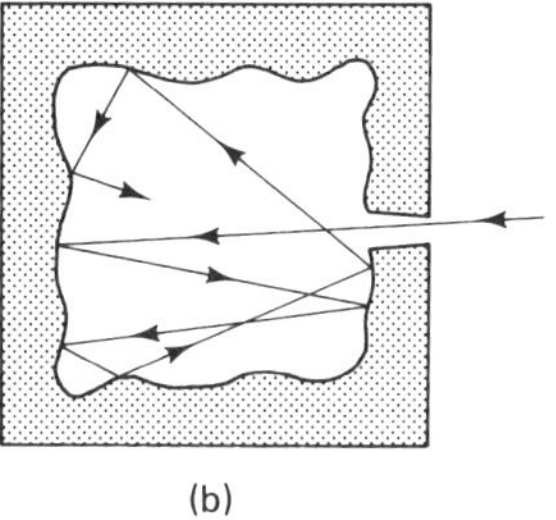

Figure 14-7 A small opening into an interior cavity approximates a blackbody, as (a) a perfect emitter and (b) a perfect absorber.

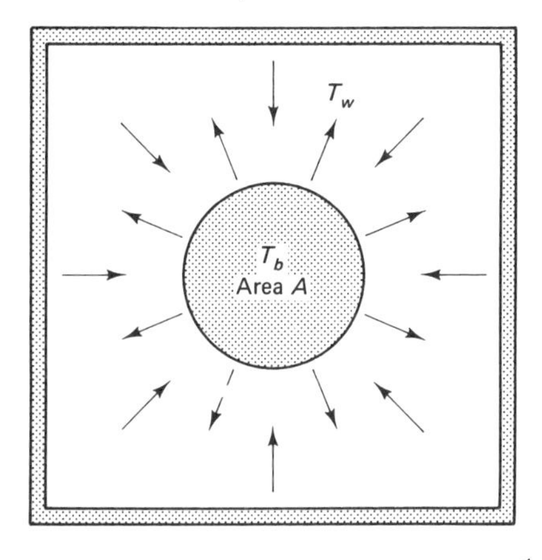

Figure 14-8 A body within an enclosure radiates heat to the walls and, at the same time, absorbs heat from the walls.

(*heat radiated*) $$Q_r = A\epsilon\sigma T_b^4(t)$$

(*heat absorbed*) $$Q_a = A\epsilon\sigma T_w^4(t)$$

The net heat transfer to or from the body is the difference in the heat radiated and the heat absorbed:

(*net heat transfer*) $$Q = Q_r - Q_a$$
$$= A\epsilon\sigma(T_b^4 - T_w^4)(t) \qquad (14\text{-}6)$$

If the body is hotter than the walls, the body will give off more heat than it absorbs. But if the body is colder than the walls, the body will absorb more heat than it gives off. When the net heat transfer is *to* the body, the temperature of the body will rise. But if the body radiates more energy than it absorbs, the temperature of the body will fall. From this we can see that the temperatures of the body and of the walls will eventually become equal.

This process is not limited to a special body within a special enclosure. A book lying on a desk in a room behaves in a similar fashion. The size of the enclosure is not important as long as the object is completely surrounded.

EXAMPLE 14-3 Suppose a sphere 1 m in diameter is located in interstellar space. If the temperature of the sphere is 20°C and if its emissivity is 0.6, calculate the energy radiated per second from the sphere.

Solution. First we convert 20°C to 293 K. Then we use Eq. (14-5) to compute the energy radiated per unit area:

$$I = \epsilon\sigma T^4$$
$$= (0.6)\left(5.67 \times 10^{-8} \frac{\text{J}}{\text{s} \cdot \text{m}^2 \cdot \text{K}^4}\right)(293\ \text{K})^4$$
$$= 250.7\ \text{J/s} \cdot \text{m}^2$$

Multiplying by the total surface area of the sphere (0.785 m^2), we get the total energy radiated per second:

$$Q/t = A \cdot I$$
$$= (0.785\ \text{m}^2)(250.7\ \text{J/s} \cdot \text{m}^2)$$
$$= 196.8\ \text{J/s}$$
$$= 47.0\ \text{cal/s}$$

Because the sphere is in interstallar space, we assume the radiation incident on the sphere is negligible.

Wien's displacement law

It is well known that as an object is heated to higher temperatures, the color of the emitted radiation changes from a dull red to a yellowish red to a brilliant white. Larger quantities of high-frequency radiation are given off at high temperatures. An object is just visible in a darkened room at about 600°C. At 700°C it is dark red; it becomes a bright red at about 900°C. At 1100°C, the object assumes a yellowish tint, which tends toward white at 1300°C. The temperature of "white heat" is on the order of 1500°C.

For a body at a given temperature, the emitted energy is not equally distributed across the frequency spectrum. For each temperature there is a particular frequency (or wavelength) at which the emitted energy will be a maximum. Moreover, as the temperature is increased, the maximum shifts (or is *displaced*) to a higher frequency. The location of the maximum is given by the following relationship, known as *Wien's displacement law*:

$$\lambda_m T = 0.290\ \text{cm} \cdot \text{K} \tag{14-7}$$

In terms of frequency, Wien's law becomes

$$T = f_m(9.70 \times 10^{-12}\ \text{s} \cdot \text{K}) \tag{14-8}$$
$$f_m = T(1.03 \times 10^{11}\ \text{s}^{-1} \cdot \text{K}^{-1}) \tag{14-9}$$

A plot of the spectral energy distribution for several temperatures is given in Fig. 14-9. A given frequency may be related to the color of visible light by referring to Fig. 14-6.

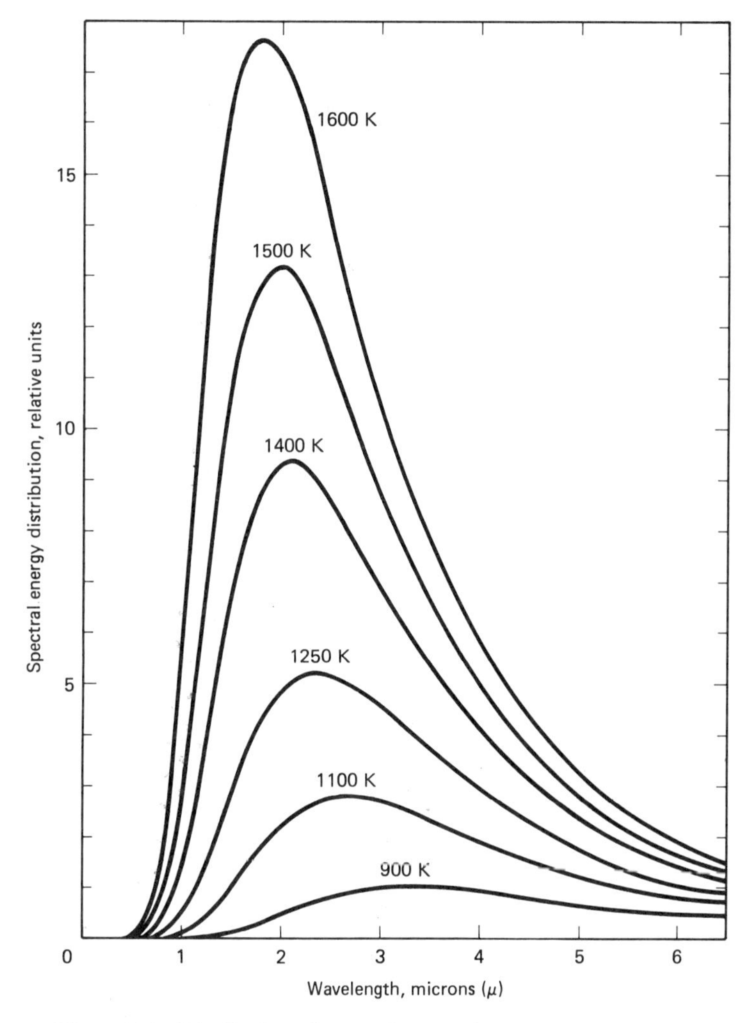

Figure 14-9 Distribution of spectral energy for several temperatures.

Wien's displacement law is put to practical use in a temperature-measuring device called an *optical pyrometer*, illustrated in Fig. 14-10. A red filter within the instrument isolates a narrow portion of the visible spectrum for examination. The object whose temperature is to be measured is viewed through the telescope. Superimposed on the field of view is a lamp filament whose temperature can be varied by carefully calibrated control circuitry. By matching the brightness of the filament to the brightness of the object (so that the filament seems to disappear), the filament is set to the same temperature as the object. The filament temperature (and object temperature) is then read directly from the temperature calibrations on the filament-brightness control.

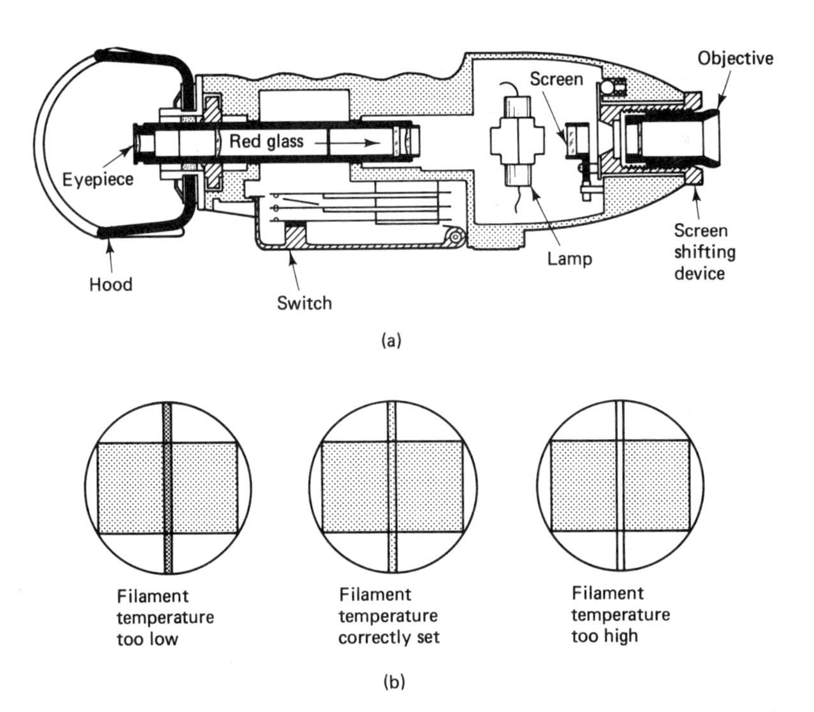

Figure 14-10 Optical pyrometer head. The filament temperature is correctly set when the color and brightness of the filament matches that of the object whose temperature is being determined.

EXAMPLE 14-4 At what wavelength will the maximum energy be radiated from an object that is white hot—for example, 2400°F?

Solution. Convert 2400°F to 1315°C and then to 1588 K. Then use Wien's displacement law.

$$\lambda_m T = 0.290 \text{ cm} \cdot \text{K}$$

$$\lambda_m = \frac{0.290 \text{ cm} \cdot \text{K}}{1588 \text{ K}} = 1.83 \times 10^{-4} \text{ cm}$$

Converting to nanometers (1 cm = 10^7 nm),

$$\lambda_m = 1826 \text{ nm}$$

This lies in the infrared portion of the spectrum, even though the object is white hot.

Radiation and convection in residential heating

There is a very noticeable difference in the sensation of warmth produced by standing in front of a fireplace as opposed to standing in a stream of warm air blown from a hot-air heating system. The difference lies in the relative proportion of radiant and convective heat transfer. The heat derived directly from a fireplace is almost entirely radiant, whereas the heat derived from the hot-air vent is almost purely convective. Radiant energy passes essentially undiminished through the air directly to the skin, but the warm air must flow over the skin in order to transfer the heat. Recall that a boundary layer develops when a fluid flows past a stationary object, and in the case of warm air flowing over a person's skin, the boundary layer acts as a layer of insulation, which impedes heat transfer. People coming in from the cold outside much prefer the radiant heat of a fireplace or a stove to warm air for getting warm quickly.

The reason that a fireplace gives off primarily radiant heat is that the warm air produced by a fireplace goes up the chimney along with most of the smoke. It is true that the surrounding brickwork gets warm and produces a convective heat transfer, but this is a relatively minor part of the total heat. Various systems of piping and ductwork can be installed in a fireplace in order to circulate air around the fire and bring it back into the room. While these devices enhance the convective heat transfer, a fireplace is still a rather inefficient source of heat.

The heat transfer from a free-standing stove in a room is both radiant and convective. The amount of radiant heat given off depends upon the temperature of the surface of the stove. If the stove is constructed of thin metal so that parts of the surface get very hot (perhaps even cherry red), the heat given off by radiation is much greater than from a stove constructed of heavier metal that heats up more uniformly.

It is interesting to observe the various forms of electric heaters on the market and speculate on whether the primary type of heat transfer is radiant or convective. If the exposed heating element becomes red-hot in operation, chances are that the primary heat transfer is radiant. But if the heating coils become a dull red while a fan blows air over the coils, the heat transfer is primarily convective. Some economy-model electric heaters include a fan whose purpose seems more to cool the thin metal of which the heater is made than to blow air over the heating coils. Whichever is best depends upon whether you wish the heat to shine forth as radiation or to appear more gently in the form of warm air.

14-4 HEAT FLOW THROUGH BUILDING STRUCTURES

Heat flow through walls, partitions, floors, roofs, and plane structures in general is commonly calculated according to the following expression:

$$Q = UA(\Delta T)t$$

where

$$A = \text{area in square feet}$$

$$\Delta T = \text{temp difference in degrees Fahrenheit}$$

$$t = \text{time in hours}$$

Here, U is the *heat-transmission coefficient*, usually expressed in the units of BTU/ft^2-h-°F. The numerical value of U gives the heat flow in British thermal units per hour per square foot of area per degree Fahrenheit of temperature difference between the hot and cold surfaces of the structure. U depends upon several factors, such as the thermal conductivity of the parts of the structure, the thickness of the structure, and the type of construction. Values of U are tabulated for various types of structures. (See the reference at the end of the chapter.) Typical values for frame walls and partitions range from about 0.15 to 0.5, with the most common value being around 0.25.

The reciprocal of the thermal transmission coefficient is the *thermal resistance*, R. It is the total resistance per unit area of the structure to the flow of heat. This is often called the *R-value* when speaking of home insulating materials. Thus,

$$R = \frac{1}{U} \qquad [R] = \frac{\text{ft}^2\text{-h-°F}}{\text{BTU}} \tag{14-10}$$

If the R-value is known for a given wall, the heat flow through the wall is given by

$$Q = A\frac{(\Delta T)}{R}t \tag{14-11}$$

In practical terms, the R-value gives the number of degrees Fahrenheit that must be established across a wall section in order to transmit through the wall 1 BTU of heat per hour for every square foot of area of the wall. For example, if insulation is rated as R 11, a temperature difference of 11°F will cause a heat transmission of one BTU every hour for each square foot of area.

The advantage of using R-values is that for a wall made up of several layers, such as sheathing, insulation, and interior paneling, the R-values of each layer may be simply added to give an overall R-value for the wall. R-values have been tabulated for a wide variety of building and insulating materials.

EXAMPLE 14-5 An external wall of a room is 9 ft high and 20 ft long. It is insulated to an R-value of 9. How much heat is lost through the wall per hour when the inside and outside temperatures are 70°F and 15°F, respectively?

Solution. The area of the wall is 180 ft^2 and T = 55°F. Using Eq. (14-11),

$$\frac{Q}{t} = A\frac{\Delta T}{R}$$

$$= (180\ \text{ft}^2)\frac{55\text{°F}}{[9\ \text{ft}^2\text{-h-°F/BTU}]}$$

$$= 1100\ \text{BTU/h}$$

Reference

Handbook of Air Conditioning System Design. Carrier Air Conditioning Company. New York: McGraw-Hill, Inc., 1965. Contains extensive data for making practical calculations of heat losses from building structures.

Questions

1. In the temperature range from 32°F to about 38°F, water becomes more dense as it is warmed. Suppose a small electric heater is centrally located in a tank of water initially

at a uniform temperature of 33°F. Describe the unusual feature of the convection currents produced when the heater is turned on.

2. What effect does the wind have upon the surface film near a warm body?
3. Why do the ungloved hands of a motorcyclist get so cold even when the temperature is a moderate 40°F?
4. Does the wind affect the reading of a thermometer?
5. Suppose a thermometer hanging outside on a cold day is wrapped in a woolen sock. What will be the effect on the indicated temperature 1 h later?
6. Explain how a typical vacuum thermos bottle minimizes heat transfer by conduction, convection, and radiation.
7. An inventor claims to have produced a perfect blackbody comparable in size to a bowling ball. Describe the appearance you would expect this object to exhibit.
8. Why do soot or ashes on snow make it melt faster?
9. How does a house gain thermal energy in the summertime? Will the things done to conserve heat in winter cause the house to be hotter in the summer?
10. People sometimes store potatoes in the woodshed. Under what circumstances will wrapping the potatoes in a blanket keep them from freezing?
11. When leaving the house to go to work, should you turn off the heat in order to save energy? Discuss. What is it that determines the heat loss from a house besides insulation, caulking, and other structural features?

Problems

1. A copper rod 1 m long and 1 cm in diameter is carefully insulated so that no heat escapes through the sides of the rod. The temperatures of the hot and cold ends of the rod are maintained at 100°C and 0°C, respectively. Compute the quantity of heat that flows down the rod: **(a)** in calories per second; **(b)** in joules per second; **(c)** in BTU per hour.
2. Repeat Problem 1, but assume the rod is made of brass instead of copper.
3. Suppose cold water flows through a continuous-flow calorimeter at a steady rate of 1 qt/min. If the water enters at 18°C and leaves at 21°C, calculate the heat carried away by the water: **(a)** in calories per second; **(b)** in joules per second; **(c)** in BTU per hour.
4. The floor of a certain building is a slab of concrete 10 ft wide, 12 ft long, and 4 in. thick. The bottom surface temperature of the slab is 52°F, and the top surface temperature is 64°F. Compute the heat conducted through the slab per hour.
5. Steam at 100°C is circulated about the hot end of the sample rod in a continuous-flow calorimeter, as in Fig. 14-2. The cross-sectional area of the rod is 4 cm^2 and the distance between thermometers T_h and T_c is 25 cm. T_h and T_c are 80°C and 50°C, respectively. At the cool end of the rod, water flows through the calorimeter at a rate of 1.25 g/s, and the temperature of the water is raised by 1°C in passing through the calorimeter. Compute the thermal conductivity of the sample rod.
6. Suppose a spherical object 1 m in diameter with emissivity $\epsilon = 0.6$ is situated in interstellar space where the radiation from distant stars in negligible. Compute the energy radiated from the object per second when its temperature is: **(a)** 1000 K; **(b)** 300 K; **(c)** 100 K; **(d)** 20 K; **(e)** 4 K.
7. If the spherical object in the problem above were solid brass, it would have a mass of 4.5×10^6 g and a heat capacity of about 1.71×10^6 J/°C.
   (a) Use Eq. (13-3) to calculate the decrease in temperature of the brass ball in 1 s occurring when its temperature is near 1000 K.
   (b) As the body cools, what happens to its rate of cooling?

8. Verify the mass and heat capacity of the brass ball in Problem 7.

9. A steel ball whose surface is oxidized is suspended in a cubical enclosure 3 m on each side. The diameter of the ball is 20 cm. The temperature of the ball is 900 K, and the temperature of the walls of the enclosure is nearly room temperature, 300 K.
   (a) Compute the rate at which energy is radiated from the ball.
   (b) Compute the rate at which the ball absorbs energy from the walls of the enclosure.
   (c) What is the net energy loss from the ball per second?

10. In the preceding problem, while the temperature of the ball is near 900 K, what is the rate of cooling of the steel ball in degrees Celsius per second?

11. The surface temperature of a 6-in. OD horizontal pipe is 125°F on a day when the air temperature is 75°F. Compute the convective heat loss per hour from a 25-ft-long section of the pipe.

12. A horizontal surface of area 14 ft^2 is heated to 225°F on a day when the ambient temperature is 75°F. Compute the heat transfer from the surface per hour if the surfaces are facing: **(a)** upward; **(b)** downward.

13. A 10-ft-long vertical plate is heated to 180°F in a room where the ambient temperature is 80°F. Compute the heat transfer per hour by convection if the plate is : **(a)** 2 ft high; **(b)** 4 ft high.

14. The filament of a certain lamp is heated to a temperature of 2400°F. Use Wien's displacement law to calculate the wavelength at which maximum emission of energy occurs.

15. The heat transmission coefficient for a particular wall of surface area 120 ft^2 is 0.10 BTU/ft^2-h-°F. Compute the heat conduction through the wall per hour for a temperature difference of 15°F.

16. Compute the R-value for the wall in Problem 15.

17. What temperature difference must be established across the wall of Problems 16 and 17 to produce a total heat conduction through the wall of 120 BTU/h?

18. Compute the R-value of a layer of asbestos 1 in. thick.

19. Compute the R-value of a concrete wall 6 in. thick.

20. Compute the R-value of a steel plate 1 in. thick.

CHAPTER 15

IDEAL GAS

This chapter is devoted to the properties of gases. In earlier chapters we give various properties—such as density, viscosity, and elastic moduli—of solids and liquids, but it is only after we study heat and temperature that the properties of gases can be meaningfully addressed.

The concept of an ideal gas is very useful because it lends itself to a fairly simple mathematical analysis and because real gases closely approximate the behavior of an ideal gas. In the first part of this chapter we present the equation of state of an ideal gas and then obtain Boyle's law and Charles' law as a consequence of the equation of state.

In the section on kinetic theory, we consider an ideal gas on the molecular level. This allows us to relate the temperature of a gas to the average kinetic energy per molecule of the gas. The Maxwell speed-distribution function for calculating molecular speeds is also given.

The last sections of this chapter deal with real gases. We present the van der Waal equation of state and note the differences between it and the equation of state of an ideal gas. Finally, we investigate the conditions under which a gas may be liquefied and see how it can be done in practice.

15-1 DESCRIPTION OF AN IDEAL GAS

An *ideal gas* is a gas whose molecules are hard, rigid spheres like miniature billiard balls. The super-subminiature spheres are assumed not to attract or repel each other until they touch; when they do touch, the resulting collision is perfectly elastic. Collisions with the walls of the containing vessel are perfectly elastic also. The molecules are assumed not to rotate. Therefore, all the kinetic energy that a particular sphere may have is due to its translational motion.

No real gas, such as hydrogen, helium, or oxygen, meets the requirements of an ideal gas because the molecules of all real gases either attract or repel each other, depending upon their separation. At a low-enough temperature, the attraction results in the condensation of the gas to a liquid. But as long as only moderate pressures are exerted on a real gas and as long as the temperature remains significantly above the condensation (liquefaction) temperature of the gas, the behavior of real gases is described to a good approximation by treating the real gas as if it were an ideal gas.

To picture an ideal gas, let us consider a cubical box filled with ideal gas at room temperature (20°C) and at atmospheric pressure. If the box is 10 cm on each side, its volume will be 1000 cm^3, which is 1 L. A simple calculation reveals that the box contains about 2.5×10^{22} molecules. If the molecules are assumed to be regularly spaced throughout the volume of the box, about 2.92×10^7 molecules would appear on a line drawn along each edge of the box. This corresponds to a distance between molecules on the order of 34×10^{-8} cm (or 34 Å, 1 Å being 1×10^{-8} cm). Thus, the spacing between the molecules is very small. This does not imply that the molecules are packed tightly together, however, because the average diameter of the molecules is only about one-tenth or so of the average spacing between the molecules. Consequently, the picture we should form of an ideal gas is that of an enormous number of Ping-Pong balls floating in space with an average distance between them of about a foot.

In our picture of an ideal gas, we have thus far ignored the fact that the molecules are moving at high speeds. The molecules are in motion because the sample of gas contains thermal energy. We have seen already that the molecules are incredibly small, and we might expect their average velocities to be small also. But such is not the case. If the mass of the hard-sphere molecules is assumed to be the same as the mass of an oxygen molecule, the average speed of a molecule in the box at 68°F and atmospheric pressure is on the order of 450 m/s. This is about 1000 mi/h!

If we now consider that our 1-L box of ideal gas contains a few thousand billion billions of molecules, each of which is traveling at supersonic speeds, it is not hard to imagine that there will be a great number of collisions between the molecules themselves and also between the molecules and the walls of the box. As a matter of fact, each molecule will experience several billion collisions per second, and each square centimeter of the walls of the box must endure about 10^{25} collisions per second.

The distance a molecule travels between successive collisions is called a *free path*, and the average of all the free paths is called the *mean free path*. The mean free path depends upon the molecular diameter and upon the number of molecules per unit volume. For oxygen and nitrogen at room temperature and atmospheric pressure, the mean free path is about 10^{-5} cm. In the near-vacuum conditions at the beginning of space about 100 mi above the surface of the earth, the mean free path may be on the order of miles.

Brownian motion

Convincing evidence for the kinetic nature of a gas is obtained from *Brownian motion* [after Robert Brown (1773–1858), English botanist] which is the zigzagging, erratic motion of small particles (such as smoke particles) floating freely in a fluid (such as air). One form of a Brownian motion apparatus is shown schemati-

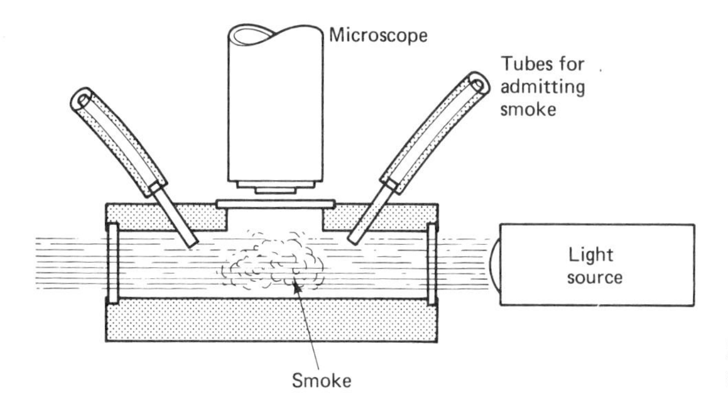

Figure 15-1 Brownian motion apparatus.

cally in Fig. 15-1. An intense beam of light illuminates smoke particles from a cigarette, and if the smoke is not too dense, individual particles show up as tiny points of light. However, the tiny points of light seemingly cannot be still; they constantly jiggle back and forth as they are bombarded billions of times per second by the air molecules. Large-scale, observable motion of a particle—a zig or a zag—occurs when an unbalance of molecular bombardment occurs from one side of the smoke particle to the other.

Macroscopic properties of a gas

Now that we have a good mental picture of a gas, we can account for several macroscopic properties, such as pressure, temperature, and the fact that the temperature of a gas increases when it is compressed. First of all, what causes a gas to exert a pressure on the walls of the containing vessel? This force arises from the collision of the molecules with the wall. Each time a molecule strikes the wall, the molecule rebounds, and the wall experiences a reaction force that is related to the change in momentum of the molecule hitting the wall. A baloon remains blown up, extended, and stretched because of molecular collisions with the inside surface of the balloon.

In a following section on kinetic theory, we write a mathematical expression that relates the temperature of a gas to the average kinetic energy per molecule. Thus, the temperature is directly related to molecular speeds. Anything that causes the molecules to be speeded up causes the temperature of the gas to rise. If a quantity of gas is enclosed in a cylinder with a movable piston, as in Fig. 15-2,

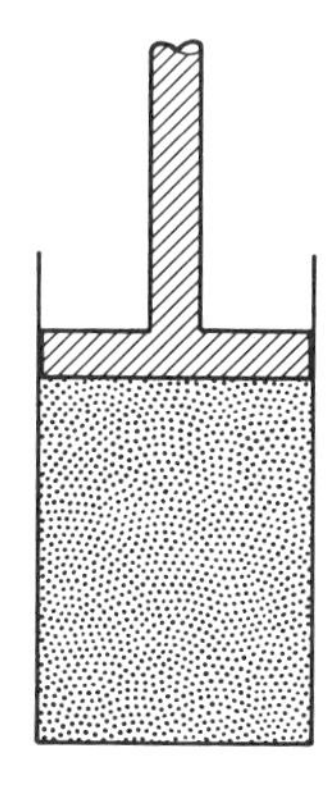

Figure 15-2 A cylinder and freely movable piston for compressing a quantity of gas.

the gas may be compressed by pushing the piston downward. Molecules colliding with the moving piston will rebound with increased velocity; the molecular speed will be increased and the temperature will rise. We may liken the molecules to baseballs and the moving piston to a bat—a ball that strikes a moving bat rebounds with increased velocity. Thus, the temperature of a gas is increased when it is compressed. If the piston moves in the opposite direction, the molecules will be slowed down as they collide with the receding piston. Consequently, the temperature of a gas is lowered when it expands against a moving piston.

15-2 MOLES; AVOGADRO'S NUMBER

We have already described a *mole* in Sect. 11-1 as a measure of mass of a particular element or compound. We recall that the advantage of measuring quantities of materials in terms of moles is that 1 mol of any substance contains the same number of molecules as 1 mol of any other substance. The number of molecules in a mole is Avogadro's number. The definition and properties of a mole are summarized in the following:

1. A mole is an appropriate unit only for elements or compounds that have a definite molecular formula, such as H_2, He, CO_2, and H_2SO_4.
2. A gram-mole (g · mol) is a mass in grams of a substance numerically equal to the molecular weight of the substance. Similarly, a kilogram-mole (kg · mol) is a mass in kilograms of a substance numerically equal to the molecular weight.
3. One gram-mole of a substance contains a number of molecules given by Avogadro's number, 6.02×10^{23} molecules/g · mol. In terms of kilogram-moles, Avogadro's number is 6.02×10^{26} molecules/kg · mol.
4. The number of moles n of a particular quantity of a substance is obtained by dividing the mass of the quantity of the substance (in grams or kilograms, as appropriate) by the molecular weight:

$$N = \frac{\text{mass of substance at hand}}{\text{molecular weight of the substance}}$$

With Avogadro's number at our disposal, we can calculate the number of atoms or molecules in common items such as an iron nail or a glass of water if we know the mass of the item. We use the following formula, which is obviously the number of moles of material times Avogadro's number

$$\text{Number of molecules} = \frac{\text{mass of material}}{\text{molecular weight}} \times \text{Avogadro's number} \qquad (15\text{-}1)$$

$$N = n \times N_a$$

EXAMPLE 15-1 An 8-oz glass of water has a mass of about 250 g. How many water molecules are present?

Solution. First, calculate the gram-moles of water in the glass. (The molecular weight of water is 18.)

$$n = \frac{250 \text{ g}}{18} = 13.89 \text{ g} \cdot \text{mol}$$

Then using Avogadro's number,

$$\begin{aligned} N &= n \times N_a \\ &= (13.89)(6.02 \times 10^{23}) \\ &= 8.36 \times 10^{24} \text{ molecules} \end{aligned}$$

15-3 EQUATION OF STATE OF AN IDEAL GAS

Suppose a sealed container is filled with an ideal gas. We may describe the state of the gas within the container by giving its volume, pressure, temperature, and the number of moles of gas that are present. These quantities are related; it is impossible to change one quantity without also affecting the others. The equation that expresses the mathematical relationship between these parameters is called the *equation of state* of the gas.

It has been determined experimentally and theoretically that the equation of state of an ideal gas is

$$PV = nRT \tag{15-2}$$

where

P = absolute pressure

V = volume

n = moles of gas in the volume V

R = the universal gas constant

T = absolute temperature

A variety of units are in common use for P and V in this equation. This presents no problem as long as they are consistent with the units of the gas constant R. Here are typical possibilities for R:

$$R = 8314 \frac{\text{J}}{\text{kg} \cdot \text{mol} \cdot \text{K}} \qquad (P = \text{N/m}^2;\ V = \text{m}^3)$$

$$R = 0.0821 \frac{\text{L-atm}}{\text{g} \cdot \text{mol} \cdot \text{K}} \qquad (P = \text{atm};\ V = \text{L})$$

$$R = 8.314 \frac{\text{J}}{\text{g} \cdot \text{mol} \cdot \text{K}} \qquad (P = \text{N/m}^2;\ V = \text{m}^3)$$

$$R = 0.0237 \frac{\text{psi-ft}^3}{\text{g} \cdot \text{mol} \cdot {}^\circ\text{R}} \qquad (P = \text{lb/in}^2;\ V = \text{ft}^3;\ T = {}^\circ\text{R})$$

The last entry is an attempt to put R in units that may be more familiar to everyday experience. Use the following to convert gram-moles of mass to the equivalent weight in pounds (where g has the standard value of 9.80665 m/s^2):

$$\text{lb} = \frac{\text{g} \cdot \text{mol} \times \text{molecular weight}}{453.5 \text{ g/lb}}$$

Furthermore, recall that absolute temperatures Rankine are obtained from degrees Fahrenheit by

$$R = F + 460$$

The following examples illustrate the usefulness of the ideal gas law [Eq. (15-2)]. We assume the behavior of the real gases involved closely approximates that of an ideal gas.

EXAMPLE 15-2 Calculate the volume occupied by 1 g-mole of an ideal gas at standard conditions: a pressure of 1 atm and a temperature of 273 K (0°C).

Solution. Solve the ideal gas equation for V

$$V = \frac{nRT}{P}$$

Then make straightforward substitutions:

$$V = \frac{(1 \text{ g} \cdot \text{mol})(0.0821 \text{ l} \cdot \text{atm/g} \cdot \text{mol} \cdot \text{K})(273 \text{ K})}{1 \text{ atm}}$$

$$= 22.4 \text{ L}$$

EXAMPLE 15-3 A 2-L flask at room temperature (20°C = 293 K) contains hydrogen, H_2, at a pressure of 2.3 atm.

(a) Compute the number of gram-moles of gas that are present.
(b) Calculate the number of hydrogen molecules in the flask.

Solution. (a) Solve for n in $PV = nRT$ and substitute:

$$n = \frac{PV}{RT}$$

$$= \frac{(2.3 \text{ atm})(2 \text{ L})}{(0.0821)(293 \text{ K})} = 0.191 \text{ g} \cdot \text{mol}$$

The units of R are L · atm/g · mol · K.
(b) Using Avogadro's number, as per Eq. (15-1),

$$\text{Molecules present} = n \times N_a$$

$$= 0.191 \times 6.02 \times 10^{23}$$

$$= 1.15 \times 10^{23} \text{ molecules}$$

EXAMPLE 15-4 Suppose a cylindrical tank whose volume is 4 ft^3 is initially evacuated and subsequently filled with oxygen to a gauge pressure of 2000 psi at a temperature of

68°F. By how many pounds does the weight of the cylinder increase as it is filled with oxygen?

Solution. First, we find the number of moles in the tank. We convert 68°F to 528°R and the gauge pressure of 2000 psi to an absolute pressure of 2014.7 psi. Hence,

$$n = \frac{(2014.7 \text{ psi})(4 \text{ ft}^3)}{(0.0237)(528°\text{R})} = 644 \text{ g} \cdot \text{mol}$$

The units of R are psi-ft^3/g · mol · °R.
The mass in grams of oxygen in the tank is

$$\text{Mass} = \text{g} \cdot \text{mol} \times \text{molecular weight of } O_2$$

$$= 644 \times 32 = 20{,}608 \text{ g}$$

Dividing this by 453.5 to convert to pounds gives 45.4 lb as the increase in weight of the tank as it is filled with oxygen.

Alternative pressure units

Sometimes pressures are conveniently expressed in units such as centimeters of mercury or millimeters of mercury (torr). Because the pressure P in the ideal gas law must be expressed in units consistent with the units of the gas constant R, it is usually most convenient to convert the centimeters or millimeters of mercury to atmospheres. The conversion is simply

$$\text{atm} = \frac{\text{cm Hg}}{76} \quad \text{or} \quad \text{atm} = \frac{\text{mm Hg}}{760} \tag{15-3}$$

Conversely, the conversion *from* atmospheres is

$$\text{cm Hg} = \text{atm} \times 76 \quad \text{or} \quad \text{mm Hg} = \text{atm} \times 760 \tag{15-4}$$

The various units of pressure are given in Sect. 11-3.

Constant volume gas thermometer

If a closed system filled with an ideal gas is designed so that its volume remains constant even as the pressure and temperature may vary, we may treat the volume V as a constant in the ideal gas equation. Therefore, we can write

$$P = \left[\frac{nR}{V}\right]T \tag{15-5}$$

where the quantity in the bracket, nR/V, is a constant. Consequently, a direct proportion exists between the pressure and the temperature which can also be expressed as

$$T = \left[\frac{V}{nR}\right]P \tag{15-6}$$

This equation forms the basis of an ideal gas thermometer, which measures

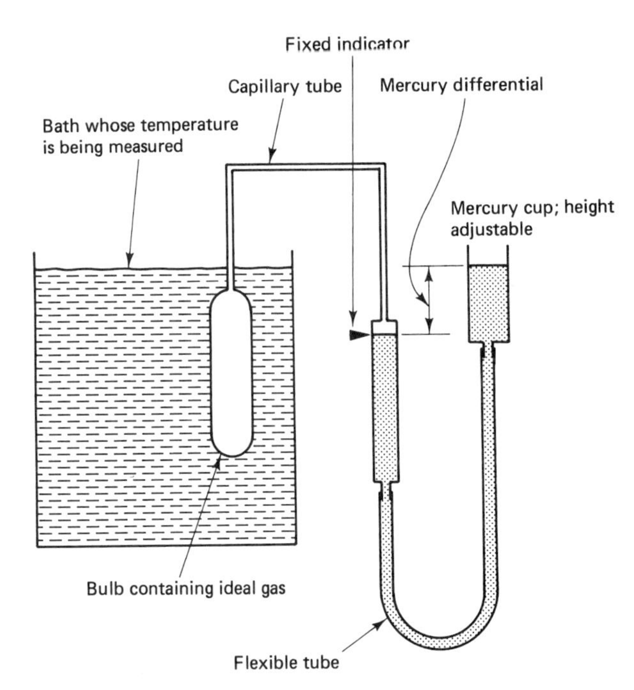

Figure 15-3 An ideal gas thermometer. The height of the mercury cup is adjusted until the mercury level reaches the fixed indicator. The mercury differential then is proportional to the pressure inside the bulb. The barometric pressure must be added to the mercury differential to get the absolute pressure in the bulb.

absolute temperatures (since T in the ideal gas equation is an absolute temperature). Such a thermometer is shown in Fig. 15-3.

In order to calculate the scale for a particular gas thermometer, the volume V and the number of moles n of gas contained in the thermometer bulb must be known very accurately. However, one of the fundamental properties of physics—the temperature of absolute zero—can be determined without knowing either V or n, provided both remain constant during the course of the experiment.

The procedure may be inferred from the graph of Fig. 15-4. Beginning at a high temperature, the pressure is accurately measured for several successively

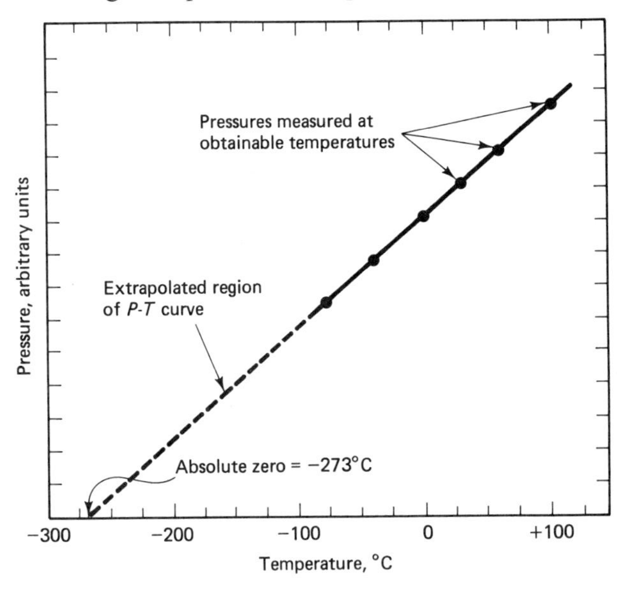

Figure 15-4 Determination of absolute zero of temperature.

lower temperatures until the lowest temperature achievable with available apparatus is reached. The data points will lie on a straight line in accordance with the linear relationship of Eqs. (15-5) and (15-6). By extrapolating (extending) the line further to the left toward lower and lower temperatures, the corresponding pressure decreases proportionally until it finally becomes zero. The temperature of this point is reasoned to be the temperature of absolute zero.

This is not to infer that any real gas could exist as a gas at zero pressure. Long before a temperature of absolute zero could be reached, the gas would cease to act as an ideal gas. Indeed, any real gas would condense to a liquid, as may be inferred from the data of Table 15-1.

TABLE 15-1 CONDENSATION TEMPERATURE OF SELECTED GASES

Gas	Temp *K*
Argon	83.8
Helium	4.22
Hydrogen	20.3
Nitrogen	77.3
Oxygen	90.2

15-4 IDEAL GAS PROCESSES

If we write the ideal gas law as

$$\frac{PV}{nT} = R = \text{constant}$$

it is evident that the quantity PV/nT is a constant because R is a constant. Any of the four parameters P, V, n, or T may vary, but the ratio PV/nT will always be the same. For example, Fig. 15-5 shows two configurations of a cylinder and piston in which all four parameters change in going from one configuration to the other. The temperature of the water bath is increased, a bit of gas is exhausted through the valve, the volume is decreased by lowering the piston, and the pressure varies in a

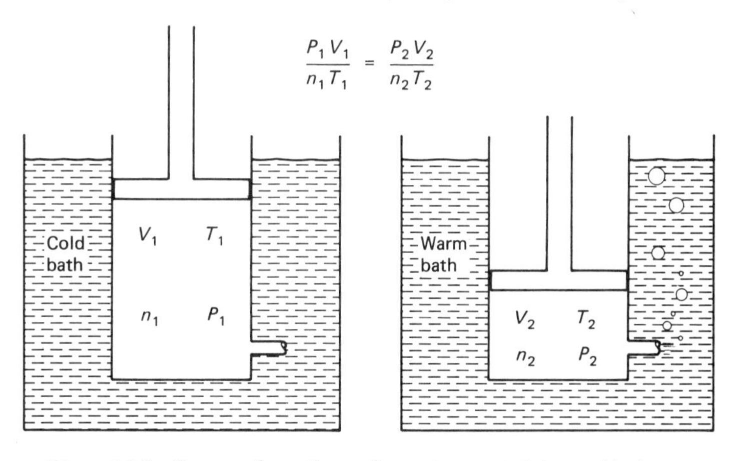

Figure 15-5 Two configurations of a system containing an ideal gas.

manner determined by the variation of the other parameters. If we denote the two configurations by subscripts 1 and 2, respectively, then it follows that

$$\frac{P_1V_1}{n_1T_1} = \frac{P_2V_2}{n_2T_2} \tag{15-7}$$

This is a very useful relationship, and it is easy to use because of the flexibility allowed with units. Any unit of measure may be used for any parameter as long as the same unit is used on both sides of the equation. However, the pressure and the temperature must be expressed on an absolute scale (not gauge pressure and not degrees Celsius or degrees Fahrenheit). This is the most general of the relationships between various configurations of an ideal gas. If one or more of the parameters are held constant, the relationship can be simplified, as we see in the following.

A sealed system

If no gas is allowed to escape from the system, the number of moles does not change, and $n_1 = n_2$. Therefore, n_1 and n_2 may be canceled to give

$$\frac{P_1V_1}{T_1} = \frac{P_2V_2}{T_2} \tag{15-8}$$

EXAMPLE 15-5 Suppose an ideal gas is in a cylinder with a movable piston at an initial temperature of 20°C and a pressure (absolute) of 160 cm Hg. The initial volume of the gas is 0.5 ft^3. What will be the pressure in the cylinder if the gas is compressed to a final volume of 0.25 ft^3 and if the cylinder and gas are cooled to 0°C?

Solution. First, we convert the temperatures of 20°C and 0°C to the absolute temperatures of 293 K and 273 K. Then, we rearrange the earlier equation and substitute to obtain

$$\begin{aligned} P_2 &= \left(\frac{T_2}{T_1}\right)\left(\frac{V_1}{V_2}\right)P_1 \\ &= \left(\frac{273\text{ K}}{293\text{ K}}\right)\left(\frac{0.5\text{ ft}^3}{0.25\text{ ft}^3}\right)(160\text{ cm Hg}) \\ &= 298\text{ cm Hg} \end{aligned}$$

Note that is is *not* necessary to convert the pressure units from cm Hg, and we can use the cubic-feet volume unit as is. In the statement of the problem the pressure was stated to be absolute.

Boyle's law

If a system is sealed and also is held at constant temperature, both n and T can be canceled from Eq. (15-7). This leaves the following expression, known as *Boyle's law*:

$$P_1V_1 = P_2V_2 \quad \text{or} \quad P_2 = \left(\frac{V_1}{V_2}\right)P_1 \tag{15-9}$$

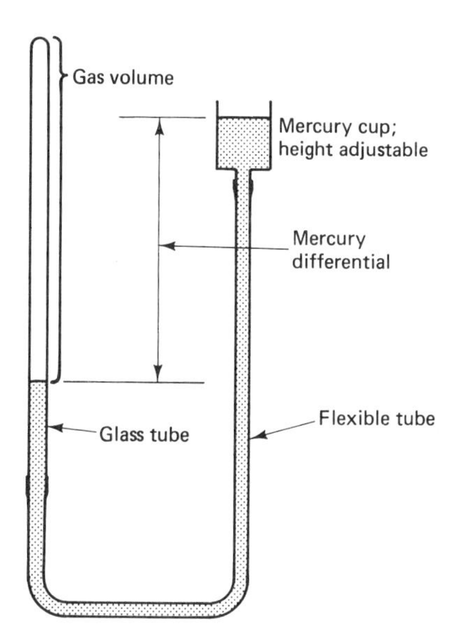

Figure 15-6 Boyle's law apparatus.

Actually, this relationship—that the product of the volume and the absolute pressure of a gas is constant—was the first property of a gas to be stated in mathematical terms. It was put forth by Robert Boyle (1627–1691) about 1662. A simple apparatus for demonstrating Boyle's law is shown in Fig. 15-6.

Charles' law

If a system is sealed and maintained at a constant pressure, n and P may be canceled out of Eq. (15-7) to give Charles' law [after Jacques Charles (1747–1823), French scientist]:

$$\frac{V_1}{T_1} = \frac{V_2}{T_2} \quad \text{or} \quad V_2 = \left(\frac{T_2}{T_1}\right)V_1 \tag{15-10}$$

An apparatus for demonstrating this relationship, shown in Fig. 15-7, consists of a cylinder whose piston is subjected to a constant applied force. The piston will rise and fall as T_2 becomes greater than or less than the reference temperature T_1.

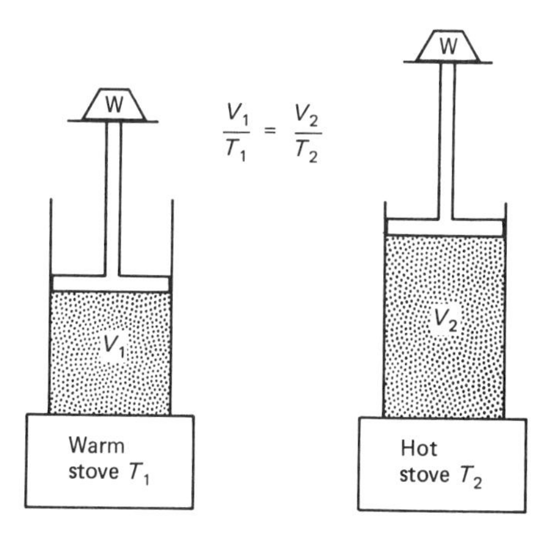

Figure 15-7 A gas expands when heated at constant pressure.

Gay-Lussac's law

If a sealed system is held at constant volume, the pressure and temperature are related by

$$\frac{P_1}{T_1} = \frac{P_2}{T_2} \quad \text{or} \quad P_2 = \left(\frac{T_2}{T_1}\right)P_1 \tag{15-11}$$

which indicates that the pressure rises as the temperature increases. [This law is named for Joseph Gay-Lussac (1778–1850), a French scientist.]

EXAMPLE 15-6 An oxygen cylinder of volume 2 ft^3 is filled to an absolute pressure of 2000 psi. If the gas is released into a light gas bag at atmospheric pressure, compute the volume of gas in the bag. (Assume the temperature remains constant.)

Solution. A pressure of 2000 psi is the equivalent of 136 atm. Therefore, using Eq. (15-10),

$$V_2 = \left(\frac{P_1}{P_2}\right)V_1$$

$$= \left(\frac{136 \text{ atm}}{1 \text{ atm}}\right)(2 \text{ ft}^3) = 272 \text{ ft}^3$$

This volume is equivalent to that of a spherical balloon very nearly 8 ft in diameter.

EXAMPLE 15-7 A 2-ft^3 oxygen cylinder is filled to an absolute pressure of 2000 psi in early morning when the air temperature is 64°F. The cylinder is then stored in direct sunlight, and the temperature rises to 94°F. Compute the pressure in the tank at 94°F.

Solution. First, we convert 64°F and 94°F to the absolute temperatures of 524°R and 554°R. Then, using Eq. (15-11), we obtain

$$P_2 = \left(\frac{T_2}{T_1}\right)P_1$$

$$= \left(\frac{554°\text{R}}{524°\text{R}}\right)(2000 \text{ psi})$$

$$= 2114.5 \text{ psi}$$

Observe that a 30°F change in temperature caused the pressure to increase by about 6%.

15-5 KINETIC THEORY

Historically, the ideal gas law and the properties of gases derived from it were deduced from experiments designed to determine the relationship between P, V, n, and T. Somewhat later, the microscopic, molecular nature of a gas became apparent. The approach of kinetic theory is exactly the opposite. Kinetic theory begins with assumptions made in regard to the molecular nature of a gas. The objective is to relate molecular behavior to macroscopically observable properties, such as pressure and temperature.

An initial assumption is that a container of gas has in it a very large (about 10^{23}) number of molecules, each of which is very small. Further, the molecules are assumed to be in a state of molecular agitation and are assumed to collide perfectly elastically with each other and with the walls of the container.

When a molecule collides with a wall, it experiences a change in momentum. The wall experiences a reaction force during each collision, and because of the exceedingly large number of collisions that occur per second, the force averages out to a (macroscopically) steady value. The force per unit area of the wall is what we perceive as pressure. Thus, the pressure that a gas exerts on the walls of a container arises from molecular collisions with the wall.

The volume of the container in part determines the number of collisions with the walls that occur per second. (A blind bug flying in a small room suffers more collisions than when flying in a very large room.) Since the pressure is related to the number of collisions, we see that a quantity of gas exerts a greater pressure in a small container than the same quantity of gas will exert in a larger container. These ideas are illustrated in Fig. 15-8.

In a real gas, molecular collisions insure that velocities in all directions are equally probable, but we cannot realistically expect all molecules to be traveling at the same speed. Thus, we must deal with molecular velocities on a statistical basis.

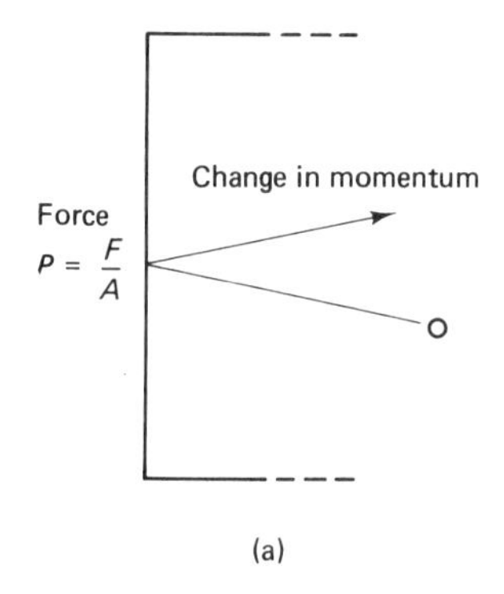

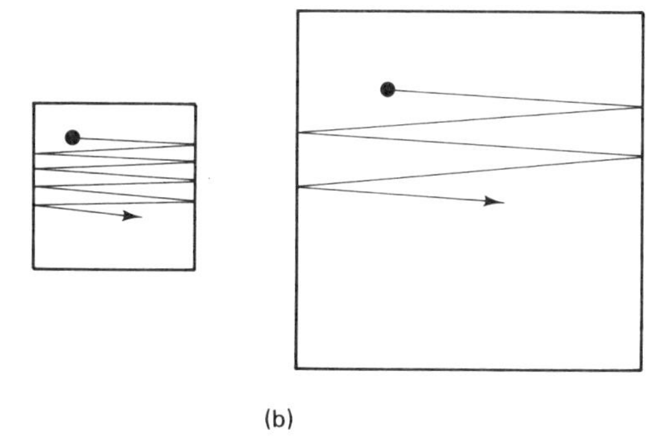

Figure 15-8 (a) Changes in molecular momenta at the wall produce a force, and therefore a pressure.
(b) At a given temperature, more collisions with the wall occur in a container of smaller volume.

Out of the momentum-pressure analysis alluded to earlier, there arise products of mass and velocity square. These products obviously are proportional to the molecular kinetic energy, and a bit of averaging over all the molecules leads to the expression

$$PV = \tfrac{2}{3}N(\overline{\mathrm{KE}}) \tag{15-12}$$

where $(\overline{\mathrm{KE}})$ is the average kinetic energy per molecule and N is the total number of molecules in the container. This equation is significant because it bridges the gap between the macroscopic parameters P and V and the microscopic parameters of N and $(\overline{\mathrm{KE}})$.

We now make use of the fact that the ideal gas law is already known, having been determined from experiments. We compare $PV = nRT$ with Eq. (15-12) and realize that for both to be valid, the right-hand sides of the two equations must be equal:

$$\tfrac{2}{3}N(\overline{\mathrm{KE}}) = nRT \tag{15-13}$$

From the definition of Avogadro's number, $N = nN_a$. When this is substituted into the left-hand side of the preceding expression, the n's cancel to leave

$$\tfrac{2}{3}N_a(\overline{\mathrm{KE}}) = RT$$

Solving for $(\overline{\mathrm{KE}})$ gives

$$(\overline{\mathrm{KE}}) = \frac{3}{2}\left(\frac{R}{N_a}\right)T \tag{15-14}$$

The quantity R/N_a is called *Boltzmann's constant*, denoted by k:

(*Boltzmann's constant*)
$$\begin{aligned} k &= \frac{R}{N_a} \\ &= 1.3805 \times 10^{-23}\ \mathrm{J/K} \\ &= 1.3805 \times 10^{-16}\ \mathrm{ergs/K} \end{aligned}$$

In terms of Boltzmann's constant, Eq. (15-14) may be written as

$$(\overline{\mathrm{KE}}) = \tfrac{3}{2}kT \tag{15-15}$$

This expression shows the direct relationship between the absolute temperature T and the average kinetic energy per molecule $(\overline{\mathrm{KE}})$.

If $(\overline{\mathrm{KE}})$ is written out explicitly, we obtain

$$(\overline{\mathrm{KE}}) = \tfrac{1}{2}m\langle v^2\rangle$$

where m is the mass of each molecule and where $\langle v^2\rangle$ is the average of the squares of all the molecular speeds. This and Eq. (15-15) give us

$$\langle v^2\rangle = \frac{3kT}{m}$$

The square root of $\langle v^2 \rangle$ is called the root-mean-square (RMS) velocity, v_{rms}:

$$v_{rms} = \sqrt{\langle v^2 \rangle}$$

$$v_{rms} = \sqrt{\frac{3kT}{m}} \qquad (15\text{-}16)$$

This expression bridges the gap between molecular velocities and absolute temperature. Thus, kinetic theory shows us the relationship between pressure, volume, and temperature, on the one hand, and molecular velocities, masses, momenta, and kinetic energies on the other.

Maxwell speed distribution

It is highly unlikely (but not impossible) that all the molecules in an observable sample of an ideal gas would ever be traveling at exactly the same speed. It is reasonable to expect some to be traveling much slower than the average molecular speed, while another fraction might be traveling much faster. A speed *distribution function* shows the portion of molecules having speeds within various segments of the total range of speeds, from very slow to extremely fast. In 1859, James Clerk Maxwell addressed himself to the problem of the distribution of molecular speeds in an ideal gas. By making only a few simple assumptions, he was able to derive the *Maxwell distribution function*.

The first part of Maxwell's treatment was largely a mathematical exercise only remotely connected to the physics of an ideal gas. He assumed (1) a large number (about 10^{23}) of molecules are being considered; (2) the directions of all molecular velocities are equally probable; and (3) the sum total of all molecules having any velocity is N, the number of molecules in the sample. From these assumptions, he deduced the mathematical *form* of the speed-distribution function. Furthermore, he was able to derive an expression for the RMS speed, and this expression served to bridge the gap between a mathematical construction, on the one hand, and a mathematical description of an ideal gas on the other.

Maxwell's second step was to compare *his* expression for v_{rms} with another expression for the same thing which was obtained from the kinetic theory of an ideal gas [our Eq. (15-16)]. By so doing, Maxwell was able to attach physical significance to the symbols in his equation.

Maxwell's distribution function has now been verified experimentally. It is given, along with an explanation of the symbols, in Fig. 15-9(a). A plot of the distribution for oxygen at various temperatures is given in Fig. 15-9(b). Note that the distribution broadens at higher temperatures.

Three characteristic speeds are commonly given for a speed-distribution function. These are (1) the most probable speed, v_{mp}; (2) the arithmetic mean (or average) speed, $\bar{v}$; and (3) the RMS speed, v_{rms}. The most probable speed is the speed that more molecules have in comparison to any higher or lower speed. It is the speed at which the peak of the distribution function occurs. The arithmetic mean speed is just the arithmetic average. The RMS speed is the square root of the

Maxwell Distribution of Molecular Speeds

$$\Delta N = (\Delta V)\,\frac{4N}{\sqrt{\pi}}\left(\frac{m}{2kT}\right)^{3/2} V^2 \exp\left[-\frac{mV^2}{2kT}\right]$$

ΔN = number of molecules having speeds within the small segment V located at the approximate speed V

ΔV = small segment of the total velocity spectrum, centered at V, as illustrated below

N = total number of molecules in the sample of gas being considered

V = central speed (centimeter per second) of the segment ΔV to which N is relevant

m = mass of the molecule (grams)

k = Boltzmann's constant; $k = 1.3805 \times 10^{-16}$ erg/K

T = absolute temperature (K)

Note: $\exp[x]$ is the same as e^x.

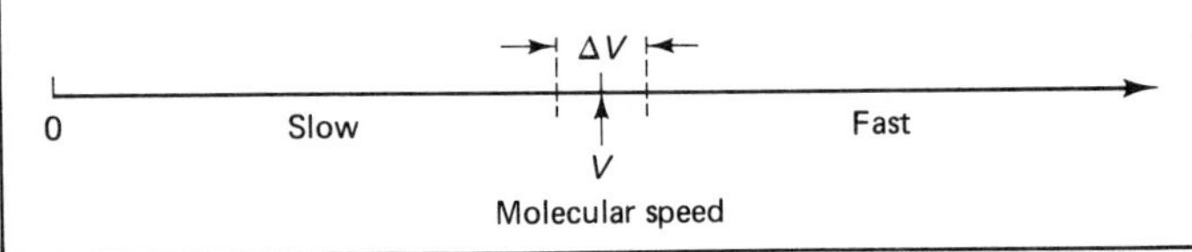

(a)

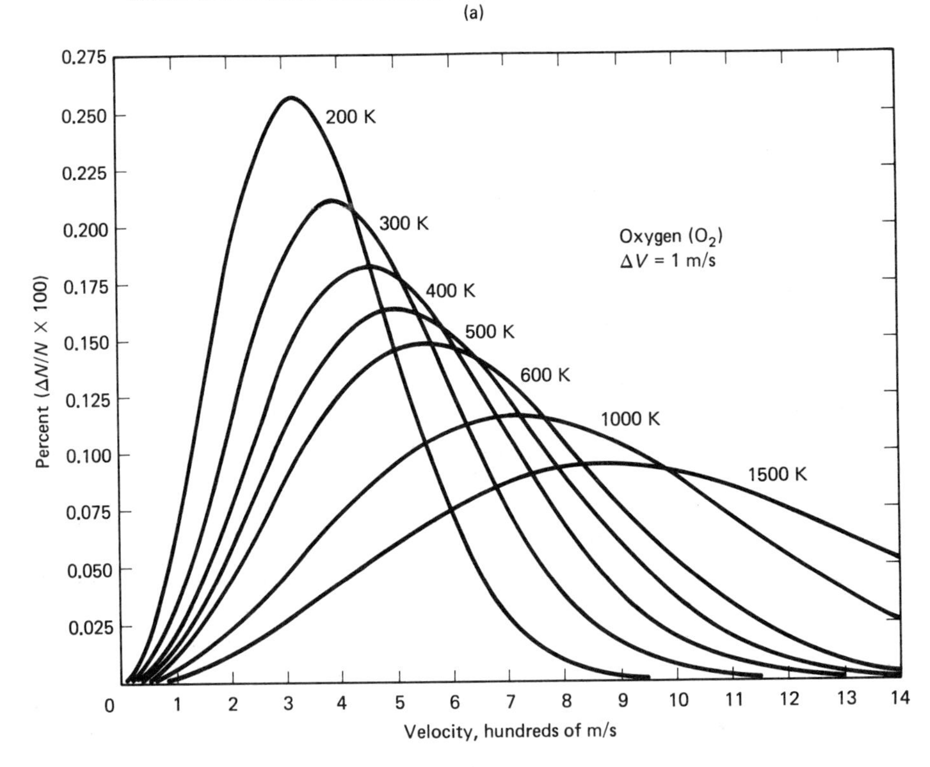

(*Example*: In a sample of oxygen at 300 K, about 0.21% of the molecules have velocities in the range from 399.5 to 400.5 m/s.)

(b)

Figure 15-9 Maxwell distribution of molecular speeds.
(a) The speed distribution function.
(b) Graph of the function for molecular oxygen at various temperatures.

average of the squares of the speeds, as described earlier. For a Maxwell speed distribution,

$$v_{mp} = \sqrt{\frac{2kT}{m}} \quad \text{(most probable speed)} \tag{15-17}$$

$$\bar{v} = \sqrt{\frac{8kT}{\pi m}} \quad \text{(arithmetic mean speed)} \tag{15-18}$$

$$v_{rms} = \sqrt{\frac{3kT}{m}} \quad \text{(root-mean-square speed)} \tag{15-19}$$

$$v_{mp}\colon \bar{v}\colon v_{rms} = 1\colon 1.128\colon 1.1224 \quad \text{(ratio)}$$

In these formulas, k is Boltzmann's constant, m is the mass of the molecule, and T is the absolute temperature.

Mean free path

If a hard-sphere, billiard-ball-type molecule has a radius ρ , its collision cross section σ is the projected area of the sphere, $4\pi\rho^2$. If the average speed of the molecule is $\bar{v}$, the molecule will sweep through a cylindrical volume given by $4\pi\rho^2\bar{v}$ per unit time. If other molecules occupy the volume swept through, the moving molecule is assumed to collide with each molecule in the volume. Hence, by computing the number n' of molecules in the volume, we can determine the average number of collisions experienced by the moving molecule per unit time. Then, by dividing the velocity $\bar{v}$ by the number n' (because the velocity is the distance traveled by the molecule per unit time), we can obtain an expression for the average distance traveled between collisions—the *mean free path*. Thus, the mean free path λ is given by

$$\lambda = \frac{1}{4\pi\rho^2 n'} = \frac{1}{\sigma n'} \tag{15-20}$$

To derive this expression, we tacitly assumed that all molecules were fixed in space except the one moving molecule. When account is taken of the fact that all molecules are moving (as in the Maxwell distribution function), the above result is reduced by $\sqrt{2}$, so that 0.707 instead of 1.0 appears in the numerator.

This expression is inconvenient for computation because the number of molecules per unit volume n' is typically not explicitly given. However, we can use the ideal gas law to convert from n' to the more commonly stated parameters P and T. The result is

$$\lambda = \frac{0.707\, kT}{4\pi\rho^2 P} \tag{15-21}$$

This expression shows how λ depends upon T and P, but it can be made even more convenient by consolidating constants and performing units conversions. The final result is

$$\text{(mean free path)} \qquad \lambda = (7.66 \times 10^{-10}) \frac{T}{\rho^2 P} \qquad (15\text{-}22)$$

where

λ = mean free path in meters

T = absolute temperature in degrees Kelvin

P = pressure in atmospheres

ρ = molecular radius in angstroms (10^{-8} cm)

TABLE 15-2 APPROXIMATE MOLECULAR DIAMETERS AND MASSES OF SELECTED GASES

Gas	Diameter, cm	Mass, kg
Argon	2.9×10^{-8}	66.3×10^{-27}
Helium	1.9	6.64
Hydrogen	2.4	3.34
Nitrogen	3.1	46.5
Oxygen	2.9	53.1
Carbon dioxide	3.3	73.0

As an example, the radius of an oxygen (O_2) molecule is on the order of 1.8 Å. Therefore, at conditions of standard temperature and pressure (STP is 0°C and 1 atm), the mean free path is

$$\lambda = (7.66 \times 10^{-10}) \frac{273 \text{ K}}{(3.24)(1 \text{ atm})} = 6.45 \times 10^{-8} \text{ m}$$

$$= 6.45 \times 10^{-6} \text{ cm} = 645 \times 10^{-8} \text{ cm} = 645 \text{ Å}$$

Thus, because the molecular diameter is about 3.6 Å, the mean free path is on the order of 180 molecular diameters.

15-6 REAL GASES

In considering how a real gas differs from an ideal gas, we should recall the two assumptions made in regard to an ideal gas, namely, that the molecules do not attract or repel each other (except in a hard-sphere, elastic collision) and that the molecules are so small that the molecular volume is negligible. Neither of these assumptions is strictly valid for real gases.

All real molecules have volumes that, although very small, are large enough to be significant in regard to the observed relationship between pressure, volume, and temperature (the equation of state). Simple molecules have diameters on the order of a few angstroms. The effect of the molecular volume is to reduce the

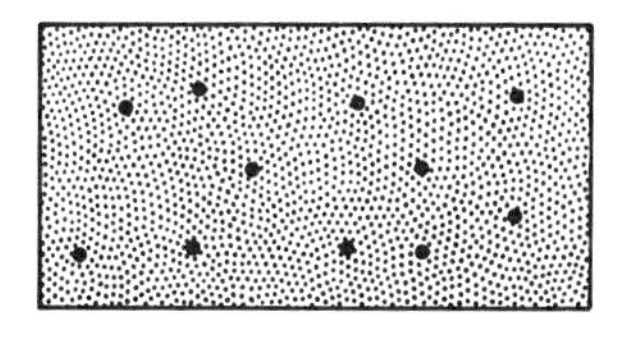
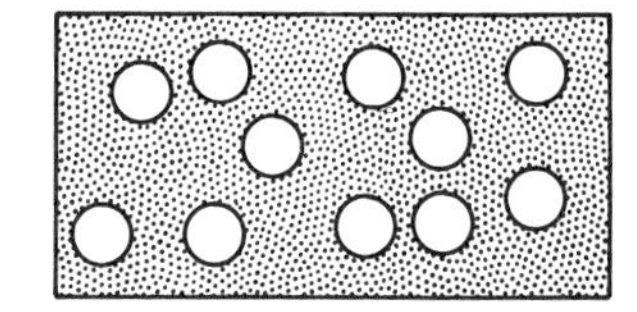

(a)

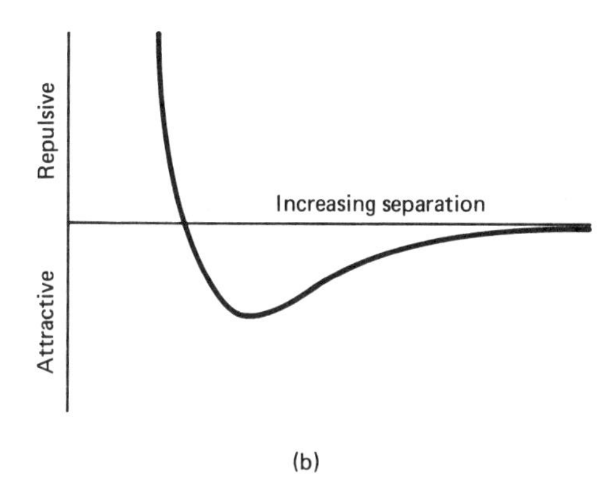

(b)

Figure 15-10 (a) The finite volume of the molecules reduces the free volume in a container of a van der Waals' gas. (b) When widely separated, molecules attract each other. The attraction changes to a strong repulsion when the molecules come close together.

amount of space between the molecules, effectively reducing the volume of the container holding the gas. This is illustrated in Fig. 15-10(a).

Real molecules either attract or repel each other, depending upon the distance between them, as illustrated in Fig. 15-10(b). When widely separated, the force between molecules is attractive but is very small. The attraction increases as they are brought closer together until, within the span of a short distance, the attraction becomes zero and then turns to a strong repulsion. At the pressures commonly exerted on gases in typical environments, the average force between molecules is attractive. Thus, the gas tends to pull itself together much as the effect of surface tension in a liquid. But of course, a gas at temperatures significantly above the condensation temperature does *not* pull itself together. The effect is merely to reduce, slightly, the pressure exerted on the walls of the container.

The origin of the intermolecular forces is electrical—the force between electrical charges (electrons and protons). Attractive forces stem from the attraction of the electrons of one atom to the nucleus of the other, and vice versa. Repulsive forces arise when the molecules are very close together because the electrons of one repel the electrons of the other, and the nucleus of one repels the nucleus of the other. These factors are called *van der Waals' forces*, and they are quite complex when treated in mathematical detail.

Van der Waals' equation of state

Many equations of state of various forms have been derived mathematically or obtained empirically (from experimental data) to describe the properties of real gases somewhat better than the ideal gas relationship, $PV = nRT$. The best-known of these, although not the most sophisticated or elaborate, is the van der Waals' equation of state, which may be written as

$$\left(P + \frac{n^2 a}{V^2}\right)(V - nb) = nRT \tag{15-23}$$

The constants a and b are determined experimentally for a particular gas and may be found in handbooks of physical and chemical data. Representative values are given in Table 15-3.

TABLE 15-3 VAN DER WAALS' CONSTANTS

Gas	a	b
Argon, A	1.345	0.0322
Carbon dioxide, CO_2	3.592	0.0427
Helium, He	0.0341	0.0237
Hydrogen, H_2	0.244	0.0266
Neon, Ne	0.211	0.0171
Nitrogen, N_2	1.390	0.0391
Oxygen, O_2	1.360	0.0318
Propane, C_3H_8	8.664	0.0844

Units of a: $L^2 \cdot atm/(g \cdot mol)^2$

Units of b: $L/g \cdot mol$

Universal gas constant, R: 0.0821 $L \cdot atm/g \cdot mol \cdot K$

The constant a stems from the attraction of the molecules for each other and tends to reduce the pressure. This can be seen more clearly if the equation is rearranged as follows:

$$P = \frac{nRT}{(V - nb)} - \frac{n^2 a}{V^2} \tag{15-24}$$

On the other hand, the constant b arises from the molecular volume, and its effect is to increase the pressure. If both a and b are set equal to zero, the van der Waals' equation reverts to the ideal gas law.

EXAMPLE 15-8 Compute the pressure that results from placing 3 moles of nitrogen into a volume of 2 L. Assume the temperature to be 20°C (293 K), and treat nitrogen as a van der Waals' gas.

Solution. From Table 15-3 we obtain the van der Waals' constants for nitrogen: $a = 1.390$ and $b = 0.0391$. A direct substitution into Eq. (15-24) yields

$$\begin{aligned} P &= \frac{nRT}{(V - nb)} - \frac{n^2 a}{V^2} \\ &= \frac{3(0.0821)(293)}{[2 - 3(0.0391)]} - \frac{(3^2)(1.390)}{2^2} \\ &= 38.33 - 3.13 = 35.20 \text{ atm} \end{aligned}$$

For an ideal gas under the same conditions, the calculated pressure is 36.08 atm. Thus, if nitrogen were treated as an ideal gas for the conditions of this example, the computed pressure would be too high by about 2.5%.

15-7 THE LIQUEFACTION OF GASES

Free expansion of gases

Suppose a sealed container is divided into two parts connected by a valve, as shown in Fig. 15-11. Section A is assumed to contain a gas at moderate pressure,

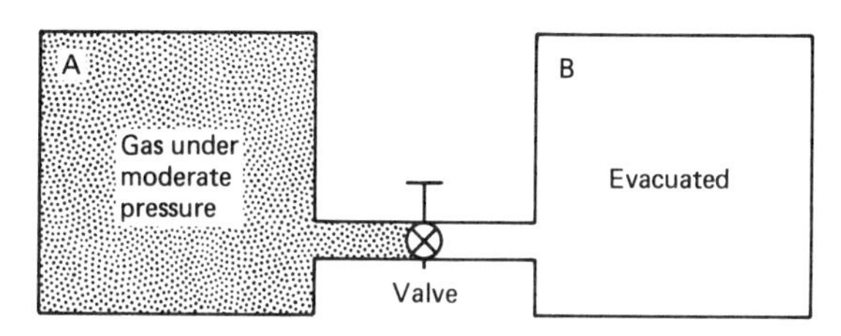

Figure 15-11 Does the temperature of the gas change as a result of the free expansion of the gas in A into B?

whereas section B is completely evacuated. If the valve is suddenly opened, the gas will quickly expand and fill section B until the pressures in A and B become equal. The gas is said to undergo a *free expansion* into section B. The point of interest is this: Does the temperature of the gas change during the free expansion? An equivalent question is: Does the speed of the molecules change as a result of the free expansion? If so, what causes the change? Historically, extensive investigation, both theoretical and experimental, has been directed to these questions.

Suppose the opening between the two sections is so small that only one molecule at a time can pass through. Let us consider what forces might act on the molecule as it passes through the opening. It is not necessary to concern ourselves with collisions with the sides of the opening because any such collision will be perfectly elastic and will not change the speed of the molecule. Consequently, any force acting to change the speed of the molecule must come from within the gas itself, arising from the intermolecular forces that the molecules exert on each other over large distances. There are three cases to consider, corresponding to attractive, repulsive, and zero intermolecular forces.

First of all, the molecules of an ideal gas exert forces on each other only when they collide. Therefore, no force will be exerted on an ideal gas molecule as it passes through the opening, and no change in speed will occur. Consequently, the temperature of an ideal gas does not change during a free expansion. On the other hand, the molecules of a van der Waals gas exhibit an attraction for each other that tends to pull peripheral molecules back toward the center of the gas. This produces a net rearward force on the molecule passing through the opening from A to B that tends to slow it down. Thus, it arrives in section B with reduced speed, corresponding to a lower temperature. We therefore conclude that if the intermolecular forces of a gas are attractive, a free expansion will produce a decrease in temperature of the gas.

Under conditions of high temperature, high pressure, or a combination of the two, molecules of real gases tend to repel each other so that a molecule escaping (via free expansion) from such a gas is speeded up as it departs. The molecules tend to "spring apart" when the constraining forces are relieved, and an increase in temperature occurs. In a free expansion, any real gas may exhibit either an increase, decrease, or no change in temperature, depending upon the initial conditions of temperature and pressure and the extent of the expansion that is allowed to occur.

In practice, the change in temperature of a gas undergoing a free expansion is difficult to measure because the temperature changes are exceedingly small. While the temperature of the gas may change by as much as several degrees, heat is quickly lost from the gas to the containing vessel whose heat capacity is typically hundreds of times as great as the heat capacity of the gas. Therefore, the resulting measurable temperature changes are on the order of only hundredths or even thousandths of a degree. Thus, free-expansion experiments are difficult to perform. A related experiment is the porous-plug experiment described in the following section.

The Joule-Kelvin effect

An experiment founded on the same principles as the free-expansion experiment, but which avoids the major technical difficulties, is the Joule-Kelvin *porous-plug* experiment illustrated in Fig. 15-12. High pressure (P_1) gas at temperature T_1 is

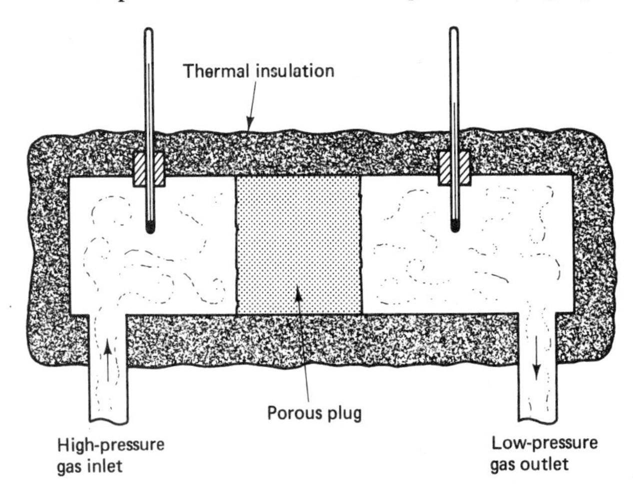

Figure 15-12 Apparatus for the Joule-Kelvin porous-plug experiment.

admitted to a chamber containing a constriction in the form of a porous plug. The gas seeps through the microscopic passages of the plug and emerges on the other side of the plug at a lower pressure P_2 and at temperature T_2. Depending upon the type of gas and upon the values of P_1 and T_1, the temperature T_2 may be either higher or lower than the initial temperature T_1. Heat loss to the surroundings is minimized by good thermal insulation. Heat transfer from the gas to the walls of the containing vessel becomes negligible after the apparatus has been in operation long enough for a steady-state condition to develop.

The expansion of the gas as it emerges from the porous plug is not a free expansion because it has to do work against the gas already present on the low-pressure side of the plug. However, this work can be computed so that the change in temperature resulting solely from the expansion of the gas can be determined. Consequently, the porous-plug experiment yields the same information as obtained from the more difficult free-expansion experiment. The passage of the gas through the porous plug is called a *throttling process*.

Analysis of throttling processes reveals that the combination $U + PV$ is not changed by the process. (Here, U is the internal energy of the gas, closely related to the temperature.) This combination is called the *enthalpy*, and it has the units of energy. The enthalpy depends upon the temperature and pressure of a gas. By performing a large number of throttling-process experiments, a chart can be constructed, as in Fig. 15-13, from which the change in temperature arising from a

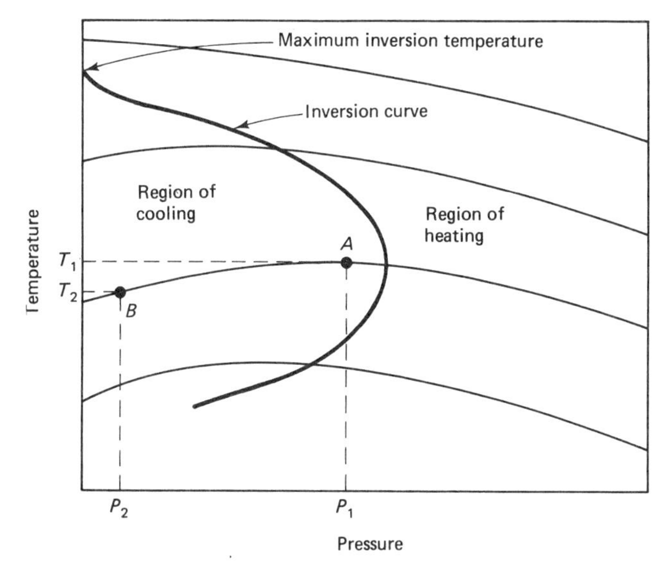

Figure 15-13 Curves of constant enthalpy, or isenthalps.

given throttling process can be determined. The curves are curves of constant enthalpy, called *isenthalps*. The initial temperature and pressure (T_1, P_1) determine the particular isenthalp relevant to a given throttling experiment (point A in Fig. 15-13). The final pressure P_2 of the throttling process locates the point (point B) on the isenthalp corresponding to the final conditions of the process. The final temperature T_2 is found from this point, and the temperature change produced by the process is readily determined.

Because throttling processes naturally proceed from a high to a lower pressure, it is evident that the isenthalp must slope upward to the right (positive slope) in the vicinity of a throttling process that is to produce a decrease in temperature. Figure 15-13 is representative of common gases, and it is seen that the isenthalps turn downward and exhibit a negative slope at high temperatures or at high pressures and lower temperatures. The curve that connects the points of zero slope on the various isenthalps is called the *inversion curve*, and it separates the region of cooling from the region of heating. Note that the intersection of the inversion curve and the temperature axis represents the highest temperature for which a throttling process will produce a decrease in temperature. This temperature is called the *maximum inversion temperature*, and a gas must be initially cooled to a lower temperature than this if further cooling is to be achieved by a throttling process. Maximum inversion temperatures for several gases are given in Table 15-4.

TABLE 15-4 MAXIMUM INVERSION TEMPERATURES OF SELECTED GASES

Gas	Temperature, K
Argon	723
Helium	40
Hydrogen	202
Nitrogen	621
Carbon dioxide	1500

Liquefaction of gases

Any real gas will condense to a liquid when cooled to a sufficiently low temperature (see Table 15-1). Thus far, we have described two processes by which a gas may be cooled. The first process, described briefly in Sect. 15-1, is the cooling of a gas by letting it expand against a moving piston. The internal energy (and temperature) of the gas is reduced as it does work by pushing against the moving piston. (If no heat is added to or removed from the gas during the expansion, the expansion is called an *adiabatic* expansion.) The second process for cooling a gas is the throttling process just described.

To produce cooling by using a throttling process, the initial temperature of the gas must be lower than the maximum inversion temperature (Table 15-4). On

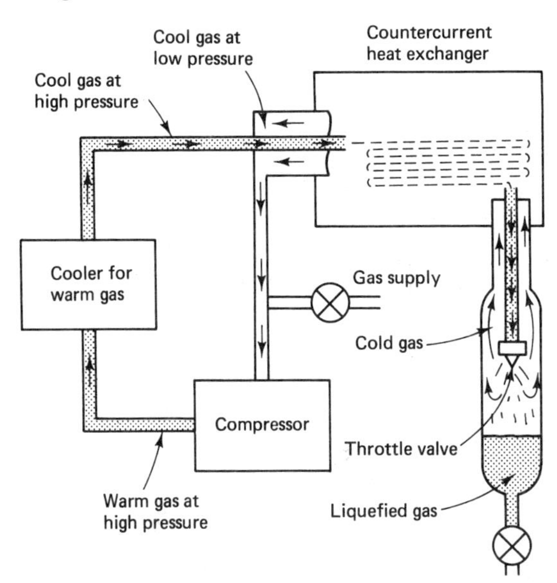

Figure 15-14 A unit to liquefy a gas by means of the Joule-Kelvin effect.

the other hand, an adiabatic expansion *always* results in cooling of the gas. Consequently, practical gas-liquefaction units typically employ both processes, each in its own optimum region of temperature and pressure. A disadvantage of adiabatic expansion is that a mechanical mechanism of moving parts must be utilized and must be lubricated at low temperatures. The disadvantage of the throttling process is the extensive precooling required to bring the gas temperature below the maximum inversion temperature.

An important component in the practical utilization of a throttling process in the liquefaction of gases is the *countercurrent heat exchanger*. The cooler gas returning from the throttling chamber is used to cool the warmer gas enroute to the throttling valve. Lower and lower temperatures are achieved by recirculating the cold gas until a steady-state condition is attained in which a fraction of the gas liquefies after throttling. A diagram of a gas liquefaction unit is shown in Fig. 15-14.

Except for their low temperatures, liquefied gases behave as any other liquids. Liquid nitrogen (77 K) can be carried around in a bucket, but it continually

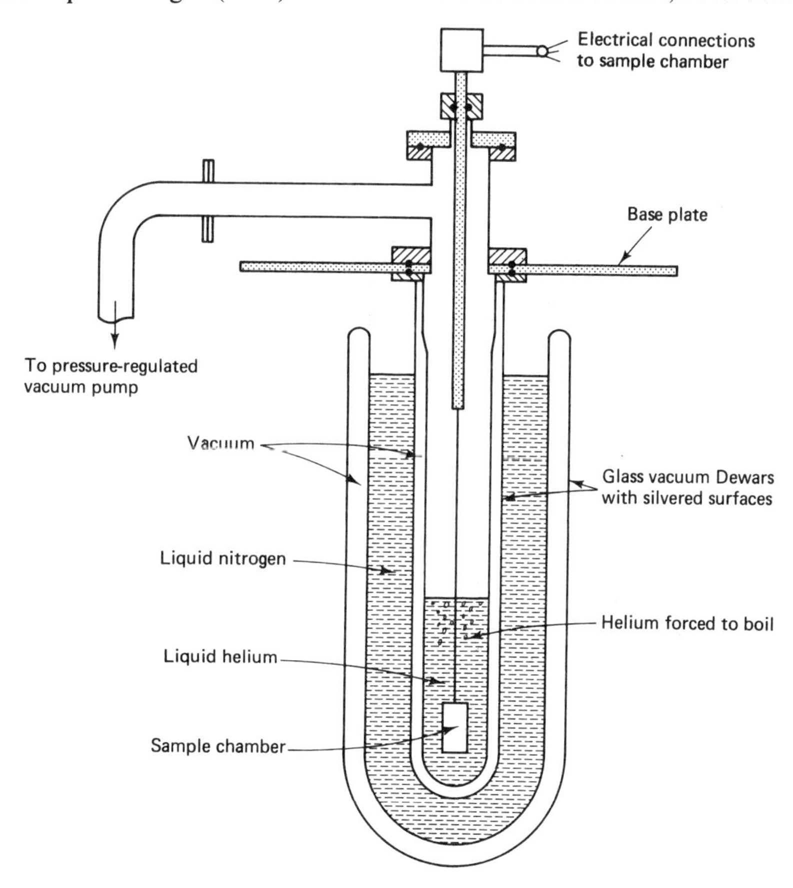

Figure 15-15 A liquid helium cryostat for conducting experiments in the temperature range from 4.2 K down to about 2.0 K. The temperature of the helium bath is controlled by controlling the pressure above the helium bath.

boils due to the heat it absorbs from the surrounding air. On the other hand, liquid helium (4.2 K) must always be protected from the "fervent heat" of air at room temperature or else it will very quickly boil away and be gone. Special, thermally insulated containers called *Dewars* [after Sir James Dewar (1842–1923)] are used for storing and transporting liquid gases.

The temperature of a thermally insulated sample of liquid gas can be lowered below its normal boiling point by using a vacuum pump to lower the pressure above the gas. This causes an increase in the rate of evaporation; this, due to the heat of vaporization, causes the temperature of the liquid to decrease. By controlling the pressure (by the rate of pumping), the temperature can be controlled. A typical arrangement is shown in Fig. 15-15.

When liquid helium is "pumped down" to about 2.2 K (the lambda point), a dramatic change occurs in the properties of the liquid. It becomes a superfluid, with zero viscosity and an infinite thermal conductivity. The phase change to superfluid helium occurs because of quantum mechanical properties of the interatomic forces acting between the helium atoms. At the lambda point, visible boiling ceases. The infinite thermal conductivity causes further evaporation to occur only at the top surface of the liquid. Furthermore, if there is even a tiny leak in the system (even one so small that normal helium cannot escape), the zero-viscosity superfluid will pass through it freely, often contaminating the surrounding vacuum that provides essential thermal insulation.

To Go Further

Read in an encyclopedia or other source about:

Robert Boyle (1627–1691), English chemist
Jacques Charles (1747–1823), French physicist
Joseph Gay-Lussac (1778–1850), French chemist
James Clerk Maxwell (1831–1879), Scottish physicist
Ludwig Boltzmann (1844–1906), German physicist

Questions

1. How does a real gas differ from an ideal gas? Under what conditions can a real gas be approximated by an ideal gas?
2. How many molecules are contained in one g-mole of mysterium (a fictitious compound)?
3. In your own language, state Boyle's law, Charles' law, the law of Gay-Lussac, and the ideal gas law.
4. At a given temperature, why does a quantity of gas exert a greater pressure when confined to a smaller container?
5. In terms of microscopic entities, why does a gas expand when heated at constant pressure?
6. In microscopic terms, why does the temperature rise when a gas is compressed?
7. What effects does van der Waals' equation of state consider that the equation of state of an ideal gas does not?
8. Describe the Joule-Kelvin effect and the porous-plug experiment.
9. Under what conditions will a real gas experience a decrease in temperature when the gas undergoes a throttling process?
10. Suppose hydrogen (H_2) and nitrogen (N_2) are mixed at room temperature. Which type molecule will move with greater speed?

Problems

1. A 3-L vacuum chamber is evacuated to a pressure of 1×10^{-8} mm Hg (a high vacuum) and is at a temperature of 294 K.
   (a) Convert the pressure from millimeters of mercury to atmospheres.
   (b) Compute the moles of gas remaining in the system.
   (c) Calculate the total number of gas molecules remaining in the system.
   (d) How many molecules are in each cubic centimeter of the vacuum chamber?
2. Interstellar gas, mostly hydrogen, near very hot stars may be ionized by the ultraviolet radiation from those stars, causing the gas to glow, which produces an emission nebula easily seen with binoculars. The density of such a nebula is typically 1000 atoms/cm^3, and the temperature is on the order of 10,000 K.
   (a) Assuming the gas to be entirely hydrogen (of molecular weight equal to 1), compute the moles per liter of the gas.
   (b) Compute the pressure of the gas in atmospheres.
   (c) Express the pressure in millimeters of mercury and compare with the pressure in the vacuum system of the problem above.
3. Compute the number of molecules in 1 g of water.
4. How many grams of the following gases does it take to contain a number of molecules equal to Avogadro's number? **(a)** CO_2 **(b)** CH_3 **(c)** He **(d)** H_2
5. A mass of 16 g of each of the following gases is stored in 8-L flasks at 294 K. Compute the pressure in each flask. **(a)** H_2 **(b)** He **(c)** N_2 **(d)** CO_2
6. Suppose we wish to prepare a mixture of helium and argon such that there is one atom of argon for each atom of helium.
   (a) Prior to mixing, how would the volumes of the respective gases compare, if at the same temperature and pressure?
   (b) How would the masses of the two gases compare?
7. Two identical steel cylinders of volume 4 ft^3 are filled with gas to a pressure of 2000 psi at a temperature of 72°F. One is filled with argon and the other is filled with helium. Unfortunately, the two tanks become confused, but the argon tank will obviously be heavier. Compute the difference in weight of the two tanks.
8. The absolute pressure exerted by the gas in the bulb of an ideal gas thermometer is 790 mm Hg when the bulb is surrounded by an ice-water mixture at 0°C. The bulb is then immersed in a bath of mineral oil, and the pressure increases to 850 mm Hg. What is the temperature of the oil: **(a)** in degrees Kelvin **(b)** in degrees Celsius?
9. Refer to the constant-volume gas thermometer of Fig. 15-3. On a day when the temperature is 32°C and the barometric pressure is 764 mm Hg, the difference in heights of the two mercury columns is 54 mm. What is the temperature on the following day if the mercury differential is 48 mm Hg while the barometric pressure is 759 mm Hg?
10. Four moles of an ideal gas are enclosed in a cylinder-piston arrangement, as in Fig. 15-5. The initial pressure, volume, and temperature are 40 psi (gauge), 1 ft^3, and 20°C, respectively. Over a period of time, 1 mole of gas leaks out, the temperature rises 5°C, and the volume is decreased to 0.4 ft^3. What is the final pressure?

*11. A spherical balloon 8 ft in diameter is to be filled with helium to a pressure of 1.002 atm on a day when the air temperature is 70°F.
   (a) How many moles of helium are required to fill the balloon?
   (b) If the helium is shipped in 2-ft^3 steel cylinders under a gauge pressure of 2000 psi at 70°F, how many moles are contained in each cylinder?
   (c) What will be the final gauge pressure indicated on the cylinder when the balloon is inflated to the proper pressure?

*12. What would be the absolute pressure (in atm) in the balloon above: **(a)** if the temperature rose to 75°F; **(b)** if the temperature fell to 65°F? **(c)** Assuming the material of which the balloon is made does not stretch or contract, what would be the condition of the balloon in part **(b)**?

13. In a Boyle's law apparatus as in Fig. 15-6, the gas column is 20 cm long when the mercury differential is 50 mm on a day when the barometric pressure is 766 mm Hg. What will be the length of the gas column when the mercury differential is increased slowly to 60 mm?

***14.** Two identical pressure vessels containing an ideal gas are connected by a small tube fitted with a valve. Each vessel has a volume of 10 L. Initially, with the valve closed, the absolute pressure in one vessel is 6 atm while the pressure in the other is 2 atm, both being at the same temperature of 300 K.

(a) Compute the moles of gas in each vessel.

(b) What will be the final pressure in each vessel after the valve is opened and the gas comes to equilibrium?

15. In the extremely rarefied regions of interstellar space where the density of particles is on the order of 100–1000 per cubic centimeter, the temperature is typically on the order of 10,000 K. Assuming the Maxwell distribution function to be applicable, compute the RMS speed of: **(a)** a hydrogen atom; **(b)** a helium atom at this temperature. **(c)** Traveling at this velocity, what period of time would be required for each particle to travel from the sun to the earth, a distance of 93 million miles?

16. Compute the most probable speed, the arithmetic mean speed, and the RMS speed of the following molecules at a temperature of 300 K. **(a)** H_2 **(b)** O_2 **(c)** CO_2 **(d)** H_2O.

17. Compute the pressure produced when 4 moles of CO_2 are placed in a volume of 3 L. Assume the temperature to be 293 K, and treat CO_2 as a van der Waals' gas.

18. Repeat Problem 17 for neon (Ne) instead of CO_2.

***19.** Suppose 12 moles of the following gases are contained in a volume of 10 L. Use the van der Waals' equation of state to compute the pressure exerted by each gas: **(a)** helium; **(b)** neon; **(c)** carbon dioxide, CO_2; **(d)** propane, C_3H_8. Assume $T = 20°C$.

20. A 10-gal water storage tank is arranged so that water enters from the bottom of the tank, trapping the air above. If the air pressure is 20 psi (gauge) when the tank contains only 1 gal of water, how many gallons of water will be in the tank when additional water is pumped in and the pressure rises to 40 psi?

21. Compute the average kinetic energy per molecule in an ideal gas at room temperature, 293 K.

22. Compute and compare the RMS velocities of an oxygen and a nitrogen molecule in a sample of air at 300 K. Which molecule is more massive and which travels faster?

***23.** **(a)** Compute the total kinetic energy of all the atoms in 4 g of helium gas at 300 K.

(b) If a 4-g bullet were given this amount of energy in the form of translational energy, what would be the velocity of the bullet?

24. Compute the mean free path of an oxygen molecule in a vacuum system at 300 K where the pressure is 1×10^{-6} mm Hg.

CHAPTER 16

PRINCIPLES AND PROCESSES OF THERMODYNAMICS

After having studied heat, temperature, and the properties of an ideal gas in the two preceding chapters, we are ready to step lightly into thermodynamics, both applied and practical. In this chapter we present the concepts that enable us to construct and analyze a heat engine, an engine that runs on a difference in temperature rather than fuel. Moreover, real engines are based on the same principles, so this chapter can lead to a much better understanding of the various types of engines we commonly encounter.

Much of this chapter is devoted to cyclic thermodynamic processes, the theoretical basis of engines of all types. After an initial description of individual thermodynamic processes, the processes are combined to form cyclic processes, such as the Carnot cycle, the Diesel cycle, and the Otto cycle. These cycles demonstrate that a quantity of heat can be converted to mechanical energy, albeit with low efficiency.

The first and second laws of thermodynamics are presented, and a brief discussion of entropy is given. The final sections deal with refrigeration, heat pumps, and practical engines, such as the Wankel rotary, the gas turbine, and the turbojet.

The material is somewhat more difficult than that in earlier chapters. An effort has been made to keep it simple, but the subject requires a higher level of algebra, and the computations become tedious at times. Even so, we barely scatch the surface, but more than that is not required of us at this point. We must leave something for the specialists.

This chapter concludes our study of heat and thermodynamics.

16-1 WORK AND MOLAR HEAT CAPACITY

Suppose a quantity of gas at pressure P is enclosed in a cylinder with a movable piston of area A, as shown in Fig. 16-1. A force $F = PA$ will be exerted on the piston by the gas, and some external mechanism must hold the piston in place. If

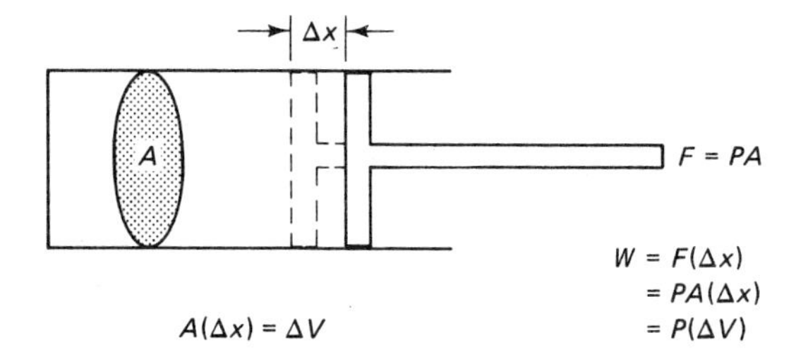

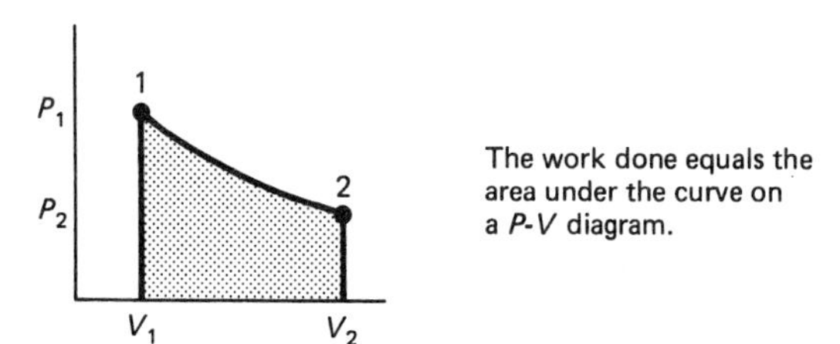

Figure 16-1 Work is done when a gas expands against a movable piston.

the piston is allowed to move a small distance Δx, allowing the gas to expand slightly, the piston will do work W on the external mechanism. The quantity of work done is given by the product of the force F and the displacement Δx. As shown in the figure, the work done by the expanding gas is given by

$$W = P(\Delta V) \tag{16-1}$$

Let us call the gas, cylinder, and piston *the system*. Then, we say that the system does work *on the external mechanism* when the piston moves to let the gas expand. On the other hand, if the external mechanism moves the piston in the opposite direction and compresses the gas, we say that work is done *on the system* by the external mechanism.

Generally speaking, a movement of the piston produces a change in the pressure, but the exact manner in which the pressure varies depends upon whether heat is added to or taken from the system while the piston is moving. A pressure-volume (P-V) diagram is useful in showing how the pressure varies as the volume changes. Three possible variations of pressure with volume are shown in Fig. 16-2. These are described in detail later.

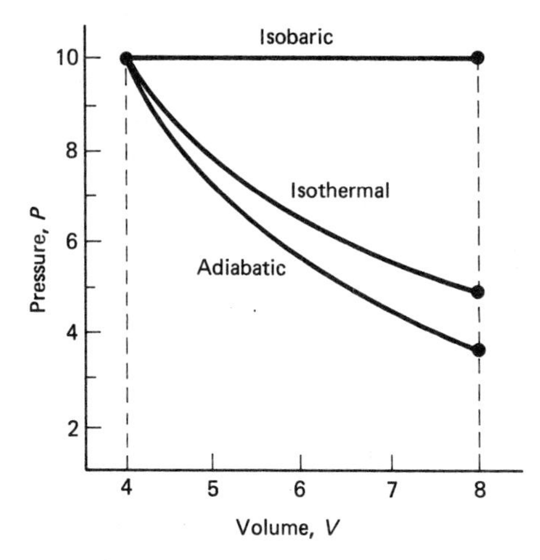

Figure 16-2 A comparison of three processes on a P-V diagram.

Because the pressure is not constant, the use of Eq. (16-1) in calculating quantities of work must be restricted to changes in volume ΔV sufficiently small so that the pressure may be considered constant. However, the work may be computed using calculus even when both the pressure and volume vary. A useful result, for conceptual purposes, is that the work done by a system in expanding from state 1 to state 2 is equal to the area under the curve connecting the two states on a P-V diagram.

When the system expands and does work on the external mechanism, energy is transferred from the system to the mechanism. Unless an equal quantity of energy in the form of heat is returned to the system, the temperature of the gas in the cylinder will decrease. But in most cases, a transfer of heat will occur during the expansion so that we cannot state categorically whether the gas temperature will rise or fall. If the heat energy added to the system is greater than the work done by the system, the temperature will rise. But if the heat added is less than the work done, the temperature will fall. Several methods for introducing heat into a system are shown in Fig. 16-3.

EXAMPLE 16-1

A cylinder 10 cm in diameter is equipped with a freely movable piston and contains gas at a pressure of 4 atm.

(a) What force in newtons is exerted on the piston?

(b) How much work is done by the expanding gas if the piston moves in the cylinder a distance of 1 mm? (Assume the pressure remains constant.)

Solution. (a) Using $\pi D^2/4$, the cross-sectional area of the piston is computed to be 7.85×10^{-3} m^2. Then, using the fact that a pressure of 1 atm is the equivalent of

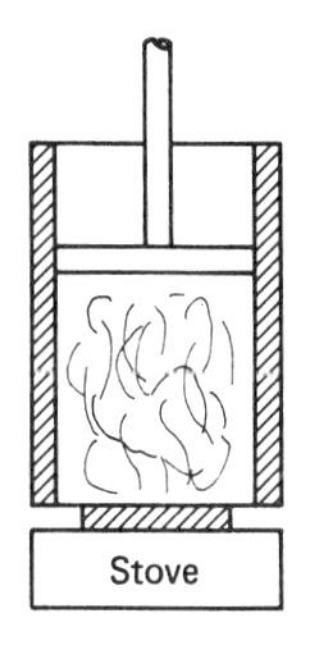

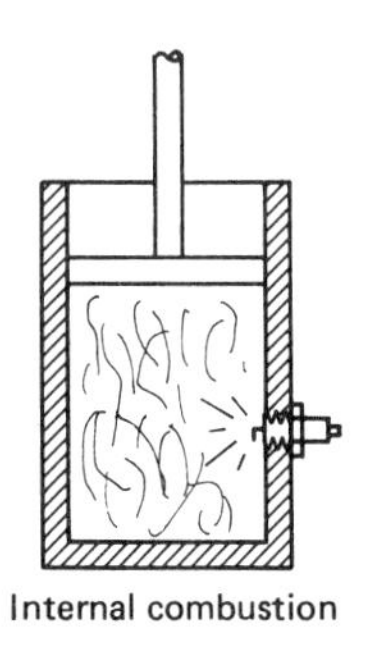

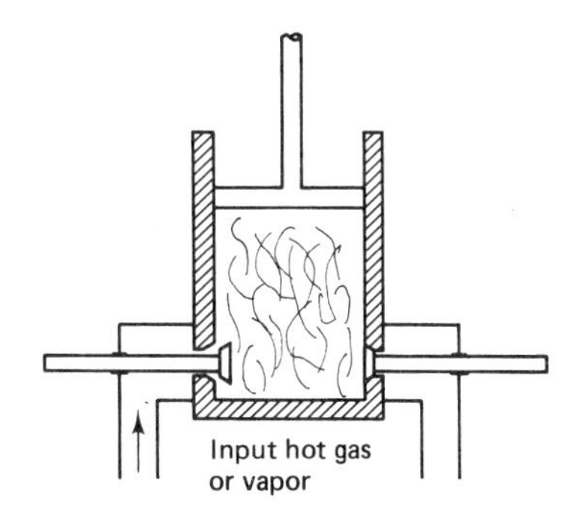

Figure 16-3 Ways to introduce heat into a system.

101.3×10^3 N/m^2, the pressure exerted on the piston is 4.05×10^5 N/m^2. Pressure times area gives the force:

$$F = PA$$
$$= (4.05 \times 10^5)(7.85 \times 10^{-3})$$
$$= 3179 \text{ N}$$

(b) The work done by the gas on the piston is the product of the force exerted on the piston and the displacement of the piston:

$$W = F(\Delta x)$$
$$= (3179 \text{ N})(1 \times 10^{-3} \text{ m})$$
$$= 3.179 \text{ J}$$

If the piston were pushed back to its original position by the external mechanism, exactly this same amount of work would be done by the piston on the gas as the gas is compressed.

Molar heat capacity, C_p and C_v

The molar heat capacity of a gas is the heat required to raise the temperature of one mole of gas 1°. But when a gas is heated, it tends to expand; if it expands, it does work, which tends to lessen the increase in temperature of the gas. Therefore, we must specify the conditions under which heat is added to the gas if we are to obtain a meaningful definition for the heat capacity of a gas. If a gas is allowed to expand during the heating process, more heat must be added to raise its temperature by 1° than if it does not expand.

The two conditions customarily specified for heating a gas are illustrated in Fig. 16-4. In part (a), the gas is not allowed to expand since the volume of the container does not change. The gas is heated at constant volume, and the appropriate molar heat capacity is denoted by C_v. If n moles of gas are present, the heat absorbed by the gas as its temperature changes from T_1 to T_2 is given by

$$\textit{(constant volume)} \qquad Q = nC_v(T_2 - T_1) \qquad (16\text{-}2)$$

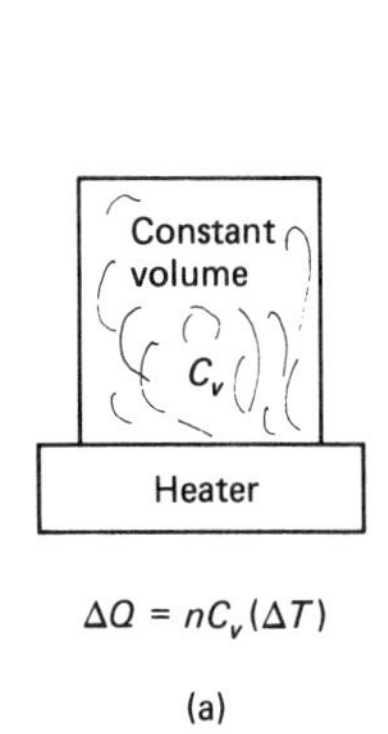

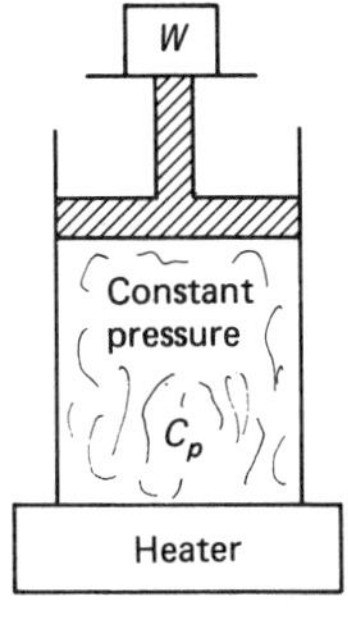

Figure 16-4 The specific heat of a gas depends upon whether the volume or the pressure is held constant during the change in temperature.

Date Due Slip

B Tech Community College
/27/12 06:52PM

* * * * * * * * * * * * *
* * * * *

TRON: 23312001114558

EM: 33312003007531
E DATE: 10/11/12
chnical physics /
g Q
LL NO. QC 21.2 .G728 1984

In Fig. 16-4(b), a constant force equal to the weight of the mass is applied to the piston, so that the pressure remains constant. The gas will expand as its temperature rises, and it will contract as its temperature falls. Because the gas does work in lifting the weight as its temperature rises, a greater quantity of heat must be added to the system in order to provide energy to (1) heat the gas, and (2) lift the weight. The expression for the heat added is

$$\text{(constant pressure)} \qquad Q = nC_p(T_2 - T_1) \tag{16-3}$$

When the gas is held at constant volume, all the energy added goes to heat the gas. But at constant pressure, the energy added to the system is divided between lifting the weight and raising the temperature of the gas. Since the net energy absorbed by the gas is the same in the two instances, the additional heat required to produce the given temperature change at constant pressure is greater than that at constant volume by an amount equal to the work done in lifting the weight. Hence, C_p is always greater than C_v. In fact, for an ideal gas, the difference between the two equals the gas constant R:

$$C_p - C_v = R \tag{16-4}$$

The preceding equations indicate that the units of C_p and C_v are the same as the units of the gas constant R. (Refer to Sect. 15-3, page 329). While several combinations of units are possible, we confine ourselves to the following units of C_v and C_p:

$$\frac{\text{J}}{\text{g} \cdot \text{mol} \cdot \text{K}} = \frac{\text{J}}{\text{g} \cdot \text{mol} \cdot {}^\circ\text{C}}$$

Other texts and handbooks may use the calorie instead of the joule as the heat unit, but since 1 cal = 4.181 J, the conversion from calories to joules is accomplished simply by multiplying by 4.181. The molar heat capacities of several gases are given in Table 16-1.

An important thermodynamic quantity is the ratio of the molar heat capacities, called the *specific heat ratio* and denoted by γ:

$$\gamma = \frac{C_p}{C_v} \tag{16-5}$$

TABLE 16-1 APPROXIMATE SPECIFIC HEATS OF SOME COMMON GASES (J/g · mol · K)

Gas	C_p	C_v	γ
Air	29.3	21.0	1.40
Argon, A	21.0	12.5	1.67
Chlorine, Cl_2	17.0	12.5	1.36
Carbon dioxide, CO_2	36.6	28.0	1.30
Carbon monoxide, CO	29.0	20.7	1.40
Helium, He	21.0	12.6	1.66
Hydrogen, H_2	28.4	20.1	1.41
Nitrogen, N_2	29.1	20.7	1.40
Oxygen, O_2	29.3	20.8	1.40
Methane, CH_4	35.5	27.1	1.31
Ammonia, NH_3	37.3	28.4	1.31

Since C_p is always greater than C_v, γ is always greater than 1.0. Typical values of γ are given in Table 16-1. Note that for monatomic gases, γ is nearly 1.67, while for diatomic gases, γ is about 1.4. Because γ is a ratio of like quantities, it has no units. It is a pure number.

EXAMPLE 16-2 A piston and cylinder are arranged so that a constant pressure of 3 atm is maintained on the enclosed 0.25 g · mol of nitrogen gas. The initial temperature of the gas is 300 K. Treating nitrogen as an ideal gas, the initial volume is found to be 2.053 L. Molar heat capacities of nitrogen are $C_v = 20.74$ and $C_p = 29.06$ J/mol · K. Heat is added to the gas causing its temperature to rise to 320 K.

(a) Compute the volume occupied by the gas at 320 K.
(b) Calculate the work done by the gas as it expands.
(c) How much heat was added to the gas?
(d) If the gas had been heated at constant volume instead of at constant pressure, how much heat would have been required to produce the same rise in temperature?
(e) Add the results of parts (b) and (d) above and compare the sum with the result of part (c).

Solution. (a) The volume at 320 K is obtained from the ideal gas law:

$$V = \frac{nRT}{P} = \frac{(0.25)(0.0821)(320)}{3} = 2.189 \text{ L}$$

(b) Using Eq. (16-1), $W = P(\Delta V)$:

$$\Delta V = 2.189 - 2.053 = 0.136 \text{ L}$$

$$W = P(\Delta V) = (3 \text{ atm})(0.136 \text{ L})$$

$$= 0.408 \text{ L} \cdot \text{atm}$$

To convert the units of work from L · atm to joules, use the fact that

$$1 \text{ L} \cdot \text{atm} = 101.3 \text{ J}$$

and then multiply by 101.3 to obtain the work done by the expanding gas:

$$W = 0.408(101.3) = 41.33 \text{ J}$$

(c) The heat added to the gas is found using Eq. (16-3):

$$Q = nC_p(T_2 - T_1)$$

$$= (0.25 \text{ mol})\left(29.06 \frac{\text{J}}{\text{mol} \cdot \text{K}}\right)(320 - 300 \text{ K})$$

$$= 145.30 \text{ J}$$

(d) Equation (16-2) applies to a gas heated at constant volume:

$$Q = nC_v(T_2 - T_1)$$

$$= (0.25 \text{ mol})\left(20.74 \frac{\text{J}}{\text{mol} \cdot \text{K}}\right)(320 - 300 \text{ K})$$

$$= 103.70 \text{ J}$$

(e) The sum of parts (b) and (d) is 41.33 plus 103.70, or 145.03 J. Note that this is the same as the result of part (c). (The small difference is probably due to uncertainties in the values of C_p and C_v and to rounding error in doing the arithmetic; it is not significant.) Thus, of the 145.3 J added to the gas at constant pressure, 103.7 J went to change the temperature of the gas while the remaining 41.33 J provided energy for doing work during the expansion:

Heat added = work done + heat required to change the temperature of the gas

This is in accordance with the law of the conservation of energy and anticipates the first law of thermodynamics, which is the topic of a later section.

16-2 ADIABATIC EXPANSION AND COMPRESSION

If a cylinder and piston are thermally insulated so that no heat enters or leaves the system, the expansion or compression is called *adiabatic*. In practice, no cylinder or piston can be perfectly insulated, but if the expansion or compression occurs quickly, as in the cylinder of an internal combustion engine, the process will be very nearly adiabatic because little heat can escape in the short time interval involved. Thus, adiabatic expansions are important to the thermodynamic analysis of reciprocating engines.

An adiabatic compression always produces an increase in both temperature and pressure. Conversely, the temperature and pressure decrease in an adiabatic expansion. The relationship fundamental to an adiabatic process is

$$PV^{\gamma} = \text{constant} \tag{16-6}$$

and it follows from this that when a gas is allowed to expand (or is compressed) adiabatically from state 1 to state 2,

$$P_1V_1^{\gamma} = P_2V_2^{\gamma} \tag{16-7}$$

The equation of state of an ideal gas can be used to obtain similar expressions that involve the temperature. The result is

$$T_1V_1^{\gamma-1} = T_2V_2^{\gamma-1} \tag{16-8}$$

and

$$T_1P_1^{(1-\gamma)/\gamma} = T_2P_2^{(1-\gamma)/\gamma} \tag{16-9}$$

These equations are summarized in Fig. 16-5 in a form convenient for computation. Note that the expressions have been rearranged to avoid negative exponents.

In an adiabatic expansion from V_1 to V_2, the system does work on the external mechanism. Energy for doing the work is taken from the gas, causing its temperature to fall. The amount of work done is equal to the area under the curve on the *P-V* diagram and is given by

$$W = \frac{P_1V_1 - P_2V_2}{\gamma - 1} \tag{16-10}$$

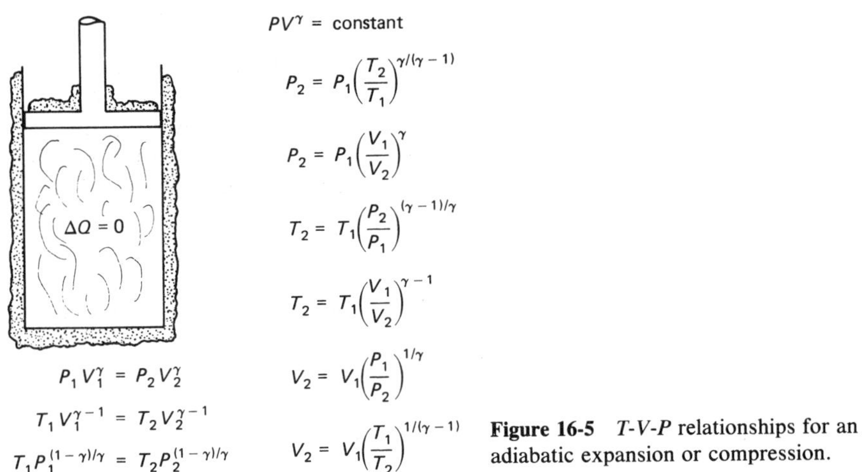

Figure 16-5 *T-V-P* relationships for an adiabatic expansion or compression.

A different expression for the work done in an adiabatic expansion may be obtained by considering the decrease in energy of the gas. The expression is

$$W = nC_v(T_1 - T_2) \tag{16-11}$$

which allows the work to be calculated in terms of different variables.

If either of the two preceding formulas are applied to an adiabatic *compression*, a *negative* quantity is obtained for the work done. This means simply that work was done *on* the system rather than *by* the system. However, when we consider adiabatic compressions as segments of thermodynamic cycles in later sections of this chapter, we arrange the formulas so that the work done is obtained as a positive number. This avoids the addition of negative numbers (as opposed to straightforward subtraction) in computing the net work done during the cycle. In each instance it is made clear whether the system is doing work or is being worked on so that no ambiguity results.

EXAMPLE 16-3 One liter of nitrogen (for which $\gamma = 1.40$) at 300 K and atmospheric pressure is compressed adiabatically to one-tenth its original volume, to 0.1 L. Compute: (a) the final pressure; (b) the final temperature; and (c) the work done on the gas during the compression.

Solution. We use the relationships of Fig. 16-4, together with Eqs. (16-10) and (16-11).

(a)

$$P_2 = P_1\left(\frac{V_1}{V_2}\right)^{\gamma}$$

$$= (1 \text{ atm})\left(\frac{1 \text{ L}}{0.1 \text{ L}}\right)^{1.4}$$

$$= (1)(10)^{1.4} = 25.12 \text{ atm}$$

(b)

$$T_2 = T_1\left(\frac{V_1}{V_2}\right)^{\gamma-1}$$

$$= (300 \text{ K})(10)^{1.4-1}$$

$$= 300(10)^{0.4} = 300(2.512) = 753.6 \text{ K}$$

(c) The work done is given by Eq. (16-10):

$$W = \frac{P_2V_2 - P_1V_1}{\gamma - 1}$$

$$= \frac{(25.12)(0.1) - (1)(1)}{1.4 - 1}$$

$$= 3.750 \text{ L} \cdot \text{atm}$$

In terms of joules (multiplying by 101.3 J/L · atm);

$$W = 3.750(101.3) = 379.9 \text{ J}$$

Alternatively, the work done in compressing the gas can be computed using Eq. (16-11), but first we must compute the number of moles of gas in the system. Using the initial conditions and the ideal gas law gives

$$n = \frac{PV}{RT} = \frac{(1 \text{ L})(1 \text{ atm})}{(0.0821)(300 \text{ K})}$$

$$n = 0.0406 \text{ g} \cdot \text{mol}$$

Then,

$$W = nC_v(T_2 - T_1)$$

$$= (0.0406)(20.74)(753.6 - 300) \tag{16-11}$$

$$= 381.95 \text{ J}$$

Compare with 379.9 J obtained earlier. The two values are close, but they are not the same even though ideally they should be. Eq. (16-11) makes use of the *measured* quantity C_v whereas Eq. 16-10 uses only *calculated* or *given* quantities that are assumed infinitely precise. Therefore, we should not expect perfect agreement because the result of any experimental measurement is going to be imprecise to some degree. On the other hand, we should be delighted to have *two* methods of calculating the work done, because the precision obtained is perfectly adequate for any practical application.

16-3 ISOTHERMAL EXPANSION

An *isothermal* expansion is an expansion that occurs at constant temperature. The expanding gas does work because there is an increase in volume. Therefore, heat must be added to the system in order to hold the temperature constant. If heat were not added, the temperature would fall as in an adiabatic expansion.

On a *P-V* diagram, the curve representing an isothermal expansion does not exhibit as great a negative slope as that of an adiabatic expansion (see Fig. 16-2). That is, the pressure does not fall as rapidly with increasing volume as it does when the expansion is adiabatic. This is because the heat added to maintain the temperature constant also gives rise to a less-abrupt decrease in pressure.

Because the temperature remains constant, there is no change in the internal energy of the gas. Consequently, the heat added to the system exactly equals the work done by the system. This is a consequence of the conservation of energy. With calculus it is easy to show that the work done in an isothermal expansion is given by

$$W = nRT \cdot \ln\left(\frac{V_2}{V_1}\right) \tag{16-12}$$

where the symbol ln denotes the natural logarithm. Using $PV = nRT$ and the fact that in an isothermal expansion $V_2/V_1 = P_1/P_2$, we also may write Eq. (16-12) in the following forms:

$$W = P_1V_1 \cdot \ln\left(\frac{V_2}{V_1}\right) \tag{16-13}$$

$$W = P_1V_1 \cdot \ln\left(\frac{P_1}{P_2}\right) \tag{16-14}$$

The important relationships for an isothermal expansion are summarized in Fig. 16-6(b).

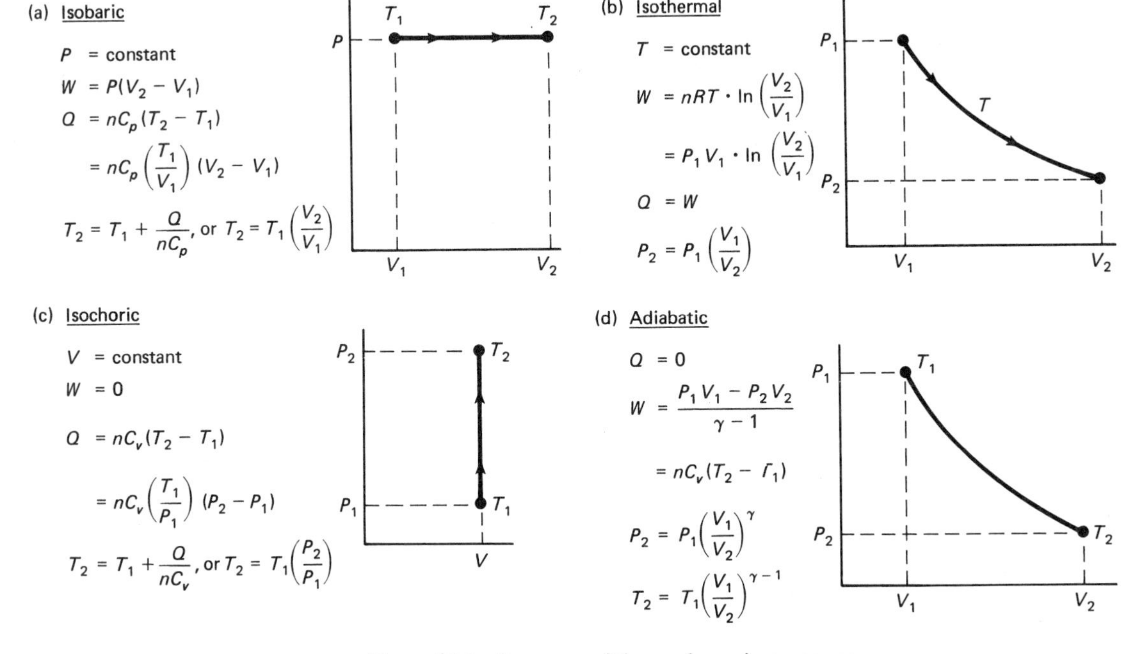

Figure 16-6 Summary of thermodynamic processes.

16-4 ISOCHORIC PROCESS

An isochoric process is one that occurs at constant volume. Because there is no change in volume, no work is done. Consequently, any heat added to the system goes entirely into the internal energy of the gas and serves to increase the temperature and pressure.

The conditions pertaining to an isochoric process are exactly those for which C_v, the molar heat capacity at constant volume, was defined. From Eq. (16-2), then, we obtain the heat added to a system during an isochoric process:

$$Q = nC_v(T_2 - T_1) \tag{16-15}$$

In terms of the pressures and temperature T_1,

$$Q = nC_v\left(\frac{T_1}{P_1}\right)(P_2 - P_1) \tag{16-16}$$

The important results for an isochoric process are illustrated in Fig. 16-6(c).

16-5 ISOBARIC EXPANSION

The last of the four fundamental processes is an *isobaric process*, one that proceeds at a constant pressure. Work is done because there is a change in volume, and as work is done, heat must be added to the system in order to maintain the pressure constant. Furthermore, the temperature of the gas increases in proportion to the increase in volume of an isobaric expansion.

Since the pressure remains constant during the expansion, the work done is easy to calculate using Eq. 16-1:

$$W = P(V_2 - V_1) \tag{16-17}$$

Since the conditions are suitable for use of C_p, the molar heat capacity at constant pressure, the heat added during an isobaric expansion is given by

$$Q = nC_p(T_2 - T_1) \tag{16-18}$$

which also may be written as

$$Q = nC_p\left(\frac{T_1}{V_1}\right)(V_2 - V_1) \tag{16-19}$$

In terms of the initial temperature T_1 and the heat added, the final temperature T_2 is given by

$$T_2 = T_1 + \frac{Q}{nC_p} \tag{16-20}$$

The important relationships for an isobaric process are given in Fig. 16-6(a).

16-6 CYCLIC PROCESSES; A HEAT ENGINE

Now that we are familiar with the four basic thermodynamic processes, we can consider cyclic processes and see how it is possible to build an engine that can take a quantity of thermal energy and convert a portion of it to mechanical work. A *cyclic process* is one that may be repeated periodically, with each cycle being identical to the previous cycle. A steam engine and an internal-combustion engine provide good examples of cyclic processes in which gases are alternately compressed and expanded as a piston moves back and forth in a cylinder. On a P-V diagram, a cyclic process forms a closed figure with each side (or segment) of the figure consisting of one of the basic processes. Several cyclic processes are illustrated in the following sections (see Figs. 16-8, 16-11, 16-15).

The concept of a *heat reservoir* is useful in discussions of heat engines. A heat reservoir is a body of known temperature with an extremely large heat capacity, so that quantities of heat may be extracted from the body without producing a noticeable change in temperature of the body. For example, a swimming pool may contain 50,000 gal of water at 75°F. If we place an ordinary ice cube in a small tin cup and then hold the cup in contact with the surface of the water, heat will be extracted from the pool as the ice melts. But the temperature of the water in the pool will not change enough to measure even with the most sensitive instruments. Thus, for our purposes with the ice cube, the pool forms a good heat reservoir. But if our objective is to melt a 5-ton block of ice, the pool could not be considered to be a good heat reservoir because the large quantity of ice would produce a change in temperature of the water in the pool by about 3°F. In general, we do not concern ourselves with the technical details of heat reservoirs; the important point is that the temperature of the reservoir does not change when quantities of heat are extracted.

A heat reservoir of very low temperature is called a *cold reservoir*. Cold reservoirs are used to absorb heat from systems or objects that are hotter. Whether a given reservoir is a hot or a cold reservoir depends upon the reservoir with which it is compared. Ocean water used to absorb waste heat from a nuclear reactor acts as a cold reservoir, but the same ocean may be considered a hot reservoir when compared to the cold mass of a polar ice cap.

In the following section we see that if hot and cold reservoirs are available, an engine can be constructed that will utilize the temperature difference between the two reservoirs to do work, such as generating electricity. Such an engine would require no fuel, and it would operate indefinitely . . . as long as the heat reservoirs are maintained.

A heat engine

The mechanism of a simple heat engine is shown in Fig. 16-7 and consists of a cylinder-piston-crankshaft arrangement that converts the reciprocating motion of the piston to rotational motion of the crankshaft and flywheel. We assume hot and cold reservoirs are available that alternately are placed in thermal contact with the gas inside the cylinder. The gas in the cylinder is called the *working substance*, and it remains in the cylinder indefinitely. No exhaust or intake valves are needed.

The basic heat engine cycle may be divided into two parts, a high-pressure expansion (the *power stroke*) followed by a low-pressure *compression stroke*. The gas does work on the external mechanism during the power stroke, but during the compression stroke, work is done on the gas by the external mechanism.

Part (a) of Fig. 16-7 shows the system at the beginning of the power stroke. The gas had been compressed to minimum volume, and the cylinder is in contact with the hot reservoir. Heat is being added to the system, causing the pressure in

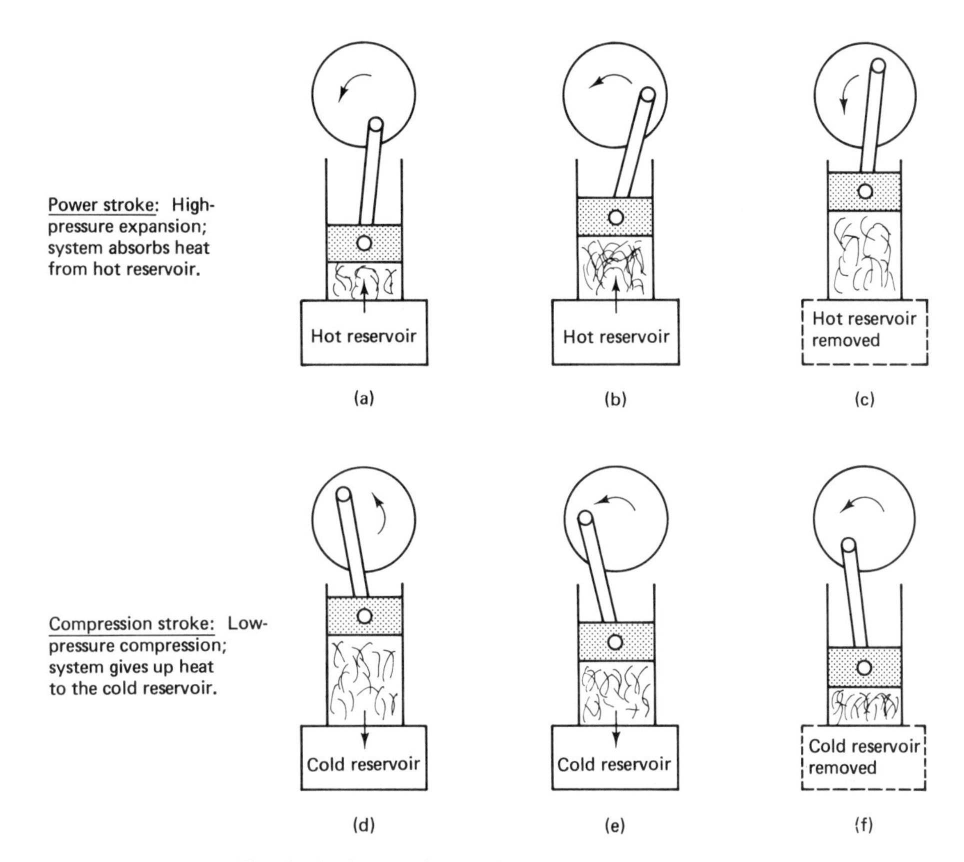

Figure 16-7 The basic heat-engine cycle. The hot and cold reservoirs are alternately placed in thermal contact with the system.

the cylinder to increase. In part (b), halfway through the power stroke, the hot reservoir continues to input heat to the system as the expanding gas does work on the moving piston. The hot reservoir is removed from the system near the end of the power stroke, as shown in part (c), and the cold reservoir is placed in contact with the cylinder as the compression stroke begins in part (d).

Since the cold reservoir absorbs heat from the system, the temperature and pressure of the gas are lower during the compression stroke. Therefore, the low-pressure gas is relatively easy to compress. However, the compression stroke requires that work be done by the external mechanism. If a large flywheel is connected to the crankshaft, the angular momentum imparted to the flywheel during the power stroke will carry the crankshaft to the end of the relatively easy compression stroke, parts (e) and (f) of the figure. The heat reservoirs then are exchanged, and another power stroke begins as the newly heated gas expands forcefully against the moving piston.

Because the system does more work during the high-pressure expansion than is required for the subsequent low-pressure compression, power may be taken from a shaft or pulley attached to the flywheel and may be used to drive an electrical generator or other device. Thus, we have an engine capable of producing electrical power that uses no fuel; its source of power is the difference in temperature between the hot and cold heat reservoirs. But this scheme is not without its shortcomings, because we note that heat is continually removed from the hot reservoir. Ultimately, for any practical reservoir, its temperature must decrease. Likewise, heat is continually added (although in lesser amounts) to the cold reservoir, so its temperature must ultimately rise. Eventually, the temperature of any two practical reservoirs will become the same, and at that point, our heat engine will cease to function. In practical engines, an expenditure of fuel is required in order to produce and maintain a high-temperature source of heat to serve as a high-temperature reservoir. The ambient air serves as the cold reservoir for a typical internal combustion engine because the air enters the cylinders at very nearly the ambient temperature.

16-7 A BASIC CYCLIC PROCESS

In this and the following sections we describe several thermodynamic cycles in quantitative terms. The cycles consist of various combinations of the four basic processes (isothermal, isobaric, isochoric, and adiabatic) described earlier and summarized in Fig. 16-6. Each process is assumed to occur within a cyclinder equipped with a reciprocating piston, and the working substance is assumed to be an ideal gas. In a practical engine, heat might be added to the system by burning fuel and might be exhausted by expelling the hot gas and replacing it with an equal quantity of cooler gas. In our description, however, we simply say that heat is obtained from a hot reservoir and heat is exhausted to a cold reservoir, as in our previous description of a heat engine. We concern ourselves neither with the mechanical details of the engine nor with the details of the combustion process. These matters are for the specialist. Our objective is to become acquainted with the thermodynamic principles involved and to become aware of some of the limitations inherent in the design of heat engines.

In light of the amount of calculation involved for each cycle, a systematic approach has been taken, so each cycle differs from the others only in detail. Only one system of units is used. Pressure is measured in atmospheres, volume in liters, and temperature in degrees Kelvin. Heat is measured in energy units of joules. When work is obtained in the units of $P \cdot V$, namely, liter-atmospheres, it is conveniently converted to joules by multiplying by 101.3. Thus, all quantities of energy are expressed in joules, and all quantities of work and heat are obtained as positive numbers.

The first cycle we describe involves only isobaric and isochoric processes and appears on the P-V diagram as a rectangle (Fig. 16-8). Point A is taken as the starting point, and the system will return to the same point when one cycle is completed.

In isobaric process AB (the compression stroke), the gas is cooled so that it may be compressed without producing an increase in pressure. The lowest temperature of the cycle occurs at point B, a point on the cycle where the volume is a minimum. Heat in the amount of Q_{ab} is removed from the system, and work W_{ab} is done on the gas as the piston compresses the gas.

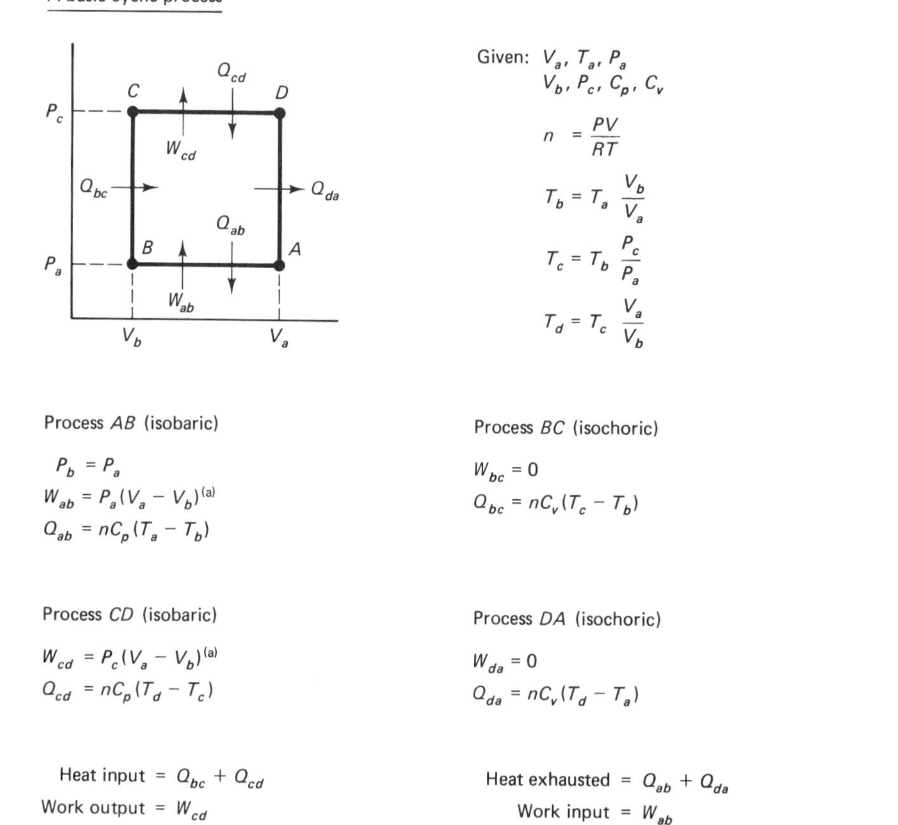

[(a)] multiply by 101.3 J/l · atm to convert energy units to joules.

Figure 16-8 A basic cyclic process.

While the piston is at the position of minimum volume V_b, heat Q_{bc} is quickly added to the gas, perhaps by burning fuel ignited by a spark plug, and the pressure rises to P_c. No work is done during this process since the volume does not change. When the gas reaches maximum pressure P_c, the piston begins the power stroke CD. Heat continues to be added to the gas, which prevents a decrease in pressure as the gas expands and does work W_{cd} on the piston. All the while, the temperature of the gas increases until it reaches the maximum temperature at point D, the end of the power stroke.

Process DA occurs while the piston is at the position of maximum volume V_a. Heat Q_{da} is removed from the gas in order to reduce the temperature and pressure of the gas to the values initially specified for point A.

In the numerical example of Fig. 16-9, the pressure, volume, and tempera-

Numerical example, basic cyclic process

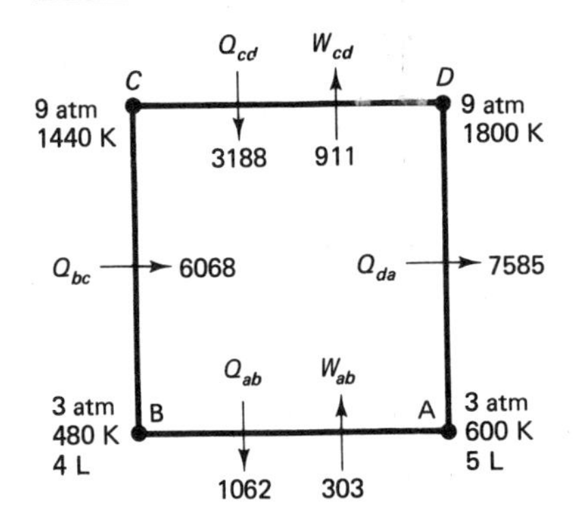

Given: $V_a = 5$ L
$V_b = 4$ L
$T_a = 600$ K
$P_a = 3$ atm
$P_c = 9$ atm
$C_p = 29.09$ J/mol · K
$C_v = 20.76$ J/mol · K

$$n = \frac{PV}{RT} = \frac{(3\text{ atm})(5\text{ L})}{(0.0821)(600\text{ K})} = 0.3045\text{ mol}$$

$T_b = 480$ K $\quad T_c = 1440$ K $\quad T_d = 1800$ K

Process AB (isobaric)

$$\begin{aligned} W_{ab} &= P_a(V_a - V_b) \\ &= (3)(5-4) \times 101.3 \\ &= 303.9\text{ J} \\ Q_{ab} &= nC_p(T_a - T_b) \\ &= (0.3045)(29.09)(600-480) \\ &= 1062.95\text{ J} \end{aligned}$$

Process BC (isochoric)

$$\begin{aligned} W_{bc} &= 0 \\ Q_{bc} &= nC_v(T_c - T_b) \\ &= (0.3045)(20.76)(1440-480) \\ &= 6068.56\text{ J} \end{aligned}$$

Process CD (isobaric)

$W_{cd} = 911.70$ J
$Q_{cd} = 3188.85$ J

Process DA (isochoric)

$W_{da} = 0$
$Q_{da} = 7585.7$ J

Heat input $= Q_{bc} + Q_{cd} = 9257.41$ J
Work output $= W_{cd} = 911.7$ J

Heat exhausted $= Q_{ab} + Q_{da} = 8648.65$ J
Work input $= W_{ab} = 303.9$ J

Net work $= 911.7 - 303.9 = 607.8$ J
Net heat $= 9257.41 - 8648.65 = 608.76$ J
} Compare.

$$\text{Efficiency} = \left(\frac{607.8}{9257.41}\right) \times 100 = 6.57\%$$

$$\text{Carnot efficiency} = \left(1 - \frac{480\text{ K}}{1800\text{ K}}\right) \times 100 = 73\%$$

Figure 16-9 Numerical example of the basic cyclic process.

ture at point A are given, which allows determination of the number of moles of gas used as the working substance. Also, V_b is given; along with V_a, V_b determines the compression ratio (V_a/V_b) of the cycle. The complete cycle is defined by additionally specifying the maximum pressure P_c. The temperatures occurring at points B, C, and D are readily computed from the formulas given in

Fig. 16-8. The quantity of heat transferred and the amount of work done for each segment are also readily computed.

The last part of the calculation amounts to an energy balance for the cycle. That is, in light of the energy conservation law, the energy entering the system during one cycle must equal the energy leaving the system during the same cycle. Consequently, the net heat and the net work must be equal. The energy balance is illustrated diagrammatically in Fig. 16-10.

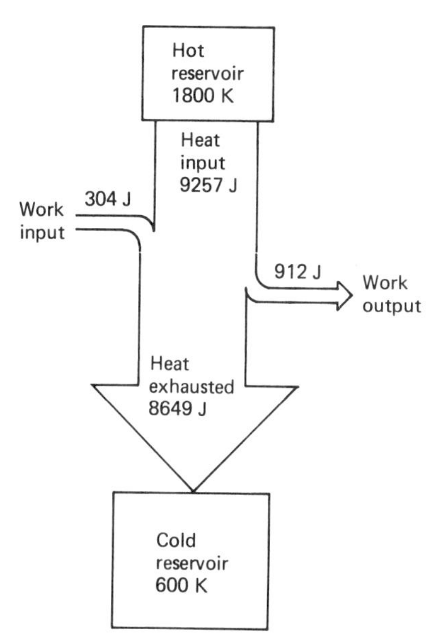

Figure 16-10 Energy balance for the cycle of Fig. 16-9.

Note that the efficiency of the cycle is a rather low 6.57%. This means that most of the heat taken in from the hot reservoir is exhausted as waste heat to the cold reservoir rather than being converted to usable mechanical work. For purposes of comparison, if a Carnot engine were operated between the same temperature limits (a high temperature of 1800 K and a low of 480 K), the efficiency of the Carnot cycle would be 73%. Thus, we must conclude that our basic cycle is not very efficient.

16-8 THE CARNOT CYCLE

The Carnot (pronounced *car-no*) cycle is of considerable theoretical importance because advanced thermodynamics shows that a Carnot cycle is the most efficient cycle that can be operated as a heat engine operating between given hot and cold reservoirs. The Carnot cycle consists of two isothermal processes bounded by two adiabatic segments of greater slope, as shown in Fig. 16-11. The efficiency of a Carnot engine is determined by the ratio of the minimum and maximum temperatures (absolute) of the cycle.

$$\text{Carnot efficiency} = \left(1 - \frac{T_{\text{min}}}{T_{\text{max}}}\right) \times 100\% \qquad (16\text{-}21)$$

Carnot cycle

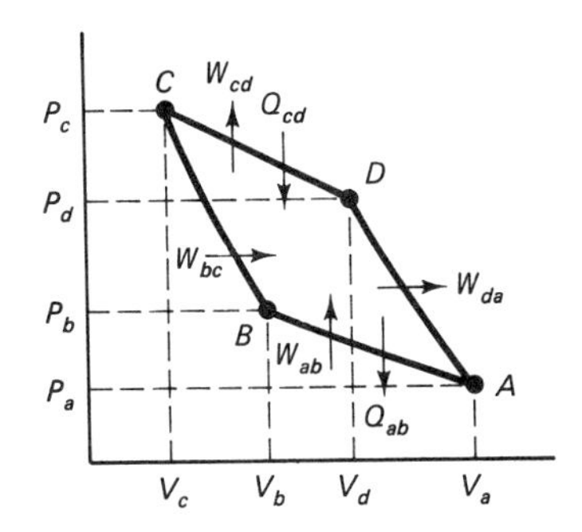

Given: $V_a, P_a, T_a, V_c, T_c, \gamma$

$\gamma - 1 = 0.4$ $\quad\quad \dfrac{1}{\gamma - 1} = 2.5$

$\dfrac{\gamma}{\gamma - 1} = 3.5$

$T_b = T_a$ $\quad\quad V_b = V_c \left(\dfrac{T_c}{T_a}\right)^{1/(\gamma - 1)}$

$T_c = T_a \left(\dfrac{V_b}{V_c}\right)^{\gamma - 1}$ $\quad\quad V_d = \dfrac{V_c V_a}{V_b}$

$T_d = T_c$

Process AB (isothermal)

$$P_a = P_a \left(\frac{V_a}{V_b}\right)$$

$$W_{ab} = P_a V_a \ln \left(\frac{V_a}{V_b}\right)^{(a)}$$

$$Q_{ab} = W_{ab}$$

Process BC (adiabatic)

$$P_c = P_b \left(\frac{V_b}{V_c}\right)^{\gamma}$$

$$W_{bc} = \frac{P_c V_c - P_b V_b}{\gamma - 1}^{(a)}$$

$$Q_{bc} = 0$$

Process CD (isothermal)

$$P_d = P_c \left(\frac{V_c}{V_d}\right)$$

$$W_{cd} = P_c V_c \ln \left(\frac{V_d}{V_c}\right)^{(a)}$$

$$Q_{cd} = W_{cd}$$

Process DA (adiabatic)

P_a = (given)

$$W_{da} = \frac{P_d V_d - P_a V_a}{\gamma - 1}^{(a)}$$

$$Q_{da} = 0$$

Heat input = $Q_{cd} = W_{cd}$

Work output = $W_{cd} + W_{da}$

Net work = work output − work input

Heat exhausted = $Q_{ab} = W_{ab}$

Work input = $W_{ab} + W_{bc}$

Net heat = heat input − heat exhausted

$$\text{Efficiency} = \left(\frac{\text{net work}}{\text{heat input}}\right) \times 100\% = \left(1 - \frac{T_a}{T_c}\right) \times 100\%$$

[a]multiply by 101.3 J/L · atm to convert energy units to joules.

Figure 16-11 The Carnot cycle.

Referring to Fig. 16-11, process AB is an isothermal compression and takes place with the system in contact with a cold reservoir. Heat is removed during the process in order to prevent a rise in temperature as the gas is compressed. At point B, the cold reservoir is removed and process BC occurs adiabatically. The temperature rises from T_a to T_c during the adiabatic compression. Processes AB and BC constitute the compression stroke.

The volume at point B is not arbitrary; it is the intersection of isotherm AB and adiabatic BC, and it must be calculated. A formula for V_b may be obtained by considering the adiabatic process BC relative to point C for which the volume and temperature are known:

$$T_c V_c^{\gamma-1} = T_a V_b^{\gamma-1}$$

Solving for V_b yields

$$V_b = V_c \left(\frac{T_c}{T_a} \right)^{1/(\gamma-1)} \tag{16-22}$$

A similar consideration of adiabatic DA gives an expression for V_d in terms of V_a, V_b, and V_c:

$$V_d = V_c \left(\frac{V_a}{V_b} \right) \tag{16-23}$$

The power stroke begins with an isothermal process CD with the system in contact with the hot reservoir. Heat Q_{cd} is added to the system to maintain the temperature constant as the system does work W_{cd}. At point D, the hot reservoir is removed in anticipation of the adiabatic process DA. The temperature falls from T_c to T_a as the system does work W_{da}, ending the power stroke at point A.

A numerical example of a Carnot cycle is given in Fig. 16-12. The computations are somewhat long, but they are straightforward. The maximum temperature of the cycle is a moderate 570 K, but even so, the efficiency is 47.4%. This may be contrasted with the basic cycle of the previous section, which involved a maximum temperature of 1800 K but achieved an efficiency of only 6.57%. Clearly, the Carnot cycle is a more efficient cycle.

We must recall, however, that a Carnot engine is an idealized engine representing the most efficient engine that can be operated between two given heat reservoirs. Practical engines typically achieve efficiencies between 30% and 40%. This means that for every unit of heat that is converted to work, roughly two units of heat must be exhausted as waste heat. This accounts for the fact that nuclear reactor installations dump large quantities of heat either into the atmosphere or into a large body of water.

16-9 THE OTTO CYCLE

A cycle that approximates the operation of a common gasoline engine is the Otto cycle, shown in Fig. 16-13. It consists of two adiabatics bounded by two constant volume (isochoric) processes. Volumes V_a and V_b are given and determine the compression ratio of the cycle. The moles of gas in the system may be computed from V_a, P_a, and T_a using the ideal gas law.

In an actual engine, the heat input to the cycle Q_{bc} is controlled by the quantity of gasoline admitted to the cylinder, and this, in turn, is controlled by the throttle (or accelerator) setting. Therefore, we take Q_{bc} as one of the quantities given in order to specify the cycle. The pressure P_c and temperature T_c are determined by the value selected for Q_{bc}.

Because of the adiabatic compression and expansion, no heat is transferred to or from the system either during the compression stroke AB or the power stroke

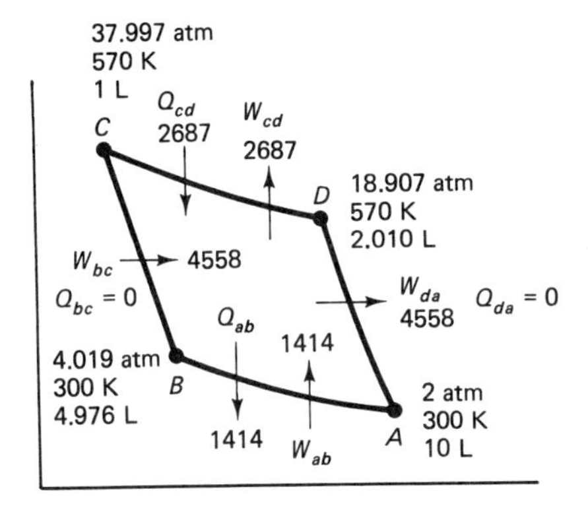

Given: $V_a = 10$ L

$P_a = 2$ atm

$T_a = 300$ K

$T_c = 570$ K

$V_c = 1$L

$\gamma = 1.4$

$$V_b = V_c \left(\frac{T_c}{T_a}\right)^{1/(\gamma - 1)}$$

$$= (1)\left(\frac{570}{300}\right)^{2.5}$$

$$= 4.976 \text{ L}$$

$$V_d = \frac{V_c V_a}{V_b} = 2.010 \text{ L}$$

$T_b = T_a = 300$ K (given)

$T_d = T_c = 570$ K (given)

Process AB (isothermal)

$$P_b = P_a\left(\frac{V_a}{V_b}\right) = 4.019 \text{ atm}$$

$$W_{ab} = P_a V_a \ln\left(\frac{V_a}{V_b}\right) \times 101.3$$

$$= (2)(10)\ln\left(\frac{10}{4.976}\right) \times 101.3$$

$$= 20(0.698)(101.3) = 1414 \text{ J}$$

$$Q_{ab} = W_{ab} = 1414 \text{ J}$$

Process BC (adiabatic)

$$P_c = P_b\left(\frac{V_b}{V_c}\right)^{\gamma} = 4.019\left(\frac{4.976}{1}\right)^{1.4}$$

$$= 37.997 \text{ atm}$$

$$W_{bc} = \frac{P_c V_c - P_b V_b}{\gamma - 1} \times 101.3$$

$$= \frac{38.00 - 20.00}{0.4} \times 101.3$$

$$= 4558 \text{ J}$$

$$Q_{bc} = 0$$

Process CD (isothermal)

$$P_d = P_c\left(\frac{V_c}{V_d}\right) = 18.907 \text{ atm}$$

$$W_{cd} = P_c V_c \ln\left(\frac{V_d}{V_c}\right) \times 101.3$$

$$= 37.997(0.698)(101.3)$$

$$= 2687 \text{ J}$$

$$Q_{cd} = W_{cd} = 2687 \text{ J}$$

Process DA (adiabatic)

$P_a = 2$ atm (given)

$$W_{da} = \frac{P_d V_d - P_a V_a}{\gamma - 1} \times 101.3$$

$$= \frac{38.00 - 20.00}{0.4} \times 101.3$$

$$= 4558 \text{ J}$$

$$Q_{da} = 0$$

Heat input = 2687 J

Work output = $W_{cd} + W_{da}$ = 7245 J

Net work = 1273 J

$$\text{Efficiency} = \left(\frac{\text{net work}}{\text{heat input}}\right) \times 100\% = \left(\frac{1273}{2687}\right) \times 100\%$$

$$= 47.4\%$$

Heat exhausted = 1414 J

Work input = $W_{ab} + W_{bc}$ = 5972 J

Net heat = 1273 J

Figure 16-12 Numerical example of a Carnot cycle.

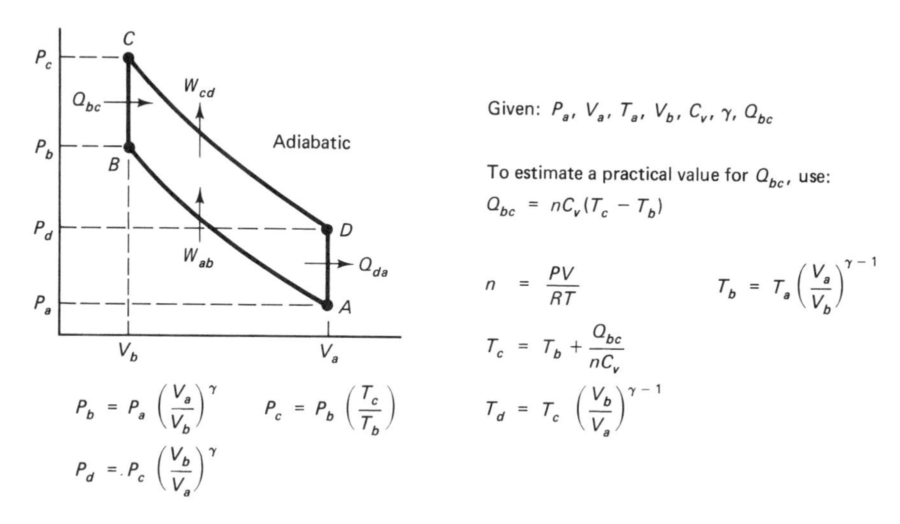

Figure 16-13 The Otto cycle.

CD. Conceptually, the system is placed in contact with the hot reservoir only after completing the compression stroke when the piston has reached the position of minimum volume. Then, following the adiabatic power stroke *CD*, with the piston at maximum volume, the system is placed in contact with the cold reservoir for removal of heat Q_{da} during the isochoric process *DA*.

16-10 THE DIESEL CYCLE

In an ordinary gasoline engine, the fuel and air are mixed in the proper proportion, and then the mixture is compressed adiabatically to high temperature and pressure. At the proper instant, a spark of the spark plug ignites the mixture and initiates the combustion process. In engines with large compression ratios, the temperature produced by the adiabatic compression frequently causes the fuel-air mixture to ignite prematurely, before the spark occurs at the plug. When this *preignition* occurs during the compression stroke, a large pressure is generated in the cylinder which places undue stress on engine components and also causes a reduced power output of the engine. The effect, called *pinging* because of the sound produced when it occurs, can be avoided by using high-octane fuels which contain additives to make the spontaneous ignition less likely.

A diesel engine avoids the problem by *not* mixing the fuel and air until it is time for combustion to occur. The air alone is drawn (or forced) into the cylinder and is compressed adiabatically with a large compression ratio to produce a high temperature. When the piston reaches an optimum point near the end of the compression stroke, the fuel under high pressure is injected into the cylinder where it ignites spontaneously. Because preignition is not a concern, lower octane (and less costly) fuels can be used, and higher compression ratios can be employed.

A thermodynamic cycle closely approximating the operation of a diesel engine is shown in Fig. 16-14. It consists of two adiabatic processes bounded by

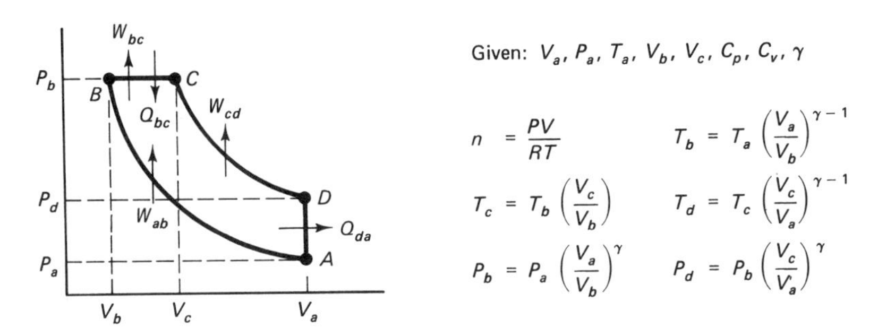

Figure 16-14 Diesel cycle.

an isobaric process at high pressure and an isochoric process at low pressure. The process BC proceeds at very nearly constant pressure because the diesel fuel is slow-burning and also because the fuel injection does not occur instantaneously. At point C, the fuel is expended, and the process CD occurs adiabatically.

Typically, the compression ratio for diesel engines is about 15. Further, the volume V_c at the end of the isobaric expansion is about one-fifth the maximum volume V_a. Hence, the *expansion ratio* V_a/V_c for a diesel engine is typically on the order of 5.

16-11 THE RANKINE AND STIRLING CYCLES

Historically, the significance of the Rankine cycle lies in the fact that it closely approximates the operation of a condensing type of reciprocating steam engine. The Rankine cycle is shown in Fig. 16-15. Water is used as the working substance in an actual steam engine, and a meaningful analysis quickly becomes involved with the thermodynamic properties of water and steam.

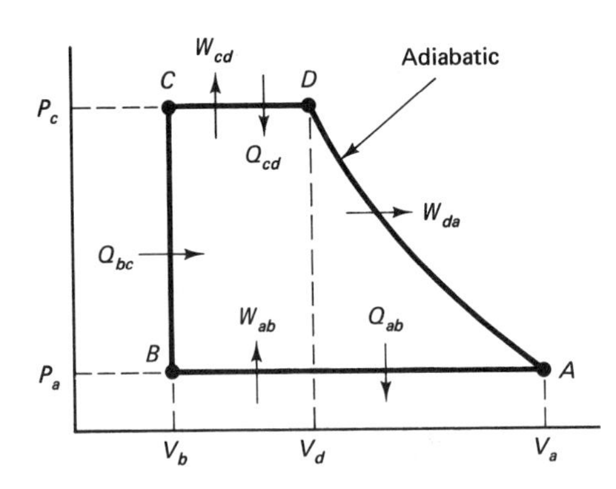

Figure 16-15 Rankine cycle.

The last cycle to be presented is the Stirling cycle, shown in Fig. 16-16. It is similar to the Otto cycle, the difference being that isothermal processes are used instead of adiabatic. Analysis of the cycle is straightforward.

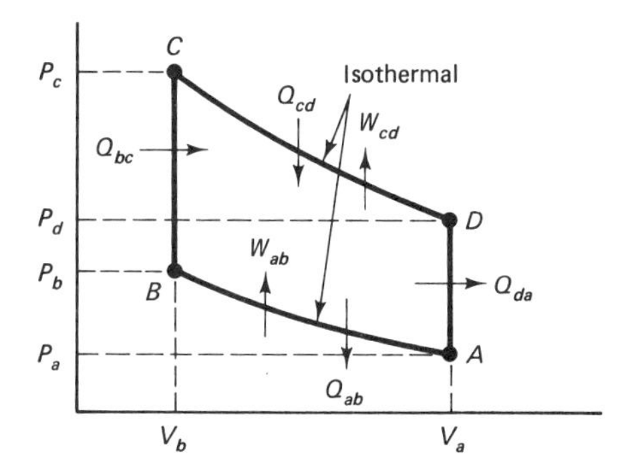

Figure 16-16 Stirling cycle.

16-12 THE FIRST LAW OF THERMODYNAMICS

We now state explicitly the first law of thermodynamics, which amounts to a generalization of the law of the conservation of energy to include heat. The first law applies to any system, but we may limit ourselves to considering a cylinder and piston enclosing a volume of ideal gas in order to keep our discussion as definite as possible.

First of all, the *internal energy* of an ideal gas is directly related to its absolute temperature. Any change in the internal energy produces a change in temperature, and if the temperature changes, we may say that the internal energy has changed. This is easily understood when we recall that the internal energy of an ideal gas is the sum total of the molecular kinetic energies, and the absolute temperature is a measure of the average kinetic energy per molecule.

In the preceding sections, we saw that a gas may do work on a piston by expanding against the piston. But if the piston moves to compress the gas, the piston does work on the gas. Equivalently, we may say the gas does a *negative* amount of work when it is compressed.

With the ideas of work W and internal energy (denoted by U) firmly in mind, we state the first law of thermodynamics:

$$\begin{array}{ccccc} \text{heat added to} & = & \text{change in internal} & + & \text{work done} \\ \text{a system} & & \text{energy of the system} & & \text{by the system} \\ Q & = & \Delta U & + & W \end{array}$$

In an isothermal process, the temperature does not change, and therefore the internal energy does not change. Consequently, $\Delta U = 0$, and the first law yields $Q = W$. As an example of this relationship, refer to the Carnot cycle of Fig. 16-12 and note that for the isothermal processes, the heat transfer equals the work done. Furthermore, observe that when work is done *on* the system, heat is removed *from* the system.

In an isochoric process (constant volume), no work is done because the gas neither expands nor contracts. Therefore, $W = 0$, and the first law reduces to $Q = \Delta U$. In this case, the heat transfer is numerically equivalent to the change in internal energy of the gas.

In an adiabatic process, no heat transfer occurs and $Q = 0$. Hence, $\Delta U = -W$ is obtained from the first law. The negative sign indicates that when the system does work, its internal energy decreases.

Our initial reference to the first law is in Example 16-2 in regard to the heating of a gas at constant pressure. It would be well to review that example with the first law in mind. Moreover, because all thermodynamic processes must obey the first law, each segment of each of the thermodynamic cycles presented earlier may be considered as an example of the application of the first law.

16-13 THE SECOND LAW; ENTROPY

All possible thermodynamic processes and physical events must take place in accordance with the first law of thermodynamics, the energy conservation law. However, there are many physical processes meeting the requirements of the first law that do not occur. Consider the following events.

A drop of ink dripped into a glass of water soon diffuses throughout the entire volume of water. A cube of ice placed in a glass of warm water melts and lowers the temperature of the water. An archer, of a culinary sort, shoots an arrow into a large ball of soft dough. These events are familiar, but the events are never observed to occur in the reverse direction. A glass of inky water never spontaneously separates to the point where the ink is again concentrated into a drop. A glass of cool water never spontaneously grows warmer with the simultaneous formation of a cube of ice at the top. And it is hard to imagine that an arrow would ever come flying backwards out of a lump of dough, thereby causing the dough to become slightly cooler. It is only in the movies, where the film is run backwards, that these processes occur, even though none violates the principle of the conservation of energy. The second law of thermodynamics deals with the *direction* in which thermodynamic processes such as those mentioned are likely to proceed.

Strictly speaking, we should not say that an arrow will *never* be ejected spontaneously from the target into which it has been fired. Specialists in a branch of physics called *statistical mechanics* can compute the probability that such an event will occur, and the probability is *not* zero. However, the odds are so greatly opposed to such an event that for all practical purposes we can say that it will *never* happen.

By introducing the concepts of order and disorder, we can establish a common ground for analyzing a wide variety of physical events. *Order* implies a lack of randomness, an organized arrangement. *Disorder* is the opposite of order. For example, two containers of water, one hot and the other cold, represent a more orderly situation than one larger container filled with an equivalent volume of warm water. Initially, we have a segregation of hot and cold molecules, and this segregation represents order. In a similar manner, a 2-L box containing a dry mixture of sugar and salt is more disorderly than two 1-L boxes with one containing sugar and the other salt. In general, a mixing process produces greater disorder. If equal quantities of an ideal gas (at the same temperature) are contained in unequal volumes, the gas enclosed in the smaller volume is more orderly because there is less uncertainty in the position of any molecule in the smaller volume.

The disorder of a system is measured by a thermodynamic property called *entropy* which increases as the system becomes more disordered. Therefore, an increase in entropy occurs when hot and cold water are mixed, when ink diffuses throughout a volume of water, and when an arrow strikes a target. Another property of entropy is illustrated in the following.

Suppose two large tanks of water are available, one boiling at 212°F while the other is an ice-cold 32°F. In principle, it is possible to build a heat engine to utilize the two tanks of water as hot and cold reservoirs. But if the water of the tanks is mixed so that both tanks come to be at the same temperature, the operation of the heat engine becomes impossible because there is no difference in temperature between the hot and cold reservoirs.

Mixing the hot and cold water causes a certain quantity of energy to become unavailable for doing work even though the total energy of the two tanks remains the same. The entropy increase that occurs during the mixing process provides a measure of the energy that becomes unavailable. When a quantity of energy appears in a form such that the energy can no longer be used, the energy is said to be *degraded*. Hence, an entropy increase represents a *degradation of energy*.

The second law

The essence of the second law of thermodynamics is embodied in the statement that physical processes tend to occur in the direction that increases the disorder of the system. Equivalently, the direction taken by any thermodynamic process will be such that entropy increases as a result of the process.

Discussions of entropy changes typically involve either or both of two regions of concern: (1) the local system, and (2) the universe. A local system might consist of two containers of water and the interior of a physics laboratory. On the other hand, the *universe* refers to the entirety of the physical universe, consisting of the solar system and untold billions of stars. This distinction becomes important to the understanding of the second law when it is realized that physical processes may reduce the entropy of a local system. For example, nothing prevents the use of a heater for heating one-half a portion of water to a high temperature while a refrigerator cools the remaining portion to a low temperature. The entropy of the local system will be reduced by the action of the stove and refrigerator. However, a very large increase in entropy might have occurred at a point far away where a lump of coal was burned in order to produce electrical power for the heater and refrigerator.

When *all* entropy changes are taken into account in order to compute the entropy change of the universe, the second law says that for all real processes, the change in entropy of the universe will be positive. Consequently, the entropy of the universe is forever increasing. Any localized decrease in entropy will be offset by a larger *increase* at a distant point so that an increase in the entropy of the universe is always the result of any real physical process.

More complete treatments of thermodynamics deal extensively with *reversible* processes, some of which may occur without producing a change in entropy. Such processes are idealizations, however, that ignore frictional effects and other dissipative phenomena. A reversible adiabatic expansion is an example of such an *isentropic* (constant entropy) process. All real processes are irreversible and produce an increase in the entropy of the universe.

The second law has been stated in many forms, all of which address the same physical principle. Here are three possible statements:

Heat will not flow from a cool body to a warmer body unless mechanical work is done to make it do so.

It is impossible to devise a heat engine that will take a given quantity of heat from a hot reservoir and convert it entirely to mechanical work without exhausting a portion of the heat to a cold reservoir.

The occurrence of any natural process produces an increase in the entropy of the universe.

The first of these is in accord with the knowledge that any refrigerator requires a source of power in order to make heat flow "the wrong way." The natural tendency is for heat to flow from regions of high temperature to regions of lower temperature.

If the second statement were not true, it would be possible to construct an oceangoing vessel that would derive its motive power from the heat of the ocean itself. But we have seen that any heat engine requires a cold reservoir, in addition to a hot reservoir. Therefore, the construction of such a vessel is impossible.

The third statement above implies that any natural process causes a certain quantity of energy to become unavailable for further use. If we assume that the total energy contained in the universe is very large but yet finite, it follows that any physical process "uses up" some of the energy that is still available for doing work. We then may foretell a time when all the available energy of the universe will have been degraded. At that time, all physical processes will cease, time will no longer have meaning, all forms of life will have long since disappeared, and the universe will have attained a state of complete uniformity. This so-called heat death of the universe lies countless billions of years in the future, however.

16-14 REFRIGERATION

An examination of each process of a Carnot cycle (as one example) reveals that the cycle may be operated in reverse. This means, in Fig. 16-17, that the cycle would proceed from point A to D to C to B and then back to point A in a counterclockwise traversal of the P-V diagram. As a consequence of the reversal, the Carnot engine (the mechanism executing the Carnot cycle) acts as a *refrigerator* rather than as an engine capable of delivering power to something else. Moreover, a source of external power is required in order to operate the cycle in reverse; the engine must be *driven* in reverse. The net effect of the operation is

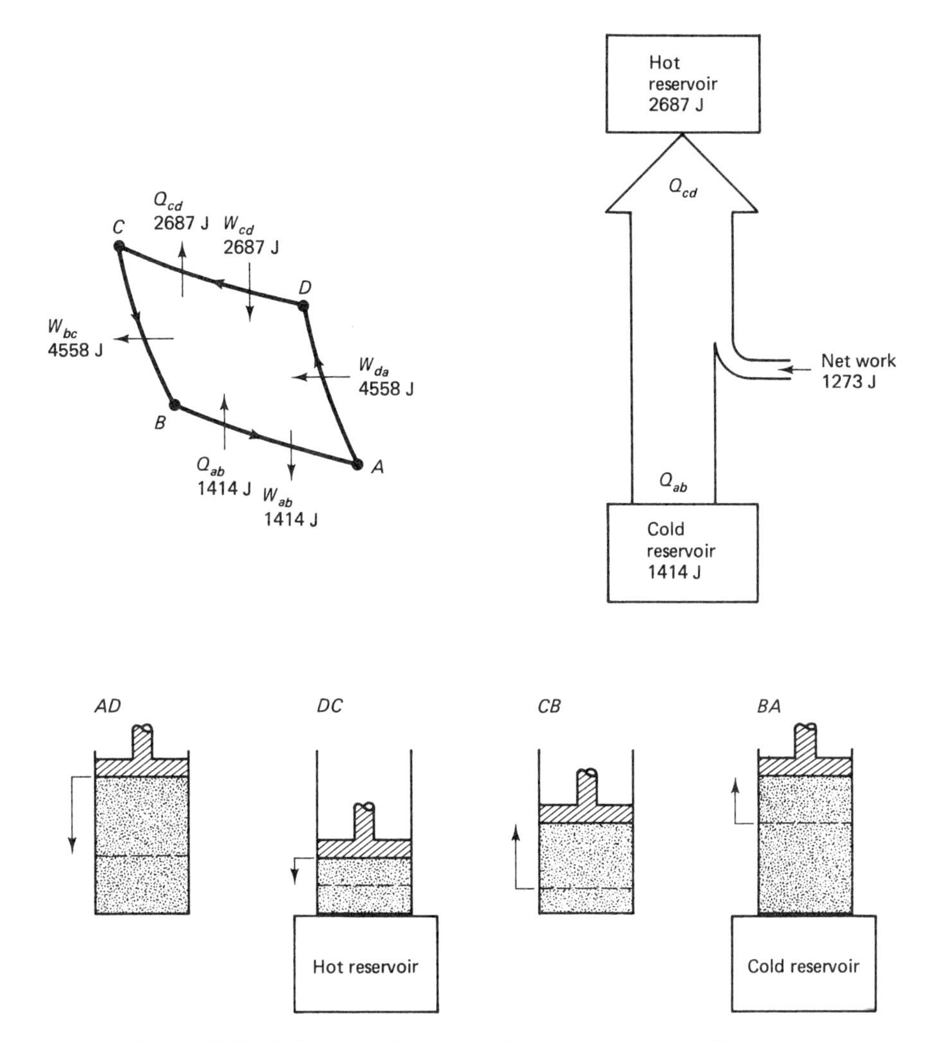

Figure 16-17 A Carnot cycle operated in reverse acts as a refrigerator.

that mechanical work done on the system results in a transfer of heat from the cold to the hot reservoir. Thus, the cold reservoir becomes cooler and the hot reservoir becomes warmer. Indeed, the cycle can be initiated with the hot and cold reservoirs even at the same temperature so that hot and cold reservoirs are actually generated. This is the action of a refrigerator.

To provide a more detailed description of a Carnot cycle operated backwards, consider Fig. 16-17, where we begin at point A and initiate an adiabatic compression until we reach point D. During this process, the temperature of the gas in the cylinder rises to the high temperature T_c of the cycle. Work is being done on the system, so that W_{da} now represents an input of energy. At point D, the system is placed in contact with the hot reservoir, but now the temperature is (must be) greater than that of the hot reservoir as the *high-pressure compression* continues isothermally to point C.

At point C, the hot reservoir is disconnected and an adiabatic expansion carries the system to point B. During this process, the system temperature falls to a temperature lower than that of the cold reservoir. When the cold reservoir is connected at point B, heat flows *from* the cold reservoir to the system as the low-pressure expansion progresses from point B to point A. Quantities of work W_{bc} and W_{ab} represent work done by the system, and heat Q_{ab} is the heat input to the system from the cold reservoir. This completes the cycle.

The objective of running a heat engine in reverse is to achieve refrigeration, the extraction of heat from the cold reservoir. Furthermore, the expense of operating a system in reverse lies in the net work that must be done on the system to carry it around the cycle. The best refrigerator is the one that extracts the most heat from the cold reservoir for a given net amount of mechanical work input. This is expressed by the *coefficient of performance* of a refrigerator, defined as

$$\begin{matrix}\text{coefficient of}\\ \text{performance}\end{matrix} = \frac{\text{heat extracted from cold reservoir}}{\text{net mechanical work input}} \tag{16-24}$$

To illustrate this, we refer to the numerical example of Fig. 16-12, but this time we assume the cycle is operated in reverse. This necessitates a change in labeling because what was work input now becomes work output, and so forth. For convenience, here are the data with new labels, taken from Fig. 16-12:

$$\begin{aligned}\text{Heat exhausted to hot reservoir} &= 2687\text{ J}\\ \text{Heat extracted from cold reservoir} &= 1414\text{ J}\\ \text{Work input during high-pressure compression} &= 7245\text{ J}\\ \text{Work output during low-pressure expansion} &= 5972\text{ J}\\ \text{Net mechanical work input} &= 7245 - 5972 = 1273\text{ J}\end{aligned}$$

We now observe that the heat extracted from the cold reservoir is greater than the net mechanical work input. Therefore, the coefficient of performance (COP) is greater than 1:

$$\text{COP} = \frac{1414}{1273} = 1.11$$

At the same time, the heat exhausted to the hot reservoir is the sum of the net work input and the heat extracted from the cold reservoir. This cycle is illustrated in Fig. 16-17.

Earlier we stated that a Carnot cycle gives rise to the most efficient heat engine that can be realized when operated between reservoirs at given temperatures. Hence, a Carnot refrigerator represents an *ideal* refrigerator in that it produces the theoretical maximum COP. All practical refrigerators fall short of the ideal because of frictional effects and other unavoidable inefficiencies, but the concept of an ideal refrigerator is still quite useful. If heat is to be removed from a cold reservoir at temperature T_{cold} and is to be exhausted to a hot reservoir at temperature T_{hot}, the COP of an ideal refrigerator can be shown to be

$$\text{COP}_{\text{ideal}} = \frac{T_{\text{cold}}}{T_{\text{hot}} - T_{\text{cold}}} \tag{16-25}$$

where the temperatures must be absolute. From this we discern that as T_{hot} increases or as T_{cold} decreases, the coefficient of performance goes down.

If the COP of a given refrigerator is known, the energy required to extract a certain quantity of heat from a cold reservoir can be computed. From Eq. (16-24), the relationship is

$$\text{Energy required} = \frac{\text{heat extracted}}{\text{COP}} \tag{16-26}$$

Thus it is possible to compute the cost of air conditioning and refrigeration of various types.

A practical refrigerator

Figure 16-18 is a diagram of a refrigeration system similar to the type used in home refrigerators. It clearly differs from the simple heat engines we have discussed, but many principles are the same. We can identify the hot and cold reservoirs and

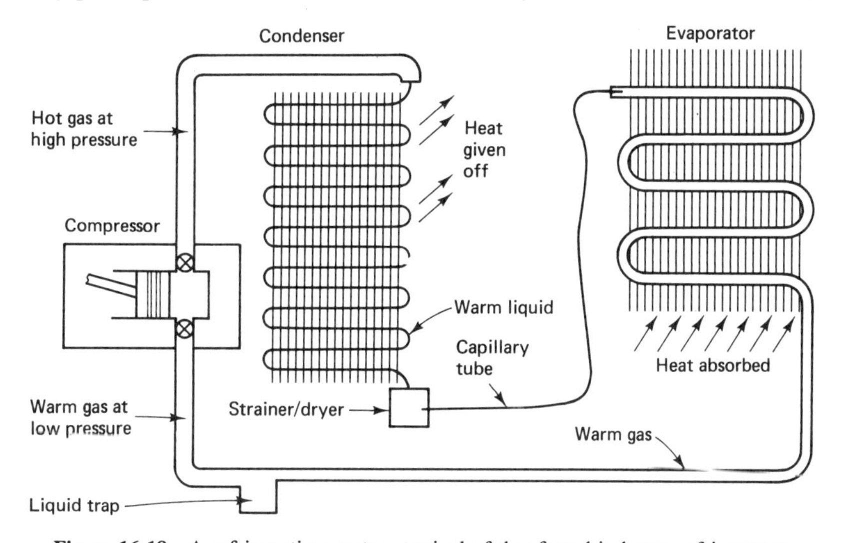

Figure 16-18 A refrigeration system typical of that found in home refrigerators.

the function of the compressor can be anticipated, but the use of a *refrigerant* that is alternately liquefied and vaporized has not been previously described. In broad terms, any gas that can be readily liquefied and vaporized can serve as a refrigerant, but common practice is to use various formulations of Freon gas, such as R12 and R22.

Recall that a liquid must absorb its heat of vaporization in order to evaporate to the gaseous (or vapor) state, and a vapor must give up an equal amount of heat in order to condense to a liquid. In a home refrigerator, the refrigerant is caused to evaporate and absorb heat in coils surrounding the cooling compartment, and to give off the heat and condense in coils that usually are located on the back of the cabinet. Thus, heat is removed from the cooling compartment (the cold reservoir) and is delivered to the room air (the hot reservoir).

The refrigerant is circulated continually by the compressor which creates the pressure differential required by the system. The refrigerant in the form of a hot gas under high pressure comes from the compressor and passes to the condensing coils on the back of the refrigerator. Here the hot gas cools and condenses to a liquid, giving up the heat of vaporization as it does. The liquid, after further cooling by the condenser coils, then passes through a strainer/dryer as it approaches the capillary tube. The strainer/dryer prevents any foreign matter from reaching and possibly obstructing the capillary, and it also absorbs any small amount of water that might be in the system as a contaminant.

A capillary tube is a small tube several feet long with a very small inside diameter. Its purpose is to limit severely the rate at which refrigerant flows from the condenser to the evaporator, thereby causing a difference in pressure to exist between the two. The size and length of the capillary controls the pressure difference.

The refrigerant enters the low-pressure region of the evaporator as a cool liquid. Evaporation of the liquid occurs and is enhanced by the low pressure. Heat is absorbed from the evaporator coils which, in turn, absorb heat from the contents of the freezer compartment. The low pressure in the evaporator is maintained by the "suction" generated at the input of the compressor. Prior to again passing through the compressor, the refrigerant passes through a liquid trap to ensure that no refrigerant in liquid form is admitted to the compressor.

Reviewing the cycle with an eye toward fundamental processes, we see that the compressor does mechanical work in compressing the gas to a high pressure and to a temperature higher than that of the air surrounding the condenser coils. The high pressure also makes the refrigerant condense more readily. A low-pressure expansion occurs in the evaporator in conjunction with evaporation. Thus, we have a high-pressure compression followed by a low-pressure expansion. This much is reminiscent of the reversed Carnot cycle, but the actual cycle of the practical refrigerator differs greatly from the idealized Carnot cycle.

Heat pump

Any refrigeration system may be called a *heat pump* because it forces heat to move from a cold to a hot region. However, the term is customarily used in reference to a type of residential heating and air-conditioning unit that uses a refrigeration system to move heat from inside the home to the outside during the summer, and to move heat from outside to inside during the winter. A single unit is able to perform both functions by virtue of a system of valves that effectively interchanges the evaporator and condenser.

In summer, the outside heat exchanger serves as the condenser and the inside heat exchanger serves as the evaporator. Consequently, heat is removed from the interior of the home, and the unit serves as an air conditioner. In winter, the outside exchanger serves as the evaporator while the inside exchanger acts as condenser. In this mode, heat is extracted from the outside air and is transferred to the interior of the home.

The advantage of a heat pump for winter heating stems from the fact that the coefficient of performance is greater than unity. If a certain quantity of electrical energy is expended in driving the compressor of the refrigeration system, an amount of heat equal to COP times that quantity is delivered to the interior of the home. A further advantage is that the same unit serves both functions of air conditioning and heating.

The primary disadvantage of a heat pump is in conjunction with winter heating. As the outside temperature falls to low levels, the COP decreases, and the heat-pumping capability of the system decreases. Thus, when heat is needed most, less heat is delivered per unit energy expended, and the rate at which the system can deliver heat is also diminished. For this reason, most heat pump installations are equipped with electrical backup heaters that switch on automatically when the situation warrants.

16-15 PRACTICAL ENGINES

Ordinary reciprocating engines are divided into two broad categories depending upon the number of piston strokes (two or four) required for the completion of one engine cycle. Each category is further divided between the familiar spark-ignited engines and the compression-ignited, diesel engines.

A *stroke* is the movement of a piston from one extreme position in the cylinder to the other, such as from the highest to the lowest point. Each stroke requires the crankshaft to turn through one-half revolution. Consequently, a four-stroke engine takes two revolutions of the crankshaft to complete one engine cycle. A two-stroke engine completes one cycle for each revolution of the crankshaft.

Engines with four strokes per cycle are commonly called *four-cycle* engines, and engines with two strokes per cycle are called *two-cycle* engines. This terminology seems to omit the *strokes per* portion of the designation *four (or two) strokes per cycle*, thereby confusing the distinction between the stroke and cycle. It is of no consequence, however, because the meaning of the shortened version is well known.

Figure 16-19 illustrates the four strokes of a four-cycle engine. The first

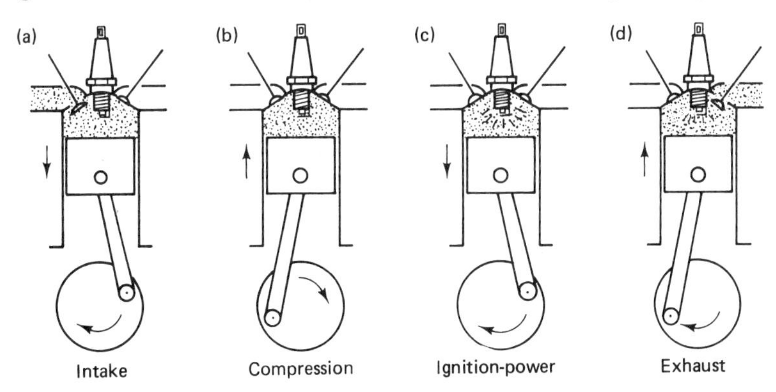

Figure 16-19 The four strokes of a four-cycle engine. One power stroke occurs for every two revolutions of the crankshaft.

stroke (intake) draws the fuel-air mixture into the cylinder. The second stroke (compression) compresses the mixture, with ignition occurring at the end of the stroke. The hot, expanding gases cause the third stroke to be the power stroke, which is then followed by the exhaust stroke. This completes the cycle. The intake valve and the exhaust valve are operated by a camshaft which rotates at one-half the crankshaft speed.

One version of a two-stroke cycle is illustrated in Fig. 16-20. In part (a) of the

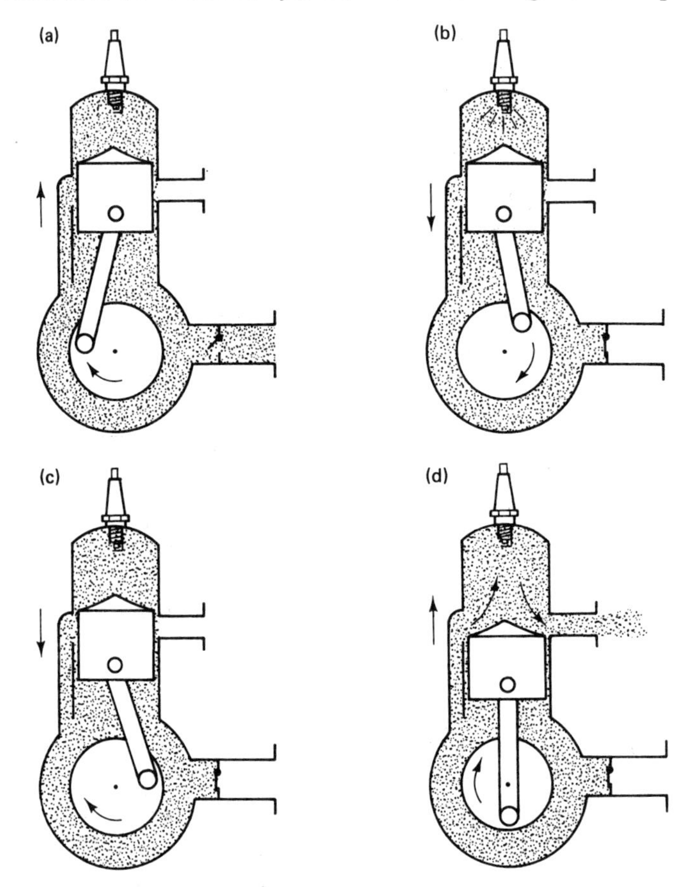

Figure 16-20 A two-stroke engine cycle.
(a) Compression and intake of new charge of fuel-air mixture through reed valve into crankcase.
(b) Ignition and beginning of power stroke.
(c) Near the end of power stroke; charge in crankcase is compressed slightly.
(d) Combination intake and exhaust; a new charge of fuel enters the cylinder as the burned fuel is exhausted.

figure, the piston is compressing the fuel-air mixture above the piston and is simultaneously drawing a charge of fuel and air into the crankcase through the reed valve. The beginning of the power stroke is shown in part (b); the spark plug has fired, the reed valve has closed, and the piston is being pushed downward. In part (c), the power stroke is nearly complete. The new charge in the crankcase is now under moderate pressure, and the intake and exhaust ports in the wall of the cylinder are about to be opened by the descending piston. In part (d), the ports are open. The combustion products leave the cylinder through the exhaust port as new fuel-air mixture enters via the intake port. The shape of the top surface of the piston minimizes undesirable mixing of the exhaust and intake gases. This process is followed by another combination intake-compression stroke, as shown in part (a), which represents the beginning of a new cycle.

A two-stroke cycle is completed for each turn of the crankshaft, and each turn of the crankshaft involves a power stroke. Recall that the four-stroke cycle requires two turns of the crankshaft for each power stroke delivered. Hence, the power output per cylinder of a two-cycle engine is greater than that of an equivalent four-cycle engine, but this greater output power is obtained at the expense of reduced fuel efficiency and increased production of pollutants.

Two-cycle engines do not require cam-operated valves and therefore may be much simpler mechanically than four-cycle engines. However, larger, more sophisticated two-cycle engines may incorporate air pumps or compressors to force the intake air into the cylinder, and *scavenge pumps* may be employed to suck the exhaust gases from the cylinder.

Both four-cycle and two-cycle engines may be designed to operate as diesel engines. In such case, the spark plug ignition system is replaced by a fuel-injection system, and the compression ratio is increased from around 10 to about 15. The large compression ratio causes the air within the cylinder to reach high temperatures so that the fuel ignites spontaneously when injected into the cylinder.

The smallest reciprocating engines probably are the tiny, fractional-horsepower, two-cycle engines used to power model airplanes. A glow plug is used to augment compression ignition so that these tiny engines require no permanent electrical system. The glow plug contains a small electrical heating element that helps to elevate the temperature in the cylinder to the combustion temperature of the fuel. Once the engine is started, the heat produced by combustion maintains the temperature of the heating element of the glow plug, and electrical power may be disconnected.

On the other hand, the larger aircraft engines may have 14, 18, or as many as 28 cylinders, and the larger of these engines may develop over 3000 hp at takeoff. Engines for marine or stationary applications may be even larger, and a single unit may produce more than 10,000 hp. The typical piston diameter of a large engine may be on the order of 13 in. with a stroke length of about 16 in. Obviously, these large engines must be designed to operate at low RPM, typically about 500.

Wankel rotary engine

The operation of a Wankel rotary engine is illustrated in Fig. 16-21. A triangular rotor replaces the pistons of the reciprocating engines and avoids the mechanical complexity of such features as the crankshaft, connecting rods, or camshafts. An inside gear on the rotor turns the central, power-output shaft. The motion of the rotor and the attendant meshing of the gears is hard to visualize, and diagrams are of little help. Only observation of a model or an actual engine allows a full

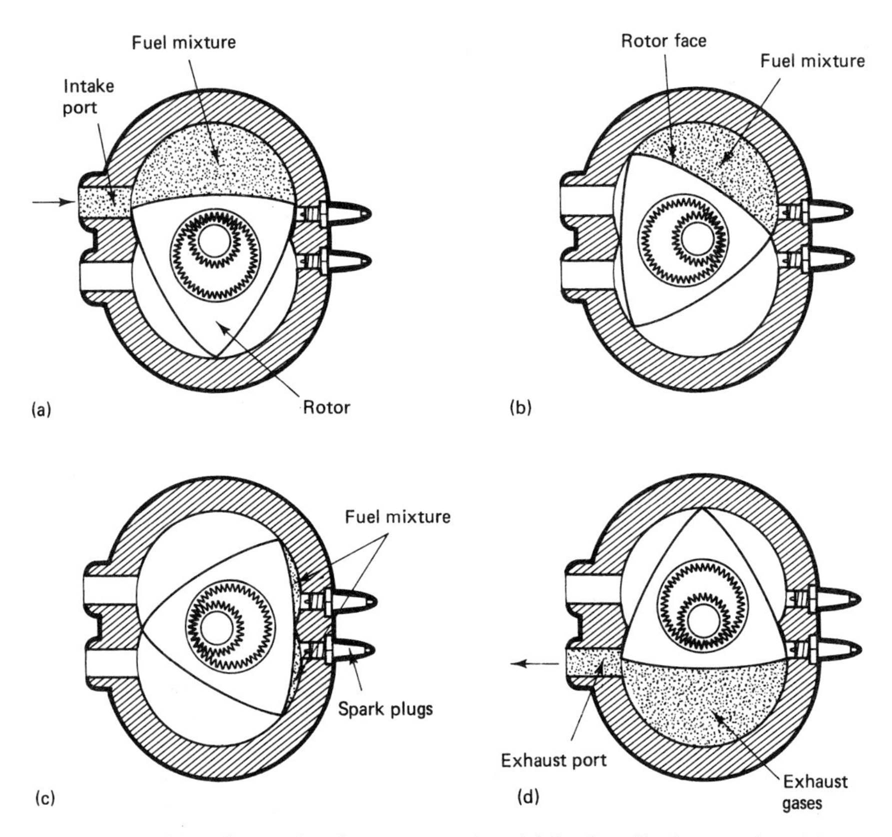

Figure 16-21 One cycle of a rotary engine. (a) Intake. (b) Compression. (c) Ignition. (d) Exhaust. Similar processes occur simultaneously in the other two chambers.

appreciation of this clever mechanism developed by the German mathematician Felix Wankel in 1956.

Advantages of a rotary engine include simplicity of design and a smooth operation due to the absence of reciprocating parts. Rotary engines are lighter and smaller for a given horsepower than a reciprocating engine, and effective power is produced over a wide range of engine speeds. A major design consideration is the seals at each apex and on the sides of the rotor, and progress continues to be made in this initially troublesome aspect. Because rotary engines have been introduced into the family car, texts on auto mechanics now describe this rather unique engine in detail.

Gas turbines

Another broad class of internal combustion engines is the *gas turbine*. These engines appear in many forms and in many sizes and are used to power a variety of vehicles including aircraft, tanks, and marine vessels. Although gas turbines have been adapted to automobiles, the strong competition from reciprocating engines and the Wankel rotary has thus far precluded the mass production of a turbine-powered passenger car. Power outputs of gas turbines range from 100–200 hp for engines suitable for light vehicles to more than 18,000 hp for engines intended for stationary or marine applications. Eventually, thermal efficiencies of gas turbine engines are expected to approach 40%, whereas conventional engines under optimum conditions achieve efficiencies on the order of 26%.

One form of a gas turbine engine is shown in Fig. 16-22. No reciprocating parts are present, and the engine operates as a continuous-flow device. The intake

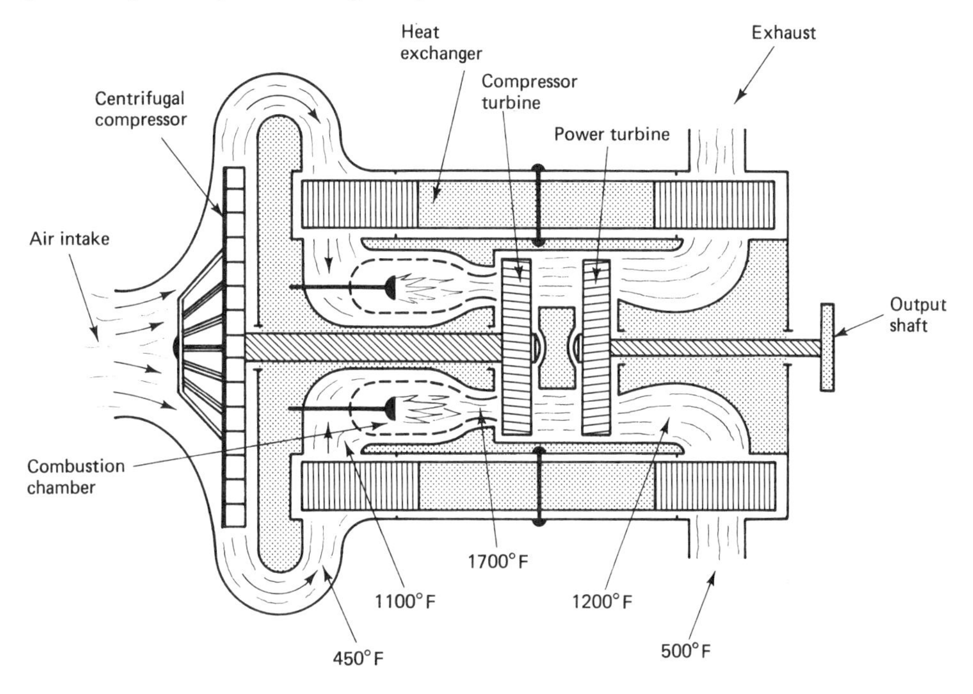

Figure 16-22 A gas turbine engine.

air is compressed by a centrifugal compressor whose impeller may turn at 22,000 RPM or higher. The air, having been heated by the compressor to about 450°F, then passes through a regenerator-type heat exchanger where heat derived from the exhaust gases raises the air temperature to about 1100°F. Fuel and air are mixed in the combustion chamber where combustion occurs as a continuous process. The hot, expanding gases leave the combustion chamber at about 1700°F and impinge upon the turbine that supplies power to the centrifugal compressor. Adjacent to the compressor turbine is the power turbine that supplies power to the output shaft. After passing through the power turbine, the gases at about 1200°F are ducted through the heat exchanger and are then exhausted at about 500°F.

A noteworthy feature of the gas turbine of Fig. 16-22 is the regenerator heat exchanger. The principle of operation is shown in Fig. 16-23. The central element is a rotating drum of open construction so that gases flow freely through the structure. The portion of the drum heated by the hot gases is continuously rotated into the stream of cool gases, causing heat to be transferred from the hot to the cool gases. Gas turbines that utilize heat from the exhaust to preheat the

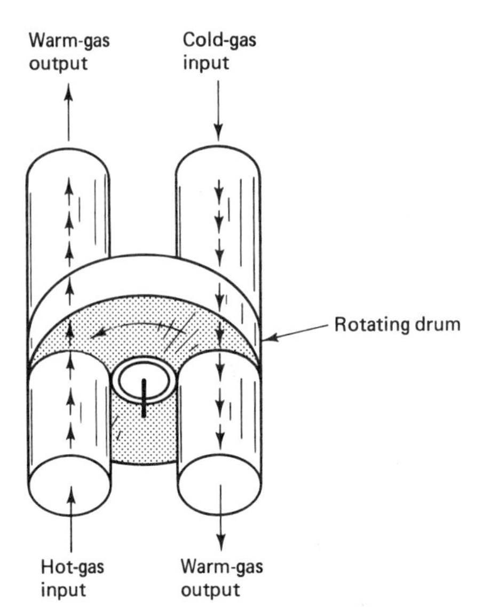

Figure 16-23 A regenerative-type heat exchanger.

combustion air are called *regenerative* gas turbines. Many gas turbines are *nonregenerative*.

Aircraft engines known as *turboprop* engines are gas turbines in which the output shaft is brought to the front of the engine to a speed-reduction gearbox that drives the propeller.

Turbojet engine

One form of a turbojet engine is shown in Fig. 16-24. A jet engine derives its forward thrust by ejecting high-speed exhaust gases to the rear, a practical utilization of Newton's action-reaction principle. The thrust is the force in the forward direction produced by the engine and is typically given in pounds. A typical turbojet may weigh 7000 lb and produce 20,000 lb of thrust. Propeller-driven aircraft are more efficient than jets at low speed (200 mi/h) and low

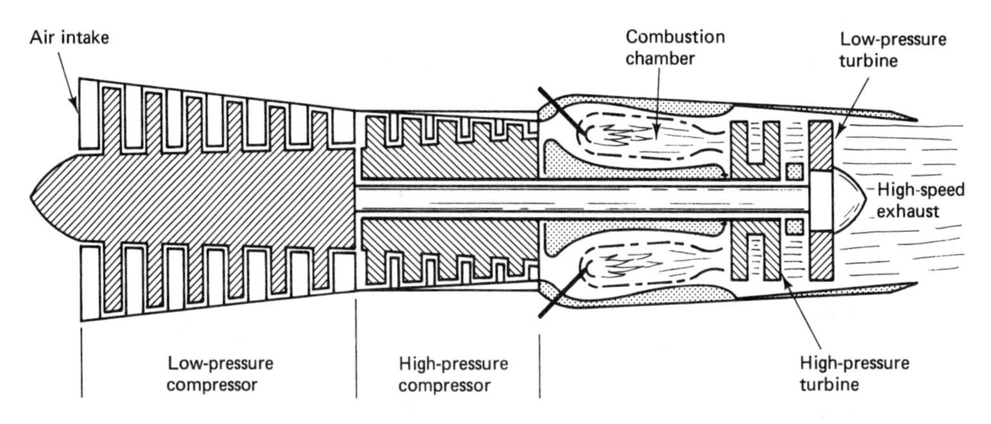

Figure 16-24 A turbojet engine.

A cut-away view of the two-stage compressor of a jet engine that is on display at the Smithsonian Institution in Washington, D.C.

altitudes, but for high altitudes and speeds above 500 mi/h, jet engines are far more efficient.

The engine of Fig. 16-24 utilizes two separate compressors to compress the intake air prior to mixing it with the fuel in the combustion chamber. The combustion occurs as a continuous process. Power to drive the compressors is derived from turbine wheels situated in the high-speed exhaust.

Questions

1. A gas expands and does work on an external system. Does the temperature of the gas increase or decrease?
2. Why are there two molar heat capacities? Which is greater, c_p or c_v? Why?
3. What is the definition of γ? What is its value for monatomic gases? For diatomic gases? For air?
4. What is the defining condition for an adiabatic expansion?
5. What is the defining condition for an isothermal expansion?
6. What is an isochoric process? Why is there never any work done during an isochoric process?
7. Describe an isobaric expansion.
8. Describe the simplest heat engine you can imagine.
9. Describe the Carnot cycle. Sketch a Carnot cycle on a P-V diagram.
10. What determines the efficiency of a Carnot cycle?
11. With what is entropy and the second law of thermodynamics concerned?
12. Describe how a Carnot cycle can be run backwards to form a refrigerator.
13. Describe the overall design of a heat pump. Why are heat pumps least effective at low temperatures: What auxiliary apparatus is customarily provided?
14. Briefly describe the operation of a gas turbine engine.
15. Describe and explain the operation of a turbojet engine.

Problems

1. (a) Perform a unit conversion to show that as a unit of energy; 1 L · atm is the equivalent of 101.3 J.
   (b) What number of calories is the equivalent of 1 L · atm?
2. A cyclinder 8 cm in diameter equipped with a freely movable piston contains a gas at a pressure of 6 atm.
   (a) What force is exerted on the piston?
   (b) Assuming the pressure to remain constant, compute the work done on the piston if it moves a distance of 0.3 cm.
   (c) If the expansion occurs adiabatically, would you expect the temperature of the gas to increase, decrease, or not change at all?
3. How much work is done when the piston of a hydraulic cylinder 6 cm in diameter moves a distance of 30 cm as fluid is pumped into the cylinder under a pressure of 6 atm?
4. How much heat must be added to 3 g · mol of carbon dioxide (CO_2) in order to raise its temperature by 10°C if the volume of the containing vessel is assumed not to change?
5. Compute the heat required to raise the temperature of 3 mol of hydrogen (H_2) from 0°C to 100°C: (a) at constant volume; (b) at constant pressure.

* 6. Work Example 16-2 assuming the gas is argon instead of nitrogen. Compare results.

7. One liter of oxygen (O_2) in an insulated cylinder at a temperature of 300 K and a pressure of 3 atm is suddenly compressed to one-half its original volume. Compute: (a) the final temperature of the gas; (b) the final pressure; (c) the work done in compressing the gas.
8. Repeat Problem 7 above for hydrogen (H_2) instead of oxygen.

* 9. Work Example 16-3 for argon instead of nitrogen and compare the results.

10. A 2-L cylinder contains helium at a pressure of 4 atm and a temperature of 293 K.
    (a) How much heat must be added to the gas to increase the pressure to 6 atm?
    (b) What will be the temperature of the gas when the pressure is 6 atm?
11. Repeat Problem 10 using hydrogen (H_2) instead of helium.
12. Suppose 4 L of carbon dioxide (CO_2) at a pressure of 10 atm and a temperature of 373 K is allowed to expand to double the volume, 8 L. Compute the final pressure: (a) if the expansion is isothermal; (b) if the expansion is adiabatic. (See Fig. 16-5.)
13. Repeat Problem 12 for carbon monoxide (CO) and note the differences.

**14. Repeat the analysis of the basic cyclic process of Fig. 16-9 with the following changes, keeping the other given quantities the same.
   (a) Reduce the starting temperature at T_a from 600 K to 500 K.
   (b) With T_a = 600 K, reduce P_c from 9 atm to 7 atm.

   For each calculation, note the effect of the change upon the efficiency of the cycle and upon the highest temperature attained in the cycle.

15. Compute the efficiency of a Carnot engine that operates between reservoirs at 0°C and 100°C.

**16. For the Carnot cycle of Fig. 16-12, investigate the effect upon efficiency of changing the compression ratio V_a/V_c. Perform an analysis as in the figure, but: (a) change V_c to 1.25 L; (b) change V_c to 0.8 L. Note the effect of these changes upon the other parameters of the cycle.

**17. For the Carnot cycle of Fig. 16-12, change the maximum cycle temperature to 700 K and see how this change affects the overall cycle.

18. Suppose an upper layer of ocean water is at a temperature of 23°C, while a lower layer is at 5°C. Compute the efficiency of a Carnot engine constructed to extract mechanical energy from this temperature difference.

**19. The Carnot cycle analysis of Fig. 16-12 pertains to a working substance with $\gamma = 1.40$. Repeat the analysis for a working substance with $\gamma = 1.67$.

**20. Perform an analysis of the Otto cycle of Fig. 16-13 using the following: $P_a = 1$ atm; $V_a = 1$ L; $T_a = 300$ K; $V_b = 0.2$ L;

$c_v = 20.76$ J/mol-K; $\gamma = 1.4$; $Q_{bc} = 350$ J. First calculate n using the ideal gas law and the data given for point A (answer: $n = 0.0406$ mol). Next, determine T and P for points B, C, and D. Then do the computations for each segment. Finally, do a heat balance and calculate the efficiency (47.5%). Compare with the Carnot efficiency.

**21. Repeat the analysis of the Otto cycle with the data given in Problem 20, but take Q_{bc} to be 400 J instead of 350 J. What happens to the efficiency and to the maximum temperature of the cycle?

**22. Perform an analysis of the Diesel cycle of Fig. 16-14 using the following data: $V_a = 1$ L; $V_b = 0.0667$ L; $V_c = 0.200$ L; $P_a = 1$ atm; $T_a = 300$ K; $c_p = 29.09$, $c_v = 20.76$ J/mol · K; $\gamma = 1.4$.

**23. Repeat the calculation for the Diesel cycle of Problem 22, but change V_c to 0.2667 L. Note any effect upon the efficiency of the cycle.

24. What is the ideal COP of a heat pump when the inside and outside temperatures, respectively, are: **(a)** 65°F and 55°F; **(b)** 65°F and 15°F?

25. The freezing compartment of a refrigerator is at a temperature of 28°F at the same time the air temperature near the condenser coils is 82°F. Calculate the COP of an ideal refrigerator operated between these temperature limits.

26. If electrical energy costs 5¢ per kilowatt-hour, what would be the cost of supplying 10 kW · h of heat to the interior of a home if the overall COP of the heat pump is 5?

27. Verify that the first law of thermodynamics is satisfied for each segment of the basic cyclic process of Fig. 16-9.

28. Recall that the power delivered by a force acting on a moving object is given by the product of the force and the velocity of the object. With this in mind, calculate the effective horsepower of a turbojet engine which produces 10,000 lb of thrust while the aircraft is traveling at 600 mi/h.

29. Compute the effective horsepower of a rocket engine that produces 100 lb of thrust while traveling at 18,000 mi/h.

*30. A certain turbine wheel rotates at 24,000 rpm. One of the smaller blades on the turbine has a mass of 300 g and travels around a circular path 40 cm in diameter.

(a) Calculate the angular velocity of the turbine wheel in radians per second.

(b) Determine the centrifugal force (in pounds) acting on the turbine blade.

*31. To produce a combustible mixture, gasoline must be mixed with air in the ratio of 1 part gasoline to 600 parts air, by volume. If a certain car gets 30 mi/gal, what volume of exhaust gases at 20°C are produced by driving the car a distance of 1 mi?

CHAPTER 17

SIMPLE HARMONIC MOTION

Simple harmonic motion is the simplest, smoothest, least-abrupt type of motion possible for a vibrating object. This chapter describes simple harmonic motion and several devices whose vibration closely approximates simple harmonic motion. These devices include the simple pendulum, the physical pendulum, the torsion pendulum, and the mass balance.

The last portion of the chapter deals with a two-dimensional pendulum, coupled pendulums, a driven oscillator, damping and critical damping, and the stability of a system that might tend to go into oscillation.

This chapter is a prelude to the next chapter which deals with wave motion of various types.

17-1 THE NATURE OF VIBRATING SYSTEMS

The world is full of vibrating systems. The pendulum of a grandfather clock provides one of the more serene examples of vibration as its steady to-and-fro motion keeps time by regulating the motion of the escapement wheel. A guitar string vibrates and produces sound as the vibration is communicated by longitudinal vibrations of the air. Indeed, all musical instruments rely on a vibrating member—such as the vibrating head of a drum, the vibrating strings of the wide assortment of stringed instruments, the vibrating reed(s) of a clarinet, saxophone, accordion, or harmonica, the vibrating column of air in a flute, pipe organ or piccolo, and even the vibrating speaker of an electronic organ or music synthesizer.

All time-keeping devices depend upon periodic motions that typically assume the form of a vibration. The most apparent of these is the grandfather clock mentioned earlier. Common clocks and watches utilize the rotational vibration of a balance wheel. Electronic watches employ the vibration of a tiny tuning fork or, more commonly, the vibration of a crystal of quartz. Atomic clocks, found in standards laboratories, make use of the vibrations of the outermost electron of cesium 133 atoms to achieve typical accuracies on the order of 1 s in 30,000 years. Periodic motions that are *not* vibratory are the uniform circular motions of the synchronous motors of common electric clocks and the motions of celestial bodies. A sundial derives its time-keeping abilities from the daily rotation of the earth.

Not all vibrations are useful or desirable. Noise is the unmusical vibration of the air resulting from a myriad of sources. Undesired vibration can degrade the strength of building structures, produce metal fatigue in aircraft, and result in excessive wear and premature failure of mechanical components. Hence, the study of vibration and its effects is an important aspect of almost all fields of engineering.

Refer to the simple pendulum of Fig 17-1 as we consider a simple vibration in more detail to become aware of certain features common to all vibrations. First,

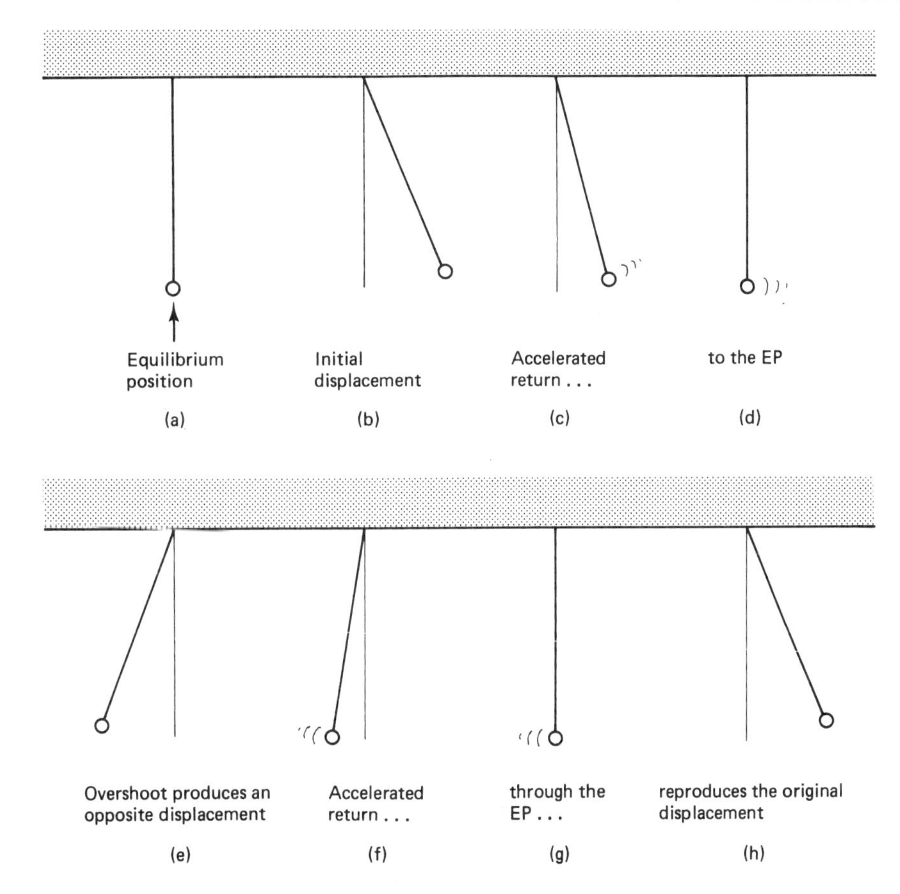

Figure 17-1 Component parts of a vibration.

we note that the *equilibrium position* (EP) is the position of the pendulum bob at rest, the point where the potential energy of the pendulum is a minimum. The pendulum bob will remain at the EP until a force acts to displace it from that position. It is only at the EP that no net force acts on the pendulum bob.

Furthermore, when the bob is displaced from the EP, a force is developed that tends to return it to the EP.

Figure 17-1(b) shows an initial displacement that must occur in order for the vibration (or *oscillation*) to begin. A momentary external force is required to produce the initial displacement. Once the initial displacement is achieved, an internal force that stems from gravity acts to return the bob to the EP. The bob accelerates toward the EP, as shown in parts (c) and (d). However, the bob arrives at the EP with considerable velocity and momentum which causes it to overshoot the EP and continue on to the opposite side until an opposite displacement is attained equal to the original displacement, as shown in part (e). Another accelerated return then occurs which carries the bob *to* and *through* the EP. Consequently, the bob reappears at the starting point, and the cycle will then be repeated.

Any practical pendulum will "run down" after a short while rather than continue to swing indefinitely. This is because the energy initially given to the system during the initial displacement is soon dissipated by frictional losses due to air friction and a combination of friction at the pivot point and bending of the supporting string. Therefore, a pendulum that is to swing indefinitely must have energy returned to it as is done by a spring or system of weights in a pendulum clock.

While this discussion is directed at a simple pendulum, a very similar sequence of events occurs for any mechanical vibration. Many systems vibrate in modes that are extremely complex with motion occurring perhaps in all three dimensions and combining translational and torsional modes. In the following section, we consider the simplest, smoothest, least-jerky type of vibration possible—that of simple harmonic motion (SHM).

17-2 HOOKE'S LAW AND SIMPLE HARMONIC MOTION

A particular type of vibratory motion occurs when the force tending to restore the vibrating member to the EP is directly proportional to the displacement of the member from the EP. This condition is approximately true for a simple pendulum. In Sect. 11-7, the force required to stretch a spring a distance x is given by

$$F = -kx \tag{17-1}$$

where k is the spring constant. That is, the force required to keep a spring extended is directly proportional to the amount the spring is stretched. This is Hooke's law. Therefore, we expect that a mass suspended by a spring, as in Fig. 17-2, will execute SHM when set in motion by an initial displacement.

In Fig. 17-2(a), the mass is stationary in the EP. The level of the mass at that point becomes the reference level, the central point of the impending vibration. When the mass is then lifted a distance x, a net downward force given by $-kx$ comes to bear on the mass due to an imbalance arising between the downward pull of gravity and the upward pull of the spring. This is shown in part (b). On the other hand, when the mass is pulled below the EP, a net upward force, $+kx$, is exerted on the mass. Thus, the restoring force is proportional to the displacement x, and SHM results when the mass is set in motion, as in part (d).

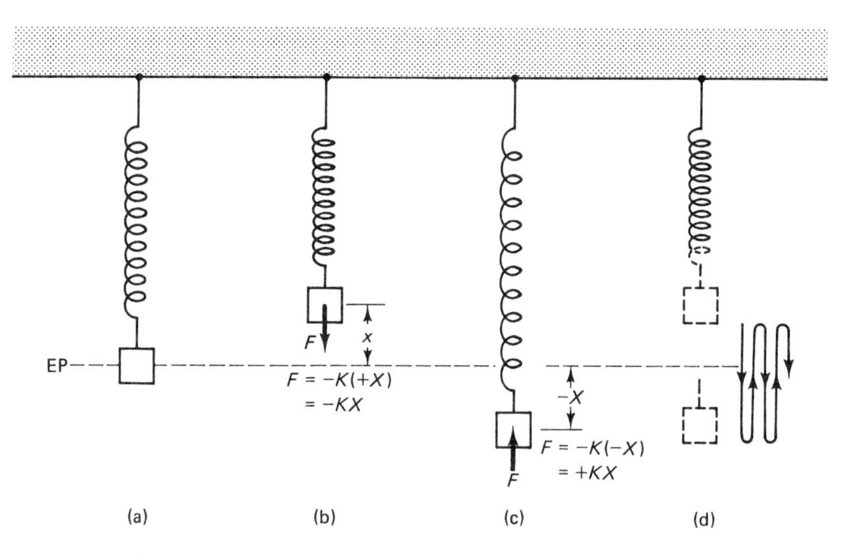

Figure 17-2 A mass suspended by a spring executes simple harmonic motion.

One complete excursion of the mass, up and down through all possible positions, is called a *cycle*. Hence, one cycle might consist of the motion of the mass from the highest point to the lowest and back to the highest. Or if the cycle is presumed to begin at the EP, one cycle includes the upward motion to the highest point followed by a descent to the lowest point and then a final ascent back to the EP. There is no unique point agreed upon for the beginning of a cycle, but a given cycle always ends at the point at which it was declared to have begun.

The *frequency* is the number of cycles that occur per second, usually expressed in hertz. One *hertz* (Hz) is the equivalent of one cycle per second. This unit is named in honor of Heinrich Hertz (1857–1894), a pioneer experimenter with electromagnetic waves. Familiar multiples of the basic unit are the kilohertz (kHz), megahertz (MHz), and gigahertz (GHz), which are 10^3, 10^6, and 10^9 hertz, respectively. Other time units, such as *cycles per minute* or *cycles per hour*, are sometimes used. These are given no special names, however.

The *period* of a vibration is the amount of time required for the execution of one cycle. The typical unit is the second, but minutes, hours, or other time units may be more convenient for particular applications. Mathematically, the period T is the reciprocal of the frequency f:

$$T = \frac{1}{f} \qquad f = \frac{1}{T} \tag{17-2}$$

Thus, a vibration whose frequency is 5 Hz has a period of

$$T = \frac{1}{f} = \frac{1}{5} = 0.2 \text{ s}$$

Going in the other direction, a vibration whose period is 0.5 s has a frequency of 2 Hz.

Because the frequency is the inverse of the period, the rigorous unit of frequency is *inverse seconds*, s^{-1}. Thus, 1 Hz is the equivalent of 1 s^{-1}. If the unit in a problem is s^{-1}, this should be expressed simply as hertz.

The *amplitude* of a vibration refers to the maximum displacement from the EP. It is measured in any convenient unit of length such as meters, inches, or feet. Consequently, the distance from the highest point to the lowest point of the vibration of Fig. 17-2 is *twice* the amplitude. Furthermore, the amplitude equals the initial displacement at the start of a vibration, assuming the vibrating member is started from rest and that the vibration has not "run down" due to frictional effects.

The reference circle

Figure 17-3 shows a rotating wheel positioned to the side of a vibrating mass suspended by a spring. Attached to the rim of the wheel is a round ball that casts a

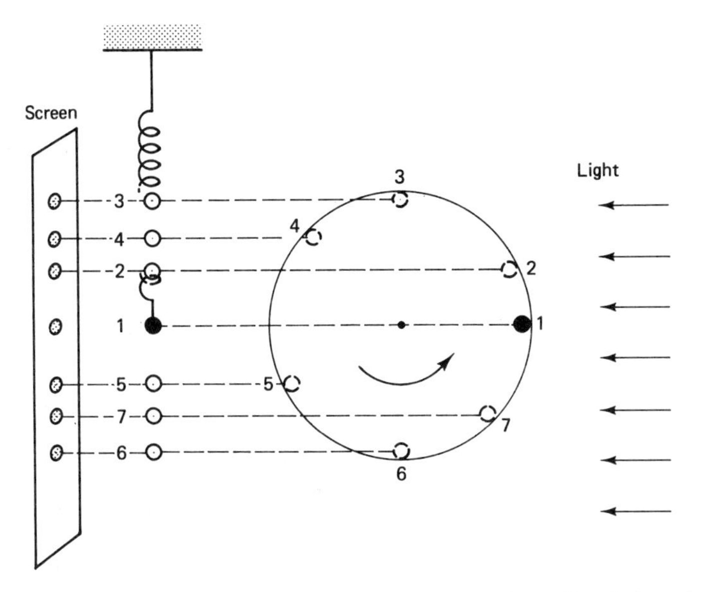

Figure 17-3 The projection (shadow) of a particle traveling uniformly in a circle executes simple harmonic motion.

shadow to the left because of the beam of light incident from the right. As the wheel turns, the ball follows a circular path around the wheel, but the shadow of the ball projected onto the screen simply travels up and down. The circular mass on the spring casts a similar shadow on the screen, and it will travel up and down the screen when the mass is set in motion. If the rotational speed of the wheel is adjustable and if the amplitude of the vibrating mass is set equal to the radius of the wheel, considerable skill will allow the two shadows on the screen to be superimposed, appearing as one shadow traveling up and down the screen.

Now, because we know that the mass executes SHM, we conclude that the shadow from the ball on the wheel does also. That is, the projection of a particle in uniform circular motion executes SHM. This fact can also be obtained mathematically. In this experiment, the amplitude of the vibration exactly matches the radius of the wheel, and the speed of rotation of the wheel exactly matches the frequency of vibration of the mass.

Figure 17-4 is a diagram showing a particle in uniform circular motion about a circle. A radius of the circle is A, and a line drawn between the center of the

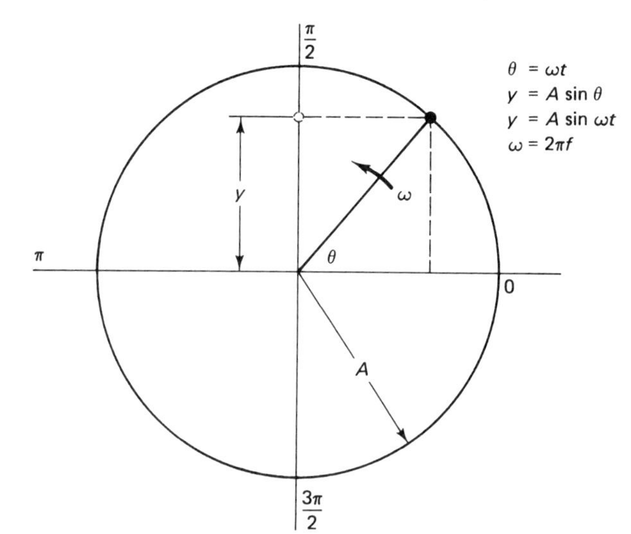

Figure 17-4 Reference circle showing the projection onto the y-axis of uniform circular motion.

circle and the particle makes an angle θ with the x-axis. Observe that the position of the particle is projected onto the y-axis, and y denotes the displacement of the particle from the x-axis. By elementary trigonometry,

$$y = A \sin \theta \qquad (17\text{-}3)$$

If the particle is moving, the angle θ will not be a fixed quantity, but will be constantly increasing. This can be represented mathematically if we let

$$\theta = \omega t \qquad (17\text{-}4)$$

where t represents time and ω is the *angular velocity*, measured in radians per second (rad/s). In terms of the frequency f,

$$\omega = 2\pi f \qquad (17\text{-}5)$$

The angle ωt is called the *phase* of the motion. The phase determines the location on the cycle of the particle at any time.

The projection of the particle (the shadow) executes SHM, so we conclude that the position of a particle executing SHM is given by

$$y = A \sin \omega t \qquad (17\text{-}6)$$

where A is the amplitude of the vibrations. Recall that the range of the sine function is from -1 to $+1$ so that the range of y in Eq. (17-6) is from $-A$ to $+A$. The nature of the variation of $\sin \omega t$ is illustrated in Fig. 17-5. The curve is called a *sine wave*, and the nature of the variation is *sinusoidal*. This type of mathematical function is called a *periodic* function because it periodically traverses the same range of values.

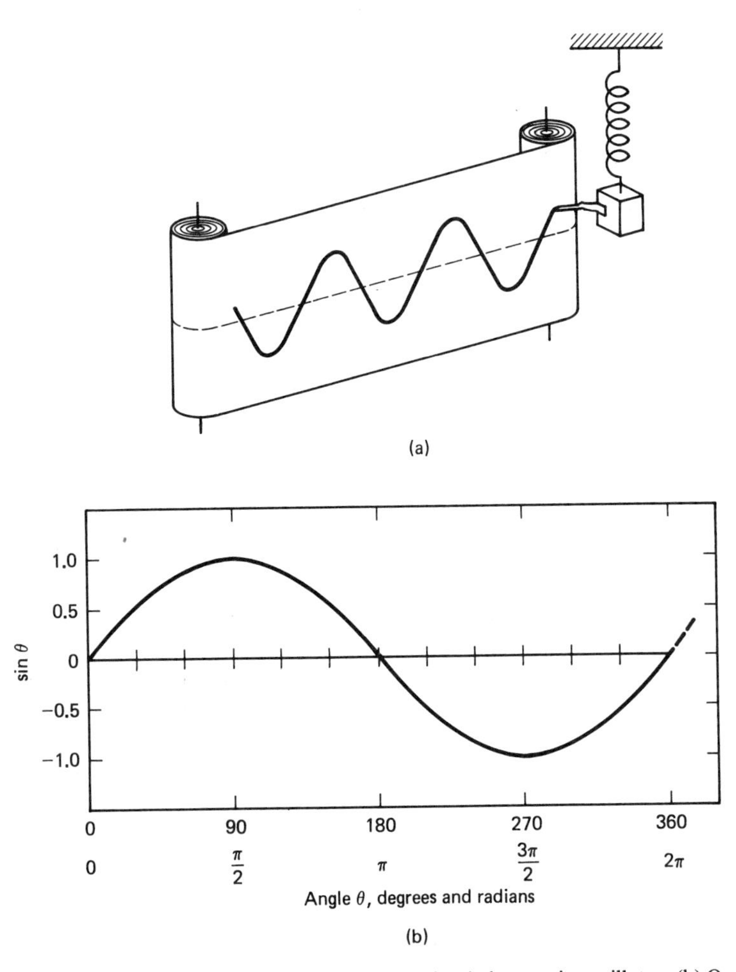

Figure 17-5 (a) A sine wave is produced by a simple harmonic oscillator. (b) One cycle of a sine wave.

EXAMPLE 17-1 A mass is executing SHM with an amplitude of 15 cm and an angular velocity of 4 rad/s. Use Eq. (17-6) to compute the position of the particle when: (a) $t = 0$ s; (b) $t = 0.2$ s; (c) $t = 1.25$ s.

Solution.

(a) $t = 0$; $t = 4(0) = 0$

$$\begin{aligned} y &= A \sin \omega t \\ &= 15 \sin(0) \\ &= 0 \end{aligned}$$

Thus, at $t = 0$ the mass is at the equilibrium position.

(b) $t = 0.2$; $t = 4(0.2) = 0.8$ rad

$$\begin{aligned} y &= A \sin(0.8) \\ &= 15(0.717) \\ &= 10.76 \text{ cm} \end{aligned}$$

At $t = 0.2$ s, the mass is 10.76 cm above the equilibrium position.

(c) $t = 1.25$; $t = 4(1.25) = 5$ rad

$$\begin{aligned} y &= 15 \sin(5) \\ &= 15(-0.959) \\ &= -14.38 \text{ cm} \end{aligned}$$

When $t = 1.25$ s, the mass is 14.38 cm below the equilibrium position. Because the amplitude is 15 cm, this is near the lowest point of the cycle.

In Example 17-1, we note that the particle was at the equilibrium position when t was zero, the instant at which we started the stopwatch used to measure t. This is the type of motion described by Eq. (17-6). In earlier discussions, however, the motion was started from the position of maximum displacement. Eq. 17-6 needs to be modified so that the motion can be initiated at any point on the cycle. The modification consists of adding an *initial phase angle* δ to the quantity ωt of Eq. (17-6) to yield

$$y = A \sin(\omega t + \delta) \tag{17-7}$$

Thus, by assigning an appropriate angle (in radians) to δ, the particle can be positioned anywhere on the cycle at $t = 0$. To initiate the motion at the point of maximum displacement, set $\delta = \pi/2$ rad.

EXAMPLE 17-2

A mass is executing SHM with an amplitude of 15 cm with an angular velocity of 4 rad/s. The timing device is started as the mass is released from the point of maximum displacement. Calculate the position of the particle when: (a) $t = 0$ s; (b) $t = 0.2$ s; (c) $t = 1.25$ s.

Solution. The similarity of this example to the previous example should be noted. We set $\delta = \pi/2 = 1.571$ rad as advised above since the motion begins from maximum displacement.

(a) $t = 0$; $\omega t = 0$; $\omega t + \delta = 0 + 1.571 - 1.571$ rad

$$\begin{aligned} y &= A \sin(\omega t + \delta) \\ &= 15 \sin(1.571) \\ &= 15(1) = 15 \text{ cm} \end{aligned}$$

This is the point of maximum displacement.

(b) $t = 0.2$; $\omega t + \delta = 4(0.2) + 1.571 = 2.371$ rad

$$\begin{aligned} y &= 15 \sin(2.371) \\ &= 15(0.697) \\ &= 10.448 \text{ cm} \end{aligned}$$

Thus, after 0.2 s, the object is nearer the equilibrium position.

(c) $t = 1.25$; $\omega t + \delta = 5 + 1.571 = 6.571$ rad

$$\begin{aligned} y &= 15 \sin(6.571) \\ &= 15(0.284) \\ &= 4.258 \text{ cm} \end{aligned}$$

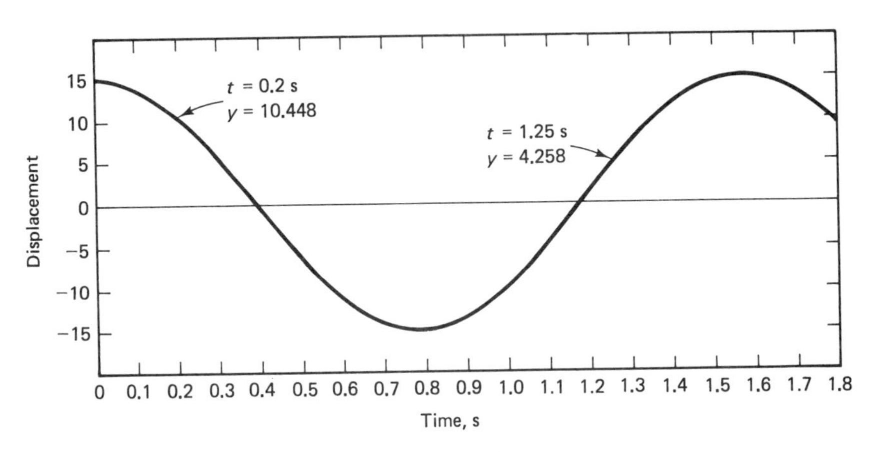

Figure 17-6

At $t = 1.25$ s, the mass is upward bound after having reached the low point (-15 cm) of the cycle. A graph of this motion is given in Fig. 17-6.

17-3 VELOCITY AND ACCELERATION IN SHM

Thus far, nothing has been said about the velocity or the acceleration of a particle executing SHM. Obviously, the velocity is constantly changing as the mass oscillates back and forth. Moreover, the acceleration also undergoes constant change. For this reason, the equations of uniformly accelerated motion developed in Chap. 3 cannot be applied to SHM. Recall, however, the fundamental relationships between position, velocity, and acceleration. Velocity is the rate of change of position; acceleration is the rate of change of velocity. We make use of these in order to obtain mathematical expressions for the velocity and acceleration in SHM.

We often allude to the use of calculus in dealing with quantities whose values are constantly changing. In particular, calculus provides a mathematical technique called *differentiation* for computing the rate of change of a mathematical function. For SHM we have obtained an expression [Eq. (17-7)] for the position of the mass at any time. If we use calculus to calculate the rate at which the position function changes with time, we obtain an expression for the velocity in SHM. Furthermore, we can then compute the rate of change of the velocity with respect to time and obtain an expression for the acceleration. This looks very appealing, but we obviously cannot delve deeply into mathematical details. Therefore, we simply indicate the process and write the result.

First of all, the notation

$$\frac{d}{dt}\text{(function)}$$

represents the rate of change with respect to time of whatever mathematical function might be enclosed in the brackets. It is called *the first derivative with respect to time*. Hence, as we have stated before,

$$\frac{d}{dt}\text{(position)} = \text{velocity}$$

and

$$\frac{d}{dt}\text{(velocity)} = \text{acceleration}$$

The first semester of a course in calculus reveals that the first derivative of the sine function is the cosine function, and the first derivative of the cosine is the negative of the sine. Omitting details, we now apply this to SHM.

For a particle executing SHM, its position is given by

$$\text{Position} = A\sin(\omega t + \delta)$$

Therefore,

$$\frac{d}{dt}\text{(position)} = \text{velocity}$$

$$\frac{d}{dt}[A\sin(\omega t + \delta)] = A\omega\cos(\omega t + \delta)$$

Furthermore,

$$\frac{d}{dt}\text{(velocity)} = \text{acceleration}$$

$$\frac{d}{dt}(A\omega\cos(\omega t + \delta)) = -A\omega^2\sin(\omega t + \delta)$$

Thus, for SHM, the velocity v and the acceleration a are given by

$$v = A\omega\cos(\omega t + \delta) \tag{17-8}$$

$$a = -A\omega^2\sin(\omega t + \delta) \tag{17-9}$$

which are the expressions we are seeking. The relationships between displacement, velocity, and acceleration are shown graphically in Fig. 17-7.

It is often of interest to know the maximum value of the velocity or acceleration for a given vibration. Because the absolute value of the sine or cosine is never greater than 1.0, the maximum value for v and a will occur when the trigonometric terms in Eqs. (17-8) and (17-9) equal 1.0. Consequently, by setting those terms equal to 1, we obtain

$$\text{Maximum displacement} = A$$

$$\text{Maximum velocity} = A\omega \tag{17-10}$$

$$\text{Maximum acceleration} = A\omega^2$$

We have ignored the negative sign in the acceleration formula because it only indicates the direction of the acceleration.

Calculating the period and frequency of SHM

One of the major concerns in dealing with vibration is the period or frequency with which a given system will oscillate—the *natural frequency*. A formula for ω, the angular frequency, can be obtained by combining Newton's third law ($F = ma$)

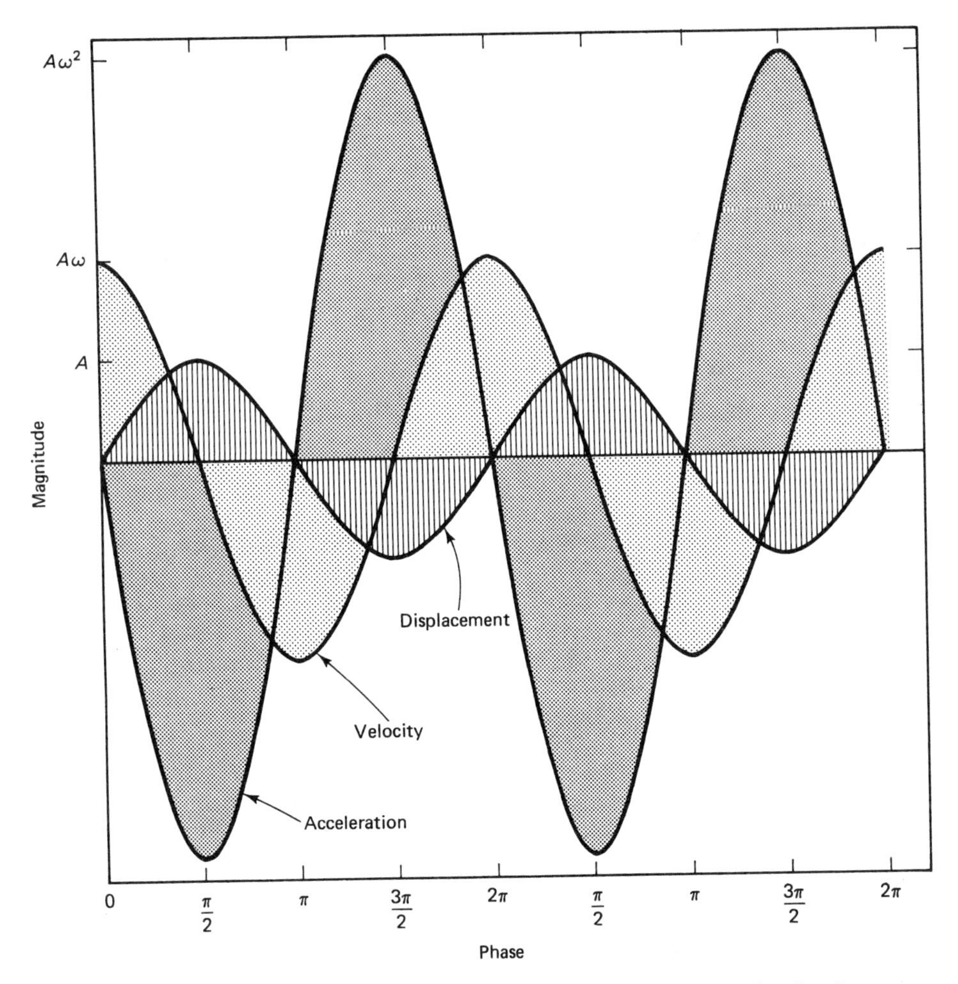

Figure 17-7 Relationships between displacement, velocity, and acceleration for simple harmonic motion with $A = 1$ and $\omega = 2$.

with Hooke's law ($F = -ky$). At any point in a cycle of SHM, the force arising from the displacement from the EP must necessarily be the same force that causes the mass to accelerate. Therefore, we may set the forces equal:

$$-ky = ma$$

Substituting Eq. (17-7) and Eq. (17-9) for y and a gives

$$-k[A \sin(\omega t + \delta)] = m[-A\omega^2 \sin(\omega t + \delta)]$$

The amplitude A and the sine term appear on both sides of the equation and therefore drop out to leave us with

$$k = m\omega^2$$

Solving for ω,

$$\omega = \sqrt{\frac{k}{m}} \tag{17-11}$$

This expression relates the angular velocity to the mass and force constant k. Using Eqs. (17-5) and (17-2) to convert to f and T gives two useful formulas.

$$f = \frac{\omega}{2\pi} \qquad f = \frac{1}{2\pi}\sqrt{\frac{k}{m}} \tag{17-12}$$

$$T = \frac{1}{f} \qquad T = 2\pi\sqrt{\frac{m}{k}} \tag{17-13}$$

Note that the amplitude A is absent from these expressions. The period and frequency of SHM do not change when the amplitude changes. This is the *isochronous* property of SHM, and it is this property that is important in the construction of time-keeping devices such as pendulum clocks.

EXAMPLE 17-3 A mass of 0.2 kg is suspended by a spring whose force constant k is 18 N/m. Compute: (a) ω; (b) f; and (c) T for the system when it executes SHM.

Solution.

(a)
$$\omega = \sqrt{\frac{k}{m}} = \sqrt{\frac{18 \text{ N/m}}{0.2 \text{ kg}}} = 9.49 \text{ rad/s}$$

(b)
$$f = \frac{\omega}{2\pi} = \frac{9.49}{2\pi} = 1.51 \text{ Hz}$$

(c)
$$T = \frac{1}{f} = \frac{1}{1.51} = 0.662 \text{ s}$$

EXAMPLE 17-4 A 0.6-kg mass executes SHM with an amplitude of 12 cm and a frequency of 1 Hz. Compute: (a) the maximum velocity of the mass; (b) the maximum acceleration of thc mass; (c) the maximum force exerted on the mass.

Solution. Anticipating the use of Eq. (17-10), we first compute the angular frequency:

$$\omega = 2\pi f; \ \omega = 2\pi(1 \text{ Hz}) = 6.28 \text{ rad/s}$$

(a)
$$\begin{aligned} \text{Maximum velocity} &= A\omega \\ &= (12 \text{ cm})(6.28 \text{ s}^{-1}) \\ &= 75.36 \text{ cm/s} = 0.7536 \text{ m/s} \end{aligned}$$

(b)
$$\begin{aligned} \text{Maximum acceleration} &= A\omega^2 \\ &= (12 \text{ cm})(6.28 \text{ s}^{-1})^2 \\ &= 473.26 \text{ cm/s}^2 = 4.73 \text{ m/s}^2 \end{aligned}$$

(c) Use $F = ma$ in conjunction with the maximum acceleration to compute the maximum force:

$$\begin{aligned} F_{\text{max}} &= ma_{\text{max}} \\ &= (0.6 \text{ kg})(4.73 \text{ m/s}^2) \\ &= 2.84 \text{ N} \end{aligned}$$

17-4 THE SIMPLE PENDULUM

A simple pendulum consists, ideally, of a point mass suspended by a light unstretchable rod or string that is caused to swing in a vertical plane. The oscillatory motion of a simple pendulum closely approximates SHM, provided the amplitude is kept small. Consequently, for small amplitudes, the period is independent of the amplitude. The period increases significantly when the pendulum is caused to swing through large amplitudes. Our objective at this point is to obtain a mathematical expression for the period of a simple pendulum.

In Fig. 17-8(a), a pendulum of length l is shown displaced from the EP until

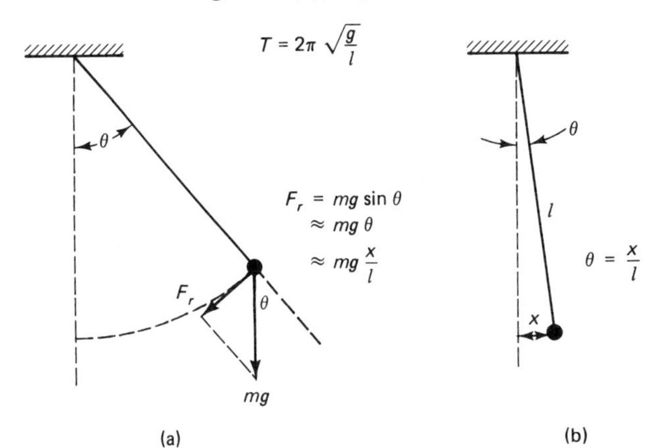

Figure 17-8 Geometry of the simple pendulum.

the supporting string makes an angle θ with the vertical. A restoring force F_r arises because of the weight of the pendulum bob.

$$F_r = -mg \sin\theta \tag{17-14}$$

From part (b), we see that the angle θ (in radians) equals the ratio of x to l, where x is the displacement from the EP measured along the arc. Hence,

$$F_r = -mg \sin\left(\frac{x}{l}\right) \tag{17-15}$$

The restoring force is not directly proportional to the displacement x, but rather is proportional to sin (x/l). Therefore, strictly speaking, a simple pendulum does not execute SHM. However, for small angles, the following approximation is good:

(for small angles)
$$\theta \approx \sin\theta \tag{17-16}$$

This is readily verified with a scientific calculator. (Angles must be in radians.) The error is on the order of 1% for angles as large as 15° (0.262 rad). Applying this approximation to Eq. (17-15) yields

$$F_r = -mg\frac{x}{l} \tag{17-17}$$

and if we identify k as mg/l, the restoring force can be written as

$$F_r = -kx$$

which is of the form required for SHM. Thus, we conclude that a simple pendulum executes SHM if the amplitude is kept small. Furthermore, the force constant k is mg/l.

We can obtain expressions for the frequency and period of a simple pendulum by substituting the preceding expression for k into the general formulas for SHM. Hence, using Eqs. (17-11), (17-12), and (17-13),

For simple pendulum

$$\omega = \sqrt{\frac{g}{l}} \qquad f = \frac{1}{2\pi}\sqrt{\frac{g}{l}} \qquad T = 2\pi\sqrt{\frac{l}{g}} \tag{17-18}$$

Observe, finally, that the frequency and period are independent of the mass of the pendulum bob.

EXAMPLE 17-5 Calculate the period of a pendulum 1 m long at a point on the earth where $g = 9.8$ m/s^2.

Solution. A direct application of Eq. (7-18) yields

$$T = 2\pi\sqrt{\frac{l}{g}} = 2\pi\sqrt{\frac{1\text{ m}}{9.8\text{ m/s}^2}} = 2\pi\sqrt{0.102} = 2.007\text{ s}$$

This is very nearly the length of the pendulum in a grandfather clock that produces a tick per second each time it passes through the EP.

Determination of g with a pendulum

If the length of a simple pendulum is carefully measured and if the period is accurately determined, the value of the acceleration of gravity can be obtained by rearranging the formula for the period:

$$g = \frac{4\pi^2 l}{T^2} \tag{17-19}$$

EXAMPLE 17-6 A simple pendulum 80.0 cm long is observed to swing through 50 cycles in 89.8 s. What is the value of g in the vicinity of the pendulum?

Solution. We first compute the period from the experimental data:

$$\text{Period} = \frac{\text{elapsed time}}{\text{number of cycles}} = \frac{89.8}{50} = 1.796\text{ s}$$

Then,

$$g = \frac{4\pi^2}{T^2} = \frac{4\pi^2(80\text{ cm})}{(1.796\text{ s})^2}$$

$$= 979\text{ cm/s}^2$$

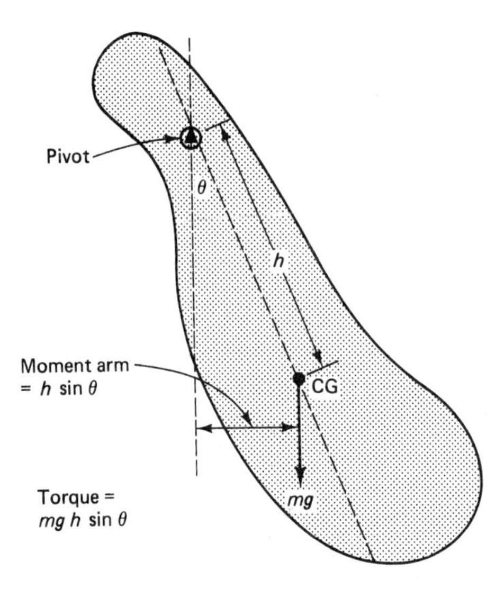

Figure 17-9 A physical pendulum.

17-5 THE PHYSICAL PENDULUM

Basically, a *physical pendulum* is any pendulum that does not meet the rather stringent qualifications of a simple pendulum. It is understood, however, that the oscillatory motion occurs in a vertical plane. A simple pendulum that is allowed to traverse an elliptical or circular path is called a *spherical pendulum*, and the mathematical analysis is considerably more involved. An example of a physical pendulum of the type we consider is shown in Fig. 17-9. It consists of an irregular body suspended on a pivot located a distance h from the center of gravity (CG) of the body.

Because the motion of a physical pendulum occurs as a to-and-fro rotation about the pivot point, the analysis must involve the moment of inertia (MOI) I that the body has when it is rotated about an axis through the pivot point. Furthermore, we must deal with a restoring *torque* rather than a restoring force. This results in no significant complication, however, and the motion turns out to be SHM if the amplitude of the oscillation is kept small.

When the object is displaced from the EP, a restoring torque L is developed that is proportional to sin θ, where θ is the angle of displacement. By limiting the oscillation to small amplitudes, we can use the approximation $\theta \approx \sin\theta$ to obtain

$$L = -mgh\theta \tag{17-20}$$

This meets the requirement for SHM because L is directly proportional to θ. If we let $k = mgh$, then

$$L = -k\theta \tag{17-21}$$

and by constructing a rotational analogy of Eq. (17-13), we obtain the following expression for the period of oscillation of a physical pendulum:

$$T = 2\pi\sqrt{\frac{I}{k}} \quad \text{or} \quad T = 2\pi\sqrt{\frac{I}{mgh}} \tag{17-22}$$

The moment of inertia I is the moment of inertia about an axis running through the pivot point perpendicular to the plane of oscillation. By using the parallel-axis theorem (Fig. 8-15), I can be related to the moment of inertia about the center of gravity, I_{CG}. Hence,

$$I = I_{CG} + mh^2 \tag{17-23}$$

and Eq. (17-22) can be written

$$T = 2\pi\sqrt{\frac{I_{CG} + mh^2}{mgh}} \tag{17-24}$$

Values of I_{CG} for several objects are given in Fig. 8-14. Therefore, if we know the mass m of the object, we can compute the period of oscillation of the object when pivoted at any point a distance h from the center of gravity. Or, we may proceed in the other direction to determine I_{CG} from an experiment designed to determine T. That is, if T, m, and h are known, we can compute I_{CG} by rearranging Eq. 17-24:

$$I_{CG} = \left(\frac{T}{2\pi}\right)^2(mgh) - mh^2 \tag{17-25}$$

Incidentally, the center of gravity of an irregular planar object can be located by balancing, as described in Sect. 5-10.

EXAMPLE 17-7 A physical pendulum is made by pivoting a thin uniform rod at one end. The rod is 1 m long, and it has a mass of 0.7 kg. Determine the period of the pendulum.

Solution. First, calculate the moment of inertia about the pivot point. The MOI of a thin rod pivoted at one end is $\frac{1}{3}mL^2$. Therefore,

$$I = \tfrac{1}{3}mL^2 = \tfrac{1}{3}(0.7\text{ kg})(1\text{ m})^2 = 0.233\text{ kg}\cdot\text{m}^2$$

Because the rod is uniform, the center of gravity is at the center of the rod. Hence, the distance h from the CG to the pivot is 0.5 m. Then, by Eq. (17-22),

$$T = 2\pi\sqrt{\frac{I}{mgh}} = 2\pi\sqrt{\frac{0.233}{(0.7)(9.8)(0.5)}}$$
$$= 1.638\text{ s}$$

By way of comparison, the period of a simple pendulum 1 m long is 2.01 s.

EXAMPLE 17-8 Use the data in Example 17-7 to compute the MOI about the center of gravity for a uniform rod 1 m long whose mass is 0.7 kg.

Solution. This requires only a straightforward application of Eq. (17-25).

$$I_{CG} = \left(\frac{T}{2\pi}\right)^2(mgh) - mh^2$$
$$= (1.638)^2(0.7)(9.8)(0.5) - (0.7)(0.5)^2$$
$$= 0.233 - 0.175$$
$$= 0.058\text{ kg}\cdot\text{m}^2$$

This agrees with the value obtained from the formula $(\frac{1}{12})mL^2$ for the MOI of a rod pivoted about its center.

Center of oscillation

It is always possible to compute the length of a simple pendulum that has the same period as a particular physical pendulum. The length of the simple pendulum is found by equating expressions for the periods of the two types, Eq. (17-18) and Eq. (17-22):

$$2\pi\sqrt{\frac{l}{g}} = 2\pi\sqrt{\frac{I}{mgh}}$$

Solving for l,

$$l = \frac{I}{mh} \tag{17-26}$$

The *center of oscillation* of a physical pendulum is the point located on the line extending from the pivot through the center of gravity at a distance l from the pivot, as illustrated in Fig. 17-10. The center of oscillation has two interesting properties. If the physical pendulum is turned over and pivoted at the center of oscillation, the oscillation will exhibit the same period as it did when suspended at

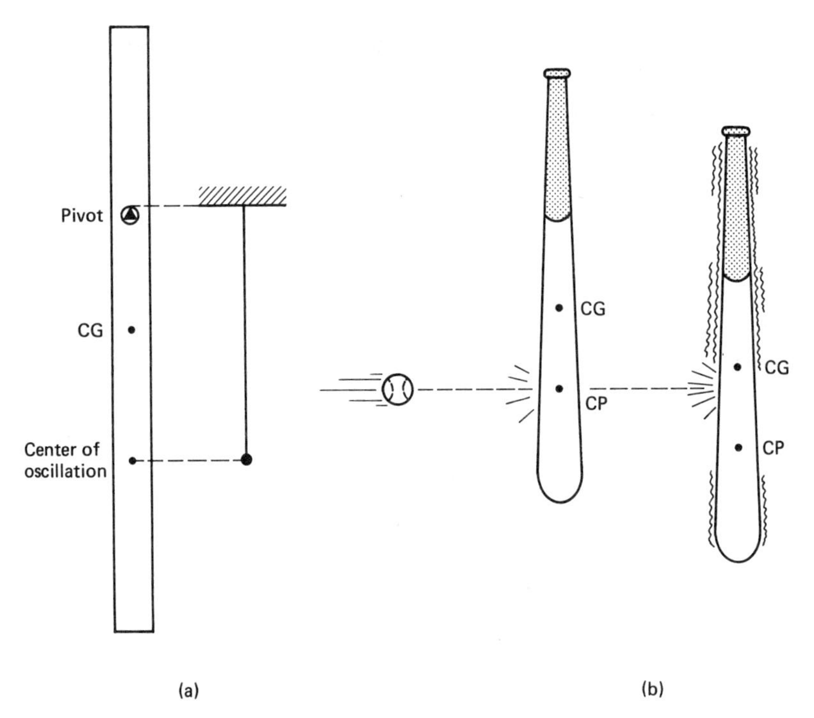

Figure 17-10 (a) Equivalent simple pendulum and the center of oscillation. (b) A stinging vibration results when the ball is hit at a point other than at the center of percussion.

the original pivot. Secondly, when the physical pendulum is struck by an object traveling on a line passing through the center of oscillation, no instantaneous reaction force is exerted on the pivot. In this regard, the center of oscillation is called the *center of percussion*. The stinging vibration sometimes felt when a baseball is hit with a bat arises from having hit the ball at a point other than at the center of percussion. This is illustrated in Fig. 17-10.

EXAMPLE 17-9 Compute the center of oscillation for the physical pendulum of Example 17-7.

Solution. Using Eq. (17-26),

$$l = \frac{I}{mh} = \frac{0.233 \text{ kg} \cdot \text{m}^2}{(0.7 \text{ kg})(0.5 \text{ m})}$$

$$= 0.666 \ m$$

This is the distance measured from the pivot to the center of oscillation. Therefore, the center of oscillation lies 16.6 cm below the center of the rod.

17-6 TORSION PENDULUM

A torsion pendulum is made by suspending an object on a thin rod, fiber, or ribbon that exerts a restoring torque when the object is rotated in the horizontal plane. When the object is given an initial angular displacement and let go, an angular oscillation results that closely approximates SHM for small amplitudes. A torsion pendulum is illustrated in Fig. 17-11.

We describe the twisting of thin rods in Sect. 11-9 where we find that a direct proportion exists between the angle of twist and the applied torque. Hence,

$$L = -k\theta \tag{17-27}$$

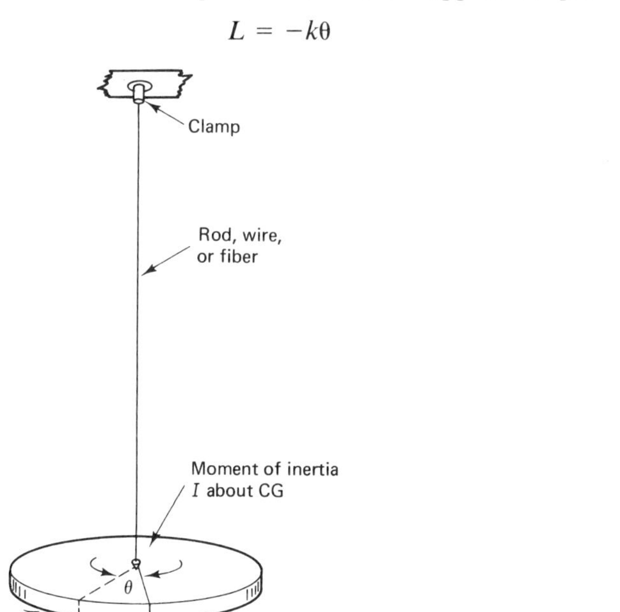

Figure 17-11 A torsion pendulum.

where k is called the *moment of torsion*, or simply the *torsional constant*, for the rod. The period of a torsion pendulum is given by

$$T = 2\pi\sqrt{\frac{I}{k}} \qquad (17\text{-}28)$$

where k is the torsional constant of the rod, fiber, or ribbon supporting the oscillating body, and I is the moment of inertia of the body about the axis of rotation.

A torsion pendulum has several applications. The torsional constant k of the supporting member can be determined if the period T and moment of inertia I are known:

$$k = \frac{4\pi^2 I}{T^2} \qquad (17\text{-}29)$$

This fact was used by Lord Cavendish in conjunction with the experiment to measure the gravitational constant G (Sect. 9-2). Another possible application is implied by the following rearrangement of the equation above:

$$I = \frac{kT^2}{4\pi^2} \qquad (17\text{-}30)$$

That is, the moment of inertia can be determined if k and I are known.

In Sect. 11-9 we obtain a relationship between the torsion constant k and the *shear modulus* n of the material of which a rod is made. For a rod of radius r and length l, the relationship is [Eq. (11-32)]

$$k = \frac{\pi r^4 n}{2l} \qquad (17\text{-}31)$$

We can solve this for the shear modulus and obtain another method for determining the shear modulus for a given material:

$$n = \frac{2lk}{\pi r^4} \qquad (17\text{-}32)$$

The method is illustrated in one of the following examples.

EXAMPLE 17-10 A brass disk is 20 cm in diameter and 1 cm thick. Accordingly, the mass of the disk is 2.733 kg. The moment of inertia of a disk about the center of gravity is given by $\frac{1}{2}mR^2$, and this gives a value of 0.0137 kg · m^2 for the disk at hand. A torsion pendulum is constructed by suspending the disk on a thin rod whose torsional constant k is 0.376 N · m (per radian). Compute the period of oscillation of the disk.

Solution. We use Eq. (17-28) directly, paying close attention to units.

$$T = 2\pi\sqrt{\frac{I}{k}} = 2\pi\sqrt{\frac{0.0137\ \text{kg}\cdot\text{m}^2}{0.376\ \text{N}\cdot\text{m}}}$$

$$= 2\pi(0.1909) = 1.20\ \text{s}$$

To reconcile the units, observe

$$\sqrt{\frac{\text{kg} \cdot \text{m}^2}{\text{N} \cdot \text{m}}} = \sqrt{\frac{\text{kg} \cdot \text{m}}{\text{N}}} = \sqrt{\frac{\text{kg} \cdot \text{m}}{\text{kg} \cdot \text{m/s}^2}} = \sqrt{s^2} = s$$

Note that the *per radian* part of the units of k does not get carried through the computation. This aspect of the radian is discussed in Sect. 8-3.

EXAMPLE 17-11 Suppose the brass disk of Example 17-10 is suspended by a copper wire 2 mm in diameter and 2 m long. Compute the period of the oscillation of the resulting torsion pendulum.

Solution. We first use Eq. (17-31) to compute the torsional constant k (n for copper is 4.2×10^{10} N/m^2). Hence,

$$k = \frac{\pi r^4 n}{2l} = \frac{(0.001 \text{ m})^4(4.2 \times 10^{10} \text{ N/m}^2)}{2(2 \text{ m})}$$

$$= 0.0330 \text{ N} \cdot \text{m} \quad \text{(per radian)}$$

Then using Eq. (17-28) and the value of I obtained previously,

$$T = 2\pi\sqrt{\frac{I}{k}} = 2\pi\sqrt{\frac{0.0137}{0.0330}} = 4.05 \text{ s}$$

EXAMPLE 17-12 The brass disk of Examples 17-10 and 17-11 ($I = 0.0137$ kg · m^2) is attached to a steel rod 1.5 m long and 3.6 mm in diameter. The resulting torsional pendulum is observed to complete 50 oscillations in 39.2 s. Compute the shear modulus of steel from this information.

Solution.

$$T = \frac{39.2 \text{ s}}{50} = 0.784 \text{ s}$$

$$k = \frac{4\pi^2 I}{T^2} = \frac{4\pi^2(0.0137)}{(0.784)^2} = 0.884 \text{ N} \cdot \text{m}$$

$$n = \frac{2lk}{\pi r^4} = \frac{2(1.5)(0.884)}{(1.8 \times 10^{-3})^4}$$

$$= 8.04 \times 10^{10} \text{ N/m}^2$$

17-7 MASS BALANCE

The device shown in Fig. 17-12 allows us to determine the mass of an object independently of any gravitational phenomena. Whereas scales and balances depend upon the mutual gravitational attraction of the earth and the mass being measured, the *mass balance* responds to the inertial properties of a mass. The mass balance would function just as well in the gravity-free environment of interstellar space as it does near the surface of the earth, but common scales and balances would be useless in the same environment.

The mass to be determined is placed on the platform which is then caused to

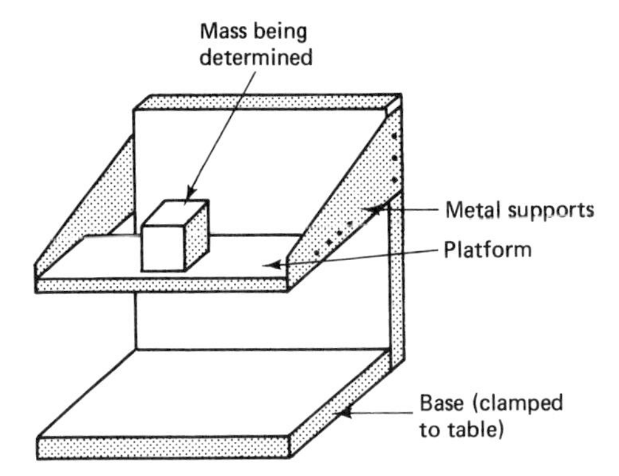

Figure 17-12 Mass balance.

execute SHM in a side-to-side motion. The spring constant of the springy metal supports is accurately known, and the period of the oscillation is determined by measuring the time required for a given number of cycles. The standard formulas for SHM are applicable, and the mass is readily computed. The working formula is a rearrangement of Eq. (17-13).

$$m = \frac{kT^2}{4\pi^2} \tag{17-33}$$

EXAMPLE 17-13 The unloaded platform of a mass balance has an effective mass of 0.20 kg. A steady sideways force of 1.48 N is sufficient to displace the platform laterally by 3.00 cm. When a certain object is attached to the platform, the period of the resulting oscillation is found to be 0.60 s. Calculate the mass of the object.

Solution. We first compute the force constant of the supports:

$$k = \frac{\text{applied force}}{\text{displacement}} = \frac{1.48 \text{ N}}{0.03 \text{ m}}$$

$$= 49.33 \text{ N/m}$$

Then, by Eq. 17-33,

$$m = \frac{(49.33 \text{ N/m})(0.60 \text{ s})^2}{4\pi^2} = 0.45 \text{ kg}$$

This mass, however, is the total mass of the platform and the object. We must subtract the mass of the platform to obtain the mass of the object:

$$\text{Mass of the object} = 0.45 - 0.20$$

$$= 0.25 \text{ kg} = 250 \text{ g}$$

17-8 TWO INTERESTING EXPERIMENTS

Figure 17-13 shows a pendulum intended to oscillate in both the x- and y-directions simultaneously, with a different period in each direction. As shown, the knot can move back and forth only in the y-direction so that the y-pendulum is effectively much longer than the x-pendulum. Consequently, the x-pendulum

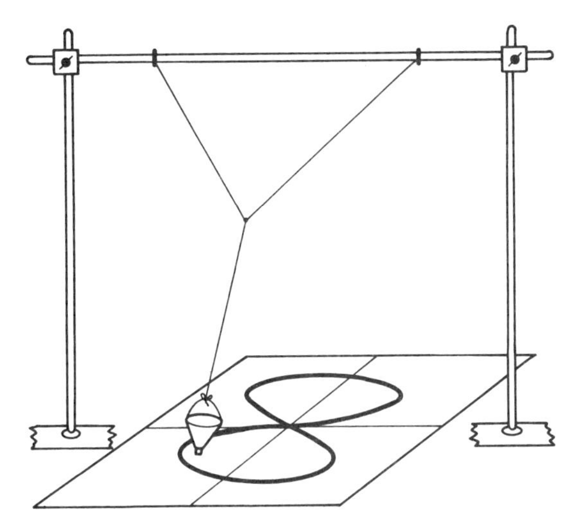

Figure 17-13 A two-dimensional pendulum. String lengths must be adjusted very carefully to produce this figure.

oscillates with the shorter period and higher frequency. The resultant displacement of the bob is the combination of the two motions.

With careful adjustment of the various string lengths, the x-frequency can be set to a multiple of the y-frequency so that a closed figure results. Sand dropping onto the paper traces the path of the bob. These figures are called *Lissajous patterns*, and a wide variety can be obtained. Several are illustrated in Fig. 17-14, where the ratios of frequencies producing the pattern are also given. Different

A Lissajous pattern displayed on an oscilloscope. The frequency ratio is 4:1 (vertical:horizontal).

patterns result from the same ratio because of differences in the phases of the two motions.

The exact pattern obtained depends not only on the frequency ratio but also upon the relative phase of the two motions. Two pendulums of the same frequency are said to be *in phase* if they reach the point of maximum displacement simultaneously. Moreover, two pendulums in phase will pass through the equilibrium position at the same time and while traveling in the same direction. When two pendulums are 180° *out of phase*, the motion of one is in the direction exactly opposite to that of the other. One reaches the maximum positive displacement at the same time the other reaches the maximum negative displacement. They pass

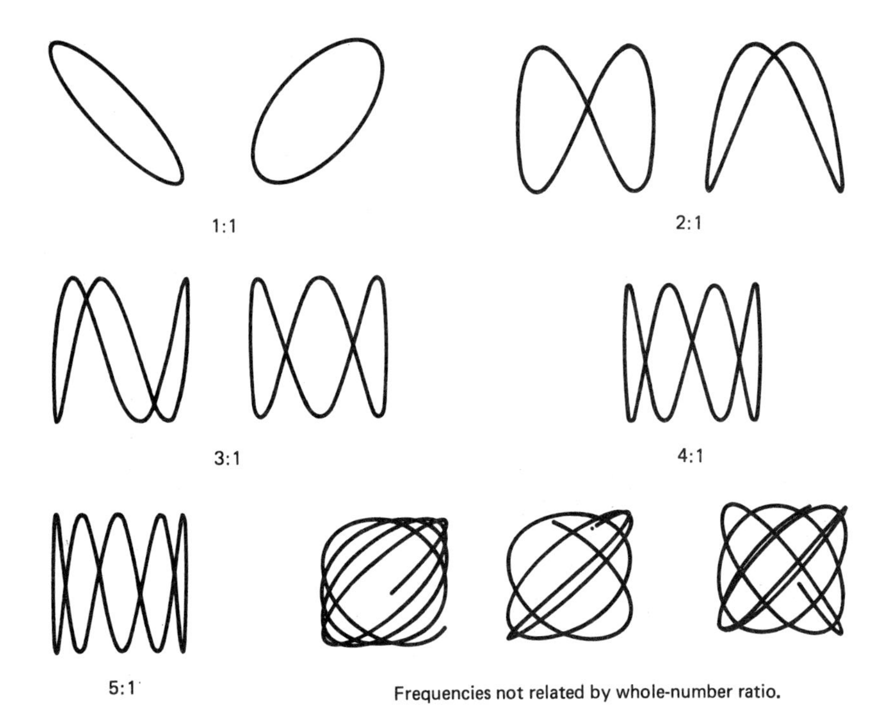

Figure 17-14 Lissajous patterns of various frequency ratios.

through the equilibrium position simultaneously, but traveling in opposite directions. A 90° phase difference produces a one-quarter-cycle offset between the two motions. One reaches a maximum displacement while the other is at the equilibrium position, and so on. The phase and amplitude are not related. Two motions can be in phase even though their amplitudes are unequal. Sine waves out of phase by various amounts are shown in Fig. 17-15.

Coupled pendulums

If two pendulums of nearly the same length are attached to a loosely strung supporting string, as shown in Fig. 17-16, the resulting motion when one pendulum is started is quite surprising. Suppose pendulum A is pulled aside and released, while B remains at or near the equilibrium position. As A swings back and forth, we note that B seems to hesitate for a moment and then begins to oscillate also. The amplitude of B is small at first, but it gradually increases while the amplitude of A becomes smaller. Soon their amplitudes become equal, but the amplitude of B continues to increase. After a while, A almost comes to rest; at the same time, B oscillates with the full amplitude originally given to A. The process then reverses, with the amplitude of B gradually diminishing to zero while the amplitude of A builds up to its original value. The cycle then repeats until the pendulums finally run down.

Because the motion of one pendulum affects the motion of the other, the pendulums are said to be *coupled*. The degree of coupling may be lessened by

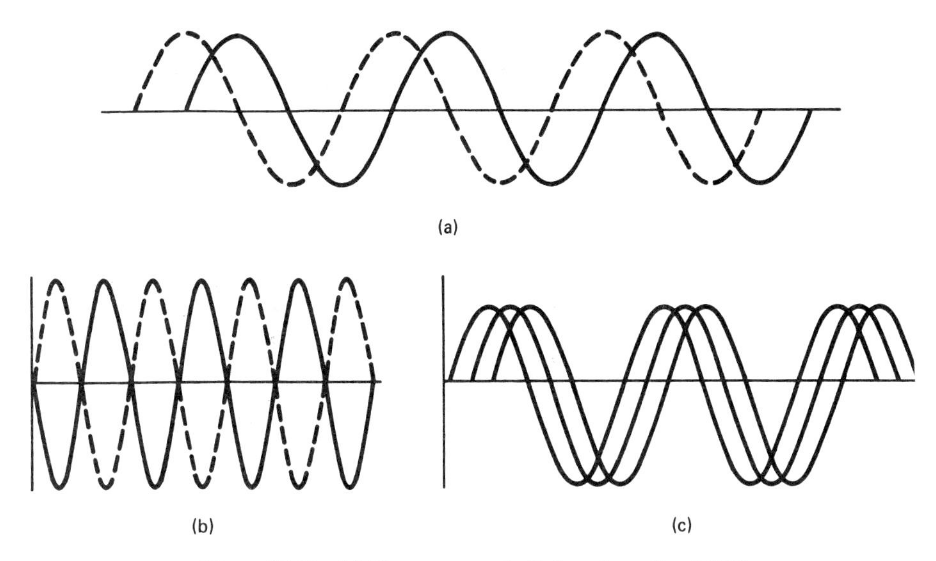

Figure 17-15 (a) Sine waves out of phase by $\pi/2$ rad (90°).
(b) Sine waves out of phase by π rad (180°).
(c) Three sine waves differing in phase by 1 rad (57.3°).

positioning the pendulums farther apart on the supporting string, and it may be increased by putting them closer together. We say that A and B are *loosely coupled* or *tightly coupled*, as the case may be. The analysis of coupled systems is an important part of physics and engineering.

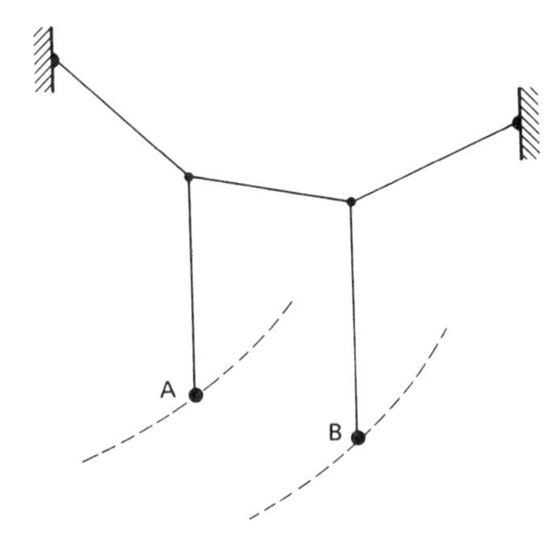

Figure 17-16 Coupled pendulums.

17-9 A DRIVEN OSCILLATOR; DAMPING

All real vibrating systems (oscillators) are subject to frictional losses that eventually rob the system of its energy and cause the vibration to cease. Consequently, energy must be continually supplied to a system if the vibration is to continue at constant amplitude. The process of reducing the amplitude through frictional losses is called *damping*, and the resulting sinusoidal waveform of decreasing amplitude is a *damped sine wave*, or a *damped oscillation*, illustrated in Fig. 17-17(a).

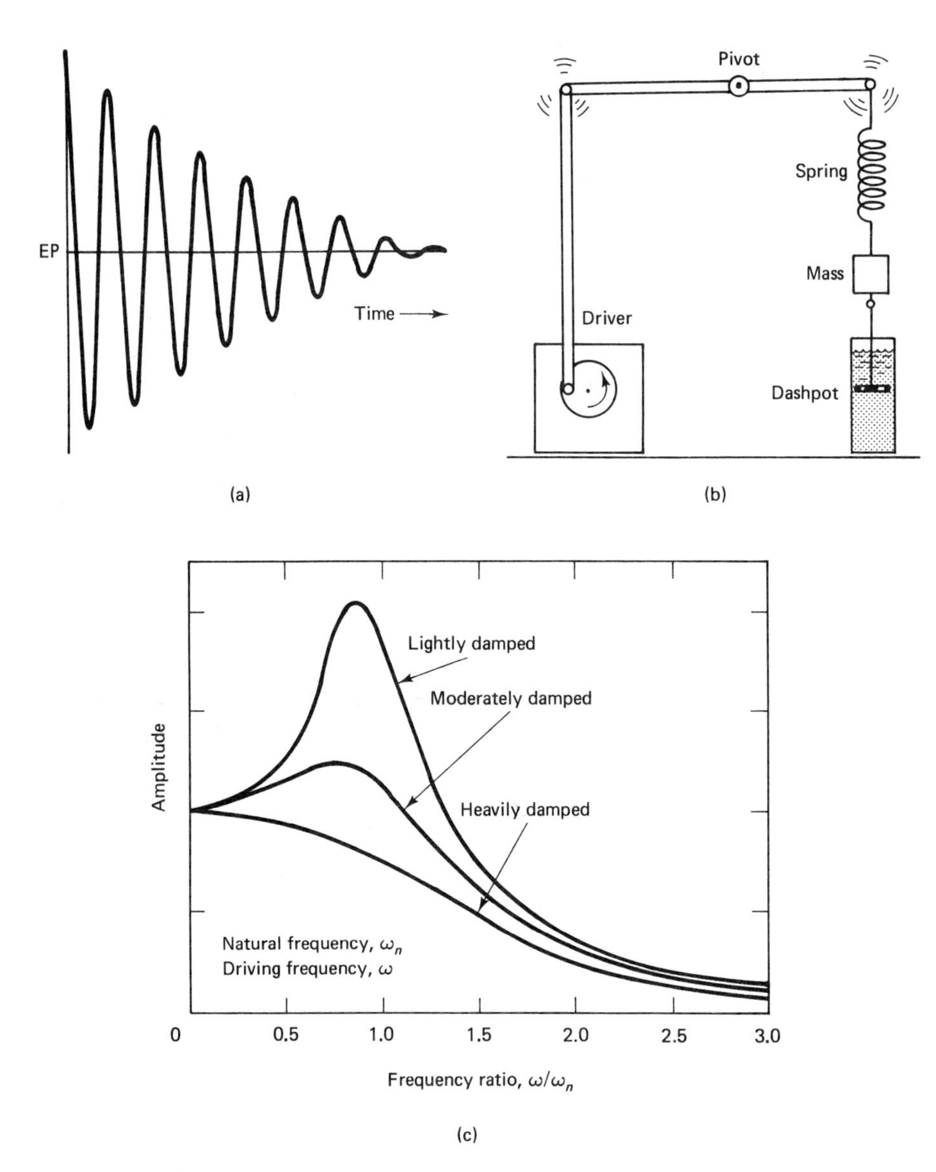

Figure 17-17 (a) A damped oscillation.
(b) A driven oscillating system with damping.
(c) Amplitude vs. frequency for three cases of damping. The same driving force was used for all three curves.

Sometimes it is desirable to provide for the damping of a vibration. The shock absorbers on a car serve this function. An experimental system in a laboratory may incorporate a *dashpot* to provide damping, as shown in the system of Fig. 17-17(b). A dashpot consists of a perforated piston that moves inside a cylinder filled with a viscous fluid.

In Fig. 17-17(b), a variable-speed motor drives a rocker arm, which supports the vibrating system. The movement of the arm produces the driving force for the spring-mass-dashpot system, and the frequency of the driving force can be altered by changing the speed of the motor. Changes in damping can be achieved by using a dashpot of different design. With this system we can see how the vibrational amplitude of the mass depends upon the driving frequency and the degree of damping.

With the dashpot disconnected and the motor turned off, the spring and mass execute SHM at the *natural frequency* of the system (assuming an initial displacement was provided). However, with the dashpot connected and the motor turned on, the mass will no longer vibrate at the natural SHM frequency, but rather it will vibrate with the same frequency as that of the driving force. Moreover, if the driving frequency changes, so does the frequency of vibration of the mass.

If the vibrational amplitude of the mass is recorded for a range of driving frequencies from very low to very high, it is found that large amplitudes occur only for driving frequencies fairly close to the natural frequency of the system. If the driving frequency is swept across the natural frequency, the amplitude increases very sharply to a maximum at the natural frequency, and it then decreases sharply as the sweep of the driving frequency continues in the same direction. This is illustrated in the graph of Fig. 17-17(c) for three degrees of damping.

The condition existing when the driving frequency equals the natural frequency of the system is called *resonance*, and the natural frequency is often called the *resonant frequency*. Hence, the maximum response of the system occurs at the resonant frequency. Observe that the greatest response occurs when the system is damped the least. Heavily damped systems exhibit a very broad response with a very moderate increase in amplitude at the resonant frequency. Systems with minimal damping exhibit a greater overall response and a dramatic increase in amplitude at the resonant frequency.

Mechanical and architectural structures of all kinds are often subjected to vibrational forces from a variety of natural and man-made sources. Moreover, all structures have natural frequencies of vibration. It is exceedingly important in the design of any structure—a bridge, a tall building, an aircraft—for the natural frequencies of the structure to be widely separated in frequency from that of any vibrational forces that might come to bear on the structure. An unexpected resonance may result in vibrational amplitudes sufficiently large to cause catastrophic failure of the structures—sometimes with tragic consequences.

17-10 CRITICAL DAMPING; STABILITY

In Fig. 17-18, a mass is held above the equilibrium position by a string, which we propose to cut. The mass will then descend to the EP where it will be supported entirely by the springs. Our interest is in the motion of the mass as it makes the transition from one position to the other. The nature of the motion will be determined by the degree of damping provided by the dashpot.

Let us consider two extremes, one for which the dashpot is removed entirely and the other with the dashpot connected and filled with molasses. It is clear that

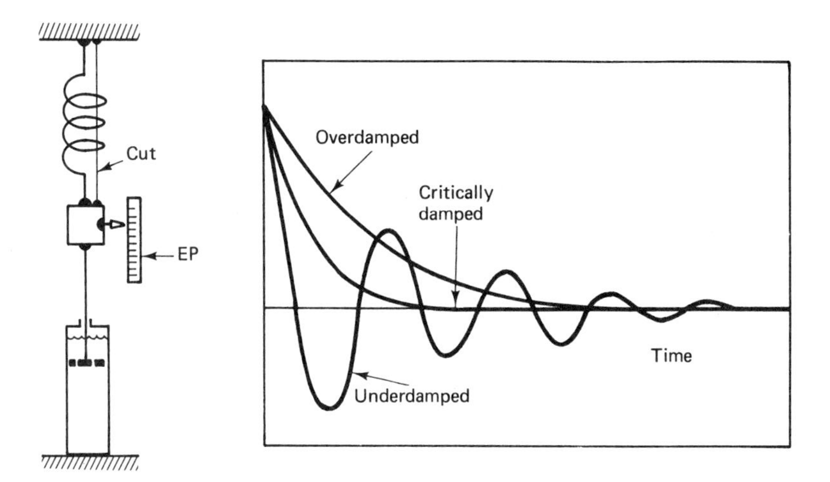

Figure 17-18 The type of motion that occurs after the string is cut depends upon the degree of damping provided by the dashpot.

without the dashpot the mass will overshoot the EP and perform a damped oscillation. It will come to rest at the EP only after the oscillation runs down, and this might take a long time. In this configuration, the system is said to be *underdamped.* On the other hand, the molasses-filled dashpot will allow the mass to settle to the EP only very slowly and without oscillation. The system is then said to be *overdamped.* No overshoot occurs when the system is overdamped. Between these extremes lies the degree of damping, called *critical damping*, that causes the mass to reach the new position in the minimum time, without overshoot and without oscillation. The system is then *critically damped.*

While this discussion relates to the mechanical system illustrated in the figure, the same effect occurs in any area of physics or engineering where a parameter having the property of inertia must change from one configuration to another. For example, the variation of voltage in an electronic circuit containing inductance and capacitance may exhibit exactly the same type of behavior as shown in Fig. 17-18. This occurs because once a voltage in such a circuit begins to change, it tends to keep on changing in the same direction (the property of inertia).

Suppose a car traveling at high speed encounters a gust of wind that makes it suddenly bear slightly to the right. What will happen if the driver panics, screams, and places both hands over his or her eyes, leaving the car to take its own course? The *desired response* of the car to such a disturbance would be for the vehicle to resume a stable path in a straight line, giving the driver time to regain composure and then bring the vehicle back to its original course with only a minor movement of the steering wheel. *Undesirable responses* include the possibility that the curve to the right will become increasingly sharp, or an oscillation of increasing amplitude might develop, or the vehicle could possibly overcompensate the gust toward the right and enter an uncontrolled swerve to the left. The possibilities are illustrated in Fig. 17-19.

The ability of a system to withstand and recover from impulsive disturbances is referred to as the *stability* of the system. Stability is a major concern of

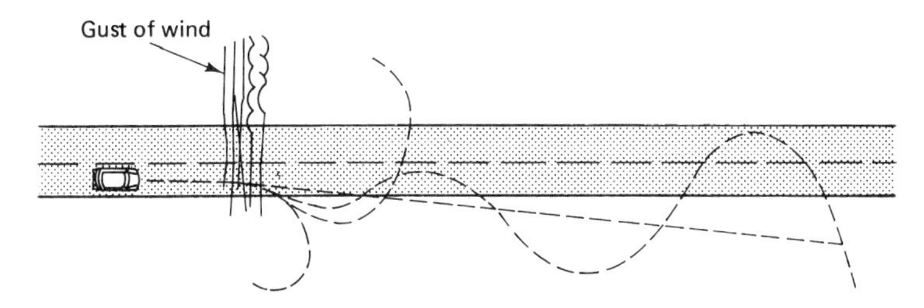

Figure 17-19 Several possible responses of a vehicle to a sudden gust of wind from the left, assuming the driver takes no corrective action.

those who design automatic controls, automobiles, aircraft, nuclear reactors, and heating systems, among *many* others.

Questions

1. On which portion of the swing of a pendulum are the velocity and acceleration directed: **(a)** in the same direction; **(b)** in opposite directions?
2. Suppose the amplitude of a simple harmonic oscillator is doubled. How is each of the following affected?
   (a) The maximum displacement.
   (b) The maximum velocity.
   (c) The maximum acceleration.
   (d) The maximum force exerted on the vibrating member.
3. Suppose the length of a simple pendulum is doubled. How does this change the **(a)** frequency; **(b)** period?
4. If a grandfather clock were taken to the moon where the acceleration of gravity is only about one-sixth its value on earth, would the clock run too fast or too slowly?
5. If the angular frequency ω of a simple harmonic oscillator is doubled, how are the following quantities affected?
   (a) Period.
   (b) Maximum velocity.
   (c) Maximum acceleration.
   (d) Maximum force exerted on the vibrating member.
6. At what point or points on the swing of a pendulum is the potential energy: **(a)** a maximum; **(b)** a minimum? **(c)** At what point is the kinetic energy a maximum?
7. How does a physical pendulum differ from a simple pendulum? In actuality, are not all pendulums physical pendulums to some extent? Why?
8. Define the center of oscillation of a physical pendulum. How is it related to the center of percussion?
9. As the pivot point of a physical pendulum is moved closer to the center of gravity, does the period increase or decrease?
10. How does a torsion pendulum differ from a physical pendulum?
11. Give a step-by-step procedure for determining the shear modulus of copper using a torsion pendulum. What equipment or instruments would be required?
12. What was the importance of a torsion pendulum to the Cavendish experiment for determining G?
13. Early Wright brothers' aircraft had little if any natural stability, but the craft did have an effective means of control (wing warping, rudders, elevators). Describe how the lack of stability increased the work load of the pilot flying the craft.
14. When the engine in a small plane is throttled back from cruise power to idle while in level flight, should the nose of the plane tend naturally to rise or fall in order for the plane to fly by itself in a stable manner?
15. The wings on most aircraft have positive dihedral (they point upward along their length when viewed from the cockpit). How does this provide natural stability against rolling tendencies? (This is a vector problem.)

Problems

1. At 60 mi/h, the crankshaft of a typical 6-cylinder engine rotates at 3000 rpm.
   - **(a)** What is the frequency of rotation in hertz?
   - **(b)** What is the angular frequency ω?
   - **(c)** What is the period?
2. What is the period of a vibrating speaker whose frequency is: **(a)** 30 Hz; **(b)** 100 Hz; **(c)** 1000 Hz; **(d)** 10 kHz?
3. What is the frequency of a vibration whose period is: **(a)** 0.01 s; **(b)** 1 ms; **(c)** 1 μs?
4. The angular frequency of a vibration is 2 rad/s. The displacement from the EP is zero when $t = 0$. What is the phase of the motion after: **(a)** 0.1 s; **(b)** 0.5 s; **(c)** 1.0 s; **(d)** π s?
5. Suppose the angular frequency is 2 rad/s as in the previous problem, but let the displacement be a maximum when $t = 0$. What is the phase of the motion after: **(a)** 0.1 s; **(b)** 0.5 s; **(c)** 1.0 s; **(d)** π s?
6. The angular frequency of a pendulum is 2 rad/s. Its amplitude of vibration is 10 cm, and the bob passes through the EP when $t = 0$. Calculate the position of the bob when the time t is: **(a)** 0.1 s; **(b)** 0.5 s; **(c)** 1.0 s; **(d)** π s. (Refer to Problem 4.)
7. Suppose a simple pendulum vibrates with an amplitude of 10 cm according to the conditions of Problem 5. Calculate the position of the bob after: **(a)** 0.1 s; **(b)** 0.5 s; **(c)** 1.0 s; **(d)** π s.
8. A tall pendulum in a science museum is 10 m long and swings with an amplitude of 0.6 m.
   - **(a)** Compute the period of the pendulum.
   - **(b)** Compute the angular frequency of the pendulum.
   - **(c)** How many swings will the pendulum make per day?
9. For the pendulum in Problem 8, suppose a stopwatch is started at 0 when the pendulum bob reaches maximum displacement near the observer. When the stopwatch reads 8 s, what will be: **(a)** the position of the pendulum; **(b)** the velocity of the pendulum bob; **(c)** the acceleration of the pendulum bob?
10. What will be the maximum **(a)** velocity and **(b)** acceleration of the pendulum bob of Problem 8?
11. When a 100-g mass is hung on a certain spring, the spring stretches 5 cm.
    - **(a)** Compute the force constant of the spring in newtons per meter.
    - **(b)** Compute the period of the SHM when the mass is set in motion.
12. In an industrial application, a platform on bearings is to be vibrated back and forth at a frequency of 2 Hz with an amplitude of 10 cm. The mass of the platform is 50 kg. If the platform executes SHM, compute the maximum force in pounds that must be exerted on the platform.
13. Repeat Problem 12 if the frequency is to be 3 Hz instead of 2 Hz.
14. Repeat Problem 12 for an amplitude of 15 cm instead of 10 cm.
15. A uniform brass rod 1 cm in diameter is 2 m long. It is pivoted 50 cm above its center.
    - **(a)** Compute the mass of the rod. (Density of brass is 8.7 g/cm^3.)
    - **(b)** Compute the moment of inertia about the center of gravity.
    - **(c)** Compute the moment of inertia about the pivot point.
    - **(d)** Compute the period of the resulting physical pendulum.
16. If the rod described in the previous problem is pivoted closer to the CG, the period will be increased. Compute the period of the rod if it is pivoted: **(a)** 25 cm above the pivot; **(b)** 10 cm above the pivot.
17. **(a)** Compute the length of the equivalent simple pendulum for the physical pendulum of Problem 15.
    (b) Locate the center of oscillation of the physical pendulum.
18. Locate the center of oscillation for the physical pendulum as described in **(a)** and **(b)** of Problem 16.
19. Suppose the rod of Examples 17-7 and 17-9 is pivoted at the center of oscillation as determined in Example 17-9. Compute the period of the resulting physical pendulum,

and compare with the result of Example 17-7.

20. A solid sphere made of aluminum is 10 cm in diameter and has a mass of 1.414 kg. It is suspended on a wire whose torsion constant is 0.25 N · m/rad.

(a) Compute the period of oscillation of this torsion pendulum.

(b) Compute the time required for 50 oscillations to occur.

21. Two lead balls, each of mass 200 g, are mounted on a very light rod 1 m long. The resulting dumbbell is suspended on a quartz fiber, and the resulting torsional pendulum has a period of 2 min.

(a) Compute the moment of torsion of the fiber.

(b) What torque must be exerted on the fiber to twist the dumbbell through an angle of 0.02 rad?

(c) Compute the distance each mass must move to produce the angular displacement of 0.02 rad.

(d) What force, applied at an angle perpendicular to the rod, must be exerted on each mass in order to maintain the angular displacement?

(e) Explain the relevance of this problem to the Cavendish balance used to measure G (Sect. 9-2).

22. The force constant of a mass balance is 60 N/m, and the effective mass of the platform is 0.3 kg. A mass of 1 kg is placed on the platform. Compute the time required for 50 cycles of the loaded platform to occur.

23. The force constant of a mass balance is known to be 78.9 N/m. When set in motion, the unloaded platform oscillates 50 times in 22.37 s. When an unknown mass is attached to the platform, the platform oscillates 50 times in 38.74 s. Determine the unknown mass.

24. Strictly speaking, a simple pendulum is not isochronous; its period varies slightly with amplitude. The exact expression for the period is given by

$$T = 2\pi \sqrt{\frac{l}{g}} \left(1 + \frac{1}{2^2} \sin^2 \frac{\theta}{2} + \frac{1 \cdot 3^2}{2^2 \cdot 4^2} \sin^4 \frac{\theta}{2} + \cdots\right)$$

where θ is the maximum angular displacement. Calculate the period of a simple pendulum one meter long for the following values of θ: **(a)** 15°; **(b)** 30°; **(c)** 60°; **(d)** 90°. Compare with the result of Example 17-5.

CHAPTER 18

WAVES

Waves and wave motion are familiar. We are surrounded by waves: sound waves, light waves, water waves, and waves on the strings of a guitar. Each type of wave may be characterized by a frequency (or period), a wavelength, a velocity, an amplitude, and intensity.

In this chapter we are concerned with these properties of waves and others. We find that standing waves are responsible for the tuning characteristics of guitar strings and organ pipes. We investigate the interference and diffraction of waves and we define the intensity of a wave.

18-1 INTRODUCTION

Everyone is familiar with the waves that occur on the surface of water whether the water is in a pond, lake, ocean, or bathtub. If we look down from the end of a fishing pier, we can observe the crests of waves as they pass underneath the pier enroute to shore. We may observe that the wave crests are regularly spaced and that they travel at a uniform velocity. It might at first appear that the bulk of the water underneath the pier is moving toward the shore, but this is only an illusion, as evidenced by the fact that objects floating on the water do not move toward the shore with the waves. Floating objects rise and fall with only a slight horizontal movement as a wave passes. It is only near the shore where the waves lap up on the beach that an appreciable horizontal movement of the water occurs.

If we toss a pebble into a pond that is otherwise calm and peaceful, a pattern of circular waves is formed which spreads out evenly in all directions from the point where the pebble hit the water. Again, the uniform wave velocity is evident, and small objects bob up and down as the waves pass. We infer that wave motion represents a transport of energy because, when the waves reach the shore, the water is stirred up and caused to slosh minutely. This requires an expenditure of energy. Kinetic energy originally possessed by the pebble is carried to shore by the waves. Indeed, a transport of energy is associated with all types of wave motion.

In addition to the wave velocity, the wavelength and frequency are important parameters useful in describing a wave. The *wavelength* is the distance from a given point on one wave to the corresponding point on the following wave. For example, the distance from the crest of one water wave to the crest of the following wave is one wavelength. Equivalently, the distance from the trough of one wave to the trough of the next is one wavelength, where the *trough* of a wave is the low point occurring between two crests.

The *frequency* is the number of waves that pass a given point in a given interval of time, or, if the source of the waves is in evidence, the frequency is the number of waves produced in a given time interval. The *period* of a wave is the time required for the passage of one complete wave and is the reciprocal of the frequency.

The *amplitude* of a wave refers to the maximum departure of the medium from the equilibrium position; the larger the amplitude, the larger the wave. Surfers prefer waves of large amplitude whereas fishermen prefer the waves to have as little amplitude as possible. Figure 18-1 shows several aspects of a wave.

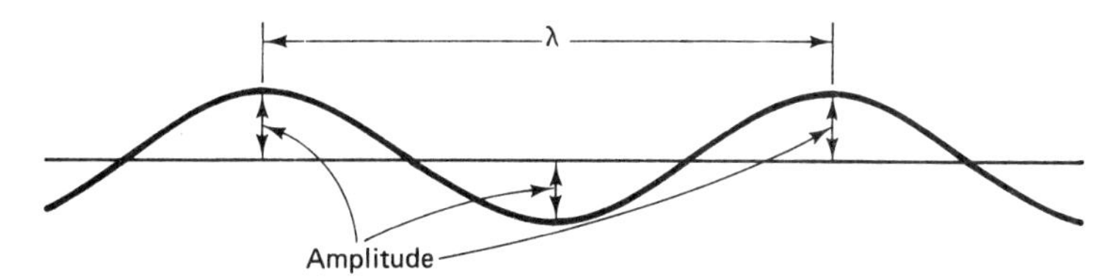

Figure 18-1 Amplitude and wavelength of a wave.

A mathematical relationship exists between the frequency f, wavelength λ, and wave velocity v:

$$f\lambda = v \quad \text{or} \quad f = \frac{v}{\lambda} \quad \text{or} \quad \lambda = \frac{v}{f} \tag{18-1}$$

For a given medium, the wave velocity tends to be a constant, so that an inverse relationship exists between the frequency and the wavelength. High frequencies give short wavelengths, and vice versa.

The crest of an ocean wave forms a straight line at right angles to the direction in which the wave is traveling. Such a wave is called a *plane* wave because the crest of a similar wave in three dimensions (a sound wave, for example) forms a plane (flat surface) at right angles to the line of the travel of the wave. On the other hand, circular waves are produced when we toss a pebble into a pond, and the three-dimensional counterpart of circular waves is *spherical* waves. Spherical sound waves are produced by an explosion occurring in midair. As spherical or circular waves travel out from the source, they become more and more like plane waves as the curvature of any small section of the wave front becomes less and less. (A *wave front* is any recognizable portion of a wave, such as the crest or trough of a water wave or the compression peak of a sound wave.) Hence, we conclude that the wave fronts of plane waves form a succession of planes, and the wave fronts of spherical waves form a set of concentric spheres as the waves travel outward from the source.

A line drawn perpendicular to a wavefront and in the direction of travel of the wave is called a *ray*. Rays are often useful in depicting the paths of waves.

18-2 PULSE ON A STRETCHED STRING

If one end of a stretched string or heavy cord is suddenly displaced sideways, the entire length of the string does not instantaneously assume the new position. The disturbance travels down the length of the string at a certain velocity and requires a definite amount of time to reach the far end. Furthermore, when the disturbance encounters the support at the far end, it will be reflected and will then travel back toward the end that was displaced. The disturbance will gradually die out, however, and leave the string in the new position. This is illustrated in Fig. 18-2.

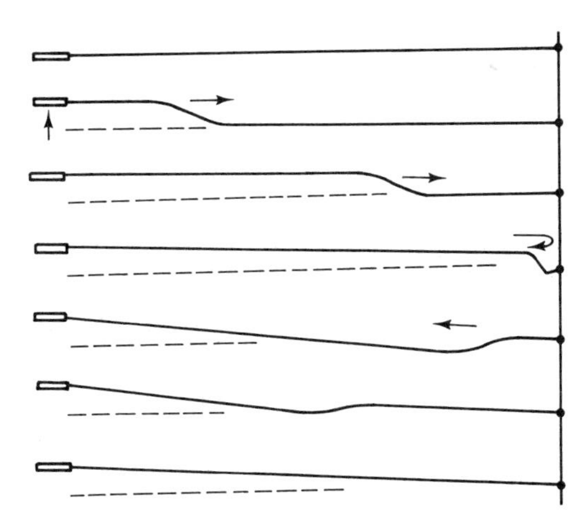

Figure 18-2 When one end of a string is suddenly displaced, the entire string does not assume the new position instantaneously. A disturbance travels down the length of the string and is reflected, perhaps several times, before it finally dies out.

A *pulse* can be produced in the string by displacing one end of the string sharply and then quickly returning the end to its original position. The pulse is then observed to travel the length of the string and to be reflected back to the point where it was produced; if the pulse has not been dissipated, it may once again be reflected. This is shown in Fig. 18-3.

If a series of experiments is performed with a variety of strings, cords, and springs, each of which is subjected to various tensions, it is found that the velocity of the pulse depends upon the tension in the string and upon its mass per unit length. The pulse moves faster if the string is stretched very tightly, but the pulse moves slower (for the same tension) along heavier cords or springs. The velocity of a pulse on a stretched string is given by

$$v = \sqrt{\frac{T}{m/L}} = \sqrt{\frac{T}{\rho}} \tag{18-2}$$

where T is the tension in the string and $\rho = m/L$ is the mass per unit length of the string.

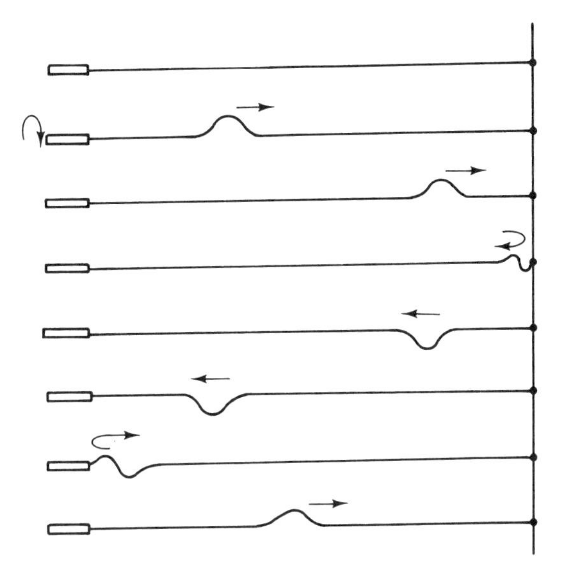

Figure 18-3 Production and reflection of a pulse on a stretched string.

EXAMPLE 18-1 A laboratory spring of mass 0.26 kg is stretched to a total length of 4 m by an applied force of 2.45 N. Calculate the velocity of a pulse traveling along the spring.

Solution. First we compute the mass per unit length of the spring:

$$\rho = \frac{m}{L} = \frac{0.26\ \text{kg}}{4\ \text{m}} = 0.065\ \text{kg/m}$$

Then, using Eq. (18-2) with $T = 2.45$ N, we obtain

$$v = \sqrt{\frac{T}{\rho}} = \sqrt{\frac{2.45}{0.065}} = 6.14\ \text{m/s}$$

Reflection of a pulse

We mentioned earlier that a pulse is reflected when it encounters the support at the far end of a stretched string. Such is the case, but the pulse is *inverted* in the process if the point of attachment of the string to the support cannot move. That is, if an upright pulse approaches the support and is reflected, the reflected pulse will be identical in shape to the incoming pulse but it will be upside down. Anticipating the following discussion of continuous waves, we say that a phase change of 180° occurs when a wave is reflected from an immovable support. This is illustrated in Fig. 18-4(a).

Quite a different thing occurs if the far end of the string is completely free to move rather than being tied to a fixed support. An incoming pulse will be reflected, but it will not be inverted. An incoming upright pulse will be reflected as an upright pulse. Hence, no phase change occurs when a wave is reflected from the freely movable end of a stretched string. This is shown in Fig. 18-4(b).

A practical method of obtaining a freely movable end of a stretched string is illustrated in Fig. 18-5. A laboratory demonstration spring serves as the wave

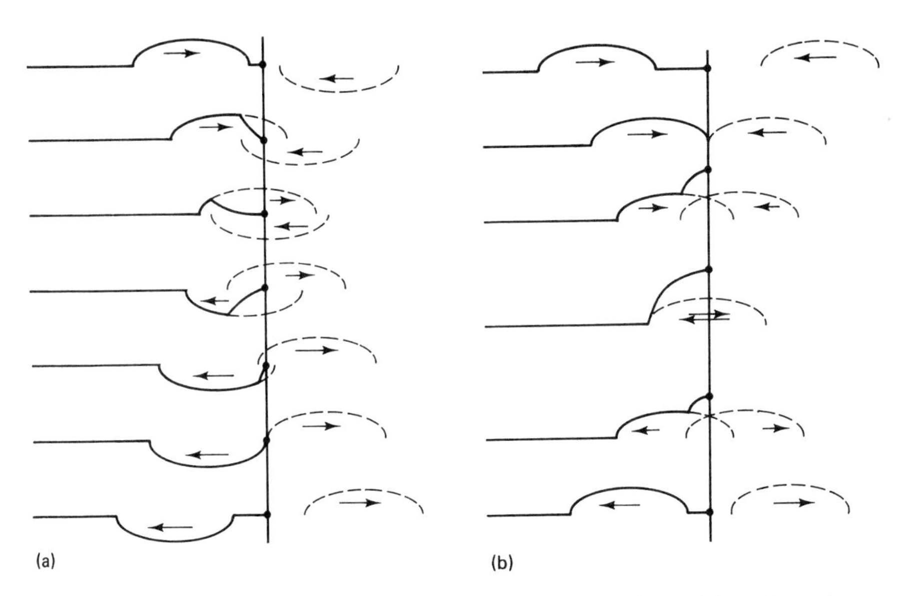

Figure 18-4 The reflection of a pulse occurs as a superposition of the pulse and its mirror image.
(a) When the end of the string is fixed, the mirror image is inverted and the reflected pulse is inverted.
(b) When the end of the string is not fixed, the mirror image is not inverted.

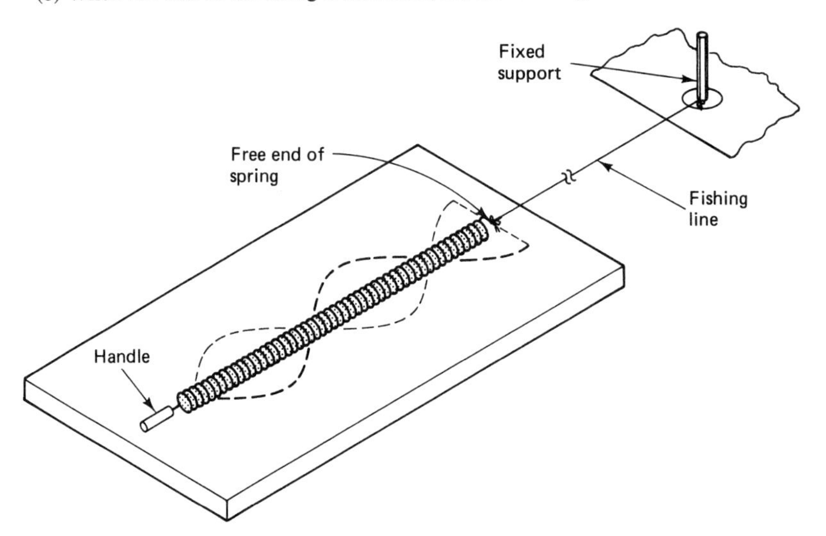

Figure 18-5 A practical method of maintaining the tension in a laboratory demonstration spring without restricting the movement of the free end of the spring.

medium, and the end of the spring connects to a fishing line that extends about another 15 ft to a fixed support. The fishing line maintains the tension in the spring without impeding the side-to-side motion of the end of the spring.

Suppose a pulse proceeding to the right encounters a second pulse on the same string proceeding to the left. What happens as the pulses collide? Before the collision, while the pulses are yet apart, each proceeds as if the other did not exist. But when the pulses begin to overlap, and for as long as there is overlap, the displacement of the string in the region of overlap is the algebraic sum of the individual displacements of the two pulses. The end result is that the two pulses appear to pass through each other without suffering any change in shape, velocity, or direction of travel.

Reflection at a discontinuity

Generally speaking, a reflection will occur whenever a pulse or wave encounters a discontinuity in the medium in which it is propagating. One possible discontinuity is shown in Fig. 18-6 in which a pulse is approaching the junction of a light string

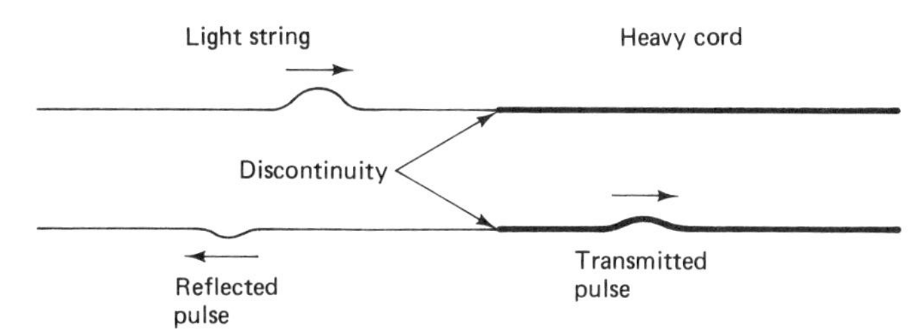

Figure 18-6 When an incoming pulse in the light string encounters the junction of the string and cord, a reflection occurs at the same time a pulse of smaller amplitude is transmitted into the heavy cord.

and a heavier cord. The incoming pulse will be reflected at the junction and will be inverted as if reflected from a fixed support. At the same time, however, an upright pulse will be produced in the heavy cord which will continue toward the right. Because the original energy of the incoming pulse is divided between the reflected and transmitted pulses, the amplitudes of both pulses will be less than that of the original pulse.

The velocity of the pulse transmitted into the heavy cord will be less than that of the reflected pulse. This causes the wavelength of the transmitted pulse to be shorter than the original because both strings are under the same tension and because, for a given tension, the pulse velocity is lower where the mass per unit length is greater. A similar thing occurs in optics when the light waves pass from air into water. Light waves travel slower in water and the wavelength is foreshortened because the frequency remains constant.

18-3 CONTINUOUS WAVES

The standard equipment of most physics laboratories includes a mechanical *wave generator* for producing a continuous train of pulses on a string or cord. One form consists of a heavy weighted rod that vibrates back and forth in the horizontal plane. The vibration is sustained by electrical pulses delivered to an electromagnet alongside the rod.

If a *very long* string is attached to the wave generator, any pulses reflected from the far end will be damped out long before they return to the vicinity of the generator. Consequently, the motion of the string near the generator will not be complicated by any returning pulses, and a steady progression of waves will travel away from the generator. If the generator executes simple harmonic motion, the shape of the *continuous wave* will be that of a sine wave. The crests and troughs will move along the string in the same manner that water waves pass under a fishing pier.

The velocity of the waves is determined by the tension in the string and the mass per unit length of the string, as in Eq. (18-2). The frequency is controlled by the generator, and the wavelength depends upon the velocity and frequency according to Eq. (18-1). The wavelength can be made shorter either by increasing the frequency or by reducing the tension in the string.

Each point on the string executes SHM from side to side as the waves travel along the length of the string. But the vibrating points are not all in phase because the relative phase varies uniformly with distance from the generator. Only points separated by a distance of an integral number of wavelengths will be in phase. If the phase of the generator is taken as a reference, a point on the string a distance of one wavelength from the generator will vibrate in phase with the generator. On the other hand, a point only one-half wavelength from the generator will be exactly out of phase (180°) with the generator, and a point one-quarter wavelength from the generator will be out of phase by 90°. Thus, distance from the generator translates directly to phase difference. Points differing in phase by 360° (one complete cycle) are in phase. Figure 18-7 illustrates the variation of phase with distance.

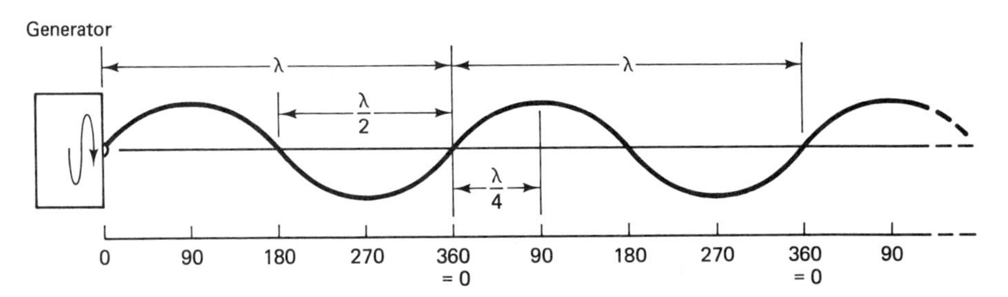

Figure 18-7 The phase of a point on the string lags the phase of the generator an amount proportional to the distance of the point from the generator. A distance of one wavelength corresponds to a phase difference of 360°, or 2π rad.

Mathematically, the variation of phase with distance x from the generator is given by

$$\Phi = 2\pi \frac{x}{\lambda} \tag{18-3}$$

where Φ, the relative phase, is in radians. For sinusoidal waves, the displacement y of a point on the string at time t is given by

$$y = A \sin\left[\frac{2\pi x}{\lambda} - \omega t\right] \tag{18-4}$$

where A is the amplitude of the wave and $\omega = 2\pi f$. This is the equation of a wave traveling along the x-axis in the positive direction.

EXAMPLE 18-2 The frequency of the waves on a light cord is 8 Hz, and the wave velocity is 24 m/s. Compute the relative phase of two points on the cord a distance of 1.5 m apart.

Solution. The wavelength is 3.0 m because $\lambda = v/f$. Therefore, using Eq. (18-3), the phase difference is

$$\Phi = 2\pi\frac{x}{\lambda} = 2\pi\left(\frac{1.5}{3.0}\right)$$

$$= \pi \text{ rad} = 180°$$

EXAMPLE 18-3 If the waves in Example 18-2 have an amplitude of 5 cm, compute the instantaneous displacement of a point on the cord 7 m from the generator when $t = 2.0$ s.

Solution. Anticipating the use of Eq. (18-4), we first calculate

$$\frac{2\pi x}{\lambda} = 2\pi\left(\frac{7}{3}\right) = 14.66 \text{ rad} \qquad \omega = 2\pi(8) = 50.27 \text{ rad/s}$$

Then,

$$y = A \sin\left[\frac{2\pi x}{\lambda} - \omega t\right]$$

$$= (5)\sin[14.66 - 50.27(2.0)]$$

$$= (5)\sin[-85.87]$$

$$= 5(0.866) = 4.33 \text{ cm}$$

18-4 STANDING WAVES

When a continuous wave encounters the fixed support at the far end of the string, the reflected wave will be a continuous wave that travels back toward the generator. The actual motion of the string will then be the resultant, or *superposition*, of two individual waves, namely, the direct wave coming from the generator and the wave reflected from the end support. The motion of the string will no longer depict a simple movement of waves as it did in the absence of reflections. At points where the direct and reflected waves are in phase, their effects will be additive resulting in a motion of the string that is much greater than if only one wave were present. But where the two are out of phase, one will tend to cancel the other so that little motion of the string results.

Whether the waves will be in phase at a particular point depends upon the distance of the point (in wavelengths) from the end support. Consider point A, which is one-quarter wavelength ($\lambda/4$) from the end, as shown in Fig. 18-8. Let us determine the relative phase of the reflected wave at that point. The phase of the direct wave changes by 90° in the region from A to the end. Further, we recall that a reflection from a fixed point produces a phase change of 180°. Finally, the phase

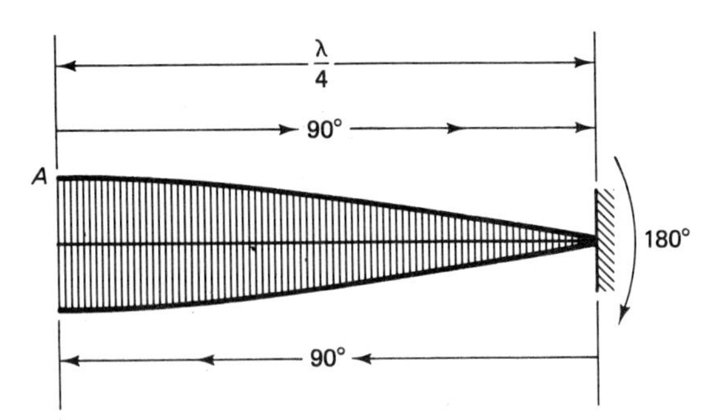

Figure 18-8 The phase difference between incident and reflected waves at a distance $\lambda/4$ from the fixed end support is 360°. This is equivalent to being in phase, so a large motion of the string is expected at that point.

of the reflected wave changes by 90° in the region from the end to A. The total phase difference is 360°. But this is equivalent to no phase difference at all, so the direct and reflected waves will be in phase a distance of $\lambda/4$ from the end support. Consequently, we expect a large motion of the string at that point.

Similar reasoning for a point a distance of $\lambda/2$ from the end shows that the two waves will be 180° out of phase. Hence, little motion of the string is expected at that point. At a distance of $3\lambda/4$ from the end, the waves are again in phase, but they are out of phase at a point one wavelength λ from the end.

If the total length of the string from generator to fixed support is an integral number of half-wavelengths, the string will vibrate in a stable pattern, as shown in Fig. 18-9. No wave motion along the length of the string will be apparent.

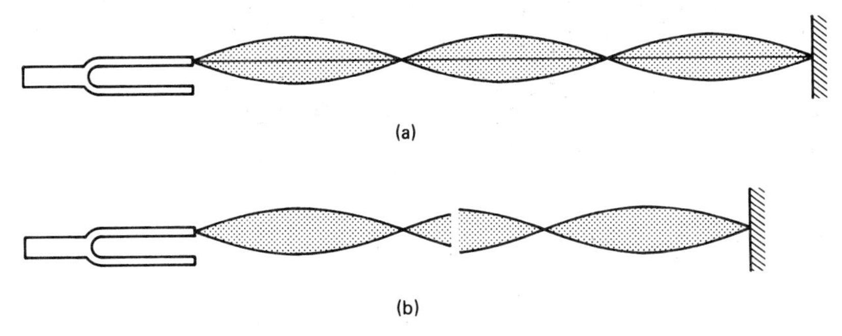

Figure 18-9 The total length of the string must be an integral number of half-wavelengths for a stable pattern of nodes and antinodes to be formed.

Therefore, the pattern is called a *standing wave*. The points of minimum amplitude are called *nodes*, and the points of maximum amplitude are called *antinodes*. If the string length is not an integral number of half-wavelengths, a stable pattern of nodes and antinodes will not be formed for the following reason.

Obviously, the far end of the string cannot move since it is attached to the support. Therefore, a node appears at the fixed, far end. Likewise, a node appears at (or very near) the generator because the generator amplitude is small in comparison to the amplitude at an antinode. We must now realize that reflections occur at the generator end of the string in the same manner as at the other end, and with the same result. Incident and reflected waves are in phase at distances of $\lambda/4$, $3\lambda/4$, . . . from the generator, but are out of phase at distances of $\lambda/2$, λ, $3\lambda/2$, ... Thus, a pattern of phases is developed at each end of the string. For a stable standing wave to be formed, these patterns must match up where they meet at the middle of the string. Furthermore, the patterns will match up when the total length of the string is an integral number of half-wavelengths.

Recall that the wavelength can be altered by adjusting the tension in the string or the frequency of the generator. Thus, a stable standing wave can usually be achieved by adjusting the length of the string, the tension in the string, or the frequency of the generator.

Allowed frequencies for standing waves

Suppose a particular string of length L is under a fixed tension T. We wish to find the frequencies of vibration that will produce a stable standing wave. This problem is more significant than it might first appear because when a stretched string—a guitar string, for example—is plucked, picked, or bowed (in the case of a violin), the resulting frequencies produced are the same as those producing stable standing waves. Consequently, the next calculation amounts to finding the natural frequency of vibration of a string on any of a multitude of stringed musical instruments.

Combining Eqs. (18-1) and (18-2) gives the following expression for λ:

$$\lambda = \frac{v}{f} \qquad v = \sqrt{\frac{T}{\rho}} \qquad \lambda = \frac{1}{f}\sqrt{\frac{T}{\rho}} \tag{18-5}$$

For a stable standing wave, the length L of the string must be an integral number of half-wavelengths. In mathematical terms, this means that

$$N\left(\frac{\lambda}{2}\right) = L \tag{18-6}$$

where it is understood that the value of N can be any positive integer.

Substituting Eq. (18-5) into Eq. (18-6) to eliminate λ gives the following expression for the resonant frequencies of a stretched string fixed at both ends:

$$f_n = N\left(\frac{1}{2L}\right)\sqrt{\frac{T}{\rho}} \qquad N = 1, 2, 3, \ldots \tag{18-7}$$

From the equation we note that the frequency increases with an increase in tension of the string and with a decrease in length of the string. This is in accordance with our experience in tuning and fretting a string on a guitar.

The physical significance of the various frequencies is shown in Fig. 18-10. The lowest frequency, called the *fundamental* and denoted f_1, is obtained when the string vibrates in one part. The next higher frequency f_2 occurs when the string vibrates in two parts, and so on for $f_3, f_4, \ldots$. The fundamental is also known as the *first harmonic*, f_2 is the *second harmonic*, f_3 is the *third harmonic*, and so forth. Furthermore, the various harmonics are *overtones*; the *second* harmonic is the *first* overtone. Likewise, the *third* harmonic is the *second* overtone, and the *fourth* harmonic is the *third* overtone, and so on. The number of the overtone is one less than that of the harmonic.

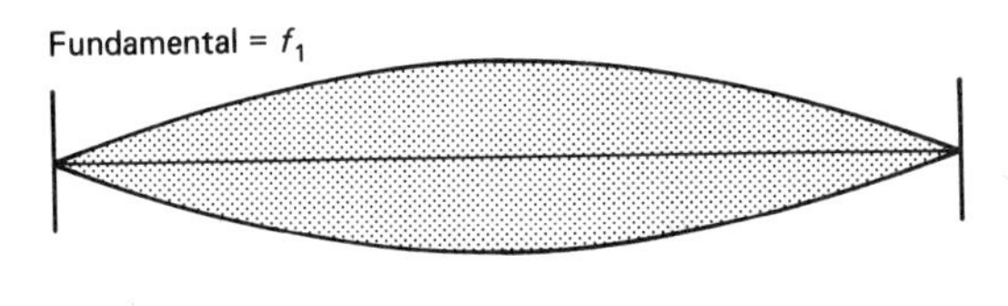

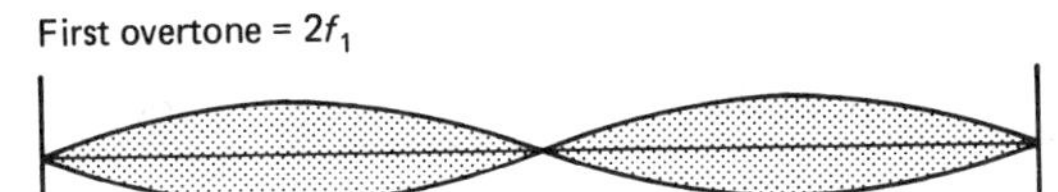

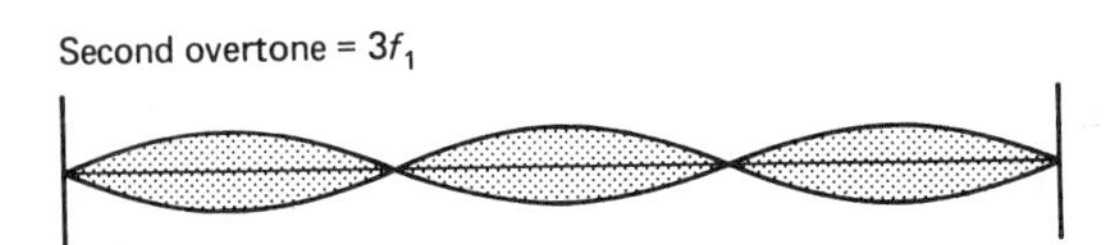

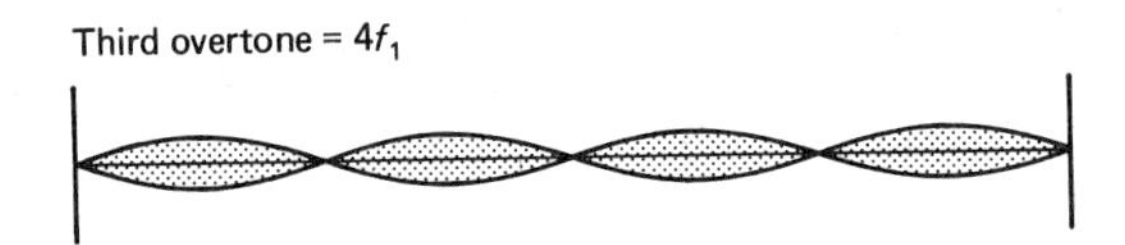

Figure 18-10 The first four modes of vibration of a string fixed at both ends.

The distinction between harmonics and overtones is that harmonics are always overtones, but overtones are called *harmonics* only if the overtone frequencies are integral multiples of the fundamental frequency. Metal bars, flat plates, and drumheads, for example, vibrate in modes such that the overtones are not integral multiples of the fundamental frequency. We say that these overtones are *not harmonically related.*

A string can vibrate in many of the allowed frequencies at the same time. The complex vibratory motion of a guitar string is the superposition of various amplitudes of the harmonics. The harmonic content, the relative amplitudes of the higher harmonics, gives a string its tonal quality. A guitar string picked near the center of the string sounds different than a string picked near the end, close to the bridge. More harmonics are excited by picking near the bridge, and a large harmonic content produces a tinny sound due to the high frequencies of the higher order harmonics.

EXAMPLE 18-4

The string of highest frequency on a certain guitar is 65 cm long and has a mass per unit length of 4.9×10^{-4} kg/m. The tension in the string is 90.2 N. Calculate: (a) the fundamental frequency; (b) the frequency of the third harmonic.

Solution. (a) A straightforward application of Eq. (18-7) is all that is required, with $N = 1$:

$$f_n = N\left(\frac{1}{2L}\right)\sqrt{\frac{T}{\rho}}$$

$$f_1 = 1\left[\frac{1}{2(0.65)}\right]\sqrt{\frac{90.2}{4.9 \times 10^{-4}}}$$

$$= 330.0 \text{ Hz}$$

(b) The frequency of the third harmonic is simply three times the fundamental:

$$f_3 = 3f_1 = 3(330.0)$$
$$= 990 \text{ Hz}$$

18-5 LONGITUDINAL WAVES

Until now our discussion has been addressed to *transverse* waves in which particle displacements occur at right angles to the direction of motion of the waves. We now describe *longitudinal* waves in which particle motions occur back and forth in line with the direction of wave travel. Whereas lateral displacements characterize transverse waves, longitudinal waves are characterized by regions of *compression* (sometimes called *condensation*) and *rarefaction*. A longitudinal wave in a stretched spring is illustrated in Fig. 18-11. A sudden movement of the

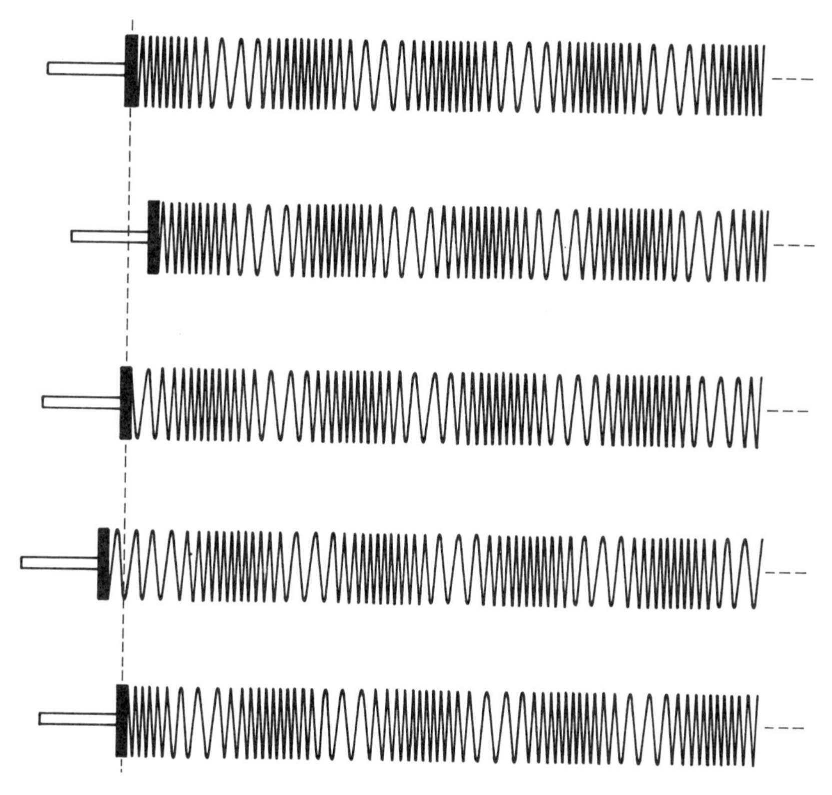

Figure 18-11 A longitudinal wave in a stretched string.

plunger to the right causes the spring to be compressed immediately ahead of the plunger. This region of compression then moves away and travels down the length of the spring. As the plunger reverses its motion, a region of the spring adjacent to the plunger is stretched, and the stretched region will also travel down the length of the spring. The stretched region may be called a region of rarefaction. Thus, a

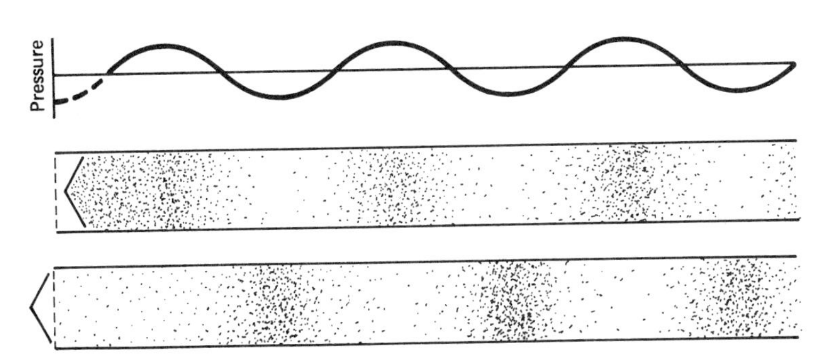

Figure 18-12 Regions of compression and rarefaction in a sound wave. The density variation is much exaggerated for purposes of illustration. The pressure variation for the upper tube is indicated.

longitudinal wave is produced and travels along the length of the spring, and the wave consists of alternating regions of compression and rarefaction.

A sound wave is a longitudinal wave that consists of alternating regions of high and low pressure, as shown in Fig. 18-12. Forward movement of the speaker cone produces a region of compression in which the pressure is slightly higher than normal, and backward movement of the cone produces a rarefied region in which the pressure is slightly lower. In the figure, a transverse sine curve is drawn to indicate the wavelength of the longitudinal wave, but this is not to imply that the sound wave has any transverse characteristics.

The *pressure amplitude* in even the loudest sounds is extremely small, amounting to about 2.8×10^{-4} atm. If the frequency of the sound is 1 kHz, the maximum displacement of an air molecule from its equilibrium position is about 10^{-5} m. For faint sounds close to the threshold of hearing, pressure amplitudes are on the order of 10^{-9} atm and molecular displacements may be only a fraction of one molecular diameter. This points out the extraordinary sensitivity of the ear in being able to respond to such small quantities.

At 20°C, the speed of sound in air is about 344 m/s (1128 ft/s). A sound wave of frequency 1 kHz will consequently have a wavelength of

$$\lambda = \frac{v}{f} = \frac{344}{1000} = 34.4 \text{ cm}$$

This is about 13.5 in., which indicates that even though the pressure amplitudes and displacements of a sound wave are very small, the wavelengths are of ordinary dimensions. At 20°C, for frequencies of 20 Hz and 20,000 Hz (the textbook limits of the audio range), the wavelengths are 56.4 ft and 0.68 in., respectively.

Longitudinal standing waves

Longitudinal standing waves may be produced in a variety of enclosures ranging from musical instruments—flute, clarinet, organ pipe—to long columns of air inside a pipe, and even in large rooms such as an auditorium or a cathedral. The primary requirement is that the waves must be reflected efficiently from both ends

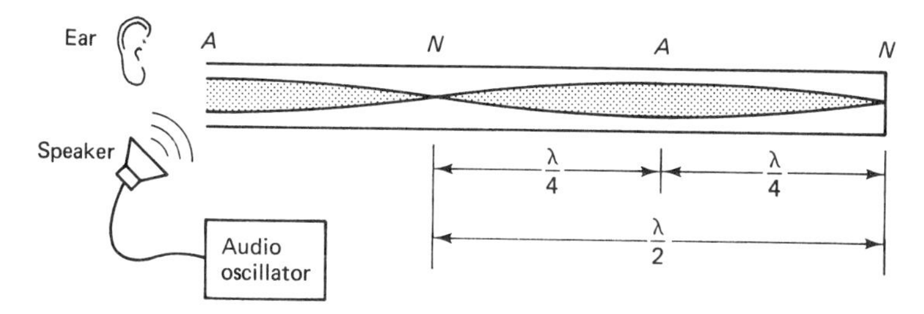

Figure 18-13 An increase in volume will occur when a stable standing wave is established within the tube.

of the enclosure. The frequency of possible standing waves depends upon the dimensions of the enclosure.

A longitudinal standing wave can be demonstrated with the apparatus shown in Fig. 18-13. A small speaker driven by a variable frequency audio generator is held near the open end of the tube. As the frequency is varied, a noticeable increase in volume will be observed at certain frequencies. This *resonance* indicates that a standing wave has been established within the tube.

As for transverse standing waves, nodes and antinodes will be set up in the tube. Air molecules exhibit maximum longitudinal vibration at the antinodes and minimal vibration at the nodes. A node always appears at the closed end of the tube because the end cap prevents longitudinal movement of the molecules. Moreover, when a standing wave is present, an antinode always appears at (or near) the open end of the tube. Between the two ends, it is possible for several nodes and antinodes to appear, and when present, a node always appears between two antinodes, and so forth. The allowed configurations of nodes and antinodes allow us to determine the resonant frequencies of the tube, the resonant frequencies being those frequencies for which stable standing waves are possible.

From the figure we see that for resonance to occur, the length L of the tube must be an odd number of quarter-wavelengths:

$$L = N\left(\frac{\lambda}{4}\right) \qquad N = 1, 3, 5, \ldots$$

If v is the speed of sound and f is the frequency, we may substitute v/f for λ to obtain

$$(\textit{closed column}) \quad L = N\left(\frac{v}{4f}\right) \quad \text{or} \quad f_n = N\left(\frac{v}{4L}\right) \qquad N = 1, 3, 5, \ldots \tag{18-8}$$

From the equation we note that longer tubes resonate at lower frequencies. Also, we note that the frequency of the first overtone, f_3, is three times the fundamental f_1. The second harmonic is missing, as is true for all even harmonics.

Resonance in an open column

Standing waves in columns open at both ends must produce antinodes at both ends with at least one node in between. This requires the length of the tube to be an even number of quarter-wavelengths, as may be surmised from Fig. 18-14. Hence, as above,

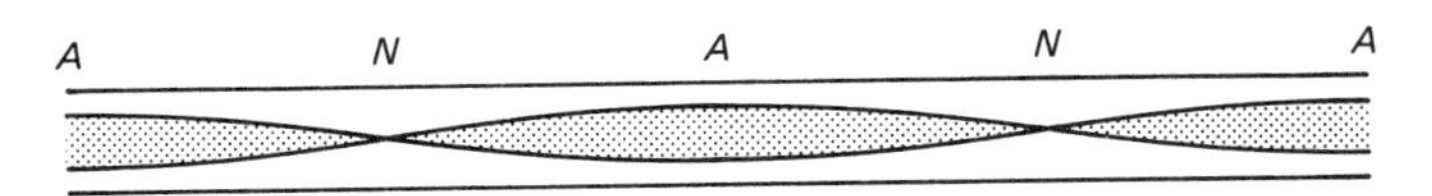

Figure 18-14 Antinodes occur at each end of an open tube when a standing wave is excited within the tube.

$$L = N\left(\frac{\lambda}{4}\right) \qquad N = 2, 4, 6, \ldots$$

and

$$(open\ column) \qquad f_n = N^*\left(\frac{v}{2L}\right) \qquad N^* = 1, 2, 3, \ldots \tag{18-9}$$

In this case the overtones are integral multiples of the fundamental so that none of the harmonics are missing.

The difference in overtone structure between open and closed columns causes the character of the sound obtained from the two to be somewhat different. Indeed, it is the overtone structure that causes a flute to sound different from a clarinet, even when both play the same note; for the same reason, the sound of a clarinet is different from a saxophone, and the sound of an oboe is distinctly different from that of a bassoon. Musicians may praise one violin over another because of its superior tone quality. Tone quality stems, in part, from the overtone structure.

EXAMPLE 18-5 Compute the allowed frequencies of vibration of the column of air inside a glass tube 120 cm long and 5 cm in diameter. Assume the tube is closed at one end and that the speed of sound is 344 m/s.

Solution. Direct application of Eq. (18-8), with $N = 1$, gives

$$f_1 = 1\left(\frac{v}{4L}\right)$$

$$= 1\left[\frac{344\ \text{m/s}}{4(1.2\ \text{m})}\right] = 71.67\ \text{Hz}$$

The frequencies of the overtones are odd multiples of this value:

$$f_3 = 3f_1 = 3(71.67) = 215.0\ \text{Hz}$$
$$f_5 = 5f_1 = 358.4\ \text{Hz}$$
$$f_7 = 7f_1 = 501.7\ \text{Hz}$$
$$\vdots$$

EXAMPLE 18-6 Assume the tube in Example 18-5 is open at both ends. Calculate the resonant frequencies and note the effect of the end cap upon the frequencies.

Solution. Using Eq. (18-9) with $N^* = 1$ yields

$$f_1 = 1\left(\frac{v}{2L}\right)$$

$$= 1\left(\frac{344\ \text{m/s}}{2(1.2)}\right) = 143.33\ \text{Hz}$$

For an open tube, the overtones are integral multiples of f_1:

$$f_2 = 2f_1 = 286.6 \text{ Hz}$$
$$f_3 = 3f_1 = 430.0 \text{ Hz}$$
$$f_4 = 4f_1 = 573.3 \text{ Hz}$$

Removing the end cap doubles the fundamental frequency and changes the harmonic structure so that all harmonics are present.

18-6 INTERFERENCE AND DIFFRACTION OF WAVES

Suppose two speakers generating sound waves of the same frequency are set up side by side, as shown in Fig. 18-15. Even if the speakers are in phase, the sound waves arriving at a listener will not be in phase unless the listener happens to be an

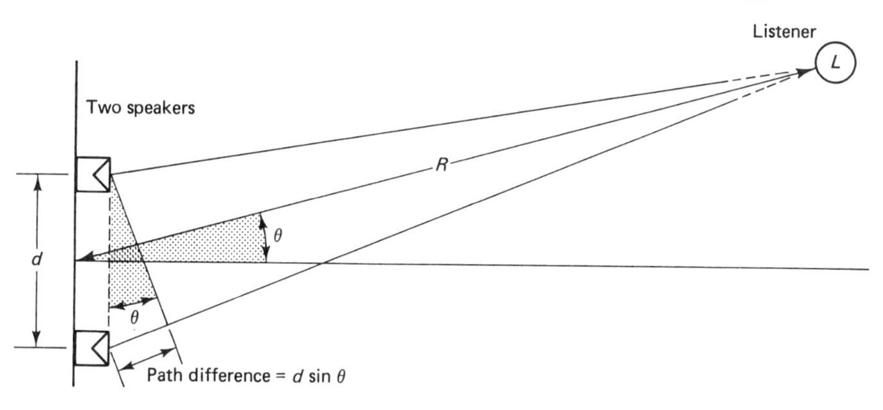

Figure 18-15 When R is large compared to d, the path difference is approximately $d \sin \theta$.

equal distance from each speaker. Therefore, unless the listener positions himself or herself on a line passing between the speakers, the listener will receive the two waves out of phase by a certain amount.

The principle of superposition states that when two waves arrive at the same point, the resultant is the algebraic (or vector) sum of the two waves, each acting as if the other were not present. (We made use of this earlier in our consideration of standing waves.) Therefore, the listener will hear the *resultant* of the individual waves from each speaker. If he or she moves to a point where they are in phase, the sound will appear louder. But if he or she moves to a point where they are out of phase, the sound level will be significantly reduced.

We established earlier that the relative phase of a wave varies uniformly along its length. It follows that the relative phase of a wave arriving at a listener depends upon the distance of the listener from the speaker that produced the wave. In particular, the difference in phase of the two waves depends directly upon the difference in distance that each wave must travel to reach the listener. We call this difference in distance the *path difference*.

When the path difference is zero, the waves will be in phase and a loud sound will be heard. Furthermore, when the path difference equals an integral number of complete wavelengths, the waves will be effectively in phase so that the sound will be loud. But when the path difference is an odd multiple of half-wavelengths, the waves will be exactly out of phase and minimal sound will be heard. In other words, an *interference pattern* is formed in which the two waves may produce either *constructive* interference or *destructive* interference, corresponding to areas of loud or soft sound levels, respectively.

All types of wave motion are capable of interference effects. Ripple tanks demonstrate the interference of water waves, as shown in Fig. 18-16. The interference of light waves is readily demonstrated, as we shall see in a later chapter.

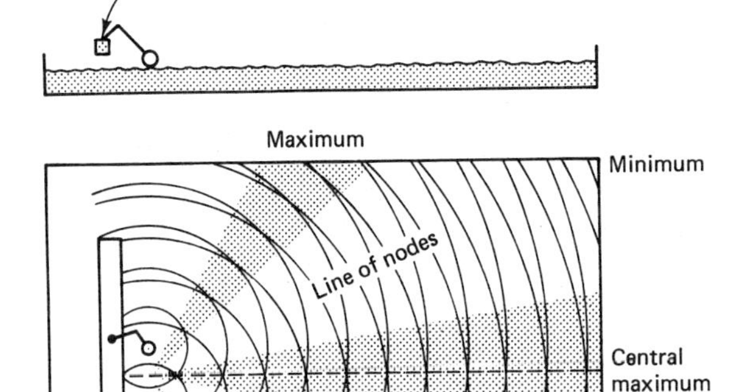

Figure 18-16 The interference of water waves demonstrated in a ripple tank. Lines of nodes separate the lines along which the wave amplitude is a maximum.

In Fig. 18-15, the path difference is shown to be $d \sin \theta$ when R is large compared to d. Setting this equal to $N(\lambda/2)$ gives an expression for the angular direction in which only minimal sound will be heard.

$$(\textit{destructive interference}) \qquad \theta = \sin^{-1}\left(\frac{N\lambda}{2d}\right) \qquad N = 1, 3, 5, \ldots \tag{18-10}$$

Of course, $N\lambda/2d$ must be less than or equal to 1.0 because the sine of an angle can never be greater than unity. Physically, this means that no destructive interference will occur unless the speaker separation is greater than at least *one* half-wavelength:

$$d \geqslant \frac{N\lambda}{2} \tag{18-11}$$

By setting the path difference equal to $N\lambda$, with $N = 0, 1, 2, 3, \ldots$, the following is obtained for *constructive* interference.

$$(\textit{constructive interference}) \qquad \theta = \sin^{-1}\left(\frac{N\lambda}{d}\right) \tag{18-12}$$

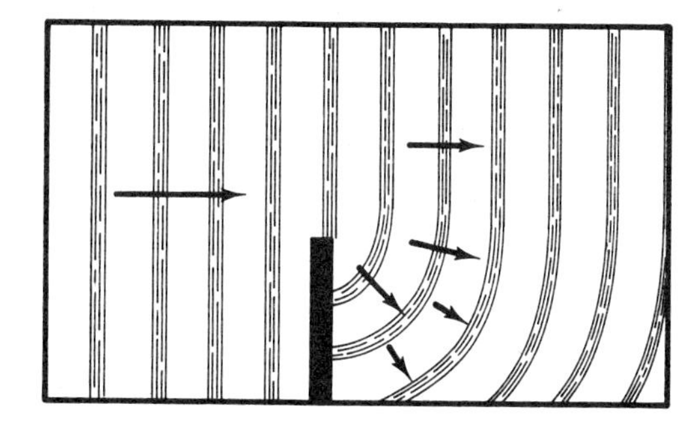

Figure 18-17 Diffraction causes plane waves incident upon an obstacle in a ripple tank to bend around into the region behind the obstacle. The amplitude of the waves behind the obstacle is considerably reduced, however.

The interference effects described here are not readily observed with speakers set up in an ordinary room because of the large amount of sound the listener receives through reflection from the walls. A carpeted auditorium or an outdoor setting is much preferred.

EXAMPLE 18-7 Two speakers 4 ft (1.22 m) apart, as in Fig. 18-15, are driven in phase by a 440-Hz tone. Calculate the angle relative to the center line that gives the direction of the first region of destructive interference. Take the speed of sound to be 344 m/s.

Solution. We first calculate the wavelength of the 440 Hz sound wave to be 0.782 m. Then, Eq. (18-10) with $N = 1$ gives

$$\theta = \sin^{-1}\left(\frac{N\lambda}{2d}\right)$$

$$= \sin^{-1}\left[\frac{1(0.782)}{2(1.22)}\right] = \sin^{-1}[0.320]$$

$$= 0.326 \text{ rad} = 18.7°$$

Regions of destructive interference are symmetrically located on both sides of the center line. Setting $N = 2$ gives $\theta = 0.694$ rad, or 39.8°, for the second region on either side of the center line.

Diffraction

The phenomenon responsible for the spreading of waves around corners or protruding walls is called *diffraction*. It can be readily observed in water waves and in light waves with the proper apparatus, but it is more difficult to demonstrate with sound simply because we cannot see or photograph sound waves. The diffraction of water waves is illustrated in Fig. 18-17.

Diffraction arises because of a property of wave motion known as *Huygens' principle*: Each point on a wavefront acts as a source of circular (or spherical) waves which propagate in the forward direction and determine the subsequent path and shape of the wave front. The application of Huygens' principle is illustrated in Fig. 18-18. Experimental evidence of Huygens' principle is illustrated in Fig. 18-19. Plane water waves incident upon two small openings in a barrier produce circular waves on the other side of the barrier as if each opening were a source of waves. Moreover, the circular waves from the two openings will produce an interference pattern, as described earlier.

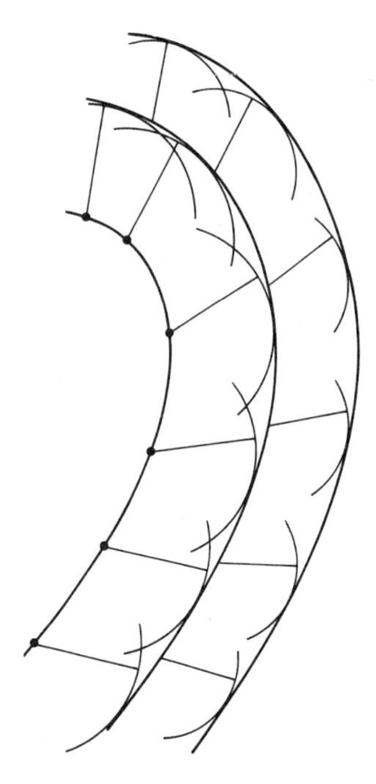

Figure 18-18 Determining the progress of a wavefront by Huygens' principle. The new wavefront is drawn tangent to the small circles drawn with centers located on the old wavefront. Only the most-forward point of the small circles is considered.

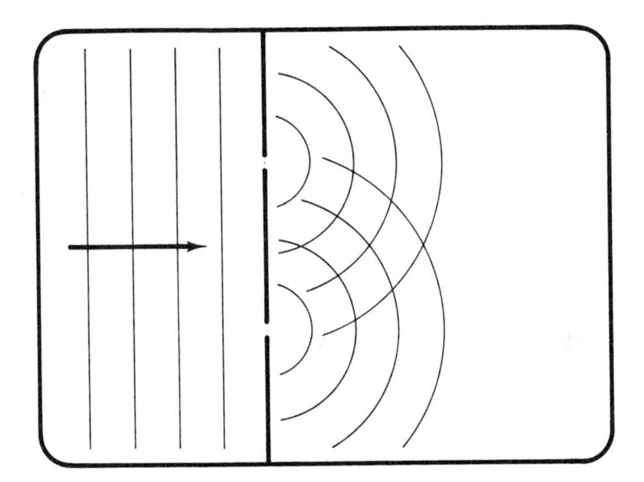

Figure 18-19 Experimental evidence of the validity of Huygens' principle. Plane waves striking small openings in the barrier placed across a ripple tank cause each opening to act as a secondary source of waves.

18-7 THE INTENSITY OF A WAVE

In the introduction to this chapter we reason from water waves that wave motion (except standing waves) involves a transport of energy in the direction of motion of the wave. The intensity of a wave is the average rate at which energy is transported across a surface perpendicular to the direction of travel of the wave. Power is the energy transferred per unit time, so the intensity of a wave is the average power transmitted per unit area of the perpendicular surface. The units of wave intensity are power per unit area, typically watts per square meter.

To make this more definite, let us consider a source of sound waves positioned high in the air, and let us assume that acoustical energy is transmitted equally in all directions. Such an omnidirectional source is called an *isotropic* source. The wavefronts form spheres of ever-increasing radius about an isotropic source. If we imagine a spherical surface of radius R_1 to be centered on the source, the wave intensity at a distance R_1 from the source is given by

$$I_1 = \frac{\text{energy/time}}{\text{surface area}} = \frac{E/t}{4\pi R_1^2} \tag{18-13}$$

If a larger sphere of radius R_2 is imagined to be concentric with the first, we see that the same energy per unit time E/t passes through the larger sphere as through the smaller. Hence, the intensity of the wave at a distance R_2 from the source is

$$I_2 = \frac{E/t}{4\pi R_2^2} \tag{18-14}$$

Dividing one of the preceding equations by the other gives

$$\frac{I_1}{I_2} = \frac{R_2^2}{R_1^2} \quad \text{or} \quad I_2 = I_1\left(\frac{R_1}{R_2}\right)^2 \tag{18-15}$$

and we see that the wave intensity is inversely proportional to the square of the distance from an isotropic source. That is, if we double the distance from the source, the wave intensity decreases by a factor of 4. This is true for all types of wave motion, whether mechanical or electromagnetic.

Another universal property of wave motion is that the intensity is proportional to the square of the amplitude of the wave. It follows that the amplitude of a wave radiated from an isotropic source is inversely proportional to the first power of the distance from the source. If the distance is doubled, the amplitude is cut in half.

If the waves are concentrated into a beam, as light waves from a searchlight or laser beam, the intensity decreases far more slowly with distance because the waves do not spread out as much. The spreading of a laser beam is very slight indeed, so the decrease in intensity of a laser beam is primarily due to absorption or scattering of the light by the atmosphere. A laser beam sent from the earth to the moon (about 238,000 mi) may be no more than a mile in diameter upon reaching the moon.

If we know the intensity I of waves crossing a particular surface of area A, the power transmitted across the surface is given by

$$\text{Power} = IA \tag{18-16}$$

EXAMPLE 18-8 An isotropic source of sound waves has an acoustical power output of 10 W. What is the intensity of the sound waves 5 m from the source?

Solution. All power transmitted must pass through an imaginary sphere of radius 5 m surrounding the source. By definition, the intensity of the wave at 5 m is the power per unit area transmitted across this sphere. Hence,

$$I = \frac{\text{power}}{\text{area}}$$

$$= \frac{10 \text{ W}}{4\pi(5)^2 \text{ m}^2}$$

$$= 3.18 \times 10^{-2} \text{ W/m}^2$$

This corresponds to a sound level of about 100 dB, which is comparable to the sound level produced by a freight train passing at close range.

18-8 COMPLEX WAVES

Not all continuous waves exhibit a sinusoidal wave form. Remember that a sinusoidal wave form results only if the wave generator executes simple harmonic motion. An important principle in wave analysis is that any periodic wave form may be considered as the superposition of a multitude of sine waves of proper amplitude and frequency. If the symbol W represents the complex wave form, we may represent this principle symbolically as

$$W = A_1\sin(1\omega t) + A_2\sin(2\omega t) + A_3\sin(3\omega t) + \cdots$$

where A_1, A_2, A_3, . . . are amplitudes, not all equal; some of the values may be zero. Typically, the lower-numbered amplitudes are larger than higher-numbered amplitudes, but the decrease in amplitude with increasing number is not necessarily *monotonic*. (A sequence of numbers or a mathematical function is said to decrease *monotonically* if each number or value of the function is less than or equal to the preceding value.) Observe that the angular frequencies of the sine waves are integral multiples of the fundamental frequency ω, where ω is the fundamental frequency of the complex wave form. Thus, the sine waves form a *harmonic series*. Unless the wave form passes through zero at $t = 0$, the series may require modification to include cosine terms as well as sine terms. A branch of mathematics that deals in part with the resolution of complex waves into sinusoidal components is *Fourier analysis*, and the series of sinusoids is known as a *Fourier series*.

One application of the above principle is in loudspeaker design. The ability of a high-fidelity speaker to reproduce a complex wave in a passage of music depends upon its ability to accurately reproduce each of the composite sine waves with the proper amplitude and phase. Conversely, if a speaker system accurately reproduces sine waves of all (audio) frequencies and amplitudes, it can reproduce a complex wave form.

18-9 SYMPATHETIC VIBRATION; BEAT NOTES

In our study of resonance in Chap. 17, we found that a vibrating system exhibits maximum response to a driving force when the driving frequency is equal to the natural, resonant frequency of the system. This idea carries over into the realm of vibrating strings, tuning forks, and musical gadgets of many kinds. Suppose two identical tuning forks are available, and that we strike one and hold it near (but not touching) the other. After a short time, we find the second fork to be vibrating also, having been excited by the sound from the first. This phenomenon is called *sympathetic vibration*; the second fork vibrates out of sympathy for the first.

A similar thing occurs when a tuning fork is held near a guitar string tuned to the same frequency. A faint sound can be heard from the guitar after the tuning fork is held near and then removed. This is good, but more impressive is the following.

A piano contains strings tuned to at least 88 different frequencies. Pressing the rightmost pedal lifts the damper pads from the strings so they may vibrate

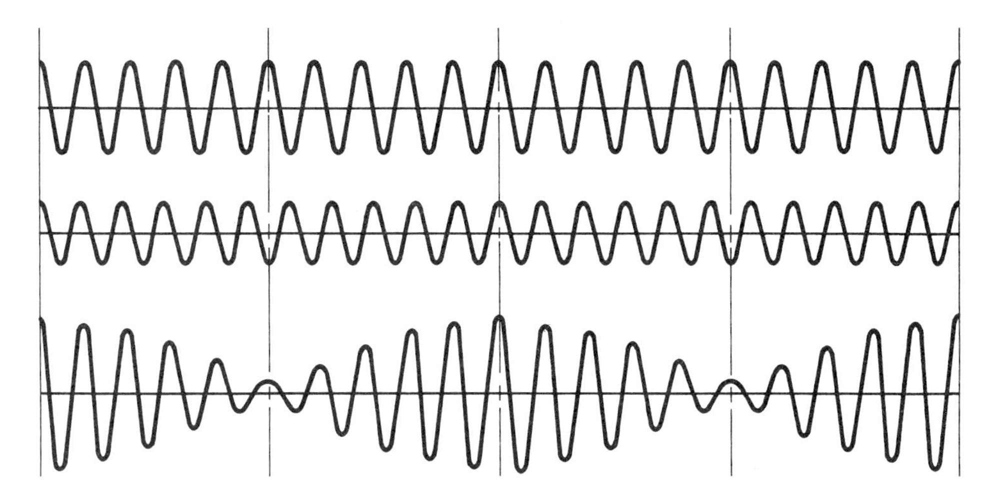

Figure 18-20 When two waves of slightly different frequency are combined, the resulting wave varies in amplitude as the two waves alternately cancel and reinforce each other. The amplitude variation may be heard as a beat note.

freely. If one opens the cabinet and whistles a note directly to the strings, the strings will vibrate sympathetically and return the whistle. If a singer sings a note to the strings, the piano will mimic the sound of the singer. And, of course, the piano will duplicate, to some degree, the sound of other musical instruments played into the strings.

Beat notes

If two strings on a guitar are tuned to very nearly the same frequency—for example, 299 and 300 Hz—and are then plucked at the same time, a slow wavering or undulation of the tone is heard that is called a *beat note*. The frequency of the beat equals the difference in frequency of the two strings—in this case, 1 Hz. As the strings are tuned closer together, the beat becomes slower, so that several seconds may be required for one complete undulation to occur. The beat note is caused by the gradual variation in phase between the two strings, as shown in Fig. 18-20.

Beats also occur when tones widely separated in frequency are sounded together, but the beat may appear as a third tone rather than as a slow variation of volume. For example, a 400-Hz tone sounded with a 600-Hz tone will produce a beat note of 200 Hz. Beat notes may be produced electronically by adding the signals from two audio generators and applying the result to a speaker.

Questions

1. What factors determine the speed of a pulse on a stretched string?
2. Describe the reflection of a pulse on a stretched string: **(a)** if the end of the string is fixed; **(b)** if the end is free to move laterally.
3. What is the phase difference represented by one wavelength?
4. What gives rise to standing waves? Why do standing waves occur only at particular frequencies?

5. What distance in wavelengths separates a node and antinode (assuming they are adjacent)?
6. What is the difference between harmonics and overtones?
7. What factors determine the pitch of a guitar string?
8. What is the difference between interference and diffraction?
9. What causes the intensity of spherical waves to decrease with the square of the distance from the source?
10. Does the intensity of an ideal beam of light decrease with distance?
11. Describe a physical situation that will give rise to the production of a beat note.

Problems

1. An experiment with sound waves in air determines the wavelength of a 440-Hz sound wave to be 0.8 m. Compute the velocity of sound from this information.
2. What is the wavelength of a 10,000-Hz sound wave when the speed of sound is 344 m/s?
3. Certain military radio transmitters operate at the extremely low frequency of 10,000 Hz using antennas measured in miles. Compute the wavelength of a radio wave of this frequency.
4. Compute the wavelength of the following radio waves: **(a)** a 1-MHz AM wave; **(b)** a 27-MHz CB wave; **(c)** a 100-MHz FM wave; **(d)** an 800-MHz TV wave; **(e)** a 10-GHz radar wave.
5. Compute the velocity of a pulse on a stretched string if the tension in the string is 8 N and if a 5 m length of the string has a mass of 1.4 g.
6. How is the velocity of a wave on a stretched string affected by doubling the tension in the string?
7. What mass must be hung on the end of a light string in order to produce a tension in the string of 13.72 N?
8. The wave frequency of a light cord is 12 Hz and the wave velocity is 30 m/s.
   (a) What is the wavelength of the wave?
   (b) Compute the relative phase of two points 1.25 m apart.
   (c) If a standing wave is set up on the cord, what will be the distance between nodes?
   (d) What is the shortest length of cord that will support a 12-Hz, 30 m/s standing wave having only one node between fixed end supports?
9. The frequency of the waves on a light cord is 6 Hz and the wave velocity is 24 m/s. The waves have an amplitude of 10 cm, and the generator passes through the equilibrium position when $t = 0$. Compute the instantaneous displacement of a point on the cord 5 m from the generator when t is: **(a)** 1.00 s; **(b)** 1.02 s; **(c)** 1.04 s; **(d)** 1.06 s; **(e)** 1.08 s.
10. A string with a mass of 0.8 g/m is stretched between two fixed supports 1 m apart with a tension of 12 N. Compute the three lowest resonant frequencies of the string.
11. Calculate the velocity of the waves on a guitar string 65 cm long tuned to 330 Hz.
12. In playing a guitar, the length of a string is changed by pressing the string against a fret. How long should the string of Problem 11 be if the string is to resonate at 349.6 Hz?
13. A section of 4-in. diameter plastic drain pipe is 8 ft long and closed at one end. Compute the first three resonant frequencies of the pipe.
14. Repeat Problem 13, but assume the pipe is open at both ends.
15. What should be the length of an organ pipe open at both ends if it is to resonate at 27 Hz?
16. A glass cylinder borrowed from the chemistry lab is 5 cm in diameter and is 40 cm tall (inside dimensions). A small speaker driven by a variable frequency generator is held near the top of the cylinder while the fre-

quency is varied. At what frequencies should resonance occur?

17. Two small speakers located 1 m apart, as in Fig. 18-15, are driven in phase by a 512-Hz tone. Calculate the angle relative to the center line of the first region of destructive interference.

18. Plane water waves in a ripple tank are incident upon two small openings in a barrier. The openings are 5 cm apart, and the wavelength of the water waves is 3 cm.
  (a) Compute the angle between the center line and the first line of nodes.
  (b) How many lines of nodes appear on each side of the center line?

19. The acoustical power output of an isotropic source of sound waves is 1 W. Calculate the wave intensity at the following distances from the source: **(a)** 1 m; **(b)** 2 m; **(c)** 4 m.

20. The wave intensity 10 m from an isotropic source of sound waves is 2×10^{-3} W/m^2. What is the acoustical power output of the source?

21. The wave intensity at a point 12 m from an isotropic source is 4×10^{-4} W/m^2. What is the wave intensity at a point 16 m from the same source?

22. Two tuning forks are tuned to 440 and 442 Hz, respectively. What is the frequency of the beat note produced?

23. The signals of two signal generators are added electronically, amplified, and fed to a speaker. The generators are set to 1200 Hz and 1300 Hz, respectively. What is the frequency of the beat note produced?

CHAPTER 19

SOUND

This chapter considers sound waves in more detail than the preceding chapter. We begin by giving a method for determining the speed of sound in air. This is followed by treatments of sound wave intensity, amplitude of vibration, and pressure amplitude. Decibels are defined and their use in expressing the intensity of a sound wave is illustrated.

The frequency response of the average human ear is given via the Fletcher-Munson contours of equal loudness. Other subjective aspects of sound include pitch, timbre, and quality. The physical cues that permit a listener to determine the direction of a sound source are also described. Other topics include room acoustics, reverberation, and the Doppler effect.

19-1 THE SPEED OF SOUND

Firm evidence that sound waves travel at a finite velocity is provided by the time lapse between a flash of lightning and the resulting clap of thunder. Similarly, a distant explosion is seen long before the sound is heard, and a noticeable time interval is required for an echo to return. The speed of sound in air at 68°F is 344 m/s, or about 1128 ft/s. Consequently, a sound wave travels 1 mi in about 5 s.

One method of determining the speed of sound is to measure the time required for a sound to travel a known distance. This can be achieved electronically, as shown in Fig. 19-1, where two identical microphones detect the passing of an impulsive sound. The signal from each microphone can be *conditioned* electronically so that a sharp pulse is produced when the sound passes. The time of travel between the microphones is then determined by an electronic timer that is started by one pulse and stopped by the other.

Another method, which avoids the use of sophisticated equipment, is shown in Fig. 19-2. The water level is adjusted in the glass tube until resonance of the air column within the tube is obtained. The top surface of the water then represents the position of a node. Lowering the liquid level to obtain the next resonance

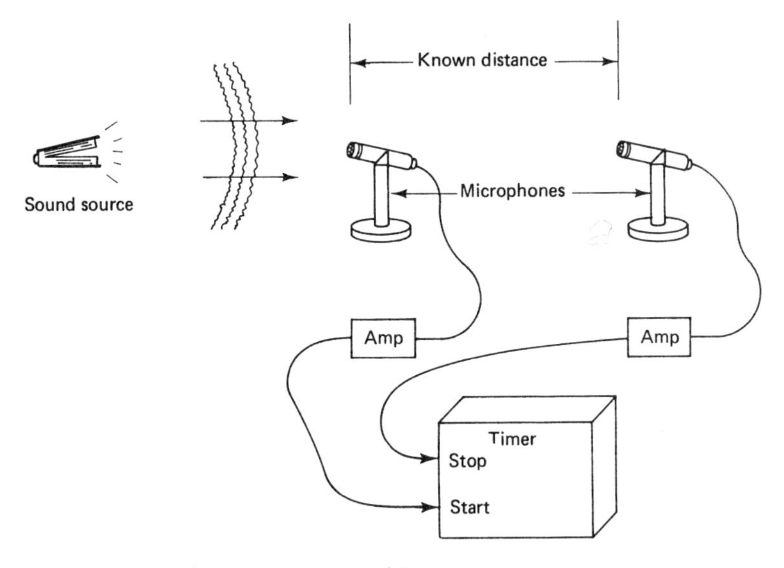

Figure 19-1 Determining the velocity of sound by electronically measuring the time for a sound wave to travel a known distance between two microphones.

locates the next node. The distance between successive nodes is one-half wavelength, and since the frequency of the tuning fork is known, the speed of sound within the column can be computed.

The speed of sound in various media

Formulas for computing the speed of sound (or longitudinal mechanical waves) in various media involve the elastic constants given in Table 11-5. Recall that Y is

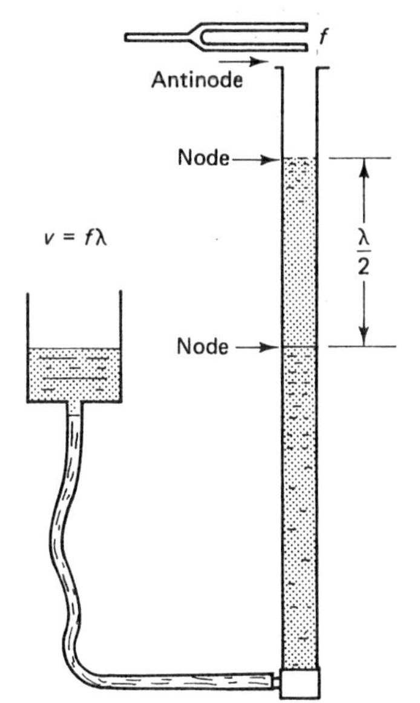

Figure 19-2 Resonance column method of measuring the velocity of sound in air.

TABLE 19-1 SPEED OF SOUND IN SELECTED MATERIALS

Gases (0°C)	Speed, m/s	
Air	331	
Argon	308	
Carbon dioxide	259	
Helium	965	
Hydrogen	1284	
Nitrogen	334	
Oxygen	316	
Liquids (25°C)		
Acetone	1174	
Kerosene	1324	
Mercury	1450	
Water, pure	1498	
Water, sea	1531	
Solids	**Bulk**	**Rod**
Aluminum	6420	5000
Brass	4760	3810
Glass, pyrex	5640	5170
Lucite	2680	1840
Steel	5854	5070

Young's modulus, S is the shear modulus, and B is the bulk modulus. The speed of sound in several materials is given in Table 19-1. The formulas for the velocity v of longitudinal waves are:

(in a rod)

$$v = \sqrt{\frac{Y}{\rho}} \tag{19-1}$$

(in an extended solid)

$$v = \sqrt{\frac{B + \frac{4}{3}S}{\rho}} \tag{19-2}$$

(in a fluid)

$$v = \sqrt{\frac{B}{\rho}} \tag{19-3}$$

In each formula, ρ is the mass density of the material.

EXAMPLE 19-1 Compute the velocity of sound in a steel rod.

Solution. Young's modulus for steel is 20×10^{10} N/m^2, and the density of steel is 7800 kg/m^3. Then, according to Eq. (19-1),

$$v = \sqrt{\frac{Y}{\rho}} = \sqrt{\frac{20 \times 10^{10}}{7800}}$$
$$= 5064 \text{ m/s}$$

EXAMPLE 19-2 Calculate the velocity of sound in water.

Solution. The bulk modulus for water is 0.2×10^{10} N/m^2 and the density is 1000 kg/m^3. By Eq. (19-3),

$$v = \sqrt{\frac{B}{\rho}} = \sqrt{\frac{0.2 \times 10^{10}}{1000}}$$
$$= 1414 \text{ m/s}$$

To obtain a theoretical expression for the speed of sound in air, which is a fluid, we look to Eq. (19-3) and then consider the bulk modulus for air. Recall, however, that the compressibility of a gas depends upon whether the compression occurs isothermally or adiabatically. That is, a given pressure increase produces a greater compression if the process occurs isothermally rather than adiabatically.

A longitudinal wave contains alternating regions of compression and rarefaction. And we expect the temperature to be elevated in the regions of compression and to be lowered in the regions of rarefaction. We might expect heat to flow from the regions of compression to the regions of rarefaction, but the wave motion occurs so rapidly that only a negligible heat transfer is possible in the brief time available. We therefore conclude that the compression and rarefaction of a gas by longitudinal waves occurs adiabatically. Moreover, the *adiabatic* compressibility of a gas is the proper quantity for Eq. (19-3).

More-advanced texts derive the following expression for the adiabatic compressibility of an ideal gas:

(adiabatic compressibility of an ideal gas)
$$B = \gamma P \tag{19-4}$$

where γ is the specific heat ratio, 1.4 for air. Substituting for B gives the following expression for the velocity of sound in a gas:

$$v = \sqrt{\frac{\gamma P}{\rho}} \tag{19-5}$$

This can be put into another form by using the ideal gas law to eliminate P and ρ. We use the relationships

$$P = \frac{n}{V}RT \qquad n = \frac{m}{M} = \frac{\rho V}{M} \qquad P = \frac{\rho RT}{M}$$

to obtain an expression for the speed of sound in terms of the absolute temperature T and the molecular weight M:

$$v = \sqrt{\frac{\gamma RT}{M}} \tag{19-6}$$

This shows that the velocity of sound in a gas depends upon the temperature. This is not surprising because the transfer of energy in a longitudinal wave in a gas is related to the molecular velocities, and we recall from our study of kinetic theory that molecular velocities are closely related to the absolute temperature.

Incidentally, the average molecular weight of air is 28.97.

EXAMPLE 19-3

Calculate the velocity of sound in air at 0°C.

Solution. We may use Eq. (19-6) directly, but we must convert 0°C to 273 K. Hence, for $\gamma = 1.40$ and

$$M = 28.97 \frac{\text{kg}}{\text{kg} \cdot \text{mol}}:$$

$$R = 8314 \text{ J/kg} \cdot \text{mol} \cdot \text{K}$$

$$v = \sqrt{\frac{\gamma RT}{M}}$$

$$= \sqrt{\frac{(1.40)(8314)(273)}{28.97}}$$

$$= 331 \text{ m/s}$$

This is in excellent agreement with experiment.

If velocity v_0 corresponds to temperature T_0 and v corresponds to T, it follows from Eq. (19-6) that

$$\frac{v}{v_0} = \sqrt{\frac{T}{T_0}} \quad \text{or} \quad v = v_0\sqrt{\frac{T}{T_0}} \tag{19-7}$$

Thus, if the velocity of sound in a gas (not necessarily air) is known at one temperature, the velocity at a different temperature is easily computed.

For temperatures in the 0 to 100°F (−17 to 38°C) range, the speed of sound is given approximately by

$$\begin{aligned} v &= 331 + 0.65C \\ &= 320 + 0.36F \end{aligned} \tag{19-8}$$

where v is in meters per second and C and F are Celsius and Fahrenheit temperatures, respectively.

19-2 SOUND WAVE INTENSITY AND AMPLITUDE

Sound waves passing through a region of air cause the molecules of the air to be displaced from their respective equilibrium positions to form the regions of compression and rarefaction. A particular molecule will execute simple harmonic motion about its equilibrium position with a frequency and period equal to that of the wave and with an amplitude determined by the intensity of the wave. The actual motion of a molecule is the superposition of the molecular motion due to the wave and the motion due to the thermal agitation of the air. For now, we ignore the thermal motion and consider the displacement amplitude and maximum velocity imparted to the molecules by the sound waves.

Advanced texts derive the following relationship for the intensity I of a sound wave in terms of the displacement amplitude A:

$$I = 2\pi^2 f^2 A^2 \rho v \tag{19-9}$$

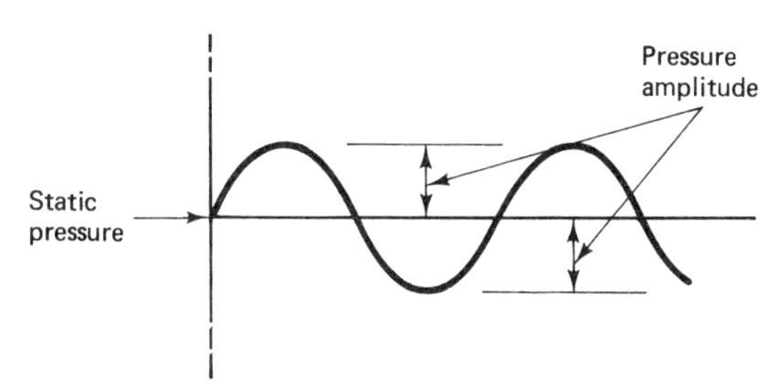

Figure 19-3 Pressure amplitude is the maximum departure of the pressure from the static value.

where ρ is the density of the medium and v is the velocity of the waves through the medium. Observe that the wave intensity is proportional to the square of the amplitude. We may solve for the amplitude in terms of the intensity to obtain

$$A = \left(\frac{I}{2\pi^2 f^2 \rho v}\right)^{1/2} \tag{19-10}$$

which allows us to calculate the displacement amplitude if the wave intensity is known.

If we know the amplitude and frequency with which a particle is executing simple harmonic motion, the maximum velocity of the particle is given by

$$\text{Maximum particle velocity} = 2\pi f A \tag{19-11}$$

which makes use of Eq. (17-10). This velocity must not be confused with the wave velocity.

The pressure increase in a region of compression is directly related to the displacement amplitude A. The *pressure amplitude* P_a is the maximum departure of the pressure from the normal static pressure, as illustrated in Fig. 19-3, and is given by

$$P_a = 2\pi f \rho A v \tag{19-12}$$

If A is eliminated between Eqs. (19-9) and (19-12), the following relationships are obtained:

$$I = \frac{P_a^2}{2\rho v} \quad \text{or} \quad P_a = \sqrt{2\rho v I} \tag{19-13}$$

For air at 20°C (68°F), $\rho = 1.20\ \text{kg/m}^3$ and $v = 344$ m/s.

EXAMPLE 19-4 Calculate the displacement amplitude of a 1000-Hz sound wave whose intensity is $1 \times 10^{-6}\ \text{W/m}^2$, a moderate intensity.

Solution. We use Eq. (19-10), but we prefer to do the calculation in two steps; MKS units are used.

$$A = \left(\frac{I}{2\pi^2 f^2 \rho v}\right)^{1/2}$$

$$2\pi f^2 \rho v = 2\pi^2(1000)^2(1.20)(344) = 8.15 \times 10^9$$

$$A = \sqrt{\frac{1 \times 10^{-6}}{8.15 \times 10^9}} = 1.11 \times 10^{-8}\ \text{m}$$

The diameter of a molecule is about 10^{-10} m, so the preceding displacement amplitude is on the order of 100 molecular diameters, a very small displacement.

Decibels

The typical human ear responds to sound waves over an intensity range of more than 100 billion to one. This points out that the ear is extraordinarily sensitive and that it is also capable of dealing with very loud sounds. We say that the *dynamic range* of the ear is very great.

When a power level or intensity varies over a wide range, relative values of the parameter may be compared by giving the number of *decibels* that one value is greater or less than another. If one particular value is established as the *reference level*, then absolute magnitudes can be expressed in terms of decibels. The defining equation is

$$\text{dB} = 10 \log \frac{I}{I_0} \tag{19-14}$$

where I_0 is the reference level and I is the value being compared to the reference level. The logarithm is to the base 10. It is easy to verify that when $I = I_0$, dB $= 0$ because the logarithm of 1.0 is zero. When I is 10 times I_0, dB $= 10$ because the logarithm of 10 is 1.0. And when I is twice I_0, dB $= 3.01$; we say that an increase of 3 dB represents a *doubling* (very nearly) of the parameter. When I is less than I_0, the logarithm term will be negative; negative dB values are frequently encountered in many areas, sound recording in particular.

If the dB and reference level, I_0, are given, the value of I can be found by the following expression obtained from Eq. (19-14):

$$I = I_0 \cdot 10^x \quad \text{where} \quad x = \frac{\text{dB}}{10} \tag{19-15}$$

For expressing the intensity of sound waves, the reference level I_0 is agreed to be

$$I_0 = 10^{-16}\ \text{W/cm}^2 \quad \text{or} \quad I_0 = 10^{-12}\ \text{W/m}^2 \tag{19-16}$$

This level corresponds to a sound that can just be heard at a frequency of 1000 Hz. Note that the first expression above involves a mixture of MKS and CGS units. This mixture is commonly encountered in the study of acoustics, but the mixed form must *not* be used in calculations. To convert from watts per square centimeter to watts per square meter, simply multiply by 10^4. The dB values of several common sounds are illustrated in Fig. 19-4.

EXAMPLE 19-5 Calculate the sound level in dB of a sound wave whose intensity is 4×10^{-8} W/cm^2.

Solution. We use Eq. (19-14) and the reference level for sound waves given earlier.

$$\begin{aligned} \text{dB} &= 10 \log \frac{I}{I_0} \\ &= 10 \log \left[\frac{4 \times 10^{-8}}{10^{-16}}\right] = 10 \log (4 \times 10^8) \\ &= 10(8.6) = 86\ \text{dB} \end{aligned}$$

Noise level in dB	Typical source
120	Threshold of pain, nearby jet engine
110	Indoor rock concert
100	Chain saw or power mower at close range
90	Inside subway train
80	Average factory; inside car in heavy traffic
70	Busy street traffic
60	Normal conversation
50	Average office
40	Average home
30	Quiet office
20	Typical whisper from a distance of 1 m
10	Gentle rustle of leaves
0	Threshold of hearing

Intensity level, dB	Intensity		Pressure amplitude	
	W/cm^2	W/m^2	dyn/cm^2	N/m^2
0	10^{-16}	10^{-12}	0.0002	2×10^{-5}
20	10^{-14}	10^{-10}	0.002	2×10^{-4}
40	10^{-12}	10^{-8}	0.02	2×10^{-3}
60	10^{-10}	10^{-6}	0.2	2×10^{-2}
80	10^{-8}	10^{-4}	2.0	2×10^{-1}
100	10^{-6}	10^{-2}	20	2
120	10^{-4}	1	200	20

Figure 19-4 Several common sound levels, intensities, and pressure amplitudes.

EXAMPLE 19-6 Calculate the intensity of a sound wave that has a sound level of 96 dB.

Solution. Using Eq. (19-15) we first divide 96 dB by 10 to get $x = 9.6$. Then,

$$\begin{aligned} I &= I_0(10^x) \\ &= (10^{-16}\ \text{W/cm}^2)(10^{9.6}) \\ &= (10^{-16})(3.98 \times 10^9) \\ &= 3.98 \times 10^{-7}\ \text{W/cm}^2 \end{aligned}$$

A scientific calculator may be used to evaluate $10^{9.6}$.

Sound levels are frequently measured by instruments that respond to the pressure amplitude of sound waves. The readings of such sound-level meters typically are in dB-SPL, where SPL denotes *sound pressure level*. The reference pressure P_0 for dB-SPL calculations is 2×10^{-4} dyn/cm^2 (2×10^{-5} N/m^2), a very small pressure that is near the threshold of hearing at a frequency of 1000 Hz. Because the intensity of a sound wave is proportional to the square of the pressure amplitude, P_a, we may express Eq. (19-14) in terms of pressure amplitudes as follows:

$$\begin{aligned} \text{dB-SPL} &= 10 \log \frac{P_a^2}{P_0^2} = 10 \log \left(\frac{P_a}{P_0}\right)^2 \\ &= 20 \log \frac{P_a}{P_0} \end{aligned} \tag{19-17}$$

where P_0 is the reference pressure, 2×10^{-4} dyn/cm^2. Note that when pressure amplitudes are being considered, the coefficient of the logarithm term is 20 instead of 10. For a given dB-SPL, the pressure amplitude may be computed using

$$P_a = P_0(10^x) \quad \text{where} \quad x = \frac{\text{dB-SPL}}{20} \tag{19-18}$$

19-3 SUBJECTIVE ASPECTS OF SOUND

Fletcher-Munson contours

We have seen already that the ear is remarkably sensitive in that it can respond to sound intensities as small as 10^{-16} W/cm^2 and can respond to sound over a loudness range of almost 120 dB. The ear, however, is not equally sensitive to all frequencies nor is it linear in its response to sounds of increasing intensity. The average ear is most sensitive at a frequency of about 4 kHz.

The *loudness* of a sound is the ear's response to the intensity. It is a *subjective* aspect of sound, meaning that the perceived loudness of a given sound depends upon the listener. Furthermore, two listeners may not agree that two sounds of different frequencies are equally loud.

Fletcher and Munson studied the hearing responses of a large group of people with normal hearing and obtained the response of the *average* ear to sounds of various frequencies and intensities. The result is the Fletcher-Munson contours of equal loudness, shown in Fig. 19-5. Each contour, or curve, is called a *phon*, and the reference frequency for the tests was 1 kHz. Therefore, at 1 kHz the

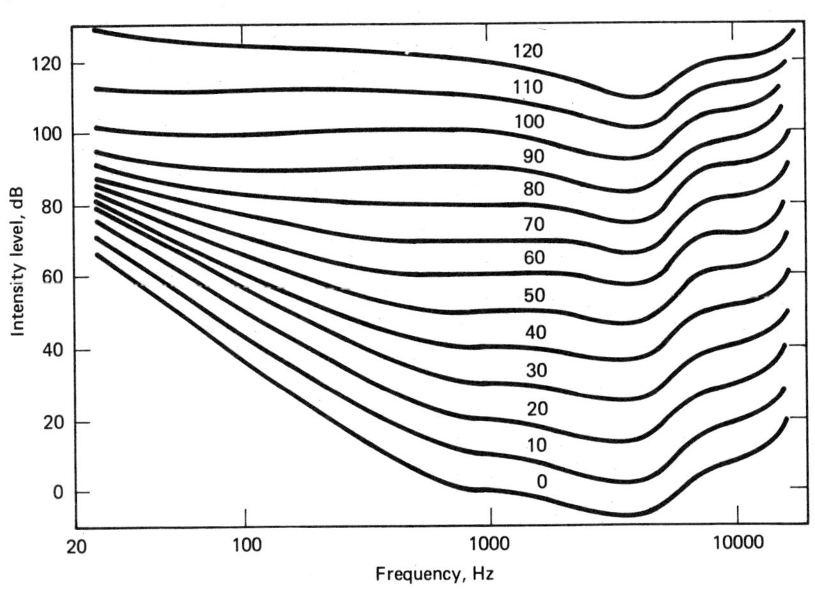

Figure 19-5 The Fletcher-Munson contours of equal loudness. The 0-dB phon represents the threshold of hearing; the 120-dB phon represents the threshold of pain.

30-phon loudness level corresponds to the 30-dB intensity level, and the same is true for the other phons at 1 kHz. But at 10 kHz, the intensity level must be about 40 dB to produce a perceived loudness level of 30 phons. In the low-frequency region at 30 Hz, a 70-dB intensity level is required in order to produce the same loudness level that a 10-dB intensity produces at 1 kHz. This points out the relative insensitivity of the ear to low frequencies at low-intensity levels. At higher levels, the response of the ear becomes more uniform with frequency.

Many stereo amplifiers have a *loudness* control (in addition to the customary tone controls) that "boosts the bass" at low volume settings. This is done to prevent the low frequencies from dropping out as the volume is reduced.

Pitch; timbre and quality

The pitch of a musical note is the characteristic that enables us to determine whether it is a low note or a high note. Obviously, the pitch is related to the frequency of the note, but the perceived pitch depends minutely upon the loudness of the sound as well as upon the frequency. The pitch of a pure tone (one without harmonics) becomes somewhat lower as the loudness is increased. The presence of harmonics seems to remove the dependence of pitch upon loudness, and because most musical instruments produce sounds rich in harmonics, the pitch, for all practical purposes, is unambiguously related to the fundamental frequency of a note.

A remarkable property of the ear is that when a group of harmonically related frequencies is presented, the perceived pitch is that of the fundamental frequency. Furthermore, this is the case even when the fundamental is totally missing; the ear seems to supply the fundamental of its own accord. This property of the ear causes a small speaker with a poor bass response to sound better than it really should, a fact used to advantage in the design of small radios.

We established earlier that the quality, or *timbre*, of a musical sound is related to the number of overtones present and to the relative intensity of each overtone. The overtones are said to be *harmonically related* when each overtone is an integral number times the fundamental frequency, and the sound is apt to be pleasing. But when the overtones are not harmonically related, the ear has difficulty in ascribing a pitch to the sound, and the sound may not be pleasant. The clanging sound of metal bars, steel plates, snare drums, and thin-metal bells arises from overtones that are not harmonically related to the fundamental frequency.

Localization of sound sources

Let us now consider the physical cues that enable a binaural (two ears) listener to determine the direction of a source of sound. In broad terms, differences in phase, amplitude, or a combination of the two are thought to be responsible for localization acuity. The relative importance of each cue is frequency dependent.

In physical terms, the ears are separated by several inches, with a head interposed. At low frequencies, up to about 800 Hz, diffraction of sound waves around the head causes the amplitude to each ear to be practically the same. Consequently, the phase difference is the important cue; the sound waves arrive

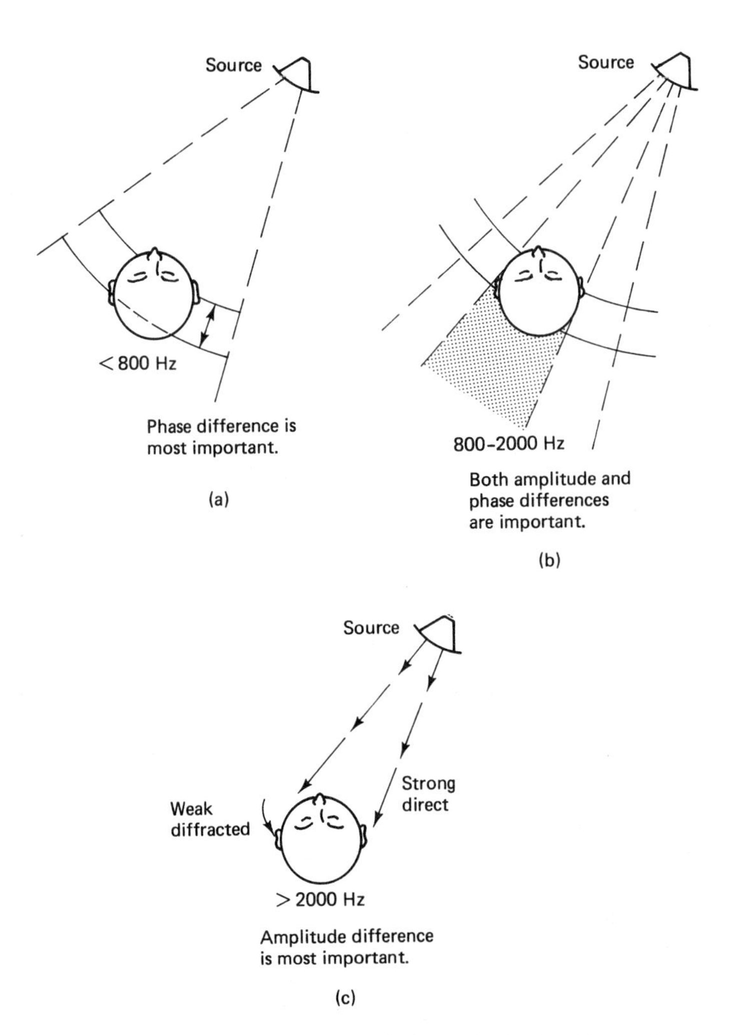

Figure 19-6 The important physical cues for localization acuity.

at each ear slightly out of phase and the brain processes this phase difference to yield dirctional information. Diffraction diminishes at higher frequencies so that the head tends to shadow one ear from the waves. This produces an amplitude difference in addition to the phase difference. Both phase and amplitude cues are operative in the transition region from about 800 to 2000 Hz. At the highest frequencies, amplitude cues are most important. These factors, illustrated in Fig. 19-6, are important to the production of stereophonic sound in which directional effects are preserved.

Two speakers symmetrically located in front of a listener and driven by the same source will cause the sound to appear to come from a point midway between the speakers. If one speaker plays slightly louder than the other, the sound will appear to come from a point nearer the louder speaker. Another effect, called the *precedence effect*, comes into play if the signal to one speaker is delayed from 5 to 25 ms relative to the other. In this case, the apparent direction tends toward the earlier speaker. Even if the sound from one speaker is a full 10 dB louder than the sound from the other, a central *image* can be maintained by delaying the louder signal by about 15 ms. The precedence effect is illustrated in Fig. 19-7.

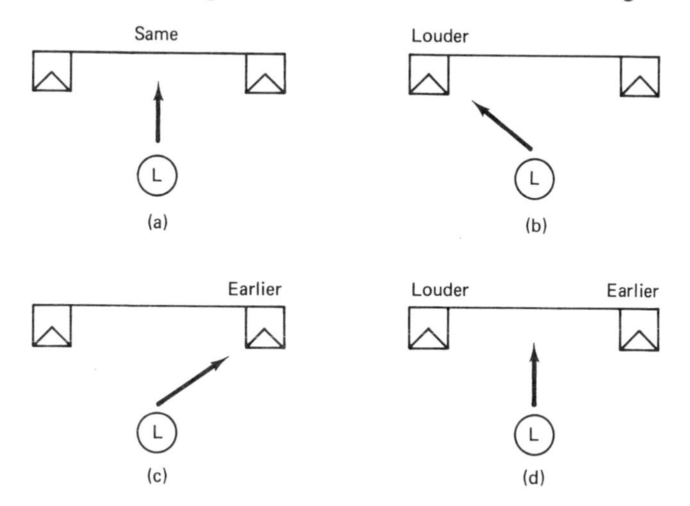

Figure 19-7 A listener, L, in front of two speakers driven with the same signal but with one speaker receiving the signal slightly louder or slightly earlier than the other will perceive the direction of the sound as indicated. The direction tends toward the earlier speaker and toward the loud speaker.

If, for one speaker of a pair driven equally loud, the phase of sound waves of all frequencies are shifted an equal amount—for example, by 60°—the apparent source of sound seems to spread out to become wider. As the phase difference is increased, the image continues to spread until, at a phase difference of about 135°, it will seem to split apart. If the phase difference is increased to 180°, as by reversing the electrical connection to one speaker, the direction of the sound becomes totally ambiguous. For this reason, it is important that stereo speakers be connected *inphase*.

19-4 ROOM ACOUSTICS; REVERBERATION

It is well known that large rooms such as auditoriums and concert halls sound different than small rooms. The characteristic that makes an auditorium sound spacious is *reverberation*, which arises from repeated reflections of sound waves back and forth in all directions within the enclosed space, as shown in Fig. 19-8. The recognizable attribute of a *reverberant soundfield* is that a sound lingers, only gradually decaying after the cause of the sound is abruptly terminated. The last note of an organ recital in a cathedral may linger for 2 s or more before finally giving way to silence.

A measure of the reverberation of a room is its *reverberation time*, RT, the time required for the sound from a steady source to decrease by 60 dB after the source is turned off. The RT is directly proportional to the volume of a room, but it is strongly influenced by the rate at which sound is absorbed by the walls and

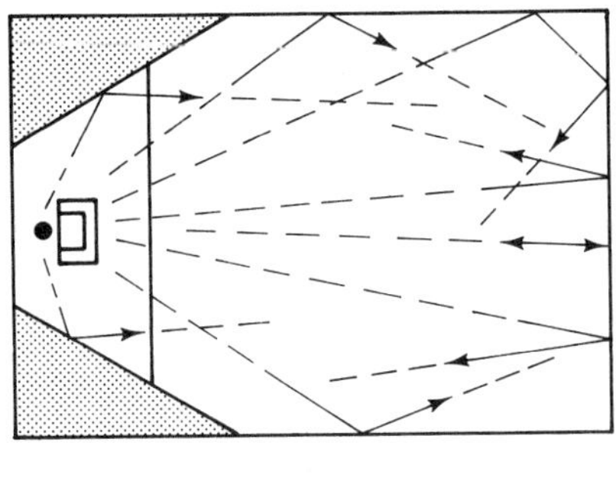

(a) Top view

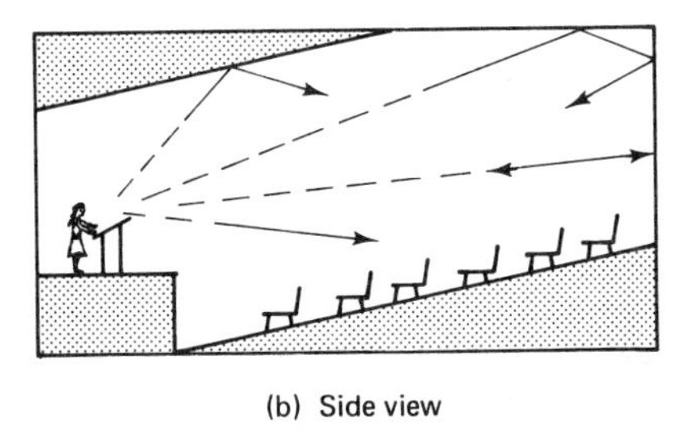

(b) Side view

Figure 19-8 Reflections of sound waves in an auditorium give rise to the reverberant field.

objects within the room. A room of a particular volume is said to be *live* if it has a long RT, and it is *dead* if the RT is short in relation to its volume. The presence of sound-absorbing surfaces in a room shortens the RT and decreases the relative strength of the reverberant field.

To examine reverberation in more detail, let us consider the sound reaching a listener in a concert hall from a piano at center stage. The first sound to reach the listener is the *direct* sound that travels straight from the piano to the listener. Then, after perhaps 25 ms, the *early group* of reflected sound waves arrive, having been reflected from surfaces fairly close to the piano. The early group, shown in Fig. 19-9, is then followed by reflections from more distant surfaces that give rise to the *diffuse reverberant field.* Thus, the listener receives sound waves that may be part of the *early sound field* (direct plus early reflections) or of the reverberant field.

An important physical cue that enables a blindfolded listener to estimate the size of a room lies in the early sound field. It is the initial time delay between the direct sound and the onset of the reverberant field. The delay will be greater in larger rooms. Also important to the overall acoustical effect is the relative strengths of the direct sound and the reverberant field; we call this the *direct-to-reverberant* ratio D/R. If the listener is fairly close to the source in a large room, D/R will be greater than it will in a smaller room having the same RT.

The relative strength of the reverberant field is greater in rooms having a long RT and is inversely proportional to the volume of the room. Therefore, an acceptable D/R ratio may be achieved in a small room having a fairly short RT, whereas a larger room requires a longer RT in order to maintain the same ratio. If D/R becomes too small, as in highly reverberant rooms where the listener is far

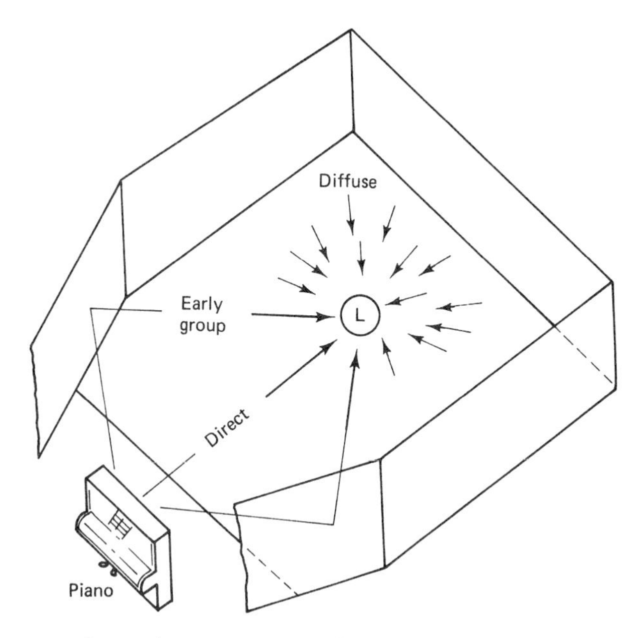

Figure 19-9 The sound reaching a listener, L, may be classified as direct sound, sound from the early group of reflections, or as sound comprising the diffuse reverberant field.

from the source, music and speech will not be clear and crisp, and the overall effect will not be pleasing.

Warmth, intimacy, and liveness are subjective attributes of a concert hall. *Warmth* is achieved via long RT's at frequencies up to about 250 Hz. *Liveness* results from comparatively long RT's in the range 500 Hz to 2 kHz, with the additional requirement that the early sound field must have the characteristics of a relatively small room. *Intimacy* refers to the perceived closeness of the source and listener and depends upon the time delay between the direct sound and the early group of reflections. The delay must be no more than about 15–20 ms for intimacy to be preserved.

19-5 THE DOPPLER EFFECT

When there is relative motion between a source of sound waves and a listener, the listener will hear the sounds at a frequency different from the frequency at which the source produces the sounds. Sounds are shifted to higher frequencies when the source and listener approach each other, and the frequency is lowered when the two are moving apart. This phenomenon is the *Doppler effect*, and it occurs for water waves and light waves as well as for sound waves. Not only does the change in frequency depend upon the relative velocity of the source and listener, but it also depends upon the individual velocities of the two. If there is a wind blowing, that makes a difference also. Examples of the Doppler effect are common. The sound of approaching boats, trains, cars, and airplanes are heard at a higher frequency than the sound of the same vehicles when departing; and when a vehicle passes by, the change in frequency is quite noticeable, especially if it is traveling at high speed.

As is illustrated in Fig. 19-10, the wavelength of sound waves in front of a moving source is shorter than the wavelength behind the source. Because the *speed* of sound is *not* affected by the motion of the source, a sound wave spreads out uniformly in a circle (or sphere) centered on the point where the wave was produced. If the source is moving, the circular waves will not be centered on the source. Expressions for the wavelength in front and back of the source are given in Fig. 19-10.

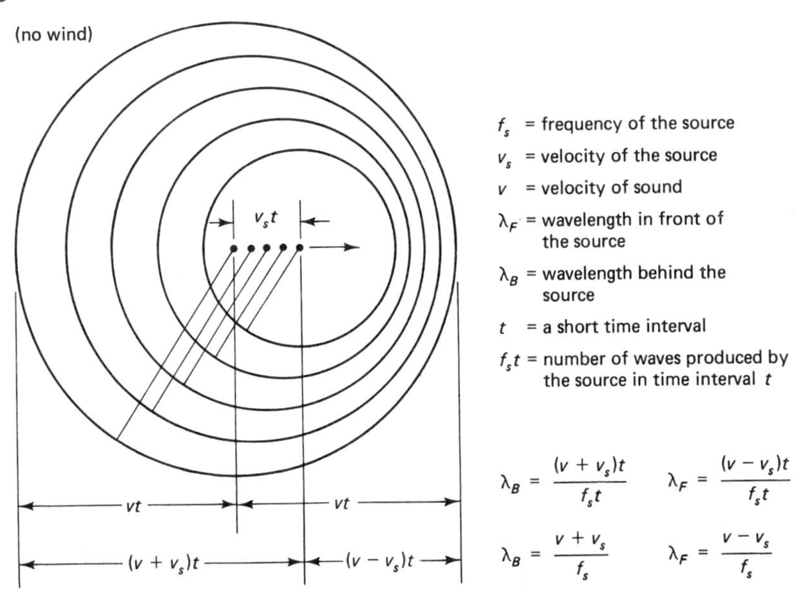

Figure 19-10 Wavelengths in front of a moving source are shorter than normal, while wavelengths in back of the source are longer.

We now derive a general formula for the frequency heard by a moving listener from a moving source in the presence of wind. For purposes of simplification, we assume that the source and listener move only on a line connecting the two, and we further assume that the wind direction is parallel to this line. The physical situation is shown in Fig. 19-11 where we may note the directions in which the source, wind, and listener are moving. The velocity of the source, v_s, is positive when the source moves toward the listener. The velocity of the listener, v_L, is positive when the listener moves toward the source. The velocity of the wind, v_w, is positive when the wind blows away from the source (toward the listener). The velocity of sound, v, is always positive. Whenever the velocity of the source, wind, or listener is in the opposite direction, a *negative* value is assigned to the velocity.

Note the position of the source when $t = 0$. We assume a wave is emitted at that instant. After an interval of time t, the wave will have traveled a distance $(v + v_w)t$ toward the listener. Observe that the total velocity of the wave is the sum of the velocity of sound in still air and the velocity of the wind. During the same interval of time, the source will have moved a distance v_st and will have emitted a total number of waves given by f_st, where f_s is the frequency of the source.

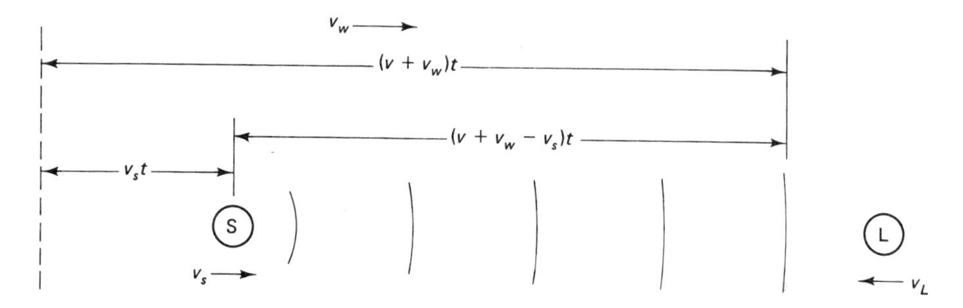

Source was here
at time $t = 0$

Number of waves $= f_s t$

$$\lambda = \frac{(v + v_w - v_s)t}{f_s t} = \frac{v + v_w - v_s}{f_s}$$

Relative velocity between listener and wave crests $= v + v_w + v_L$

$$f_L = \frac{v + v_w + v_L}{\lambda} \qquad f_L = \frac{(v + v_w + v_L)}{(v + v_w - v_s)} f_s$$

Figure 19-11 Derivation of the Doppler formula for the frequency heard by the listener.

At time t, the wave emitted at $t = 0$ is a distance from the source given by

$$(v + v_w - v_s)t$$

and a number of waves equal to $f_s t$ appear in this distance. Dividing the distance by the number of waves gives the distance per wave, the wavelength, in front of the source:

$$\lambda = \frac{v + v_w - v_s}{f_s} \tag{19-19}$$

This formula can also be used to calculate the wavelength behind the source, provided the velocity of the wind is inserted as a negative quantity (because in the region behind the source, the wind blows *toward* the source).

The frequency heard by the listener f_L equals the number of wavefronts encountered by the listener per second. To determine this, we note first that the relative velocity between the wavefronts and the listener is

$$v + v_w + v_L$$

When this velocity is divided by the wavelength λ just given, we obtain

$$f_L = \frac{v + v_w + v_L}{\lambda}$$

$$= \left(\frac{v + v_w + v_L}{v + v_w - v_s}\right) f_s \tag{19-20}$$

This is often written in the following form:

$$\frac{f_L}{v + v_w + v_L} = \frac{f_s}{v + v_w - v_s} \tag{19-21}$$

This formula is applicable to many situations, as illustrated by the following examples.

EXAMPLE 19-7

A horn on a train sounds with a frequency of 100 Hz. What frequency will be heard in each case?

(a) The train moves at 60 mi/h directly toward a stationary listener.
(b) The listener moves at 60 mi/h directly toward the stationary train.
(c) Both the train and listener move toward each other, each with a velocity of 30 mi/h.

Assume the wind to be calm and the speed of sound to be 1128 ft/s.

Solution. For each part, we substitute into Eq. (19-20). Recall that 60 mi/h is equivalent to 88 ft/s.

(a) $v_L = 0$; $v_w = 0$

(moving train, stationary listener)

$$f_L = \left(\frac{v + v_w + v_L}{v + v_w - v_s}\right) f_s$$

$$= \left(\frac{1128 + 0 + 0}{1128 + 0 - 88}\right)(100 \text{ Hz})$$

$$= 108.46 \text{ Hz}$$

(b) $v_s = 0$; $v_w = 0$

(stationary train, moving listener)

$$f_L = \left(\frac{1128 + 0 + 88}{1128 + 0 - 0}\right)(100 \text{ Hz})$$

$$= 107.80 \text{ Hz}$$

(c) $v_s = 44$; $v_L = 44$; $v_w = 0$

(both train and listener move)

$$f_L = \left(\frac{1128 + 0 + 44}{1128 + 0 - 44}\right)(100 \text{ Hz})$$

$$= 108.12 \text{ Hz}$$

If a wind blows at 30 mi/h from the train toward the listener, the result of each calculation above will be changed by the effect of the wind to 108.12, 107.51, and 107.80 Hz, respectively.

EXAMPLE 19-8

On a day when the speed of sound is 1080 ft/s, an aircraft travels at mach 0.5 (540 ft/s) and emits a loud sound at a frequency of 400 Hz. Calculate the wavelength of the sound: (a) in front of the plane; (b) in back of the plane. Assume the wind is calm.

Solution. We may use Eq. (19-19) with $v_w = 0$, or we may use the expressions given in Fig. 19-10. Using Eq. (19-19) gives the following.

(a) $v_w = 0$; $v_s = +540$

(in front)

$$\lambda = \frac{v + v_w - v_s}{f_s}$$

$$= \frac{1080 + 0 - 540}{400}$$

$$= 1.35 \text{ ft}$$

(b) When considering a point behind the source, v_s must go in as a negative quantity. Hence,

$$\lambda = \frac{1080 + 0 - (-540)}{400} \quad (in\ back)$$

$$= \frac{1620}{400} = 4.05 \text{ ft}$$

If the plane is flying at low altitude in line with a stationary listener, the observed frequencies with the plane approaching and departing are 800 Hz and 266.7 Hz, respectively.

Doppler effect for reflected waves

The Doppler effect is widely used in speed-measuring devices, the most familiar being the traffic radar used by law enforcement officers to determine the speed of passing cars. In such cases, waves of known frequency are transmitted toward a reflecting object which reflects the waves back to a receiver located at the source. Relative motion between the source and reflector causes a frequency change to occur between the transmitted and reflected waves, and this difference in frequency can be used to compute the unknown speed. We shall consider two cases, one where a moving source-receiver approaches a stationary reflector, and the other where a stationary source-receiver is approached by a moving reflector (as in traffic radar).

The results of analyses similar to that done earlier are given in Fig. 19-12. The formulas are based upon sound waves, and we have included the effects of the wind. Note that the wind velocity is considered positive when it blows from the source-listener toward the reflector.

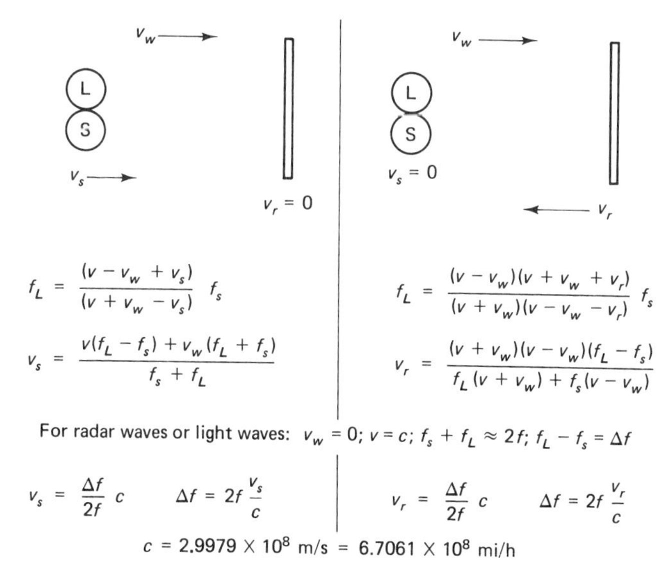

Figure 19-12 Working formulas for the Doppler effect for reflected waves.

Radar waves are electromagnetic waves, which travel at the speed of light (2.9979×10^8 m/s) and are not affected by any phenomenon corresponding to the wind. To use the formulas given in the figure for radar wave calculations, we set $v = c$, the speed of light, and we set $v_w = 0$. Furthermore, the frequency difference between f_s and f_L will be very small, so that we can make the following approximation:

$$f_s \approx f; \qquad f_L \approx f; \qquad f_s + f_L \approx 2f$$

where f is the operating frequency of the radar transmitter. Working formulas are given in the figure.

EXAMPLE 19-9 A person in a convertible drives at 60 mi/h on a street that passes between two large buildings. The horn on the car operates at 200 Hz, and the wind is blowing toward the buildings at 10 mi/h. Assuming the driver can hear the echo of the horn from the buildings, compute the frequency of the echo heard by the driver if the speed of sound is 770 mi/h.

Solution. We substitute directly into the formula given in Fig. 19-12 for a stationary reflector:

$$f_L = \frac{(v - v_W + v_s)}{(v + v_W - v_s)} f_s$$

$$= \frac{(770 - 10 + 60)}{(770 + 10 - 60)}(200)$$

$$= 227.78 \text{ Hz}$$

If there had been no wind, the echo frequency would have been 233.80 Hz, but if the wind had been blowing at 10 mi/h away from the buildings, the frequency would have been 240.00 Hz.

EXAMPLE 19-10 A stationary police officer using a radar unit operating at 20 GHz (20×10^9 Hz) aims the device at a motorist approaching at 60 mi/h. Calculate the frequency difference between the transmitted and reflected waves.

Solution. From Fig. 19-12, we have

$$f = 2f\frac{v_r}{c}$$

$$= 2(20 \times 10^9 \text{ Hz})\frac{60 \text{ mi/h}}{6.70 \times 10^8 \text{ mi/h}}$$

$$= 3582 \text{ Hz}$$

To Go Further

Consult an encyclopedia for information about:

Christian Doppler (1803–1853), Austrian scientist

In a text on astronomy, find how the Doppler shift for light enables astronomers to determine radial velocities of celestial objects. Look under *red shift*, or perhaps even *blue shift*.

Questions

1. Suggest a method for determining the speed of sound in the ocean at a depth of 500 ft.
2. Molecules of H_2 are only about $\frac{1}{16}$ as massive as molecules of O_2. Based upon kinetic theory, which type molecules will have the greatest speed at a given temperature? Does sound travel faster in H_2 or O_2?
3. Tell how the intensity of a sound wave depends upon (or varies with) the: **(a)** frequency; **(b)** amplitude; **(c)** density of the medium; **(d)** the wave velocity.
4. Explain the meaning of pressure amplitude. How is the pressure amplitude related to the intensity of a sound wave?
5. How does loudness differ from intensity?
6. What physical aspects of a sound wave determine: **(a)** its pitch; **(b)** its timbre or quality; **(c)** its loudness?
7. Is the perceived pitch of a sine-wave tone determined by frequency alone? Does the pitch increase or decrease as the loudness is increased?
8. What effect do sound absorbing surfaces have on the reverberation time of a room of a given volume?
9. Why does a large auditorium sound differently when it is filled with people than when it is empty, assuming the people are perfectly quiet?
10. Describe the sound of a concert given in an auditorium with a very long reverberation time. What happens to passages of music that are performed very rapidly?
11. What physical characteristics are responsible for the subjective attributes of warmth, liveness, and intimacy in regard to the acoustics of a concert hall?
12. Will a hobo riding near the rear of a train detect a change in pitch of the whistle as the train accelerates from rest to 80 mi/h?
13. In a certain small town, a whistle atop the courthouse blows each day at noon. Does the direction or speed of the wind have any effect upon the pitch heard by the citizens of the town as they have lunch?
14. How is it possible for astronomers to determine whether a star is moving toward or away from the earth?

Problems

1. Calculate the velocity of sound: **(a)** in an aluminum rod; **(b)** in a large rectangular block of aluminum.
2. Calculate the velocity of sound in a brass rod.
3. Calculate the velocity of sound in mercury.
4. Use Eq. (19-6) to calculate the velocity of sound in air at 27°C.
5. Use Eq. (19-8) to compute the velocity of sound in air at 27°C and compare with the result of the preceding problem.
6. Use Eq. (19-6) to calculate the velocity of sound at 20°C in : **(a)** H_2; **(b)** O_2.
7. Given that the speed of sound in air is 331 m/s at 0°C, use Eq. (19-7) to calculate the speed of sound at 15°C.
8. Compute the intensity of a 500-Hz sound wave in air at 20°C if the displacement amplitude is: **(a)** 2×10^{-8} m; **(b)** 4×10^{-8} m.
9. Calculate the maximum particle velocity for the two cases in the previous problem. How does this compare with the speed of sound?
10. One source of sound produces an intensity at a certain point that is 700 times as great as another. Express this intensity ratio in decibels.
11. Express the following power ratios in decibels: **(a)** 2 : 1; **(b)** 5 : 1; **(c)** 10 : 1; **(d)** 20 : 1; **(e)** 50 : 1; **(f)** 100 : 1.
12. What ratio of intensities is represented by the following values? **(a)** 0 dB **(b)** 20 dB **(c)** 40 dB **(d)** 55 dB.
13. A good ear can perceive an intensity level change as small as 1 dB at 1000 Hz. What intensity ratio does this minimum perceptible change represent?
14. A very small audio amplifier may produce an electrical power output of 0.1 W. If we adopt 0.1 W as our reference level, what would be the output in dB of amplifiers that have the follow-

ing electrical outputs in watts? **(a)** 0.5 W **(b)** 1 W **(c)** 10 W **(d)** 50 W **(e)** 100 W **(f)** 300 W

15. Refer to the Fletcher-Munson curves of Fig. 19-5 and determine the approximate loudness level of a sound wave with an intensity level of 50 dB at a frequency of: **(a)** 1000 Hz; **(b)** 10,000 Hz; **(c)** 200 Hz; **(d)** 30 Hz.

16. **(a)** What approximate intensity level in dB corresponds to the threshold of hearing at 50 Hz?
(b) What is the intensity of this sound in watts per square meter?

17. Sound intensities from multiple sources can be added to get the total intensity only if the intensities are expressed in watts per square meter or similar units. Intensities expressed in dB cannot be added directly. With this in mind, calculate the total intensity in dB resulting from the combination of: **(a)** two intensities of 4×10^{-8} W/cm^2; **(b)** an intensity of 86 dB and 96 dB. (See Examples 19-5 and 19-6.)

18. Calculate the pressure amplitude corresponding to a sound level of 80 dB-SPL.

19. A 1-kHz sound wave has a displacement amplitude of 3×10^{-8} m in air at 20°C. Calculate: **(a)** the pressure amplitude in N/m^2; **(b)** the intensity in W/m^2; **(c)** the intensity level in dB; **(d)** the sound pressure level in dB-SPL; **(e)** the approximate loudness level in phons (use the Fletcher-Munson curves).

20. Repeat Problem 19, but assume the frequency to be 100 Hz instead of 1 kHz.

21. Verify that the units of Eq. 19-10 are indeed that of a length.

22. Verify the units for Eq. 19-12.

23. If one ear is 4 in. farther from a source of sound that the other ear, compute the time difference between the arrival of a sound wave at the near ear and the more distant ear. (Use v = 344 m/s.)

24. **(a)** Calculate the additional distance a reflected wave must travel in order to arrive at a listener 15 ms after the direct wave.
(b) How far does a sound wave travel (in inches) in 1 ms when the temperature is 20°C?

25. If a Doppler radar operates at 10 GHz and is able to discern a change in frequency Δf as small as only 1 Hz, what is the minimum velocity that the unit can measure?

26. A car whose horn operates at 250 Hz approaches a stationary listener at 50 mi/h on a calm day.
(a) What frequency is heard by the listener?
(b) What frequency is heard after the car passes by as it travels away from the listener?
(c) What is the total frequency change heard by the listener as the car passes?

27. A jumper with unopened chute approaches the ground at 120 mi/h while screaming at a frequency of 400 Hz. At what frequency will a person on the ground hear the scream? (Take v = 344 m/s.)

28. Two cars approach each other, each traveling at 60 mi/h. If one car sounds its horn at 300 Hz, what frequency is heard by the driver of the other car: **(a)** before they pass, and **(b)** after they pass? (Note v = 344 m/s.)

29. A horn mounted on the hood of a convertible operates at 400 Hz. What frequency will the driver hear from the horn while traveling at 60 mi/h?

30. The prominent H_α line in the optical spectrum of hydrogen occurs at a wavelength of 6563 Å. This corresponds to a frequency of 4.568×10^{14} Hz. Compute: **(a)** the frequency; **(b)** the wavelength of this line in the spectrum of a star moving away from the earth at a speed of 22 km/s. (Assume the earth to be stationary in space.)

31. Suppose the H_α line of the hydrogen spectrum of a certain star is shifted from 6563 to 6565 Å. Calculate the speed of recession of the star.

32. The sonar of a stationary submarine sends out a short burst (a *ping*) of ultrasonic sound waves of frequency 32,000 Hz toward an approaching ship. The ping returns 1.62 s later with a frequency of 32,172 Hz. **(a)** How far is the ship from the submarine, and **(b)** with what velocity is the ship approaching? (The speed of sound in sea water is 1480 m/s.)

CHAPTER 20

LIGHT AND COLOR

This chapter marks the beginning of our study of light and optics. We begin with an investigation of the atomic origins of light, considering the spectrum of hydrogen in particular. We find that the line spectra of the elements are directly related to the electronic structure of the atoms.

Light is said to consist of photons. Moreover, photons have mass and momentum. This brings us to the inevitable question: Is light a wave phenomenon or a phenomenon involving small particles? It appears to be both; thus we are faced with the wave-particle duality that seems to be a property indigenous to all matter.

On a less theoretical level, we consider the principles of light intensity and illumination. All in all, this chapter should provide the principles and concepts for understanding a wide variety of nonoptical properties and applications of light.

20-1 ATOMIC ORIGINS OF LIGHT

Almost all light we encounter originates in the outer portion of atoms and arises from transitions made by electrons in going from one allowed orbit to another of lower energy.[1] Therefore, to understand the atomic origin of light, we must first become more familiar with the electronic structure of atoms; in particular, we must be concerned with the energy level associated with each electron orbit. In this section we present only the most basic concepts for the simplest atom, hydrogen. A more detailed presentation of atomic structure is given in Chap. 31.

[1]Two sources of light that do not involve electron transitions in atoms are *Cherenkov radiation* and *synchrotron radiation*. Cherenkov radiation occurs when high-speed particles pass through a medium at a speed greater than the speed of light in the medium; it is most commonly observed in the vicinity of nuclear reactors that are immersed in water (swimming pool reactors). Synchrotron radiation occurs when charged particles are caused to travel in curved paths, in keeping with the principle that accelerated charges radiate energy; it was first observed in conjunction with the operation of synchrotron particle accelerators, but it has subsequently been identified in radiation from astronomical entities such as the Crab nebula.

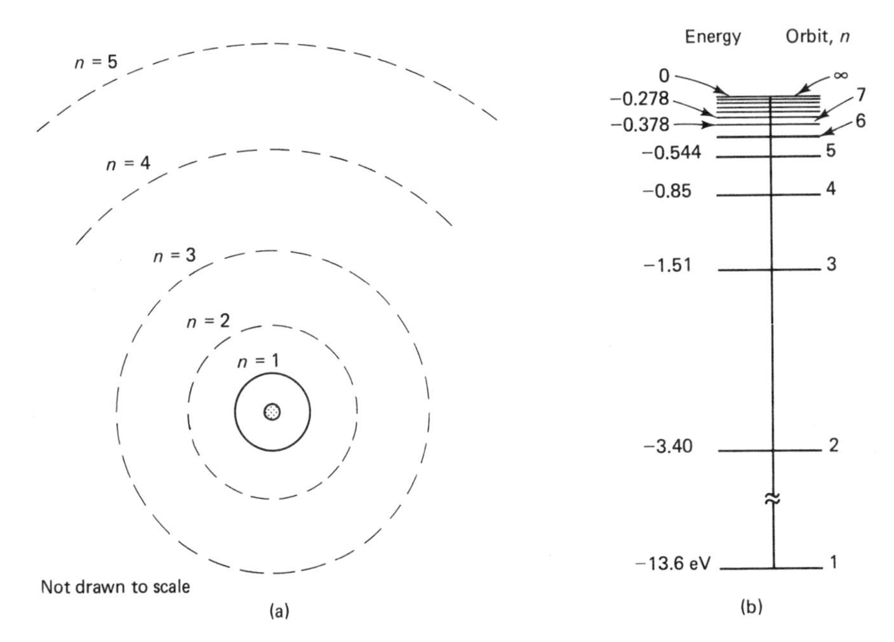

Figure 20-1 A diagrammatic representation of a hydrogen atom. (a) Possible orbits for the electron of a hydrogen atom. (b) Approximate energy levels of a hydrogen atom.

A representation of a hydrogen atom is shown in Fig. 20-1(a). The nucleus consists of one proton, and a single electron orbits the nucleus in one of many allowed orbits. Only the orbit nearest the nucleus is shown as an unbroken line; this orbit is special because it is the orbit of lowest energy, and the electron of an *unexcited* hydrogen atom will reside in this orbit. The other orbits are higher-energy orbits, and the electron must be given additional energy in order for it to be *excited* to one of the higher orbits. A hydrogen atom whose electron occupies a higher orbit is said to be *excited*. If, however, the electron is in the lowest orbit, the atom is said to be in the *ground state*.

Each orbit is designated by a number n, where $n = 1, 2, 3, \ldots$, as shown in the figure. For hydrogen, the approximate radius of each orbit is given in angstroms (10^{-10} m) by $0.53n^2$. Note that the diameter of the lowest orbit is very nearly *one* (1.06) angstrom.

A simplified energy level diagram for hydrogen is shown in part (b) of Fig. 20-1. Energies are given in units of *electron volts* (eV), 1 eV being the energy acquired by an electron in being accelerated through an electrostatic potential difference of 1 V. In terms of joules, $1 \text{ eV} = 1.602 \times 10^{-19}$ J. Observe also that the tabulated energies are negative. This is done to conform to the standard practice of assigning *zero* energy to an electron that is completely free. Hence, an electron in the lowest orbit of hydrogen has 13.6 eV less energy than a free electron. Lower orbits have *less* energy than higher orbits even though the magnitudes assigned are numerically greater. An approximate formula for computing the energy level of the nth orbit of a hydrogen atom is

$$E_n = -\frac{13.6 \text{ eV}}{n^2} \tag{20-1}$$

Electron transitions

An electron excited to a high orbit cannot remain there indefinitely; it will soon fall back to a lower orbit, giving up energy in the process. Furthermore, the energy that an electron gives up during a downward transition appears as a *photon*, which is a basic bundle of energy. Therefore, light is produced when electrons in atoms undergo downward transitions.

An electron may be excited to a state of higher energy (a higher orbit) as a result of a collision between atoms, a collision between a free electron and an orbiting electron, or by the absorption of a photon. A photon of the proper energy (frequency) can give up its energy to excite an electron from a lower state to a higher state. In so doing, the photon disappears. This process is the exact opposite of the process in which a photon is produced.

In an ordinary incandescent lamp, an electric current is caused to flow through a tungsten filament, heating the filament to a high temperature (to perhaps 2800 K). At high temperatures, atomic collisions become sufficiently energetic to excite electrons within the atoms to higher energy levels. Light is given off as the electrons fall back to the lower levels.

We now consider the production of a photon in more detail, recognizing that the fundamental processes involved are not yet known to science. Suppose an electron in a high orbit of energy, E_h, falls to a lower orbit of energy, E_l. The energy difference, $E_h - E_l$, will be the energy carried away by the single photon resulting from the transition. A simple and direct relationship exists between the energy E of a photon and the frequency ν of the waves of which it is composed. The relationship is

$$E = h\nu \tag{20-2}$$

where h is Planck's constant, named in honor of the German physicist, Max Planck (1858–1947). The numerical value of h is

$$h = 6.625 \times 10^{-34}\ \mathrm{J \cdot s}$$

With this fundamental relationship, we can calculate the frequency of the photon emitted during an electron transition.

$$\nu = \frac{E_h - E_l}{h} \quad \text{or} \quad \nu = \frac{\Delta E}{h} \tag{20-3}$$

where ΔE is the energy difference, $E_h - E_l$, between the two orbits. Thus we see that high-energy photons also have a higher frequency than less energetic photons.

EXAMPLE 20-1 Suppose an electron falls from the third orbit ($n = 3$) of hydrogen to the second orbit ($n = 2$). Use the information of Fig. 20-1 to calculate the frequency and wavelength of the emitted photon.

Solution. From the figure [or Eq. (20-1)], we see that $E_3 = -1.51$ eV and $E_2 = -3.40$ eV. The energy difference between these two levels is 1.80 eV.

At this point, in order to reconcile electron volts with joules, let us do a unit conversion and express Planck's constant in terms of eV-seconds instead of joule-seconds. You may verify that the result is

$$h = 4.135 \times 10^{-15} \text{ eV} \cdot \text{s}$$

We use this in Eq. (21-3) to calculate the frequency ν of the photon:

$$\nu = \frac{\Delta E}{h} = \frac{1.89 \text{ eV}}{4.135 \times 10^{-15} \text{ eV} \cdot \text{s}}$$

$$= 4.57 \times 10^{14} \text{ Hz}$$

(Recall that the unit s^{-1} is the equivalent of hertz.)

We now go one step further and calculate the wavelength of the photon using the familiar relationship between frequency, velocity, and wavelength, $f\lambda = v$. Using the symbols of this section, this is expressed as

$$\nu\lambda = c$$

from which we calculate

$$\lambda = \frac{c}{\nu}$$

$$= \frac{2.997 \times 10^{8} \text{ m/s}}{4.57 \times 10^{14} \text{ Hz}}$$

$$= 6.56 \times 10^{-7} \text{ m} = 656 \text{ nm}$$

$$= 6560 \text{ Å}$$

This is the well-known H_α line in the spectrum of hydrogen. A more accurate value of its wavelength is 6562.8 angstroms.

Carbon arc lamp

An intense source of light is the carbon arc lamp, shown in Fig. 20-2(a). An electric arc is formed between the carbon electrodes when they are brought into momentary contact and then separated a few millimeters. The tips of the electrodes reach temperatures on the order of 3600 K and thereby form an intense source of light. A comparatively small quantity of light comes from the carbon arc itself because carbon vapor is a very poor radiator. A high-wattage, current-limiting resistor is connected in series with the arc to limit the current to about 5 A.

Gas discharge tube

A gas such as hydrogen may be caused to emit light by enclosing the gas at a pressure of about 20 mm Hg in a gas discharge tube (spectral tube) such as that shown in Fig. 20-2(b). A high-voltage source (10,000 V or more) connected to electrodes at opposite ends of the tube accelerates free electrons within the gas to very high velocities. Collisions of these speeding electrons with the atoms of the gas excites the atoms and causes the emission of light as they return to the ground state.

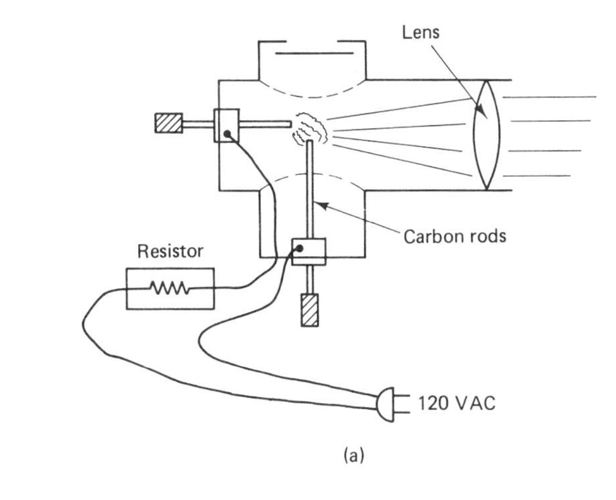

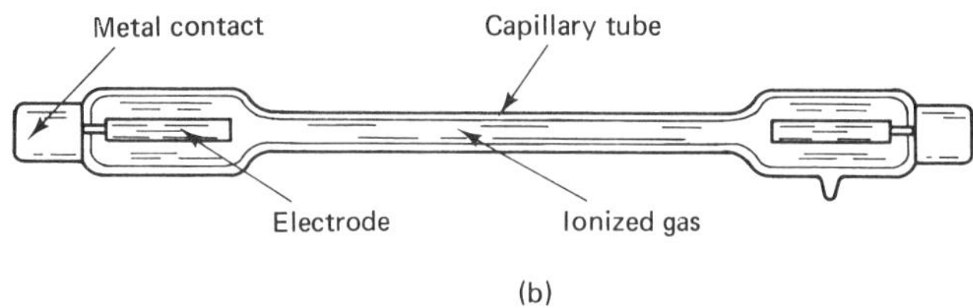

Figure 20-2 (a) A carbon arc light source.
(b) Spectral tube for obtaining the line spectrum of a gas.

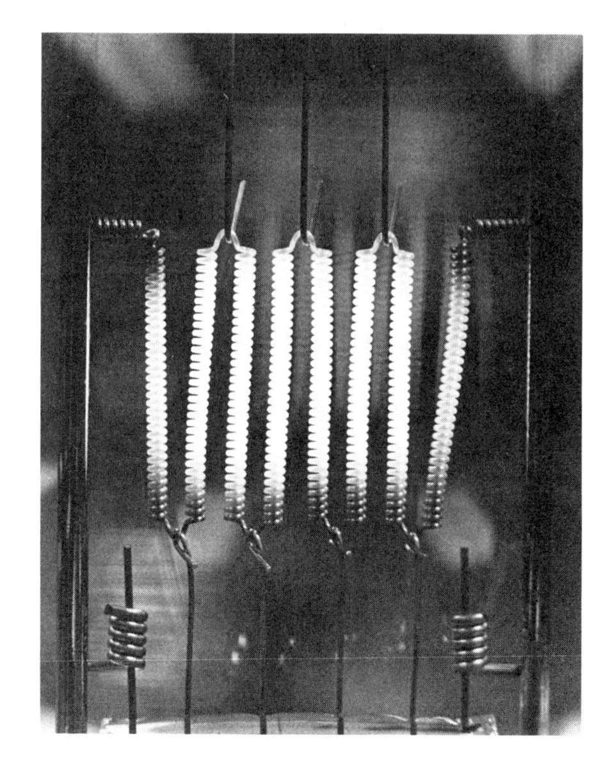

The tungsten filament of a large incandescent lamp.

The color of light emitted depends upon the type of gas within the tube. The familiar neon sign is a gas discharge tube containing neon, which emits a fiery orange red light.

20-2 PROPERTIES OF PHOTONS

We have seen already that a photon may be characterized by its energy. The energy determines its frequency and the frequency determines its wavelength. All photons travel at the same velocity, the speed of light (c).

Within the spectrum of visible light, the frequency of a photon determines its color. More precisely, we should say that the human eye-brain complex assigns the attribute of color according to the frequency of photons. Photons that produce the sensation of red have a lower frequency than those that produce blue. The relationship of color, frequency, and wavelength is shown in Table 20-1 for several colors in the visible region of the electromagnetic spectrum.

TABLE 20-1 COLOR AND WAVELENGTH OF VISIBLE LIGHT

Color	Wavelength, Å	Frequency, Hz	Photon energy, eV
Red	6600	4.541×10^{14}	1.878
Orange	6100	4.913	2.032
Yellow	5800	5.167	2.137
Green	5500	5.449	2.253
Blue	4700	6.337	2.637
Violet	4100	7.310	3.023

The human eye responds only to a very narrow range of photon frequencies. Photons in the frequency range just above the visible constitute *ultraviolet* light, and those in the range just below the visible constitute *infrared* light. *Black light*, used to stimulate the fluorescence of fluorescent paints in displays and posters, is actually ultraviolet light. There is no dramatic difference between visible photons and photons of the infrared or ultraviolet; only the energy is different, and the energy difference mandates that the frequency and wavelength must be different also.

Photons have mass and momentum. We can obtain a formula for the mass of a photon by using the Einstein mass-energy relationship, $E = mc^2$. Setting the energy $h\nu$ of a photon equal to mc^2 gives

$$E = h\nu = m_p c^2$$

(*mass of a photon*)
$$m_p = \frac{h\nu}{c^2} \quad \text{or} \quad m_p = \frac{h}{c\lambda} \tag{20-4}$$

Photons can exist only while traveling at the speed of light. If a photon is stopped (brought to rest), it ceases to exist. We say that a photon is a particle that has *zero rest mass*.

The momentum of a photon is the product of its mass and velocity. Hence, from above,

$$\text{Momentum of a photon} = m_p c = \frac{h}{\lambda} \tag{20-5}$$

Experimental verification of the momentum of a photon is provided by the Compton effect which deals with the collision between a photon and a free electron. In short, the electron recoils from the collision as if it had collided with a particle having momentum, and the photon loses energy and is deflected as if it were a particle having momentum.

Wave-particle duality of light

We have mentioned light briefly in earlier chapters, treating it as a wave phenomenon without making any significant reference to photons. The foregoing discussion implies that light consists of particles having energy, mass, and momentum, particles that somehow are imbued with the properties of waves. We therefore must face the question as to whether light consists of waves or of particles.

Ordinarily we would look to laboratory experiment for the answer, but in this case we find direct and compelling evidence for both sides of the question. Experiments concerned with interference and diffraction indicate light to be a wave phenomenon because particles cannot produce destructive interference. On the other hand, the Compton effect and the photoelectric effect (described in a later chapter) indicate that light has the attributes of particles. Thus, experimental evidence only reinforces our dilemma. We are forced to conclude that light exhibits a dual character. In certain experiments it acts as waves; in others it acts as particles. However, it always acts as one *or* the other, never both at the same time. The character exhibited depends upon the particular experiment.

A similar duality is found for particles. Electrons, which we normally regard as particles, can be diffracted by firing them through the regular array of a crystal lattice, and diffraction is definitely a wave phenomenon. Indeed, the wave-particle duality of light and matter now seems to be a fundamental precept of physics. The dilemma it presents has challenged some of the best minds this world has seen.

The two extremes of the electromagnetic spectrum are represented by radio waves at the low-frequency end and by gamma rays at the other. It is difficult to picture a radio wave as a particle, and it is equally difficult, experimentally, to demonstrate the wavelike attributes of a high-energy gamma ray. Thus, the relative prominence of wave characteristics steadily gives way to particle characteristics as we proceed through the electromagnetic spectrum in the direction of higher frequencies.

In the design of an experiment involving electromagnetic radiation, it is possible to predict whether the wave or particle aspect will be presented. If the experiment involves interactions between electromagnetic radiation and charged particles, the radiation will manifest itself as photons. But if interactions occur only between various portions of the radiation itself (as in interference and diffraction experiments), the radiation will appear as waves. Indeed, one concept of a photon is that a photon represents the discrete bundle, or *quantum*, of energy and momentum that is exchanged when an electromagnetic wave interacts with a charged particle. Moreover, the energy and momentum exchanged in the process are determined by the frequency of the electromagnetic wave. This concept avoids the apparent necessity of trying to picture an electromagnetic wave as an assemblage of photons.

It is inevitable that we should try to form a mental image, a model of a photon. While such imaging may not be strictly justified, we offer the following in order to lend a bit of substance to a mental image. We must not carry this too far, however, nor take it too seriously because our concept of a photon is only a representation of physical reality.

A photon may be envisioned as a skinny, needle-shaped packet of waves with a typical packet of the visible spectrum containing about 10^5 waves. A typical length of the wave train is a few centimeters, but it may be a meter or more for photons from highly monochromatic (one color) sources of light. The lateral extent is thought to be very small in comparison with its length because photons interact with matter at very small points (as in the photoelectric effect or in a photographic emulsion) rather than over an appreciable area. Individual photons do not spread out as do the waves from a point source.

20-3 SPECTRA

When a beam of white light is passed through a prism, as in Fig. 20-3, a band of colors forming the visible spectrum is projected onto the screen. This experiment, first performed by Sir Isaac Newton, demonstrates that white light is a mixture of all colors.

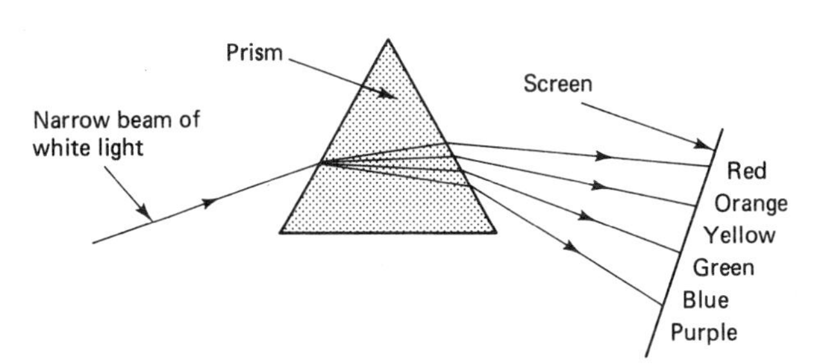

Figure 20-3 A narrow beam of white light sent through a prism is broken into its constituent colors; a continuous spectrum is produced.

If all colors are present in the spectrum of a particular source of light (that is, if none are missing), the spectrum is said to be a *continuous spectrum*. Continuous spectra are produced by light emitted from solids heated to high temperature. The carbon arc and an incandescent lamp produce continuous spectra. On the other hand, the light from a gas discharge tube forms a *line spectrum*, also called an *emission spectrum*. The spectrum of hydrogen, which we now consider, is an example of a line spectrum.

The hydrogen spectrum

In Sect. 20-1 we calculated the wavelength of light associated with a selected electron transition in hydrogen. It is clear that many other transitions are possible, as shown on the energy diagram of Fig. 20-4(a). Moreover, each transition produces a photon whose frequency is characteristic of that particular transition,

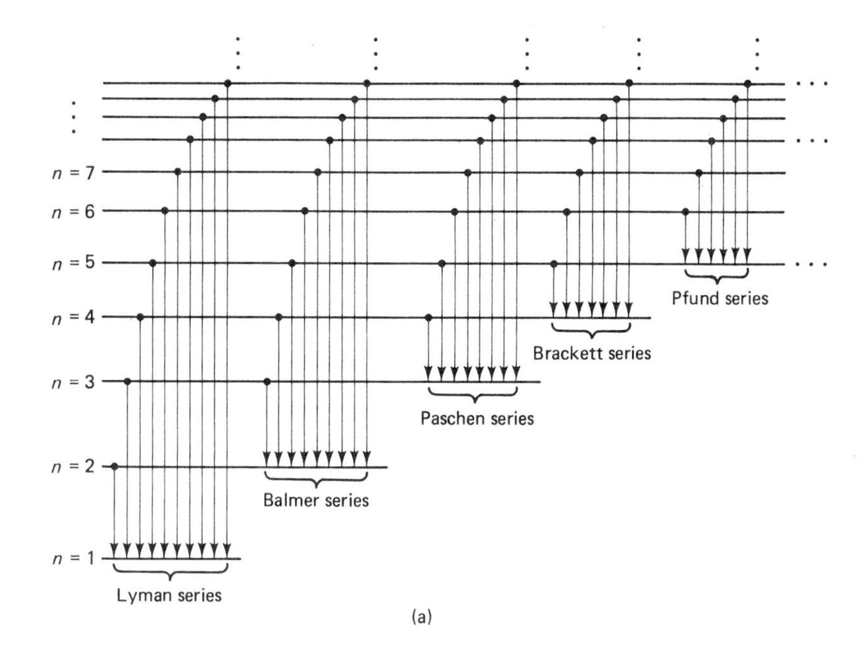

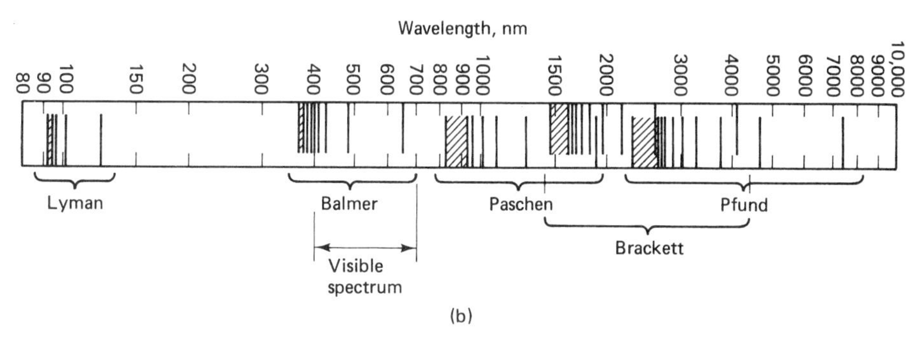

Figure 20-4 (a) Energy level diagram of hydrogen showing the transitions that comprise the various series.
(b) A schematic representation of the hydrogen spectrum drawn on a logarithmic scale. Overlapping series are distinguished by a small vertical separation.

and each discrete frequency produces a line in the spectrum. Hence, the spectrum of hydrogen consists of a series of lines, as shown in Fig. 20-4(b).

The transitions are arranged into groups according to the final level of the transition. Each group of transitions produces a *series* of spectral lines. For example, transitions ending on the lowest level ($n = 1$) form the Lyman series. Transitions ending on the second level ($n = 2$) form the Balmer series, and so on for the Paschen, Brackett, and Pfund series, which correspond to the levels where n equals 3, 4, and 5, respectively. Note that only lines of the Balmer series fall within the visible spectrum. The Lyman series lies in the ultraviolet, and the Paschen and other series lie in the infrared.

The spacing between the lines of a given series becomes smaller in proceeding toward the violet. This arises because the energy levels within the

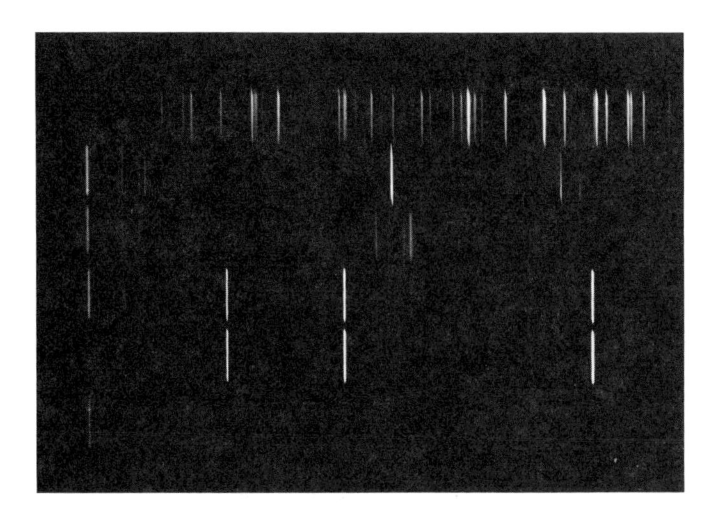

A small section of a spectrum plate showing spectral lines of several different elements.

atom are closer together for larger values of n. There exists a *series limit* for each series, and this limit, representing the minimum wavelength for each series, corresponds to the downward transition (capture, if you prefer) of an electron that was initially free ($n = \infty$) and at rest relative to the atom. Beyond the series limit lies a region of a faint continuous spectrum that arises from the capture of electrons with an initial kinetic energy.

Uniqueness of spectra

It is not surprising that atoms of each element should possess an energy-level diagram unique to that element. Consequently, because the characteristics of the emission spectrum are determined by the energy levels of the element, the spectrum of each element is unique. This provides a method for identifying elements and does not involve a chemical procedure. The spectrum of an element serves as a "fingerprint" by which the identity of the element may be determined.

One consequence of this is that we can determine the chemical composition of the sun by examining the light from the sun. The same holds true for stars. We have determined that stars thousands or even millions of light-years away are made of the same type of atoms as we have all around us here on earth.

Continuous spectra

Now that we have seen that the emission spectrum of each element is unique, we might wonder why the hot tip of a solid carbon electrode of an arc lamp gives off a continuous spectrum instead of a spectrum of carbon. Why does a block of iron heated to incandescence emit a continuous spectrum rather than an emission spectrum of iron? The answer has to do with the proximity of the atoms to each other.

When one atom is brought near another, the presence of the second atom *perturbs* the energy levels of the first. In particular, each energy level of the two atoms will be split into two levels of slightly different energy. If three atoms are brought close together, as in a solid, each level of all three atoms will be split into three components. If N atoms are brought together, each energy level splits into N components—one component for each atom in the assembly.

Obviously, for a solid specimen of appreciable size, N is very large. The result is that the original energy levels of individual atoms become *energy bands* when the atoms are assembled into a solid, and these bands typically widen and overlap to form a *continuum* of energy levels. Hence, electron transitions are possible for essentially every value of energy, and a continuous spectrum is the result.

Absorption spectra

We stated earlier that an electron in an atom can be excited to a higher orbit by absorbing the energy of an incident photon. When this happens, the photon disappears.

A constraint placed on this process is that the photon absorbed must have exactly the same amount of energy as that required by the electrons to make the upward transition. If a photon has a bit too much or too little energy, it will not be absorbed and the excitation of the electron will not occur. Therefore, the wavelengths of photons absorbed in upward transitions are exactly equal to the wavelengths of photons emitted in downward transitions.

If light from a carbon arc (which has a continuous spectrum) is sent through sodium vapor, as in Fig. 20-5, the sodium vapor will absorb a portion of the photons that have wavelengths corresponding to the spectral lines of sodium. The initially continuous spectrum will then exhibit dark lines having the characteristics of a sodium spectrum. Such a spectrum is an *absorption spectrum*.

The excited electrons eventually return to their initial states, emitting photons of sodium wavelengths as they do. These photons, however, are emitted in all directions, whereas the photons absorbed were a part of the light beam from the arc lamp. Thus, the net result of the process of absorption and reemission is

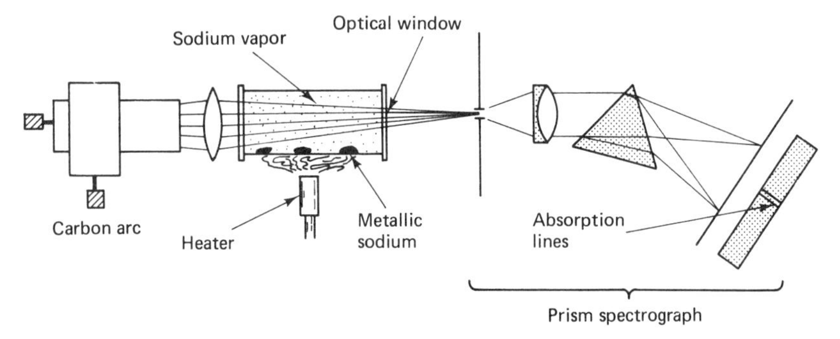

Figure 20-5 An apparatus for producing an absorption spectrum of sodium. A continuous spectrum (white light) enters the region of sodium vapor, where the wavelengths corresponding to sodium are absorbed. The spectrograph presents a continuous spectrum with gaps corresponding to the sodium lines.

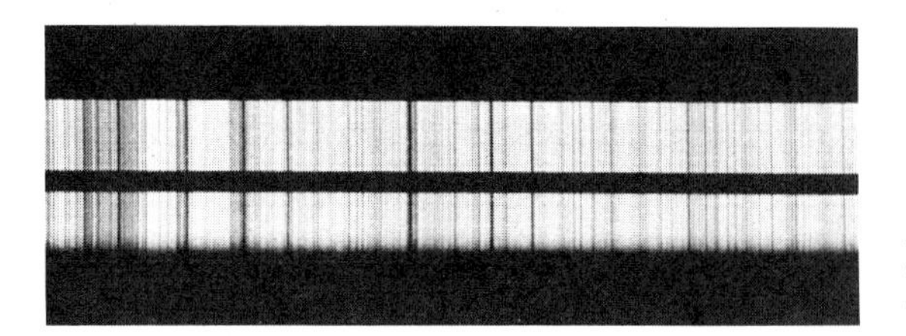

A high-resolution photo of a portion of the solar spectrum (two exposures). The Fraunhoffer lines are prominent.

that photons of sodium wavelengths are removed from the beam and are scattered in all other directions. As a result of the scattering, the path of the beam through the sodium vapor becomes visible, giving off a pale yellow light that is characteristic of sodium.

Light emitted from the surface of the sun forms a continuous spectrum, but the light must pass through the outer regions of the solar atmosphere enroute to the earth. If the spectrum of sunlight is examined, it is found to be interspersed by thousands of dark absorption lines, which are called *Fraunhofer lines* in honor of Joseph von Fraunhofer (1787–1826). The dark lines are produced by the selective absorption of light by the solar atmosphere. The most prominent lines are designated by letters of the alphabet and are shown in Table 20-2, along with the element responsible for the absorption.

TABLE 20-2 PROMINENT FRAUNHOFER LINES

Designation	Absorbing element	Wavelength[a]
A	Oxygen	7594
B	Oxygen	6870
C	Hydrogen	6562
D	Sodium	5893
E	Iron	5270
F	Hydrogen	4861
G	Iron	4308
H	Calcium	3969
K	Calcium	3935

[a]In angstroms; for nanometers, divide by 10.

Historically, certain Fraunhofer lines were determined to correspond to wavelengths calculated for an element with atomic number 2 and an atomic weight of 4. In this way, helium was discovered to exist on the sun prior to its discovery on the earth (hence its name—after *helios*, meaning *sun*).

Monochromatic sources of light

A *monochromatic* source emits light of only one frequency and wavelength whereas most sources of light emit either a continuous spectrum or a spectrum having a multitude of lines. Monochromatic sources are useful in the laboratory in experiments involving interference and diffraction.

The spectrum of sodium contains a pair of lines (a doublet) at a wavelength of about 589 nm that are far brighter than any other lines in the spectrum. Consequently, the light from a sodium vapor lamp can be used as a practical monochromatic source for lab demonstrations and other nonexacting applications. The light from a sodium lamp appears yellow.

The spectrum of mercury has a strong green line at about 546 nm. A mercury vapor lamp can be used as a monochromatic source if a green filter is placed over the lamp to absorb other undesired wavelengths. The red light from a helium-neon laser is monochromatic with a wavelength of 6328 Å.

20-4 LIGHT RAYS; DIFFRACTION EFFECTS

A *ray* of light is imagined to be the thinnest-possible, pencil-like beam of light. At one time light rays were thought to be a physical reality, but it turns out to be impossible to isolate a single ray of light. We now regard light rays as being simply lines drawn to show the path taken by light waves, and for this purpose, light rays are very useful. The rays are shown as lines drawn perpendicular to the wave fronts, and arrowheads are often superimposed upon the lines to indicate the direction of travel of the light waves.

Figure 20-6 shows an experiment designed to isolate a single ray of light. A beam of light is incident upon an adjustable slit that is made progressively

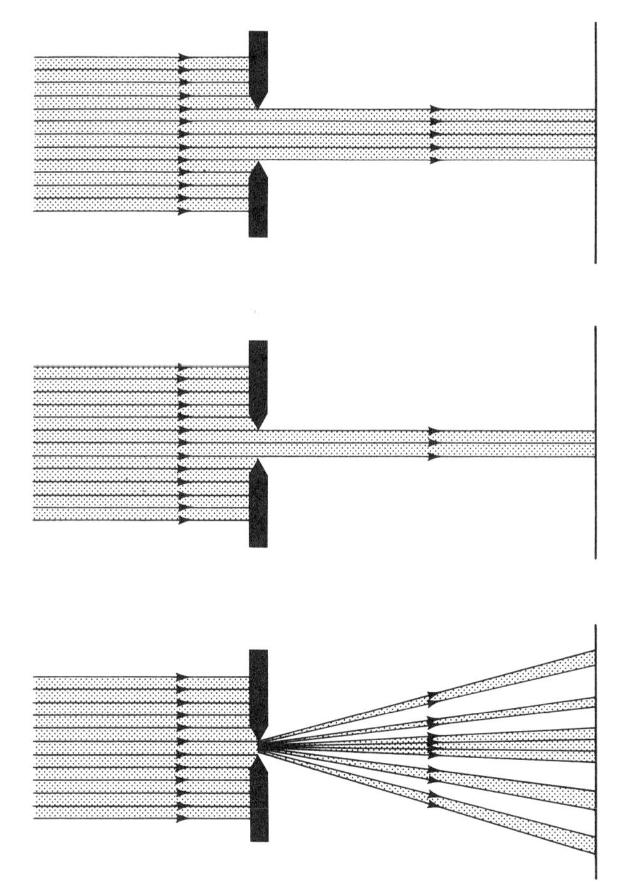

Figure 20-6 An attempt to isolate a single ray of light ends in failure due to the effects of diffraction.

narrower in an effort to permit only a single ray of light to pass through. It is expected that when only one ray passes through the slit, only a single, very small point of light will appear on the screen.

What actually happens as the width of the slit approaches the same order of magnitude as the wavelength of light is that diffraction occurs as predicted by Huygens' principle (see Sect. 18-6). Consequently, the beam of light spreads out in the region to the right of the slit. If the slit is made narrower still, the beam spreads out even more. We therefore conclude that it is impossible to isolate a single ray of light.

Diffraction at an edge

Suppose a beam of light is aimed across an object with a sharp edge, as shown in Fig. 20-7(a). The portion of the screen behind the object will be in shadow, and the other portion of the screen will be illuminated. Our interest is in the transition region between the shadow and lighted area. Will the edge of the shadow be perfectly sharp, with the level of illumination changing abruptly from completely dark to fully lighted? Or will the transition from dark to light be gradual, passing through all shades of gray in going from dark to light?

If light behaved as rays that always travel in straight lines, we would expect the shadow to be perfectly sharp. If the edge somehow deviated the rays passing closest to it, we would expect a *gradual* transition from dark to light. What actually happens, however, is that a *diffraction pattern* is formed in the transition region; a series of light and dark *fringes* mark the transition region, as shown in

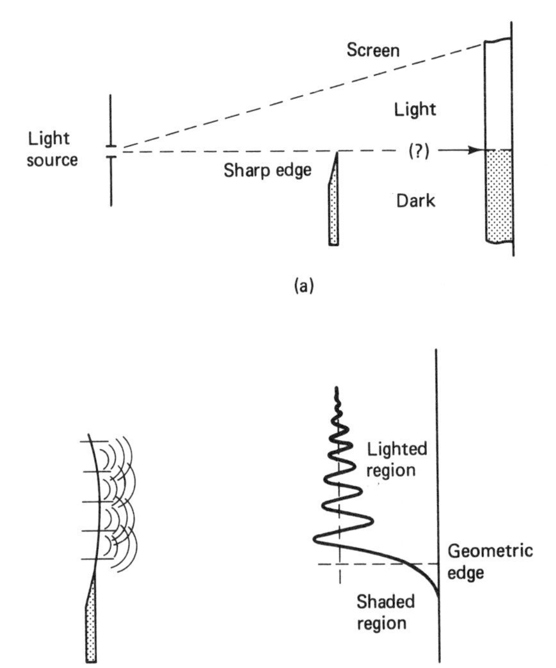

Figure 20-7 (a) A sharp edge obstructs light falling on a screen. (b) The wavefront passing the edge may be considered to be divided into many small sections, each of which emits Huygens' wavelets. (c) The intensity of light on the screen near the transition region; a series of fringes is produced.

Fig. 20-7(c). This is a consequence of Huygens' principle, and this phenomenon is evidence that light consists of waves.

To understand the origin of the fringes, consider the wave fronts passing near the edge to be divided into sections, as shown in Fig. 20-7(b). Each section acts as a source of secondary waves that determine the subsequent path of the wave front (Huygen's principle). A point on the screen will not be an equal distance from all wave front sections near the edge, and the small differences in distance will translate directly into phase differences. At some points on the screen, most of the Huygens' wavelets will arrive in phase and interfere constructively to produce a bright region. At adjacent points, most of the wavelets will arrive out of phase, interfere destructively, and produce a dark region. Thus, a system of alternating light and dark fringes is produced.

An interesting and novel consequence of the theory of diffraction is that diffraction occurring at the edge of a circular object (a penny, for example) placed in a beam of light should produce a bright spot at the center of the shadow. This prediction arises as a consequence of the wavelike character of light and was at first thought to represent a difficulty for the wave theory because no such bright spot had ever been observed. Subsequent laboratory investigations turned up a remarkable bit of fact: A bright spot appears at the center of the shadow of a circular object. This seemed to prove the wave-nature of light, once and for all.

20-5 LIGHT INTENSITY AND ILLUMINATION

Solid angle; steradians

In Sect. 8-1 we define the radian as a unit of angular measure, a rad being the angle subtended by an arc length of R on a circle of radius R. Hence, there are 2π rad in a complete circle because the circumference of a circle is 2π.

To state the obvious, an *angle* is a parameter for describing the intersection of two lines in a plane; a *plane* is a flat, *two-dimensional* surface. In the following we define a different kind of angle, a *solid angle*, that has a three-dimensional character. Whereas an ordinary angle consists of two lines that radiate outward from a point, a solid angle is like a cone that radiates outward from a point, the point being at the apex of the cone. Long, skinny cones with small apex angles represent small solid angles. Short, fat cones with large apex angles represent large solid angles. Even so, the apex angle does not enter directly into the definition of a solid angle.

A solid angle is defined in relation to a sphere in terms of an area marked off within a boundary on the surface of the sphere. If an area equal to R^2 is contained within the boundary on a sphere of radius R, the solid angle subtended by the area is one steradian (sr). This is illustrated in Fig. 20-8(a).

We now see that the solid angle will be shaped like a cone only if the boundary of the surface area is a circle. The shape of the boundary is not of primary importance; square, rectangular, or irregular boundaries are equally valid.

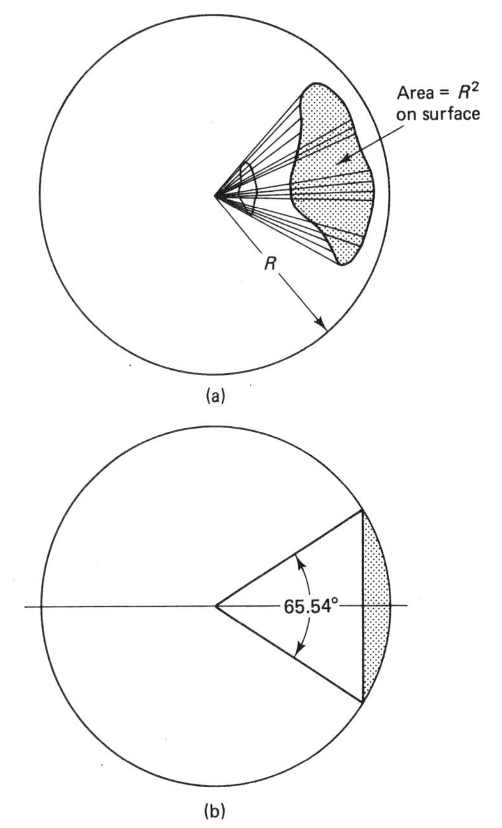

Figure 20-8 (a) An area of R^2 within a boundary on the surface of a sphere of radius R subtends a solid angle of 1 sr. (b) If the boundary is circular so that the solid angle is in the form of a cone, the angle at the apex of the cone is 65.54°.

If the area A is given, the solid angle subtended by the area is calculated by

$$\Omega = \text{solid angle (steradians)} = \frac{A}{R^2} \tag{20-6}$$

Because the surface area of a complete sphere is $4\pi R^2$, there are 4π sr in a sphere ($4\pi R^2/R^2 = 4\pi$). This is analogous to the fact that 2π rad make a complete circle. A solid angle of 1 sr in the shape of a cone gives rise to a cone whose apex angle is 65.54°. This is shown in Fig. 20-8(b).

Luminous flux, *F*

The quantity of light that either falls upon a surface or passes through a surface is called the *luminous flux*. The terms *light* and *luminous* as used here refer to the visible portion of the electromagnetic spectrum. If, for purposes of visualization, light is assumed to consist of discrete, countable *rays*, the luminous flux is the total number of rays that are present. The unit of luminous flux is the *lumen*. We can imagine a lumen to be a certain "number of rays" of light, but it is impossible to give the number because individual rays of light have no physical significance. (A similar situation exists for the number of lines of force in a magnetic field.) The lumen is defined in terms of the intensity of a point source in the following section.

Luminous intensity, I

The *luminous intensity* of a source is what we perceive as its brightness. Originally, the standard of intensity was a standard candle that burned a special whale oil at the rate of 120 g/h. The technical problems associated with the use of a flickering open flame as a standard source of light are easily imagined, so the unit of luminous intensity is now defined in terms of a blackbody at the solidification temperature of platinum, 2045 K. The SI unit of luminous intensity is the *candela* (cd). A blackbody radiator at 2045 K has a luminous intensity (in a direction normal to the surface) of 1 cd for each 1/60 cm^2 of surface area of the radiator.

For conceptual purposes we may refer to a *standard candle* (which has a luminous intensity of 1 cd) even though a candle is not used as a standard. Thus, the intensity of a source of light may be given in terms of candela or *candlepower* (cp).

We can now define the lumen (lm), the unit of luminous flux. By definition, a point source having a luminous intensity of 1 cd emits luminous flux in the amount of 1 lm/sr of solid angle, as shown in Fig. 20-9. This means that the total luminous

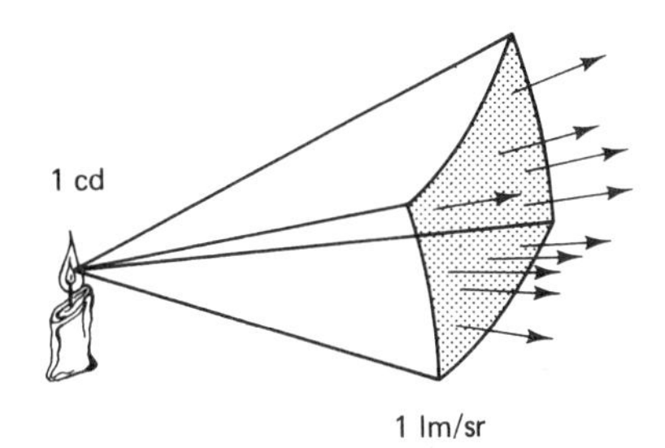

Figure 20-9 A source with a luminous intensity of 1 cd emits luminous flux in the amount of 1 lm/sr.

flux of a 1-cd point source is 4π (about 12.57) lm. The relationship between the flux F and the intensity I of an isotropic point source is

$$F = 4\pi I \tag{20-7}$$

Based on the definition of the lumen, a candela is equivalent to an intensity of one lumen per steradian:

$$1 \text{ candela (cd)} = \frac{1 \text{ lumen (lm)}}{\text{steradian (sr)}} \tag{20-8}$$

Illuminance

Illuminance refers to the luminous flux per unit area falling on a surface such as the top of a desk or the page of a book. The illuminance E is the luminous flux F per unit area:

$$E = \frac{F}{A} \tag{20-9}$$

The SI unit of illuminance is the lm/m^2, called the *lux* (lx). An older unit is the

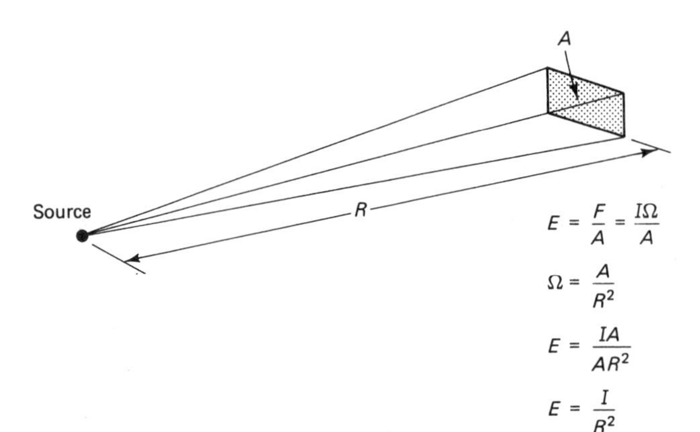

Figure 20-10 The illuminance E of a point source shining on a flat surface a distance R from the source.

footcandle (fc) which corresponds to 1 lm/ft^2. The footcandle is related to the lux:

$$1 \text{ fc} = 10.764 \text{ lx} \tag{20-10}$$

The 10.764 arises from the number of square feet in 1 m^2.

Let us now derive an expression for the illuminance of a flat surface of area A located a distance R from a point source, as shown in Fig. 20-10. The intensity I of the source is given in candela (or candlepower). We assume the flat surface area to be small in comparison with R^2 so that we may use Eq. (20-6) to compute the solid angle subtended by the surface without introducing appreciable error. (Strictly speaking, in order to use Eq. (20-6) the surface should be a portion of the surface of a sphere of radius R.) The light source lies on the normal passing through the center of the surface. The luminous flux F falling on A is the product of the source intensity I and the solid angle Ω subtended by the area,

$$F = I\Omega \tag{20-11}$$

The illuminance E is the luminous flux per unit area falling on A. Hence,

$$E = \frac{F}{A} = \frac{I\Omega}{A} \tag{20-12}$$

and it follows that

$$E = \frac{I}{R^2} \tag{20-13}$$

as shown in the figure.

If the flat surface is not aimed directly at the source, the projected area must be used in computing the solid angle Ω. If the line from the surface to the source makes an angle α with the normal to the surface, the projected area is $A \cos \alpha$ so that

$$\Omega = \frac{A \cos \alpha}{R^2} \tag{20-14}$$

The illuminance for a surface lighted obliquely then becomes

$$E = \frac{I \cos \alpha}{R^2} \tag{20-15}$$

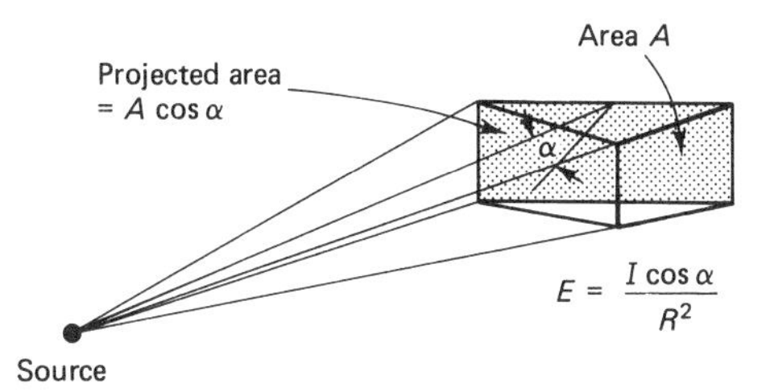

Figure 20-11 When the area A is oriented at an angle α to the source, the illuminance is reduced by the factor cos α.

When $\alpha = 0$, $\cos\alpha = 1$ and this formula reverts to Eq. (20-15). The geometry is shown in Fig. 20-11.

EXAMPLE 20-2 A flat rectangular surface of dimensions 0.2 m by 0.3 m is located 3 m from a 60-cd bulb located on the normal of the surface.

(a) Compute the solid angle subtended by the surface.
(b) Compute the luminous flux incident on the surface.
(c) Compute the illuminance at the surface.

Solution. (a) The area of the surface is 0.06 m^2. Thus, by Eq. (20-6), the solid angle subtended is

$$\Omega = \frac{A}{R^2}$$

$$= \frac{0.06 \text{ m}^2}{(3 \text{ m})^2} = 6.67 \times 10^{-3} \text{ sr}$$

(b) Using Eq. (20-11), the luminous flux incident on the surface is

$$F = I\Omega$$

$$= (60 \text{ cd})(6.67 \times 10^{-3} \text{ sr})$$

$$= 0.40 \text{ lm}$$

(c) The illuminance is computed using Eq. (20-13):

$$E = \frac{I}{R^2} = \frac{60 \text{ cd}}{(3 \text{ m})^2}$$

$$= 6.67 \text{ lm/m}^2 = 6.67 \text{ lx}$$

In part (c), the illuminance could have been obtained directly from the definition $E = F/A$ [Eq. (20-6)].

EXAMPLE 20-3 If the surface of Example 20-2 is rotated so that the angle between the normal and the source is 45°, compute: (a) the solid angle subtended; (b) the illuminance of the surface.

Solution. We use Eqs. (20-14) and (20-15) directly.

(a)
$$\Omega = \frac{A \cos\alpha}{R^2}$$

$$= \frac{(0.06 \text{ m}^2)\cos 45}{(3 \text{ m})^2}$$

$$= 4.714 \times 10^{-3} \text{ sr}$$

(b)

$$E = \frac{I \cos \alpha}{R^2}$$

$$= \frac{(60 \text{ cd})(\cos 45)}{(3 \text{ m})^2}$$

$$= 4.71 \text{ lx}$$

Luminous efficiency

The *luminous efficiency* of a source is a measure of its ability to convert energy into visible light. Because most sources of light are electrical and because the unit of electrical power is the watt, the luminous efficiency η is given as lumens per watt. For example, a typical 75-W incandescent lamp produces 1150 lm. Its luminous efficiency is

$$\eta = \frac{\text{luminous flux emitted}}{\text{input power, watts}} \tag{20-16}$$

$$= \frac{1150 \text{ lm}}{75 \text{ W}} = 15.33 \text{ lm/W}$$

This is typical of incandescent bulbs. Fluorescent lamps, on the other hand, have typical luminous efficiencies of about 60 lm/W.

TABLE 20-3 RECOMMENDED LEVELS OF ILLUMINATION

Situation	Footcandles[a]
Reading, printed matter	30
pencil writing	70
Assembly of extra-fine components	1000
Artwork, dark colors	500
medium colors	100
Beauty parlor	100
Lobby of bank	70
Dance hall	5
Elevators	20
Kitchen sink	70
Living room	10
Basketball	50
Baseball, infield	150
outfield	100
Tennis	30

[a]Multiply by 10.764 to convert to lux.

Photometric radiation equivalent

Luminous flux is a subjective phenomenon that depends upon the properties of the human eye in the same manner that the loudness of a sound depends upon the properties of the ear. (Recall the Fletcher-Munson contours of equal loudness, Fig. 19-5.) On the other hand, *radiant flux* is *not* subjective because it represents the energy per unit time radiated as electromagnetic waves without regard to the response of the eye to the radiation.

Because the eye is not equally sensitive to all portions of the visible spectrum, the same amount of monochromatic radiant flux will produce differing quantities of luminous flux if the wavelength of the radiant flux is varied. For example, 1 W of radiant flux having a wavelength of 555 nm (5550 Å) produces 680 lm. But at a wavelength of 500 nm, 1 W of radiant flux produces only about 220 lm. The conversion factor between radiant flux (watts) and luminous flux (lumens) for monochromatic radiation is called the *photometric radiation equivalent* K, and its numerical value varies with the wavelength of the radiation as shown in Fig. 20-12. The units of K are lumens per watt.

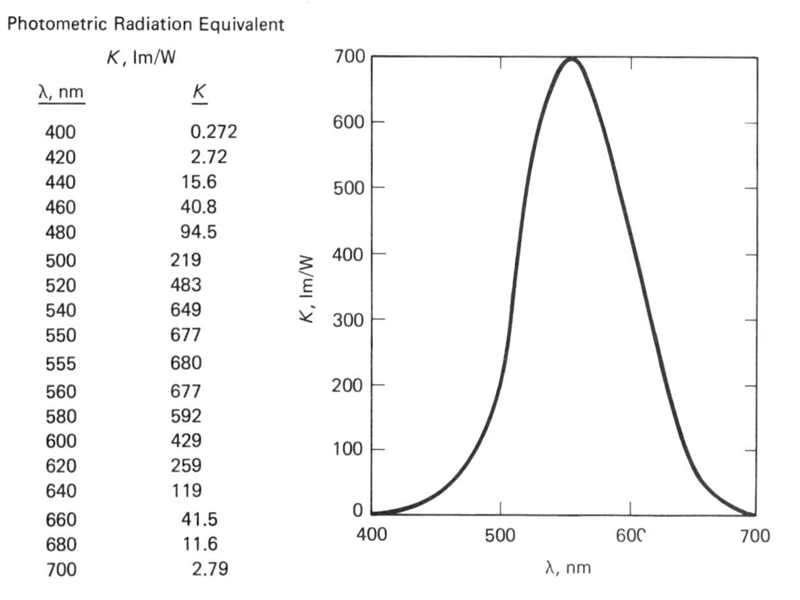

Photometric Radiation Equivalent

K, lm/W

λ, nm	K
400	0.272
420	2.72
440	15.6
460	40.8
480	94.5
500	219
520	483
540	649
550	677
555	680
560	677
580	592
600	429
620	259
640	119
660	41.5
680	11.6
700	2.79

Figure 20-12 Photometric radiation equivalent.

Transparent, translucent, and opaque

These terms characterize the light-transmitting properties of materials. Light passes essentially unaltered through materials that are *transparent*. The glass of a picture window is transparent. Light also passes through materials that are *translucent*, but we cannot "see through" such materials. They often appear milky white, as does the translucent mixture of water and a smaller portion of milk. *Opaque* materials block light completely and consequently appear dark.

Bunsen photometer

A *photometer* is a device for determining the intensity of a source of light. A Bunsen grease-spot photometer is shown in Fig. 20-13. With it the intensities of two sources can be compared, and if one is a standard, the other intensity can be computed.

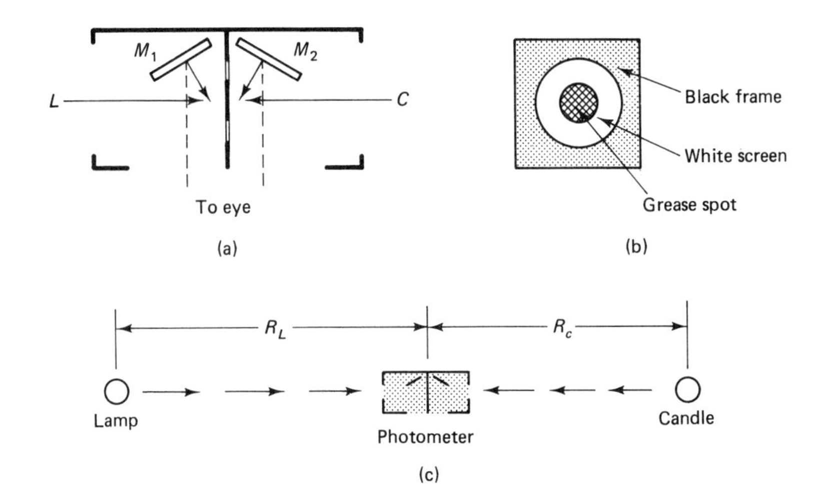

Figure 20-13 Bunsen grease-spot photometer.
(a) Arrangement of grease spot and mirrors. (b) Grease spot is situated within a white screen held by a frame painted black. (c) Distances R_l and R_c are adjusted until the grease spot appears equally bright in both mirrors.

A drop of oil or grease applied to a piece of paper causes the paper to become translucent; the grease spot resembles the wax paper found in kitchens. Such a grease spot is used in a photometer as shown in Fig. 20-13.

The grease spot is illuminated on one side by lamp at L and on the other by a candle at C. When the illuminance of L is greater than that from C, the image of the grease spot seen in mirror M_2 will appear brighter than the similar image visible in M_1. By adjusting R_c and R_l, the two images can be made equally bright, so that the illuminances E_c and E_l become equal.

Using Eq. (20-13) and setting the illuminances equal gives

$$\frac{I_c}{R_c^2} = \frac{I_l}{R_l^2} \tag{20-17}$$

which may be rearranged to give

$$I_l = \left(\frac{R_l}{R_c}\right)^2 I_c \tag{20-18}$$

Thus, if the intensity of the candle (or equivalent standard source) is known, the intensity of the lamp can be computed.

To Go Further

Read about early investigators of spectra:

Joseph von Fraunhofer (1787–1826), Bavarian optician
Robert Bunsen (1811–1899), German scientist of Bunsen burner fame
Gustav Kirchhoff (1824–1887), German physicist
Sir Norman Lockyer (1836–1920), discoverer of helium in the sun

Questions

1. Explain three ways that an electron in an atom can obtain sufficient energy to undergo an upward transition.
2. What determines the color of the light emitted by a gas discharge tube? Why do the elements have unique spectra?
3. Describe the Lyman, Balmer, and Paschen series of the hydrogen spectrum. What is the origin and location in the spectrum of each series?
4. What causes Fraunhofer lines and in what spectrum do they appear?
5. Explain how it was possible to discover helium on the sun before it was discovered on earth.
6. What happens when one attempts to isolate one ray of light?
7. What is a solid angle? A steradian? How many steradians represent a complete sphere?
8. Describe the relationship between luminous flux, luminous intensity, and illuminance.
9. In the Northern Hemisphere, the earth is actually closer to the sun during the winter months. Then why is it colder in winter?
10. Luminous flux and luminous intensity refer to the visible portion of the electromagnetic spectrum. How is radiant flux related to luminous flux?
11. Does the intensity of a laser beam decrease according to the inverse square law?

Problems

1. How many angstroms does it take to make 1 nm?
2. Use the relationship $R = 0.53n^2$ to compute the radius of each of the first six orbits of hydrogen. Make a sketch (to scale) of the orbits.
3. Compute the approximate energy of each of the following orbits of the hydrogen atom. **(a)** $n = 4$ **(b)** $n - 8$ **(c)** $n - 12$ **(d)** $n = 100$
4. Compute the wavelength of light emitted when an electron makes the following transitions.
   (a) From $n = 8$ to $n = 4$. **(b)** From $n = 12$ to $n = 4$. **(c)** From $n = 100$ to $n = 4$.
5. Compute the wavelength of light emitted when an electron makes the following transitions.
   (a) From $n = 12$ to $n = 4$. **(b)** From $n = 12$ to $n = 3$. **(c)** From $n = 12$ to $n = 2$. **(d)** From $n = 12$ to $n = 1$.
6. Suppose an electron in a hydrogen atom is excited from the $n = 2$ level to the $n = 5$ level.
   (a) What energy must be delivered to the electron in order for it to make the transition?
   (b) What is the wavelength of the photon capable of delivering the energy to the electron?
7. Suppose an electron transition in an atom requires 4×10^{-9} s to occur. Compute the maximum length of the emitted photon.
8. Calculate the mass of a 589-nm photon from a sodium vapor lamp and compare it to the rest mass of an electron.
9. Compute the momentum of a 550-nm photon.
10. Determine the velocity an electron must have in order to have the same momentum as a 550-nm photon.
11. What solid angle is subtended by an area of 8 cm^2 on the surface of a sphere whose radius is 8 cm.
12. A small lamp bulb is located 6 ft above the top of a desk. A rectangular sheet of paper (8.5×11 in.) lies on the desk. If the sheet of paper is imagined to form part of a sphere centered on the bulb, what solid angle is subtended by the sheet of paper?
13. Determine the intensity in candela that the bulb of Problem 12 must have in order to provide an illuminance of 300 lx, the mini-

mum recommended for reading printed matter.

14. **(a)** Assuming the bulb in Problems 12–13 to be an isotropic source, determine the total lumens that it should provide.
(b) Based upon the typical luminous efficiency of incandescent bulbs, what should the wattage rating of the bulb be? Comment.

15. What is the total luminous flux emitted by an isotropic point source whose luminous intensity is: **(a)** 1 cd; **(b)** 1 cp; **(c)** 100 cd?

16. How many lumens per steradian of solid angle are emitted by a source having a luminous intensity of 40 cd?

17. The luminous flux of a certain 100-W bulb is 1750 lm. Consider the bulb to be an isotropic point source and: **(a)** calculate the intensity of the bulb in candela; **(b)** determine the luminous efficiency of the bulb; **(c)** compute the illuminance produced by the bulb on a surface located 8 ft from the bulb.

18. Long-life incandescent bulbs have a larger than normal filament that operates at a lower temperature in order to last longer. A certain 100-W long-life bulb produces 1100 lm. Compute: **(a)** the intensity of the bulb; **(b)** the luminous efficiency; **(c)** the illuminance produced by the bulb on a surface located 8 ft from the bulb. Compare with the previous problem.

19. A flat surface of area 0.05 m^2 is 4 m from a 50-cd bulb located on the normal to the surface.
(a) What solid angle is subtended by the surface?
(b) Determine the luminous flux incident upon the surface.
(c) Calculate the illuminance at the surface.

20. Repeat Problem 20 assuming the surface is tilted so that the normal to the surface makes an angle of 30° to the line from the surface to the bulb.

21. How many lumens will be produced by a monochromatic source of each wavelength if the radiated power is 1 W?
(a) 400 nm **(b)** 500 nm **(c)** 555 nm

22. The grease spot of a Bunsen photometer appears equally bright on both sides when a candle and an unknown lamp are being compared. The distance from the grease spot to the candle is 40 cm; the distance to the lamp is 3.0 m. If the candle has an intensity of 1 cd, what is the intensity of the lamp?

23. Two sources, A and B, are being compared using a Bunsen grease-spot photometer. Source A is four times as bright as source B. How must the photometer distances compare in order for the grease spot to be lighted equally on both sides?

CHAPTER 21

THE REFLECTION OF LIGHT

Light is reflected by the surface of a mirror, by the smooth surface of a quiet pond, more generally, from the surface of any object you can see. In this chapter we study the properties of the reflection of light and several practical applications of mirrors, both plane and curved.

Curved mirrors are used as image-forming elements in many optical instruments. On the largest scale, parabolic reflectors constitute the objective element of the world's largest optical and radio telescopes. Smaller reflectors (mirrors) are used by amateur astronomers in backyard telescopes, many of which are of excellent quality. More down-to-earth is the magnifying personal mirror, a concave mirror that helps us to see ourselves as others do not. In the following sections, we present the major principles of image formation by curved mirrors.

21-1 BASIC PROPERTIES OF REFLECTION

The law of reflection

When a beam of light is incident upon a surface, the electric and magnetic fields of the light waves interact with the charged particles (electrons) within the surface material. Forces are exerted on the electrons, and they vibrate minutely back and forth in response to the oscillating forces applied by the incident electromagnetic waves. The result is that the vibrating charges produce a secondary beam of light which travels away from the surface. This secondary beam is said to be *reflected* from the surface.

Figure 21-1 illustrates the law of reflection, which relates the directions of the incident and reflected rays. Quite simply, the law of reflection is that the angle of incidence ϕ_i equals the angle of reflection ϕ_r, with ϕ_i and ϕ_r being measured from the normal to the surface. Furthermore, the incident ray, the normal to the surface, and the reflected ray all lie in the same plane.

If the reflecting surface is highly polished so that the surface is flat (on a

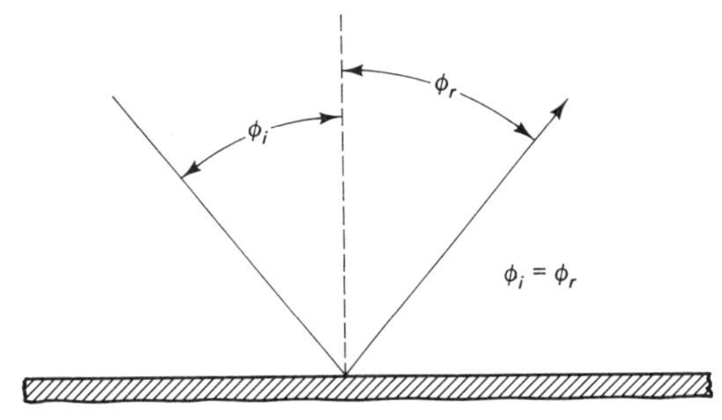

Figure 21-1 The law of reflection.

microscopic level), the reflection of light will be such that reflected images of nearby objects can be seen. The surface of smooth glass, still water, silver spoons, and polished metal produce this type of reflection, called *specular reflection*. If, on the other hand, the surface is rough and uneven on the microscopic level, the surface will not appear shiny and no reflected images will be visible. Such a surface is said to have a *matte* finish, and the reflection is *diffuse*. The difference between specular and diffuse reflection is illustrated in Fig. 21-2.

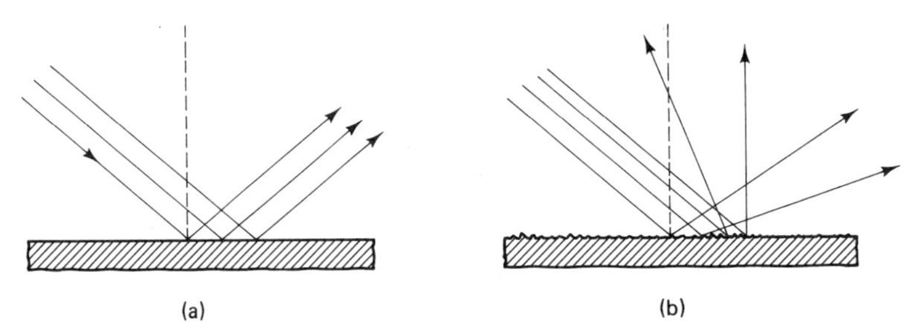

Figure 21-2 (a) Specular reflection. (b) Diffuse reflection.

All surfaces are reflective to some degree whether or not the surfaces are specially prepared. The reflection may be diffuse or specular, depending upon the nature of the reflecting surface. The ratio of reflected light to incident light that occurs in a given situation depends upon the nature of the surface, the material involved, and the angle of incidence. Almost all surfaces become highly reflective if the angle of incidence is very nearly equal to 90°, as shown in Fig. 21-3. If the

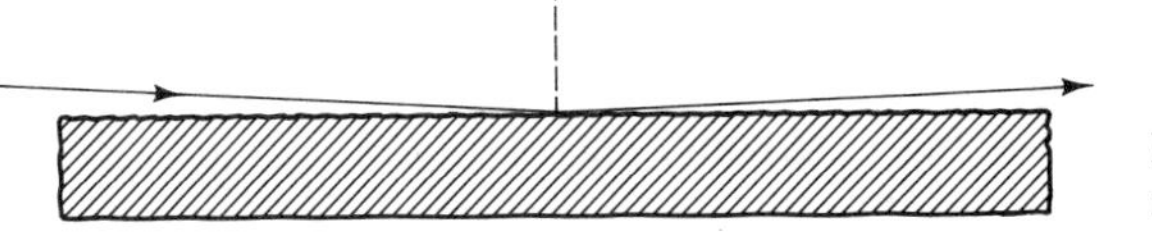

Figure 21-3 Surfaces become highly reflective at the grazing angle.

incident light travels nearly parallel to the reflecting surface, the reflection is said to be at the *grazing angle*.

21-2 PLANE MIRRORS

Mirrors are highly polished surfaces specially prepared to be efficient specular reflectors. Mirrors may be made either of solid metal or of glass to which a reflective coating of metal (usually silver or aluminum) is applied. Glass mirrors

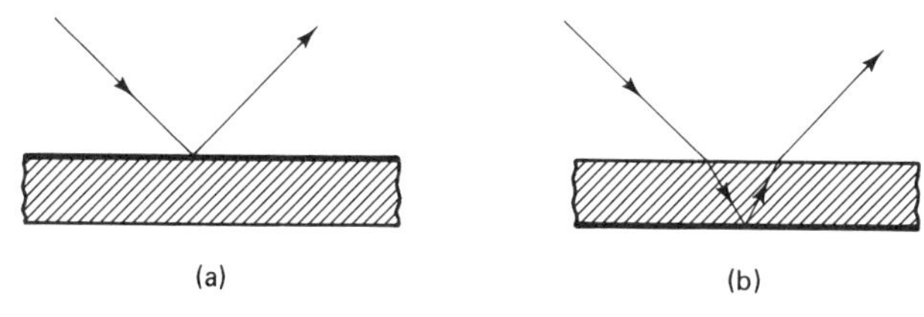

Figure 21-4 (a) Front-surface mirror. (b) Back-surface mirror.

may be either *front-surface* or *back-surface* mirrors, as illustrated in Fig. 21-4. Metal mirrors and front-surface glass mirrors may be *overcoated* with a thin transparent material to prevent contamination of the reflective surface. Back-surface glass mirrors usually have an opaque paint applied as a protective layer. Ordinary mirrors found in the home are back-surface mirrors. Front surface mirrors are typically used in laboratory situations or in optical instruments where it is undesirable for the light to pass through the glass.

A mirror whose surface is flat is called a *plane mirror*. However, no surface of any mirror is absolutely flat in the strictest sense. The flatness of a mirror is usually expressed in terms of wavelengths of light (550 nm is typical). The high-quality plane mirrors used as diagonals in reflecting telescopes (see Fig. 21-18) are typically flat to $\frac{1}{8}$ or $\frac{1}{10}$ wave. Reflective surfaces specially prepared to be *flat* are called *optical flats*. Ordinary mirrors are considerably less flat than those used in optical instruments (telescopes, bombsights, or range finders).

If the reflective coating of a glass mirror is very thin, the mirror will be partially reflecting. Only a part of the incident light will be reflected; the rest will be transmitted through the mirror. Partially reflecting mirrors (sometimes called *half-silvered* mirrors) may be used as beam splitters, as illustrated in Fig. 21-5, or

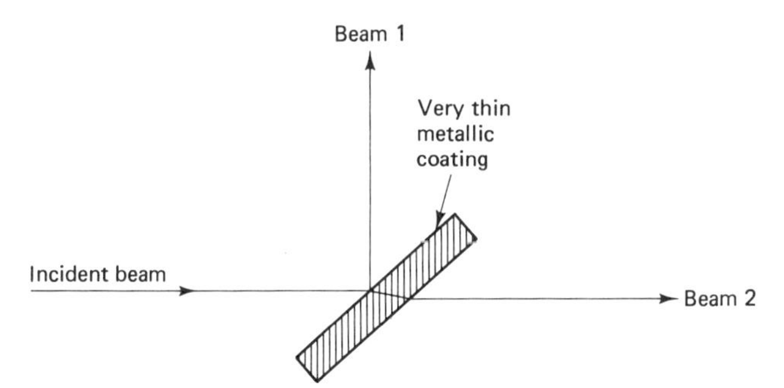

Figure 21-5 A partially reflecting mirror used as a beam splitter.

as one-way mirrors, as in Fig. 21-6. Partially reflecting mirrors can also be used to create special effects, as in having a candle burn underwater (Fig. 21-7).

Image in a plane mirror

The origin of the image in a plane mirror is illustrated in Fig. 21-8. The image is located behind the mirror at a distance equal to the distance from the object to the mirror, and the image is the same size as the object. Each ray of light that travels from the object to the eye obeys the law of reflection at the surface of the mirror.

In actuality, there is no light behind the mirror at the point where the image appears to be. Therefore, the image is a *virtual* image. We encounter *real* images in the following sections; light *is* actually present at the location of a real image.

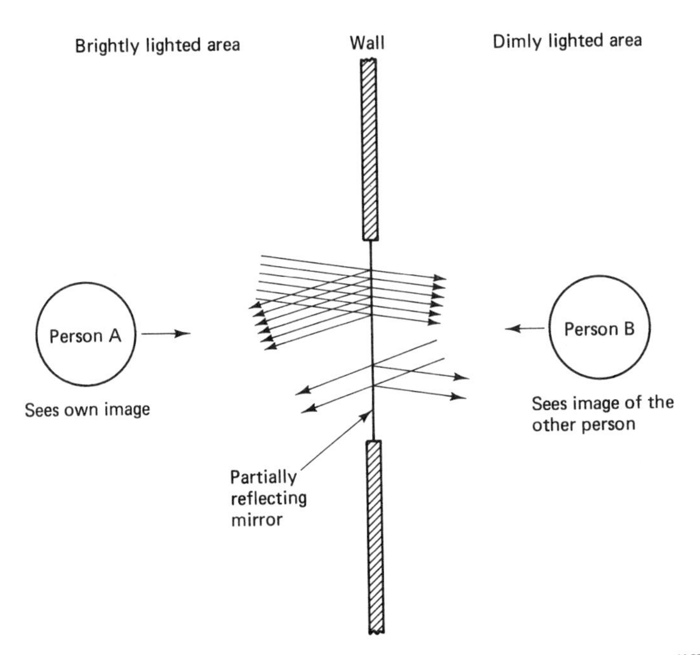

Figure 21-6 A partially reflecting mirror used as a one-way mirror. The difference in the level of illumination on the two sides of the mirror gives a partially reflecting mirror its one-way properties. The feeble illumination of the darker side is swamped by the greater illumination from the brighter side.

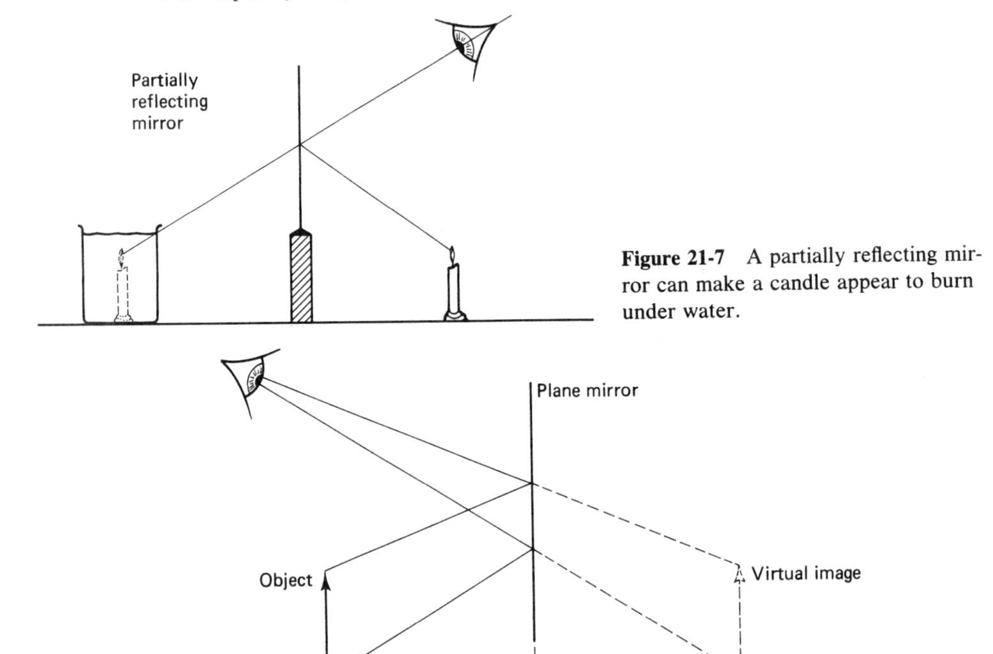

Figure 21-7 A partially reflecting mirror can make a candle appear to burn under water.

Figure 21-8 Object and virtual image of a plane mirror. No light is physically present at a virtual image.

21-3 APPLICATIONS OF PLANE MIRRORS

Optical lever

Plane mirrors are often useful in measuring small angles of rotation by devising an optical lever, as shown in Fig. 21-9(a). A small, high-quality mirror is attached to the rotating member, and an intense but very narrow beam of light is reflected off

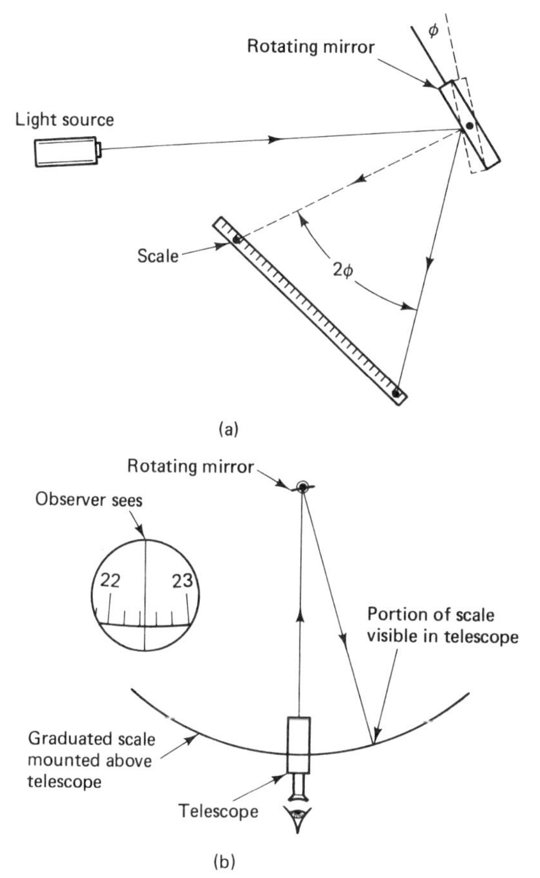

Figure 21-9 (a) The principle of the optical lever.
(b) One version of a practical optical lever.

the mirror onto a scale. As the mirror rotates through an angle ϕ, the reflected beam rotates through an angle twice as great, 2ϕ. Thus, very small angular deflections of the mirror can be observed, especially if the distance between the mirror and the scale is fairly large. This technique was used in the Cavendish experiment (Fig. 9-4) to determine the gravitational constant G.

Figure 21-9(b) shows a variation of the optical lever which uses the line of sight of a telescope rather than a beam of light. The observer looks through the telescope and sees a small portion of the semicircular scale. As the mirror rotates, the numbered scale seems to run by the cross hair of the telescope. With a good telescope and an accurately calibrated scale, very fine measurements can be made in this manner.

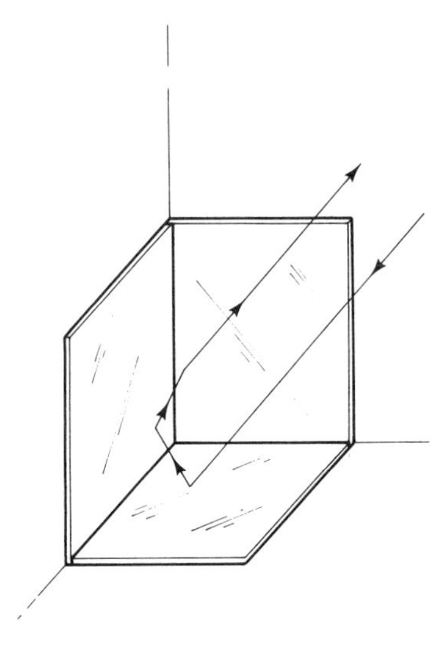

Figure 21-10 An incident ray of light reflected from a corner reflector will be reflected back in the direction of its source.

Corner reflector

Three plane mirrors set at 90° to each other to form a corner will reflect an incident ray of light in the direction from which it came. Three mirrors so arranged form a *corner reflector*, illustrated in Fig. 21-10. In practice, the mirrors (reflecting surfaces) are ground on the surface of a solid piece of glass to form a *corner prism* that has the same property of returning an incident beam to its source. An array of corner prisms left on the moon by the lunar astronauts reflects a laser beam back to its source on the earth. The transit time of the beam is used to calculate the distance from the earth to the moon.

Range finder

A diagram of an optical range finder is shown in Fig. 21-11. Light rays from the object reach the eye by two different routes, the direct route through a split mirror

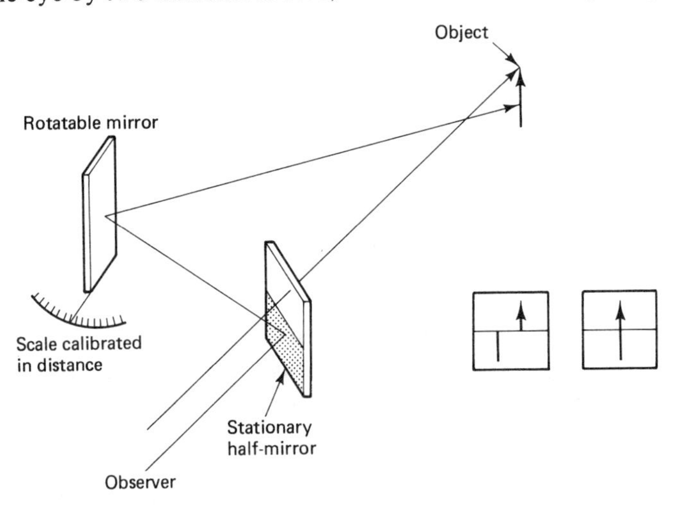

Figure 21-11 The principle of a split-image range finder.

and the indirect route using a mirror set on a rotating platform. Two distinct images of the object will be seen until the mirror is rotated to the position shown, at which time the two images overlap. The angular position of the mirror is related to the distance to the object. Such range finders are incorporated into the focusing mechanism of one type of 35-mm camera.

21-4 CONCAVE SPHERICAL MIRRORS

A curved mirror whose reflecting surface forms a segment of the surface of a sphere is called a *spherical mirror*. The *radius of curvature* of the mirror is the radius, R, of the sphere, and the *center of curvature*, C, of the mirror lies at the center of the hypothetical sphere. This is illustrated in Fig. 21-12. A spherical

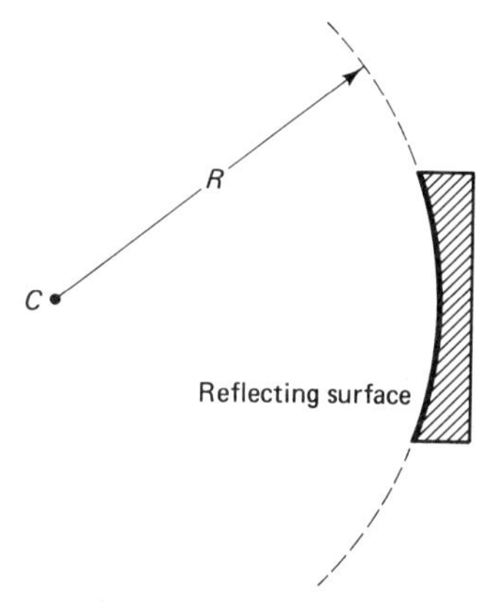

Figure 21-12 The radius of curvature R and the center of curvature C of a spherical mirror.

mirror may be either concave or convex, depending upon whether the inside (for concave) or the outside surface of the spherical segment is the reflecting surface.

A concave spherical mirror has the ability to focus incoming rays of light, as shown in Fig. 21-13. When the incident rays are parallel, the focal point F will lie

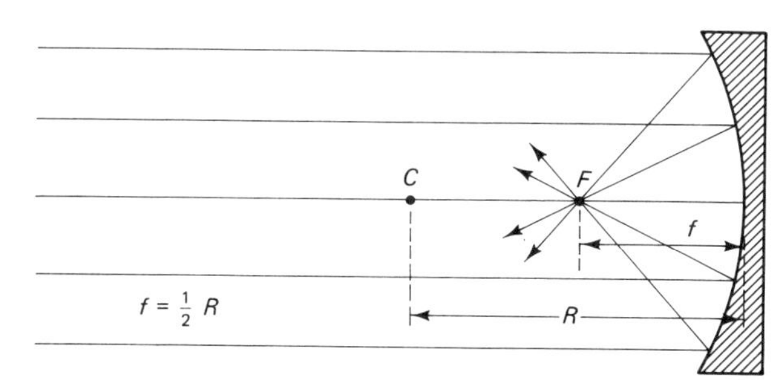

Figure 21-13 The focusing action of a concave mirror.

halfway between the center of curvature and the mirror surface. If the light source (or object) is located far from the mirror, the incident rays will be (practically) parallel. Thus, rays from the sun or stars are parallel.

When the incident light rays are parallel, the focal point F is called the *primary focal point*, and the distance, f, from the mirror to the primary focal point is called the *primary focal length*. Thus, from the figure it is apparent that

$$f = \tfrac{1}{2}R \tag{21-1}$$

where R is the radius of curvature of the mirror. As a consequence of the focusing action of a concave mirror, an image of the source of light is produced at the focal point. If a small screen or photographic film is located at the focal point (for parallel light only), the image can be seen or recorded photographically, as desired.

We may gain a better understanding of the image-forming properties of a concave mirror by constructing a ray-tracing diagram as illustrated in Fig. 21-14.

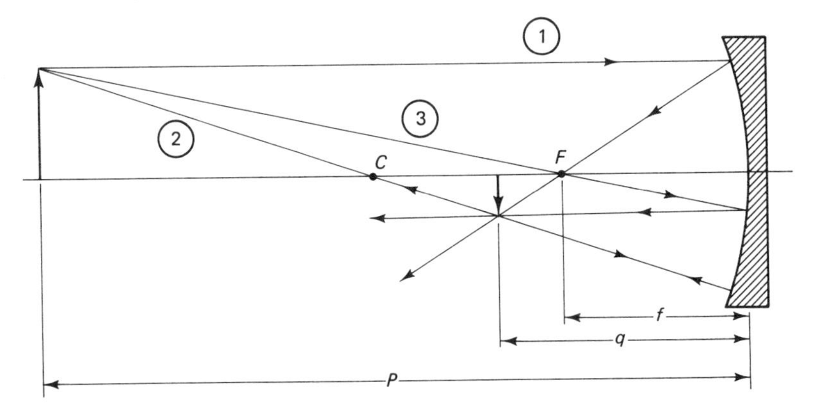

Figure 21-14 Ray-tracing diagram for a concave mirror.

The position of the mirror, center of curvature, and focal point are accurately drawn to scale, and the object is drawn as shown.

Each and every point on the object acts as a source of light rays, whether or not the object is self-luminous. Points on nonluminous objects emit rays of light by virtue of diffuse reflection of incident light whereas luminous objects such as the filament of an incandescent lamp generate the light they emit.

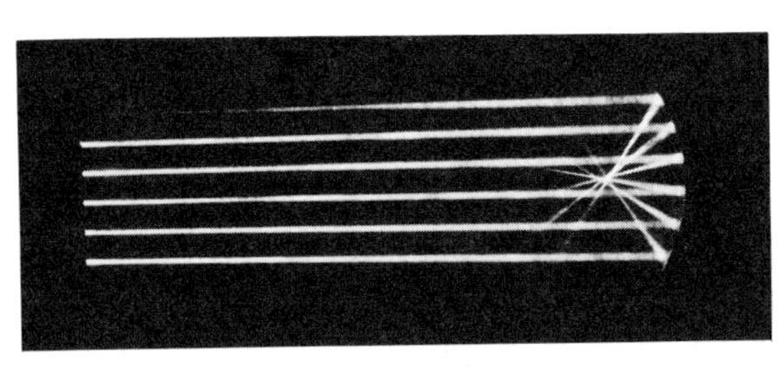

Parallel rays incident on a concave mirror are brought to a focus. Spherical aberration causes the focus to be imperfect.

To locate the image formed by a concave mirror, we use the following properties of rays emitted by the object in three different directions:

1. The incident ray that travels parallel to the axis will be reflected through the primary focal point F.
2. The incident ray that passes through the center of curvature C of the mirror will be reflected on itself so as to again pass through C.
3. The incident ray that passes through the primary focal point F will be reflected parallel to the axis.

These three rays are drawn in Fig. 21-14. The image is formed where the three rays come together. The image is *not* located at the primary focal point. Also, the image is inverted.

The distance p from the object to the reflecting surface is called the *object distance*. The distance from the reflecting surface to the image is called the *image distance*, denoted by q. A simple relationship exists between the object distance, image distance, and primary focal length. It is called the *mirror equation*:

$$\frac{1}{p} + \frac{1}{q} = \frac{1}{f} \tag{21-2}$$

This equation[1] permits the calculation of any one of p, q, or f if the other two are known.

EXAMPLE 21-1 A certain concave mirror brings the rays of the sun to a focus at a point 20 cm from the mirror. Compute the primary focal length of the mirror.

Solution. Since the sun is very distant in comparison with 20 cm, we set $p = \infty$ in Eq. (21-2). Hence, with $q = 20$ cm,

$$\frac{1}{\infty} + \frac{1}{20} = \frac{1}{f}$$

$$0 + 0.05 = \frac{1}{f}$$

$$f = 20 \text{ cm}$$

Thus, the primary focal length equals the image distance when the object is at infinity.

EXAMPLE 21-2 Calculate the image distance when an object is located 100 cm in front of a mirror having a primary focal length of 20 cm.

Solution. The mirror equation [Eq. (21-2)] yields

$$\frac{1}{p} + \frac{1}{q} = \frac{1}{f}$$

$$\frac{1}{100} + \frac{1}{q} = \frac{1}{20}$$

$$q = 25 \text{ cm}$$

It is helpful to solve the mirror equation for each unknown in terms of the other two, which gives

$$f = \frac{pq}{p + q} \qquad q = \frac{fp}{p - f} \qquad p = \frac{fq}{q - f} \tag{21-3}$$

These expressions make the arithmetic a bit more direct in the course of using the mirror equation.

[1]The mirror equation is identical to the lens equation derived in Sect. 22-4. For brevity, we derive it only once.

Sign convention

Thus far we have encountered only positive values for the image and object distances in the use of the mirror equation. However, if the object is closer to the mirror than a distance equal to the primary focal length, the image distance will be negative. For example, if a spherical concave mirror has a focal length of 20 cm and if an object is 12 cm from the mirror, the image distance will be $q = -30$ cm. What is the meaning of the negative sign?

This situation is shown in Fig. 21-15. With the object close to the mirror, the rays incident on the mirror are so highly divergent that the converging power of

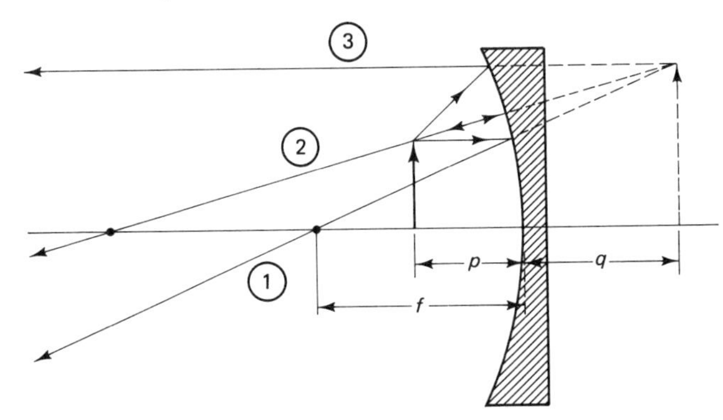

Figure 21-15 When the object is closer to the mirror than a distance equal to the primary focal length f, the image will be virtual and the image distance q will be negative.

the concave mirror is not great enough to bring the rays back together. The effect of the mirror is to reduce the divergence of the rays, and, of course, to reverse the overall direction of travel of the rays by 180°. To an observer, the rays appear to come from a point behind the mirror; we say that a *virtual image* is formed behind the mirror. The distance from the mirror to the virtual image is given by the image distance q, with q being negative. Thus, when q turns out to be negative, it means that the image is virtual and that it is located behind the mirror. Later we shall see that it is possible for an object distance to be negative also.

With the physical arrangement as in Fig. 21-14, the following sign convention is followed:

1. The primary focal length f of the concave mirror is *positive*.
2. The object distance p is *positive* when a real object is placed to the left of the mirror. (A negative object distance is used when a virtual object exists in the region behind the mirror.)
3. A positive image distance indicates the formation of a real image in front of the mirror. A negative image distance signifies the formation of a virtual image behind the mirror.

To clarify the distinction between a real and a virtual image, note that a real image is formed by rays of light traveling toward and converging to the image. A real image will appear on a screen placed in the image plane, and a real image will expose a photographic film. On the other hand, the rays appear to come *from* a virtual image, and, even more noteworthy, no light is present at the location of a virtual image. Recall that no light can pass through the mirror to the region behind the mirror. Consequently, a virtual image cannot appear on a screen nor expose a photographic film.

Magnification

The magnification produced by a mirror or lens is defined as

$$\text{Magnification} = M = -\frac{\text{image size}}{\text{object size}} \tag{21-4}$$

It can be shown that the magnification is given by

$$M = -\frac{\text{image distance}}{\text{object distance}} \tag{21-5}$$

$$= -\frac{q}{p}$$

When the magnification is a negative quantity, the image is *inverted* relative to the object. When the magnification is positive, the image is *erect*.

EXAMPLE 21-3 A personal mirror that magnifies is a slightly concave mirror that has a focal length of about 75 in. If the mirror is held 15 in. from a person's face, calculate: (a) the image distance q; (b) the magnification M. (c) Is the image erect or inverted?

Solution. (a) We use one of Eqs. (21-3) to calculate the image distance:

$$q = \frac{fp}{p - f}$$

$$= \frac{(75)(15)}{15 - 75}$$

$$= -18.75 \text{ in.}$$

The image appears to be 18.75 in. behind the mirror.
(b) The magnification is found from Eq. (21-5):

$$M = -\frac{q}{p}$$

$$= -\frac{(-18.75)}{15}$$

$$= +1.25$$

(c) Because the magnification is positive, we conclude that the image is erect rather than inverted. This agrees with our experience. The image is 25% larger than the object.

EXAMPLE 21-4 A concave mirror with a focal length of 1 ft is aimed at a tree that is 300 ft away. Calculate: (a) the position of the image of the tree; (b) the magnification of the image. (c) If the tree is 25 ft tall, compute the size of the image.

Solution. (a) We calculate the image distance:

$$q = \frac{fp}{p - f}$$

$$= \frac{(1)(300)}{300 - 1}$$

$$= 1.003 \text{ ft}$$

Thus, the image is formed very nearly at the primary focal point since the tree is a relatively great distance away.

(b) The magnification is

$$M = -\frac{q}{p}$$

$$= -\frac{1.003}{300}$$

$$= -0.0033$$

(c) The size of the image is the product of the magnification and the size of the object:

$$\text{Image size} = M \times \text{object size}$$

$$= 0.0033 \times 25 \text{ ft}$$

$$= 0.0825 \text{ ft} = 0.99 \text{ in.}$$

The image is about 1 in tall and is inverted (as determined by the negative value obtained for the magnification).

Spherical aberration

The *aperture* of a spherical mirror is the straight-line distance from one edge of the mirror to the other; it is the diameter of the mirror. If a portion of the mirror is covered by a ring of opaque material, the aperture is the diameter of the active portion of the mirror.

Incoming rays are brought to an acceptable focus by a concave spherical mirror only when the aperture is small in comparison to the radius of curvature of the mirror. When the aperture is large, rays near the periphery of the mirror are not reflected through the focal point. This effect gives rise to an image that is slightly out of focus, an effect known as *spherical aberration*. It is illustrated in Fig. 21-16.

Parabolic reflector

The ideal shape for a concave mirror is that of a *paraboloid*, the surface obtained by rotating a parabola about its axis of symmetry. All rays that approach the mirror (parallel to the axis) are reflected through the focus irrespective of the size of the aperture. Conversely, if a point source of light is placed at the focus of a

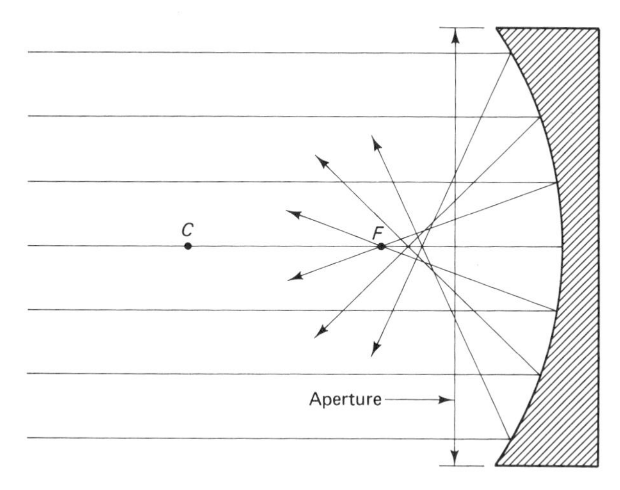

Figure 21-16 When the aperture of a spherical mirror is large in comparison with the radius of curvature, the focus becomes imperfect due to spherical aberration.

parabolic reflector, the transmitted beam will be perfectly parallel. The mirrors used in large reflecting telescopes are parabolic for the same reason. The paraboloid eliminates spherical aberration.

Microwaves may be transmitted as beams just as beams of visible light.

21-5 CONVEX SPHERICAL MIRRORS

A ray-tracing diagram for a convex spherical mirror is shown in Fig. 21-17. Note that the center of curvature and the primary focal point are to the right of the mirror. For real objects to the left of the mirror, the rays will be reflected so that

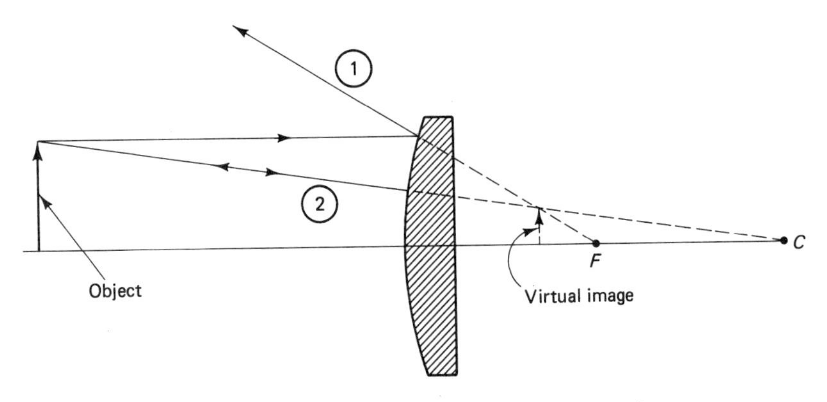

Figure 21-17 Ray-tracing diagram for a convex mirror.

they appear to come from an erect virtual image to the right of the mirror. The image can be located by the intersection of two rays with the following properties:

1. An incoming ray parallel to the axis is reflected as if it had come from the primary focal point, F.
2. The ray that proceeds toward the center of curvature is reflected on itself.

The mirror equation also applies to a convex mirror. However, because the primary focal point is to the right of the mirror, the focal length f is taken to be a negative quantity. The sign convention for the image and object distances is the same as for the concave mirror.

EXAMPLE 21-5 An object 3 cm tall is located 25 cm in front of a convex mirror with a radius of curvature of 10 cm. Determine: (a) the location of the image; (b) the magnification; and (c) the size of the image.

Solution. (a) The primary focal length is one-half the radius of curvature, and we assign to it a negative value. Hence, $f = -5$ cm. We then use the mirror equation to compute the image distance:

$$q = \frac{pf}{p - f}$$

$$= \frac{(25)(-5)}{25 - (-5)}$$

$$= -6.25 \text{ cm}$$

Thus, we find the image to be 6.25 cm to the right of the mirror.
(b) As before, the magnification is the negative of the ratio of the image and object distances:

$$M = -\frac{q}{p}$$

$$= -\frac{-6.25}{25}$$

$$= +0.25$$

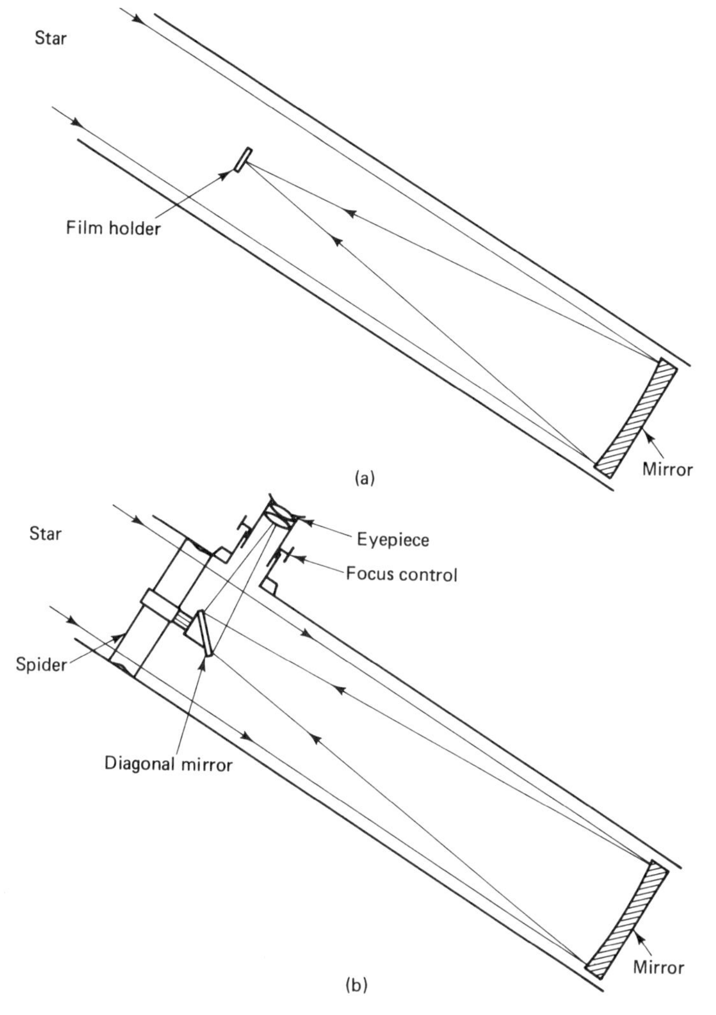

Figure 21-18 (a) Large Newtonian reflectors may have the film holder located within the tube where the light is focused to form the image.
(b) In the realm of amateur astronomy, a diagonal mirror deflects the image to the outside of the telescope tube.

The positive sign of the magnification indicates that the image is erect.
(c) The size of the image is the product of the object size and the magnification:

$$\begin{aligned}\text{Image size} &= M \times \text{object size}\\ &= 0.25 \times 3 \text{ cm}\\ &= 0.75 \text{ cm}\end{aligned}$$

21-6 REFLECTING TELESCOPE

One application of concave mirrors (ground to the shape of a paraboloid) is in the design of an astronomical telescope. The basic design of a Newtonian reflector is shown in Fig. 21-18(a). Parallel rays from a celestial object strike the mirror and are brought to a focus at the primary focal length f where a real, inverted image is formed. A photographic film may be used to capture the image, as shown in the figure.

Mirrors for such telescopes may be quite large. The largest in existence is a Soviet telescope which has a mirror 6 m (236 in.) in diameter. The next largest is the 200-in. telescope on Palomar Mountain in southern California. An observer using one of these large telescopes may observe the image directly while riding in a cage located within the "tube" of the telescope at the position of the image.

Smaller instruments used by amateur astronomers usually are like that shown in Fig. 21-18(b). A plane, front-surfaced mirror, called a *diagonal*, deflects the image out of the side of the tube to a point where it is accessible to the observer. The image may then be photographed or viewed directly with the aid of an eyepiece. The eyepiece serves as a magnifier to make the image appear larger, and it enables a person to bring his or her eye close to the image while still being able to focus on the image. We describe this in more detail in our discussion of lenses.

Newtonian reflectors used by amateur astronomers range in (mirror) diameter from about 3 in. to about 18 in. Relatively few amateur astronomers have mirrors 12 in. or larger, but the new Dobsonian reflectors make it possible for amateurs to afford mirrors as large as 17 or 18 in. Advertisements in popular astronomy magazines may be consulted for more information.

The advantage of a large mirror is twofold. A large mirror provides a greater *light-gathering ability* making it possible to see fainter objects. At the same time, the larger aperture increases the theoretical *resolving power* of the telescope. Resolving power refers to the ability of a telescope to separate two stars that appear very close together. Consequently, a double star might appear as a single point of light in a 4-in. telescope, whereas an 8-in. telescope might resolve the double into two individual stars.

The smallest angular separation of two objects that can be resolved by an instrument having a circular aperture d is given by the following expression which stems from the Rayleigh criterion:

$$\phi_{min} = 1.22\frac{\lambda}{d} \tag{21-6}$$

where λ is the wavelength of the light by which the objects are observed and ϕ_{min} is in radians. This expression arises from consideration of the diffraction patterns produced by circular apertures and from the criterion for resolution proposed by the British scientist, Lord Rayleigh.

The *magnification*, or *power*, of a telescope is determined by the combination of the mirror and eyepiece:

$$\text{Power} = \frac{\text{focal length of mirror}}{\text{focal length of eyepiece}} \tag{21-7}$$

It is not necessarily true that the greater the power, the more you see because the use of higher power reduces the brightness of the image. One may see more detail in a smaller, brighter image than in a larger, dimmer image.

Furthermore, the use of higher powers makes atmospheric disturbances more objectionable, and high power magnifies any defects in the objective mirror. Under optimum conditions, the greatest power that is practically useful is about 50X for each inch of mirror diameter. Generally, the optimum power to use is far less than this, depending, or course, upon the quality of the equipment used and upon the nature of the object being viewed.

The *f-number* of a telescope mirror (or lens) is the ratio of the focal length to the mirror diameter:

$$f\text{-number} = \frac{\text{focal length}}{\text{mirror diameter}} \tag{21-8}$$

The *f*-number relates to the brightness of the image and provides a method of comparing the brightness of the images produced by mirrors and lenses of different designs. The images of the same object produced by mirrors or lenses of different designs will be equally bright if the *f*-number of each system is the same. Most Newtonian reflector telescopes used by amateur astronomers fall into the range from *f*/4 to *f*/8.

EXAMPLE 21-6 The focal length of the 6-in. mirror of a Newtonian reflector is 48 in. The eyepiece used has a focal length of 28 mm.

(a) What is the power of the telescope?

(b) What is the *f*-number of the telescope mirror?

Solution. (a) The power is the ratio of the focal length of the mirror to the focal length of the eyepiece. We convert the focal length to millimeters: 48 in. = 1219 mm. Hence,

$$\text{Power} = \frac{1219 \text{ mm}}{28 \text{ mm}}$$

$$= 43.5$$

(b) Using Eq. (21-8), we find the *f*-number of the mirror to be

$$f\text{-number} = \frac{48 \text{ in.}}{6 \text{ in.}}$$

$$= 8$$

We say the telescope is an *f*/8 Newtonian reflector.

EXAMPLE 21-7 How much more light is "gathered" by a 6-in. mirror in comparison with a 4-in. mirror?

Solution. The light gathered by a mirror is proportional to its area. Thus, the ratio of the light-gathering ability of two mirrors is the same as the ratio of their areas, and the ratio of the areas is equivalent to the square of the ratio of the diameters. Therefore, the ratio of the light-gathering ability of a 4-in. and a 6-in. mirror is

$$(\tfrac{6}{4})^2 = 2.25$$

The 6-in. mirror gathers 2.25 times as much light as the 4-in. mirror.

EXAMPLE 21-8 The Arecibo radiotelescope in Puerto Rico is 305 m in diameter (the world's largest). If it receives electromagnetic waves of wavelength 21 cm from interstellar hydrogen, calculate the smallest angular separation ϕ_{min} of objects that may be resolved.

Solution. Use Eq. (21-6) in a straightforward manner:

$$\phi_{min} = 1.22\frac{\lambda}{d}$$

$$= (1.22)\frac{0.21 \text{ m}}{305 \text{ m}}$$

$$= 8.4 \times 10^{-4} \text{ rad}$$

$$\approx 3' \text{ of arc}$$

To Go Further

Find out about the development of the telescope and the contributions made by Isaac Newton, Galileo, and the American astronomer George Ellery Hale.

References

LOVELL, SIR BERNARD, *The Story of Jodrell Bank*. New York: Harper & Row, 1968. Tells the story of the conception, design, and construction of one of the largest fully steerable radiotelescopes.

Questions

1. What properties of a surface determine whether reflection from the surface will be specular or diffuse?

2. What is an optical flat? How is the flatness of a mirror usually expressed?

3. How is a partially reflecting mirror made? What are some applications of partially reflecting mirrors?

4. What is the shortest mirror in which a 6-ft-tall person can see his or her full-length image?

5. What is the special property of a corner prism. Why were such prisms left on the moon?

6. What will the image do when a person in front of a mirror: **(a)** parts hair on left; **(b)** raises left hand; **(c)** rotates counterclockwise; **(d)** approaches mirror?

7. Describe (or sketch) the three rays used in a ray-tracing diagram for a concave mirror.

8. What is the advantage of a parabolic reflector over a spherical reflector?

9. Give the procedure for constructing the ray-tracing diagram for a convex spherical mirror.

10. Give a step-by-step procedure for determining the primary focal length of a concave mirror: **(a)** on a sunny day; **(b)** on a cloudy day or at night.

11. How can a parabolic reflector be used to transmit a beam of light or, in the case of radio telescopes, to transmit a narrow beam of electromagnetic radiation into space?

12. Why should a solar telescope (one used for observing the sun) have a large aperture even though light-gathering ability is not an advantage?

Problems

1. A diagonal mirror to be used in a Newtonian reflector telescope is flat to $\frac{1}{10}$ wavelength of 550-nm light. Express this distance in inches.
2. A small child is 4 ft in front of a plane mirror. How far from the child is the image?
3. The horizontal beam from a small laser is directed at a two-sided mirror that stands vertically at the center of a turntable of a phonograph. If the rate of rotation of the turntable is 33.3 rpm, what will be the rate at which the reflected laser beam will sweep about the room?
4. Phonograph turntables rotate in the clockwise direction. If one views a turntable in a mirror, which way will the turntable appear to rotate?
5. An optical lever as shown in Fig. 21-9(a) is used to measure small rotations of a mirror attached to a thermal expansion apparatus. A laser beam is reflected from the mirror onto a scale mounted on a wall 10 ft from the mirror. Calculate the angle through which the mirror must rotate in order to move the beam 0.1 in. across the scale. Express the answer in: **(a)** radians; **(b)** degrees, minutes, and seconds.
6. The scale of the optical lever of Fig. 21-9(b) is graduated in millimeters and is 1 m from the mirror. Through what angle must the mirror rotate in order for the observer to see the scale move 1 mm past the cross hairs in the telescope?
7. What is the focal length of a spherical mirror whose radius of curvature is 18 cm if: **(a)** the mirror is concave; **(b)** if the mirror is convex?
8. The focal length of a concave mirror is 44 in. What is the radius of curvature of the mirror?
9. * On a sheet of graph paper, construct a ray-tracing diagram for a concave mirror whose radius of curvature is 20 units and whose primary focal length is, therefore, 10 units. Locate the object 40 units from the mirror, and make the object 5 units tall. Locate the image and determine its size as accurately as possible. (*Hint*: Represent the concave mirror by a straight vertical line even though in actuality it is slightly curved.)
10. Use the mirror equation to verify the results of the ray-tracing diagram of Problem 9. That is: **(a)** calculate the image distance; **(b)** compute the magnification; **(c)** compute the size of the image and tell whether it is erect or inverted.
11. A certain concave mirror has a primary focal length of 20 cm. Calculate the image distance and the magnification when an object is located at the following distances from the mirror.
    (a) 100 cm **(b)** 40 cm **(c)** 21 cm **(d)** 19 cm
    (e) 10 cm **(f)** 2 cm
12. When an object is placed 75 cm in front of a concave mirror, a real image is formed 19 cm in front of the mirror. Calculate the focal length of the mirror.
13. A concave mirror taken from a Newtonian reflector telescope is 6 in. in diameter and has a primary focal length of 24 in. By aiming the mirror at a screen 20 ft away and by placing a small clear-glass incandescent lamp between the mirror and the screen, it is possible to project an image of the lamp filament onto the screen.
    (a) How far in front of the mirror should the lamp be placed so that the image will be in focus at the screen 20 ft away?
    (b) If the filament of the lamp is 1 cm long, what will be the length of the filament projected onto the screen?
14. The sun is about 93 million miles from the earth, and it is about 800,000 mi in diameter. Use this information to calculate the size of the image of the sun produced by a concave mirror having a primary focal length of 48 in.
15. An object 5 cm tall is located 30 cm in front of a convex mirror having a radius of curvature of 12 cm. Determine: **(a)** the location of the image; **(b)** the magnification; **(c)** the size of the image. **(d)** Is the image erect or inverted?
16. *On a sheet of graph paper, construct a ray-tracing diagram for convex mirror whose

radius of curvature is 20 units and whose primary focal length is 10 units. Assume the object is 30 units from the mirror, and assume the object to be 5 units tall. Locate the image, determine its size, and note whether it is erect or inverted.

17. Use the mirror equation to verify the results of the previous ray-tracing problem. That is, calculate: **(a)** the image distance; **(b)** the magnification; **(c)** the size of the image. **(d)** Determine whether the image is erect or inverted.

18. What is the diameter in feet of the largest radio telescope?

19. The Palomar telescope is 200 in. in diameter. How many times more light will be gathered by the Palomar telescope than a 6-in. telescope operated by an amateur astronomer?

20. A telescope having a focal length of 28 in. is operated with eyepieces having focal lengths of: **(a)** 28 mm; **(b)** 20 mm; **(c)** 15 mm. Calculate the power of the telescope when operated with each eyepiece.

21. A certain telescope is advertised as a 6″ *f*/5.
- **(a)** What is the focal length of the telescope?
- **(b)** What should be the focal length of an eyepiece that will provide a power of about 90 when used with the telescope?

22. What is the *f*-number of a telescope mirror that is 8 in. in diameter and has a focal length of 56 in.?

23. In an experiment to determine the focal length of a telescope mirror, a candle is set up 5 m from the mirror in a darkened room. The image is found at a distance of 1 m from the mirror. What is the focal length of the mirror?

***24.** Verify mathematically that the minimum height required of a mirror in which a person can see himself or herself full length is one-half the height of the person. (*Note*: The distance from the person to the mirror is not important.)

***25.** A certain telescope has an angular field of view of 3°. Calculate the time required for a star to cross the field of view due to the rotation of the earth when the telescope is aimed perpendicular to the axis of rotation of the earth.

***26.** Using a technique called aperture syntheses, radio astronomers can utilize two or more antenna dishes widely separated to achieve an effective aperture equal to the separation of the dishes. Near Socorro, New Mexico, the VLA (very large array) utilizes 27 movable 85-ft dishes arranged in a Y to achieve an aperture of about 27 km. Compute ϕ_{min} for the array when operated at a wavelength of 6 cm. Compare this with the 1 arc-second resolutions obtained with typical large-diameter optical telescopes.

CHAPTER 22

THE REFRACTION OF LIGHT

When light passes from one medium to another, its speed changes. This change in speed causes the direction of the light to change as it crosses the interface between the two media. This effort is called *refraction* and it is responsible for the fact that a straight stick inserted in water appears to be bent.

Refraction makes possible the design of lenses that are capable of forming images, and such lenses make feasible photography, television, slide and movie projectors, telescopes, microscopes, and corrective lenses for human eyes. A major portion of this chapter is devoted to the image-forming characteristics of lenses. The final sections are devoted to the telescope and microscope.

The material of this chapter is very practical; applications of the principles are found in a wide variety of commonly encountered scientific instruments, in equipment for entertainment, and in everyday experience.

22-1 REFRACTION AT PLANE SURFACES

The general situation occurring when a beam of light is incident upon a plane surface of transparent material is shown in Fig. 22-1. The incident beam is split into two components. One component forms the *reflected* beam and the other forms the *transmitted* beam. The direction taken by the reflected beam is determined by the law of reflection, as described in the preceding chapter. The direction taken by the transmitted beam is determined by the physical properties of the medium above and below the surface. Generally speaking, a change in direction occurs as the beam passes through the surface. The incident beam is said to be *refracted* at the surface, and the process involving the change in direction is called *refraction*.

It is an experimental fact that light travels slower through a medium such as water, glass, or air than it travels through a vacuum. Put differently, light travels at its maximum velocity through empty space, and this maximum velocity is the speed of light, denoted by c. The speed of light in a certain medium is given

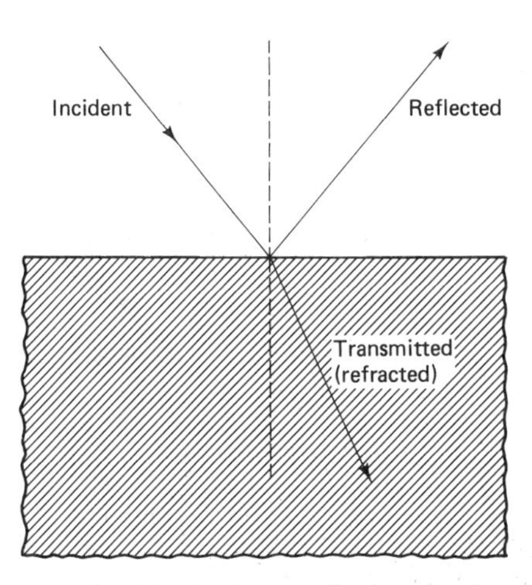

Figure 22-1 A beam of light incident upon a surface separating two media is split into a reflected and a refracted beam.

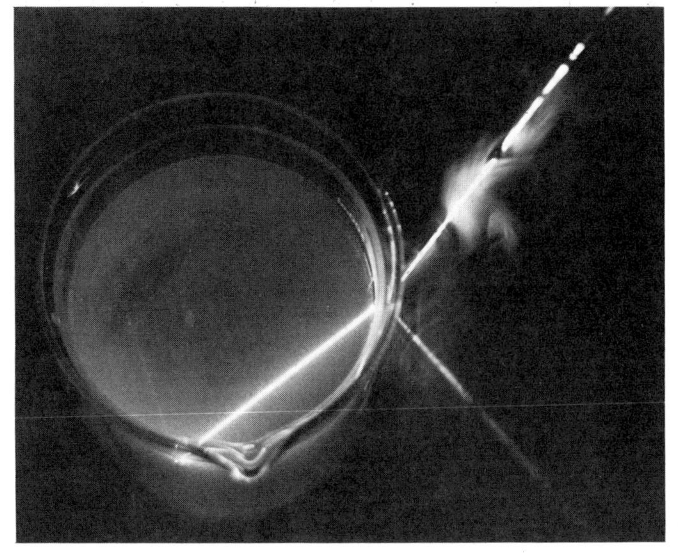

A laser beam incident upon a beaker is both reflected and refracted. Smoke in the air and coffee creamer in the water make the beam visible.

indirectly by the *index of refraction* of the medium. The index of refraction n is defined by

$$n = \frac{c}{v} \qquad (c = 2.9979 \times 10^8 \text{ m/s}) \tag{22-1}$$

where v is the speed of light in the medium and where c is the speed of light in a vacuum. Rearranging the above equation gives

$$v = \frac{c}{n} \tag{22-2}$$

The index of refraction of water is 1.33. It follows from Eq. (22-2) that the speed of light in water is 2.25×10^8 m/s, which is only about three-fourths its

TABLE 22-1 INDICES OF REFRACTION OF SELECTED MATERIALS

Material	$n = c/v$
Air (0°C, at 1 atm)	1.0003
Water, at 0°C	1.334
at 20°C	1.333
at 40°C	1.331
at 60°C	1.327
Glass, crown	1.52
Light flint	1.58
Dense flint	1.65
Fused quartz	1.46
Ethyl alcohol	1.36
Plexiglas	1.51
Diamond	2.42

velocity in a vacuum. Table 22-1 gives the indices of refraction of selected materials.

Refraction is caused by the difference in speed with which light travels in two different materials. The difference in speed gives rise to differing indices of refraction; we may say, equivalently, that refraction is caused by differences in the index of refraction. Refraction may occur abruptly, as at the surface between two media having different indices of refraction, or it may occur gradually, as when a beam of light passes through a nonuniform medium.

Several examples of refraction are illustrated in Fig. 22-2. Because light passes more slowly through the more dense portions of the atmosphere, the observed image of the setting sun is higher than the actual position of the sun. Atmospheric refraction also alters the positions of stars near the horizon in a

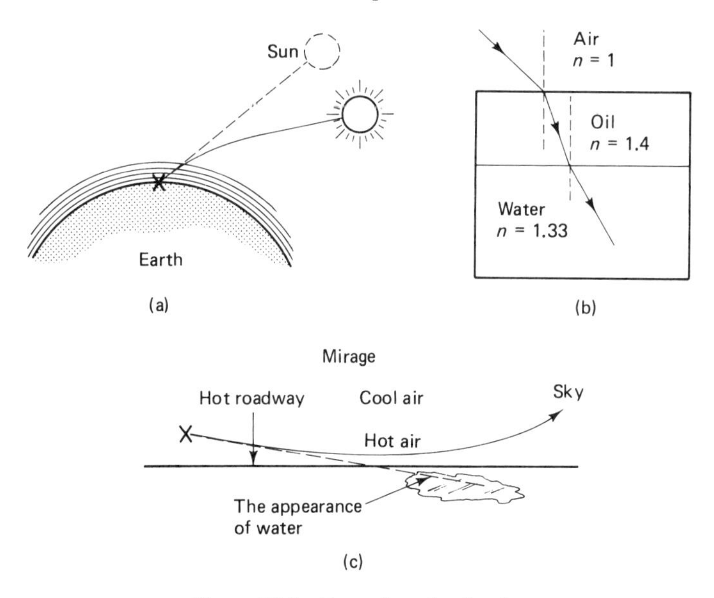

Figure 22-2 Examples of refraction.

similar manner. Part (b) shows a beam of light passing successively from air through a layer of oil and then into a layer of water. The ray of light is refracted toward the normal in passing from the air into the oil, but the ray is refracted away from the normal in passing from the oil into the water. Oil has a greater index of refraction than air or water.

The cause of a mirage is shown in Fig. 22-2(c). A roadway heated by the sun heats the air immediately above the surface to a higher temperature than the air above. Then, because light travels faster through hot air, which is less dense, the line-of-sight is curved upward toward the sky. The observer at X sees an image of the sky where the road surface should be, and the image might mistakenly be attributed to water on the surface of the road. A similar thing happens over hot sands of a desert, causing a thirsty traveler to see images of water when actually looking at the sky in a roundabout way.

22-2 SNELL'S LAW

Snell's law provides the means of calculating the direction taken by a refracted beam of light when it passes through a surface from one medium to a second medium having an index of refraction different from that of the first. We can derive Snell's law with the aid of Fig. 22-3.

Plane waves are incident upon the surface between the two media. The waves are incident at an angle ϕ_i measured relative to the normal to the surface. The refracted waves travel at an angle ϕ_r to the normal. We now consider the wave front section AB as it passes through the surface to the position CD.

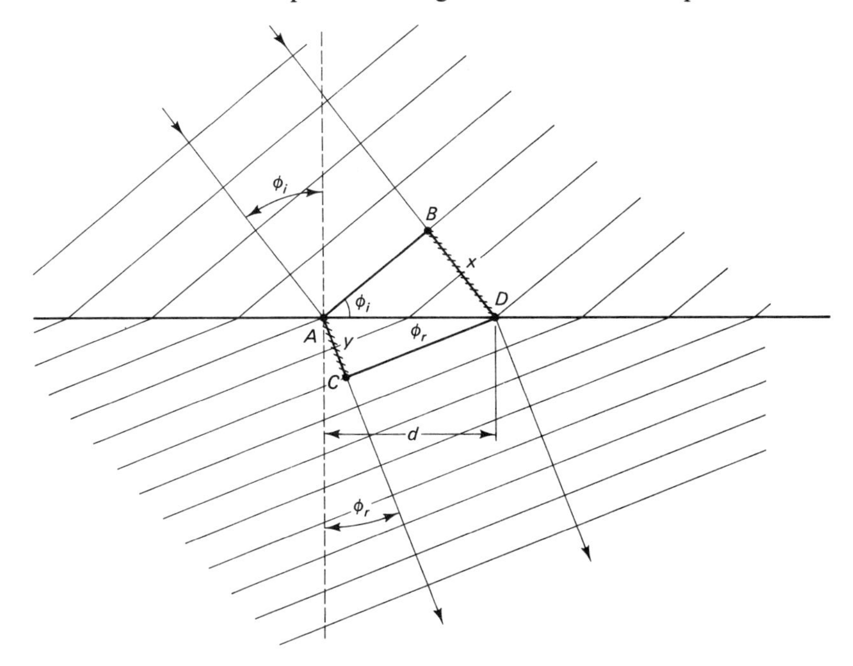

Figure 22-3 Derivation of Snell's law.

From the diagram we see that the portion of the wave near A will travel the distance y from A to C in the same amount of time that the wave near B travels the distance x from B to D. The actual times, which we denote by t_x and t_y, are given by

$$t_x = \frac{x}{c/n_i} = \frac{n_i x}{c} \quad \text{and} \quad t_y = \frac{y}{c/n_r} = \frac{n_r y}{c} \tag{22-3}$$

Setting the expressions for t_x and t_y equal and rearranging gives

$$\frac{x}{y} = \frac{n_r}{n_i} \tag{22-4}$$

From the diagram we note that

$$x = d \sin \phi_i$$

$$y = d \sin \phi_r \tag{22-5}$$

Dividing one of these equations by the other gives another expression for x/y:

$$\frac{x}{y} = \frac{\sin \phi_i}{\sin \phi_r} \tag{22-6}$$

Setting the two expressions for x/y equal to each other and simplifying yields

$$\frac{n_r}{n_i} = \frac{\sin \phi_i}{\sin \phi_r} \tag{22-7}$$

which may be rearranged to give

$$n_r \sin \phi_r = n_i \sin \phi_i \tag{22-8}$$

This is the mathematical statement of Snell's law.

EXAMPLE 22-1 In Fig. 22-4 a beam of light traveling in water ($n = 1.333$) is incident upon a cube of glass whose index of refraction is 1.52. The incident ray (in water) makes an angle of 40° to the normal. Calculate the angle of refraction, ϕ_r.

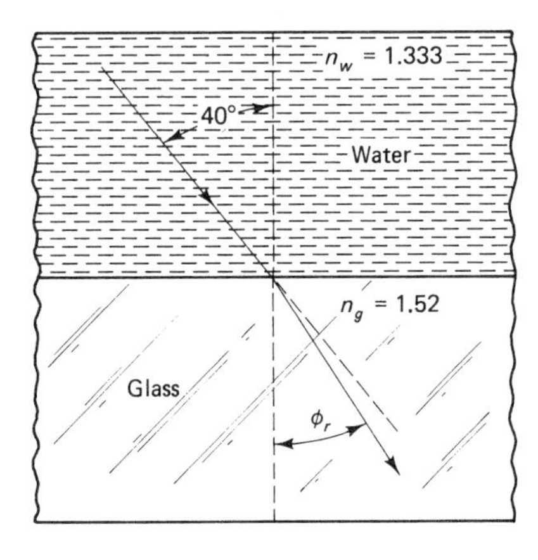

Figure 22-4

Solution. Direct application of Snell's law to water and glass gives

$$n_g \sin \phi_r = n_w \sin \phi_i$$
$$(1.52)\sin \phi_r = (1.333)\sin 40$$
$$\sin \phi_r = 0.564$$
$$\phi_r = 34.3°$$

We note that the ray is refracted toward the normal because the index of refraction of glass is greater than that of water.

EXAMPLE 22-2 Snell's law may be used to predict the path of a ray of light through a prism, as shown in Fig. 22-5. The ray passes through an equilateral prism of dense flint glass having an

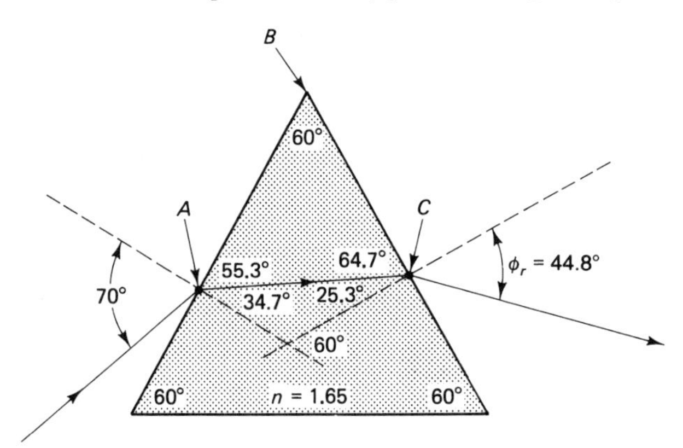

Figure 22-5 The path taken by a ray of light in passing through a prism is computed by using Snell's law.

index of refraction of 1.65. The angle of incidence at point *A* is 70°. Straightforward application of Snell's law gives the angle of refraction at point *A* to be 34.7°.

Angle *BAC* is the complement of 34.7°, 55.3°. The angle at point *B* is 60°. Because the angles of the triangle *ABC* must have a sum of 180°, the angle *ACB* at point *C* must be 64.7°. It follows that the angle of incidence (relative to the normal) of the ray at point *C* is 25.3°. We now calculate the angle of refraction at point *C*.

Solution. We use Snell's law as follows:

$$n_g \sin (25.3) = n_{air} \sin \phi_r$$
$$(1.65)(0.427) = (1.00)\sin \phi_r$$
$$\sin \phi_r = 0.705 \quad \phi_r = 44.8°$$

At point *C*, the ray is refracted away from the normal because the ray is passing from an optically dense medium into air whose approximate index of refraction is 1.00.

Total internal reflection

Suppose a ray of light in an optically dense medium is incident upon a surface that contacts a less-dense medium, as in Fig. 22-6. As ϕ_i is gradually increased from a small value, the angle of refraction ϕ_r also increases and eventually approaches 90°. At exactly 90°, the refracted ray grazes the surface. Recall from Fig. 22-1, however, that a reflected ray is associated with the refracted ray. As ϕ_i is

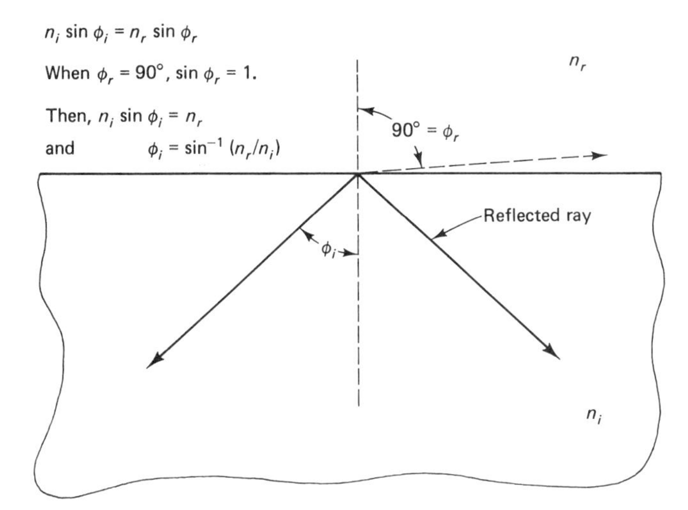

Figure 22-6 Total internal reflection occurs whenever ϕ_r becomes 90°.

increased until ϕ_r reaches 90°, the reflected ray increases in intensity until, finally, all the incident light is reflected back into the dense medium. The refracted ray disappears, and the reflection becomes perfect. This phenomenon is called *total internal reflection*, and the angle ϕ_i at which total internal reflection first occurs is called the *critical angle*. A formula for computing the critical angle for any two media is given in the figure. Total internal reflection occurs also for values of ϕ_i greater than the critical angle.

Use is often made of total internal reflection in optical instruments to achieve perfectly reflecting surfaces. The *porro prisms* in prism binoculars make use of total internal reflection. A *fiber-optics light pipe* utilizes total internal reflection to channel the reflected light along the fiber.

As an illustration, you may verify that the critical angle for a beam of light incident upon a water-air interface is 48.7°.

22-3 CONVERGING (POSITIVE) LENSES

It is well known that lenses are capable of focusing the rays of light from an object or a scene to form an image. Mirrors function by means of reflection, while lenses function via refraction. We have already seen that a ray of light is deviated in passing through a prism. The shape of a lens is such that rays striking the lens at different points are deviated by differing amounts so as to be brought to a focus. The analogy between a lens and an array of prisms is illustrated in Fig. 22-7.

Lenses are classified as positive or negative depending upon whether the net effect of the lens is to produce greater convergence or divergence, respectively, of rays passing through the lens. If a lens is thicker at its center than at the edges, it will be a positive lens. A lens thinner at the center is a negative lens. The surfaces of most lenses are spherical, and the converging (or diverging) power of the lens is related to the radius of curvature of the surfaces. Various types of positive and negative lenses are illustrated in Fig. 22-8.

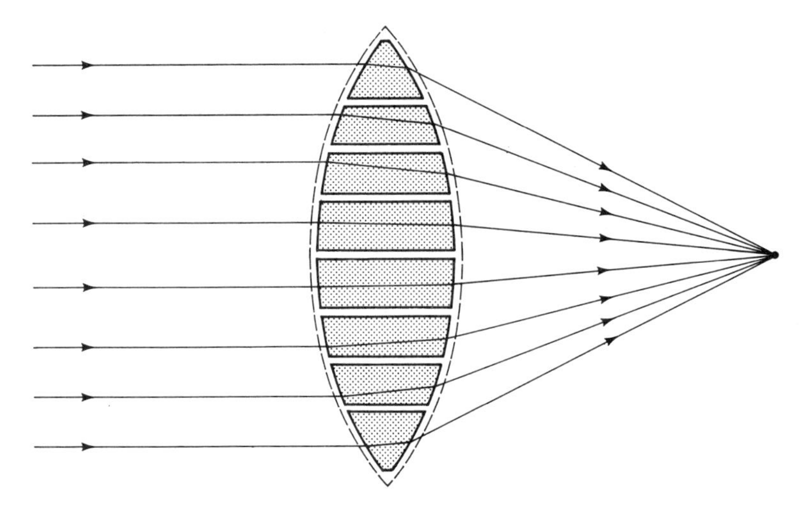

Figure 22-7 The analogy between a lens and an array of prisms.

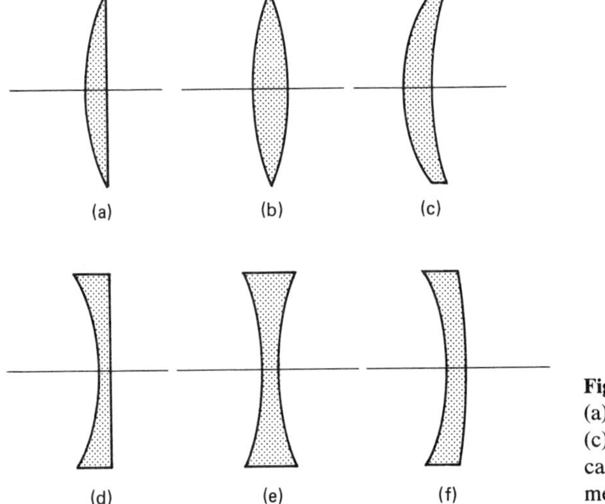

Figure 22-8 Types of lenses. (a) Plano convex. (b) Double convex. (c) Positive meniscus. (d) Plano concave. (e) Double concave. (f) Negative meniscus.

Image formation

We now examine the formation of an image in more detail, beginning with a question. When a screen is held near a luminous object, why does an image of the object *not* appear on the screen? Why is a lens or concave mirror required in order to produce an image on the screen?

To understand this, recall that each and every point on the object emits rays of light, and the rays are emitted in all directions from each point. Therefore, a single point on the screen will receive light from every point of the object. The result is that at the screen, the rays get all mixed up. Rays from the top of the object spread out all over the screen, and these rays are joined by rays from the bottom of the object which also spread out all over the screen. The result is that the screen is evenly illuminated (more or less) by the object, and no image is formed.

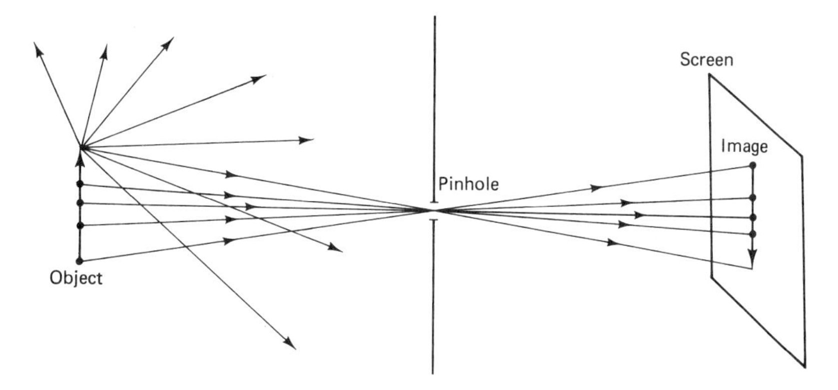

Figure 22-9 Image formation by a pinhole.

In order to form an image on the screen, some way must be found to keep the rays from getting all mixed up. One way to do this is to insert a pinhole between the object and screen, as shown in Fig. 22-9. Then, rays from a point at the top of the object can reach the screen only by passing through the pinhole, and the rays will illuminate the screen only at a single point. Thus, the pinhole limits the light reaching the screen so that a *one-to-one correspondence* is established between points of the object and points of the image. This one-to-one correspondence is the essential prerequisite of image formation.

When a converging lens is placed in front of an object, a multitude of rays from a point on the object strike the lens, and the rays are more or less evenly distributed over the entire aperture of the lens. Each of these rays, however, is deviated by the lens so that the rays converge and intersect after passing through the lens. The point of intersection represents a point of the image. Adjacent image points are formed in a similar manner. The image may be observed by placing a screen at the location of the image or by using a magnifier (eyepiece) to view the image, as shown in Fig. 22-10.

An important, but not necessarily obvious, point is that the image is *there* even if no screen or eyepiece is present to allow it to be observed. However, if no screen is present, the rays of light do not stop at the image. They proceed as if their intersection at the image had never occurred. Light rays have the peculiar ability to pass through each other without any effect being produced by one ray upon another.

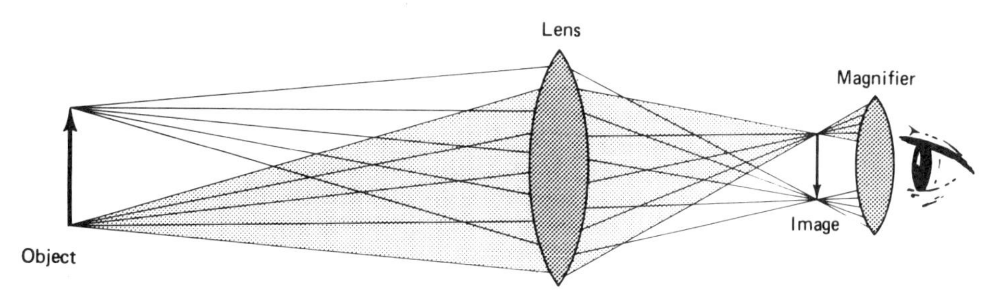

Figure 22-10 Image formation by a positive lens.

Primary focal length of a converging lens

The focal length resulting when the object is at a great distance (when the object is at infinity) from the lens is called the *primary focal length*. It follows that incoming parallel rays will be brought to focus at the primary focal point F. Because light may pass through a lens in either direction, a lens may be envisioned as having two focal points, one on each side of the lens. These are denoted by F and F' in Fig. 22-11 and are called *conjugate foci*.

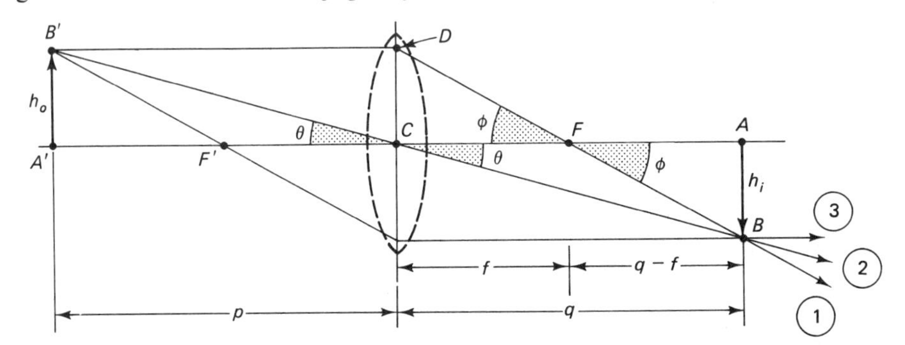

Figure 22-11 Ray-tracing diagram for a positive lens. The lens is represented by a straight vertical line.

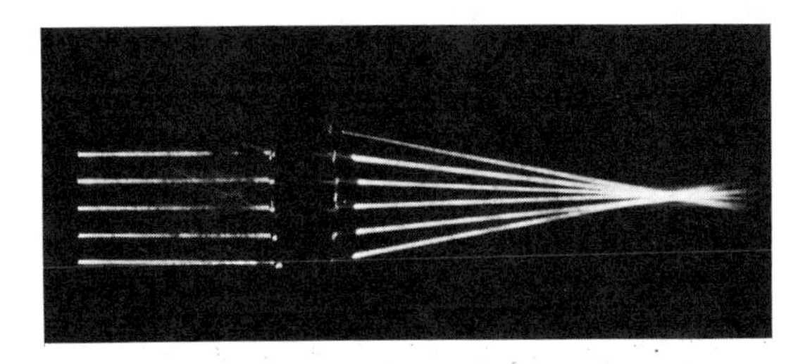

Parallel rays focused by a positive lens.

A familiar demonstration with a converging lens an inch or so in diameter is to allow the rays of the sun to pass through the lens and be focused at a point on a flammable material, which will soon be ignited. The distance from the lens to the burning point is the primary focal length.

Ray tracing

A ray-tracing diagram for a converging lens is shown in Fig. 22-11. Three rays are used to locate the position of the image (although any two are sufficient). The properties of the rays are:

1. The ray parallel to the axis of the lens is refracted through the primary focal point F.
2. The ray that passes through the center of the lens is undeviated.
3. The incident ray that passes through the conjugate focal point F′ is refracted so that it subsequently travels parallel to the axis.

In the next section we derive the mathematical relationship between the object distance, image distance, and primary focal length of a lens.

22-4 THE LENS EQUATION

We can use the ray-tracing diagram of Fig. 22-11 to derive the lens equation, which relates p, q, and f. First, we note that angle $A'CB'$ (θ) is equal to angle ACB and angle DFC (ϕ) is equal to angle AFB. We then make use of similar triangles $A'CB'$ and ACB to write

$$\tan\phi = \frac{h_o}{p} = \frac{h_i}{q} \rightarrow \frac{h_i}{h_o} = \frac{q}{p} \tag{22-9}$$

In a similar manner we use triangles DFC and AFB to write

$$\tan\phi = \frac{h_o}{f} = \frac{h_i}{q-f} \rightarrow \frac{h_i}{h_o} = \frac{q-f}{f} \tag{22-10}$$

At this point we have two different expressions for h_i/h_o. Setting them equal gives

$$\frac{q}{p} = \frac{q-f}{f} \quad \text{or} \quad qf = pq - pf \tag{22-11}$$

which may be rearranged as

$$(p+q)f = pq \quad \text{or} \quad \frac{p+q}{pq} = \frac{1}{f} \quad \text{or} \quad \frac{p}{pq} + \frac{q}{pq} = \frac{1}{f} \tag{22-12}$$

We finally obtain the lens equation:

$$\frac{1}{p} + \frac{1}{q} = \frac{1}{f} \tag{22-13}$$

This equation is identical to the mirror equation given without proof in Chap. 21.

Sign convention

We assume that rays normally pass through the lens from left to right so objects normally appear on the left and images normally appear on the right. Hence, object distances to the left of the lens are positive; image distances to the right of the lens are positive; the focal length of a converging lens is positive. We shall see that both the object and the image may be virtual as indicated by negative values of p and q.

Magnification

The magnification M relates the size of the image h_i to the size of the object h_o. It is given by

$$M = -\frac{q}{p} \tag{22-14}$$

as justified by Eq. (22-9). The negative sign is appended in order to distinguish erect and inverted images; a negative magnification means the image is inverted.

EXAMPLE 22-3 An object 5 cm tall is placed 60 cm in front of a converging lens whose focal length is 12 cm.

(a) Calculate the image distance.

(b) Compute the size of the image and tell whether the image is erect or inverted.

Solution. (a) We make direct application of the lens equation:

$$\frac{1}{p} + \frac{1}{q} = \frac{1}{f}$$

$$\frac{1}{60} + \frac{1}{q} = \frac{1}{12}$$

$$q = 15 \text{ cm}$$

(b) The size of the image h_i is the product of the magnification (without regard to sign) and the size of the object h_o. Hence,

$$h_i = Mh_o$$

where

$$M = \frac{q}{p} = -\frac{15}{60} = -0.25$$

It follows that

$$h_i = (0.25)(5 \text{ cm})$$

$$= 1.25 \text{ cm}$$

Because the magnification is negative we conclude that the image is inverted.

EXAMPLE 22-4 A positive (converging) lens has a focal length of 50 cm. An object 5 cm tall is placed in front of the lens 30 cm from the lens.

(a) Find the location of the image.

(b) Describe the image.

Solution. (a) We use the lens equation and solve for q in terms of p and f:

$$q = \frac{pf}{p - f}$$

$$= \frac{(30)(50)}{30 - 50}$$

$$= -75 \text{ cm}$$

Because q is negative, the image is virtual and is located to the left of the lens.

(b) The magnification is

$$M = -\frac{q}{p} = -\frac{(-75)}{30}$$

$$= +2.5$$

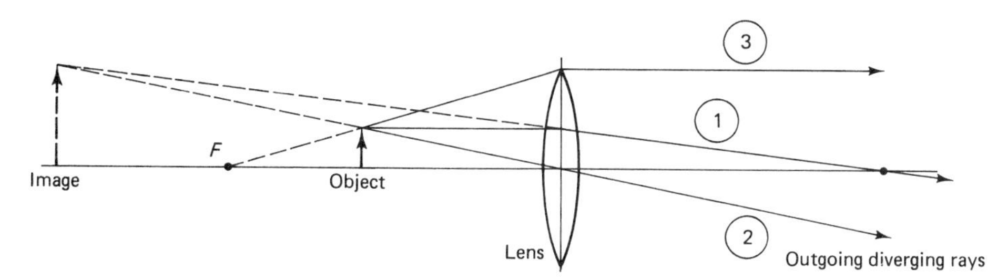

Figure 22-12 The object is closer to the lens than is the primary focal point. This causes the image to be virtual; the lens does not bring the rays back together to form a real focus.

Thus, the image is virtual, erect, and enlarged. Its size is

$$h_i = 2.5(5 \text{ cm})$$
$$= 12.5 \text{ cm}$$

A ray diagram of this example is shown in Fig. 22-12.

Magnifier

Everyone is familiar with a *magnifying glass*, a positive lens of short focal length that magnifies an object when the object is viewed through the glass. Example 22-4 demonstrates that when the object is closer to a positive lens than a distance equal to the primary focal length of the lens, the image appears to be on the object side of the lens and is erect and magnified. This is the principle on which the magnifier is based.

To be seen most distinctly, an object must be held a distance of about 25 cm from the normal, unaided eye. This distance, which we denote by N, serves to locate the *near point* of the eye, or the distance of most distinct vision. A more distant object may be seen clearly, but the image formed on the retina will be smaller and less distinct so that less object detail is visible. If an object is held closer to the eye than the distance N, the lens of the eye is not able to form a sharply focused image on the retina.

One way to use a magnifier is shown in Fig. 22-13. The magnifier is held near the eye, and the object is positioned so that a virtual image is formed at the point

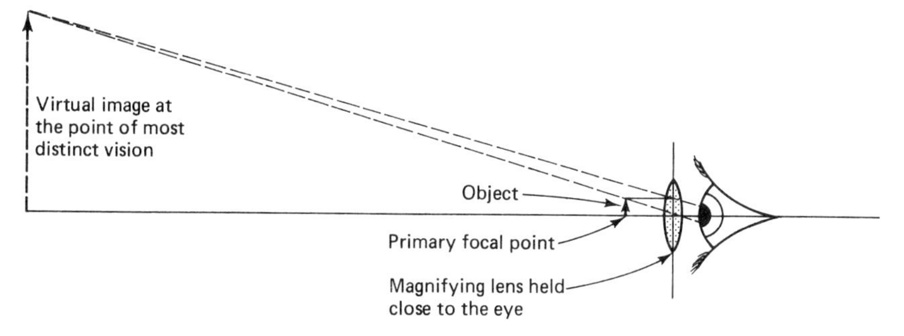

Figure 22-13 With the object located just inside the primary focal distance, an erect, virtual image will be formed at the point of most distinct vision.

of most distinct vision. Therefore, the image distance is $-N$, and the corresponding object distance is

$$p = \frac{Nf}{N + f} \tag{22-15}$$

The magnification is $-q/p$, and substitution of the above quantities gives the magnification of a magnifier:

$$M = \frac{N}{f} + 1 \tag{22-16}$$

If f is small in comparison with N, the magnification is given approximately by

$$M = \frac{N}{f} \tag{22-17}$$

As an example, a positive lens having a focal length of 2 cm used as shown in Fig. 22-13 will produce a magnification of 13.5 according to Eq. (22-16) or a magnification of 12.5 according to Eq. (22-17).

Applications

The projection lenses of slide projectors, movie projectors, photographic enlargers, and other projection apparatus are highly corrected, positive-lens systems of relatively short focal length. The object (a slide, film, or negative) is located relatively near the primary focal point so that the image distance is comparatively large—the image is projected. Example 22-5 illustrates this.

EXAMPLE 22-5

The projection lens of a 35-mm slide projector has a focal length of 100 mm. The slide is positioned (upside down and reversed from left to right) so that the object distance is 102 mm.

(a) Compute the image distance, which is the required distance from the projector to the screen.

(b) If the object is 24 mm tall, compute the size of the image on the screen.

Solution. (a) Use a rearrangement of the lens equation:

$$q = \frac{pf}{p - f}$$

$$q = \frac{(102 \text{ mm})(100 \text{ mm})}{102 - 100 \text{ mm}}$$

$$= 5100 \text{ mm} = 5.1 \text{ m} = 16.7 \text{ ft}$$

(b) The magnification is

$$M = -\frac{5100}{102} = -50$$

Because the object height is 24 mm, the image is $50 \times 24 = 1200$ mm tall. This is nearly 4 ft.

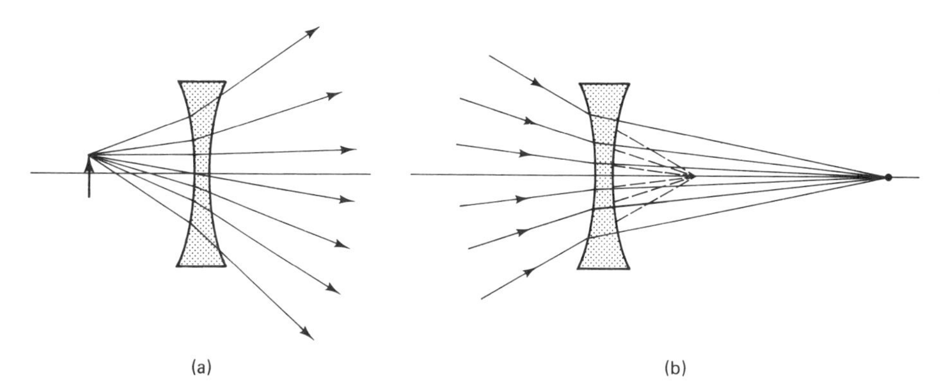

Figure 22-14 The effect of a diverging (negative) lens.
(a) The divergence of the rays from an object is increased.
(b) Rays converging to a focus are caused to converge less rapidly by the diverging lens.

22-5 DIVERGING (NEGATIVE) LENS

A diverging lens increases the divergence of rays emanating from an object, or, if the rays have been converged by a positive lens, a diverging lens reduces the convergence of the rays. The basic action of a diverging lens is shown in Fig. 22-14.

The sign convention for a negative lens is the same as for a positive lens except the focal length of a negative lens is taken to be a negative quantity. The lens equation is applicable to negative lenses.

With the object at infinity, the incoming rays to a negative lens will be parallel. After passing through the lens, the rays will be diverging and will appear

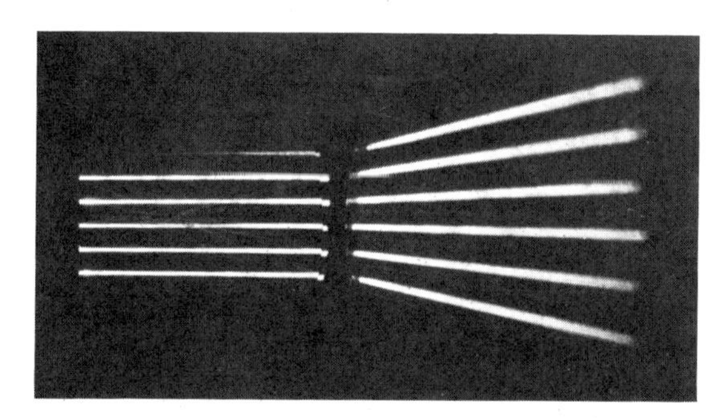

Parallel rays passing through a negative lens are caused to diverge.

to have originated at a point on the object side of the lens. The distance from the lens to the apparent point of origin of the rays is the *primary focal distance*, and the point is called the *primary focal point*. Because light may pass in either direction through the lens, the lens has conjugate foci, one on either side of the lens.

Ray tracing

A ray-tracing diagram for a negative lens is shown in Fig. 22-15. Two rays are used to locate the image:

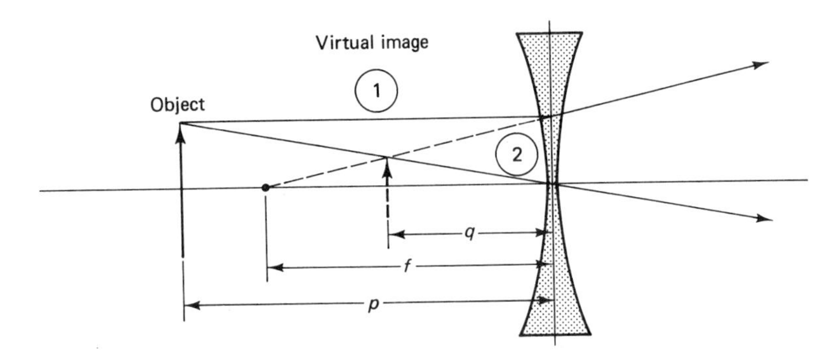

Figure 22-15 Ray-tracing diagram for a negative lens.

1. The incident ray traveling parallel to the axis is deviated by the lens so that it appears to have come from the primary focal point.
2. The ray passing through the center of the lens is undeviated.

For any positive object distance, the image will be erect, virtual, and reduced in size.

EXAMPLE 22-6 An experimenter holds a negative lens between his eye and an object. The focal length of the lens is −20 cm and the object is 10 cm from the lens.
(a) What is the location of the image as seen by the observer?
(b) If the object is 2 cm in diameter, what is the diameter of the image?

Solution. (a) We use the lens equation to compute the image distance:

$$q = \frac{pf}{p - f}$$

$$= \frac{(10 \text{ cm})(-20 \text{ cm})}{(10) - (-20) \text{ cm}}$$

$$= -6.67 \text{ cm}$$

The image appears to be 6.67 cm from the lens on the object side of the lens.
(b) The magnification is

$$M = -\frac{q}{p}$$

$$= -\frac{-6.667 \text{ cm}}{10 \text{ cm}} = +0.667$$

The magnification is positive, which means the image is erect. The size of the image is 2(0.667) = 1.33 cm. The image is two-thirds as large as the object when the lens is 10 cm from the object. At other object distances, however, the position and size of the image will be different.

22-6 LENS COMBINATIONS

Lenses are often used in combinations that comprise a *lens system* in which the rays of light pass first through one lens and then through the other lenses in succession. Indeed, most lenses are actually systems of lenses rather than simple single lenses. Typical of lens systems are lenses of cameras of all types, projectors of all types, photographic enlargers, and eyepieces of binoculars, microscopes, and telescopes. In real applications, a lens that is a single glass element is more unusual than systems of lenses.

The principal idea in computing the location (and other parameters) of the image is this: The image formed by the first lens of the system becomes the object for the second lens of the system. The image formed by the second lens becomes the object for the third, and so on for other elements of the system. The lens equation is used repeatedly in a straightforward manner, but due care must be given to sign conventions. The following examples illustrate the calculation procedure for several configurations.

EXAMPLE 22-7 Two identical positive lenses, each having a focal length of 20 cm, are separated a distance of 10 cm, as shown in Fig. 22-16. An object 3 cm tall is 65 cm to the left of the first lens. Find the location and the size of the image formed by the system.

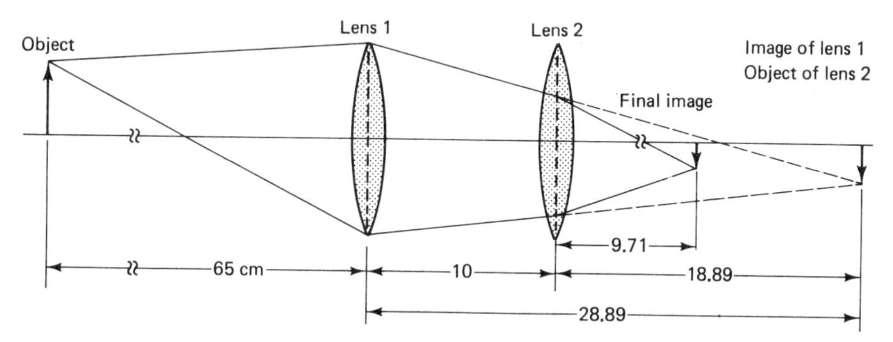

Figure 22-16 When dealing with a combination of lenses, the image formed by lens 1 becomes the object for lens 2.

Solution. We first apply the lens equation to the first lens, treating it as if the second lens were not present:

$$q_1 = \frac{p_1 f_1}{p_1 - f_1}$$

$$= \frac{(65\text{ cm})(20\text{ cm})}{65 - 20\text{ cm}} = 28.89\text{ cm}$$

$$M_1 = -\frac{q_1}{p_1} = -\frac{28.89\text{ cm}}{65\text{ cm}} = -0.444$$

Thus, the image formed by the first lens is inverted and, in the absence of the second lens, would be located 28.89 cm to the right of the first lens.

The first image now becomes the second object. Note, however, that the object is to the right of the second lens, and that the incident rays are already converging when they arrive at the second lens. Rays of light diverge from real objects; in this case we have a virtual object. To take this into account, we assign a negative value to the object distance for the second lens.

From the figure we see that the numerical value of the object distance of the second lens is 18.89 cm. A second application of the lens equation gives q_2:

$$q_2 = \frac{p_2 f_2}{p_2 - f_2} = \frac{(-18.89)(20)}{(-18.89 - 20)} = 9.71 \text{ cm}$$

$$M_2 = -\frac{q_2}{p_2} = -\frac{9.71}{(-18.89)} = +0.514$$

We see that the final image is formed 9.71 cm to the right of the second lens. A positive value of M_2 indicates that the second lens does not invert the image a second time.

The magnification of the system M_s is the product of the individual magnifications. Hence,

$$M_s = M_1 M_2$$
$$= (-0.444)(0.514) = -0.226$$

The final image is inverted relative to the original object, as is indicated by the negative value of M_s. The size of the final image h_f is

$$h_f = M_s h_o$$
$$= (0.226)(3 \text{ cm})$$
$$= 0.678 \text{ cm}$$

EXAMPLE 22-8 Repeat Example 22-7, but assume the second lens is a negative (diverging) lens whose focal length is -40 cm.

Solution. The computation for the first lens is the same as before; the object distance for the second lens is -18.89 cm. We then apply the lens equation to the second lens:

$$q_2 = \frac{p_2 f_2}{p_2 - f_2}$$
$$= \frac{(-18.89)(-40)}{(-18.89) - (-40)}$$
$$= \frac{755.6}{21.11} = 35.79 \text{ cm}$$

$$M_2 = -\frac{q_2}{p_2} = -\frac{35.79}{(-18.89)} = 1.89$$

$$M_s = M_1 M_2 = (-0.444)(1.89) = -0.841$$

$$h_f = M_s h_o = (0.841)(3 \text{ cm}) = 2.52 \text{ cm}$$

The effect of the negative lens is to extend the image distance of the first lens. The diverging lens cancels part of the convergence of the rays produced by the first lens.

Lenses in contact

Suppose two thin lenses are placed in contact. We calculate the location of the final image as for any other two-lens system, but in this case, the image distance q_1 becomes the object distance p_2 without any numerical modification. The quantity p_2 is taken as a negative quantity because the object for the second lens is virtual. The two-lens combination acts as an equivalent single lens; the focal length of the equivalent single lens is given by

$$\frac{1}{f_e} = \frac{1}{f_1} + \frac{1}{f_2} \quad \text{or} \quad f_e = \frac{f_1 f_2}{f_1 + f_2} \tag{22-18}$$

The lens equation for two thin lenses in contact can be written as

$$\frac{1}{p} + \frac{1}{q} = \frac{1}{f_1} + \frac{1}{f_2} \tag{22-19}$$

Power of a lens; diopters

Lenses that have a short focal length are said to be high-power lenses whereas long-focal-length lenses are low power. The greater the power of the lens, the shorter is its focal length. If the focal length is expressed in meters, the power of a lens in diopters is given by

$$D = \frac{1}{f} \qquad (f \text{ in meters, } D \text{ in diopters}) \tag{22-20}$$

It follows that a 1-diopter lens has a focal length of 1 m, or 100 cm. A 2-diopter lens has a focal length of 0.5 m, or 50 cm. Diverging lenses are said to have negative powers because the focal length of a diverging lens is negative.

For lenses in contact, as described above, the powers of the component lenses may be added together to obtain the power of the equivalent single lens. We may write

$$D_e = D_1 + D_2 \tag{22-21}$$

where D_e is the power of the single-lens equivalent and D_1 and D_2 are the powers of the lenses in contact. The lens equation for the two lenses in contact is

$$\frac{1}{p} + \frac{1}{q} = D_e \quad \text{or} \quad \frac{1}{p} + \frac{1}{q} = D_1 + D_2 \tag{22-22}$$

and it follows that p and q for lenses in contact are given by

$$p = \frac{q}{D_e q - 1} \qquad q = \frac{p}{D_e p - 1} \tag{22-23}$$

Corrective lenses; eye defects

Two types of vision defects, *nearsightedness* and *farsightedness*, are illustrated in Fig. 22-17. These defects are known also as *myopia* and *hyperopia*, respectively. In a myopic eye the lens is too strong; the image is formed in front of the retina. In a hyperopic eye, the lens is too weak; the image is formed behind the retina. Myopic (nearsighted) persons can focus clearly on near objects but cannot see distant objects clearly. Hyperopic (farsighted) persons see distant objects clearly but cannot see near objects clearly.

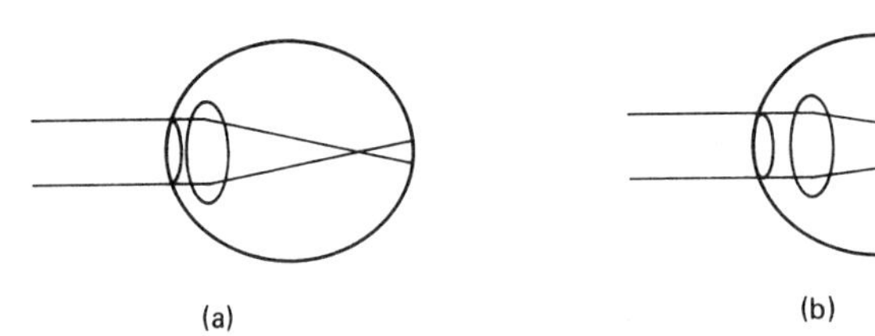

Figure 22-17 Two types of visual defects (a) Myopia. (b) Hyperopia.

Myopia is corrected by placing a diverging lens in front of the eye in order to counteract the excessive power of the lens of the eye. Hyperopia is corrected by placing a converging lens in front of the eye in order to add to the power of the lens of the eye. The corrective lenses may be conventional lenses or they may be contact lenses.

The ability of the eye to focus on objects at different distances is called *accommodation*. As a person gets older, the eye becomes less able to focus on near objects until a condition similar to hyperopia—called *presbyopia*—is attained. The lack of accommodation may be corrected by *bifocal* lenses, lenses in which the top and bottom parts of the lens have different focal lengths. The top part is used for viewing distant objects, and the power of the top portion of the lens is less than the bottom portion which is used for viewing near objects.

Astigmatism is another common eye defect. It is caused by the optical surfaces of the eye being slightly out-of-round. An effect of astigmatism is that horizontal and vertical lines are not focused equally sharply on the retina.

A cylindrical lens provides an extreme example of astigmatism. Focusing occurs only in one plane. In Fig. 22-18, an object that is a point source is focused to a line by a cylindrical lens.

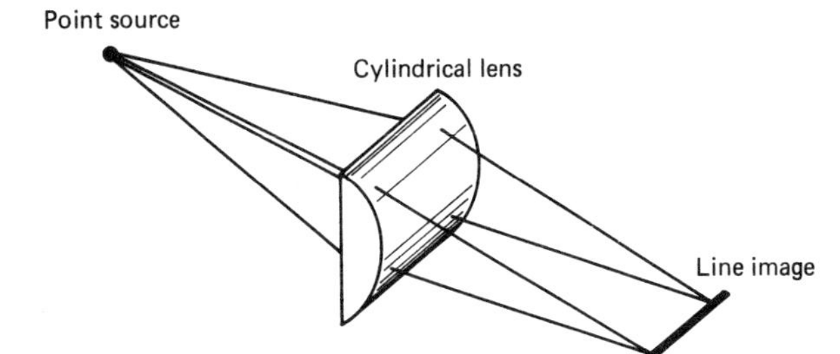

Figure 22-18 A cylindrical lens focuses a point source to a line. This may be considered an extreme form of astigmatism.

22-7 THE LENSMAKER'S FORMULA

The focal length of a lens is determined by the radius of curvature of each of its two surfaces, the index of refraction of the material from which the lens is made and the index of refraction of the medium surrounding the lens. A lensmaker must select the type of glass and the radii of curvature for the surfaces of a lens in order to produce a given focal length. For a thin lens that is to be used in air, the following formula is applicable:

$$\frac{1}{f} = (n_L - 1)\left(\frac{1}{R_1} + \frac{1}{R_2}\right) \tag{22-24}$$

where n_L is the index of refraction of the lens material and R_1 and R_2 are the radii of curvature, as shown in Fig. 22-19.

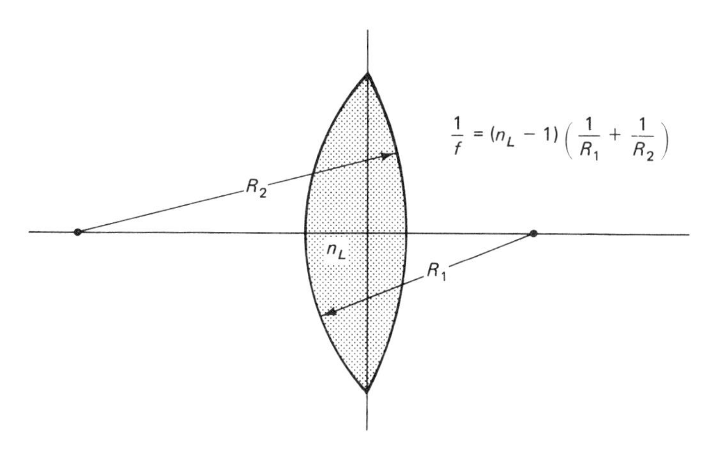

Figure 22-19 The lensmaker's formula relates the focal length of a lens to the radii of curvature of the lens' surfaces and the index of refraction of the lens' material.

The above formula may be used for both converging and diverging lenses, and the lens surfaces may be either concave or convex. A convex surface is assumed to have a positive radius of curvature; a concave surface is assumed to have a negative radius of curvature.

When a lens is to be used in a medium other than air or vacuum, the following formula applies:

$$\frac{n_m}{p} + \frac{n_m}{q} = (n_L - n_m)\left(\frac{1}{R_1} + \frac{1}{R_2}\right) \tag{22-25}$$

where p and q are the object and image distances, n_L is the index of refraction of the lens material, and n_m is the index of refraction of the material surrounding the lens. If a lens has a focal length in air of f_{air}, the same lens will have a focal length of

$$f_{med} = \frac{n_m(n_L - 1)}{n_L - n_m} f_{air} \tag{22-26}$$

when immersed in a medium such as water. When a glass lens is immersed in water, the focal length is increased by a factor of more than three.

It frequently happens that the medium is not the same on both sides of a lens, as illustrated in Fig. 22-20. The following formula relates the indices of refraction, the radii of curvature, and the image and object distances:

$$\frac{n_o}{p} + \frac{n_i}{q} = \frac{n_L - n_o}{R_1} + \frac{n_L - n_i}{R_2} \tag{22-27}$$

where

n_o = n for medium on object side of lens
n_i = n for medium on image side of lens
n_L = n for the material of the lens
R_1 = radius of curvature of lens surface nearest the object

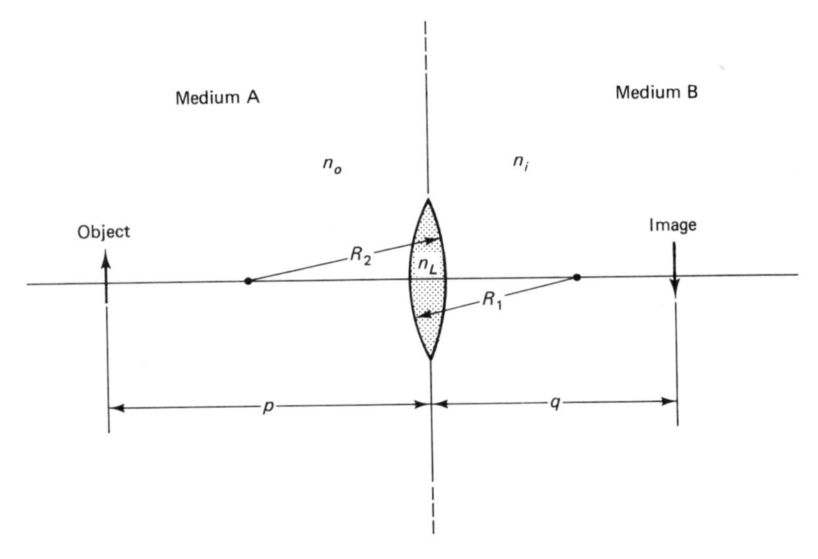

Figure 22-20 A lens located at the interface between two optically different media. [See Eq. (22-27).]

R_2 = radius of curvature of lens surface nearest the image

p, q = object and image distances, respectively

EXAMPLE 22-9 A double-convex lens (n_L = 1.52) is sealed into the side of a container of water (n = 1.33) as shown in Fig. 22-20. The air side of the lens has a radius of curvature of 20 cm; the water side has a radius of curvature of 15 cm. A small lamp is located 80 cm from the lens. Where in the water will the image of the lamp be formed?

Solution. We substitute directly into Eq. (22-27):

$$\frac{1.00}{80 \text{ cm}} + \frac{1.33}{q} = \frac{1.52 - 1.00}{20 \text{ cm}} + \frac{1.52 - 1.33}{15 \text{ cm}}$$

$$\frac{1.33}{q} = 0.0262$$

$$q = 50.83 \text{ cm}$$

The image is formed 50.83 cm from the lens in the water. Note that both R_1 and R_2 are taken as positive quantities because they are convex surfaces of the lens.

22-8 THE REFRACTING TELESCOPE

The optics of a simple refracting telescope are shown in Fig. 22-21. A long-focal-length objective lens forms a real inverted image of the object being viewed, and an eyepiece (shown as a simple positive lens) serves as a magnifier for viewing the image formed by the objective lens. Analysis of this system follows the lines of the calculations for combinations of lenses, so we merely give the results as needed in order to derive an expression for the angular magnification of a telescope.

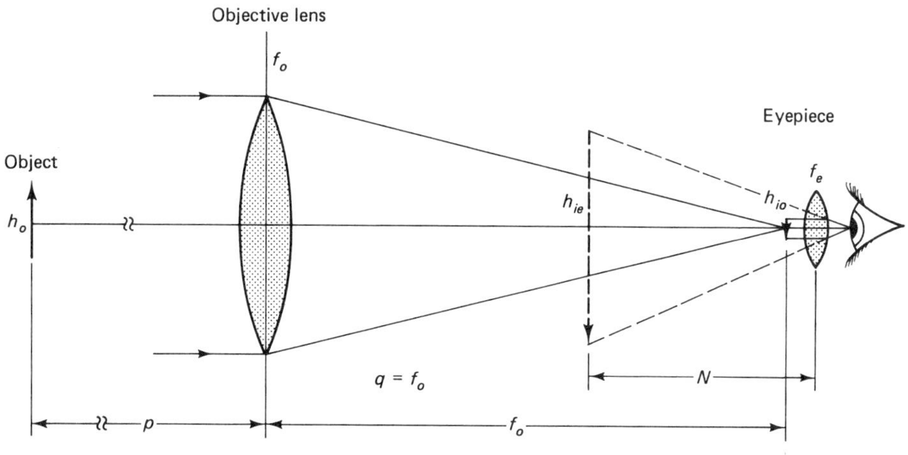

Figure 22-21 The optics of a simple refracting telescope. The objective lens has a very long focal length; in practice it will be an achromatic doublet so that chromatic aberration will be reduced.

Suppose an object h_o units tall is located a large distance p (perhaps several hundred meters) from the telescope. The image will be formed very close to the primary focal point of the objective lens, and the size of the image will be

$$h_{io} = \frac{q}{p} h_o = \frac{f_o}{p} h_o \tag{22-28}$$

where f_o is the focal length of the objective lens.

If the position of the eyepiece is adjusted so that the image formed by the eyepiece is a distance N from the eyepiece, the size h_{ie} of the image formed by the eyepiece is given by

$$h_{ie} = \frac{(N - f_e)}{f_e} h_{io} \tag{22-29}$$

The human eye can focus on an object (or, in this case, a virtual image) if the object is farther than a certain minimum distance from the eye, typically 25 cm. Therefore, the distance N in this expression must be 25 cm or greater. The minimum distance at which the eye can focus is called the *near point* of the eye.

We can now calculate the angular magnification of the telescope. We begin by computing the angles subtended by the object and the image as seen by the observer *with* and *without* the telescope.

The object, whose height is h_o, is located a distance p from the observer. The angle subtended by the object is given (in radians) by

$$\theta_{\text{obs}} = \frac{h_o}{p} \tag{22-30}$$

The image seen in the eyepiece appears to be a distance N away, and its height is given by Eq. (22-29). Hence, the angle θ_e subtended by the image in the eyepiece is given by

$$\theta_e = \frac{h_{ie}}{N} = \frac{(N - f_e)}{f_e N} h_{io}$$

$$= \frac{(N - f_e) f_o h_o}{f_e N p} \quad (22\text{-}31)$$

where Eq. (22-28) has been substituted for h_{io}.

The angular magnification is defined as

$$M_\theta = \frac{\theta_e}{\theta_{obs}} \quad (22\text{-}32)$$

When the expressions for θ_{obs} and θ_e are substituted, the following formula for the angular magnification of a refracting telescope is obtained:

$$M_\theta = \frac{f_o}{f_e}\left(1 - \frac{f_e}{N}\right) \quad (22\text{-}33)$$

If N is much larger than f_e, which is ordinarily the case as f_e is typically only 2 or 3 cm, the angular magnification is given approximately by

$$M_\theta = \frac{f_o}{f_e} \quad (22\text{-}34)$$

This is simply the ratio of the focal length of the objective lens to the focal length of the eyepiece.

22-9 THE MICROSCOPE

For viewing very large, distant objects, we construct a telescope by using a long-focal-length objective lens and an eyepiece whose focal length is 1 or 2 cm. For viewing very small objects located extremely close to the objective lens, we can construct a microscope by using an eyepiece and an objective lens whose focal length is extremely short. A diagram of a microscope is shown in Fig. 22-22. We have shown the objective lens and eyepiece as a single lens even though, in practice, both must be complicated systems of lenses to minimize image distortion and aberrations.

In general, the distance L between the eyepiece and objective is fixed. In use, the object distance p_o is adjusted so that the image formed by the objective falls very near the focal point of the eyepiece. More precisely, the objective image is positioned (by varying p_o) so that the virtual image formed by the eyepiece is located a distance N from the eyepiece; N must be greater than about 25 cm, the minimum focusing distance of the eye.

The total magnification is the product of the magnifications of the objective and of the eyepiece. To a good approximation, it is given by

$$M = \frac{NL}{f_e f_o} \quad (22\text{-}35)$$

where f_e and f_o are the focal lengths of the eyepiece and objective lens, L is the

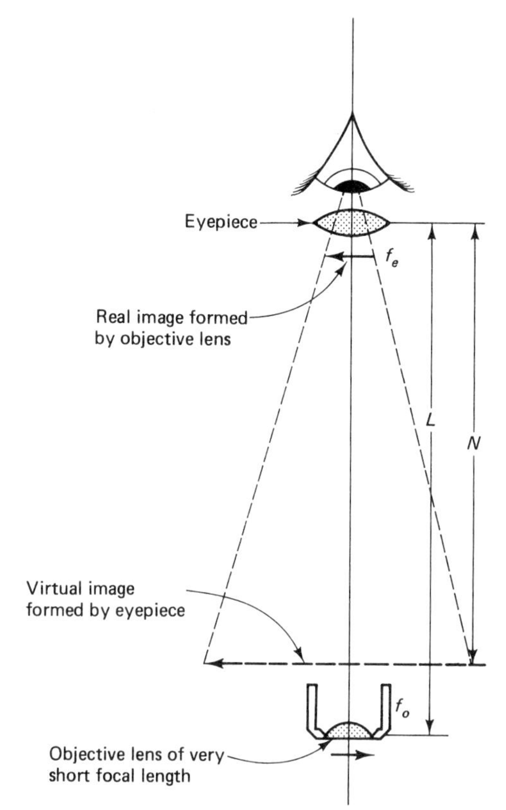

Figure 22-22 A microscope uses an objective lens with a very short focal length.

distance between the lenses, and N is the near-point distance of the eye, about 25 cm.

As an example, suppose a microscope has an objective lens of focal length 2.8 mm and an eyepiece of focal length 18 mm. The distance L is 12 cm, and N is taken as 25 cm. You may use Eq. (22-35) to verify that the total magnification is about 595.

To Go Further

Read in an encyclopedia about:

Willebrord Snell (1591–1626), Dutch physicist

The 40-in. (world's largest) refracting telescope at Yerkes Observatory in Williams Bay, Wisconsin

Questions

1. In simple terms, what causes refraction?
2. In regard to refraction at a surface, is it possible to have a transmitted beam without having also a reflected beam? Under what circumstances can you obtain a reflected beam without having a transmitted beam?
3. Describe the circumstances that give rise to total internal reflection.
4. If a white sheet of paper is held near a candle, why does an image of the candle not appear on the paper? What is essential for image formation?

5. Give a quick method for determining the focal length of a positive lens: **(a)** on a clear day; **(b)** at night. What equipment is required?

6. Give a step-by-step procedure for constructing a ray-tracing diagram for a positive lens.

7. What is the significance of a negative magnification?

8. Under what conditions will a positive lens produce a virtual image?

9. Explain the use and principle of a simple magnifier. What property of the eye determines the magnification of a particular magnifier?

10. How is the focal point (or the primary focal length) of a negative lens defined?

11. Give a step-by-step procedure for constructing a ray-tracing diagram for a negative lens.

12. How is the power of a lens defined? In what units must the focal length be expressed in order for the power to appear in diopters?

13. Can a person see clearly under water without goggles, with the water in direct contact with the opened eyes?

14. How does the objective element of a microscope differ from that of a telescope?

15. Will a contact lens still work if it is held near the eye rather than being inserted into the eye?

Problems

1. Compute the speed of light in the following materials.
   (a) Fused quartz. **(b)** Crown glass.
   (c) Ethyl alcohol. **(d)** Diamond.

2. By what percentage, relative to its velocity in water at 20°C, does light travel faster in water at 60°C?

* 3. Suppose two pulses of light begin at the same point at the same time and travel in parallel paths for a distance of 1 mi. One pulse travels in air; the other travels through water at 20°C.
   (a) Compute the time required for the air pulse to travel 1 mi.
   (b) How far will the water pulse have traveled when the air pulse passes the 1-mi marker?

4. A ray of light incident upon the surface of water at 20°C makes an angle of 50° to the normal. Compute the angle of refraction of the refracted (transmitted) ray.

5. Repeat Problem 4, but assume the temperature of the water is 60°C. How much does the direction of the refracted ray change as the water is warmed from 20°C to 60°C? (Answer in minutes and seconds of arc.)

6. In Fig. 22-4, suppose the ray had been incident upon the top surface of the water at an angle of 40° to the normal.
   (a) Calculate the angle of refraction of the transmitted ray as it travels through the layer of water.
   (b) Compute the angle of refraction of the refracted ray that travels through the glass.

7. Extend Problem 6 by supposing the refracted ray in the glass leaves the glass by way of the bottom surface. Calculate the direction taken by the ray that leaves the glass.

* 8. Redo the calculation of Example 22-2, but assume the prism is made of fused quartz (n = 1.46). (See Fig. 21-5.)

9. Compute the critical angle for total internal reflection if the following materials are surrounded by air.
   (a) Crown glass. **(b)** Ethyl alcohol. **(c)** Diamond.

10. A student finds a plano-convex lens in a far corner of the lab, but its focal length is not known. Outside, the lens forms a very bright image of the sun (the burning point) a distance of 24 cm behind the lens. What is the primary focal length of the lens?

11. An object 10 cm tall is located 150 cm from a positive lens of focal length 30 cm.
    (a) How far from the lens will the image be formed?
    (b) Will the image be real or virtual, erect or inverted?

(c) What will be the magnification produced?

(d) What will be the height of the image?

12. Repeat Problem 11, but assume the lens has a focal length of 40 cm.

13. In an experiment to determine the focal length of a positive lens, a small candle is placed 100 cm from the lens, and a sharply focused image is formed on a small white card held 60 cm from the lens on the opposite side. What is the focal length of the lens?

***14.** A certain camera utilizes a 55-mm lens (primary focal length) to focus the image of the scene being photographed onto the film. If a 6 ft tall person stands 15 ft from the camera, compute the height of the person's image as it appears on the film. Give the answer in millimeters.

15. A positive lens being used as a magnifier is held near the eye, and the object is positioned so that the virtual and erect image appears 25 cm from the lens. The focal length of the lens is 4 cm.

(a) What is the magnification produced by the magnifier?

(b) How far from the lens should a rare coin be held in order to view the coin most distinctly?

16. The distance of most distinct vision for a certain nearsighted person is 12.5 cm (without corrective lenses). What magnification will be obtained with a magnifier having a focal length of 4 cm?

17. A penny is located 5 cm behind a negative lens with a focal length of -15 cm.

(a) How far behind the lens does the image appear?

(b) Will the image be erect or inverted, real or virtual?

(c) Will the image of the penny appear larger or smaller than the penny without the lens?

***18.** In an experiment to determine the focal length of a negative lens, a candle is set up in front of a *positive* lens which projects an image onto a small screen located 30 cm on the other side of the positive lens. The negative lens is then inserted between the positive lens and the screen, the negative lens being located 10 cm from the positive lens. To regain the focus of the image on the screen, the screen must be moved back 10 cm so that it is 30 cm from the negative lens. From this information, compute the focal length of the negative lens.

***19.** Two positive lenses of 15-cm focal length are separated a distance of 5 cm. An object 5 cm tall is placed 40 cm to the left of the first lens.

(a) Where is the image formed by the first lens located relative to the first lens?

(b) What is the object distance for the second lens?

(c) What is the location relative to the second lens of the image formed by the system?

(d) What is the size of the image formed by the system?

***20.** A negative lens of focal length -20 cm is located 3 cm to the right of a positive lens whose focal length is 12 cm. An object 4 cm tall is located 100 cm from the positive lens.

(a) What is the location of the image formed by the positive lens?

(b) What is the object distance for the negative lens?

(c) Where is the system image, relative to the negative lens?

(d) What is the size of the final image?

21. Two thin positive lenses of focal length 20 cm are in contact. An object is located 30 cm to the left of the combination.

(a) What is the location of the image?

(b) What is the effective focal length of the two-lens combination?

22. A positive lens of focal length 5 cm is in contact with a negative lens whose focal length is -8 cm.

(a) What is the power of each lens?

(b) What is the power of the combination?

(c) What is the effective focal length of the combination?

23. Suppose a positive lens of focal length $+10$ cm is in contact with a negative lens of -10 cm. What will be: **(a)** the effective power of the combination; **(b)** the effective focal length of the combination?

24. A negative lens ($f = -10$ cm) is located 4 cm to the right of a positive lens of the same focal length ($f = +10$ cm). An object is located 50 cm to the left of the positive lens.

(a) Compute the location of the image formed by the system.

(b) What is the power in diopters of each lens?

(c) What is the effective power of the system?

25. A symmetrical convex lens has surfaces of radius of curvature of 5 cm. The index of refraction of the glass is 1.52.

(a) What is the focal length of the lens in air?

(b) What is the focal length of the lens in water?

26. A plano-convex lens is to be made of dense flint glass and is to have a focal length of 10 cm. What should be the radius of curvature of the convex surface? [The flat side of the lens has a radius of curvature of ∞.]

27. A negative lens is made of glass having an index of refraction of 1.52. The concave surfaces of the lens have radii of curvature of -8 cm and -12 cm. What is the focal length of the lens?

28. A simple refracting telescope utilizes an objective lens that has a focal length of 80 cm. What angular magnification will the telescope have when used with eyepieces having the following focal lengths?

(a) 40 mm **(b)** 20 mm **(c)** 16 mm **(d)** 10 mm

29. The distance between the lenses of a microscope is 18 cm. The focal length of the objective lens is 0.4 cm, and the focal length of the eyepiece is 2.4 cm. What is the magnification of the microscope?

30. Suppose the distance between the lenses of Problem 30 is shortened to 15 cm. What is the new magnification of the microscope?

CHAPTER 23

CONCEPTS OF PHYSICAL OPTICS

Physical optics deals more with the fundamental properties of light than with applications to image formation that is the realm of geometrical optics. In geometrical optics, light is often represented as rays, with little concern for the fundamental nature of light. The ray-tracing diagrams of the previous chapter are typical of the methods of geometrical optics.

Dispersion, polarization, interference, and diffraction are four broad topics addressed by physical optics. Each of these phenomena requires that the wave properties of light be considered in any meaningful treatment of the topic. We have already encountered interference and diffraction in conjunction with sound waves and water waves, so these topics already should be familiar. However, dispersion and polarization are new; these phenomena are not often mentioned except in conjunction with electromagnetic waves.

We begin this chapter with a discussion of dispersion, and we give a fairly detailed (for this level of physics) discussion of the properties and uses of a prism. Dispersion allows us to understand the rainbow, and polarization and scattering allow us to understand why the sky is blue and why the sunsets are red. The Michelson interferometer is described in the section on interference; the final section is addressed to the diffraction grating which plays a major role in high-resolution spectroscopy. And, we determine the origin of the colors in soap bubbles and oil slicks.

23-1 DISPERSION

Until now we have treated the index of refraction of a particular material as if it were a constant. Actually, the index of refraction depends somewhat upon the wavelength of the light being refracted. The variation of the index with wavelength for four types of optical glass is shown in Fig. 23-1, and quantitative values are given in Table 23-1. Note that the index increases as the wavelength becomes shorter.

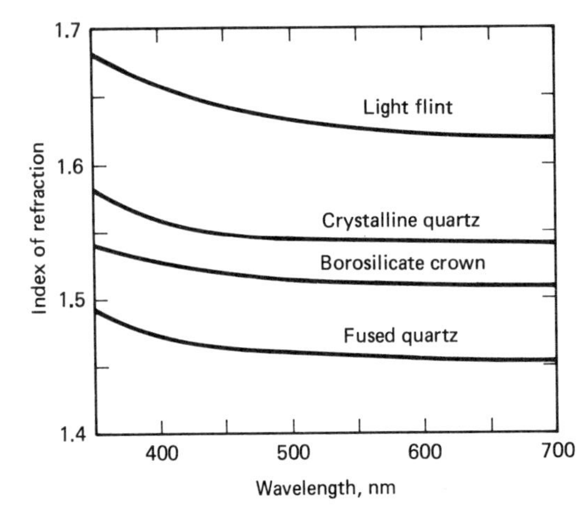

Figure 23-1 The variation of the index of refraction with wavelength of four materials.

TABLE 23-1 VARIATION OF THE INDEX OF REFRACTION WITH WAVELENGTH

	Red 660 nm	Yellow 580 nm	Green 550 nm	Blue 470 nm
Crown glass	1.5200	1.5225	1.5260	1.5310
Light flint	1.5850	1.5875	1.5910	1.5960
Dense flint	1.6620	1.6670	1.6738	1.6836
Fused quartz	1.5420	1.5438	1.5468	1.5510

When a beam of light is refracted at a surface, the angle of refraction depends, in part, upon the index of refraction. Because the index varies with wavelength, the angle of refraction will vary with wavelength also. Thus, if a beam of white light is incident upon a surface, as shown in Fig. 23-2, the component colors of the beam will be refracted at different angles. This phenomenon is called *dispersion*. Even though the effect is slight, dispersion is responsible for the ability

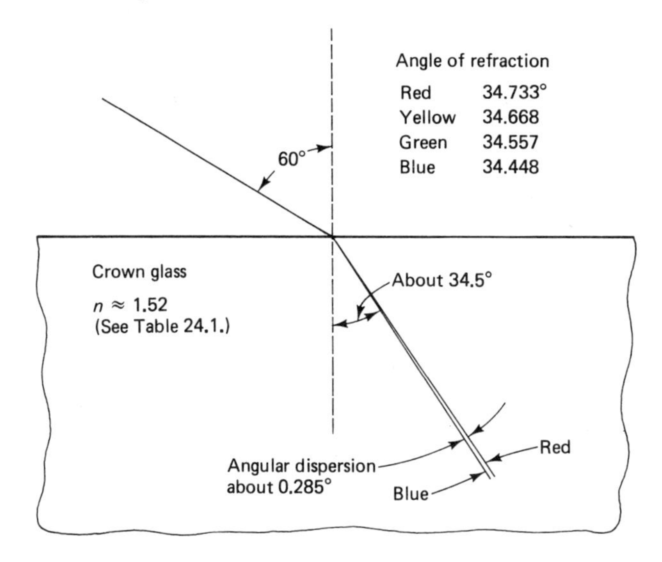

Figure 23-2 An incident beam of white light is separated into its component colors by dispersion.

of a prism to produce a spectrum, and dispersion in tiny droplets of water is responsible for the formation of a rainbow.

A medium that produces dispersion is called a *dispersive medium*. It follows from the definition of the index of refraction that the speed of light in a dispersive medium depends upon the color (wavelength) of the light. In the glasses listed in Table 23-1, blue light travels slower than red light.

No net dispersion results when a beam of light is sent completely through a flat piece of glass (or other dispersive medium) with parallel surfaces. The dispersion introduced at the point of incidence is nullified by an equal and opposite dispersion occurring at the point of emergence. However, if the surfaces are not parallel, a net dispersion will occur; the component colors will emerge in slightly different directions. This is the property of a prism that allows it to form a spectrum from an incident beam of white light.

The dispersion that occurs at the two surfaces of a prism is additive. This accounts for the ability of a prism to produce sufficient angular separation of colors to produce a visible spectrum. In Example 22-2 (Fig. 22-5), we computed the path taken by a ray of light in passing through a prism. We now examine that process in more detail.

Prism deviation

A prism obviously changes the direction of a beam of light that passes through it. The angular change in direction is called the *deviation* and is the angle δ in Fig. 23-3. The deviation depends upon the prism angle α, the index of refraction of the prism n_p, and it also depends upon the angle of incidence ϕ_{i1}.

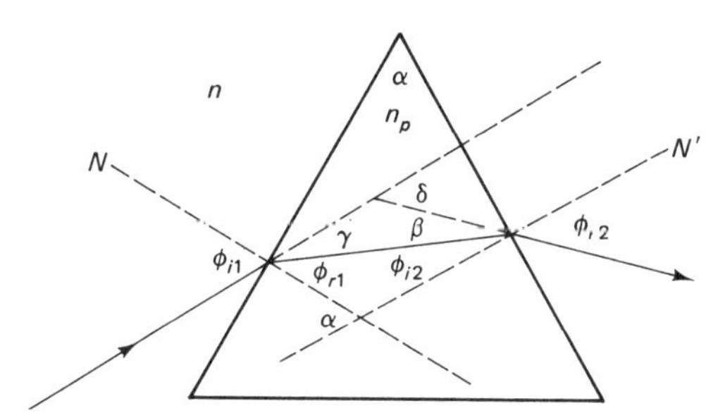

Figure 23-3 The deviation of a ray of light by a prism.

The following relationships are true for the angles of Fig. 23-3:

1. $\gamma = \phi_{i1} - \phi_{r1}$
2. $\beta = \phi_{r2} - \phi_{i2}$
3. $\alpha = \phi_{ri} + \phi_{i2}$
4. $\delta = \gamma + \beta$

By substituting (1) and (2) into (4) and using (3) to eliminate ϕ_{r1} and ϕ_{i2}, we obtain an expression for the angle of deviation:

$$\delta = \phi_{i1} + \phi_{r2} - \alpha \tag{23-1}$$

This expression is useful only if ϕ_{r2} is known. The angle ϕ_{r2} can be calculated as in

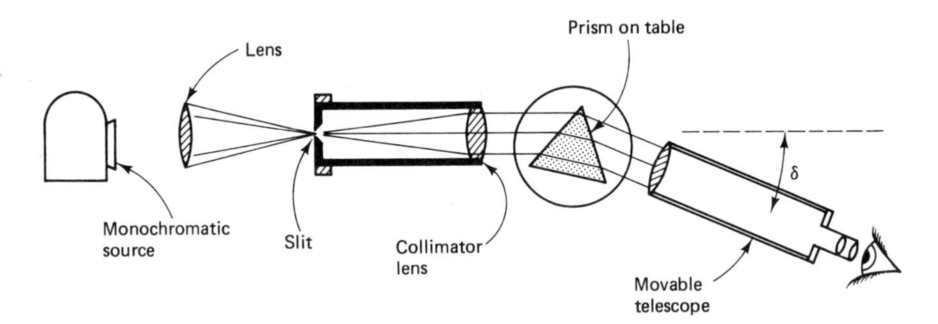

Figure 23-4 Measuring the angle of deviation of light by a prism.

Example 22-2 by two applications of Snell's law. In purely algebraic terms, the angle is given by

$$\phi_{r2} = \sin^{-1}\left\{n_p \sin\left[\alpha - \sin^{-1}\left(\frac{\sin \phi_{i1}}{n_p}\right)\right]\right\} \tag{23-2}$$

where n_p is the index of refraction of the prism. Hence, the angle of deviation can be computed using Eqs. (23-1) and (23-2).

The deviation angle can be measured with the experimental arrangement of Fig. 23-4. Light from a monochromatic source is focused on a narrow slit, and the rays that enter the slit are rendered parallel by the collimator lens. The rays are deviated by the prism, and the angle of deviation is measured by the movable telescope which is attached to an accurately calibrated scale. (If a monochromatic source is not used, the observer will see a portion of a spectrum rather than a sharp image of the slit.)

EXAMPLE 23-1 A beam of light is incident at an angle ϕ_{i1} of 50° upon one face of a prism. The prism angle is 60°, and the index of refraction of the prism is 1.67.
(a) Calculate the angle of refraction ϕ_{r2} at the second surface.
(b) Compute the angle of deviation.

Solution. (a) We use Eq. 23-2 in a straightforward manner:

$$\begin{aligned}\phi_{r2} &= \sin^{-1}\left\{n_p \sin\left[\alpha - \sin^{-1}\left(\frac{\sin \phi_{i1}}{n_p}\right)\right]\right\}\\ &= \sin^{-1}\left\{1.67 \sin\left[60 - \sin^{-1}\left(\frac{\sin 50}{1.67}\right)\right]\right\}\\ &= \sin^{-1}\{1.67 \sin[60 - \sin^{-1}(0.459)]\}\\ &= \sin^{-1}[1.67 \sin(60 - 27.30)]\\ &= \sin^{-1}(1.67 \sin 32.7)\\ &= \sin^{-1}(0.902)\\ &= 64.45^\circ\end{aligned}$$

(b) The deviation is given by Eq. (23-1):

$$\begin{aligned}\delta &= \phi_{i1} + \phi_{r2} - \alpha\\ &= 50 + 64.45 - 60\\ &= 54.45^\circ\end{aligned}$$

Minimum deviation

If, for a particular prism, the deviation is calculated and plotted as a function of the angle of incidence ϕ_{i1}, it is found that a certain angle of incidence produces a minimum value of the deviation. This is illustrated in Fig. 23-5. This angle of

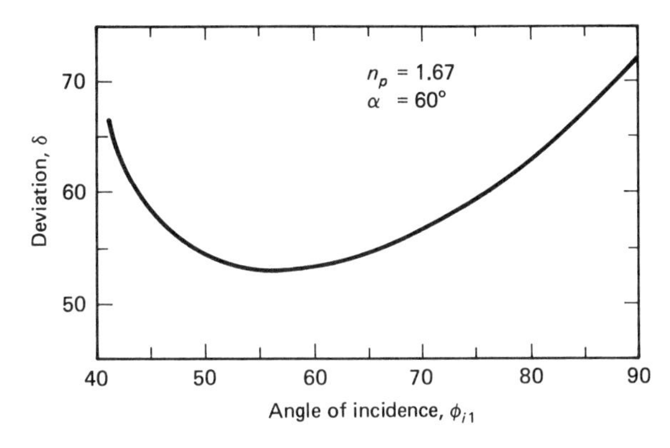

Figure 23-5 The deviation of a ray of light by a prism as a function of the angle of incidence.

deviation is called the *minimum angle of deviation*, δ_m. In the spectroscope of Fig. 23-4, the angle of incidence can be varied simply by rotating the platform on which the prism rests. When the platform is positioned so that the deviation δ is a minimum, the prism is said to be set at minimum deviation.

The geometry of a light ray traversing a prism set at minimum deviation is shown in Fig. 23-6. The angle of incidence, ϕ_{i1}, and the angle of refraction, ϕ_{r2},

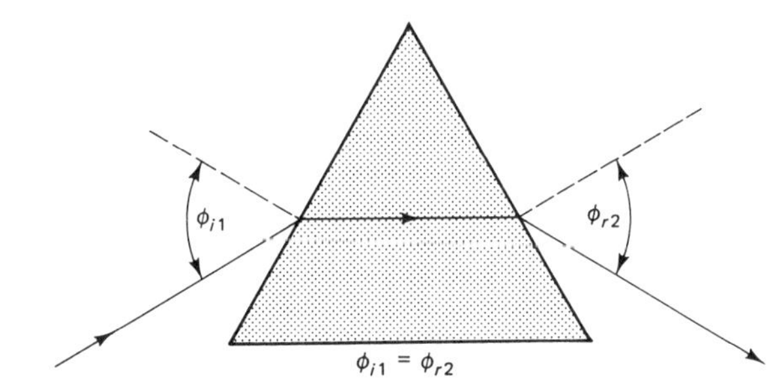

Figure 23-6 A light ray traversing a prism at the angle of minimum deviation.

are equal, and if the prism has sides of equal length, the ray will traverse the prism traveling parallel to the base of the prism. In prism spectroscopes, the prism is always set as nearly as possible to minimum deviation.

More advanced texts on optics give the following formula, which relates the indices of refraction of the prism, n_p, and the surrounding medium, n_m, the prism angle, α, and the angle of minimum deviation, δ_m:

$$\frac{n_p}{n_m} = \frac{\sin \frac{1}{2}(\alpha + \delta_m)}{\sin \frac{1}{2} \alpha} \tag{23-3}$$

If the surrounding medium is air, n_m may be taken as unity. This expression proves a convenient method for determining the index of refraction of a material of which a prism is made. The prism angle α and the minimum deviation are easily determined with a spectroscope, and n_p can then be calculated using the expression above.

Solving the expression above for δ_m gives

$$\delta_m = 2 \sin^{-1}\left[\frac{n_p}{n_m} \sin \frac{1}{2} \alpha\right] - \alpha \tag{24-4}$$

which allows the angle of minimum deviation to be calculated. From Eq. (23-1) it follows that the angle of incidence, ϕ_{i1}, required for minimum deviation is given by

$$\phi_{i1} = \tfrac{1}{2}(\alpha + \delta_m) \tag{23-5}$$

We make use of this expression in order to compute the angle of incidence required for a prism to be set at minimum deviation. We are now ready to investigate the dispersive properties of a prism.

Dispersion by a prism

Suppose an equilateral prism ($\alpha = 60°$) is made of crown glass. Let us assume the prism is set at minimum deviation in a spectroscope. We wish to calculate the deviation the prism produces for red, yellow, green, and blue rays of light.

From Table 23-1 we can see that the nominal value of the index of refraction for crown glass is 1.52. We use this in Eq. (23-4) to calculate the angle of minimum deviation:

$$\delta_m = 2 \sin^{-1}\left[\frac{1.52}{1} \sin \frac{1}{2} (60)\right] - 60$$

$$= 38.93°$$

We then use Eq. (23-5) to compute the angle of incidence required for minimum deviation:

$$\phi_{i1} = \tfrac{1}{2}(60 + 38.93)$$

$$= 49.46°$$

This is the angle of incidence for all colors. We are now ready to use Eqs. (23-2) and (23-1) (in that order) to calculate the deviation for light rays of each wavelength. These calculations proceed in the same manner as Example 23-1. Using the data of Table 23-1 for crown glass, we obtain the results tabulated in Table 23-2. Note that the difference in deviation between red and blue light is only

TABLE 23-2 CALCULATED DEVIATIONS FOR AN EQUILATERAL PRISM OF CROWN GLASS SET AT MINIMUM DEVIATION

Color	Wavelength, nm	Index	Deviation, °
Red	660	1.5200	38.928
Yellow	580	1.5225	39.149
Green	550	1.5260	39.460
Blue	470	1.5310	39.907

Difference in deviation: blue − red = 0.979°.

0.979°, and the blue light is deviated more than the red light. This is a rather small angular difference, but recall that the telescope has considerable angular magnification so that the resulting image of the spectrum may subtend a visual angle of 30° or more.

The rainbow

A *rainbow* is apt to be observed when the sun shines in one part of the sky while rain is falling in the opposite part. The display of color—a *spectrum*—results from the dispersion of sunlight by spherical droplets of water as the rays of sunlight are refracted, internally reflected, and again refracted by the droplets. The most commonly observed rainbow is the *primary rainbow*, with violet on the inside and red on the outside. Another rainbow may appear outside of but concentric with the primary rainbow. This less-frequently observed, dimmer rainbow is called the *secondary rainbow*. The colors of the secondary rainbow are reversed; red appears on the inside and violet appears on the outside. Furthermore, the secondary rainbow is twice as wide as the primary.

The mechanisms of refraction and internal reflection responsible for the primary and secondary rainbows are shown as part of Fig. 23-7. Dispersion causes

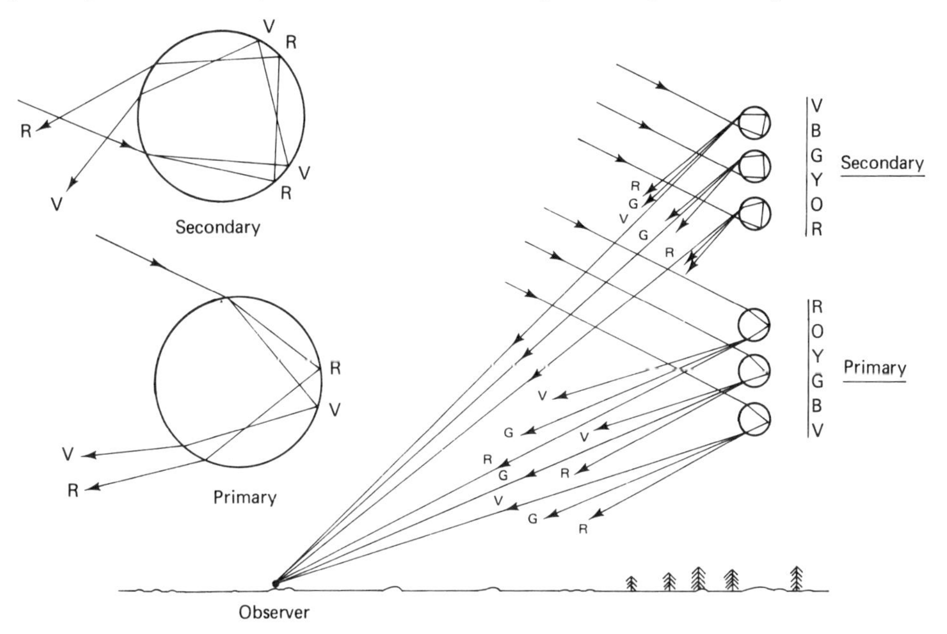

Figure 23-7 The origin of the primary and secondary rainbow.

the violet and the red rays to leave the droplets at different angles relative to the incident rays of sunlight. Therefore, the observer must look in slightly different directions to see the various colors. This angular separation of color constitutes a spectrum—a rainbow.

Any droplet that appears red in the primary rainbow must be located relative to the observer so that the line from the droplets to the observer makes an angle of about 42° to the rays of the sun (in the direction of the observer's shadow). This condition is satisfied by droplets lying on the arc of a circle, and, therefore, the rainbow appears as the arc of a circle. Flyers often observe rainbows that form complete circles, with the center of the circle lying in the direction of the observer's shadow. For the violet portion of the primary rainbow, the angle of interest is 40°. For the red portion of the secondary rainbow, the angle is 50°, and for the violet of the secondary, the angle is 54°. Thus, the primary rainbow is 2° wide while the width of the secondary is about 4°. The two rainbows are separated by 8°.

Chromatic aberration

Because the power of a lens depends upon the index of refraction and because the index of refraction varies with the wavelength (color) of light, we might expect the focal length (and the resulting image distance) of a given lens will also be color dependent. This is indeed the case, and the result is a lens defect—an aberration—called *chromatic aberration*. Chromatic aberration for a positive lens is illustrated in Fig. 23-8(a). The image composed of violet light is formed nearer the lens than

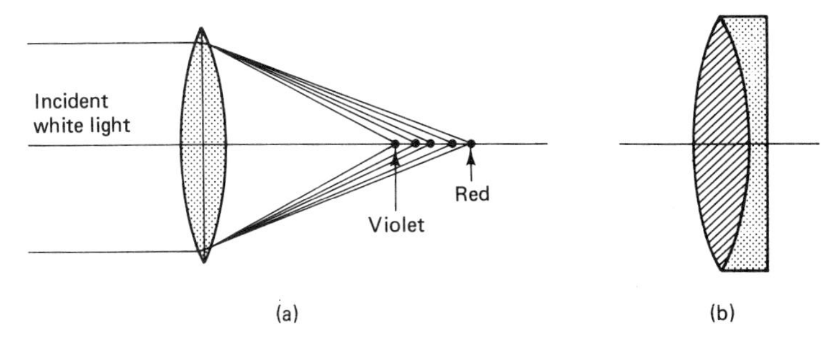

Figure 23-8 (a) Chromatic aberration. The focal length of a lens is slightly shorter for violet than for red light.
(b) An achromatic doublet for minimizing chromatic aberration.

the image composed of red light. This causes the image to appear ringed with color to the observer, and it is quite objectionable.

Fortunately, it is possible to construct a lens *doublet* that is virtually free of chromatic aberration. By combining a positive and a negative lens, the dispersion of one can be made equal and opposite to that of the other so that no net dispersion (and chromatic aberration) results. This is possible because glasses are available that have different indices of refraction while having nearly the same dispersion characteristics. By using a glass of high index for the positive lens and a glass of lower index for the negative lens, a net positive power can be achieved at the same time the dispersions cancel. The resulting lens is called an *achromatic lens*, an *achromatic doublet*, or simply an *achromat*. One type is illustrated in Fig. 23-8(b).

We have already seen that astronomical telescopes may be either reflecting or refracting. When dealing with large aperture telescopes (anything larger than about 3 in.), the reflecting telescopes have the distinct advantage of being totally free of chromatic aberration. Larger diameter achromatic lenses are very expensive.

23-2 POLARIZATION

A light wave is a transverse electromagnetic wave that consists of an oscillating electric field and an associated oscillating magnetic field. At any point, the directions of the two fields and the velocity of the wave are mutually perpendicular. The *polarization* of a light wave refers to the orientation of the electric field. If the electric field of a light wave at a point in space first points up, becomes zero, then points down, and so forth in an oscillating manner, the wave is said to be *vertically polarized*. Similarly, a wave may be *horizontally polarized*. These two cases are special cases of *plane polarization*, sometimes called *linear polarization*. In plane-polarized waves, the electric field is always directed parallel to a stationary plane. A representation of a plane-polarized electromagnetic wave is shown in Fig. 23-9.

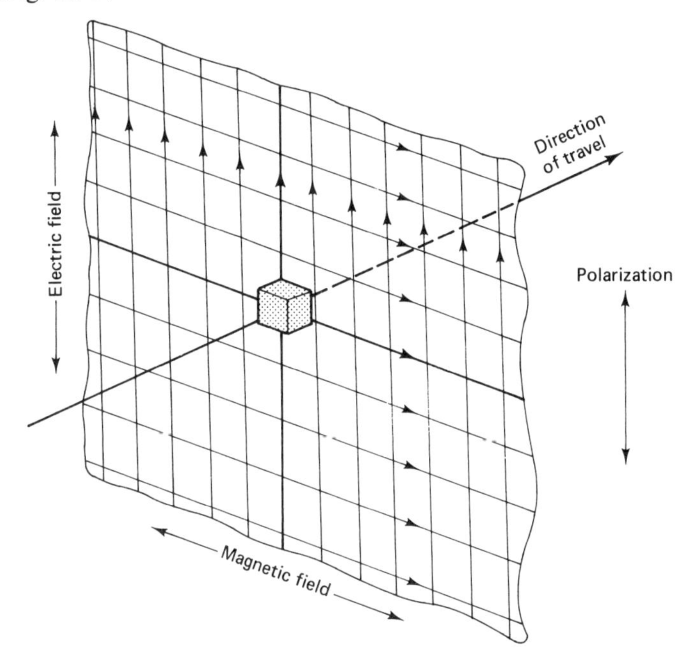

Figure 23-9 A representation of a section of a plane, plane-polarized electromagnetic wave. The fields oscillate sinusoidally, in phase. Because the electric field is vertical, we say the wave is vertically polarized. (The tiny box is to aid visualization.)

A convenient mechanical analogy is provided by the mechanical vibrations (transverse) of a string or cord. The displacement of a point of the cord corresponds to the strength and direction of the electric field of an electromagnetic wave. The analogy is shown in Fig. 23-10.

In real situations, we always deal with beams of light rather than with

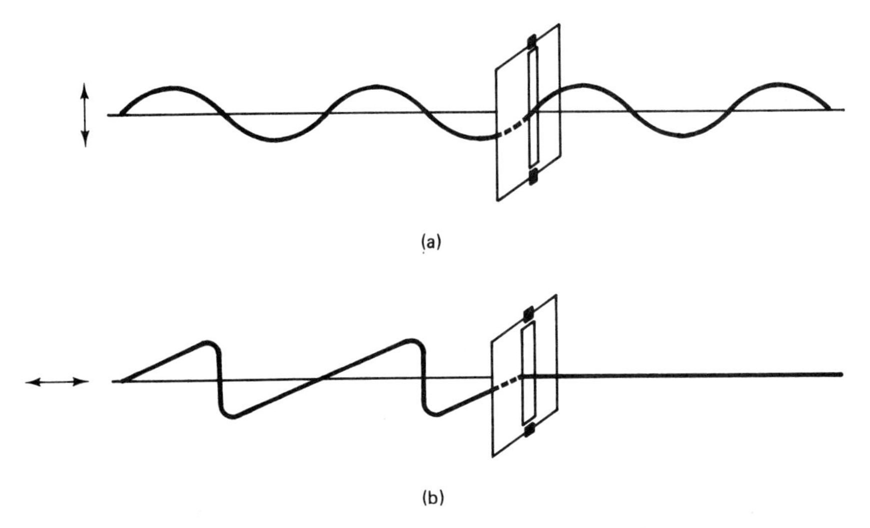

Figure 23-10 (a) The vertical slot allows vertical vibrations to pass. (b) The vertical slot attenuates the horizontal vibrations.

individual waves. With light beams, we deal with extremely large numbers of individual waves (photons) each of which may be randomly polarized. Therefore, for a beam of light to exhibit a distinguishable polarization, all the waves of the beam must exhibit the same polarization. A beam of light (as from a flashlight) whose individual waves are randomly polarized is said to be *unpolarized*.

A light wave may be *circularly polarized*. In this case, at a given point in space, the direction of the electric field rotates in a plane perpendicular to the direction of travel of the wave. Circularly polarized light may be either right-hand or left-hand polarized. An analogy is that of stretched string, the end of which is rapidly rotated in a circle so that the string assumes the shape of a spiral or corkscrew.

Actually, plane polarized light and circularly polarized light are special cases of *elliptically polarized light*. In elliptical polarization, the direction of the electric field rotates, but the strength of the field varies as the direction changes. Thus, circular polarization is the special case of elliptical polarization in which the strength of the field is the same for all directions taken by the field. Plane polarization represents the special case in which the elliptical polarization degenerates to a linear polarization. Recall that the extremes of ellipses of differing eccentricities are a circle on the one hand and a straight line on the other.

To form a better mental picture of circularly polarized light, consider the following. Suppose we are looking in the direction in which a right-hand, circularly polarized beam of light is traveling past a tiny red dot that serves as a reference point. As the circularly polarized wave passes the dot, an electric field will exist at the location of the dot. We can represent this field by a vector of constant length whose tail is always attached to the dot. As the circularly polarized wave passes the dot, the vector rotates clockwise uniformly, pivoting around the dot in a plane perpendicular to the line of travel of the wave. However, if the beam is elliptically polarized, the length of the rotating vector will not

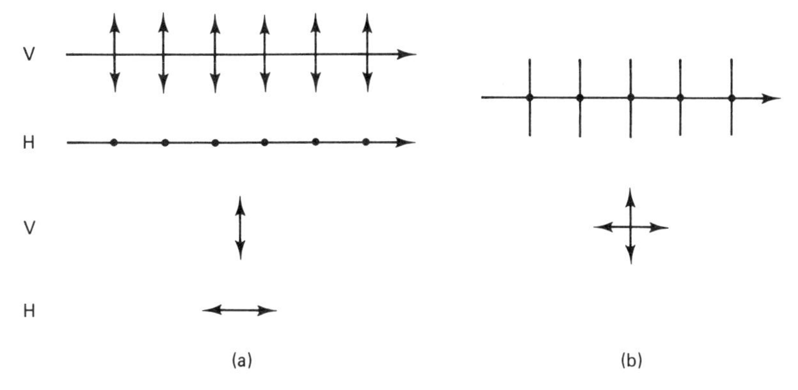

Figure 23-11 (a) Polarized light.
(b) Unpolarized light.

remain constant as it rotates through each cycle. For an elliptically polarized beam whose major axis is vertical, the vector will attain a maximum length as it passes through the vertical, and it will attain a minimum length as it passes through the horizontal.

Chapter 17 shows that two vibrations at right angles to each other can be combined to produce an elliptical vibration. (See Figs. 17-13 and 17-14 regarding Lissajous patterns.) The converse is also true. An elliptical vibration can be resolved into two linear components perpendicular to each other. Therefore, we may regard a beam of polarized light as consisting of two plane-polarized components, one vertical and one horizontal. Hence, the two components are of equal strength in circularly polarized light (and 90° out of phase) and are of unequal strength in elliptically polarized light; one component is of zero strength in plane-polarized light.

Because of the extent of the subject, we shall consider neither elliptical nor circular polarization in any greater detail. We limit further discussions, for the most part, to plane-polarized light. Various means of illustrating polarized and unpolarized light are illustrated in Fig. 23-11.

Polaroid sheets

A beam of plane polarized light is easily obtained by passing unpolarized light through a substance that absorbs one component of the unpolarized light so that the transmitted beam is plane polarized. Because unpolarized light can be resolved into two perpendicular components, the intensity of the beam is reduced by one-half in passing through the polarizing material. (This ignores any absorption of light by the polarizing medium. Any real polarizer will reduce the intensity slightly more than one half.)

Today, the most common polarizing medium is the Polaroid material produced in the form of thin sheets or films that may be sandwiched between protective layers of glass or plastic. One type of Polaroid film consists of thin sheets of nitrocellulose that are densely packed with needle-shaped microscopic crystals of iodosulfate of quinine with the long axes of the crystals all parallel. The tiny crystals almost completely absorb one component of the polarization.

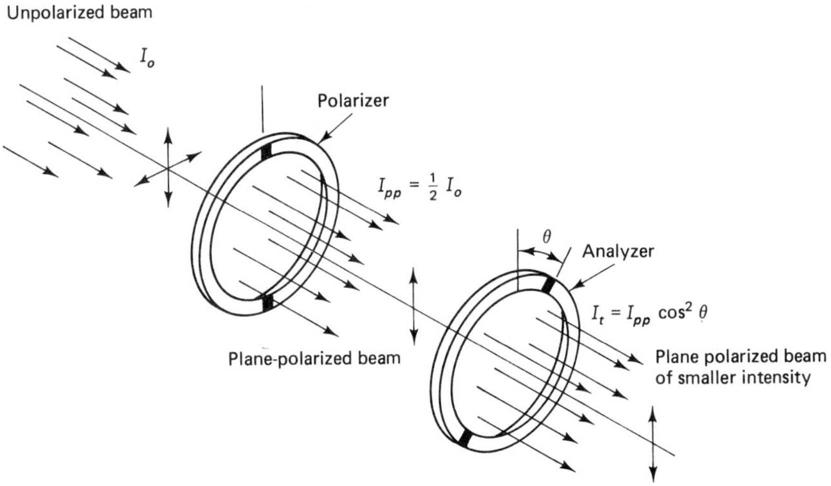

Unpolarized beam

I_o

Polarizer

$I_{pp} = \frac{1}{2} I_o$

θ

Analyzer

$I_t = I_{pp} \cos^2 \theta$

Plane-polarized beam

Plane polarized beam of smaller intensity

Figure 23-12 When the polarizing axes of the polarizer and analyzer are set at an angle relative to each other, the intensity of the transmitted beam is given by the law of Malus, Eq. (23-6).

Figure 23-12 shows an arrangement in which a beam of light is passed through two polarizing disks in succession. (Each disk contains a Polaroid film.) In this configuration, the disks are identified as the *polarizer* and *analyzer*, as indicated. When the analyzer is rotated relative to the polarizer, the intensity of the transmitted beam can be made to vary from (nearly) zero to one-half the intensity of the original unpolarized beam. Recall that one-half the intensity is lost at the polarizer in conjunction with the initial polarization process.

If a beam of plane polarized light is passed through an analyzer whose polarization axis is rotated an angle θ relative to the plane of polarization of the incident light, the intensity of the transmitted beam is given by

$$I_t = I_{pp}\cos^2\theta \qquad (23\text{-}6)$$

where I_{pp} is the intensity of the plane-polarized beam. This relationship is known as the *law of Malus*, and the basis of the law is indicated in Fig. 23-13.

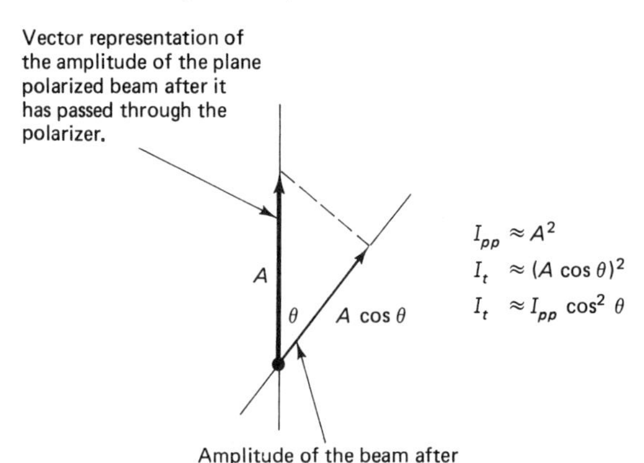

$I_{pp} \approx A^2$

$I_t \approx (A\cos\theta)^2$

$I_t \approx I_{pp}\cos^2\theta$

Figure 23-13 The basis of the law of Malus.

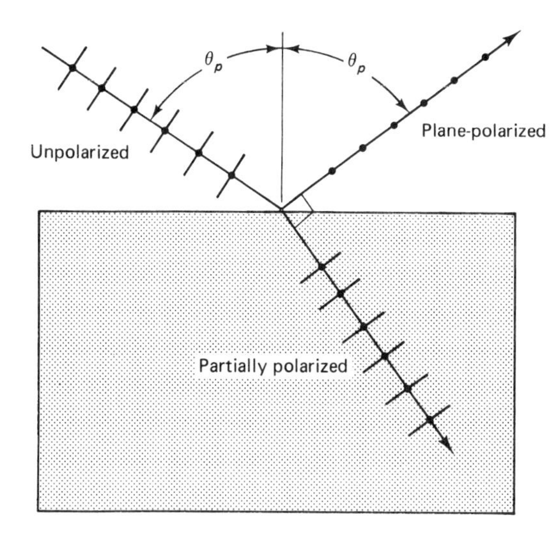

Figure 23-14 The reflected ray is completely plane-polarized when the angle of incidence equals the Brewster angle θ_p given by Eq. (23-7).

The implication of this law is this: If we examine a beam of light with an analyzer and if rotation of the analyzer reduces the intensity of the transmitted beam to zero, we can conclude that the beam being examined is plane-polarized. Furthermore, if the polarization of the analyzer is known, we can determine from it the plane of polarization of the beam. If, however, no variation of intensity occurs as the analyzer is rotated, we may conclude that the beam is either unpolarized or circularly polarized. Yet another possibility is that the intensity would vary but would never approach zero. This would indicate that the beam is a mixture of polarized light and unpolarized light, or that the beam is elliptically polarized.

Polarization by reflection

When a beam of light is reflected from a nonmetallic surface (such as glass, lucite, or water) at an angle other than normal to the surface, the reflected beam becomes polarized in a direction perpendicular to the plane of incidence. The polarization is only partial for all angles of incidence except one—called the *polarizing angle*, or *Brewster angle*—at which the polarization is complete. When the beam is incident at the polarizing angle, the reflected and refracted rays are exactly perpendicular to each other, as shown in Fig. 23-14. The refracted ray becomes partially polarized parallel to the plane of incidence, but it never becomes completely plane-polarized for any angle of incidence of the incoming beam.

The perpendicularity of the reflected and refracted beams mandates the following relationship between the indices of refraction and the polarizing angle:

$$\tan \theta_p = \frac{n_r}{n_i} \tag{23-7}$$

This relationship is known as *Brewster's law* in honor of Sir David Brewster who discovered the relationship in 1812. It is obtained directly from Snell's law, and it allows us to calculate the polarizing angle θ_p by simply computing the inverse tangent of the ratio of the indices of refraction of the two media.

Light reflected from a glass mirror (coated with silver or aluminum) is actually reflected from a metallic surface so that the polarization effect does not occur. In general, a beam of light shows no tendency toward polarization when it is reflected from a metallic surface. In fact, when a beam of plane-polarized light is reflected from a metallic surface, it is not reflected as plane-polarized light; the reflected beam will be elliptically polarized.

Because reflected light tends to be polarized, sunglasses made of polarizing material will be quite effective in eliminating reflected light if the polarization axis of the glasses is at right angles to that of the reflected light. By orienting the axis of the glasses vertically, most reflections, which tend to be horizontally polarized, can be significantly reduced. Polaroid sunglasses are preferred by those who fish for their ability to eliminate reflections from the surface of the water so that fish and other objects underneath the surface of the water can be more easily observed.

It is easy to demonstrate that Polaroid sunglasses are made of polarizing material by holding a lens of one pair in front of and at right angles to a lens of a second pair. With the polarization axes at right angles to each other, no light will be transmitted through the overlapping lenses.

The scattering of light

When a light wave encounters a charged particle, the oscillating electric field of the wave exerts an oscillating force upon the charged particle and causes the particle to execute a forced vibration. The vibration occurs at the same frequency as that of the light wave. Because a light wave is transverse, the vibration of the particle will be at right angles to the velocity of the wave.

The sinusoidal motion of the vibrating charged particles involves accelerations. It is well established both experimentally and theoretically, that an accelerated electrical charge will radiate energy. Therefore, the vibrating particle will emit light of the same frequency as that of the original beam, but the light will be emitted in directions at right angles to the original direction of the beam. Because the radiated energy must come from the original beam, the beam is weakened by the radiation by the charged particle. This phenomenon is called *scattering* because light originally contained in the beam is effectively removed from the beam and scattered in directions perpendicular to the beam.

If, for example, a beam of unpolarized light is directed along the z-axis of a rectangular coordinate system, as shown in Fig. 23-15, any charged particles present will vibrate in a plane parallel to the x- and y-axes. If an observer is located on the x-axis, only the y-component of the charged particles' motion will radiate light in the direction of the observer. This is because light is a transverse wave, and the motion of the particles toward or away from the observer cannot radiate light in the direction of the observer. Consequently, the scattered light is polarized. It will be plane-polarized in directions at right angles to the beam and elliptically, or partially, polarized in other directions.

Atoms and molecules of the earth's atmosphere contain electrons and protons which scatter light from the rays of the sun. Thus, skylight (not from

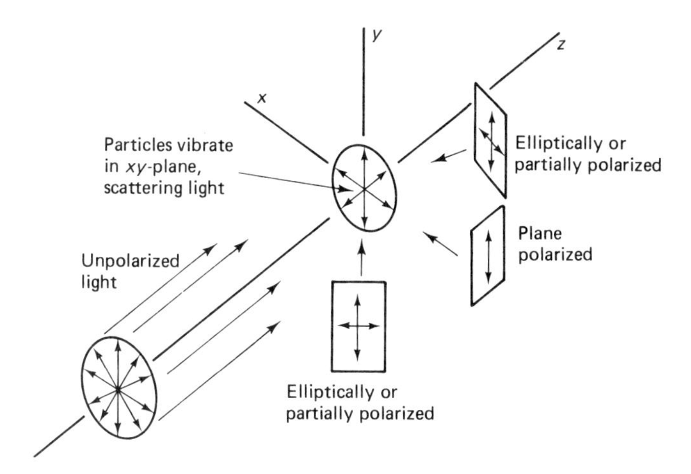

Figure 23-15 A beam of unpolarized light incident on charged particles causes the particles to vibrate in the *xy*-plane. In turn, the particles radiate light that is plane polarized when viewed parallel to the *xy*-axis. In other directions, it is only partially polarized or elliptically polarized.

clouds) tends to be polarized, the greatest polarization being in the direction perpendicular to the sun's rays. Use is often made of this fact in photography when it is desired to darken the sky in a photograph for artistic purposes. A polarizer attached to the camera lens (and rotated appropriately) will reduce the light from the sky and will also reduce the amount of reflected light entering the camera from objects on the ground. This increases the saturation of the colors and produces a more dramatic photograph.

A system undergoing forced vibrations will vibrate with greater amplitude when the driving force is close to the natural resonant frequency of the system. The natural frequencies of the electrons in the atoms and molecules of the air correspond to the frequencies of ultraviolet light. Therefore, the blue end of the sun's spectrum produces a greater amplitude of vibration than the red end, and the greater amplitude produces a correspondingly larger proportion of blue in the scattered light. Hence, blue light is scattered more than red, and it is because of this that the sky appears blue.

At sunset, the rays of the sun travel obliquely through the atmosphere and therefore travel a much larger distance through the atmosphere than they do at midday. Consequently, more blue light is removed (by scattering) so that the red end of the spectrum becomes progressively more prominent. When these rays fall on scattered clouds, dust, and smoke near the earth's surface, a beautiful orange red sunset is apt to be observed.

Optical activity

Many substances possess the peculiar ability to rotate the plane of polarization of a beam of polarized light as it passes through the substance. This property is known as *optical activity*. Optically active substances include crystalline quartz, turpentine, sugar crystals, and a solution of sugar in water. Optical activity stems

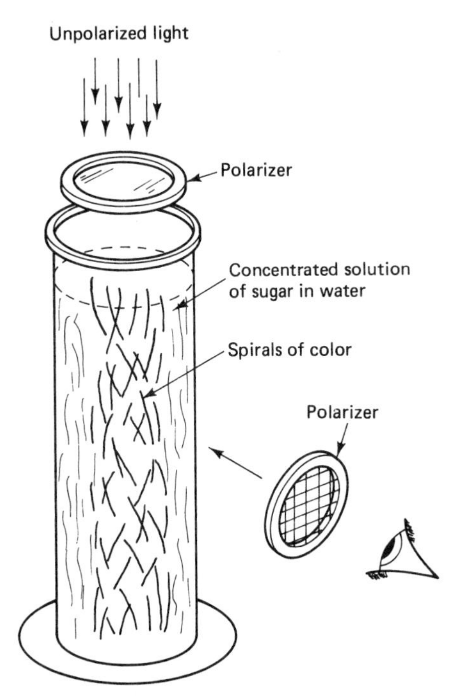

Figure 23-16 Optical activity.

from spiral-shaped characteristics of complex molecules or crystals and may produce either right-handed or left-handed rotation, depending upon the material. An interesting demonstration of optical activity is illustrated in Fig. 23-16, in which a barber pole of colors may be observed because the optical activity is strongly dependent upon the wavelength of light.

23-3 INTERFERENCE OF LIGHT

We have already presented the basic concepts of the interference of waves in Sect. 18-6 in conjunction with sound waves and water waves. Therefore, the concept of constructive and destructive interference should be familiar. In this section, and in the remainder of this chapter, we address several aspects of the interference of light.

First of all, the wavelength of light is very short in comparison to the dimensions of objects we normally encounter. For example, it takes slightly more than 45,700 waves of green light (555 nm) to span a distance of 1 in. Because of the shortness of light waves, interference phenomena for light waves occur on a very small scale. Most of the time a magnifier or a microscope or some other means of enlargement is required in order to observe them directly. It is for this reason that the interference of light is as unfamiliar in everyday experience as it is.

Coherent light

We know that most light stems from electron transitions in atoms, and a typical source of light consists of billions of billions of atoms, at least. Within a particular source, each atom radiates more or less independently of the others so that the beam of light emitted from the source is composed of a large number of individual waves. Moreover, there is no definite phase relationship between the individual waves because each "atomic emitter" turns on, radiates, and then turns off without regard for the other atoms around it. The light produced by such a source is said to be *incoherent*. All the common sources of light (incandescent, gas discharge, carbon arc, and so on) produce incoherent light. However, the light of a laser beam is *coherent*. (The laser is the only common source of coherent light. It is described in Sect. 31-8.)

In order for two beams of light to interfere and produce an observable interference pattern, the phases of the waves comprising the two beams must be related. That is, the beams must be coherent, and the beams must therefore originate from *coherent sources*. We are now ready to see how coherent sources can be achieved in practice.

Young's double-slit experiment

The interference of light can be demonstrated with the experimental arrangement of Fig. 23-17. Moreover, measurements made using the apparatus can be used to determine the wavelength of light.

Light of one wavelength (from a monochromatic source) is focused on the entrance slit S_e, which then acts as a point source of light for the rest of the apparatus. Waves from S_e then reach the double slit composed of S_1 and S_2, and the wave fronts arrive at S_1 and S_2 simultaneously. By Huygens' principle, S_1 and S_2 act as sources and appear to emit a new wavelet each time a wave front arrives from S_e. Thus, the emissions of wavelets by S_1 and S_2 are related, and the two slits serve as coherent sources. An interference pattern will be produced on the screen.

Bright fringes will appear on the screen at points located such that the path difference W is an integral number of wavelengths. Thus, for bright fringes,

$$W = n\lambda \qquad n = 0, 1, 2, 3, \ldots$$

and in terms of the slit separation d,

$$W = d \sin \theta$$

Equating these expressions for W gives

$$d \sin \theta = n\lambda \quad \text{or} \quad \sin \theta = \frac{n\lambda}{d} \tag{23-8}$$

Slit separations d are typically no less than 10^{-4} m, so θ is a very small angle. Hence, we can approximate $\sin \theta$ by x/R to obtain

$$d \frac{x}{R} = n\lambda \quad \text{or} \quad x = \frac{nR\lambda}{d} \tag{23-9}$$

which allows us to calculate the lateral positions of the bright fringes on the screen. If Δx is the lateral distance on the screen from one bright fringe to the next, we can write the above expression in terms of Δx to yield

$$\lambda = \frac{d(\Delta x)}{R} \tag{23-10}$$

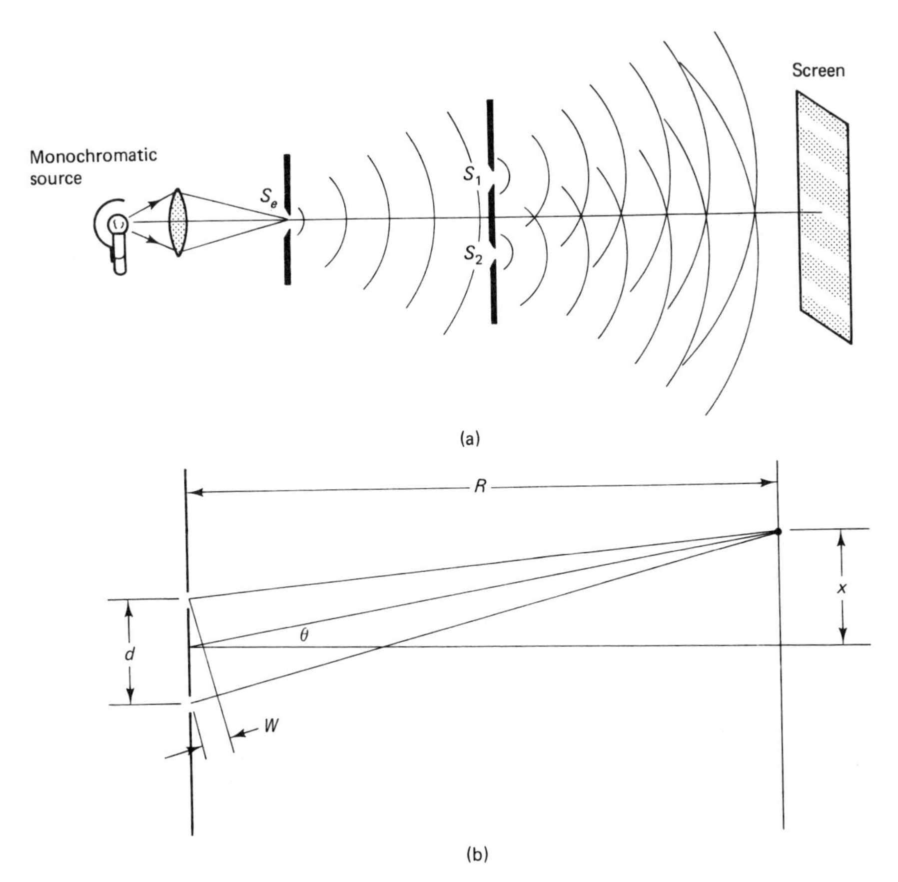

Figure 23-17 The interference of light through a double slit. (a) Experimental arrangement. (b) Geometry.

which may be used to determine the wavelength of light. In practice, a photographic film is substituted for the screen to facilitate the determination of Δx.

EXAMPLE 23-2 In a particular double-slit experiment, the slit separation d is 10^{-4} m and the distance from the slit to the screen is 20 cm. A monochromatic source of wavelength 650 nm is used.

(a) Calculate the angle for the first bright fringe (not the central fringe).
(b) Determine the lateral displacement of the first fringe.

Solution. (a) Use Eq. (23-8).

$$\sin \theta = \frac{n\lambda}{d}$$

$$= \frac{1(650 \times 10^{-9}\text{ m})}{10^{-4}\text{ m}}$$

$$= 6.5 \times 10^{-3}$$

$$\theta = \sin^{-1}(6.5 \times 10^{-3}) = 0.372°$$

(b) The lateral displacement is easily determined by converting 0.372° to 6.5×10^{-3} rad and multiplying by R:

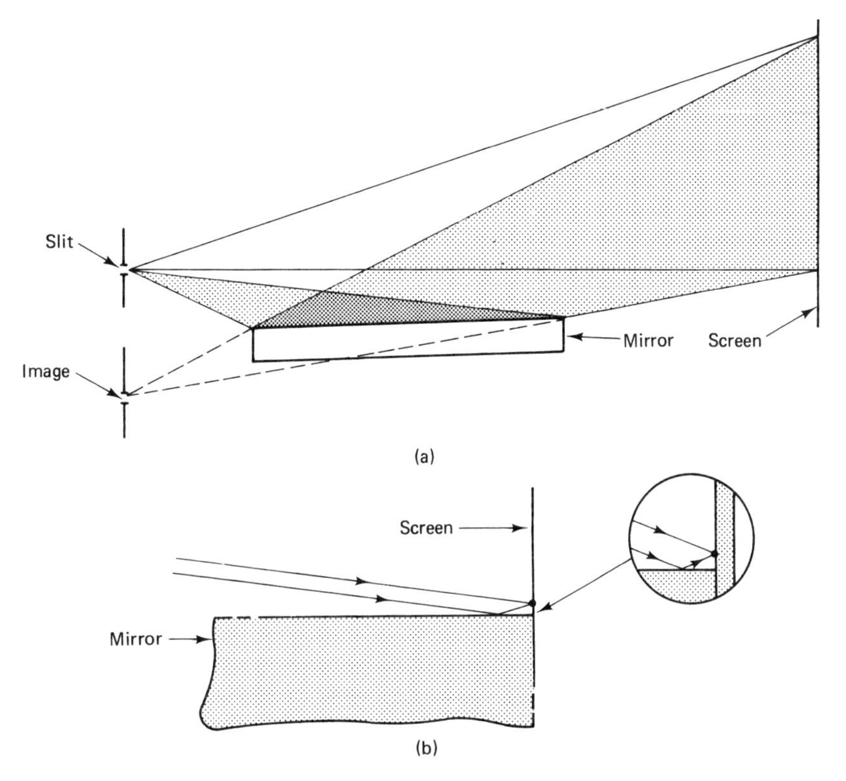

Figure 23-18 Lloyd's mirror.
(a) Experimental arrangement.
(b) Screen is moved to the edge of the mirror. A dark band forms at the line of contact between screen and mirror.

$$x = R\theta = (20 \text{ cm})(6.5 \times 10^{-3})$$
$$= 0.13 \text{ cm} = 1.3 \text{ mm}$$

EXAMPLE 23-3 Monochromatic light of unknown wavelength falls on a double slit whose spacing is 10^{-4} m. A photographic film is located 20 cm from the slit, and the separations of the resulting fringes are measured to be 1.18 mm. Calculate the wavelength of the light.

Solution. Use Eq. (23-10).

$$\lambda = \frac{d(\Delta x)}{R}$$
$$= \frac{(10^{-4} \text{ m})(1.18 \times 10^{-3} \text{ m})}{0.20 \text{ m}}$$
$$= 5.90 \times 10^{-7} \text{ m} = 590 \text{ nm}$$

Lloyd's mirror

The arrangement shown in Fig. 23-18(a) is known as *Lloyd's mirror*. The slit and its image in the mirror constitute the double source, and an interference pattern is formed on the screen.

Two beams of light reach the screen; one is direct, the other is reflected. If the screen (or a photographic film) is placed in contact with the end of the mirror, as shown in part (b), a dark fringe (instead of a bright fringe) is formed at the surface of the mirror. This is surprising because the rays reaching the screen at that point have traveled over almost identical paths to get to that point. The only difference is that one ray was reflected at the very last instant of its journey.

A dark fringe implies destructive interference, and destructive interference occurs only between waves that are out of phase. Therefore, we must conclude that one ray experienced a phase change of 180°. Because the only significant difference between the rays is the reflection of one, we conclude that the reflection introduced a phase change of 180°. This is an important result.

When a ray traveling in one medium is reflected from the surface of another medium of higher index of refraction, the reflected ray experiences a phase change of 180°. This is illustrated in Fig. 23-19. This bit of information gives special significance to the Lloyd's mirror experiment.

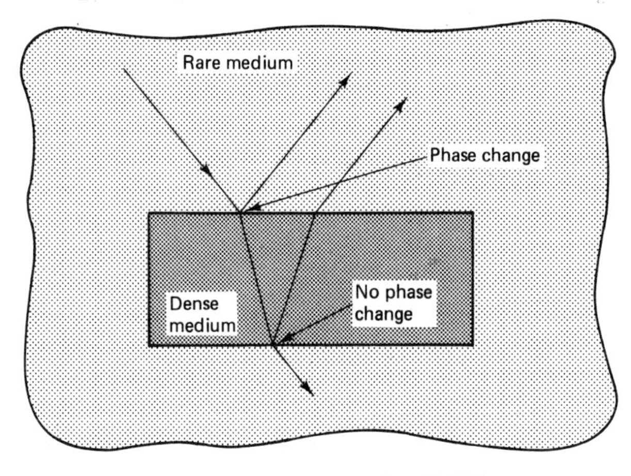

Figure 23-19 A phase change occurs when a ray is reflected from a dense medium, but no phase change occurs when a ray is reflected from the rare medium back into the dense medium.

23-4 THE MICHELSON INTERFEROMETER

The preceding experiments produced interference by bringing together two different portions of a wave front. This process is known as *wave front division*. We shall now consider the Michelson interferometer, which produces interference by a process known as *amplitude division* in which the energy within the same section of a wave front is divided into two beams that are subsequently brought back together to form an interference pattern.

The physical configuration of a Michelson interferometer is shown in Fig. 23-20. Light from an extended monochromatic source (one having considerable area, as opposed to a point source or a slit) is incident upon a partially silvered mirror B that serves as a beam splitter. Thus, the incident beam is divided into two beams of approximately the same intensity. The two beams are then reflected back to the beam splitter by mirrors M_1 and M_2, and a portion of each beam is reflected to the eye of the observer. The observer will see interference fringes either as straight lines, curved lines, or as concentric circles, depending upon how the instrument is adjusted. Glass plate C is a corrector plate of the same thickness

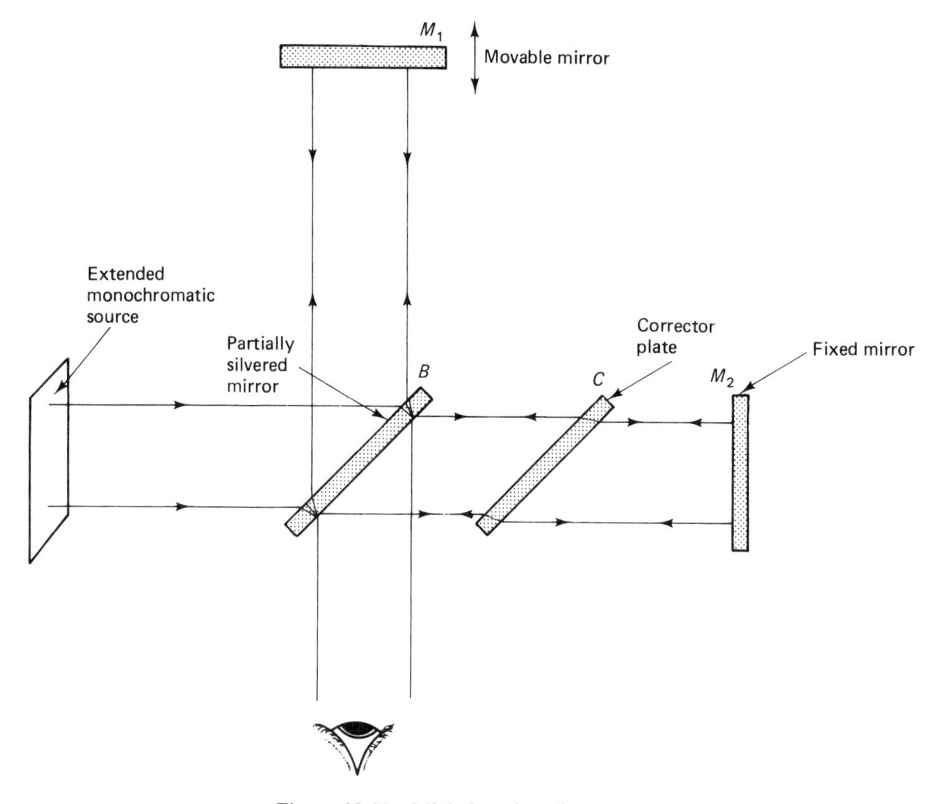

Figure 23-20 Michelson interferometer.

and type of glass as the beam splitter. It is included so that both beams travel through the same thickness of glass in order to keep the optical paths of the two beams very nearly the same.

When the instrument is adjusted for circular fringes, the central fringe—the "bull's eye"—may appear either light or dark. Furthermore, moving the movable mirror causes the fringe pattern to expand or contract (depending upon the direction of movement of the mirror), and one fringe will appear or disappear at the center each time the mirror moves a distance equal to one-half the wavelength of the monochromatic light being used. Thus, very small movements can be discerned and measured to a precision of a fraction of a wavelength of light. By connecting the movable mirror to an external mover, many exotic experiments can be performed, such as measuring the bending of a heavy steel beam caused by the weight of a butterfly.

The Michelson interferometer has played an important role in the development of physics. In the famous Michelson-Morley experiment, an advanced interferometer was used in an effort to detect the motion of the earth relative to the ether—that mysterious substance imagined to pervade all space so that light waves may have a medium in which to propagate. The null result of that experiment provided initial support for Albert Einstein's special theory of relativity and to the general conclusion that if there is an ether, we shall never be able to detect it.

On a level that is more applied, a modification of a Michelson interferometer is used in defining the standard of length (the standard meter) in terms of a large number of wavelengths of monochromatic light.

23-5 INTERFERENCE IN THIN FILMS

Everyone has seen the colors that appear in a soap bubble and the array of colors when a film of oil covers the surface of water. This display of color in otherwise colorless materials is a consequence of interference, namely, interference in *thin films*. Before we plunge into this topic, however, we need to define the concept of optical path.

Optical path

Light travels slower in any material medium than it does in a vacuum, and so a greater phase change occurs per unit length in a medium than in a vacuum. The *optical path* is the distance in vacuum required for the occurrence of an equivalent phase change as that which occurs in a given distance in a medium. If the distance in the medium is d_m, the optical path is given by

$$\text{Optical path} = n \cdot d_m \qquad (23\text{-}11)$$

where n is the index of refraction. When considering changes of phase of light waves in various media, we must see the optical path rather than the simple geometric distance between the two points under consideration.

Reflections in a thin film

When a ray of light is incident upon a thin film, as shown in Fig. 23-21, a reflection occurs at the top surface and also at the bottom surface. The ray reflected from

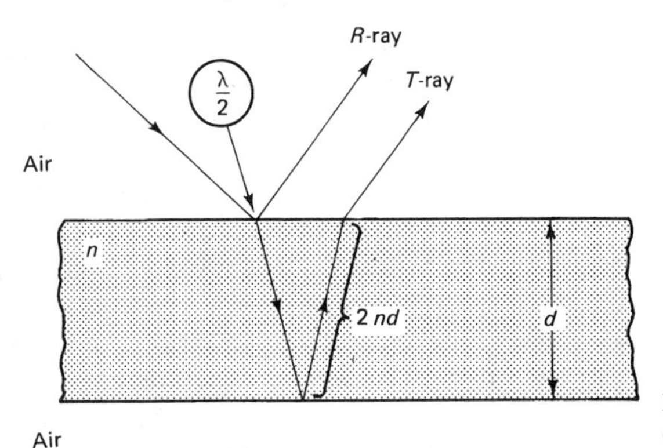

Figure 23-21 Geometry for interference in thin films.

the bottom surface is subsequently refracted at the top surface and leaves the film traveling parallel to the ray reflected from the top surface. Either constructive or destructive interference can occur between the two rays, depending upon their relative phases. If *destructive* interference occurs, no light will be reflected from that point. If *constructive* interference occurs, a greater than normal amount of reflection will occur.

For convenience, we speak of phase changes in terms of wavelengths. A phase change of one wavelength λ is equivalent to 360°, or 2π rad. A phase change of 180° is equivalent to $\lambda/2$. Also, for the sake of simplicity, we consider reflections only for beams that are incident nearly normal to the surface.

Referring to Fig. 23-21, we see that the T-ray traverses the thickness of the film two times, traveling a distance of $2d$ within the film. The corresponding optical path is $2nd$. This represents the difference in optical path traversed by the two rays. We must now consider any phase changes introduced in the process of reflection itself. A phase change of $\lambda/2$ is imparted to the R-ray, but no phase change is imparted to the T-ray in the reflection at the bottom of the film.

For constructive interference, the R-ray and the T-ray must be in phase as they leave the film. This requires the optical path difference of the T-ray to be given by

$$2nd = m\lambda + \tfrac{1}{2}\lambda \tag{23-12}$$

where m is a positive integer or zero. (Recall that a phase change of an integral number of wavelengths is equivalent to no phase change at all.) Thus, for constructive interference in a film bounded by air on both sides,

$$2nd = (m + \tfrac{1}{2})\lambda \tag{23-13}$$

This expression can also be written as

$$d = \frac{(m + \frac{1}{2})\lambda}{2n} \quad \text{or} \quad \lambda = \frac{2nd}{(m + \frac{1}{2})} \qquad m = 0, 1, 2, 3, \ldots \tag{23-14}$$

For destructive interference, the T-ray and the R-ray must differ in phase by $\lambda/2$. This requires that

$$2nd = m\lambda \qquad m = 1, 2, 3, \ldots \tag{23-15}$$

When conditions for destructive interference are met, the amount of light reflected from the top surface is reduced.

When white light shines on a soap bubble, the thickness of the soap film might be such that constructive interference occurs for red light at the same time that destructive interference occurs for blue light. In this case, more red would be reflected than blue, and the bubble would appear red. The opposite case is equally likely, and it is not unusual for the top and bottom parts of a bubble to exhibit different colors. This indicates that the film of the bubble is thicker on the bottom.

Newton's rings

When a convex lens is placed on a plane surface, as shown in Fig. 23-22, a thin film of air is created between the two surfaces of glass. Light reflected from the top and bottom of the film will interfere, and an interference pattern in the form of concentric circles will be visible. This interference pattern is known as *Newton's rings* (even though the original discovery is now attributed to Robert Hooke).

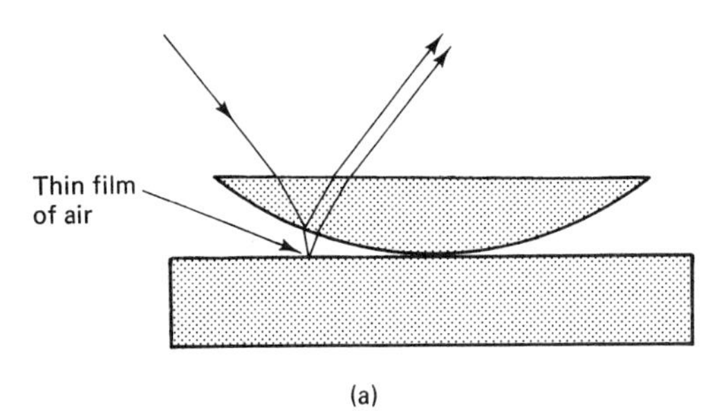

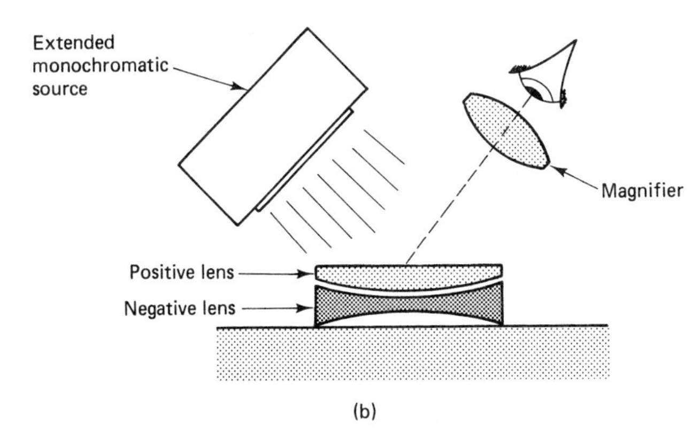

Figure 23-22 (a) Reflections from opposite sides of the air film give rise to Newton's rings.
(b) A practical setup for observing the rings.

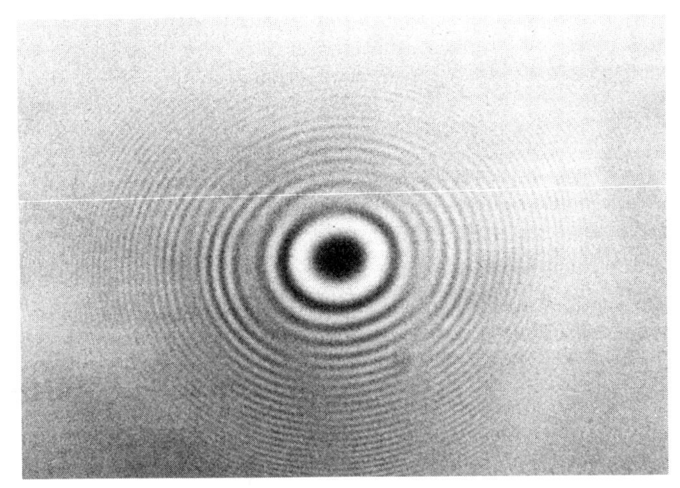

Newton's rings. The actual size of this pattern is less than one millimeter in diameter.

23-6 THE DIFFRACTION GRATING

Diffraction has been discussed or at least mentioned in two chapters already. In Sect. 18-6 we see that diffraction arises because of Huygens' principle (Figs. 18-18 and 18-19), which can be demonstrated with water waves: Each point of a wave front acts as a source of wavelets which together determine the future progress of the wave. We ran into a diffraction phenomenon again in Sect. 20-4 in our attempt

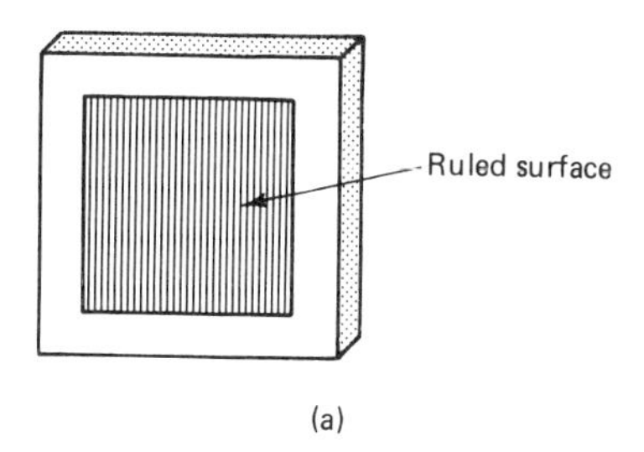

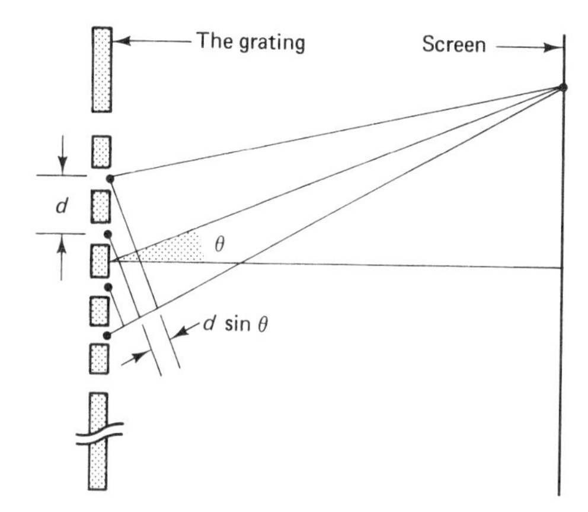

Figure 23-23 (a) Appearance of a diffraction grating (except the lines are quite invisible on a real grating). (b) Geometry for obtaining the grating equation.

to isolate an individual ray of light (Fig. 20-6). Diffraction produces light and dark fringes at the edge of a shadow, as shown in Fig. 20-7.

Edge diffraction occurs because the wave front near the edge acts as if it were divided into small sections. This occurs even when nothing physical divides the wave front, but the phenomenon is confined to a region near the edge. We now propose to approach the matter directly and physically divide the wave front into sections. This can be done by ruling (scribing) a large number of fine straight lines on a flat piece of optical glass. When a wave front passes through the glass and through the array of lines, it will be divided as we proposed, and the diffraction phenomenon will be achieved in a useful form. The ruled piece of glass is called a *diffraction grating*. A typical grating may be ruled with as many as 1200 lines/mm, and the grooved area may be on the order of 1 in.^{2} for a small grating.

A diffraction grating acts like an enhanced double-slit experiment. An interference pattern is formed, and the angular deviations are dependent upon the wavelength of the incident light. But a grating is far more effective than a double slit in dispersing light of different colors. Whereas a double slit produces a bright fringe [for a given value of n in Eq. (23-8)], a grating will produce an entire spectrum, assuming white light is incident upon the grating. Hence, the value of a grating lies in its dispersion characteristics—in its ability to form a spectrum comparable to that formed by a prism.

The appearance of a grating is shown in Fig. 23-23(a) and the theoretical representation is shown in Fig. 23-23(b). Reasoning as for the double slit leads to the grating equation:

$$d \sin \theta = m\lambda \qquad m = 0, 1, 2, 3, \ldots \tag{23-16}$$

The *grating spacing*, d, is the distance between grooves (lines), and m is the *order* of the diffraction. A complete spectrum is produced for each value of m except zero, and the angular width of each spectrum is proportional to m. The second-order spectrum is twice as wide as the first, and so on. The use of a grating in a simple spectroscope is illustrated in Fig. 23-24.

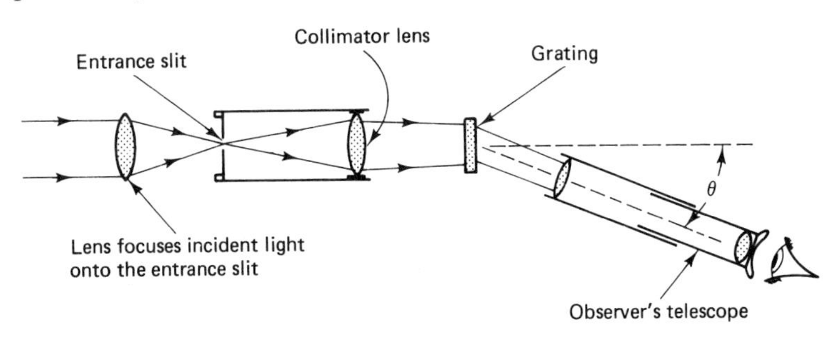

Figure 23-24 A simple spectroscope that uses a diffraction grating rather than a prism.

The smaller the grating spacing, the wider the spectrum of each order. Further, the greater the total number of lines on the grating, the greater the *resolving power* of the grating. Thus, by constructing a grating with a very large number of very closely spaced lines, spectral lines separated in wavelength by only 0.01 Å (0.001 nm) can be resolved.

The primary advantage of a grating over a prism lies in the high resolving power in conjunction with a large dispersion. A lesser advantage is that the spectrum produced by a grating is linear (a *normal* spectrum) whereas the spectrum produced by a prism tends to be compressed on the red end. A disadvantage of a grating is that the incident light energy is divided among several spectra (one for each order), but methods have been developed for concentrating more than 90% of the incident light into the spectrum of a particular order. Also, portions of adjacent spectra may overlap for orders greater than 4 or 5.

Several types of gratings may be distinguished according to the method of manufacture or optical properties. A *transmission grating* is one that transmits light through the glass on which the rulings are made. A *reflection grating* serves as a mirror on which the lines have been ruled. Obviously, all *metal gratings* must be reflection gratings. A *concave grating* is a reflection grating that serves also as a concave mirror in order to bring the spectrum to a focus without intervening optical components. A *replica grating* is a reproduction of a grating, made by using a ruled grating as a mold in order to duplicate the rulings in a thin film of plastic material that subsequently is mounted on a clear-glass slide.

EXAMPLE 23-4 Light of wavelength 589 nm is incident upon a grating ruled with 1200 lines/mm. Compute the angle of the first-order diffraction.

Solution. We must first compute the grating spacing, d:

$$d = \left(\frac{1200 \text{ lines}}{1 \text{ mm}}\right)^{-1} = \frac{1 \times 10^{-3} \text{ m}}{1200 \text{ lines}}$$

$$= 8.33 \times 10^{-7} \text{ m/line}$$

Then use Eq. (23-16). For $m = 1$:

$$\sin \theta = \frac{m\lambda}{d}$$

$$= \frac{1(589 \times 10^{-9} \text{ m})}{8.33 \times 10^{-7} \text{ m}}$$

$$= 0.707$$

$$\theta = \sin^{-1}(0.707) = 45.0^\circ$$

To Go Further

Find out about the lives and contributions to optics made by the following:

A. J. Angstrom (1814–1874), professor of physics at Uppsala, Sweden
Thomas Young (1773–1829), English physician and physicist
Sir David Brewster (1781–1868), English physicist
Augustin Fresnel (1788–1827), French contributor to the theory of light
A. A. Michelson (1852–1931), American physicist
Etienne Malus (1775–1812), French engineer

Questions

1. Dispersion is caused by what basic physical phenomenon?
2. What accounts for the fact that a prism is able to form a spectrum?
3. What causes chromatic aberration in a lens? How is it minimized?
4. Explain the law of Malus. How is the intensity of the transmitted beam related to the vector component of the amplitude of the transmitted beam?
5. Why is a polarizing material often incorporated into sunglasses? Is the polarization axis of Polaroid sunglasses vertical or horizontal? Why?
6. What particular angular relationship exists between the transmitted and reflected rays when light is incident on a glass surface at the Brewster angle? What is significant about the reflected light?
7. Explain the blueness of the sky in terms of forced vibrations, scattering, and natural resonant frequencies.
8. Why is a polarizing disk that is attached to a camera lens able to darken the blue sky and also increase the color saturation of objects on the ground?
9. A student asks a perceptive old physicist, "Why do two beams of ordinary incoherent light not interfere and produce an interference pattern?" The physicist answers, "Well, they do; we're just not fast enough to see it." Explain.
10. What property of reflection was discovered from the Lloyd's mirror experiment?
11. Of the interference experiments in this chapter, which use wave front division and which use amplitude division?
12. What causes the colors to appear in a soap bubble and in an oil slick on calm water where the layer of oil is very thin?

*13. How can a flat piece of glass be tested for flatness if an optical flat is available for purposes of comparison?

14. What accounts for the reddish purple color of the lenses in binoculars, small telescopes, and cameras?

*15. Light reflected from a black phonograph record often appears tinged with color. From where do the colors come?

*16. In a spectroscope, what is the purpose of the collimator lens?

*17. The dispersion of a prism is maximum when the prism is set at minimum deviation. Give a practical method for setting the prism of a small spectroscope at minimum deviation.

*18. Can interference phenomena be explained in terms of photons? What attributes must be assigned to the photons in order to do so?

Problems

1. Use the data of Table 23-1 to compute the velocity of red and of blue light through: **(a)** crown glass; **(b)** fused quartz.
2. Use Snell's law to compute the angle of refraction of a red and of a blue ray incident upon fused quartz at an angle of 60°.
3. Repeat Problem 2 for an angle of incidence of 40°.
4. Use Eqs. (23-1) and (23-2) to compute the deviation of **(a)** a red ray and **(b)** a blue ray through an equilateral prism made of crown glass. Assume ϕ_{i1} is 50°, and use the data of Table 23-1.
5. Repeat Problem 4 for fused quartz.
6. Compute the angle of minimum deviation for an equilateral prism whose index for green light is 1.67.
7. Repeat Problem 6 for a prism having an index of 1.55.

* 8. A symmetric positive lens made of crown glass has surfaces whose radius of curvature is 15 cm. Use the lensmaker's formula to compute the primary focal length of the lens for: **(a)** red light; **(b)** blue light.

* 9. Repeat Problem 8 assuming the lens is made of fused quartz.

10. An unpolarized beam of light passes through a single polarizing disk. What fraction of the original intensity does the transmitted plane-polarized beam possess? How do the amplitudes of the incident and transmitted beams compare?
11. A vertically polarized beam of light is passed through an analyzer whose axis is rotated from the vertical by 40°. What fraction of the intensity of the incident beam is transmitted?
12. Compute the angle of reflection at which light reflected from water will be completely plane polarized.
13. Suppose the slits of Young's double-slit experiment are separated by 10^{-4} m. Green light of wavelength 550 nm is incident upon the slits.
    (a) Compute the angle at which the first bright fringe ($n = 1$) will be located.
    (b) If a photographic film is located 30 cm behind the slits, compute the linear separation between successive bright fringes as they appear on the film.
14. A double-slit experiment is performed to determine the wavelength of a certain monochromatic light source. The slit separation is 0.1 mm, and a photographic film is located 50 cm from the slits. When the film is developed, the bright fringes are found to be separated by 2.9 mm. Compute the wavelength of the light.
15. Calculate the optical path of 1 cm of glass whose index of refraction is 1.55.
16. Compute the minimum thickness of a film of water (in a soap bubble) that will produce constructive interference for red light of wavelength 660 nm.
17. A nonreflective coating is to be applied to a glass lens whose index of refraction is 1.58.
    (a) What should be the index of refraction of the film?
    (b) If the reflection is to be minimized at a wavelength of 550 nm, what should be the thickness of the film?
18. A diffraction grating is ruled with 600 lines/mm and is illuminated with yellow light of wavelength of 589 nm.

(a) Compute the grating spacing d in meters.

(b) Calculate the angles at which first-order ($n = 1$) and second-order ($n = 2$) diffraction occurs.

19. A grating ruled with 1200 lines/mm is used to determine the wavelength of a monochromatic source of light. First-order diffraction occurs at an angle of 38.2°. Calculate the wavelength of the light.

20. The movable mirror of a Michelson interferometer is driven by a screw thread having 40 threads/in. If 589-nm monochromatic light is used by the interferometer, how many fringes will "sink" into the bull's eye when the drive screw is turned through one turn?

21. The free end of a steel rod 60 cm long connects to the movable mirror of a Michelson interferometer. The rod expands as it is heated and causes the mirror to move. Calculate the number of fringes that "sink" per second as the temperature of the rod increases at the rate of 1°C/min. (Assume $\lambda = 589$ nm.)

22. An air prism is made by cementing together thin sheets of clear plastic to form an equilateral prism whose interior is filled with air. It is then immersed in a shallow tray filled with water. A laser beam is incident on one face at an angle to the normal of 10°. Use Snell's law directly to calculate the deviation of the beam as it passes through the prism. Make a sketch.

CHAPTER 24

ELECTROSTATICS

This chapter marks the beginning of our study of electricity, magnetism, and electronics. In this chapter we study the basic properties of electrical charge, the electrostatic field, and electrostatic potential, all under static (stationary) conditions. The concept of capacitance is introduced and the static properties of capacitance are presented.

We do not consider batteries and electric current explicitly in this chapter. These topics appear in Chap. 25. However, we do use a battery to charge a capacitor. We trust everyone is aware that a battery is an "electron pump" and an electric current is a "flow of electrons." If this is the case, there should be no difficulty experienced with the material presented in this chapter.

24-1 ELECTRIC CHARGE

There are two types of electric charge. One type is *positive*; the other is *negative*. The basic unit of positive charge is carried by *protons*; the basic unit of negative charge is carried by *electrons*. All protons carry exactly the same amount of positive charge, and all electrons carry exactly the same amount of negative charge. The amount of positive charge carried by the proton is equal in magnitude to the amount of negative charge carried by the electron. Consequently, one unit of negative charge can combine with and neutralize one unit of positive charge.

A hydrogen atom consists of one proton that is orbited by one electron. Hence, the atom contains one positive and one negative charge so that the atom, overall, is electrically neutral. A neutron is a very tightly bound combination of an electron and a proton, that, consequently, is electrically neutral. Any assembly of particles that contains an equal number of positive and negative charges will be, overall, electrically neutral.

Electrical charges have the peculiar ability to exert forces on each other. Like charges repel each other, and unlike charges attract each other. Thus, electrons repel other electrons, and protons repel other protons. On the other hand, electrons and protons attract each other. The magnitude of the force is given by Coulomb's law, described later.

The fundamental nature of electric charge is not presently understood. How an electron *carries* its charge is unknown, just as the fundamental structure of the electron is unknown (if such a thing as a "fundamental structure" even has meaning.) Nevertheless, without knowing the details, we can describe electrical phenomena in terms of the behavior and properties of electrons and protons. Other subatomic particles are imbued with electrical charge, but we need not concern ourselves with that here. In any event, no single discrete charge has ever been observed that is smaller in magnitude than the charge on an electron.

To observe the effects of electric charge, some means must be found for segregating the negative and positive charges. One way to achieve this is to rub a glass rod vigorously with a silk cloth. The rod loses electrons to the cloth, and this disturbs the normal balance of charges on the rod. The rod becomes positively charged because of its loss of negative electrons that originally were paired with positive protons contained in the rod. In a similar manner, a rubber or a plastic rod will gain electrons and become negatively charged when rubbed with a piece of fur. This is illustrated in Fig. 24-1.

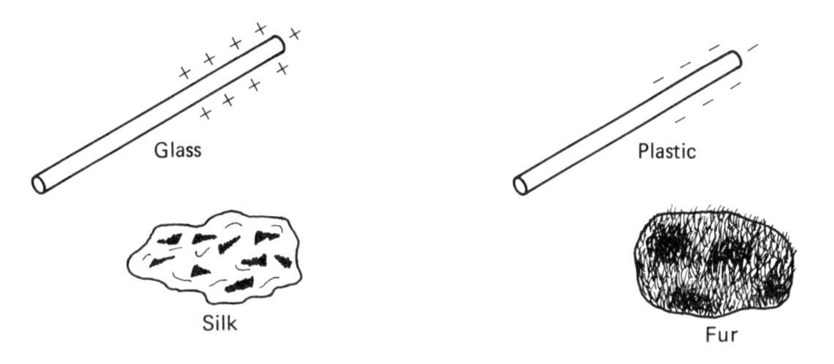

Figure 24-1 A glass rod rubbed with silk becomes positively charged. A plastic rod rubbed with fur becomes negatively charged.

Conductors and insulators

At this point we need to clarify how electrical charge is able to move through various materials. First of all, when we speak of the movement of charge, we are really speaking of the movement of electrons or protons because these particles carry the charge. Furthermore, electrons move much more freely than protons. Protons are heavy particles found in the nucleus of an atom, whereas electrons are very light particles that orbit the nucleus. Consequently, almost all movement of charge in typical experiments with static electricity is associated with the movement of the negatively charged electrons.

Atoms of metals typically have one or more electrons in orbits on the outer periphery of the atom. These outermost, *valence* electrons are not as tightly bound to the nucleus as the inner electrons. As a matter of fact, when a large number of copper atoms, for example, are joined together as in a tiny chunk of copper, the valence electrons forsake the individual atoms and become a part of the "electronic system" of the metal. The valence electrons no longer traverse orbits around single atoms; they travel in much larger paths that weave complex patterns which might extend throughout the entire chunk of metal of which they are a part. Thus, the electrons in a metal are extemely mobile; for this reason, electrons travel through metals with ease, and we say that metals are good *conductors*.

In nonmetals, however, the electrons remain bound to their individual host atoms and are not free to move. These materials, through which electrons move with great difficulty, are called *insulators*. Typical insulators are glass, plastics, rubber, wood, and chalk. The electronic structures of a metal and of an insulating material are illustrated in Fig. 24-2.

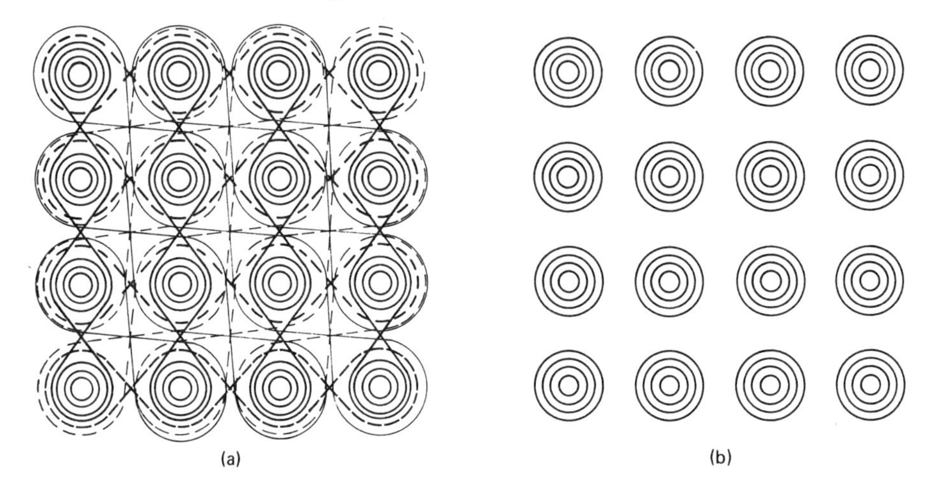

Figure 24-2 Representations of the electronic structures of: (a) a metal; (b) an insulator.

The ease with which electrons move through a material is given by the *conductivity* of the material, which we define later. The inverse of conductivity is the *resistivity*. The actual conductivity of a material depends upon many factors and is not as simple as the explanation above would lead us to believe. Indeed, conductivities of various materials span a range greater than 10^{24}. At this point, however, we need realize only that electrons move quite easily through conductors but are essentially immobile in insulators.

Static electricity

Because the word *static* means *stationary*, we might infer, correctly, that *static electricity* refers to electrical charges that are not moving. This situation is exactly opposite to that in which an *electric current* flows through a wire. An electric current is the *movement* of electric charge. We study electric current in more detail in a later chapter.

The static nature of static electricity can be demonstrated by touching a negatively charged rod to a sheet of Teflon, as shown in Fig. 24-3. It is possible to establish several "piles" of electrons on the Teflon sheet by touching the negatively charged rod at various points. Moreover, the piles of charge will

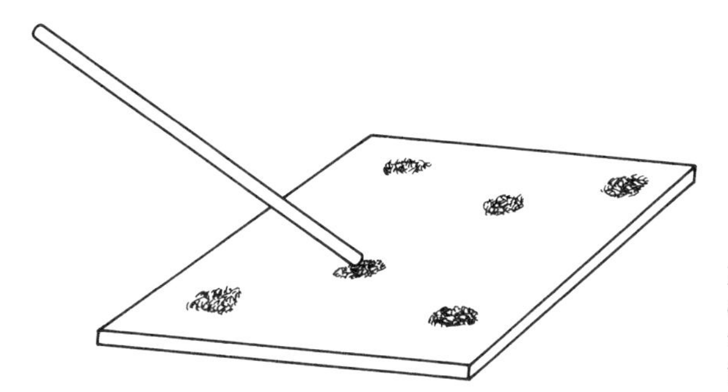

Figure 24-3 Piles of electrons may be established on a sheet of Teflon or plastic by touching a negatively charged rod to the sheet at various points.

remain for a considerable time because the electrons cannot flow through or across the Teflon, which is a very good insulator. Exactly how long the piles stay put will depend upon the cleanliness of the Teflon surface and upon the humidity of the air. Generally, static electricity experiments do not work very well on humid days because the electrons tend to leak off into the air.

How do we detect a static charge? How can we tell if we have a pile of electrons on the end of a rod or on the surface of a sheet of Teflon? In the traditional manner, we might use an *electroscope*, as shown in Fig. 24-4. On the

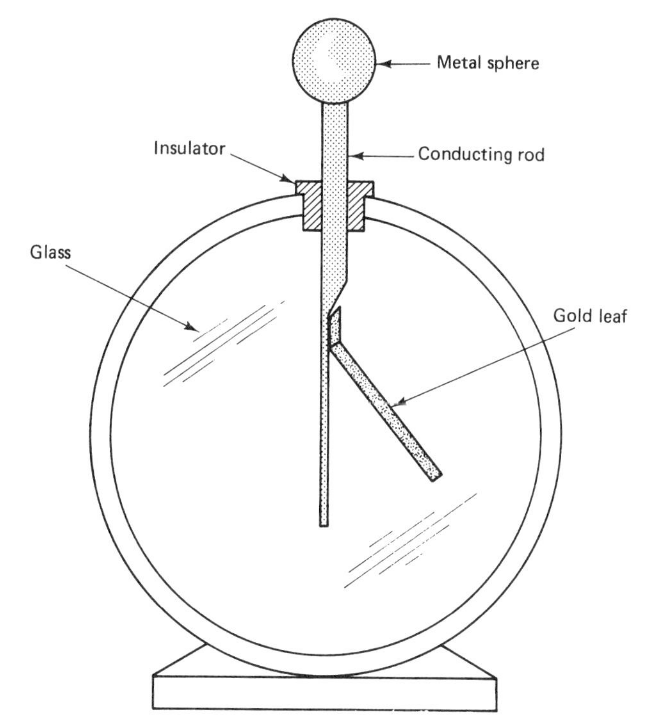

Figure 24-4 A gold-leaf electroscope.

other hand, modern *electrometers* are available that detect and measure the intensity of static charges when a small probe is held near or is touched to the region of the charge.

The *gold-leaf electroscope* makes use of the fact that like charges repel each other. When a charge is imparted to the gold leaf and its metal support, the

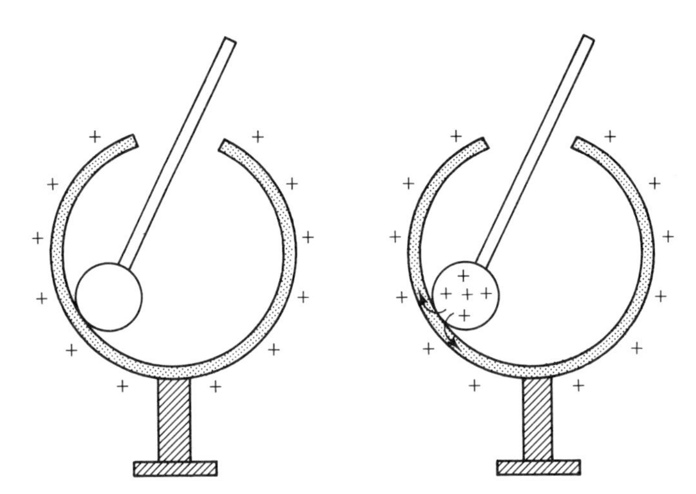

Figure 24-5 The Faraday ice pail experiment demonstrates that an electric charge resides on the surface of a conductor.
(a) A small disk touched to the inside of a charged sphere does not become charged.
(b) Any charge introduced to the interior of the sphere quickly flows to the outer surface.

charged leaf is repelled, and it rises to an angle that indicates the magnitude of the charge. Other types of electroscopes are available that work on the same principle, but the gold-leaf electroscope is the most sensitive.

In use, with the electroscope initially uncharged, a charge is transferred to the electroscope via a small metal disk attached to an insulating handle. The disk is first touched to the charged area and is then touched to the electroscope. The charge picked up by the disk is transferred to the electroscope, and the gold leaf deflects accordingly.

Charged metal sphere

A static charge may be imparted to a metal sphere by touching the sphere with a charged rod, and the charge may be either positive or negative. Because like charges repel each other, the charge imparted to a sphere will reside totally on the surface of the sphere. This results from the attempt of each electron (in the case of a negative charge) to get as far away from all the other electrons as possible. The interior of the metal sphere remains oblivious to the charge on the surface; the sphere can even be hollow without affecting the behavior of the charges on its surface.

This behavior was discovered by Michael Faraday (1791–1867) in conjunction with the ice pail experiment, a modern version of which is illustrated in Fig. 24-5.

Charging by induction

A negative rod can be used to induce a positive charge on a metal sphere, as shown in Fig. 24-6. The negative rod, when held near one side of the sphere, forces electrons to the opposite side. Touching the sphere with a finger or a wire

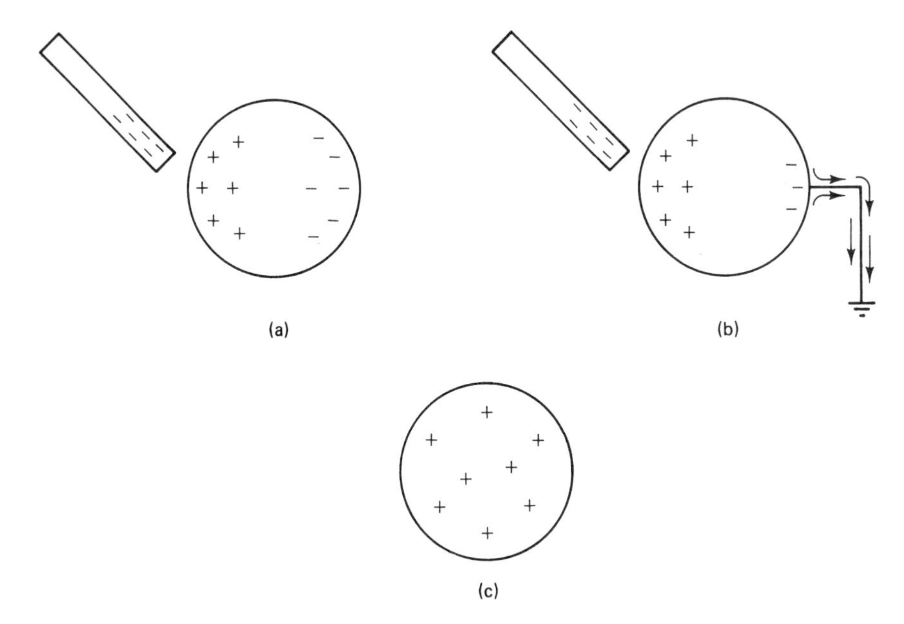

Figure 24-6 Charging a sphere by induction.
(a) A negative rod held near the sphere causes a separation of charge to occur.
(b) Electrons escape to ground when the sphere is touched by a conductor.
(c) The sphere is left with a positive charge when the rod is removed.

connected to ground then allows the electrons to escape from the sphere. When the negatively charged rod is removed, the sphere is left with deficiency of electrons, a net positive charge that becomes uniformly distributed over the surface of the sphere.

24-2 COULOMB'S LAW

In the laboratory, electrostatic charges are conveniently handled by placing the charges on metal spheres. A charge on a sphere may be cut in half by touching the sphere with an identical sphere that is initially uncharged. Thus, it is possible to manipulate charges without having sophisticated equipment, and this makes it possible to determine the quantitative nature of the force that one charge exerts on another.

In 1784, Charles Augustin de Coulomb (1736–1805) announced what is now known as Coulomb's law:

> The magnitude of the force existing between two electrostatic charges is directly proportional to the product of the magnitudes of the charges and is inversely proportional to the square of the distance between the charges. The forces lie on a line connecting the two charges and are repulsive if the charges are of the same polarity and are attractive if the charges are of opposite polarity.

In mathematical terms, Coulomb's law is

$$F = k\frac{q_1 q_2}{R^2} \tag{24-1}$$

where q_1 and q_2 are the magnitudes of the charges, R is the distance between the charges, and k is a constant of proportionality. The numerical value of k depends upon the system of units used. We now must investigate the units used to measure electrical charge. We shall use the SI system.

The SI unit of electrical charge is the coulomb (C). One coulomb of charge is the total charge carried by about 6.24192×10^{18} electrons (or protons). Hence, the charge e carried by one electron or by one proton is

$$e = 1.60207 \times 10^{-19}\ \text{C} \tag{24-2}$$

If the force F is measured in newtons while the distance R is measured in meters, the proportionality constant k turns out to have the value

$$\begin{aligned} k &= 8.98742 \times 10^9\ \text{N} \cdot \text{m}^2/\text{C}^2 \\ &\approx 9 \times 10^9\ \text{N} \cdot \text{m}^2/\text{C}^2 \end{aligned} \tag{24-3}$$

This value was determined by experiment using a torsion balance in a system very similar to that used by Lord Cavendish to determine the gravitational constant G. Of course, charged metal spheres were used instead of the large masses used in the Cavendish apparatus. Incidentally, note the similarity between Coulomb's law and the law of gravitation [Eq. (9-1)]. Both involve a direct proportion to a product and an inverse proportion to the square of a distance. However, we must not surmise that the similarity of the two laws is in any way connected to the similarity of the experimental apparatus used to verify the laws.

In order to simplify the proportionality constants found in many equations encountered in advanced electrostatics, Coulomb's law is often written as

$$F = \frac{1}{4\pi\epsilon_0} \cdot \frac{q_1 q_2}{R^2} \tag{24-4}$$

where k has been replaced by

$$k = \frac{1}{4\pi\epsilon_0} \tag{24-5}$$

and where ϵ_0 is called the *permittivity of free space*. Its value is

$$\epsilon_0 = \frac{1}{4\pi k} = 8.85418 \times 10^{-12}\ \text{C}^2/\text{N} \cdot \text{m}^2 \tag{24-6}$$

Inclusion of the 4π into the definition of ϵ_0 is called *rationalization*; hence, we have the *rationalized MKS system* of units, the previous name of the SI system.

EXAMPLE 24-1 Two small metal spheres, separated by a distance of 1 m, carry 1 C of negative charge each. Compute the repulsive force that each sphere exerts on the other.

Solution. Direct application of Coulomb's law gives

$$\begin{aligned} F &= k\frac{q_1 q_2}{R^2} \\ &= (9 \times 10^9\ \text{N} \cdot \text{m}^2/\text{C}^2)\frac{(1\ \text{C})(1\ \text{C})}{(1\ \text{m})^2} \\ &= 9 \times 10^9\ \text{N} = 9 \text{ billion newtons!} \end{aligned}$$

The force computed in the preceding example is an extremely large force, and we realize that such an experiment could not be performed in the lab. Our conclusion is twofold: First, electrostatic forces are very strong; second, a coulomb of charge is a very large amount of charge. A microcoulomb (μC) is a more practical unit for laboratory use. By computing the gravitational force and the electrostatic force between an electron and a proton separated an arbitrary distance, it is found that the electrical force is about 10^{39} times as strong as the gravitational force. Even so, electrical forces are not the strongest forces found in nature; short-range nuclear forces are even stronger. Otherwise, how would atomic nuclei with more than one proton keep from flying apart?

24-3 THE ELECTRIC FIELD

Suppose a positive charge Q of physically small diameter is located in a region that is otherwise free of electrical charge except for one very much smaller positive test charge q_t. We assume the test charge to be movable to any point in the region. When it is brought near the larger charge, a repulsive force is exerted on it by the larger charge. We can represent the magnitude and direction of the force on q_t by a vector located at the same point whose length is proportional to the magnitude of the force and which points in the direction of the force. If we place the test charge at various points in the vicinity of the larger charge and draw the vector for each point, we might obtain an array of vectors surrounding the larger charge like the one shown in Fig. 24-7. This array of vectors describes the *electric field* surrounding the larger charge.

If a large number of vectors are drawn so that near the head of each lies the tail of another, the vectors will appear to form an array of lines surrounding the charge. The lines will indicate the direction of the force exerted on the test charge. They are called *lines of force*, and they provide a convenient method for representing an electric field, as shown in Fig. 24-7.[1]

The magnitude, or strength, of the electric field E is defined to be the ratio of F_t to q_t, where F_t is the force exerted on the test charge whose magnitude is q_t. That is,

$$E = \frac{F_t}{q_t} \tag{24-7}$$

where F_t is given by Coulomb's law—

$$F_t = k\frac{Qq_t}{R^2} \tag{24-8}$$

and R is the distance from the center of charge Q to the test charge q_t. When Eqs. (24-7) and (24-8) are combined, we obtain a formula for the magnitude of the electric field surrounding a point charge:

[1]The concept of using lines of force to represent an electric (or magnetic) field was originated by Michael Faraday, an English scientist. A victim of extreme poverty, Faraday became an apprentice to a bookbinder at age 14. This proximity to books led to his self-education and an interest in science. He spent most of his life at the Royal Institution, founded by Count Rumford, and made many important discoveries in electromagnetism and chemistry.

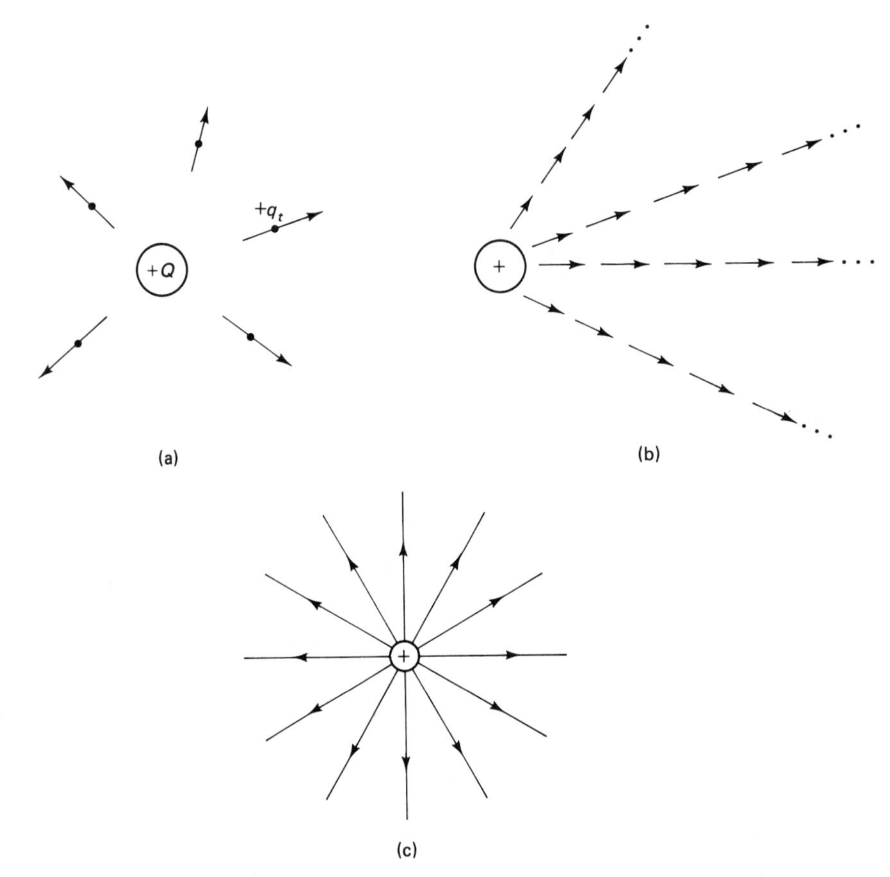

Figure 24-7 (a) A test charge q_t is acted upon by a force represented by the vector.
(b) A large number of vectors can be drawn to represent the forces exerted upon q_t at various locations.
(c) The arrows (vectors) may be connected to form a representation of the electric field around Q.

$$E = \frac{kQ}{R^2} \tag{24-9}$$

Observe that q_t, the magnitude of the test charge, does not appear in the above equation, and note the inverse square dependence upon R. Even though we obtained this formula by considering a point charge, the formula is equally applicable to charged metal spheres or any spherically symmetric distribution of charge. When so applied, R is the distance from the center of the spherical distribution of charges to the point of observation.

EXAMPLE 24-2 A metal-coated Ping-Pong ball suspended by an insulating thread is given a charge of 1 μC. Compute the electric field a distance of 1 m from the center of the ball.

Solution. Direct application of Eq. (24-9) yields

$$E = k\frac{Q}{R^2}$$

$$= (9 \times 10^9 \text{ N} \cdot \text{m}^2/\text{C}^2) \left(\frac{1 \times 10^{-6} \text{ C}}{(1 \text{ m})^2}\right)$$

$$= 9 \times 10^3 \text{ N/C}$$

E as force per unit charge

Suppose the electric field at a certain point is known to have a certain magnitude. We can then readily calculate the force that will be exerted on a charge placed at that point. The formula is

$$F = qE \tag{24-10}$$

This relationship stems from Eq. (24-7), which implies that the electric field strength E at a certain point is the electrostatic force exerted on a unit charge placed at the point in question.

Note from Example 24-2 that the electric field is given in units of newtons per coulomb. This is reasonable because the electric field is defined in terms of force per unit charge in Eq. (24-7).

The usefulness of the electric field lines for describing an electric field may be surmised from Fig. 24-8, in which the electric field resulting from several configurations is illustrated.

EXAMPLE 24-3 A charge of 8 μC is located at a point where the electric field is 4×10^3 N/C. Calculate the force exerted on the charge.

Solution. Direct application of Eq. (24-10) gives

$$F = qE = (8 \times 10^{-6} \text{ C})(4 \times 10^3 \text{ N/C})$$

$$= 32 \times 10^{-3} \text{ N}$$

Vector nature of the electric field

The electric field is a vector quantity because force is a vector quantity. Therefore, if two or more charges contribute to the electric field at a certain point, the direction of the force associated with each charge must be taken into account and the forces must be combined vectorially.

To illustrate this, consider the two charges shown on the x-axis in Fig. 24-9. A positive charge of 2 μC is located at $x = -3$ m, and a negative charge of -1 μC is located at $x = 4$ m. Let us find the magnitude and direction of the electric field at a point A on the y-axis, at $y = 6$ m.

First, we compute the magnitudes of E_1 and E_2 using Eq. (24-9):

$$E_1 = k\frac{Q_1}{R_1^2} = \left(9 \times 10^9 \frac{\text{N} \cdot \text{m}^2}{\text{C}^2}\right) \left[\frac{1 \times 10^{-6} \text{ C}}{(7.21 \text{ m})^2}\right] = 173 \text{ N/C}$$

$$E_2 = k\frac{Q_2}{R_2^2} = \left(9 \times 10^9 \frac{\text{N} \cdot \text{m}^2}{\text{C}^2}\right) \left[\frac{2 \times 10^{-6} \text{ C}}{(6.71 \text{ m})^2}\right] = 400 \text{ N/C}$$

The distances $R_1 = 7.21$ m and $R_2 = 6.71$ m were obtained by applying the Pythagorean theorem to the geometry indicated in the figure. Further, the angles

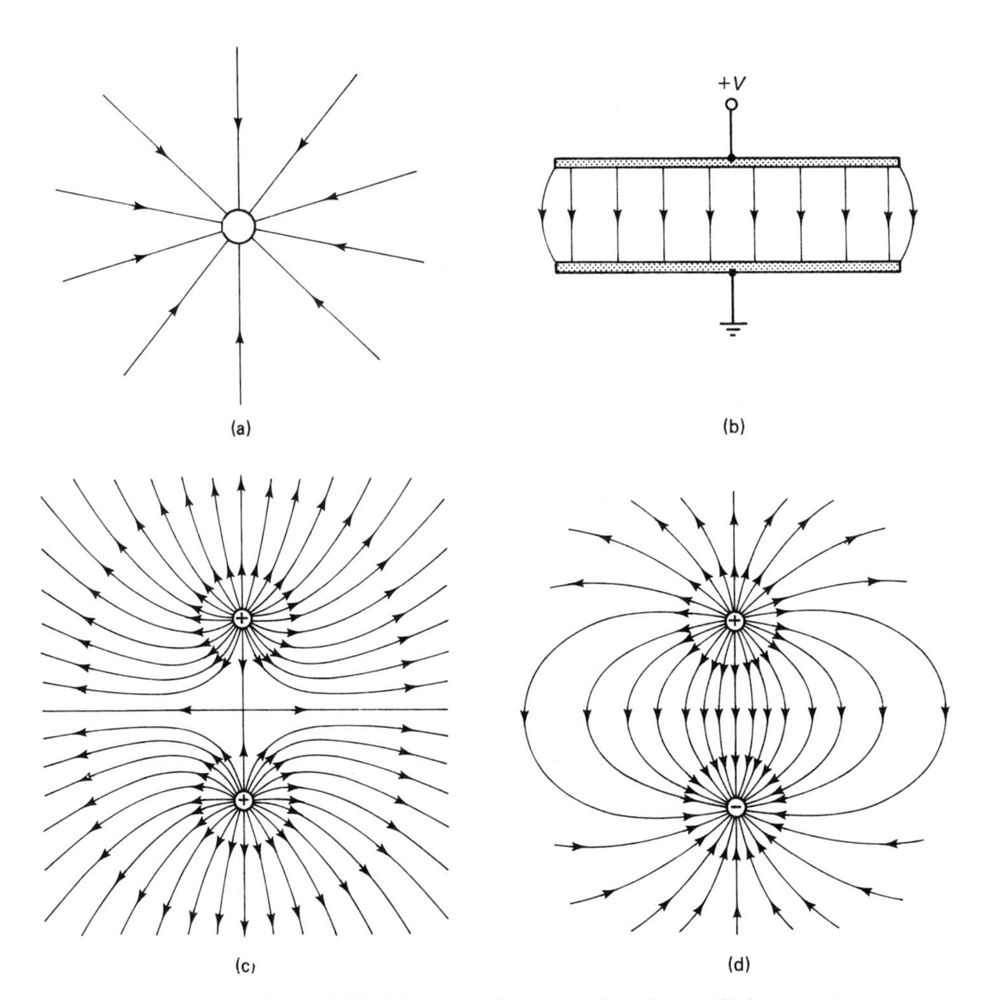

Figure 24-8 Electric field: (a) surrounding a negative charge; (b) between two parallel plates; (c) surrounding two like charges; (d) surrounding two unlike charges.

are determined using simple trigonometry and are shown in the figure. Observe that E_2 is directed away from Q_2 (which is positive), while E_1 is directed toward Q_1 (which is negative).

The electric field at A is the vector sum of E_1 and E_2. Summing components gives

$$\begin{aligned} E_x &= E_1\cos 56.3 + E_2\cos 60 \\ &= (173)(0.55) + (400)(0.5) = 295 \text{ N/C} \\ E_y &= E_1\sin 56.3 + E_2\sin 60 \\ &= -(173)(0.83) + (400)(0.87) = 204 \text{ N/C} \end{aligned}$$

Then,

$$E = \sqrt{E_x^2 + E_y^2} = 359 \text{ N/C}$$

$$\theta = \text{Tan}^{-1}\left(\frac{E_y}{E_x}\right) = 34.7°$$

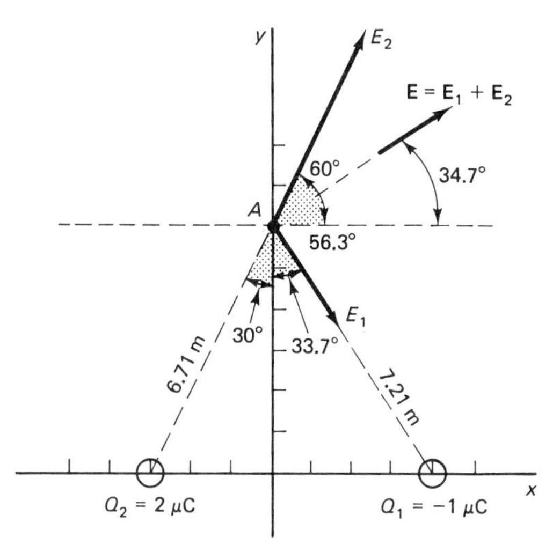

Figure 24-9 Geometric illustration of the vector nature of the electric field.

Thus, the electric field at point *A* has a strength of 359 N/C and is directed at an angle of 34.7° to the *x*-axis.

24-4 ELECTROSTATIC POTENTIAL

In our study of gravity we saw that when a body of mass *m* is lifted a distance *h*, work is done in the amount of *mgh* by the external agent in lifting the body. Subsequently, the object has potential energy equal to the work done in lifting the body. A similar principle is applicable in electrostatics.

Suppose a small test charge, $+q_t$, is located a very great distance from a larger charge, $+Q$. The force of repulsion between the two charges will be negligible because of their great separation. But if q_t is acted upon by an external agent and is brought closer to Q, the repulsive force increases in magnitude; q_t has to be pushed in the direction of Q. Consequently, work is done in forcing q_t toward Q, and q_t gains energy in the process.

When q_t is located close to Q, q_t has electrostatic potential energy because of the work that was done on it in bringing it to its location near Q. If q_t is released, the force of repulsion will cause q_t to move away from Q, accelerating as it goes. As q_t gains speed, the electrostatic potential energy it had initially is converted to kinetic energy.

If calculus is used to sum up the work done on a test charge q_t in bringing it from infinity to a point a distance R from another charge Q, we obtain the following expression for the electrostatic potential energy (denoted U) possessed by the two charges:

$$U = k\frac{Q \cdot q_t}{R} \qquad (24\text{-}11)$$

With this expression at hand we can go on to define the electrostatic potential.

Though Eq. (24-11) was obtained by considering two point charges, it is not limited in application to point charges. It may be applied to any pair of spherically symmetric charge distributions with R then being the distance from the center of one to the center of the other. This fact arises from the dependence of Coulomb's law on $1/R^2$, and it parallels a similar phenomenon for gravity discussed in Sect. 9-1.

The *electrostatic potential* at a point is defined as the electrostatic potential energy that one unit of charge has when placed at that point. The electric potential V a distance R from the center of a spherically symmetric distribution of charges is given by

$$V = k\frac{Q}{R} \tag{24-12}$$

where V is the electrostatic potential and where Q is the total charge in the distribution. The distance R is assumed to be greater than the radius of the charge distribution so that only points outside the region are considered.

Observe that all points equidistant from the center of a spherical distribution of charge have the same electrostatic potential. The locus of points having the same potential form *equipotential surfaces* in the form of spheres concentric with the charge distribution. In two dimensions, the *equipotentials* form circles, as shown in Fig. 24-10. Note, as is always the case, that the equipotential lines intersect the lines of the electric field at right angles.

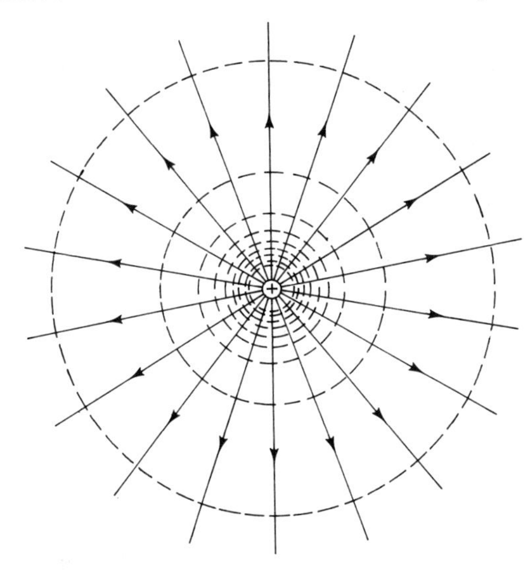

Figure 24-10 A two-dimensional representation of the equipotential surfaces (dashed lines) surrounding a charged sphere.

If a charge q is located at a point where the electrostatic potential is V, the potential energy possessed by the charge is given by

$$PE = qV \tag{24-13}$$

The unit of electrostatic potential is joules per coulomb, which is in the form of energy per unit charge. For brevity, 1 J/C is called a volt: 1 V = 1 J/C.

EXAMPLE 24-4 Suppose a small brass sphere 3 cm in diameter carries a positive charge of 0.05 μC. Calculate the magnitude of: (a) the electric field, (b) the electrostatic potential 9 cm from the center of the charge.

Solution. We use Eqs. (24-9) and (24-12) in a straightforward manner:

(a)
$$E = k\frac{Q}{R^2}$$
$$= (9 \times 10^9 \text{ N} \cdot \text{m}^2/\text{C}^2)\frac{0.05 \times 10^{-6} \text{ C}}{(0.09 \text{ m})^2}$$
$$= 5.56 \times 10^4 \text{ N/C}$$

(b)
$$V = k\frac{Q}{R}$$
$$= (9 \times 10^9 \text{ N} \cdot \text{m}^2/\text{C}^2)\frac{0.05 \times 10^{-6} \text{ C}}{0.09 \text{ m}}$$
$$= 5.00 \times 10^3 \text{ N} \cdot \text{m/C} = 5 \times 10^3 \text{ J/C}$$
$$= 5 \times 10^3 \text{ V}$$

EXAMPLE 24-5 An electron-volt (eV) is a unit of energy equivalent to the energy possessed by an electron at a point where the electrostatic potential is 1 V. Express this amount of energy in joules.

Solution. We use Eq. (24-13) written in terms of the electronic charge e:

$$PE = eV \qquad (e = 1.60219 \times 10^{-19} \text{ C})$$
$$= (1.60219 \times 10^{-19} \text{ C})(1 \text{ V})$$

Hence,

$$1 \text{ eV} = 1.60219 \times 10^{-19} \text{ J}$$

Potential difference

In practical electricity and electronics, the potential difference (PD) between points in the circuit is perhaps the most often measured parameter. The PD is measured in volts and often is simply called the *voltage*. The PD existing between two points is the difference in electrostatic potential that exists between the two points.

In more formal language, we say that the potential difference between points A and B is 1 V if 1 J of energy is required to move 1 C of charge from one point to the other.

In practical circuits, A and B will be represented by solder joints, connecting posts, or some other type of conductor. A *voltmeter* connected between the two points will indicate the potential difference. The two points, however, do not have to be part of a circuit; they may be points in space near a charged sphere, for example. The concept of potential difference is still a valid concept even when A and B are in empty space.

EXAMPLE 24-6 An ordinary 12-V battery develops a potential difference of approximately 12 V between its two terminals. Compute the energy required to move 4 C of negative charge (electrons) from the positive terminal to the negative terminal.

Solution. We use the following relationship obtained from Eq. (24-13):

$$\text{Energy} = \text{charge} \times \text{PD}$$
$$= (4\ \text{C})(12\ \text{V}) = 36\ \text{C} \cdot \text{V}$$
$$= 36\ \text{J}$$

The path taken by the charge in traveling from one terminal to the other is unimportant as far as this calculation is concerned.

EXAMPLE 24-7 A small metal sphere is negatively charged; the amount of charge on the sphere is 0.001 μC. Calculate the electrostatic potential: (a) 50 cm from the sphere; (b) 25 cm from the sphere. (c) Compute the energy required to move an electron from the 50-cm point to the 25-cm point.

Solution. For parts (a) and (b), we use Eq. 24-11: $V = k(Q/R)$.

(a) $$V_a = (9 \times 10^9)\frac{0.001 \times 10^{-6}\ \text{C}}{0.5\ \text{m}} = 18\ \text{V}$$

(b) $$V_b = (9 \times 10^9)\frac{0.001 \times 10^{-6}\ \text{C}}{0.25\ \text{m}} = 36\ \text{V}$$

(c) The potential difference between the two points is 18 V. Therefore, the energy required to move the electron is given by

$$\text{Energy} = e \times \text{PD}$$
$$= (1.602 \times 10^{-19}\ \text{C})(18\ \text{V})$$
$$= 2.884 \times 10^{-18}\ \text{J} = 18\ \text{eV}$$

Note that this is 18 eV. Whenever an electron moves through a potential difference of 1 V, the energy it gains or loses (depending upon the direction) is one eV, by definition.

Absolute potential; ground

Recall that we determine the electrostatic potential at a point (conceptually) by bringing a small test charge from infinity to the point in question, summing the work done on the charge as it is moved. This is quite impossible to do, practically speaking. Consequently, the *absolute potential* at a point is difficult—if not impossible—to determine. For example, how could we determine the absolute potential of the earth?

In practical applications and in lab demonstrations and experiments, the important parameter is potential difference. We may have a metal sphere charged to a potential of 50,000 V, but we cannot be sure that the absolute potential of the sphere is 50,000 V. We can say only that the potential of the sphere is 50,000 V relative to the rest of the lab. For all we know, the entire lab and the earth on which it sits may be charged to 100,000 V, so that the absolute potential of the sphere is 150,000 V. However, only the difference in potential is important and observable to the people working in the lab. The sphere will behave as if its absolute potential were 50,000 V and as if the absolute potential of the walls, floor, and fixtures of the lab were zero.

Airplanes in flight pick up static charges of thousands of volts due to air friction, but the passengers inside the plane remain unaware of their high electrostatic potential. Short, resistive wires attached to the trailing edge of the wings help to drain the charge back to the air so that sporadic sparking—which might otherwise interfere with radio communications—is minimized.

The reference potential in practical electricity, electrostatics, and electronics is called *ground*. Frequently, the ground is achieved by a physical connection to the earth, forming an *earth ground*. In other situations, the ground may be the potential of the metal frame of an automobile or plane, the metal chassis of a radio receiver, or a conducting foil on the printed circuit board of an electronic circuit. In each case, potentials (voltages) are measured relative to ground. Grounds that are not connected to earth ground are said to be *floating grounds*.

24-5 CAPACITANCE

Suppose we have two metal spheres, one of radius 10 cm and the other of radius 20 cm. Further, let us suppose we place a charge of 0.001 μC on each sphere. Our point of concern is the electrostatic potential that the same charge produces on spheres of unequal radius. Will the same charge produce the same potential on a large sphere as on a small sphere?

First, we must consider an important detail. Faraday's ice pail experiment (Fig. 24-5) demonstrates that the charge imparted to a conducting sphere resides totally on the surface of the sphere. Now we propose to calculate the electrostatic potential of that same surface. More advanced texts show that it is valid to consider all the free surface charge to be concentrated at the center of the sphere, so we may then treat the real surface as an equipotential surface. This allows us to use Eq. (24-12) to calculate the potential obtained by a conducting sphere when it is given a charge Q. We now return to the proposal of imparting the same charge to spheres of different diameter.

A simple calculation using Eq. (24-12) indicates that when 0.001 μC is introduced onto each sphere, the large sphere obtains a potential of 45 V, but the smaller sphere obtains a potential of 90 V. Thus, the smaller sphere suffers the greater rise in potential when the same charge is placed upon it.

Let us now approach this situation from the opposite point of view. Suppose the two spheres are both charged to a potential of 1000 V. What amount of charge must be placed on each sphere to produce this rise in potential?

By rearranging Eq. (24-11) and solving for Q in terms of V, we obtain, for the charge Q contained on a metal sphere of radius R and potential V,

$$Q = \frac{RV}{k} \tag{24-14}$$

Using this formula, we find that a charge of 0.0111 μC will cause a rise in potential of 1000 V on a sphere of radius 10 cm. On a sphere of radius 20 cm, a charge of 0.0222 μC is required in order to produce the same rise in potential.

From this we see that the abilities of the two spheres to hold a charge are different. This leads us to the concept of capacitance. *Capacitance* refers to the ability of a conductor to accept a charge without suffering a large increase in potential. More specifically, the capacitance C is the ratio of the charge placed on a conductor to the rise in potential V that the charge produces. Put differently, it is the charge transferred per unit increase in potential. Hence, by definition of capacitance,

$$C = \frac{Q}{V} \tag{24-15}$$

From this definition we discern that the unit of capacitance is coulombs per volt. This combination is called a farad (F), in honor of Michael Faraday: 1 F = 1 C/V. Thus, we say that the capacitance of a conductor, such as a sphere, is 1 F if a charge of 1 C placed on the conductor produces a rise in potential of 1 V. The farad is a very large unit of capacitance, as we shall see in a following example.

By applying the definition of capacitance to a sphere, we can obtain a formula for the capacitance of a conducting sphere isolated in space. When a charge Q is placed on the sphere, a potential given by Eq. (24-12) results. Hence,

$$C = \frac{Q}{V} = \frac{Q}{kQ/R} \tag{24-16}$$

$$= \frac{R}{k} \qquad \text{(for a conducting sphere)} \tag{24-17}$$

where C is in farads, R is in meters, and $k \approx 9 \times 10^9 \text{ N} \cdot \text{m}^2/\text{C}^2$.

If k is written in terms of ϵ_0 as in Eq. (24-5), the formula for the capacitance of a sphere becomes

$$C = 4\pi\epsilon_0 R \qquad \epsilon_0 = 8.85 \times 10^{-12} \text{ C}^2/\text{N} \cdot \text{m}^2 \tag{24-18}$$

EXAMPLE 24-8

Compute the radius of a metal sphere whose capacitance is 1 F.

Solution. From Eq. (24-17), $R = kC$. Hence, for $C = 1$ F,

$$R = (1 \text{ F})(9 \times 10^9 \text{ N} \cdot \text{m}^2/\text{C}^2)$$

$$= 9 \times 10^9 \text{ m}, \quad \text{or} \quad 9 \text{ billion meters!}$$

This is a very large sphere indeed. It is almost 13 times as large in diameter as the sun!

24-6 THE PARALLEL-PLATE CAPACITOR

In the previous section we introduced the concept of capacitance by considering an isolated conducting sphere. We now consider the capacitance that exists between two conductors separated by some form of electrical insulation. The insulation is called the *dielectric*, and the overall assembly is called a *capacitor*. Capacitors are widely used in electricity and electronics. (Capacitors were once called *condensers*, but this term is now obsolete.)

One of the simplest types of capacitors is the *parallel-plate capacitor* shown in Fig. 24-11. The dielectric may consist of a variety of insulating materials,

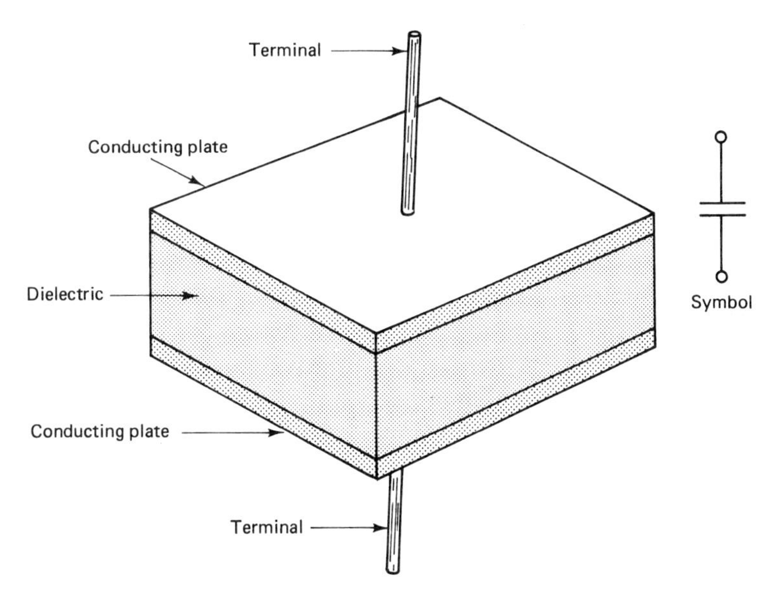

Figure 24-11 A parallel-plate capacitor and the symbol used to denote a capacitor in a electronics circuit. The lateral dimensions of the plates are assumed to be very large in comparison with the thickness of the dielectric.

including air or vacuum. When the dielectric is a vacuum, the capacitance is given by

$$C = \epsilon_0 \frac{A}{d} \tag{24-19}$$

where A is the inside area of one of the plates, d is the separation of the plates, and ϵ_0 is the permittivity of free space, given earlier.

A capacitor is *charged* by transferring electrons from one conducting plate to the other. One way to charge a capacitor is to use a battery, as shown in Fig. 24-12. The battery pulls electrons from one plate and forces them onto the other plate. The plate from which the electrons are removed becomes positively charged, and the plate receiving the electrons becomes negatively charged. The *charge on the capacitor* is the quantity of electrical charge carried by the electrons that are transferred. In other words, it is the charge on *one* of the plates.

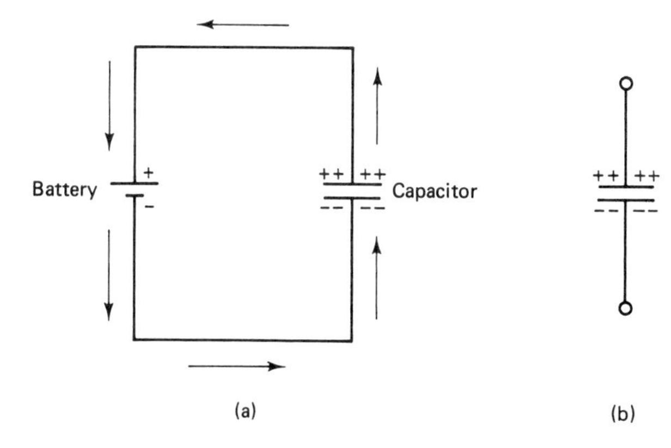

Figure 24-12 Using a battery to charge a capacitor. The arrows in (a) indicate the direction of electron flow during the charging process. A capacitor remains charged even after the battery is disconnected, as shown in (b). Note the symbol used to denote a battery, and observe how it differs from the symbol used to depict a capacitor.

Whenever a capacitor is given a charge Q, a voltage V develops between the plates of the capacitor. The relationship that relates Q, V, and C is

$$Q = CV \tag{24-20}$$

where C is the capacitance of the capacitor. Of course, Q is measured in coulombs, C in farads, and V in volts.

EXAMPLE 24-9 An 8-μF capacitor is charged until a PD of 100 V exists between its plates. Compute the quantity of charge that is transferred from one plate to the other during the charging process.

Solution. We use Eq. (24-20).

$$\begin{aligned} Q &= CV \\ &= (8 \times 10^{-6} \text{ F})(100 \text{ V}) \\ &= 800 \times 10^{-6} \text{ C} \\ &= 800 \ \mu\text{C} \end{aligned}$$

E-field inside a parallel-plate capacitor

Generally speaking, an electric field arises whenever positive and negative charges become segregated. A parallel-plate capacitor (with vacuum dielectric, for simplicity) provides a good example of this: An electric field is established between the plates when the capacitor is charged. The strength of the field is proportional to the charge on the capacitor; therefore, the field is proportional to the voltage V to which the capacitor is charged. [See Fig. 24-8(b).]

If the distance d between the plates is small in comparison to their lateral dimensions, the field between the plates will be uniform (except near the edges where fringing occurs). With this bit of knowledge, we can obtain an expression for the field strength E by considering the work done in moving a small, positive test charge q_t from the negative plate across the dielectric to the positive plate.

According to Eq. (24-13), the change in potential energy of q_t in being moved across the dielectric from one plate to the other is

$$PE = q_t \cdot V \tag{24-21}$$

While q_t is between the plates, a force $q_t \cdot E$ acts on q_t according to Eq. (24-10). Consequently, an external agent must do work against this force in moving q_t. The work is force exerted times distance moved; hence, the work W is given by

$$W = q_t \cdot E \cdot d \tag{24-22}$$

By conservation of energy, $W = PE$. Therefore,

$$q_t \cdot E \cdot d = q_t \cdot V$$

from which we may cancel q_t to obtain the formula for the field strength between the plates of a parallel-plate (vacuum) capacitor:

$$E = \frac{V}{d} \tag{24-23}$$

This relationship reveals that the electric field strength may be expressed in units of volts per meter as well as in newtons per coulomb. (The field strength of a radio wave is given by radio engineers in volts per meter.)

Energy storage in a capacitor

The electric field between the plates of a charged capacitor contains energy, and this is a general property of electric fields. Indeed, the energy stored by a capacitor is stored in the electric field, but such considerations are beyond our needs at this point. An easier way to obtain an expression for the energy stored in a capacitor is to sum the work done by the source in transferring the charge from one plate to the other, but even this requires calculus (because of the changing voltage), which we wish to avoid. Therefore, we simply write the result: The energy stored by a capacitor of capacitance C charged to a voltage V is given by

$$\text{Energy} = \tfrac{1}{2}CV^2 \qquad (24\text{-}24)$$

Note that the energy stored is directly proportional to the capacitance C and is proportional to the square of the voltage.

EXAMPLE 24-10 Calculate the energy stored in an 8-μF capacitor that is charged to a PD of 100 V.

Solution. Direct application of Eq. (24-24) gives

$$\begin{aligned} \text{Energy} &= \tfrac{1}{2}CV^2 \\ &= (0.5)(8 \times 10^{-6}\ \text{F})(100\ \text{V})^2 \\ &= 0.04\ \text{J} \end{aligned}$$

The dielectric constant

If the region between the plates of a parallel-plate capacitor is filled with a physical dielectric such as glass, mica, or Teflon, the resulting capacitor has a greater capacitance than the original vacuum capacitor. That is, more charge must be transferred between the plates of the capacitor to establish the same potential difference between the plates.

The reason for this can be discerned from Fig. 24-13. As the charging of the

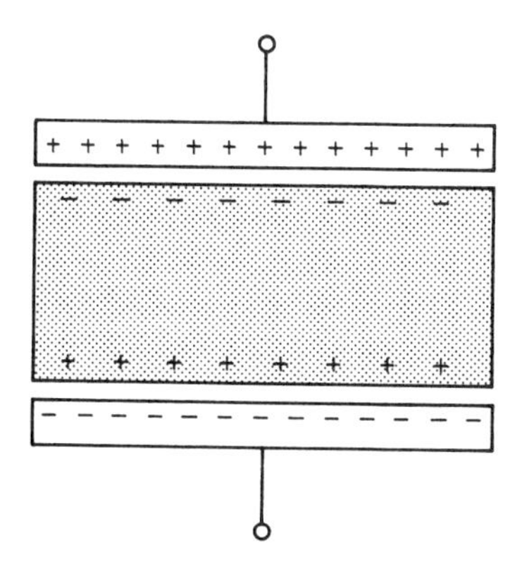

Figure 24-13 The polarization charge on the surface of the dielectric effectively reduces the net separation of charge so that the voltage appearing between the capacitor plates is reduced. This translates to an increased capacitance.

capacitor begins, an electric field is established in the region of the dielectric. The action of this field is to exert forces on the positive and negative charges (protons and electrons) contained in the dielectric. The result is that the atoms and molecules of the dielectric become *polarized*; the negative charges move slightly in one direction, and the positive charges move slightly in the other. This gives rise to *polarization charges* on the surface of the dielectric near the capacitor plates. Furthermore, the polarization charge is of opposite polarity to the charge on the capacitor plate.

The voltage existing between the plates of a capacitor arises ultimately from the separation of positive and negative charge that was effected by charging the capacitor. We can now see that the polarization charge effectively reduces the net separation of positive and negative charge so that a smaller voltage results for a given transfer of charge between the capacitor plates. This translates, using Eq. (24-15), into a larger capacitance. Therefore, the effect of the dielectric is to increase the capacitance of the original vacuum capacitor.

The *dielectric constant* of a particular dielectric is the factor by which the capacitance is increased when the dielectric is inserted into a vacuum capacitor. Hence, if C_{vac} is the capacitance of a vacuum capacitor and if C_{die} is the capacitance of a similar capacitor having the dielectric present, the dielectric constant K is defined as

$$K = \frac{C_{die}}{C_{vac}} \tag{24-25}$$

We can then define the permittivity of the dielectric to be

$$\epsilon = K\epsilon_0 \tag{24-26}$$

and the capacitance of a parallel-plate capacitor with a physical dielectric is given by

$$C = K\epsilon_0\frac{A}{d} \quad \text{or} \quad C = \epsilon\frac{A}{d} \tag{24-27}$$

Dielectric strength

A particular thickness of a given dielectric cannot withstand the application of an arbitrarily large potential difference. As the applied voltage is gradually increased, a point will be reached at which the dielectric will break down and conduct. The ability of a dielectric to resist this breaking down is reflected in the *dielectric strength* of the material. It is the maximum voltage per unit thickness of the dielectric that the dielectric can safely withstand. The dielectric constant and the dielectric strength of several dielectric materials are given in Table 24-1.

For a dielectric of given thickness, the voltage at which the dielectric may be expected to fail is given by

$$\text{Breakdown voltage} = \text{dielectric strength} \times \text{thickness} \tag{24-28}$$

Because of the limitations imposed by the dielectric strength, commercially available capacitors are rated according to the maximum voltage that may be safely applied to the capacitor.

TABLE 24-1 DIELECTRIC PROPERTIES OF SELECTED MATERIALS

Material	Dielectric constant, *K*	Dielectric strength, *V/m*
Vacuum	1	—
Air	1.0006	3×10^6
Glass	6	30×10^6
Mica	6	200×10^6
Mylar	3	—
Rubber	3	21×10^6
Polystyrene	2.5	24×10^6
Aluminum oxide	7	8×10^6
Tantalum oxide	11	4×10^6
Water (distilled)	78	—

Values are only approximate.

24-7 COMBINATIONS OF CAPACITORS

Capacitors in parallel

When two or more capacitors are connected as shown in Fig. 24-14, the capacitors are said to be connected in *parallel*. For comparison, note the capacitors in Fig. 24-15, which are connected in *series*.

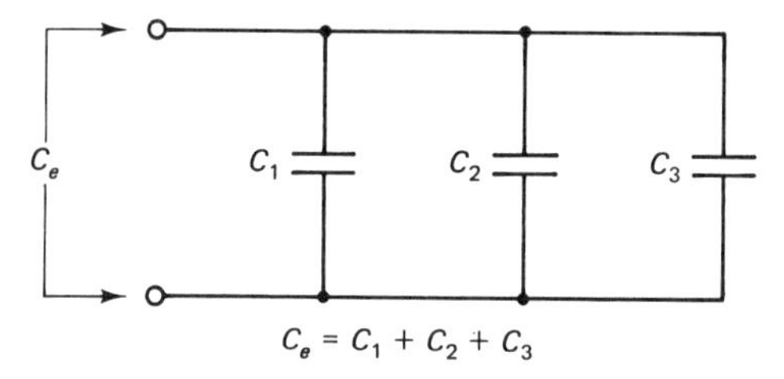

Figure 24-14 A parallel combination of three capacitors.

Any combination of capacitors will act as a single capacitor, but the effective capacitance of the equivalent single capacitor will depend upon the capacitances of the capacitors in the combination and will also depend upon the manner in which the capacitors are connected together. We now consider the parallel combination of three capacitors C_1, C_2, and C_3, as shown in Fig. 24-14.

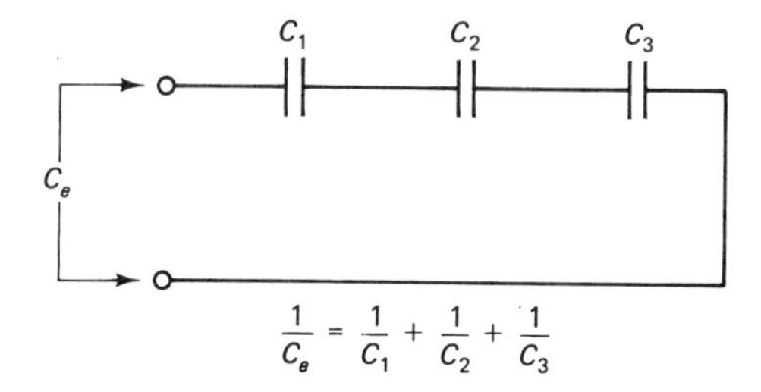

Figure 24-15 A series combination of capacitors.

It is apparent that the same potential difference appears across capacitors connected in parallel, and it is apparent that the total charge held by the combination is the sum of all the individual charges. Therefore, we may write

$$Q_e = Q_1 + Q_2 + Q_3 \tag{24-29}$$

$$V_e = V_1 = V_2 = V_3 \tag{24-30}$$

For any capacitor, the relationship $Q = CV$ always holds. Therefore, we can rewrite Eq. (24-29) as

$$C_eV_e = C_1V_1 + C_2V_2 + C_3V_3 \tag{24-31}$$

In light of Eq. (24-30), the V's may be canceled from the above equation to give a formula for the equivalent capacitance for capacitors in parallel:

$$C_e = C_1 + C_2 + C_3 \tag{24-32}$$

That is, the equivalent capacitance is simply the sum of the individual capacitances. The result holds for any number of capacitors connected in parallel.

Capacitors in series

When capacitors are connected in series, as in Fig. 24-15, the transfer of electrons does not occur from a plate of a given capacitor to its other plate. Instead, electrons are transferred from a plate of one capacitor to a plate of the adjacent capacitor, as shown in Fig. 24-16. Thus, all capacitors in a series combination will

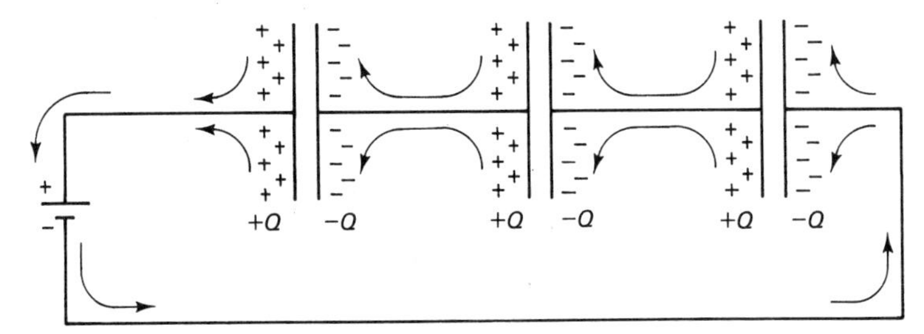

Figure 24-16 The effective charge of a series combination of capacitors is the same as the charge of each individual capacitor.

develop the same charge, but the same voltage will not be developed across each individual capacitor unless all individual capacitors have the same capacitance. Generally speaking, each capacitor will develop a different voltage, but the sum of the individual voltages will equal the voltage applied to the combination.

In light of the above, we may write, for capacitors in series,

$$Q_e = Q_1 = Q_2 = Q_3 \tag{24-33}$$

$$V_e = V_1 + V_2 + V_3 \tag{24-34}$$

Using $V = Q/C$ for each capacitor, we write Eq. (24-31) as

$$\frac{Q_e}{C_e} = \frac{Q_1}{C_1} + \frac{Q_2}{C_2} + \frac{Q_3}{C_3} \tag{24-35}$$

Then, using Eq. (24-33), we cancel the Q's to obtain a formula for the equivalent capacitance for a combination of capacitors connected in series:

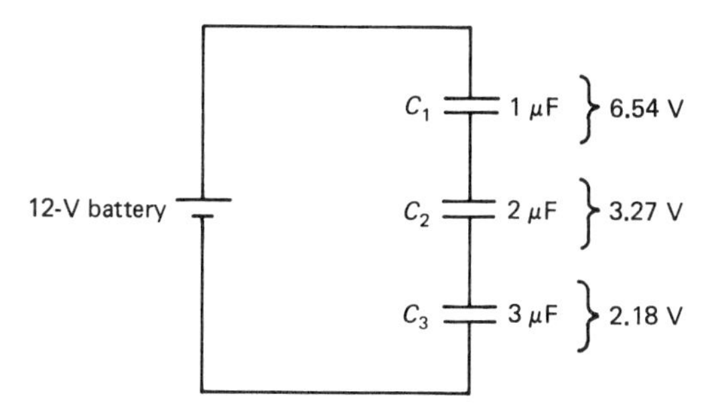

Figure 24-17

$$\frac{1}{C_e} = \frac{1}{C_1} + \frac{1}{C_2} + \frac{1}{C_3} \tag{24-36}$$

This result may be extended to more than three capacitors if desired. If only two capacitors are connected in series, Eq. (24-36) can be converted to the form of a product over a sum:

$$C_e = \frac{C_1C_2}{C_1 + C_2} \tag{24-37}$$

EXAMPLE 24-11 A 1-μF, a 2-μF, and a 3-μF capacitor are connected in series, and the series combination is connected to a 12-V battery as shown in Fig. 24-17.
(a) Calculate the equivalent capacitance of the combination.
(b) Compute the charge on each capacitor.
(c) Calculate the voltage appearing across each capacitor.

Solution. (a) We use Eq. (24-36) to calculate the equivalent capacitance:

$$\frac{1}{C_e} = \frac{1}{C_1} + \frac{1}{C_2} + \frac{1}{C_3}$$

$$= \frac{1}{1} + \frac{1}{2} + \frac{1}{3}$$

$$C_e = 0.545\ \mu\text{F}$$

(b) The charge on each individual capacitor is the same as the charge of the combination. We use the formula $Q = CV$, using C_e for C:

$$Q = C_eV$$
$$= (0.545\ \mu\text{F})(12\ \text{V})$$
$$= 6.54\ \mu\text{C}$$

Note that the product of microfarads and volts gives microcoulombs.
(c) Each capacitor is considered separately to calculate the voltage appearing across each capacitor:

For C_1, with $C = 1\ \mu\text{F}$ and $Q = 6.54\ \mu\text{C}$, $V_1 = \dfrac{6.54\ \mu\text{C}}{1\ \mu\text{F}} = 6.54\ \text{V}$

For C_2, with $C = 2\ \mu\text{F}$ and $Q = 6.54\ \mu\text{C}$, $V_2 = \dfrac{6.54\ \mu\text{C}}{2\ \mu\text{F}} = 3.27\ \text{V}$

For C_3, with $C = 3\ \mu\text{F}$ and $Q = 6.54\ \mu\text{C}$, $V_3 = \dfrac{6.54\ \mu\text{C}}{3\ \mu\text{F}} = 2.18\ \text{V}$

You may verify that $V_1 + V_2 + V_3$ equals 12 V, the voltage of the battery connected to the series combination. (A slight discrepancy is caused by rounding.)

Commercial capacitors

Capacitors, along with resistors and inductors, are widely used in practical electronic circuits and are found in many different shapes and sizes. The smallest discrete capacitors have capacitances on the order of a picofarad (pF); 1 pF is 10^{-12} F. On the other extreme, the largest capacitors may have capacitances

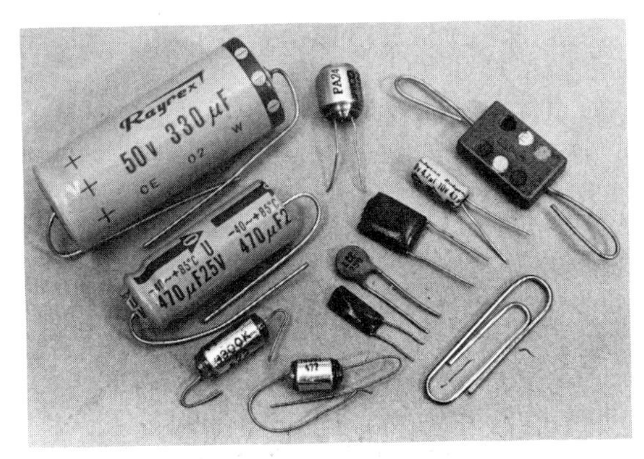

Assorted capacitors used in electronics.

approaching 1 F. Most capacitors used in electronics are found to lie in the range from a few picofarads to a few hundred microfarads.

Capacitors that must withstand high voltages must use dielectrics that are thicker than their low-voltage counterparts. Consequently, high-voltage capacitors are physically larger (and more expensive) than low-voltage capacitors.

One type of capacitor uses a chemically formed layer as a dielectric and provides a large capacitance in a comparatively small volume due to the very small thickness of the layer. Such capacitors are called *electrolytic capacitors* because of the chemical *electrolyte* used to form the insulating layer between the conductors. Electrolytic capacitors are usually *polarized*, which, in this context, means that one terminal is designated *positive* while the other is designated *negative*. Conventional capacitors are not polarized.

Materials used as dielectrics include oil-impregnated paper, mica, mylar, polystyrene, and ceramic materials. The ability to select the best capacitor for a given application is part of the stock-in-trade of an electronics-design engineer.

To Go Further

Investigate these names in electricity and electrostatics.

Charles A. de Coulomb (1736–1806), French physicist
Michael Faraday (1791–1867), English scientist
Alessandro Volta (1745–1827), Italian scientist
Benjamin Franklin (1706–1790), of Philadelphia
Luigi Galvani (1737–1810), Italian professor of anatomy
Henry Cavendish (1731–1810), English physicist
Robert A. Millikan (1868–1953), American physicist

Questions

1. Why is static electricity a good name?
2. Why does a static charge of electrons placed upon a conducting sphere reside totally on the surface of the sphere?
3. How may one use a plastic rod and a piece of fur to impart a positive charge to a conducting sphere?
4. Describe the similarity between Coulomb's law and the law of gravitation.
5. Suppose an electric field in a certain room is directed from west to east.
   (a) In what direction would positive charges within the room tend to move?
   (b) In what direction would negative charges within the room tend to move?
6. What evidence can you cite to support the contention that an electric field is a vector quantity?
7. Explain the difference between absolute potential and potential difference.
8. Describe the concept of capacitance as it relates to an isolated metal sphere and to a parallel-plate capacitor.

*9. Suppose a parallel-plate capacitor is given a charge Q. If the plates are then isolated so that Q remains constant, what happens to the voltage across the capacitor as the plates are physically moved farther from each other? What happens to the energy stored in the capacitor? Does work have to be done in pulling the plates apart?

*10. If the plates of a charged parallel-plate capacitor (vacuum) are pulled apart while the charge is held constant, does the electric field between the plates become stronger, weaker, or does it stay the same?

*11. A parallel-plate capacitor is given a charge Q, and the plates are then isolated.
   (a) If a dielectric material is then inserted between the plates, what happens to the voltage across the capacitor?
   (b) Does the energy stored by the capacitor change?
   (c) Does the capacitor return to its initial state of charge if the dielectric is removed?

*12. Suppose we have two identical capacitors, one charged, the other uncharged. If the two are then connected together in parallel, the charge redistributes itself equally between the two. Calculation reveals that energy is lost in the process. Visualize this process, and determine, if you can, where the "lost" energy goes.

Problems

1. Calculate the number of electrons represented by 1 μC of electrical charge.
2. A copper sphere 1 cm in diameter contains about 4.4×10^{22} atoms. If the tiny sphere is given a positive charge of 1×10^{-9} C, what fraction of the copper atoms give up an electron in the process? (*Answer*: 1 electron per 7×10^{12} atoms.)
3. Suppose an electron and a proton are 1 cm apart. Use Coulomb's law to calculate the force that each exerts upon the other.
4. Compute the acceleration that an electron will have toward a proton when it is 1 cm from the proton. (Use the result of Problem 3.)
5. Compute the acceleration that a proton will have toward an electron when it is 1 cm from the electron. Compare with the result of Problem 4.
6. A metal sphere 10 cm in diameter is given a charge of 1 μC.
   (a) Calculate the potential of the sphere. (*Hint*: Use Eq. (24-12) with $R = 0.05$ m.)
   (b) Calculate the electric field at a point in space 10 cm from the surface of the sphere. (*Hint*: Use $R = 0.15$ m.)
   (c) Compute the electrostatic potential 10 cm from the surface of the sphere.
7. Show that the combination of units of volts

per meter is the equivalent of newtons per coulomb so that volts per meter is a valid unit for expressing the magnitude of an electric field.

8. Positive charges of 0.1 C are located on the x-axis at $x = 4$ m and at $x = -2$ m. Determine the magnitude and direction of the electric field at the origin. Make a sketch.

9. A square centered on the origin is 4 m on each side. Positive charges of 0.01 C are located at the left-hand corners and negative charges of the same magnitude are located at the right-hand corners. Determine the magnitude and direction of the electric field: **(a)** at the origin; **(b)** at the point $x = 0$, $y = 2$.

10. Suppose two positive point charges of 0.01 μC are separated a distance of 1 m.
- **(a)** Calculate the electrostatic energy of the two charges.
- **(b)** Given that electrostatic potential obeys a scalar superposition principle, determine the potential at a point midway between the two charges.

11. Calculate the magnitude of the electric field 0.53 Å away from a proton.

12. An electron is located 1 m from the center of a sphere 10 cm in diameter. The sphere carries a positive charge of 0.1 μC.
- **(a)** Determine the strength of the electric field of the sphere at a distance of 1 m from the sphere.
- **(b)** Compute the force of attraction between the electron and the sphere using Eq. (24-10).
- **(c)** Determine the mutual attractive force between the sphere and the electron by using Coulomb's law and compare to the result of part (b).

13. Calculate the velocity an electron must have in order to have a kinetic energy of 1 eV.

14. Assume the earth to be an isolated conducting sphere. What is the capacitance of the earth?

15. Suppose a metal sphere 10 cm in diameter is charged to an absolute potential of 1000 V.
- **(a)** Calculate the charge Q that must be imparted to the sphere in order to raise its potential to 1000 V.
- **(b)** Determine the capacitance of the sphere.
- **(c)** Use the result of part (b) and the relation $Q = CV$ to check the result you obtained for part (a).

16. What amount of energy is expended when 1 C of charge moves from a point where the potential is 70 V to a point where the potential is 30 V?

17. Demonstrate that the units of Eq. (24-16), $C = R/k$, are consistent with the definition of a farad as 1 C/V.

18. A parallel-plate capacitor is made of two square plates 1 m square. Determine the capacitance of the capacitor if the separation of the plates is: **(a)** 1 cm; **(b)** 0.1 cm. (Assume the dielectric is air.)

19. For the capacitor of Problem 18, assume the dielectric is made of glass with a dielectric constant of 6 and recalculate the capacitance.

20. The dielectric strength of glass is about 30×10^6 V/m. What potential difference could a sheet of glass 4 mm thick withstand?

21. A parallel-plate capacitor has a removable dielectric whose dielectric constant is 6. With the dielectric in place, the capacitor is charged to 100 V and the power supply is then disconnected. The dielectric is then pulled from between the plates. Assuming no charge leaks from the plates, determine the voltage across the capacitor after the dielectric is removed.

22. What capacitance is required for a capacitor to hold 1 C of charge when it is charged to a potential of 10 V?

23. Suppose 8-μF and 12-μF capacitors are connected in parallel and the parallel combination is charged to a potential of 40 V.
- **(a)** What charge appears on each capacitor?
- **(b)** What is the total charge held by the parallel combination?
- **(c)** What is the equivalent capacitance of the combination?
- **(d)** When an equivalent capacitor is charged to 40 V, how does the resulting charge compare with the result of part (b)?

24. Suppose 8-μF and 12-μF capacitors are connected in series, and the series combination is charged to a potential of 40 V.

(a) What is the equivalent capacitance of the combination?

(b) What charge appears on each capacitor?

(c) What voltage appears across each capacitor?

25. Suppose a 1-μF capacitor is charged to a potential of 20,000 V. Calculate the energy stored in the capacitor.

26. Suppose a 10-μF capacitor is charged to 100 V and is then connected to another identical capacitor that is initially uncharged.

(a) Compute the charge of the capacitor that is charged to 100 V.

(b) What charge will the parallel combination have after the two capacitors are connected together?

(c) What voltage will appear across the parallel combination?

27. For Problem 26 above, compute the energy contained in the capacitor charged to 100 V, and then compute the energy contained in the parallel combination. Can you account for the difference in energy?

28. Show that the equivalent capacitance of three identical capacitors connected in series is one-third the capacitance of one of the capacitors. Is this a rule that holds for two capacitors? for four capacitors?

29. To determine the capacitance C_x of an unknown capacitor, a physics professor first charges C_x to a test voltage V_t. The professor then disconnects C_x from the power supply and quickly connects to C_x, in parallel, another capacitor C_s whose capacitance is accurately known. The voltage across the combination is then quickly measured to be V_m.

(a) Use the fundamental relationships for capacitors to show that C_x is given by $C_x = C_s(V_t/V_m - 1)$.

(b) Determine C_x if $C_s = 1\ \mu F$, $V_t = 10$ V, and $V_m = 2.5$ V.

CHAPTER 25

ELECTRIC CURRENT

Our society would be much different if it were not for the fact that energy can be transported from one place to another by an electric current. Suppose energy could be transported only by rotating shafts—can you imagine a network of shafts that could deliver energy to homes and factories in a manner similar to that of the electric power distribution system?

In this chapter we investigate the fundamental principles of electric current. We consider simple DC circuits, series and parallel resistances, power dissipation in resistances, resistivity, the internal resistance of batteries, and Kirchhoff's laws. It is a full chapter, but most of the material is not difficult at all. In the next chapter we study the magnetic field and its properties and relationship to electric current.

25-1 ELECTRIC CURRENT FUNDAMENTALS

Definition of current

An *electric current* is the *flow* of *electrical charge*. Most often the flow of charge occurs by way of the movement of electrons through metallic conductors, but this is not the only way in which an electric current can occur. The electron gun inside a cathode ray tube (a CRT, such as a television picture tube) fires a stream of electrons through a vacuum to the screen at the front of the tube. The electron beam represents an electric current even though no conductor is physically present.

Electrons are very much more mobile than protons, but many situations exist where positive, as well as negative, charges may move. When salt, NaCl, is dissolved in water, the Na and Cl atoms become dissociated to form positive Na^+ and negative Cl^- ions. These ions are free to move so that when an electric field is established within the solution, the positive ions move in one direction while the negative ions move in the other. This movement of charged particles represents an

An elegant alternative to the transport of energy via rotating shafts.

electric current; in this case, both positive and negative charges participate in the movement.

In the following discussion, we assume that we are speaking of electron flow through a conductor in order to be as specific as possible. Hence, we can consider an electric current to be the flow of electrons through a conductor.

At this point, we need to recall the nature of a metal in order to see how it is possible for electrons to travel through a copper wire, for example. A copper wire consists of an exceedingly large number of copper atoms joined together, and each copper atom contributes a valence electron (sometimes called a *free electron*) to the electronic structure of the metal. Thus, a metal contains an abundance of free electrons which are very mobile.

The nucleus of an atom is extremely small in comparison to the overall atomic diameter. Therefore, an atom is mostly empty space. Because of this, things made of atoms are mostly empty space. More specifically, a copper wire—even though it appears solid under examination by the naked eye—contains plenty of empty space in which the free electrons may move.

The ampere

The SI unit of electric current is the ampere (A), named in honor of Andre Marie Ampere (1775–1836). This unit is often called simply an amp, and the smaller units milliamps (mA), microamps (μA), and nanoamps (nA) are commonly encountered. The ampere is defined in terms of the quantity of charge that flows past a point on a conductor in a time interval of one second:

A conductor is said to carry a current of 1 A if a quantity of electrical charge equal to 1 C flows past a point on the conductor in 1 s.

This definition is illustrated in Fig. 25-1. Because 1 C is represented by about 6.242×10^{18} electrons, this number of electrons will flow past any point on a conductor each second if the conductor is carrying a current of one amp.[1]

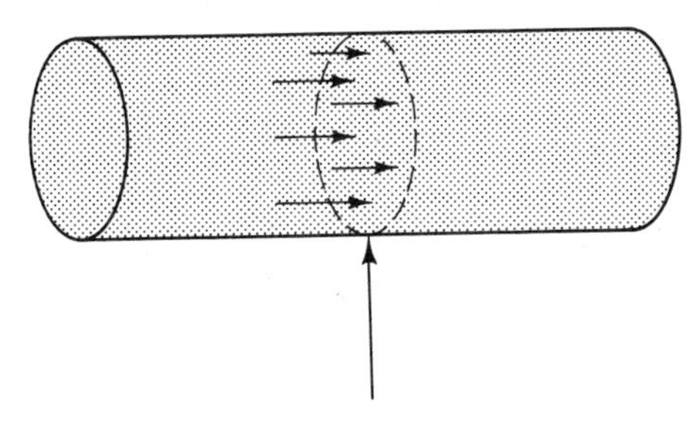

Figure 25-1 A conductor is said to carry a current of 1 A if 1 C of charge per second passes through a cross section adjacent to any point on the conductor.

In this text we consider the *direction* of current flow to be the direction in which the electrons move. However, many texts take the direction of the current as the direction in which positive charges move. In that system, current flows from the positive toward the negative and is called *conventional current*. But in actuality the electrons do most of the moving; thus, for our purposes it seems more direct to speak of current as being based upon and in the direction of electron flow.

In mathematical terms, the current I is given by

$$I = \frac{Q}{t} \tag{25-1}$$

where Q is the electrical charge that passes a given cross section of a conductor in a time period of t seconds. This may be rearranged to give a formula for computing the quantity of charge that passes a point on a conductor carrying current I in a time interval of t seconds:

$$Q = I \cdot t \tag{25-2}$$

From Eq. (25-1) we can infer that the units of a current are charge per unit time. Hence, 1 A = 1 C/s.

EXAMPLE 25-1 A copper wire carries a current of 5 A. A spot of red paint marks a point on the surface of the wire. Compute the quantity of charge that flows past the red spot in 8 s.

[1]In the SI system of units, the ampere is in fact taken as the fundamental unit and the coulomb is derived from the ampere. This stems from the fact that it is easier to measure a current accurately than to measure a quantity of electrical charge to the same accuracy. The standard ampere is defined in terms of the mutual force exerted on parallel current-carrying conductors. Such conductors are described in Sect. 26-7.

Solution. We use Eq. (25-2).

$$Q = I \cdot t = (5\ \text{A})(8\ \text{s}) = 40\ \text{C}$$

Even though a current of 1 A represents the flow of a very large number of electrons, we should keep in mind that the total number of electrons in a wire is far greater. If we could look inside a wire that is carrying a current of 1 A, we would see that only a very small fraction of the free electrons seem to be moving as part of the current. The rest of the electrons would not exhibit any unusual behavior.

Electromotive force

To establish a current in a conductor, the electrons in the conductor must experience a force that tends to make them move. The *electromotive force* (emf), as it is called, may arise from the influence of either a magnetic field or an electric field. We speak of *sources of emf*, by which we mean batteries, electromechanical generators, fuel cells, solar cells, and so on. When a source of emf is connected to a conductor, a current will flow through the conductor.

In terms of a water-pump analogy, we can think of a battery as being an *electron pump*. Batteries have two terminals which we can liken to an input port and an output port of a pump. The electron-pumping action stems from chemical reactions within the battery, and the electrical pressure (the voltage) developed by a particular battery depends upon the chemistry used by that particular type of battery. Any good college-level general chemistry text will provide the details of the electron-pumping action.

Another view is that a battery provides a potential difference in a package. The voltage of the battery is a measure of the potential difference that the battery produces. When the potential difference is applied to a conductor, an electric field is set up (on a microscopic level) in the conductor, and the electric field, in turn, causes the electrons to move which constitutes the current.

The chemical processes inside a battery remove electrons from the positive terminal and transfer the electrons to the negative terminal. Thus, inside a battery, electron flow occurs from the positive to the negative terminal. When an external conductor is connected to the terminals of the battery, electrons flow from the negative terminal through the external conductor to the positive terminal. The external conductor is said to complete the *circuit*, or path of continuous electron flow. This is illustrated in Fig. 25-2.

Electrical resistance

Electrons are not able to move through a conductor with perfect ease. They collide with other electrons, some of which may be bound to the atoms comprising the conductor, so that the path taken by an electron through the conductor may be very zigzagging. The trouble that electrons have in flowing through a conductor is related to the *electrical resistance* of the conductor.

When a particular voltage is applied to a conductor such as a long copper wire, the electrical resistance of the wire limits the current that will flow.

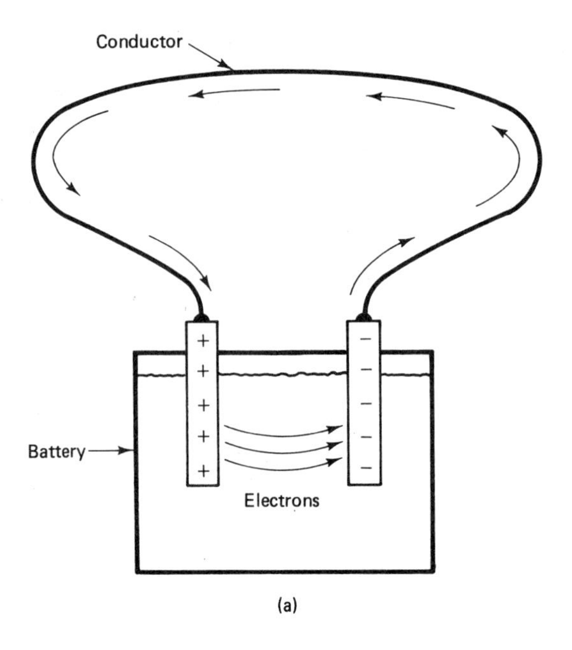

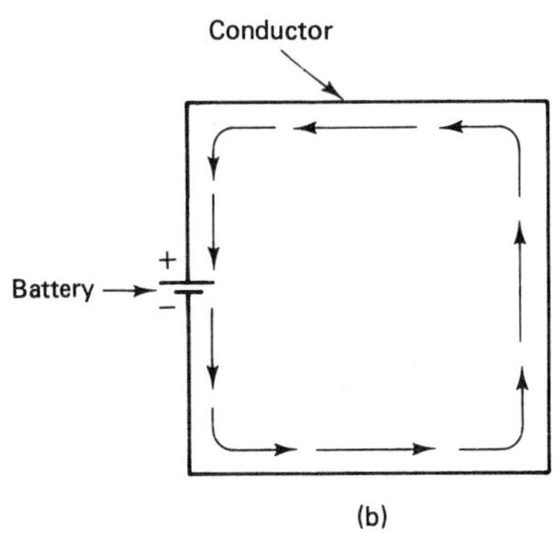

Figure 25-2 (a) A pictorial representation of the electron flow (denoted by arrows) inside and outside a battery whose terminals are connected by a conductor.
(b) Schematic representation of the same circuit.

Electrical resistance is measured in ohms, denoted by the capital Greek letter omega (Ω).

We now have described three important parameters of an electrical circuit. These are: (1) the source of *electromotive force* whose effect is described by giving its voltage; (2) the *resistance* of the conductors that comprise the circuit, measured in ohms; and (3) the *current*, measured in amperes, which is determined by the applied voltage and the electrical resistance. In the following section we describe the mathematical relationship between these three parameters, a famous relationship known as Ohm's law.

25-2 OHM'S LAW

When an emf of V volts is applied to a conductor, the current I that flows through the conductor is directly proportional to the applied emf:

$$I \propto V \tag{25-3}$$

This proportionality was discovered in 1827 by Georg Simon Ohm, a German high-school teacher. If the proportionality is expressed in terms of the resistance R, we obtain the relationship commonly known as *Ohm's law*:

$$I = \frac{V}{R} \tag{25-4}$$

Thus, the current varies in direct proportion to the applied voltage and inversely as the resistance.

The electronics industry makes wide use of *resistors*. A resistor is a small component (typically) that exhibits a fixed amount of electrical resistance. They are manufactured in a wide range of ohmic values which span from a few tenths of an ohm to perhaps a hundred megohms.

A circuit that can be used to confirm the validity of Ohm's law is shown in Fig. 25-3. A battery is connected to a resistor, but note that a voltmeter is included to measure the voltage developed across the resistor. Also, note that an ammeter is connected in series with the resistor to indicate the current that flows through the resistor. We make several assumptions in order to simplify the analysis of such circuits. First, we assume that the wires used to connect the battery to the resistor, ammeter, and so on are perfect conductors. This will be a good assumption if the resistance R is very large in comparison to the resistance of the wires, which is usually the case. Secondly, we assume the ammeter has zero internal resistance; quality ammeters exhibit very low resistance and an ideal ammeter would have zero resistance. Finally, we assume that the voltmeter has a very high internal resistance so that a negligible current will flow through the voltmeter. While this may seem like a lot of assumptions, they are good assumptions for most practical circuits.

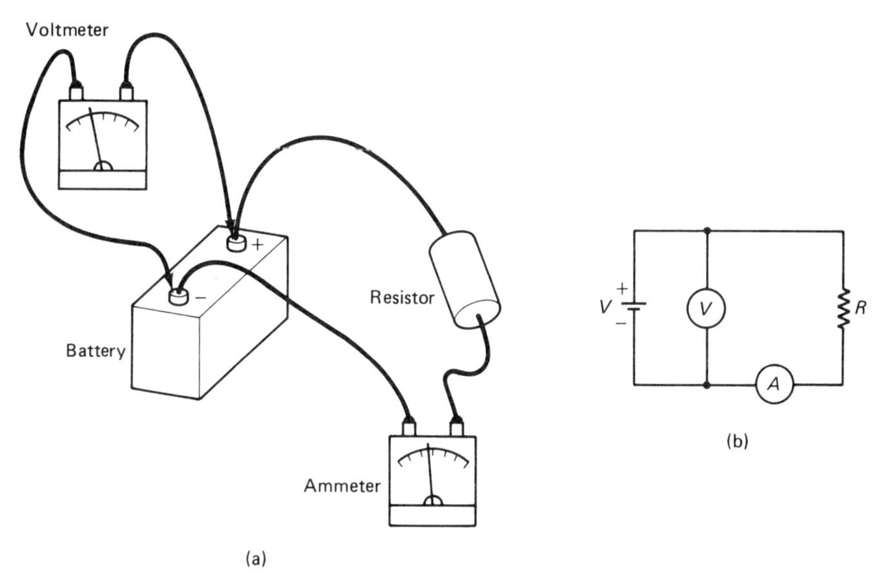

Figure 25-3 A circuit for checking the validity of Ohm's law. (a) Pictorial. (b) Schematic.

EXAMPLE 25-2 A 12-V battery is connected to a small incandescent lamp whose filament has a resistance of 8 Ω. What current flows through the lamp?

Solution. By Ohm's law, $I = V/R = 12\ \text{V}/8\ \Omega = 1.5\ \text{A}$.

EXAMPLE 25-3 A current of 4 A is known to be flowing through a resistor whose resistance is 6 Ω. What voltage is developed across the resistor?

Solution. Using Ohm's law, $V = IR = (4\ \text{A})(6\ \Omega) = 24\ \text{V}$.

EXAMPLE 25-4 When a voltage of 120 V is applied to an ordinary 100-W incandescent lamp, 0.833 A of current flows through the lamp. Calculate the electrical resistance of the lamp.

Solution. Use Ohm's law again:

$$R = \frac{V}{I} = \frac{120\ \text{V}}{0.833\ \text{A}} = 144\ \Omega$$

25-3 POWER DISSIPATION IN A RESISTANCE

We see that a potential difference V must be applied to a resistance in order to establish a current through the resistance. The current is, by definition, charge transported per unit time, Q/t. Hence, as current flows through a resistance, electric charge traverses the potential difference established across the resistance.

In Chap. 24 we find that a transformation of energy occurs whenever a charge Q moves through a potential difference V; the energy transformed is QV. Therefore, energy is transformed in a resistor as current flows through it. The power dissipation is the energy transformed per unit time:

$$\text{Power} = \frac{\text{energy}}{\text{time}} = \frac{QV}{t}$$

But Q/t equals the current I. It follows that the power dissipation in a resistance is given by

$$\text{Power} = I \cdot V \tag{25-5}$$

where I is the current flowing *through* the resistance and V is the voltage applied to or developed *across* the resistance. If I is given in amps and V in volts, the unit of power is the watt, W.

Ohm's law, $V = IR$, can be used to convert Eq. (25-5) to

$$\text{Power} = I^2R = \frac{V^2}{R} = I \cdot V \tag{25-6}$$

A simple example is given in Fig. 25-4.

When current flows through a resistance, the electrical energy given up is converted to heat. The resistor gets hot. This makes sense when we recall that heat is molecular motion and when we further realize that electron collisions with the atoms of a conductor will cause those atoms to move. Thus, the heating of a current-carrying conductor (or resistance, if you prefer) is due to collisions

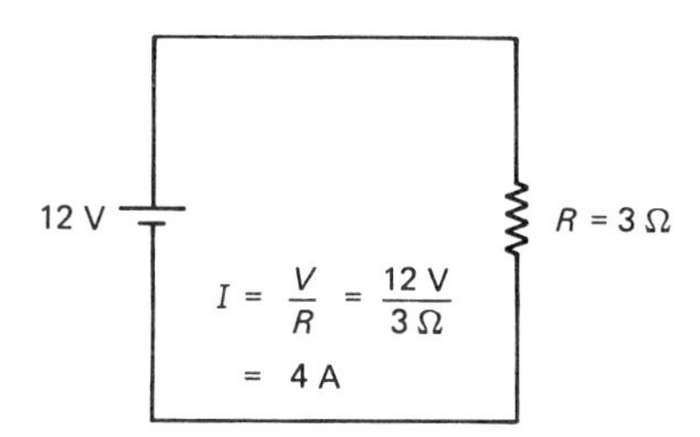

$$\text{Power dissipation} = I \cdot V = \frac{V^2}{R} = I^2 R$$
$$= (4\text{ A})(12\text{ V}) = 48\text{ W}$$

Figure 25-4 Formulas for computing the power dissipation in a resistor, and a sample calculation.

between electrons comprising the current and the atoms of the material of which the conductor is made.

25-4 RESISTORS IN SERIES

In Fig. 25-5 two resistors are connected in series. Electrons coming from the negative terminal of the battery flow through one resistor and then flow through the other. Therefore, the same current must flow through both resistors. Moreover, the total circuit resistance is the sum of the two resistances.

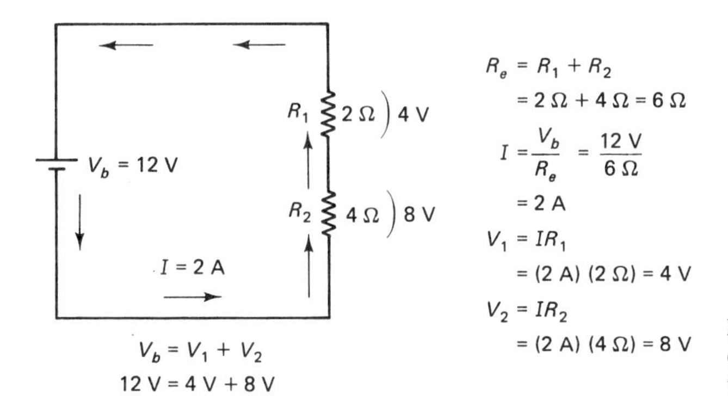

$$R_e = R_1 + R_2 = 2\,\Omega + 4\,\Omega = 6\,\Omega$$
$$I = \frac{V_b}{R_e} = \frac{12\text{ V}}{6\,\Omega} = 2\text{ A}$$
$$V_1 = IR_1 = (2\text{ A})\,(2\,\Omega) = 4\text{ V}$$
$$V_2 = IR_2 = (2\text{ A})\,(4\,\Omega) = 8\text{ V}$$

Figure 25-5 A circuit with resistors connected in series. A sample calculation is provided.

We can find an equivalent resistance R_e so that when the equivalent resistance is connected to the battery, the same current will flow through the battery as when the series combination of R_1 and R_2 was connected. Obviously, the equivalent resistance is given by

$$R_e = R_1 + R_2 \tag{25-7}$$

If more than two resistors are connected in series, the equivalent resistance is the sum of all the resistances in the series combination.

The current that flows in a series circuit, such as that in Fig. 25-5, is given by

$$I = \frac{V}{R_e} \tag{25-8}$$

It should be clear that the same current flows at all points in a series circuit. Therefore, the current calculated from Eq. (25-8) is the current that flows through the battery, and it is also the current that flows through each resistor.

To calculate the voltage that appears across each resistor in a series combination, we apply Ohm's law to each resistor. The voltage appearing across R_1 is given by

$$V_1 = IR_1 \tag{25-9}$$

and the voltage appearing across R_2 is given by

$$V_2 = IR_2 \tag{25-10}$$

This procedure is continued if more than two resistors are included in the series combination.

It is easy to verify that the sum of the voltages appearing across the individual resistors in a series combination equals the voltage of the battery (or other source) that causes current to flow in the series combination. If V_b is the battery voltage,

$$V_b = V_1 + V_2 \tag{25-11}$$

Batteries connected in series

When batteries are connected in series, as shown in Fig. 25-6, the equivalent voltage of the combination of batteries is the sum of the separate battery voltages:

$$V_e = V_{b1} + V_{b2} + \cdots \tag{25-12}$$

This assumes that all batteries are connected in the same direction, that they are connected *series adding*. If any battery is connected backwards, the voltage of that battery must be subtracted instead of added when evaluating Eq. (25-12), as illustrated in Fig. 25-7. Batteries of any voltage may be connected in series, but we shall see later that only batteries of the same voltage may be connected in parallel.

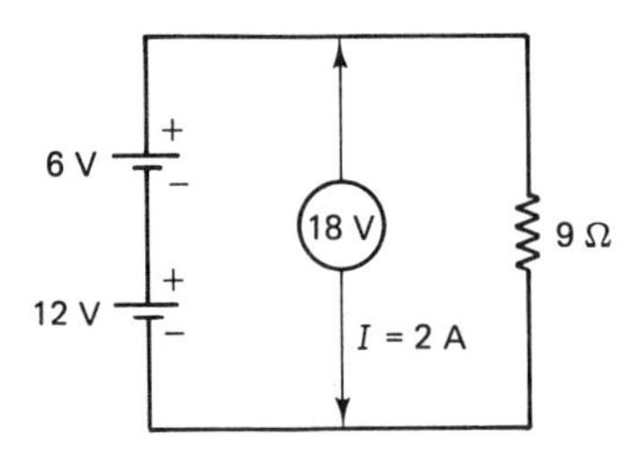

Figure 25-6 The equivalent voltage of batteries connected in series adding is the sum of their separate voltages.

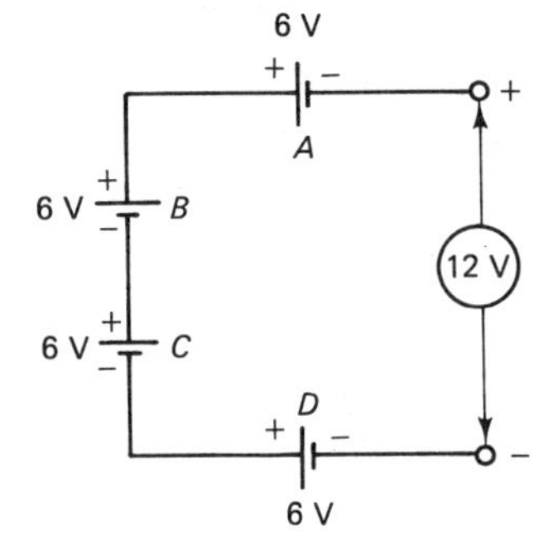

Figure 25-7 *A* is connected so that its emf opposes that of the other batteries. It is connected in *series opposition*.

Batteries connected in parallel

Only batteries having the same terminal voltage may be connected in parallel. When this is the case, as in Fig. 25-8, the voltage of the combination is the same as that of any one battery. The advantage of connecting batteries in parallel is that the batteries share the current load and therefore will be capable of operating for a longer period of time before being recharged (or discarded).

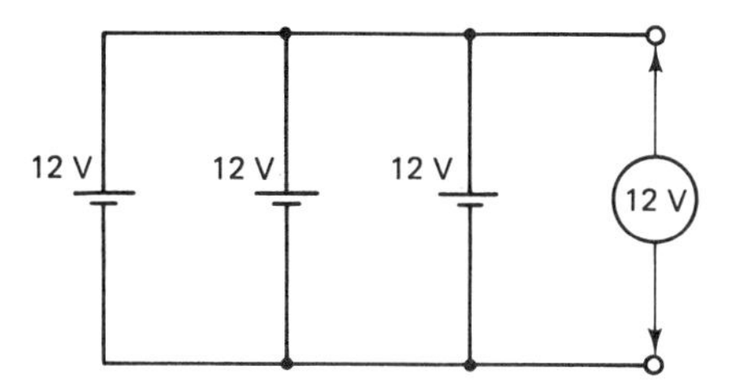

Figure 25-8 Batteries that have the same terminal voltage may be connected in parallel. The output voltage of the combination is the same as that of any one battery.

To see why batteries that have unequal terminal voltages must not be connected in parallel, examine Fig. 25-9, in which a 12-V battery is connected in parallel with a 6-V batttery. The circuit has been redrawn in part (b) of the figure to illustrate that the circuit can be considered as a series circuit in which the batteries are series opposing. Clearly, a net voltage of 6 V exists, and a current will flow through the conductors connecting the batteries. In most cases, the resistance of the conductors is very small so that very large currents will flow. If the batteries are large, as in the case of automobile batteries, the conductors will get hot very quickly, and the large dissipation of energy inside the batteries is apt to cause one or both batteries to explode.

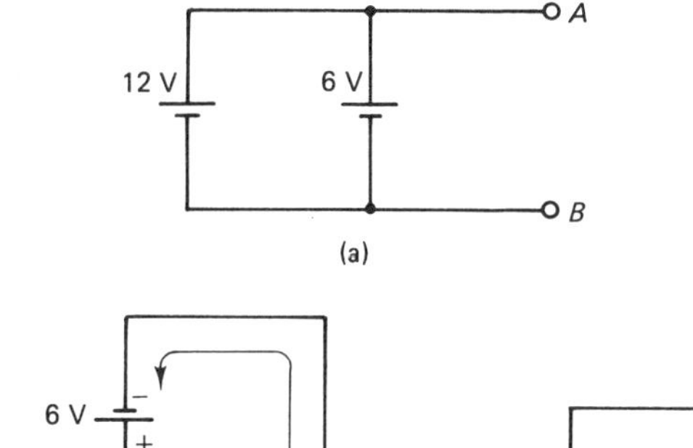

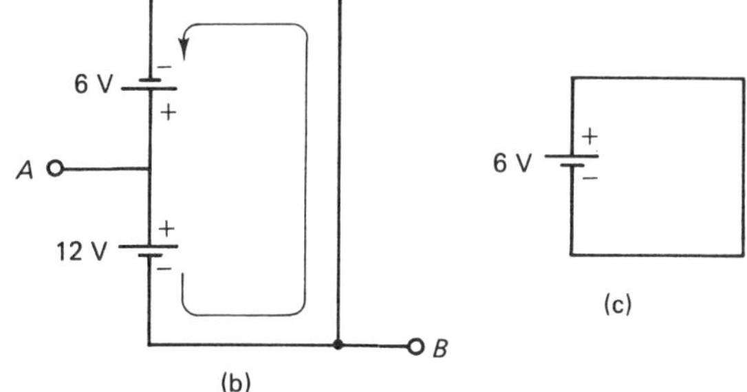

Figure 25-9 (a) Batteries of unequal terminal voltage connected in parallel. (b) The effect is to create a series circuit in which a very large current will flow.
(c) Equivalent circuit.

The common cold-weather practice of using jumper cables to jump-start automobiles involves connecting the batteries of the two vehicles in parallel. Therefore, one should never use a 12-V battery to jump-start a vehicle that utilizes a 6-V battery. It is an extremely dangerous practice.

When connecting batteries in parallel, it is essential that the connection be made with proper polarity. The positive of one battery connects to the positive of the other, and the negative of one connects to the negative of the other. If a reversal occurs, the two batteries will be connected together in series adding, and some sort of explosion is certain. (Hydrogen gas produced within the battery is heated and ignites explosively. This often blows the battery apart and scatters sulfuric acid violently. Automobile batteries deserve to be handled with care.)

25-5 RESISTORS IN PARALLEL

When resistors are connected in parallel, as shown in Fig. 25-10, the same voltage is applied to the terminals of each resistor. Hence, the current that flows through each resistor will be determined by the resistance of that particular resistor. The current through the battery will be the sum of the currents flowing through the separate resistors.

In mathematical terms, for the three resistors of Fig. 25-10,

$$I_1 = \frac{V}{R_1} \qquad I_2 = \frac{V}{R_2} \qquad I_3 = \frac{V}{R_3} \tag{25-13}$$

(*current through battery*) $$I_b = I_1 + I_2 + I_3 \tag{25-14}$$

Equivalent resistance of resistors in parallel

For a parallel combination of resistors, we can find an equivalent resistance such that when the equivalent resistance is connected to the battery, the same current will flow through the battery as when the parallel combination was connected to the battery. We make use of the two equations given above:

$$I_b = I_1 + I_2 + I_3$$

$$\frac{V}{R_e} = \frac{V}{R_1} + \frac{V}{R_2} + \frac{V}{R_3}$$

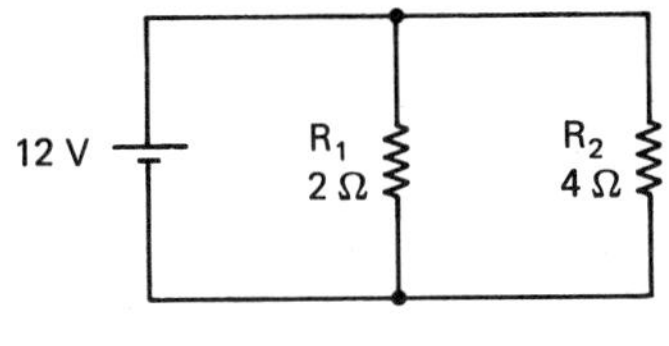

12 V — R_1 2 Ω — R_2 4 Ω

(a) $I_1 = \dfrac{V}{R_1} = \dfrac{12\text{ V}}{2\ \Omega} = 6\text{ A}$

$I_2 = \dfrac{V}{R_2} = \dfrac{12\text{ V}}{4\ \Omega} = 3\text{ A}$

(b) $I_b = I_1 + I_2 = 6\text{ A} + 3\text{ A} = 9\text{ A}$

(c) Power $= I \cdot V$

$P_1 = I_1 V = (6\text{ A})(12\text{ V}) = 72\text{ W}$

$P_2 = I_2 V = (4\text{ A})(12\text{ V}) = 48\text{ W}$

(d) $R_e = \dfrac{R_1 R_2}{R_1 + R_2} = \dfrac{(2\ \Omega)(4\ \Omega)}{2\ \Omega + 4\ \Omega} = 1.333\ \Omega$

Figure 25-10 The smallest resistance in a parallel combination dissipates the greatest amount of energy.

Because V, the voltage of the battery, is the same as that applied to all the resistors, we may cancel the V's to obtain a formula for the equivalent resistance for resistors connected in parallel:

$$\frac{1}{R_e} = \frac{1}{R_1} + \frac{1}{R_2} + \frac{1}{R_3} \tag{25-15}$$

For the special case in which only two resistors are connected in parallel, Eq. 25-15 may be rewritten as a product over a sum.

$$R_e = \frac{R_1 R_2}{R_1 + R_2} \tag{25-16}$$

EXAMPLE 25-5 A 12-V battery is connected to the parallel combination of a 2-Ω and a 4-Ω resistor, as shown in Fig. 25-10. Calculate: (a) the current through each resistor; (b) the current that flows through the battery; (c) the power dissipation in each resistor; (d) the equivalent resistance of the parallel combination of resistors.

Solution. The solution is given as part of the figure.

25-6 POTENTIALS AND POTENTIAL DIFFERENCES IN PRACTICAL CIRCUITS

Schematic diagrams of electronic devices such as radios, television sets, and so forth frequently give the voltage that normally appears at various test points within the circuit. Also indicated is the *circuit ground*, which is the conductor relative to which all other voltage measurements are made.

A voltmeter has two leads. One lead, generally colored black, is called the *common*; the other, generally colored red, is called either the *high side* or the *hot lead*, as local jargon dictates. One of the first steps performed in the process of doing a voltage analysis of a circuit is to connect the common lead of the voltmeter to the ground conductor of the circuit under test. This assures that all voltages subsequently measured will be measured relative to ground. In most circuits, the conductor chosen to be ground is somewhat arbitrary, but once the ground is chosen, all other voltages of the circuit become fixed relative to that ground. It is this feature of circuits and grounds that we now wish to investigate.

Examine the series circuit of Fig. 25-11. Three resistors are shown, and the voltage appearing across each is given. Note the points marked A, B, C, and D, and observe the ground symbol connected to the negative terminal of the battery. You may wish to verify that a current of 2 mA flows in the circuit and that the voltage appearing across each resistor is as indicated.

We now consider the voltages appearing at various points. First, note that the voltage at D is 0 V. This must be the case because point D connects directly to ground. Next, observe that 6 V appears across R_3. This implies that the voltage at C is 6 V higher than the voltage at D. Consequently, the voltage at C is +6 V. We reason that the polarity at point C must be positive because point C is (electrically speaking) closer to the positive terminal of the battery than point D.

Continuing, we note that a potential difference of 9 V appears across R_2. This implies that point B is 9 V higher in potential than point C. Thus, the potential

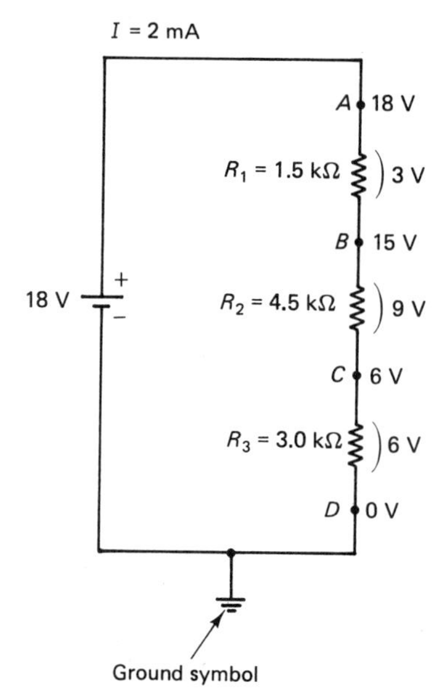

Figure 25-11 Potentials and potential differences in a circuit where ground is taken as the most negative point in the circuit.

(or voltage) at point B is 15 V. In the same manner we can determine that the voltage at point A is 18 V.

The foregoing illustrates the procedure for determining the voltages at various points in a circuit. We now consider the *same* circuit, with the exception that the ground is now connected to point C instead of to point D. This circuit is shown in Fig. 25-12.

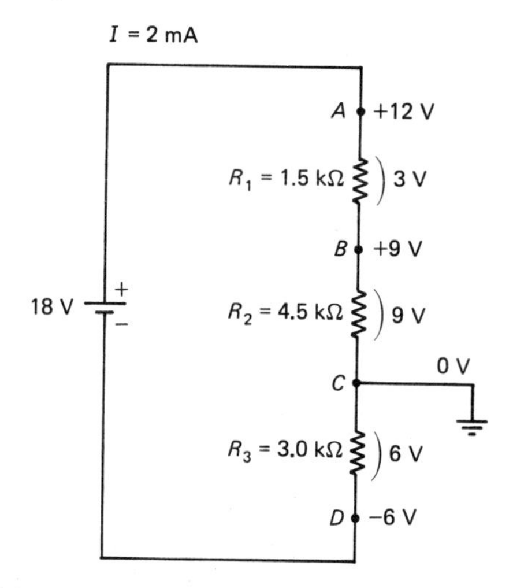

Figure 25-12 The circuit of Fig. 25-11 with the ground point shifted to point C.

To determine the voltages at the various points, we begin at the ground symbol which denotes the point at 0 V. Thus, the voltage at point C is 0 V. Observe that moving the ground point does not affect the voltages appearing across each of the resistors, nor does it affect the current flowing in the circuit.

Because of the 6-V potential difference appearing across R_3, we know that the potential at point D must be -6 V, *negative* because point D is electrically closer to the negative terminal of the battery.

Going in the other direction and reasoning along the same lines as before, we conclude that the voltage at point B is $+9$ V, and the voltage at point A is $+12$ V.

After observing that point A is connected directly to the positive terminal of the 18-V battery, we might wonder if the voltage at A should not be $+18$ V. This is not the case, however, because the negative terminal of the battery is connected directly to point D, which is at a potential of -6 V. This brings up the following important point.

The rated voltage of a battery is not the potential of either the positive or the negative terminal of the battery. It is the *difference in potential* that the battery produces from one of its terminals to the other. Indeed, in Fig. 25-12 we can determine that point A is 18 V more positive than point D, which is as it should be because these points are connected directly to the battery terminals.

To carry this example even further, if the ground point is shifted to point A, the potential at points B, C, and D become -3 V, -12 V, and -18 V, respectively. In this case, the positive terminal of the battery is at 0 V while the negative terminal is at -18 V.

Another circuit that demonstrates the effect upon measured voltages of shifting the ground point is given in Fig. 25-13. The circuit is easy to analyze by following the procedure given above.

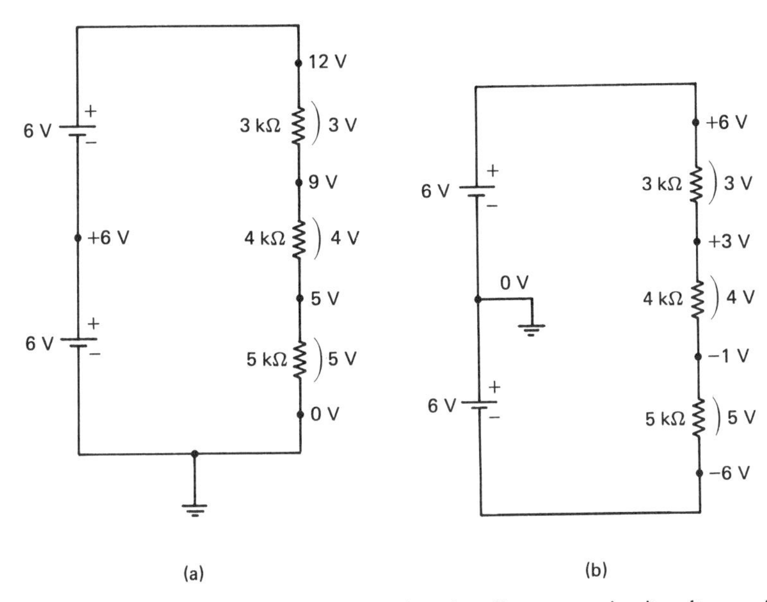

Figure 25-13 A second example showing the effect upon circuit voltages of shifting the ground point.

If we examine the voltage at points A, B, C, and D in Fig. 25-11, we observe that the voltage becomes less positive as we move farther from the positive terminal of the battery. Moreover, it is the resistors that produce the drop in voltage from one point to the next. Therefore, we often speak of the voltage drop that occurs across a resistor, and, in light of the formula for computing the voltage drop, it is often referred to as the *IR drop*.

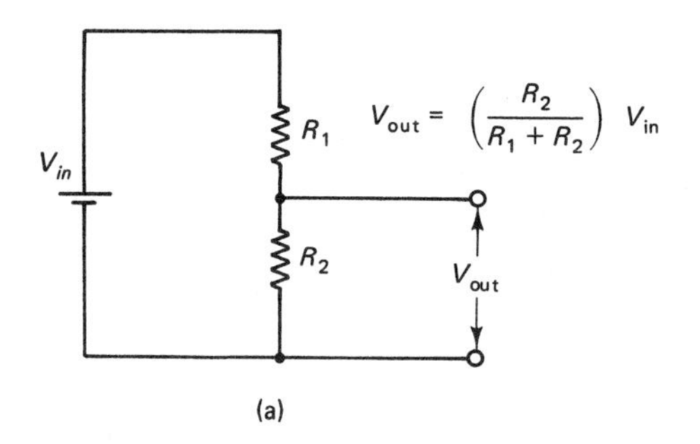

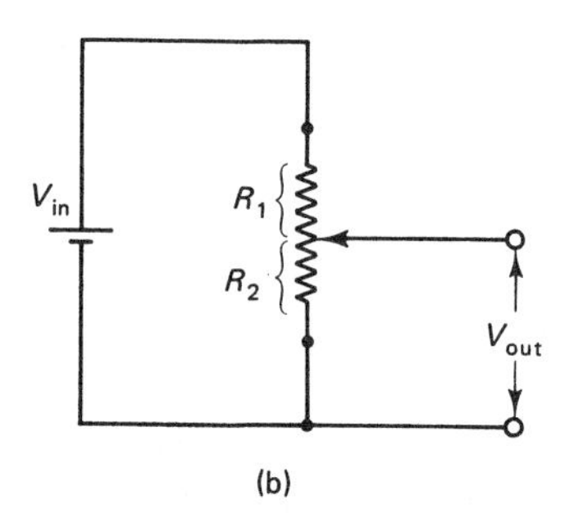

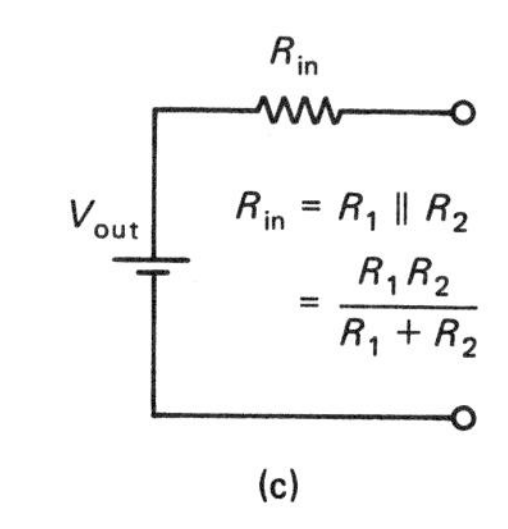

Figure 25-14 Two forms of a voltage divider.
(a) Using fixed resistors.
(b) Using a variable resistor.
(c) Equivalent circuit.

Voltage divider

Figure 25-14 shows a method for producing a voltage that is a certain fraction of the battery voltage. The circuit is called a voltage divider and is used extensively, in various forms, in practical electronics. It is a simple matter to derive the formula for the output voltage.

The battery and the two resistors are connected in series. The current that flows in this series circuit is

$$I = \frac{V_{in}}{R_1 + R_2} \tag{25-17}$$

The output voltage, V_{out}, is the voltage developed across R_2 by the current I flowing through it. By Ohm's law, this voltage is IR_2, where I is given by the equation above. Hence,

$$V_{out} = \left(\frac{R_2}{R_1 + R_2}\right) V_{in} \tag{25-18}$$

This is often called the *voltage-divider formula*.

A disadvantage of the voltage divider of Fig. 25-14 is that current flowing in a load connected to the output must flow through the resistance of the divider circuit. This causes a simple voltage divider to have a large internal resistance. The equivalent circuit is shown in Fig. 25-14(c). The internal resistance R_{in}, is the parallel equivalent of R_1 and R_2.

The circuit of Fig. 25-14(b) utilizes a variable resistor called a *potentiometer*, or *pot*, in order to produce an output voltage that is continually adjustable from zero volts to a maximum equal to the battery voltage. A *slider* is moved by turning a knob to various positions along the resistance element, and the output voltage is determined by the slider position. A typical pot is illustrated in Fig. 25-15.

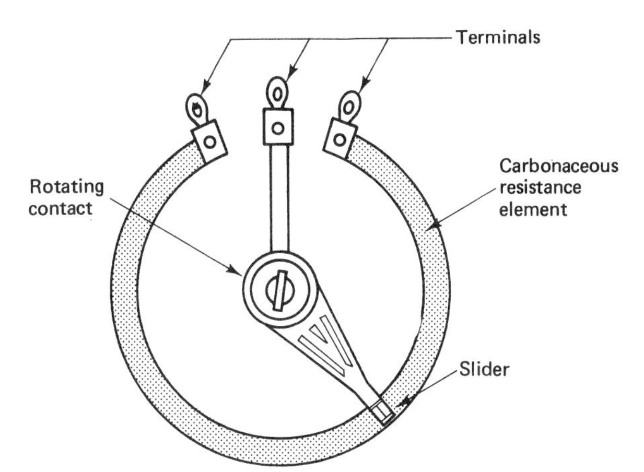

Figure 25-15 The construction of a three-terminal variable resistor.

25-7 RESISTIVITY AND RESISTANCE OF CONDUCTORS

As we state in an earlier section, various materials differ in their abilities to conduct an electric current. For example, a long wire made of iron will exhibit a greater resistance to the flow of a current than an identical wire made of copper. The physical parameter that describes the ability of a material to conduct a current is its *conductivity*, σ. Good conductors have numerically large conductivities. A related parameter that is more commonly encountered is the *resistivity*, ρ. Mathematically, the resistivity is the reciprocal of the conductivity:

$$\rho = \frac{1}{\sigma} \tag{25-19}$$

Good conductors have numerically small resistivities. The following discussion utilizes the resistivity rather than the conductivity.

Experiments indicate that, for a given material, a long wire has a greater resistance than a short wire, and a wire of large diameter exhibits a smaller resistance than a wire of smaller diameter if the length of the two wires is the same. Therefore, it is not surprising that the following formula yields the resistance of a conductor of length L and cross-sectional area A:

$$R = \rho\frac{L}{A} \tag{25-20}$$

In this formula, ρ is the resistivity of the material of which the conductor is made. Resistivities of selected materials are given in Table 25-1.

An experimental arrangement for determining the resistivity of a given material is shown in Fig. 25-16. A suitable source of current such as a power supply or battery sends a current through a measured length of a conductor made of the material whose resistivity is being determined. The voltage developed across the measured length is determined by a sensitive voltmeter, and the resistance R is then calculated using Ohm's law. The cross-sectional area of the conductor is determined, and the following rearrangement of Eq. (25-20) is used to calculate the resistivity:

$$\rho = \frac{RA}{L} \tag{25-21}$$

TABLE 25-1 RESISTIVITY AND TEMPERATURE COEFFICIENT OF SELECTED CONDUCTING MATERIALS AT 20°C[(a)]

Material	ρ, $\Omega \cdot$ m	α, °C^{-1}
Aluminum	2.8×10^{-8}	3.8×10^{-3}
Brass	7	2.0
Copper	1.8	3.9
Gold	2.3	3.8
Iron	9.5	6.5
Nichrome	100	0.13
Silver	1.6	3.6
Tungsten	5.6	4.8

[(a)]Values vary depending upon the state of the material: annealed, work-hardened, and so forth.

An examination of the units of the quantities involved in the above equation reveals that the units of resistivity are ohm-meters, provided the length of the conductor is expressed in meters and the cross-sectional area is given in square meters. Other units are possible and are frequently encountered.

Once the resistivity of a material, such as copper, is known, it is a simple matter to use Eq. (25-20) to compute the resistance of a given conductor made of copper.

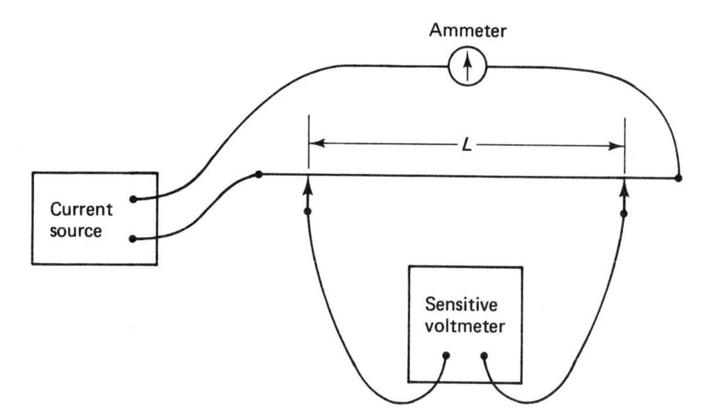

Figure 25-16 Experimental setup for determining the resistivity of a material in the form of a long wire.

EXAMPLE 25-6

Calculate the resistance of a copper wire that is 1 mm in diameter and 50 m long.

Solution. We first compute the cross-sectional area A of the wire, using a radius of 5×10^{-4} m, which is one-half of 1 mm. Using $A = \pi R^2$, we find the area to be 7.85×10^{-7} m^2. Equation (25-20) is then used to compute the resistance:

$$R = \rho \frac{L}{A}$$

$$= (1.79 \times 10^{-8}\ \Omega \cdot \text{m}) \frac{50\ \text{m}}{7.85 \times 10^{-7}\ \text{m}^2}$$

$$= 1.14\ \Omega$$

The resistivity of metals increases with temperature. For this reason, the tungsten filament of an incandescent lamp will exhibit a greater resistance when the filament is hot than when the filament is cool. This accounts for the fact that failure of most incandescent lamps occurs when power is first applied. The application of a high voltage to the cool filament results in a large current surge which can cause failure of a small section of the filament before the entire filament has time to heat up, increase in resistance, and limit the current to a smaller value.

Temperature coefficient of resistivity

The variation of resistivity with temperature is described by the temperature coefficient of resistivity, α. The resistivity at a given temperature T is given by

$$\rho = \rho_0 + \rho_0\alpha(T - 20°\text{C}) \qquad (25\text{-}22)$$

where ρ_0 is the resistivity at the reference temperature of 20°C. The units of α are $°\text{C}^{-1}$; values of α for selected materials are given in Table 25-1.

EXAMPLE 25-7 Compute the fractional increase in resistivity of copper as its temperature rises from 20°C to 100°C.

Solution. Rearrange Eq. (25-21) to obtain the fractional increase $(\rho - \rho_0)/\rho_0$:

$$\frac{\rho - \rho_0}{\rho} = \alpha(T - 20°\text{C})$$

Then,

$$\begin{aligned}\text{Fractional increase} &= (3.9 \times 10^{-3}\,°\text{C}^{-1})(100 - 20°\text{C}) \\ &= 0.31 = 31\%\end{aligned}$$

25-8 BATTERIES: TERMINAL VOLTAGE AND AMP-HOUR RATING

Terminal voltage

The electromotive force produced by a battery is determined by the chemistry of the materials making up the cell. However, the voltage that appears at the battery terminals is not always the same as the emf of the cell. This is because the current-carrying conductors inside the battery exhibit electrical resistance, and a part of the emf of the battery is lost in the form of an IR drop across the internal resistance.

A useful way to think of a real battery is to consider that a real battery is made up of an ideal battery (one without internal resistance) in series with a resistance equal to the internal resistance of the real battery. This model is illustrated in Fig. 25-17.

The voltage appearing at the terminals, the terminal voltage, equals the emf minus the voltage that is developed across the internal resistance. The terminal voltage V_t is given by

$$V_t = V_0 - IR_{\text{in}} \qquad (25\text{-}23)$$

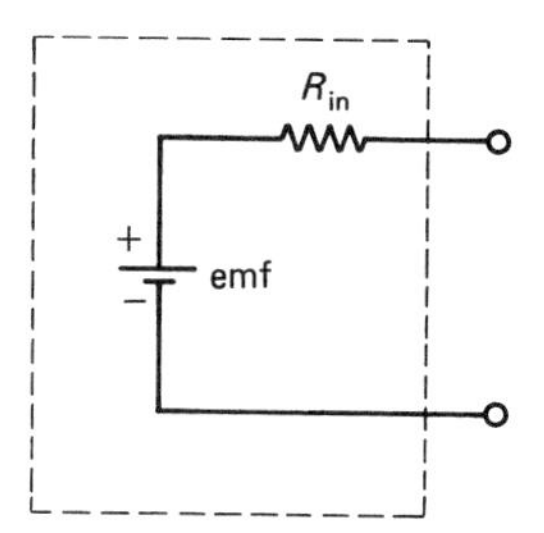

Figure 25-17 A more-realistic model of a battery.

where I is the current flowing through the battery and V_0 is the open-circuit (i.e., when $I = 0$) voltage that is identical to the emf. From Eq. (25-23) we can see that as the current flow through the battery increases, the product IR_{in} increases and V_t becomes smaller.

EXAMPLE 25-8 A sensitive electronic voltmeter measures the open-circuit voltage at the terminals of a battery to be exactly 12.00 V. If the internal resistance of the battery is 0.15 Ω, calculate what the terminal voltage will be when the battery delivers a current of 4 A to an external circuit.

Solution. We assume the sensitive voltmeter draws negligible current from the battery, so 12.00 V is the open-circuit voltage of the battery. Then, using Eq. (25-23),

$$\begin{aligned} V_t &= V_0 - IR_{in} \\ &= 12.00\ \text{V} - (4\ \text{A})(0.15\ \Omega) \\ &= 11.40\ \text{V} \end{aligned}$$

Amp-hour rating of a battery

The amp-hour rating of a battery is a measure of the energy contained in the battery when it is fully charged. This rating is used to determine the period of time the battery will function under various current loads. For example, if a certain battery is rated at 4 A · h, the battery will deliver a current of 4 A for 1 h, or it will deliver a current of 1 A for 4 h. The following relationships apply:

$$\text{Current load} \times \text{time period} = \text{amp-hours}$$

or

$$\text{Time period} = \frac{\text{amp-hours}}{\text{current load}}$$

As another example, a 12-A · h battery can be expected to deliver a current of $\frac{1}{2}$ A for a time period of 24 h without a significant decrease in terminal voltage of the battery.

The amp-hour capacity of a battery is in some sense proportional to the physical size of the battery. Small batteries may be rated in milliamp-hours instead of amp-hours.

25-9 THE WHEATSTONE BRIDGE

In this section we describe a simple circuit that can be used to determine the ohmic value of an unknown resistor provided the ohmic value of three other resistors in the circuit are accurately known. The circuit is called a *Wheatstone bridge*. It is often encountered in various forms in modern electronics circuits.

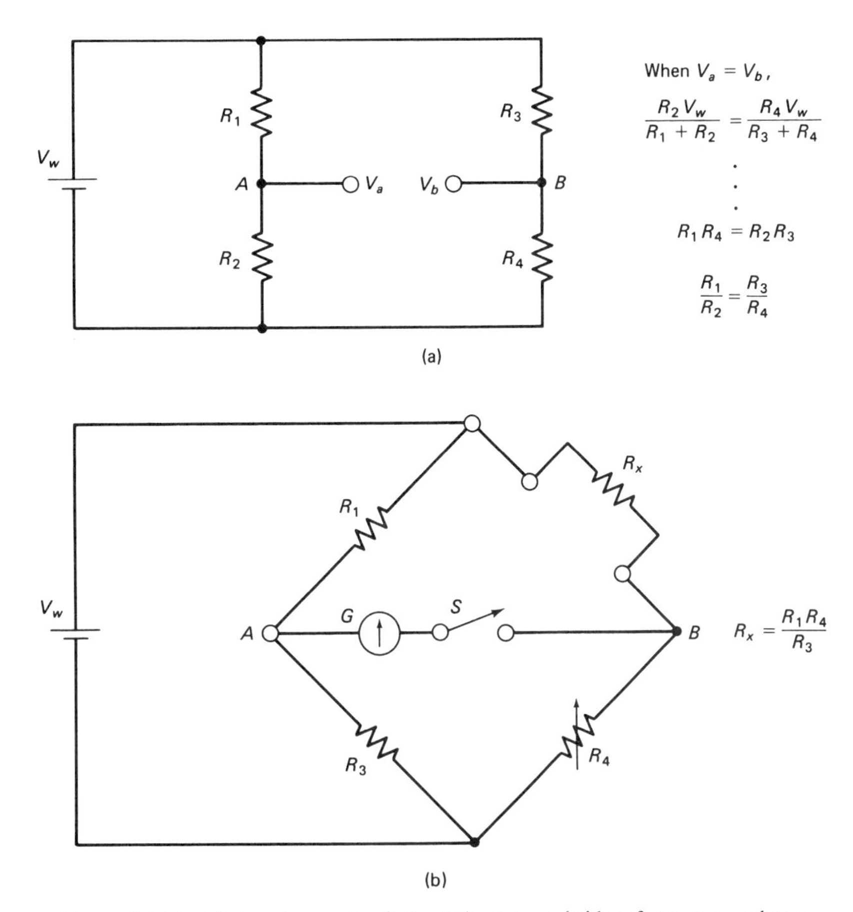

Figure 25-18 (a) Development of the Wheatstone bridge from two voltage dividers.
(b) Practical implementation of a Wheatstone bridge.

The development of the circuit is indicated in Fig. 25-18(a). A working battery of voltage V_w is shown connected to two voltage dividers that appear in parallel. The voltages V_a and V_b may be calculated using the voltage divider formula, Eq. (25-18). By adjusting the resistor values, the voltage V_a can be made equal to voltage V_b. When this is the case, R_1 and R_2 will be in the same ratio as R_3 and R_4.

Part (b) of Fig. 25-18 shows a practical rendition of the same circuit. However, R_3 is now the unknown resistor, denoted by R_x, and R_4 is shown as a variable resistor. We assume that a means is available for determining the ohmic value of R_4; we can assume it is equipped with a calibrated dial. Also shown is a sensitive *galvanometer*, *G*, and a momentary contact switch, *S*, that are connected across the bridge from point *A* to point *B*. The galvanometer indicator will deflect whenever the switch is depressed while the voltage at point *A* is different from the voltage at point *B*. Whenever the two voltages are the same, the galvanometer will not deflect when the switch is depressed, and we then say that the bridge is *balanced*. When the bridge is balanced, the resistance ratios are then the same, and the formula given in the figure can be used to determine R_x.

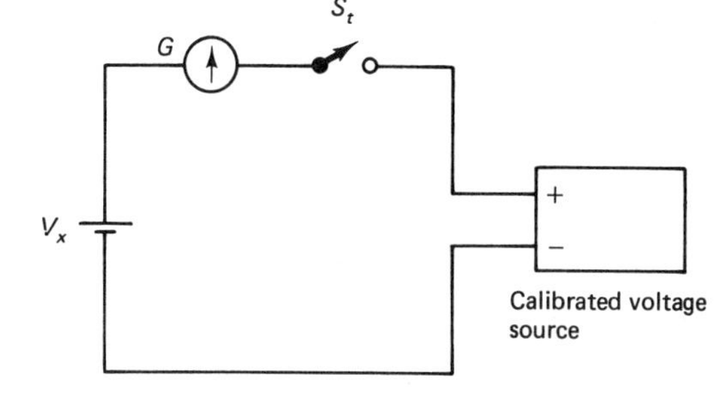

Figure 25-19 Determining the voltage V_x by a potentiometric means so that no current flows through V_x during the measurement process. When the source voltage equals V_x, no deflection will occur on the galvanometer when S_t is depressed.

A good Wheatstone bridge is capable of extreme accuracy. Observe that neither a current- nor voltage-measuring device is required. The ability of the galvanometer to respond to very small currents is essential to the accuracy of the Wheatstone bridge, but the galvanometer does not have to be calibrated. Only the three resistors must be precision components.

25-10 SLIDE-WIRE POTENTIOMETER

It is possible to determine the emf of a battery without drawing a current from the battery during the measurement process. The principle is illustrated in Fig. 25-19, where an unknown voltage source V_x is connected in series opposition with a calibrated voltage source whose output voltage is accurately known and which is variable. When the calibrated voltage is made equal to V_x, no current will flow when S_t is depressed, and no deflection will occur on the galvanometer. The unknown voltage V_x is then the same as whatever voltage is indicated on the calibrated source.

The technique of balancing one voltage against another for the purpose of measuring one of the voltages is called *potentiometric* voltage measurement. Instruments are available commercially that perform this function automatically; they are called *potentiometric voltmeters* or *potentiometric microvoltmeters*, as the case might be.

A simple system for performing potentiometric measurements of small voltages is shown in Fig. 25-20. This system requires only two precision

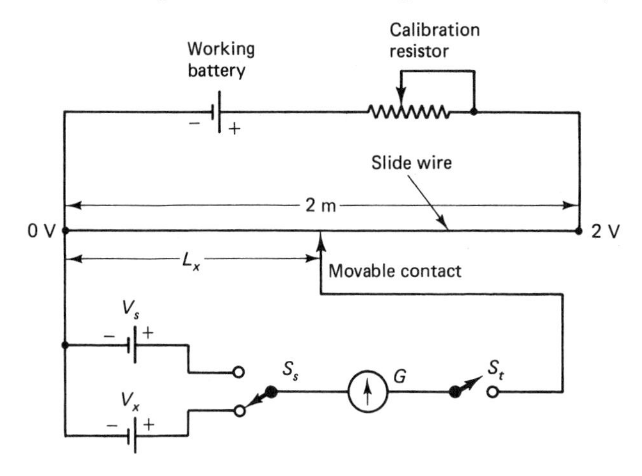

Figure 25-20 A slide-wire potentiometer for determining the emf of a battery without drawing a current from the battery whose emf is being measured.

components: the *standard cell*, V_s, whose emf must be accurately known, and the slide wire, which must be uniform along its length. The system of Fig. 25-20 is often called a *slide-wire potentiometer* and it is commonly found in sophomore physics laboratories.

The working battery sends a current through the slide wire and is adjusted using the calibration resistor so that exactly 2 V is developed across the 2-m length of the slide wire. This establishes a correspondence between length along the wire and potential difference; a change in position of the movable contact of 1 mm corresponds to a change in potential of 1 mV. That is, the potential is distributed along the wire at the rate of 1 V/m.

The standard cell is used in adjusting the current through the slide wire to give exactly 2 V across its entire length. One type of standard cell has an emf of 1.018 V. With switch S_s set to the standard cell, the movable contact is moved until the distance L_x in the figure is 1.018 m. Switch S_t is then depressed, and any deflection of the galvanometer is noted. The calibration resistor is then varied until no deflection of the galvanometer occurs when S_t is depressed. When this situation is obtained, 1.018 V is developed across the 1.018 m of the slide wire, and this ensures that 2 V will be developed across the entire length of the slide wire.

After the slide wire is calibrated as described above, switch S_s is switched over to the unknown cell. The movable contact is then adjusted until no galvanometer deflection occurs when S_t is depressed. The distance L_x then indicates the emf of the unknown cell. For example, a distance L_x of 1.573 m converts directly to a V_x of 1.573 V.

Incidentally, the term *potentiometer* was used earlier in conjunction with a variable resistor (Fig. 25-15) and we use it here to denote a device for measuring potential. The name for the variable resistor probably stems from the use of the slide wire and movable contact in the potentiometer just described.

25-11 KIRCHHOFF'S LAWS

The German physicist Gustav Kirchhoff (1824–1887) developed two principles that are useful in analyzing circuits that are more complicated than the simple series and parallel circuits we consider earlier in this chapter. The two principles are known as *Kirchhoff's voltage law* (KVL) and *Kirchhoff's current law* (KCL). First we apply the laws to simple circuits and then consider circuits of somewhat greater complexity.

Kirchhoff's current law amounts to a statement of the conservation of electric charge. It relates to a junction (or node) in a circuit, to a point where two or more conductors are joined. In simple terms:

The sum of the currents flowing toward a junction must equal the currents flowing away from the junction. Mathematically, $\Sigma I_{in} = \Sigma I_{out}$.

This is equivalent to saying that current can neither be created nor destroyed at a junction. KCL is illustrated in Fig. 25-21.

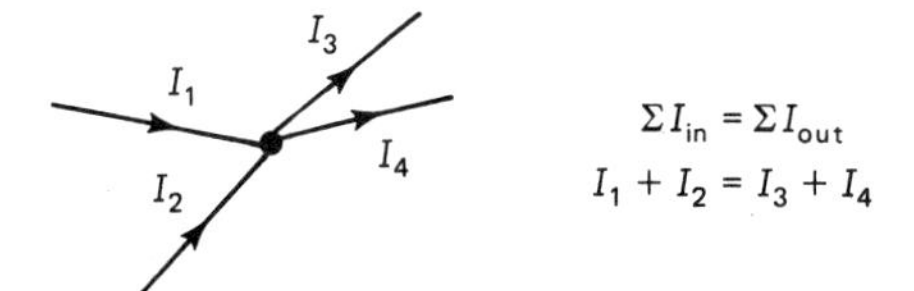

Figure 25-21 An illustration of Kirchhoff's current law.

The voltage law (KVL) deals with closed loops within electrical circuits. A loop is any closed path within a circuit that begins with a particular junction and ends at the same junction. If we limit our discussion to circuits in which only batteries and resistors appear, KVL may be stated as follows:

The sum of the battery voltages encountered in traversing a loop equals the sum of the voltages developed across the resistors in the loop. That is, $\Sigma V_b = \Sigma V_r$ *where* $V_r = I \cdot R$.

You may wish to ascertain that KVL is satisfied by the one-loop circuits of Figs. 25-5, 25-6, and 25-11. Also, verify KCL for Fig. 25-10.

For more complex circuits, KVL and KCL are used to obtain as many simultaneous equations as there are unknowns in the circuit. Loops and nodes are selected for consideration, and each loop and each node contributes an equation. The following is a procedure for writing an equation for a loop.

1. On a diagram of the circuit, label each unknown current and assign it an arbitrary direction.
2. Identify the loop to be traversed, and arbitrarily pick a direction in which to proceed around the loop.
3. Follow the sign convention:
   (a) Battery voltages are positive when the battery is traversed from + to −.
   (b) An IR drop across a resistor is positive when the assumed direction for the current through it is the same as the direction in which the loop is being traversed.
4. Go once around the loop looking for batteries and writing terms; this is for the left-hand side of the equation. Go again around the loop looking for resistors to obtain the right-hand side of the equation.
5. Rearrange (and simplify if possible) the resulting equation to put it into a form suitable for including in a system of simultaneous equations.

To obtain an equation from a node using KCL:

1. Identify the currents flowing (in) toward the node and those flowing (out) away from the node. Then form an equation according to $\Sigma I_{in} = \Sigma I_{out}$.
2. Rearrange the equation as necessary for including into the system of equations.

An example

We wish to analyze the circuit of Fig. 25-22(a). Currents I_1, I_2, and I_3 have been arbitrarily assigned a direction. One loop has already been identified. We write the equation in two steps. First, look for batteries. We find two. Hence,

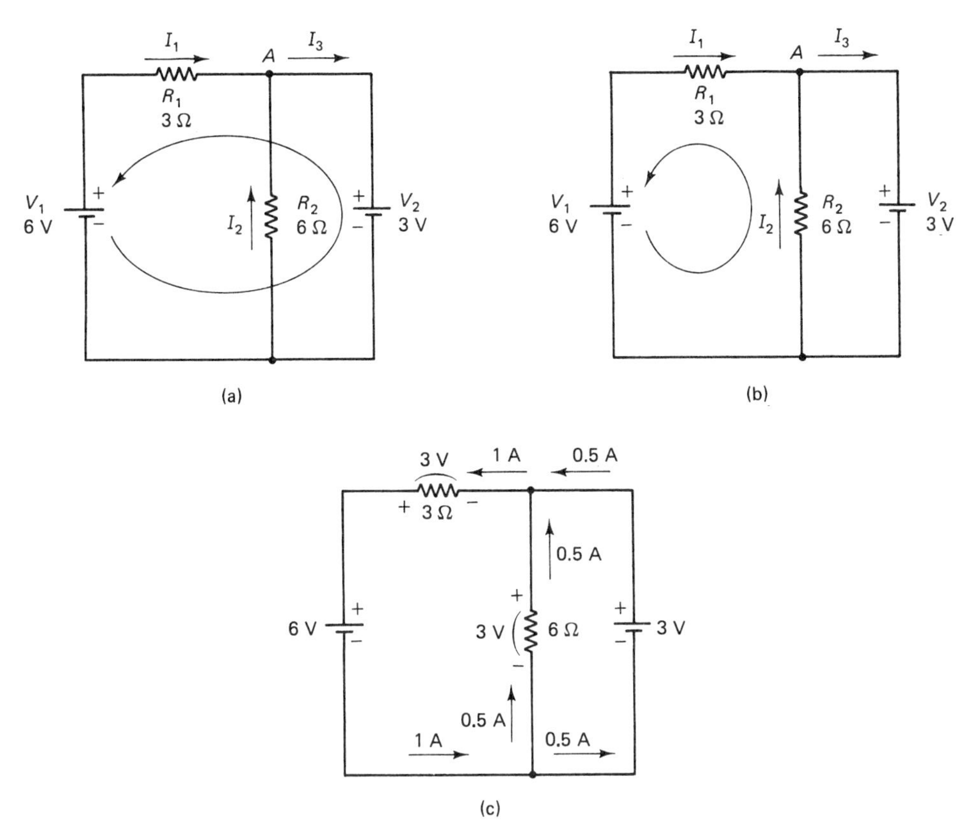

Figure 25-22 Circuit analyzed by using Kirchhoff's laws.
(a) The given circuit after labeling currents and identifying a loop.
(b) A second loop.
(c) The final result.

$$6\text{ V} - 3\text{ V} =$$

Note that V_2 is negative according to the sign convention. Next, look for resistors. We find one. Therefore,

$$6\text{ V} - 3\text{ V} = -3I_1$$

Rearranging, we obtain,

$$-3I_1 = 3\text{ V} \qquad (25\text{-}24)$$

Next, we consider the loop shown in part (b) of the figure. We find only one battery:

$$6\text{ V} =$$

We find two resistors:

$$6\text{ V} = 6I_2 - 3I_1$$

Rearranging,

$$-3I_1 + 6I_2 = 6\text{ V} \qquad (25\text{-}25)$$

Now apply KCL to the node at A. Currents *in* are I_1 and I_2; the current *out* is I_3. Therefore,

$$\Sigma I_{\text{in}} = \Sigma I_{\text{out}}$$

$$I_1 + I_2 = I_3$$

Rearranging,

$$I_1 + I_2 - I_3 = 0 \tag{25-26}$$

We now have as many equations as there are unknowns. Form Eqs. (25-24), (25-25), and (25-26) into a system of simultaneous equations:

$$\begin{aligned} -3I_1 + 0I_2 + 0I_3 &= 3 \\ -3I_1 + 6I_2 + 0I_3 &= 6 \\ 1I_1 + 1I_2 - 1I_3 &= 0 \end{aligned}$$

We apply Cramer's rule (Appendix A-8) to solve the system. The matrix of coefficients is

$$A = \begin{pmatrix} -3 & 0 & 0 \\ -3 & 6 & 0 \\ 1 & 1 & -1 \end{pmatrix} \quad \text{and} \quad \det A = 18$$

The constant matrices and their determinants are as follows.

for I_1:

$$B = \begin{pmatrix} 3 & 0 & 0 \\ 6 & 6 & 0 \\ 0 & 1 & -1 \end{pmatrix}$$

$$\det B = -18$$

for I_2:

$$C = \begin{pmatrix} -3 & 3 & 0 \\ -3 & 6 & 0 \\ 1 & 0 & -1 \end{pmatrix}$$

$$\det C = 9$$

for I_3:

$$D = \begin{pmatrix} -3 & 0 & 3 \\ -3 & 6 & 6 \\ 1 & 1 & 0 \end{pmatrix}$$

$$\det D = -9$$

And finally,

$$I_1 = \frac{\det B}{\det A} = \frac{-18}{18} = -1 \text{ A} \qquad I_2 = \frac{\det C}{\det A} = \frac{9}{18} = 0.5 \text{ A} \qquad I_3 = \frac{\det D}{\det A} = \frac{-9}{18} = -0.5 \text{ A}$$

We note that currents I_1 and I_3 are negative. This means that the direction initially assumed for each of these is just opposite of what it should be, and we should turn them around.

Once the currents are known, the voltage appearing across each resistor is easily calculated. Moreover, the polarity of the voltage can be determined by remembering that current (electron flow) flows through a resistor from the negative to the positive end. The final result is shown in Fig. 25-22(c).

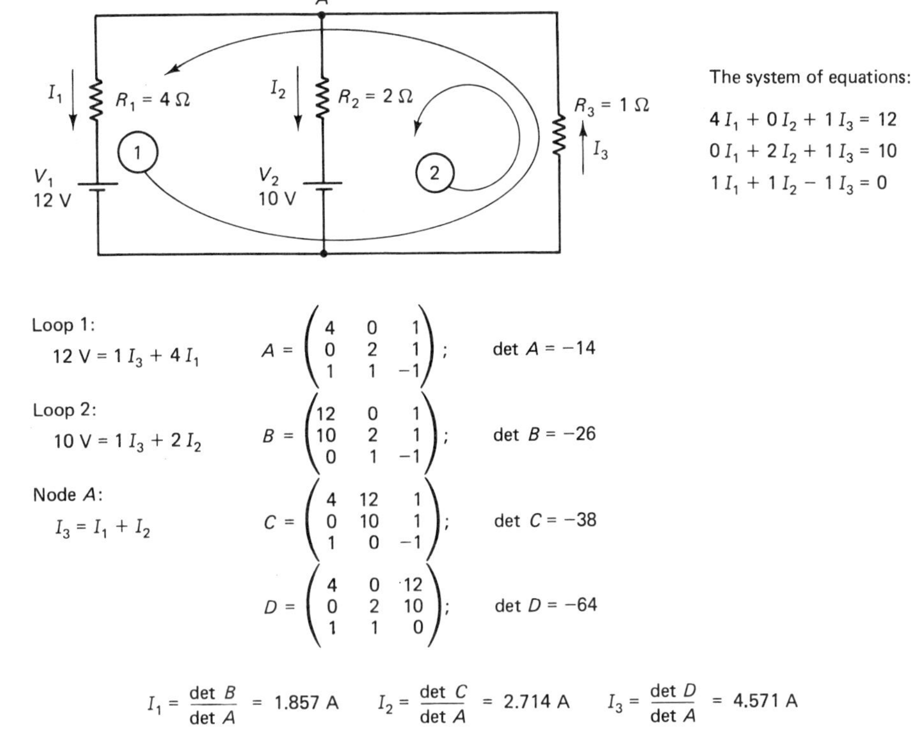

Figure 25-23 Solution of a complex network by Kirchhoff's laws.

Another example

Another circuit is shown in Fig. 25-23, along with the heart of the analysis. Details are omitted for brevity. This example is significant because it parallels the problem that would arise if two real batteries of unequal terminal voltage were connected in parallel to drive a load. In this example, the battery having the least voltage ($V_2 = 10$ V) actually carries the most of the current load. This arises because of the relative values of R_1 and R_2.

Final comment

Complex circuit analysis is a large part of an electrical engineer's stock-in-trade. In addition to Kirchhoff's laws, there are many other theorems and principles available that facilitate circuit analysis. Among these are the superposition theorem, Thevenin's theorem, Norton's theorem, Millman's theorem, and many variations and combinations. These topics are not unduly complicated, but they are best left to those who choose to specialize in electricity and electronics.

To Go Further

Find out about the lives and work of:

- Georg Simon Ohm (1789–1854), German physicist
- Andre Ampere (1775–1836), French scientist
- Gustav Kirchhoff (1824–1887), German physicist

References

The following popular texts treat electrical principles and network theory comprehensively at the introductory level:

BOYLESTAD, ROBERT L., *Introductory Circuit Analysis*. Columbus, Ohio: Charles E. Merrill Publishing Company, 1977.

HARTER, JAMES H., PAUL Y. LIN, *Essentials of Electric Circuits*. Reston, Va.: Reston Publishing Co., 1982.

JACKSON, HERBERT W., *Introduction to Electric Circuits*, 5th ed. Englewood Cliffs, N. J.: Prentice-Hall, Inc., 1981.

MOTTERSHEAD, ALLEN, *Introduction to Electricity and Electronics*. New York: John Wiley & Sons, Inc., 1982.

Questions

1. Many texts define the direction of an electrical current to be in the direction of movement of positive charge.
   (a) How is this definition somewhat at variance with what actually occurs in metallic conductors?
   (b) In what sense is a flow of negative charge to the left equivalent to the flow of positive charge to the right?
2. Explain the analogy between a battery and a water pump. How is a potential difference analogous to a difference in hydrostatic pressure?
3. Students often speak of "voltage flowing from a battery to a lamp." Is it proper to speak of voltage as flowing?
4. We defined current, for our purposes, as the *flow* of electrons. Is it redundant to say that current flows? Is this the same as saying that a flow flows? Is this a problem?
5. Why is it improper to say that resistance slows down the current? Why is it better to say that the resistance *limits* the current?
6. Someone said that electrical resistance is "the trouble electrons have in getting through a conductor." Explain how this idea makes it easier to remember that resistors in series are added to obtain the equivalent resistance.
7. What is the advantage of connecting batteries in parallel?
8. * Draw the schematic diagram of a circuit that can be used to determine the power dissipation in an incandescent lamp. In particular, show how the ammeter and the voltmeter are connected.
9. Suppose a 3-Ω and a 4-Ω resistor are connected in series. If the combination is connected to a 12-V battery, which resistor will dissipate the most power?
10. Give a step-by-step procedure for determining the resistance of a long wire made of copper. Draw a schematic diagram of the circuit used.
11. *A fan is directed at an electric heater in order to get the most heat out of the heater. With the fan turned on, the current drawn by the heater is greater than when the fan is off. Explain how this observation is consistent with the principle of the conservation of energy.
12. Would you expect a simple DC motor to draw more current when it is driving a heavy mechanical load or when it is running free? Why?
13. *Give a step-by-step procedure for determining the internal resistance of a battery. What equipment is required? (*Caution*: Do not use an ohmmeter.)
14. How does a potentiometric voltage measurement differ from a voltage measurement made using an ordinary voltmeter? If the voltage of a battery is made potentiometrically and also with an ordinary voltmeter, which reading will be higher?

15. Devise an analogy between cars approaching and leaving an intersection and Kirchhoff's current law.

*16. Devise an analogy using a water pump and a series of dams to illustrate Kirchhoff's voltage law.

Problems

1. What current is flowing in a wire if 40 C of charge pass a point on the wire in 8 s?
2. How many coulombs of charge pass a point on a wire per minute if the wire is conducting a current of 10 A?
3. A copper sphere 1 cm in diameter contains about 6.65×10^{22} free electrons (one per atom).
   (a) What charge in coulombs does this represent?
   (b) If a certain wire carries a current of 1 A, how long would it take for this number of electrons to flow past a point on the wire?
4. What resistance connected to a 12-V battery will produce a current of 6 A through the battery?
5. Compute the resistance of incandescent lamps of the following wattage ratings if the lamps are designed to operate at 120 V.
   (a) 100 W **(b)** 60 W **(c)** 25 W **(d)** 7 W
6. A popular type of small indicator lamp extensively used in electronic apparatus of all sorts is the *light-emitting diode* (LED). A typical LED requires 15 mA of current at about 2 V to operate at its rated brilliance. Calculate the power dissipation of the LED under these conditions.
7. For the circuit of Fig. 25-24, calculate:

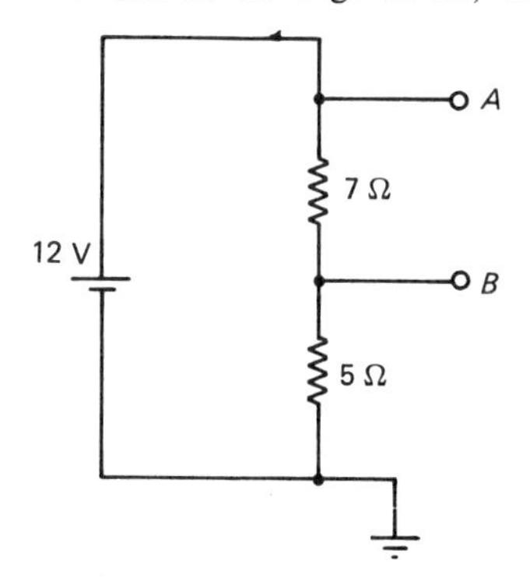

Figure 25-24

(a) the current that flows in the circuit; **(b)** the voltage at point A; **(c)** the voltage at point B; **(d)** the power dissipation in the 7-Ω resistor.

8. **(a)** Determine the current that flows through each resistor in Fig. 25-25.
   (b) Compute the power dissipation in each resistor.

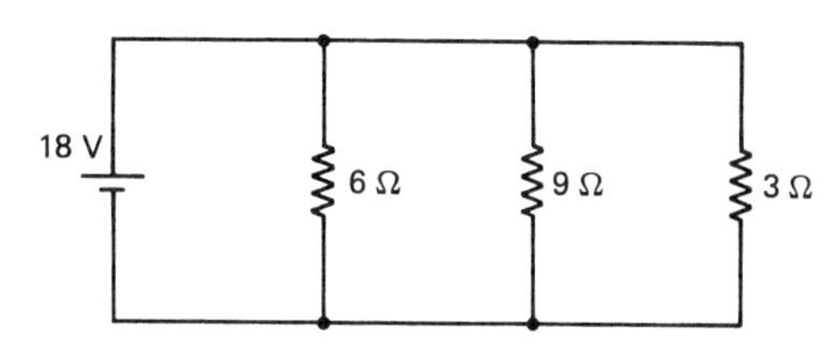

Figure 25-25

9. Refer to the circuit of Fig. 25-26 and determine: **(a)** the current that flows through the 12-V battery; **(b)** the current that flows through the 6-V battery; **(c)** the voltage drop across the 6-kΩ resistor; **(d)** the voltage drop across the 3-kΩ resistor; **(e)** the potential difference between points A and B.

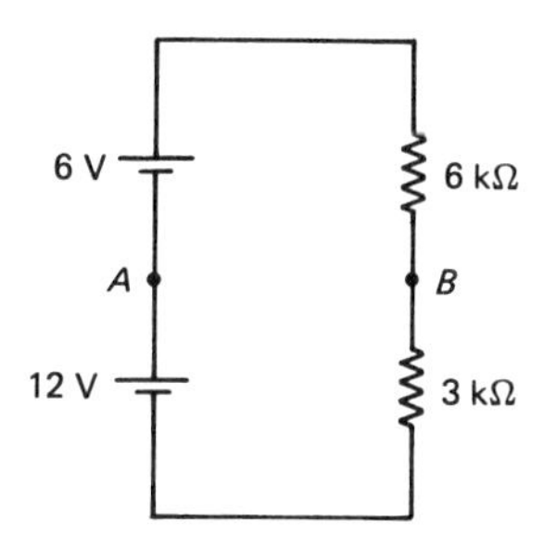

Figure 25-26

10. The emf of an ordinary flashlight battery is about 1.5 V, and a typical flashlight uses two cells connected in series. A typical flashlight bulb requires 250 mA at 3 V to operate at normal brilliance.
    (a) Find the resistance of a flashlight bulb while it is lighted.

(b) Calculate the power dissipation of a flashlight bulb.

11. What current flows through the battery of Fig. 25-27?

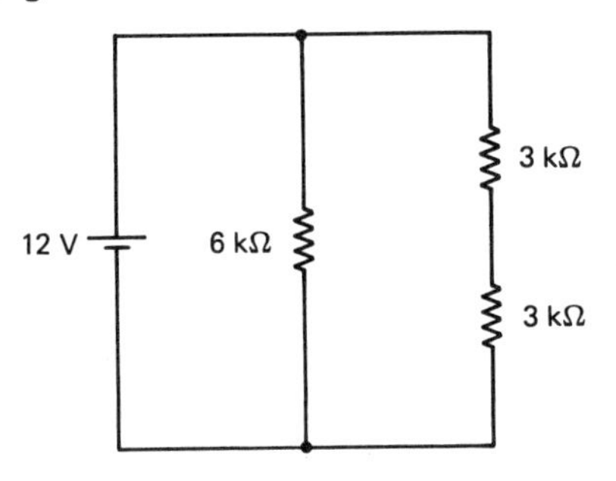

Figure 25-27

12. Determine the current that flows through each resistor and through the battery in Fig. 25-28.

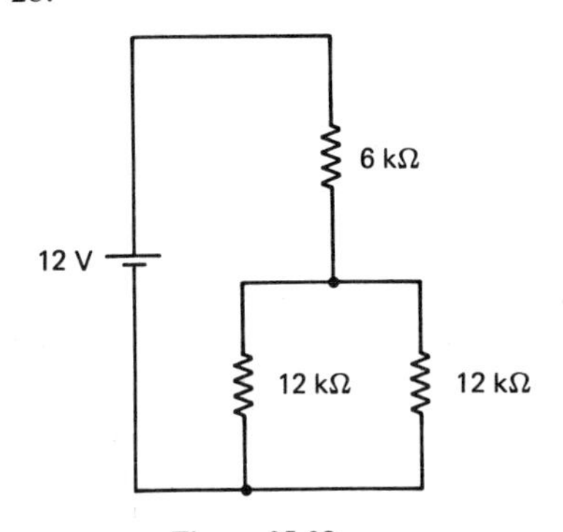

Figure 25-28

13. What is the output voltage of the voltage divider of Fig. 25-29?

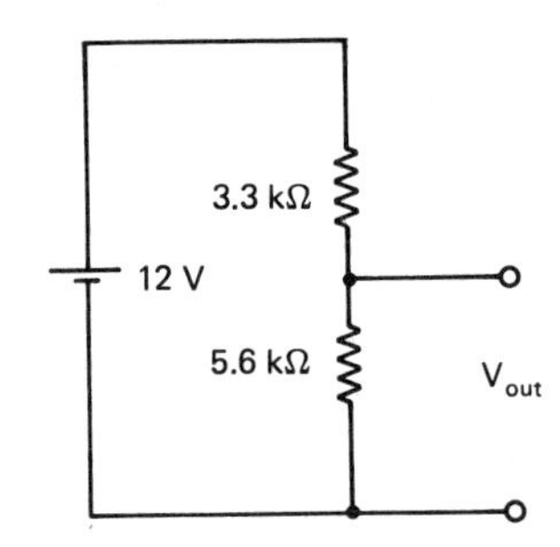

Figure 25-29

*14. In the circuit of Fig. 25-29, suppose a 1-kΩ resistor is connected between points *A* and *B*.

(a) What voltage will be applied to the resistor?

(b) What current will flow through the resistor?

What current flows through the 3.3-kΩ resistor: (c) before; (d) after the 1-kΩ resistor is connected?

15. A small motor is designed to operate with a terminal voltage of 6 V while drawing a current of 2 A. It is desired to operate the motor from a 12-V battery.

(a) Determine the ohmic value of a resistor connected in series with the motor that will permit it to operate with 6 V across its terminals.

(b) What will be the power dissipation in the resistor?

16. Commercial resistors are rated according to ohmic value and maximum power dissipation. Determine (a) the maximum voltage and (b) the maximum current that can be applied to or conducted by a resistor rated 10 Ω, 5 W.

17. Determine the resistance of a copper wire 3 m long, whose diameter is 0.5 mm.

18. Suppose the wire of Problem 17 is made of nichrome instead of copper. What is its resistance?

19. What voltage will be developed across a 1-m length of aluminum wire if the wire is 1.2 mm in diameter and if 0.5 A flows through the wire?

*20. A circular-mil is a unit of area equal to the area of a circle 1 mil (0.001 in.) in diameter.

(a) Obtain a conversion factor for converting resistivities in ohm-meters to units of ohm-circular-mils/foot.

(b) What is the resistivity of copper in these units?

*21. Use the data of Table 25-2 to determine the resistivity of annealed solid copper.

22. Electric power is run from a house to a water pump 100 yd away. AWG 10-gauge wire is used. (See Table 25-2.) The pump motor draws 10 A of current. If the line voltage is 120 V at the house, what voltage will be applied to the motor? [*Note*: Two conductors are required.]

TABLE 25-2 AMERICAN WIRE GAUGE (AWG) (PARTIAL LISTING) (ANNEALED SOLID COPPER, 20°C)

Gauge number	Diameter, mils	Ω/1000 ft
0000	460	0.0490
0	325	0.0983
4	204	0.249
8	128	0.628
10	102	0.999
12	81	1.59
14	64	2.53
16	51	4.02
18	40	6.39
20	32	10.2
24	20.1	25.7
28	12.6	64.9
32	8.0	164
36	5.0	415
40	3.1	1049

Note: 1 mil = 0.001 in.
1 mm = 39.37 mils

23. Repeat the preceding problem but assume AWG 14-gauge wire is used.

24. The open-circuit voltage of a certain battery is 13.24 V. When a current load of 8 A is placed on the battery, the terminal voltage drops to 12.46 V. What is the internal resistance of the battery?

25. A 12-V battery whose internal resistance is 0.05 Ω is accidentally short-circuited. That is, a large conductor of negligible resistance comes in direct contact with the terminals of the battery. Compute the current that flows through the battery while it is short-circuited.

***26.** Suppose a battery has an emf of 12 volts and an internal resistance of 24 Ω, which is rather large. The battery is connected, one at a time, to resistors having the following values: **(a)** 48 Ω; **(b)** 36 Ω; **(c)** 30 Ω; **(d)** 24 Ω; **(e)** 18 Ω; and **(f)** 12 Ω. Calculate the current that flows in the circuit for each case.

***27.** Using the results of Problem 26, determine the power dissipation **(a)** in the battery and **(b)** in the external resistor for each case. For which case is the power dissipation in the external resistor a maximum? For which case is the power dissipation inside the battery a maximum? For which case is the total power dissipation a maximum?

28. What would be the amp-hour rating of a battery that can deliver a current of 0.25 A for 40 h before becoming discharged?

29. For what period of time can a 12-A · h battery deliver a current of 25 mA?

30. Assuming the bridge of Fig. 25-30 is balanced, compute the value of R_x.

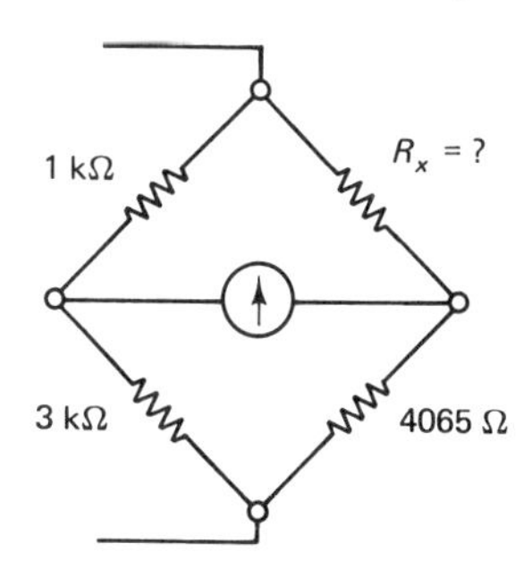

Figure 25-30

31. If the working battery supplying current to the bridge in Fig. 25-30 has a terminal voltage of 12 V, calculate the current that flows through the battery when the bridge is balanced.

*32. Suppose the slide wire of the potentiometer of Fig. 25-20 is made of nichrome of diameter 0.5 mm.
   (a) Compute the resistance of a 2-m length of the wire.
   (b) What current must flow through the slide wire to produce a voltage across the wire of 2 V?
   (c) If the working battery has a terminal voltage of 6 V, what value should the calibration resistor be in order to establish 2 V across the slide wire?
   (d) What will be the power dissipation in the calibration resistor?

*33. Two batteries having an open-circuit voltage of 12.00 V are connected in parallel to a 1-Ω load. The internal resistance of battery A is 0.1 Ω and of battery B is 0.2 Ω. Use Kirchhoff's laws to determine the current flow through each battery.

*34. A new car battery has an emf of 13.6 V and an internal resistance of 0.05 Ω. It is connected in parallel with an older battery having an emf of 12.4 V and an internal resistance of 0.1 Ω. The parallel combination is connected to a load resistance of 0.5 Ω. Use Kirchhoff's laws to find the current flow through each battery and through the load.

*35. Determine the current that flows through each resistor in the circuit of Fig. 25-31. What is the voltage at point A?

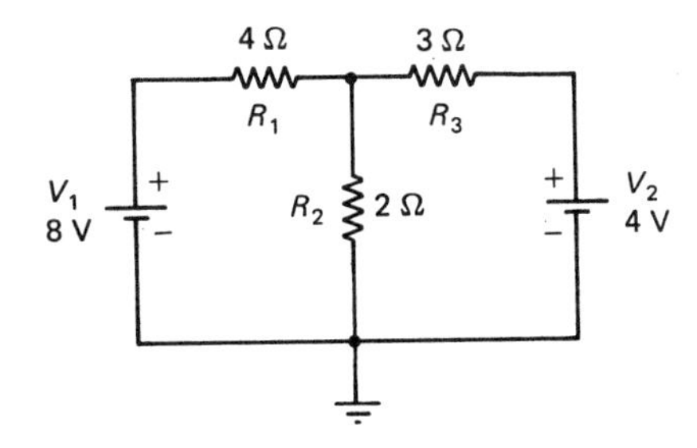

Figure 25-31

*36. Reverse the polarity of V_2 in Fig. 25-31 and then find the current in each resistor and the voltage at point A.

CHAPTER 26

THE MAGNETIC FIELD

The discovery that a fundamental relationship exists between an electric current and a magnetic field represented a major advance in the theory of electricity and magnetism. Today the interaction between electric currents and magnetic fields is used in inductors, transformers, motors, generators, television deflection systems, meter movements, and on and on. In this chapter we investigate the basic properties of magnetic fields. In the last sections of this chapter we describe the D'Arsonval galvanometer and show how a meter movement can be used to make any of a voltmeter, ammeter, or ohmmeter.

26-1 THE MAGNETIC FIELD OF PERMANENT MAGNETS

Almost everyone has had the opportunity to play with magnets. We are initially surprised by the fact that a magnet can attract a paper clip, a nail, or another magnet without touching the object. There seems to be something in the space surrounding a magnet that is capable of exerting a force on certain objects. We call this "something" a magnetic field.

The magnetic field of a magnet can be pictured by drawing magnetic lines of force around the magnet. The magnetic field of a bar magnet is shown in Fig. 26-1. Notice that the lines are closer together near the ends of the magnet where the magnetic field is strongest. The end of a bar magnet constitutes what we call a *magnetic pole*. One end is a *north pole* and the other end is a *south pole*.

The earth is surrounded by a magnetic field whose lines of force run, generally speaking, in a north-south direction. A long bar magnet, freely suspended so that it can rotate in a horizontal plane, will align itself parallel to the magnetic field of the earth. The end of the bar magnet that points north is the north pole of the magnet. An ordinary compass consists of a needle-shaped bar magnet mounted on a pivot so that it can align itself with the magnetic field of the earth.

If two bar magnets are brought together end-to-end, one will exert a force on the other. The force will be attractive if the two approaching poles are of opposite

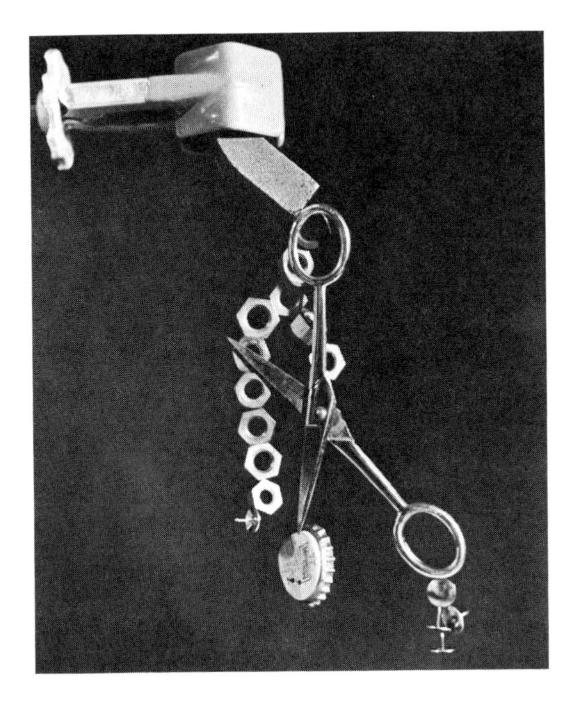

Induced magnetism causes objects to cling to a magnet and to each other.

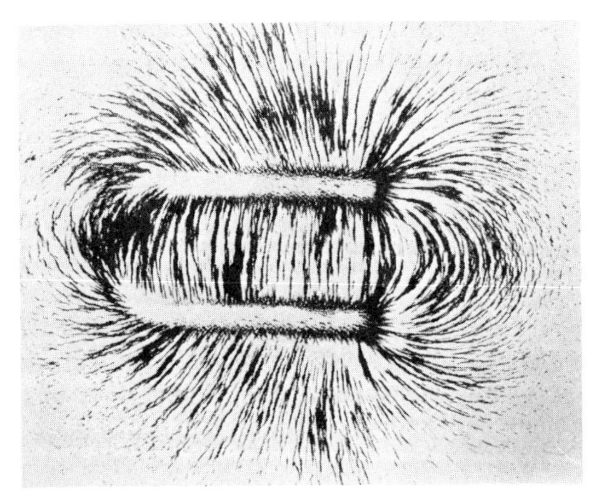

Iron filings sprinkled on a paper above a horseshow magnet indicate the shape of the magentic field.

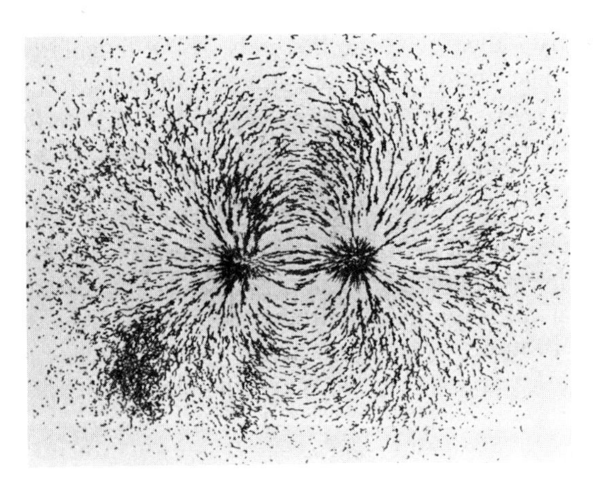

Iron filings sprinkled over a bar magnet.

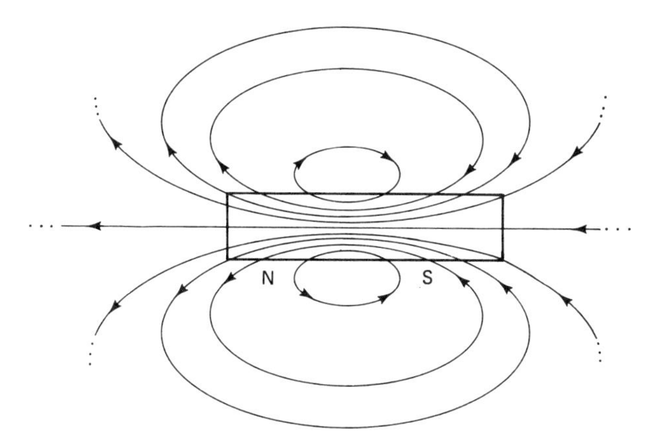

Figure 26-1 The magnetic field around a bar magnet. Each magnetic line forms a continuous loop.

type, but it will be repulsive if the poles are of the same type. If a small compass is located near the north pole of a magnet, the north pole of the compass needle will be repelled and the south pole of the compass will be attracted. Hence, the compass needle will rotate and align itself parallel to the direction of the magnetic lines at the location of the compass.

The direction of a magnetic field is defined to be that direction in which a north pole tends to move when placed in the field. Thus, the lines of force around a bar magnet are directed from its north pole to the south pole, as shown in Fig. 26-1.

Recall that lines are drawn to indicate the nature of an electric field surrounding electrostatic charges. The lines of an electric field may be imagined to begin on positive charges and end on negative charges. However, magnetic lines of force have no beginning and no end. This is because there is no such thing as an isolated north magnetic pole or an isolated south magnetic pole. North and south poles always occur in pairs. An isolated pole is called a *magnetic monopole*, but none has ever been observed.

If an attempt is made to isolate a magnetic pole by bisecting a bar magnet, as shown in Fig. 26-2, the two halves of the original magnet are themselves imbued with both a south pole and a north pole. Further division of the magnet always results in smaller magnets, each of which has both a north and a south pole.

In light of this, we conclude that magnetic lines of force can exist only as complete, closed loops. In Fig. 26-1, each line can be traced around a complete circuit. Any attempt to establish a magnetic field in a region must provide for the magnetic field to exist as a myriad of lines of force, each of which exists as a closed loop.

Magnetic lines cannot intersect. If they did, a magnetic field would have two different directions at the point of intersection. This, of course, is impossible.

Magnetic lines behave as if each line were under a tension along its length

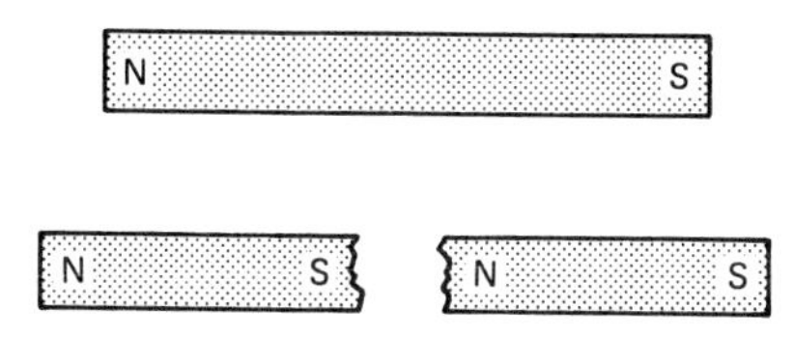

Figure 26-2 When a magnet is broken, each part develops a north and a south pole.

that tends to pull the line to its shortest possible length. Moreover, magnetic lines appear to resist being crowded together.

We must emphasize that magnetic lines only *represent* a magnetic field. A magnetic field does not exist in the form of lines; if it did, we might be tempted to ask what lies between the lines. Furthermore, nothing flows along the lines, and we may be amused by the idea of looking at a magnetic line of force through a microscope. Nevertheless, magnetic lines are very useful as an aid to visualizing a magnetic field. The method is due to Michael Faraday.

Magnetic domains

If the magnetic character of the interior of a permanent magnet is examined, the volume of the metal is found to be divided into regions of irregular shape that behave as tiny magnets. Each region is called a magnetic *domain*, and each domain has a definite magnetic orientation. The size of magnetic domains are typically only a fraction of a millimeter along the longest dimension.

In an unmagnetized specimen of iron, the magnetic domains are randomly oriented so that no net magnetic field is exhibited on the outside of the metal. However, the specimen may be magnetized by stroking it along one pole of a strong magnet and causing the domains to become aligned. An external field will cause the magnetic direction of the domains to rotate, or it will cause part of the domains to shrink in size while those of the preferred direction become larger. Whether a piece of iron is magnetized or unmagnetized is determined by the state of the domains. If they are oriented at random, no external magnetic field will be observed. But if the domains are aligned, the iron will act as a magnet. This is illustrated in Fig. 26-3.

A permanent magnet can be demagnetized by any process that tends to randomize the domains. Beating on a magnet with a hammer will eventually destroy the magnetism; heating the magnet to a high temperature will destroy the magnetism due to increased thermal agitation.

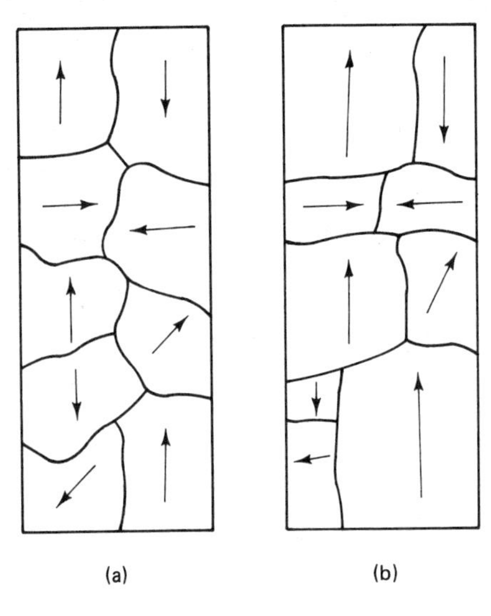

Figure 26-3 (a) In an unmagnetized specimen of iron, the domains are randomly oriented.
(b) When the domains are preferentially aligned, the specimen will exhibit an external magnetic field and will act as a permanent magnet.

If an unmagnetized piece of iron is placed in a magnetic field, the object becomes temporarily slightly magnetized. An unmagnetized nail held near a bar magnet will be attracted to the bar magnet because magnetic poles will be induced on the nail, as shown in Fig. 26-4.

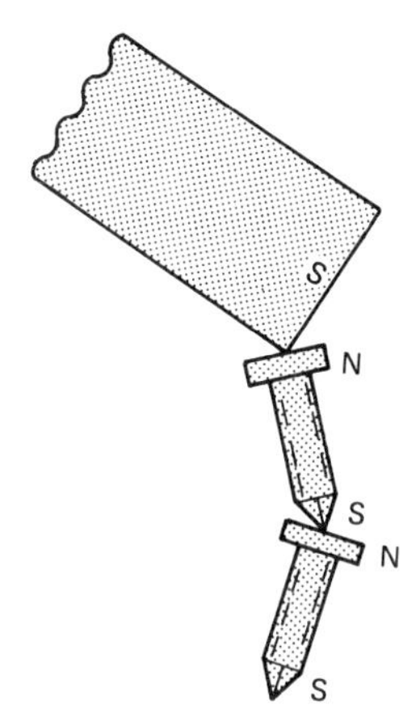

Figure 26-4 A bar magnet temporarily induces magnetic poles in objects that were originally unmagnetized.

26-2 MAGNETIC FLUX

Suppose we wish to express the *quantity* of magnetic field that comes out of the end of a certain bar magnet. How can we do this? First of all, it is convenient to imagine the magnetic field to consist of a number of discrete lines that can be counted. Then, we can express the quantity of magnetic field by giving the number of lines that come out of the end of the magnet.

The term *magnetic flux* refers to the number of lines in a particular magnetic field. The SI unit of magnetic flux is the weber (Wb), and the symbol for the magnetic flux is the Greek letter phi (Φ). The question that naturally arises at this point is how many lines does it take to make a weber? This is an appropriate question, but it is one that physicists never answer because, we must remember, the lines have no actual physical existence. They simply provide a good way for us to think about a magnetic field. There is another way to define the amount of magnetic flux contained in 1 Wb without having to refer to magnetic lines. We see that in a later section, but for now we can simply say that a weber represents a certain number of magnetic lines without worrying about what that number actually is.

Magnetic flux density

The physical quantity that we relate to the strength of a magnetic field is the number of lines that cross a unit area constructed perpendicular to the direction of the lines. We call this quantity the *magnetic flux density*, B. It is also known as the *magnetic induction*. The flux density is a vector quantity because the magnetic field has a direction as well as a magnitude. The magnitude of the flux density is given by

$$B = \frac{\Phi}{A} \tag{26-1}$$

The SI unit of flux density is the tesla (T). A tesla is equal to a flux density of one weber per square meter:

$$1 \text{ tesla} = \frac{1 \text{ weber}}{1 \text{ m}^2}$$

An illustration of the definition of flux density is given in Fig. 26-5.

An older unit of flux density that is often encountered is the gauss (G). It is a much smaller quantity of flux than the tesla:

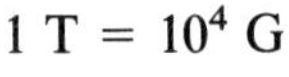

$$1 \text{ T} = 10^4 \text{ G}$$

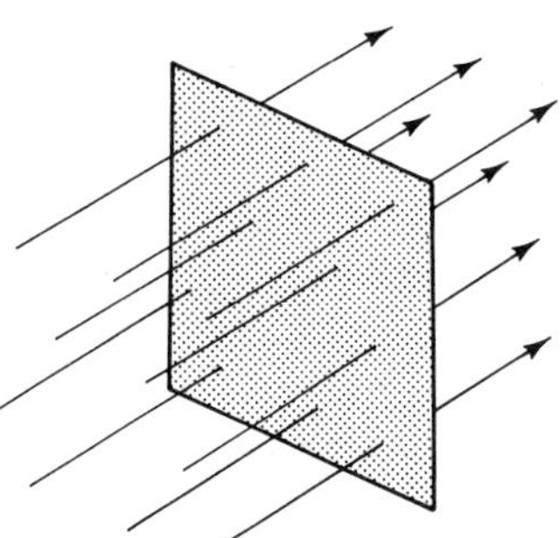

Figure 26-5 When a quantity of flux equal to 1 Wb crosses an area of 1 m^2, the flux density is 1 Wb/m^2, which is defined to be 1 T.

26-3 ELECTRIC CURRENT AND MAGNETISM

If a current-carrying wire is suspended vertically, as shown in Fig. 26-6, a compass needle will orient itself tangent to a circle drawn around the wire. When the current is turned off, the compass will again assume a north-south orientation. This simple experiment demonstrates that an electric current produces a magnetic field. Furthermore, the presence of iron is not essential to the production of a magnetic field. It is now known that all magnetism, including the magnetism of the earth and of permanent magnets, results from electric current.

More advanced texts derive the following formula for the magnetic flux density, B, a distance R from a long wire carrying a current I:

$$B = \frac{\mu_0 I}{2\pi R} \tag{26-2}$$

The constant μ_0 is called the *permeability of free space*. Its value, in SI units, is exactly

$$\mu_0 = 4\pi \times 10^{-7} \text{ W/A} \cdot \text{m}$$

The permeability of a material, or of free space, is a measure of the ease with which magnetic flux may be established in the material.

Iron inserted into a current-carrying coil becomes magnetized and attracts iron filings.

A current-carrying coil attracts iron filings even when no other iron is present.

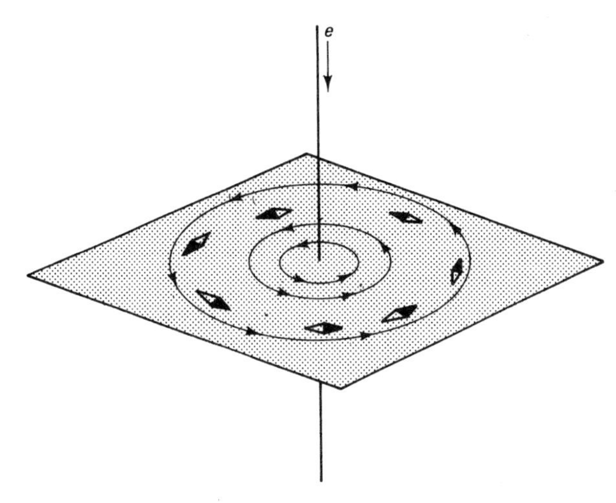

Figure 26-6 A magnetic field surrounds a current-carrying wire.

If a current-carrying wire is formed into a circular loop, as shown in Fig. 26-7, the magnetic field will be more concentrated within the loop because more lines of flux are caused to flow across a smaller area. For a loop of radius R carrying a current I, the flux density at the center of the loop is given by

$$B = \frac{\mu_0 I}{2R} \tag{26-3}$$

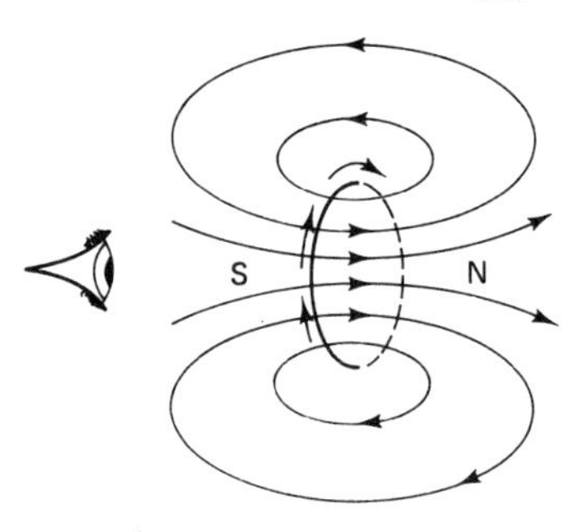

Figure 26-7 The magnetic field of a current loop. Electron flow is counterclockwise when the loop is viewed as shown.

The direction of the magnetic field around a long, straight wire and around a current loop can be remembered by the *left-hand rule*, illustrated in Fig. 26-8. If the conductor is gripped with the left hand so that the left thumb points in the direction of the electron flow, the fingers will extend in the direction of the magnetic field.

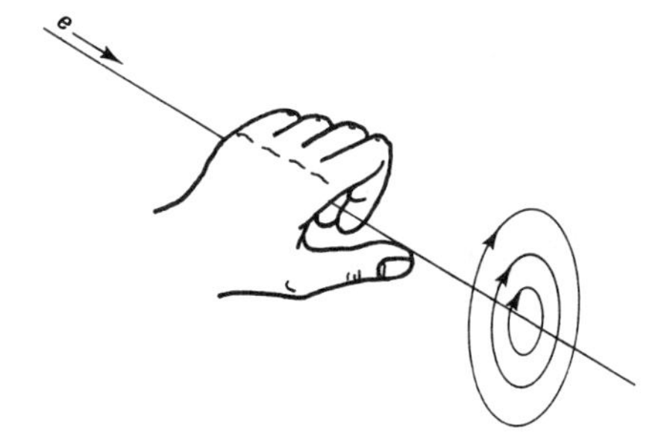

Figure 26-8 The left-hand rule.

Many texts use a right-hand rule to obtain this direction, but those texts define the direction of an electric current to be in the direction of the flow of positive charge. In this text we deal more directly with electron flow, and therefore we have to use the left hand instead of the right hand.

EXAMPLE 26-1 A long wire carries a current of 10 A. Calculate the magnetic flux density at a point 5 cm from the axis of the wire.

Solution. We use Eq. (26-2):

$$B = \frac{\mu_0 I}{2\pi R}$$

$$= \frac{(4\pi \times 10^{-7})(10\text{ A})}{2\pi\,(0.05\text{ m})}$$

$$= 4.00 \times 10^{-5}\text{ T}$$

$$= 0.4\text{ G}$$

EXAMPLE 26-2 A wire carrying a current of 10 A is formed into a loop of radius 5 cm. Determine the magnetic flux density at the center of the loop.

Solution. Using Eq. (26-3) we obtain

$$B = \frac{\mu_0 I}{2R}$$

$$= \frac{(4\pi \times 10^{-7})(10\text{ A})}{2(0.05\text{ m})}$$

$$= 12.6 \times 10^{-5}\text{ T}$$

$$= 1.26\text{ G}$$

Inspection reveals that the magnetic field of a current loop resembles that of a bar magnet, and the shape of the field is that of a magnetic *dipole*. The dipole field results whenever a north and a south pole are slightly separated in space that is otherwise free of materials that would alter the shape of the field. The field around a long bar magnet closely approximates a dipole field.

It is impossible to locate the north and south poles of a current loop. We can identify a region as being the north side of the loop, but we cannot point to a particular spot and identify it as a magnetic pole.

The concept of magnetic poles is useful mainly when dealing with permanent magnets or electromagnets that have iron pole-pieces. In a toroid, we shall see that it is impossible to locate poles even approximately. But this is no problem because poles are not essential to magnetic fields.

It is often convenient to speak of the *magnetic moment* of a current loop. The magnetic moment is defined as the product of the current flowing around the loop and the area enclosed by the loop. Thus, we can refer to the magnetic moment of an electron in orbit around a hydrogen nucleus; by this we refer to the magnetic field produced by the very rapid circular motion of the electron.

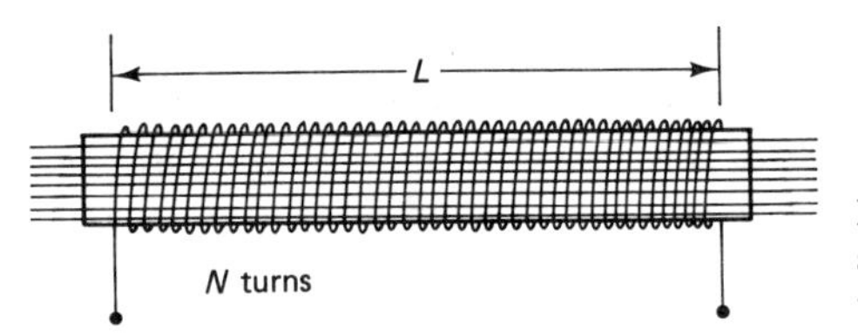

Figure 26-9 The magnetic field inside a solenoid is uniform and runs parallel to the axis.

The solenoid

A *solenoid* is a coil or wire wound in the form of a circular cylinder, as shown in Fig. 26-9. The magnetic field of each turn of wire adds vectorially with the field of the other turns to produce a field inside the solenoid that is uniform and runs parallel to the axis. The magnetic flux density inside a solenoid of length L which has a total of N turns is given by

$$B = \frac{\mu_0 NI}{L} \tag{26-4}$$

where I is the current flowing through the solenoid. We assume the length of the solenoid to be at least 10 times its diameter; otherwise the calculation of the magnetic field inside the solenoid is much more complex.

EXAMPLE 26-3 A solenoid 1 m long is 4 cm in diameter. It is wound uniformly, one turn per millimeter, and has a total of 1000 turns. Determine the magnetic flux density inside the solenoid if a current of 1 A flows through the winding.

Solution. Use Eq. (26-4):

$$\begin{aligned} B &= \frac{\mu_0 NI}{L} \\ &= \frac{(4\pi \times 10^{-7}\ \text{Wb/A}\cdot\text{m})(1000\ \text{turns})(1\ \text{A})}{1\ \text{m}} \\ &= 12.57 \times 10^{-4}\ \text{T} = 12.57\ \text{G} \end{aligned}$$

The toroid

A *toroid* is a coil wound in the shape of a doughnut, as shown in Fig. 26-10. The magnetic lines run parallel to the circumference of the toroid, and *all* the magnetic lines are contained within the interior of the coil [in the shaded region of Fig. 26-10(b)]. No magnetic flux exists in the hole of the doughnut nor in the region surrounding the doughnut. This is a peculiar feature of a toroid.

The magnetic field within a toroid is not uniform as it is within a long solenoid. The field is given by

$$B = \frac{\mu_0 NI}{2\pi R} \tag{26-5}$$

where N is the number of turns wound on the toroid and where R is the distance from the center of the toroid as shown in the figure. We discern from this formula

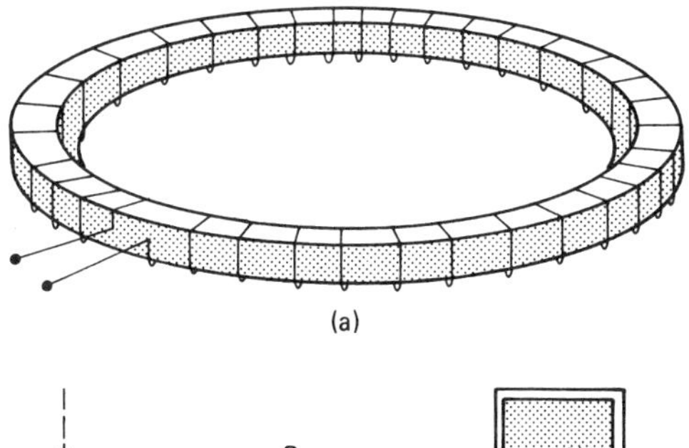

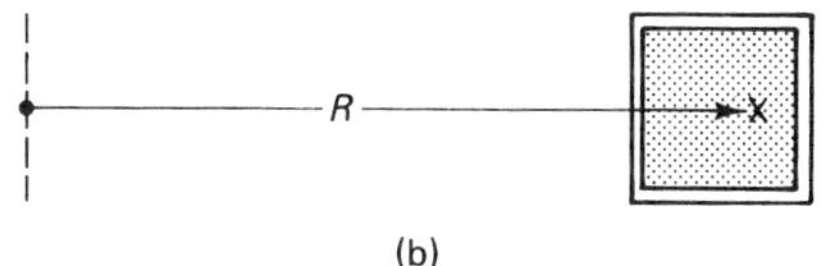

Figure 26-10 (a) A toroid. All the magnetic flux is contained within the core of the toroid.
(b) Cross section of the core showing the dimension R.

that the magnetic flux density varies with R, but if the cross-sectional area of the toroid is small in comparison with its diameter, we can ignore this variation. If we consider the toroid to be a long solenoid that has been bent into a circle, the quantity $2\pi R$ equals the length L of the solenoid, and Eq. (26-5) becomes the same as Eq. (26-4) for the flux density inside a long solenoid.

Atomic magnetism

We have seen that a current flowing in a loop of wire produces a magnetic field. Within atoms, electrons orbiting the nucleus act as tiny current loops within the atom. These atomic current loops create magnetic fields (magnetic moments) in the same way that electrons do when flowing around a loop of wire. Additionally, an electron in orbit about a nucleus seems to spin on its axis. A crude (classical) analogy is the daily rotation of the earth as it revolves about the sun once each 365 days. The *spin* of the electron also produces a magnetic moment. Thus, magnetism in atoms results from the orbital motion of the electrons and also from the electron spin.

In atoms of most materials the magnetic moments cancel so that the atom as a whole does not exhibit a net magnetic moment. In particular, any completely filled electron shell will have zero magnetic moment. In some atoms, however, the moments are not exactly balanced, and the atom will exhibit a net magnetic moment. Such materials are known as *paramagnetic* materials.

Atoms of the elements iron, nickel, and cobalt have very large magnetic moments due to a large number of uncancelled magnetic moments within the atom. In these atoms, electrons occupy the fourth shell, while the third electronic shell is not completely filled. This, in part, is responsible for the large magnetic moments exhibited by atoms of these materials, which are known as *ferromagnetic* materials.

The thing that sets the ferromagnetic materials apart from the paramagnetic materials is a complex exchange interaction that causes the atomic magnets in a region to line up in exactly the same direction to form a magnetic domain, which we described earlier. The magnetic effects of ferromagnetic materials are on the order of 100,000 times as strong as those of paramagnetic materials.

26-4 A SIMPLE MAGNETIC CIRCUIT

Because lines of flux must exist as closed loops, it is common practice to refer to the curved path taken by a given line, or a number of lines, as a *magnetic circuit.* Further, we can imagine the lines to "flow" along the circuit, and we can even say that the force that makes the lines flow is *magnetomotive force*, F_{mm}. Moreover, a magnetic circuit will offer a resistance to the flow of the magnetic lines; we call this resistance *reluctance*, R_m. Our purpose is to strike an analogy with Ohm's law and a circuit for an electric current.

In an electrical circuit, an electromotive force acts against a resistance to produce a current. In magnetic circuits, a magnetomotive force acts against a reluctance to produce a quantity of magnetic flux, Φ. Hence, we make the following analogies:

$$I \rightarrow \Phi \qquad V \rightarrow F_{mm} \qquad R \rightarrow R_m$$

In direct analogy with Ohm's law, we write the following relationship for a magnetic circuit.

$$\Phi = \frac{F_{mm}}{R_m} \tag{26-6}$$

Perhaps the simplest magnetic circuit is that of a toroid whose cross-sectional area is small in comparison with its diameter. Such a toroid is illustrated in Fig. 26-10. We know already that the flux is established by the flow of current through the many turns of wire wrapped around the toroid. Thus, the magnetomotive force is related to the current and to the number of turns of wire wrapped around the toroid. Specifically, the magnetomotive force F_{mm} equals the product of the two:

$$F_{mm} = NI \tag{26-7}$$

The practical unit of magnetomotive force is the ampere-turn. A current of 2 A, for example, flowing through 10 turns creates a magnetomotive force of 20 A-turns.

We now consider the reluctance, R_m, and we recall the formula of Eq. (25-20) for the resistance of a conductor:

$$R = \rho\frac{L}{A}$$

In terms of the conductivity σ rather than the resistivity ρ, this becomes

$$R = \frac{1}{\sigma}\frac{L}{A}$$

We use this formula to write the following for the magnetic circuit:

$$R_m = \frac{1}{\mu}\frac{L}{A} \tag{26-8}$$

where μ is the *permeability* of the material through which the magnetic lines must flow.

We now substitute Eq. (26-7) and Eq. (26-8) into Eq. (26-6) to obtain

$$\Phi = \frac{F_{mm}}{R_m} = \frac{NI}{\frac{1}{\mu}\frac{L}{A}} = \frac{\mu NIA}{L} \tag{26-9}$$

By definition, the flux density B is Φ/A. Therefore, after dividing the preceding equation by A, we obtain an expression for the magnetic field inside a large, tightly wound toroid whose cross-sectional area A is small in comparison with its diameter:

$$B = \frac{\Phi}{A} = \frac{\mu NIA}{AL}; \qquad B = \frac{\mu NI}{L} \tag{26-10}$$

It is no coincidence that this is the same equation as was given earlier for the field inside a solenoid.

The magnetic field intensity

We now introduce a magnetic quantity that is analogous to the electric field. It is helpful to recall that the electric field between the plates of a parallel-plate capacitor is given by V/d, where V is the potential difference between the plates and d is the distance by which the plates are separated. Thus, the electric field is expressible as a voltage per unit length.

We now introduce a quantity that is magnetomotive force per unit length, where the length is measured along the path of a magnetic circuit:

$$H = \frac{F_{mm}}{L} \tag{26-11}$$

The name of this quantity is *magnetic field intensity*. This is immediately confusing, because field intensity seems to refer to the *strength* of the magnetic field. The parameter that we perceive as the strength of a magnetic field is B, the flux per unit area. We can think of H as being the *cause* of the magnetic field because it is related to the magnetomotive force.

Going back to the toroid, we can substitute Eq. (26-7) into Eq. (26-11) to obtain the magnetic field intensity inside the toroid:

$$H = \frac{NI}{L} \tag{26-12}$$

When this is substituted into Eq. (26-10), we obtain the following relationship between B and H:

$$B = \mu H \tag{26-13}$$

This is a general result, not limited in application just to a toroid. The quantity μ is a property of the material in which the magnetic field is being established, and, generally speaking, its value varies considerably for different values of H.

Consequently, B and H are not strictly proportional, as Eq. (26-13) would at first glance lead one to believe.

26-5 MAGNETIC PROPERTIES OF MATERIALS

Suppose we have a "dissectable" toroid, one that can be taken apart so that we can change the core material. Further, let us assume we have a means of very accurately measuring the flux density inside the toroid. We wish to investigate the flux density that results when different materials are inserted for the core. We keep the current through the toroidal winding the same at all times.

With a vacuum as the core, we might obtain a flux density of 1 G. With an aluminum core, we will then obtain a flux density of 1.00002 G. A copper core gives a flux density of 0.99999 G. When an alloy of iron is inserted for the core, we obtain a flux density of 1982 G.

This hypothetical experiment illustrates that some materials (copper) reduce the flux density slightly while other materials (aluminum) increase it slightly, relative to vacuum. But when an alloy of iron is introduced into the core, the flux density is increased more than a thousandfold!

Materials such as bismuth, copper, lead, and mercury, which reduce the flux density when introduced into a magnetic field are called *diamagnetic* materials. The atoms of diamagnetic materials initially have no magnetic moment, but the external magnetic field induces minute magnetic moments in these atoms, and the induced magnetic moments oppose the external field. Actually, all atoms exhibit the property of diamagnetism, but the slight diamagnetic effect is swamped by the more pronounced effects of paramagnetism and ferromagnetism.

Paramagnetic materials increase the flux density slightly due to alignment of the atomic magnetic moments (not induced) with the external field. Aluminum is paramagnetic. However, the paramagnetic effect is essentially negligible in comparison to the dramatic increase in flux density produced by the ferromagnetic alloy of iron. When an external field is applied to a ferromagnetic material, the magnetic domains align themselves with the field and increase the flux density greatly.

It is because of the extreme effect of ferromagnetic materials that iron, and alloys of iron, cobalt, and nickel, are widely used in electromagnets associated with motors, generators, transformers, inductors and all types of devices that ultilize magnetic fields in their operation.

Relative permeability

The permeability of a material is conveniently given in terms of the relative permeability k_m, which is defined by

$$\mu = k_m \mu_0 \tag{26-14}$$

$$k_m = \frac{\mu}{\mu_0} \tag{26-15}$$

The relative permeability gives the effectiveness of a material when the material is used as a core material for a magnetic circuit. For example, if B_0 is the flux density resulting when a certain toroid utilizes a vacuum core and if B is the flux resulting when a material core is used, then

$$k_m = \frac{B}{B_0} \tag{26-16}$$

The relative permeability is unitless because it is the ratio of like quantities.

The *magnetic susceptibility* is defined as

$$X_m = k_m - 1 \tag{26-17}$$

The magnetic susceptibility has the property that diamagnetic materials have small negative values, paramagnetic materials have small positive values, and ferromagnetic materials have large positive values. The relative permeabilities and susceptibilities of selected materials are given in Table 26-1.

TABLE 26-1 SELECTED MAGNETIC PROPERTIES

Material	Relative susceptibility
Air	3.6×10^{-7}
Aluminum	2.1×10^{-5}
Carbon	-1.4×10^{-5}
Lead	-1.6×10^{-5}
Tungsten	7.8×10^{-5}
	Relative permeability at 20 gauss
Iron (sheet)	200
Mu metal (sheet)	20,000
Supermalloy	100,000

Magnetic hysteresis

Suppose a coil is wound on a ferromagnetic toroidal core, as shown in Fig. 26-11, and suppose a variable current source is connected to the coil. We assume a means is available for determining the magnetic flux density, B, within the core. We wish to see how the flux density varies as we increase the current through the coil. Because the magnetic field strength H is proportional to the current through the coil, we can make a graph of B vs. H to show the result of the experiment. Such a graph is shown in Fig. 26-11(b).

We start with the core unmagnetized. As the current is increased from zero, domains oriented along the direction of the field increase in volume while the other domains become smaller. Hence, the flux density increases. With a stronger current and larger value of H, the magnetization directions of intially unaligned domains begin to rotate and align themselves with the external field. At this point, B increases with greater slope. At a fairly high current level, nearly all the domains will have rotated to the preferred direction, and the slope of the B-H curve decreases. Finally, after all domains have rotated, the value of B does not

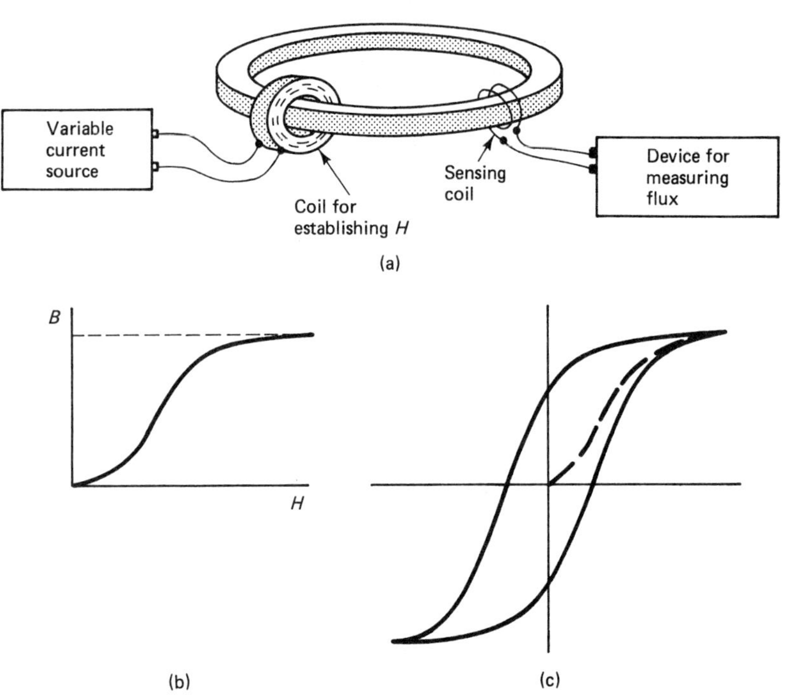

Figure 26-11 (a) Experimental arrangement for investigating magnetic hysteresis.
(b) The initial magnetization curve.
(c) A hysteresis loop.

significantly increase with increasing current levels. The core is then said to be *saturated.*

As the current is decreased, the domains do not automatically rotate back to their original position. It is found that an appreciable flux density will remain even after the current is reduced to zero. That is, the core has been permanently magnetized.

As the current level is increased in the opposite direction, the flux density is gradually reduced to zero and then it reverses direction. A further increase in current will carry the specimen to saturation in the opposite direction. Subsequent cycling of the direction and strength of the current will give rise to the hysteresis loop shown in Fig. 26-11(c).

Energy is required in order to rotate the domains, and the area enclosed by the hysteresis loop provides a measure of the energy absorbed by the core each time the loop is traversed. Materials having tall, skinny loops are suitable for use as transformer cores and electric motor or generator pole-pieces, but materials having shorter, fatter loops are best for permanent magnets.

26-6 MAGNETIC FORCE ON ELECTRIC CHARGE

A stationary magnet does not exert a force on a stationary electric charge. An electric charge must be moving relative to the magnetic field in order for the magnetic field to exert a force on the charge. Then, the magnitude of the force is directly proportional to the velocity with which the charge moves through the field, the magnitude of the charge, and the flux density B.

Because of the vector nature of the flux density and the velocity of the charge, the direction of the charge's velocity relative to the direction of the magnetic field also affects the magnitude of the force exerted on the moving charge. For example, when the charge moves parallel to the lines of the magnetic field, no force is exerted on the charge. The force is a maximum when the charge moves at right angles to the field.

Mathematically, the force F on a charge q moving with velocity v through a magnetic field at an angle θ to the flux density B is given by

$$F = qvB \sin \theta \tag{26-18}$$

The direction of the force F is perpendicular to both B and v, as shown in Fig. 26-12. Consequently, the magnetic force can never do work on the moving charge such as to either speed it up or slow it down. The force on a negative charge is oppositely directed to that exerted on a positive charge.

When the velocity of the charged particle is at right angles to the direction of the field, $\sin \theta = 1$, and the equation above simplifies to

$$F = qvB \tag{26-19}$$

A charged particle traveling at right angles to a magnetic field tends to travel in a circular path whose radius R is such that the centripetal force exactly balances the

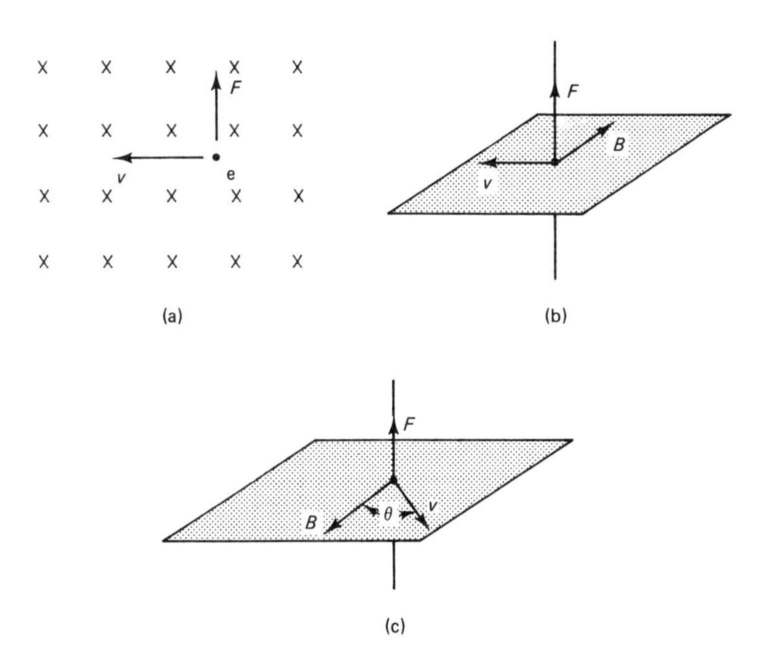

Figure 26-12 (a) With the magnetic field directed into the page, the force exerted on an electron moving to the left is directed upward.
(b) Vector relationships.
(c) The electron velocity may be at an angle to the magnetic field.

A beam of electrons deflected upward by a magnetic field.

magnetic force. The *centripetal force* is given by mv^2/R, where m is the mass of the particle. Equating the magnetic and centripetal forces yields

$$\frac{mv^2}{R} = qvB \tag{26-20}$$

which may be solved for R:

$$R = \frac{mv}{qB} \tag{26-21}$$

The circular motion of a charged particle in a magnetic field is illustrated in Fig. 26-13. If the particle makes an angle to the magnetic field, the resulting motion is a spiral.

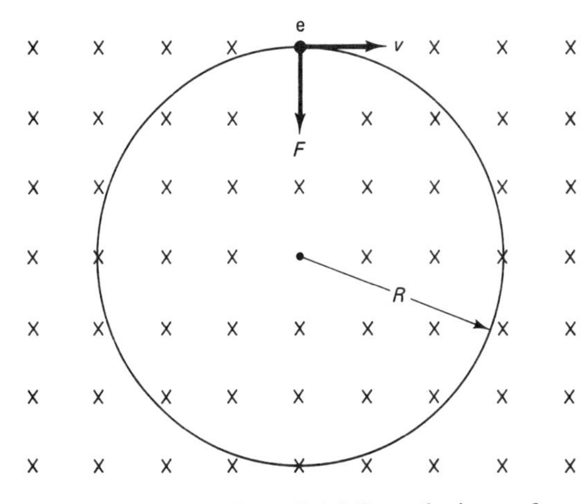

Figure 26-13 An electron traveling at right angles to a magnetic field will travel in either a circle or a spiral.

We can use Eq. (26-21) to derive a formula for the revolutions per second that a particle in a magnetic field will traverse. This frequency is called the *cyclotron frequency*. It is given by $v/2\pi R$, where R is obtained from the above equation. The result is

$$f = \frac{qB}{2\pi m} \tag{26-22}$$

Note that the cyclotron frequency is independent of the particle velocity and the radius of the circular path.

Units

The relationship $F = qv\mathrm{B}$ [Eq. (26-19)] provides a convenient means for expressing the units of B (teslas) in fundamental SI units. We have

$$\text{newton} = \text{coulomb} \cdot \frac{\text{meter}}{\text{second}} \cdot \text{tesla}$$

But *coulombs per second* are amperes, so that

$$\text{newton} = \text{ampere} \cdot \text{meter} \cdot \text{tesla}$$

and

$$\text{tesla} = \text{newtons/ampere-meter.}$$

Because 1 tesla = 1 weber/meter^2, it follows that

$$\text{weber} = \text{newton-meter/ampere.}$$

Frequently, we encounter the combination, coulomb-tesla. From the above we see that

$$\text{coulomb-tesla} = \text{newton-second/meter}$$

which facilitates the reconciliation of units in many problems.

EXAMPLE 26-4 An electron travels at a speed of 1×10^6 m/s at an angle of 50° to a magnetic field of 2000 G. Determine the magnitude of the force on the electron.

Solution. Using Eq. (26-18), $F = qvB \sin \theta$, with $e = 1.6 \times 10^{-19}$ C, and $B = 0.2$ T:

$$F = (1.6 \times 10^{-19}\ \text{C})(1 \times 10^6\ \text{m/s})(0.2\ \text{T})\sin 50$$
$$- 2.45 \times 10^{-14}\ \text{N}$$

The direction of the force may be determined from Fig. 26-12. The resulting motion of the electron will be a clockwise spiral when viewed along the direction of B.

EXAMPLE 26-5 Determine the cyclotron frequency of the electron in the preceding example. (For an electron, m is 9.1×10^{-31} kg.)

Solution. Substituting directly into Eq. (26-22) gives

$$f = \frac{qB}{2\pi m} = \frac{(1.6 \times 10^{-19}\ \text{C})(0.2\ \text{T})}{2\ (9.1 \times 10^{-31}\ \text{kg})}$$
$$= 5.6 \times 10^9\ \text{Hz} = 5.6\ \text{GHz}$$

To justify units,

$$\frac{\text{CT}}{\text{kg}} = \frac{\text{N} \cdot \text{s/m}}{\text{kg}} = \text{s}^{-1} \qquad (\text{because kg} = \text{N} \cdot \text{s}^2/\text{m})$$

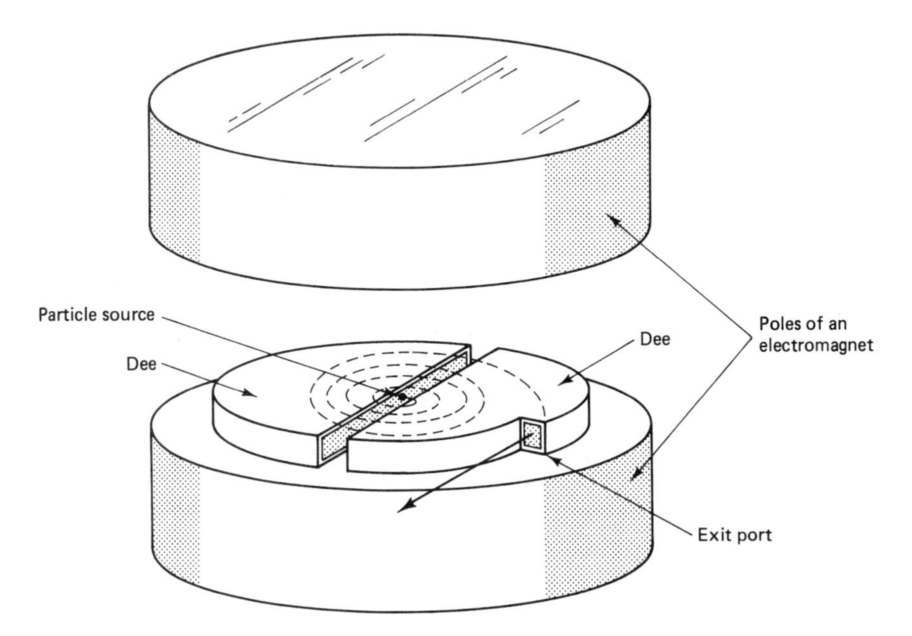

Figure 26-14 Schematic representation of a cyclotron.

The cyclotron

A device for accelerating charged particles, such as protons and electrons, to high velocities is the *cyclotron*, the first of which was constructed in 1932 by E. Lawrence and M. Stanley. It makes use of the fact that charged particles traveling at right angles to a magnetic field travel in a circle. The main features of a cyclotron are shown in Fig. 26-14.

A large electromagnet produces a uniform magnetic field directed vertically in the region occupied by two semicircular *dees* that are connected to a source of high-frequency alternating voltage. The dees are hollow on the inside and are made of a metal that does not affect the magnetic field. A strong electric field is established in the gap between the two dees, and the direction of this field changes in step with the high-frequency voltage supplied to the dees.

The particles to be accelerated are introduced at the center of the dees and will be attracted toward one of them. The magnetic force exerted on the moving particles causes them to traverse a circular path within the hollow chamber of the dee so they once again approach the gap between the dees. By the time the particle reaches the gap, the polarity of the electric field across the gap will have changed, and the particle will once again be accelerated across the gap. The particles gain energy each time they cross the gap, and as the particle velocity increases so does the radius of the circular path in accordance with Eq. (26-21). The particles spiral outward toward the outer periphery of the dees until they encounter a small deflecting magnetic field which causes them to leave the chamber.

By solving Eq. (26-21) for v and substituting into the formula for kinetic energy ($\frac{1}{2}mv^2$), it is easy to show that the maximum energy obtainable is

$$KE_{max} = \frac{1}{2} \frac{(qBR)^2}{m} \tag{26-23}$$

where R is the maximum radius accommodated by the dees. The frequency of the voltage supplied to the dees is the cyclotron frequency given by Eq. (26-22). [*Note*: If the particle velocity approaches the speed of light, the mass of the particle increases due to relativistic effects and Eq. (26-23) will no longer be applicable.]

26-7 MAGNETIC FORCE ON CURRENT-CARRYING CONDUCTORS

Because an electric current is comprised of moving electric charges, it is not surprising that a current-carrying conductor is acted upon by a magnetic force when the conductor is placed in a magnetic field. The force is proportional to the current I, the length L of the conductor, and the magnetic flux density B. Also, the magnitude and direction of the force depends upon the orientation of the wire in the magnetic field.

In analogy with Eq. (26-18), the magnetic force on a current-carrying wire is given by

$$F = IBL \sin \theta \tag{26-24}$$

where θ is the angle the conductor makes with the magnetic field, as shown in Fig. 26-15. If a wire is oriented perpendicular to a magnetic field, the force exerted on the wire will be at right angles to both the wire and the magnetic field.

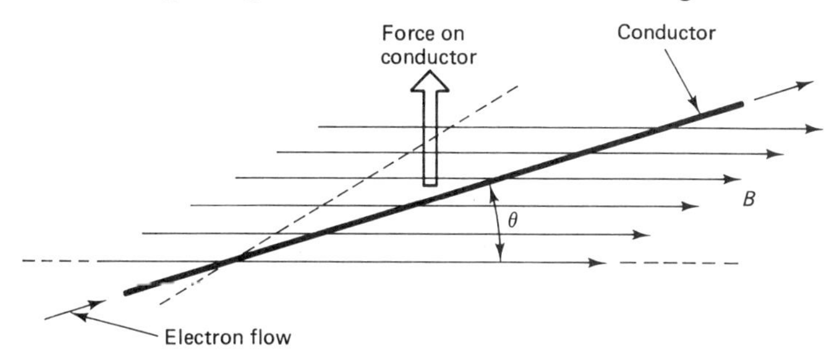

Figure 26-15 Force on a conductor in a magnetic field.

Another way to consider the force exerted on a current-carrying wire is as an interaction between the external field and the magnetic field produced around the wire by the current flowing through it. The two fields add vectorially and reinforce on one side of the wire and cancel on the other. The direction of the force is toward the side where the cancellation occurs. This is shown in Fig. 26-16.

Force between two parallel conductors

We saw earlier that a magnetic field surrounds a long straight wire in which a current is flowing. In the last section we found that a force is exerted on a current-carrying conductor placed in a magnetic field. It follows that if two current-

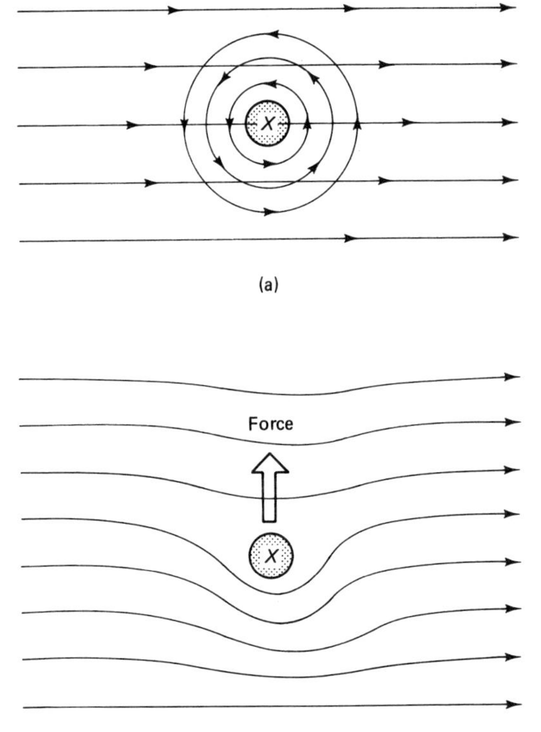

Figure 26-16 (a) The field of a conductor (carrying electrons into the page) embedded in a uniform field produces a field like that shown in (b).

carrying conductors are placed side by side, a mutual force will be exerted on the conductors because the second conductor lies in the magnetic field of the first, and vice versa. Let wires 1 and 2 be separated a distance d. If wire 1 carries a current I_1, the magnetic field produced by wire 1 at the location of wire 2 is given by [using Eq. (26-3)]

$$B_1 = \frac{\mu_0 I_1}{2\pi d} \tag{26-25}$$

Because the wires are assumed to be parallel, wire 2 is oriented perpendicular to the magnetic field produced by wire 1. Hence, $\sin\theta = \sin 90° = 1.0$ and the term $\sin\theta$ effectively drops out of Eq. (26-24), which we now use to compute the force on wire 2 due to the magnetic field produced by wire 1. Therefore,

$$F_2 = I_2 B_1 L \tag{26-26}$$

where B_1 is given by Eq. (26-25). Making the substitution gives the mutual force between two parallel conductors:

$$F = \frac{\mu_0 I_1 I_2 L}{2\pi d} \tag{26-27}$$

We dropped the subscript from F because $F_1 = F_2$, as dictated by Newton's third law. The force between the wires will be attractive if the current flows in the same direction in both wires. If the current flows in opposite directions, as shown in Fig. 26-17, the force will be repulsive.

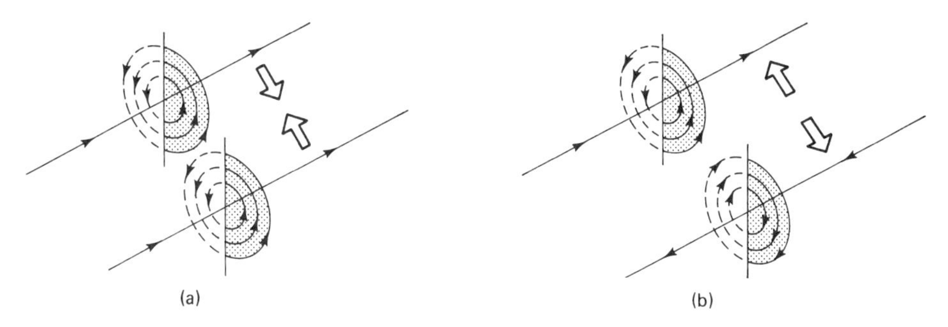

Figure 26-17 (a) With electron flow in the same direction in two parallel conductors, the force between the conductors will be attractive.
(b) The force will be repulsive if the electron flow is in opposite directions.

26-8 MAGNETIC TORQUE ON A CURRENT LOOP

Let us now consider a current-carrying conductor formed in the shape of a rectangular loop, as shown in Fig. 26-18. The axis of the loop is assumed to be perpendicular to the direction of the magnetic field, and the loop is oriented at an angle to the field. Each side of the loop will experience a magnetic force. We wish to calculate the torque on the loop that tends to cause it to rotate about its axis.

First of all, the magnetic forces acting on sides 3 and 4 will be oppositely directed and will cancel each other. Only the forces acting on sides 1 and 2 produce a torque about the axis of the loop. If the length of the loop is L, the force exerted on sides 1 and 2 is given by Eq. (26-24) with the angle equal to 90°: $F = IBL$. The force on side 1 is down; the force on side 2 is up. These forces tend to cause the loop to rotate about its axis. For each force the moment arm is $R \sin \theta$. Therefore, the torque acting on the loop is

$$\begin{aligned} \tau &= 2(FR \sin \theta) \\ &= 2IBLR \sin \theta \end{aligned} \qquad (26\text{-}28)$$

Dees of an early heavy-particle cyclotron on display at the Smithsonian Institution in Washington, D.C.

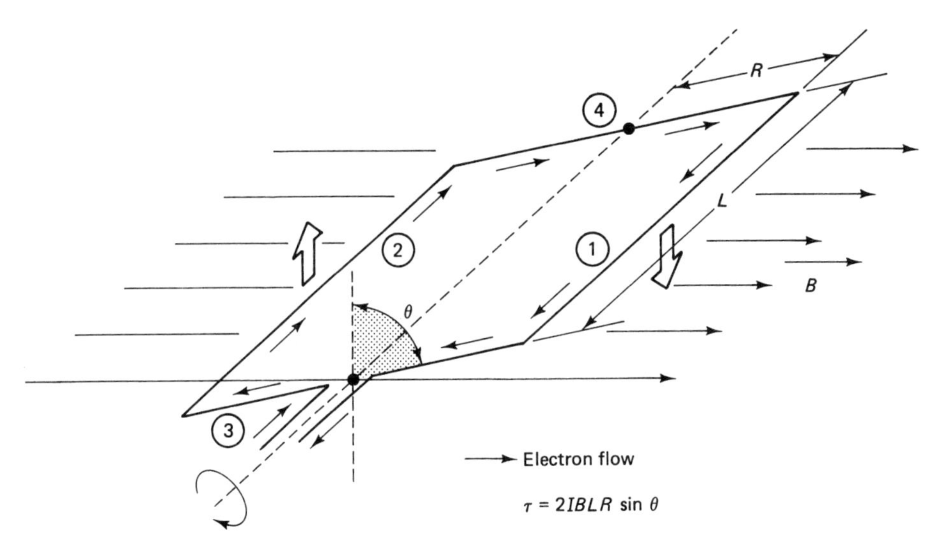

Figure 26-18 The torque on a current loop in a magnetic field.

But the quantity $2LR$ equals the area A enclosed by the loop, so we may write

$$\tau = IAB \sin \theta \tag{26-29}$$

The magnetic moment μ of a current loop is the product of the current I and the area A enclosed by the loop. Hence, the torque exerted on the loop is given by

$$\tau = \mu B \sin \theta \tag{26-30}$$

If a loop comprises N turns instead of just one, the magnetic moment is given by NIA, and the torque on the loop is effectively multiplied by N. (Recall that the symbol μ was used to denote permeability. Do not confuse magnetic moments and permeabilities.)

EXAMPLE 26-6

A single-turn loop 12 cm long and 8 cm wide is located in a magnetic field of 400 G. The loop is oriented as in Fig. 26-18 with $\theta = 30°$. A current of 5 A flows through the loop. Calculate the torque exerted on the loop by the magnetic field.

Solution. Since 10^4 G = 1 T, $B = 0.04$ T. The area enclosed by the loop is 9.6×10^{-3} m^2. We now use Eq. (26-29):

$$\begin{aligned}\tau &= IAB \sin \theta \\ &= (5 \text{ A})(9.6 \times 10^{-3} \text{ m}^2)(0.04 \text{ T})\sin 30° \\ &= 9.6 \times 10^{-4} \text{ N} \cdot \text{m}\end{aligned}$$

The units are reconciled by using the following:

$$1 \text{ A} = 1 \frac{\text{N} \cdot \text{m}}{\text{Wb}} \qquad 1 \text{ T} = 1 \frac{\text{Wb}}{\text{m}^2}$$

The first of these is from Eq. (26-26), and the second is by definition.

Electric motors

The rotation of a current loop in a magnetic field is fundamental to the operation of electric motors. The magnetic field in which the loop rotates may be produced either by a permanent magnet or by an electromagnet comprised of a field coil and pole pieces. The rotating loop is a part of the armature, and current is typically conducted to the rotating loop via a brush and slip-ring assembly. An elementary motor is shown in Fig. 26-19(a), and the operation of the motor is illustrated in Fig. 26-19(b).

When current is first applied, the loop tends to rotate until the plane of the

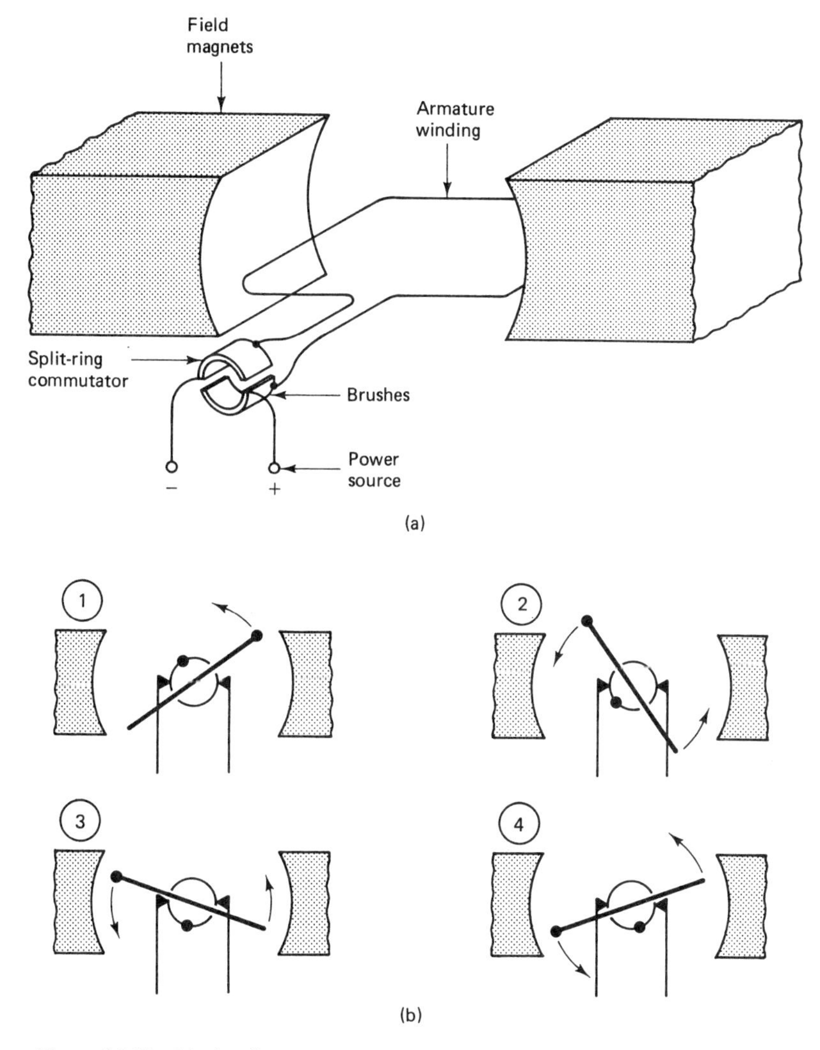

Figure 26-19 (a) An elementary motor that uses permanent magnets and a DC power source.
(b) A sequence showing how the brush–slip-ring assembly reverses the direction of current flow in the armature.

An elementary motor showing the armature, field magnets, and brush assembly.

loop is vertical, at which point it would normally stop. However, current flows to the loop by way of a split-ring commutator. The commutator assembly reverses the direction of the current flow around the loop just as the loop becomes vertical, and a torque is again applied to the loop. Inertia causes the loop to continue its motion in the same direction. This process is repeated again and again, and the loop rotates smoothly as the armature of an electric motor.

Basic meter movement; the galvanometer

The fact that a torque is exerted on a current-carrying loop in a magnetic field provides the basis of operation of a widely used current measuring device known as the *D'Arsonval galvanometer*. The mechanism is illustrated in Fig. 26-20. A

A D'Arsonval meter movement. The large permanent magnet is located behind the faceplate.

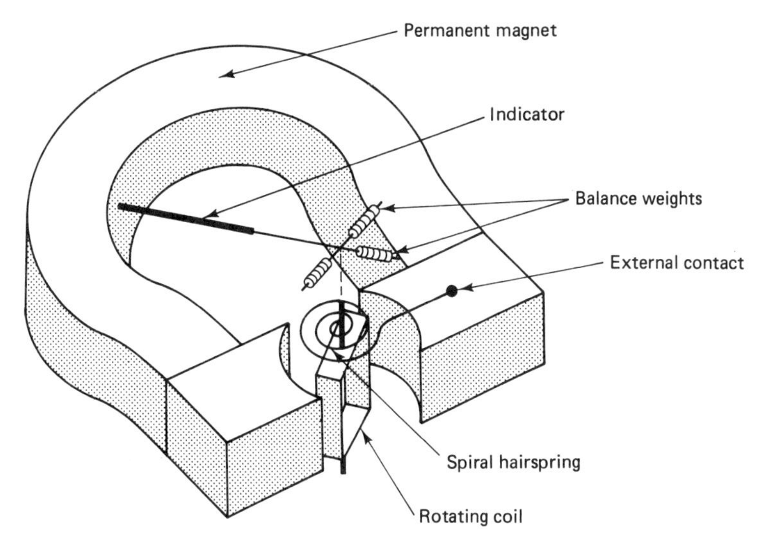

Figure 26-20 Essential features of a D'Arsonval galvanometer. (The pivots, central iron slug, and second hairspring and contact are not shown.)

permanent magnet creates a strong magnetic field in the gap between the pole pieces, and the iron core situated in the gap causes the magnetic lines to cross the gap in a radial direction. This, in effect, removes the term sin θ from Eq. (26-29) so that the torque on the coil is proportional to the current flowing through it:

$$\begin{aligned} \tau &= 2IBLR \\ &\approx I \end{aligned} \tag{26-31}$$

The coil consists of perhaps several hundred turns of very fine wire, wound on a very light, nonmagnetic bobbin to which an indicator needle is attached. The rotating assembly, including appropriate counterweights, is mounted in jewels to minimize friction. An opposing torque is provided by two spiral *hairsprings*, which also conduct the current to and from the coil. A carefully designed movement will exhibit a needle deflection that is directly proportional to the current flowing through the coil.

Meter movements are characterized by the current required to drive the needle to full-scale deflection. The full-scale deflection current is denoted by I_{fs}. Also, the electrical resistance of the meter coil R_m is often given. For inexpensive movements, I_{fs} is typically 100 μA, or perhaps 50 μA. More sensitive and more expensive movements have values of I_{fs} as low as 10 μA. Typical coil resistances (R_m) are on the order of several hundred ohms or less.

26-9 BASIC ELECTRIC MEASURING INSTRUMENTS

The voltmeter

Only a very small voltage (about 0.01 V) applied to the terminals of a meter movement will drive the meter to full scale. To construct a voltmeter capable of measuring voltages ranging to the hundreds or thousands of volts, a large

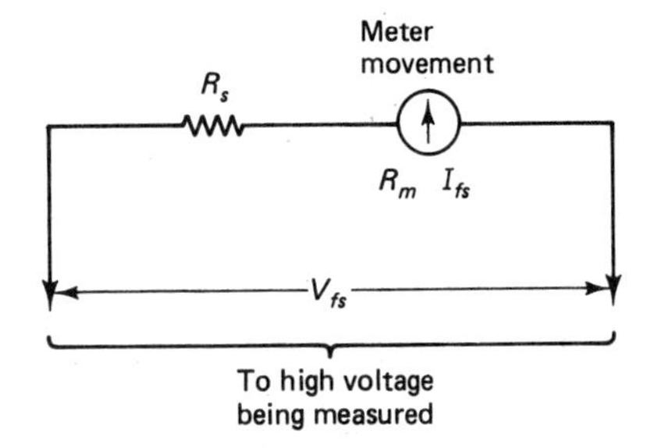

Figure 26-21 To construct a voltmeter from a basic meter movement, a large resistance, R_s, is connected in series with the movement.

resistance, R_s, must be connected in series with the movement, as shown in Fig. 26-21.

Let us denote the full-scale voltage being measured as V_{fs}. The following relationship applies:

$$I_{fs} = \frac{V_{fs}}{R_s + R_m} \tag{26-32}$$

from which we obtain

$$R_s = \frac{V_{fs}}{I_{fs}} - R_m \tag{26-33}$$

This allows us to compute the ohmic value of R_s required for the movement to deflect to full scale when voltage V_{fs} is applied to the terminals of the voltmeter.

The sensitivity of a voltmeter is given by the *ohms per volt* rating of the voltmeter. It is given by

$$\Omega/\text{V} = \frac{R_s + R_m}{V_{fs}} \tag{26-34}$$

and is determined by the I_{fs} rating of the meter movement. For typical voltmeters that are not electronically amplified, the sensitivity is 20,000 Ω/V or 50,000 Ω/V.

The ammeter

The current-measuring range of a meter movement may be extended by connecting a resistor (a *shunt*) in parallel with the movement, as shown in Fig. 26-22. If the shunt resistance R_{sh} is small in comparison with the resistance of the movement R_m, most of the current being measured will flow through the shunt resistor.

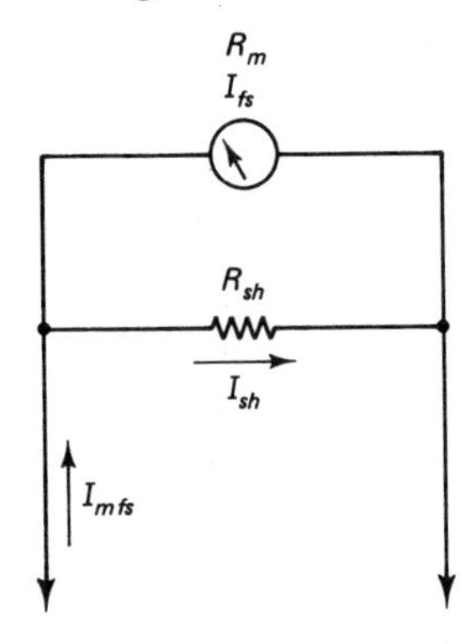

Figure 26-22 An ammeter is constructed by connecting a shunt resistor, R_{sh}, in parallel with the basic meter movement.

Suppose we wish to design an ammeter that will read full scale when a large current I_{mfs} flows through the ammeter. We may write

$$I_{mfs} = I_{fs} + I_{sh} \tag{26-35}$$

where I_{fs} is the full-scale deflection current of the movement and I_{sh} is the current that flows through the shunt under full-scale conditions. Because the voltage across the shunt and across the movement are the same, we may also write

$$I_{fs}R_m = I_{sh}R_{sh} \tag{26-36}$$

These two equations may be combined to eliminate I_{sh} and solve for R_{sh}:

$$R_{sh} = \frac{I_{fs}R_m}{I_{mfs} - I_{fs}} \tag{26-37}$$

EXAMPLE 26-7

For a certain meter movement, $R_m = 200\ \Omega$ and $I_{fs} = 100\ \mu\text{A}$. What value of shunt resistance R_{sh} is required so that the resulting ammeter will read full scale when a current of 1 A flows through the meter?

Solution. Equation (26-37) yields

$$R_{sh} = \frac{(100 \times 10^{-6}\ \text{A})(200\ \Omega)}{1\ \text{A} - (100 \times 10^{-6}\ \text{A})}$$

$$= 0.0200\ \Omega$$

This is a very small resistance. Resistances of this magnitude are formed by winding a short length of wire into a coil that is attached directly to the terminals of the meter movement.

The ohmmeter

The schematic diagram of a simple ohmmeter is shown in Fig. 26-23. A battery is included that causes a current to flow in the series circuit consisting of R_v, R_m, and the unknown R_x. Maximum current will flow when the ohmmeter test leads are connected directly together ($R_x = 0$), and R_v is adjusted so that when $R_x = 0$, full deflection of the meter is obtained. Under these circumstances, the following relationship holds:

$$I_{fs} = \frac{V}{R_v + R_m} \tag{26-38}$$

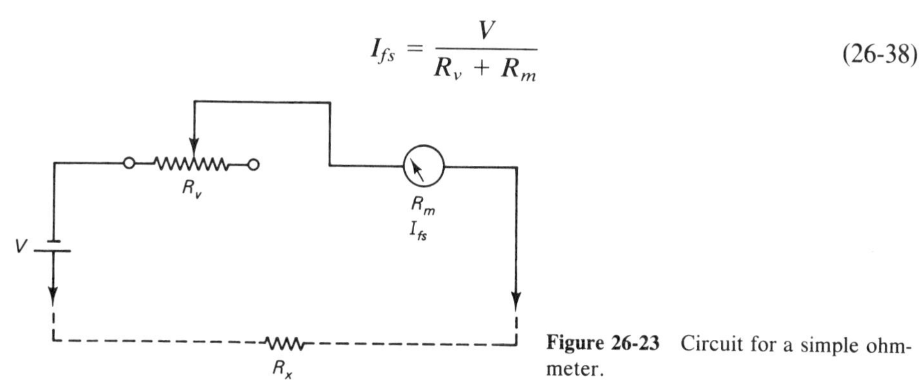

Figure 26-23 Circuit for a simple ohmmeter.

This may be solved for the variable resistance R_v:

$$R_v = \frac{V}{I_{fs}} - R_m \tag{26-39}$$

Once R_v is set, the ohmmeter may be used to determine an unknown resistor R_x. When the leads are connected to R_x, a current I_x will flow around the circuit and through the meter movement given by

$$I_x = \frac{V}{R_v + R_m + R_x} \tag{26-40}$$

The fraction of full scale that the meter will be deflected is given by

$$\text{def} = \frac{I_x}{I_{fs}} = \frac{R_v + R_m}{R_v + R_m + R_x} \tag{26-41}$$

This expression is obtained from Eqs. (26-38) and (26-40).

For an ohmmeter of this type, 0 Ω corresponds to full-scale deflection; the scale is reversed, and it is nonlinear. Designing a practical, useful ohmmeter is somewhat more involved than the above analysis indicates, but the principles are the same. Electronics texts provide details for interested readers.

EXAMPLE 26-8 An ohmmeter is to use a meter movement having I_{fs} = 100 μA and R_m of 200 Ω. The battery is a standard 1.5-V D-cell. Calculate the value of R_v required to set the full-scale deflection. Also, compute the deflection that occurs when an unknown resistance of 20,000 Ω is connected to the ohmmeter leads.

Solution. Equation (26-39) gives a value of R_v of 14,800 Ω. We then substitute this into Eq. (26-41):

$$\text{def} = \frac{14{,}800\ \Omega + 200\ \Omega}{14{,}800\ \Omega + 200\ \Omega + R_x}$$

When 20,000 Ω is substituted for R_x, we obtain

$$\text{def} = \frac{(14{,}800 + 200)\ \Omega}{(14{,}800 + 200 + 20{,}000)\ \Omega}$$

$$= 0.43$$

which is 43% of full scale.

To Go Further

Investigate the lives and work of:

Karl Friedrich Gauss (1777–1855), a great mathematician
Hans Christian Oersted (1777–1851), Danish physicist
Nikola Tesla (1856–1943), Yugoslavian-born American engineer

Questions

1. Suppose two bar magnets are located near each other so that their fields overlap. Do the fields *really* overlap? Do the lines of force of one magnet cross over or intersect those of the other? Explain.
2. How is the direction of a magnetic line defined?
3. Explain why the magnetic pole near the North Pole is actually a *south* magnetic pole.

4. Suppose we have two identical cylindrical pieces of iron, one of which is magnetized. Without additional apparatus, how can we determine which is the magnet?
5. What is the difference between a specimen of iron that is magnetized and one that is not, as far as the domains are concerned?
6. Why does a permanent magnet attract an iron nail that is not permanently magnetized?
7. Why do physicists not try to determine the number of magnetic lines in a weber?
8. Explain the left-hand rule for determining the direction of the magnetic field around a current-carrying wire.
9. Describe the peculiar feature of a toroid in regard to its magnetic field.
10. Explain the difference in behavior of a diamagnetic and a paramagnetic material.
11. What is the relationship between susceptibility and the relative permeability?
12. Explain the distinction between flux density B and magnetic field intensity H.
13. Describe the relationship between magnetomotive force, reluctance, and magnetic flux.
14. If an electron headed west is traveling through a magnetic field pointing north, what will be the direction of the force exerted on the electron?
15. Under what circumstances can an electrical charge travel through a magnetic field without experiencing a force?
16. Why can a static magnetic field do no work on a charge moving through it?
17. Explain how a galvanometer and an electric motor are similar.
18. Explain the function of: **(a)** the large series resistance in a volmeter; **(b)** the low resistance shunt in an ammeter; **(c)** the battery in an ohmmeter.

*19. The permeability of free space is μ_0. By definition, free space is a region devoid of matter, a region that is perfectly empty. Can a perfectly empty region have physical properties?

Problems

1. Magnetic flux in the amount of 0.04 Wb passes across an area of 2 m^2. Calculate the flux density B.
2. Determine the magnetic flux that passes through a 1-cm^2 area between the poles of an electromagnet where the flux density is 0.1 T.
3. A long wire carries a current of 1 A. Compute the magnetic flux density at points **(a)** 1 cm, **(b)** 2 cm, and **(c)** 10 cm from the wire. Express the results in teslas and in gauss.
4. A coil of wire 20 cm in diameter consists of 50 turns wound closely together. A current of 2 A flows through the coil. What is the flux density of the center of the coil?
5. A coil of wire 10 cm in diameter has 10 turns wound closely together. What current must flow through the coil in order to produce a flux density of 1 G at the center of the coil?
6. What is the magnetic moment of the coil described in Problem 4?
7. A solenoid consists of 1000 turns of wire wound on a coil-form 50 cm long and 4 cm in diameter. A current of 10 A flows through the solenoid. Determine the magnetic flux density on the inside of the solenoid.
8. For the solenoid of Problem 7, determine the total amount of magnetic flux that passes across an interior cross section.
9. The core of a certain toroid has a rectangular cross section and is made of iron having a relative permeability of 900. The inside diameter of the core is 15 cm; the outside diameter is 25 cm. Four hundred turns are wound on the core, and a current of 2 A passes through the winding.
   (a) Determine B near the inner edge of the core where R = 7.5 cm.
   (b) Determine B at the center of the core where R = 10 cm.
   (c) Determine B at the outer edge where R = 12.5 cm.
10. A toroid whose diameter is 12 cm has a cross-sectional area of 1 cm^2. The core

material has a relative permeability of 400. A current of 0.5 A flows through the 150-turn winding.
(a) What magnetomotive force is produced by the winding?
(b) What is the magnetic field intensity within the core?
(c) What is the flux density within the core, assuming it to be uniform?

11. Use the toroidal core of Problem 10.
(a) What is the reluctance of the core?
(b) What is the total flux established within the core?
(c) Compute the flux density B using Eq. (26-10) and compare the result with part (c) of Problem 10.

12. Use Eq. 26-6, $\Phi = F_{mm}/R_m$, to determine the units of reluctance R_m.

13. A toroidal core of small cross section has a ring diameter of 30 cm. It is wound with 500 turns, and a current of 4 A flows through the winding.
(a) Determine the magnetic field intensity H inside the core.
(b) Find the flux density B if the relative permeability of the core material is 5400, which is typical of ordinary transformer steel.

14. An electron travels with a velocity of 6×10^6 m/s at right angles to a magnetic field of 300 G.
(a) Calculate the force exerted on the electron.
(b) What acceleration of the electron will this force produce?
(c) Use Eq. (26-21) to determine the radius of the circular path in which the electron will move.

15. Repeat Problem 14 for a proton traveling at the same velocity. Calculate the cyclotron frequency for (a) an electron and (b) a proton if the magnetic flux density is 1 T.

16. A magnetic field is used to make charged particles traverse a circular path of radius 1 m. The particles travel at a velocity of 1×10^5 m/s. What value of B is required if the particles are: (a) protons; (b) singly ionized atoms of helium; (c) doubly ionized helium atoms (helium nuclei)?

17. A magnetic field of $B = 0.4$ T between the poles of an electromagnet extends over a region 10 cm in length. A copper wire carrying a current of 5 A passes through the region. What force will be exerted on the wire?

18. According to the theory of relativity, the mass of particles increases as their velocities approach the speed of light. The mass m_v of a particle traveling at velocity v is given by

$$m_v = \frac{m_0}{\sqrt{1 - v^2/c^2}}$$

where m_0 is the *rest mass* of the particle and c is the velocity of light.
(a) Compute the mass of an electron traveling at one-half the speed of light.
(b) Use $\frac{1}{2}mv^2$ to calculate the kinetic energy of the electron.
(c) For a value B of 6 G, compute the cyclotron frequency for an electron traveling at low speed.
(d) Repeat (c) for a speed of $\frac{1}{2}c$. What is the practical implication of this change in frequency?

19. Repeat Problem 18 for a proton, but use a value of B of 2000 G.

20. Two parallel conductors 1 m long carry currents of 1 A in the same direction. Compute the mutual force exerted by one conductor upon the other if the distance between the conductors is: (a) 1 cm; (b) 2 cm; (c) 10 cm.

21. Two parallel wires separated a distance of 1 m carry equal currents of 1 A in the same direction. Demonstrate that the force of attraction per unit length of the wires is 2×10^{-7} N/m. (This principle is used to provide a standard for definition of the ampere.)

22. A one-turn loop of wire carrying a current of 3 A is located in a magnetic field of 1000 G. The loop is a square 8 cm on each side and is pivoted about an axis through the center of the loop. Calculate the torque on the loop when the plane of the loop makes the following angles to the direction of the magnetic field.
(a) 0° (b) 30° (c) 60° (d) 90

23. Compute the magnetic moment of the loop in Problem 23 and use Eq. (26-30) to verify the results obtained in that problem.

24. A certain D'Arsonval meter movement has an I_{fs} of 200 μA and a meter resistance R_m of 100 Ω. Calculate the resistance R_s that must be connected in series with the movement in order to construct a voltmeter having a full-scale voltage of: **(a)** 1 V; **(b)** 10 V; **(c)** 100 V; **(d)** 1000 V.

(e) Calculate the ohm-per-volt sensitivity of each voltmeter.

25. Repeat Problem 24 for a meter movement having an I_{fs} of 50 μA and an R_m of 400 Ω.

26. A certain meter movement has an I_{fs} of 1 mA and R_m of 100 Ω.

(a) What should be the resistance R_{sh} of the shunt to make an ammeter that reads 10 A full scale?

(b) What voltage will be developed across the shunt when the meter reads full scale?

27. A meter movement having I_{fs} of 1 mA and $R_m = 200\ \Omega$ is used in conjunction with a 1.5-V battery to construct an ohmmeter.

(a) What value of R_v, the calibration resistor, is required to set the full-scale deflection?

(b) What percentage deflections will result for the following unknown resistances R_x when measured by the ohmmeter: 100 Ω; 1000 Ω; 1500 Ω; 10,000 Ω?

28. Repeat Problem 27 for a meter movement having $I_{fs} = 10$ mA and $R_m = 200\ \Omega$.

CHAPTER 27

MAGNETIC INDUCTION

Chapter 26 gives the basic properties of magnetic fields. In this chapter we investigate the manner in which a changing magnetic field is able to induce a current in a coil of wire. Faraday's law of induction relates the emf produced in a loop to the rate of change of magnetic flux passing through the loop. Lenz's law provides a basis for determining the directions of emf's, forces, and reactions, and a method is given for determining the quantity of magnetic flux in a weber without having to count the lines. The concept of inductance is introduced, and the properties of inductors are presented.

On the practical side, we describe the emf produced when a loop is rotated in a magnetic field. This is the basis of operation of electrical generators and alternators. The fundamentals of transformers are given; the turns-ratio formula is obtained; and the method of determining the primary and secondary currents is described.

In the last section we describe Maxwell's equations. The mathematics of the equations is above our level, but the principles on which the equations are based are described and illustrated. These equations tie together the theories of electricity and magnetism and further predict the existence of electromagnetic waves.

27-1 BASIC PRINCIPLES OF MAGNETIC INDUCTION

Suppose a loop of wire is connected to a galvanometer, as shown in Fig. 27-1, and that a permanent magnet is positioned so that the magnetic flux from the magnet passes through the loop. Experiments indicate that no electric current will flow through the galvanometer unless there is relative motion between the loop and the permanent magnet. That is, no electromotive force is produced in the loop unless the quantity of flux passing through the loop is changing. A change in the quantity of flux passing through the loop may be produced by moving the magnet and/or the loop, as shown in Fig. 27-1, or by rotating either the magnet or the loop, as shown in Fig. 27-2.

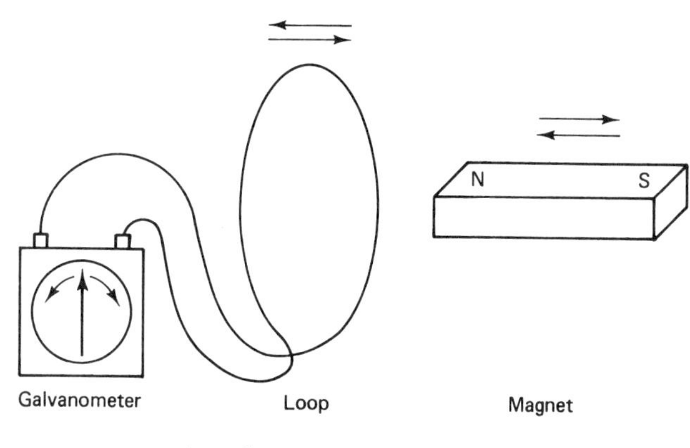

Figure 27-1 When the magnet moves toward the loop or when the loop moves toward the magnet, a current will be induced in the loop and the galvanometer will deflect.

A galvanometer deflection will occur in one direction when the magnet moves toward the loop, and a deflection in the opposite direction will occur when the magnet moves away from the loop. The magnitude of the deflection is proportional to the speed with which the relative motion occurs.

Uniform rotation of either the magnet or the loop, as shown in Fig. 27-2, produces alternating deflections of the galvanometer due to the changing direction of the flux through the loop.

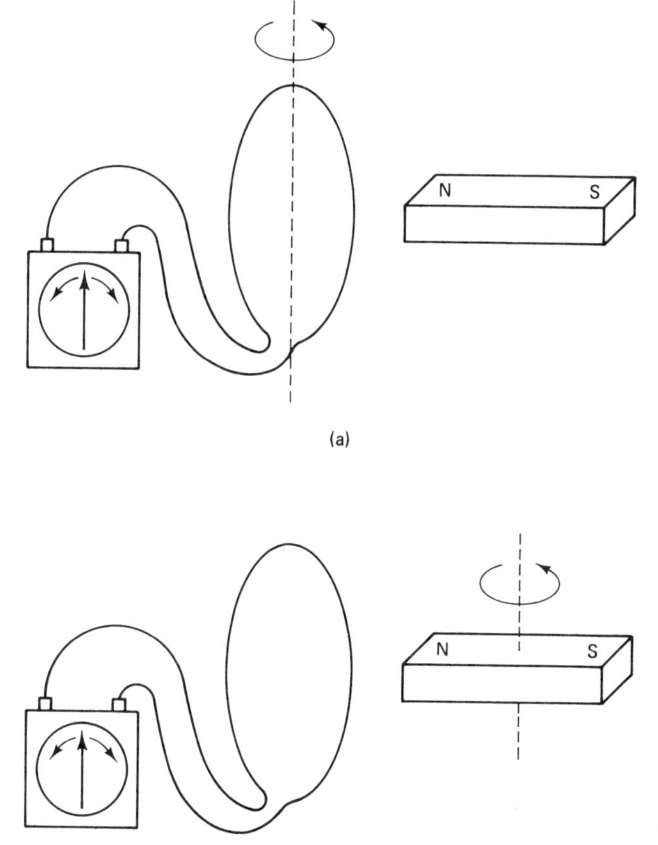

Figure 27-2 Rotation of either (a) the loop or (b) the magnet produces alternating deflections of the galvanometer.

In 1831, Michael Faraday put forth the mathematical relationship we now know as *Faraday's law*. If the magnetic flux passing through a coil of N turns changes by an amount $\Delta\Phi$, in a time interval Δt, the electromotive force induced in the coil is given by

$$\mathscr{E} = -N\frac{\Delta\Phi}{\Delta t} \tag{27-1}$$

The induced emf, $\mathscr{E}$, is given in volts if the rate of change of flux is expressed in webers per second and Δt in seconds. The negative sign is associated with the direction or polarity of the induced emf.

Another means of establishing a magnetic flux through a coil is shown in Fig. 27-3(a). When switch S is closed, current flows through the coil connected to the battery (which we shall call the primary coil) and the current sets up a magnetic field around the coil. A portion of the flux will pass through the other coil (which we call the secondary).

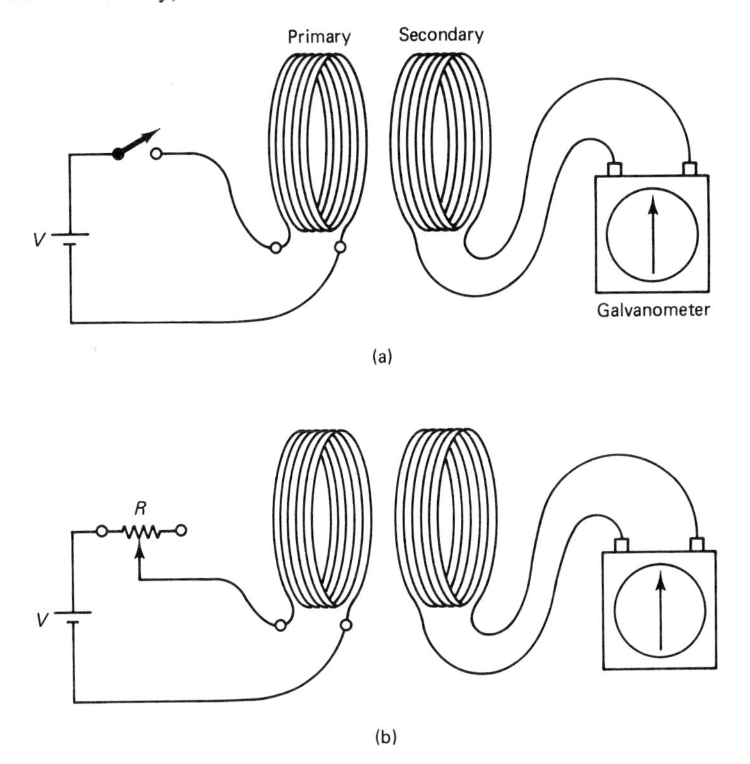

Figure 27-3 (a) The galvanometer will deflect when the switch is either closed or opened.
(b) When resistance R is varied, the galvanometer will deflect.

When the switch is closed, deflection of the galvanometer will be observed in one direction but the needle will quickly return to zero deflection. When the switch is subsequently opened, deflection in the opposite direction will occur, and the needle will again quickly return to zero.

A modification is shown in Fig. 27-3(b), which allows the magnitude of the magnetic field around the primary to be slowly varied. A change in resistance R causes a change in current through the primary coil, and this results in a change in the magnetic flux.

Coils arranged as in Fig. 27-3 are said to be *linked* by the magnetic flux, and the flux linkage is proportional to the magnitude of the flux and to the number of turns in the coils.

Definition of the weber

In Chap. 26 we refrained from specifying the number of magnetic lines contained in one weber because of the physical nonexistence of the lines. Eq. 27-1 provides a method for specifying that quantity of magnetic flux without having to count the lines. Hence, a weber is defined as follows. We assume that a variable source of magnetic flux is located so that the flux passes through a single loop of wire. If the flux passing through the loop initially is zero, and if it increases uniformly at a rate such that exactly 1 V of emf is induced in the loop, then at the end of 1 s the amount of flux passing through the loop is exactly 1 Wb.

Lenz's law

When a permanent magnet approaches a conducting loop, as shown in Fig. 27-4, a current is established in the loop and this current sets up its own magnetic field around the loop. Furthermore, the induced magnetic field will interact with the

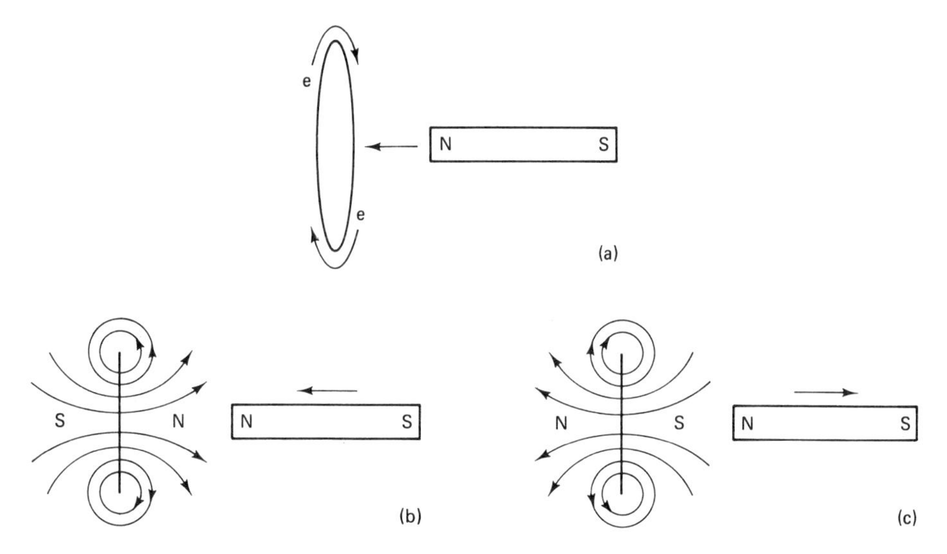

Figure 27-4 (a) A north pole approaching a loop sets up a clockwise flow of electrons in the loop, as seen from the position of the magnet.
(b) The induced current establishes a magnetic field that tends to repel the approaching magnet.
(c) When the magnet recedes from the loop, the induced field tends to attract the magnet back toward the loop.

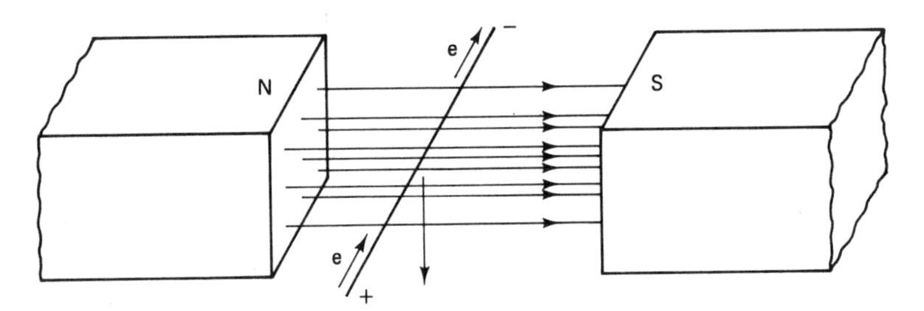

Figure 27-5 When the conductor moves downward in the magnetic field, the electrons move as shown.

field of the permanent magnet and will either attract or repel the permanent magnet. According to *Lenz's law*, the induced emf and resulting current will always be of such polarity that it *opposes* the effect that produced it. Consequently, a magnet approaching a loop will experience a force that tends to oppose the motion of the magnet toward the loop. The converse is also true. If a magnet is withdrawn from a loop, the current induced in the loop will set up a magnetic field that tends to pull the magnet back toward the loop.

That Lenz's law is a physical necessity can be understood by considering the conservation of energy as a magnet approaches a conducting loop. The loop exhibits a small electrical resistance, and energy will be dissipated in the resistance as the induced current flows through it. Consequently, the loop will be heated slightly. The energy that appears in the loop as heat must come from the work done in pushing the magnet toward the loop. An external force must be exerted on the moving magnet in order to move it toward the loop at constant velocity. It follows that the loop appears to repel the approaching magnet.

Emf produced in a moving conductor

In Chap. 26 we see that an electron moving in a magnetic field experiences a force. The same holds true even if the electron is part of a conductor and is being carried through the field by the motion of the conductor. Thus, it is not surprising to find that when a wire passes through a magnetic field an emf is developed in the wire. This is illustrated in Fig. 27-5.

The magnitude of the induced emf is easily determined by applying Eq. (27-1) to the situation shown in Fig. 27-6. A moving conductor forms one side of a rectangular loop whose plane is perpendicular to a magnetic field **B.** As the conductor moves, the area of the loop increases, and because the flux contained within the loop is proportional to the area, the flux also increases. Therefore, an emf is produced that is proportional to the length L of the conductor, to the velocity v with which it moves, and to the magnetic flux density B. The relationship is

$$\mathscr{E} = BLv \tag{27-2}$$

If the velocity of the conductor makes an angle θ to the lines of the magnetic field, a term $\sin\theta$ must be appended to Eq. (27-2) to give

$$\mathscr{E} = BLv \sin\theta \tag{27-3}$$

which is similar to Eq. (26-18).

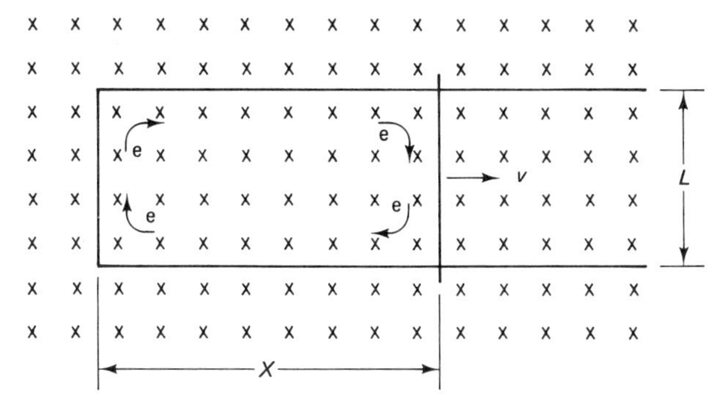

$$\mathcal{E} = -N\frac{\Delta\Phi}{\Delta t} = -\frac{\Delta\Phi}{\Delta t} = -\frac{B\Delta A}{\Delta t} = -\frac{BL\Delta X}{\Delta t}$$
$$= -BLv$$

Figure 27-6 A conductor of length L slides along a rectangular loop situated in a magnetic field directed into the page. An emf is directed around the loop that causes electrons to move as shown.

EXAMPLE 27-1 A conductor 15 cm long is part of the armature of a generator. Determine the emf induced in the conductor when it moves at 8 m/s perpendicular to the magnetic field of 1000 G.

Solution. Recall that 1000 G = 0.1 T. Then by Eq. (27-2):

$$\begin{aligned} \mathcal{E} &= -BLv \\ &= -(0.1\ \text{T})(0.15\ \text{m})(8\ \text{m/s}) \\ &= -0.12\ \text{T}\cdot\text{m}^2/\text{s} = -0.12\ \text{V} \end{aligned}$$
(minus sign indicates direction only)

To justify units,

$$\text{T} = \text{N/A}\cdot\text{m}; \qquad \text{A} = \text{C/s}$$

$$\text{T-m}^2/\text{s} = \text{N}\cdot\text{m}^2/\text{A}\cdot\text{m}\cdot\text{s} = \text{N}\cdot\text{m/(C/s)}\cdot\text{s} = \text{N}\cdot\text{m/C} = \text{J/C} = \text{V}$$

27-2 THE EMF OF A ROTATING LOOP

Suppose a rectangular loop of length L and width W is rotated in a uniform magnetic field B, as shown in Fig. 27-7. An emf will be produced in conductors ab and cd because these conductors move laterally relative to the field. Furthermore, the emf produced in ab will be series adding with the emf produced in cd because the two sections always move in opposite directions relative to the field. We can use Eq. (27-3) to compute the net emf produced in the rotating loop. This, of course, provides the basis for practical electrical power generation.

If the angular velocity of the loop is ω, the tangential velocity of conductors ab and cd is $\omega W/2$. Because $\omega = 2\pi f$, where f is the revolutions per second of the loop, the tangential velocity may be written as

$$v = 2\pi f W/2 = \pi f W \tag{27-4}$$

As the loop rotates, the direction of the velocity of ab and cd constantly changes relative to the magnetic field. The result is that the two conductors do not cut across the magnetic lines at a uniform rate. For example, in Fig. 27-7, the conductors will travel exactly at right angles to the field when θ is 90°, and they

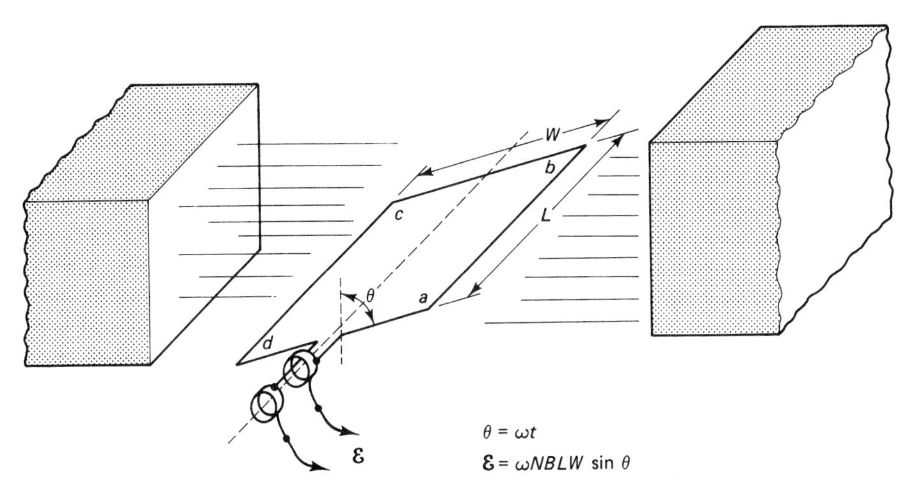

Figure 27-7 An emf is induced in a loop that rotates in a magnetic field.

will simultaneously experience the greatest induced emf. When the plane of the loop is vertical ($\theta = 0°$), the conductors will be traveling parallel to the magnetic field and will experience no induced emf.

Substitution of Eq. (27-4) into Eq. (27-3) gives the emf in one of the conductors:

$$\mathscr{E} = BLv \sin \theta$$
$$\mathscr{E}_1 = BL\pi fW \sin \theta \qquad (27\text{-}5)$$

The angle θ varies uniformly with time and is given by $\theta = \omega t$. Therefore, we may write

$$\mathscr{E}_1 = BL\pi fW \sin \omega t$$

and because there are two conductors (*ab* and *cd*) involved, we multiply the preceding expression by 2 to obtain the total emf induced in the rotating loop:

$$\mathscr{E} = 2BL\pi fW \sin \omega t$$

Because $\omega = 2\pi f$, this equation can be written in the following two forms:

$$\mathscr{E} = 2\pi fNBLW \sin(2\pi f)t \qquad (27\text{-}6)$$

$$\mathscr{E} = \omega NBLW \sin \omega t \qquad (27\text{-}7)$$

where the factor N has been included to apply to a loop having N turns rather than a single turn of wire.

For a particular loop rotating at a uniform and constant speed in a nonvarying magnetic field, the product of ω, N, B, L, and W will be a constant, which we may denote as $\mathscr{E}_0$:

$$\mathscr{E}_0 = \omega NBLW \qquad (27\text{-}8)$$

Thus, we may write Eq. (27-7) in the following form:

$$\mathscr{E} = \mathscr{E}_0 \sin \omega t \tag{27-9}$$

This demonstrates the sinusoidal variation of the emf produced in the loop. Because the algebraic sign of the sine function changes polarity in a periodic, cyclical fashion, the polarity of the induced emf varies in a similar manner. If the circuit is completed by connecting a lamp to the terminals of the rotating loop, the current flow through the lamp will vary sinusoidally, flowing first in one direction and then the other. Such a current is called an *alternating current.*

The induced emf will be a maximum when $\sin \omega t = 1.0$, and from Eq. (27-9), we see that the maximum value will be $\mathscr{E}_0$. The quantity $\mathscr{E}_0$, the maximum value, is called the *amplitude* of the alternating emf.

EXAMPLE 27-2 A rectangular coil 10 cm long and 8 cm wide is wound with 200 turns of wire. It rotates 30 times per second in a uniform magnetic field of 1000 G (0.1 T). Compute the amplitude of the alternating emf induced in the coil.

Solution. We first calculate ω for the rotating coil:

$$\begin{aligned} \omega &= 2\pi f \\ &= 2\pi(30\ \text{s}^{-1}) = 188.5\ \text{s}^{-1} \end{aligned}$$

We then substitute directly into Eq. (27-8):

$$\begin{aligned} &= \omega NBLW \\ &= (188.5\ \text{s}^{-1})(200)(0.1\ \text{T})(0.1\ \text{m})(0.08\ \text{m}) \\ &= 30.16\ \text{V} \end{aligned}$$

The units were reconciled to give volts as follows. A tesla is 1 Wb/m^2. Further, a weber is equivalent to a volt-second, from the definition of a weber. Thus,

$$1\ \text{T} = 1\ \text{Wb/m}^2 = 1\ \text{V} \cdot \text{s/m}^2$$

When this result is substituted for T in the preceding, all units except volts cancel.

Properties of a sine wave

One complete cycle of alternating current (AC) is produced each time the loop of Fig. 27-7 makes one complete revolution. The output voltage is plotted as a function of time in the graph of Fig. 27-8. The result is a sine wave, in accordance with Eq. (27-9). Two complete cycles are shown, and each cycle is divided into 360°.

The *frequency* of an alternating current is the number of complete cycles that occur each second. For an AC generator consisting of a simple loop, the frequency is the same as the number of revolutions per second that the loop rotates. The unit of frequency is the hertz, which corresponds to one cycle per second (s^{-1}).

The *period* T is the time required for one complete cycle to occur. The period and frequency are reciprocals:

$$T = \frac{1}{f} \qquad f = \frac{1}{T} \tag{27-10}$$

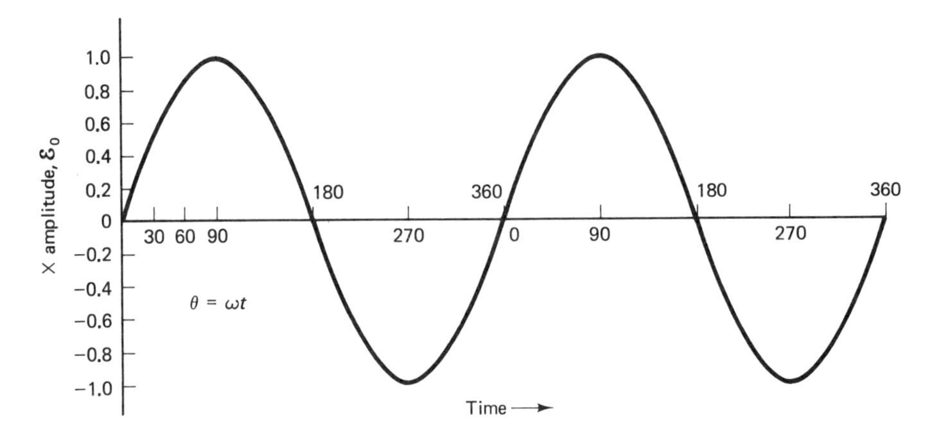

Figure 27-8 A graph of output voltage (emf) vs. time for a loop rotating in a magnetic field.

The *amplitude* $\mathscr{E}_0$ is one-half the total excursion of the wave form. That is, it is measured from the reference line to the maximum (or peak) value assumed by the wave form.

A *cathode ray oscilloscope* is a device used extensively by electronics technicians to display wave forms appearing in electronics circuits. It displays voltage (vertical) vs. time (horizontal) in the manner of the graph of Fig. 27-8. An oscilloscope is described in more detail in a later section.

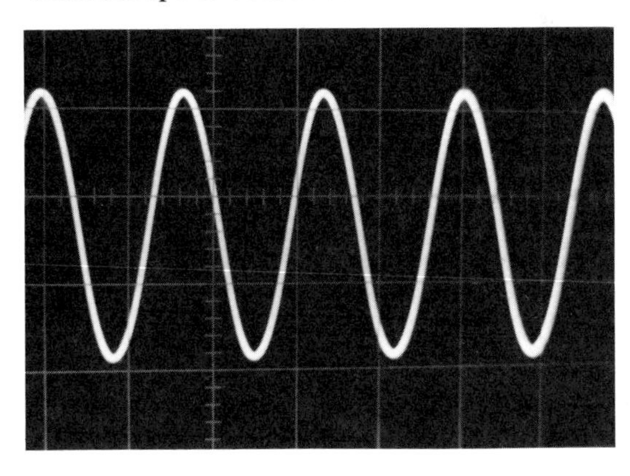

A sine wave displayed on an oscilloscope.

27-3 INDUCTANCE

Self-inductance of a coil

We now consider another one of the basic components of electrical and electronics circuits—inductors. Basically, an *inductor* is a coil of wire, and the coil may be wound on a core of iron or other material. Two types of inductors are shown in Fig. 27-9.

Refer to the two circuits of Fig. 27-10. In part (a), the current rises virtually instantaneously to its final value when switch S is closed, and the current falls to zero almost instantaneously when the switch is opened.

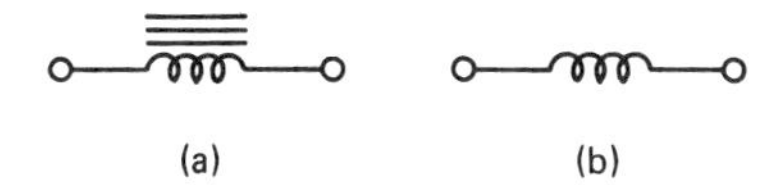

Figure 27-9 Symbols for inductors: (a) iron core; (b) air core.

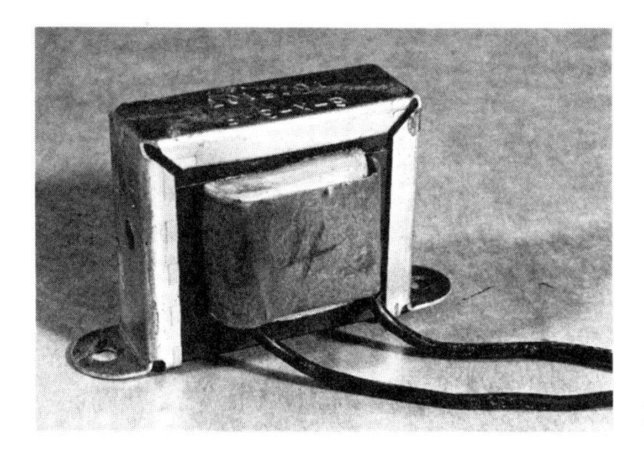

A small iron-core inductor used in electronics applications.

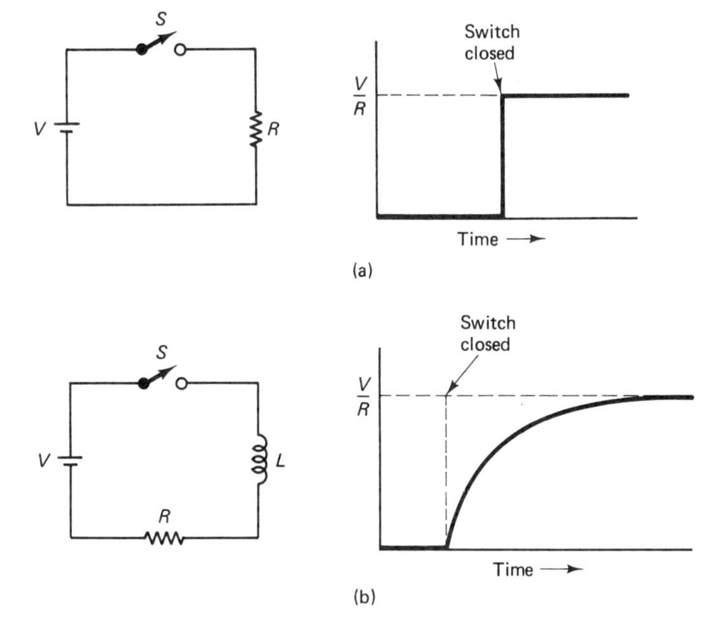

Figure 27-10 (a) In a noninductive circuit, the current rises almost instantaneously to its final value.
(b) In an inductive circuit, the back-emf of the inductor causes the current to rise more gradually.

In part (b) of the figure, a coil of wire (an inductor) appears in series with the resistor. When the switch is closed, a small current begins to flow and a magnetic field begins to expand outward around the inductor. The expanding field induces an emf in the windings of the inductor, and the emf, in accordance with Lenz's law, opposes the initial flow of current through the inductor. The result is that the current flow through the inductor (and also through the series resistor) builds up only gradually. We say that the inductor generates a *back-emf* that initially opposes the current flow. After a short time, however, the back emf disappears and the current level reaches its final value which is determined by V and R in accordance with Ohm's law.

This magnetic property of a coil is called *inductance*, and the ability of a coil to induce an emf in itself is called *self-inductance*. Inductance L is measured in units called *henries* (H) after Joseph Henry. The functional property of an inductor is that it generates a back emf that tends to resist any change in the amount of current flowing through it. Accordingly, an inductor is said to have an inductance of 1 H if a current changing at the rate of 1 A/s causes a back emf of 1 V to appear across the inductor. Hence,

$$1\text{ H} = 1\,\frac{\text{V}}{\text{A/s}} = 1\,\frac{\text{V}\cdot\text{s}}{\text{A}}$$

It is not difficult to calculate the instantaneous current flowing in an inductive circuit such as that shown in Fig. 27-10(b). The formula, obtained using simple calculus, is

$$I = \frac{V}{R}(1 - e^{-t/\tau}) \tag{27-11}$$

where $\tau = L/R$. In this formula, t is the time (seconds) elapsed after the switch S is closed, V is the battery voltage, R is the resistance in series with the inductance, and τ is the time constant for the particular circuit. Of course, e is the base of the natural logarithms. A graph of this formula is shown in Fig. 27-11.

The rapidity with which the instantaneous current approaches the final equilibrium value is governed by the ratio L/R, which we have defined as the time constant τ of the circuit. It is a property of the exponential function that in an interval of time equal to τ, the current rises to about 63% of its final value. After 5τ, the current will have risen to more than 99% of its final value.

Once a current is established in an inductor, the inductor opposes any attempt to change the magnitude of the current. If switch S is opened in Fig. 27-10(b), the current in the circuit will suddenly drop to zero and the magnetic field around the inductor will collapse. The collapsing field will generate a high voltage across the inductor, with the polarity being such that it tries to maintain the current at its original level. This voltage is applied to the opening switch contacts and causes a heavy spark to occur between the contacts as they open. The voltage produced by the collapsing field may be many times greater than the source voltage that caused the field to be established in the first place. For example, a 12-volt battery connected to an inductance of several henries easily produces an "inductive kick" of more than 100 V when the circuit is broken.

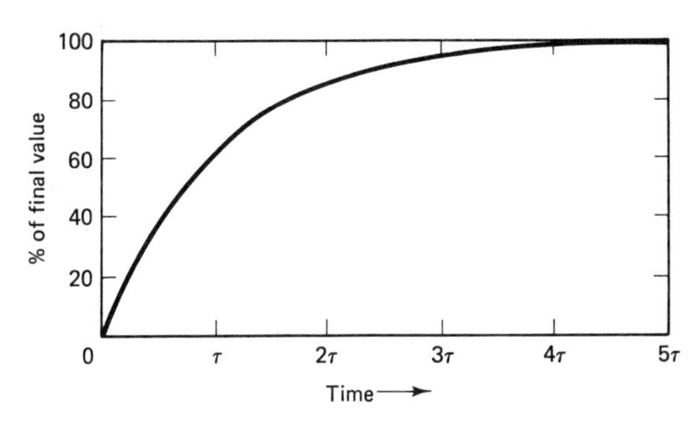

Figure 27-11 The rise of current in an inductive current in terms of the time constant τ. The level of 100% corresponds to V/R.

To avoid excessive arcing at the contacts of a switch that must operate in an inductive circuit, a capacitor is often connected in parallel with the switch. The capacitor prevents the excessive rise in voltage and makes the arcing less severe. This is the function of the *condenser* (same as capacitor) across the breaker points in the (nonelectronic) ignition system of an automobile engine.

Electric motors contain coils of wire (used as electromagnets) and therefore exhibit considerable inductance. Large motors are often equipped with special *zero-crossing* switches that open or close only when the instantaneous AC voltage is near zero volts. Consequently, the circuit is never broken when a large magnetic field surrounds the coils, and severe arcing at the switch contacts is avoided.

EXAMPLE 27-3 A 12-V battery is connected through a switch to a 10-Ω resistor in series with an 8-H inductor [as in Fig. 27-10(b)].

(a) Compute the time constant of the circuit.

(b) Calculate the instantaneous current in the circuit 1 s after the switch is closed.

Solution. (a) The time constant τ is given by

$$\tau = \frac{L}{R}$$

$$= \frac{8\ \text{H}}{10\ \Omega} = 0.8\ \text{s}$$

(b) We use Eq. (27-11):

$$I = \frac{V}{R}(1 - e^{-t/\tau})$$

$$= \left(\frac{12\ \text{V}}{10\ \Omega}\right)(1 - e^{-1.0/0.8})$$

$$= (1.2\ \text{A})(1 - e^{-1.25})$$

$$= 0.856\ \text{A}$$

Energy stored in an inductor

When a current I flows through an inductance L, the energy stored in the field of the inductor is

$$\text{Energy} = \tfrac{1}{2}LI^2 \qquad (27\text{-}12)$$

If L is expressed in henries and I in amperes, the units of energy will be joules. As an example, if a current of 10 A flows through an inductance of 1 H, the energy stored in the magnetic field surrounding the inductor is 50 J.

Mutual inductance

When two coils are positioned so that a portion of the magnetic flux produced by one coil passes through the other, the coils are said to be *magnetically coupled*. A pair of magnetically coupled coils are shown in Fig. 27-3. The degree of coupling is given by the coefficient of coupling, k, defined by

$$k = \frac{\Phi_m}{\Phi_p} \tag{27-13}$$

where Φ_p is the flux produced by the primary coil and Φ_m, called the *mutual flux*, is the flux produced by the primary that also passes through the secondary coil. Values of k range from 1.0 for *tightly coupled* coils, to 0 for coils that are *loosely coupled*. For $k = 1.0$, all the primary flux passes through the secondary, and for $k = 0$, none of the primary flux passes through the secondary.

The *mutual inductance* of a pair of coils is a measure of the effectiveness of one coil in inducing a voltage in the other. It is measured in henries, the unit of inductance. Two coils have a mutual inductance of 1 H when a current change of 1 A/s in the primary coil induces an emf of 1 V in the secondary.

Suppose two coils having self inductances of L_1 and L_2 are magnetically coupled. The mutual inductance M between the two coils is given by

$$M = k\sqrt{L_1 L_2} \tag{27-14}$$

where k is the coefficient of coupling. From this we see that the mutual inductance M is related to the degree of coupling (k).

Inductors connected in series

When resistors are connected in series, the total (or equivalent) resistance is the sum of the individual resistances. When inductors are connected in series, the equivalent inductance depends upon whether there is magnetic coupling between the series-connected inductors. More specialized texts derive the following formula for the equivalent inductance of two inductors connected in series:

$$L_T = L_1 + L_2 \pm 2M \tag{27-15}$$

We see that the mutual inductance M appears in the formula, but the term $2M$ may be either added or subtracted to the sum of L_1 and L_2. This is because the inductors may be connected *series adding* or *series opposing*, as illustrated in Fig. 27-12.

If the series-connected inductors are located sufficiently far apart so that the coefficient of coupling is zero, M will be zero, and the equivalent inductance will simply be the sum of the individual inductances.

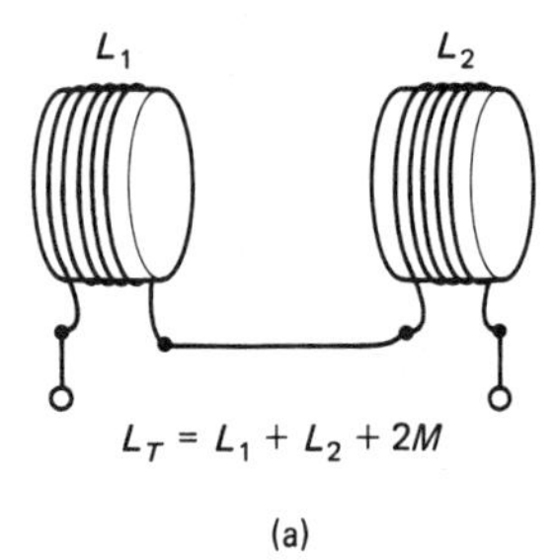

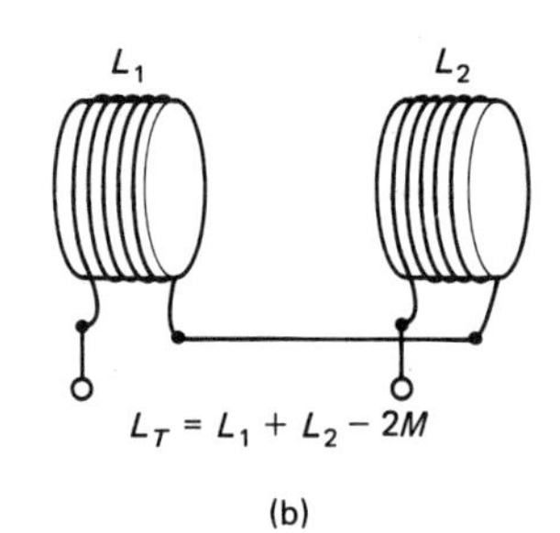

Figure 27-12 Inductors connected: (a) series adding; (b) series opposing.

A transformer for stepping line voltages down to the levels of residential use.

EXAMPLE 27-4 A 3-mH inductor is positioned next to a 5-mH inductor so that the coefficient of coupling k is 0.3.

(a) What is the mutual inductance M?

(b) If the two are connected series adding, what will be the total inductance?

(c) What will be the total inductance if the two are connected in series opposition?

Solution. (a) Using Eq. (27-14):

$$\begin{aligned} M &= k\sqrt{L_1 L_2} \\ &= (0.3)\sqrt{(3 \text{ mH})(5 \text{ mH})} \\ &= 1.16 \text{ mH} \end{aligned}$$

(b) Use Eq. (27-15) with the plus sign:

$$\begin{aligned} L_T &= L_1 + L_2 + 2M \\ &= 3 + 5 + 2(1.16) \\ &= 10.32 \text{ mH} \end{aligned}$$

(c) Use Eq. (27-15) with the negative sign:

$$\begin{aligned} L_T &= L_1 + L_2 - 2M \\ &= 3 + 5 - 2(1.16) \\ &= 5.68 \text{ mH} \end{aligned}$$

Inductors connected in parallel

The equivalent inductance of inductors connected in parallel can be computed by a simple calculation only if there is no mutual inductance between the inductors. For this simplified case, the inductors obey the same rule as for resistors connected in parallel. Therefore,

$$\frac{1}{L_T} = \frac{1}{L_1} + \frac{1}{L_2} + \cdots \tag{27-16}$$

If only two inductors are involved, Eq. (27-16) may be written in the form of a product over a sum. Hence,

$$L_T = \frac{L_1 L_2}{L_1 + L_2} \tag{27-17}$$

27-4 TRANSFORMERS

It is often desirable either to step up or step down the voltage of a particular alternating current source. A *transformer* provides a convenient means of doing this efficiently. In this section we consider only iron-core transformers, one type of which is illustrated in Fig. 27-13. In such transformers, we can assume the coefficient of coupling between the primary and secondary coils is unity so that all the flux generated by the primary passes through the secondary. This ideal can be closely approached in practical transformers.

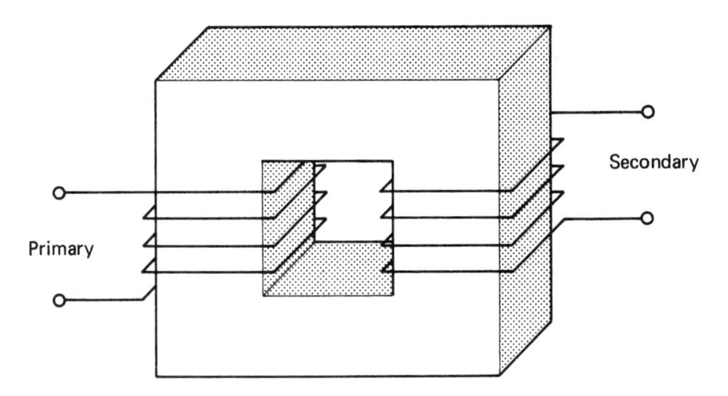

Figure 27-13 One type of iron-core transformer.

To obtain the relationship between the primary and secondary voltages, we assume that the flux in the core is changing at the rate $\Delta\Phi/\Delta t$. This flux change will induce voltages in the primary and secondary coils given by

$$V_p = N_p \frac{\Delta\Phi}{\Delta t} \qquad V_s = N_s \frac{\Delta\Phi}{\Delta t} \tag{27-18}$$

where N_p and N_s are the number of turns on the respective coils. Because $\Delta\Phi/\Delta t$ is the same for both coils, we can write

$$\frac{V_p}{N_p} = \frac{V_s}{N_s} \tag{27-19}$$

which may be rearranged to give

$$V_s = \left(\frac{N_s}{N_p}\right) V_p \tag{27-20}$$

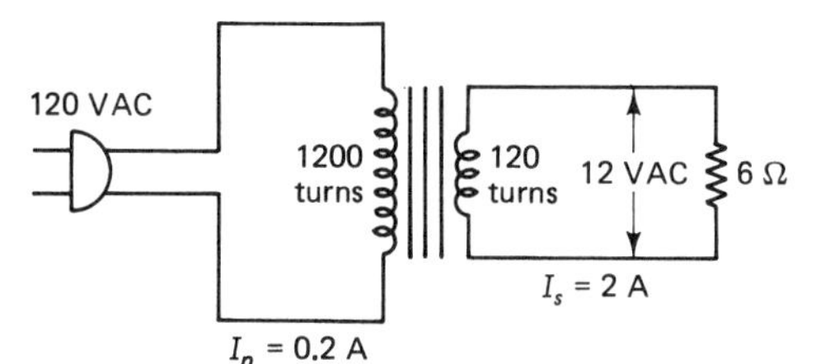

Figure 27-14

Thus, the *turns ratio*, N_s/N_p, becomes the proportionality constant relating the primary and secondary voltages.

In actuality, the voltage V_p applied to the primary causes the flux to be established within the core. Nevertheless, Eq. (27-18) is still valid even though both parts were obtained by considering the matter from a different point of view.

A *step-up* transformer has more turns on its secondary than on its primary, so N_s/N_p is greater than unity. Consequently, the secondary voltage is greater than the primary voltage. The converse is true for a *step-down* transformer.

Primary and secondary current

Practical transformers operate very efficiently. Larger units may achieve an efficiency of 99%. Therefore, it is a good assumption that all the energy that enters a transformer through the primary will leave through the secondary. (This is a characteristic of an ideal transformer.) It follows that the power level at the primary and secondary will be the same.

The power (energy/time) delivered to the primary is the product of the primary voltage and the primary current. It is customary to express this product in units of volt-amps (V · A). Hence,

$$\text{Power (primary)} = V_p I_p \tag{27-21}$$

In a similar manner, the power that flows out of the secondary is

$$\text{Power (secondary)} = V_s I_s \tag{27-22}$$

Setting these two equations equal gives

$$V_p I_p = V_s I_s \tag{27-23}$$

This relationship may be used to calculate the primary current when the secondary current is known.

EXAMPLE 27-5 Figure 27-14 shows a step-down transformer connected to a resistance of 6 Ω. The primary winding consists of 1200 turns and the secondary has 120 turns. The primary is powered by 120 VAC (volts alternating current) obtained from a conventional household wall receptacle. Calculate: (a) the secondary voltage; (b) the secondary current; (c) the primary current.

Solution. (a) We use the turns ratio formula, Eq. (27-20):

$$V_s = \left(\frac{N_s}{N_p}\right) V_p$$

$$= \left(\frac{120 \text{ turns}}{1200 \text{ turns}}\right)(120 \text{ VAC})$$

$$= 12 \text{ VAC}$$

A Tesla coil is a step-up transformer that operates at high frequency. The resulting fields cause nearby fluorescent bulbs to light up.

(b) The secondary voltage of 12 VAC is applied to a 6-Ω resistor. We make use of Ohm's law:

$$I_s = \frac{V_s}{R}$$

$$= \frac{12 \text{ VAC}}{6\ \Omega} = 2 \text{ A}$$

(c) A rearrangement of Eq. (27-23) allows us to calculate the primary current:

$$I_p = \left(\frac{V_s}{V_p}\right) I_s$$

$$= \left(\frac{12 \text{ VAC}}{120 \text{ VAC}}\right)(2 \text{ A}) = 0.2 \text{ A}$$

We observe from this example that a larger current flows in the secondary than flows in the primary, even though the transformer is a step-down transformer. This indicates that a step *down* in voltage is accompanied by a step *up* in current. This is true in general, and the converse is also true for a step-up transformer. When the voltage is stepped *up*, the current is stepped *down*.

Here is a noteworthy point: Suppose the resistor is disconnected from the secondary in Fig. 27-14, so the secondary circuit is *open*. The secondary current I_s will then be zero, and according to Eq. (27-23), the primary current will be zero also. Thus, we see that the primary current depends upon the condition of the secondary circuit. When no current is drawn from the secondary of an ideal transformer, no current will flow in the primary, and no power will be dissipated.

In actual transformers, however, a small magnetization current will flow in the primary even when no current flows in the secondary. This is due to the properties of the iron core.

Multiple secondaries

A transformer may have more than one secondary winding, as shown schematically in Fig. 27-15. In such case, the secondaries act independently of each other, and a transformer may be both step-up and step-down at the same time.

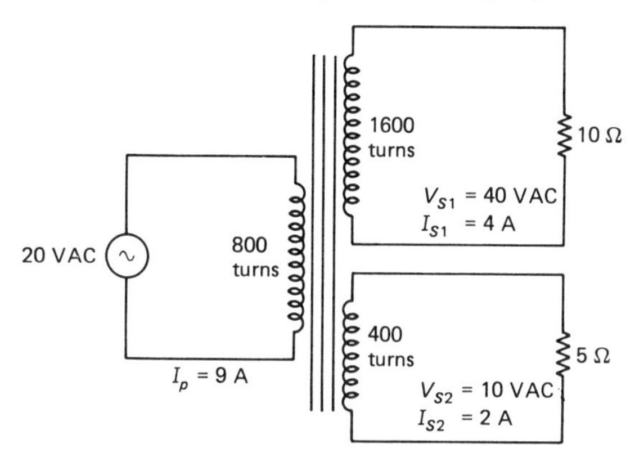

Figure 27-15 A transformer with multiple secondaries.

The secondary voltages are calculated by applying Eq. (27-20) to each secondary. Further, the secondary currents will generally not be the same.

The primary current of a multiple-secondary transformer is computed by extending Eq. (27-23). Because the power delivered to the primary must equal the sum of the powers of all the secondaries, we may write

$$V_p I_p = V_{s1} I_{s1} + V_{s2} I_{s2} + \cdots \tag{27-24}$$

where a $V_s I_s$ term appears for each secondary. Solving for I_p gives the primary current for a multiple-secondary transformer:

$$I_p = \frac{V_{s1} I_{s1} + V_{s2} I_{s2} + \cdots}{V_p} \tag{27-25}$$

The current and voltage levels of Fig. 27-15 may be verified using Eqs. (27-20) and (27-25) (and Ohm's law, of course).

27-5 EDDY CURRENTS

We have seen that a magnet approaching a conducting loop causes a current to flow in the loop (Figs. 27-1 and 27-4). In Fig. 27-16, a magnet is shown approaching a solid conducting disk, and, not surprisingly, electric currents are induced in the disk because of the changing magnetic field. These currents are called *eddy currents*. It is helpful to imagine the disk as being comprised of a large number of conducting rings. The production of eddy currents as the magnet approaches is then easily understood.

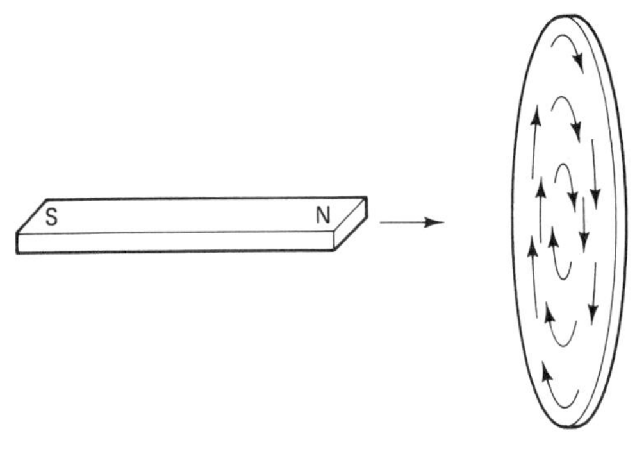

Figure 27-16 When a magnet approaches a stationary conducting disk, eddy currents are induced in the disk.

Eddy currents are also formed when a conducting disk rotates through a magnetic field, as in Fig. 27-17. The eddy currents cause the disk to be heated due to the fact that heat is produced any time a current flows through a resistance. Thus, by the principle of conservation of energy, we conclude that a retarding force is exerted on the disk which will cause it to slow down unless a torque is constantly applied. The work done by the applied torque appears in the disk as heat. We could have arrived at this result by considering Lenz's law.

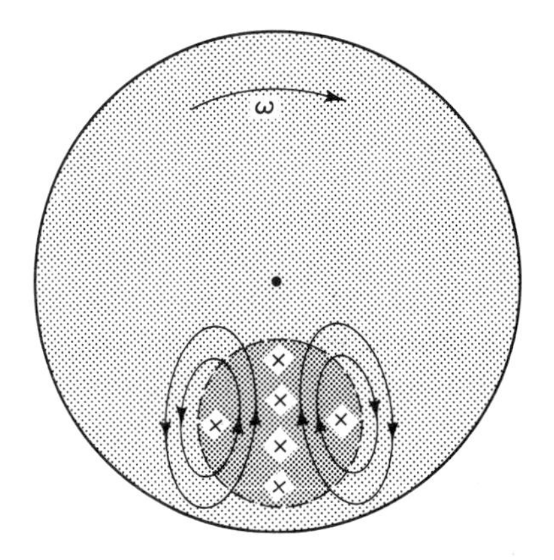

Figure 27-17 Eddy currents are produced when a conducting disk rotates through a magnetic field.

Another demonstration of the effects of eddy currents is shown in Fig. 27-18(a). A pendulum whose bob is a conducting disk swings between the poles of a strong magnet. The pendulum is made of nonmagnetic materials (such as aluminum) so that it is not attracted to the magnet. However, eddy currents are produced when the disk passes through the magnetic field, and the retarding force very rapidly reduces the amplitude with which the pendulum swings.

In Fig. 27-18(b), an identical disk has been slotted in order to break up the conducting loops. The retarding force is reduced dramatically, and the pendulum will swing for a much greater time before coming to rest. Thus, it is possible to reduce the effects of eddy currents by reducing the dimensions of the conducting loops.

Laminated iron cores

If a core of solid iron is used in a transformer or inductor, eddy-current heating of the core makes the device very inefficient and seriously limits its ability to operate satisfactorily. A typical eddy current loop is shown in the cutaway drawing of Fig.

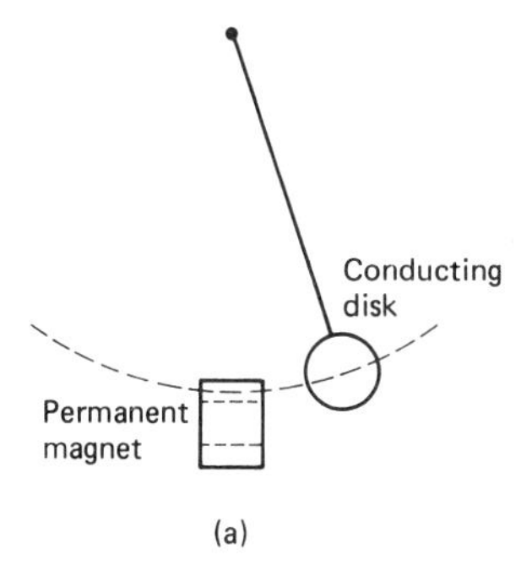

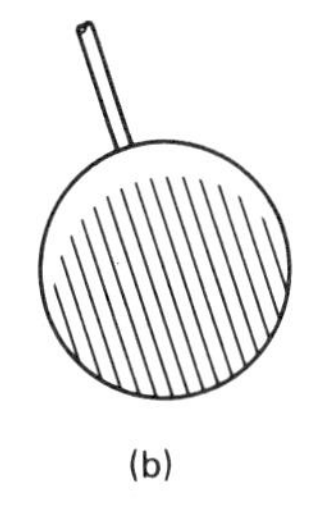

Figure 27-18 (a) When a conducting disk swings as a pendulum through a strong magnetic field, the motion of the pendulum is rapidly reduced due to eddy currents generated in the disk. (b) Slotting the disk reduces the effects of the eddy currents.

27-19, and we can envision the core as a very large number of one-turn secondaries through which the eddy currents flow. Clearly, this is a less than optimum situation.

The iron cores of practical inductors, transformers, motors, and generators are always *laminated* (built up in layers), and each lamination is coated with an insulating layer of varnish or oxide. Consequently, the conducting loops are broken up, and eddy current losses are greatly reduced.

Inductors and tuned transformers used in high-frequency electronics applications frequently utilize iron cores consisting of *powered iron* that has been molded into the desired shape. Each particle of iron is insulated from adjacent particles by the binder material so that significant eddy-current loops are avoided.

Induction heating

Eddy-current heating is used to advantage in a technique for heating objects without bringing the objects into contact with a flame or hot surface. The object to be heated is located within or near a coil energized by a high-frequency alternating

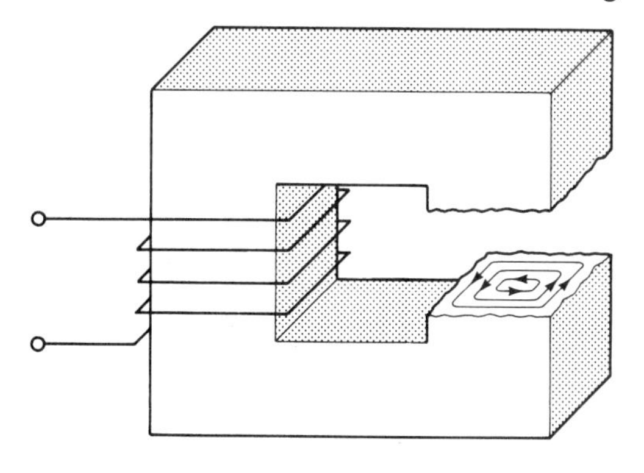

Figure 27-19 Conducting loops within a solid iron core act as one-turn secondaries through which large currents flow.

current. The resulting high-frequency magnetic field establishes large eddy currents within the object, and the object may reach a high temperature very quickly. The high-frequency current is provided by a power oscillator which in many respects resembles a radio transmitter.

27-6 MAXWELL'S EQUATIONS

In 1865 James Clerk Maxwell obtained a set of four equations that unified the theories of electricity and magnetism. Maxwell's equations, as they are known, provide the mathematical foundation for all electromagnetic phenomena. When the equations for the force exerted on a charge by electric and magnetic fields plus Newton's second law are added to these equations, we obtain a complete basis for all of classical electrodynamics. It is only in the realm of relativistic (very fast) and/or quantum (very small) electrodynamics that the equations require modification or extension. The scope of the equations is truly vast, and their elegance is widely appreciated.

At the time of Maxwell, the theory of electricity and magnetism was already well founded. The work of Coulomb, Oersted, Faraday, Henry, Ampere, and others paved the way for Maxwell by providing *individual* laws that were to be *unified* by the equations of Maxwell. In this section we describe the physical principles embodied in the equations without treating the mathematics in detail. An in-depth treatment requires a knowledge of calculus and vector analysis that is usually not obtained until the third year of undergraduate study.

Here are Maxwell's equations, written in the modern notation of vector analysis:

$$\nabla \cdot \mathbf{D} = \rho \quad \text{(M-1)} \qquad \nabla \times \mathbf{H} = \mathbf{J} + \frac{\partial \mathbf{D}}{\partial t} \quad \text{(M-3)}$$

$$\nabla \cdot \mathbf{B} = 0 \quad \text{(M-2)} \qquad \nabla \times \mathbf{E} = -\frac{\partial \mathbf{B}}{\partial t} \quad \text{(M-4)}$$

First we note the symbols $\nabla \cdot$ and $\nabla \times$. These are the *divergence* and the *curl*, respectively, both differential operators. Secondly, we observe a quantity **D**, called the *displacement*, and the *current density*, **J**. We begin our discussion with a brief look at the displacement.

The displacement D

In Sect. 24-6 we consider a parallel-plate capacitor with a dielectric material between the plates. Further, we see that the dielectric becomes polarized when the capacitor is charged, and layers of polarization charge are formed near the plates of the capacitor. The effect of the polarization charge is to reduce the magnitude of the electric field within the dielectric and thereby increase the capacitance of the capacitor.

As shown in Fig. 27-20, the conducting plates of the capacitor contain charge $+Q$ and $-Q$. We call this the *free charge*, and it is the *charge* of the capacitor.

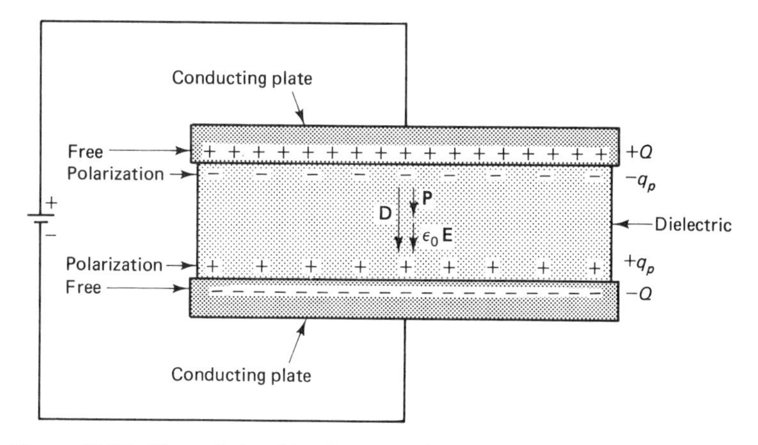

Figure 27-20 The relationships between free and polarization charges and between the electric vectors **D**, **P**, and **E**.

Adjacent to each plate is a layer of polarization charge. The charge of the layers are $+q_p$ and $-q_p$ as shown. Inside the dielectric, the magnitude of **E** depends upon both the free charge Q and the polarization charge q_p; the effect of q_p is to reduce the effect of Q in establishing **E** within the dielectric.

By virtue of its definition, the displacement **D** depends upon only the free charge Q. For the parallel-plate capacitor, its magnitude is the free charge per unit area of the plates, Q/A, and it has the units of coulombs per square meter. It is a vector quantity; its direction is the same as that of **E**. The displacement can be considered as a vector field in the same manner as **E**, but it is not the same as **E**.

The polarization of the dielectric can be represented as a vector, **P**, whose magnitude equals the polarization charge per unit area, q_p/A. The units of **P** are coulombs per square meter. The direction of **P**, by definition, is from the negative toward the positive polarization charge. For isotropic (uniform in regard to direction) dielectrics, **P**, **D**, and **E** will all have the same direction.

By definition,

$$\mathbf{D} = \epsilon_0\mathbf{E} + \mathbf{P}$$

The term $\epsilon_0\mathbf{E}$ has the units of coulombs per square meter, as also do **D** and **P**. The magnitude $\epsilon_0\mathbf{E}$ is that portion of the free charge density required to establish **E** in the absence of the dielectric. The magnitude **P** is the polarization charge density, q_p/A. Thus, the above equation amounts to a relationship between charge densities. (The polarities of q_p and Q on or near one plate are opposite, but this does not affect the algebraic sign of the magnitudes of **E** or **P**. The actual value of **P** depends upon the electric susceptibility of the medium.)

Current density

The vector quantity **J** is the current density in units of amperes per square meter, and it has the direction of flow of positive charge. In terms of magnitudes, $I = J \cdot A$, where A is the cross sectional area of the conductor or medium in which the current I is established. We are now ready to consider Maxwell's four equations.

(M-1) and (M-2)

We can consider the divergence ($\nabla \cdot$) of a vector field in terms of lines of flux of the field entering or leaving an arbitrary volume in which the field is established. If the divergence is zero, exactly as many lines *leave* as *enter* the region. Moreover, a zero divergence implies that the lines are continuous.

According to (M-2), the divergence of **B** is zero: $\nabla \cdot \mathbf{B} = 0$. This implies that magnetic lines are continuous; they have no ends. This is in accord with the fact that magnetic monopoles have never been observed.

Equation (M-1) tells us that the divergence of **D** is *not* zero. Therefore, lines of **D** may have ends; the lines are imagined to begin at positive charges and end at negative charges. Figure 27-21 illustrates the divergence of **D**. If more lines leave a volume than enter it, then the volume must contain a net positive charge to act as a "source" for the extra lines.

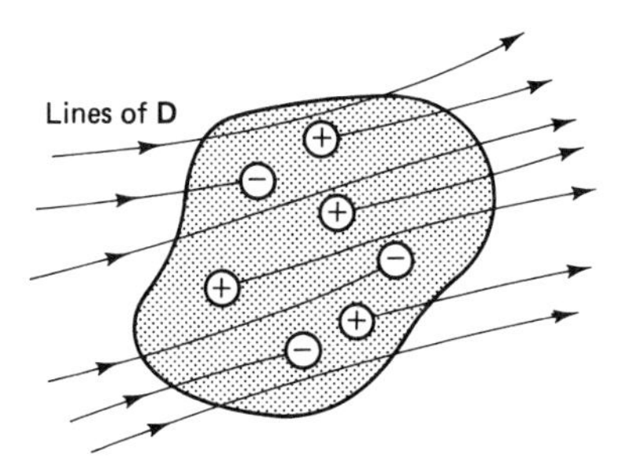

Leaving :	7	Positive charges :	4
Entering:	−6	Negative charges:	−3
	+1		+1

Figure 27-21 An illustration of the divergence. The difference in lines entering and lines leaving equals the net positive charge within the volume.

Faraday's law (M-4)

Recall that Faraday's law relates the emf induced in a loop to the rate of change of magnetic flux passing through the loop. The equation of Maxwell embodying this principle is

$$\nabla \times \mathbf{E} = -\frac{\partial \mathbf{B}}{\partial t} \qquad \text{(M-4)}$$

An implication of this equation is illustrated in Fig. 27-22, in which a vertical magnetic field is steadily increasing. The changing magnetic field produces an electric field in the form of a continuum of infinitesimal circlets lying in horizontal planes. A vector field having such a character is said to have *curl*.

If a closed path is drawn in a horizontal plane, some of the circlets will contact the path so that a component of the electric field will be tangent to the path. If we go completely around the path and sum the components tangent to the path, we obtain a result in the form of $E \cdot L$ that is not zero. The product of E and L (path length) is equivalent to an emf. Thus, the changing magnetic field induces an emf around the loop, and if the loop is constituted by a conductor, a current will flow.

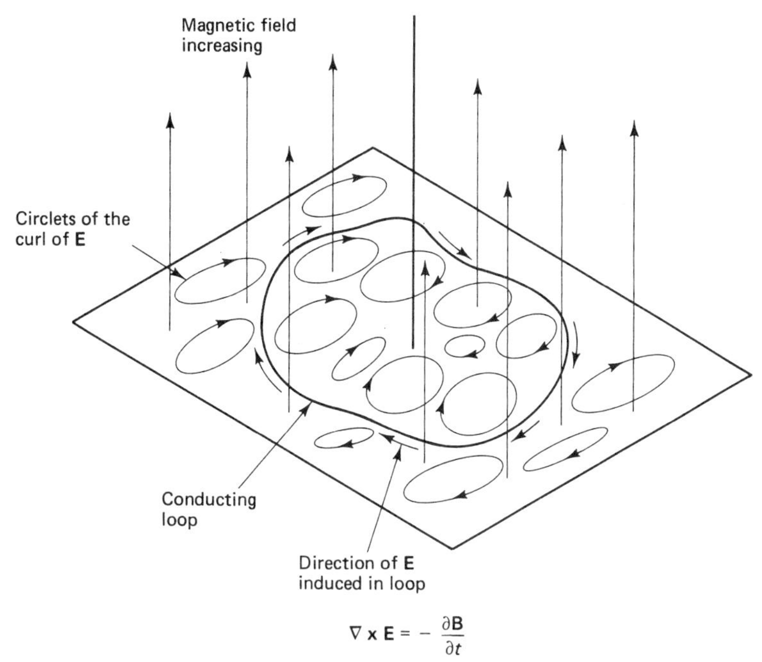

Figure 27-22 A changing magnetic field produces an electric field that gives rise to an emf in a conducting loop.

The left-hand side of (M-4), $\nabla \times \mathbf{E}$, is the curl of the electric field, in the notation of vector analysis. The right-hand side, $-\partial \mathbf{B}/\partial t$, is the rate of change of the magnetic field, in the notation of differential calculus.

Ampere's law (M-3)

In broad terms, Ampere's law relates magnetic fields to electric currents. An early form of the relationship was

$$\nabla \times \mathbf{H} = \mathbf{J}$$

which equates the curl of the magnetic field intensity $\mathbf{H}$ to the current density $\mathbf{J}$. Maxwell, however, recognized that this equation was incomplete. To the right-hand side he added the term $\partial \mathbf{D}/\partial t$, the rate of change of the displacement $\mathbf{D}$. The result is

$$\nabla \times \mathbf{H} = \mathbf{J} + \frac{\partial \mathbf{D}}{\partial t} \qquad \text{(M-3)}$$

An implication of this equation is illustrated in Fig. 27-23, in which we consider the charging of a parallel-plate capacitor. In the wires flowing to the capacitor, a current $I = J \cdot A$ flows as the capacitor charges. Between the plates, however, no real current flows (because of the insulating properties of the dielectric). However, when we consider the displacement field established within the dielectric, it is not steady; it is changing, and $\partial \mathbf{D}/\partial t$ is not zero.

According to (M-3), this produces an $\mathbf{H}$-field that has curl. Therefore, if we form a loop in a horizontal plane, a net magnetomotive force will be developed

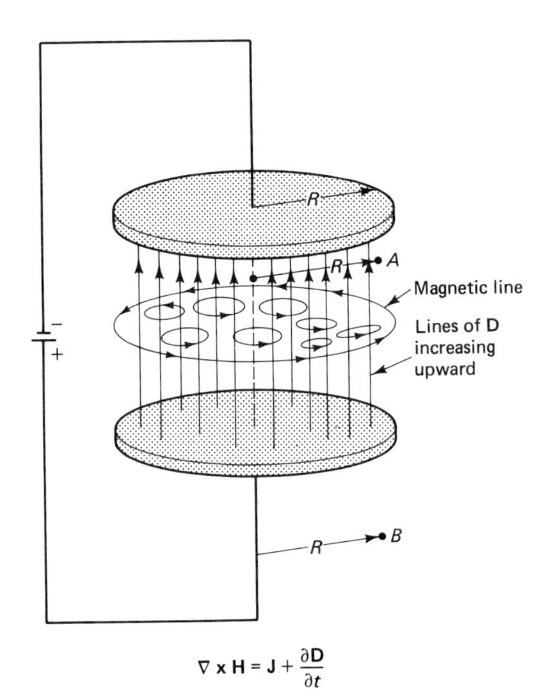

Figure 27-23 A changing **D**-field (or **E**-field) produces a magnetic field. The magnitude of the **B**-field at points A and B is the same.

around the loop, and this magnetomotive force will give rise to a magnetic field within the volume of the dielectric. Thus, a changing **D**-field (or **E**-field, since **D** is related to **E**) gives rise to a magnetic field.

If the capacitor plates are circular with radius R, the magnetic field at points A and B in Fig. 27-23 will be identical.

Symmetry between E and B

Equations (M-3) and (M-4) point out a fundamental relationship between **E** and **B** (recall $\mathbf{B} = \mu\mathbf{H}$). That is, one results from a change in the other. A changing electric field produces a magnetic field, and a changing magnetic field produces an electric field.

When (M-3) and (M-4) are combined and use is made of (M-1) and (M-2), a *wave equation* results whose solution is a wave comprised of electric and magnetic components. Furthermore, the speed of these waves in vacuum is calculated to be

$$\text{Speed of waves} = \frac{1}{\sqrt{\epsilon_0\mu_0}}$$

where μ_0 and ϵ_0 are the permeability and permittivity, respectively, of free space. The numerical value of the speed is the same as the speed of light!

The predicted waves were soon demonstrated experimentally, and they are now accepted as a common aspect of everyday life—radio waves, television waves, microwaves, and so forth, and even light itself. All are manifestations of the same physical phenomenon. Even so, the waves were unknown in 1865 when Maxwell formulated the equations that now bear his name.

To Go Further

Investigate the following:

Joseph Henry (1797–1878), American physicist
Heinrich Lenz (1804–1865), German physicist
Heinrich Hertz (1857–1894), German physicist

Questions

1. Suppose a magnet is held vertically above a horizontal coil of wire that is connected to a galvanometer. Describe the action of the galvanometer when the magnet is released and allowed to fall through the coil.
2. In what manner does the amplitude of the emf generated by a rotating loop depend upon the rotational speed of the loop?
3. What determines the frequency of the emf produced by a simple rotating loop?
4. Suppose the axis of a rotating loop is aligned parallel to the magnetic field. Will an emf be produced in the loop as it rotates?
5. The output terminals of a small hand-driven generator are connected to an incandescent lamp. Why is the generator harder to turn when the lamp is switched on?
6. Approximately how many time constants must elapse for the current in a series *L-R* circuit to reach its final value?
7. What is responsible for the heavy arcing at the contacts of a switch in an inductive circuit?
8. The switch connected in series with an electric motor is constructed so that its contacts are visible. Most of the time, a heavy arc is observed when the switch is opened, but occasionally, little, if any, arc is observed. Explain.
9. When two coils are oriented exactly at right angles to each other, their coefficient of coupling is zero even though they may be very close together. Why?
10. Does a long, straight wire have inductance?
11. Is it possible to reverse a step-down transformer and operate it as a step-up transformer?
12. Under heavy load (large currents), practical transformers sometimes get hot. Can you identify two sources of heat in a transformer?
13. An aluminum plate is attached to the end of the beam of a triple-beam balance used in a physics lab, and the plate is located between the poles of a permanent magnet. Why?

Problems

1. The magnetic flux passing through a 25-turn coil increases from 0.2 Wb to 0.8 Wb in 3 s. What is the average emf induced in the coil?
2. Two coils are located in close proximity, as in Fig. 27-3. A variable current source is connected to the primary, and a sensitive voltmeter is connected to the secondary, which consists of 100 turns. Starting from zero, the primary current is increased at a uniform rate so that 5 V is induced in the secondary. How much flux will pass through the secondary coil at the end of 20 s?
3. A conductor 10 cm long moves at right angles to a uniform magnetic field of flux density of 0.5 T. If the velocity of the conductor is 0.6 m/s, what emf is induced in the moving conductor?
4. If the conductor of the preceding problem moves at an angle of 45° to the field, what emf will be induced?
5. A square loop 1 m on a side and which contains 100 turns is rotated uniformly at 1 rev/s about a vertical axis in an open region where the magnetic field of the earth (assumed to be horizontal) is 1 G (10^{-4} T). What emf is induced in the coil?

6. A device for measuring magnetic flux density **B** incorporates a small, motor-driven coil on the end of a long probe that may be inserted into the field. The dimensions of the coil are $L = 2$ cm and $W = 1$ cm. It rotates at 2400 RPM, and contains 50 turns. What voltage will be induced in the coil when **B** is: **(a)** 100 G; **(b)** 1000 G; **(c)** 1 T; **(d)** 5 T?
7. Electrical power supplied to a motor causes the conducting loops of the armature to rotate in the magnetic field produced by the field coils of the motor. In so doing, a back-emf is induced in the windings of the armature that opposes the applied emf. The motor also acts as a generator! Compute the back-emf generated in an armature winding of 10 turns when the armature rotates at 1750 rpm. The magnetic flux density is 0.2 T, and the armature winding is a rectangular coil 5 cm by 10 cm.
8. What are **(a)** the frequency and **(b)** the period of the emf produced by a simple generator rotating at 1750 rpm?
9. Household current in the United States is AC at a frequency of 60 Hz. What is its period?
10. An inductance of 8 H is connected in series with a resistance of 20 Ω as shown in Fig. 27-10(b). The battery voltage is 12 V.
    (a) Calculate the *L-R* time constant for this circuit.
    (b) What will be the final value of the current after the switch is closed?
11. For the circuit of Problem 10, compute the current at the following times after the switch is closed.
    (a) 0.2 s **(b)** 0.4 s **(c)** 0.8 s **(d)** 2.0 s
12. Repeat Problems 10 and 11, but change the resistance to 40 Ω instead of 20 Ω.
13. Calculate the energy stored in a 10 H inductor when the following currents flow through it.
    (a) 1 A **(b)** 2 A **(c)** 3 A **(d)** 4 A
14. The mutual inductance between a pair of coils is 0.4 H. Current changes in the first coil at the rate of 2 A/s. What emf is produced in the second coil?
15. Two coils are magnetically coupled with $k = 0.2$. The self-inductances are 15 mH and 25 mH. Calculate the mutual inductance.
16. For the inductors of Problem 15, compute the equivalent inductances obtained when the inductors are connected: **(a)** series adding; **(b)** series opposing.
17. Suppose a 6-H and a 3-H inductor are connected in parallel (with $k = 0$). What is the equivalent inductance?
18. A certain transformer has 1200 turns on the primary and 800 turns on the secondary. A 30-VAC source is connected to the primary.
    (a) Calculate the secondary voltage.
    (b) What current will flow in the secondary when it is connected to a 40-Ω resistor?
    (c) What will be the primary current when the resistor is connected to the secondary.
19. Refer to Fig. 27-24.
    (a) Compute each secondary voltage.
    (b) Compute each secondary current.
    (c) What is the total volt-amps delivered by the secondaries?
    (d) Compute the primary current.

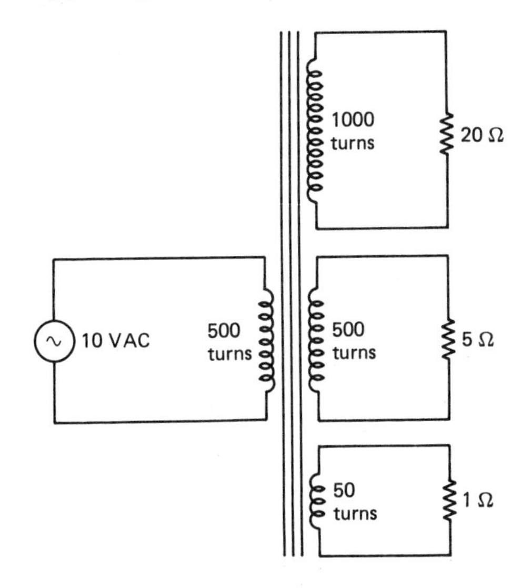

Figure 27-24

20. Use the I^2R formula to calculate the power dissipation (in watts) in each resistor of the circuit shown in Fig. 27-24. Compare the result with the volt-amps delivered by each secondary.

CHAPTER 28

ALTERNATING CURRENT CIRCUITS

An earlier chapter considers the properties of electrical currents flowing in circuits; in particular we studied direct-current (DC) circuits. However, practically all power distribution systems use alternating current (AC), and a major portion of all electronics systems use alternating current in some fashion. In this chapter we study the principles of alternating current and AC circuits.

In the realm of alternating current we encounter the phenomena of reactance, phase shifts, and resonance, so AC circuits are somewhat more complex than their DC counterparts.

28-1 PHASOR REPRESENTATION OF SINE WAVES

Frequently, in electrical and electronics applications, two or more sine waves must be compared, considering both the amplitudes and phases of the sine waves. Sine waves, or sinusoidally varying voltages or currents, can be conveniently represented by rotating vectors called *phasors*. Fig. 28-1 shows that the projection

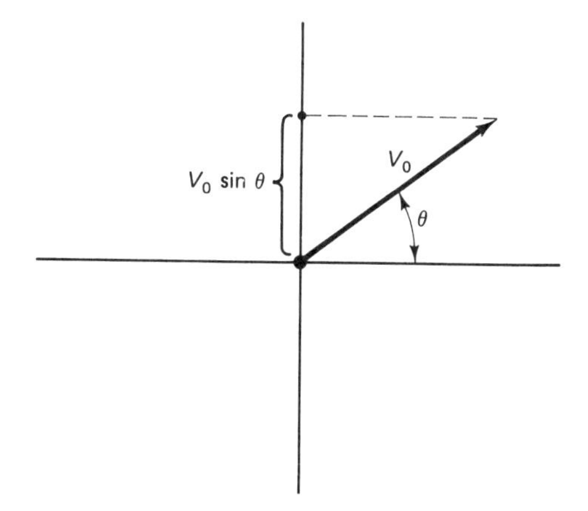

Figure 28-1 A phasor is a rotating vector that is used to represent a sine-wave voltage or current. The length of the phasor is proportional to the amplitude of the sine wave.

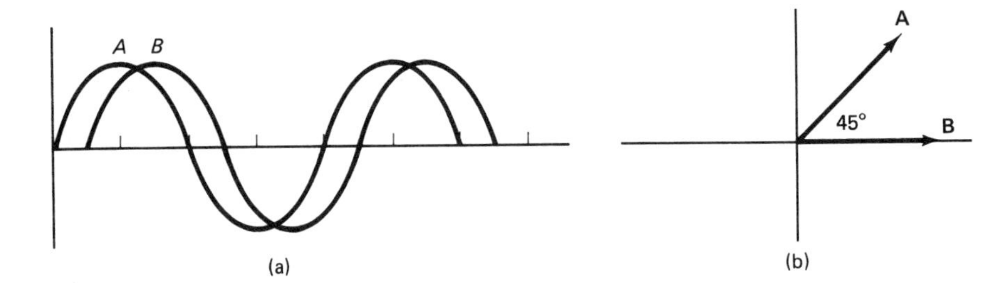

Figure 28-2 (a) Two sine waves 45° out of phase. (b) Phasor representation of two sine waves 45° out of phase.

of the phasor onto the vertical axis is proportional to sin θ, and the projection will vary sinusoidally as the phasor rotates uniformly according to $\theta = \omega t$. Thus, we may recognize the correspondence between the projection of the phasor and the instantaneous value of a voltage or current in an AC circuit. The length of the phasor represents the amplitude of the sine wave, and the rotational speed is determined by the frequency. The phasor executes one complete revolution per cycle of the sine wave that it represents.

Phasors are particularly useful in comparing relative phases of sine waves, as shown in Fig. 28-2. For the relative phase to be constant, the frequency of both sine waves must be the same, and this implies that the two phasors must rotate at the same angular velocity.

Phasors obey the rules of vector addition: They may be added graphically by placing the tail of one at the tip of the other, as shown in Fig. 28-3. It follows that when two sine waves of the same frequency are added, the resultant is a sine wave of the same frequency but generally of different amplitude and phase. If two sine waves of different frequency are added, the resultant will not be of constant amplitude and the resulting wave form will not be a simple sine wave.

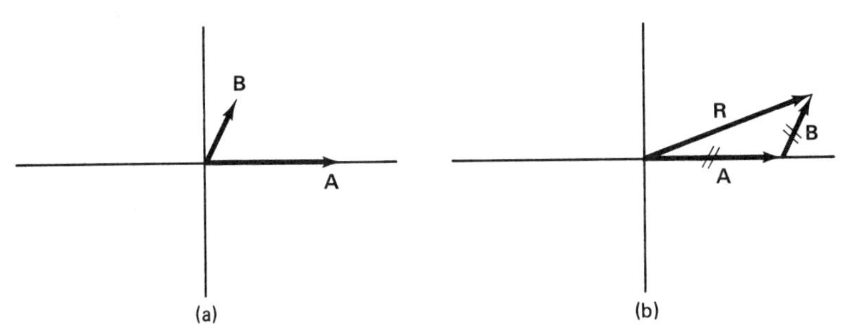

Figure 28-3 Phasors are added according to the rules of vector addition. The resultant phasor is also assumed to rotate.

28-2 AC IN RESISTIVE CIRCUITS

We have seen that the emf produced by a rotating loop or another source of AC can be expressed in the form

$$V_i = V_0 \sin \omega t \tag{28-1}$$

where V_i is the instantaneous value of the emf and where V_0 is the amplitude. If

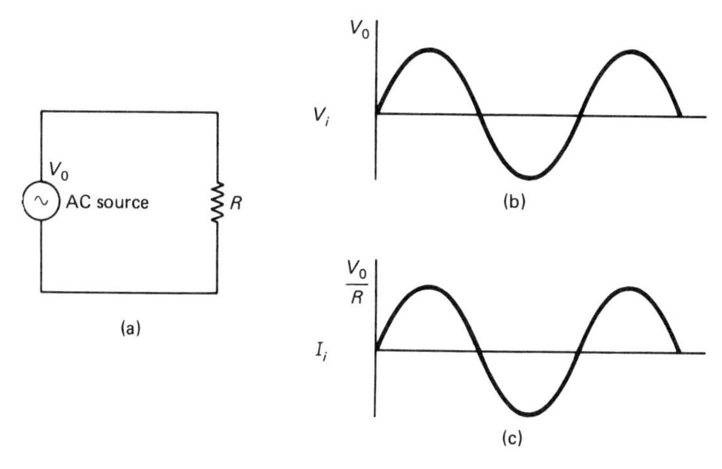

Figure 28-4 (a) A source of AC connected to a resistor. (b) Variation of the voltage applied to the resistor. (c) The resulting current flow through the resistor.

such an emf is applied to a resistor, as shown in Fig. 28-4, the resulting current flow through the resistor will vary in a similar manner:

$$I_i = I_0 \sin \omega t \tag{28-2}$$

Thus, the current, as well as the emf, alternates. By Ohm's law, $I_0 = V_0/R$ so that

$$I_i = \frac{V_0}{R} \sin \omega t \tag{28-3}$$

Power dissipation in a resistor

The power dissipation in a resistor is given by I^2R, and this is true for AC as well as DC circuits. However, in an AC circuit, the current level is not constant; it varies sinusoidally. Hence, using Eq. (28-2),

$$\begin{aligned} \text{Power} &= I^2R = (I_0 \sin \omega t)^2 R \\ &= I_0^2 R \sin^2 \omega t \end{aligned} \tag{28-4}$$

A graph of this function is shown in Fig. 28-5, and we see that the power

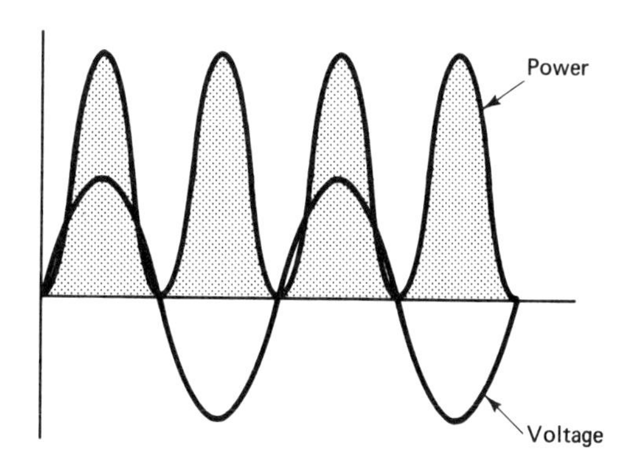

Figure 28-5 Power dissipation in a resistor when a sine-wave voltage is applied to the resistor.

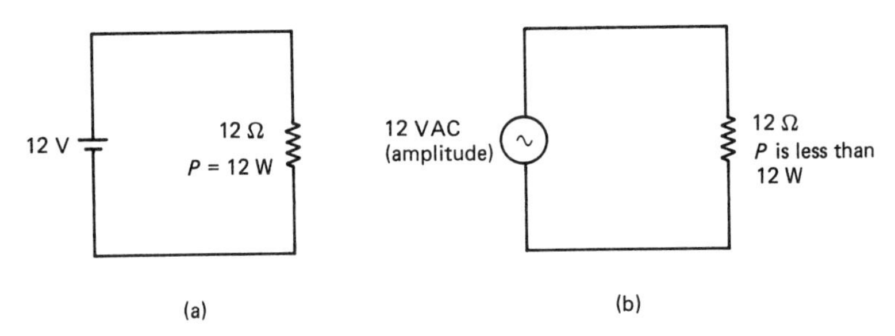

Figure 28-6 (a) A 12-V DC source produces a power dissipation of 12 W in the resistor.
(b) A 12-V (amplitude) AC source dissipates less than 12 W.

dissipation occurs in spurts, with two spurts occurring for each cycle of the applied voltage.

Let us now compare an AC to a similar DC circuit, shown in Fig. 28-6. In the DC circuit, a steady current of 1 A produces a steady power dissipation of 12 W in the resistor. But in the AC circuit, the power dissipation is not steady; it reaches a maximum of 12 W two times each cycle, but at other times, it is less than 12 W. Twice per cycle, the instantaneous power dissipation is zero. Clearly, the DC circuit is more effective in producing a power dissipation in the resistor.

In practice, it would be convenient for a 12-V AC source to produce the same power dissipation as a 12-V DC source. But for this to be the case, we must revise our method of expressing the magnitude of AC voltages. In the above, we tacitly assume the AC voltage to be 12 V in amplitude. In practice, the amplitude of an AC voltage is not customarily given. Rather, AC voltages are expressed in RMS (root-mean-square) values. The *RMS voltage* is defined in relation to the amplitude, so that a 12-V RMS source will produce the same time-averaged power dissipation in a resistor as a 12-V DC source.

Using calculus, it is possible to show that the RMS voltage is related to the amplitude according to

$$\text{RMS} = \frac{\text{amplitude}}{\sqrt{2}} \tag{28-5}$$

Thus, to obtain an RMS voltage of 12 V, the amplitude must be 16.968 V.

Voltages: peak, peak-to-peak, and RMS

These voltages are shown relative to the sine wave of Fig. 28-7. The peak voltage, V_p, is the same as the amplitude, and the peak-to-peak voltage, $V_{p\text{-}p}$, is twice the amplitude. The same definitions may be applied to current; we often speak of either peak current or RMS current.

Obviously there is room for considerable confusion unless everyone is consistent in expressing AC voltages in terms of RMS values. To preclude such confusion, AC voltmeters and ammeters are calibrated in RMS values. Ordinary AC voltmeters read RMS voltages. Peak-reading voltmeters are available, but they are more specialized instruments.

The AC voltage appearing at the wall socket of a house is about 120 V, as read on an ordinary AC voltmeter. Accordingly, the peak voltage is about 169 V.

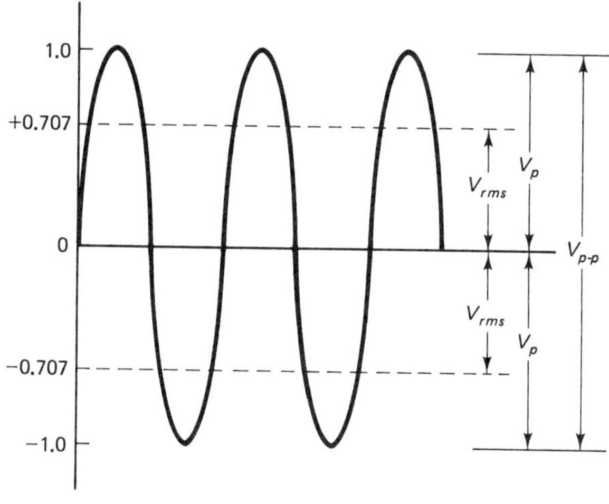

Figure 28-7 Definitions of $V_{p\text{-}p}$, V_p, and V_{rms}.

28-3 AC IN AN INDUCTIVE CIRCUIT

In Fig. 27-10 we see that the action of an inductor is to oppose any change in the current flowing through it. When a sinusoidal voltage is applied to an inductor, the back-emf generated causes the inductor to oppose the flow of current through it. In other words, the current is limited by the inductive effect. We say the inductor exhibits a *reactance* that opposes the flow of current through it. Reactance is somewhat analogous to resistance, but the two have many different properties.

When an AC voltage V is applied to an inductance L, a current given by

$$I = \frac{V}{X_L} \tag{28-6}$$

will flow through the inductor, where X_L is the *inductive reactance*. The value of X_L depends upon the frequency f of the applied voltage and upon the size of the inductance L:

$$X_L = 2\pi fL \tag{28-7}$$

We note that inductive reactance is greater at high frequencies, and larger inductors exhibit proportionally greater reactances at a given frequency.

EXAMPLE 28-1 A 12-V source of 400 Hz alternating current is applied directly to an inductance of 0.2 H.

(a) Calculate the inductive reactance X_L.
(b) Compute the current that flows through the inductor.
See Fig. 28-8.

Solution. (a)

$$\begin{aligned} X_L &= 2\pi fL \\ &= 2\pi(400\text{ Hz})(0.2\text{ H}) \\ &= 502.6\ \Omega \end{aligned}$$

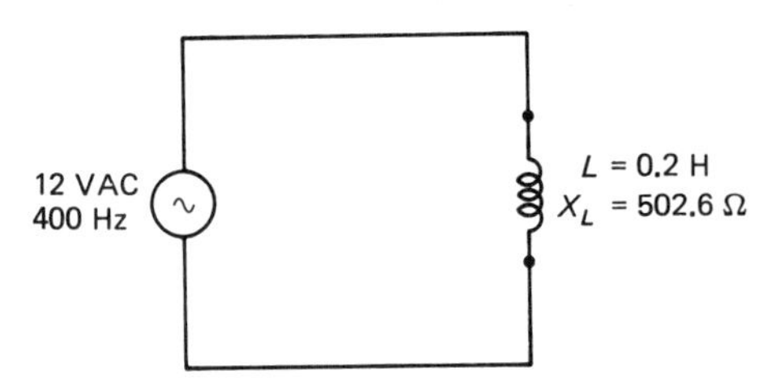

Figure 28-8

(b)

$$I = \frac{V}{X_L}$$

$$= \frac{12 \text{ V}}{502.6\ \Omega} = 0.0239 \text{ A}$$

$$= 23.9 \text{ mA}$$

Because the type of voltage is not specified, we assume that RMS voltages are intended.

R and *L* in series

An AC voltage source is connected to a series combination of R and L in Fig. 28-9. We wish to find the current that flows in the circuit. As for DC circuits, the current is the same at all points in a series circuit, even when AC voltages are applied.

Both the resistance R and inductive reactance X_L oppose the flow of current, but the total opposition is not simply the sum of R and X_L as we might expect. To see why this is so, consider a sinusoidal current to be flowing in the circuit. The resistance develops the maximum *opposing voltage* when the current is a maximum, at the peak of the sine wave. However, the inductor develops its greatest opposing voltage (back-emf) when the current is changing most rapidly. And, the current changes most rapidly when it is instantaneously zero. Thus, the individual oppositions to the flow of current presented by the resistance and inductance reach a maximum at different times, at points on the sine wave differing in phase by 90°.

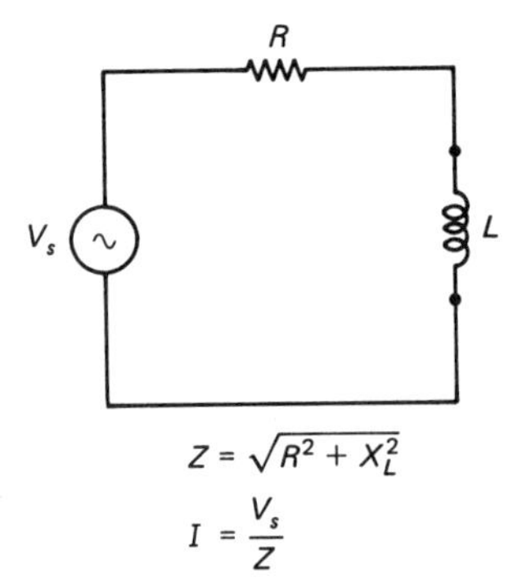

Figure 28-9 A series circuit consisting of an inductor and a resistor.

To get the total opposition (the *impedance*) to the flow of current, the resistance and reactance must be added vectorially. We do this by constructing an impedance diagram, as shown in Fig. 28-10. The impedance Z is the hypotenuse of the triangle and is consequently given by

$$Z = \sqrt{R^2 + X_L^2} \tag{28-8}$$

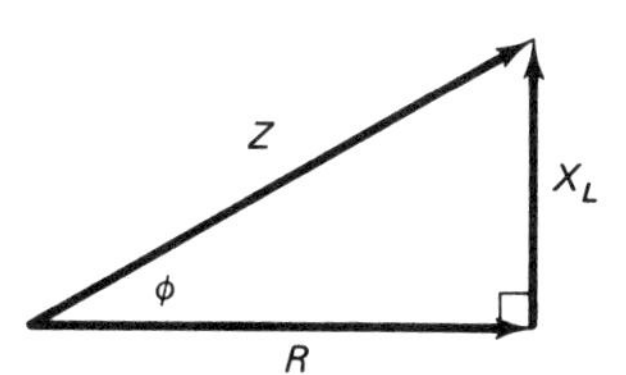

Figure 28-10 Impedance diagram (triangle) for a series combination of R and L.

The current that flows through L, R, and the source is given by

$$I = \frac{V_s}{Z} \tag{28-9}$$

which is the AC counterpart of Ohm's law.

The current causes voltages to be developed across both the inductor and the resistor, but these voltages are not in phase. The voltage across the inductor ($V_L = IX_L$) reaches a peak 90° of the cycle before the voltage across the resistor ($V_R = IR$). This is represented by the phasor diagram of Fig. 28-11(a), which gives rise to the voltage triangle of Fig. 28-11(b). Note that the voltages appearing across the resistor and inductor (as measured by the voltmeter) do not add to give the source voltage. The source voltage, V_s, forms the hypotenuse of the triangle, and so,

$$V_s = \sqrt{V_R^2 + V_L^2} \tag{28-10}$$

as dictated by the Pythagorean theorem.

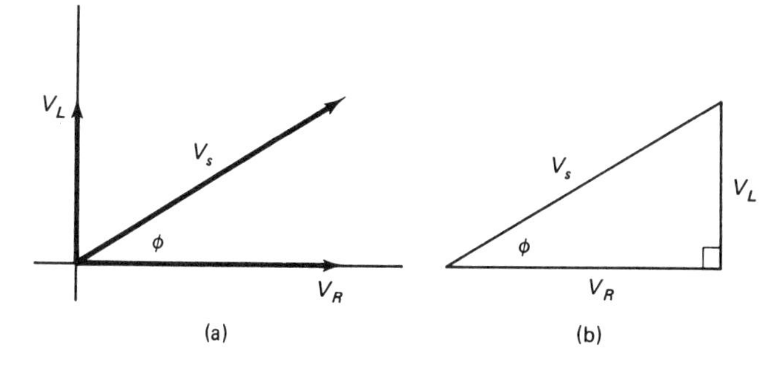

Figure 28-11 (a) Phasor diagram for series L-R circuit.
(b) Voltage triangle for series L-R circuit.

Phase angle

The current flowing in a series L-R circuit has the same phase as the voltage across the resistor. From the voltage triangle, we see that the phase of the current differs from the phase of the applied voltage (the source voltage) by the angle ϕ, called the *phase angle*. In an L-R series circuit, the current *lags* the applied voltage an amount equal to the phase angle.

The phase angle can be calculated from either the impedance triangle or the voltage triangle. We recognize the following relationships:

$$\tan \phi = \frac{X_L}{R} \qquad \phi = \tan^{-1}\frac{X_L}{R} \tag{28-11}$$

$$\tan \phi = \frac{V_L}{V_R} \qquad \phi = \tan^{-1}\frac{V_L}{V_R} \tag{28-12}$$

An example illustrating the solution of a series L-R circuit is given in Fig. 28-12.

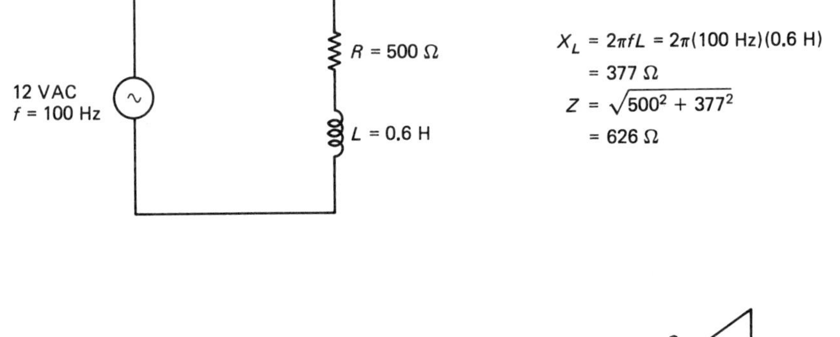

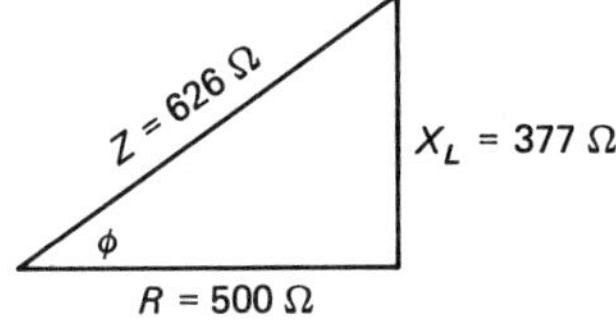

$$I = \frac{V_s}{Z} = \frac{12\text{ V}}{626} = 0.0192\text{ A}$$

$$V_R = IR = (0.0192\text{ A})(500\ \Omega) = 9.58\text{ V}$$

$$V_L = IX_L = (0.0192\text{ A})(377\ \Omega) = 7.24\text{ V}$$

$$\phi = \tan^{-1}\left(\frac{X_L}{R}\right) = \tan^{-1}\left(\frac{377}{500}\right)$$

$$= \tan^{-1}(0.754) = 37.0°$$

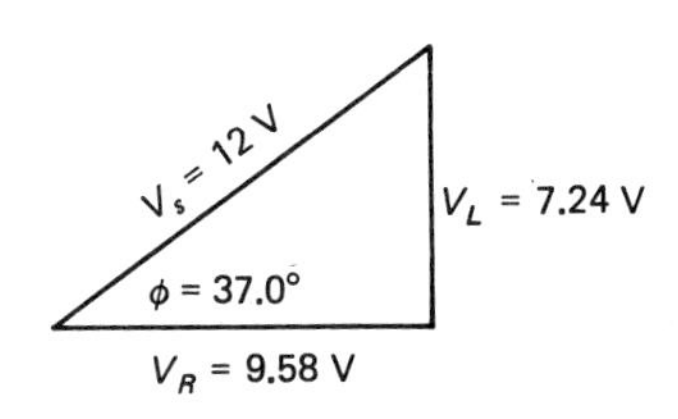

Figure 28-12 An example illustrating the solution of a series L-R circuit.

Power dissipation; reactive power

The power dissipation in a resistor is I^2R, and we might expect the power dissipation in an inductor to be I^2X_L. This is almost true, but we must examine the inductive power in more detail.

First of all, let us recall that the instantaneous power dissipation in any device equals the instantaneous current flowing through it multiplied by the instantaneous voltage developed across it. Secondly, the voltage developed across an inductor stems from its back-emf and is out of phase by 90° with the current flowing through it. This means that if V_L is proportional to sin ωt, then I_L

will be proportional to cos ωt because of the 90° phase difference between the sine and cosine functions. Furthermore, for angles within certain ranges, the sine and cosine have different algebraic signs (from 90° to 180° and 270° to 360°). It follows that the product of I_L and V_L (the instantaneous power dissipation in an inductor) will be *negative* in these intervals. A negative power dissipation implies that energy is delivered from the inductor back to the circuit of which it is a part.

A *pure* inductance (one without a resistance associated with the windings) does not dissipate power in the same sense as a resistor. An inductor absorbs energy as its magnetic field builds up around it, but this energy is given back to the circuit when the magnetic field collapses. Therefore, an inductor does not dissipate *true power*. We say that an inductor dissipates *reactive power*. Reactive power Q is measured in vars (volt-amps reactive) and is given by

$$Q = I_L V_L = I_L^2 X_L = \frac{V_L^2}{X_L} \tag{28-13}$$

Reactive power is a measure of the energy that an inductor cyclically absorbs and gives back to the circuit.

Apparent power

The apparent power S delivered by an AC source to a circuit is the product of the source voltage V_s and the source current I_s:

$$S = V_s I_s \tag{28-14}$$

But we have seen that the power dissipation in an AC circuit may not be all true power—a portion may be reactive power. The relationship between apparent power, S, true power, P, and reactive power, Q, is reflected in the power triangle, shown in Fig. 28-13. Note that apparent power is measured in volt-amps, reactive power is measured in vars, and true power is measured in watts. The Pythagorean theorem gives

$$S = \sqrt{P^2 + Q^2} \qquad P = \sqrt{S^2 - Q^2} \qquad Q = \sqrt{S^2 - P^2} \tag{28-15}$$

Power factor

Observe that the same angle, ϕ, the phase angle, appears in the impedance triangle, the voltage triangle, and the power triangle. Furthermore, from the power triangle, note that the true power P is related to the apparent power S by

$$P = S \cos \phi \tag{28-16}$$

The term $\cos \phi$ is called the *power factor* of the circuit. The following relationships are evident for a series *L-R* circuit:

$$\cos \phi = \frac{R}{Z} \qquad \cos \phi = \frac{V_R}{V_s} \qquad \cos \phi = \frac{P}{S} \tag{28-17}$$

The power factor provides an indication of the phase relationship between

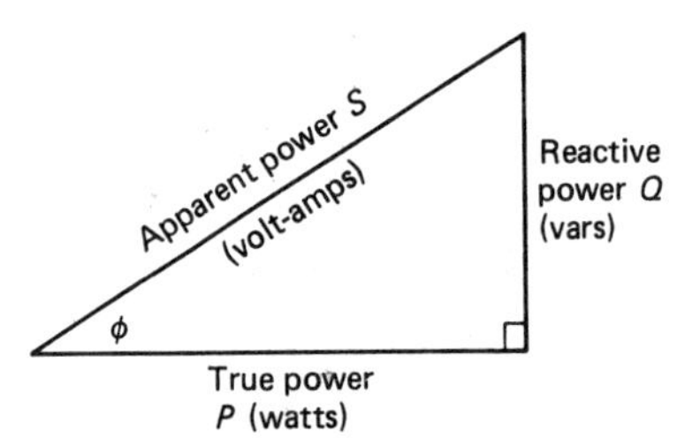

Figure 28-13 Power triangle for series *L-R* circuit.

the current and voltage in an AC circuit. It is desirable in power distribution systems to keep the power factor as close to 1.0 as possible so that the line current will be as small as possible for a given true power dissipation.

An example of the power relationships in an *L-R* circuit is given in Fig. 28-14.

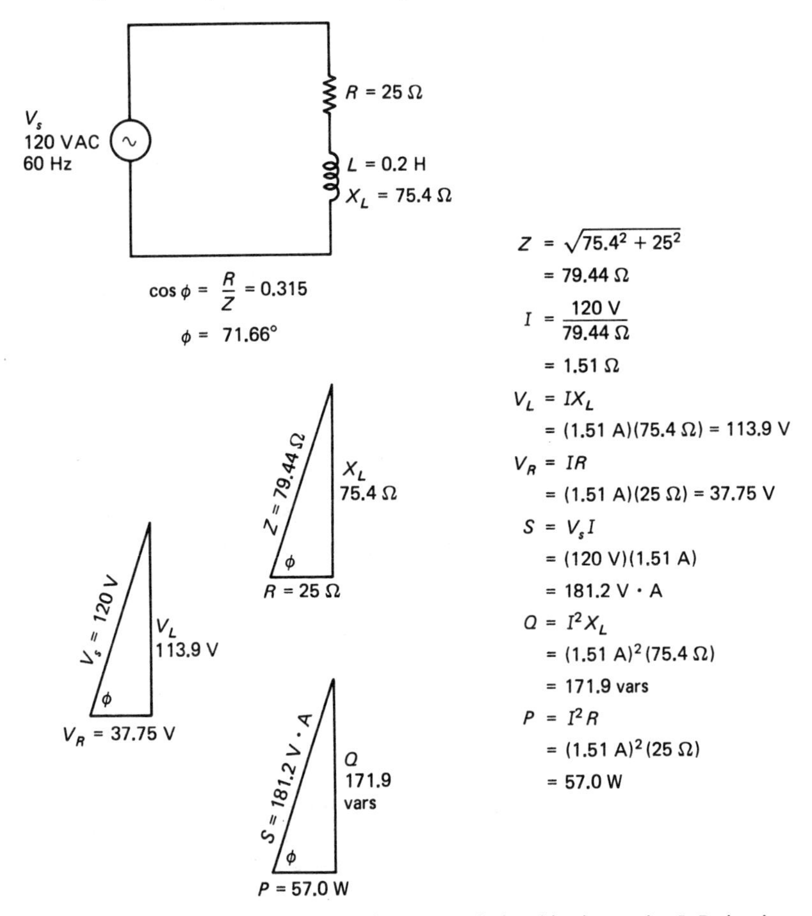

Figure 28-14 An example showing the power relationships in a series *L-R* circuit.

28-4 CAPACITIVE CIRCUITS

We first encountered capacitance in Sect. 24-5 in our discussion of electrostatics, but we have yet to describe the action of capacitors in either DC or AC circuits. We recall that a voltage is developed across a capacitor due to electrons being transferred from one plate of the capacitor to the other. Further, in order for the voltage across the capacitor to change, electrons must flow through the external circuit from one plate to the other. Because this process takes a finite amount of time, we say that the voltage across the capacitor cannot change quickly.

The charging of a capacitor is demonstrated in the circuit of Fig. 28-15. The capacitor is initially uncharged so the voltage across it is zero. When the switch is closed, the battery draws electrons from one side of the capacitor and forces them onto the other side, and a voltage is thereby developed across the capacitor.

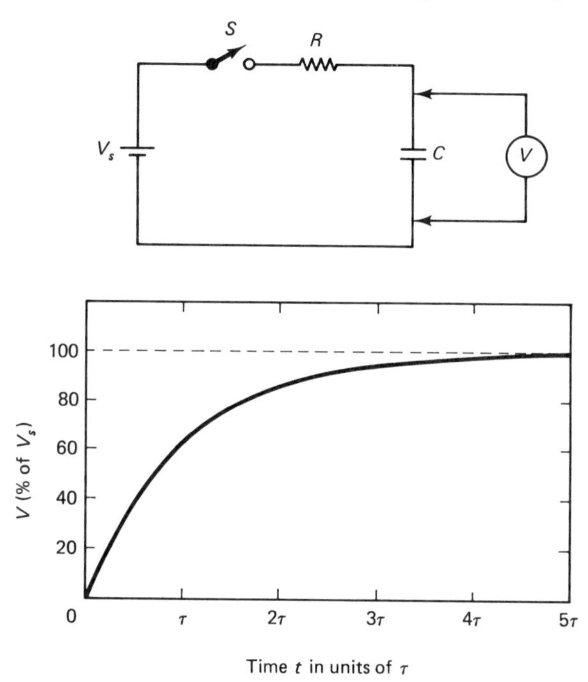

Figure 28-15 Charging of a capacitor, according to Eq. (28-19).

The series resistor R limits the rate at which the transfer of electrons can occur. This, together with the size of the capacitor, determines the time constant of the R-C circuit. The time constant is given by

$$\tau = RC \tag{28-18}$$

and the voltage across the capacitor t seconds after the switch is closed is given by

$$V = V_s(1 - e^{-t/\tau}) \tag{28-19}$$

This equation is of the same form as the equation for the rising current in an inductor [Eq. (27-11)]. Hence, in a time interval of one time constant, the voltage across the capacitor rises to about 63% of its final value.

Capacitive reactance

A direct current cannot flow through a capacitor because electrons cannot cross the dielectric. However, for a short period of time, a current can appear to flow due to the charging and/or discharging of the capacitor. But the charging of a capacitor produces a voltage across the capacitor that opposes the voltage applied to the circuit by the source.

Suppose a capacitor is connected to an AC source so the voltage applied to the capacitor continually changes polarity in a sinusoidal manner. The capacitor will be repeatedly charged and discharged and a current will appear to flow through the capacitor. The capacitor, however, will exhibit an opposition to the flow of current through it; this opposition is called *capacitive reactance*, X_C.

It is inversely proportional to the frequency f and to the size of the capacitance C:

$$X_C = \frac{1}{2\pi f C} \tag{28-20}$$

If an AC source of frequency f and voltage V_s is connected to a capacitor, a current, given by

$$I_C = \frac{V_s}{X_C} \tag{28-21}$$

will flow through the capacitor.

R and *C* in series

As for inductive reactance, capacitive reactance must be combined with R via an impedance triangle to obtain the total opposition to the flow of current in a series R-C circuit. However, the capacitive reactance is drawn downward, opposite to the direction in which inductive reactance is drawn. This stems from the generally opposite characteristics of capacitors and inductors. An impedance triangle for an R-C circuit is shown in Fig. 28-16. The Pythagorean theorem is used to obtain Z:

$$Z = \sqrt{R^2 + X_C^2} \tag{28-22}$$

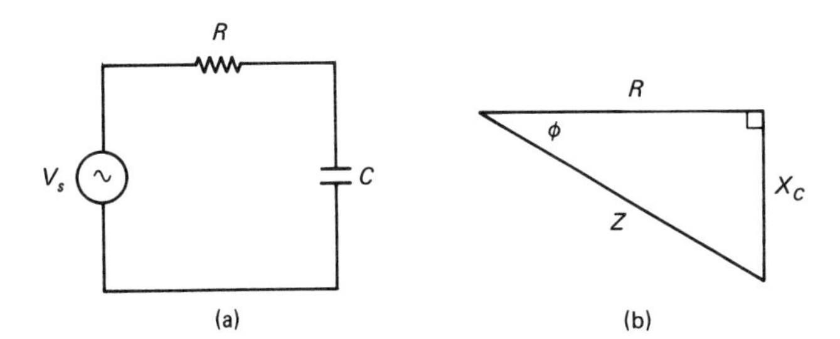

Figure 28-16 (a) A series R-C circuit. (b) Impedance triangle.

and the voltage across the resistor (V_R) and capacitor (V_C) are given by

$$V_R = IR \qquad V_C = IX_C \tag{28-23}$$

where I for the circuit is given by V_s/Z, as for the inductive circuit. The voltage triangle is shown in Fig. 28-17(a).

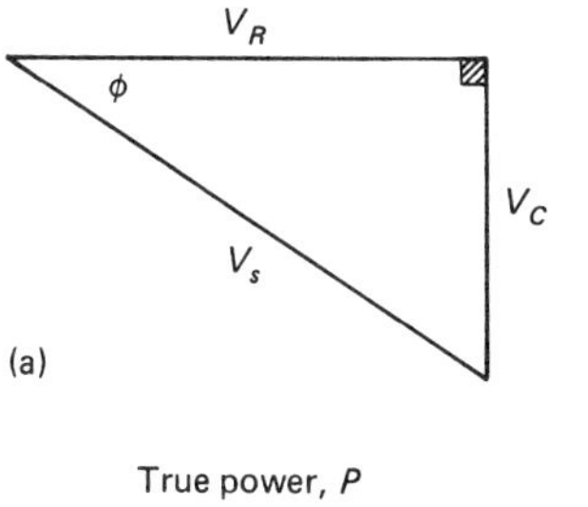

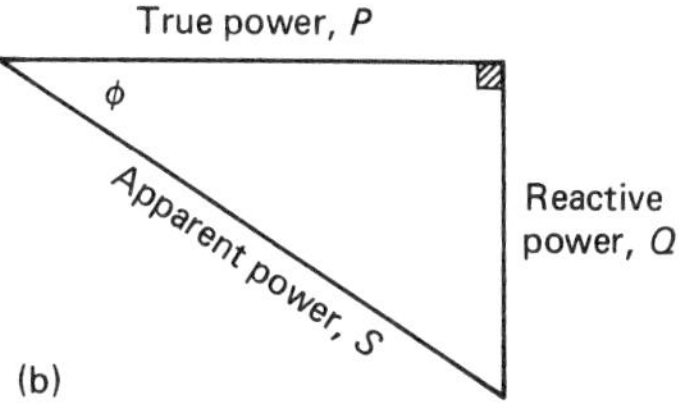

Figure 28-17 (a) Voltage triangle and (b) power triangle for a series *R-C* circuit.

The power dissipation in a capacitor is reactive power, measured in vars. It is given by

$$Q = V_C I_C = I_C^2 X_C = \frac{V_C^2}{X_C} \tag{28-24}$$

Also, a power triangle may be constructed as in Fig. 28-17(b).

Note that in a capacitive circuit, the current leads the voltage by the angle ϕ, whereas in an inductive circuit, the current lags the applied voltage. An analysis of a capacitive circuit is given as an example in Fig. 28-18.

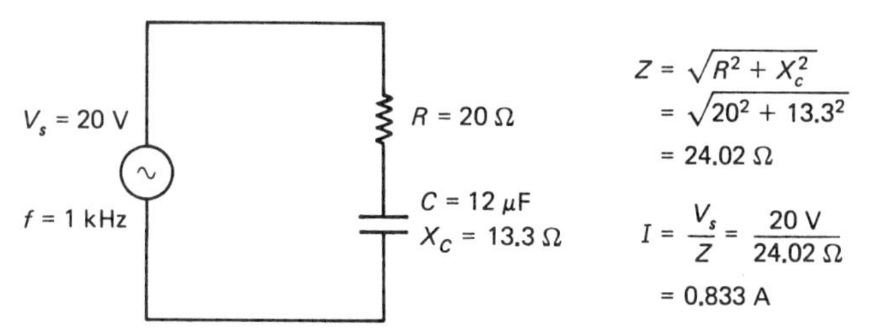

$$V_R = IR = (0.833\text{ A})(20\ \Omega) = 16.65\text{ V}$$

$$V_C = IX_C = (0.833\text{ A})(13.3\ \Omega) = 11.07\text{ V}$$

$$Q = I^2X_C = (0.833\text{ A})^2(13.3\ \Omega) = 9.23\text{ vars}$$

$$P = I^2R = (0.833\text{ A})^2(20\ \Omega) = 13.88\text{ W}$$

$$S = IV_s = (0.833\text{ A})(20\text{ V}) = 16.66\text{ V}\cdot\text{A}$$

$$\phi = \tan^{-1}\left(\frac{X_C}{R}\right) = \tan^{-1}\left(\frac{13.3}{20}\right) = 33.6^\circ$$

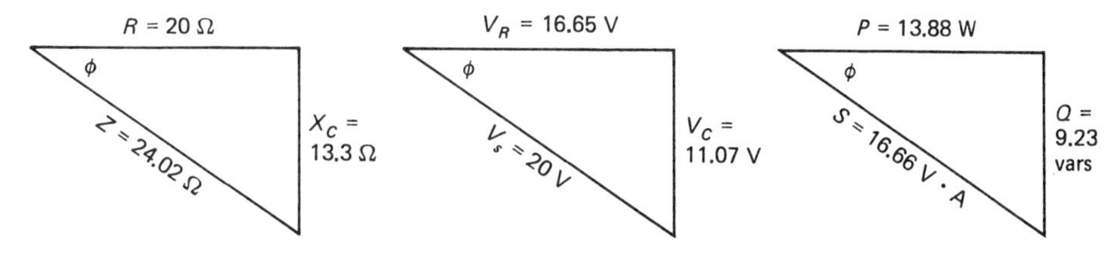

Figure 28-18 Analysis of a capacitive circuit.

28-5 *L, R,* AND *C* IN SERIES

We have now seen that the effect of an inductance is to cause the current to lag the applied voltage, whereas a capacitor causes the current to lead the applied voltage. These two effects oppose each other when both an inductor and a capacitor are included in the same circuit.

A series *L-R-C* circuit is shown in Fig. 28-19(a) and the impedance diagram is shown in Fig. 28-19 (b). Note that the capacitive reactance is drawn downward from the tip of the inductive reactance vector. The impedance Z is given by

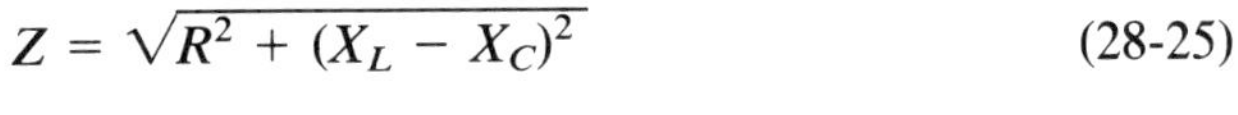

$$Z = \sqrt{R^2 + (X_L - X_C)^2} \qquad (28\text{-}25)$$

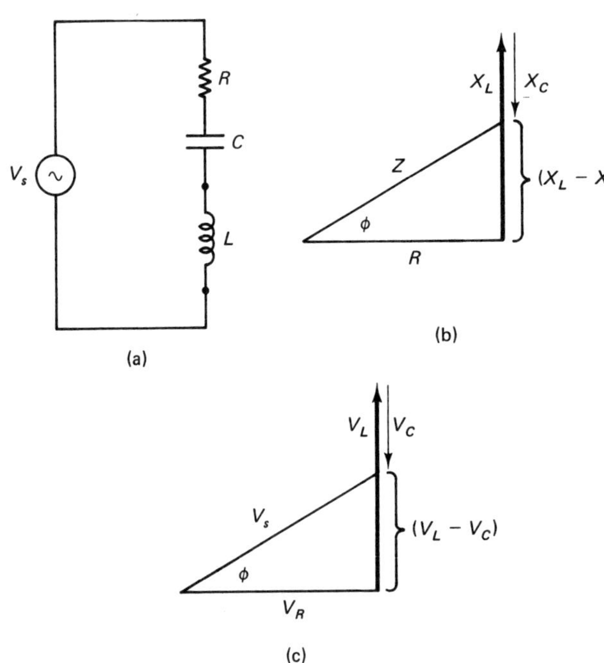

Figure 28-19 (a) A series *L-R-C* circuit. (b) Impedance triangle. (c) Voltage triangle.

and the current in the circuit is found using

$$I = \frac{V_s}{Z} \qquad (28\text{-}26)$$

Once the current is known, the voltage across each component can be computed:

$$V_R = IR \qquad V_C = IX_C \qquad V_L = IX_L \qquad (28\text{-}27)$$

The voltage triangle is shown in Fig. 28-19(c). Note that the voltage V_C is drawn opposite in direction to V_L. The voltages developed across L and C are 180° out of phase and tend to cancel each other.

The phase angle ϕ is calculated using

$$\phi = \tan^{-1}\left(\frac{X_L - X_C}{R}\right) \qquad (28\text{-}28)$$

A power capacitor used on a distribution line to cancel the inductance of the wires.

Observe that if X_C is greater than X_L, ϕ will be negative, which implies that the current will lead the applied voltage. As before, the power factor is the cosine of the phase angle, $\cos \phi$.

The power triangle is constructed using the net reactive power:

$$Q_{\text{net}} = Q_L - Q_C \tag{28-29}$$

The difference of the reactive powers, rather than the sum, is used in the power triangle (to determine the apparent power) because a periodic, cyclic transfer of energy occurs between the capacitor and inductor. The capacitor absorbs energy as the inductor gives it up, and the inductor absorbs energy as the capacitor gives it up. We shall see in the following that when X_C and X_L are equal, the net reactive power becomes zero and the inductive and capacitive effects cancel out even though large quantities of energy are cyclically exchanged between the inductor and capacitor.

EXAMPLE 28-2

A series *L-R-C* circuit consists of an inductance of 0.75 H, a resistance of 50 Ω, and a capacitance of 12 μF. The series combination is connected to a 120-V, 60-Hz source.

(a) Compute the current that flows in the circuit.
(b) Compute the voltage appearing across each component.
(c) Compute the power dissipation in each component.
(d) Calculate the power factor.
(e) Sketch the impedance, voltage, and power triangles.

The circuit is shown in Fig. 28-20(a).

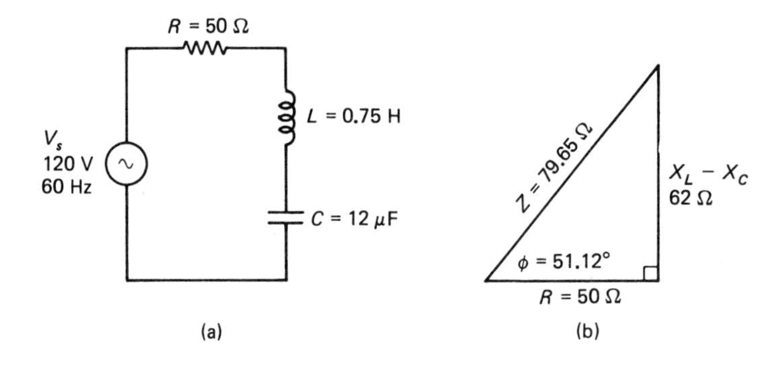

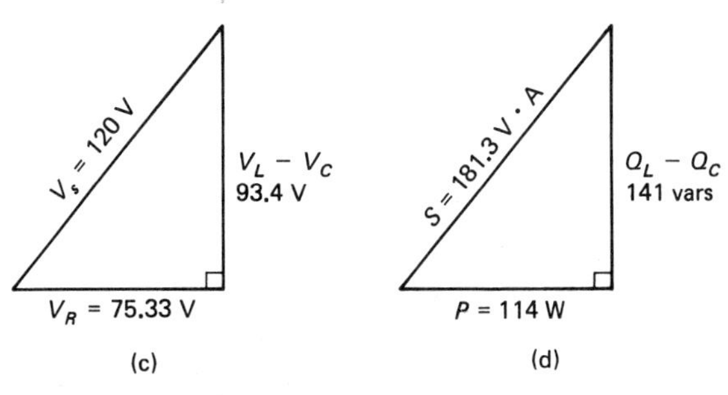

Figure 28-20 (a) Circuit.
(b) Impedance triangle.
(c) Voltage triangle.
(d) Power triangle.

Solution. First we calculate the inductive and capacitive reactances. The result: $X_L = 283\ \Omega$; $X_C = 221\ \Omega$. We then calculate the impedance:

$$Z = \sqrt{R^2 + (X_L - X_C)^2}$$

$$= \sqrt{50^2 + (283 - 221)^2}$$

$$= 79.65\ \Omega$$

(a) Once the impedance is known, the current can be calculated using the AC-form of Ohm's law:

$$I = \frac{V_s}{Z} = \frac{120\text{ V}}{79.65\ \Omega} = 1.51\text{ A}$$

(b) The voltage across each component may then be found:

$$V_R = IR = (1.51\text{ A})(50\ \Omega) = 75.33\text{ V}$$
$$V_C = IX_C = (1.51\text{ A})(221\ \Omega) = 333.0\text{ V}$$
$$V_L = IX_L = (1.51\text{ A})(283\ \Omega) = 426.4\text{ V}$$

(c) The power dissipation in each component is:

$$P = I^2R = (1.51\text{ A})^2(50\ \Omega) = 114\text{ W}$$
$$Q_L = I^2X_L = (1.51\text{ A})^2(283\ \Omega) = 645\text{ vars}$$
$$Q_C = I^2X_C = (1.51\text{ A})^2(221\ \Omega) = 504\text{ vars}$$

(d) Before we can compute the power factor, we must first determine the phase angle:

$$\phi = \tan^{-1}\left(\frac{X_L - X_C}{R}\right)$$

$$= \tan^{-1}\left(\frac{283 - 221}{50}\right)$$

$$= 51.12°$$

The power factor, $\cos \phi$, is then

$$\cos \phi = \cos 51.12° = 0.63$$

(e) The triangles are shown in Fig. 28-20(b), (c), and (d).

Series resonance

When the frequency of the applied voltage and sizes of L and C are properly selected, X_L and X_C can be made equal. In this situation, called *resonance*, the inductive and capacitive reactances cancel so that only the resistance R opposes the flow of current through the circuit. To find the relationship between the resonant frequency f_r, L, and C, we set X_L and X_C equal:

$$X_L = X_C$$

$$2\pi f_r L = \frac{1}{2\pi f_r C}$$

Solving for f_r gives the resonant frequency of a series circuit:

$$f_r = \frac{1}{2\pi\sqrt{LC}} \qquad (28\text{-}30)$$

Inspection of Eq. (28-25) reveals that when X_L and X_C are equal, $Z = R$, and the current flowing through the circuit is

$$I_r = \frac{V_s}{R} \qquad (28\text{-}31)$$

where the subscript r denotes the current at resonance. Furthermore, the phase angle will be zero, implying that the current and applied voltage are in phase. It follows that the power factor will be unity.

EXAMPLE 28-3 A 0.1-μF capacitor, a 150-mH inductor, and a 100-Ω resistor are connected in series.
(a) Calculate the resonant frequency of the circuit.
(b) If a 12-V source, operating at the resonant frequency, is applied to the circuit, what current will flow?
(c) What voltage will appear across each component?

Solution. (a) Use Eq. (28-30):

$$f_r = \frac{1}{2\pi\sqrt{LC}}$$

$$= \frac{1}{2\pi\sqrt{(150 \times 10^{-3})(0.1 \times 10^{-6})}}$$

$$= 1299 \text{ Hz}$$

(b) According to Eq. (28-31):

$$I_r = \frac{V_s}{R} = \frac{12 \text{ V}}{100 \ \Omega} = 0.12 \text{ A}$$

(c) The voltage across each component is

$$V_R = I_r R = (0.12 \text{ A})(100 \ \Omega) = 12 \text{ V}$$
$$V_C = I_r X_C = (0.12 \text{ A})(1225 \ \Omega) = 147 \text{ V}$$
$$V_L = I_r X_L = (0.12 \text{ A})(1225 \ \Omega) = 147 \text{ V}$$

(The computation of X_L and X_C for $f_r = 1299$ Hz was not shown.)

Resonant rise in voltage

Observe that with only 12 V applied to the circuit of Example 28-3, 147 V appears across the inductor and across the capacitor. And, at the same time, a voltage equal to the applied voltage appears across the resistor. This is a feature of the resonance of a series circuit.

If we treat the frequency as an independent variable and sweep the frequency through the resonant frequency, we observe a pronounced rise in voltage across the inductor and across the capacitor at the resonant frequency. We refer to this as the *resonant rise in voltage*, which occurs when the applied frequency is near the resonant frequency. A plot of the voltage is shown in Fig. 28-21.

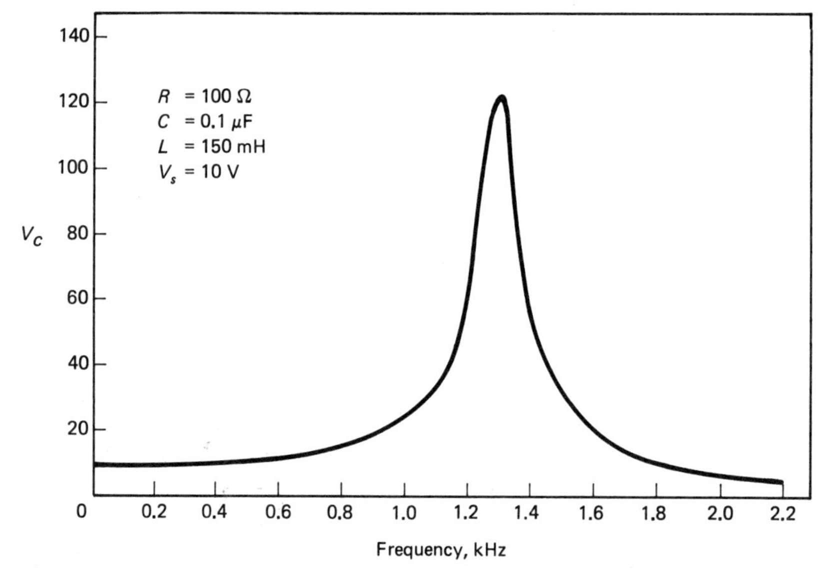

Figure 28-21 A resonant rise in voltage occurs across the capacitor and across the inductor of a series *L-R-C* circuit when the applied signal is near the resonant frequency.

Electronics texts define the *quality factor*, Q_f (we use the subscript f to avoid confusion with reactive power), of a resonant circuit in order to express the sharpness of the resonance. For a series *L-R-C* circuit,

$$Q_f = \frac{1}{R}\sqrt{\frac{L}{C}} \tag{28-32}$$

Also, at the resonant frequency, Q_f is given by

$$Q_f = \frac{X_L}{R} = \frac{2\pi f_r L}{R} \tag{28-33}$$

One use of Q_f is in computing the voltage V_C or V_L at resonance. In terms of V_s the applied voltage, the voltage across L or C at resonance is

$$V_L = V_C = Q_f V_s \tag{28-34}$$

Clearly, the resonant rise of voltage is associated with the principle of *tuning*, which is extremely important in electronics.

The quality factor has many other properties, such as determining the selectivity of a tuned circuit. A more complete description of resonance and resonant circuits may be found in texts devoted to electrical and electronics circuits.

28-6 PARALLEL RESONANT CIRCUIT

A parallel connection of R, L, and C is shown in Fig. 28-22. The parallel combination is shown in series with a large resistance R_s that serves to isolate, somewhat, the parallel network from the signal source. The parallel network is capable of resonance and tuning, but the characteristics are, in many respects, opposite to those of the series resonant circuit described earlier. In this discussion we assume the inductor to be an "ideal" inductor that has no effective series resistance.

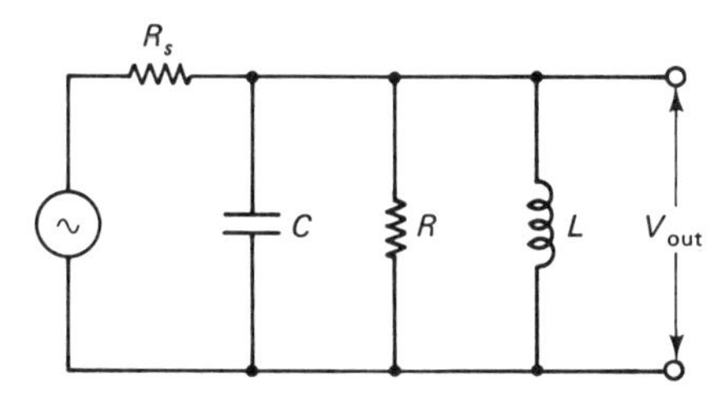

Figure 28-22 A parallel resonant circuit consisting of L, R, and C. Resistor R_s isolates the resonant circuit from the source so that the variation in impedance of the resonant circuit may be observed.

When the capacitive and inductive reactances are equal, the capacitor and inductor engage in a cyclic exchange of energy due to the charging and discharging of the capacitor in concert with the building up and collapsing magnetic field of the inductor. The frequency of the resulting voltage developed across the resonant circuit is given by Eq. (28-30).

For series resonant circuits, the current is a maximum at resonance. In a parallel resonant circuit, however, the current is a minimum at resonance. The

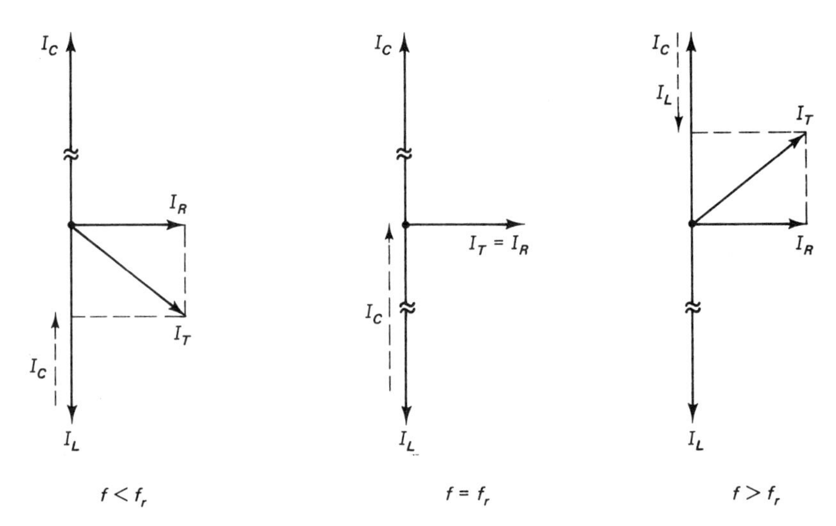

Figure 28-23 The total current I_T flowing in a parallel resonant circuit is the vector sum of I_L, I_C and I_R. It is a minimum when $f = f_r$.

reason for this is illustrated in Fig. 28-23, where the currents through R, L, and C are shown via phasors for applied frequencies below, equal to, and above the resonant frequency. Note that I_R is independent of the frequency, but for low frequencies, $I_L > I_C$, and for high frequencies, $I_C > I_L$. The total current is the vector sum of the three components.

The fact that the total current reaches a *minimum* at resonance indicates that the impedance of the parallel network reaches a *maximum* at resonance (opposite to the series circuit). Accordingly, an output voltage taken from across the network will attain a maximum at resonance, because the network impedance becomes large relative to the series resistance R_s. A plot of output voltage vs. frequency is shown in Fig. 28-24 for the circuit of Fig. 28-22.

Even though the total current (or *line* current) reaches a minimum at resonance, the component currents I_L and I_C attain a pronounced *maximum* at resonance. I_L and I_C can easily be from 10 to 50 times as large as the total current.

A meaningful analysis of parallel resonant circuits is far more involved and extensive than we treat here. Many texts on electronic circuit theory provide excellent discussions for those wishing to pursue the subject in more detail.

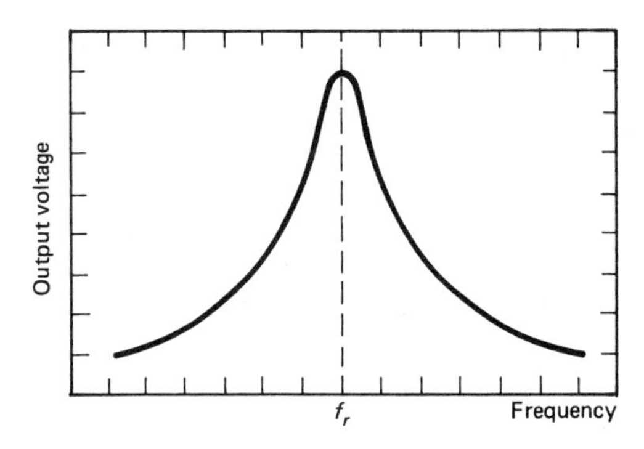

Figure 28-24 Output voltage vs. frequency for a parallel resonant circuit.

28-7 THREE-PHASE AC SYSTEMS

In its most basic form, a three-phase AC generator has three loops (or coils) oriented at 120° to each other, as shown in Fig. 28-25(a). The induced emf's are therefore 120° out of phase with each other, as shown in Fig. 28-25(b) and (c).

A *four-wire, wye-connected* system is shown in Fig. 28-26. When the load is balanced ($R_a = R_b = R_c$), no current will flow in the neutral line because the vector sum of the phase currents is zero. Current flows in the neutral only when the loads are unbalanced.

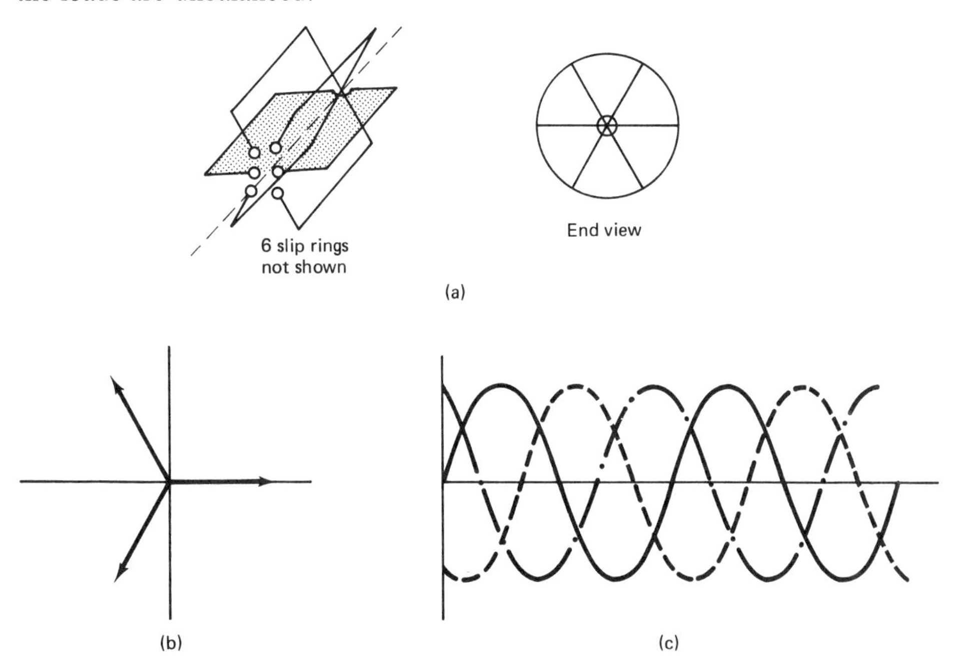

Figure 28-25 (a) The coils in a three-phase generator are oriented 120° relative to each other, and each coil is connected to a pair of slip rings.
(b) Phasor diagram for a three-phase generator.
(c) Sine waves 120° out of phase.

Each phase of the four-wire, wye-connected system may be treated as a single-phase line when used in conjunction with the neutral. Hence, both three-phase and single-phase motors (for example) may be operated within an establishment supplied only by a three-phase system.

A three-phase system may involve transformers for stepping the line voltages up or down. Either three single-phase transformers or a special three-phase transformer may be used in a given application according to the dictates of good engineering practice.

Polyphase systems exhibit two main advantages over single-phase systems: (1) The power output or dissipation by a polyphase generator or load occurs at a constant rate. Recall that the power dissipation in a single-phase system occurs in pulses according to the sine-square waveform. (2) Smaller line currents are required to effect a given power transfer at a specified voltage. This translates to a

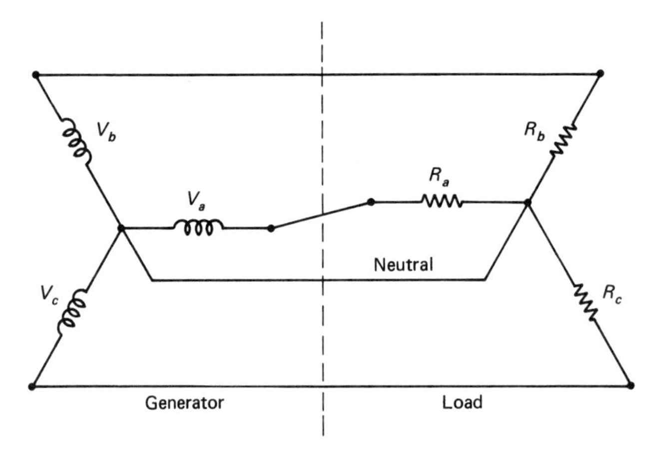

Figure 28-26 A four-wire, wye-connected, three-phase system.

saving in conductor material (copper or aluminum). Less conductor material per unit power delivered is required in polyphase systems.

For purposes of reference, a *three-wire, delta-connected* system is shown in Fig. 28-27. Each system (wye and delta) have respective advantages at various points in particular power distribution applications, and it is possible, with proper design, to connect wye to delta and vice versa. This is in the realm of the power distribution specialist, however.

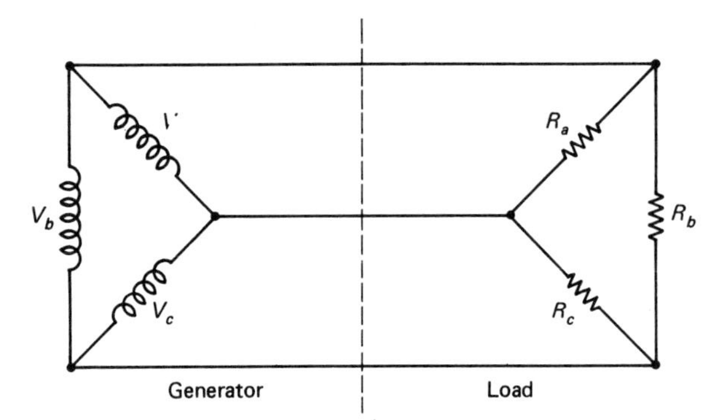

Figure 28-27 A three-wire, delta-connected, three-phase system.

Line losses

Suppose power is delivered from a power source to a load located a considerable distance from the source, as shown in Fig. 28-28. The conductor connecting the load to the source is assumed to have a total resistance R. The source voltage is

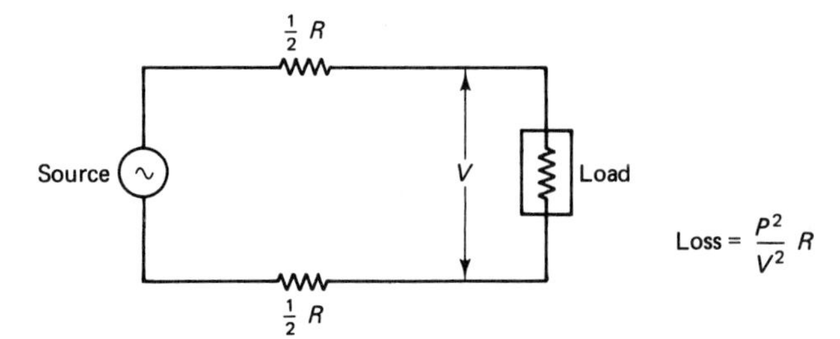

$$\text{Loss} = \frac{P^2}{V^2}R$$

Figure 28-28 The conductors carrying power to a load dissipate power.

such that voltage V is applied to the load, and current I flows through the load. We wish to calculate the power dissipated in the conductor (which we call *loss*) connecting the source to the load.

The power P delivered to the load is given by IV. The loss in the conductor is I^2R. Hence,

$$P = I \cdot V \qquad \text{Loss} = I^2R$$

Therefore, since $I = P/V$, it follows that

$$\text{Loss} = \frac{P^2}{V^2}R \tag{28-35}$$

Here we see that for a given power level P and conductor resistance R, the line loss is inversely proportional to the square of the voltage. It is for this reason that electrical power is transferred long distances over high-tension lines that operate at hundreds of thousands of volts. Transformers at each end of the line step the voltage up and down appropriately.

Questions

1. What type of wave form results when two sine waves of the same frequency and amplitude are added, even if the two sine waves are out of phase?
2. When a sinusoidal voltage is applied to a resistor, is the power dissipation also sinusoidal?
3. Are ordinary AC voltmeters calibrated in peak volts, RMS volts, or peak-to-peak volts?
4. How do capacitive reactance and inductive reactance vary with frequency?
5. In an inductive circuit, does the current lag or lead the applied voltage?
6. What is the justification for drawing X_C and X_L in opposite directions on an impedance diagram?
7. What is the significance of a negative instantaneous power dissipation?
8. What is apparent power, and how is it calculated?
9. For a series circuit at resonance, tell whether the following reach a maximum or a minimum: impedance; current; voltage across capacitor; voltage across inductor; current through inductor; reactive power; phase angle.
10. For a parallel circuit at resonance, tell whether the following reach a maximum or a minimum: impedance; line current; voltage across the capacitor; current through the inductor; power factor.

Problems

1. Two sine waves of amplitude 10 V are 50° out of phase. What will be the amplitude of the resulting sine wave obtained by adding the two?
2. A sine-wave voltage of amplitude 10 V is applied to a 5-Ω resistor. Make a sketch of the current wave form that flows through the resistor. What is the maximum current that will flow?
3. In Fig. 28-6(b), a 12-V (amplitude) sine-wave voltage is applied to a 12-Ω resistor. Calculate the power dissipation in the resistor.
4. A sine wave has a peak voltage of 20 V.
   (a) What is the RMS voltage?
   (b) What is the peak-to-peak voltage?
5. What is the RMS value of a sinusoidal voltage whose peak-to-peak value is 30 V?

6. What DC voltage applied to a 10-Ω resistor will produce the same power dissipation as a sine wave whose amplitude is 10 V?
7. An AC voltage of 12 V_{rms} is applied directly to an inductance of 1 H. What current will flow through the inductor if the frequency of the applied voltage is: **(a)** 60 Hz; **(b)** 120 Hz; **(c)** 1000 Hz?
8. A 30-V source of 400 Hz AC is applied to the series combination of a 0.5-H inductor and a 2200-Ω resistor.
   (a) What is the impedance of the circuit?
   (b) What current will flow?
   (c) What voltage will appear across the resistor?
   (d) What voltage will appear across the inductor?
   (e) Compute the phase angle and sketch the voltage and impedance triangles.
9. For Problem 8, calculate: **(a)** the true power dissipation; **(b)** the reactive power: **(c)** the apparent power; **(d)** the power factor. Sketch the power triangle.
10. For the circuit shown in Fig. 28-12, calculate: **(a)** the true power; **(b)** the reactive power; **(c)** the power factor; **(d)** the apparent power. **(e)** Show that true power equals the power factor times apparent power.
11. Suppose, in the circuit of Fig. 28-15, that $C = 100\ \mu F$, $R = 10{,}000\ \Omega$, and $V_s = 10$ V.
   (a) Compute the time constant for the circuit.
   (b) What will be the instantaneous voltage across the capacitor 2 s after the switch is closed?
12. Compute the capacitive reactance of a 1-μF capacitor at frequencies of **(a)** 1 Hz; **(b)** 10 Hz; **(c)** 100 Hz; **(d)** 1000 Hz.
13. A 10-V, 600-Hz source is connected to a series combination of a 200-Ω resistor and a 2-μF capacitor.
   (a) Calculate the impedance of the circuit.
   (b) What current will flow?
   (c) What voltage will appear across the capacitor?
   (d) What voltage will appear across the resistor?
   (e) Compute the phase angle and power factor for the circuit.
14. For Problem 13, calculate: **(a)** the true power; **(b)** reactive power; **(c)** apparent power.
15. What is the resonant frequency of a 0.1-H inductor connected in series with a 0.01-μF capacitor?
16. An AC source produces 50 V output at a frequency of 400 Hz. The source is connected to a series circuit in which $R = 400\ \Omega$, $L = 0.6$ H, and $C = 1\ \mu F$. Compute: **(a)** the impedance of the circuit and **(b)** the current that flows. **(c)** What is the phase angle? Does the current lead or lag the applied voltage? **(d)** What voltage appears across the capacitor?
17. Repeat Problem 16, but assume the source frequency to be 200 Hz instead of 400 Hz.
18. A resistance of $R = 1000\ \Omega$ is connected in series with an inductance of 15 mH and a capacitance of 0.001 μF.
   (a) Compute the resonant frequency of the circuit.
   (b) Compute X_L amd X_C at the resonant frequency and construct the impedance diagram.
   (c) Do the necessary calculations to construct the impedance diagram for frequencies of 30 kHz and **(d)** 50 kHz.
19. **(a)** For the circuit of Problem 18, compute the quality factor Q_f.
   (b) What voltage will appear across the capacitor at resonance if the applied voltage is 10 V?
20. Suppose a 1-HP motor (746 W) is located a considerable distance from the source of electrical power, and suppose the conductors carrying the power to the motor have a total resistance of 1 Ω.
   (a) Compute the power lost in the line if the motor is designed to operate at 120 V.
   (b) Compute the power lost if the motor is connected so that it operates at 240 V.

CHAPTER 29

SEMICONDUCTOR FUNDAMENTALS

Since the development of the first transistor in 1948, advances in solid-state physics have completely revolutionized the field of electronics. Whereas vacuum tubes are palm sized, discrete transistors are usually no larger than the eraser of a pencil, and in integrated circuits they are truly microscopic. Thus, the small size and reduced power requirements of transistors have paved the way for electronic miniaturization. Computers that once would have filled a room now fit conveniently on the corner of a desk, and pocket calculators now perform functions that once were done only on large computers. Because more circuitry can now be put in a smaller volume, test instruments and electronic apparatus in general have become more compact and more sophisticated. In only three decades a revolution has occurred, and it appears that we have just begun.

In this chapter we present a few of the basic concepts of semiconductor physics and show how a solid-state diode and transistor work. We describe a practical transistor amplifier that is easily constructed. We limit our discussions to bipolar-junction transistors, omitting, for lack of space, any reference to field-effect devices. Many books are available for readers who wish to go further. Our objective is just to provide an introduction and to indicate the flavor of solid-state electronics.

29-1 INTRINSIC SEMICONDUCTORS

The two most important semiconductor materials are silicon and germanium. The properties of the two are similar because both have four valence electrons. Considering that the nucleus and the inner electrons play only passive roles in determining the electronic properties of these materials, it is convenient to speak of a silicon *core* or a germanium *core* when referring to the nucleus and inner electrons. Hence, we may visualize a silicon or germanium atom as a core surrounded by four electrons. This picture emphasizes the importance of the four outer electrons.

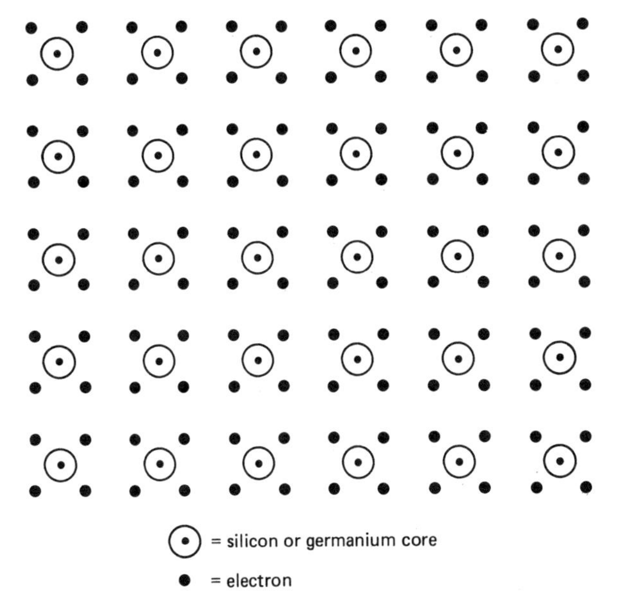

Figure 29-1 A two-dimensional representation of a silicon or germanium crystal. The electronic structure easily accommodates four electrons near each core.

The crystal structure of silicon and germanium easily accommodates four electrons in the vicinity of each atomic core. However, the electrons that occupy these positions belong more to the electronic structure of the crystal than to individual atoms, even though the electrons were carried into the structure by individual atoms. Each core, taken by itself, contains four positive charges (protons) that are not balanced with negative electrons. Hence, four electrons per core are required to establish electrical neutrality, and these electrons are provided by the electronic structure of the crystal. Figure 29-1 is a two-dimensional representation of a silicon or germanium crystal.

In the following discussion we refer only to silicon, but it is to be understood that the principles involved apply also to germanium.

Electron-hole pairs

To understand the electrical properties of an *intrinsic* (pure) semiconductor, we must consider the effects of thermal agitation because, even at room temperature, these effects are quite important. The valence electrons participate in the thermal motion. When an electron accumulates about 1.1 eV of thermal energy (for silicon), it is able to break away from its core and become free. We say the electron is *excited* from the *valence band* (of energy) into the *conduction band*. This produces a free electron (the one that was excited) and also a vacancy in the electronic structure where the electron originally resided. The vacancy is called a *hole*. Obviously, a hole is produced each time a free electron is produced, and we therefore speak of *thermally induced electron-hole pairs*. The word *pair*, however, does not imply that the free electron and hole stay together or even near each other after they are produced.

For temperatures above absolute zero, there will exist an equilibrium number of electron-hole pairs in an intrinsic semiconductor. At higher temperatures, the number increases. The presence of the free electrons and holes enables the semiconductor to conduct an electric current because mobile, charged particles are available to form the current. Furthermore, because more electron-hole pairs are present at higher temperatures, the electrical conductivity increases with temperature; warm semiconductors conduct better than cool semiconductors.

Hole conduction

It is easy to understand that the presence of free electrons in a silicon crystal will increase its conductivity, but it is somewhat surprising to find that the holes also increase the conductivity. That is, a hole can act as a charge carrier.

Recall that a hole is a vacancy, a place in the electronic structure where an electron could be but is not. But a hole in conjunction with a silicon core represents an imbalance of electrical charge. The hole-core combination appears positive because of the electron deficiency represented by the hole. The positive charge actually resides in the nucleus of the silicon atom, but it is convenient (and only slightly misleading) to speak of a hole as being positively charged an amount equal to the charge of one proton.

The mechanism of hole conduction is illustrated in Fig. 29-2. A hole moves from one core to an adjacent core via an electron jump from one core to the other in the direction opposite to the motion of the hole. Motion of the hole to the left, from core 10 to core 1, is accomplished by successive electron jumps from 9 to 10, then from 8 to 9, and so on. The net result is a transfer of one negative charge from core 1 to core 10. Far less energy is required for an electron to move from a given core to a hole on an adjacent core than is required to excite the same electron to the realm of free electrons (into the conduction band).

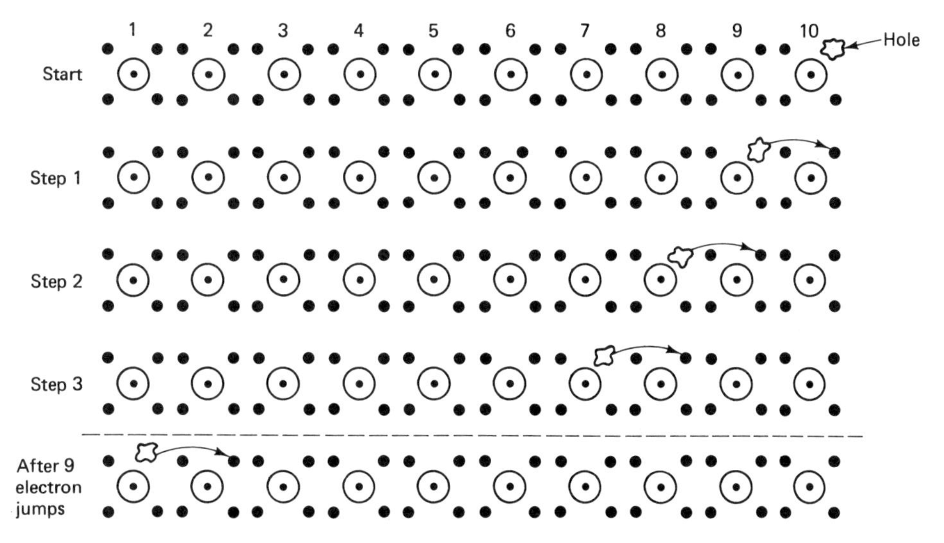

Figure 29-2 An illustration of hole motion along a line of silicon atoms. The hole moves to the left as electrons jump to the right.

29-2 DOPING—ADDING IMPURITIES TO SEMICONDUCTORS

Pure (intrinsic) semiconductors as such are not very useful because they are rather good insulators. The electrical conductivity of semiconductors can be increased dramatically, however, by the addition of minute quantities of selected impurities to the semiconductor crystal. This process is called *doping*, and it serves to introduce either free electrons or free holes into the crystal, depending upon which dopant is used. An impurity concentration on the order of one part per million can result in a millionfold increase in the electrical conductivity of the material.

N-type material

The elements phosphorus, arsenic, antimony, and bismuth consist of atoms that have five electrons in the outer shell. If one of these elements is used to dope a silicon crystal, a small number of the silicon atoms will be replaced by the impurity atoms. Since the impurity atoms have five electrons in the outer shell and since the electronic structure of the crystal easily accommodates only four electrons per core, the extra electron does not fit conveniently into the electronic structure of the crystal. But it must accompany the impurity atom into the crystal in order for the impurity atom to remain electrically neutral. The result is that the fifth electron is not as tightly bound to the atomic cores as are the four electrons that are easily accommodated. The fifth electron becomes a free electron (at ordinary temperatures) that can move under the influence of an electric field and form an electric current.

Semiconductors doped with impurities that contribute extra electrons are called *N-type semiconductors* because the polarity of the mobile charge carriers (the free electrons) is negative. Likewise, the impurities that produce *N*-type semiconductors are known as *N-type impurities*.

P-type material

The elements aluminum, boron, gallium, and indium have only three electrons in the outer shell. When these elements are added as impurities to a semiconductor, each impurity atom contributes only three electrons instead of four to the electronic structure of the crystal. Thus, a hole exists at the site of each impurity atom. At ordinary temperatures, the holes become free holes and serve to increase the conductivity of the crystal.

Semiconductors doped with impurities that produce holes are called *P-type semiconductors*, and the impurities are called *P-type impurities*. The polarity of charge carriers in *P*-type materials (holes) is positive. A symbolic representation of *N*-type and *P*-type semiconductors is shown in Fig. 29-3.

Current conduction in semiconductors

We have now described three types of semiconductors, intrinsic, *N*-type, and *P*-type. In intrinsic (undoped) semiconductors, current flow results from the motion of both electrons and holes that stem from thermally induced electron-hole pairs.

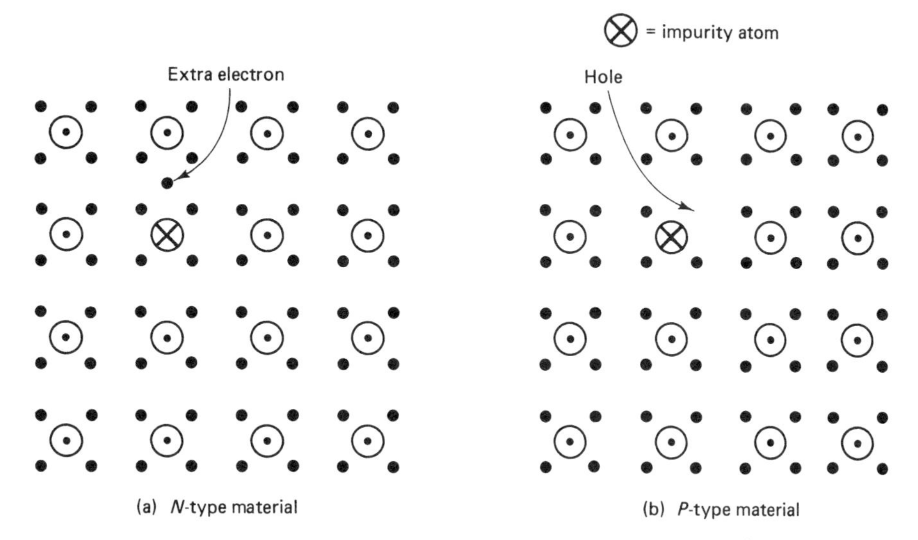

Figure 29-3 A representation of *N*- and *P*-type materials. The *N*-type impurity atom contributes an extra electron in *N*-type material, while the *P*-type impurity atom contributes a hole in the *P*-type material.

Holes move in one direction and electrons move in the other. In comparison with doped semiconductors, the conductivity of intrinsic material is very small.

In *N*-type material, the impurity atoms contribute large numbers of free electrons in addition to the thermally induced electron-hole pairs. Consequently, current conduction is due primarily to the movement of free electrons. The thermally generated holes also contribute to the total current, but their effect is very small in comparison to that of the free electrons. In *N*-type material, the free electrons are the *majority carriers* while the holes are the *minority carriers*.

In *P*-type materials, free holes are the majority carriers. This is to say that current flow occurs primarily via the process of hole conduction. The thermally generated free electrons constitute the minority carriers.

We should always be mindful of the "nothingness" of holes. Hole conduction is actually caused by the movement of electrons in the opposite direction.

29-3 THE *PN* JUNCTION

A *PN junction* is formed where a region of *N*-type material is in intimate contact with a region of *P*-type material with the continuity of the crystal structure maintained at the interface between the two regions. Mechanically speaking, a dramatic change does not occur at the junction. The thing that changes, in crossing the junction, is the type of impurity atoms that are substituted for an occasional atom of the host material. Since the impurity concentrations are small, even this change is hardly noticeable, but the effect upon the electronic properties of the crystal is dramatic.

Suppose a crystal of *N*-type material is suddenly joined to a crystal of *P*-type material, with the contact being such that the geometrical structures of the two crystals match up at the point of contact. The *N*-type material has an abundance

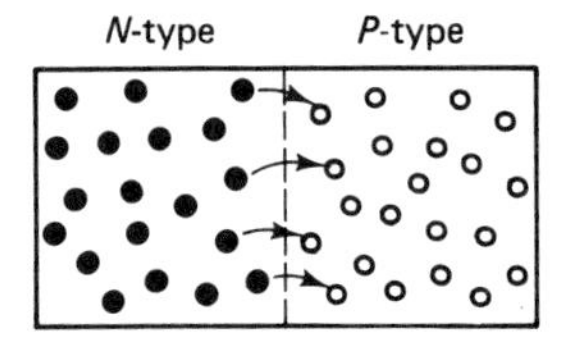

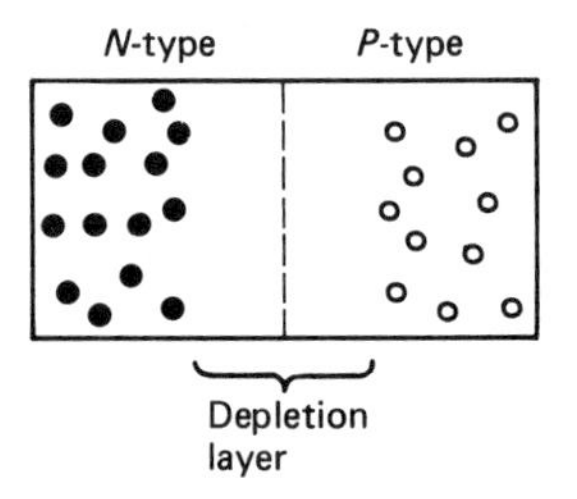

Figure 29-4 Formation of the depletion layer at the interface of a *PN* junction. Electrons cross the interface from the *N*-material to fill holes near the interface in the *P*-material.

of free electrons whereas the *P*-type material has an abundance of holes. When the two crystals are joined, what happens at the interface between the two types of material?

The obvious occurs. Free electrons near the interface on the *N*-side cross the interface and fill holes located near the interface on the *P*-side. The result is that free electrons of the *N*-type material disappear from the vicinity of the interface, and the holes near the interface on the *P*-side are filled by electrons coming over from the *N*-side. Thus, all charge carriers near the interface are *depleted*, and the region surrounding the interface, where there are neither electrons nor holes, is called a *depletion layer*. This is illustrated in Fig. 29-4.

The migration of free electrons across the interface does not continue until all free electrons in the *N*-material have gone over to fill all the holes in the *P*-material. When an electron crosses from the *N*-side to the *P*-side, it leaves an unbalanced positive charge behind in the *N*-material and it contributes an extra negative charge to the *P*-material. A layer of positive charge develops near the interface in the *N*-material, and a layer of negative charge develops near the interface in the *P*-material. These layers establish an electric field across the interface that tends to oppose further migration of electrons. This limits the depletion layer thickness to values that are extremely thin in comparison to the overall dimensions of the *N*- and *P*-regions.

The depletion layer has no free electrons or holes, which makes it a good insulator. Thus, the depletion layer represents a layer of insulation that naturally develops between the *N*- and *P*-regions. In the following, however, we shall see that the thickness of this layer may be altered by applying an external voltage to the *N*- and *P*-regions. We shall see that electrons will flow across the junction in one direction but not the other. That is, a *PN* junction exhibits the properties of a *rectifier*.

Forward-biased *PN* junction

In Fig. 29-5(a), a voltage source is connected to a *PN* junction with the positive voltage applied to the *P*-type material. Holes are repelled from the positive terminal through the *P*-region toward the interface. On the other side, electrons are repelled from the negative terminal through the *N*-region toward the interface. The result is that the depletion layer becomes narrower, and when the applied voltage reaches about 0.6 V, the depletion layer practically disappears and a comparatively large current will flow across the junction and through the external circuit. Under this condition, the junction is said to be *forward-biased*.

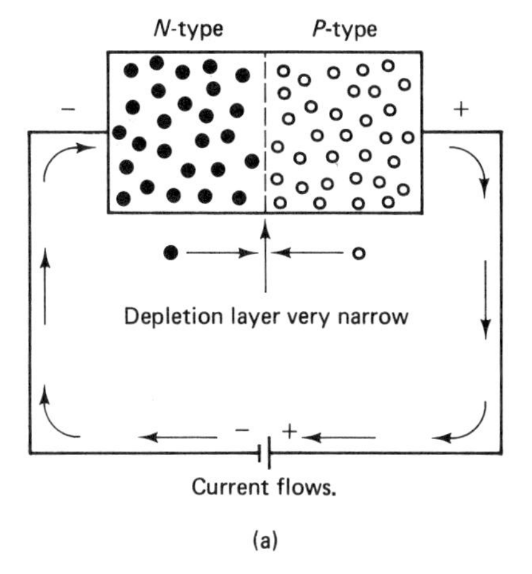

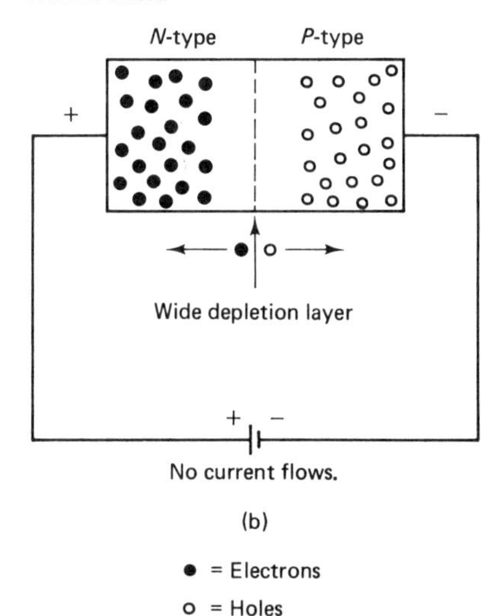

Figure 29-5 (a) Forward-biased and (b) reverse-biased *PN* junction.

Electron flow through the *N*-material of a forward-biased *PN* junction is by the movement of free electrons through the crystal toward the interface. At the interface, the electrons combine with holes arriving at the interface from the other direction. In the *P*-material, conduction is by the movement of holes in the manner depicted in Fig. 29-2.

Reverse-biased *PN* junction

In Fig. 29-5(b), the external voltage is connected with the positive voltage applied to the *N*-type material. Free electrons are attracted to the positive terminal and are drawn through the *N*-material away from the interface. On the other side, holes are attracted to the negative terminal and are also drawn away from the interface. The result is that the depletion layer becomes wider; no current flows across the junction due to the absence of mobile charge carriers in the depletion layer. Under this condition, the junction is said to be *reverse-biased.*

Conduction characteristics of a *PN* junction

The forward-and-reverse conduction characteristic of a *PN* junction is shown in Fig. 29-6. Current flow is plotted along the vertical axis and the applied voltage is plotted along the horizontal axis. The right-hand side of the graph is the region of forward bias; the reverse-bias region is on the left.

As the applied voltage is increased from zero in the direction of forward bias, the current increases very slowly (actually exponentially) until a forward bias of about 0.5 V is reached. The current then increases more rapidly, and the plot turns upward sharply at about 0.65 V. The region where the plot turns upward is called the *knee* of the curve. Beyond about 0.7 V, the current increases dramatically as the applied voltage is increased.

The reverse characteristic is obtained by reversing the polarity of the voltage applied to the junction. As the reverse voltage is increased, a very small *saturation current*, I_s, begins to flow that remains essentially constant until the reverse voltage becomes quite large, perhaps several hundred volts. This saturation current is on the order of 10^{-12} A and is therefore negligible in most practical applications, because the forward current is typically on the order of 10^{-3} A or larger. As the reverse voltage is further increased, the *breakdown region* of the device is approached, and it begins to conduct in the reverse direction. In

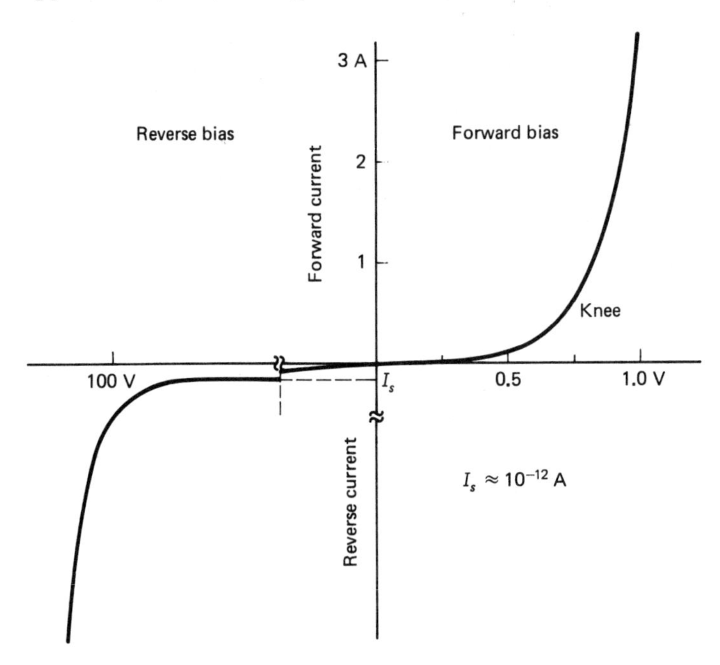

Figure 29-6 The forward-and-reverse conduction characteristic of a *PN* junction. Note that the scales are different in the forward and reverse regions.

practical applications, such as a rectifier, a *PN* junction (a *diode*) must not be allowed to enter the reverse breakdown region.

For voltages greater than about 100 mV applied in the forward direction, the current that will flow through a diode is given to a good approximation by

$$I = I_s e^{V_d/26} \qquad (29\text{-}1)$$

for a silicon diode at room temperature. In this equation, V_d is the voltage applied to the diode (expressed in millivolts) and I_s is the saturation current. You may verify that when V_d is 700 mV and I_s is 10^{-12} A, the current I is 0.49 A.

Solid-state diodes

PN junctions are widely used in electronics and are known as *solid-state diodes* (to distinguish them from the now obsolete vacuum diodes). They serve many functions, one of which is as rectifying elements in power supplies that convert alternating current to direct current for use in TVs, radios, phonographs, and so forth.

Ordinary diodes intended for use in power supplies are characterized by two important parameters. One is the *maximum forward current*, which is the maximum current that the diode can safely conduct in the forward direction. The other is the *peak inverse voltage* (PIV), which is closely related to the reverse breakdown voltage of the junction. Inexpensive diodes may safely handle a current of 10 A while larger, more expensive units may handle currents in the hundreds of amps. PIVs may range from 50 V to 1000 V or more for diodes commonly encountered.

The schematic symbol for a solid-state diode and the direction of electron flow is shown in Fig. 29-7.

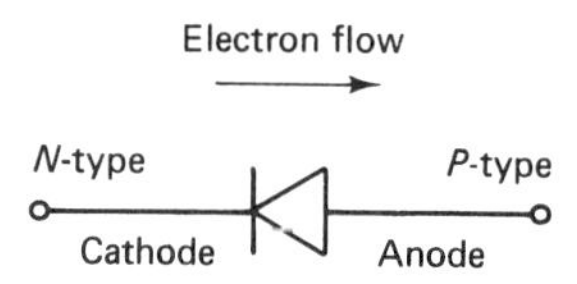

Figure 29-7 Symbol for a solid-state diode. Electrons flow from the *cathode* to the *anode* (terminology held over from the era of vacuum tubes).

29-4 POWER SUPPLY CIRCUITS

Every electronic device that handles audio, video, or digital information requires DC voltages for use by the amplifiers and other signal-processing stages contained in the device. Because electrical power is distributed as AC, a power supply within the device must convert the AC from the power source (a wall socket, in most cases) to DC for use by the amplifiers. In this section we describe two circuits to illustrate how the conversion from AC to DC is achieved.

Half-wave rectifier

Figure 29-8 shows a simple half-wave rectifier circuit. A step-down transformer steps the line voltage (120 VAC) down to 12 VAC, a typical voltage. One side of the transformer secondary is connected to ground, which is the point relative to

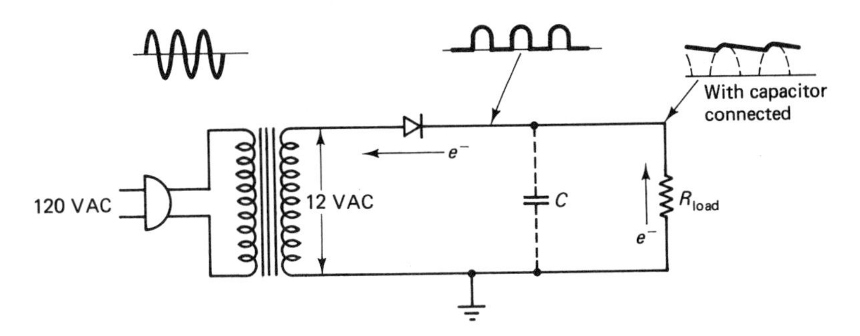

Figure 29-8 A simple half-wave rectifier circuit.

which voltages are measured, and the other side connects to the rectifier diode. The diode permits electrons to flow through it and through the load resistor in only one direction. Thus, the voltage waveform applied to the resistor is one-half of a sine wave (hence the name *half-wave rectifier*) and is therefore *pulsating* DC. Pulsating DC is not a suitable power source for amplifiers, however.

Note that a capacitor may be connected across the load resistor. This capacitor, called a *filter capacitor*, charges during the intervals of diode conduction and maintains the voltage across the load when the diode is not conducting. Thus, the capacitor smooths the pulsating DC wave form. A bit of *ripple* remains even with the capacitor connected, the amount being dependent upon the size of the capacitor and upon the ohmic value of the load resistor. With proper design, the ripple can be made sufficiently small for noncritical applications, but usually a somewhat more elaborate filter circuit is used.

Full-wave bridge rectifier

A full-wave bridge rectifier is shown in Fig. 29-9. Four diodes are used, but only two are active at a given time. We must consider the two polarities of the input voltage from the transformer secondary, as shown in parts (b) and (c) of the figure. Note that for both polarities, electron flow through the load is in the same direction. Further, because the bridge conducts on both the positive and negative portions of the sine wave, the rectifier is called a *full-wave rectifier*.

Observe that two capacitors and a filter resistor are connected between the bridge rectifier and load resistor. This forms a two-section filter, which is more effective in reducing the ripple voltage.

Zener diode voltage regulator

A special class of solid-state diodes called *zener diodes* are fabricated so that the reverse-breakdown characteristic is very sharp and occurs at low voltages. These diodes are used as voltage regulators because the voltage developed across the diode in the reverse-breakdown region is essentially independent of the current flowing through the diode. The reverse-conduction characteristic of a zener diode

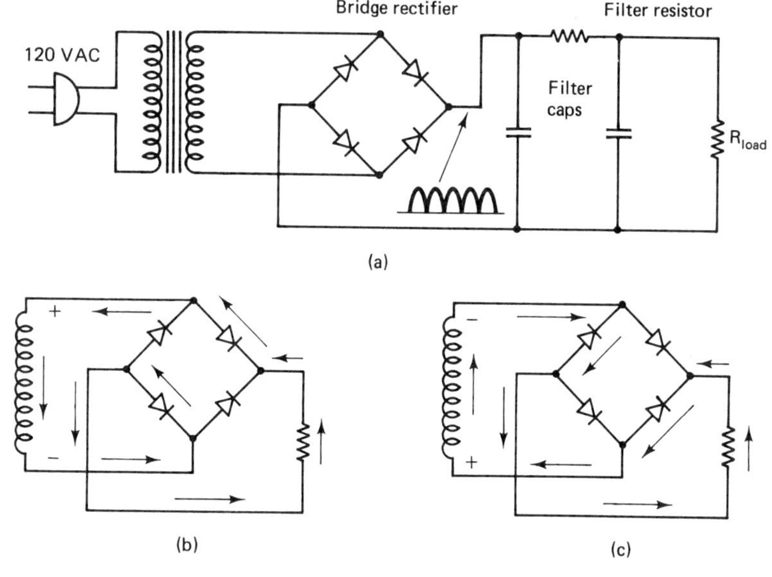

Figure 29-9 (a) A full-wave bridge rectifier circuit with two-section filter. (b) and (c) Conduction patterns for the two polarities of transformer voltage.

is shown in Fig. 29-10(a) and a simple regulator circuit is shown in Fig. 29-10(b). As voltage regulators, zeners are always operated in the reverse breakdown region. Zeners are characterized by their zener voltage V_z, and values of V_z range from about 3 V to 100 V or more.

As the source voltage in Fig. 29-10(b) is increased from zero, the voltage

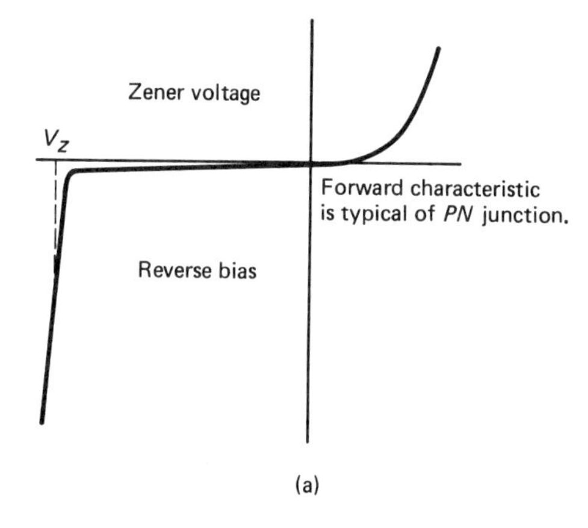

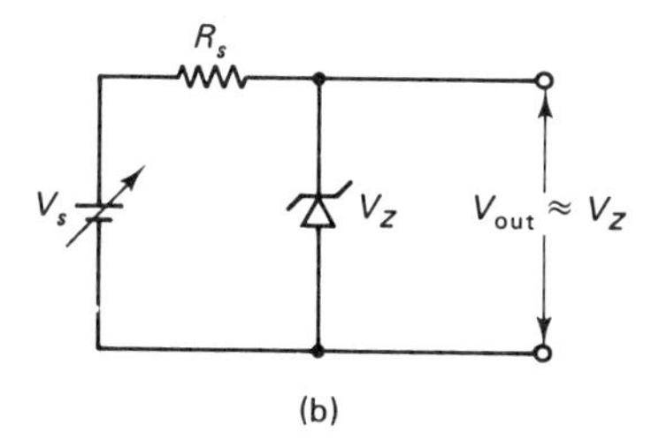

Figure 29-10 (a) The reverse breakdown of a zener diode is very sharp. (b) A simple zener regulator circuit. Source V_s is variable, but as long as V_s is greater than V_z, the output voltage will be very nearly V_z.

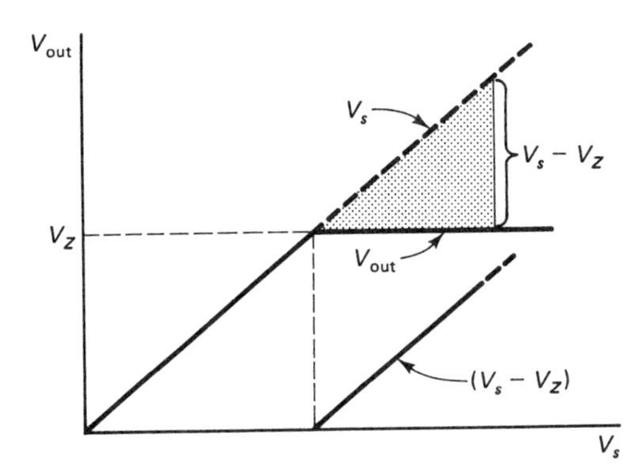

Figure 29-11 The output voltage varies in direct proportion to V_s until V_s reaches V_Z. Thereafter, V_{out} remains nearly constant at V_Z.

across the zener increases linearly until V_z is approached. No current (practically speaking) flows through the zener until the vicinity of V_z is reached, and then the zener current increases sharply. At this point, voltage $V_s - V_z$ appears across the series resistor R_S. A further increase in V_s causes the voltage across the zener to increase only slightly, but the additional current increases the *IR* drop across R_s so that the voltage across the zener remains nearly equal to V_Z. This action is shown in the graph of Fig. 29-11.

29-5 BIPOLAR JUNCTION TRANSISTORS

We now consider a solid-state device that is capable of amplification, the *bipolar junction transistor* (BJT). It is *bipolar* because both *N*- and *P*-type materials are used, and it is a *junction* transistor because two *PN* junctions form the heart of the device.

A BJT is a semiconductor sandwich consisting of a thin layer of *N*- or *P*-material sandwiched between two regions of material of the opposite type, as shown in Fig. 29-12. Two possible configurations give rise to *NPN* and *PNP* transistors as shown. The terminals are identified as the emitter E, base B, and collector C, and we identify the junctions as the base-emitter (BE) junction and the base-collector (BC) junction. We shall describe BJT operation in terms of an *NPN* silicon device.

Examine the circuit of Fig. 29-13, and note that battery *A* forward biases the BE junction. Observe that switch *S* is open so that no external voltage is applied to the collector. When the voltage of *A* reaches the knee voltage of the junction, electrons will flow from the *N*-type emitter region into the *P*-type base region and

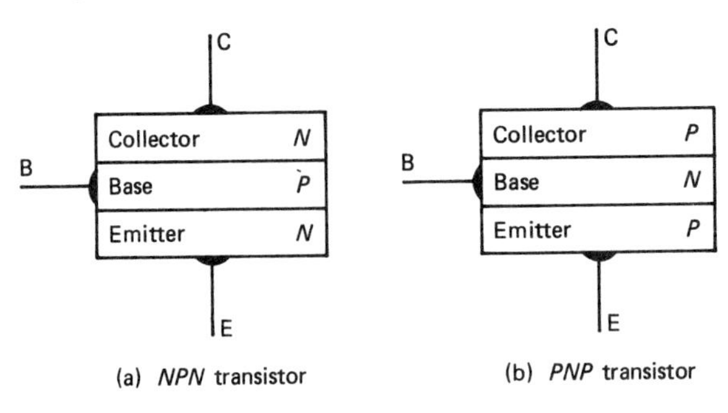

Figure 29-12 *NPN* and *PNP* bipolar junction transistors. Actual device geometries differ greatly from this simple pictorial representation.

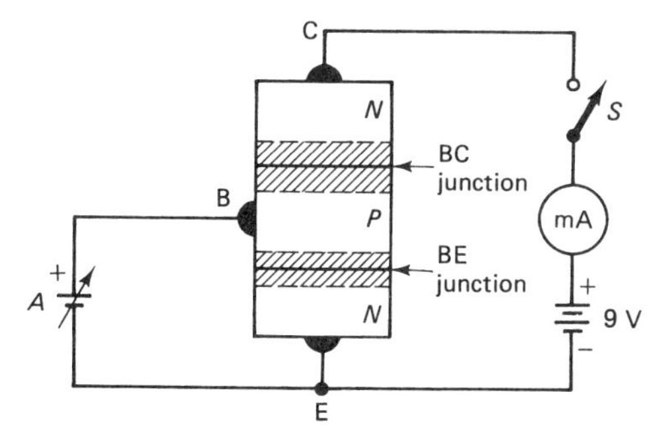

Figure 29-13 Battery *A* forward biases the BE junction. When Switch *S* is closed, a collector current flows even though the BC junction is reverse biased.

out into the external circuit. The magnitude of the current depends upon the voltage applied to the junction by battery *A*, and a graph of the current vs. the applied voltage would resemble the forward conduction characteristic that is typical of *PN* junctions. When the voltage of *A* drops below the knee voltage (about 0.6 V for silicon), electron flow into the base drops to very low levels. If the voltage of *A* were increased to about 0.9 V, a very large current would flow that in many transistors might destroy the BE junction. Electron flow from the emitter to the base is controlled by the voltage applied to the BE junction.

With switch *S* open, no voltage is applied to the collector, and the collector does not affect the operation of the BE junction. When *S* is closed, however, a large positive voltage (9 V in this case) is applied to the collector making the collector region of the transistor much more positive than the base region. This places a reverse bias on the BC junction, and the BC depletion layer becomes wide. We do not expect a current to flow across the BC junction because of the scarcity of charge carriers in the BC depletion layer. We find, however, that a large current does flow across this junction. Why?

The base region is physically very thin and is lightly doped with impurities in comparison with the emitter and collector. These factors reduce the total number of holes in the base region. When the BE junction is forward biased, more electrons cross the BE interface than there are holes in the *P*-type base region for them to drop into. Electrons "pile up" at the base edge of the BE depletion layer, and this pile of electrons extends across the thin base region to the base edge of the BC depletion layer. But the pile of electrons is not of equal depth across the base region.

In other words, the positive voltage applied to the base pulls electrons from the emitter into the base region. A small part of these electrons combine with holes in the *P*-type base material, but most remain in the base region as free electrons. The electron concentration is greatest near the BE junction; it becomes smaller with increasing distance from the BE junction and becomes nearly zero at the BC depletion layer. This is illustrated in Fig. 29-14.

The effect of this electron concentration *gradient* is to produce a *diffusion current* of electrons from the region of greatest electron concentration to the region of lower concentration. This produces a current flow from the emitter side to the collector side of the base region.

Recall that a wide depletion layer appears between the base and the collector regions that acts as an insulator due to the scarcity of charge carriers within the layer. The diffusion current serves to inject free electrons into the BC depletion

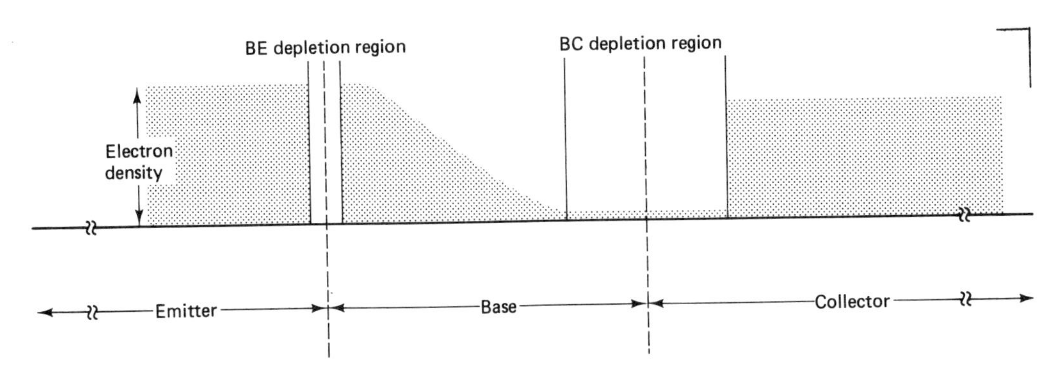

Figure 29-14 Electron densities in an *NPN* transistor. Electrons are most dense in the emitter, and somewhat less dense in the collector. The uneven density in the base region produces a diffusion current that introduces electrons into the depletion layer between the base and collector.

layer, and the injected electrons are readily drawn across the region to the collector. These electrons ultimately form the collector current in the external circuit.

Thus, a large current flows from the emitter to the collector of the transistor, and this current is under the control of the voltage applied to the base, or equivalently, the collector current is determined by the forward bias applied to the BE junction. When the forward bias is increased, more electrons are drawn into the base region, the electron concentration gradient is increased, the diffusion current increases, and the collector current increases. The opposite effects occur when the forward bias of the BE junction is decreased.

If the forward bias is removed from the base, no electrons will be drawn into the base region and no diffusion current will occur. Consequently, the collector current will be zero. Even if the collector is operated at a high voltage (less than the breakdown voltage of the transistor), no appreciable collector current will flow because the collector is not able to extract electrons from the emitter without assistance from the base.

Current gain from base to collector

Of the total number of electrons entering the base from the emitter, a small fraction (about 1%) will combine with holes in the base and be removed from the diffusion current. It is these electrons that combine with holes that ultimately constitute the current that flows from the base terminal of the transistor. In practice, it is desirable for the base current to be minimized.

Here is the principle that makes the transistor useful as an amplifier. A small base current, resulting from a voltage applied to the base, can effectively control a much larger current that flows from the emitter to the collector. In terms of energy, the base uses a small amount of energy to control a much greater expenditure of energy in the collector circuit. This is the principle on which amplification is based.

For a given transistor, there is an almost constant relationship between the base current I_b and collector current I_c. To a good approximation, the collector current is a constant multiple of the base current:

$$I_c = \beta I_b \tag{29-2}$$

The factor β is called the *beta* of the transistor, defined as the ratio of collector current to base current:

$$\beta = \frac{I_c}{I_b} \tag{29-3}$$

Typical values of β range from about 20 to 200 for transistors commonly encountered. The division of emitter current between the base and collector is illustrated in Fig. 29-15.

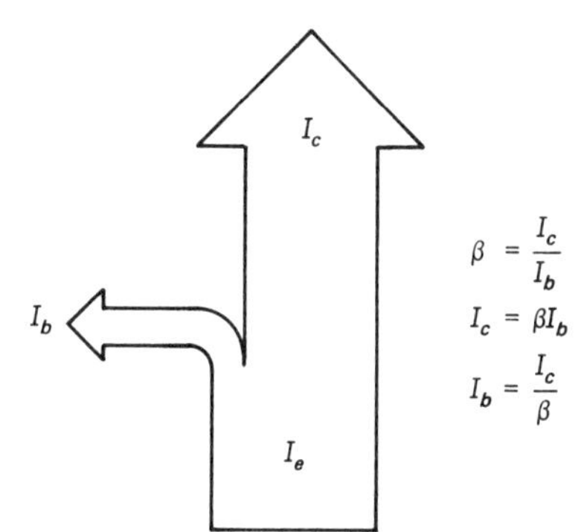

Figure 29-15 The division of current between the base and collector.

PNP transistors

If we substitute the term *holes* for *electrons* in the preceding description of an *NPN* transistor, the result will hold true for a *PNP* transistor. Additionally, the voltages applied to the transistor terminals must be reversed: The collector of a *PNP* transistor is made negative whereas the collector of an *NPN* transistor is positive. For both *NPN* and *PNP* transistors, the base is biased toward the polarity of the collector to obtain conduction through the transistor. *PNP* transistors are used less extensively than *NPN* devices, but this is not to imply that *PNP* transistors are rarely encountered.

29-6 A PRACTICAL VOLTAGE AMPLIFIER

In this section we describe two amplifier circuits in order to give an idea of how a BJT can be caused to amplify a signal from a microphone, phonograph cartridge, or other signal source. These amplifiers are called *voltage amplifiers* to distinguish them from *power amplifiers*. A voltage amplifier increases the amplitude of a signal with little concern for delivering power to a load, whereas a power amplifier is mainly concerned with driving a load such as a loudspeaker.

One of the simplest transistor amplifiers is shown in Fig. 29-16. To provide the small positive voltage required by the base in order to bias the BE junction

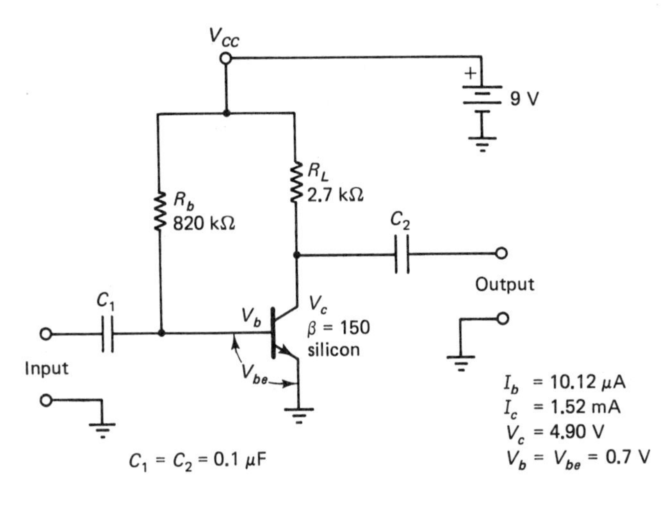

Figure 29-16 A simple transistor amplifier. Almost any general purpose *NPN* silicon transistor can be used.

into conduction, resistor R_b, called the base resistor, is connected from the base to the positive terminal of the power supply. With the BE junction biased into conduction, V_{be}, the voltage between the base and emitter, will be on the order of 0.7 V (for silicon; 0.3 V for germanium). The voltage across R_b is $V_{cc} - V_{be}$, and the current through R_b (which is the base current I_b) is given by

$$I_b = \frac{V_{cc} - V_{be}}{R_b} \tag{29-4}$$

Once the base current is known, since we assume the β of the transistor is also known for the transistor being used, we can compute the collector current I_c using

$$I_c = \beta I_b \tag{29-5}$$

Furthermore, we can now calculate the collector voltage V_c by observing that V_c is the power supply voltage V_{cc} minus the *IR* drop across the load resistor R_L:

$$V_c = V_{cc} - I_c R_L \tag{29-6}$$

By inspection we determine that the emitter voltage V_e is zero, and the base voltage is simply V_{be}. We now have mathematical expressions for all the currents and voltages involved. Sample calculations are given in the figure.

Let us now see how the amplifier amplifies a signal. First of all, the signal is assumed to be an AC wave form such as a voice wave obtained from a microphone. The signal will pass with little opposition from the input through C_1 to the base, but the DC voltage on the base is prevented by C_1 from flowing back into the signal source. Capacitor C_1 is used as a *coupling* capacitor; it blocks DC but passes AC.

The input signal voltage is small in comparison to V_{be}, yet it causes the base voltage to vary because the total voltage at the base is V_{be} plus the signal voltage. Variations in the base voltage produce corresponding variations in the collector current. Thus, the input signal applied to the base causes the collector current to vary accordingly.

Note that the collector current must flow through load resistor R_L, and recall that the collector voltage is determined, in part, by the collector current due to the

IR drop across R_L. A variation of the collector current produces a variation in the *IR* drop across R_L, and this causes the collector voltage to vary in a similar manner. Moreover, the variation of voltage at the collector constitutes the output signal of the amplifier. Coupling capacitor C_2 prevents the DC voltage at the collector from appearing at the output terminals of the amplifier.

The output signal is larger in amplitude than the input signal because of the current-amplifying properties of the transistor. Resistor R_L transforms the variations in collector current to variations in collector voltage, which represents the output signal.

We should point out that the voltage gain is *not* the same as the current gain β of the transistor. The voltage gain is defined in terms of signal amplitudes V_{in} and V_{out}:

$$\text{Voltage gain} = \frac{V_{\text{out}}}{V_{\text{in}}} \tag{29-7}$$

To calculate the voltage gain of this stage, we must know the characteristics of the source of the signal (microphone, or whatever), which we have not given. Therefore, for the sake of brevity, we simply state that the voltage gain of this stage will be no more than about 4 or 5 for a typical microphone.

A better amplifier

A disadvantage of the preceding amplifier is that the performance of the circuit is quite sensitive to small variations of the transistor parameters. A better circuit is shown in Fig. 29-17. This circuit is commonly encountered in commercial electronics devices.

A voltage divider consisting of R_{hb} and R_{lb} delivers a fraction of the power supply voltage to the base in order to forward bias the BE junction. Due to the presence of R_e in the emitter lead, the emitter voltage V_e will not be zero as it was

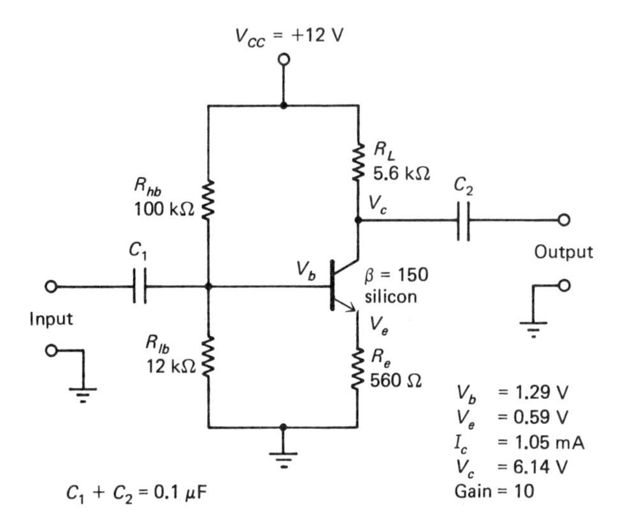

Figure 29-17 An improved version of a transistor amplifier. This circuit is more nearly independent of the transistor parameters than the circuit of Fig. 29-16.

in the previous circuit. Rather, V_e will pull up to within V_{be} of the base voltage. Thus,

$$V_b = \left(\frac{R_{hb}}{R_{lb} + R_{hb}}\right)V_{cc} \tag{29-8}$$

$$V_e = V_b - V_{be} \tag{29-9}$$

Because the base current is small in comparison with the collector current, we can say that the emitter and collector currents are approximately equal. Hence, because

$$V_e = I_c R_e \tag{29-10}$$

we may compute the collector current using

$$I_c = \frac{V_e}{R_e} \tag{29-11}$$

As for the preceding circuit, the collector voltage V_c is given by

$$V_c = V_{cc} - I_c R_L \tag{29-12}$$

The base current may be computed using $I_b = I_c/\beta$, which stems from the definition of β.

To a good approximation, the gain of this stage is given by

$$\text{Voltage gain} = \frac{R_L}{R_e} \tag{29-13}$$

which we state without justification.

Obviously the description of the amplifiers given here is sketchy at best. For more information on transistor amplifiers, consult the reference at the end of this chapter.

29-7 THE CATHODE RAY OSCILLOSCOPE

An oscilloscope is perhaps the most useful instrument to an electronics technician. A "scope" allows one to see the wave form existing at various points in an electronic device, to measure amplitudes fairly accurately, and to determine frequencies of signals to a good approximation. By examining the wave form at the output of an amplifier and comparing it with the input wave form, the performance of the amplifier can be quickly determined.

The heart of a scope is the *cathode ray tube* (CRT), the general features of which are shown in Fig. 29-18. An electron gun at the back of the tube fires a sharply focused beam of electrons (cathode rays) toward a phosphor-coated screen at the front of the tube. The electron beam produces a small but intense dot of light where it strikes the screen.

In traveling from the electron gun to the screen, the beam passes between two sets of deflection plates that are capable of deflecting the beam both vertically and horizontally. Thus, the dot on the screen can be moved to any location on the

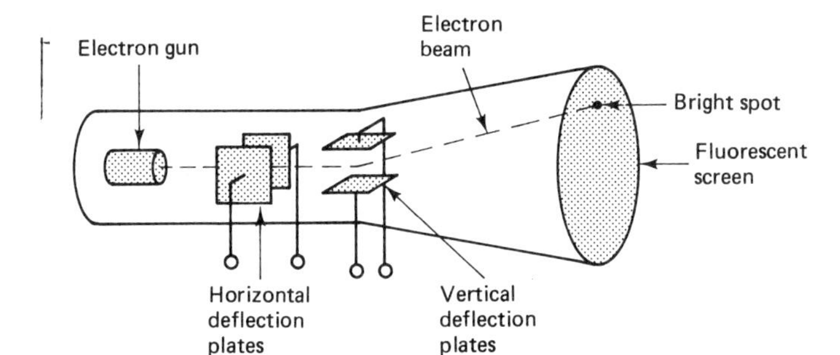

Figure 29-18 A cathode ray tube (CRT) is the heart of an oscilloscope.

screen by applying the appropriate voltages to the vertical and to the horizontal deflection plates.

In use, the dot is caused to sweep repeatedly across the screen with uniform velocity from left to right by a time-base generator contained in the instrument. The speed with which the dot moves across the screen is determined by the settings of front-panel controls. When the dot reaches the right-hand side of the screen, it is turned off and quickly returned for another pass across the screen from left to right. At high sweep rates, the dot is perceived as a line due to the persistence of the phosphors on the screen and due also to the persistence of vision of the eye. Nevertheless, the dot is still a dot.

The voltage whose wave form is to be investigated is applied to the input of the vertical section of the scope, which moves the dot vertically on the screen in direct proportion to the instantaneous value of the input voltage. As the input becomes more positive, the dot moves upward, and so forth. If the input is a sine wave, the dot will move vertically in a sinusoidal manner.

The result of the combined vertical and horizontal motions is to plot a trace on the screen of input voltage (vertical axis) vs. time (horizontal axis). Thus, a sinusoidal voltage appears as a sine wave on the screen.

In order for the trace to form a stationary pattern on the screen, the dot must not begin its horizontal sweep until the input voltage reaches the same point on its cycle as it did when the previous horizontal sweep was begun. Synchronization circuits within the instrument delay the start of each horizontal sweep until the input signal reaches a predetermined point on its cycle. A group of front-panel controls are devoted to this synchronizing function.

Reference

MALVINO, ALBERT P., *Transistor Circuit Approximations*. 3rd ed. New York: Gregg Division, McGraw-Hill Book Company, 1980. An excellent text on the practical design of simple transistor amplifiers.

Questions

1. What is an electron-hole pair? Why do free electrons and holes occur in pairs in intrinsic semiconductors? What is responsible for the generation of electron-hole pairs?
2. Describe the mechanism by which the presence of electron-hole pairs increases the electrical conductivity of a semiconductor.
3. Why does a hole appear to have a positive

charge even though a hole is really nothing?

4. Explain why the presence of holes in a *P*-type semiconductor increases the electrical conductivity of the material.
5. What are the majority and the minority carriers in *N*-type semiconductor material?
6. What causes a depletion layer to form at the interface between the *N* and *P* regions of a *PN* junction?
7. What happens to the depletion layer when a *PN* junction is forward biased?
8. When a *PN* junction made of silicon is forward-biased, what approximate voltage appears across the junction under conditions of moderate conduction?
9. Describe the reverse-breakdown characteristic of a zener diode. What is the primary application of zener diodes?
10. In normal operation of an *NPN* transistor, what voltage polarities are applied to the base and collector regions? Which is more positive?
11. Why is it necessary to forward-bias the BE junction?
12. In normal operation the BC junction of a BJT is reverse-biased, yet a large current easily flows across the junction. What is responsible for this effect?
13. Why do most of the electrons entering the base region go on to the collector?
14. A transistor is often called a current multiplier. What current is multiplied? What is the beta of a transistor?
15. How do the emitter and collector currents compare? How does the base current compare with the collector current?
16. In normal operation of a BJT, the collector current is controlled by the voltage existing between which two elements of the transistor?
17. Describe the major features of a cathode ray tube (CRT).
18. Explain the functions performed by the vertical and horizontal sections of an oscilloscope.

Problems

1. The atomic density of silicon in the solid state is about 10^{22} atoms/cm^3. At 300 K, intrinsic silicon contains about 10^{10} electron-hole pairs per cm^3. Hence, one electron-hole pair will exist for every ___?___ silicon atoms.
2. Suppose 10^{16} atoms of phosphorus are added as a dopant to 1 cm^3 of silicon. What is the ratio of silicon atoms to impurity atoms in such material?
3. With an impurity concentration of 10^{16} atoms/cm^3 (as for Problem 2), how many impurity atoms will be found in a tiny cube of silicon having 100 silicon atoms on each edge of the cube?
4. If 10^{16} *N*-type impurity atoms are added per cubic centimeter of silicon, what will be the ratio of free electrons contributed by the impurity to free electrons arising from thermally induced electron-hole pairs at 300 K? (See Problem 1.)
5. The reverse saturation current, I_s, for a particular diode is 1.5×10^{-12} A. What forward current will flow through the diode when the following voltages are applied to the diode?
   (a) 100 mV **(b)** 400 mV **(c)** 600 mV **(d)** 800 mV
6. Repeat Problem 5, but use $I_s = 3 \times 10^{-12}$ A instead of 1.5×10^{-12} A.
7. For a certain diode, $I_s = 2 \times 10^{-12}$ A. What voltage V_d must be applied to the diode in order to produce a forward current of 10 A?
8. The numerical constant (26 mV) in Eq. (29-1), the diode equation, is actually kT/q, where k is Boltzman's constant (1.38×10^{-23} J/K), T is the absolute temperature, and q is the electronic charge (1.6×10^{-19} C). Verify that the constant is very nearly 26 mV when $T = 300$ K.
9. In the circuit of Fig. 29-10(b), the zener voltage V_z is 9 V, R_s is 100 Ω, and V_s is 12 V.
   (a) What *IR* drop appears across R_s?
   (b) What is the current that flows through R_s?
   (c) If no load is connected to the output,

what current will flow through the zener diode?

10. The beta β of a certain transistor is 150. Compute the collector current when I_b is **(a)** 10 μA; **(b)** 1 mA.

11. The collector current of a certain transistor is 3 A when the base current is 50 mA. What is the beta of the transistor?

12. What base current must flow in order to produce a collector current of 1 mA in a transistor whose beta is 170?

13. Verify the values of I_b, I_c, V_c, and V_b for the circuit of Fig. 29-16.

14. For the circuit of Fig. 29-16, compute the value of V_c that will result when the beta increases from 150 to 160.

15. Verify the values of I_c and V_c listed in Fig. 29-17 for the circuit shown in the figure.

16. What will be the output voltage V_{out} if a sine wave of amplitude 100 mV is applied to the input of the circuit of Fig. 29-17?

CHAPTER 30

PRINCIPLES OF ELECTRONICS

In this chapter we show how the physics of electromagnetic waves and electronic circuits can be applied to practical communications systems such as AM and FM radio, television, tape systems, and phonographs. This chapter provides a rather cursory glimpse of what is involved in the communications area of electronics technology.

Not too many years ago, a computer was something associated with research laboratories or large record-keeping institutions. Now computers are everywhere, and we are probably yet seeing only the tip of the iceberg.

In this chapter we also present the principles upon which digital computers are based. We show how electronic circuits are able to handle numbers in digital form and how basic logic gates may be combined to implement logic expressions represented in terms of Boolean algebra. Flip-flops form the basis for electronic counters, frequency dividers, and storage registers in the memory of a computer. With the basic gates it is possible to make adders and subtracters, and only a slight increase in complexity gives us electrically alterable circuits that form the heart of a digital computer.

Interested readers may find many specialized books written on the topics that are mentioned in this chapter.

30-1 TRANSMISSION AND RECEPTION OF ELECTROMAGNETIC WAVES

In electronics, extensive use is made of circuits called *oscillators* that are capable of producing high-frequency alternating currents. Typically, an oscillator consists of an amplifier and a feedback network that couples a fraction of the output signal back to the input of the amplifier, as shown in Fig. 30-1. The feedback network may contain tuning elements (inductors and capacitors) for setting the frequency of oscillation or the tuning elements may be incorporated within the amplifier.

Oscillators of various types are available that will generate sine waves and

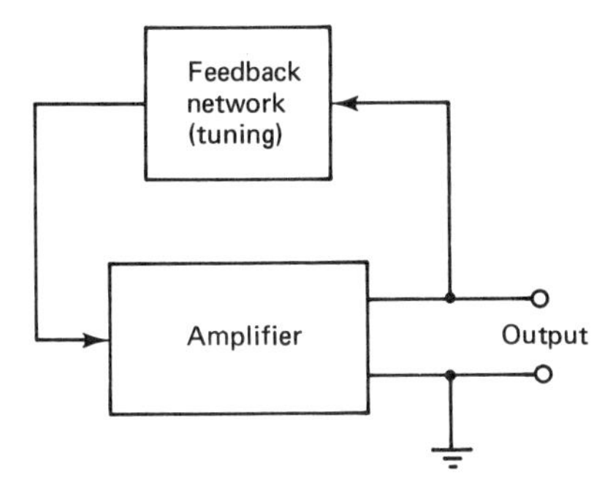

Figure 30-1 Generalized block diagram of an oscillator. The frequency of the output may range from 0.001 Hz to over 10^9 Hz.

other wave forms over the frequency range from about 0.001 Hz to over 10^9 Hz. *Audio oscillators* cover the range from about 20–20,000 Hz; *radio frequency* (RF) oscillators operate in the range above the audio.

Suppose an oscillator is coupled to a dipole antenna, as shown in Fig. 30-2. At first we might expect *zero current* to flow in the transmission line and in the antenna sections because the ends of the antenna are "open." However, the alternating voltage from the oscillator causes electrons to surge from end to end in the dipole so that a significant current does flow. The surges of electrons give rise to charge accumulations at the ends of the dipole, whose magnitude and polarity vary sinusoidally at the oscillator frequency.

Whenever positive and negative charges are separated, an electric field (*E*-field) is developed in the region between the charges. Hence, an electric field exists around the dipole antenna whenever nonzero charges appear at the ends, and the lines of the *E*-field lie in planes parallel to and including the antenna. Furthermore, the current flow in the antenna produces a magnetic field around the antenna whose lines lie in planes perpendicular to the length of the antenna. Thus, both an electric and a magnetic field exist in the region surrounding the antenna.

The production of an electromagnetic wave can be envisioned as follows, in terms of the electric field. When the field reaches a maximum, it immediately begins to decrease in magnitude and ultimately to change polarity. We can imagine the electric field to be "snapped off" due to the rapidity of the electron surges driven by the oscillator, and once the field becomes detached from the

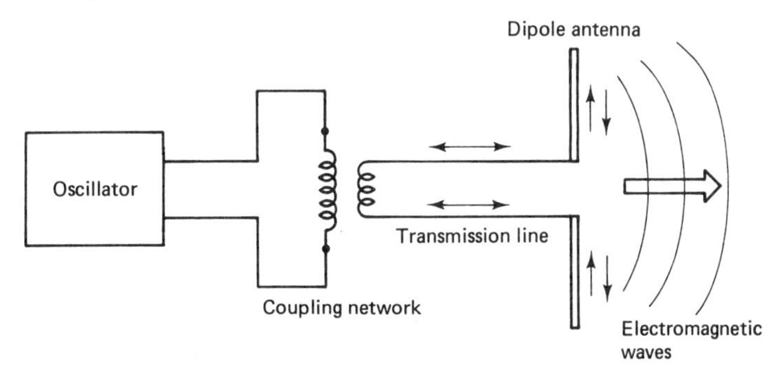

Figure 30-2 One means of producing electromagnetic waves. The high-frequency signal produced by the oscillator sets up currents in the dipole antenna which give rise to the electromagnetic waves.

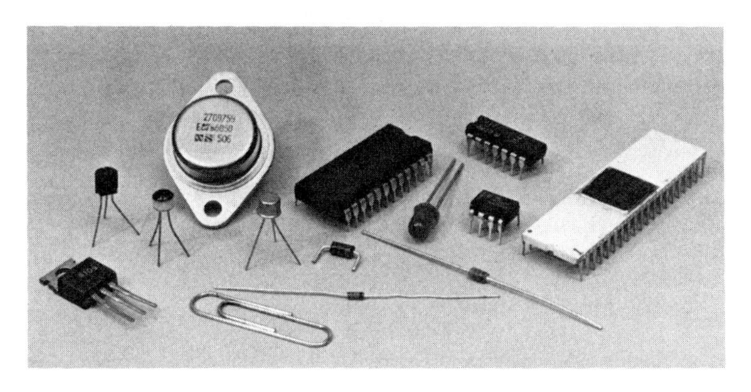

Assorted semiconductor devices. The ceramic (white) unit is a 40-pin microprocessor chip.

antenna, it travels away from the antenna at the speed of light c. A similar thing happens for the magnetic field so that the electromagnetic wave consists of both an electric and a magnetic field. This, of course, is in accord with the predictions of Maxwell's equations.

The frequency of electromagnetic waves produced by an antenna is the same as that of the oscillator driving the antenna. Moreover, the magnitude of the wave, its strength, is proportional to the magnitude of the driving voltage. As for other types of wave motion, the following relationships hold for electromagnetic waves:

$$f = \frac{c}{\lambda} \qquad \lambda f = c \qquad \lambda = \frac{c}{f} \tag{30-1}$$

where $c = 2.9979 \times 10^8$ m/s, f is the frequency in hertz Hz, and λ is the wavelength in meters. The radio portion of the electromagnetic spectrum is shown in Fig. 30-3.

Name	Frequency range
Subaudio	Less than 20 Hz
Audio	20 Hz to 20 kHz
Extremely low frequencies (ELF)	30–300 Hz
Superlow frequencies (SLF)	300 Hz to 3 kHz
Very low frequencies (VLF)	3–30 kHz
Low frequencies (LF)	30–300 kHz
Medium frequencies (MF)	300 kHz to 3 MHz
High frequencies (HF)	3–30 MHz
Very high frequencies (VHF)	30–300 MHz
Ultrahigh frequencies (UHF)	300 MHz to 3 GHz
Superhigh frequencies (SHF)	3–30 GHz
Extremely high frequencies (EHF)	30–300 GHz

Figure 30-3 Radio portion of the electromagnetic spectrum.

Electromagnetic wave generator at the local airport.

Reception of electromagnetic waves

Electromagnetic waves would be of little practical consequence if they could not be received by a receiving antenna. The simplest receiving antenna is a long wire oriented (for optimum efficiency) parallel to the E-field of the wave being received. When the EM wave passes across the receiving antenna, a current is induced in the antenna. This current may be coupled into a tuning circuit as shown in Fig. 30-4, and with proper design, the output voltage V_{out} will be a sine wave whose frequency equals that of the incident wave and whose amplitude is proportional to the strength of the incident wave. The tuning circuit exhibits the greatest response to the EM waves of the tuned frequency and tends to reject waves of other frequencies.

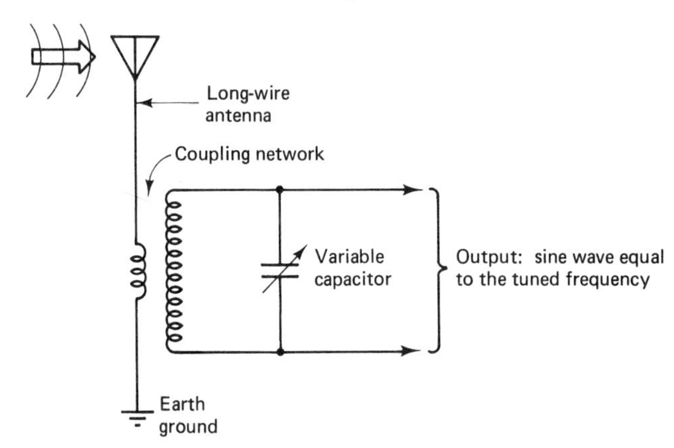

Figure 30-4 A simple circuit for detecting electromagnetic waves of a particular frequency. The output signal strength depends upon the strength of the incident electromagnetic waves.

30-2 INTERRUPTED CARRIER TELEGRAPHY

One system for transmitting information by the medium of EM waves is illustrated in Fig. 30-5. The transmitter is equipped with a key (a telegraph key) so that EM waves are transmitted only when the key is depressed. Letters of the alphabet and numbers may be sent by utilizing Morse code.

The train of waves transmitted is called the *carrier wave*, or simply the *carrier*. Hence, the carrier is interrupted by the key in order to encode the characters being sent, and we have *interrupted carrier telegraphy*.

The carrier is picked up at the receiver by an antenna coupled into a tuned circuit. An RF amplifier increases the amplitude of the radio-frequency sine wave appearing at the output of the tuned circuit. The output of the RF amplifier is applied to one of the two inputs to the *mixer*.

The frequency of the signal at the output of the RF amplifier is much higher than that to which the ear can respond. Therefore, we must somehow convert the interrupted carrier into an audible tone so that we may hear the dots and dashes of the tone. Toward this end, a beat-frequency oscillator (BFO) is incorporated into the receiver, and the BFO generates a continuous high-frequency sine wave

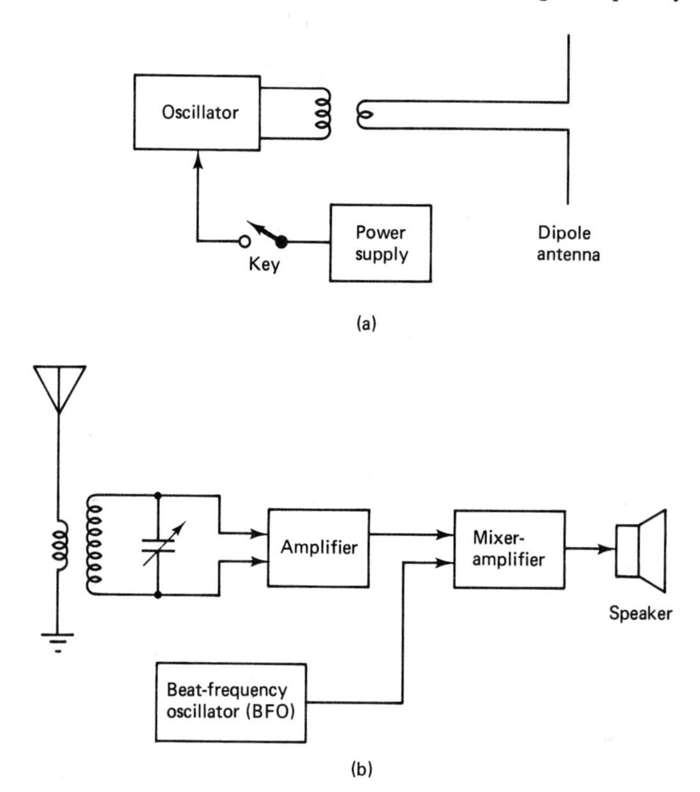

Figure 30-5 A simple system for interrupted-carrier telegraphy.
(a) The oscillator is keyed to produce Morse code.
(b) The receiver mixes the transmitter signal with a signal from a beat-frequency oscillator to produce an audible tone.

whose frequency is nearly equal to that of the carrier being received. The BFO output is applied to the second input to the mixer.

The mixer causes a beat note to be formed between the antenna signal and the signal from the BFO. The frequency of the beat note equals the *difference* in frequency of the antenna and BFO signals:

$$f_{bn} = f_{ant} - f_{bfo} \tag{30-2}$$

The beat note constitutes the output of the mixer and will be audible (through earphones or speakers) provided f_{bn} lies in the audio spectrum. The interruptions of the carrier (corresponding to Morse code) will give rise to corresponding interruptions of the beat note because the beat note occurs only when the carrier signal is present.

The frequency of the audible tone may be altered by making slight adjustments to the BFO frequency relative to that of the carrier.

30-3 AMPLITUDE MODULATION (AM)

The transmission of voice or music via EM waves requires a more sophisticated system than that previously described. One such system is the system of *amplitude modulation*.

To transmit voice or music requires that the audio wave form of the music be reproduced at the receiver. Somehow, the audio information must be added to the carrier produced by the transmitter. This is accomplished by causing the instantaneous amplitude of the carrier to be directly proportional to the instantaneous amplitude of the audio wave form. Accordingly, the carrier amplitude becomes a maximum when the audio wave form reaches an instantaneous positive maximum, and the carrier amplitude reaches a minimum when the audio wave form reaches a negative minimum. This process is illustrated in Fig. 30-6 and is called *amplitude modulation*.

A simplified block diagram of an AM transmitter is shown in Fig. 30-7. The oscillator stage generates the high-frequency sine wave that ultimately becomes the carrier. Amplifier stages following the oscillator increase the amplitude of the oscillator signal. The audio signal from the source (microphone, turntable, tapes,

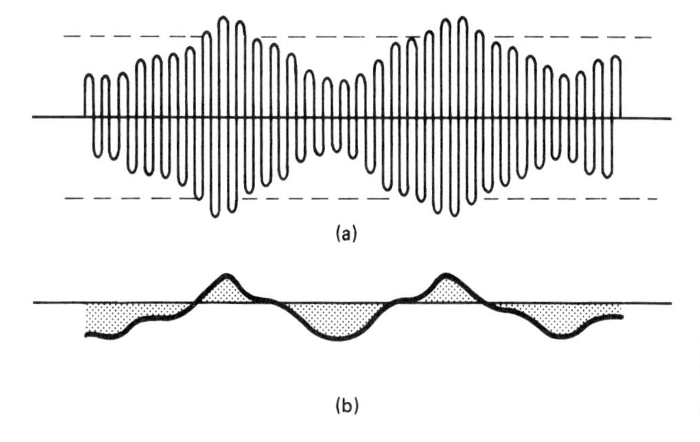

Figure 30-6 Amplitude modulation. (a) The AM waveform produced by the modulating signal of (b).

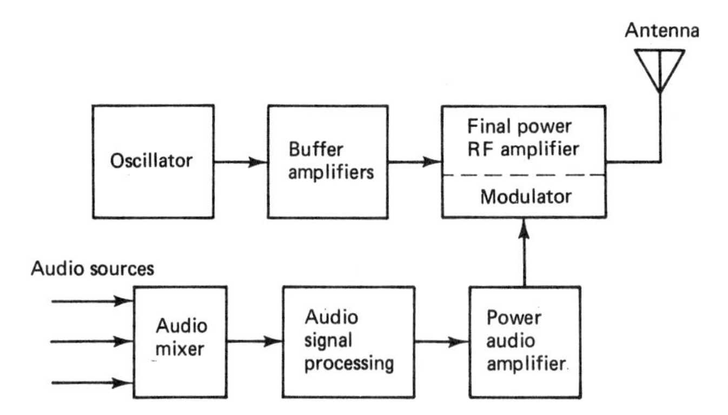

Figure 30-7 Block diagram of an AM radio transmitter.

and so on) is amplified and then applied to the *modulator*. The modulator works in conjunction with the final RF power amplifier to cause the output signal amplitude of the RF amplifier to vary in accordance with the audio signal.

AM receiver

An AM receiver in its simplest form is shown in Fig. 30-8. A long-wire antenna is magnetically (transformer) coupled to a tuning circuit consisting of L and C. The AM wave form of the transmitter to which the receiver is tuned appears at the output of the tuned circuit. A solid-state diode D performs the function of a half-wave rectifier and removes the bottom half of the AM wave form applied to the diode. Capacitor C_d smoothes the pulses to reconstruct the original audio wave form. The reconstructed (demodulated) audio signal is delivered to the earphones which produce the sound for the listener.

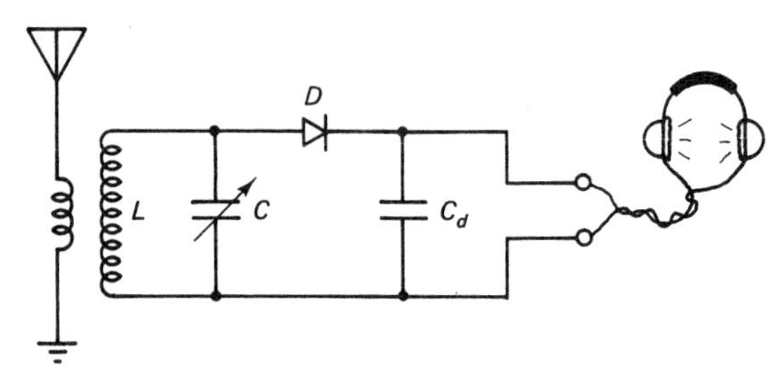

Figure 30-8 A simple AM receiver.

This simple receiver contains no amplifiers to increase the amplitude of the signal. Consequently, only very strong stations will be received, and a long-wire, external antenna is required. Modern receivers contain elaborate amplifiers and signal-processing stages to increase the sensitivity of the receiver and to provide sufficient power output to drive a system of speakers rather than earphones. Radio receivers are described in detail in the references given at the end of this chapter.

30-4 FREQUENCY MODULATION (FM)

In an FM system, voice, music, or other information to be transmitted is used to vary the *frequency* of the carrier rather than its amplitude. The audio signal causes the carrier frequency to vary over specific limits, while the amplitude of the carrier remains constant. The principle of frequency modulation is illustrated in Fig. 30-9, where an audio sine wave causes the carrier to deviate above and below the *center frequency* as the audio wave form goes positive and negative. A comparison of AM and FM is given in Fig. 30-9(a).

A major advantage of FM over AM is that FM is less sensitive to electromagnetic noise. Electromagnetic waves are produced by electrical discharges (sparking) that occur in electrical machinery (motor brushes), automobile ignition systems, and so forth, including electrical discharges in the air (lightning). These spurious waves can be very large in amplitude and may produce loud clicks and pops or static in an AM receiver. On the other hand, because the frequency of a signal is independent of its amplitude, an FM receiver incorporates an amplitude *limiter* which limits the signal amplitude to a predetermined design value. Hence, any amplitude variation of the FM carrier is removed, and noise pulses do not reach the speaker. The action of a limiter is illustrated in Fig. 30-10.

The transmissions of commercial FM are of higher fidelity than the transmissions of commercial AM. One reason for this is that the noise immunity of the FM system effectively eliminates the annoying static that often accompanies AM reception. Moreover, the commercial FM band is located in a comparatively quiet portion of the spectrum, from 88 MHz to 108 MHz. Most atmospheric noise occurs below 30 MHz. The AM broadcast band extends from 0.54 MHz to 1.6 MHz, a relatively noisy portion of the spectrum.

The fidelity of a transmission depends also upon the *bandwidth* of the transmitted audio signal. Stations in the AM broadcast band do not transmit audio frequencies higher than 5 kHz, but commercial FM stations transmit audio

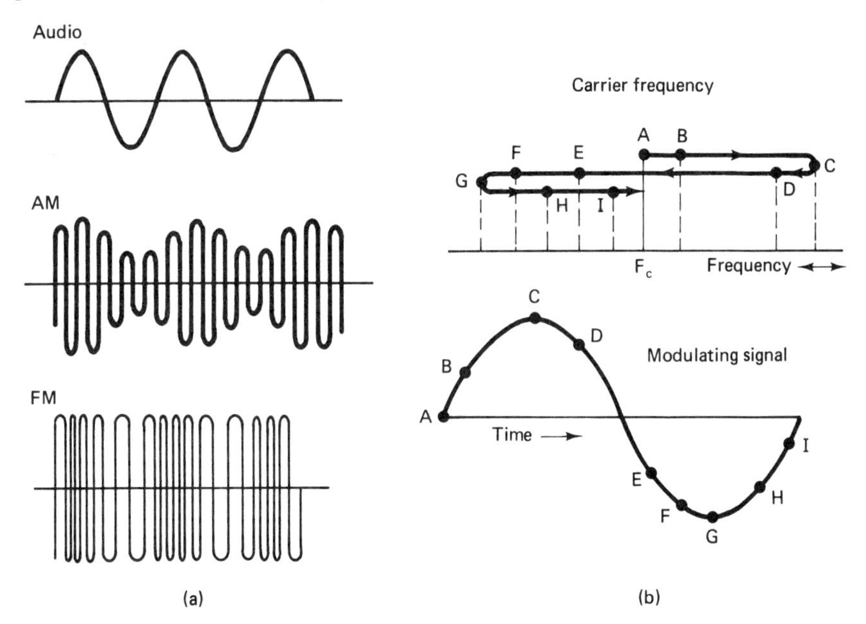

Figure 30-9 (a) Comparison of AM and FM waveforms.
(b) The frequency varies above and below the center frequency F_c as the modulating signal rises and falls above and below the zero reference.

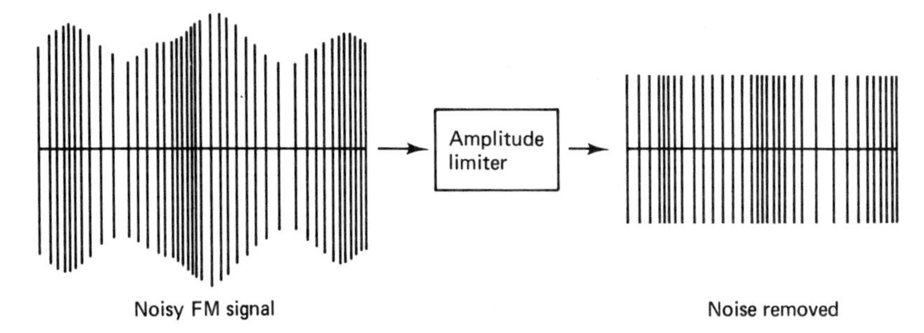

Figure 30-10 Noise can be removed from an FM signal by an amplitude limiter. Note that the frequency variations remain.

frequencies up to 15 kHz and therefore achieve greater fidelity. These bandwidths are mandated by the Federal Communications Commission and do not represent inherent limitations in either the AM or FM system.

30-5 TELEVISION

A television system can be divided into four broad components for purposes of discussion. The television *camera* converts the scene to be televised to electronic signals. The *transmission system* transmits the electronic signals to the receiver. The television *picture tube* converts the electronic signals back to pictorial form for viewing. The *audio subsystem* transmits the sounds accompanying the scene being televised. In this section we briefly describe the physical principles employed in each of these components.

TV picture tube

A TV picture tube is a CRT, similar to that found in an oscilloscope, which has been adapted for viewing for entertainment. An electron gun in the "neck" of the tube fires a sharply focused electron beam toward a phosphor-coated screen. A tiny spot of illumination appears where the beam strikes the screen. The brightness of the spot may be varied from completely dark to maximum brightness by varying the intensity of the electron beam, and the beam intensity is easily controlled by the voltage applied to an electrode that is part of the electron gun.

A full-screen picture is formed on the TV screen by sweeping the electron beam rapidly horizontally and slowly vertically so that the single spot is swept over the entire picture area. As the spot moves along, the *video signal* delivered to the electron gun varies the intensity of the spot so that light and dark areas of the scene are reproduced. In the United States, the screen is divided into 525 horizontal lines (not all of which are visible), and 30 complete pictures (*frames*) are transmitted per second.

To reduce the flicker that would be apparent at a frame speed of only 30 per second, a technique of *interlaced scanning* is used that effectively doubles the frame speed. Each frame is divided into two *fields*; one field consists of even-numbered horizontal lines, while the other consists of odd-numbered lines. Thus, the spot traverses the 262.5 lines of a field in 1/60 s and the eye is tricked into thinking that 60 complete pictures are transmitted per second. The interlaced scanning pattern is shown in Fig. 30-11.

In television, the electron beam is swept across the screen by *magnetic deflection*. Horizontal and vertical deflection coils placed over the neck of the picture tube set up magnetic fields inside the tube that deflect the beam appropriately. In an oscilloscope, the beam is deflected via electrostatic deflection, implemented through the horizontal and vertical deflection plates located inside the tube. The *deflection yoke* (as the horizontal and vertical deflection coil assembly is known) is an important adjunct of a TV picture tube.

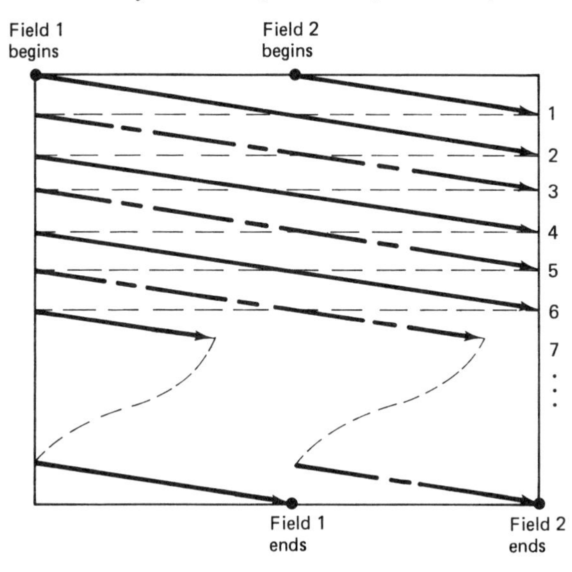

Figure 30-11 System of interlaced scanning used to avoid flicker of a TV screen. A total of 525 lines are provided in the U.S. system, but not all lines are active. The vertical retrace is not shown in this illustration. Several horizontal lines occur in the vertical retrace interval during which the beam is blanked.

TV camera tube

The TV camera tube and associated circuitry must convert an optical image into a video signal that may be transmitted to the TV receiver. A popular type is the *vidicon*, shown in Fig. 30-12. An image of the scene to be televised is optically focused onto the flat target area on the faceplate of the tube. The target consists of two active layers, a transparent conductive layer adjacent to the faceplate and a layer of photoconductive material in contact with the conductive layer. The photoconductive layer is in the form of a mosaic of cells. The conductivity of each cell is determined by the amount of light striking the cell.

Each cell acts as a tiny capacitor. The conductive layer forms one plate that is common to all cells. The opposite surface of each cell acts as the other plate, which appears electrically open. Thus, in effect, the target is an array of tiny capacitors that leak an amount determined by the light striking the cell.

An electron gun situated at the other end of the tube provides a beam that is scanned over the target area by deflection coils external to the tube. As the beam scans across a particular cell, the open plate of the capacitor is effectively grounded so the capacitor is discharged. The discharge current flows in the

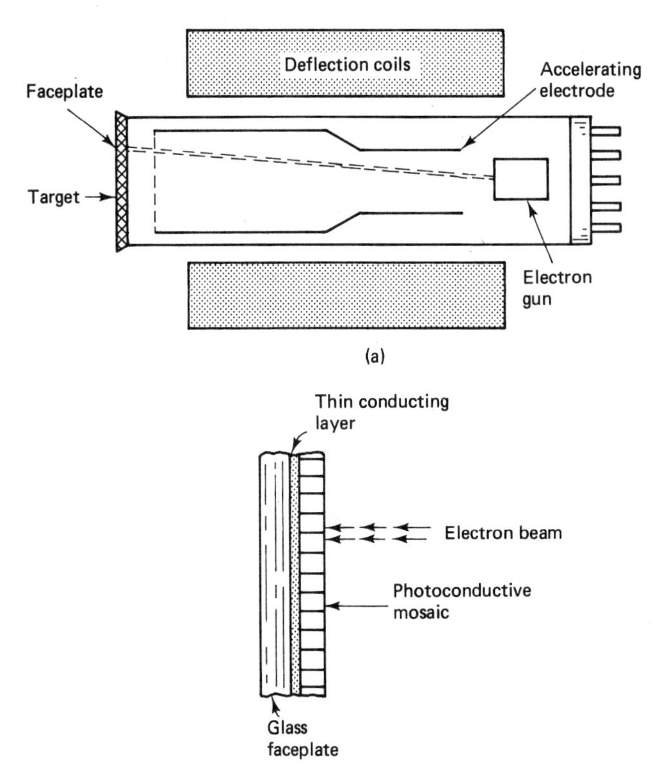

Figure 30-12 Major features of a vidicon camera tube.
(a) Overall arrangement.
(b) Cross section of the target.

external circuitry connected to the conductive layer (the common plate). Hence, because the charge on each cell is determined by the level of illumination reaching the cell, the variations of the discharge current will correspond to the variations of light intensity in the scene focused on the vidicon target. The discharge current represents the video-signal output of the tube.

Clearly, the electron beam in the picture tube of the TV receiver must be synchronized perfectly with the beam in the vidicon tube in the TV camera. Indeed, a considerable fraction of the circuitry of a TV system is devoted to the task of synchronizing the beam of the TV receiver with the beam of the camera tube.

TV transmission system

Once the video signal is obtained from the TV camera, it is used to amplitude-modulate the *video carrier* that is transmitted from the TV station. Simultaneously, the audio is obtained from microphones or other audio sources and is used to frequency modulate a separate *audio carrier*. Thus, a TV station transmitter incorporates both an AM and an FM transmitter; a unit called a *diplexer* permits both transmitters to transmit from the same antenna. Synchronizing pulses for synchronizing the picture tube beam are included as part of the video information. A block diagram of a TV system is given in Fig. 30-13.

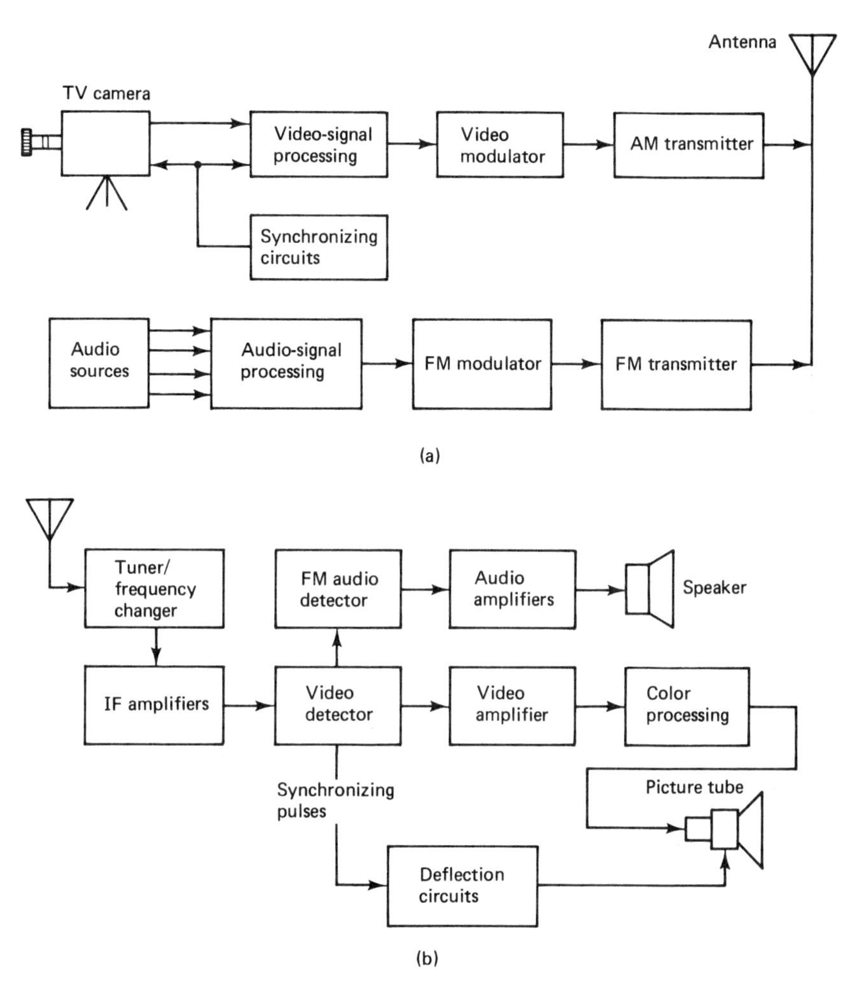

Figure 30-13 Block diagram of: (a) a TV transmitter; (b) a TV receiver.

Color TV

Color TV is based on the principle that white light can be resolved into three primary colors, and any color can be obtained by combining appropriate amounts of three primary colors. The colors adopted as primaries for color TV are red, green and blue.

A color TV camera includes an optical system that resolves the scene into three primary images, a red, a green, and a blue. Each image is applied to a separate vidicon, so red, green, and blue color signals are produced. These signals are multiplexed onto the video carrier in such a manner that black-and-white TV sets are able to receive color transmissions in black and white while color TV sets are able to reproduce the scenes in color.

The picture tube of a color TV is designed to reproduce and combine the red, green, and blue images present at the color camera. Consequently, the screen incorporates red, green, and blue phosphors in a definite pattern so that each color is controlled independently of the others. One type of color picture tube uses three different electron guns (one for each color) arranged so that the beam for each gun strikes phosphors of only one color. Thus, the relative quantities of red, green, and blue produced on the screen can be varied by altering the beam intensity of the respective electron gun. The beams scan the screen in unison.

All in all, the TV system is quite complex and sophisticated. A color television set is one of the most complicated electronic devices we are likely to encounter. It is a tribute to the processes of mass production that such devices can be manufactured at prices affordable by the general population.

30-6 RADAR

Radar is an acronym for *radio detection and ranging*. It provides a means of detecting a target, determining the distance (the range) of the target, and, perhaps, determining the velocity with which the target is moving.

A radar transmitter sends out a short burst of electromagnetic waves, a pulse, in the direction of the target. When the waves strike the target, the waves are reflected and a small portion of the incident energy will be reflected back toward the radar installation, where the radar receiver will detect the echo. By measuring the time interval between transmission of the pulse and reception of the echo, the distance to the target can be determined.

Radar waves travel at the velocity of light, $c = 2.99 \times 10^8$ m/s. In 1 μS a wave will travel 299 m, or about 981 ft. Because the waves must travel to and from the target, about 490 ft of range corresponds to each microsecond of elapsed time between pulse transmission and echo reception.

Doppler radar utilizes the Doppler effect to determine the radial velocity of a moving target (the velocity toward or away from the radar set). An approaching target returns waves of slightly higher frequency; a receding target returns waves of slightly lower frequency. The frequency difference is directly related to the target velocity.

Radar is used extensively in air-traffic control, in particular near airports that handle a large volume of traffic. Radar service is facilitated by the use of radar *transponders* located in the aircraft. The pilot is given a transponder code which is entered into the transponder. The ground-based radar then transmits that code to the airborne transponder, and the transponder transmits encoded information back to the ground-based station. This avoids the dependence upon echos and also makes aircraft identification more reliable.

30-7 ELECTRONIC REPRESENTATION OF NUMBERS

Digital electronics is the study of electric circuits designed to deal with numbers in digital form. Not only has digital electronics made computers possible as we know them today, but it is also widespread in electronics devices that have nothing to do with computers, such as TV sets, citizens' band radios, and industrial controls.

As we consider the available options for representing numbers electronically, a *proportional system*, in which numbers are represented by proportional voltages, comes to mind. For example, the number 5.42 might be represented by 5.42 V on a certain conductor; the number 25.4 might be represented by 25.4 V. Even though this system is feasible in principle (and is used in analog computers), practical difficulties soon become apparent. Precision (and therefore expensive) electronic components are required to yield even moderate accuracy, and voltage levels must be precisely controlled. Consider the precision required to represent the value of π to 6 decimal places: 3.141593. A more convenient system has been developed that uses the *binary number system*.

Only two digits (0 and 1) are used in the binary system. Hence, only two voltage levels need be distinguished on a given conductor to determine if the conductor carries a 0 or a 1. A 1 is usually represented by a *high* voltage and a 0 is represented by a *low* voltage. Further, considerable variation of the high (or low) voltage may be allowed without introducing any ambiguity as to whether the conductor carries a 1 or 0. The voltage levels chosen to represent 0s or 1s in a given system are known as *logic levels*. The logic levels used by a popular family of logic-circuit components [the transistor-transistor logic (TTL) family] are shown in Fig. 30-14. Any voltage greater than 2.0 V is interpreted as a 1; any voltage less than 0.8 V is interpreted as a 0.

Number systems

In addition to the decimal system, three other number systems are used extensively in the literature regarding digital computers. These systems are the *binary*, *octal*, and *hexadecimal* systems, based on the numbers 2, 8, and 16, respectively. Each system uses the concept of *place value*, which we now describe.

In the decimal system (base 10), each digit of a number is the multiplier of a power of 10. For example, the number 349.24 actually represents

$$3 \times 10^2 + 4 \times 10^1 + 9 \times 10^0 + 2 \times 10^{-1} + 4 \times 10^{-2}$$

The position of a digit relative to the decimal point determines the corresponding power of 10:

$$\ldots\ \underline{10^3}\ \underline{10^2}\ \underline{10^1}\ \underline{10^0}\ .\ \underline{10^{-1}}\ \underline{10^{-2}}\ \underline{10^{-3}}\ \ldots$$

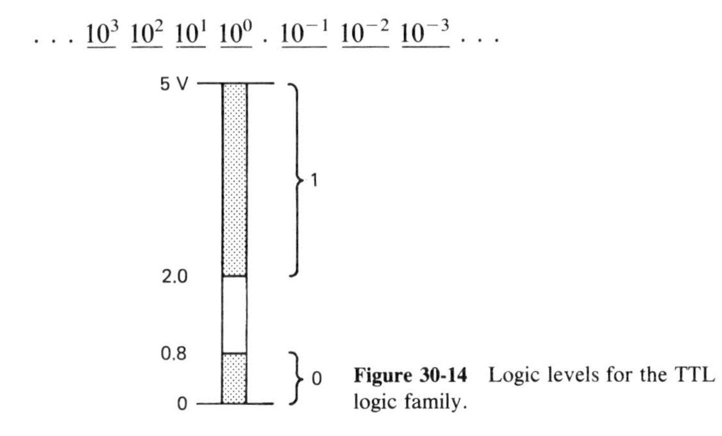

Figure 30-14 Logic levels for the TTL logic family.

This gives rise to units, tens, hundreds, thousands, and so forth, as we are taught at an early age.

In a similar manner, the binary, octal, and hexadecimal systems are based on powers of 2, 8, and 16, respectively:

(*binary*) $\ldots \underline{2^3}\ \underline{2^2}\ \underline{2^1}\ \underline{2^0}\ .\ \underline{2^{-1}}\ \underline{2^{-2}}\ \underline{2^{-3}} \ldots$

(*octal*) $\ldots \underline{8^3}\ \underline{8^2}\ \underline{8^1}\ \underline{8^0}\ .\ \underline{8^{-1}}\ \underline{8^{-2}}\ \underline{8^{-3}} \ldots$

(*hexadecimal*) $\ldots \underline{16^3}\ \underline{16^2}\ \underline{16^1}\ \underline{16^0}\ .\ \underline{16^{-1}}\ \underline{16^{-2}}\ \underline{16^{-3}} \ldots$

The decimal system utilizes ten different digits: 0–9. The binary system uses only two, 0 and 1. The octal system uses eight: 0–7. The hexadecimal system uses 16, as follows:

0 1 2 3 4 5 6 7 8 9 *A* *B* *C* *D* *E* *F*

Letters are used in lieu of introducing special symbols. Thus, in hexadecimal, there are *C* eggs in a dozen and there are *D* cookies in a baker's dozen (13). The first 18 numbers of the four systems are listed in Fig. 30-15.

EXAMPLE 30-1 Find the decimal equivalent of $8EC_{16}$. (The subscript indicates that 8*EC* is a hexadecimal number.)

Solution. In terms of place value,

$$8EC_{16} = 8 \times 16^3 + E \times 16^2 + C \times 16^0$$
$$= 8 \times 4096 + 14 \times 256 + 12 \times 1$$
$$= 36{,}364$$

Thus,

$$8EC_{16} = 36{,}364_{10}$$

Decimal	Binary	Octal	Hexadecimal
0	0	0	0
1	1	1	1
2	10	2	2
3	11	3	3
4	100	4	4
5	101	5	5
6	110	6	6
7	111	7	7
8	1000	10	8
9	1001	11	9
10	1010	12	*A*
11	1011	13	*B*
12	1100	14	*C*
13	1101	15	*D*
14	1110	16	*E*
15	1111	17	*F*
16	10000	20	10
17	10001	21	11

Figure 30-15 The first 18 numbers in the four systems commonly encountered in the field of computers.

Data bus

In a digital system, a single data line (a conductor) can carry only a 0 or a 1. To represent larger numbers, more data lines are required, one wire for each *bit* (binary digit). For example, to represent 31_{10} requires 5 bits; to represent 33_{10} requires 6 bits. Digital systems employ groups of data lines called *buses* to convey large numbers in binary form. Each conductor carries a single bit, and the conductors must be organized according to the place value of the bit. The *least significant bit* (LSB) and the *most significant bit* (MSB) of a bus must be identified, and intermediate bits must not be confused.

A nomenclature has evolved regarding the grouping of bits of a data bus. The smallest unit is a *bit*, a binary digit carried by a single data line. A group of bits, usually four, is called a *nybble*. A *byte* usually consists of 8 bits, which is the equivalent of two nybbles. A *word* consists of 2 or more bytes. Large computers may be organized to handle 32-bit words, each word consisting of four 8-bit bytes. This is illustrated in Fig. 30-16.

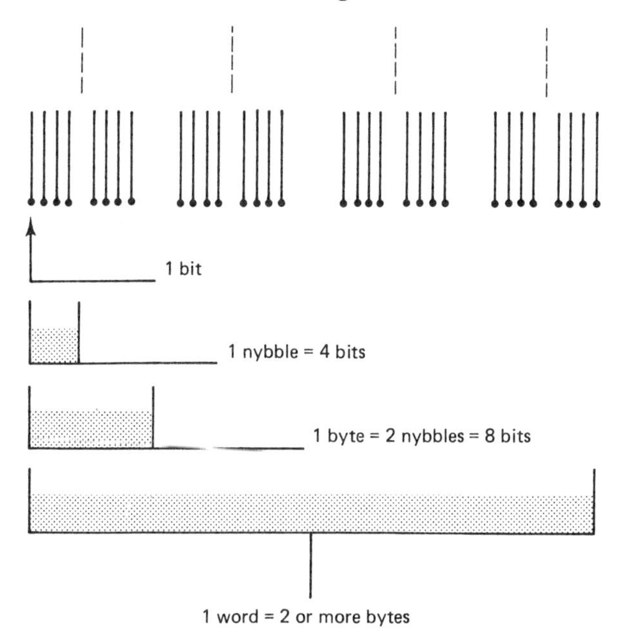

Figure 30-16 The groupings of bits in a data bus.

30-8 LOGIC GATES; BOOLEAN ALGEBRA

The basic building blocks of a digital logic circuit are *logic gates*, of which there are six basic types. The basic gates are AND, OR, INVERTER, NAND, NOR, and EXCLUSIVE-OR (EX-OR). Actually, the NAND, NOR, and EX-OR can be made using AND's, OR's, and INVERTER's, so only three may be considered basic. The function performed by each gate is conveniently shown in a *truth table*, which is a listing of the output x for all possible combinations of inputs. The gates and truth tables are shown in Fig. 30-17.

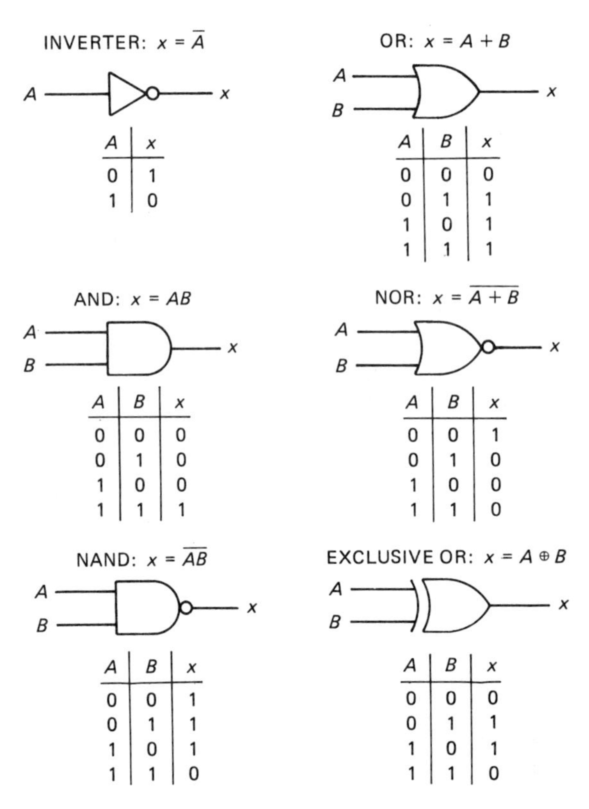

INVERTER:

A	x
0	1
1	0

OR:

A	B	x
0	0	0
0	1	1
1	0	1
1	1	1

AND:

A	B	x
0	0	0
0	1	0
1	0	0
1	1	1

NOR:

A	B	x
0	0	1
0	1	0
1	0	0
1	1	0

NAND:

A	B	x
0	0	1
0	1	1
1	0	1
1	1	0

EXCLUSIVE OR:

A	B	x
0	0	0
0	1	1
1	0	1
1	1	0

Figure 30-17 Basic logic gates.

Because these gates are used only in digital systems, the inputs and outputs will be either 0's (low voltages) or 1's (high voltages). Thus, each input variable can be either a 0 or a 1, and any output is either a 0 or a 1 as determined by the truth table.

The INVERTER is the simplest gate, having only one input. If a 1 is applied to the input, a 0 appears at the output. Conversely, if a 0 is applied to the input, a 1 appears at the output. This may be discerned from the truth table for an INVERTER.

An OR gate produces a 1 at the output whenever a 1 is applied to *either* or *both* inputs. An EX-OR gate produces a 1 at the output when a 1 is applied to *either* input but *not* to *both* inputs.

An AND gate produces a 1 at the output only when 1's are applied to *both* inputs. A NAND gate is the equivalent of an AND followed by an INVERTER, and a NOR gate is the equivalent of an OR followed by an INVERTER.

With the exception of the inverter, all the basic logic gates may have more than two input variables.

Boolean algebra

A system of algebra called *Boolean algebra*, developed by Irish mathematician George Boole in the 1850s, is particularly suited to the design and analysis of logic circuits. Variables may assume values of only 0 or 1, and the basic operations parallel the operation of the basic logic gates. Boolean addition corresponds to the

OR operation, and Boolean multiplication corresponds to the AND operation. As such, there is no Boolean subtraction or division, but another operation called *complementation* corresponds to the action of an INVERTER. If A is a Boolean variable, the complement of A is denoted as $\bar{A}$ (A-*bar*). When $A = 0$, $\bar{A} = 1$, and when $A = 1$, $\bar{A} = 0$. The operations are indicated as shown in Fig. 30-17. (*Note*: $A + B$ is read as A OR B, not as A plus B.)

Many properties of Boolean algebra may be discerned from Fig. 30-18, in which several Boolean identities are listed. Some, such as $1 + 1 = 1$, may appear unusual at first. This is the OR operation; the left-hand side of the equation represents the input side of an OR gate while the right-hand side is the output. To justify a Boolean identity, resort to the truth table for the operation involved.

OR	AND	Inverter
$0 + 0 = 0$	$0 \cdot 0 = 0$	$\bar{0} = 1$
$1 + 0 = 1$	$0 \cdot 1 = 0$	$\bar{1} = 0$
$0 + 1 = 1$	$1 \cdot 0 = 0$	$\bar{\bar{A}} = A$
$1 + 1 = 1$	$1 \cdot 1 = 1$	

Commutative property:

$A \cdot B = B \cdot A \qquad A + B = B + A$

Distributive property:

$A(B + C) = AB + AC$

De Morgan's laws:

$\overline{A + B} = \bar{A} \cdot \bar{B} \qquad \overline{A \cdot B} = \bar{A} + \bar{B}$

Miscellaneous:

$A + AB = A \qquad A + \bar{A}B = A + B$

Figure 30-18 Properties of Boolean algebra.

EXAMPLE 30-2 Evaluate the expression $A\bar{B}(C + 1)$ for the case where $A = 1$, $B = 0$, and $C = 0$.

Solution. We may write

$$x = A\bar{B}(C + 1)$$

Note that when $B = 0$, $\bar{B} = 1$. Substituting:

$$x = 1(1)(0 + 1)$$
$$= 1(1)(1) = 1$$

The TTL logic family

Ready-made logic gates are commercially available in *integrated-circuit* (IC) form, and one chip may contain several identical gates. Moreover, complicated logic circuits containing a hundred or more gates may be constructed in one IC package. In the realm of large-scale integration, a 40-pin IC package may contain thousands of logic gates (a microprocessor chip, for example).

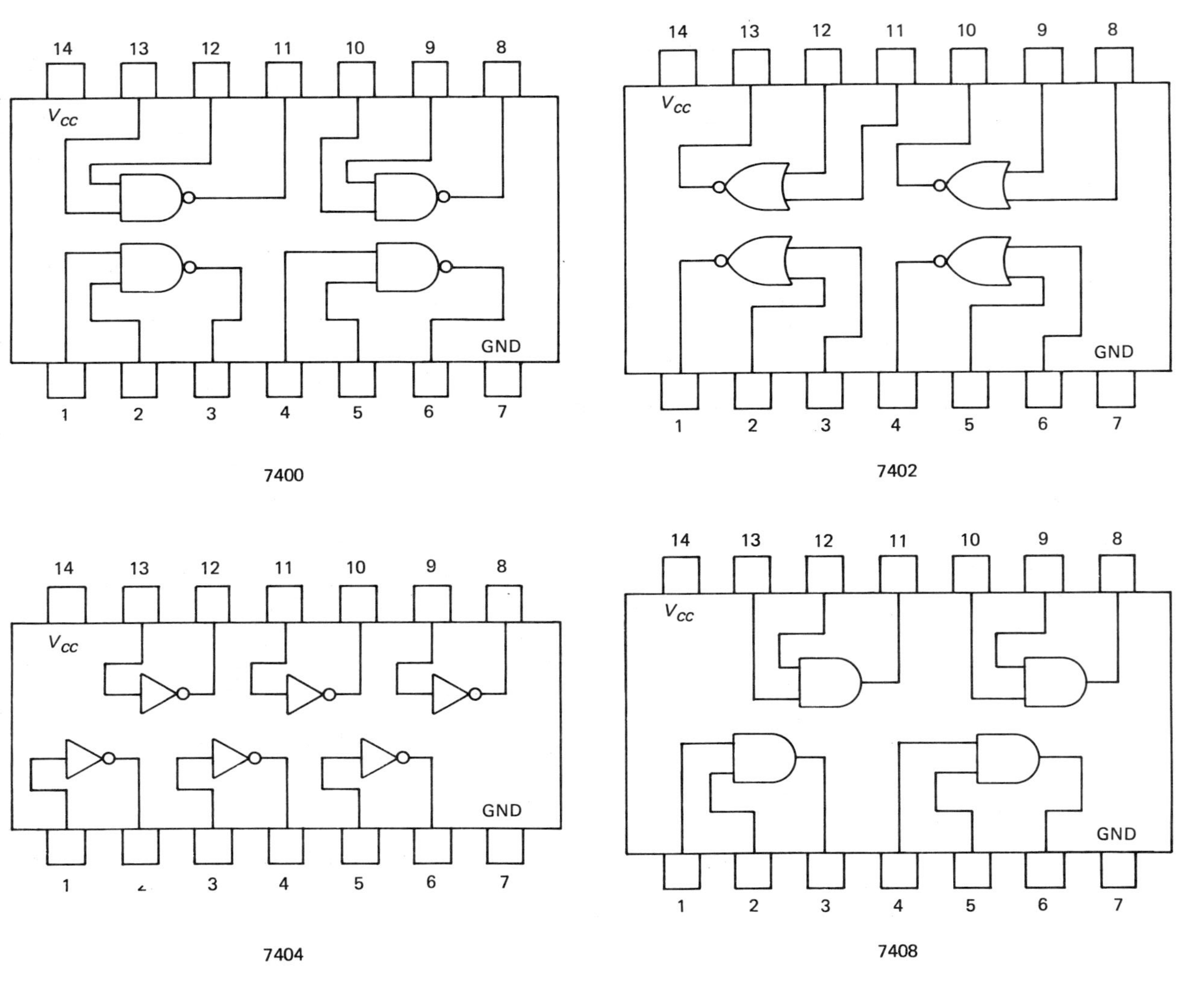

Figure 30-19 Typical 7400-series logic gates.

The TTL family is widely used by professionals and hobbyists alike for implementing digital circuits. The TTL family consists of perhaps 300 different chips containing a wide variety of logic circuits ranging from basic gates to circuits containing the equivalent of a hundred gates or more. The output(s) of one chip may be connected directly to the inputs of another so that complete circuits may be constructed with a minimum of discrete components. TTL chips are identified by a numbering system carrying the 74 prefix; typical examples are the 7400, 7404, 7490, 74193, and so on, some of which are shown in Fig. 30-19. Additionally, there are several TTL subfamilies that bear similar numbers, such as 74L00, 74H00, 74LS00, ... Another logic family, the C-MOS family, carries 7400-series numbers with a C, such as 74C00, 74C193, . . . , even though this is not a TTL family.

An extensive knowledge of electronics fundamentals is not required for a person to experiment with digital circuits. Practical details are provided in the references at the end of this chapter. A circuit for demonstrating the operation of basic gates is shown in Fig. 30-20.

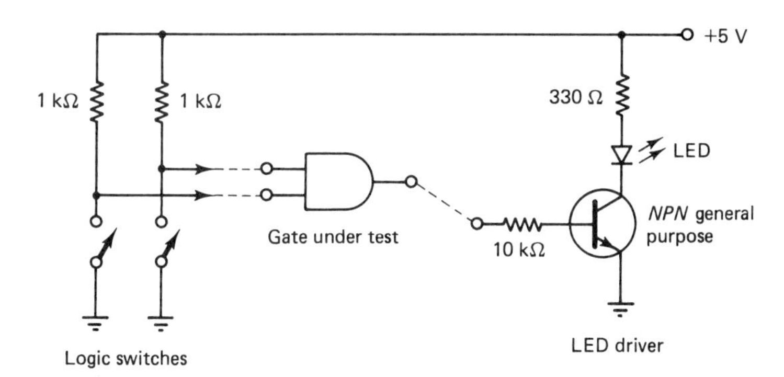

Figure 30-20 A circuit for demonstrating or testing basic TTL logic gates. The logic switches provide the input signals (corresponding to logic 1 or 0) and the LED driver serves as the indicator for the output-logic level.

30-9 COMBINATIONAL LOGIC CIRCUITS

With the aid of Boolean algebra, complicated logic circuits can be expressed in mathematical form. Then, by making use of various Boolean identities, the mathematical representation can be simplified, giving rise to a simpler logic circuit. Several Boolean identities are given in Fig. 30-18.

Two Boolean expressions are equal (or equivalent) if they yield the same truth table. Thus, to verify that a Boolean identity is valid, a truth table should be constructed for each side of the equation. If identical truth tables are obtained, the identity is valid. Alternatively, a Boolean identity may be verified by using other identities to simplify the identity in question to a previously proven identity.

The correspondence between Boolean expressions and logic circuits is demonstrated in Fig. 30-21.

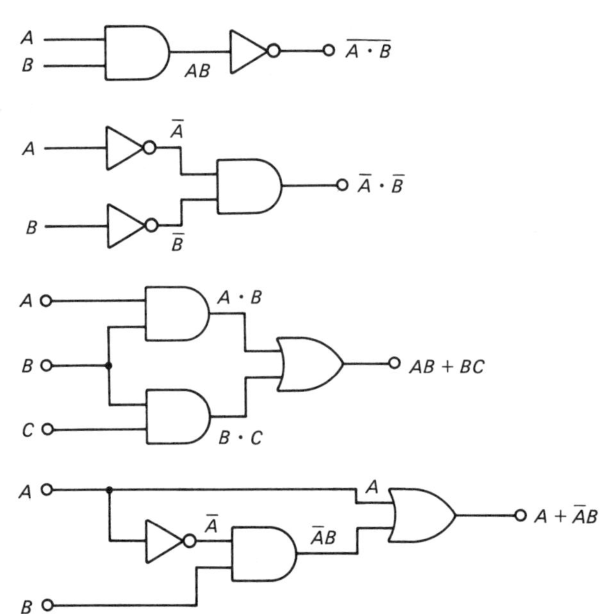

Figure 30-21 Circuits demonstrating the correspondence between Boolean expressions and the logic circuits.

30-10 FLIP-FLOPS AND COUNTERS

A flip-flop (FF) is a one-bit data-storage device that has two distinct states (SET and RESET). A FF has two outputs, typically labeled Q and $\bar{Q}$; one is the complement of the other. When the FF is *set*, $Q = 1$ and $\bar{Q} = 0$. When the FF is *reset*, $Q = 0$ and $\bar{Q} = 1$. When a FF changes states, it is said to *toggle*.

The logic diagram of a typical FF is shown in Fig. 30-22(a). The input lines S, R, and C are the SET, RESET, and CLOCK inputs, respectively. A positive pulse applied to the S input will cause the FF to toggle to the SET state ($Q = 1$; $\bar{Q} = 0$) or to remain in the SET state if it is already set. A positive pulse applied to the R input will cause the FF to toggle to the RESET state ($Q = 0$; $\bar{Q} = 1$) or to remain in the RESET state if it is already reset.

A pulse applied to the CLOCK input causes the FF to toggle—on the *falling edge* of the pulse. That is, each pulse applied to the CLOCK input causes the FF to change states: If it is initially reset, it will become set; if it is initially set, it will become reset. (The rising edge of the clock pulse has no effect.) The operation of the FF is illustrated in the *timing diagram* of Fig. 30-22(b).

Several variations of FF's are commercially available in IC form. They differ primarily in the configuration of the inputs and in whether the FF toggles on the rising or falling edge of the clock pulse. More information is provided in the references at the end of this chapter.

Binary ripple counter

An electronic circuit that can count pulses is shown in Fig. 30-23. It consists of FF's cascaded together so that the output of one provides the input to the next. Each FF represents one binary bit. The counter shown consists of four FF's and

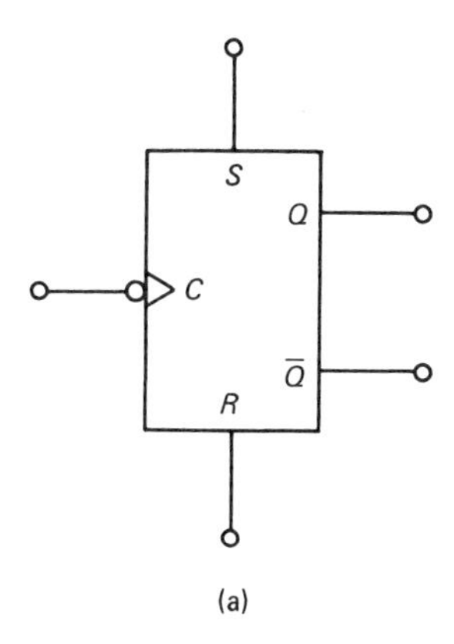

(a)

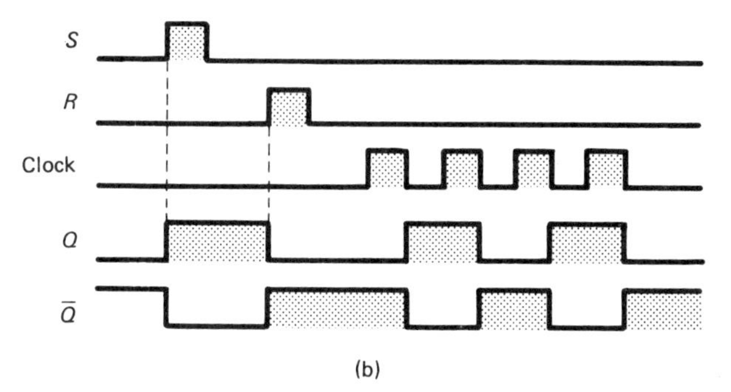

(b)

Figure 30-22 (a) Logic diagram for a flip-flop. S, C, and R are inputs; Q and $\bar{Q}$ are outputs.
(b) Timing diagram showing the response of the FF to S, R, and clock signals.

therefore has 16 (= 2^4) distinct states. More FF's may be added in order to increase the number of states. If N FF's are cascaded, the resulting counter has 2^N distinct states.

The timing diagram of Fig. 30-23 indicates that the period of the output of each FF is twice that of the wave form at the input to the same FF. Hence, each FF acts as a *frequency divider* which divides the signal frequency by two. In music, tones differing in frequency by a factor of two differ in pitch by one octave. Consequently, in an electronic organ or music synthesizer, tones of lower octaves may be obtained by successive division by two of the tones of the top octave.

30-11 ARITHMETIC CIRCUITS

Binary adder

Four possibilities arise when two 1-bit binary numbers are added:

$$\begin{array}{cccc} 0 & 0 & 1 & 1 \\ +0 & +1 & +0 & +1 \\ \hline 0 & 1 & 1 & 1\ 0 \end{array}$$

(Carry) (Sum)

It is easy to construct a digital circuit to perform the addition of two bits, which we

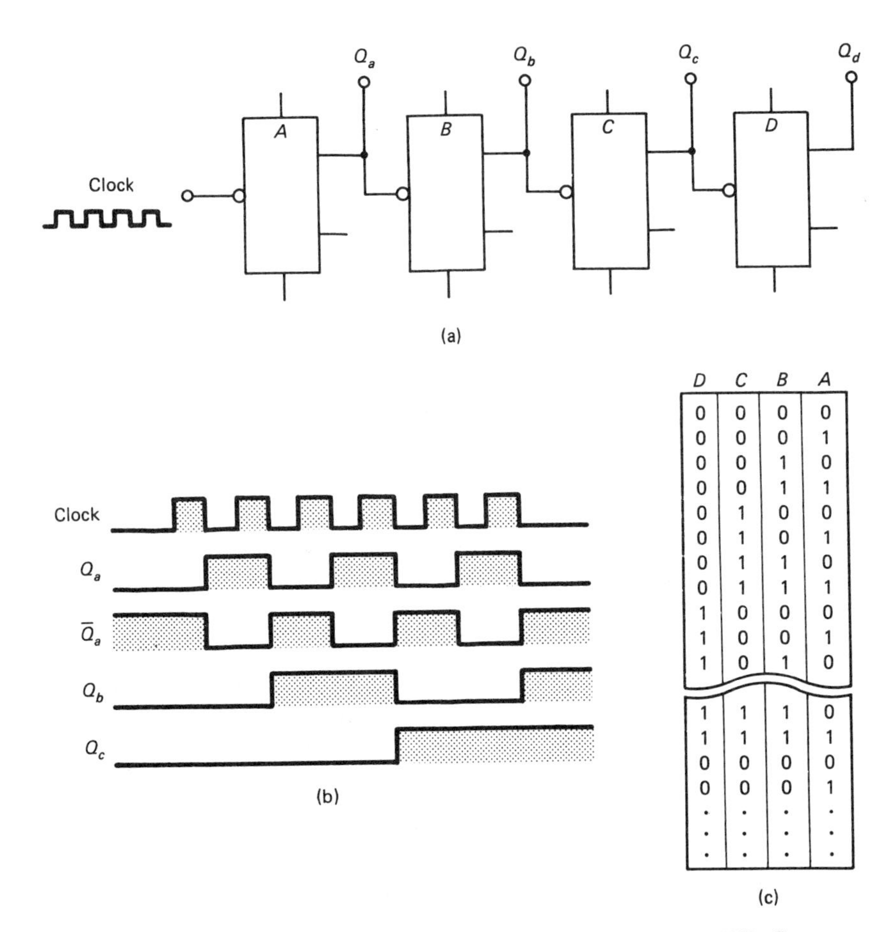

D	C	B	A
0	0	0	0
0	0	0	1
0	0	1	0
0	0	1	1
0	1	0	0
0	1	0	1
0	1	1	0
0	1	1	1
1	0	0	0
1	0	0	1
1	0	1	0
1	1	1	0
1	1	1	1
0	0	0	0
0	0	0	1
⋮	⋮	⋮	⋮

Figure 30-23 (a) Logic diagram of an electronic counter that uses *T* flip-flops. (b) Partial timing diagram. (c) Partial listing of the sequence of states.

may denote by *A* and *B*. In terms of Boolean algebra, the sum and carry are given by

$$\text{Sum} = A \oplus B$$
$$\text{Carry} = A \cdot B$$

The corresponding logic diagram is shown in Fig. 30-24.

When binary numbers of more than one bit are being added, the possibility exists that a carry will be generated in a certain column of bits that must be added into the next higher column. Thus, an adder must be designed that can add three

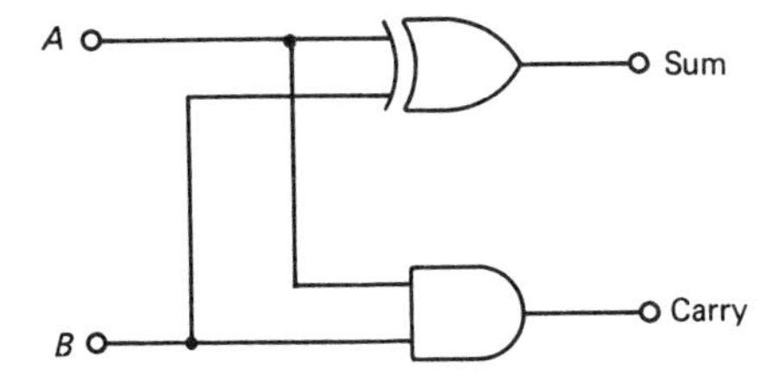

Figure 30-24 Logic diagram of a half-adder.

binary bits. A 3-bit adder is called a *full-adder*; a 2-bit adder is a *half-adder*. A full-adder composed of two half-adders is shown in Fig. 30-25.

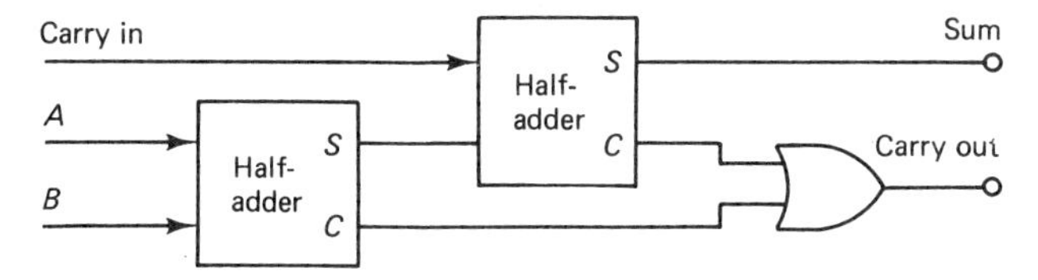

Figure 30-25 A full-adder composed of two half-adders.

Binary subtractor

As for addition, four possibilities arise for binary subtraction:

$$\begin{array}{cccc} & & & \text{(borrow)} \downarrow \\ 0 & 1 & 1 & 10 \\ -0 & -0 & -1 & -1 \\ \hline 0 & 1 & 0 & 1 \leftarrow \text{(difference)} \end{array}$$

Boolean expressions for the *difference* D and the *borrow* B_o are

$$D = A \oplus B$$
$$B_o = \bar{A} \cdot B$$

where it is understood that bit B is being subtracted from bit A. These expressions represent a *half-subtractor*; the corresponding logic diagram is shown in Fig. 30-26.

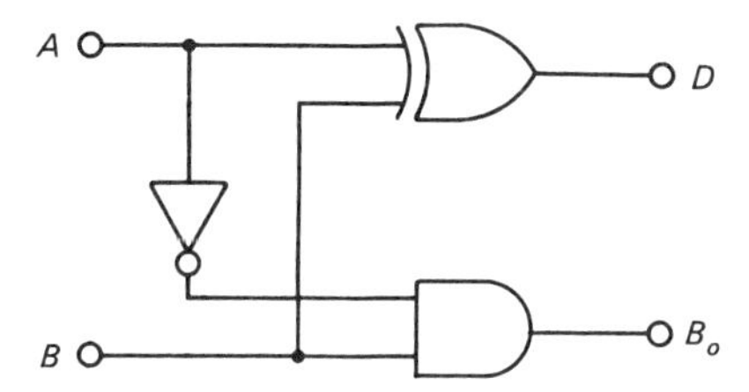

Figure 30-26 Logic diagram of a half-subtractor.

A *full-subtractor* provides an additional input so that borrows from lower-order bit columns can be accommodated. The Boolean expressions for a full-subtractor are

$$D = (A \oplus B) \oplus B_i$$
$$B_o = \bar{A}B + \overline{(A \oplus B)}B_i$$

where B_i and B_o are the input and output borrows, respectively.

Comparator

A *comparator* compares two binary numbers and determines whether the numbers are equal or, alternatively, which of the two is larger. A circuit for detecting the equality of two 4-bit numbers is shown in Fig. 30-27. The output goes

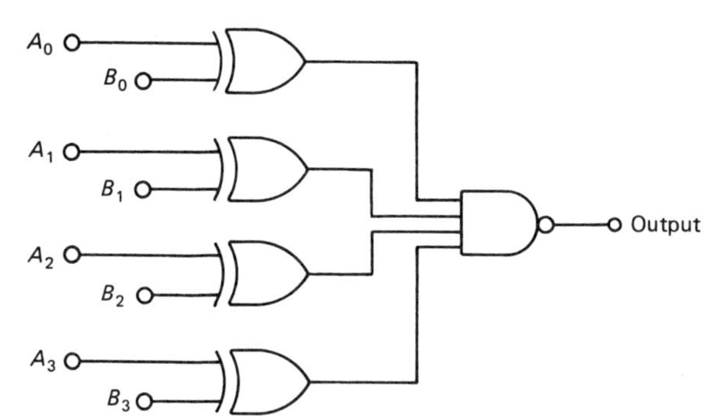

Figure 30-27 A 4-bit comparator for detecting the equality of two numbers.

high when the numbers are equal. Circuitry for detecting $A < B$ and $A > B$ is only slightly more complex and may be found in the references.

30-12 DATA STORAGE; MEMORY

We have seen already that a flip-flop is capable of storing one bit of data, a 1 or a 0. By extension, then, a group of FF's can store a group of bits, and if the FF's are organized according to the significance of the bits, a group of FF's can store a binary number. Such a group of FF's is called a *register*. Registers in digital systems usually have numbers of bits corresponding to the number of bits contained in a nybble, byte, or word as defined for a particular system.

One of the major subsystems of a digital computer is its *memory*, an array of hundreds, thousands, or perhaps millions of addressable storage registers used to store both data and instructions (the computer program). Each register is assigned a number (its address), and each register may be 8 or 16 bits wide, depending upon the size and organization of the memory. Each bit of each register is represented by a flip-flop (or the equivalent), so large memories involve a very large number of storage cells. Computer memories store information only in binary form. Alphabetic characters and special symbols are encoded into binary form prior to storage.

Solid-state memory technology has advanced so dramatically in the past decade that computer memory is now relatively inexpensive and compact. A typical and commonplace memory IC may contain 4096 individually addressable bits in a 16-pin DIP package.

Electronic data storage systems may be divided into two broad classifications according to the manner in which the data is accessible. In a *random-access memory* (RAM), any data byte can be retrieved in the same amount of time as any other; all bytes have the same *access time*. In a *sequentially accessible* memory, the data bytes are stored in a serial or sequential manner. Examples of sequential storage systems are magnetic tape, rotating drum, and floppy disk. In sequential systems, the access time depends upon the location of the data relative to the read head when the request for the data is made. Considerable movement of the tape, drum, or disk may be required in order to access the desired data.

Another classification of computer memory distinguishes between *read/*

write memory and *read-only memory* (ROM). Data are permanently stored in a ROM and cannot be changed once the ROM is *programmed.* On the other hand, data can be stored in and read from a read/write memory by the computer of which the memory is a part. (In current usage, the term RAM implies read/write as well as random-access memory.) A typical computer may utilize both RAM and ROM and also mass storage such as magnetic tape or floppy disks.

30-13 DIGITAL COMPUTERS

We are all well aware of the fantastic things that digital computers can do. On the most basic level, however, the elemental tasks that a computer can perform are very simple and comparatively few. Digital computers operate with binary numbers; all information or data used by a computer must be stored in binary form. Thus, a computer is a manipulator of binary numbers. It can move a binary number from one place to another within the computer or to or from an input-output device such as a printer or video screen; it can compare two binary numbers; it can add or subtract binary numbers (only larger machines have direct provision for multiplication and division); and, a computer can perform the basic logic operations, AND, OR, and EX-OR. Also, most computers have the facilities to examine various bits of a binary number to determine if the bit is a 1 or 0. With a few exceptions, perhaps, these few things are what a computer can do. But these simple operations can be combined in ingenious ways to perform tasks of enormous complexity.

The concept of electrically alterable circuits is fundamental to the design of a digital computer. In Fig. 30-28(a), an EX-OR is used as a *controlled inverter.* When a logic 1 is applied to the control input, the gate acts as an inverter. But when a 0 is applied to the control input, no inversion occurs. In part (b) of the figure, the control line determines whether data from point A or point B will be delivered to point C. In part (c), the control line determines whether the circuit is a half-adder or a half-subtractor. Thus, circuits can be caused to perform different functions by changing the logic levels at various control points.

A computer performs its elemental tasks according to a list of instructions (the *program*) stored in the computer memory in the form of binary numbers. The central processing unit (CPU) *fetches* an instruction from memory to the instruction decoder. The decoder circuitry examines each bit of the instruction and generates corresponding control signals that "set up" the arithmetic logic unit (ALU) to *execute* the instruction. If, for example, the instruction is to add, subtract, or compare two numbers, the ALU is configured as an adder, subtracter, or comparator, respectively. A pulse generator, known as the *master clock*, provides timing pulses that ensure that all actions are properly sequenced. After each instruction is executed, the next instruction in the program is fetched, and the process is repeated until a *halt* instruction is encountered. Typical small computers require only several microseconds to fetch and execute an instruction.

The internal organization of a computer is called its *architecture*, and many variations are found. Nevertheless, most computer architectures have several features in common. The *program counter* is a register that contains the address

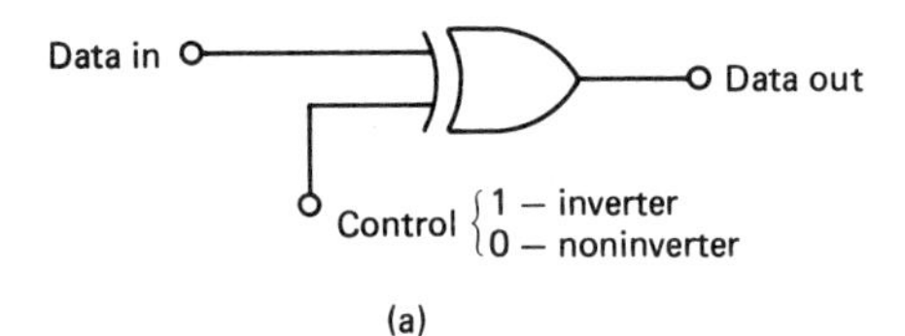

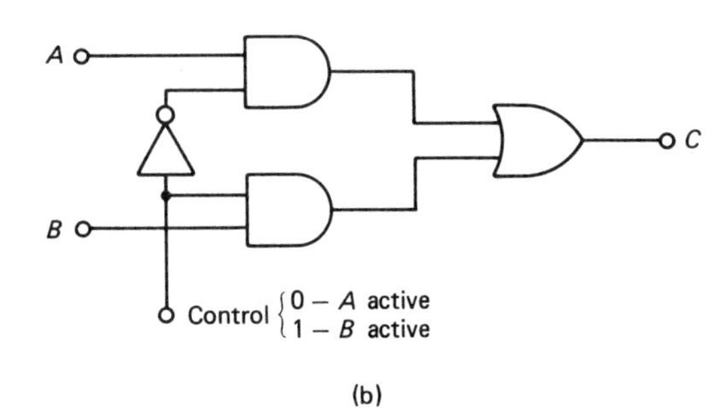

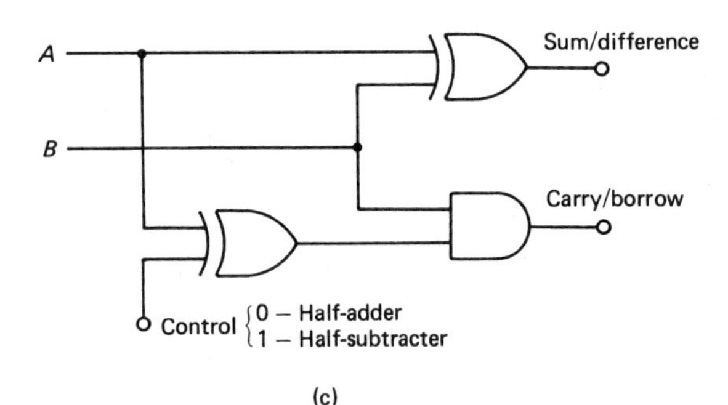

Figure 30-28 Electrically alterable circuits.

of the next instruction to be fetched. It is incremented automatically each time an instruction is fetched. The *accumulator* is a register associated with an instruction, such as *add*. After the operation is performed, the result of the operation is returned to the accumulator. Several *scratch-pad registers* may be included as part of the CPU for temporary storage of data, and *index registers* may be included for added flexibility in addressing data stored in the memory.

A group of *conditional branch* instructions give a digital computer decision-making capability. A branch instruction changes the contents of the program counter so that subsequent instructions are fetched sequentially from another portion of the memory. Conditions upon which branches *are* or *are not* taken include (1) the algebraic sign of the number in the accumulator; (2) whether the number in the accumulator is zero; (3) whether a carry or borrow *flag* is set (by a prior operation); and (4) for many computers, the logic level of special sense lines that connect to external devices. With the ability to manipulate numbers, make decisions, accept data from external devices, and deliver data to output devices, a digital computer becomes a very powerful and useful device.

Clearly, the capabilities of a computer depend upon the ingenuity of the human (computer programmer) who prepares the list of instructions (the program) for the computer to execute. A computer by itself has no innate intelligence; its every move, its ability to make the simplest decision must be thought out and programmed in advance by a human being.

30-14 COMPUTER LANGUAGES

The only language a computer understands is that of binary numbers and logic levels. Consequently, a program written in binary numbers is often called a *machine-language* program. Unfortunately, human beings find binary numbers rather tedious and confusing, so ways to avoid the binary have been developed. An 8-bit binary number is conveniently represented by two hexadecimal digits simply by dividing the 8 bits into two groups of 4 bits each and expressing each 4-bit group as a hexadecimal digit, as shown here:

(*binary*)	$\underbrace{0111}\underbrace{0110}$	$\underbrace{1010}\underbrace{0011}$	$\underbrace{1111}\underbrace{0000}$
(*hexadecimal*)	7 6	*A* 3	*F* 0

Thus, it is common practice to write machine-language programs is hexadecimal form. The hex numbers are electronically converted to binary as they are keyed into a hexadecimal keypad.

Even two-digit hexadecimal numbers are not easily remembered. Accordingly, a system of mnemonics is developed for the *instruction set* of each computer, so each instruction is uniquely represented by a two- or three-letter *mnemonic*. For example, the instruction "branch if the accumulator is not zero" is more easily remembered as BNZ than as 3A or as 00111010. But a computer cannot accept the mnemonics directly; a computer understands only binary. It is necessary to convert the program in mnemonics to a machine-language program in binary form before delivering it to the computer for execution.

This in itself would be a tedious task, but computers themselves are ideally suited to performing such chores. Fairly advanced computer programs, called *assemblers*, are used to generate the binary equivalent of the program written in mnemonics. The programmer writes the *source program* in *assembly language* (in mnemonics) and the assembler converts it to the *object program* (machine-language program) in binary form.

The state of the art is now such that several high-level computer languages have been developed that completely relieve the programmer of any concern for binary code or even the architectural details of the computer being programmed. Examples are FORTRAN, BASIC, ALGOL, PL-1, COBOL, and RPG. These languages are standardized codes that programmers use to communicate with the corresponding *compiler*. The compiler is a computer program that converts the source program (written, for instance, in FORTRAN) to a binary form specifically designed for the computer at hand. High-level languages are very powerful in that quite extensive machine-level programs can be generated with only a minimal expenditure of energy by the programmer.

In most computer systems, the compiler is stored permanently in the computer memory, and the object program generated by the compiler is placed directly into the memory so that the programmer sees neither. Thus, attention can be devoted entirely to the source program.

References

DUNLAP, ORRIN E. JR., *Marconi: The Man and his Wireless*. New York: Arno Press and the *New York Times*, 1971.

GREEN, CLARENCE R. and ROBERT M. BOURQUE, *The Theory and Servicing of AM, FM, and FM Stereo Receivers*. Englewood Cliffs, N. J.: Prentice-Hall, Inc., 1980.

LIFF, ALVIN A., *Color and Black & White Television Theory and Servicing*. Englewood Cliffs, N. J.: Prentice-Hall, Inc., 1979.

SHRADER, ROBERT L., *Electronic Communication*. 4th ed. New York: Gregg Division, McGraw-Hill Book Company, 1980.

TOCCI, RONALD J., *Digital Systems Principles and Applications*. Englewood Cliffs, N. J.: Prentice-Hall, Inc., 1979.

Questions

1. How is it possible for current to flow in a dipole antenna even though the ends of the antenna are open?
2. What electronic components are used to tune the antenna circuit of a receiver of electromagnetic waves?
3. Briefly explain the concept of amplitude modulation. What is the difference between modulation and demodulation?
4. Briefly describe the system of frequency modulation.
5. Give two reasons why the commercial FM band is capable of higher-fidelity transmissions than the AM broadcast band.
6. Describe the major components of a TV picture tube.
7. What phenomenon is effectively avoided by the use of an interlaced scanning system?
8. Why is it necessary to have synchronization circuits in a TV receiver?
9. Why does a TV station transmit two carriers?
10. Briefly describe the principle on which radar is based.
11. What is the purpose of the high-frequency bias signal that is added to an audio signal being recorded on magnetic tape?
12. What is the primary advantage of binary numbers for representing numbers in electronic systems?
13. In Boolean algebra, what are the only two values a Boolean variable may assume?
14. What function is performed by a comparator?
15. What is the difference between a random-access memory and a sequentially accessed memory?
16. Explain how an EXCLUSIVE-OR gate can be used as a controlled inverter.
17. Explain five elemental tasks that a computer can perform.
18. Describe a machine-language program. Why do humans not like machine-language programs very much?
19. Describe and tell the function performed by an assembler and by a compiler.
20. What is the primary objective or advantage of a high-level computer langauge?

Problems

1. How much time is required for a radio wave to travel from the earth to the moon (a distance of about 238,000 miles)?
2. A typical distance from Earth to Saturn is 0.8 billion miles. Compute the time required for a radio wave to travel from the earth to Saturn.
3. Compute the wavelength in meters of
   (a) an AM broadcast band radio wave of frequency 1 MHz

   (b) an FM broadcast band wave of frequency 100 MHz
   (c) microwaves of frequency 1 GHz (1 GHz = 10^9 Hz)

4. Radio astronomers frequently receive radio waves of 3-cm wavelength. Calculate the frequency of these waves.
5. Two signals having frequencies of 140 kHz and 145 kHz are applied to the inputs of a mixer. What is the frequency of the beat note produced?
6. Signals of frequency 1100 kHz and 1555 kHz are applied to the inputs of a mixer. What frequencies appear at the output of the mixer?
7. In the United States television system, a frame consists of 525 horizontal lines. Thirty frames are transmitted per second.
   (a) How many horizontal lines are transmitted per second?
   (b) What is the total time (trace and retrace) required for the transmission of one horizontal line?
8. A radar echo is received 118 μs after the pulse is transmitted. What is the range of the object that returned the echo?
9. At a time when the planet Venus is 30 million miles from the earth, how much time is required for a radar pulse to travel from Earth to Venus and back?
10. Write the decimal number 37 in the binary, octal, and hexadecimal systems.
11. Determine the decimal equivalent of the following binary numbers:
    (a) 1011
    (b) 100.01
    (c) 11111111.
12. Determine the decimal equivalent of the following octal numbers:
    (a) 765
    (b) 76.5
    (c) 0.001
13. Determine the decimal equivalent of the following hexadecimal numbers:
    (a) 97
    (b) FE
    (c) 1A.A
    (d) C.0A
14. Write the decimal 1000 in binary, octal, and hexadecimal.
15. Given that $A = 1, B = 1$, and $C = 1$, evaluate the following Boolean expressions:
    (a) $x = AB + C$ (b) $x = \overline{AB} + \overline{C}$
    (c) $x = \overline{AB + C}$ (d) $x = \overline{AB\overline{C}}$
    (e) $x = \overline{A + B} + \overline{C}$ (f) $x = \overline{AB} + \overline{\overline{B}\,\overline{C}}$
16. Evaluate the preceding expressions for $A = 0, B = 1$, and $C = 1$.
17. Draw the logic diagrams corresponding to each Boolean expression in problem 15.
18. Simplify the following Boolean expressions:
    (a) $x = AB + A\bar{B}$ (b) $x = A + \bar{A}B$
    (c) $x = ABC + AB\bar{C}$ (d) $x = AB(A + BC)$
19. A binary ripple counter consists of six flip-flops in cascade.
    (a) How many distinct states does the counter have?
    (b) What is the largest number (decimal equivalent) the counter can hold?

CHAPTER 31

FUNDAMENTALS OF ATOMIC PHYSICS

An atom is an extremely small system—so small, in fact, that its components and structure cannot be observed directly. Thus, it is not immediately apparent exactly what an atom is or how it is put together. The determination of the structure of the atom represents one of the noteworthy accomplishments of the first decades of the twentieth century.

As it turned out, a whole new physical theory had to be developed in order to describe the behavior of the orbital electrons in an atom. The classical theories failed to account for the observed properties of atoms. The new theory, quantum mechanics, has been able to account for all observed properties of the electronic structure of atoms, and it therefore represents a major intellectual achievement.

Quantum mechanics presented physicists with at least two radical concepts. One is that there is no clear dividing line between particles and waves; an electron, for example, seems somehow to be both. The other is that in quantum mechanics we must deal with probabilities and statistical manifestations rather than with the hard, firm predictions of classical physics.

In this chapter we present the fundamental concepts of atomic physics and of quantum mechanics. Obviously, one chapter can only give a slight indication of what the subject is about, but we have tried to capture the key ideas on which the subject is based.

31-1 EARLY MODELS OF THE ATOM

In 1897, Sir J. J. Thomson (1856–1940) determined the ratio of charge to mass, e/m, for electrons, which then were known as "cathode rays" or "cathode corpuscles." Thus, the electron was identified as a negatively charged particle, and evidence suggested that electrons were a constituent of atoms. Similar experiments for "positive rays" led to the proton as being the massive, positive component of a hydrogen atom even though the nature of the nucleus was unknown at the time.

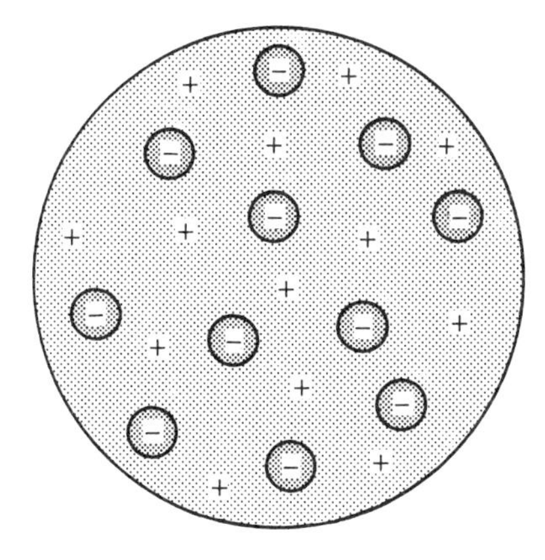

Figure 31-1 The plum-pudding model of the atom envisioned the electrons being embedded in a heavy sphere of positive charge.

With atoms believed to be composed of both positive and negative charges—in equal amounts since atoms are electrically neutral—it was only natural to speculate about the structure of an atom. Thomson proposed the first model of an atom, the so-called plum-pudding model. This model pictured the atom as a heavy sphere of positive charge in which was embedded enough electrons (like plums in a pudding) to make it electrically neutral. A plum-pudding atom is shown in Fig. 31-1.

In 1911, Ernest Rutherford (1871–1937) performed an experiment to test the plum-pudding model. A collimated beam of alpha particles emitted from a small quantity of radium was directed at a thin gold foil. It was reasoned that collisions between the alpha particles (helium nuclei) and the atoms of the foil would cause the incident beam to be scattered as it passed through the foil. The beam was detected by a scintillation technique. A small screen of zinc sulfide was viewed through a low-power microscope. When an alpha particle hit the screen, a tiny flash of light was produced. By counting the flashes, the number of alpha particles hitting the screen could be determined. The experimental arrangement is shown in Fig. 31-2. By rotating the arm carrying the foil and alpha source, scattering angles could be investigated ranging from a few degrees to very nearly 180°.

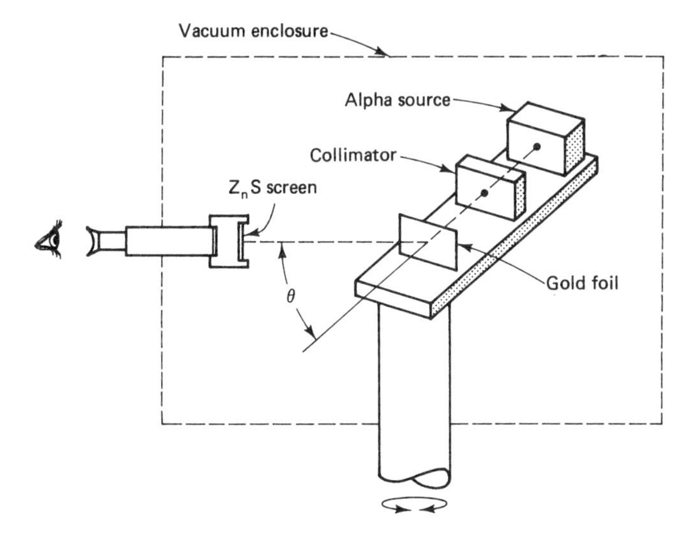

Figure 31-2 Experimental arrangement used for the Rutherford scattering experiment. Alpha particles were scattered through the angle by the nuclei of the metal foil.

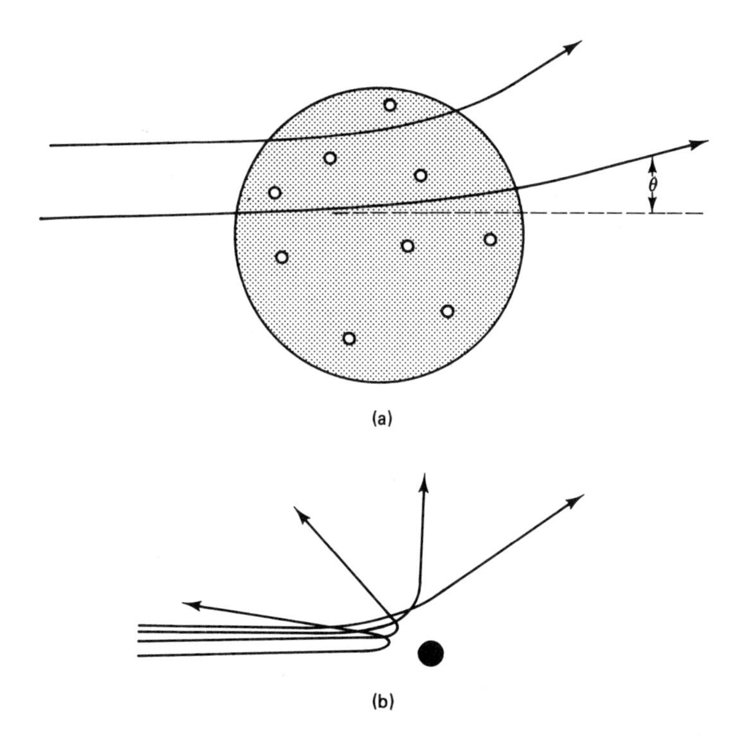

Figure 31-3 The plum-pudding model is not expected to produce large scattering angles, as shown in (a). In (b) however, where the scattering center is very small, large angles are expected.

Figure 31-3(a) indicates that plum-pudding atoms should not be capable of deflecting the comparatively massive positively charged alpha particles through large angles. Because electrons are very light, their effect can be ignored; only the heavier positive charge should produce significant deflections. Furthermore, only a very small deflection should result when an alpha particle passes through the sphere of positive charge.

Data from the experiment showed that a large portion of the alpha particles either passed straight through the foil or were scattered by very small amounts. On the other hand, a few were scattered through very large angles, even approaching 180°. The plum-pudding model could not account for such large scattering angles.

In an effort to explain his data, Rutherford proposed a planetary model for an atom. He assumed the positive, massive portion of an atom to be concentrated in a very small region at the atom's center. The electrons were imagined to orbit the central core in orbits whose diameters were very large in comparison to the diameter of the *nucleus* (as it is now called). This model accounts for the experimental data as follows.

First of all, the atom is mostly empty space and, as before, the electrons are unable to produce significant scattering. Thus, most of the alpha particles pass through the foil without being scattered through a significant angle. Occasionally, however, a particle will pass through the electron cloud and approach the nucleus very closely. Electrostatic forces of repulsion become very great between the two positive particles, so the path of the alpha particle is changed greatly. In a perfect head-on collision, the alpha particle reverses its direction completely. Thus, the nuclear or planetary model of Rutherford can explain the result of the scattering experiment.

While the Rutherford model made no pretense of being complete, difficulties with the model quickly became apparent. Electromagnetic theory predicted that an electron in a planetlike orbit should radiate energy as light, and that a continuous spectrum should be produced. Moreover, the orbits should decay as energy is lost so that the electrons should spiral into the nucleus in a very brief time, about 10^{-11} s. These effects are in direct contradiction to experience. First, an atom does not emit radiation unless it is first excited, and when an atom does emit, a line spectrum rather than a continuous spectrum is produced. Secondly, atoms are stable; if there were not, we would not exist. An improved atomic model was put forth by Neils Bohr in 1913 which, in part, resolved the difficulties of the Rutherford model. The Bohr atom is described in a following section.

The Balmer formula

In 1884, J. J. Balmer, a Swiss high-school mathematics teacher, discovered a simple formula that would give the wavelengths of the four visible lines of the hydrogen spectrum. Other investigators generalized the *Balmer formula* so that it could be applied to the other hydrogen series (Lyman, Paschen, and so on). The generalized formula may be written as

$$\frac{1}{\lambda} = (0.01097)\left(\frac{1}{m^2} - \frac{1}{n^2}\right) \tag{31-1}$$

$$n = m + 1, m + 2, m + 3, \ldots$$

where m is the number of the orbit to which an electron makes a transition from a higher orbit of number n. We have taken the liberty to do a unit conversion so that λ is given in nanometers. To obtain the Balmer series, set $m = 2$ and let n take on the values 3, 4, 5, ... You may verify that the first four lines of the Balmer series have wavelengths of 656, 486, 434, and 410 nm. To obtain the Lyman series, set $m = 1$ and let n take on the values 2, 3, 4, ... (Refer to Sect. 20-3 for a description of the hydrogen spectrum.)

Even though the Balmer formula was obtained empirically, it played an important role in the development of atomic theory. It essentially gave Niels Bohr the final result in the problem we now call the Bohr atom.

31-2 THE BOHR ATOM

In 1913, the Danish physicist Niels Bohr (1885–1962) succeeded in deriving the Balmer formula as a product of his model of a hydrogen atom. Bohr assumed that a hydrogen atom consists of a heavy positive nucleus (a proton) about which revolves a negative electron in a circular orbit. The electrostatic attraction between the proton and electron provides the centripetal force required to keep the electron in its orbit. Additionally, he made three important assumptions:

1. An electron can exist in certain allowed orbits without radiating energy (which would cause the orbit to decay).
2. Allowed orbits are those for which the angular momentum of the electron is an integral multiple of $h/2\pi$ (h is Planck's constant).
3. An electron may make a transition from one allowed orbit to another by emitting or absorbing a quantum of energy in the form of a photon.

Proceeding with a mathematical treatment, we equate the centripetal force of an electron of mass m and charge e in an orbit of radius R to the electrostatic force of attraction between a charge e and another charge Ze separated by a distance R:

$$\frac{mv^2}{R} = \frac{Ze^2}{4\pi\epsilon_0 R^2} \tag{31-2}$$

[Refer to Eqs. (24-1) and (24-5).] Solving for v^2 gives

$$v^2 = \frac{Ze^2}{4\pi\epsilon_0 mR} \tag{31-3}$$

We now consider the quantization of the angular momentum, mvR, and write, by virtue of assumption (2):

$$mvR = \frac{nh}{2\pi} \tag{31-4}$$

where n is an integer: $n = 1, 2, 3, \ldots$.

Solving for v and squaring gives

$$v^2 = \frac{n^2h^2}{(2\pi)^2 m^2 R^2} \tag{31-5}$$

This expression for v^2 may be set equal to Eq. (31-3) and subsequently solved for R. The result is

$$R = \frac{n^2h^2\epsilon_0}{\pi m e^2 Z} \tag{31-6}$$

where

$$h = 6.62 \times 10^{-34}\ \text{J} \cdot \text{s}$$

$$\epsilon_0 = 8.85 \times 10^{-12}\ \text{C}^2/\text{N} \cdot \text{m}^2$$

$$m = 9.11 \times 10^{-31}\ \text{kg}$$

$$e = 1.60 \times 10^{-19}\ \text{C}$$

$$Z = 1\ \text{(for hydrogen)}$$

By substituting Eq. (31-6) into Eq. (31-4) to eliminate R, the following expression for v is obtained:

$$v = \frac{Ze^2}{2\epsilon_0 nh} \tag{31-7}$$

Thus, we have expressions for the radius of each Bohr orbit and for the velocity v of the electron in the orbit. You may verify that for $n = 1$, $R = 5.29 \times 10^{-11}$ m and $v = 2.18 \times 10^6$ m/s.

Energy of the orbit

The total energy of an electron in a particular orbit is the sum of the electron's kinetic energy and the electrostatic potential energy of the electron-proton combination. The kinetic energy is $\frac{1}{2}mv^2$, where Eq. (31-7) is substituted for v. The electrostatic potential energy is given by

$$PE = -\frac{Ze^2}{4\pi\epsilon_0 R} \tag{31-8}$$

obtained from Eqs. (24-11) and (24-12) [plus Eq. (24-5)]. It is negative because the charges are of opposite sign. Equation (31-6) may be substituted in order to eliminate R. The final result, after a bit of algebra, is

$$E_n = -\frac{Z^2e^4m}{8\epsilon_0^2\, n^2\, h^2} \tag{31-9}$$

This can be written as

$$E_n = -\frac{B}{n^2} \tag{31-10}$$

if we define B (for convenience) as

$$B = \frac{Z^2e^4m}{8\epsilon_0^2\, h^2} \tag{31-11}$$

You may verify that the value of B is 2.179×10^{-18} J, which is the equivalent of 13.6 eV because 1 eV equals 1.6×10^{-19} J.

Electron transitions

According to the third assumption of Bohr, a photon is emitted when an electron makes a transition from a higher to a lower orbit. If E is the difference in energy between the orbits, we may write

$$E = E_i - E_f \tag{31-12}$$

where the subscripts i and f denote initial and final orbits. Substitution of Eq. (31-10) gives

$$E = -\frac{B}{n_i^2} + \frac{B}{n_f^2} = B\left(\frac{1}{n_f^2} - \frac{1}{n_i^2}\right) \tag{31-13}$$

Recalling that a photon's energy and frequency are related by $E = hf$, which may be written as $E = hc/\lambda$, we obtain

$$\frac{hc}{\lambda} = B\left(\frac{1}{n_f^2} - \frac{1}{n_i^2}\right) \quad \text{or} \quad \frac{1}{\lambda} = \frac{B}{hc}\left(\frac{1}{n_f^2} - \frac{1}{n_i^2}\right) \tag{31-14}$$

The constant B/hc has the value $1.097 \times 10^7\ \text{m}^{-1}$. If λ is expressed in nanometers, the constant becomes $0.01097\ \text{nm}^{-1}$ so that (with a change in notation: $n_f = m$ and $n_i = n$):

$$\frac{1}{\lambda} = (0.01097)\left(\frac{1}{m^2} - \frac{1}{n^2}\right) \tag{31-15}$$

This is identical to the Balmer formula [Eq. (31-1)], which was obtained empirically almost 30 years earlier.

The Lyman series of hydrogen is produced by electrons falling from higher orbits to the lowest orbit. Hence, $m = 1$ in Eq. (31-15), and $n = 2, 3, 4, \ldots$. The Lyman series limit is obtained when $n = \infty$.

Even though the Bohr theory was able to predict the hydrogen spectrum with a high degree of accuracy, it was not able to account for *all* aspects of the hydrogen spectrum. High-resolution spectroscopy revealed that many of the hydrogen lines consisted of two or more separate lines located very close together. The Bohr theory could not explain this *fine structure* of the lines, nor could it account for the relative intensity of the different spectral lines. Furthermore, the Bohr theory could only treat one-electron atoms such as hydrogen, singly ionized helium, doubly ionized lithium, and so on. In short, the Bohr model was not a general solution to the problem of atomic structure. Nevertheless, the Bohr model represented a major step forward in the understanding of the atom, and it set the stage for the more comprehensive quantum theory.

31-3 WAVES AND PARTICLES

The photoelectric effect

We now briefly describe an experiment that verifies the energy-frequency relationship for a photon, $E = hf$. Further, this experiment provides a means of determining the value of h (Planck's constant) and it provides the first indication that light may exhibit properties both of waves and of particles.

The photoelectric effect, in broad terms, refers to the fact that electrons are ejected from a metal when light of sufficiently high frequency strikes the metal. An experimental arrangement for investigating this phenomenon is indicated in Fig. 31-4. Two electrodes, an emitting and a collecting electrode, are sealed in an evacuated bulb provided with a window through which light is beamed at the emitting electrode. Electrons are ejected from the emitting electrode when it is irradiated with light, and these electrons are collected by the collecting electrode. A low-voltage power supply (about 10 V) applies a small positive (or negative) voltage to the collecting electrode, and a sensitive ammeter measures the electron

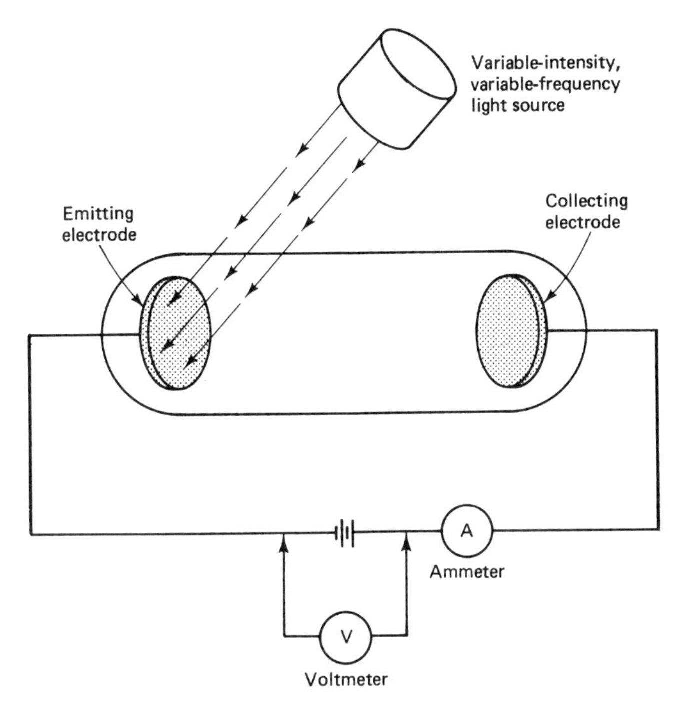

Figure 31-4 Schematic representation of the apparatus for performing the photoelectric-effect experiment.

current flowing from the collecting electrode. The light source is variable both in intensity and wavelength (or frequency). Several aspects of the photoemission process can be investigated by performing the experiment in different ways.

1. For light of a given wavelength and with a positive voltage applied to the collector, the photocurrent varies in direct proportion to the intensity of the light, as one would expect. Intense light produces more photoelectrons than dim light.
2. By making the collector slightly negative relative to the emitter, it is possible to reduce the photocurrent to zero because the electrons must overcome the negative potential in order to reach the collector. This makes it possible to determine the maximum kinetic energy, KE_{max}, with which electrons are ejected from the emitting electrode. It is found that KE_{max} depends upon the wavelength of the light and upon the material of which the target is made, but not upon the intensity of the light.
3. With the collector slightly positive, and with the intensity of the light held constant, photoemission ceases when the frequency of the light falls below a certain *threshold frequency*. For frequencies below the threshold, neither an increase in light intensity nor an increase in collector voltage causes a resumption of photoemission.
4. Finally, measurement of KE_{max} when *extremely* low intensity light is incident upon the emitter reveals that the value of KE_{max} is the same as for greater intensities. Further, there is essentially no time delay between the time the light is applied and the time of ejection of the electrons.

Interpretation of the photoelectric effect

The results of (2), (3), and (4) cannot be explained if light is considered to be a classical electromagnetic wave. For example, in (2) we would expect KE_{max} to depend upon the *intensity* rather than the *frequency* because the energy content of a classical wave is proportional to its intensity. In (3), even electromagnetic waves of very low frequency should be capable of ejecting electrons, provided the wave intensity is sufficiently great. In (4), we would expect the electrons ejected by low-intensity light to have far less KE_{max} than electrons ejected by intense beams. Also, we would expect a time delay between application of the dim light and ejection of the electrons, the delay resulting from a gradual buildup of the electron energies. Thus, the photoelectric effect presents formidable problems for the classical concept of light as an electromagnetic wave.

In 1905 Albert Einstein satisfactorily explained the photoelectric effect by extending a proposal made by Max Planck around 1900. Planck assumed that the "oscillators" that were responsible for emitting radiation from a heated object could exist only in states for which the energy was an integral multiple of hf. Einstein postulated that a beam of light consists of small bundles (*quanta*) of energy, which we now call *photons*, and that the energy of a quantum of light is given by hf. This was a radical concept.

Once the concept of a *photon* is at hand, we can explain the photoelectric effect in simple terms. First, a particular metal is characterized by its *work function*, W, which is the minimum energy required to remove an electron from its surface. Secondly, when a photon transfers its energy to an electron, the electron gets all the photon's energy, and the photon disappears. If the photon energy exceeds the work function, W, the electron will be ejected from the metal with the excess energy appearing as kinetic energy of the electron. Thus, Einstein's photoelectric equation is:

$$KE_{max} = hf - W \tag{31-16}$$

The appropriate kinetic energy is the *maximum* because an ejected electron may have less KE due to collisions that occur after its absorption of the photon.

It is now clear that when hf is less than W, the electron will not be ejected. This accounts for the existence of a threshold frequency f_0, given by

$$f_0 = \frac{W}{h} \tag{31-17}$$

obtained by setting $KE_{max} = 0$ in the equation above.

With an irradiating beam of very low intensity, only a few photons hit the emitter electrode per second. Consequently, only a few electrons are ejected, but each ejected electron receives the full quota of energy from the photon with which it interacts. Thus, KE_{max} depends upon f but not upon the intensity.

The fact that an incident photon reacts with only a single electron indicates that the energy of a photon is concentrated at a point (the location of the electron) rather than to be distributed over a large area. A classical magnetic wave would tend to spread out over the entire surface of the emitter. This aspect of a photon is more characteristic of a *particle* than of a *wave*. Thus, we must conclude that light has the characteristics of particles and of waves.

A photon can exist only while traveling at the speed of light. If it slows down, it ceases to exist. Hence, we say its *rest mass* is zero. But a moving photon has mass because of its energy. This is in accordance with Einstein's famous result from the special theory of relativity:

$$E = mc^2 \tag{31-18}$$

The energy of a photon is hf, and we may write

$$hf = m_p c^2 \tag{31-19}$$

and solve for the photon mass m_p:

$$m_p = \frac{hf}{c^2} \quad \text{or} \quad m_p = \frac{h}{\lambda c} \tag{31-20}$$

We may continue this line of reasoning to arrive at an expression for the momentum of a photon. The momentum is the product of the mass and the velocity (as for a classical particle). Hence, the photon momentum is given by

$$p = m_p c = \left(\frac{h}{\lambda c}\right) c = \frac{h}{\lambda} \tag{31-21}$$

At this point, we have obtained the mass and momentum of a photon, two parameters normally associated with particles. We now turn to the opposite situation in which particles such as electrons, protons, and so forth are observed to have certain characteristics of waves.

Matter waves

In 1924, Louis de Broglie made a bold proposition that material particles, even large particles such as a grain of sand, have the characteristics of waves, namely a frequency and a wavelength and the capability of producing the effects of interference and diffraction. As radical as such a proposal may seem, it is in fact true. Electrons, neutrons, and even atoms have been diffracted, and diffraction is definitely a wave phenomenon.

The frequency and wavelength of a particle are given by expressions analogous to those for a photon:

$$f = \frac{E}{h} \quad \text{and} \quad \lambda = \frac{h}{p} = \frac{h}{mv} \tag{31-22}$$

When an electron is accelerated through a potential difference (voltage) V, the electrostatic potential energy, eV, is converted to kinetic energy, $\frac{1}{2}mv^2$. Setting the two equal and solving for v gives

$$v = \sqrt{\frac{2eV}{m}} = \sqrt{\frac{2e}{m}} \cdot \sqrt{V} \tag{31-23}$$
$$v = (5.93 \times 10^5) \sqrt{V}$$

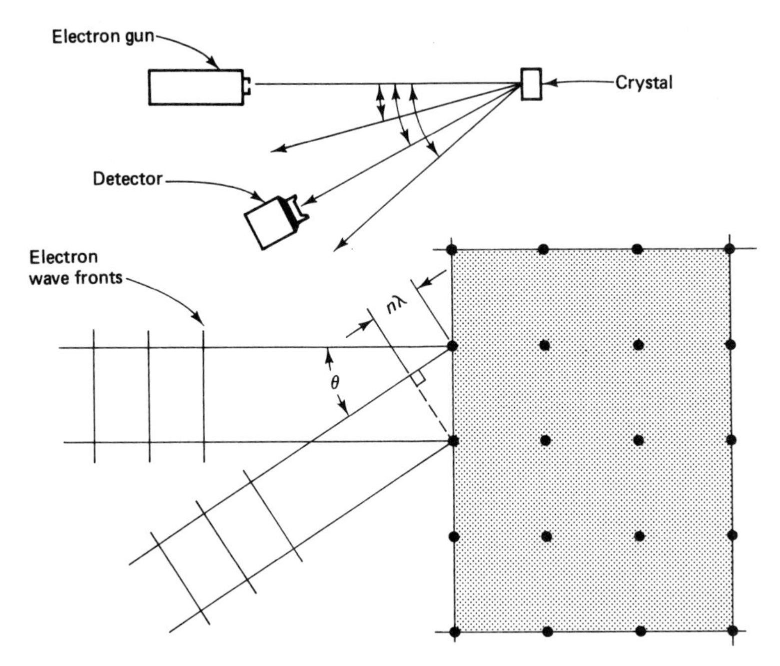

Figure 31-5 Electron diffraction by a crystal demonstrates the wave properties of electrons.

where V is in volts and v is the velocity in m/s. Substituting this into Eq. (31-22) gives the following formula for the wavelength of electrons (in nanometers) that have been accelerated through a potential of V volts:

$$\lambda = \frac{1.23}{\sqrt{V}} \text{ nm} \tag{31-24}$$

Thus, electrons that have been accelerated through a potential of 1 V have a wavelength of 1.23 nm. In contrast, the visible Balmer line of highest frequency has a wavelength of 410 nm.

Electron diffraction

The wavelength of electrons accelerated even through moderate potentials is far too short to be diffracted by even the finest optical diffraction grating. However, the atomic planes of a large single crystal serve the same function as the rulings of a grating, so electron diffraction may be demonstrated by firing a beam of electrons at a crystal, as shown in Fig. 31-5. Electrons exit the crystal along preferential angles rather than in a continuous distribution and thereby verify that diffraction has occurred. Diffraction of X-rays (which are photons) and of electrons can be done in such a manner so as to produce similar diffraction patterns. This confirms the wavelike properties of the electron.

Wave-particle duality

The dual character of material particles and of photons is now accepted as a fundamental aspect of nature. Consequently, there is no simple answer to: Is it a particle or is it a wave? Someone proposed the concept of "wavicles," partly in jest, but with a hint of reason. Our imaginations balk at the proposition of constructing a mental image of something that is both a particle and a wave.

As it turns out, a wavicle (particle or photon) will never manifest both types of behavior at the same time in a given situation. Either wave properties (electron diffraction) or particle properties (photoelectric effect) will dominate, depending upon the physical conditions. Thus, the electron beam in a TV picture tube acts as a stream of particles, and the electron beam of an electron microscope acts as a wave. Only one aspect—waves or particles—is exhibited at a time.

31-4 QUANTUM MECHANICS

It is now well established that subatomic particles such as electrons and protons do not obey the classical laws of mechanics as set forth by Newton. Put differently, Newtonian mechanics does not have the capability of dealing with electrons in orbit about the nucleus of an atom, mainly because Newtonian mechanics is addressed to the motions of particles, and we have now seen that small particles are as much waves as particles. Consequently, a new system of mechanics is called for that takes into account the wave-particle duality inherent in nature. A highly successful theory that does this is known as *quantum mechanics*.

There is no sharp dividing line between the realm of quantum mechanics and Newtonian mechanics. In the region of overlap, where the wave properties of a particle are less significant, quantum mechanics and Newtonian mechanics give very nearly the same results. For example, an electron in a high-numbered atomic orbital (large n) may be described reasonably well by Newtonian mechanics. The *correspondence principle* tells us that in the limit, the predictions of quantum mechanics and classical mechanics will correspond.

Quantum mechanics is a wave theory. We no longer picture an electron as a particle, a tiny speck of concentrated matter moving in a definite orbit about the nucleus. Rather, we speak of the *quantum states* that electrons in atoms may occupy, and these states are described in mathematical terms by *wave functions*. The wave function for a particular state embodies all the information that can be known about an electron in that state. For example, the energy and angular momentum are characteristics of quantum states and are determined using mathematical operations upon the wave functions.

The wave functions are obtained as solutions to the *Schrödinger wave equation*, which has the general form

$$(\mathbf{K} + \mathbf{V})\psi = \mathbf{E}\psi \qquad (31\text{-}25)$$

where **K**, **V**, and **E** are the kinetic, potential and total energies of the system under consideration, and where ψ is the wave function. This equation is not an ordinary equation; the factors **K**, **V**, and **E** represent differential *operators* that operate on ψ so that the Schrödinger equation is a wave equation. Solutions to the equation, the wave functions, describe the de Broglie waves of the particle.

Interpretation of the wave functions is not immediately obvious and has been the subject of much discussion. The accepted interpretation is that the wave function is a *probability amplitude*, and the *square* of the probability amplitude represents the *probability* that a particle will be at a certain location. (Recall that in classical wave theory, the intensity is the square of the amplitude.) For example, the wave function for an electron in the ground state of the hydrogen atom has a maximum amplitude at a radial distance corresponding to the radius of the first Bohr orbit, 0.53 Å. Hence, the electron is most likely to be found at that distance from the nucleus. But there is a significant probability that the electron (in the ground state) will be found at other distances from the nucleus. The *probability density* is the probability per unit volume that a particle will be found in the region of interest. A plot of the probability density for the ground state of hydrogen is shown in Fig. 31-6.

A characteristic of waves is that they are spread out in space, whereas a particle is localized. Because the probability waves are not localized, QM does not predict that electrons in atoms travel in distinct and definite orbits about the nucleus. Quantum mechanics tells us nothing about the path taken by an electron as it orbits the nucleus (or even if it does indeed orbit the nucleus). It only provides the probability that an electron will be found at certain locations. This gives rise to the mental image of a cloud of probability density surrounding the nucleus; where the cloud is most dense, the electron is most likely to be found. Thus, the planetary model of atoms must be modified. We can no longer envision an atom as a miniature solar system with the electrons traveling in distinct and definite orbits.

The ideas of quantum mechanics, the concept of probability amplitudes, and so forth, are radical, to say the least. However, experimental data from a wide variety of experiments support the theory. Quantum mechanics provides a unified theory for dealing with the wave and particle characteristics of matter.

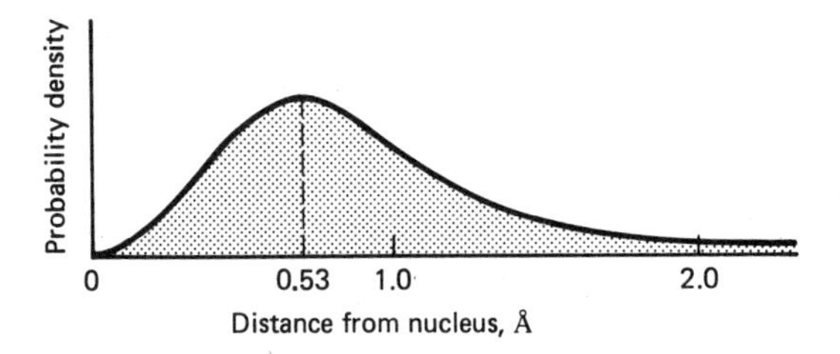

Figure 31-6 Graph of the probability density as a function of distance from the nucleus for the ground state of hydrogen.

31-5 THE HEISENBERG UNCERTAINTY PRINCIPLE

Any measurement is subject to error or uncertainty due to the limitations of the instruments used to make the measurement. But QM predicts that even with ideal instruments there is a limit to the accuracy or precision with which measurements can be made. This limitation is inherent in nature and arises from two considerations: One is the wave-particle duality of matter and the other is the unavoidable interaction between the measuring instrument and the thing being measured.

Suppose we suspect an electron to be located at a certain point in space.

How could we ascertain that it is indeed there? Obviously, by looking at it. But this implies an interaction between at least one photon and the electron; the photon must collide with the electron in order for us to "see" it. The photon will impart momentum (velocity) to the electron so that when we do see it, we are not seeing where it is, but where it was. In determining its position, we create an uncertainty in its velocity.

This situation is similar to that of a blindfolded person in a room trying to locate a balloon by probing with a meterstick. The balloon moves when the searcher hits it with the stick.

In order to reduce the effect of the photon, we may try a photon of lower energy. But photons of lower energy have longer wavelengths, so we suffer a loss of optical resolution. We reduce the uncertainty in the electron's velocity, but we incur a loss of accuracy in its position. Conversely, if we use a photon of higher energy (and shorter wavelength) to reduce the uncertainty in its position, we increase the uncertainty in its velocity.

In 1927 Werner Heisenberg (1901–1976) stated the *Heisenberg uncertainty principle*: It is impossible simultaneously to determine the exact position and velocity of a particle. In mathematical terms, the product of the uncertainty in position (Δx) and the uncertainty in momentum (Δp) will always be equal to or greater than $h/2\pi$:

$$(\Delta x)(\Delta p) \geqslant \frac{h}{2\pi} \tag{31-26}$$

Because $\Delta p = m(\Delta v)$, we can write (31-26) as

$$(\Delta x)(\Delta v) \geqslant \frac{h}{2\pi m} \tag{31-27}$$

This implies that the uncertainties for particles of larger mass are smaller than for particles of small mass. For macroscopic particles, the uncertainties are negligible.

Another form of the uncertainty relationship exists for energy and time. The relationship is

$$(\Delta E)(\Delta t) \geqslant \frac{h}{2\pi} \tag{31-28}$$

This means, essentially, that an energy level cannot be determined exactly unless an infinite amount of time is taken for the determination. When an electron is excited to a higher level in an atom, it remains in the higher level only about 10^{-8} s before giving up its energy and falling back to the ground state. This gives rise to an uncertainty in energy of the high state of

$$\begin{aligned}\Delta E &= \frac{h}{2\pi(\Delta t)} = \frac{6.62 \times 10^{-34}\ \text{J}\cdot\text{s}}{2\pi(10^{-8}\ \text{s})} \\ &= 1.06 \times 10^{-27}\ \text{J}\end{aligned} \tag{31-29}$$

This translates to an uncertainty in frequency of the emitted photon given by

$$\begin{aligned}\Delta f &= \frac{\Delta E}{h} \\ &= \frac{1.06 \times 10^{-27}\ \text{J}}{6.62 \times 10^{-34}\ \text{J} \cdot \text{s}} = 1.59 \times 10^{7}\ \text{Hz}\end{aligned} \tag{31-30}$$

This is quite small in comparison with the frequency of visible light (about 10^{14} Hz) so that spectral lines are very sharp even though their width is easily observed. This mechanism of *spectral line broadening* is called *natural broadening*. Another type, *Doppler broadening*, is caused by the motion of atoms toward or away from the observer; it is an atomic manifestation of the Doppler frequency shift described for sound in an earlier chapter.

In classical mechanics, it is possible, in principle, to determine exactly the instantaneous position and velocity of a particle. Then, if the forces acting on the particle are known, its position and velocity at all later times can be calculated. We find, however, that QM does not provide such determinacy, such definiteness. The Heisenberg uncertainty principle forbids exact knowledge of initial conditions, and the foundation stones of quantum mechanics are probability amplitudes, the wave functions. The probabilistic character of QM has been widely debated but is now generally accepted as reflecting an inherent aspect of nature.

31-6 QUANTUM NUMBERS

Even though the wavefunctions for an atom are mathematically rather complex, many qualitative aspects can be described in terms of four quantum numbers. These quantum numbers may be grouped according to certain rules, and each combination represents a particular quantum state that an electron in an atom may occupy. The principle quantum number n is the same as in the Bohr theory of the hydrogen atom. The values of n range from 1 to ∞.

The orbital quantum number, l, specifies the angular momentum of the electron and can take on values ranging from 0 to $n - 1$. The magnitude of the angular momentum is given by

$$L = \sqrt{l(l+1)}\left(\frac{h}{2\pi}\right) \qquad l = 0, 1, 2, \ldots, (n-1) \tag{31-31}$$

The orbital quantum number l is related to the shape of the cloud of probability density corresponding to the state. In atoms having more than one electron, the energy depends upon l as well as n.

For convenience, values of l are usually denoted by the letters s, p, d, f, g, h, . . . , which correspond to values of l of 0, 1, 2, 3, 4, ... The choice of the first four of these letters stems from historical considerations, namely, the sharp, principle, diffuse, and fundamental series in the realm of spectroscopy.

The magnetic quantum number m_l relates to the projection of the angular momentum onto an axis parallel to the direction of an externally applied magnetic

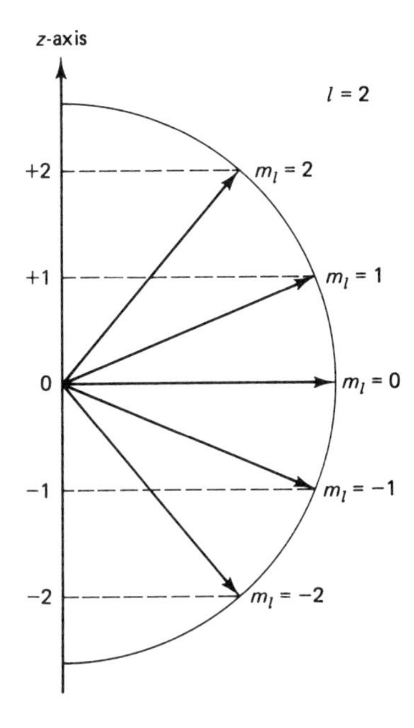

Figure 31-7 Spatial quantization of angular momentum. An externally applied magnetic field is assumed parallel to the z-axis.

field. The moving electron produces a magnetic moment directed parallel to the angular momentum, and the orientation of the magnetic moment relative to the external field is important in determining the total energy of the state. Allowed values of m_l range from $-l$ to $+l$ in integral steps, including zero. Thus, for a given l, there are $2l + 1$ possible values for m_l. The spatial quantization of angular momentum is illustrated in Fig. 31-7.

The *Zeeman effect* stems directly from the spatial quantization of angular momentum (and the magnetic moment). It refers to the splitting of spectral lines into two, three, or more closely spaced components when the source is placed in a strong magnetic field. With no external field applied, the orientation of the angular momentum is arbitrary because nothing is present that will distinguish one direction from any other. When the field is turned on, the angular momentum vectors of the myriad of atoms of the source distribute themselves among the possible states according to the allowed values of m_l. The states have slightly different energies, and electron transitions from the states to lower levels produce spectral lines of slightly different frequencies. This is illustrated in Fig. 31-8.

The remaining quantum number, m_s, is called the *spin quantum number*, and it describes the spin magnetic moment of the electron. Values of m_s are $+\frac{1}{2}$ and $-\frac{1}{2}$. The classical view of electron spin pictures the electron spinning on its axis as

TABLE 31-1 QUANTUM NUMBERS AND PERMITTED VALUES FOR AN ELECTRON IN AN ATOM

Name	Symbol	Permitted Values
Principle	n	$1, 2, 3, 4, \ldots$
Orbital	l	$0, 1, 2, \ldots, n-1$
Magnetic	m_l	$-l \ldots -1, 0, +1, \ldots, +l$
Spin	m_s	$-\frac{1}{2}, +\frac{1}{2}$

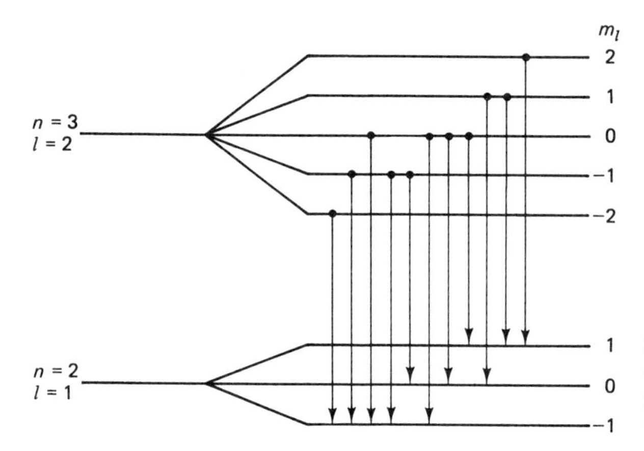

Figure 31-8 An external magnetic field causes energy levels to split. This gives rise to many more spectral lines than would otherwise occur.

it revolves about the nucleus in a manner similar to the spinning earth in orbit about the sun. This is not a legitimate analogy, however, because the electron is not a localized object capable of spinning. It is a wavicle! The spin is an inherent property of the electron. The four quantum numbers are summarized in Table 31-1.

Figure 31-9 illustrates the possible states for principle quantum number $n = 3$. Vertical separations on such a diagram represent differences in energy. (No attempt has been made to draw the vertical separations to scale.) It is seen that 18 different states are possible for $n = 3$.

Because the hydrogen atom has only one electron, only one state at a time will be occupied. The many other states described by the allowed combinations of quantum numbers represent states into which the electron may be excited from the ground state.

Atoms having many electrons have a corresponding number of states occupied. In many-electron atoms, electrons fill the states of lowest energy first. The *Pauli exclusion principle* states that in a particular atom (or other system), no two electrons can have the same set of quantum numbers. That is, only one

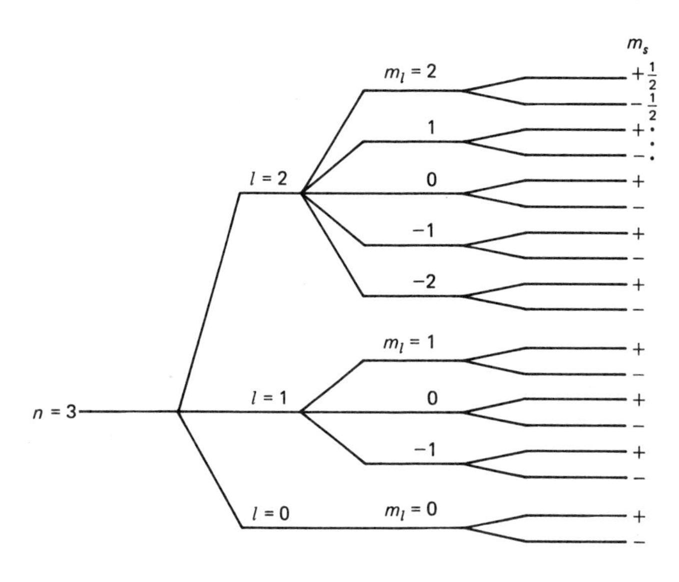

Figure 31-9 Quantum states for principle quantum number $n = 3$.

electron is allowed per quantum state. Thus, a maximum of two electrons can exist in the first shell ($n = 1$); a maximum of eight electrons can exist in the second shell ($n = 2$); and a maximum of 18 electrons can exist in the third shell ($n = 3$).

Symbolic representation of atomic quantum states

It is convenient to represent atomic quantum states in the form nx^a, where n is the principle quantum number, x is a letter corresponding to the orbital quantum number (s, p, d, f, g, h, . . .), and a is the number of electrons in the group of states. For example, $2s^2$ represents the states for $n = 2$ and $l = 0$. The superscript 2 signifies that two electrons are in the $2s$ states. This notation does not distinguish the different states corresponding to the possible values of m_l and m_s, so the notation $2p$ actually represents six different states. The notation $2p^4$ signifies that 4 electrons are in the p subshell corresponding to the principle quantum number $n = 2$. The state for which $n = 4$ and $l = 3$ is designated by $4d$.

Electronic structure of complex atoms

We can now conveniently depict the electronic structure of many-electron atoms. The structure of oxygen, for example, is

$$1s^2 2s^2 2p^4$$

which indicates that two electrons occupy the subshell $n = 1$ and $l = 0$, two electrons occupy the $n = 2$, $l = 0$ subshell, and four electrons occupy the $n = 2$, $l = 1$ subshell.

A mnemonic diagram is shown in Fig. 31-10 for determining the order in which electrons fill the subshells. Each subshell is filled to a maximum before any

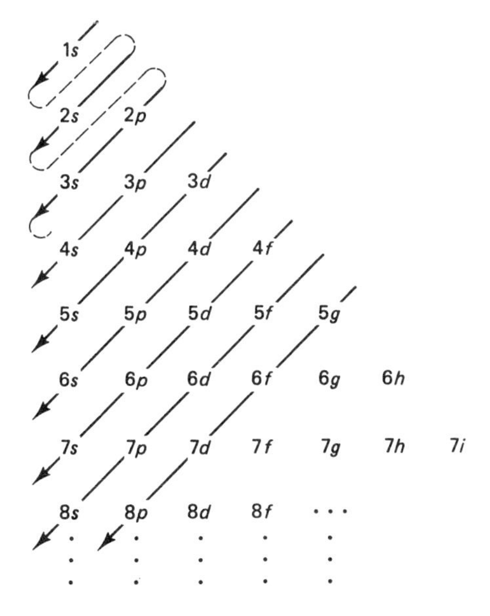

Figure 31-10 A mnemonic diagram for the ordering of energy levels in multi-electron atoms.

electrons occupy the next higher subshell. Note that the $4s$ subshell is filled before the $3d$ subshell, so energy-ordering does not strictly follow the sequence of the principle quantum number n. The maximum occupancy of the subshells is given by $2(2l + 1)$; hence, for an f subshell ($l = 3$) the maximum occupancy is $2[2(3) + 1] = 14$.

As an example, iron has 26 electrons. The electronic configuration is therefore

$$1s^2 2s^2 2p^6 3s^2 3p^6 3d^6 4s^2$$

Note that the s subshell is occupied even though the $3d$ subshell is not completely filled.

Complex optical spectra

Each quantum state is characterized by an energy level, although more than one quantum state may have the same energy (*degeneracy*). The energy levels may be plotted on an energy-level diagram, as shown in Fig. 31-11 for sodium. In light of the large number of energy levels, it is not surprising that the line spectrum of a many-electron atom contains a large number of lines, with each line correspond-

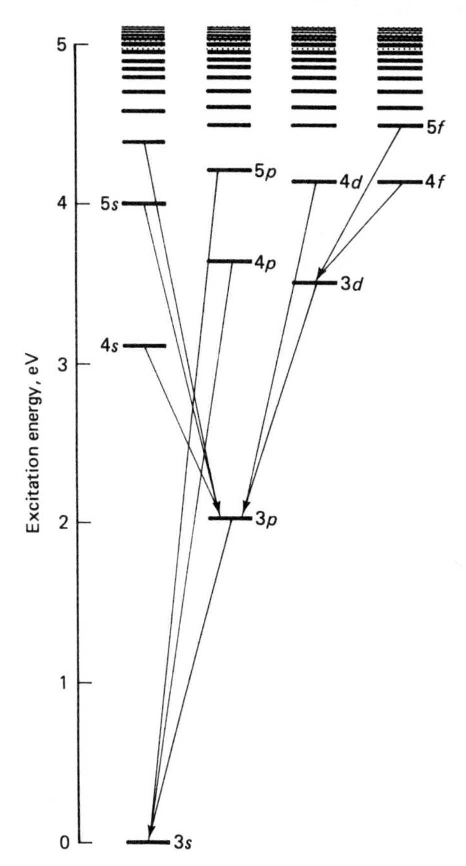

Figure 31-11 A portion of the energy-level diagram for sodium. A few electron transitions are shown.

ing to an electron transition from one level to another. However, fewer lines appear in the spectrum than one might at first expect because selection rules do not allow certain transitions. The only transitions that can occur are those for which the orbital quantum number changes by +1 or −1 and the change in the magnetic quantum number m_l is +1, −1, or 0. That is,

$$\begin{aligned} \Delta l &= \pm 1 \\ \Delta m_l &= 0, \pm 1 \end{aligned} \qquad \text{(selection rules)}$$

No restrictions are placed on the principle quantum number n or upon the spin quantum number m_s.

The existence of selection rules arises from the conservation of angular momentum. When an atom emits (or absorbs) a photon, the photon carries angular momentum away from the atom, so the orbital quantum number l must change accordingly in order to conserve the angular momentum.

Not all lines of the line spectrum of an element are equally intense. This means that certain electron transitions are more probable than others. Quantum mechanics provides a means for computing the relative intensities of the spectral lines. In short, quantum mechanics can account for all known features of atomic structure. This provides compelling evidence in support of the nonclassical ideas upon which the theory is based.

31-7 X-RAYS

X-rays are high-energy photons having wavelengths in the range from about 0.1 to 100 Å. This corresponds to photon energies ranging from about 10^2 to 10^5 eV. In comparison, the photon energies of visible light range from about 2 to 4 eV. The penetrating power of X-rays is well known, as is the use of X-rays in medicine.

X-rays may be produced by bombarding a metal target with high-energy electrons. A typical X-ray tube is shown in Fig. 31-12. A high voltage is established between the cathode and target, which accelerates the electrons to very high velocities. X-rays are produced when the electrons strike the target, and the X-rays easily pass through the glass to the outside world. Typical target materials are rhenium, tungsten, and molybdenum. Because most electrons

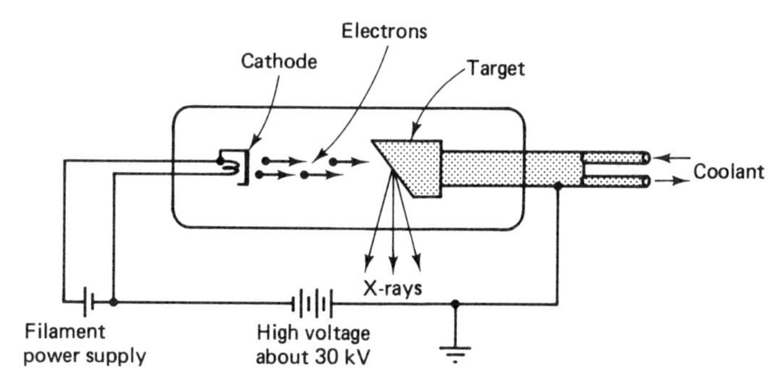

Figure 31-12 A schematic representation of a typical X-ray tube.

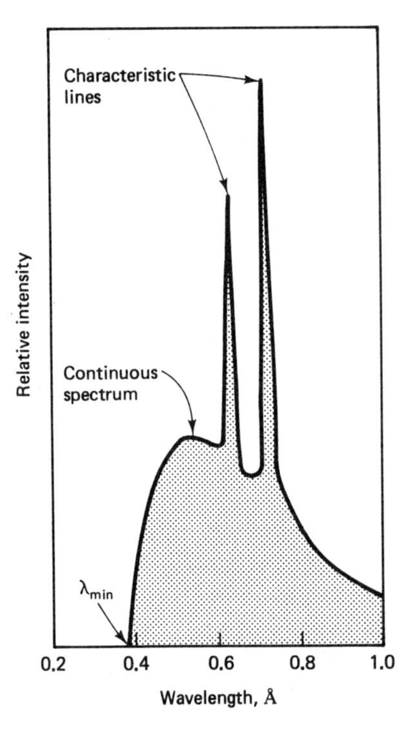

Figure 31-13 A typical X-ray spectrum consists of characteristic lines superimposed upon a continuous spectrum.

produce only heat in the target rather than X-rays, the target is provided with a means of cooling which may consist of a system for circulating water through the target assembly.

The production of X-rays occurs via two mechanisms. A continuous spectrum of X-rays is produced by the small fraction of incident electrons that lose their kinetic energy in a single collision with an atom of the target. The rapid deceleration of the electron causes a photon to be produced in keeping with classical electromagnetic theory, which predicts that electromagnetic radiation will be produced whenever a charge is accelerated. Radiation so produced is called *bremsstrahlung*, which means *braking radiation*. The other mechanism of X-ray production involves electron transitions in the target atoms.

The energy of the quantum states in an atom is proportional to Z^2, where Z is the atomic number of the atom [see Eq. (31-9)]. Thus, the energy of the ground state in copper ($Z = 29$) is about $29^2 = 841$ times as great as the ground state of hydrogen (13.6 eV). Thus, about 10^4 eV of energy is required to eject a 1s electron from a copper atom. When the vacated 1s state is refilled by a free electron, a photon is produced whose energy is about 10^4 eV, and this photon is, by definition, an X-ray.

A typical X-ray spectrum, shown in Fig. 31-13, is a composite of the *continuous* spectrum and the *characteristic* spectrum. The continuous spectrum is independent of the target material, but because the characteristic spectrum arises from electron transitions, it is *dependent* upon the target material. The characteristic spectrum is actually part of the line spectrum of the target atoms. The lines are very sharp and may have intensities 10,000 times greater than that of the continuous spectrum.

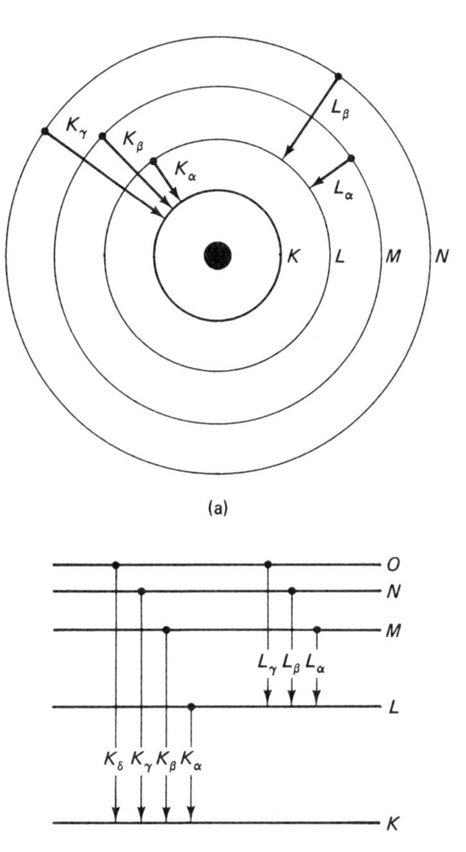

Figure 31-14 (a) Electron shells are denoted by *K, L, M,* . . . in X-ray spectroscopy.
(b) X-ray series are designated by the level to which the electrons fall.

In X-ray spectroscopy it is customary to designate atomic orbitals as *K*, *L*, *M*, . . . , corresponding to principle quantum numbers $n = 1, 2, 3, \ldots$. Hence, characteristic X-rays arising from electron transitions to the *K* shell form the *K-series*, and electron transitions to the *L* shell produce the *L-series*, and so on. The longest wavelength of a series is denoted by α, the next longest by β, and so forth, as shown in Fig. 31-14.

The continuous spectrum cuts off abruptly at λ_{min}, as shown in Fig. 31-13. This threshold wavelength represents the X-rays of maximum energy, and the maximum energy is the same as the maximum kinetic energy of the electrons striking the target. Therefore, λ_{min} depends upon the accelerating voltage V applied to the tube and is given by

$$\lambda_{\text{min}} = \frac{hc}{eV} \qquad (31\text{-}32)$$

If λ_{min} is in angstroms, then

$$\lambda_{\text{min}} = \frac{12{,}400}{V}\,\text{Å} \qquad (31\text{-}33)$$

It follows that a voltage of 12,400 V is required to produce a λ_{min} of 1 Å.

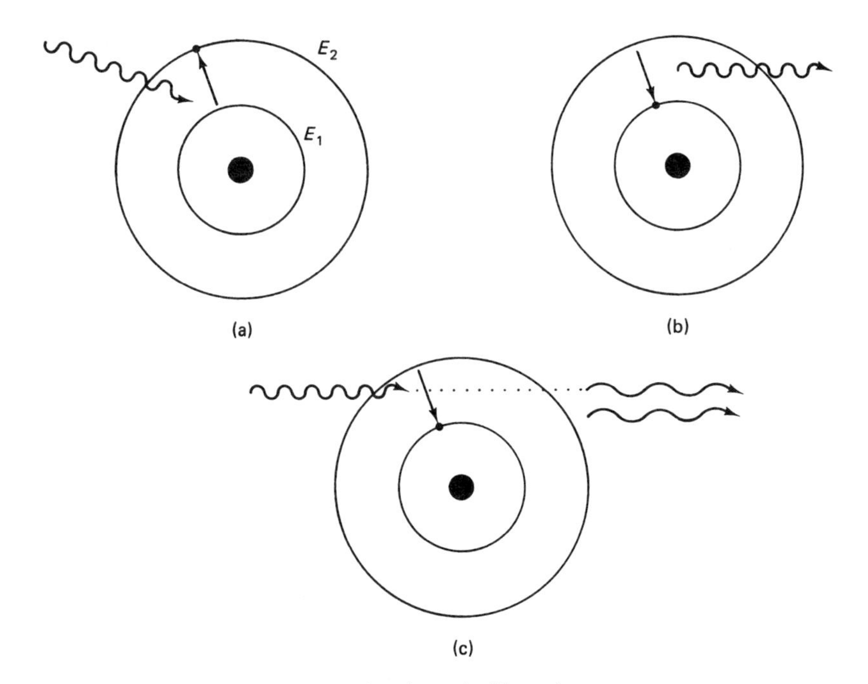

Figure 31-15 (a) An electron absorbs an incident photon.
(b) An electron causes a photon to be emitted as it spontaneously falls to a lower level.
(c) Stimulated emission: An incident photon causes the downward transition and a second photon is produced that is in phase with the first.

31-8 THE LASER

The development of the laser represents one of the most important technological advances of the 1960s. For theoretical work on lasers and masers, Charles H. Townes was awarded a Nobel Prize in 1964. Lasers have a wide range of applications in communications, medicine, the military, and industry, and new applications are continually being discovered.

A *laser* (Light Amplification by the Stimulated Emission of Radiation) produces an intense beam of monochromatic and coherent light that has virtually zero angular divergence. Because the light of a laser beam is almost perfectly parallel, it can be focused to an extremely small point to produce energy concentrations as high as 10^{12} W/cm^2.

To understand the operation of a laser, we must first be aware of three types of interactions between photons and electrons in an atom. Two of these are already familiar. (1) An electron may absorb a photon of proper energy and be excited from energy level E_1 to a higher level E_2. (2) An electron may spontaneously emit a photon as it makes a transition from a high-energy state, E_2, to a lower state, E_1. The third process is new and was suggested by Albert Einstein in 1916. (3) An incident photon may *stimulate* or *induce* a downward transition of an electron from state E_2 to E_1, and a second photon will be produced that will be exactly in phase with the photon that induced the transition. These processes are illustrated in Fig. 31-15.

When an electron is excited to a higher level in an atom, the electron remains in the excited state only for about 10^{-8} s before it falls back to the lower level, producing a photon in the process. However, *metastable states* exist in certain atoms in which an electron may remain for a much greater time—for periods on the order of 10^{-3} s. The existence of metastable states is essential to the operation of a laser. Another crucial point is this: A downward transition from a metastable state can be stimulated by a photon. That is, the fact that a state is metastable does not preclude the possibility of stimulated emission.

Let us now consider the fate of a photon as it passes through an assembly of atoms about which we make various assumptions. First, if none of the atoms are excited (all atoms are in the states of lowest possible energy), there is no possibility for a stimulated emission to occur. The photon can only be absorbed (or it may pass through the assembly without interacting at all), and its absorption will result in exciting one electron of one atom to a higher level.

Assume now that *all* the atoms of the assembly were initially excited to the high state. The incident photon cannot be absorbed because no upward transition of an electron is available to absorb it. However, the photon will very likely stimulate the emission of a second photon that matches the incident photon in phase and frequency, so—after the stimulated emission occurs—there will be *two* photons present instead of *one*. These two, in turn, can trigger other stimulated emissions; a chain reaction sets in and produces an avalanche of photons, all in phase, that have arisen because of the arrival of the first photon. We might say that the incident photon has been *amplified*.

Finally, let us assume that the assembly of atoms is perfectly balanced between excited and unexcited atoms. An incident photon will then have a 50% chance of being absorbed or of stimulating an emission. In this case, no overall increase in the number of photons present would be expected because of the equal probability of absorption and stimulated emission.

In most assemblies of atoms, only a very small fraction of the atoms are excited at a given time, far less than 50%. Even when energy is supplied to the system, electrons remain in the excited state only for about 10^{-8} s before spontaneously falling back to the state of lower energy. But we have seen that more than 50% of the atoms must be excited in order for an overall photon multiplication to occur. The existence of metastable states provides a means of achieving a *population inversion*, a situation in which more electrons occupy a higher level of energy, E_2, than occupy a lower level, E_1.

Figure 31-16 shows a portion of an energy level diagram of a hypothetical atom in which state E_2 is metastable. By supplying energy to a collection of such atoms via optical or electrical means (a process called *pumping*), electrons may first be excited to level E_3, from which they subsequently fall in about 10^{-8} s to metastable state E_2. Because E_2 is metastable, the electrons remain there for much longer times (about 10^{-3} s), so it is possible to obtain a condition where more electrons exist in state E_2 than E_1; a population inversion may be established. Once this is achieved, the stage is set for *lasing action* to occur. An incident photon of energy $E_2 - E_1$ can initiate the induced transition from E_2 to E_1 and the stimulated emission and avalanche process will be set in motion and will continue as long as the population inversion is maintained.

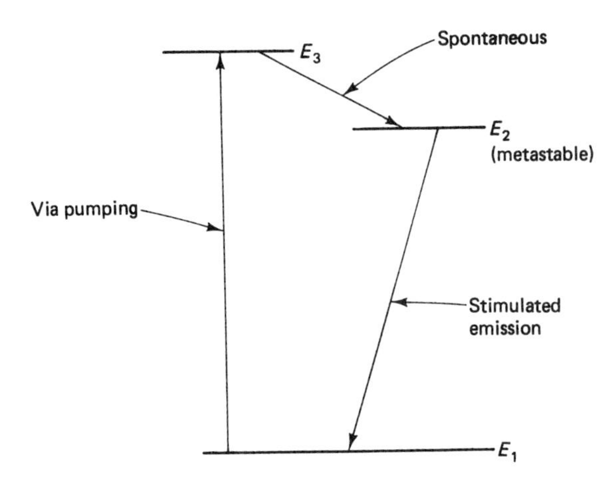

Figure 31-16 Electrons pumped from ground state E_1 to state E_3 fall spontaneously to metastable state E_2, where they remain for comparatively long periods. However, the downward transition from E_2 to E_1 may be stimulated by incident photons of the proper frequency.

The ruby laser

The essential features of a ruby laser are shown in Fig. 31-17(a). A ruby is a crystal of aluminum oxide, Al_2O_3, in which chromium ions (Cr^{3+}) replace a small portion of the aluminum (Al^{3+}) ions. The ends of the ruby are highly polished and are parallel; one end is fully "silvered," while the other is only partially silvered and is partially transparent. A xenon flash tube surrounds the ruby rod for the purpose of optical pumping. The length of the ruby rod is made equal to an integral number of half-wavelengths, so an optical standing wave may be established within the rod.

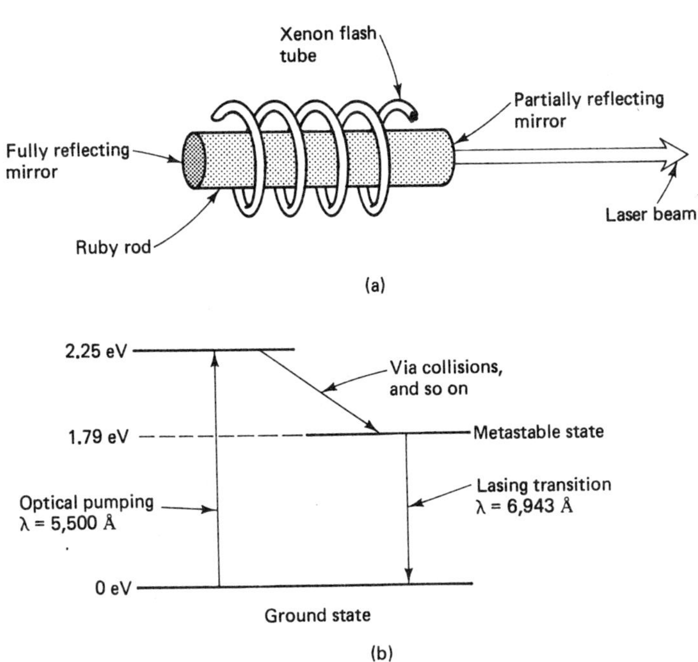

Figure 31-17 (a) Physical configuration of the ruby laser. A xenon flash tube provides optical pumping.
(b) A portion of the energy-level diagram for chromium.

Figure 31-17(b) is the energy level diagram for chromium. The flash tube excites electrons to the 2.25-eV level; the electrons then lose energy to other ions in the crystal by collisions and fall to the 1.79-eV metastable state. A few photons will be produced by the spontaneous emission by a few of the Cr^{3+} ions, and these will travel back and forth through the rod between the mirrored ends, stimulating other Cr^{3+} ions to emit in the process. After a few microseconds, an intense pulse of coherent light at 6943 Å will be emitted from the partially transparent end of the ruby rod. Optical pumping through the flash tube must then occur to reestablish the population inversion in order for the process to be repeated.

The helium-neon laser

The tube of a helium-neon laser contains a mixture of about 7 parts helium to 1 part neon at a pressure of about 1 mm Hg. The tube is provided with parallel mirrors, one of which is partially transparent. The gas within the tube is excited by a high-frequency voltage applied to electrodes that may be external to the tube. Excitation of electrons to the metastable state of helium (20.61 eV) occurs via electron collisions, and collisions between the excited helium atoms cause electrons to be excited to the metastable state of neon at 20.66 eV. Thus, a population inversion is created in the neon atoms, due, in part, to the presence of the helium gas. Once the population inversion is achieved, lasing action follows.

In neon, the final state of the laser transition (the 18.70-eV level) is above the ground state, so electrons do not remain in the state after arriving there via the laser transition. They undergo spontaneous emission to lower levels or make the transition to lower levels via energy losses due to collisions. Hence, few electrons are available to absorb photons at the laser frequency of 6328 Å. Therefore, lasing action can occur when considerably less than 50% of the electrons involved are in the metastable state. Further, because the collisions that excite electrons to the metastable state occur continuously, the lasing action of a helium-neon laser is continuous.

Holography

The coherent light from a laser can be used to produce a truly three-dimensional image by a process called *holography*, as shown in Fig. 31-18. First of all, lenses are used to cause the laser beam to diverge, giving it considerable width. This coherent beam is then split into a reference beam and an object beam by a partially reflecting mirror. Light reflected from the object combines with the reference beam to form a complicated interference pattern on a photographic plate. This plate, when developed, is called a *hologram*. The details of the hologram are determined by the characteristics of the object.

To view the three-dimensional image, light from a diverging laser beam is directed at the hologram and wave fronts are reconstructed that are replicas of those that produced the hologram. The resulting image is truly three-dimensional. By changing the angle from which the image is viewed, it is possible to "look

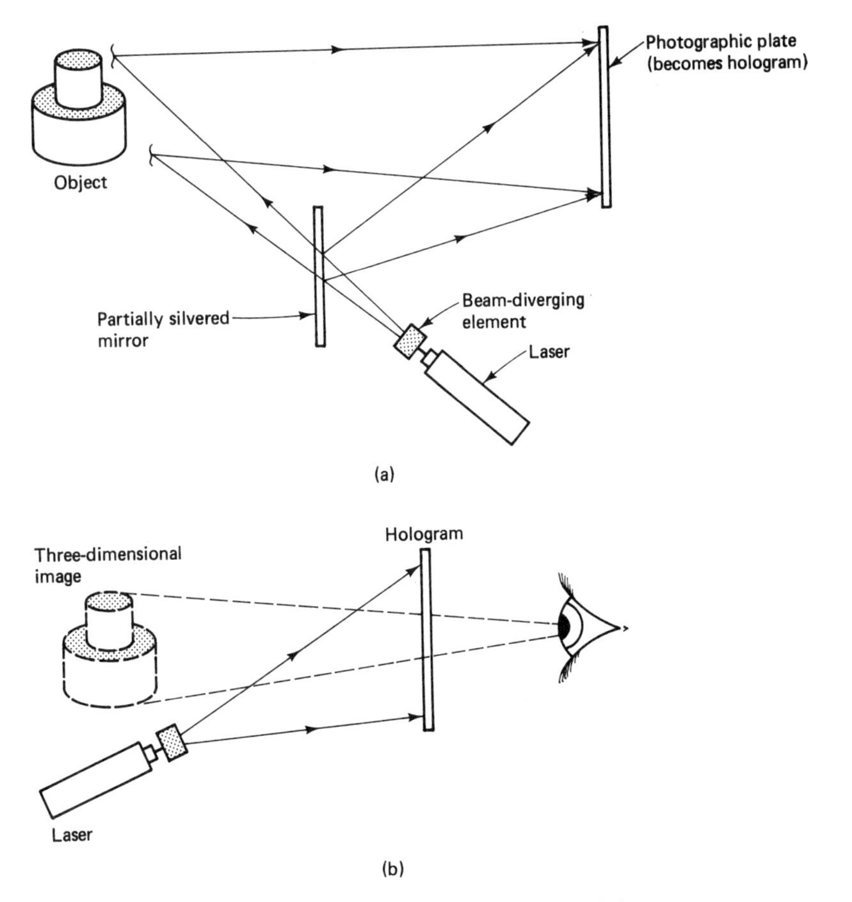

Figure 31-18 (a) Making a hologram. (b) Viewing a hologram.

behind'' objects in front of the image. Circular holograms have been produced that enable complete inspection of the image, from any direction.

To Go Further

Read about the contributions to the understanding of the atom made by the following:

Johann Balmer (1825–1892), Swiss mathematics teacher
Sir J. J. Thomson (1856–1940), English physicist
Max Planck (1858–1947), German physicist
Ernest Rutherford (1871–1937), British physicist
Niels Bohr (1885–1962), Danish physicist
Erwin Schrödinger (1887–1961), Austrian-born physicist
Louis de Broglie (1892–), French physicist
Werner Heisenberg (1901–1976), German physicist
Wolfgang Pauli (1900–1958), Austrian physicist

References

ANDRADE, E. N. DA C., *Rutherford and the Nature of the Atom*. New York: Anchor Books, Doubleday & Company, Inc., 1964.

MOORE, RUTH, *Niels Bohr*. New York: Alfred A. Knopf, Inc. 1966.

Questions

1. How did the Rutherford scattering experiment indicate that the plum-pudding model was not accurate?
2. The planetary model of Rutherford predicted that atoms should "decay" in a very short time. What phenomenon is responsible for the predicted decay?
3. Describe the assumptions made by Bohr in his analysis of the atom.
4. Briefly describe the photoelectric effect and its implications regarding the nature of photons.
5. What aspects of the photoelectric effect cannot be explained by classical concepts?
6. Give two experiments that might be performed to demonstrate that a certain type of physical entity occurs in the form of particles.
7. How can we demonstrate that a physical entity occurs in the form of waves?
8. Discuss the physical reality of the abstraction we call a probability wave. Does the fact that quantum mechanics predicts the right answers lend physical significance to the waves?
9. What is the wavelength of an electron at rest?
10. To say that an electron is at rest implies that its velocity is precisely known. In light of the answer to Question 9, what can then be said about the electron's position?
11. What is responsible for the natural line broadening of spectral lines?
12. What conservation law is responsible for the existence of selection rules governing the electron transitions within an atom?
13. Give two mechanisms for the production of X-rays within an X-ray tube.
14. Describe the three possible interactions between photons and electrons in an atom.
15. What is distinctive about the stimulating and the emitted photon when an incident photon induces the downward transition of an electron?
16. Why is the existence of metastable states crucial to the action of a laser?
17. How is a population inversion achieved in a ruby laser?
18. How are electrons excited into a metastable state in a helium-neon laser?

Problems

1. Use the Balmer formula [Eq. (31-1)] to compute the wavelengths of the first four lines of the Lyman series.
2. What is the conversion factor for converting wavelengths given in nanometers to angstroms?
3. Compute the series limit of: **(a)** the Lyman series; **(b)** the Balmer series; and **(c)** the Paschen series.
4. Compute the radius of the first four Bohr orbits of hydrogen. Make a sketch of the orbits, roughly to scale.
5. Compute the velocity of the electron in the first and second Bohr orbits.
6. Calculate the energy levels for the first three Bohr orbits, corresponding to $n = 1$, 2, and 3. These three energy levels provide for photons of threc different wavelengths. Calculate the wavelengths.
7. Compute **(a)** the radius, **(b)** the electron velocity, and **(c)** the energy level corresponding to the lowest orbit ($n = 1$) of singly ionized helium.
8. What is the threshold frequency for gold, whose work function is 4.82 eV?
9. Ultraviolet light of wavelength 2300 Å ejects electrons from gold. What is the maximum kinetic energy with which an electron may leave the gold?
10. Compute **(a)** the mass and **(b)** the momentum of a photon of wavelength 589 nm.
11. Compute **(a)** the mass and **(b)** the momentum of an X-ray of wavelength 1 Å.
12. Compute the velocity and wavelength [use

Eq. (31-22)] of an electron in the third Bohr orbit for hydrogen. Then show that the circumference of the orbit is such that the orbit contains an integral number of electron wavelengths.

13. Through what voltage must an electron be accelerated in order to have a wavelength of 1 Å?

14. Suppose the uncertainty in the position of an electron is 4×10^{-10} m. Calculate the minimum uncertainty in its velocity.

15. Repeat Problem 14 for a proton instead of an electron.

16. Suppose the average lifetime of an electron in an excited state of an atom is 5×10^{-8} s before it falls to the ground state, where it remains forever.

(a) Compute the uncertainty of the energy of the excited state.

(b) Convert this to an uncertainty in frequency of the emitted photon.

(c) If the electron produces a photon of nominal energy 2.5 eV during the downward transition, what is the fractional uncertainty $\Delta f/f$ of the frequency of the photon?

17. Demonstrate that the units of Planck's constant h are those of angular momentum.

18. Use the formula mvR to compute the angular momentum of an electron in the first Bohr orbit.

19. Calculate the angular momentum of a quantum state having orbital quantum number $l = 1$ and compare with the result of the previous problem.

20. List all possible combinations of quantum numbers for principle quantum number $n = 4$. What is the maximum number of electrons that can occupy the electron shell corresponding to $n = 4$?

21. Repeat Problem 20 for $n = 1$, 2, and 3.

22. Determine the electronic configuration for copper in the ground state. (The atomic number of copper is 29.)

23. Make use of the selection rules to determine the maximum number of transitions that can occur from the $n = 3$ shell to the $n = 2$ shell.

24. Repeat Problem 23 for transitions from $n = 4$ to $n = 3$.

25. Determine the minimum wavelength of X-rays produced by an X-ray tube to which is applied a voltage of 50,000 V.

26. Repeat Problem 25 for a voltage of 100,000 V.

CHAPTER 32

THE NUCLEUS

In Chap. 31 we investigated the outer reaches of the atom and found that quantum mechanics provides the solution to the electronic structure. In this chapter we consider several aspects of the nucleus. Unfortunately, no general theory of the nucleus has yet been developed that satisfactorily encompasses all the many nuclear properties. Instead, several models have been proposed with each model having certain advantages and disadvantages.

Several nuclear phenomena can be understood on a practical level without the benefit of a detailed theory. Radioactivity and nuclear fission are two examples, and both have many practical applications. Radioactive dating of ancient artifacts, medical use of radioactive isotopes, nuclear power, and, of course, nuclear weapons are possible because of our understanding of the nucleus.

32-1 GENERAL PROPERTIES OF THE NUCLEUS

Even though more than 99.9% of the mass of an atom is contained in the nucleus, the nucleus is extremely small in comparison to the size of the electron shells. For example, the diameter of a proton (hydrogen nucleus) is about 1×10^{-15} m. The diameter of the ground-state electron orbit of hydrogen is about 1×10^{-10} m, roughly 10^5 times as large as the nucleus. If the proton were enlarged to a diameter of 1 in., the diameter of the electron orbit would be well over 1 mi!

With the exception of hydrogen, whose nucleus consists of only a single proton, atomic nuclei contain neutrons and protons, to which we refer collectively as *nucleons*. Protons and neutrons have approximately the same mass, 1.67×10^{-27} kg. The neutron is neutral, as its name implies, but the proton carries a positive charge equal in magnitude to the charge of an electron, 1.602×10^{-19} C.

The atomic number Z is the number of protons contained in a nucleus, and the total number of nucleons (protons and neutrons) is called the *atomic mass number A*. The *neutron number*, N, equals $A - Z$. Nuclei of different types are often called *nuclides*.

The *proton number* Z distinguishes one element from another. However, the neutron number N is not unique to an element, as is the proton number. Nuclides of the same element that have different numbers of neutrons are called *isotopes*. Deuterium and tritium are isotopes of hydrogen, having one and two neutrons, respectively. Isotopes are conveniently represented in the form ${}^{A}_{Z}X$, where X is the chemical symbol for the element and where Z and A are as defined earlier. To illustrate, six isotopes of uranium are

$$ {}^{232}_{92}\text{U} \qquad {}^{233}_{92}\text{U} \qquad {}^{235}_{92}\text{U} \qquad {}^{236}_{92}\text{U} \qquad {}^{238}_{92}\text{U} \quad {}^{239}_{92}\text{U} $$

Nuclear masses are expressed in terms of *unified atomic mass units*, u. A mass of 1 u is defined as exactly $\frac{1}{12}$ the mass of a neutral ${}^{12}_{6}\text{C}$ atom, the most abundant isotope of carbon. Consequently, the mass of a proton or neutron is approximately, but not exactly, 1 u, and 1 u corresponds to a mass of about 1.660×10^{-27} kg.

Because modern physics is often concerned with the conversion of mass to energy and the fundamental equivalence of the two, nuclear masses are often expressed in terms of energy according to $m = E/c^2$, a rearrangement of $E = mc^2$. Thus, a mass of 1 u corresponds to an energy of about 931.5 MeV:

$$ 1 \text{ u} = 1.660 \times 10^{-27} \text{ kg} = 931.5 \text{ MeV/c}^2 \qquad (1 \text{ eV} = 1.602 \times 10^{-19} \text{ J}) $$

TABLE 32-1 REST MASS OF SELECTED PARTICLES AND ATOMS

Electron	0.00055 u
Proton	1.00728 u
Neutron	1.00867 u
${}^{1}_{1}\text{H}$ atom	1.00783 u
${}^{2}_{1}\text{H}$ atom	2.01410 u
${}^{3}_{1}\text{H}$ atom	3.01605 u
${}^{3}_{2}\text{He}$ atom	3.01603 u
${}^{4}_{2}\text{He}$ atom	4.00260 u

$1 \text{ u} = 1.660 \times 10^{-27}$ kg

Mass and energy are dimensionally not the same. Hence, it is not entirely proper to write a mass as a number of megaelectron-volts, even though it is commonly done. The c^2 in the expression MeV/c^2 is included for the sake of dimensional consistency.

Scattering experiments using high-speed electrons indicate that nuclei are nearly spherical in shape with a radius given approximately by

$$ R = (1.2 \times 10^{-15} \text{ m})A^{\frac{1}{3}} \tag{32-1} $$

Thus, the volume of a nucleus is directly proportional to the number of nucleons A, and this suggests that the nucleons combine as if they were impenetrable spheres (see Fig. 32-1).

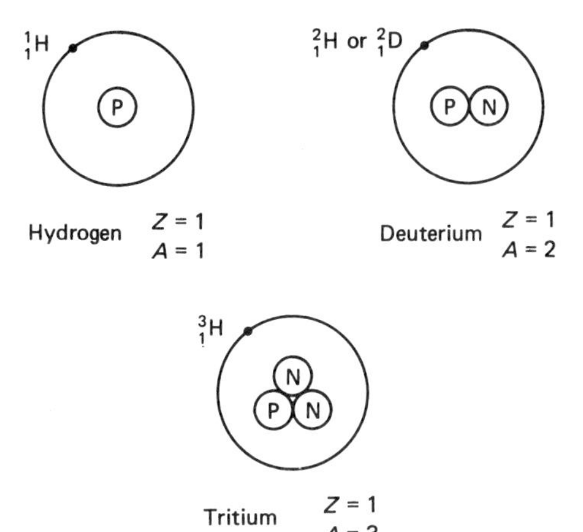

Figure 32-1 The three isotopes of hydrogen.

Binding energy

At first thought it would appear that a nucleus should fly apart due to electrostatic repulsion of the positive protons. However, when nucleons approach each other, a much stronger force, a *nuclear force*, comes into play; it is strongly attractive when the particle separation is less than about 10^{-14} m. For greater separations, the nuclear force is essentially zero, so it is a very short-range force. The fundamental nature of the nuclear force is yet unknown, but it is thought to act between any two nucleons (proton-proton and proton-neutron) independently of the electrical charge of the nucleons involved.

The mass of a nucleus is always less than the sum of the masses of its individual protons and neutrons. The difference is called the *mass defect*, and we may wonder where the "defecting" mass has gone. The answer is that a small portion of the total mass has been converted into *binding energy* which holds the nucleus together. For a helium nucleus, the mass defect is 0.0305 u which corresponds to a binding energy of 28.4 MeV. To tear a helium nucleus completely apart, this much energy must be added to the nucleus.

An important consideration pertaining to nuclear stability is the *average binding energy per nucleon*, which is the total binding energy divided by *A*, the number of nucleons in the nucleus. This provides a rough indication of how tightly a given type of nucleus is held together. A plot of average binding energy per nucleon as a function of mass number *A* is shown in Fig. 32-2. We may discern that heavy nuclei (large *A*) are held together less tightly than those having intermediate values of *A*.

32-2 RADIOACTIVE DECAY

Radioactivity refers to the emission by radioactive isotopes of *nuclear radiations*. We now know that radioactivity stems from the decay of unstable nuclides into daughter nuclides that may or may not be stable. Further, nuclear radiations consist primarily of three types, *alpha*, *beta*, and *gamma*. An alpha particle is a

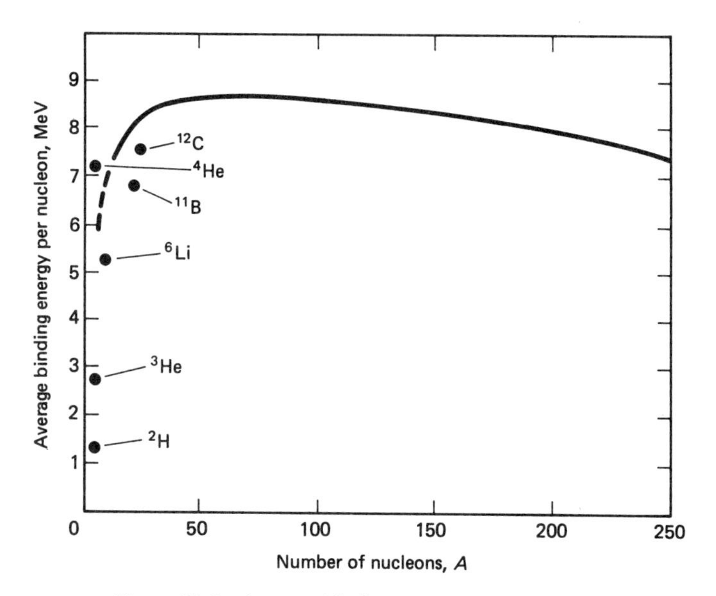

Figure 32-2 Average binding energy per nucleon.

helium nucleus, ${}^4_2$He; a beta particle may be either an electron or its antimatter counterpart, a positron; and gamma radiation is high-energy photons. Figure 32-3 illustrates an experiment that shows that nuclear radiation consists of at least three types.

Initial work in the area of radioactivity was done around 1898 by Henri Becquerel (who is credited with its discovery) and by Pierre and Marie (Madame) Curie. The history of the discovery of radioactivity and subsequent investigations is very interesting and can be found in many books.

Radioactivity is independent of macroscopic entities such as temperature, pressure, or chemical compounds formed by the radioactive substance. This stems from the fact that radioactive decay is a nuclear phenomenon. Moreover, when a particular nucleus will decay is unpredictable. In a collection of identical nuclei, any nucleus is as apt as any other to decay at a given time, and it is not true that older nuclei decay first. The *radioactive decay law*, which describes

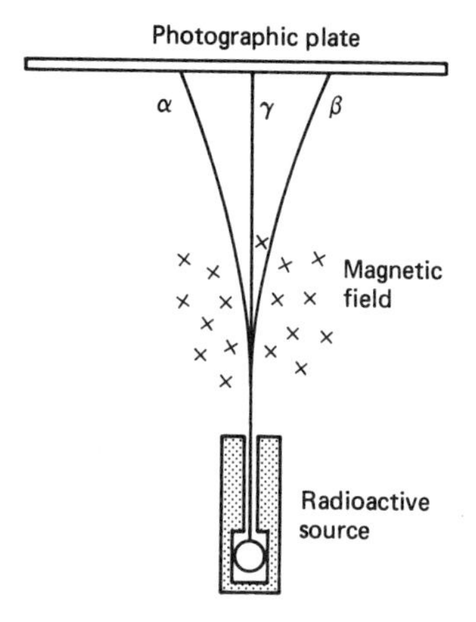

Figure 32-3 An experiment that demonstrates that nuclear radiation consists of at least three types.

statistically the rate at which disintegrations occur, is described in a following section. Radioactive decay occurs by three primary processes, which we now describe. Examples are shown in Fig. 32-4.

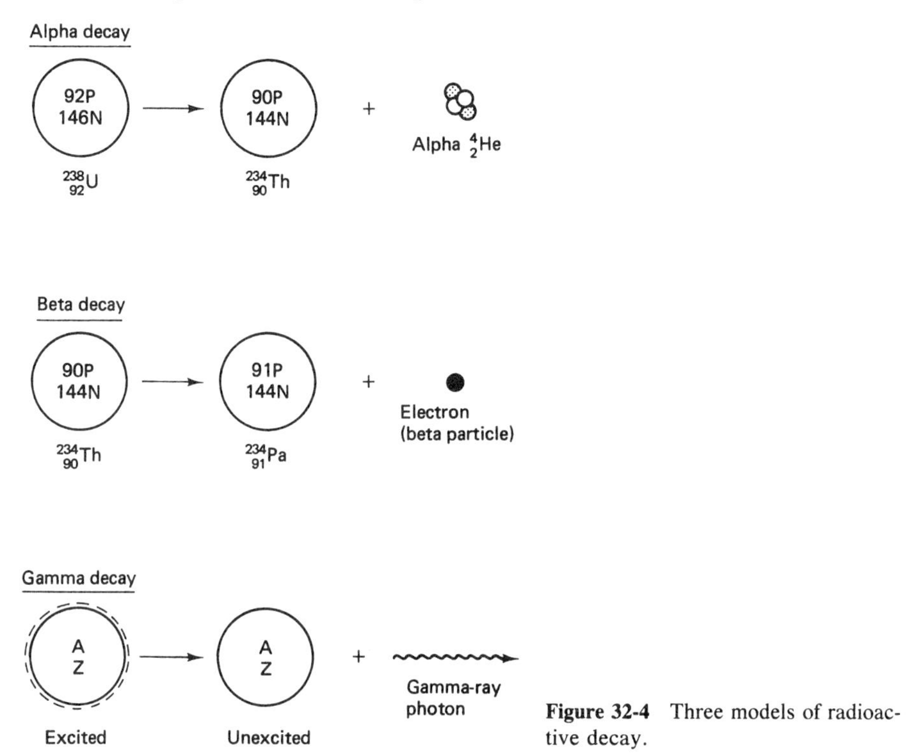

Figure 32-4 Three models of radioactive decay.

Alpha decay

Alpha decay occurs when a nucleus ejects an alpha particle, a ${}^{4}_{2}He$ nucleus. Consequently, the parent nucleus loses two protons and two neutrons; its proton number Z becomes $Z - 2$, and the mass number A becomes $A - 4$. An example is the alpha decay of uranium-238 into thorium-234:

$${}^{238}_{92}U \rightarrow {}^{234}_{90}Th + {}^{4}_{2}He \qquad 4.5 \times 10^{9} \text{ years} \qquad \text{(half-life)}$$

The alpha particle will be ejected with an energy of about 4.2 MeV. Other examples of alpha decay are:

$$\begin{array}{ll} {}^{234}_{92}U \rightarrow {}^{230}_{90}Th + {}^{4}_{2}He & 2.5 \times 10^{5} \text{ years} \\ {}^{230}_{90}Th \rightarrow {}^{226}_{88}Ra + {}^{4}_{2}He & 8 \times 10^{4} \text{ years} \\ {}^{226}_{88}Ra \rightarrow {}^{222}_{86}Rn + {}^{4}_{2}He & 1622 \text{ years} \\ {}^{210}_{84}Po \rightarrow {}^{206}_{82}Pb + {}^{4}_{2}He & 138 \text{ days} \end{array}$$

In these examples, all daughter nuclides except ${}^{206}_{82}Pb$ are also unstable and will subsequently decay either by alpha or beta decay.

Beta decay

When a nucleus has more neutrons than required for its stability, it is apt to emit a high-speed electron (beta particle) and thereby convert one of its neutrons into a proton. This process is called *beta decay*. The proton number Z increases by one; the neutron number N decreases by one; the mass number A does not change. Examples of beta decay are:

$$^{234}_{90}\text{Th} \rightarrow {}^{234}_{91}\text{Pa} + {}^{0}_{-1}\text{e} \qquad 24 \text{ days}$$

$$^{234}_{91}\text{Pa} \rightarrow {}^{234}_{92}\text{U} + {}^{0}_{-1}\text{e} \qquad 6.8 \text{ h}$$

$$^{214}_{83}\text{Bi} \rightarrow {}^{214}_{84}\text{Po} + {}^{0}_{-1}\text{e} \qquad 5 \text{ days}$$

The electron ejected from the nucleus as a beta particle is "created" within the nucleus as a neutron decays into a proton and an electron.

Beta decay is hardly as forthright as the above examples indicate. Initial investigations seemed to indicate that, in beta-decay processes, the conservation laws of energy, angular momentum, and linear momentum were violated. The difficulty was resolved, however, by realizing that another type of particle (the *neutrino*) was involved in beta decay processes that carried away both energy and momentum. That this particle had not been observed in cloud-chamber photographs of nuclear reactions is due to its charge neutrality, zero (probably) rest mass, and to the fact that it reacts only very slightly with matter (and therefore leaves no trails in the photographs). The existence of the neutrino, now firmly established, was first proposed by Wolfgang Pauli in 1930, and the details of beta decay were developed in 1934 by Enrico Fermi.

Actually, it is not the neutrino, ν, but rather is the antineutrino, $\bar{\nu}$, that is involved in ordinary beta decay. Thus, a beta-decay process may be written to include the antineutrino as

$$^{14}_{6}\text{C} \rightarrow {}^{14}_{7}\text{N} + {}^{0}_{-1}\text{e} + \bar{\nu}$$

If a nucleus is unstable by virtue of having an excess of protons relative to the number of neutrons, it is likely to decay by β^+ emission (as opposed to β^- emission) in which a positron (positive electron) is ejected from the nucleus. Effectively, a proton is converted to a neutron so that a more stable nucleus is obtained. An example of β^+ decay, complete with the neutrino, is

$$^{19}_{10}\text{Ne} \rightarrow {}^{19}_{9}\text{F} + {}^{0}_{+1}\text{e} + \nu$$

Gamma decay

A gamma ray is a photon having energies on the order of a few kiloelectron-volts up to several megaelectron-volts. Hence, it has no rest mass and no electrical charge. Consequently, the emission of a gamma ray from an *excited* nucleus does not alter A, Z, or N. A transmutation of nuclides does not occur for gamma emission as it does for alpha and beta decay. The emission of a γ-ray by an excited nucleus is analogous to the emission of a photon by an excited atom as an orbital electron falls to a level of lower energy.

A nucleus may come to be in an excited state as a result of a previous

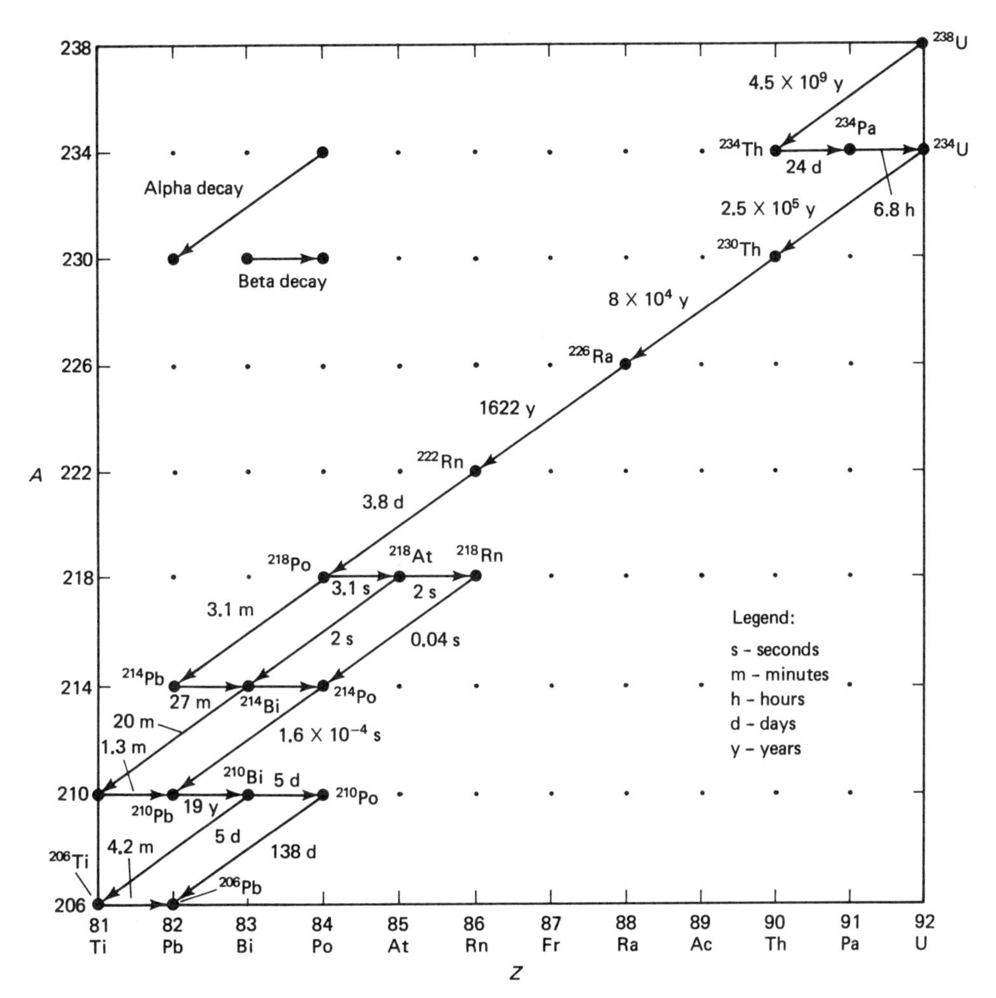

Figure 32-5 The uranium decay series.

radioactive decay or as a result of a collision with another particle. Also, the two daughter nuclei resulting from the fission (splitting) of a nucleus may be formed in excited states from which they decay and emit gamma rays. This accounts for a portion of the gamma radiation at the site of a nuclear blast and for that associated with nuclear reactors.

Radioactive decay series

We have seen that when a radioactive isotope decays to a daughter isotope, the daughter might also be radioactive and decay to yet another isotope which might, in turn, be radioactive. This possibility leads to a *decay series* in which a parent isotope, such as $^{238}_{92}U$, undergoes a succession of decays until a stable isotope is reached ($^{206}_{82}Pb$, in the case of $^{238}_{92}U$).

There are four different radioactive series, named according to the parent nuclide: thorium, neptunium, uranium, and actinium. The end product is an isotope of lead with the exception of the neptunium series whose end product is an isotope of bismuth. Details of the uranium series is shown in Fig. 32-5.

Radioactive decay rate

Suppose a sample of N_o radioactive atoms is observed for a brief time interval Δt. During the interval, a portion ΔN of the atoms will decay. Experimentally, it is found that ΔN is related to N_o and Δt by

$$\Delta N = -\lambda N_o(\Delta t) \quad (32\text{-}2)$$

The negative sign is affixed because ΔN represents a decrease in the number of radioactive atoms present; λ is the decay constant for the particular type of radioactive atoms being observed.

With calculus it is easy to show that if N_o atoms are initially present, after a time t there will be N atoms remaining, where N is given by

$$N = N_o e^{-\lambda t} \quad (32\text{-}3)$$

It is customary, rather than to specify λ for a particular material, to give its *half-life* $T_{1/2}$. This is the period of time required for one-half of the original atoms to decay. In terms of λ, $T_{1/2}$ is

$$T_{1/2} = \frac{0.693}{\lambda} \quad \text{or} \quad \lambda = \frac{0.693}{T_{1/2}} \quad (32\text{-}4)$$

TABLE 32-2 HALF-LIVES OF SELECTED RADIOISOTOPES

Carbon-14	5730 years
Calcium-45	165 days
Cesium-137	30 years
Cobalt-60	5.26 years
Iodine-131	8.0 days
Iron-59	45.6 days
Gold-198	2.7 days
Phosphorus-32	14.3 days
Plutonium-239	2.41×10^4 years
Radium-226	1600 years
Sodium-24	15 hours
Strontium-90	28 years
Thorium-230	8×10^4 years
Tritium, ${}^3_1H$	12.3 years
Uranium-235	7×10^8 years
Uranium-238	4.5×10^9 years
Uranium-239	23.5 minutes
Technetium-99	6.0 hours

Substituting into the previous equation gives

$$N = N_o exp(-0.693t/T_{1/2}) \quad (32\text{-}5)$$

Because the mass of a sample is proportional to N, we can use Eq. (32-5) to compute in terms of masses rather than in terms of the number of atoms. A graph of Eq. (32-5) is shown in Fig. 32-6.

EXAMPLE 32-1 The half-life of ${}^{210}_{83}Bi$ is 5 days for the beta decay to ${}^{210}_{84}Po$. If a sample initially contains 100 g of ${}^{210}_{83}Bi$, how much will remain after (a) 5 days; (b) 10 days; (c) 15 days?

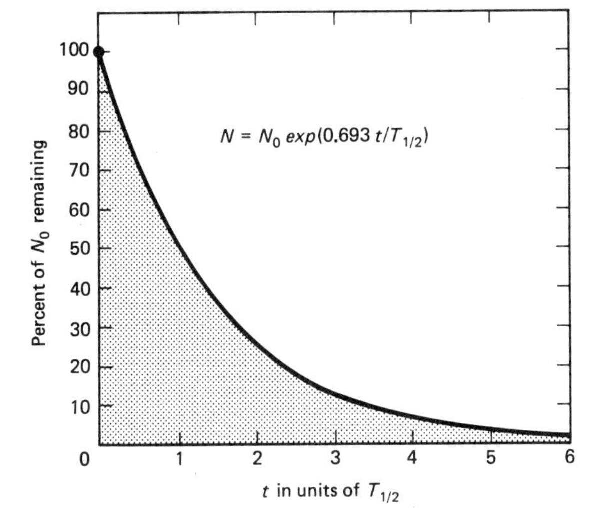

Figure 32-6 Percent of a radioactive sample that remains after a time t.

Solution. By definition of the half-life, 50 g will remain after 5 days ($T_{1/2}$); 25 g will remain after 10 days ($2T_{1/2}$) and 12.5 g will remain after 15 days ($3T_{1/2}$).

EXAMPLE 32-2

Refer to Example 32-1. How much ${}^{210}_{83}Bi$ will remain after 12 days?

Solution. Since 12 days is not an integral number of half-lives, we must use Eq. (32-5) (using M for N):

$$M = M_o exp(-0.693t/T_{1/2})$$

$$= (100\text{ g})exp[(0.693)(12)/5]$$

$$= (100\text{ g})exp(-1.66) = 18.95\text{ g}$$

32-3 RADIOACTIVE DATING

The age of once-living matter can be estimated by carefully measuring the natural radioactivity of the carbon atoms in the object whose age is to be determined. This is possible because of the radioactivity of ${}^{14}_{6}C$, an isotope that occurs in the atmosphere at the rate of one per 10^{12} atoms of ordinary carbon, ${}^{12}_{6}C$. The half-life of ${}^{14}_{6}C$ is about 5730 yr. Carbon-14 is continually produced in the upper atmosphere by the bombardment of nitrogen, ${}^{14}_{7}N$, with neutrons produced by cosmic rays. The reaction is

$${}^{14}_{7}N + {}^{1}_{0}n \rightarrow {}^{14}_{6}C + {}^{1}_{1}H$$

This carbon isotope then reacts chemically with oxygen to form CO_2, which is absorbed directly by plants and indirectly by other forms of life. By assuming that the ratio of ${}^{14}_{6}C$ to ${}^{12}_{6}C$ has remained constant over a long period of time, we can assert that the same ratio existed in the organisms that lived perhaps 20,000 years ago.

When a tree, for example, is living, it exchanges carbon rather freely with its environment. But when the tree dies, the exchange stops. Whatever ${}^{14}_{6}C$ atoms the tree contains are trapped in the wood, as are the ordinary atoms of carbon. If, then, a piece of wood is preserved for thousands of years, the ratio of $^{14}_{6}C$ to $^{12}_{6}C$ will slowly decrease due to the ratioactive decay of the $^{14}_{6}C$. In 5730 years ($T_{1/2}$ for $^{14}_{6}C$), the ratio will drop to one-half its original value; it will drop to one-fourth its original value in 11,460 years. Thus, by measuring the radioactivity of a sample of carbon taken from a once-living organism, the ratio $^{14}_{6}C/^{12}_{6}C$ can be determined and the age of the organism inferred. The decay of $^{14}_{6}C$ is via beta decay:

$$^{14}_{6}C \rightarrow {}^{14}_{7}N + {}_{-1}^{0}e$$

Enough $^{14}_{6}C$ is present in 1 g of carbon from a recently deceased organism to yield about 16 beta emissions/min. Thus, after 5730 years, the count will decrease to 8/min, and after 11,460 years the count will be about 4/min. It is not difficult to show that the age of a specimen is given by

$$\text{Age} = -1.44\ T_{1/2} \ln R \tag{32-6}$$

where R is the ratio of the present rate of decay to the initial rate. A graph of this equation is shown in Fig. 32-7.

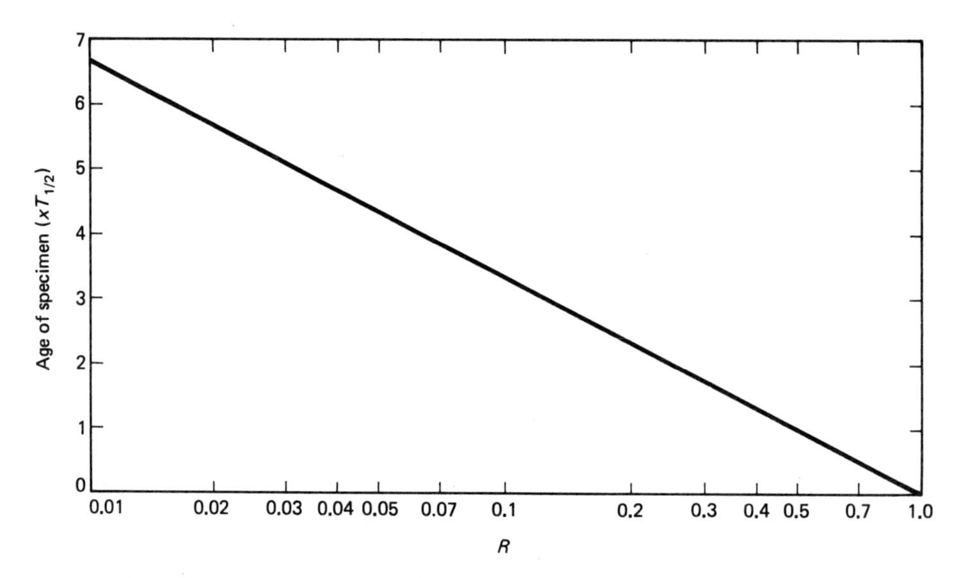

Figure 32-7 A semilog graph of Eq. (32-6) for determining the age of a specimen by radioactive dating. R is the ratio of the present rate to the initial rate.

EXAMPLE 32-3 One gram of carbon from an ancient artifact yields 6 beta emissions per minute. Estimate the age of the artifact.

Solution. Using Eq. (32-6),

$$\begin{aligned}\text{Age} &= -1.44\ T_{1/2} \ln R \\ &= -1.44\ (5730 \text{ years}) \ln \tfrac{6}{16} \\ &= 8093 \text{ years}\end{aligned}$$

32-4 THE DETECTION AND MEASUREMENT OF RADIATION

Probably the most familiar radiation detector is the *Geiger counter*, shown schematically in Fig. 32-8. A cylindrical metal tube with a thin window is filled with a gas, perhaps argon, at a pressure of about 0.1 atm. A high voltage (about 1000 V) is maintained between a central electrode and the metal tube, but the voltage is not high enough to ionize the gas directly.

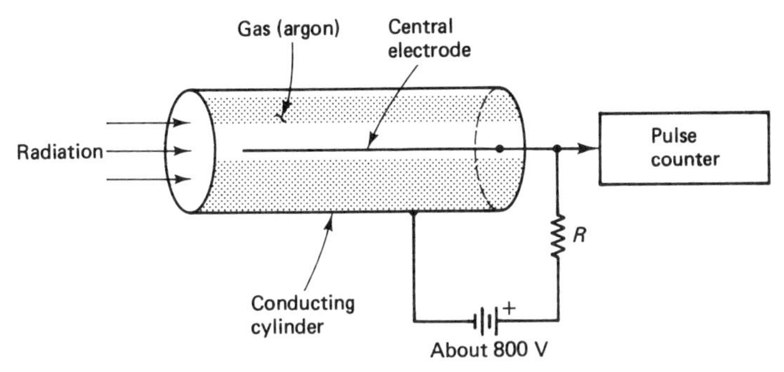

Figure 32-8 Major features of a Geiger counter.

When an energetic particle enters the tube, it is likely to ionize one or more gas molecules, producing free electrons. These few free electrons are then accelerated toward the central electrode and soon gain enough energy to produce further ionization of the gas. A multiplication process is set in motion and an *avalanche discharge* occurs between the cylinder wall and central electrode. This creates a pulse in the external circuit which may be amplified and sent to a pulse counter, a metering circuit, or to a speaker to produce an audible click.

A *scintillation counter* utilizes the fact that energetic particles incident upon certain materials produce tiny flashes of light. The scintillation material may be incorporated into a photomultiplier tube that senses the flashes and converts them to electrical pulses. The overall features of a scintillation counter tube are shown in Fig. 32-9.

When an incoming particle strikes the scintillation material, a flash of light is produced, and photons from the flash fall on a photocathode, which emits electrons as in the photoelectric effect. The photoelectrons are then accelerated

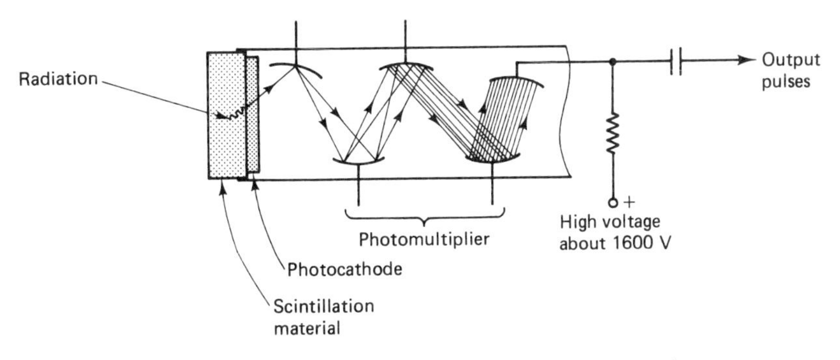

Figure 32-9 Main features of a scintillation counter tube.

toward a nearby electrode, called a *dynode*, which emits from two to five electrons for every one that is incident on it. These secondary electrons then are accelerated toward another dynode at a higher potential, and another multiplication process takes place. By using several dynodes in succession, a very large pulse can be delivered for even a very tiny flash of light (one photon, perhaps).

Several means are available for recording the track of a charged particle, the simplest of which is a photographic film. The passage of the particle through the emulsion layer of the film exposes the grains of silver halide so that when the film is developed, the resulting particles of silver reveal the path of the particle.

A *cloud chamber* was one of the early devices used to study the particles of radioactivity. Its chamber is filled with a supercooled gas, a gas cooled to a temperature slightly below its condensation point. The gas will, therefore, readily condense on any ions that may be present, and the passage of an energetic particle through the chamber produces a trail of ions along its path. Thus, the track of the particle is revealed by the trail of droplets that form. The track may be photographed to provide a permanent record.

In a *bubble chamber*, a liquid (perhaps liquid hydrogen) is heated to a temperature slightly above its normal boiling point. The ion trail of a particle causes bubbles (boiling) to be initiated along the trail which makes the trail visible.

Units of measurement

The strength, or *activity*, of a radioactive source is the number of disintegrations of "events" that occur per second. Accordingly, the SI unit of radioactive activity is the becquerel (Bq) which is defined as one event per second:

$$1 \text{ Bq} = 1 \text{ event per second}$$

An earlier unit, the curie (Ci) was based on the activity of 1 g of radium. One curie is defined as exactly 3.7×10^{10} events per second, even though precise measurements indicate that the activity of 1 g of radium is slightly different from this. Hence,

$$1 \text{ Ci} = 3.7 \times 10^{10} \text{ Bq}$$

Smaller units, the millicurie ($1 \text{ mCi} = 10^{-3}$ Ci) and microcurie ($1\ \mu\text{Ci} = 10^{-6}$ Ci) are popular.

EXAMPLE 32-4 A 10-g sample of radioactive material produces 7.4×10^4 events per second. What is the activity of the sample in microcuries and in becquerels?

Solution. By definition of the becquerel, the activity of the sample is 7.4×10^4 Bq. A unit conversion gives the equivalent in curies, 2 μCi.

Activity calculation

The activity of a particular sample can be calculated if the mass m of the sample is given and if the half-life is known. The formula is

$$\text{Activity (Bq)} = nN_a\, \lambda e^{-\lambda t} \qquad (32\text{-}7)$$

where

$$n = \text{g-mol of material} = \text{mass/atomic weight}$$
$$N_a = \text{Avogadro's number, } 6.02 \times 10^{23} \text{ molecules/g} \cdot \text{mol}$$
$$t = \text{elapsed time}$$
$$\lambda = \text{decay constant} = 0.693/T_{1/2}$$

EXAMPLE 32-5

Determine the activity of 1 g of radium, $^{226}_{88}Ra$, whose half-life is 1620 years.

Solution. Anticipating the use of Eq. (32-7), we first compute n and λ:

$$n = \frac{1 \text{ g}}{226 \text{ g/g} \cdot \text{mol}} = 4.42 \times 10^{-3} \text{ g} \cdot \text{mol}$$

$$\lambda = \frac{0.693}{T_{1/2}} = \frac{0.693}{5.11 \times 10^{10} \text{ s}}$$
$$= 1.36 \times 10^{-11} \text{ s}^{-1}$$

Note that the half-life of 1620 years was converted to seconds. Then, using Eq. (32-7) with $t = 0$ (the time when the mass was 1 g),

$$\text{Activity (Bq)} = (4.42 \times 10^{-3})(6.02 \times 10^{23})(1.36 \times 10^{-11})\, e^{0}$$
$$= 3.62 \times 10^{10} \text{ s}^{-1}$$
$$= 3.62 \times 10^{10} \text{ Bq}$$
$$\cong 1 \text{ Ci}$$

EXAMPLE 32-6

Determine the activity of the sample in Problem 32-5 after 800 years.

Solution. We can use the result of the example to simplify the application of Eq. 32-7:

$$\text{Activity (Bq)} = (3.62 \times 10^{10} \text{ Bq})e^{-\lambda t}$$
$$= (3.62 \times 10^{10} \text{ Bq}) \exp[-(0.693)(800)/1620]$$
$$= 2.57 \times 10^{10} \text{ Bq}$$

Biological effects

An important consideration pertaining to the biological effects of radiation is the ability of the radiation to ionize atoms and molecules within the cells of an organism. The ions (electrically charged atoms or molecules) or free radicals (chemical fragments) produced may interfere with the normal chemistry of the cell to the extent that the cell may die. More seriously, the reproductive capacity of the cell may be altered so that the defective cell reproduces to form other defective cells at a very rapid rate. This condition—cancer—often threatens the life of the entire organism.

DNA molecules embody the patterns, or blueprints, according to which an organism is constructed. If a *gene* is altered by radiation, a mutation occurs, and most mutations are harmful; mutations may be passed on to future generations.

Alpha and beta rays are charged particles that produce ions by electrostatic interactions. Due to the high energies of these particles, a single particle can cause literally thousands of ionizations. X-rays and gamma rays (photons) produce ions via the photoelectric effect and by *collisions* with orbital electrons (the *Compton effect*). Neutrons interact mainly by means of collisions with nuclei, and a nucleus may be split apart and thereby alter the molecule of which it is a part.

Dosimetry

The SI unit of *radiation dose* is the *gray* (Gy). One gray corresponds to the absorption of 1 Joule of energy per kilogram of the material absorbing the radiation. A related unit is the rad, which corresponds to the absorption of 10^{-2} Joules of energy per kilogram of target material. Hence,

$$1 \text{ Gy} = 100 \text{ rad}$$

The first unit of radiation dosage was the roentgen (R). By definition, 1 R is the amount of radiation that will produce ionization in the amount of 2.1×10^9 ion pairs per cubic centimeter of air at STP. A disadvantage of the roentgen as a unit is that it does not take into account the differences in materials, in regard to their abilities to absorb radiation, and differences in the types of radiation in regard to the tendency of the radiation to produce ions.

Some radiations are more effective than others in producing biological damage. For example, 1 rad of alpha radiation produces from 10 to 20 times as much damage as 1 rad of beta or gamma radiation. A factor that expresses this is the *relative biological effectiveness* (RBE) of the particular type of radiation. It is the number of rads of X-rays or gamma rays that produces the same biological damage as 1 rad of the radiation being considered. RBE's for the different radiations are given in Table 32-3.

TABLE 32-3 RBE VALUES OF DIFFERENT RADIATIONS

Type	Typical RBE
X-rays	1
Gamma rays	1
Low-energy beta rays	1
Slow neutrons	3–5
Fast neutrons	10
Protons	10
Alpha particles	10–20

Finally, a unit that incorporates the physical dose (in rads) and the RBE is the *rem* (for rad-equivalent man) given by

$$\text{rem} = \text{rad} \times \text{RBE} \tag{32-8}$$

In principle, 1 rem of any type of radiation produces the same amount of damage.

The effects of a large dose of radiation depends upon the period of time over which the dose is received. A 1000-rem dose in a short period is nearly always fatal. A 400-rem dose in a short period is fatal in 50% of the cases, whereas a 400-rem dose over a period of several weeks is seldom fatal even though considerable damage will result to the body.

The consensus now is that *any* radiation dose is bad. Nevertheless, acceptable levels have been established at 5 rem/year for occupational exposures and 0.5 rem/year for the general population. Cosmic rays and other sources of natural radioactivity contributes about 0.125 rem/year for the average person while diagnostic X-rays supply an additional 0.07 rem/year for the average person in the United States. Averaged over the U.S. population, nuclear power plants contribute less than 0.00001 rem/year.

32-5 NUCLEAR REACTIONS

The term *nuclear reaction* refers to the general situation in which a nucleus is bombarded by a simpler particle (alpha or beta particle, for example) to produce an interaction that produces a change in the target nucleus. The bombarding particles may be produced by a naturally radioactive material or by a particle accelerator or nuclear reactor.

Ernest Rutherford in 1919 reported the first nuclear reaction ever observed. He bombarded nitrogen ${}^{14}_{7}N$ with alpha particles from ${}^{214}_{83}Bi$ to achieve the following reaction:

$${}^{4}_{2}He + {}^{14}_{7}N \rightarrow {}^{17}_{8}O + {}^{1}_{1}H$$

In this reaction, the nitrogen nucleus absorbs the alpha particle to produce the compound nucleus ${}^{18}_{9}F$, which subsequently decays to ${}^{17}_{8}O$. Since the discovery of this reaction in 1919, a multitude of nuclear reactions, both natural and artificial, have been documented.

When beryllium in the form of a fine powder is mixed with an alpha emitter such as radium or polonium, the alphas interact with the beryllium according to

$${}^{9}_{4}Be + {}^{4}_{2}He \rightarrow {}^{12}_{6}C + {}^{1}_{0}n$$

Because one reaction product is a high-speed neutron, this reaction provides a means of fabricating a *source* of neutrons by tightly sealing the materials in a small metal capsule. Many other reactions provide neutrons as daughter products and may be used as neutron sources.

Due to the electrical neutrality of neutrons, they do not represent *ionizing radiation* and therefore produce no direct effect in ionization chambers or the Geiger-Mueller tubes of Geiger counters. Thus, the detection of neutrons is somewhat more complicated than is the detection of other forms of radiation. One method used is to fill a Geiger-Mueller tube with BF_3 gas so that the following reaction may occur when neutrons pass through the gas:

$${}^{10}_{5}B + {}^{1}_{0}n \rightarrow {}^{7}_{3}Li + {}^{4}_{2}He$$

Both the alpha particle and the lithium nucleus produce ionization within the tube that can initiate a discharge.

Induced radioactivity

A sample of material that initially is not radioactive may, in certain cases, be made radioactive by bombarding the sample with particles from another source. The radioactivity is said to be *induced* or *artificial* as opposed to *natural*. One example is the alpha bombardment of aluminum $^{27}_{13}\text{Al}$ to produce phosphorus $^{30}_{15}\text{P}$ which decays via the emission of a positron (β^+ decay) to a stable isotope of silicon $^{30}_{14}\text{Si}$:

$$^{27}_{13}\text{Al} + ^{4}_{2}\text{He} \rightarrow ^{30}_{15}\text{P} + ^{1}_{0}\text{n}$$

$$^{30}_{15}\text{P} \rightarrow ^{30}_{14}\text{Si} + _{+1}^{0}\text{e}$$

The unstable isotope of phosphorus ($T_{1/2} = 2.5$ m) accounts for the radioactivity of irradiated aluminum.

Two features of induced radioactivity are worthy of note. First, an unstable nuclide produced by bombardment of a stable nuclide usually decays to a stable nuclide in one step. Thus, induced radioactivity does not yield a radioactive series that is characteristic of natural radioactivity. Secondly, an unstable daughter of a bombardment interaction is apt to become stable by capturing one of the orbital electrons. Inside the nucleus, the captured electron combines with a proton p to form a neutron n and a neutrino ν that is emitted:

$$\text{p} + \text{e}^- \rightarrow \text{n} + \nu$$

When an orbital electron is captured, a vacancy is created in one of the inner electron orbits of the atom, and this vacancy will be filled by an electron from a higher orbit, or, more likely, by the capture of a free electron. A photon (X-ray) characteristic of the daughter nucleus will thereby be produced, and the presence of the X-ray gives evidence of the occurrence of electron capture.

Energy considerations

Nuclear reactions must obey the conservation laws of angular and linear momentum and the extended version of the law of conservation of energy, the conservation of *mass-energy*. Accordingly, energy considerations play a major role in determining the conditions under which a given nuclear reaction may occur. Nuclear reactions may result in either a net release or absorption of energy. The fission of $^{235}_{92}\text{U}$, described in the next section, is an example of a reaction that releases energy in large amounts.

32-6 NUCLEAR FISSION

Nuclear fission refers to the splitting of a heavy nucleus into two roughly equal *fission fragments*; it is accompanied by the release of tremendous amounts of energy per fission and is the basis of the atomic (fission) bomb and nuclear reactors. The phenomenon was discovered in 1939.

The best-known example of fission is that of $^{235}_{92}\text{U}$. This isotope occurs in natural uranium ore at the rate of about 0.7% relative to $^{238}_{92}\text{U}$. Three possible fission reactions are

$$^{1}_{0}n + ^{235}_{92}U \rightarrow ^{140}_{54}Xe + ^{94}_{38}Sr + 2(^{1}_{0}n)$$

$$^{1}_{0}n + ^{235}_{92}U \rightarrow ^{141}_{56}Ba + ^{92}_{36}Kr + 3(^{1}_{0}n)$$

$$^{1}_{0}n + ^{235}_{92}U \rightarrow ^{150}_{60}Nd + ^{81}_{32}Ge + 5(^{1}_{0}n)$$

For each of these reactions, on the order of 200 MeV of energy is released. In contrast, ordinary combustion reactions (the burning of coal or gas) produce only a few electron-volts per reaction, and nonfission nuclear reactions liberate at most only a few million electron-volts. Most (about 83%) of the energy released in fission goes into kinetic energy of the fission fragments and eventually appears as heat.

The liquid-drop model of nuclear fission is shown in Fig. 32-10. Absorption of a neutron produces an excited nucleus of $^{236}_{92}U$ which oscillates wildly until it breaks apart into two fission fragments plus a small number of neutrons. Once apart, the fission fragments repel each other strongly due to electrostatic repulsion and are quickly accelerated to high speeds.

Because (on the average) 2.5 free neutrons are produced per fission, the possibility of a chain reaction exists in which the neutrons collide with and are absorbed by other $^{235}_{92}U$ atoms, which subsequently fission. If all the neutrons go on to produce additional fission reactions, the number of fissions will grow according to a geometric progression to an exceedingly large number in a time period of a few microseconds; an explosion will result—as occurs in a fission bomb.

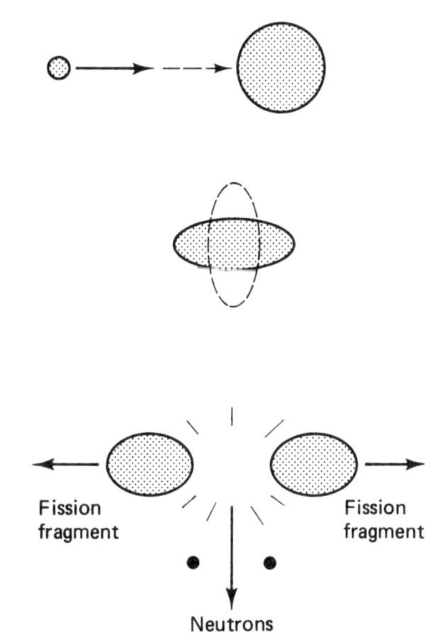

Figure 32-10 The liquid-drop model of nuclear fission.

If, on the other hand, most of the neutrons escape without inducing further fission reactions, the process will gradually diminish and the reaction will not be sustained. But if exactly one neutron per fission goes on to produce a subsequent fission, the reaction will be sustained at a steady rate. The terms *supercritical*, *subcritical*, and *critical*, respectively, are applied to these three situations.

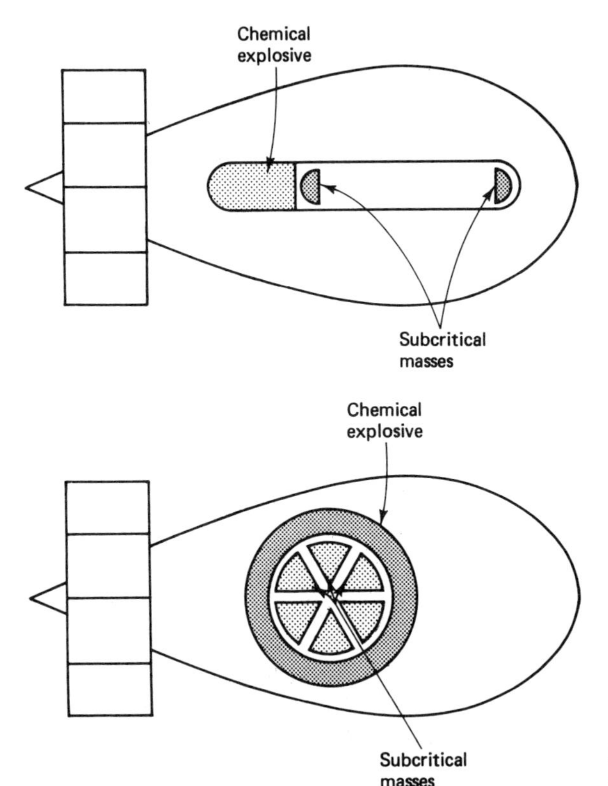

Figure 32-11 Two methods for implementing an atomic bomb.

Suppose a quantity of fissionable material is in the shape of a sphere. The number of neutrons produced will be proportional to the volume of the sphere. Therefore, the ratio of neutrons produced to neutrons escaping will be proportional to the first power of R, the radius of the sphere. That is, comparatively fewer neutrons escape from larger spheres.

If material is gradually added so that the radius is slowly increased, a point of *criticality* will be reached where an average of one neutron per fission produces a secondary fission. The mass of this sphere is called the *critical mass* of the material. A larger mass will be supercritical, and, under certain conditions, can explode as a fission bomb. The magnitude of the critical mass depends upon the type of fissionable material involved; it is on the order of 2 kg for $^{235}_{92}U$. (Precise data are not available to the general public.)

The atomic bomb

The principle of the atomic bomb is very simple. A nuclear explosion will be produced when a mass of fissionable material is brought together (in a time span of a few microseconds) to form a supercritical mass in which a chain reaction will occur. Technical details include such considerations as (1) how to provide for transporting the bomb; (2) the arming (or disarming) and safety mechanism; (3) the firing mechanism; and (4) sufficient shielding so that persons who handle the device are not exposed to large quantities of radiation.

Two possibilities for a bomb are shown in Fig. 32-11. In one, a charge of chemical explosive is fired in order to hurl a subcritical mass of fissionable material down a tube toward another subcritical mass. A supercritical mass is formed when the two come together, and the pressures generated by the chemical explosion hold the mass together long enough for the chain reaction to progress to an advanced state before the effect of the nuclear reaction begins to blow the bomb mechanism (and everything else) to pieces. The second design has the fissionable material arranged as (separated) sections of a sphere. A chemical explosion blasts the sections inward to form a solid, supercritical sphere, and a nuclear explosion results.

Nuclear fallout

A nuclear explosion produces intense radiation of all types in the vicinity of the blast, and large quantities of dust and debris are carried upward to the higher regions of the atmosphere. The intense radiation at the moment of the blast induces radioactivity in many otherwise stable materials, and the winds distribute these materials over large areas. In particular, the fission fragments are highly radioactive. These materials eventually fall back to the surface of the earth and are, therefore, called *radioactive fallout*.

The radioactivity of the fission fragments stems from the excess of neutrons with which they are formed. A nucleus of $^{235}_{92}U$ has 51 more neutrons than protons. When the nucleus fissions, only a few neutrons are released so that the fission fragments have a large excess of neutrons. (Nuclides in the $Z = 30$ to $Z = 60$ region must have roughly the same number of neutrons as protons in order to be stable.) Hence, the fragments are radioactive and decay mainly by β^- emission as the excess neutrons are converted to protons.

Fallout is particularly dangerous when it enters the food chain and is ingested and stored in the body. For example, $^{90}_{38}Sr$, because of its chemical similarity to calcium, becomes concentrated in bone and may cause bone cancer or leukemia resulting from the destruction of the bone marrow.

Nuclear reactors

In broad terms a nuclear reactor is a device for producing and controlling a self-sustaining chain reaction. To be self-sustaining, an average of one neutron per fission must go on to produce another fission. Thus, controlling a nuclear reactor amounts to controlling the neutron flux within the reactor.

First of all, the design of the reactor must provide for the neutron flux to be great enough for the reactor to *go critical*. Neutrons that escape from the structure or that are absorbed in nonfission reactions do not help to sustain the reaction. The most abundant (99.3%) isotope of uranium $^{238}_{92}U$ can absorb neutrons without producing a fission according to

$$^{1}_{0}n + ^{238}_{92}U \rightarrow ^{239}_{92}U + \gamma$$

This reaction is more likely to occur if the neutrons are *fast* (high energy). On the other hand, if the neutrons are slowed down by a *moderator*, the above reaction is less likely to occur, and also the probability of neutron absorption by $^{235}_{92}U$ is increased markedly. Therefore, the fuel for a reactor is imbedded in a material such as graphite which slows the neutrons via collisions to speeds they would acquire naturally from being in thermal equilibrium with the moderator material. Other moderator materials are *light water* (H_2O) and *heavy water* (D_2O), water formed with deuterium, $D = {}^2_1H$.

To increase the likelihood of fission reactions, the percentage of U-235 in nuclear fuel is increased relative to U-238 from 0.7% to between 2 and 3%. Such fuel is said to be *enriched.*

Control rods, made of cadmium or boron, are incorporated into the fuel-moderator matrix to provide control over the neutron flux density. Cadmium and boron readily absorb slow neutrons. To increase the neutron flux density (and the rate at which fission reactions occur), the control rods are partially withdrawn from the core so that they absorb fewer neutrons. To shut down a reactor, the control rods are fully inserted and the chain reaction gradually diminishes.

Energy is taken from a reactor (in the form of heat) by circulating water or liquid sodium through the core. The heat produces steam in a heat exchanger and the steam is used to drive a turbine, which—in turn—drives an electrical generator. Essential features of a reactor are shown in Fig. 32-12.

Because the conversion of thermal energy to rotational mechanical energy is only about 33% efficient, most of the heat produced by a reactor must be "dumped" as waste heat into the environment. Roughly twice as much energy is wasted as is used; we throw away two units for every one we keep. This inefficiency is inherent in the thermal-to-mechanical energy conversion, and automobile engines, for example, suffer the same plight. Therefore, nuclear power stations must be carefully considered from the standpoint of *thermal pollution*, the unnatural warming of a river or body of water that may be injurious to biological processes.

A reactor consumes only about 1 g of fissionable material per day per

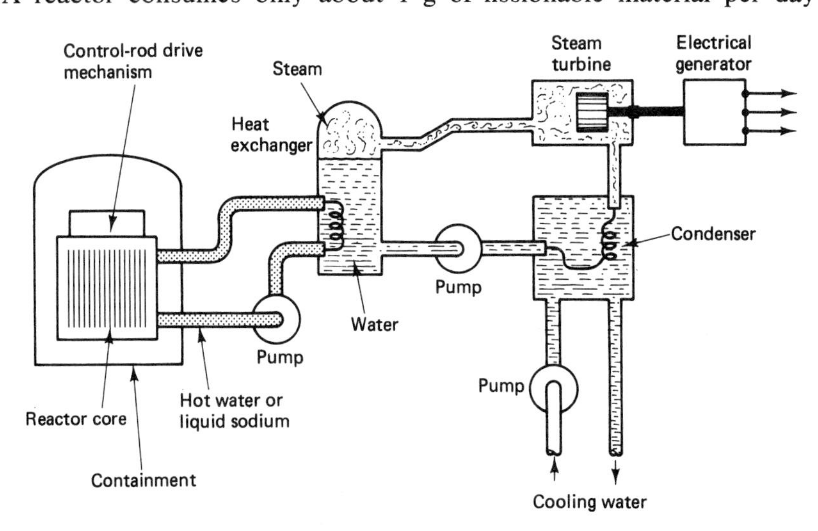

Figure 32-12 Essential features of a nuclear-reactor power station.

megawatt of power output. Nevertheless, the supply of nuclear fuel is not infinite; nuclear reactors in their present form are not the solution to the long-range energy problem.

A *breeder reactor* is actually capable of producing more fissionable material than it consumes. This is due to the fact that unfissionable $^{238}_{92}U$ and $^{232}_{90}Th$ may be transmuted to fissionable $^{239}_{94}Pu$ and $^{233}_{92}U$, respectively, according to the sequences

$$^{238}_{92}U + ^{1}_{0}n \rightarrow ^{239}_{92}U \xrightarrow[T_{1/2} = 24 \text{ m}]{\beta^- \text{ decay}} ^{239}_{93}Np \xrightarrow[T_{1/2} = 2.3 \text{ d}]{\beta^- \text{ decay}} ^{239}_{94}Pu$$

$$^{232}_{90}Th + ^{1}_{0}n \rightarrow ^{233}_{90}Th \xrightarrow[T_{1/2} = 22 \text{ m}]{\beta^- \text{ decay}} ^{233}_{91}Pa \xrightarrow[T_{1/2} = 27 \text{ d}]{\beta^- \text{decay}} ^{233}_{92}U$$

The neutrons required are provided by fissions within the breeder-reactor core: The *fertile* materials U-238 and Th-232 are simply inserted into the reactor core where they are exposed to the neutron flux. The thorium sequence is most effective with slow neutrons, but the uranium sequence works best with fast neutrons. Hence, a *fast breeder reactor* functions with *fast* rather than *slow* neutrons. Research and development of practical breeder reactors is presently under way, but not everyone agrees upon the desirability of constructing breeder reactors, or even regular nonbreeding reactors for that matter.

A major concern regarding widespread use of reactors (and breeders in particular) is that reactors can be used to produce the fissionable material required for a bomb. The isotope of plutonium $^{239}_{94}Pu$ is excellent material for making bombs, and a breeder reactor can produce it in sufficient quantities. The fear is that if everyone has a reactor, everyone will have a bomb. It follows that, in time, someone will use a bomb and trigger a sequence of events that will render this planet uninhabitable.

Another disadvantage of the use of reactors for power production is the generation of highly radioactive waste materials. The disposal of such materials is a major problem and is the topic of much discussion and the center of much controversy.

Routine concerns and problems with operating power reactors have much to do with safety mechanisms, leaks in the plumbing, and the inevitable human error. While a reactor cannot blow up in a full-fledged nuclear explosion, a far less exciting event—such as the release of radioactive materials into the environment—could have devastating consequences.

In assessing the hazards of nuclear energy, we must remember that other sources are not without disadvantages. Burning coal pollutes the air, and each year hundreds of miners lose their lives through mining accidents. Oil spills contaminate our oceans and strip mining frequently destroys our landscapes.

32-7 NUCLEAR FUSION

A loss of mass occurs whenever two or more isolated nucleons join together to form a more complex nuclide. The process is called *fusion*, and the mass that is lost is *given off* as energy. For example, the following reaction is the production of deuterium ${}^2_1H$ from the fusion of a proton and a neutron:

$${}^1_1H + {}^1_0n \rightarrow {}^2_1H + \gamma$$

Mass of hydrogen	1.007825 u	Mass of deuterium	2.014102 u
Mass of neutron	1.008665		
Total mass	2.016490		

$$\text{Mass lost} = 2.016490 - 2.014102 = 0.002388 \text{ u}$$

$$\begin{aligned}\text{Energy equivalent} &= (0.002388 \text{ u})(931.5 \text{ MeV/u})\\ &= 2.224 \text{ MeV}\end{aligned}$$

This quantity of energy is carried off by the deuterium nucleus and the gamma ray.

One of the reactions being studied in the course of research aimed at the construction of a practical fusion reactor is

$${}^2_1H + {}^2_1H \rightarrow {}^3_2He + {}^1_0n$$

In this reaction, 0.003510 u of mass is lost; the ${}^3_2He$ nucleus and neutron share about 3.27 MeV of kinetic energy, the energy equivalent of the mass lost.

In order for the preceding reaction to occur, the two deuterons must approach to about 10^{-14} m of each other to allow the attractive nuclear force to overcome the repulsive electrostatic force. At this distance, the electrostatic potential energy is on the order of 0.14 MeV. For a deuteron to acquire this amount of kinetic energy due to its thermal motion, the temperature would have to be on the order of 10^8 K. A contained sample of deuterium heated to this temperature will undergo fusion. Nuclear reactions that require such high temperatures to ignite are called *thermonuclear* reactions.

A major research effort is under way to develop a practical nuclear-fusion reactor for energy production. This is in the face of the foreseeable depletion of fossil fuels and uranium fuel for conventional (not breeder) reactors. There is an abundant supply of deuterium in the oceans and lakes of the earth (about 1 g of deuterium per gallon of water). Another advantage of the fusion process is that fewer long-lifetime nuclear wastes are produced than in the fission process.

The practical implementation of a fusion power plant appears to be many years or even decades in the future. The problem is this: how to produce the high temperatures under high pressure for a time period sufficiently long for the fusion process to occur.

Magnetic "bottles" consisting of specially shaped magnetic fields are used in an attempt to keep the hot *plasma* (ions and electrons) from contacting the walls of the containment vessel. When a charged particle approaches the converging regions of the field shown in Fig. 32-13, the particle is reflected back toward the

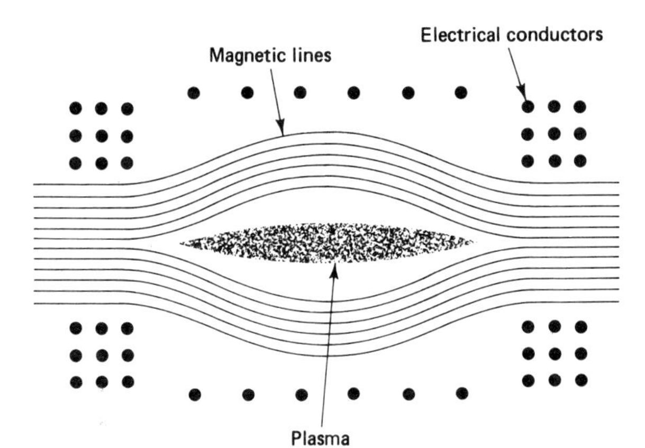

Figure 32-13 A magnetic bottle for containing hot plasmas.

center. However, such bottles have not proven completely satisfactory, and research continues.

Another experimental approach is shown in Fig. 32-14. A pellet of fusionable material is dropped into the region of convergence of the beams from several high-powered lasers. The radiation from the lasers simultaneously compresses and heats the pellet to fusion temperatures, and a brief burst of fusion-produced energy should be obtained. Thus far, lasers sufficiently powerful for this approach have not been developed.

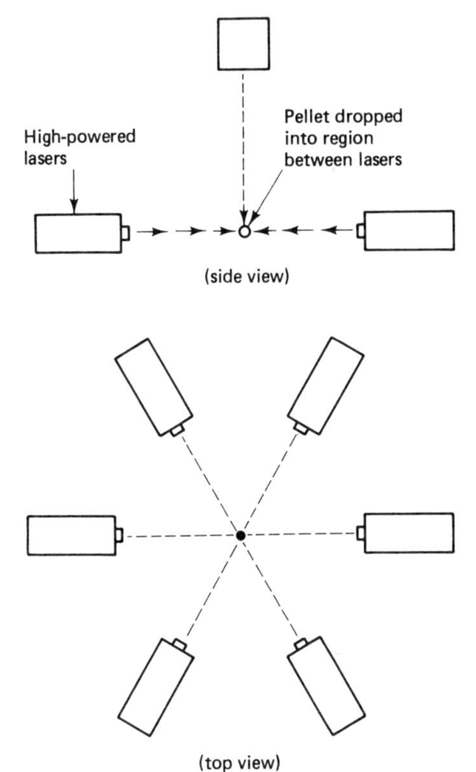

Figure 32-14 An approach to nuclear fusion that avoids the problems of confining high-temperature plasmas.

The fusion bomb

It is a comparatively simple matter to produce an uncontrolled fusion reaction, i.e., an H-bomb. An ordinary fission bomb produces temperatures on the order of 10^8 K, and this temperature is quite sufficient to ignite a fusion reaction. Thus, a small fission bomb can be used to trigger a larger fusion bomb.

An H-bomb may consist of three different types of fuel. First a charge of chemical explosive, such as TNT, is arranged to blast together a more-than-critical mass of $^{235}_{92}U$. The supercritical mass of U-235 then explodes as a result of a chain-reaction fission process. The resulting high temperatures ignite the fusion reaction in a quantity of lithium hydride, LiH, the fuel for the H-part of the bomb. Finally, there may be a quantity of U-238 included, which will undergo fission due to bombardment by fast neutrons.

The LiH used is not ordinary LiH. For the lithium, only ^6_3Li is used (as opposed to the more abundant ^7_3Li) and for hydrogen, ^2_1H (deuterium) is used rather than ordinary hydrogen. One deuterium reaction produces neutrons:

$$^2_1H + ^2_1H \rightarrow ^3_2He + ^1_0n$$

and these neutrons react with ^6_3Li to produce tritium (^3_1H) and helium:

$$^1_0n + ^6_3Li \rightarrow ^3_1H + ^4_2He$$

The tritium then reacts according to

$$^2_1H + ^3_1H \rightarrow ^4_2He + ^1_0n$$

Thus, the lithium not only serves as a carrier for the hydrogen (deuterium), but also participates in the fusion process.

H-bombs have been tested that give an explosive power equal to more than 100 million tons of TNT.

Energy source of the sun

The sun derives its energy from thermonuclear reactions in which hydrogen (primarily) is fused into helium. As the sun was forming, gravitational attraction gave rise to temperatures within the sun sufficiently high that thermonuclear fusion began. One set of reactions, called the *proton-proton cycle* is

$$^1_1H + ^1_1H \rightarrow ^2_1H + e^+ + \nu$$

$$^1_1H + ^2_1H \rightarrow ^3_2He + \gamma$$

$$^3_2He + ^3_2He \rightarrow ^4_2He + ^1_1H + ^1_1H$$

These reactions produce more than 26 MeV of kinetic energy. Many other reactions also occur, but the proton-proton cycle is most prevalent in hydrogen-rich stars like the sun. Fig. 32-15 is an illustration of the proton-proton cycle.

The first reaction produces neutrinos, which have the ability to pass through matter with little interaction. Consequently, neutrinos produced at the center of the sun readily escape and are emitted into space. An experiment to detect these

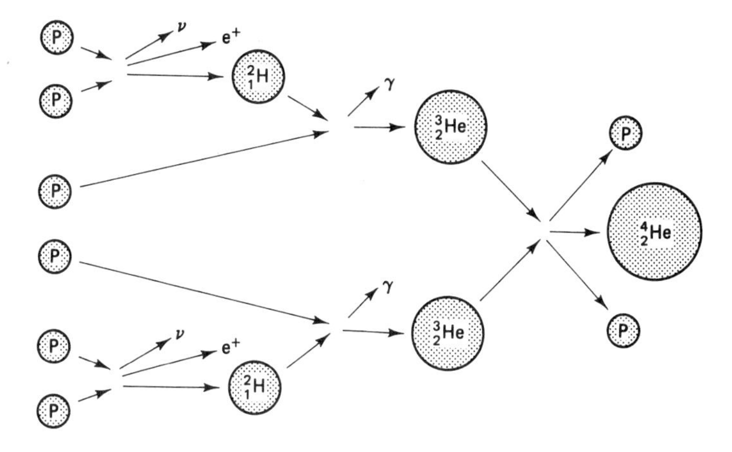

Figure 32-15 The proton-proton cycle, one mechanism of nuclear fusion that occurs in the sun.

neutrinos on earth detected less than 10% of the anticipated number of neutrinos. Hence, we must conclude that the question of solar energy production is not yet fully solved.

To Go Further

Find out about the lives and work of:

Henri Becquerel (1852–1908), French scientist
Pierre Curie (1859–1906), French physicist; husband of Marie Curie
Marie Curie [Marja Sklodowska] (1867–1934), Polish-born pioneer in the study of radioactivity
W. K. Roentgen (1845–1923), German-born scientist
Lise Meitner (1878–1968), Austrian-born physicist
Otto Hahn (1879–1968), German physicist
Louis Harold Gray, British radiobiologist
Enrico Fermi (1901–1954), Italian physicist

Questions

1. How is the atomic mass unit defined?
2. Does the binding energy represent energy that a nucleus *has* or has given up?
3. What experimental evidence indicates that radioactivity is a nuclear phenomenon?
4. Briefly describe alpha, beta, and gamma radiation.
5. What does it mean when we say that radioactive decay exhibits a statistical character?
6. Which two of the three radioactive decay modes results in the transmutation of the parent nuclide?
7. What problem in the theory of beta decay led to the discovery of the neutrino?
8. Explain the two types of transitions that occur in a radioactive decay series.
9. Explain the principle on which radioactive dating is based. What assumptions are made? How is it possible to test the procedure?
10. What are the units of measure of the activity of a radioactive source?
11. Describe the manner in which radioactivity can be induced in a material that initially was not radioactive.

12. Describe the transmutation that results from electron capture.
13. What might happen if a subcritical mass could somehow be compressed into a smaller volume?
14. Why is a nuclear reactor able to "go critical" without producing an explosion? What is the relationship between neutrons that produce fissions and the number of fissions occurring at the point of criticality?
15. Why is it impossible for a nuclear reactor to produce a blast during a malfunction that would resemble the mushroom-cloud explosion?
16. Why must nuclear reactors dump large quantities of heat into the environment? Do plants that burn fossil fuel also have to dump large quantities of heat into the environment?
17. Describe the process by which a breeder reactor can produce more nuclear fuel than it consumes. What are the difficulties or concerns associated with a breeder reactor?
18. Describe the major problem to be overcome in the development of a fusion reactor.
19. What are the advantages of fusion (as envisioned) over fission reactors?

Problems

1. Use $E = mc^2$ to show that one atomic mass unit has an energy equivalent of about 931.5 MeV.
2. Determine the approximate radius of a nucleus of: **(a)** $^{12}_{6}C$; **(b)** $^{56}_{26}Fe$; **(c)** $^{108}_{47}Ag$; **(d)** $^{238}_{92}U$.
3. Given that the mass defect of a $^{4}_{2}He$ nucleus is 0.0305 u, compute the average binding energy per nucleon.
4. Use the data of Table 32-1 to determine: **(a)** the mass defect; **(b)** binding energy; **(c)** binding energy per nucleon for deuterium, $^{2}_{1}H$.
5. Repeat Problem 4 for tritium, $^{3}_{1}H$.
6. The half-life of $^{90}_{38}Sr$ is 28 years. Compute the decay constant λ in units of: **(a)** $(\text{years})^{-1}$; **(b)** $(\text{days})^{-1}$.
7. Suppose we have an assembly of 1 million $^{90}_{28}Sr$ atoms. Use the results of the previous problem to compute the number of atoms that are expected to decay in 1 day.
8. How many disintegrations per second would be expected from a 1-g sample of $^{90}_{38}Sr$?
9. How many disintegrations per second would be expected from a 1-g sample of $^{14}_{6}C$?
10. The half-life of $^{131}_{53}I$ is 8 days. If 5 days are required to ship a 10-g sample from the point of production to the point of use, how many grams of the isotope will be received at the point of use?
11. A specimen of $^{131}_{53}I$ produces 100 counts/s on a Geiger counter. How many counts per second will the specimen produce after: **(a)** 4 days; **(b)** 8 days; **(c)** 16 days; and **(d)** 1 year?
12. Approximately what dose is represented by 0.1 rad of: **(a)** X-rays? **(b)** beta particles? **(c)** alpha particles?
13. Calculate the number of U-235 atoms that must fission in order to release 1 J of energy.
14. **(a)** Compute the energy produced by a hypothetical reactor that consumes 1 g/day of U-235.
    (b) If the thermal-electrical energy conversion is 33% efficient, what useful energy is produced?
    (c) How much thermal energy must be dumped into the environment?

APPENDIX A

MATHEMATICS

A-1 ALGEBRA

The *FOIL* (First, Outer, Inner, Last) method of multiplying two sums is

F L

$$(a + b)(c + d) = ac + ad + bc + bd$$

I O

Hence,

$$\begin{aligned}(x + 2)(x - 1) &= x^2 - x + 2x - 2 \\ &= x^2 + x - 2\end{aligned}$$

$$\begin{aligned}(a + b)^2 = (a + b)(a + b) &= a^2 + ab + ba + b^2 \\ &= a^2 + 2ab + b^2\end{aligned}$$

The *difference of two perfect squares* may be factored as

$$(x^2 - y^2) = (x - y)(x + y)$$

Hence,

$$(x^2 - 1) = (x - 1)(x + 1)$$

$$(9y^2 - 4) = (3y - 2)(3y + 2)$$

The *distributive law of multiplication over addition* is

$$a(b + c) = ab + ac$$

Hence,

$$3(x + 1) = 3x + 3$$

$$3a(x + 2) = 3ax + 6a$$

The *binomial expansion* is

$$(1 + x)^n = 1 + \frac{n}{1}x + \frac{n(n-1)}{2(1)}x^2 + \frac{n(n-1)(n-2)}{3(2)(1)}x^3 + \cdots$$

When x is small compared to 1, dropping the higher powers of x gives the following approximations:

$$(1 + x)^n \approx 1 + nx \qquad (x << 1)$$

$$(1 + x)^2 \approx 1 + 2x$$

$$(1 - x)^2 \approx 1 - 2x$$

$$\frac{1}{1 + x} = (1 + x)^{-1} \approx 1 - x$$

$$\frac{1}{1 - x} = (1 - x)^{-1} \approx 1 + x$$

$$\sqrt{1 + x} = (1 + x)^{1/2} \approx 1 + \frac{1}{2}x$$

$$\frac{1}{\sqrt{1 + x}} = (1 + x)^{-1/2} \approx \frac{1}{2}x$$

A *quadratic equation* has the form

$$ax^2 + bx + c = 0$$

where a, b, and c are constants. The *quadratic formula* provides the *two* solutions to a quadratic equation:

$$x = \frac{-b \pm \sqrt{b^2 - 4ac}}{2a}$$

The quantity $b^2 - 4ac$ is called the *discriminant*. When the discriminant is:

positive, the roots are real and unequal;
zero, the roots are real but equal;
negative, the roots are imaginary and unequal.

EXAMPLE Find the solutions to $x^2 - 7x + 12 = 0$

Solution. $a = 1$, $b = -7$, and $c = 12$.

$$x = \frac{-(-7) \pm \sqrt{(-7)^2 - 4(1)(12)}}{2(1)}$$

$$= \frac{7 \pm \sqrt{49 - 48}}{2}$$

$$= \frac{7 \pm 1}{2}$$

$$x_1 = \frac{7 + 1}{2} = 4 \qquad x_2 = \frac{7 - 1}{2} = 3$$

EXAMPLE Find the solutions to $x^2 + 2x + 5 = 0$

Solution. $a = 1$, $b = 2$, and $c = 5$.

$$x = \frac{-2 \pm \sqrt{2^2 - 4(1)(5)}}{2(1)}$$

$$= \frac{-2 \pm \sqrt{-16}}{2}$$

$$= \frac{-2 \pm 4i}{2} = -1 \pm 2i$$

$$x_1 = -1 + 2i \qquad x_2 = -1 - 2i$$

Beware! Here are three errors that are commonly made:

$$\frac{1}{a} + \frac{1}{b} \neq \frac{1}{a + b}$$

$$(a + b)^2 \neq a^2 + b^2$$

$$(a^2 + b^2) \neq (a + b)(a + b)$$

A-2 EXPONENTS

An exponent is the power to which a number is raised. When we write x^2, x^3, or x^4, 2, 3, and 4 are called *exponents*. Hence, by definition,

$$x^2 = (x)(x) \qquad x^3 = (x)(x)(x) \qquad x^4 = (x)(x)(x)(x)$$

Negative exponents denote reciprocals:

$$x^{-1} = \frac{1}{x} \qquad x^{-2} = \frac{1}{x^2} \qquad x^{-3} = \frac{1}{x^3}$$

When numbers expressed as powers are *multiplied*, the exponents are *added*.

$$(x^m)(x^n) = x^{m+n} \qquad x^2(x^3) = x^5$$

When numbers expressed as powers are *divided*, the exponent of the denominator is *subtracted* from the exponent of the numerator:

$$\frac{x^m}{x^n} = x^{m-n} \qquad \frac{x^5}{x^3} = x^2 \qquad \frac{x^2}{x^3} = x^{-1} = \frac{1}{x}$$

A number expressed as a *power* may in turn be *raised* to a *power*. In such a case, the exponents are *multiplied*:

$$(x^2)^3 = x^6 \qquad (a^3)^4 = a^{12} \qquad (x^{-2})^3 = x^{-6}$$

$$(a^2x^3)^2 = a^4x^6 \qquad \left(\frac{a}{b}\right)^n = \frac{a^n}{b^n}$$

Any number (except 0) to the *zero power* is *unity*: $x^0 = 1$; $a^0 = 1$; $(5 - x)^0 = 1$. This stems from

$$1 = \frac{x^n}{x^n} = x^{n-n} = x^0$$

Fractional exponents denote *roots* of numbers:

$$x^{1/2} = \sqrt{x} \qquad x^{1/3} = \sqrt[3]{x} \qquad x^{1/4} = \sqrt[4]{x}$$

$$(4)^{1/2} = 2 \qquad 8^{1/3} = 2 \qquad 16^{-1/4} = \frac{1}{2}$$

This arises from the fact that $(x^{1/n})^n = x$.

A-3 LOGARITHMS

Any positive number N can be written in the form $N = b^x$ provided $b > 1$. Accordingly, x is said to be the *logarithm* of N to the *base* b. Conversely, N is said to be the *antilogarithm* of x.

The number 10 is used as the base for *common* logarithms. Thus, the common logarithm of 100 is 2:

$$\log_{10}100 = 2 \quad \text{because} \quad 10^2 = 100$$

We use *log* to denote logarithms to the base 10 and omit the subscript that denotes the base. Thus,

$$\log 100 = 2 \qquad \log 1000 = 3 \qquad \log 10 = 1$$

Fractional logarithms are possible:

$$\log 200 = 2.301 \quad \text{because} \quad 10^{2.301} = 200$$

$$\log 500 = 2.699 \quad \text{because} \quad 10^{2.699} = 500$$

Tables of logarithms have been prepared but are now obsolete for everyday use because scientific calculators provide the logarithms at the touch of a button.

Properties of logarithms

The logarithm of a *product* is the *sum* of the logarithms of the factors:

$$\log(ab) = \log a + \log b$$

The logarithm of a *quotient* is the *difference* of the logarithms of the factors:

$$\log\left(\frac{a}{b}\right) = \log a - \log b$$

The logarithm of a number *raised* to a *power n* is *n* times the logarithm of the number:

$$\log a^n = n \log a$$

The following stems from the definition of logarithms:

$$N = 10^{\log N}$$

Logarithms of numbers less than 1 are *negative*. Logarithms of negative numbers do not exist.

Natural logarithms

The base of the natural logarithms is the irrational number e, which may be obtained from the following infinite series:

$$e = 1 + \frac{1}{1} + \frac{1}{2(1)} + \frac{1}{3(2)(1)} + \frac{1}{4(3)(2)(1)} + \cdots$$
$$\approx 2.71828 \ldots$$

The *symbol* used to denote natural logarithms is ln. With a calculator you may verify the following:

$$\ln 5 = 1.609 \qquad \ln 10 = 2.303$$

$$\ln 50 = 3.912 \qquad \ln 100 = 4.605$$

The natural logarithms have a different base, but they obey the same properties as the logarithms to the base 10.

The natural and common logarithms are directly proportional:

$$\ln x = 2.3026 \log x$$

Calculating roots of numbers is easy using logarithms:

$$\sqrt[n]{a} = 10^{(1/n)\log a}$$

or

$$\sqrt[n]{a} = e^{(1/n)\ln a}$$

For example,

$$\sqrt{64} = 10^{(\log 64)/2} = 10^{(1.806)/2} = 8.000$$

A-4 SOME PROPERTIES OF RADICALS

$$\sqrt{a} = a^{1/2} \qquad \sqrt{\frac{a}{b}} = \frac{\sqrt{a}}{\sqrt{b}}$$

$$\sqrt{ab} = \sqrt{a}\,\sqrt{b} \qquad \sqrt{a^2} = a$$

$$\frac{\sqrt{b}}{a} = \sqrt{\frac{b}{a^2}} \qquad \sqrt{a^2 b} = a\sqrt{b}$$

A-5 SCIENTIFIC NOTATION

Very large or very small numbers are more conveniently written in scientific notation form, as indicated by the following:

$$352 = 3.52 \times 10^2 \qquad 0.0352 = 3.52 \times 10^{-2}$$

$$3250 = 3.52 \times 10^3 \qquad 0.00352 = 3.52 \times 10^{-3}$$

$$35{,}200 = 3.52 \times 10^4 \qquad 0.000352 = 3.52 \times 10^{-4}$$

$$93{,}000{,}000 = 9.3 \times 10^7$$

$$0.0000005 = 5 \times 10^{-7}$$

Numbers may be *added* or *subtracted* only if the power of 10 is the same for both numbers:

$$\begin{array}{r} 6 \times 10^8 \\ + \; 3 \times 10^8 \\ \hline 9 \times 10^8 \end{array} \qquad \begin{array}{r} 5.40 \times 10^4 \\ 2.30 \times 10^3 \\ \hline \end{array} \rightarrow \begin{array}{r} 5.40 \times 10^4 \\ 0.23 \times 10^4 \\ \hline 5.63 \times 10^4 \end{array}$$

Multiplication is accomplished by multiplying the constants and adding the powers of 10:

$$(3 \times 10^4)(2 \times 10^5) = 6 \times 10^9$$

$$(7 \times 10^4)(5 \times 10^3) = 35 \times 10^7 = 3.5 \times 10^8$$

$$(4 \times 10^{-6})(2 \times 10^9) = 8 \times 10^3$$

Division is accomplished by dividing the constants and subtracting the exponents as follows:

$$\frac{6 \times 10^8}{3 \times 10^5} = 1.23 \times 10^3 \qquad \frac{8 \times 10^{-3}}{4 \times 10^{-5}} = 2 \times 10^2$$

$$\frac{3 \times 10^{-10}}{5 \times 10^{-7}} = 0.6 \times 10^{-3} \qquad \frac{8 \times 10^4}{2} = 4 \times 10^4$$

A number written in scientific notation is *raised* to a *power* as follows:

$$(2 \times 10^3)^2 = 2^2 \times (10^3)^2 = 4 \times 10^6$$

$$(4 \times 10^{-6})^2 = 16 \times 10^{-12} = 1.6 \times 10^{-11}$$

$$(4 \times 10^8)^3 = 64 \times 10^{24} = 6.4 \times 10^{25}$$

Scientific notation is rendered by most *computers* in a notation involving the letter E:

$$4.2 \times 10^6 = 4.2\text{E}6 \quad \text{or} \quad 4.20\text{E}06$$

$$4.2 \times 10^{-6} = 4.2\text{E}{-6} \quad \text{or} \quad 4.20\text{E}{-06}$$

The meaning is the same; most computers cannot easily reproduce subscripts or exponents.

Roots of numbers are computed as follows:

$$\sqrt{9 \times 10^6} = \sqrt{9} \times (10^6)^{1/2} = 3 \times 10^3$$
$$\sqrt{49 \times 10^8} = \sqrt{49} \times (10^8)^{1/2} = 7 \times 10^4$$
$$\sqrt{36 \times 10^{-4}} = \sqrt{36} \times (10^{-4})^{1/2} = 6 \times 10^{-2}$$
$$\sqrt[3]{8 \times 10^6} = \sqrt[3]{8} \times (10^6)^{1/3} = 2 \times 10^2$$
$$\sqrt[5]{32 \times 10^5} = \sqrt[5]{32} \times (10^5)^{1/5} = 2 \times 10^1 = 20$$

Sometimes the form of the number must be written in a different power of 10:

$$\sqrt{4 \times 10^5} = \sqrt{40 \times 10^4} = \sqrt{40} \times (10^4)^{1/2} = 6.32 \times 10^2$$
$$\sqrt[3]{3 \times 10^8} = \sqrt[3]{300 \times 10^6} = \sqrt[3]{300} \times (10^6)^{1/3} = 6.69 \times 10^2$$
$$\sqrt{5 \times 10^{-11}} = \sqrt{50 \times 10^{-12}} = \sqrt{50} \times (10^{-12})^{1/2} = 7.07 \times 10^{-6}$$

A-6 TRIGONOMETRY

Basic trigonometry deals primarily with *right triangles*, triangles in which one interior angle is 90°, as shown in Fig. A-1. The sum of the interior angles of any plane triangle equals 180° so that

$$\theta + \phi + 90 = 180$$

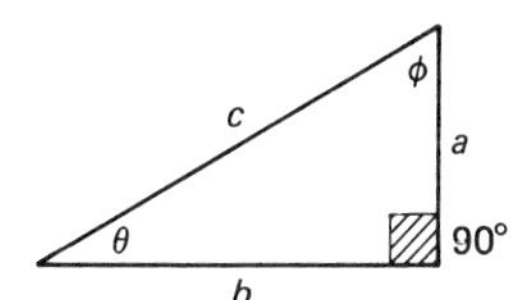

Figure A-1

and therefore,

$$\theta + \phi = 90$$

The *Pythagorean theorem* relates the lengths of the sides of a right triangle:

$$c^2 = a^2 + b^2 \quad \text{or} \quad c = \sqrt{a^2 + b^2}$$
$$b = \sqrt{c^2 - a^2} \quad \text{or} \quad a = \sqrt{c^2 - b^2}$$

The longest side of a right triangle is called the *hypotenuse*.

The *trigonometric functions* (sine, cosine, and tangent) may be defined in terms of a right triangle. Referring to Fig. A-2, the definitions are:

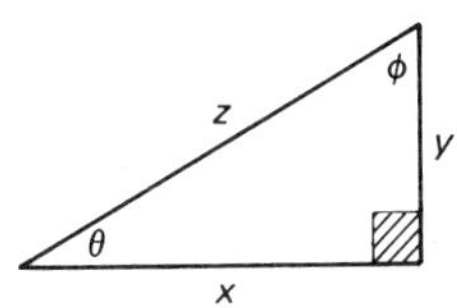

Figure A-2

$$\sin\theta = \frac{\text{side opposite }\theta}{\text{hypotenuse}}$$

$$\sin\theta = \frac{y}{z}$$

$$\cos\theta = \frac{\text{side adjacent }\theta}{\text{hypotenuse}}$$

$$\cos\theta = \frac{x}{z}$$

$$\tan\theta = \frac{\text{side opposite }\theta}{\text{side adjacent }\theta}$$

$$\tan\theta = \frac{y}{x}$$

The three other trigonometric functions (cotangent, secant, and cosecant) may be defined in terms of the previous three:

$$\cot\theta = \frac{1}{\tan\theta} = \frac{x}{y}$$

$$\sec\theta = \frac{1}{\cos\theta} = \frac{z}{x}$$

$$\csc\theta = \frac{1}{\sin\theta} = \frac{z}{y}$$

Further,

$$\tan\theta = \frac{\sin\theta}{\cos\theta}$$

and

$$\sin^2\theta + \cos^2\theta = 1$$

Referring again to Fig. A-2, the functions of the *complementary angles* θ and ϕ are related by

$$\sin\phi = \cos\theta$$
$$\cos\phi = \sin\theta$$
$$\tan\phi = \cot\theta$$

In Fig. A-3 a triangle is shown that is not a right triangle. The lengths of the sides are a, b, and c, and the interior angles opposite the sides are A, B, and C, respectively.

The *law of sines* expresses the equality of the ratios of the sine of an angle to its opposite side:

$$\frac{\sin A}{a} = \frac{\sin B}{b} = \frac{\sin C}{c}$$

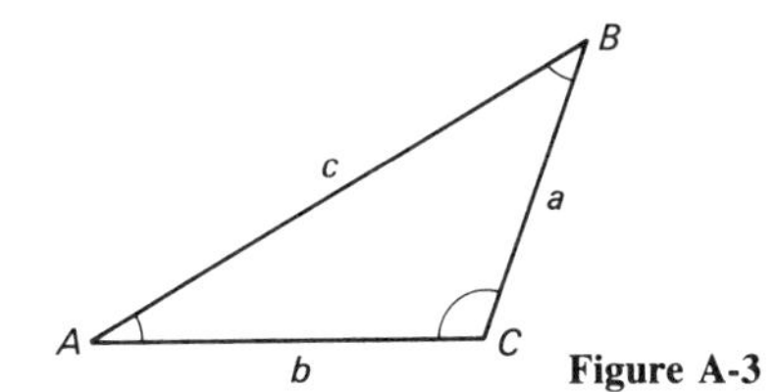

Figure A-3

The *law of cosines* is a more general case of the Pythagorean theorem and may assume three different arrangements:

$$a^2 = b^2 + c^2 - 2bc \cos A$$
$$b^2 = a^2 + c^2 - 2ac \cos B$$
$$c^2 = a^2 + b^2 - 2ab \cos C$$

Angles may be specified in radians as an alternative to degrees. Because 2π radians is the equivalent to 360°, we find that

$$1 \text{ radian} = \frac{180}{\pi} \approx 57.3°$$

A-7 FORMULAS

Surface area

Rectangle: length × width = lw

Triangle: $\frac{1}{2}$ base × altitude = $(0.5)bh$

Circle: $\pi R^2 = \dfrac{\pi D^2}{4}$ (R = radius, D = diameter)

Sphere: $4\pi R^2 = \pi D^2$

Right circular cylinder: $2\pi R^2 + 2\pi Rh = 2\pi R(R + h)$ (h = height)

Volume

Rectangular solid: length × width × height = lwh

Right circular cylinder: $\pi R^2 h$

Sphere: $\dfrac{4}{3}\pi R^3 = \dfrac{\pi D^3}{6}$

A-8 SIMULTANEOUS LINEAR EQUATIONS (CRAMER'S RULE)

A *matrix* is an array of numbers. The *determinant* of a matrix is a single number derived from the matrix. Here, in symbols, is a 3 × 3 matrix:

$$A = \begin{pmatrix} a_{11} & a_{12} & a_{13} \\ a_{21} & a_{22} & a_{23} \\ a_{31} & a_{32} & a_{33} \end{pmatrix}$$

The determinant of A, det A, is

$$\det A = \begin{vmatrix} a_{11} & a_{12} & a_{13} \\ a_{21} & a_{22} & a_{23} \\ a_{31} & a_{32} & a_{33} \end{vmatrix}$$

Determinants are defined only for square matrices. The following method can be used to evaluate the determinant of a 3×3 matrix:

1. Recopy and append the first two columns.

$$\begin{matrix} a_{11} & a_{12} & a_{13} & a_{11} & a_{12} \\ a_{21} & a_{22} & a_{23} & a_{21} & a_{22} \\ a_{31} & a_{32} & a_{33} & a_{31} & a_{32} \end{matrix}$$

2. Form the products $P_1, P_2, \ldots,$ as indicated. For example:

$$P_1 = a_{11} \cdot a_{22} \cdot a_{33}$$
$$P_5 = a_{32} \cdot a_{23} \cdot a_{11}$$
$$\vdots$$

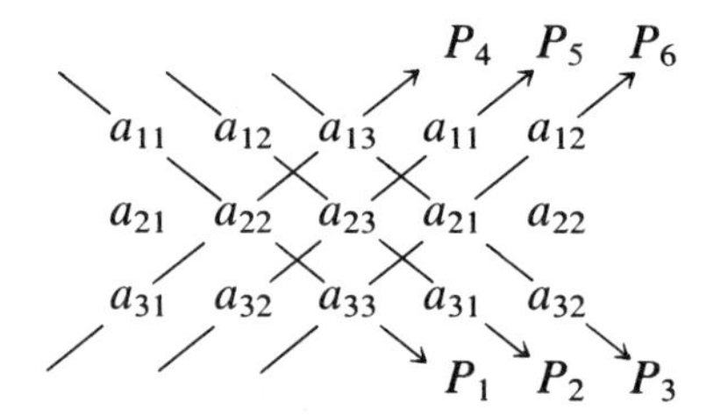

3. The determinant of A is then given by

$$\det A = P_1 + P_2 + P_3 - P_4 - P_5 - P_6$$

This procedure is illustrated below.

One application of determinants is to solving simultaneous linear equations. The following is a system of *three* simultaneous equations in three unknowns, x_1, x_2, and x_3:

$$\begin{aligned} a_{11}x_1 + a_{12}x_2 + a_{13}x_3 &= c_1 \\ a_{21}x_1 + a_{22}x_2 + a_{23}x_3 &= c_2 \\ a_{31}x_1 + a_{32}x_2 + a_{33}x_3 &= c_3 \end{aligned}$$

We wish to solve these equations for x_1, x_2, and x_3.

1. Form the matrix of coefficients and find its determinant. The matrix of coefficients is:

$$A = \begin{pmatrix} a_{11} & a_{12} & a_{13} \\ a_{21} & a_{22} & a_{23} \\ a_{31} & a_{32} & a_{33} \end{pmatrix}$$

Its determinant, det A, is evaluated as given above.

2. Form the *constant matrix* for each of x_1, x_2, and x_3. We denote the constant matrices by B, C, and D, respectively. To form the constant matrix for a

particular variable, substitute the column of constants for the column of coefficients for that variable in the matrix of coefficients. Hence,

for x_1: for x_2: for x_3:

$$B = \begin{pmatrix} c_1 & a_{12} & a_{13} \\ c_2 & a_{22} & a_{23} \\ c_3 & a_{32} & a_{33} \end{pmatrix} \quad C = \begin{pmatrix} a_{11} & c_1 & a_{13} \\ a_{21} & c_2 & a_{23} \\ a_{31} & c_3 & a_{33} \end{pmatrix} \quad D = \begin{pmatrix} a_{11} & a_{12} & c_1 \\ a_{21} & a_{22} & c_2 \\ a_{31} & a_{32} & c_3 \end{pmatrix}$$

3. Evaluate the determinant of each constant matrix. That is, find the value of det B, det C, and det D.
4. Then, x_1, x_2, and x_3 are given by:

$$x_1 = \frac{\det B}{\det A} \qquad x_2 = \frac{\det C}{\det A} \qquad x_3 = \frac{\det D}{\det A}$$

This method is known as *Cramer's rule*.

As an example, let us solve the following system:

$$\begin{aligned} 4x_1 + 5x_2 - 1x_3 &= 3 \\ 1x_1 - 2x_2 + 1x_3 &= 9 \\ 2x_1 + 5x_2 + 2x_3 &= 9 \end{aligned}$$

Form the matrix of coefficients and evaluate its determinant:

$$A = \begin{pmatrix} 4 & 5 & -1 \\ 1 & -2 & 1 \\ 2 & 5 & 2 \end{pmatrix}$$

$$\begin{array}{ccccc} & & 4 & 20 & 10 \\ 4 & 5 & -1 & 4 & 5 \\ 1 & -2 & 1 & 1 & -2 \\ 2 & 5 & 2 & 2 & 5 \\ & & -16 & 10 & -5 \end{array}$$

$$\det A = (-16) + (10) + (-5) - (4) - (20) - (10)$$
$$\det A = -45$$

The constant matrices are

for x_1: for x_2: for x_3:

$$B = \begin{pmatrix} 3 & 5 & -1 \\ 9 & -2 & 1 \\ 9 & 5 & 2 \end{pmatrix} \quad C = \begin{pmatrix} 4 & 3 & -1 \\ 1 & 9 & 1 \\ 2 & 9 & 2 \end{pmatrix} \quad D = \begin{pmatrix} 4 & 5 & 3 \\ 1 & -2 & 9 \\ 2 & 5 & 9 \end{pmatrix}$$

Omitting details, the determinants of the constant matrices are:

$$\det B = -135 \qquad \det C = 45 \qquad \det D = -180$$

Therefore,

$$x_1 = \frac{\det B}{\det A} = \frac{-135}{-45} = 3$$

$$x_2 = \frac{\det C}{\det A} = \frac{45}{-45} = -1$$

$$x_3 = \frac{\det D}{\det A} = \frac{-180}{-45} = 4$$

Substitution back into the original system verifies that these values are correct.

Comments: (1) The procedure for evaluating determinants given here does not work for determinants larger than 3×3. (2) If the determinant of the matrix of coefficients is zero, the original equations are not independent and no unique solution exists.

APPENDIX B

THE PERIODIC CHART

1 H 1.0080																	2 He 4.0026
3 Li 6.941	4 Be 9.0122											5 B 10.81	6 C 12.011	7 N 14.0067	8 O 15.9994	9 F 18.9984	10 Ne 20.179
11 Na 22.9898	12 Mg 24.305											13 Al 26.9815	14 Si 28.086	15 P 30.9738	16 S 32.06	17 Cl 35.453	18 Ar 39.948
19 K 39.102	20 Ca 40.08	21 Sc 44.956	22 Ti 47.90	23 V 50.941	24 Cr 51.996	25 Mn 54.9380	26 Fe 55.847	27 Co 58.9332	28 Ni 58.71	29 Cu 63.54	30 Zn 65.37	31 Ga 69.72	32 Ge 72.59	33 As 74.9216	34 Se 78.96	35 Br 79.909	36 Kr 83.80
37 Rb 85.467	38 Sr 87.62	39 Y 88.906	40 Zr 91.22	41 Nb 92.906	42 Mo 95.94	43 Tc (99)	44 Ru 101.07	45 Rh 102.906	46 Pd 106.4	47 Ag 107.870	48 Cd 112.40	49 In 114.82	50 Sn 118.69	51 Sb 121.75	52 Te 127.60	53 I 126.9045	54 Xe 131.30
55 Cs 132.906	56 Ba 137.34	57 La 138.906	72 Hf 178.49	73 Ta 180.948	74 W 183.85	75 Re 186.2	76 Os 190.2	77 Ir 192.2	78 Pt 195.09	79 Au 196.967	80 Hg 200.59	81 Tl 204.37	82 Pb 207.2	83 Bi 208.981	84 Po (210)	85 At (210)	86 Rn (222)
87 Fr (223)	88 Ra (226)	89 Ac (227)	104 Rf (261)	105 Ha (262)													

Lanthanide Series

58 Ce 140.12	59 Pr 140.908	60 Nd 144.24	61 Pm (147)	62 Sm 150.4	63 Eu 151.96	64 Gd 157.25	65 Tb 158.925	66 Dy 162.50	67 Ho 164.930	68 Er 167.26	69 Tm 168.934	70 Yb 173.04	71 Lu 174.97

Actinide Series

90 Th (232)	91 Pa (231)	92 U (238)	93 Np (237)	94 Pu (242)	95 Am (243)	96 Cm (248)	97 Bk (249)	98 Cf (249)	99 Es (254)	100 Fm (257)	101 Md (258)	102 No (259)	103 Lr (260)

Atomic weights of stable elements are those adopted in 1969 by the International Union of Pure and Applied Chemistry. For those elements having no stable isotope, the mass number of the "most stable" well-investigated isotope is given in parentheses.

APPENDIX C

IMPORTANT METRIC PREFIXES

Prefix	Abbreviation	Meaning	Typical Examples
tera	T	$\times 10^{12}$	1 TeV = 10^{12} eV
giga	G	$\times 10^{9}$	1 GHz = 10^{9} Hz = 10^{9} cycles/s
mega	M	$\times 10^{6}$	1 megaton = 10^{6} tons
kilo	k	$\times 10^{3}$	1 kg = 1000 g
deci	d	$\times 10^{-1}$	1 dB = 0.1 B
centi	c	$\times 10^{-2}$	1 cm = 0.01 m
milli	m	$\times 10^{-3}$	1 mA = 0.001A
micro	μ	$\times 10^{-6}$	1 μV = 10^{-6} V
nano	n	$\times 10^{-9}$	1 ns = 10^{-9} s
pico	p	$\times 10^{-12}$	1 pF = 10^{-12} F
femto	f	$\times 10^{-15}$	1 fm = 10^{-15} m

APPENDIX D

THE ELEMENTS

Symbol	Name	Atomic Number	Symbol	Name	Atomic Number	Symbol	Name	Atomic Number
Ac	Actinium	18	Ge	Germanium	32	Po	Polonium	84
Ag	Silver	47	H	Hydrogen	1	Pr	Praseodymium	59
Al	Aluminum	13	Ha	Hahnium	105	Pt	Platinum	78
Am	Americium	95	He	Helium	2	Pu	Plutonium	94
Ar	Argon	18	Hf	Hafnium	72	Ra	Radium	88
As	Arsenic	33	Hg	Mercury	80	Rb	Rubidium	37
At	Astatine	85	Ho	Holmium	67	Re	Rhenium	75
Au	Gold	79	I	Iodine	53	Rf	Rutherfordium	104
B	Boron	5	In	Indium	49	Rh	Rhodium	45
Ba	Barium	56	Ir	Iridium	77	Rn	Radon	86
Be	Beryllium	4	K	Potassium	19	Ru	Ruthenium	44
Bi	Bismuth	83	Kr	Krypton	36	S	Sulfur	16
Bk	Berkelium	97	La	Lanthanum	57	Sb	Antimony	51
Br	Bromine	35	Li	Lithium	3	Sc	Scandium	21
C	Carbon	6	Lr	Lawrencium	103	Se	Selenium	34
Ca	Calcium	20	Lu	Lutetium	71	Si	Silicon	14
Cd	Cadmium	48	Mg	Magnesium	12	Sm	Samarium	62
Ce	Cerium	58	Mn	Manganese	25	Sn	Tin	50
Cf	Californium	98	Mo	Molybdenum	42	Sr	Strontium	38
Cl	Chlorine	17	Mv	Mendelevium	101	Ta	Tantalum	73
Cm	Curium	96	N	Nitrogen	7	Tb	Terbium	65
Co	Cobalt	27	Na	Sodium	11	Tc	Technetium	43
Cr	Chromium	24	Nb	Niobium	41	Te	Tellurium	52
Cs	Cesium	55	Nd	Neodymium	60	Th	Thorium	90
Cu	Copper	29	Ne	Neon	10	Ti	Titanium	22
Dy	Dysprosium	66	Ni	Nickel	28	Tl	Thallium	81
Es	Einsteinium	99	No	Nobelium	102	Tm	Thulium	69
Er	Erbium	68	Np	Neptunium	93	U	Uranium	92
Eu	Europium	63	O	Oxygen	8	V	Vanadium	23
F	Fluorine	9	Os	Osmium	76	W	Tungsten	74
Fe	Iron	26	P	Phosphorus	15	Xe	Xenon	54
Fm	Fermium	100	Pa	Protactinium	91	Y	Yttrium	39
Fr	Francium	87	Pb	Lead	82	Yb	Ytterbium	70
Ga	Gallium	31	Pd	Palladium	46	Zn	Zinc	30
Gd	Gadolinium	64	Pm	Promethium	61	Zr	Zirconium	40

Alphabetical according to symbol.

APPENDIX E

THE GREEK ALPHABET

Α	α	alpha	Η	η	eta	Ν	ν	nu	Τ	τ	tau
Β	β	beta	Θ	θ	theta	Ξ	ξ	xi	Υ	υ	upsilon
Γ	γ	gamma	Ι	ι	iota	Ο	ο	omicron	Φ	φ	phi
Δ	δ	delta	Κ	κ	kappa	Π	π	pi	Χ	χ	chi
Ε	ϵ	epsilon	Λ	λ	lambda	Ρ	ρ	rho	Ψ	ψ	psi
Ζ	ζ	zeta	Μ	μ	mu	Σ	σ	sigma	Ω	ω	omega

APPENDIX F

PHYSICAL DATA OF THE PLANETS

Planet	Equatorial radius[a]	Mass[a]	Mean density, (g/cm^3)	Surface gravity[a]
Mercury	0.832	0.0553	5.44	0.378
Venus	0.949	0.815	5.24	0.894
Earth	1.000	1.000	5.50	1.000
Mars	0.533	0.107	3.9	0.379
Jupiter	11.19	317.9	1.3	2.54
Saturn	9.41	95.2	0.7	1.07
Uranus	4.4	14.6	1.0	0.8
Neptune	3.8	17.2	1.7	1.2
Pluto	0.4	0.02	~0.6	—

[a]in multiples of Earth value:

Equatorial radius of Earth: 6.378×10^6 m = 3963 mi

Mass of Earth: 5.98×10^{24} kg

Standard acceleration of gravity of Earth: $g = 9.80665\ m/s^2$

SOLUTIONS TO ODD-NUMBERED PROBLEMS

Chapter 2

1. (a) 91.44 m; (b) 109.36 yd **3.** (a) 1.112 N; (b) 9/10 **5.** 17.07 jiggers **7.** 52.99 L **9.** 28.2 mi/gal **11.** 1 mi/gal = 6.88 ft/tsp **13.** (a) 4800/h; (b) 115,200/day; (c) 4.2×10^7/y; (d) 2.94×10^9/70 y **15.** 2366 lb **17.** (a) 2.49 h; (b) 374 gal **19.** (a) 44.3 m; (b) 1962 m^2 **21.** 192 in.-oz **23.** (a) 5.89×10^{-5} cm; (b) 589 nm

Chapter 3

1. 15 mi/h **3.** 3.67 ft/s^2 **5.** (a) 40 ft/s; (b) 160 ft; (c) 20 ft/s **7.** 2300 mi/h **9.** 1.29 s **11.** 3760 ft/s (b) 73.4 s **13.** (a) 103 ft; (b) 342 ft **15.** (a) 88.36 ft/s; (b) 1.95 s; (c) 0.81 s **17.** 231 ft **19.** 15.5 ft **21.** (a) 30.25 ft; (b) 121 ft **23.** (a) 1.12 s; (b) 134 ft/s; (c) 139 ft/s

Chapter 4

1. 0 **3.** 3 m/s^2 **5.** 1.92 ft/s^2 **7.** 0.2248 lb **9.** 32.7 N **11.** 275 lb **13.** 15.4 ft **15.** a = 0.98 m/s^2 downward **17.** (a) 1.96 m/s^2; (b) 47 N **19.** (a) 7.84 m/s^2; (b) 17.6 N **21.** (a) 6.53 m/s^2; (b) 6.53 N; (c) 3.27 m **23.** 265 lb

Chapter 5

1. (a) (−7,2); (b) (−1,−4); (c) (6,−6) **3.** (a) 5 at 53.13°; (b) 5.39 at 21.8° **5.** (a) (7,5); (b) 35.54° **7.** (3,2) **9.** (7.01, 14.38) **11.** F_x = 36.25 lb; F_y = 16.9 lb **13.** 600.33 mi/h 1.9° south of east **15.** (a) 14.43 lb; (b) 17.68 lb; (c) 12.5 lb; (d) 143.2 lb **17.** (a) 28.87 lb; (b) 25 lb; (c) 50 lb; (d) 286.84 lb **19.** (a) 50.77 lb; 77.79 lb; (b) 32.63 lb, 67.37 lb; (c) 100 lb **21.** 30° **23.** (a) 187.9 g; (b) 68.4 g; (c) 68.4 g; (d) 0 **25.** 0.81 **27.** (a) 574.5 cm/s; (b) 482.1 cm/s; (c) 168.4 cm; (d) 1.17 s; (e) 565.3 cm **29.** (a) 200, −32 ft/s; 200, −64 ft/s; 200, −96 ft/s; (b) 3.35 s; (c) 221.8 ft/s at 25.6° below horizontal **31.** 100 lb **33.** (a) 87.5, 12.5 lb; (b) 75.0, 25 lb; (c) 50, 50 lb; (d) 25, 75 lb **35.** (a) 119.2 lb; (b) 155.6 lb **37.** 25.7 lb, 74.3 lb

Chapter 6

1. 100 ft-lb **3.** 980 ergs **5.** 1 J = 0.737 ft-lb **7.** 25 J **9.** (a) 0, 980 J; (b) 245, 735 J; (c) 490, 490 J; (d) 735, 245 J; (e) 980, 0 J **11.** 11.25 m/s **13.** (a) 2800 ft-lb/s; (b) 5.09 hp **15.** 36.4 hp **17.** (a) 7 J/s; (b) 0.056 kW-h; (c) 0.28 cents **19.** 24 hp **21.** 3.57 hp **23.** (a) 9×10^{13} J; (b) 8.33×10^6 kW-h; (c) $417,000; (d) 16.7×10^6 kW-h

Chapter 7

1. 1 kg-m/s **3.** (a) 29.4 N; (b) 3.0 kg; (c) 29.4 m/s **5.** (a) 78.4 N-s; (b) 39.2 m/s; (c) 78.4 N-s **7.** (a) 4 m/s; (b) 5760 J; (c) 19.2 J **9.** (a) 0.5 ft/s; (b) 2.5 ft/s; (c) 4.5 ft/s; (d) 0 ft/s **11.** (a) 0.4, −0.4 kg-m/s; (b) 0.8 kg-m/s; (c) 5.33 N **13.** (a) 2.44 m/s; (b) 2.44 m/s; (c) 0.098 kg-m/s; (d) 48.8 N **15.** −5.08 m/s, 7.46 m/s **17.** 0, 3 m/s **19.** (a) −2 m/s, 4 m/s; (b) 36 J, 0 J; (c) 4 J, 32 J; (d) 0.89; (e) same

Chapter 8

1. (a) $\pi/2$; (b) π; (c) 2π; (d) $\pi/4$; (e) $\pi/6$ **3.** (a) 3 cm; (b) 6 cm; (c) 2.356 cm; (d) 1.571 cm **5.** (a) 1.89×10^{-2} rad, 1.08°; (b) 184.3 ft **7.** (a) 0.089 rad/s; (b) 80 mi/h **9.** (a) 3.491 rad/s; (b) 20.9 in./s **11.** (a) −3.05 rad/s^2; (b) 873.7 rev **13.** (a) 4 N; (b) 2 N; (c) 1.33 N **15.** (a) 3750 N-m **19.** 0.05 kg-m^2; (a) 15 rad/s, 11.25 J; (b) 22.5 rad/s, 2.81 J; (c) 15 rad/s, 11.25 J; (d) 0, 20 J **21.** (a) 0.171 rad/s; (b) 3.43 J; (c) 0.2 rad/s; (d) 24 J **23.** (a) 1.352 s; (b) 1.277 J; (c) 1.277 J; (d) same **25.** (a) 1.884 s; (b) 2.25 m; (c) 0.75 m **27.** (a) difference = 0.274%; (b) relative to stars **29.** (a) 900 mi/h; (b) 735 mi/h; (c) 520 mi/h; (d) 0

Chapter 9

1. (a) 400 N; (b) 100 N; (c) 44.4 N; (d) 25 N **5.** 7.38×10^{22} kg **7.** 9.38 m/s^2 **9.** 628 kg/m^3 **11.** (a) 6450 m/s; (b) 2.6 h **13.** (a) 1.16×10^{-4} m/s; (b) 75.54 h **15.** (a) 3.76×10^7 J; (b) 1.98 **17.** 2.38×10^3 m/s **19.** 4.73×10^9 J **21.** 36×10^6 mi **23.** (a) 1.28 s; (b) 8.31 min; (c) 4.24 years **25.** 6460 mi **27.** (b) 7460 m/s **29.** (a) 9.30 vs 9.35 m/s^2; (b) 4.85 vs 6.27 m/s^2; (c) approximation fails **31.** (a) 3 in.; (b) 0.667

Chapter 10

1. 16.7 lb **3.** 5.76 **5.** (a) 5; (b) 5 ft; (c) 40 lb; (d) 200 ft-lb; (e) 200 ft-lb **7.** (a) 18; (b) 18 ft; (c) 1800 lb **9.** (a) 2.76; (b) 3; (c) 3; (d) 450 ft-lb; (e) 88.9% **11.** (a) 437.5 rpm; (b) 183.26, 45.81 rad/s; (c) 10.8 ft-lb; (d) 8.59 hp **13.** (a) 11; (b) 159 rpm **15.** 60, 100, 166167 rpm **17.** (a) 603; (b) 3.32 lb; (c) 6.63 lb **19.** (a) 83.3 lb; (b) 2500 ft-lb

Chapter 11

1. 1.55 mi **3.** (a) Si: 2,8,4; Ge: 2,8,18,4 **5.** (a) 9800 N/m^3; (b) 1000 kg/m^3; (c) 0; (d) 1000 kg/m^3 **7.** 12.42 lb/gal **9.** (a) 161.8 cm^3; (b) 2788 cm^3 **11.** (a) 866.7 lb/in.^{2}; (b) 24,500 lb **13.** (a) 3.528 mm Hg; (b) 763.5 mm Hg **15.** 6.95 lb **17.** 12.06 lb **19.** 0.82 **21.** 0.62 **23.** (a) 62.67; (b) 112.9 lb/in^2 **25.** 4.83 g **27.** 101.68 dynes/cm **29.** −1.25 mm **31.** 0.228 cm **33.** 9.05% **35.** (a) 34,300 dynes/cm; (b) 35 g/cm **37.** 0.1 m **39.** (a) 2×10^8 N/m^2; (b) 1×10^{-3} **41.** (a) 8.33×10^4 N/m^2; (b) 2.38×10^{-4}%; (c) 1.36×10^{-4} deg **43.** (a) 2.14 N-m; (b) 4.26×10^{14} N/m^2

Chapter 12

1. 9.79 gal/min **3.** 5985 gal/s **5.** (a) 2.5 s^{-1}; (b) 0.93 N **7.** 1.48 N-s/m^2 **9.** 424, 931 N/m^2 **11.** (a) 8.71 m/s; (b) 34.8 m/s; (c) 149.7 ft^3/min **13.** 2.6 lb/in^2 **15.** (a) 0.43 in.; (b) (1.73 in.; (c) 3.90 in. **17.** 11.83 ft **19.** 21.6 **21.** 4.26 **23.** for N_r = 200: 0.5 m/s **25.** (a) 17.9 lb; (b) 5.74 hp **27.** 23 m/s **29.** (a) 1.01; (b) 1.04; (c) 1.12; (d) 1.47

Chapter 13

1. −297°F; 163°R; 90 K **3.** 232°C; 910°R; 505 K **5.** 0.28°C **7.** (a) 7.74×10^5 J; (b) 1.85×10^5 cal; (c) 734 BTU **9.** 70°F 11. 77.1°F **13.** water equivalent = Mc = 11 g **15.** 13.82°C **17.** (a) 23.99 ft; (b) 11.9967 in. **19.** 0.799 in. **21.** 0.52 mm/°C **23.** 2.26 mm/°C **25.** (a) 1.38 min; (b) 16.67 min; (c) 20.83 min; (d) 112.5 min **27.** 77.33 cal/g **29.** (a) 2.54 cal/min; (b) 0.177 J/s **31.** 0.14°C **33.** 0.61°C **35.** 30.4 g/m^3 **37.** 14.2%

Chapter 14

1. (a) 0.732 cal/s; (b) 3.025 J/s; (c) 10.33 BTU/h **3.** (a) 47.25 cal/s; (b) 197.8 J/s; (c) 675 BTU/h **5.** 26.04 cal/m-°C-s **7.** (a) 0.063°C/s; (b) decreases **9.** (a) 3750 J/s; (b) 46.3 J/s; (c) 3704 J/s **11.** 1570 BTU/h **13.** (a) 1400 BTU/h; (b) 3600 BTU/h **15.** 180 BTU/h **17.** 10°F **19.** 0.5 ft^2-h-°F/BTU

Chapter 15

1. (a) 1.316×10^{-11} atm; (b) 1.635×10^{-12} g-mol; (c) 9.84×10^{11}; (d) 3.28×10^{8} **3.** 3.34×10^{22} **5.** 24.1 atm; (b) 12.1 atm; (c) 1.72 atm; (d) 1.1 atm **7.** 50.4 lb **9.** 27.8°C **11.** (a) 314.6 g-mol; (b) 318.4 g-mol; (c) 9.45 psi (gauge) **13.** 19.76 cm **15.** (a) 15,700 m/s; (b) 7900 m/s; (c) 110 days for H; 220 days for He **17.** 27.62 atm **19.** (a) 29.662 atm; (b) 29.167 atm; (c) 25.25 atm; (d) 19.64 atm **21.** 6.07×10^{-21} J **23.** (a) 3738 J; (b) 1376 m/s

Chapter 16

3. 515.5 J **5.** (a) 6030 J; (b) 8520 J **7.** (a) 395.8 K; (b) 7.92 atm; (c) 2.4 J **9.** (a) 46.77 atm; (b) 1403 K; (c) 555.9 J **11.** (a) 979.3 J; (b) 439.5 K **13.** (a) 5 atm; (b) 3.79 atm **15.** 26.8% **17.** heat in, 871.44 J; heat out, 373.47 J; work in, 7126.80 J; work out, 7624.77 J; net heat = net work = 497.97 J; Eff = 57.1% **19.** heat in, 5175.8 J; heat out, 2724.2 J; work in, 5445.6 J; work out, 7897.2 J; net heat = net work = 2451.6 J; Eff = 47.4% **21.** heat in, 400 J; heat out, 210 J; work in, 229 J; work out, 419 J; net heat = net work = 190 J; Eff = 47.5% **23.** heat in, 3137.9 J; heat out, 1507.2 J; work in, 494.7 J; work out, 2125.2 J; net heat = net work = 1630 J; Eff = 52% **25.** 9.04 **29.** 4800 hp **31.** about 20 gal due to air alone

Chapter 17

1. (a) 50 Hz; (b) 314 rad/s; (c) 0.02 s **3.** (a) 100 Hz; (b) 1 kHz; (c) 1 MHz **5.** (a) 1.771 rad; (b) 2.571 rad; (c) 3.571 rad; (d) 7.854 rad **7.** (a) 9.800 cm; (b) 5.401 cm; (c) 4.163 cm; (d) 10.00 cm **9.** (a) −3.958 cm; (b) −0.593 m/s; (c) 0.039 m/s^2 **11.** (a) 196 N/m; (b) 0.142 s **13.** 398.7 lb **15.** (a) 1.366 kg; (b) 0.456 kg-m^2; (c) 0.797 kg-m^2; (d) 2.17 s **17.** (a) 1.166 m; (b) 1.666 m from top of rod **19.** same as in example 17-7 **21.** (a) 2.742×10^{-4} N-m/rad; (b) 5.483×10^{-6} N-m; (c) 1 cm; (d) 5.483×10^{-6} N **23.** 0.8 kg

Chapter 18

1. 352 m/s **3.** 18.6 mi **5.** 169 m/s **7.** 1.4 kg **9.** (a) 10.00 cm; (b) 7.29 cm; (c) 0.63 cm; (d) −6.37 cm; (e) −9.92 cm **11.** 429 m/s **13.** 35.26 Hz; 105.78 Hz; 176.30 Hz **15.** 20.9 ft **17.** 19.6° **19.** (a) 0.080 W/m^2; (b) 0.020 W/m^2; (c) 0.005 W/m^2 **21.** 2.25×10^{-4} W/m^2 **23.** 100 Hz

Chapter 19

1. (a) 5139 m/s; (b) 6393 m/s **3.** 1434 m/s **5.** 348.5 m/s **7.** 340 m/s **9.** (a) 6.28×10^{-5} m/s; (b) 12.57×10^{-5} m/s **11.** (a) 3.01 dB; (b) 6.99 dB; (c) 10 dB; (d) 13.01 dB; (e) 16.99 dB; (f) 20 dB **13.** 1.26:1 **15.** (a) 50 dB; (b) 40 dB; (c) 40 dB; (d) <0 dB **17.** (a) 89 dB; (b) 96.4 dB **19.** (a) 7.78×10^{-2} N/m^2; (c) 68.6 dB; (d) 71.8 dB-SPL; (e) 68.6 dB **23.** 2.95×10^{-4} s **25.** 1.5×10^{-2} m/s **27.** 472.9 Hz **29.** 400 Hz **31.** 91,300 m/s

Chapter 20

1. 10 Å = 1 nm **3.** (a) −0.850 eV; (b) −0.213 eV; (c) −0.094 eV; (d) -1.36×10^{-3} eV **5.** (a) 1639 nm; (b) 874.6 nm; (c) 374.9 nm; (d) 91.7 nm **7.** 1.2 m **9.** 1.205×10^{-27} kg-m/s **11.** 0.125 sr **13.** 1005 cd **15.** (a) 4π lm; (b) same as part (a); (c) 400π lm **17.** (a) 139.3 cd; (b) 17.5 lm/W; (c) 23.4 lx **19.** (a) 3.125×10^{-3} sr; (b) 15.63 lm; (c) 312 lx **21.** (a) 0 lm; (b) 220 lm; (c) 680 lm **23.** $R_a = 2R_b$

Chapter 21

1. 2.16×10^{-6} in. **3.** 66.6 rpm **5.** (a) 1.667×10^{-3} rad; (b) 0°5′44″ **7.** (a) 9 cm; (b) −9 cm **9.** size = 1.66 units, inverted **11.** (a) 30 cm, −0.300; (b) 40 cm, −1.00; (c) 400 cm, −19.05; (d) −400 cm, 21.05; (e) −20 cm, 2.00; (f) −2 cm, 100 **13.** (a) 26.67 in.; (b) 9 cm **15.** −5 cm (behind lens); (b) 0.167; (c) 0.833 cm; (d) erect **17.** (a) −7.5 units; (b) 0.25; (c) 1.25 units; (d) erect **19.** 1111 **21.** (a) 30 in.; (b) 8.5 mm **23.** 83.3 cm **25.** 12 min

Chapter 22

1. (a) 2.05×10^8 m/s; (b) 1.97×10^8 m/s; (c) 2.21×10^8 m/s; (d) 1.24×10^8 m/s **3.** (a) 5.36×10^{-6} s; (b) 0.753 mi **5.** 35.259°; 0°10′56″ **7.** $\theta_{air} = 40°$ **9.** (a) 41.14°; (b) 47.3°; (c) 24.4° **11.** (a) 37.5 cm; (b) real, inverted; (c) −0.25; (d) 2.5 cm **13.** 37.5 cm **15.** (a) 7.25; (b) 3.44 cm **17.** (a) 3.75 cm; (b) erect, virtual; (c) smaller **19.** (a) 24 cm to right; (b) −19 cm; (c) 8.38 cm to right; (d) 1.32 cm **21.** (a) 15 cm; (b) 10 cm **23.** (a) 0; (b) infinity **25.** (a) 4.81 cm; (b) 17.82 cm **27.** −9.23 cm **29.** 469

Chapter 23

1. (a) 1.974×10^8 m/s, 1.960×10^8 m/s; (b) 1.946×10^8 m/s, 1.934×10^8 m/s **3.** 24.64°; 24.48° **5.** (a) 40.891°; (b) 41.713° **7.** 41.61° **9.** (a) 13.84 cm; (b) 13.61 cm **11.** 0.59 **13.** (a) 0.315°; (b) 1.65 mm **15.** 1.55 cm **17.** (a) $n_f = \sqrt{n_g} = 1.25$; (b) $d = \lambda/4n_f = 110$ nm **19.** 515 nm **21.** 0.374 s^{-1} or 22.4 fringes/min

Chapter 24

1. 6.242×10^{12} per μC **3.** 2.31×10^{-24} N **5.** 1.38×10^3 m/s **9.** (a) 3.182×10^7 N/C along +x-axis; (b) 4.9×10^7 N/C parallel to +x-axis **11.** 5.13×10^{11} N/C **13.** 5.98×10^5 m/s **15.** (a) 0.0111 μC; (b) 1.11×10^{-11} F; **19.** (a) 5.31×10^{-9} F; (b) 5.31×10^{-8} F **21.** 600 V **23.** (a) 320 μC, 480 μC; (b) 800 μC; (c) 20 μF; (d) same **25.** 200 J **27.** 50,000 μJ, 25,000 μJ **29.** 3 μF

Chapter 25

1. 5 A **3.** (a) 10,653 C; (b) 2 h 57 min 34 s **5.** (a) 144 Ω; (b) 240 Ω; (c) 576 Ω; (d) 2057 Ω **7.** (a) 1 A; (b) 12 V; (c) 5 V; (d) 7 W **9.** (a) 2 mA; (b) 2 mA; (c) 12 V; (d) 6 V; (e) 6 V **11.** 4 mA **13.** 7.55 V **15.** (a) 3 Ω; (b) 12 W **17.** 0.276 Ω **19.** 12.38 mV **21.** 1.72×10^{-8} Ω-m **23.** 104.8 V **25.** 240 A **27.** 48 Ω: 0.669 W, 1.339 W; 36 Ω: 0.960 W, 1.440 W; 30 Ω: 1.183 W, 1.479 W; 24 Ω: 1.500 W, 1.500 W; 18 Ω: 1.936 W, 1.472 W; 12 Ω: 2.661 W, 1.331 W **29.** 480 h **31.** 5.21 mA **33.** $I_a = 7.5$ A; $I_b = 3.75$ A **35.** $I_{R1} = 1.2308$ A; $I_{R2} = 1.5385$ A; $I_{R3} = 0.3077$ A

Chapter 26

1. 0.02 T **3.** (a) 2×10^{-5} T, 0.2 G; (b) 10^{-5} T, 0.1 G; (c) 2×10^{-6} T, 0.02 G **5.** 0.796 A **7.** 251 G **9.** (a) 1.92 T; (b) 1.44 T; (c) 1.15 T **11.** (a) 7.5×10^8 A/Wb; (b) 10^{-7} Wb; (c) 0.1 T **13.** (a) 2122 A-turns/Wb; (b) 14.4 T **15.** (a) 28 GHz; (b) 15.2 MHz **17.** 0.2 N **19.** (a) 1.92×10^{-27} kg; (b) 1.35×10^8 eV; (c) 3.05 MHz; (d) 2.65 MHz **23.** (a) 1.93×10^{-3} N-m; (b) 1.66×10^{-3} N-m; (c) 9.6×10^{-4} N-m; (d) 0 **25.** (a) 19.9 kΩ; (b) 199.9 kΩ; (c) 2 MΩ; (d) 20 MΩ **27.** (a) 1300 Ω; (b) 94%, 60%, 50%, 13%

Chapter 27

1. 5 V **3.** 0.03 V **5.** 0.0628 V **7.** 1.83 V **9.** 16.7 ms **11.** (a) 0.24 A; (b) 0.38 A; (c) 0.52 A; (d) 0.60 A **13.** (a) 5 J; (b) 20 J; (c) 45 J; (d) 80 J **15.** 3.87 mH **17.** 2 H **19.** (a) 20 V, 10 V, 1 V; (b) 1 A, 2 A, 1 A; (c) 41 V-A; (d) 4.1 A

Chapter 28

1. 18.13 V **3.** 6.00 W **5.** 10.61 V **7.** (a) 31.83 mA; (b) 15.92 mA; (c) 1.91 mA **9.** (a) 0.308 W; (b) 0.176 var; (c) 0.355 W; (d) 0.868 **11.** (a) 1 s; (b) 8.64 V **13.** (a) 240 Ω; (b) 41.67 mA; (c) 5.675 V; (d) 8.334 V; (e) 33.54°, 0.833 **15.** 5033 Hz **17.** (a) 402.2 Ω; (b) 0.124 A; (c) −5.99°, current leads voltage; (d) 98.7 V **19.** (a) 3.873; (b) 38.73 V

Chapter 29

1. 10^{12} **3.** 1 **5.** (a) 7.02×10^{-11} A; (b) 7.2×10^{-6} A; (c) 16 mA; (d) 34.6 A **7.** 760 mV **9.** (a) 3 V; (b) 30 mA; (c) 30 mA

Chapter 30

1. 1.28 s **3.** (a) 300 m; (b) 3 m; (c) 0.3 m **5.** 5 kHz **7.** (a) 15,750; (b) 63.5 μs **9.** 5.36 min **11.** (a) 11; (b) 4.25; (c) 255 **13.** (a) 151; (b) 254; (c) 26.625; (d) 12.0391 **15.** (a) 1; (b) 0; (c) 0; (d) 1; (e) 1; (f) 1 **19.** (a) 64; (b) 63

Chapter 31

1. 182 nm, 103 nm, 97.2 nm, 95.0 nm **3.** (a) 91.16 nm; (b) 364.6 nm; (c) 820 nm **5.** 2.18×10^6 m/s, 1.09×10^6 m/s **7.** (a) 0.265 A; (b) 4.36×10^6 m/s; (c) −54.5 eV **9.** $KE_{max} = 0.575$ eV **11.** (a) 2.21×10^{-32} kg; (b) 6.62×10^{-24} kg-m/s **13.** 151 V **15.** 157 m/s **19.** 1.49×10^{-34} J-s **23.** 30 **25.** 0.248 A

Chapter 32

3. 7.1 MeV **5.** 2.83 MeV/nucleon **7.** 67.8 events/day **9.** 3.61×10^{10} Bq **11.** (a) 70.7 Bq; (b) 50 Bq; (c) 25 Bq; (d) zero **13.** 3.13×10^{10} atoms

INDEX